# Mountain Sheep in North America

*Mountain Sheep in North America: Biology, Ecology, Conservation, and Management* provides a detailed account of bighorn and thinhorn sheep across their geographic ranges from Alaska and Canada to Mexico. This book examines all aspects of the sheep life history including activity; biology; conservation; diet; ecology; genetics; habitat requirements; health and disease; historical perspectives; management approaches; predation; competition with other species; natural history; social behavior; taxonomy; the changing role of agency, Indigenous, and conservation stakeholders; and the considerations for future management actions, opportunities, and conservation risks.

This book describes advancements in new technologies supported by the current science, disproving some long-held assumptions about the species and their ecology, making it the most comprehensive source of information about mountain sheep in North America today. It offers the reader a better understanding of the management of mountain sheep and the numerous challenges of a developing world including anthropogenic influences, altered habitats, species assemblages, and changing climates. Each chapter is written by internationally recognized experts and practicing mountain sheep managers from agencies and academia. New research is presented within broader shifts in conservation policy and funding, helping readers understand the practical implications of evolving science in real-world situations and offering wisdom that will help create positive conservation outcomes for mountain sheep.

This is an essential text for students of wildlife, early-career and experienced biologists, managers, policy- and decision-makers, administrators, and anyone interested in mountain sheep in our world today, including those who are simply passionate about them. This book will provide even the most knowledgeable mountain sheep enthusiast with an improved understanding of this magnificent wildlife resource.

# Mountain Sheep in North America
Biology, Ecology, Conservation, and Management

Edited by
Paul R. Krausman and Bill A. Jex

CRC Press
Taylor & Francis Group
Boca Raton London New York

CRC Press is an imprint of the
Taylor & Francis Group, an **informa** business

Designed cover images: top image: Dall's sheep (males), Sonny Parker; center row, left to right: Stone's sheep, Darryn Epp; Rocky Mountain male, Jeff Jackson; desert female and lambs, Darryn Epp; bottom image: Dall's sheep, Sonny Parker

First edition published 2026
by CRC Press
2385 NW Executive Center Drive, Suite 320, Boca Raton FL 33431

and by CRC Press
4 Park Square, Milton Park, Abingdon, Oxon, OX14 4RN

*CRC Press is an imprint of Taylor & Francis Group, LLC*

© 2026 selection and editorial matter, Paul R. Krausman and Bill A. Jex; individual chapters, the contributors

ISBN: 978-1-032-84398-8 (hbk)
ISBN: 978-1-032-85558-5 (pbk)
ISBN: 978-1-003-51868-6 (ebk)

DOI: 10.1201/9781003518686

Typeset in Times
by codeMantra

# Dedication

---

*To Rick, Martha, John, Terri, Bruce, Valerius, Ray, Manfred, and all others who have devoted their professional lives to the management and study of mountain sheep and other wildlife.*

# Contents

Foreword .................................................................................................................................................x
Preface.................................................................................................................................................. xiii
Acknowledgments.................................................................................................................................xiv
About the Editors ................................................................................................................................ xvi
List of Contributors.............................................................................................................................xvii

## SECTION 1   Biology and Ecology

**Chapter 1**   Origin, Classification, and Distribution of Mountain Sheep in North America...............................3

*Raul Valdez*

**Chapter 2**   Historical Trends in Mountain Sheep Populations and Their Habitats...........................................25

*Raul Valdez, Paul R. Krausman, and Bill A. Jex*

**Chapter 3**   Range-Wide Status of Mountain Sheep ........................................................................................47

*Daryl W. Lutz, Froylán Hernández, and Darren L. Bruning*

**Chapter 4**   Physical Characteristics and Horn Development .........................................................................55

*Kevin L. Monteith, Seth T. Rankins, Jaron T. Kolek, and Tayler N. LaSharr*

**Chapter 5**   Social Behavior and Reproductive Strategies ..............................................................................72

*Marco Festa-Bianchet and Fanie Pelletier*

**Chapter 6**   Application of Genetics to Taxonomy, Biology, Conservation, and Management...........................84

*Clinton W. Epps, Michael R. Buchalski, Joshua P. Jahner, and Zijian Sim*

**Chapter 7**   Nutritional Ecology of Mountain Sheep: From Bites to Populations............................................ 114

*Kristin Denryter and Thomas R. Stephenson*

**Chapter 8**   Diseases......................................................................................................................................... 136

*Michael W. Miller, Karen A. Fox, Margo J. Pybus, and Helen M. Schwantje*

**Chapter 9**   Relationships among Mountain Sheep and Their Predators........................................................ 160

*Brian F. Wakeling, Vernon C. Bleich, Marcus E. Blum, and Justin A. Dellinger*

**Chapter 10**  Competition among Mountain Sheep and Other Ungulates ....................................................... 175

*Jericho C. Whiting, Vernon C. Bleich, and Tom Smith*

**Chapter 11** Seasonal Migrations and Other Movements ................................................................ 191

  *Blake Lowrey*

**Chapter 12** Weather, Water, and Precipitation ............................................................................. 208

  *James W. Cain III and Madelon Van de Kerk*

## SECTION 2  Population Management

**Chapter 13** Mountain Sheep Survey Methods ............................................................................... 233

  *Vernon C. Bleich, Daryl W. Lutz, Marcus E. Blum, Danielle M. Glass, Steven G. Torres,
  and Jericho C. Whiting*

**Chapter 14** Capture and Translocation of Mountain Sheep ......................................................... 258

  *Daryl W. Lutz and Vernon C. Bleich*

**Chapter 15** Modeling Population Dynamics of Mountain Sheep .................................................. 280

  *Mark S. Boyce and Evelyn H. Merrill*

**Chapter 16** Harvest Management of Mountain Sheep in North America ....................................... 290

  *Chadwick Lehman, Todd J. Brinkman, and Brad R. Wendling*

**Chapter 17** Mountain Sheep Management in Mexico .................................................................. 305

  *Raul Valdez, Raymond M. Lee, Hugo Sotelo Gallardo, and Juan Manuel Segundo Galán*

**Chapter 18** Human Dimensions and Mountain Sheep ................................................................ 330

  *Kathryn A. Schoenecker, Nicholas W. Cole, and Saeideh Esmaeili*

## SECTION 3  Habitat, Climate Change, and Conflicts

**Chapter 19** Characteristics of Habitat for Mountain Sheep ......................................................... 347

  *Paul R. Krausman*

**Chapter 20** Habitat Management: The Key to Mountain Sheep Conservation ............................... 378

  *James C. deVos, Jr. and Raymond M. Lee*

**Chapter 21** Threats to Mountain Sheep Habitat ......................................................................... 397

  *Bill A. Jex*

**Chapter 22** Conflict Management ............................................................................................... 414

  *Brian F. Wakeling, Stephen L. Webb, Jeffrey W. Gagnon, Emily S. Almberg, and Jared T. Beaver*

# SECTION 4    Science-Related Management Opportunities

**Chapter 23** Long-Term Research on Individual Sheep.................................................................................441

   *Marco Festa-Bianchet and Fanie Pelletier*

**Chapter 24** Challenges for Mountain Sheep Management and Conservation in North America...................................455

   *Paul R. Krausman and Bill A. Jex*

**Index**.................................................................................................................................................465

# Foreword

When Dr. Paul R. Krausman asked in early Autumn 2023 if I would be willing to write the Foreword for a new mountain sheep book, my initial reaction was, "Sure!" Then, pondering the collective talent, knowledge, and pedigrees of the many authors writing these powerful chapters, my second thought was, "Oh no, what have I gotten myself into?" In my opinion, these authors are *the* North American experts on mountain sheep. I am quite pleased to offer my thoughts and perspective on what I consider a timely, comprehensive, and critically important reference text on the conservation and management of mountain sheep. As noted in the Preface for this book, these are challenging and rapidly evolving times for wild sheep globally. I am so grateful that Paul R. Krausman and Bill A. Jex have made this enormous contribution; I congratulate and thank them for this effort, and I sincerely appreciate their invitation to write this Foreword.

Having been a wildlife biologist for nearly 50 years, and having worked more than 45 years on mountain sheep conservation in both the public (i.e., agency) and private (i.e., NGO) sectors, I recognize and greatly admire the combined expertise of academic and agency-ministry contributors to this book, to look back and describe the rocky path mountain sheep have taken and to also lay out their collective vision and describe hoped-for trajectories for future wild sheep management. The List of Contributors to these chapters is a veritable "Who's Who" of mountain sheep experts, and I am awed by what they have done in their careers and what they have collectively written here. If I may be considered a colleague by these mountain sheep experts, I would be deeply humbled and most appreciative.

Other than opportunistic observations before launching my graduate studies at the University of Wyoming in the early 1980s, I hadn't spent much time looking at or studying Rocky Mountain bighorn sheep, or any other wild sheep, for that matter. Reading as much technical and popular literature as I could before launching my MS field work, I read that Teddy Roosevelt, considered by many to be the greatest conservation president our country has ever known, and an extremely well-traveled man, called the 50-mile stretch between Cody, Wyoming and Yellowstone National Park, along the North Fork of the Shoshone River, "…the most scenic 50 miles in America…." I could not believe my good fortune in securing this study area for my master's research to be located between the North Fork of the Shoshone River and the beautiful Sunlight Basin, nor could I anticipate the wild sheep path I would follow for the next 4.5 decades.

Once I hit the field in June 1982 to study the Trout Peak bighorn herd located between Cody and Yellowstone National Park, my first close encounter was with a half-curl male, which watched me from a ridgetop less than 90 m away, silhouetted and back-lit by a rising early morning sun. In the conglomerated breccia volcanic rubble of northwest Wyoming's Absaroka Range, I had lost my footing while solo backpacking beneath and on my way to Trout Peak. As I sat there, regaining my composure and my breath after nearly sliding off a steep slope into nothingness, I watched and admired this young male, instantly recognizing that, as a visitor to his mountain realm, I was no match for his sure-footedness, and I was indeed a visitor to his world. That inaugural experience modified my approach and sharpened my safety protocols; that exquisite memory has remained vivid, deepening my infatuation with mountain sheep over the past four plus decades.

I was fortunate to spend about 30 years studying and managing bighorn sheep in northwestern Wyoming's Absaroka Range, described as a stronghold for bighorn sheep by Ralph Honess and Ned Frost in their 1942 Wyoming Bighorn Sheep Study. The Absaroka Range is where the Mountain Shoshone (Tukadika) people, known as Sheep Eaters, had lived with, and relied on, bighorn sheep for millennia, with much of their food, clothing, weapons, tools, and culture derived from, and centered around, bighorn sheep. The Tukadika quarried obsidian (i.e., volcanic glass) from sites in what is now known as Yellowstone National Park, turning obsidian chips into razor-sharp arrowheads, knives, and scrapers. By soaking bighorn sheep horns in geothermal waters to soften and render them pliable, strong, durable hunting bows were made with mountain sheep horn and sinew. When paired with obsidian arrowheads, these horn bows were an effective hunting tool, used successfully for ground-hunting bighorns in alpine and subalpine terrain; these horn bows were also traded widely with other tribes by the Sheep Eaters. Rock and linear-arranged downed timber on high ridgelines is all that remains from the time of the Tukadika and the elegant drive traps they constructed to harvest mountain sheep in support of their unique way of life.

Indigenous First Nation peoples, in what is now known as the Yukon Territory, were closely tied to Dall's sheep and the tools, clothing, and sustenance wild sheep provided. Anasazi (more correctly Ancestral Puebloan) peoples, in what is now known as the southwestern United States, also relied on, and heavily used, desert bighorn sheep for their customs, culture, and sustenance. Around the western United States (including Alaska) and western Canada, Indigenous people, tribes and First Nations, and early explorers, trappers, and mountain men were deeply connected to, and often dependent on, mountain sheep. I have always admired the talent, ability, and proximity of those early human mountain-dwellers for their close ecological, and some might say spiritual, connection to mountain sheep, their habitats, and the remote landscapes they survived and thrived in.

Over the past 4.5 decades, I have had the opportunity, privilege, and good fortune to visit and learn about many mountain sheep ranges across North America, each with its history, challenges, and opportunities for wild sheep management. Even though I have not physically set foot there, I've learned about Central Asian countries, where congeneric wild sheep have been integral to cultures and communities, again for millennia.

As I ponder the history and the present status and anticipated future of mountain sheep, I cannot help but think how early naturalists and budding scientists guesstimated mountain sheep numbers in the western United States and Canada. In their epic journeys West, Lewis, and Clark took great care to document the abundance and distribution of bighorn sheep, recording observations of those then-common ungulates as they journeyed west toward the Pacific. No one today can validate Ernest Thompson Seton's 1928 estimate of 1.5–2 million bighorn sheep in the West, but, from widespread petroglyphs and pictographs, tales passed down through generations, mountain man observations, early explorer journals, and other anecdotal and quasi-scientific records, mountain sheep were widely distributed in the West before Spanish conquistadors, missionaries, and early European settlers brought their languages, beliefs, livestock, accompanying human- and animal-based pathogens, and civilization to mountain sheep and their once-remote habitats.

Alarmingly, and widely accepted, by 1960, when Helmut K. Buechner published his monograph, *The Bighorn Sheep in the United States: Its Past, Present and Future*, bighorn sheep numbers were somberly estimated at only 15,000–18,000 remaining in a dozen western states. Buechner identified multiple factors that solely, or in combination, contributed to the catastrophic decline of bighorn sheep: unregulated hunting (to provide sustenance for wagon trains, railroad workers, and early miners and settlers), competition for forage with livestock, pathogens and parasites inadvertently transmitted from domestic livestock to mountain sheep, and early western settlement that adversely influenced mountain sheep and their habitats.

The past century has seen more than 1,500 transplants conducted by wild sheep managers and conservation stakeholders to repopulate historic ranges, with nearly 25,000 individual bighorn sheep being trapped and relocated across virtually every western state (except Alaska) and southern Canadian province where bighorns were once abundant. While bighorn sheep numbers have increased threefold or more since Buechner's 1960 monograph described their population nadir, current bighorn sheep numbers and distribution can only be described as a small fraction of their historic levels. While other western big game (e.g., elk, pronghorn, mule deer) experienced dramatic population recoveries and, in some cases, significant range expansions during the conservation decades of the 1920–1970s, notably, bighorn and bison are the exceptions to that pattern of successful restoration.

For decades, northern wild sheep (i.e., Dall's, Stone's) were thought to be just fine, due to the remoteness of their habitats and a relatively light human footprint.

Recent decades have seen increased penetration into and disturbance of thinhorn sheep range, and while the geographic distribution of thinhorn sheep may be largely intact compared to Charles Sheldon's seminal mapping more than a century ago, absolute numbers of Dall's and Stone's sheep have widely declined in most jurisdictions. Transmission powerlines, cell phone towers, mining ventures, two-track roads pioneered into previously inaccessible mountain ranges and alpine regions, jet boat and ARGOS high-powered vehicular access, increased fly-in and drop-camp traffic to backcountry lakes and river systems, heli-skiing, heli-hiking, heli-biking, heli-kayaking, and other motorized human access into once-remote thinhorn ranges have influenced what was previously viewed as pristine, intact mountain sheep country.

Combined with changing climates and altered precipitation patterns, rain-on-snow events, repeated freeze-and-thaw cycles, an ascending brush line moving upslope a meter/year, biotic and abiotic changes have altered the landscapes and ecological niches Dall's and Stone's sheep evolved in. In drier habitats well south of thinhorn sheep range, chronic and acute drought have affected the availability of water for desert bighorns and other desert-dwelling wildlife, and drought effects on vegetative quality and production have adversely influenced important seasonal mountain sheep forage. Invasion of noxious weeds (e.g., star thistle, Dalmatian toadflax) and flashy annual grasses (e.g., cheatgrass) has changed historic fire regimes, while increased juniper, limber pine, pinyon pine, and other coniferous cover has further complicated management of fire-adapted vegetation communities relied on by mountain sheep.

Predation on wild sheep has been variable through time, between jurisdictions, and by land ownership. Some jurisdictions (e.g., NM) have restored desert bighorn sheep via active, targeted predator management (e.g., mountain lions) programs, while other states (e.g., CA) struggle to enact large predator management programs, due largely to societal pressure. In northern reaches like Alaska, federally protected golden eagles are a significant predation factor on Dall's sheep lamb survival, but legal control options are restricted. While predation helped shape the evolutionary and ecological attributes of mountain sheep that many people admire, there is continued, intense debate about intervention and control of large predators, from wolves and wolverines in the North, to mountain lions in the Southwest, and from coyote control on lambing ranges to aerial predation of thinhorn neonates.

In many jurisdictions and regulatory frameworks, there is little-to-no recognition that wildlife, in particular mountain sheep, are an important component of federally designated wilderness. State wildlife agency (e.g., AZ, NV, WY) primacy to capture, radio-collar, conduct pathogen surveillance, or perhaps relocate mountain sheep in federally designated wilderness is often compromised, requiring complex, lengthy minimum tools analyses that sometimes hinder important, time-sensitive decision-making by state wild sheep managers.

Forage competition from burgeoning feral horse and burro populations, inside and beyond designated Horse Management Areas, and an antiquated Appropriate Management Level for horses and burros established in 1971, coupled with increased horse and burro advocacy and significant budget restrictions on federal land management agencies that must maintain off-range and expensive private-land sanctuaries, have all compromised mountain sheep conservation and much-needed habitat management. In some jurisdictions (e.g., TX), introduced exotic ungulates (e.g., aoudad) can out-produce native desert bighorn sheep and recent research has documented transmission of lethal respiratory pneumonia bacteria from aoudad to desert bighorn sheep, compounding the influence of these exotics on mountain sheep.

Failure to achieve effective spatial and temporal separation between domestic sheep and goats and wild sheep on public-land grazing allotments, combined with increasing presence and potential contact from domestic sheep and goat farm flocks and hobby herds, continues and escalates the friction between agricultural and mountain sheep conservation factions. Another variable added in recent times to this complex equation is the growing demand for pack-goat and pack-llama recreational use in occupied wild sheep habitat; by necessity, federal land managers walk a fine line between too much or too little protection of mountain sheep ranges from otherwise well-intended recreational users of public lands in the West.

Agency and ministry responses to pneumonia outbreaks and repeated die-offs in mountain sheep are often, but not always, guided by academic research and the best available science, as political forces squabble over land-use decisions and multiple-use resource allocations. How many times have bighorn sheep been trapped and transplanted to historic, still-suitable range displaying a short-lived growth trajectory, only to be followed by some sort of pneumonia die-off? Wild sheep managers and their NGO conservation partners continue to invest sweat equity and critical funding for wild sheep management, but often, little to no return is realized on those investments.

At times, the list of challenges for mountain sheep conservation and management seems infinite. Mountain sheep conservation is not for the faint of heart, but it is such critically needed work. A long-time wild sheep conservationist colleague analogized, much like the comparison between NBA basketball and NFL football, that wild sheep conservation is a full-contact sport. This work is not for the meek. In March 1960, the legendary outdoor writer Jack O'Connor, whom many consider the grandfather of wild sheep hunting, posited:

> The sheep hunter is willing to climb until his lungs are bursting, to walk until his legs are dead and weary, to grow hungry and thirsty for great rewards. There is no halfway. After his first exposure, a man is either a sheep hunter or he isn't. He either falls under the spell of sheep hunting and sheep country, or he won't be caught dead on another sheep mountain.

Comparatively, wild sheep managers, researchers, and conservationists have fully engaged for decades. The exceptional authors of the chapters in this book have gone to great lengths and no doubt have made great sacrifices during their careers to document and describe the challenges that mountain sheep have endured over the past two centuries. I cannot recall such a collection of knowledge and experience incorporated into a single reference book, as is evident here. Over my working career, I have learned, or attempted to learn, from so many of these mountain sheep experts, and I thank them for their individual and collective accomplishments and contributions to the conservation and management of the most fascinating animals that roam the mountains, canyons, deserts, and ridgelines of the West. For the love of wild sheep, for these authors, for the legions of wild sheep conservationists, there truly is no halfway. We are all in. Willingly. Enjoy these chapters, my fellow conservationists!

**Kevin P. Hurley**
*Wild Sheep Foundation*
*Vice-President of Conservation – Emeritus*

# Preface

Over the past two decades, significant technological advances have enabled a deeper understanding of mountain sheep, their distributions, habitats, and life requirements, and the human and environmental factors that can influence population trajectories and increase vulnerability to localized extirpation. Advances in genetic research have enabled whole-genome sequencing of mountain sheep, helping to redefine subspecies distributions and evolutionary lineages. This also assists researchers in understanding the parasites and diseases of mountain sheep, which supports a significant understanding of pathogen pathways and why disease virulence in mountain sheep populations appears to vary. With scientific information becoming more current and human and mountain sheep responses to environmental stochasticity becoming better understood in recent years, widely held assumptions by managers of mountain sheep, as recently as 5 years ago, were incomplete and, in some cases, inaccurate. The danger in this time of research, knowledge evolution, policy, and management transformation is that those seeking information about mountain sheep may be faced with trying to understand and accept what seem to be contradictory views of cause-and-effect relationships and future threats.

We are also at a point where public and private investments in mountain sheep management are changing how jurisdictions consider project implementation and establish objectives. Significant financial investments in restoring species, such as desert bighorn sheep populations, to historic and new ranges, and understanding climate change and the role of humans, hunters, habitat, and changing natural predator systems, now benefit from millions of dollars in annual investments, sourced directly from non-governmental and non-academic sources (both financial and in-kind support). This reshaped investment engine has the potential to lead to important shifts in jurisdictional priority setting.

As such, the worlds of mountain sheep management, policy, and regulatory reform require an up-to-date baseline for mountain sheep that helps ensure directions are based on the best available, accurate, and reliable science appropriate for the scope and scale of the management action or decisions being considered.

The recognized races of mountain sheep (Rocky Mountain bighorn sheep [*Ovis canadensis canadensis*], California bighorn sheep [*O. canadensis californiana*], desert bighorn sheep [*O. canadensis* spp.], Dall's thinhorn sheep [*O. dalli dalli*], Stone's thinhorn sheep [*O. dalli stonei*]) are the central theme of this book; we have also referred to other races because they are still managed as different entities (e.g., Sierra Nevada bighorn sheep [*O. canadensis sierrae*]). We also use the metric system for most measurements in the book but use the Imperial system when measurements have been and are traditionally recorded in that system (e.g., horn measurements).

We were privileged to have worked with some of the most superb mountain sheep biologists in the continent on this book. They were all interested, dedicated, responsive, and a pleasure to work with in creating this exceptional volume. Our overall goal is that this book will serve as a reference resource for the conservation and management of mountain sheep in North America, benefiting students, mountain sheep biologists and managers, hunters, recreationists, other interested stakeholders, and ultimately our iconic and globally unique, mountain sheep.

**Paul R. Krausman,**
*Professor Emeritus, University of Arizona*

**Bill A. Jex,**
*RPBio., British Columbia Provincial Wild Sheep and Mountain Goat Specialist (retired)*

# Acknowledgments

Throughout our careers, we have had the opportunity and benefit of interacting with experts in the biology, management, and conservation of mountain sheep. Those interactions have continued with this book's development, and the chapters' referees and authors represent an impressive collection of experts on large mammals, including mountain sheep. We, and the authors are grateful to the referee's dedication, timely reviews, and expert advice provided by Morgan Anderson, Robert M. Anderson, Carlos A. Astorga Toro, Nayo Balderrama, S. Barbosa, Louis C. Bender, Vernon C. Bleich, R. Terry Bowyer, Melanie Bucci, Darren Bruning, D. Coltman, Steeve Côté, Gavin Cotterill, Cody Deane, Daniella Dekelaita, Jonathan Derbridge, Carlos H. Alcalá Galván, Bob Garrott, Steve M. Gordon, James R. Heffelfinger, John J. Hervert, Kevin Hurley, A. G. Jex, D. L. Jex, Heather Johnson, Krystal Kriss, Meghan Larivee, Brent Lonner, Doug McWhirter, Jerod Merkle, J. Miller, Enrique de Jesús Ruiz Mondragón, Roberto Noriega-Valdez, Elizabeth Painter-Flores, Katherine Parker, Ronald H. Pine, Kim Poole, Chris Procter, Kelly M. Proffitt, Rob Roy Ramey II, Gary Roemer, Eric Rominger, Esther Rubin, C. Q. Ruhl, Tom Ryder, Benjamin N. Sacks, Griselda Sánchez, Andrew V. Sandoval, Tim Schommer, K. George Schultze, Eahsan Shahriary, Krista Sittler, Kelley Stewart, Caeley Thacker, Raul Valdez, Andrew Walker, John Wehausen, Don Whittaker, E. J. Wilkins, Michael Wisdom, and P. L. Wolff. Chapters 5 and 23 benefited from assistance from J. T. Jorgenson, W. D. Wishart, C. Feder, A. Hubbs, J. T. Hogg, J.-M. Gaillard, D. W. Coltman, M. B. Morrissey, and K. Ruckstuhl for important contributions to long-term studies of bighorn sheep. Over 100 students and field assistants collected data in the field. The Natural Sciences and Engineering Research Council of Canada provided long-term financial support through grants and scholarships. Funding was also provided by the Canada Research Chairs, Alberta Conservation Association, the Fonds de Recherche du Québec, and the Université de Sherbrooke. Logistic support from Alberta Fish and Wildlife was key to the success of both Ram Mountain and Sheep River. Time and space were provided to the authors of Chapter 7 by the Alaska Department of Fish and Game and the California Department of Fish and Wildlife. The authors of Chapter 18 were supported by the U.S. Geological Survey, Ft. Collins Science Center, Colorado for funding the chapter, although mention of products is not an endorsement of the U.S. Government. J. P. Jahner was supported by the Modelscape Consortium with funding from the NSF (OIA-2019528). C. W. Epps was supported, in part, by the Oregon Agricultural Experiment Station with funding from the Hatch Act capacity funding program, award numbers NI25HFPXXXXXG022 and/ or NI25HMFPXXXXG029, from the USDA National Institute of Food and Agriculture. We are grateful to all the organizations, agencies, and funders that supported the authors during the preparation of chapters. Bill Morrison assisted with some of the figures. Ian Freeman, Doug McWhirter, personnel from the Wyoming Game and Fish Department, Montana State University, Yellowstone National Park, and Kevin Monteith assisted with the development of figures for Chapter 11 by providing data or assisting with the cartography. We also thank the author's families for the time taken from home. Tiffany Gasparini was invaluable for navigating the world of book publishing. The editors of CRC Press were instrumental in bringing this work to completion, especially Randy Brehm, Amelia Bashford, Shikha Garg, Tom Connelly, Matthew Shobbrook, and Michele Dimont. Proofs were prepared by Assunta Petrone and the team of copyeditors from codeMantra. The index was prepared by Melanie Bucci and editing of all chapters was completed by Paul R. Krausman and Bill A. Jex. The Wild Sheep Initiative of the Western Association of Fish and Wildlife Agencies and the Wild Sheep Foundation were instrumental with the development and distribution of books to contributors. To each of you, we owe a debt of gratitude. Many, many thanks.

## THE WILD SHEEP FOUNDATION

The purpose of the Wild Sheep Foundation is to put and keep wild sheep on the mountain.

Their mission is to enhance wild sheep populations and their habitats, promote scientific wildlife management, and educate the public and youth on sustainable use and the conservation benefits of hunting while promoting the interests of the hunter. The Wild Sheep Foundation's vision is to be the best managed, most respected, influential, and relevant conservation organization benefiting wild sheep worldwide.

## THE WESTERN ASSOCIATION OF FISH AND WILDLIFE AGENCIES AND THE WILD SHEEP INITIATIVE

The Western Association of Fish and Wildlife Agencies' goal has been to support sound resource management and build partnerships at all levels to conserve wildlife for the use and benefit of all citizens, now and in the future. Their mission is to "advance collaborative, proactive, science-based fish and wildlife conservation and management across the west" through leadership, integrity, teamwork, excellence, support, respect, and accountability. The Wild Sheep Initiative began as the Wild Sheep Working Group in 2007, when agency directors saw the need, value, and urgency to convene managers, researchers, veterinarians, and others to address the myriad challenges mountain sheep faced in North America. The Wild Sheep Initiative is committed to the restoration and conservation of mountain sheep for the benefit of future generations and their mission is: "To identify priority topics and management challenges to wild sheep in western North America; collaboratively develop solutions to those challenges; and foster strong relationships between state, provincial, territorial jurisdictions, federal land management agencies, indigenous groups, and wild sheep advocates; and engage with domestic livestock owners and users that share the same landscapes inhabited by wild sheep." This book encompasses the mission of these organizations by providing a better understanding of the status of mountain sheep and their ecology. The associations are delighted to be a part of the publication of this book.

# About the Editors

**Dr. Paul R. Krausman** was an assistant professor at Auburn University (1976–1978), professor of wildlife conservation and management at the University of Arizona (1978–2007), and the Boone and Crockett Professor at the University of Montana (2007–2015). He is currently a professor emeritus at the University of Arizona and the editor of the Conservation and Management Book Series published by Johns Hopkins University Press and The Wildlife Society. He was raised in North Africa, Europe, and Asia before returning to the United States where he earned degrees from The Ohio State University (BS, zoology), New Mexico State University (MS, wildlife management), and the University of Idaho (PhD, 1976), and he was awarded the Aldo Leopold Medal from The Wildlife Society in 2006. Dr. Krausman was also a visiting professor at the Wildlife Institute of India, Dehra Dun (1989–2000) and the Universidade de Trás-os-Montes e Alto Douro, Vila Real, Portugal (2005–2006). He was also the recipient of the Desert Ram Award from the Desert Bighorn Council (2000) and the O. C. "Charlie" Wallmo Award (1999). He has numerous journal and book publications and has been the editor of the *Desert Bighorn Council Transactions*, *The Journal of Wildlife Management*, and *Wildlife Monographs*, and has been an associate and guest editor for five other journals. Dr. Krausman is a certified Wildlife Biologist, Wildlife Fellow, and Honorary Member of the Wildlife Society, and served as faculty advisor to the student chapters of The Wildlife Society at Auburn, Arizona, and Montana. Dr. Krausman has taught an array of wildlife-related classes throughout his career, including scientific writing, big game ecology, and applied wildlife management. He has directed nearly 100 graduate students on ungulate and predator studies throughout the United States and in India. The most rewarding aspect of his career is the development of future leaders in the wildlife profession—students. He currently lives with his wife and an assemblage of animals on Whidbey Island, Washington.

**Bill A. Jex** always had a keen interest in the outdoors and started working in the natural resources field in Ontario, Canada in 1984. He graduated from Sir Sandford Fleming College in 1989 as a fish and wildlife technologist while doing some teaching along the way. Since his graduation, he has worked in the fisheries, wildlife, habitat, enforcement, and regulatory fields. Beginning in 1996, Bill Jex started employment with the province of British Columbia, and since 2018, he has represented the province as its first wild sheep and mountain goat specialist before retiring in 2025. He continues to support the province in enforcement roles as a provincial wild sheep expert, and also supports non-government organizations as a member of the Wild Sheep Foundation's Professional Resource Advisory Board, the Rocky Mountain Goat Alliance's Science and Conservation Committee, and the Northern Wild Sheep and Goat Council, and as a collaborator with the Western Association of Fish and Wildlife Agencies' Wild Sheep Initiative. He was one of the first technologists to become accepted as a registered professional biologist in British Columbia in 2001. Currently, Bill Jex works with the Wild Sheep Foundation and the Wild Sheep Society of British Columbia, and continues to support research with universities, conservation-focused organizations, and Indigenous governments on wild sheep and mountain goat issues. He also supports, edits, and authors scientific and popular articles and conducts public education workshops. Mountain sheep and goats, and especially thinhorn sheep, are truly special to Bill Jex. These animals live where few can: on the tops of the world and the edges of the earth, where the line between the land and sky is at times obscured by clouds or blinded in sunshine. His goal is to continue to work for the conservation and management of mountain ungulates so future generations can be inspired by their resilience and majesty, and the places they live. He currently lives in Smithers, British Columbia, with his lovely wife of 30 years and their two dogs, a chocolate lab and a golden retriever.

# List of Contributors

**Emily S. Almberg**
Montana Fish, Wildlife & Parks
Bozeman, Montana

**Jared T. Beaver**
Department of Animal and Range Sciences
Montana State University
Bozeman, Montana

**Vernon C. Bleich**
University of Nevada Reno
Bismarck, North Dakota

**Marcus E. Blum**
Natural Resources Institute
Texas A&M University
College Station, Texas

**Mark S. Boyce**
Biological Sciences
University of Alberta
Edmonton, AB, Canada

**Todd J. Brinkman**
University of Alaska Fairbanks
Fairbanks, Alaska

**Darren L. Bruning**
Alaska Department of Fish and Game
Delta Junction, Alaska

**Michael R. Buchalski**
Wildlife Genetics Research Unit, Wildlife Health
    Laboratory
California Department of Fish and Wildlife
Sacramento, California

**James W. Cain III**
U.S. Geological Survey
New Mexico Cooperative Fish and Wildlife Research Unit
Las Cruces, New Mexico

**Nicholas W. Cole**
U.S. Geological Survey
Fort Collins Science Center
Fort Collins, Colorado

**Justin A. Dellinger**
Wyoming Game and Fish Department
Lander, Wyoming

**Kristin Denryter**
Alaska Department of Fish and Game
Palmer, Alaska

**James C. deVos, Jr.**
Arizona Game and Fish Department
Phoenix, Arizona

**Clinton W. Epps**
Department of Fisheries, Wildlife, and Conservation
    Sciences
Oregon State University
Corvallis, Oregon

**Saeideh Esmaeili**
Colorado State University
Fort Collins, Colorado

**Marco Festa-Bianchet**
Département de biologie
Université de Sherbrook
Sherbrooke, QC, Canada

**Karen A. Fox**
Colorado State University
Fort Collins, Colorado

**Jeffrey W. Gagnon**
Arizona Game and Fish Department
Phoenix, Arizona

**Danielle M. Glass**
California Department of Fish and Wildlife
Bishop, California

**Froylán Hernández**
Texas Parks and Wildlife Department
Alpine, Texas

**Joshua P. Jahner**
Department of Biology
New Mexico Institute of Mining and Technology
Socorro, New Mexico

**Jaron T. Kolek**
University of Wyoming
Laramie, Wyoming

**Tayler N. LaSharr**
University of Wyoming
Laramie, Wyoming

**Raymond M. Lee**
Ray Lee LLC
Cody, Wyoming

**Chadwick Lehman**
South Dakota Game, Fish, and Parks
Custer, South Dakota

**Blake Lowrey**
U.S. Geological Survey
Bozeman, Montana

**Daryl W. Lutz**
Wyoming Game and Fish Department
Lander, Wyoming

**Evelyn H. Merrill**
University of Alberta
Sherwood Park, AB, Canada

**Michael W. Miller**
Veterinarian
Fort Collins, Colorado

**Kevin L. Monteith**
University of Wyoming
Laramie, Wyoming

**Fanie Pelletier**
Département de biologie
Université de Sherbrooke
Sherbrooke, QC, Canada

**Margo J. Pybus**
Alberta Fish and Wildlife Division
Lethbridge, AB, Canada

**Seth T. Rankins**
University of Wyoming
Laramie, Wyoming

**Kathryn A. Schoenecker**
U.S. Geological Survey
Fort Collins Science Center
Fort Collins, Colorado

**Helen M. Schwantje**
Veterinarian
Duncan, BC, Canada

**Juan Manuel Segundo Galán**
Independent Consultant
Hermosillo, Sonora, Mexico

**Zijian Sim**
University of Alberta
Edmonton, AB, Canada

**Tom Smith**
Brigham Young University
Provo, Utah

**Hugo Sotelo Gallardo**
Consultant
Monterrey, Nuevo León, Mexico

**Thomas R. Stephenson**
California Department of Fish and Wildlife
Bishop, California

**Steven G. Torres**
California Department of Fish and Wildlife
Granite Bay, California

**Raul Valdez**
New Mexico State University
Las Cruces, New Mexico

**Madelon Van de Kerk**
Western Colorado University
Gunnison, Colorado

**Brian F. Wakeling**
Montana Fish, Wildlife & Parks
Helena, Montana

**Stephen L. Webb**
Texas A&M University
College Station, Texas

**Brad R. Wendling**
Alaska Fish and Game, Region III
Fairbanks, Alaska

**Jericho C. Whiting**
Department of Biology
Brigham Young University-Idaho
Rexburg, Idaho

# Section 1

## Biology and Ecology

# 1 Origin, Classification, and Distribution of Mountain Sheep in North America

*Raul Valdez*

## 1.1 INTRODUCTION

Thinhorn (*Ovis dalli*; widespread in Alaska to northwestern Canada) and bighorn (*Ovis canadensis*; occurring from southwestern Canada to northern Mexico) comprise the two endemic mountain sheep species in North America, both of which evolved from an Asian ancestor that accessed North America via the Bering Land Bridge during the Pleistocene. The snow sheep (*Ovis nivicola*), a species endemic to Siberia, is the closest living relative of New World sheep. Unstable landscapes and climates of the Pleistocene and glacial-induced vicariance, hybridization, and introgression are the major influences of North American mountain sheep genetic and phenotypic evolution. The diverse geomorphology of temperate montane northwestern, western, and southwestern North America afforded colonizing mountain sheep travel corridors in the expansive mountain systems and adjoining varied terrain that facilitated their historical distribution from Alaska to Mexico. Rugged landscapes, essential for evading predators, are the unifying habitat component of thinhorn and bighorn mountain sheep environments. The large distinct white rump patch extending above the base of the tail except for the white Dall's sheep (*O. dalli dalli*), massive transversely ridged curling horns unique to adult males, and a 2n (diploid chromosome number)=54 are some of the anatomical and genetic attributes that differentiate North American mountain sheep from other American bovids. The significance of the horns of adult male mountain sheep is summarized by Seton (1927: 525): "The horns of the ram are the coveted trophy of the sportsman; for they are the creature's peculiar gift and crowning glory, his one distinguishing and imperishable ornament, the record of his size and of his life." Sexual dimorphism is evident in thinhorn and bighorn sheep, with males having more massive and longer horns and greater body mass and metrics than females. Anthropogenic habitat degradation, overexploitation, and exposure to domestic ungulate diseases were the primary factors of large-scale mountain sheep population declines and extirpations. Continued conservation concerns (e.g., disease outbreaks, habitat fragmentation, climate change) require regional and jurisdictional cross-border recovery efforts. Their widespread geographical distribution in Arctic to desert environments is characterized by ecoregions and habitats differentiated by temperature, precipitation, plant growth patterns, and regional soil and mineral attributes. Mountain sheep occur at elevations ranging from 90 m in hilly areas, cliffs, escarpments, and buttes to 4,270 m in rocky, precipitous alpine habitats. Physiological, behavioral, and ecological adaptations enable them to tolerate extreme climatic conditions including subfreezing temperatures of −51°C in northern Arctic and alpine habitats to more than 4.4°C of scorching heat in southern desert regions, in some instances without free-standing water. Nevertheless, they are vulnerable to climate change, which is expected to cause ever more restricted habitats in the future. In this chapter, I review the origins; historical basis of their discovery and description of their diversity; and the taxonomy, nomenclature, and distribution of mountain sheep of North America.

## 1.2 ORIGINS OF MOUNTAIN SHEEP

The genus *Ovis*, which originated in Asia, is a complex taxon with a tangled evolutionary and taxonomic history. It has a Holarctic distribution which encompasses Eurasia (Palearctic) and temperate northern, western, and southwestern North America (Nearctic), including northern Mexico and the Baja California Peninsula. Adult mountain sheep species differ in coat color and pelage patterns, anatomical metrics, diploid chromosome number, genomic diversity, and geographic distribution. The evolutionary history (e.g., diversification, adaptation, speciation) of mountain sheep has been characterized by complex hybridization and introgression histories that have complicated the interpretation of the phylogenetic and phylogeographic history of the genus (Schaller 1977, Geist 1985, Rezaei 2010, Groves and Grubb 2011, Damm and Franco 2014, Chapter 6).

The genus *Ovis* probably diverged from other Caprini lineages during the Pliocene, perhaps from a Middle Eastern ancestral species. Wang et al. (2016) described a Pliocene Caprini (sheep, goats, and close relatives) species (*Protovis himalayensis*) from the western Tibetan Plateau that potentially occupies a transition phase to the *Ovis* lineage 5.46 to 3.10 million years ago (MYA). The lineage that gave rise to the Siberian and American mountain sheep lineage diverged from other Eurasian lineages 5.6 to 2.4 MYA. *Ovis* species expanded their ranges east and west in the immense landscapes of Eurasia, eventually crossing the Bering Land Bridge and giving rise to populations in North America (Bunch et al. 2006, Korobitsina et al. 1974, Rezaei et al. 2010, Wang et al. 2016, Chen et al. 2021, Upadhyay et al. 2021, Feijó et al. 2022, Chapter 6).

DOI: 10.1201/9781003518686-2

Extant *Ovis* species form three distinct clades, classified as subgenera, based on their evolutionary lineages. These distinctions are supported by genetic (including chromosomal) and anatomical differences, phylogeography, and life history adaptations (Geist 1971, Valdez 1982, 2011a,c,d; Valdez et al. 1978, Nadler et al. 1973a,b; Schaller 1977, Bunch et al. 1999, Dotsev et al. 2019, Groves and Grubb 2011, Damm and Franco 2014). The three lineages include the moufloniforms (two species with 2*n*=54 or 58) of the Middle East and nearby regions, the argaliforms (one species with 2*n*=56), occurring principally in central Asia, and pachyceriforms (three species with a 2*n*=52 in the Asian species and 2*n*=54 in mountain sheep species in North America) of Siberia and North America. Species of the moufloniform and argaliform subgenera have long-legged antelopian body shapes adapted for undulating terrain in a variety of habitats (Geptner et al. 1961, Schaller 1977, Valdez 1982, Geist 1985, 1991; Damm and Franco 2014). Members of the argali lineage additionally have the greatest biomass and longest horns, with maximum recorded horn lengths of 191 cm (Fedosenko and Blank 2005). The three species composing the pachyceriforms include the snow sheep (*O. nivicola*) from eastern Siberia and Dall's thinhorn (*O. dalli*), and bighorn (*O. canadensis*) mountain sheep of North America (Rezaei 2010, Chen et al. 2021), all of which are adapted for rugged, precipitous terrain (Figures 1.1–1.3).

Based on chromosomal differences, Nadler et al. (1973b) proposed that moufloniform and argaliform species may have occurred in ringed distributions during the Pleistocene with repeated cycles of isolated and overlapping populations leading to hybridization, introgression, and genetic differentiation of populations. Other genetic studies have revealed introgression within pachyceriforms and between pachyceriforms and argali lineages, indicating these lineages probably had larger and overlapping geographical distributions during the Pleistocene (Upadhyay et al. 2021). All extant mountain sheep species can interbreed with others in captivity, and wild populations can potentially form hybrid zones (Pulling 1945, Young and Manville 1960, Gray 1972, Hoefs and Nowlan 1997, Bunch 1978, Valdez et al. 1978, Bagirov et al. 2012, Buchalski et al. 2016), which is evidence of their close phylogenetic relationships. In these hybrid zones, populations that differ genotypically and phenotypically from parental populations may form but also continue to admix through hybridization (Chapter 6). In summary, the genus *Ovis* originated in Asia where several lineages evolved and radiated across Europe, the Asian Middle East, northern India, southwest and central Asia, and Siberia, and subsequently dispersed to North America. During this evolutionary dispersal, at least six extant wild species evolved, presumably by adapting to heterogeneous environments. There was extensive hybridization among species resulting in introgression and instilling populations with additional genetic and phenotypic variation. Whole-genome sequencing reveals a complex genetic introgression history and the genetic basis of mountain sheep physiological and anatomical adaptations (Upadhyay et al. 2021). Wilson and Mittermeier (2011) and Castelló (2016) provide color illustrations of the wild sheep species and subspecies of the world.

## 1.3 AN OVERVIEW OF THE ORIGINS OF MOUNTAIN SHEEP IN NORTH AMERICA

The origin and evolution of North American mountain sheep have been a research topic of interest and conjecture. Determining genetic interrelationships and adaptations to specific environments is integral to specifying subspecies taxonomy and delineating subspecific distributions.

**FIGURE 1.1**  An adult female Dall's sheep and lambs, part of a female-lamb group, in the Kluane National Park, Yukon Territory, Canada. (Photo by Sonny Parker; used with permission.)

FIGURE 1.2 The most massive horns recorded for a North American mountain sheep, based on the scoring system of the Boone and Crockett Club (Lehr and Schwab 2023), are of this Rocky Mountain bighorn male, from Wildhorse Island, Montana. The horns retain their massiveness throughout their length. It died a natural death in 2016. (Photo by L. Victor Clark; used with permission.)

Such understanding, in turn, has conservation and management implications. Perhaps the earliest speculation of the origin of North American mountain sheep was that of Cuvier (1817), who proposed that mountain sheep evolved from an Asian argaliform (*Ovis ammon*) ancestor that reached North America over an ice bridge that connected both continents. Cowan (1940) suggested that mountain sheep evolved from a snow sheep ancestor. Glacial barriers divided mountain sheep into a northern Alaska population that evolved into Dall's sheep and a southern population in the Rocky Mountains and further south that evolved into the bighorn sheep. Furthermore, these disjunct populations may have formed sympatric populations during an interglacial event "with results impossible to determine" (Cowan 1940: 507), a prescient speculation (Figures 1.4–1.6).

Mountain sheep diversified into species and subspecies after being isolated from an Asiatic ancestor in ice-free refugia in northwestern North America during the Pleistocene. Their subsequent dispersal into southern climatically diverse habitats also provided refugia. Continental glaciers formed a barrier that isolated northern mountain sheep populations in Alaska and northwestern Canada from populations further south. Mountain sheep and other species in southern refugia expanded their distribution northward to recolonize the northern American continent (Shafer et al. 2010, Hewitt 2011, Deakins et al. 2019) after the periodic melting of glacial barriers. Molecular genetic studies have been pivotal in clarifying the temporal and geographic complexities of the processes that produced the anatomical, physiological, and genetic differentiation and adaptations in mountain sheep in North America. Cycles of isolation and admixture during glacial and interglacial periods have been influential in thinhorn subspecific relationships. The evolutionary history of bighorn sheep has also been influenced by glacial cycles and associate habitat changes in southern refugia (Chapter 6). Even though mountain sheep

FIGURE 1.3 A subadult male and female Stone's sheep photographed near Fort Nelson, British Columbia. Stone's sheep have a variable body coloration. Note the variation in rump patch size and distinct black tail. (Photo by Darryn Epp; used with permission.)

**FIGURE 1.4** A herd of four California bighorn sheep photographed near Kamloops, British Columbia. (Photo by Bill A. Jex; used with permission.)

**FIGURE 1.5** An adult male and female desert bighorn sheep in the Muddy Mountains Wilderness in southern Nevada. (Photo by L. Victor Clark; used with permission.)

FIGURE 1.6 A male group of desert bighorn sheep from the central region of the Baja California Peninsula, Mexico, near the shoreline of the Gulf of California. Note the broken right horn of the foremost male. (Photo by Enrique de Jesús Ruiz Mondragón; courtesy of the Fundación Universidad Autónoma de Baja California, A. C.)

populations in North America were isolated and, in some cases, continue to be like their Asian congeners, populations with adjacent ranges hybridize and signatures of this admixture, evident in their genes, have contributed to their evolution.

Genetic studies (Worley et al. 2004, 2006; Sim et al. 2016, 2019, Chapter 6) have clarified Dall's thinhorn sheep origins, phylogeographic structure, infraspecific relationships, and range distributions. Thinhorn sheep taxa referred to as Dall's sheep, Stone's sheep (*Ovis dalli stonei*), and Fannin's sheep are populations that have admixed genetics and diverse color expressions. Phylogenomic analyses have further supported these relationships. Stone's sheep have had episodes of ancient hybridization with bighorn sheep, and several genomic fragments show that these hybridization events involved the retention of genes associated with melanogenesis, and these genes are thought to have influenced the darker pelage observed in Stone's sheep (Santos et al. 2021). Santos et al. (2021) estimated that the mountain sheep in North America diverged from the snow sheep approximately 2.85 MYA and that the bighorn sheep diverged from the thinhorn sheep, both Dall's and Stone's, about 2.50 MYA (Figure 1.7). These divergence times, coupled with the even more recent divergence times of the subspecies of bighorn sheep, reveal a rapid radiation of sheep after colonization of North America. Perhaps human predation and anthropogenic habitat modifications before European colonization may have been factors in the process of genetic differentiation in some populations (Darimont et al. 2009, Guthrie 2017, Ålund et al. 2023, Chapter 2). Chapter 6 provides a detailed review of mountain sheep evolution in North America.

## 1.4 TAXONOMY AND CLASSIFICATION

Biological taxonomists name and describe taxa (groups of organisms recognized at some level of classification) and classify organisms in categories (levels of classification). The modern concept of animal classification is based on the Linnaean system, named for Carolus Linnaeus (Blunt 2001), a Swedish botanist who consistently applied binomials (genus and specific epithet) to designate a species beginning with the tenth edition of *Systema Naturae* (Linnaeus 1758). Taxonomists organize the seemingly confusing diversity of life into nested sets of taxa that can be placed into a comprehensible framework (Thomson et al. 2018). The recognition of taxa enables biologists to ask additional questions regarding the specifics of biodiversity and the ecological and evolutionary processes that have created and sustained it, and that clarify genetic relationships and enlarge our knowledge of evolution. Taxonomy is of fundamental importance in many aspects of biological sciences, including ecology and genetics, and has practical applications in wildlife law enforcement, wildlife disease management, medicine, agriculture, and bioeconomics. There is urgency in documenting the diversity of life because of the accelerating deleterious anthropogenic effects on the world's fauna and flora, resulting in the reduction and extirpation of wild populations.

Taxonomists classify extant and extinct organisms into a structural hierarchy, a multilevel nested categorical system in which a taxon is placed in a larger group of related organisms and then in smaller lower groupings based on common anatomical and genetic characteristics that reflect the evolutionary lineage as based on phylogenetic relationships.

**FIGURE 1.7** The snow sheep is the closest living Asian wild sheep relative of mountain sheep in North America. Note the similar coloration to bighorn sheep and similarity of horn shape to those of thinhorn sheep but differ from American forms in that the rump patch does not extend above the base of the tail. (Photo by Yuri Yarovenko; courtesy of Gerhard Damm and the CIC Caprinae Atlas of the World.)

Progressing down the taxa from higher to lower levels, there are progressively more similar diagnostic characteristics because of increasing relatedness. The hierarchical arrangement with decreasing number of species at each lower level of categories is evident in the classification of extant mountain sheep at the order to tribe categories. Wild sheep are ungulates (hoofed mammals) classified in the following categories and taxa: order Artiodactyla (~240 species); suborder Ruminantia (ruminating ungulates; 193 species), infraorder Pecora (horned [bovids] and antlered [cervids] ungulates: 184 species); family Bovidae (horned ungulates: 143 species); subfamily Caprinae (sheep, goats, aoudad, chamois, and related species: ~35 species); and tribe Caprini (sheep and goats and relatives: 15 species). Organisms classified in a genus represent closely related species with shared characters that were derived from a last common ancestor. In the case of the hierarchical arrangement of wild sheep, there is a decreasing number of species from the order to the tribe level. At the genus level, there may be one (monotypic) or more (polytypic) species. Sheep are classified in the genus *Ovis* and includes several species. The number of species and subspecies can be highly variable. The criteria for recognizing species and subspecies may include ecological, behavioral, anatomical, and genetic characteristics depending on the taxonomist's concept of species and subspecies.

part of a clade often combined in the order Cetartiodactyla. Although cetaceans and even-toed ungulates share a most recent common ancestor and are therefore monophyletic, the validity of this term has been questioned as being unnecessary and misleading. Cetaceans and even-toed ungulates are not sister taxa (the closest relative[s] of another given unit in an evolutionary tree); cetaceans are actually an embedded clade within a branch of even-toed ungulates (Prothero et al. 2022). Therefore, some zoologists place them both in Artiodactyla. Here I use the term Artiodactyla to include only the even-toed ungulates and exclude the cetaceans. Artiodactyls are the most diverse group of hoofed mammals; these include sheep and goats, cattle, deer, antelopes, giraffe (*Giraffa* spp.), okapi (*Okapi* spp.), pronghorns (*Antilocapra americana*), swine and peccaries (*Pecari* spp.), camels (*Camelus* spp.) and llamas (*Lama* spp.), hippos (*Hippopotamus* spp.), tragulids (*Tragulus* spp.; chevrotains), and moschids (*Moschus* spp.: musk deer). Artiodactyls have a cosmopolitan distribution and are the source of most domesticated ungulates, including cattle, swine, camels, goats, and sheep. In terms of biomass, artiodactyls constitute approximately 50% of the combined biomass of terrestrial wild mammals and also constitute the greatest biomass of domestic animals (Greenspoon et al. 2023).

## 1.5 ORDER ARTIODACTYLA

Wild sheep are hoofed mammals classified in the order Artiodactyla (even-toed or cloven-hoofed ungulates; >200 spp.). Artiodactyls and cetaceans (whales and dolphins) are

## 1.6 SUBORDER RUMINANTIA

The suborder Ruminantia (ruminating ungulates; 193 species) includes all of the above except camels, swine, peccaries, and hippos. Ruminants are a diverse and evolutionarily successful group; they constitute the highest number of ungulate species. Most possess sexually dimorphic head-gear (permanent horns of bovids, usually in both males and females; antlers of deer, which are deciduous and except for caribou [*Rangifer tarandus*], possessed only by males; pronghorn: horn sheaths, which are shed by males and females; and ossicones: bone processes covered with skin and hair in giraffes and okapi); and all possess a multi-chambered stomach, which facilitates the digestion of plant cellulose with the aid of protozoans and bacteria. They occur from forest to desert, tropical to Arctic, and lowland to alpine and are highly variable in mass ranging from 2 to 1,200 kg (Gentry 1990, Bubenik 1990, Etnyre et al. 2011, Castelló 2016, Wang et al. 2019).

## 1.7 INFRAORDER PECORA AND FAMILY BOVIDAE

The infraorder Pecora includes the horned (males and most female bovids) and antlered (restricted to male cervids or deer, with exception of the caribou [*R. tarandus*], and the antlerless water deer [*Hydropotes inermis*]) ungulates, although species in the families Moschidae and Tragulidae do not have cranial appendages. The family Bovidae, or hollow-horned ruminants, is the most diverse family of ungulates with 143 species (Grubb 2005), has a widespread distribution, occurring in Africa and Eurasia, and in North America in the Arctic, in boreal and temperate forests, and in grasslands and deserts. The term hollow-horn or cavi-corn refers to horns with a bony core into which the frontal sinuses often extend to form air spaces. Groves and Grubb (2011) recognized 279 species of bovids, including 20 species of wild sheep by elevating ecotypes and subspecies to species status; however, this increase in species numbers has been criticized for unjustified taxonomic inflation (Heller et al. 2013, Zachos et al. 2013, Zachos 2015). I use the basic classification, especially as applied to wild bovid species, of Schaller (1977), Shackleton and Lovari (1997), Wilson and Reeder (2005), and Damm and Franco (2014).

Bovids include cattle, bison (*Bison bison*), yak (*Bos grunniens*), antelopes, sheep, goats, and related species. They lack upper incisors and canines, lower canines are incisiform, premolars are molariform, cheek teeth usually hypsodont (high-crowned), there are hollow sinuses in the frontal bone of the skull that extend into the horn core, and strikingly sexually dimorphic (Hall and Kelson 1959, Groves and Grubb 2011, Groves and Leslie 2011). The greatest diversity of extant bovids (about 75 spp.) inhabit Africa, followed by Eurasia, but were absent in Central and South America and Australia and New Zealand before the introduction of domestic (some now in feral populations) and wild bovid species (Wilson and Mittermeier 2011, Kingdon 2018, Baker and Gynther 2023, Webb and Blincow 2024, Aulager et al. 2025). North America is depauperate in bovid species as compared to the diversity of bovids in Eurasia. There are five bovid species in North America as compared to 48 species in Eurasia including ten of the tribe Bovini (wild cattle and related species), twelve antelope species, four sheep species, eight goat and seven related species (Grubb 2005, Groves and Leslie 2011), and 3 goral (*Naemorhedus* spp) and 4 serow (*Capricornis* spp; Kays and Wilson 2009, Mori et al. 2019). Wild bovids in North America include the musk ox (*Ovibos moschatus*), Rocky Mountain goat (*Oreamnos americanus*), bison, and the two mountain sheep species (Hall 1981, Wilson and Mittermeier 2011). Wild bovids are some of the most economically important ungulates because of the high number of species that are hunted for sport, commercial, and subsistence purposes in Africa; have widespread ecotourism value; and the numerous montane bovid species of high economic value that are sport-hunted in Eurasia, including wild sheep and goats, and the two mountain sheep species in North America (Holechek and Valdez 2018, Adhikari et al. 2021, Lehr and Schwab 2023, Parker et al. 2023). Mountain sheep are the highest-priced big game trophies in North America (Chapter 17). Also, bovids are an important food source for many predator species. Most bovid species readily breed in zoos and 39 species of antelopes from Africa or Asia have been established in private ranches in the United States, principally in Texas, but also several species in Mexico. Populations of foreign bovid species have also been the source of animals for reestablishing wild populations in their native habitats (Mungall and Sheffield 1994, Valdez 2019, Mungall 2000, 2024).

## 1.8 SUBFAMILY CAPRINAE AND TRIBE CAPRINI

The subfamily Caprinae is composed of sheep, goats, aoudad (erroneously named Barbary sheep; *Ammotragus lervia*), bharal or blue sheep (*Pseudois nayaur*), and relatives (~35 species). Other species include Eurasian chamois (*Rupicapra* species), musk ox (*Ovibos* spp.) and mountain goat, which is not a true goat. Wild sheep (genus *Ovis*) and goats (genus *Capra*) attain their greatest species diversity in temperate Asia. Bharal (*P. nayaur*; $2n=54$) are restricted in distribution to northern Asia (Schaller 1977, Bunch et al. 1999). The tribe Caprini consists of wild sheep and goats and close relatives (Figure 1.8). Wild sheep have a widespread distribution in Eurasia and North America but attain their greatest species diversity in Asia (Schaller 1977, Groves and Grubb 2011, Valdez and Weinberg 2011, Damm and Franco 2014, Castelló 2016, Li et al. 2022). Wild sheep can be distinguished anatomically from other species of the tribe Caprini by the presence of preorbital (in front of the eye), inguinal (groin), and interdigital (between the hoofs) glands, and a short tail not exceeding 15.2 cm. Characteristic of all genera in the chromosomal evolution of the tribe Caprini has been a series of centric fusions resulting in a reduction in the chromosome number (Bunch et al. 1999, Chapter 6). Adult male mountain sheep horns grow in a tapering curl or spiral, with a maximum recorded

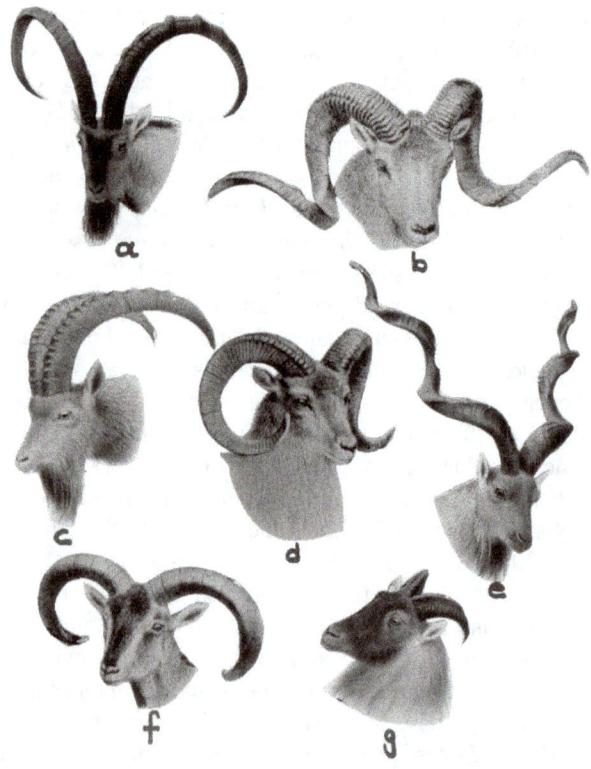

**FIGURE 1.8** Horn shape diversity of adult males of species in the tribe Caprini (sheep, goats, and relatives) from Asia: (a) wild or bezoar goat (*Capra aegagrus*), (b) Marco Polo argali (*Ovis ammon*), (c) Himalayan ibex (*Capra sibirica*), (d) urial sheep (*Ovis vignei*), (e) markhor (*Capra falconeri*), (f) bharal (*Pseudois nayaur*), and (g) Himalayan tahr (*Hemitragus jemlahicus*). (Reproduced from Lydekker 1907.)

horn length in American sheep of 131 cm in a Dall's sheep and a maximum basal circumference of 44.5 cm in a Rocky Mountain bighorn sheep (Lehr and Schwab 2023); female horns are short (rarely exceeding 30.5 cm), slender (basal circumference up to 15.2 cm), and slightly arched (Chapter 4). Wild sheep and goats are genetically closely related and have a common ancestor, having diverged into separate lineages about 5–7 MYA (Mereu et al. 2019). Wild goats (nine spp.) are endemic to Eurasia and northern Africa (Schaller 1977, Valdez 1985, Damm and Franco 2014). Goats differ from sheep in having subcaudal glands, a uniform $2n=60$, lack the forementioned glands of sheep, and adult males have a noticeable beard (i.e., goatee). Also, unlike wild sheep, wild goat species have a variety of horn shapes, from the sickle-shaped horns of the wild goat (*Capra aegagrus*) and the upright spiraling horns of markhor (*Capra falconeri*) to the supracervical (curving above the neck) horns of the east Caucasian tur (*Capra cylindricornis*; Figure 1.8; Valdez 1985, Damm and Franco 2014). Most species of wild sheep and goats occur in depleted populations and under protective national and international laws and treaties (Valdez and Weinberg 2011). Wild goat species readily interbreed in captivity. Private ranches in the United States, principally in Texas, have established populations of several

species of Eurasian wild goats and their hybrids, principally for sport hunting purposes (Mungall 2000, 2024).

Hybridization of species classified in separate genera indicates a close genetic and evolutionary relationship. Under captive conditions, domestic sheep (derived from the Eurasian mouflon [*Ovis gmelini*]) and the domestic goat can hybridize producing fertile offspring (Bunch et al. 1977, Pauciullo et al. 2016). Van Gelder (1977) proposed that goats and sheep be placed in the same genus (*Capra*) because they are interfertile; however, crosses are usually unsuccessful due to genetic incompatibility (Pauciullo et al. 2016). Also, there are no documented cases of wild sheep and goats producing hybrid offspring in the wild in Asia, where species of both genera are sympatric (overlapping in distribution). Goats, the aoudad ($2n=58$), and bharal are closely related (Hight and Nadler 1976, Schaller 1977, Valdez and Bunch 1980). The domestic goat and aoudad can hybridize in captivity (Bunch et al. 1977). The domestic goat is derived from the wild or bezoar goat (*C. aegagrus*; Colli et al. 2015) with its distribution centered in the Middle East (Pidancier et al. 2006, Weinberg and Amberli 2020). A bezoar goat population has been established in the Florida Mountains of southcentral New Mexico as a game species by the state game agency (Bavin 1975); attempts to establish a free-ranging population of Siberian ibex (*Capra sibirica*) failed (Mungall 2024). Invasive feral domestic goat populations have a cosmopolitan distribution and are of concern to wildlife biologists because of their propensity to degrade native vegetation (Hess et al. 2018) and are potential competitors of bighorn sheep and a reservoir of mountain sheep diseases (Krausman et al. 1999). Aoudad are native to North Africa (Cassinello et al. 2021) but have been introduced and have established invasive wild populations in Europe, principally in southern Spain, and in arid North America in the United States and northeastern Mexico, where they are abundant; populations are widespread and expanding their range into mountain sheep habitats (Valdez 2011b, Damm and Franco 2014, Mori et al. 2017). Aoudad populations are of particular concern to mountain sheep biologists in North America because of the potential competition for food and space resources and as a reservoir for respiratory pathogens such as *Mycoplasma ovipneumoniae* that can be transmitted to mountain sheep and result in disease-related die-offs (Wright et al. 2024, Parikh et al. 2025, Chapter 8).

## 1.9 GENUS *OVIS*

The genus *Ovis*, in which all true wild sheep are classified, is one of seven genera in the tribe Caprini (Grubb 2005). *Ovis* is a complex taxon and one of the most widespread, if not the most widespread, bovid genus in the world, occurring in eastern Europe, widespread in the Middle East, Southwest and Central Asia (including northern India), Siberia, and from Alaska to Mexico (Nadler et al. 1973b, Valdez 1982, Geist 1971, Groves and Grubb 2011, Valdez and Weinberg 2011, Damm and Franco 2014). Groves and Grubb (2011) noted that the method of fighting, facilitated by broad-based horns, allows for maximum power of impact (Chapter 4). Also, unlike other ungulates, *Ovis* adult males have enlarged basal horn circumferences on horns that curl in a spiral growth pattern. All wild sheep species are sexually dimorphic, with males having more massive and longer horns than females (Damm and Franco 2014). The genus *Ovis* in North America can be anatomically and genetically differentiated from other North American bovid genera by the massive curling horns in adult males; large, distinct, white rump patch which extends above the base of the tail in both sexes (not distinguishable in Dall's sheep because of the uniformly white body except for some individuals with a black tail), rectangular dark pupils, yellow irises, the presence of preorbital and interdigital glands, and a $2n = 54$ (Valdez 1982, Krausman and Bowyer 2003, Chapter 4).

## 1.10 WILD SHEEP SPECIES AND SUBSPECIES CONCEPTS

The species is the basic category of the taxonomic framework and a component of biodiversity (Ruedas et al. 2024). Species concepts have important implications for biodiversity conservation (Frankham et al. 2012, Sandall et al. 2023). A basic component of a species definition requires that the members of a species in the wild breed with one another and produce fertile offspring. Generally, hybrids will have reduced viability or fertility, and therefore are reproductively isolated. Naturally occurring hybrid zones do exist and can be interpreted to indicate conspecificity or to be comprised of separate species, depending on the criteria used. With the advent of molecular technology, the species concept underwent a broader interpretation, by adding a genetic and phylogenetic component. The genetic concept emphasizes genetic isolation rather than reproductive isolation. The most robust genetic and phylogenetic studies integrate morphological and genetic criteria (Wheeler 2024). Baker and Bradley (2006), De Queiroz (2007), Zachos (2016), and Bradley and Dowler (2019) provide a review of species concepts.

Mammalogists have differences of opinion regarding the number of wild sheep species recognized, varying from 1 to as many as 20 species (Lydekker 1913, Valdez 1982, Shackleton and Lovari 1997, Grubb 2005, Groves and Grubb 2011). I adopt the classification of Rezaei et al. (2010) and Damm and Franco (2014), who recognize six wild sheep species. Wild sheep species can be differentiated by pelage color and hair patterns, horn shape and metrics, cranial anatomy, genomic diversity, diploid chromosome number, and geographic distribution. The variation in the number of wild sheep species recognized is partly due to different interpretations of species concepts but also to the unique evolutionary history of the genus, which has resulted in numerous differentiated populations with high levels of genetic and phenotypic variation, in diverse habitats, and nontrivial levels of recent and earlier hybridization.

There has been a trend among American and European mammalogists since the late 20th century in recognizing two mountain sheep species in North America: bighorn and thinhorn (Cowan 1940, Hall and Kelson 1959, Hall 1981, Valdez 1982, Shackleton and Lovari 1997, Grubb 2005, Groves and Grubb 2011, Damm and Franco 2014). Dall's thinhorn and bighorn sheep are differentiated by coat color (white in Dall's except for a black tail in some individuals, shades of grizzled white to dark coloration in Stone's and shades of brown in bighorn), average greater body and cranial measurements in bighorn sheep males and females, and more massive horns and a shorter horn tip to tip spread in bighorn males (Cowan 1940, Hall 1981, Valdez and Krausman 1999, Damm and Franco 2014, Castelló 2016, Lehr and Schwab 2023, Chapter 4).

Subspecies are populations or groups of populations within a species and that are anatomically or genetically distinct and occupy one or more specific allopatric geographical distributions or are sympatric in a narrow zone or zones of intergradation with one or more subspecies. Thus, subspecies can denote distinct intraspecific phenotypic and genetic geographical variation. It is especially difficult to perceive unequivocal boundaries between most mountain sheep subspecies in North America. During the early Holocene and until the European colonization of North America, when mountain sheep populations were probably regionally more continuously distributed, phenotypic characters may have exhibited clinal variation (a gradual change in an anatomical attribute or attributes in a species' geographical range). In such a case, it would have been arbitrary to designate subspecies boundaries using morphological criteria. There are differences of opinion on the value of the criteria used to designate subspecies, which questions the usefulness of subspecies designations (Torstrom et al. 2014). The use of genetic data since the 1980s has been instrumental in detecting infraspecific variation, but has not resolved the subspecies taxonomic conundrum because of a lack of a standard method to determine subspecies boundaries. Determining mountain sheep subspecies and infraspecific genetic variation in North America is an ongoing research agenda (Chapter 6).

The mountain sheep specimens deposited in American museums, most of which were collected from 1805 to 1920 during zoological discovery in western North America, became the basis for documenting their diversity and distribution (Figure 1.9). During the early phase of documenting mountain sheep diversity, biologists described and named mountain sheep species and subspecies based on geographically restricted and small samples of specimens, in some cases based on only one specimen, which was the case with Sheldon's desert bighorn (O. sheldoni Merriam 1916). Zoological taxonomy during the 19th and early 20th centuries was based on the concept of typology; species were aggregations of individuals based on the common characteristics that agreed with the author's diagnosis. The significance of geographical anatomical variation and the spatial and temporal genetic and biotic and abiotic environmental components involved in evolution, including speciation, was not understood. Moreover, early descriptions of mountain sheep were based on body mass, skull and horn metrics,

**FIGURE 1.9**  This photograph of a mountain sheep collecting expedition for the United States National Museum to the Grapevine Mountains, Inyo County, California, was taken in November 1924. Museum specimen collectors often did not limit the number of specimens collected even though mountain sheep populations were in decline. The individual on the left is J. R. Acorn, a then employee of the California Game and Fish Department, and on the right is E. H. Ober, a California game warden. (Photo courtesy of L. Victor Clark; used with permission.)

and pelage coloration differences, and restricted to those of males, all of which are variable characters even within a population and do not necessarily define evolutionary units. Also, dimensional morphological traits that arise from genetic with environmental interaction and size-based morphometric traits can be ecophenotypic if the environment (e.g., forage quality and quantity or nutrient availability) exerts a greater role than genetics. Heredity-based variation should thus be corroborated by independent genetic information (Wehausen and Ramey II 2000), a technology for which was then not available.

## 1.11  NOMENCLATURAL AND TAXONOMIC HISTORY OF NORTH AMERICAN MOUNTAIN SHEEP SPECIES AND SUBSPECIES

The correct scientific name of bighorn sheep was a contentious issue for over a century. There was uncertainty concerning the correct scientific name for the Rocky Mountain bighorn because it was unknown which of three proposed names published in 1804 had priority (which one of the three names had been published first). The three names were used interchangeably in zoological literature into the early 20th century. According to the rules of zoological nomenclature, the first published binomial designation for a species takes priority over other published names. The bighorn was designated as a new species in 1804, based on three published names: Canadian sheep (*O. canadensis*) by George Shaw, a British zoologist, accompanied with a written description and a color illustration of a male bighorn; deer-like sheep (*Ovis cervina*) by A. G. Desmarest, a French zoologist, accompanied with an illustration; and mountain sheep (*Ovis montana*) by J. C. D. von Schreber, a German zoologist, with a written description (Osgood 1913). Osgood (1913) opined that *canadensis*, even though he did not determine dates of publication, had priority because it was the only name with both a written description and an illustration. Subsequently, Osgood (1914) determined that *canadensis* had been published earlier (early February 1804) than either *cervina* (around March 1) or *montana* (April 1). Therefore, *O. canadensis* is the correct species designation for Rocky Mountain and desert bighorn sheep. Equally confusing was the designation of the common English names for bighorn sheep (Stewart 1935), which included argali (an Asian sheep species: *O. ammon*) and ibex (common name of several Eurasian wild goat species) because of the early uncertain relationships of American and Eurasian mountain sheep and goat species. Thinhorn sheep common names included white sheep and Alaskan sheep for Dall's sheep and black, Fannin's and Liard sheep for Stone's sheep, among others (Seton 1927). In the 1900s, mountain sheep became the standard collective name for North American species and bighorn for Rocky Mountain and desert bighorn. Thinhorn sheep is the collective name specifically for Dall's and Stone's sheep.

The color illustration (Figure 1.10) accompanying Shaw's (1804) original description of the Rocky Mountain bighorn was based on the horns and a skin of an adult male specimen sent to the Natural History Museum in London, England, by Duncan McGillivray, a fur trader who collected specimens in 1800 in the Rocky Mountains near the Bow River in the vicinity of the present location of Exshaw, Alberta, Canada, which is the type locality for the species. McGillivray and his Native American guide collected five specimens, including a mature male shot by McGillivray. The first illustration of a Rocky Mountain bighorn sheep in North America was published in *The Medical Repository*, an American popular publication, in 1803, also based on the horns and skin of a male provided by McGillivray, probably the mature male shot by him (Mitchill 1803, Chapter 2).

The first naturalists to question the then prevailing classification of mountain sheep were Audubon and Bachman (1846–1853), who, using the name *O. montana* for the bighorn, synonymized *Ovis californiana*, named by David Douglas in 1829, type locality from Washington State, and *montana* (=*canadensis*). They based their conclusion on the continuous distribution of bighorn sheep in northern North America, similarity in climate of the two type localities, and they dismissed the purported differences in coat color between northern sheep populations. Hornaday (1908), after comparing body and horn measurements of seven males collected in the Pinacate region of the Sonoran Desert in northwestern Mexico with males from the northern Canadian Rocky Mountains (Hornaday 1906), concluded that desert bighorn sheep were smaller due to the lower quality desert forage and inbreeding owing to the isolation of desert populations in discontinuous mountain ranges. Hornaday (1908), acknowledging ecotypic differences, considered the names of northern (*O. canadensis* Shaw) and desert bighorn (*O. nelsoni* Merriam 1897) to be synonymous but retained the Mexican desert bighorn sheep (*Ovis mexicana*) as a separate species.

The intraspecific taxonomy of bighorn sheep has been complicated, especially by the many named taxa. Between 1897 and 1937, there were five subspecies proposed for Rocky Mountain and seven subspecies for desert bighorn sheep. Some of the names, now subsumed under subspecies, were first proposed as North American mountain sheep species. Eight of the named wild sheep taxa were determined to be synonyms of taxa previously described

**FIGURE 1.10** This painting was included in the original scientific description of the Rocky Mountain bighorn sheep by George Shaw in 1804. The illustration was based on the horns and skin of a specimen collected in the Canadian Rockies in 1800. The painter assigned a deer body conformation. Refer to the first published illustration (Chapter 2) of a mountain sheep in the United States in 1803.

(Allen 1912, Cowan 1940, Hall 1981). Most of the taxonomic confusion created from the numerous named species and subspecies was the outcome of naming taxa based on few specimens from restricted localities, a nebulous concept of species and infraspecific delimitations, and a lack of knowledge of anatomical geographical variation. Also, molecular analysis was not available to use on interspecific and infraspecific genetic variation.

Allen (1912) was the first zoologist to comprehensively review the species, subspecies, synonymies, type localities, and geographical distributions of mountain sheep in North America. Cowan (1940) conducted the first comprehensive quantitative taxonomic revision of mountain sheep in North America, basing subspecies classification on horn shape, pelage coloration, and with an emphasis on male craniometric differences (with limited statistical analysis), as summarized by Buechner (1960: Table 1.1). Cowan (1940) grouped taxa based on the classification of Allen (1912) and recognized three subspecies of Dall's sheep (*dalli, kenaiensis, stonei*), three subspecies of northern bighorn sheep (*canadensis, californiana, auduboni*), and four subspecies (*nelsoni, mexicana, cremnobates, weemsi*) of desert bighorn sheep. He acknowledged that his classification was based on inadequate sample sizes and that only one museum had a large enough sample size to statistically analyze for individual variation: "in only one or two instances are series available to illustrate differences due to sex and age" (Cowan 1940: 510); also, he did not specify how he determined age categories and probably erred in determining the age of sample specimens. He provided a map showing the probable distribution of subspecies with boundaries

that he recognized were based on unsatisfactory data, and on which he designated potential hybrid zones (Cowan 1940). Cowan's taxonomic study of mountain sheep based on comparisons of skull and horn measurements among populations was the basis for subspecies designations and their distributions for over 50 years. The initial taxonomic publications of Ramey II (1993, 1995) and Wehausen and Ramey II (1993, 2000) were the stimulus for the subsequent and ongoing mountain sheep systematic studies in North America (Chapter 6).

Refer to Hall (1981) for a detailed review of North American mountain sheep taxonomy including synonymies, type specimen localities, and specimen collecting localities of species and subspecies based on the classification of Cowan (1940); also included is a revised map based on arbitrary subspecies distributions. Nineteenth-century illustrations of bighorn sheep include those in the original bighorn description in Shaw (1804), in Godman (1826) based on specimens collected by the Lewis and Clark expedition members (Figure 1.11), and in Audubon and Bachman (1845–1848) which included a color painting of a male and female bighorn sheep based on specimens collected and mountain sheep observed in the badlands of the Upper Missouri River (Figures 1.12 and 1.13) during Audubon's western expedition in 1843 (Audubon 1897). The first color illustration of a thinhorn sheep, a painting by Carl Rungius (Whyte and Hart 1985), a celebrated North American big game painter, was included in Hornaday's (1901) original description of Fannin's thinhorn sheep (Figure 1.14). For the first published illustration of a Rocky Mountain bighorn sheep, see Chapter 2.

## TABLE 1.1

### Classification of Mountain Sheep in North America (Species, Subspecies, and Partial Synonymies)

**Species and Subspecies**

**Thinhorn Sheep:** *Ovis dalli*

*Ovis dalli dalli* Nelson 1884 (Dall's thinhorn sheep)

*Ovis dalli stonei*[a] J. A. Allen 1897 (Stone's thinhorn sheep)

  Synonym: *Ovis d. kenaiensis* J. A. Allen 1902

**Bighorn Sheep:** *Ovis canadensis*

*Ovis canadensis canadensis* Shaw 1804 (Rocky Mountain bighorn sheep)

*Ovis canadensis californiana*[b] Douglas 1829 (California bighorn sheep)

*Ovis canadensis sierrae* Grinnell 1912 (Sierra Nevada bighorn sheep)

*Ovis canadensis auduboni* Merriam 1901 (Audubon's bighorn sheep)

*Ovis canadensis nelsoni* Merriam 1897 (Nelson's desert bighorn sheep)

  Synonyms of Nelson's desert bighorn: *Ovis c. mexicana* Merriam 1901

*Ovis c. cremnobates* Elliot 1904

*Ovis c. weemsi* Goldman 1937

*Refer to Chapter 6 or a Review of the Current Knowledge of Genetic Variation in Dall's and Stone's Sheep on Which Their Classification Is Based. The Classification of Bighorn Sheep Is Based Principally on the Speculative Revision of Epps et al. (Chapter 6) Founded on Current Knowledge of Genetic Variation. Audubon's Bighorn Has Not Been Reevaluated Genetically. Refer to Hall (1981) for a Complete Synonymy.*

[a] *Ovis dalli fannini* Hornaday 1901 (Fannin's thinhorn sheep) is based on an admixed population of *O. d. dalli* and *O. d. stonei*.

[b] *Ovis canadensi califoniana* as defined by Cowan (1940) is not supported, but recent analyses (Chapter 6 and Jahner et al. [2025]) support subspecies-level distinction for a portion of the range of that taxon.

FIGURE 1.11    This painting, published in Godman (1826), is based on mountain sheep specimens collected during the Lewis and Clark expedition (1804–1806).

FIGURE 1.12    Audubon and Bachman's book, *Viviparous Quadrupeds of North America*, included a color painting of an adult male and female mountain sheep based on specimens collected in the badlands of the Upper Missouri River in habitat of Audubon's bighorn sheep. Audubon observed mountain sheep during his western expedition in 1843.

**FIGURE 1.13** An adult male Audubon's bighorn sheep shot in the badlands of Montana in January 1882. Even at that early period mountain sheep populations were greatly diminished. Note the relatively steep but low elevation habitat typical of the badlands of the Upper Missouri River. (Photo by L. A Huffman; courtesy of the Montana Historical Society, Helena.)

## 1.12 ROCKY MOUNTAIN AND DESERT BIGHORN SHEEP TAXONOMY AND CLASSIFICATION

Ramey II (1993), based on a reanalysis of Cowan's (1940) data, validated the separation of Rocky Mountain and desert bighorn sheep but noted the lack of statistical support for recognizing the several named desert bighorn sheep taxa. Wehausen and Ramey II (1993), based on a morphometric analysis of male and female skull measurements, synonymized the peninsular (*O. c. cremnobates*) with Nelson's (*O. c. nelsoni*) desert bighorn sheep. Ramey II (1995), based on mitochondrial DNA (mtDNA) data analysis, noted low levels of differentiation in southwestern desert bighorn sheep populations and that their ranges were not consistent with the subspecies boundaries proposed by Cowan (1940). Subsequently, Wehausen and Ramey II (2000) conducted a taxonomic revision of bighorn sheep using discriminant function analysis and 17 skull measurements from 694 bighorn sheep specimens from the Great Basin in the United States, north to British Columbia and Alberta in Canada. Rocky mountain bighorn sheep are characterized by phylogeographic continuity, indicating there was a lack of biogeographic barriers that would have impeded gene flow on a regional scale. Wehausen and Ramey II (2000) synonymized *auduboni* (Merriam 1901) with *canadensis* and assigned formerly classified extant and extinct California (*californiana*) bighorn populations from British Columbia and Washington to *canadensis*. They assigned the Great Basin Desert form of Nelson's (*nelsoni*) desert bighorn sheep to include extinct native populations

**FIGURE 1.14** The original description of Fannin's thinhorn sheep (Hornaday 1901) included this color painting by famed big game painter Carl Rungius. (Reproduced from the Fifth Annual Report of the New York Zoological Society.)

from northeastern California, Oregon, southwestern corner of Idaho, and northern Nevada, formerly classified as California bighorn. Also, they restricted the distribution of California bighorn sheep to populations in the Sierra Nevada, the same populations which had previously been named *sierrae* by Grinnell (1912). These sheep have an affinity with desert bighorn sheep populations but are genetically unique relative to Great Basin populations (Wehausen and Ramey 2000 II). Wehausen et al. (2005) subsequently designated the bighorn sheep population in the central and southern Sierra Nevada as Sierra Nevada bighorn sheep (*O. c. sierrae*) because that name had nomenclatural priority. Wehausen and Ramey (2000) recognized two subspecies of Rocky Mountain bighorn sheep (Rocky Mountain bighorn sheep, Sierra Nevada bighorn sheep) and two subspecies of desert bighorn sheep (Nelson's bighorn sheep, Weem's bighorn sheep [*O. c. weemsi*]) that were adapted to arid environments.

Buchalski (2016) identified three divergent bighorn sheep clades corresponding to desert, Rocky Mountain and Sierra Nevada bighorn sheep. Buchalski et al. (2016) also recognized three desert bighorn evolutionary lineages generally in agreement with Cowan's (1940) desert subspecies designations in the United States and northern Mexico: *nelsoni*, *mexicana*, and *cremnobates*. These subspecific delineations were based on clearly defined mtDNA haplogroups and strong population subdivisions evidenced from mitochondrial and nuclear $F_{ST}$ analyses. There was notable admixture between adjacent but well-defined genetic groups, including *O. c. cremnobates* × *O. c. nelsoni* and *O. c. nelsoni* × *O c. mexicana*. These analyses revealed that different populations of bighorn sheep contain unique genetic diversity that contributes to the overall gene pool of the species, and these evolutionary units should be conserved to prevent a loss of intraspecific genetic diversity (Buchalski et al. 2016). Weem's desert bighorn sheep from the southern Baja California Peninsula is the southernmost subspecies of North American mountain sheep. Genetic and skull morphometric analyses suggest that desert bighorn sheep (*O. c. cremnobates, O. c. weemsi*) from the Baja California Peninsula are not significantly differentiated from mainland Mexico populations (Chapter 6). Based on genetic and morphometric studies, desert bighorn sheep can be considered a polytypic subspecies with *mexicana*, *cremnobates,* and *weemsi* synonymized with *nelsoni*. Refer to Chapter 6 for a detailed discussion of desert bighorn sheep taxonomic relationships.

## 1.13 DALL'S AND STONE'S SHEEP TAXONOMY AND CLASSIFICATION

Dall's and Stone's sheep were described as separate species but Stone's sheep was reclassified as a subspecies of Dall's sheep by Allen (1912). Four other names for supposed species or subspecies from northwestern Canada and relevant to the discussion of *O. stonei* were later proposed, including *fannini*, which was treated and named by Hornaday (1901) based on a specimen from near Dawson City and later classified as a pelage color variant of *stonei* by Allen (1912). Based on extensive field studies of Stone's sheep color variation throughout its range, Sheldon (1911) carefully mapped the distributions of dominant color phases of each region. He concluded that *fannini* was a color variant of Stone's sheep not meriting subspecific status.

Genetic studies of thinhorn sheep have significantly improved our understanding of the *Ovis dalli* populations' infraspecific genetic structure and subspecies' range distributions. Glaciations have been the key agents leading to subspecies differentiation of Dall's sheep and Stone's sheep populations. Patterns of subspecies genetic variation are generally consistent with genetic subspecies boundaries at the species level (Sim et al. 2016). Those studies also revealed three previously unreported Stone's sheep genetic clusters in Stikine-Skeena, Cassiar, and Rocky Mountains and identified a new geographic range for Stone's sheep that is much more restricted than previously determined and is almost exclusively confined within British Columbia. Populations in the Yukon and areas in British Columbia west of Teslin Lake previously described as Fannin's or Stone's sheep are admixed. In summary, Dall's, Stone's, and Fannin's sheep consist of populations that have admixed genetics and wide-ranging color expressions (Worley et al. 2004, Loehr et al. 2006, Sim et al. 2016, 2019). See Chapter 6 for a detailed discussion of the genetic relationships and evolution of thinhorn sheep.

## 1.14 CURRENT CLASSIFICATION OF NORTH AMERICAN MOUNTAIN SHEEP

Here I adopt the mountain sheep classification in North America based on the speculative revision of Epps et al. (Chapter 6) founded on current knowledge of genetic variation (Table 1.1). Based on a review of phylogenetic relationships within the two species of North American mountain sheep, Epps et al. recognize two subspecies of thinhorn sheep: *O. d. dalli* and *O. d. stonei*. Fannin's thinhorn sheep is an admixture of Dall's and Stone's sheep. Relative to Rocky Mountain bighorn sheep, Epps et al. (Chapter 6) recognize four subspecies of bighorn sheep (Rocky Mountain bighorn, California bighorn, Sierra Nevada bighorn, Audubon's bighorn [*O. c. auduboni*]) and consolidate all desert bighorn sheep populations into one subspecies: Nelson's bighorn (Chapter 6, Figure 6.2). Audubon's bighorn has not been reevaluated genetically. This classification differs from the traditional classification of Cowan (1940) that recognized three thinhorn subspecies and three Rocky Mountain bighorn sheep subspecies (*canadensis, californiana, auduboni*) and four desert bighorn sheep subspecies (*nelsoni, mexicana, cremnobates, weemsi*) and with subspecies distributions that differ from those of Epps et al. (Chapter 6, Figure 6.2). See Chapter 6 for a detailed review of the genetic relationships of North American mountain sheep subspecies and their distributions (Figure 1.15).

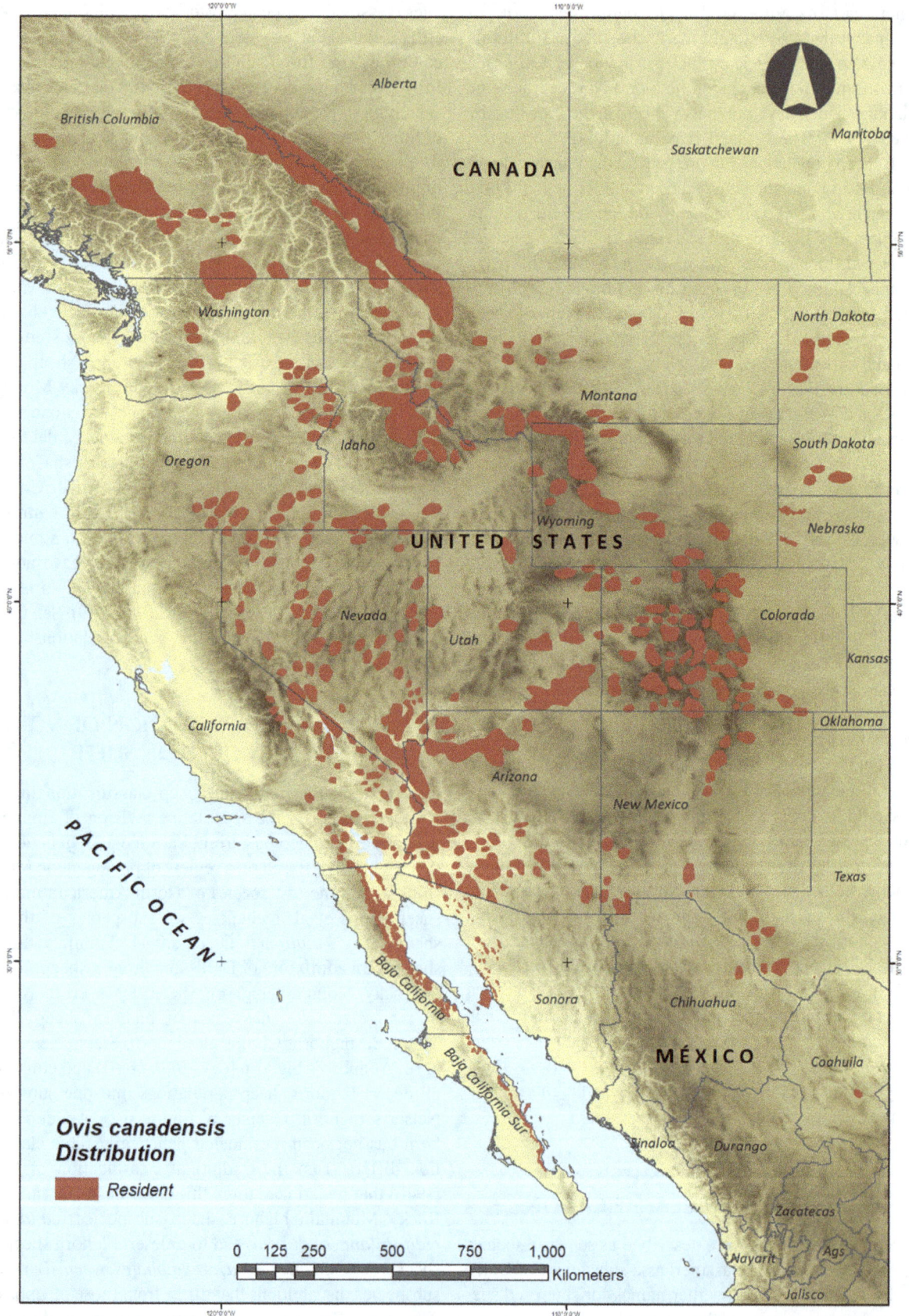

**FIGURE 1.15** Distribution of North American bighorn sheep modified from the Wild Sheep Working Group Initiative (2024). Mountain sheep in the Mexican states of Chihuahua and Coahuila were extirpated but reestablished in several populations in each state principally from captive populations. The single distribution areas noted in Chihuahua and Coahuila indicate the only free-ranging populations in each state. (Cartography by Ramiro Velázquez Rincón.)

## 1.15 DISTRIBUTION

Mountain sheep occupied the western North American mountain massifs from Alaska and Canada to northern Mexico west of the 100th meridian in the United States. Arid and semiarid environments in the United States occur west of this meridian compared to the more humid region to the east. Only Dall's sheep occur in Alaska and Dall's sheep and Stone's sheep and their hybrids occur in Canada. The distribution range of Dall's sheep extends from northern Alaska to northwestern Canada. Dall's sheep are most abundant and widespread in Alaska in eight principal mountain ranges: Brooks, White, Tanana, Alaska, Talkeetna, Wrangell, Chugach, and Kenai mountains. Dall's sheep in Canada occur west of the Mackenzie River on the Yukon-Northwest territories border, mountainous regions of the Yukon Territory, and northwestern British Columbia. Stone's sheep occur only in Canada, principally in northwest British Columbia, and extend from the Cassiar and Pelly mountains to the MacArthur Range (Festa-Bianchet 2020a, Chapter 3). For detailed maps of thinhorn sheep distributions for each Canadian and United States jurisdiction, refer to the Wild Sheep Initiative (2024).

Bighorn sheep occurred from the Canadian Rocky Mountains in western Canada primarily in southwestern Alberta and southeastern British Columbia to northern New Mexico; adjacent inhabited mountains included the Cascade Range in the states of Washington, Oregon, and northern California, Sierra Nevada Mountains in eastern California; in arid regions of the United States (Mohave, Great Basin, Chihuahuan, Sonoran deserts) in west Texas, southern New Mexico, central and southern Nevada, southern Utah, western and southern Arizona, and southeastern California. The southernmost distribution was in northern Mexico in the Chihuahuan desert (states of Chihuahua and Coahuila) and in the Sonoran desert (state of Sonora and the eastern portion of the Baja California Peninsula in the states of Baja California and Baja California Sur; Bailey and Klein 1997, Toweill and Geist 1999, Valdez and Krausman 1999, Festa-Bianchet 2020b, Brewer and Bleich 1923, Chapter 3). Extensive subsequent discontinuities in Rocky Mountain sheep historical distributions were due to population extirpations. Mountain sheep were extirpated from the badlands (Audubon's mountain sheep habitat) of the Yellowstone and Missouri rivers in eastern Montana, eastern Wyoming, western North and South Dakota, and northwestern Nebraska. Rocky Mountain or California bighorn were extirpated in Washington, Oregon, northern California, northern Nevada, and northern New Mexico. Desert bighorn sheep were extirpated in Texas, and Coahuila and Chihuahua in Mexico. Populations currently found in regions where they had previously been extirpated have been reestablished through transplant programs that restored Rocky Mountain and desert bighorn sheep to much of their former range (Toweill and Geist 1999, Valdez and Krausman 1999, Festa-Bianchet 2020b, Chapters 3 and 17).

Refer to the Wild Sheep Initiative (2024) for detailed mountain sheep distributions for each jurisdiction in Canada and the United States and Chapter 17 for distributions in Mexico.

## 1.16 SUMMARY

After having probably evolved in western Asia, wild sheep dispersed to Siberia and gave rise to snow sheep. An ancestral form of the Siberian snow sheep accessed northwestern North America via the Bering Land Bridge and became the progenitor of the two species (bighorn, thinhorn) of mountain sheep. The North American western mountain systems provided habitats, refugia, and corridors that enabled mountain sheep to occupy their present distribution extending from northern Alaska to the southwestern United States and northern Mexico, including the eastern portion of the Baja California Peninsula. Cycles of isolation and admixture during glacial and interglacial periods have been highly influential in the evolution of American species and subspecific relationships. The diverse habitats enabled localized populations to evolve genetically distinct genomes, and in some instances, formed hybrid zones and subsequent introgressed populations, further complicating the determination of genomic histories and infraspecific genetic relationships. Mountain sheep are classified in a hierarchical arrangement in the order Artiodactyla (even-toed hoofed mammals) to species and subspecies. Since the late 20th century, there has been unanimity among American and European mammalogists in recognizing two species of North American mountain sheep: thinhorn and bighorn. Currently, thinhorn sheep occupy most of their native range and, until recently, maintained close to ancestral population size. The infraspecific taxonomy of bighorn sheep has been complicated, especially owing to named taxa originally based on phenotypic criteria that are not diagnostically meaningful. The use of genetic data since the 1980s has been instrumental in determining infraspecific variation, but has not resolved the subspecies taxonomic conundrum in part because subspecies boundaries have not been clearly defined. Although the naming of subspecies was originally intended to indicate variation within species, its subjective nature often obfuscated actual infraspecific variation. The application of genetic methods (e.g., karyotyping, mtDNA sequencing, single-nucleotide polymorphisms) has enabled the identification of unique and distinct populations that represent the overall genetic variation within a species and are important conservation units.

## REFERENCES

Adhikari, L., B. Khan, S. Joshi, L. Ruijun, G. Ali, G. M. Shah, M. Ismail, K. Bono, R. Ali, G. Kahn, et al. 2021. Community-based trophy hunting programs secure biodiversity and livelihoods: learnings from Asia's high mountain communities and landscapes. *Environmental Challenges* 4:100175.

Allen, J. A. 1912. Historical and nomenclatural notes on North American wild sheep. *Bulletin of the American Museum of Natural History* 31:1–29.

Ålund, M., C. Cenzer, N. Bierne, J. W. Boughman, J. Cerca, M. S. Comeford, A. Culicchi, B. Langerhans, S. e. McFarlane, M. H. Möst, et al. 2023. Anthropogenic change and the process of speciation. *Cold Spring Harbor Perspectives in Biology* 15:a041455.

Audubon, J. J., and J. Bachman. 1846–1853. *The Viviparous Quadrupeds of North America*. Three volumes. J. J. Audubon, New York, New York, USA.

Audubon, M. R. 1897. *Audubon and His Journals, with Zoological and Other Notes by Elliot Coues*. Two volumes. Charles Scribner's Sons, New York, New York, USA.

Aulager, S., P. Haffner, A. J. Mitchell-Jones, F. Moutou, and J. Zima. 2025. *A Field Guide to the Larger Mammals of Europe, North Africa, and the Middle East*. Third edition. Bloomsbury, London, United Kingdom.

Bagirov, V. A., P. M. Klenovitskiy, B. S. Iolchiev, N. A. Zinovieva, V. V. Kalashnikov, O. V. Shilo, V. A. Soloshenko, Sh. N. Nasibov, V. P. Kononov, and A. V. Kolesnikov. 2012. Cytogenetic characteristics *of Ovis ammon ammon, O. nivicola borealis* and their hybrids. *Sel'skokhozyaistvennaya Biologiya [Agricultural Biology]* 6:43–48.

Bailey, J. A., and D. R. Klein. 1997. United States of America. Pages 307–316 *in* D. M. Shackleton, editor. *Wild Sheep and Their Relatives, Status Survey and Conservation Action Plan for Caprinae*. IUCN, Gland, Switzerland.

Baker, A. M., and I. C. Gynther, editors. 2023. *Strahan's Mammals of Australia*. Fourth edition. Bloomsbury, London, United Kingdom.

Baker, R. J., and R. D. Bradley. 2006. Speciation in mammals and the genetic species. *Journal of Mammalogy* 87:643–662.

Bavin, R. L. 1975. *Ecology and Behavior of the Persian Ibex in the Florida Mountains*. Thesis, New Mexico State University, Las Cruces, USA.

Blunt, W. 2001. *Linnaeus: The Compleat Naturalist*. Princeton University Press, Princeton, New Jersey, USA.

Bradley, R. D., and R. C. Dowler. 2019. A century of mammal research: changes in research paradigms and emphases. *Journal of Mammalogy* 100:719–732.

Brewer, C. E., and V. C. Bleich. 1923. Desert bighorn sheep: species spotlight. Pages 19–33 *in* K. M. Lehr and J. Schwab, editors. *Records of North American Big Game. A Book of the Boone and Crockett Club Containing Tabulations of Outstanding North American Big Game Trophies, Compiled from Data in the Club's Big Game Records Archives. Volume 1*. Fifteenth edition. Boone and Crockett Club, Missoula, Montana, USA.

Bubenik, A. B. 1990. Epigenetical, morphological, physiological, and behavioral aspects of evolution of horns, pronghorns, and antlers. Pages 3–113 *in* G. A. Bubenik and A. B. Bubenik, editors. *Horns, Pronghorns, and Antlers: Evolution, Morphology, Physiology, and Social Significance*. Springer, New York, New York, USA.

Buchalski, M. R., B. N. Sacks, D. A. Gille, M. C. T. Penedo, H. B. Ernest, S. A. Morrison, and W. M. Boyce. 2016. Phylogeographic and population genetic structure of bighorn sheep (*Ovis canadensis*) in North American deserts. *Journal of Mammalogy* 97:823–838.

Bunch, T. D. 1978. Fundamental karyotype in domestic and wild species of sheep: identity and ranking of autosomal acrocentrics involved in biarmed formations. *Journal of Heredity* 69:77–80.

Bunch, T. D., A. Rogers, and W. C. Foote. 1977. G-band and transferrin analysis of aoudad-goat hybrids. *Journal of Heredity* 68:210–212.

Bunch, T. D., C. Wu, Y. P. Zhang, and S. Wang. 2006. Phylogenetic analysis of snow sheep (*Ovis nivicola*) and closely related taxa. *Journal of Heredity* 97:21–30.

Bunch, T. D., and W. C. Foote. 1977. Evolution of the 2N=54 karyotype of domesticated sheep (*Ovis aries*). *Genetics Selection Evolution* 9:509–515.

Bunch, T. D., R. S. Hoffmann, and C. F. Nadler. 1999. Cytogenetics and genetics. Pages 263–276 *in* R. Valdez and P. R. Krausman, editors. *Mountain Sheep of North America*. University of Arizona Press, Tucson, Arizona, USA.

Cassinello, J., F. Bounaceur, J. C. Brito, E. Bussière, F. Cuzin, J. Gil-Sánchez, F. Herrera-Sánchez, and T. Wacher. 2021. *Ammotragus lervia*. The IUCN Red List of Threatened Species 2021: e.T1151A22149987. https://dx.doi.org/10.2305/IUCN.UK.2021-3.RLTS.T1151A22149987.en Accessed 20 November 2023.

Castelló, J. R. 2016. *Bovids of the World: Antelopes, Gazelles, Cattle, Goats, Sheep, and Relatives*. Princeton University Press, Princeton, New Jersey, USA.

Chen, Z.-H., Y.-X. Xu, X.-L. Xie, D.-F. Wang, D. A. Gómez, G.-J. Liu, X. Li, A. Esmailizadeh, V. Rezaei, J. Kantanen, et al. 2021. Whole-genome sequence analysis unveils different origins of European and Asiatic mouflon and domestication-related genes in sheep. *Communications Biology* 4:1307.

Colli, L., H. Lancioni, I. Cardinali, A. Olivieri, M. R. Capodiferro, M. Pellecchia, M. Rzepus, W. Zamani, S. Naderi, F. Gandini, et al. 2015. Whole mitochondrial genomes unveil the impact of domestication on goat matrilineal variability. *BioMed Central Genomics* 16:1115.

Cowan, I. M. 1940. Distribution and variation in the native sheep of North America. *American Midland Naturalist* 24:505–580.

Cuvier, M. 1817. Le Regne Animale. Volume 1. Deterville, Paris, France.

Damm, G. R., and N. Franco. 2014. The CIC Caprinae Atlas of the World. Two volumes. CIC International Council for Game and Wildlife Conservation, Budakeszi, Hungary in cooperation with Rowland Ward Publications, Johannesburg, Republic of South Africa.

Darimont, C. T., S. M. Carlsonc, M. T. Kinnisond, P. C. Paquete, T. E. Reimchena, and C. C. Wilmers. 2009. Human predators outpace other agents of trait change in the wild. *Proceedings of the National Academy of Sciences* 106:952–954.

Deakins, S., J. C. Gorrell, J. Kneteman, D. S. Hik, R. M. Jobin, and D. W. Coltman. 2019. Spatial genetic structure of Rocky Mountain bighorn sheep (*Ovis canadensis canadensis*) at the northern limit of their native range. *Canadian Journal of Zoology* 98:317–330.

De Queiroz, K. 2007. Species concepts and species delimitation. *Systematic Biology* 56:879–886.

Dotsev, A. V., E. Kunz, A. V. Shakhin, S. N. Petrov, O. V. Kostyunina, I. M. Okhlopkov, T. E. Deniskova, M. Barbato, V. A. Bagirov, D. G. Medvedev, et al. 2019. The first complete genome of snow sheep (*Ovis nivicola*) and thinhorn sheep (*Ovis dalli*) and their phylogenetic implications for the genus *Ovis*. *Mitochondrial DNA Part B* 4:1334–1333.

Etnyre, E., J. Lande, A. McKenna, and J. Berini. 2011. Artiodactyla. Animal Diversity Web at https://animaldiversityweb.org/accounts/Artiodactyla. Accessed 30 August 2024.

Fedosenko, A. K., and D. A. Blank. 2005. *Ovis ammon. Mammalian Species* 773:1–15.

Feijó, A., D. Ge, Z. Wen, J. Cheng, L. Xia, B. D. Patterson, and Q. Yeng. 2022. Mammalian diversification bursts and biotic turnovers are synchronous with Cenozoic geoclimatic events in Asia. *Proceedings of the National Academy of Sciences* 119:e2207845119.

Festa-Bianchet, M. 2020a. *Ovis dalli*. The IUCN Red List of Threatened Species 2020:e.T39250A22149895. https://dx.doi.org/10.2305/IUCN.UK.2020-2.RLTS.T39250A22149895.en. Accessed 12 November 2024.

Festa-Bianchet, M. 2020b. *Ovis canadensis*. The IUCN Red List of Threatened Species 2020: e.T15735A22146699. https://dx.doi.org/10.2305/IUCN.UK.2020-2.RLTS.T15735A22146699.en. Accessed 15 November 2024.

Frankham, R. J. D. Ballouc, M. R. Dudashd, M. D. B. Eldridgeb, C. B. Fensterd, R. C. Lacye, J. R. Mendelson, I. J. Portonh, K. Rallsc, and O. A. Ryder. 2012. Implications of different species concepts for conserving biodiversity. *Biological Conservation* 53:25–31.

Geist, V. 1971. *Mountain Sheep: A Study in Behavior and Evolution*. University of Chicago Press, Chicago, Illinois, USA.

Geist, V. 1985. On evolutionary patterns in the Caprinae with comments on the punctuated mode of evolution, gradualism and a general model of mammalian evolution. Pages 15–30 *in* S. Lovari, editor. *The Biology and Management of Mountain Ungulates*. Croom Helm, London, United Kingdom.

Geist, V. 1991. On the taxonomy of the giant sheep (*Ovis ammon* Linnaeus, 1766). *Canadian Journal of Zoology* 69:706–723.

Gentry, W. W. 1990. Evolution and dispersal of African Bovidae. Pages 195-227 *in* G. A. Bubenik and A. B. Bubenik, editors. *Horns, Pronghorns, and Antlers. Evolution, Morphology, Physiology, and Social Significance*. Springer, New York, New York, USA.

Geptner, V. G., A. A. Nasimovich, and G. Bannikov. 1961. *Mammals of the Soviet Union. Volume* 1. *Artiodactyla and Perissodactyla*. 1988, English translation. Smithsonian Institution Libraries and the National Science Foundation, Washington, D.C., USA.

Godman, J. D. 1826. *American Natural History. Volume* 2. *Part 1. Mastology*. H. C. Carey and I. Lea., Philadelphia, Pennsylvania, USA.

Gray, A. P. 1972. *Mammalian Hybrids*. Second edition. Commonwealth Agricultural Bureaux, Farnham Royal, United Kingdom.

Greenspoon, L., E. Kriegera, R. Sendera, Y. Rosenberg, Y. M. Bar-On, U. Morana, T. Antman, S. Meiri, U. Roli, E. Noor, et al. 2023. The global biomass of wild mammals. *Proceedings of the National Academy of Sciences USA* 120:e2204892120.

Grinnell, J. 1912. *The bighorn of the Sierra Nevada*. University of California Publications in Zoology 10:143–153.

Groves, C., and P. Grubb. 2011. *Ungulate Taxonomy*. Johns Hopkins University Press, Baltimore, Maryland, USA.

Groves, C., and D. M. Leslie Jr. 2011. Family Bovidae. Pages 444–571 *in* D. E. Wilson and R. Mittermeier, editors. *Handbook of the Mammals of the World. Volume* 2. *Hoofed Mammals*. Lynx Edicions, Barcelona, Spain.

Grubb, P. 2005. Order Artiodactyla. Pages 637–722 *in* D. E. Wilson and D. M. Reeder, editors. *Mammal Species of the World: A Taxonomic and Geographic Reference. Volume* 1. Second edition. Smithsonian Institution Press, Washington, D.C., USA.

Guthrie, R. D. 2017. Paleoecology of the Dry Creek site and its implications for early hunters. Pages 153–192 *in* T. Goebel, editor. *Dry Creek: Archaeology and Paleoecology of a Late Pleistocene Alaskan Hunting Camp*. Texas A&M University Press, College Station, Texas, USA.

Hall, E. R. 1981. *The Mammals of North America*. Two volumes. Second edition. John Wiley and Sons, New York, New York, USA.

Hall, E. R., and K. R. Kelson. 1959. *The Mammals of North America*. Two volumes. Ronald Press, New York, New York, USA.

Heller, R., P. Frandsen, E. D. Lorenzen, and H. R. Siegismund. 2013. Are there really twice as many bovid species as we thought? *Systematic Biology* 62:490–493.

Hess, S. C., D. H. Van Vuren, and G. W. Witmer. 2018. Feral goats and sheep. Pages 289–309 *in* W. C. Pitts, J. Beasley, and G. W. Witmer, editors. *Ecology and Management of Terrestrial Vertebrate Invasive Species in the United States*. CRC Press, Boca Raton, Florida, USA.

Hewitt, H. G. M. 2011. Quaternary phylogeography: the root of hybrid zones. *Genetica* 129:617–638.

Hight, M. E., and C. F. Nadler. 1976. Relationships between wild sheep and goats and the aoudad (Caprini) studied by immunodiffusion. *Comparative Biochemistry and Physiology* 54B:265–269

Hoefs, M., and U. Nowlan. 1997. Hybridization of thinhorn and bighorn sheep, *Ovis dalli* X *O. canadensis*. *Canadian Field Naturalist* 111:647–648.

Holechek, J., and R. Valdez. 2018. Wildlife conservation on the rangelands of eastern and southern Africa: past, present, and future. *Rangeland Ecology and Management* 71:245-258.

Hornaday, W. T. 1901. Notes on the mountain sheep of North America, with description of a new species. *New York Zoological Society Annual Report* 5:77–122.

Hornaday, W. T. 1906. *Camp-Fires in the Canadian Rockies*. Charles Scribner's Sons, New York, New York, USA.

Hornaday, W. T. 1908. *Camp-Fires on Desert and Lava*. Charles Scribner's Sons, New York, New York, USA.

Jahner, J.P., T. L. Parchment, M. D. Motocq, M Cox, R. S. Crowhurst, L. M. Galland, S. M. Burdo, et al. In review. The history of bighorn sheep diversification informs future management. Evolutionary Implications.

Kays, R. W., and D. E. Wilson. 2009. *Mammals of North America*. Second edition. Princeton University Press, Princeton, New Jersey, USA.

Kingdon, J. 2018. *Kingdon Field Guide to Arican Mammals*. Second edition. Bloomsbury, London, United Kingdom.

Korobitsina, D. V., C. F. Nadler, N. N. Vorontsov, and R. F. Hoffmann. 1974. Chromosomes of the Siberian snow sheep, *Ovis nivicola*, and implications concerning the origin of amphiberingian wild sheep (subgenus *Pachyceros*). *Quaternary Research* 4:325–345.

Krausman, P. R., and R. T. Bowyer. 2003. Mountain sheep. Pages 1095–1115 *in* G. A. Feldhamer, B. C. Thompson, and J. A. Chapman, editors. *Wild Mammals of North America: Biology, Management, and Conservation*. Second edition. Johns Hopkins University Press, Baltimore, Maryland, USA.

Krausman, P. R., A. V. Sandoval, and R. C. Etchberger. 1999. Natural history of desert bighorn sheep. Pages 137–191 in R. Valdez and P. R. Krausman, editors. *Mountain Sheep of North America*. University of Arizona Press, Tucson, Arizona, USA.

Lehr, K. M, and J. Schwab, editors. 2023. *Records of North American Big Game. A Book of the Boone and Crockett Club Containing Tabulations of Outstanding North American Big Game Trophies, Compiled from Data in the Club's Big Game Records Archives. Volume 2.* Fifteenth edition. Boone and Crockett Club, Missoula, Montana, USA.

Li, X., S.-G. He, W. R. Li, L. Y. Luo, Z. Yan, D.-X. Mo, X. Wan, F. H. Lv, J, Yang, Y.-X. Xu, et al. 2022. Genomic analysis of wild argali, domestic sheep, and their hybrids provide insights into chromosomal evolution, phenotypic variation, and germplasm innovation. *Genome Research* 32:1669–1684.

Linnaeus, C. 1758. *Systema Naturæ Per Regna Tria Naturæ, Secundum Classes, Ordines, Genera, Species, Cum Characteribus, Differentiis, Synonymis, Locis. Volume 1: Regnum Animale.* Tenth edition. Laurentii Salvii, Stockholm, Sweden.

Loehr, J., K. Worley, A. Grapputo, J. Carey, A. Veitch, and D. W. Coltman. 2006. Evidence for cryptic glacial refugia from North American mountain sheep mitochondrial DNA. *Journal of Evolutionary Biology* 19:419–430.

Lydekker, R. 1907. *The Game Animals of India, Burma, Malaya, and Tibet. Being a New and Revised Edition of 'The Great and Small Game of India, Burma and Tibet'.* Rowland Ward, London, United Kingdom.

Lydekker, R. 1913. *Catalogue of the Ungulate Mammals in the British Museum (Natural History). Volume 1. Artiodactyla. Family Bovidae, Subfamilies Bovinae to Ovibovinae (Cattle, Sheep, Goats, Chamois, Serows, Takin, Musk-Ox, etc.).* British Museum, London, United Kingdom.

Mereu, P., M. Pirastru, M. Barbato, V. Satta, E. Hadjisterkotis, L. Manca, S. Naitana, and G. G. Leoni. 2019. Identification of an ancestral haplotype in the mitochondrial phylogeny of the ovine haplogroup B. *PeerJ* :e7895.

Merriam, C. H. 1897. *Ovis nelsoni*, a new mountain sheep from the desert region of southern California. *Proceedings of the Biological Society of Washington* 11:217–218.

Merriam, C. H. 1901. Two new bighorns and a new antelope from Mexico and the United States. *Proceedings of the Biological Society of Washington* 24:29–32.

Merriam, C. H. 1916. *Ovis sheldoni*, a new mountain sheep from Sierra del Rosario, Sonora, Mexico. *Proceedings of the Biological Society of Washington* 29:129–132.

Mitchill, S. L. 1803. Account of the wild North-American sheep. *The Medical Repository* 6:237–240.

Mori, E., G. Mazza, L. Saggiomo, L. Sommese, and B. Essattore. 2017. Strangers coming from the Sahara: an update on the worldwide distribution, potential impacts and conservation opportunities of alien aoudad. *Annales Zoologici Fennici* 54:373–386.

Mori, E., L. Nerva, and S. Lovari. 2019. Reclassification of the serows and gorals: the end of a neverending story? *Mammal Review* 49:256–262.

Mungall, E. C. 2000. Exotics. Pages 736–764 *in* S. Demarais and P. R. Krausman, editors. *Ecology and Management of Large Mammals in North America.* Prentice Hall, Upper Saddle River, New Jersey, USA.

Mungall, E. C. 2024. *Exotic Animal Field Guide: Nonnative Hoofed Mammals in the United States.* Second edition. Texas A&M University Press, College Station, Texas, USA.

Mungall, E. C., and W. J. Sheffield. 1994. *Exotics on the Range: the Texas Example.* Texas A&M University Press, College Station, Texas, USA.

Nadler, C. F. 1971. Chromosomes of the Dall sheep, *Ovis dalli dalli* (Nelson). *Journal of Mammalogy* 52:461–463.

Nadler, C. F., R. S. Hoffmann, and W. Woolf. 1973a. G-band patterns as chromosomal markers and the interpretation of chromosomal evolution in wild sheep (*Ovis*). *Experientia* 29:17–119.

Nadler, C. F., K. V. Korobitsina, R. S. Hoffmann, and N. N. Vorontsov. 1973b. Cytogenetic differentiation, geographic distribution, and domestication in Palearctic sheep (Ovis). *Zeitschrift fur Säugetierkunde* 28:109–125.

Osgood, W. H. 1913. The name of the Rocky Mountain sheep. *Proceedings of the Biological Society of Washington* 26:57–64.

Osgood, W. H. 1914. Dates for *Ovis canadensis, Ovis cervina* and *Ovis montana. Proceedings of the Biological Society of Washington* 27:1–4.

Parikh, G. L., J. E. Etehart, R. O'Shaughnessy, L. A. Harveson, and J. W. Cain III. 2025. Feeding habits of sympatric aoudad (*Ammotragus lervia*) and desert bighorn sheep (*Ovis canadensis mexicana*) in West Texas. *Journal of Wildlife Management* 89:e7008.

Parker, B. G., M. Khanyhari, H. Ambarli, B. Bauveibaatar, M. Kabiri, G. Khanai, H. R. Mirzadeh, Y. Onon, and M. S. Faradinia. 2023. A review of the ecological and socioeconomic characters of trophy hunting across Asia. *Animal Conservation* 26:609–624.

Pauciullo, P., C. Knorr, A. Perucatti, A. Iannuzzi, L. Iannuzzi, and G. Erhardt. 2016. Characterization of a very rare case of living ewe-buck hybrid using classical and molecular cytogenetics. *Scientific Reports* 6:34781.

Pidancier, N., S. Jordan, G. Luikart, and P. Taberlet. 2006. Evolutionary history of the genus *Capra*: discordance between mitochondrial DNA and Y-chromosome phylogenies. *Molecular Phylogenetics and Evolution* 40:739–749.

Prothero, D. R., D. Domning, R. E. E. Fordyce, S. Foss, C. Janis, S. Lucas, K. L Marriott, G. Metais, D. Naish, K. Padian, et al. 2022. On the unnecessary and misleading taxon "Cetartiodactyla." *Journal of Mammalian Evolution* 29:93–97.

Pulling, A. von S. 1945. Hybridization in bighorn and domestic sheep. *Journal of Wildlife Management* 9:82–83.

Ramey, R. R. II. 1993. *Evolutionary Genetics and Systematics of North American Mountain Sheep: Implications for Conservation.* Dissertation, Cornell University, Ithaca, New York, USA.

Ramey, R. R. II. 1995. Mitochondrial DNA variation, population structure, and evolution of mountain sheep in the south-western United States and Mexico. *Molecular Ecology* 4:429–439.

Rezaei, H. R., S. Naderi, I. C. Chintauan-Marquier, et al. 2010. Evolution and taxonomy of the wild species of the genus *Ovis* (Mammalia, Artiodactyla, Bovidae). *Molecular Phylogenetics and Evolution* 54:315–326.

Ruedas, L. A., R. W. Norris, R. M. Timm. 2024. Best practices for the naming of species. *Journal of Mammalogy* 106:gyae113.

Sandall, E. L., A. A. Maureaud, R. Guralnick, M. A. McGeoch, Y. V. Sica, M. S. Rogan, D. B. Booher, R. Edwards, N. Franz, K. Ingenloff, et al. 2023. A globally integrated structure of taxonomy to support biodiversity science and conservation. *Trends in Ecology and Evolution* 38:1143–1153.

Santos, S. H. D., R. M. Peery, J. M. Miller, A. Dao, F.-H. Lyu, X. Li, M.-H. Li, and D. W. Coltman. 2021. Ancient hybridization patterns between bighorn and thinhorn sheep. *Molecular Ecology* 30:6273–6288.

Schaller, G. B. 1977. *Mountain Monarchs: Wild Sheep and Goats of the Himalaya*. University of Chicago Press, Chicago, Illinois, USA.

Seton, E. T. 1927. *Lives of Game Animals. An Account of those Land Animals in America, North of the Mexican Border, Which are Considered 'Game," Either Because They Have Held the Attention of Sportsmen, or Received the Protection of Law*. Volume 3. Doubleday, Doran and Company. Garden City, New York, New York, USA.

Shackleton, D. M., and S. Lovari. 1997. Classification adopted for the Caprinae survey. Pages 9–14 *in* D. M. Shackleton, editor. *Wild Sheep and Goats and Their Relatives: Status Survey and Conservation Action Plan for Caprinae*. IUCN, Gland, Switzerland and Cambridge, United Kingdom.

Shafer, A. B. A., C. I. Cullingham, S. D. Côté, and D. W. Coltman. 2010. Of glaciers and refugia: a decade of study sheds new light on the phylogeography of northwestern North America. *Molecular Ecology* 19:4589–4621.

Shaw, G. 1804. Ovis canadensis. *Naturalist's Miscellany* 15: Text (unpaginated) attached to plate 610.

Sheldon, W. G. 1911. *The Wilderness of the Upper Yukon*. Charles Scribner's Sons, New York, New York, USA.

Sim, Z., C. S. Corey, B. A. Jex, T. Hegel, and D. W. Coltman. 2019. Management implications of a highly resolved hierarchical population genetic structure in thinhorn sheep. *Conservation Genetics* 20:185–201.

Sim, Z., J. C. Hall, B. A. Jex, T. M. Hegel, and D. W. Coltman. 2016. Genome-wide set of SNPs reveals evidence for two glacial refugia and admixture for postglacial recolonization in an alpine ungulate. *Molecular Ecology* 25:3696–3705.

Stewart, G. R., Jr. 1935. Popular names for the mountain sheep. *American Speech* 10:283–288.

Thomson, S. A., R. L. Pyle, S. T. Ahyong, M. A. Zarazaga, J. Ammiratim, J. F. Araya, J. S. Ascher, T. L. Audisio, V. M. A. Santos, N. Bailly, et al. 2018. Taxonomy based on science is necessary for global conservation. *PLoS Biology* 16:e2005075.

Torstrom, S. M., K. L. Pangle, and B. J. Swanson. 2014. Shedding subspecies: the influence of genetics on reptile subspecies taxonomy. *Molecular Phylogenetics and Evolution* 76:143–143.

Toweill, D. E, and V. Geist. 1999. *Return to Royalty: Wild Sheep of North America*. Boone and Crockett Club and Foundation for North American Wild Sheep, Missoula, Montana, USA.

Upadhyay, M., E. Kunz, E. S. Castellanos, A. Hauser, S. Krebs, A. Graf, H. Blum, A. Dotsev, I. Okhlopkov, A. Shakhin, et al. 2021. Whole genome sequencing reveals a complex introgression history and the basis of adaptation to subarctic climate in wild sheep. *Molecular Ecology* 30:6701–6717.

Valdez, R. 1982. *Wild Sheep of the World*. Wild Sheep and Goat International, Mesilla, New Mexico, USA.

Valdez, R. 1985. *Lords of the Pinnacles: Wild Goats of the World*. Wild Sheep and Goat International, Mesilla, New Mexico, USA.

Valdez, R. 2011a. Anatolian sheep (*Ovis gmelini*). Pages 727–728 *in* D. E. Wilson and R. Mittermeier, editors. *Handbook of the Mammals of the World. Volume 2. Hoofed Mammals*. Lynx Edicions, Barcelona, Spain.

Valdez, R. 2011b. Aoudad. Pages 714–715 *in* D. E. Wilson and R. Mittermeier, editors. *Handbook of the Mammals of the World. Volume 2. Hoofed Mammals*. Lynx Edicions, Barcelona, Spain.

Valdez, R. 2011c. Snow sheep. Pages 737–738 *in* D. E. Wilson and R. Mittermeier, editors. *Handbook of the Mammals of the World. Volume 2. Hoofed Mammals*. Lynx Edicions, Barcelona, Spain.

Valdez, R. 2011d. Bighorn sheep. Pages 738–739 *in* D. E. Wilson and R. Mittermeier, editors. *Handbook of the Mammals of the World. Volume 2. Hoofed Mammals*. Lynx Edicions, Barcelona, Spain.

Valdez, R. 2019. Historical and conservation perspectives of wildlife in Mexico. Pages 1–18 *in* R. Valdez and J. A. Ortega- S., editors. *Wildlife Ecology and Management in Mexico*. Texas A&M University Press, College Station, Texas, USA.

Valdez, R., and T. D. Bunch. 1980. Systematics of the aoudad. Pages 27–29 *in* C. D. Simpson, editor. *Proceedings of the Symposium on Ecology and Management of Barbary Sheep*. Texas Tech University, Lubbock, Texas, USA.

Valdez, R., and P. R. Krausman. 1999. Description, distribution, and abundance of mountain sheep in North America. Pages 3–22 *in* R. Valdez and P. R. Krausman, editors. *Mountain Sheep of North America*. University of Arizona Press, Tucson, Arizona, USA.

Valdez, R., C. F. Nadler, and T. D. Bunch. 1978. Evolution of wild sheep in Iran. *Evolution* 32:56–72.

Valdez, R., and P. Weinberg. 2011. Tribe Caprini. Pages 712–749 *in* D. E. Wilson and R. Mittermeier, editors. *Handbook of the Mammals of the World. Volume 2. Hoofed Mammals*. Lynx Edicions, Barcelona, Spain.

Van Gelder, R. G. 1977. Mammalian hybrids and generic limits. *American Museum Novitates* 2635:1–25.

Venegas, M. 1757. Noticia de la California y Su Conquista Temporal, y Espiritual hasta Tiempo Presente. Volume 1. Emprenta de la Viuda de Manuel Fernandez. Madrid, Spain.

Wang, B., L. Chen, and W. Wang. 2019. Genomic insights into ruminant evolution: from past to future prospects. *Zoological Research* 40:476–487.

Wang, X., Q. Li, and G. T. Takeuchi. 2016. Out of Tibet: an early sheep from the Pliocene of Tibet, *Protovis himalayensis*, genus and species nov. (Bovidae, Caprini), and origin of Ice Age mountain sheep. *Journal of Vertebrate Paleontology* 2016:e1169190.

Webb, R., and J. Blincow. 2024. *A Field Guide to the Larger Mammals of South America*. Princeton University Press, Princeton, New Jersey, USA.

Wehausen, J. D., and R. R. Ramey II. 2000. Cranial morphometric and evolutionary relationships in the northern range of *Ovis canadensis*. *Journal of Mammalogy* 81:145–161.

Wehausen, J. D., V. C. Bleich, and R. R. Ramey II. 2005. Current nomenclature for Sierra Nevada bighorn sheep. *California Fish and Game* 91:216–218.

Wehausen, J. D., and R. R. Ramey II. 1993. A morphometric reevaluation of the peninsular bighorn subspecies. *Desert Bighorn Council Transactions* 37:1–10.

Weinberg, P., and H. Ambarli. 2020. *Capra aegagrus*. The IUCN Red List of Threatened Species 2020:e.T3786A22145942. https://dx.doi.org/10.2305/IUCN.UK.2020-2.RLTS.T3786A22145942.en. Accessed 10 Nov 2024.

Wheeler, Q. 2024. *Species, Science and Society: The Role of Systematic Biology*. Routledge, New York, New York, USA.

Whyte, J., and E. J. Hart. 1985. *Carl Rungius, Painter of the Western Wilderness*. Salem House, Massachusetts, USA.

Wild Sheep Initiative. 2024. *Western Association of Fish and Wildlife Agencies*. Wild Sheep Initiative, Boise, Idaho, USA.

Wilson, D. E., and R. E. Mittermeier, editors. 2011. *Handbook of the Mammals of the World. Volume 2. Hoofed Mammals.* Lynx Edicions, Barcelona, Spain.

Wilson, D. E., and D. M. Reeder, editors. 2005. *Mammal Species of the World. Volume 1.* Third edition. Johns Hopkins University Press, Baltimore, Maryland, USA.

Worley, K., C. Strobeck, S. Arthur, J. Carey, H. Schwantje, A. Veitch, and D. W. Coltman. 2004. Population genetic structure of North American thinhorn sheep (*Ovis dalli*). *Molecular Ecology* 13:2545–2556.

Wright, E. A., G. G. Brugette, K. F. Buckert, F. Hernández, J. H. Reed, S. R. Wyckoff, J. C. Taylor, K. R. Manlove, C. D. Phillips, and R. D. Bradley. 2024. Multi-locus sequence typing indicates multiple strains of Mycoplasma in desert bighorn sheep and aoudad in Texas. *Journal of Wildlife Management* 88:e2594.

Young, S. P., and R. H. Manville. 1960. Records of bighorn hybrids. *Journal of Mammalogy* 41:523–525.

Zachos, F. E. 2015. Taxonomic inflation, the phylogenetic species concept and lineages in the Tree of Life: a cautionary comment on species splitting. *Journal of Zoological Systematics and Evolutionary Research* 53:180–184.

Zachos, F. E. 2016. *Species Concepts in Biology.* Springer International, Cham, Switzerland.

Zachos F. E., M. Apollonio, E. V. Bärmann, M. F. Bianchet, U. Göhlicha, J. C. Habele, E. Haringa, L. Kruckenhausera, S. Lovari, A. D. McDevitt, et al. 2013. Species inflation and taxonomic artefacts—a critical comment on recent trends in mammalian classification. *Mammalian Biology* 78:1–6.

# 2 Historical Trends in Mountain Sheep Populations and Their Habitats

*Raul Valdez, Paul R. Krausman, and Bill A. Jex*

## 2.1 INTRODUCTION

Bighorn (*Ovis canadensis*) and thinhorn (*Ovis dalli*), the two species of American mountain sheep, were, until the 1800s, unknown to naturalists except for vague descriptions by missionaries and European explorers in the 16th to early 18th centuries. Europeans first encountered mountain sheep in Mexico when Spaniards explored and founded missions in northwestern Mexico in the 1500s, and then in the southwestern United States in the early 1600s. From their observations and written casual descriptions and illustrations, the presence of mountain sheep became known in Europe (Allen 1912). In the 1800s, European explorers and fur traders recorded bighorns in the areas that would later be known as the Canadian Rockies and the western United States, but it was not until 1804 that bighorn sheep specimens, collected in Canada, were formally described by the scientific community. European and American explorers, fur trappers, naturalists, and sport hunters at this time also added to the knowledge of bighorn sheep anatomy, distribution, habitats, and life histories (McGillivray 1803, Audubon 1897, Seton 1927, Bates 1988, Valdez 1988, Czech and Valdez 2016, Chapter 1). Mountain sheep in the western United States and Canada occurred in regions far removed from eastern North American populated areas, and thinhorn sheep occupied habitats even more remote in the climatically harsh landscapes of northwestern Canada and Alaska. Although explored by Daniel Harmon, Simon Fraser, and Alexander Mackenzie, it was not until Samuel Black (Rich and Johnson 1955 in Demarchi and Hartwig 2004) recorded his observations in 1824 that thinhorn sheep in the Finlay River became known to Europeans; the sheep were, however, familiar to the Native North Americans that inhabited the area.

It's most likely that Indigenous Peoples first encountered sheep in Alaska when they crossed from Siberia to North America via the Bering Land Bridge, about 25,000 years ago. Indigenous Peoples had thousands of years of contact in northeastern Asia with Siberian snow sheep (*Ovis nivicola*) in habitats similar to those of North American sheep. After recession of the glaciers in the post-Pleistocene period, these Native People continued their utilitarian use of mountain sheep, evidenced by hand-hewn artifacts such as ceremonial bowls, tools, and utensils dating back over 7,000 years (Banfield 1974).

Subsequent explorations and zoological expeditions, principally sponsored by governments in the United States and Canada, fur hunters and gold rush prospector accounts, and sport hunting exploits in the 1800s began to reveal the distributions of mountain sheep. Most of the public became aware of mountain sheep principally from museum (e.g., American Museum of Natural History 1885) and zoo exhibits (Hornaday 1899, 1909; Vandersommers 2016), and hunting and life history accounts in popular magazines (Phillips 1930, Biscotti 2019) and books (Phillips 1930, Heller 1997, Czech and Valdez 2016) including those of Theodore Roosevelt (1885, 1888, 1905), John Muir (1874, 1894), and W. T. Hornaday (1904, 1906, 1908), and the popularized and natural history accounts of Seton (1901, 1927). Mountain sheep sport hunting accounts emphasized the massive horns of adult males for which big game hunters developed a special attraction; the quality of sheep horns would come to have a significant influence in managing mountain sheep populations. By the late 1800s, mountain sheep in the conterminous United States were relegated to small, isolated or extirpated populations over most of their range (Buechner 1960), except for populations in Alaska and Canada, and unprotected, small desert bighorn (*O. canadensis mexicana*) populations in northwestern Mexico. When territorial and subsequent state and provincial governments in the United States and Canada began to manage populations with protective legislation, their efforts lacked effectiveness simply due to an inability to implement effective enforcement across such a vast and remote landscape. Governments were also challenged with inadequate knowledge of the status, distribution, and habitat requirements of wildlife in general, and also the extent to which subsistence and market-based hunting (that supported railroad and overland transportation route work crews and gold rush prospectors) was having. The deleterious effects of domestic livestock that overwhelmed forage supplies on the rangelands of the western United States in the early 1800s and in Mexico beginning in the 1500s also exposed mountain sheep to new and novel domestic livestock parasites and pathogens to which native mountain sheep had no defenses (Cowan 1940, Allen 1942, Buechner 1960, Krausman et al. 1996). The understanding of North American mountain sheep as immunologically naïve species was very poorly understood by those charged with the management and welfare of mountain sheep resources. Early indications that sheep were becoming infected with disease and parasites, resulting in high mortalities, alongside suspicions that these were transmitted by domestic sheep, began to be documented (Grinnell 1904, Rush 1927, Wright et al. 1933, Bailey 1936, Allen 1942, Honess and Frost 1942, Thiessen 2020). It was evident in the early 1900s that bighorn sheep were in

DOI: 10.1201/9781003518686-3

a precarious situation because of their continuous decline and regional extirpations; state wildlife agencies failed to recognize or to react adequately to this decline due to lack of trained veterinarians and biologists. At this time the Bureau of Biological Survey (Taylor 1924, 1948) and the Geological Survey of Canada (Anderson 1919) began to emphasize an ecological and behavioral emphasis on wildlife research to replace random natural history observations. Funding for mountain sheep research remained minimal because of limited hunting opportunities and insufficient funding for state wildlife management programs. State agencies initiated bighorn sheep trapping and transplanting programs in the 1940s but on a small scale (Chapter 14). Management agencies instead focused their efforts on the widespread and increasing cervid populations, including deer (*Odocoileus* spp.), elk (*Cervus canadensis*), and moose (*Alces alces*), in the north, which were most in demand by big game and subsistence hunters, and the viewing public. Most of the Dall's (*O. dalli dalli*) and Stone's (*O. d. stonei*) sheep populations were in remote, sparsely settled, rugged montane habitats with harsh climatic landscapes, and they continue to be the least anthropogenically modified today, although that is changing. Hence, most thinhorn populations occupying remote habitats did not appear to undergo similar widespread declines, extirpations, or suffer habitat fragmentation to the degree experienced by bighorn sheep populations further south.

Leopold (1933) can be credited for initiating a paradigm shift in wildlife conservation by emphasizing the importance of critical thinking and the application of ecological principles in wildlife management; however, these for the most part were initially ignored by government wildlife departments. Game agencies progressively recognized the need for long-term mountain sheep field studies: of the twelve long-term field studies conducted between 1930 and 1959, one was conducted in the 1930s, six in the 1940s, and five in the 1950s. Of these studies, two were conducted in Canada and ten in the United States, including two in Alaska. The Pittman-Robertson account, which redirected an existing tax on guns and ammunition (later expanded to include a very wide variety of outdoor use items) to state wildlife agencies, provided the multi-year funding opportunities needed to implement research and management efforts for mountain sheep populations and to concentrate studies directly related to mountain sheep. Establishment of the United States Cooperative Fish and Wildlife Research Units in 1935 and associated graduate student education programs gave further impetus to state wildlife research efforts (Organ and McCabe 2018). In Canada there is no similar user taxation funding opportunity, so provincial, territorial, and Indigenous governments and stakeholders are challenged even today with securing these types of long-term funding support.

By the 1960s, game agencies hired biologists with professional degrees who began to conduct research on a host of environmental factors affecting mountain sheep, interspecific and intraspecific ecological and behavioral relationships, including effects of the human footprint on wildlife populations and habitats, and applying technologically advanced research methods and analytics in support of the research. Foremost was the recognition of an ecosystem approach to understanding community interactions and the application of ecologically based habitat management (Bleich and Thompson 2018, Bishop and Hubbard 2018). In this chapter we present: an historical account of Native North American relationships with mountain sheep; initial European and American encounters during the missionary, exploration, and fur trade periods; and a review of the history of state and provincial sheep management programs. We also note the importance of the evolution of state and provincial management efforts, initially focused on protective laws and regulations, establishing hunting seasons with restricted game harvests and designating game refuges. As species conservation needs changed, these programs were followed by initiation of trapping and transplant (i.e., translocation) programs, establishing standardized monitoring protocols, and adoption of modern, scientifically based big game research and management methodologies.

## 2.2  NATIVE NORTH AMERICAN ENCOUNTERS WITH MOUNTAIN SHEEP (PREHISTORIC PERIOD–1500)

### 2.2.1  Indigenous North Americans: Mountain Sheep Habitat Managers

Native People were the first to encounter and hunt North American mountain sheep when their ancestral progenitors crossed the Bering Land Bridge over 20,000 years ago (Pigoti 2023), establishing permanent settlements in the New World, and encountering thinhorn sheep populations in Alaska. First Nations peoples had coexisted with North American mountain sheep for thousands of years before the European encounters with North American wildlife; humans were present in North America south of the ice sheets between 20,000 and 23,000 years ago during the Last Glacial Maximum (Pigoti et al. 2023). Native Siberians, who initially inhabited Siberia 45,000 years ago (Shichi et al. 2023), were familiar with the native sheep in the Siberian Mountains, the Asian congeners of North American mountain sheep. Siberian mountain sheep (i.e., snow sheep [*O. nivicola*]) also entered North America via the Bering Land Bridge during the Pleistocene Epoch (~2.6 mya-11,700 years ago), as did several other large wild mammals of Siberian origin (Chapter 1). The snow sheep is the closest Asiatic mountain sheep relative of North American sheep species (Bunch et al. 2006), having a widespread distribution across northcentral and eastern Siberia, and occurring in similar montane habitats to those of mountain sheep in Alaska and Canada. Siberian and North American thinhorns share similar horn morphology; however, Siberian sheep are similar in coat color to Rocky Mountain bighorn sheep, but the rump patch does not extend above the base of the tail (Bunch 2006, Geist

1993, Valdez 2011, Damm and Franco 2014). They were a source of meat protein, and the horns, bones, and skins had utilitarian value (Banfield 1974, Grant 1980). Above all, mountain sheep and other wildlife were social, subsistence, cultural, and spiritual assets upon which the livelihoods of Native People depended (Gilman 2003).

In addition to large-scale lightning-caused fire regimes, Indigenous Peoples were a major management agent in modifying habitats using fire (Lewis 1978, Lewis and Ferguson 1988), presumably to create plant communities favorable for communal settlement areas, and to attract game species, including mountain sheep. Changes in the structure and species composition of plant communities and the ecological consequences resulting from creation of a diversity of plant successional stages were the first human management interventions in wildlife habitats in North America. Human-mediated ecological processes were important in clearing forests and other tall vegetation, areas avoided by mountain sheep, thus creating, maintaining, and in some areas potentially expanding mountain sheep habitat. In turn, Native North Americans benefited by increasing the diversity and number of game species populations that were a major food source. Use of fire to maintain optimal sheep foraging habitat is a basic management practice still in use (Risenhoover and Bailey 1985, Whiting et al 2023, Chapter 20). Use of fire by Indigenous Peoples is described by Williams (2002) and Boyd (2022), and for fire use in modern forest management by Ffolliott (1996) and Bowman et al. (2016). First Nations communities also added an efficient domestic carnivore that amplified their ability to detect and kill prey: the domestic dog. The climate-predator-prey-habitat relationship of humans and North American mountain sheep has evolved over a period of thousands of years developing from ecosystem modification practices that restructured plant and animal communities (Kay 1994). This ecological relationship abruptly ended when Indigenous Peoples were forcibly removed from their ancestral homelands and the governments of the day implemented widespread wild-fire suppression programs in support of industrial and commercial interests targeting natural resource extraction. Designation of bighorn sheep by government programs as a wilderness species, implying that they were isolated from human interactions, is arbitrary because of their long-term ecological relationships with Native Americans.

### 2.2.2 Indigenous North American Hunting Methods

Indigenous North Americans used a variety of hunting methods including ground-blind construction and game drives. These involved cooperative hunting efforts in which the intended prey was driven toward hiding spearmen or archers and entrapment areas, using winged corrals (Muir 1894, Benedict 1992, 1996). Game drives, described from numerous archaeological sites, were a hunting method that had been in use by First Nations for thousands of years from the Arctic to the American southwest, but also by hunters in Eurasia and Africa (Lemke 2022). Game drives, desert kite structures, or large-scale traps consist of a rock or wood construct and were used through a group or communal effort to trap or ambush migrating artiodactyls to secure large quantities of prey animals including bighorn sheep. Lemke (2022) used the term "hunting architecture" to describe hunting structures that included any form of permanent or semi-permanent built structure used to aid hunting. Game drives could be more than a kilometer long; usually U-shaped or V-shaped, and the walls made of loosely piled rocks up to a meter high. This required concentrated efforts sometimes involving dozens of hunters who were familiar with the migratory movements of large ungulates (Benedict 1996). In the Rocky Mountains, wooden catch pens for capturing sheep extended over long distances and included pens with large logs and a ramp directing animals to the entrance (Frison et al. 1990, Frison 2004). Large-scale traps were diverse and in widespread use in the North American Great Basin region of eastern California, western Utah, and Nevada during the past 9,000 years (Hockett and Dillingham 2023, Figure 2.1). Stone structures were also used for directing or capturing sheep traveling along migration routes and across mountain passes, which combined pre-planned use of topographical features and knowledge of mountain sheep behavior (LaBelle and Pelton 2013, Schneider et al. 2014).

Pronghorn (*Antilocapra americana*), mule deer (*Odocoileus hemionus*), and bighorn sheep were the primary artiodactyls targeted by Native Americans on the Great Basin. Use of game drives in the Rocky Mountains appears to have ended after adopting a horse culture around 1650. Grinnell (1904) described the use of horses when hunting sheep that sought escape cover through use of isolated buttes in areas surrounded by prairie in Wyoming and Montana where sheep occasionally occurred in prairie habitats, sometimes in association with pronghorn. After chasing sheep to an isolated butte, mounted hunters would surround the butte, while other hunters climbed the butte and chased the sheep to the waiting hunters. Parkman (1849) observed a similar hunt when a migrating party of Sioux Indians unexpectedly encountered a herd of bighorn sheep. Other native tribes, such as the Bannocks, stalked bighorns by disguising themselves as sheep. They covered their bodies with sheep skins, placed sheep horns on their heads, and crouched to assume the appearance of a sheep (Grinnell 1904). Another group of Native Americans, the Tukudika or Sheepeaters, a band of Shohone from the Yellowstone Plateau, acquired their name because mountain sheep constituted their preferred game animal and were a fundamental element of tribal culture. Russell (1921:31), a fur trapper, described a family group he encountered as "all neatly clothed in dressed deer and sheep skins of the best quality" and from whom he obtained a "large number" of "elk, deer,

**FIGURE 2.1** Rock art is a common form of expression in North American Native cultures. This is an example that depicts what may be a game drive. (Photo by L. Victor Clark; used with permission.)

and sheep skins." He added: "Their bows were artistically made from sheep, buffalo, and elk horns, and secured with deer and elk sinews, and ornamented with porcupine quills, and generally about three feet long."

Most First Nations considered thinhorn sheep as animals that were difficult to hunt due in large part to the challenge and remoteness of their habitats. The primary hunting weapon for thinhorn sheep became the bow and arrow, having been recorded in archaeological sites in Alaska dating from the early Holocene (Maschner and Mason 2013). Use of the bow and arrow was facilitated by access to obsidian which First Nations hunters crafted (and later traded) into arrowheads, spear tips, and cutting instruments. Also widely used was the atlatl (or spear thrower), a narrow, 0.55-m-long stick used to propel a spear (Grant 1980), and snares set up in areas along traditional sheep trails and where they could be secured to boulders (Gwich'in Elders 1997). With these types of weapons capable at killing game at a distance, the use of ground blinds, often constructed with rock wall and shrub screening and positioned along game trails, stream crossing, or in areas where topography naturally funneled mountain sheep movement, proved effective. Although thinhorn sheep were harvested opportunistically, more often hunting of thinhorn sheep occurred as part of an annual hunting trip (J. John, retired guide outfitter and First Nations elder, deceased, formerly of Carcross, Yukon, Canada, Wedge and Van Bibber; from Demarchi and Hartwig 2004), timed to harvest animals while they still retained a large amount of body fat. Males were preferred before the rut and females after the rut (Gwich'in Elders 1997).

### 2.2.3 Native North American Mountain Sheep Artistry

Native North Americans are noted for the artistic portrayals of North American mountain sheep using pictographs and petroglyphs or rock art, a widespread art form in western North America and northern Mexico in present and former mountain sheep habitats (Grant et al. 1968, 1980; Murray and Espinosa-T. 2006). The bighorn was a favorite rock art motif and perhaps signified a special relationship with those early peoples (Figure 2.2). Of the 100,000 images of bighorn sheep in the Coso Range of southwestern California, about 50% are bighorn sheep (Grant 1980, Garfinkel and Austin 2011). Perhaps the most impressive bighorn pictographs are in the central Baja California Peninsula (Crosby 1984), some larger than life-size. Bighorn may have become a venerated animal deity, and rock art was a form of honoring mountain sheep to ensure their continued abundance (Grant 1980), although the purpose and meaning of rock art has several contentious interpretations (Slifer 1998, Garfinkel et al. 2010, Garfinkel and Austin 2011, Whitley 2013, 2023). Gámez-Brunswick and Rojas-Soto (2020), using ecological niche modeling and rock art as an alternative source of distributional information for desert bighorn in the past 6,000 years, revealed a recent increase in the species' distribution and responses to past climatic changes.

**FIGURE 2.2** A female mountain sheep photographed in the Muddy Mountains of southern Nevada with a petroglyph of a male mountain sheep in a rock overhang. Prehistoric rock art has been used to determine the historical distribution of mountain sheep. (Photo by L. Victor Clark; used with permission.)

### 2.2.4 CONTROVERSIES ON EFFECTS OF INDIGENOUS HUNTING

Based on the overkill hypothesis (Martin 1967), Indigenous Peoples were responsible for the widespread megafaunal extinction of North American mammals during the Pleistocene. This is a controversial hypothesis that continues to be debated without a consensus as to its validity (Lyman 2004, Nagaoka 2012, Nagaoka et al. 2018); however, the large mammal North American Holocene fauna, including mountain sheep, survived during the thousands of years of a human presence without verging on extinction, based on archaeological data (Surovell 2008). Indigenous Peoples may have numbered in the millions, acting as a keystone species that limited ungulate numbers, structured plant communities, had a greater influence on prey population dynamics than carnivores, and were an integral part of wilderness but not conservationists because they locally overharvested prey (Kay 1994, 2002a,b). In any case, there is no archaeological or historical incontrovertible evidence that Indigenous Peoples extirpated or significantly limited an ungulate species, including mountain sheep, which were abundant and widespread, before and the period of Spanish settlement in southwestern North America and the periods of early exploration and fur trade of western America (1800–1840).

Indigenous Americans had minor deleterious effects on mountain sheep populations and habitats compared to those imposed by European colonizers in North America.

Mountain sheep had not been exposed to domestic livestock pathogens and diseases before colonization. After arrival of European explorers and then settlers, wildlife in North America, and mountain sheep specifically, were subjected to gradually overwhelming numbers of domestic sheep and other livestock that were transported, raised, and foraged in mountain sheep habitats. Firearms were not available before the European arrival, that later saw expanded settlement and the human footprint and widespread industrial development. This change created a new pattern of game consumption developed through market hunting, which saw Indigenous hunters employed to provide game meat for work camps and crews, and in most areas where people settled, a general degradation of mountain sheep populations and habitats. As European interests in natural resources grew, culminating in the gold rush era of the 1800s, First Nations people saw the commodification of nature, inherently changing why Native Peoples hunted and fished, and this affected Indigenous relationships with the environment and their connection to their foods (Peyton 2011). Until European settlers imported livestock and developed agricultural trade systems, Native North Americans did not have domestic ungulates; hence, mountain sheep were not previously affected by the habitat intrusions and potential competition for forage resources by exotic, domestic ungulates. Acquisition of domesticated horses and their dispersal into the Great Plains and northern Rockies as early as the mid-17th century among native tribes (Taylor et al. 2023) greatly improved Indigenous hunting capabilities for plains game; however, horses were not efficient for hunting mountain sheep. The exception was in areas occupied by mountain sheep in less steep terrain such as the Missouri River Breaks. Widespread use of firearms by First Nations imperiled localized mountain sheep populations, but this did not occur on a widespread scale until the 1850s when firearms first became readily available and market hunting became more important. In the 15th and 16th centuries, horses and later firearms, introduced into North America by Europeans, were two of the most disruptive cultural intrusions for Native North Americans, rapidly changing the long-standing hunter-game balance between First Nations and subsistence prey species.

Indigenous peoples would often be accused of being a major obstacle of enforceable, conservative game laws. In some regions Native communities, however, had been deprived of access to their ancestral hunting grounds, marginalized in land-management decision-making, decimated by disease, and lived in poverty without an option except to participate in the market hunting economy and having to rely on subsistence hunting. In other areas lifeways changed as peoples such as the Tahltan, who occupied a large area of northwest British Columbia, were brought into the commodified hunting industry as experts and laborers; this while the government enacted new regulations changing how they could interact with the environment around them and the way that wildlife was consumed (Peyton 2011). Still, the quantity of commercially harvested game by Native American peoples, including mountain sheep, was probably overestimated, because of the prejudicial attitude of the white population who were themselves involved in illegal hunting (Warren 1997, Jacoby 2001, Valdez 2021) or who had personal or commercial interests in sport hunting that same wildlife. Even some of the early leaders of the American wildlife conservation movement participated in what today could be considered unethical hunting behavior (Taylor 2016); in any case, the effects of illegal sheep hunting by Native Americans remain speculative and mostly without context. For example, there are records of what was considered excessive harvest of wildlife by a Native individual (e.g., six sheep and several moose in winter; Peyton 2011); however, the reality is that wildlife harvest in support of subsistence provisioning in First Nation communities was a communally shared activity that supported multiple families, and numerous domesticated dogs and sled dogs, which were important as a means of transportation and food procurement.

In the north, the effects of game consumption supported through market hunting eventually led to implementation of game laws (e.g., Game Amendment Act 1905), harvest limits, and creation of dedicated enforcement officers or game wardens. This challenged the established meat procurement practices of rural and Indigenous Peoples while at the same time generating new economic opportunities in transportation, guiding, cooking, horse-wrangling, and general outfitting (Peyton 2011). Predator control and culling by some

First Nations was initiated to generate income from fur sales while also reducing consumptive competition between predators and people (inclusive of consumptive use but also trophy hunting). As prey populations responded favorably to predator removals, this created fewer restrictions on game harvesting, thus improving economic opportunities from supplying furs and game meat to markets and from commercial trophy hunting. For some Indigenous People, sale of winter-caught furs could reach a thousand dollars in the early 1900s (Peyton 2011), and near the same or possibly more could be realized from summer and fall meat harvest sales.

## 2.3 EUROPEAN EXPLORATION, FUR TRADE, AND GOVERNMENT SURVEYS (1500–1879)

### 2.3.1 MOUNTAIN SHEEP INITIAL ENCOUNTERS WITH EUROPEAN CULTURES

Europeans first became aware of the existence of mountain sheep in North America from cursory wildlife accounts by Spanish missionaries and explorers in northwestern Mexico and what would become the western United States. The earliest descriptions of desert bighorn sheep were those of missionaries in southern Baja California including those of Francis Maria Picolo and Juan Maria de Salvatierra (Allen 1912). The first illustration of a North American mountain sheep was published by Venegas (1757) based on a specimen from Baja California, Mexico; that illustration established a visual presence of mountain sheep in North America (Figure 2.3). Several Spanish explorers including Coronado (Winship 1896), Cabeza de Vaca (Adorno and Pautz 1999), Anza (Bolton 1930), and Escalante (Warner 2004) encountered mountain sheep during their exploration of western North America and recorded brief descriptions. The first illustration (Figure 2.4) of a Rocky Mountain bighorn was published in 1803 in a popular publication, The Medical Repository, based on a male specimen provided by Duncan McGillivray, a fur trader, who collected specimens in the Canadian Rockies and was drawn by E. Savage (McGillivray 1803, White 2001). This was also the specimen that Shaw (1804) used for his description of Rocky Mountain bighorn sheep (Chapter 1).

During the Lewis and Clark expedition (1804–1806), Lewis wrote a detailed description of bighorn sheep on 25 May 1805, based on two specimens collected by members of the expedition on the Missouri River Breaks or Badlands. He recorded, among other anatomical features, that the head and horns of a male weighed 12.3 kg and mentioned the disproportionately large size of the male horns, the smaller size of female horns, and the bighorn's preference for precipitous terrain to escape from predators, among other natural history observations (Burroughs 1961, Moulton 1987, Tubbs and Jenkinson 2003). One of the most famous hunting expeditions was that of Prince Maximillian

FIGURE 2.3    The first illustration of a desert mountain sheep was published in 1757 in a book authored by M. M. Venegas.

FIGURE 2.4    The first illustration of a Rocky Mountain bighorn sheep was published in 1803 based on a male collected by Duncan McGillivray in the Canadian Rockies in 1800. The body depicted is similar to that of a domestic sheep, but note the broomed horn and white rump patch.

**FIGURE 2.5** Bighorn sheep in the badlands of the Upper Missouri River were painted by Carl Bodmer in 1833 during the western North American expedition of Prince Maximillian of Wied, a German naturalist (Witte and Gallagher 2010).

of Wied, a German explorer, ethnologist, naturalist, and big game hunter, who ventured to the Upper Missouri region in 1833–1834. He hunted mountain sheep in the same general area as Lewis and Clark. He hired a 23-year-old Swiss artist, Carl Bodmer, to illustrate the expedition and who depicted sheep in the badlands of the Upper Missouri in 1833 (Witte and Gallagher 2010, Figure 2.5). Members of the Audubon expedition (Audubon 1897) also procured sheep specimens in the badlands and illustrated them in Audubon and Bachman (1846–1854, Chapter 1).

Canada and Alaska experienced a different group of explorers such as Daniel Harmon, Simon Fraser, and Alexander Mackenzie; however, it was Samuel Black who recorded his observations of thinhorn sheep in the Finlay River in 1824 (Rich and Johnson 1955 in Demarchi and Hartwig 2004) that created awareness of these new species of sheep to Europeans. This in turn fostered sport hunting interests by wealthy aristocrats and even European royals such as the Austrian Prince Adolph Von Hagen, who sought multiple specimens that included the heads of Rocky Mountain goats and sheep, with each being obtained with comparative ease in the Cassiar region, British Columbia (Peyton 2011), and again, First Nations people were enlisted as guides in support of this hunting activity (Valdez 1988). During these decades between the mid-1800s and early 1900s, most hunting parties harvested multiple sheep (along with other game species) during their multi-day expeditions (Palliser 1853; Southesk 1875; Dunraven 1876; Baillie-Grohman 1882; Roosevelt 1885, 1888, 1893,1929; Emmet 1901; Bradley 1904; Bannon 1911; Cudahy 1928), some of which were the first to publish the presence of sheep populations in previously unrecorded areas (Figures 2.6 and 2.7).

**FIGURE 2.6** The American West became a favorite big game hunting venue for European hunters beginning in the 1870s. Among these hunters was W. A. Baillie-Grohman, who hunted mountain sheep in the United States and Canada in the 1800s and pictured here surrounded by European and American big game trophies.

**FIGURE 2.7** Theodore Roosevelt's hunting exploits in the American West included bighorn sheep in the North Dakota badlands and forged his strong commitment to wildlife conservation. His extensive publications on his sport hunting ventures popularized big game hunting in the American West. Note the mountain sheep skulls and horns in the pile of elk antlers. Reproduced from Roosevelt (1905).

## 2.4 HITTING THE BOTTOM AND INITIAL PROTECTION (1880–1930)

Mountain sheep were a relatively unknown ungulate in North America, not having been formally described until 1804, and remained an unknown species to most North Americans until the early 1900s because of their scarcity relative to other North American species and distant distribution from the population centers of the eastern United States and Canada. By then, bighorn sheep were relegated to small, isolated, or extirpated populations over most of their range in the lower 48 states, except for abundant populations in Canada, and unprotected, illegally hunted, and reduced desert bighorn populations in northwestern Mexico (Cowan 1940, Allen 1942, Buechner 1960, Tapia Landeros 2008, Brown 1989, Carmony and Brown 1993). Wild thinhorn sheep populations in Alaska and northwestern Canada were the least affected. It was in this period that game departments were established and early laws enacted. Unregulated hunting in the 1800s and early 1900s included subsistence and market hunting, habitat loss through land conversion to agricultural purposes, the facilitation of transportation provided by railroads and their construction through bighorn sheep habitats, extensive habitat fragmentation due to fencing and road construction ostensibly restricting movements and gene flow, and mining and logging, all of which affected habitat function. Intensive grazing of domestic livestock, initiated before 1900 and continuing to about 1940 (Wagner 1978), had the most

deleterious effects on mountain sheep populations because of the excessive numbers of domestic livestock pastured in mountain sheep habitats and the likely transmission of disease and pathogens transmitted from livestock to mountain sheep (Miller et al. 2012).

### 2.4.1 Livestock Overgrazing on Western Rangelands: Effects on Mountain Sheep

Domestic sheep, probably derived from native Spanish breeds, were first introduced into the Americas in 1493 on Caribbean islands by Christopher Columbus, and by Hernan Cortes into Mexico in 1522 (Crosby 1972, Dunmire 2013, Campos et al. 2020). Domestic sheep, cattle, and goats were being introduced in 1540–1542, in what would become part of the United States when Francisco Coronado led an expedition through the southwestern United States to the central Great Plains (Winship 1896). Juan de Oñate also brought domestic sheep to New Mexico in 1598 when he founded the first Spanish settlement (Hammond and Rey 1953). By the 1600s, livestock ranching was established in the Baja California Peninsula and California by missionaries. Hence, domestic sheep have been present and close to bighorn sheep habitats for centuries (Gulliford 2018).

In 1890, there were an estimated 625,000 domestic sheep in Arizona (Brown 2009a). By 1880 there were 2,090,000 adult sheep in New Mexico, with the highest densities in the northern Rio Grande watershed (Bernalillo County alone had 587,000; Carlson 1969), which included prime Rocky Mountain bighorn sheep habitats (Bailey 1932). New Mexico became a domestic sheep supply center for the American West and mining operations in northern Mexico (Wildeman and Brock 2000, Gulliford 2018). By 1902, there were 3,400,000 domestic sheep in Mexico, and 4,464,000 by the 1930s (Whetten 1948). Experimental research facilities at the University of Alaska, Fairbanks, and in Kodiak, Alaska, saw the arrival of domestic sheep as early as 1907, coupled with attempts to hybridize Dall's sheep and domestic Lincoln-Cotswold sheep in hopes of creating a commercial opportunity for sheep meat production (Husby et al. 1998). In the United States, domestic sheep numbers increased from 7 million in the early 1800s to 56 million in 1945, with numbers dropping precipitously afterward (Jones 2004). Introduction of domestic sheep in the American southwest and northern Rockies caused catastrophic widespread population extirpations and declines of mountain sheep populations principally through the introduction of Old-World sheep diseases (Besser and Cassirer 2023, Chapter 8).

In Canada, the first domestic sheep appear to have been brought from France in the middle of the 17th century. Following the initial importation, there were no new arrivals for approximately the next 100 years (Spencer and MacMillan 1937) when United Empire Loyalists moved north from the eastern United States into Canada's maritime provinces, Quebec and Ontario, but it was not until the late 1800s that domestic sheep arrived on the Canadian Pacific coast, having been brought there by the Hudson's Bay Company and later the Puget Sound Agricultural Company (Spencer and MacMillan 1937). In addition to the agriculture and wool industry, domestic sheep were also transported north in large part to provide mutton and clothing during the gold rush era (Bawtree and Zabek 2011). During this time, government aid to the industry consisted of financial incentives and assistance, evolving in the early 1990s into marketing boards and wool grower associations. With these supports in place and by the 1920s, thousands of domestic sheep were being ranched and allowed to graze in alpine and montane habitats, across bighorn sheep range in southern British Columbia and into Alberta. In 1935, the two provinces hosted significant numbers of domestic sheep, with estimates approaching a million head shared between these two provinces (650,000 in Alberta and 176,000 in British Columbia; Spencer and MacMillan 1937). Some ranching operations in southern Alberta during this time ranged from 1,000 to as many as 20,000 sheep (Spencer and MacMillan 1937). The practice of large-scale sheep farming, however, appeared to be largely abandoned by the 1960s, with loss of markets and public concerns about observed linkages between domestic sheep disease and bighorn sheep die-offs. "…[T]there is a hazard there of communicable disease that doesn't apparently affect the domestic sheep to any great extent, but does affect mountain sheep" (Game Department of British Columbia 1957:87). A review of disease-caused die-offs in British Columbia's bighorn sheep indicates that since the start of the 1900s, the

likely cause of most large-scale mortality events is relatable to respiratory disease (Chapter 8); however, our understanding and ability to detect respiratory pathogens such as *Mycoplasma ovipneumoniae* were limited in early years by technology that undoubtedly hindered identification of specific causes to more general descriptions (except where samples were collected and retrospective analyses could be completed).

Feral ungulates were and continue to be abundant, especially donkeys and horses in mountain sheep habitats in the United States and Mexico, and were to become a major management challenge (Jones 1980, Petersen et al. 2023). The Bureau of Land Management (2023) reported that there were approximately 82,883 feral horses and burros requiring an expenditure to manage and care for these animals in excess of $148 million. Feral livestock consume water resources and significant amounts of natural forage resources, influencing range, grazing leases, and wildlife. Feral donkeys and horses also became a major management challenge in desert bighorn environments (Marshal et al. 2008, McKnight 1964, Jones 1980, Seegmiller and Ohmart 1981, Petersen et al. 2023, Brewer et al. 2014, Rubin et al. 2024). Feral donkeys have a historical presence in North American desert environments. Feral donkeys were observed in the early 1900s near two springs in southwestern Arizona, including an estimated 3,000 burros near Agua Dulce Spring in what is now the Cabeza Prieta National Wildlife Refuge, an area inhabited by desert bighorn sheep (Lumholtz 1912). In 1915, numerous feral donkeys were observed from a vantage point in the Sierra Pinacate in northwestern Sonora, Mexico (Carmony and Brown 1993). Although western Canada does not host the numbers of feral horses and burros that occupy the western United States, continued feral equine population growth in bighorn sheep habitats has become a significant enough issue, that both Alberta and British Columbia are developing strategies to better understand the magnitude of potential overgrazing conflicts between feral horses, burros, and native species like bighorn sheep. Intensive livestock grazing also results in habitat alteration, degradation, loss, and fragmentation (Chapter 10), and habitat abandonment due to human-caused disturbance (e.g., shepherds and their domestic dogs), and some population declines are due to subsistence hunting by shepherds (Pafici et al. 2023).

Cattle and other domestic ungulates, which were first imported into Mexico, have also deleteriously influenced mountain sheep habitats. By the late 1930s, there were 11,640,000 cattle in Mexico followed in numbers by 6,854,000 goats (Whetten 1948). Domestic goats have had particularly deleterious effects on vegetation in the arid areas of northern Mexico and southwestern United States (Castellanos-Perez et al. 2002, Mellado et al. 2005) and are of special concern in mountain sheep habitats because browse is a major component of goat and desert bighorn diets (Sandoval et al. 2019), which can result in direct competition for forage resources. Goats also can graze on steep and rough terrain, habitats which are favored by mountain sheep. Introduction of cattle and the subsequent uncontrolled increase in numbers in the western United States also had negative effects on wild herbivore habitats (Fleischner 1994, 2010; Donahue 1999, Wilshire et al. 2008, Garrison et al. 2016, Sayre 2023), with similar realities occurring in western Canada at this same time (Breen 2024). Between 1870 and 1884, cattle increased from 4.6 million to 35–40 million in the 17 western states (Wildeman and Brock 2000), perhaps bolstered by the loss of bison (*Bison bison*) populations that would have disrupted subsistence food provisioning. In 1891, in Arizona alone, cattle numbers approached 1.5 million (Brown 2009a). Similarly in western Canada, construction of the Canadian Pacific Railway, resource extraction, human population expansion, and treaties with Indigenous Peoples (Treaty 6 in Alberta and Saskatchewan 1876) required a larger and more stable food supply, leading to Alberta and British Columbia's cattle industry that saw significant expansions in and around 1850–1890 (MacLachlan 2006) across the south of the two provinces and through much of this region's bighorn sheep range. Mismanagement of cattle grazing results in soil erosion and compaction, forage depletion, encroachment of tall vegetation, and the displacement of native ungulates (Wagner 1978, Fleischner 1994, 2004; Brown et al. 2010, Krausman et al. 1996, Krausman et al. 2011, Mellink and Riojas-López 2020). More importantly, excessive removal of herbage beyond a certain critical stage will result in plants losing physiological vigor, decreased herbage production, and eventual death (Head and Child 1994). Cattle forage consumption can limit bighorn populations in areas of sympatry (Garrison et al. 2016), and the presence of cattle can influence the spatial distribution and create changes in vigilance and grazing behavior (Brown et al. 2010). The United States Forest Service, which was responsible for grazing management, failed to establish sustainable forest land grazing programs during its initial years, and did not take into consideration the forage requirements necessary to sustain native wild herbivores (Wagner 1978, Chambers and Holthausen 2000). As an example, cattle grazing in unmanaged numbers in sheep habitats in what is now Sequoia National Park, California, was initiated in about 1861 and was continued until 1900, without any consideration for the forage needs of wildlife (Vankat and Major 1978), including mountain sheep. Probably more damaging to the mountain sheep population was the subsequent intensive grazing of large numbers of domestic sheep in the same area, compounding the negative effects and habitat degradation, and introduction of fatal pathogens and diseases to mountain sheep.

### 2.4.2 State and Provincial Historical Perspectives

During the territorial years and early statehood in the late 1880s, some western states outlawed market hunting and initiated hunting seasons, bag limits, and special restrictions on the hunting of some species in response to a growing realization that wildlife populations were developing

scattered distributions and were generally declining and low in numbers in some areas (Duda et al. 2010). Game harvest regulations were generally enacted without data-supported knowledge of the status and distribution of game animal populations. So, the enactment of early game laws curtailing hunting proceeded cautiously, because city and rural populations continued to rely on game animals as a source of meat. While there must have been a reluctant understanding by the public that wildlife populations were declining, the sentiment that populations could sustain continued harvesting surely existed. In the absence of biologists to monitor populations and lack of law enforcement personnel, the new laws were not enforced. There simply were no provisions for funding professionals or law enforcement; game wardens were not provided with support personnel, adequate budgets, or equipment that would enable them to be effective, and therefore, game laws were ignored by the public in favor of food provisioning and continued harvest. State game policies from the late 1800s to 1930s were based on limiting harvest, implementing predator control (also a means to reduce competition with humans for dwindling game and prey resources), and subsidizing subsistence need for meat through provision of alternative sources supplied by domestic species (e.g., cattle, sheep). In frontier areas of the north during this same period, new game laws were also being enacted to address the effects of market hunting in support of industrial camps and the gold rush that attracted thousands of people, but also to ensure that a newly formed commercial opportunity of hunting could be realized. In 1905, the newly appointed Provincial Game Warden for British Columbia enacted a new regime of control over landscape, animals, and ecology (Peyton 2011). Similar to the United States, these new laws closed seasons, created harvest limits (often referred to as bag limits today), and introduced structured enforcement and accounting procedures in response to the changed way both Indigenous Peoples and also frontiersmen saw wildlife; hunting of wildlife was no longer solely for subsistence purposes (at least some of that need could be addressed through livestock markets) and there was significant economic gain to be made through commercialization of hunting. European and southern sportsmen would pay thousands of dollars for guided hunting expeditions in northern British Columbia and Alberta, and these in turn created employment and generated significant revenue for local economies and various levels of government (Rutledge 1989, Peyton 2011). So, new game laws that created conservation outcomes were ironically not implemented in the name of conservation, as we define it today. Still, in the United States and Canada, the implementation of controls over harvest created some of the first conservation outcomes in wildlife management (Organ and McCabe 2018).

*Thinhorn sheep.* During the territorial years in Alaska, Dall's sheep were illegally hunted for commercial purposes (Sherwood 1981). Due to the remoteness of Dall's sheep habitats, however, populations had remained relatively stable with an estimated 58,150 in 1990 and 52,100 in 2000. Also, in 1990 in Canada, there were approximately 1,500 Dall's sheep and 10,500 Stone's sheep in British Columbia, 20,000 Dall's sheep in the Northwest Territory, and 22,500 Dall's (including Fannin's) and approximately 100 Stone's sheep in the Yukon Territory for a total thinhorn sheep population estimate of 54,600 (Wild Sheep Initiative 2024). The 2023 population estimates indicate that there may be as many as 43,100 Dall's sheep and 10,610 Stone's sheep in Canada, although these estimates were generated based on trends observed in a small number of surveys and the declines may be more significant than reported. Dall's sheep in Alaska numbered 22,500 in 2023, a significant decrease from the estimate of 58,150 in 1990 (Wild Sheep Initiative 2024).

*Rocky Mountain bighorn sheep.* In territorial Colorado in 1867, the hunting season was closed between 15 January and 15 August. In 1883, Colorado, which became a state in 1876, established a big game season between 1 January and 15 September, and closed sheep hunting for the next 10 years. The first game warden was appointed in 1891, the Colorado Department of Forestry, Game and Fish was created in 1897, hunting licenses were approved in 1903, and in 1913, mountain sheep were protected until 1924 (Barrows and Holmes 1990). In Montana in 1872, when still a territory, the big game hunting season was closed from February to 15 August, including sheep hunting, and market hunting was outlawed in 1877. In 1889, the year Montana became a state, there was continued protection for bison, moose, and elk, and an open season from 15 September to 31 December on mountain sheep, deer, mountain goat (*Oreamnos americanus*) and pronghorn. In 1895, the Montana state legislature established a Board of Game Commissioners, which for the first time set bag limits of eight each of sheep, deer, mountain goats, pronghorn, and three elk with a hunting season from 1 September to 1 January. In 1897, Montana prohibited the sale of game mammals and birds. The first state game warden was appointed in 1901 and the Fish and Game Commission was created. In 1907, the open deer season was from 1 October to 1 December, with a three-deer bag limit. The bag limit on elk, sheep, and goats was one of each species. In 1911, there were 59,291 hunting and fishing licenses sold, an exceptional increase. In 1915, the bighorn season was closed and not reopened until 1953 (Couey and Schallenberger 1971). There was also a high demand for live big game animals for private collectors and zoos. One dealer in Montana in the 1890s sold over 500 live big game animals, including mountain sheep, to buyers in the eastern United States (Picton and Lonner 2008). Wyoming became a state in 1890. In 1895 the hunting season extended from 1 September to 31 November with limited out-of-state licensing. In 1899, the office of state game warden was created. That year, the bag limit was one sheep, two deer, two elk, and three pronghorns. Hunting guides required certification and were appointed as assistant game wardens. In 1911, a record of 6,393 big game licenses were sold; hunting licenses were a growing source of income (Blair 1987). In Idaho, the state created a game warden position in 1899 and game licenses were first issued in 1903. A state game warden

estimated a state population of only 300 to 400 bighorn sheep in 1906. The low bighorn population ebb occurred in the 1930s. During the 1931–1934 hunting seasons, only seven bighorn tags were sold with no reported kills. Only 26 bighorns were legally hunted during the 1935–1938 seasons. The state closed the bighorn hunting season during 1940–1945 and reopened limited hunting with 13 rams collected in 1946 and 15 rams in 1947 but closed the hunting season in 1948 due to low bighorn productivity. The state opened a general hunt in 1956 and through the 1970 hunting season, hunter success was only 8% (Thiessen 2020). In New Mexico in the 1900s, Rocky Mountain bighorn sheep had been extirpated and only small populations of desert bighorns remained in southcentral, southwestern, and southeastern areas of the state (Ligon 1927). In the late 1950s, there were 8,100–10,100 Rocky Mountain bighorn sheep in the United States (Buechner 1960); estimates increased to 10,660–12,680 in 1974 (Wishart 1975). In British Columbia, Brooks (1923) declared that bighorn sheep were decimated in the eastern part of the interior due to acquisition of rifles by First Nations and spread of domestic sheep diseases, which decimated sheep east of the Fraser River in about 1870. Also, cougar (*Puma concolor*), coyote (*Canis latrans*), and golden eagle (*Aquila chrysaetos*) predation significantly influenced declining sheep populations, slowing their recovery from disease die-off events. In 1974, there were an estimated 5,200 to 5,800 bighorns in Canada of which 3,150–4,400 were in provincial parks (Wishart 1975). In 1998, there were about 42,000–45,000 Rocky Mountain and California bighorn sheep in North America, with approximately 11,630 to 14,630 occurring in Canada (Toweill and Geist 1999). Based on 2023 estimates, the number of bighorn sheep (California, Rocky Mountain, and Sierra Nevada) numbered an estimated 47,000 in North America (11,500 in Canada and 35,500 in the United States), indicating a much more robust North American population than earlier in the century (Wild Sheep Initiative 2024).

*Desert bighorn sheep.* Management of sheep in desert environments has concentrated on translocations, disease management, water developments, competition from feral livestock and the aoudad (*Ammotragus lervia*), an exotic ungulate, and predator management (Goldstein and Rominger 2003, Wakeling et al. 2009, Tsukamoto 1987, Brewer and Bleich 2023). Historically, management differed from that of Rocky Mountain sheep in that some states established captive breeding programs, although in most states for limited periods as in Texas (Cook 1994). The New Mexico Department of Game and Fish established a permanent captive rearing facility in 1972, which has been the source of animals for augmenting and reestablishing desert bighorn populations (Toweill and Geist 1999, Goldstein and Rominger 2003). As an example of a desert bighorn population historical trend, in Arizona, sheep populations through the 1860s were widespread in rugged terrain along the Colorado River to alpine habitats in the San Francisco and Kendrick mountains in northern Arizona. By 1877, the Southern Pacific Railroad had reached the Colorado River and in 1880 it reached Tucson. The railroads initiated an influx of settlers, lumbermen, miners, and livestock producers, and the ever-present market hunters, whose activities greatly degraded wildlife habitats and reduced sheep populations. The Southern Pacific Railroad also made sheep hunting accessible to hunters from California and the eastern United States. Arizona included the protection of mountain sheep in the territorial game code of 1887, but game codes were not enforced. The 1893 amended game code strictly protected mountain sheep, although the territorial government did not follow the national trend and go as far as establishing game refuges. Mountain sheep also declined in this period and by 1912, the year Arizona became a state, the urgent necessity to close big game hunting was still not recognized. That same year however, the state legislature passed a referendum requiring a hunting license and creation of the position of a salaried game warden; the referendum was approved by popular vote. A state game department was established in 1913. Sheep had been reduced in number, having been extirpated in several areas, with extant sheep numbers being particularly low throughout the state. Sheep in small numbers still occurred in the San Francisco Peaks, White Mountains, and the Superstition, Virgin, and other mountain ranges where they would also be extirpated. Consequently, the hunting season was closed in 1913; however, sheep surveys to determine their numbers and distribution were not initiated until the early 1930s (Brown 2009, 2012). There has been an upward trend in populations, beginning from the early to mid-1980s in Arizona, with population in 2023 of 5,600, second only to Nevada with 7,100 (Wild Sheep Initiative 2024). In contrast, in New Mexico, desert bighorns have had a slower recovery. By 1927, remnant populations of desert bighorn sheep in New Mexico occurred in the Big Hatchet Mountains in the southwestern area and San Andres Mountains in the southcentral region (Ligon 1927), and up to the early 1960s continued to face a bleak forecast (Sands 1967). Desert bighorns in New Mexico began to show a steady increase in numbers with the reestablishment of sheep populations in formerly distributed areas; however, recurrent disease outbreaks limited population growth (Toweill and Geist 1999). The 2023 populations estimate in New Mexico was 1,170 with a slight decrease over the last 5 years (Wild Sheep Initiative 2024). In the United States, there were an estimated 6,700–8,100 desert sheep in the late 1950s (Buechner 1960), 9,800–11,490 in 1978 (Monson 1980), 18,550 in 1991 (Valdez and Krausman 1999), and 18,000 in 1998 (Toweill and Geist 1999). In 2023, there were an estimated 22,700 desert bighorn in the United States (Wild Sheep Initiative 2024).

In Sonora and the Baja California Peninsula in northwestern Mexico, bighorn sheep populations in the modern period were unrecorded until the late 1800s and early 1900s based on sport hunting accounts, principally those of Americans in Baja California and Sonora (Hornaday 1908, Roosevelt 1920, Cudahy 1928, O'Connor 1974), and those of Sheldon (1925) who hunted sheep in Chihuahua in the late 1800s and early 1900s and Sonora in the early 1900s (Tinker 1978, Carmony and Brown 1993, Mellink 1993, Medellín et al. 2005). By the beginning of the 1900s, sheep had greatly limited distributions in Coahuila and Chihuahua occurring in isolated populations (Chapter 1). There was widespread killing of sheep in Baja California by market hunters to supply mining camps and communities during 1905–1906 (Nelson 1921). The government of northern Baja (present state of Baja California) issued a decree in 1917 prohibiting the hunting of mountain sheep (but without enforcement). In 1922, the president of Mexico issued a decree banning the hunting of mountain sheep for 10 years; that was extended in 1933 for another 10 years and again banned in 1944. Hunting was reopened in 1963 with limited sport hunting permits (Tapia 2008), and in 1974, 41 permits were issued for the Baja California Peninsula for the price of $15,000 per permit for foreign hunters of which 30 were from the United States. Nineteen (44%) of the hunters collected a male (Alvarez 1976a,b). During the decade of 1980–1990 sheep hunting seasons, 625 hunting permits were issued with about a 75% hunter success (Lee and Mellink 1996). There was concern in the public and among biologists that hunting permits were issued without adequate knowledge of the sheep population status in the state of Baja California. In 1990, the president of Mexico issued what became a permanent decree banning sheep hunting in that state (Mellink 1993, Tapia Landeros 2008, Toweill and Geist 1999, Chapter 17). By the 1960s, sheep had been extirpated in Coahuila and Chihuahua and only occurred in small, isolated populations in northwestern Sonora and the Baja California Peninsula (Valdez 1997). In 1964, there were an estimated 5,000 desert bighorn sheep in Mexico, principally in the Baja California Peninsula (Trefethen 1975). By 1978, the estimate was increased to 5,560–8,800 in Mexico, with the highest estimates (4,560–7,800) in the Baja California Peninsula (Monson 1980); however, these estimates are unreliable because they are based on minimal, random, and inaccurate survey data. In 1991 and 1998, there were an estimated 4,500 desert sheep in Mexico (Toweill and Geist 1999, Valdez and Krausman 1999), a number more consistent with habitat conditions and other limiting factors (e.g., illegal hunting). Captive sheep herds were reestablished in Chihuahua and Coahuila in 2000 and those populations increased significantly and subsequently were the source used to establish free-ranging populations by private landowners. By 2012, there were 12,700 bighorn sheep in Mexico, with approximately 7,500 in Sonora alone, of which 3,600 were in captive herds, a major mountain sheep management strategy in Mexico (Sandoval et al. 2019, Chapter 17). In 2023, there were an estimated 19,000 desert bighorn sheep in Mexico, with increasing numbers of captive and free-ranging populations indicating successful management (Chapter 17).

### 2.4.3 RESTORATION OF MOUNTAIN SHEEP POPULATIONS: HABITAT MANAGEMENT (1930–1980)

Losses in mountain sheep populations near the turn of the century did not go unnoticed and since the early 1900s, state and provincial agencies had biological and enforcement staff managing wildlife populations, with many jurisdictions undertaking bighorn sheep translocation and reintroduction efforts. As early as the 1920s Rocky Mountain bighorn sheep were being moved from Alberta to nearly all the western states, except California, Colorado, Texas, and Wyoming, to reestablish their sheep populations. In the 1950s, British Columbia also began exporting California bighorns to states such as Washington, Oregon, Idaho, and Nevada. Also, in the 1950s Arizona began moving desert bighorns to Texas (Wild Sheep Working Group 2015). So, by the 1960s, game agencies needed to hire biologists with professional degrees who could begin to conduct wildlife research on interspecific and intraspecific ecological and behavioral relationships, applying statistical and quantitative methods to wildlife research and management. Coincidentally with reintroduction efforts, a cadre of professional wildlife biologists was developed, some who concentrated on mountain sheep ecology, behavior, and diseases. Foremost was the transition to an ecosystem approach for understanding biotic and abiotic community interactions and the application of ecologically based mountain sheep habitat management. Leopold (1933) laid the foundation for initiating this modernized era of wildlife management. His ecological tenets were adopted by federal, provincial, and state agencies during this period, which enabled the beginning of recovery of many bighorn sheep populations. In the 1930s, however, although bighorn sheep populations were recovering, they were not recovering with the same success as other big game mammals. Some prominent wildlife biologists were still concerned that mountain sheep numbers in the United States relative to other widespread and numerous ungulates were unnaturally low. Estimated big game numbers in the contiguous United States in 1937 were 3.8 million white-tailed deer (*Odocoileus virginianus*), 2.2 million mule and Columbian black-tailed deer (*Odocoileus hemionus columbianus*), and 199,380 pronghorn, but only 15,325 bighorn sheep (10,168 Rocky Mountain bighorn and 5,157 desert bighorn). By comparison, the number of big game killed by sport hunters in 1941 were 440,159 deer (*Odocoileus* spp.), 22,548 elk, and 5,260 pronghorn. The only reported bighorn sheep killed legally were in Wyoming where 25 were harvested in 1941 and 22 in 1942; data from other states was not available (Jackson 1944). Few huntable mountain sheep populations remained, and these were under threat of disease die-offs (at that time attributed to lung-worm, scabies [*Psoroptes* spp.], and generic descriptions of *pneumonia*), habitat degradation, and illegal hunting. So, while pronghorn, mule deer, and elk populations increased to huntable populations in the western United States by the 1950s (Mackie 2000), mountain sheep were slow to respond to recovery efforts in the contiguous United States; in the central Okanagan and Rocky Mountains, Canada, during this period, bighorn populations had recovered. The maximum estimated Rocky Mountain and desert bighorn populations combined numbered less than 20,000 in 1955 (Buechner 1960). By 1950, Rocky Mountain bighorn occurred only in Alberta, British Columbia, Colorado, Wyoming, Montana, and Idaho, forming the impetus for a more widespread and concerted effort across all the western United States to actively reintroduce and translocate Rocky Mountain, California, and desert bighorn sheep (Whiting et al. 2023).

Federal and state governments responded to decreasing numbers of wild ungulates by creating refuges, some specifically for mountain sheep (Brown 1989), and monitoring populations and enforcing game laws on national public lands. States responded by imposing closed seasons, requiring hunting licenses, enforcing game laws, predator management, and initiating long-term studies, and concentrating their wildlife research and management programs on huntable ungulate populations. As such, mountain sheep were often not a priority species for allocation of funding by state and provincial agencies.

With the availability of Pittman-Robertson funding, game departments were able to support studies of mountain sheep, resulting in an increasing number of mountain sheep and internal agency research publications. Between 1930 and 1961, there were 12 mountain sheep field studies spanning more than a single year that were supported by state and federal agencies, except the study of Sierra Nevada bighorns (Jones 1950). There were three studies in Canada: Green (1949) on Rocky Mountain bighorns in Alberta, and Blood (1961) and Sugden (1961) who both studied California bighorns in southern British Columbia. Those studies conducted in the conterminous United States were Couey (1950) and Ogren (1954) in Montana, Carson (1941) in Texas, Honess and Frost (1942) in Wyoming, Jones (1950) and Welles and Welles (1961) in California, Smith (1954) in Idaho, and Russo (1955) in Arizona. In Alaska there were two long-term studies: Dixon (1938) and Murie (1944). Mountain sheep research studies in Mexico, other than population surveys, were initiated in the late 1970s (Eaton-González et al. 2017, Tarango and Krausman 1997) and have progressively increased (Valdez 2019). The growing focus on mountain sheep biological studies in the United States and Canada is evident in the increasing number of technical publications published between 1950 and 1969 (Post 1971), with an accelerated increase in publications from the 1970s to the 1990s (Ayers et al. 1999). These research efforts provide scientific and technological knowledge to enhance mountain sheep populations and habitats and their management in North America (Bishop and Hubbard 2018, Bleich and Thompson 2018).

Of fundamental importance for furthering desert bighorn sheep research and its dissemination was the creation of the Desert Bighorn Council, which presented a forum for desert bighorn sheep population status, research, and management reports and a publication format: Desert Bighorn

Council Transactions (1957–present; www.desertbighorn-council.com, accessed 10 April 2025). The Northern Wild Sheep Council (1971–1976) and the expanded Northern Wild Sheep and Goat Council (1977–present; www.nwsgc.org, accessed 10 April 2025) through their meetings and proceedings provide a similar forum for northern mountain sheep and mountain goat populations. Mountain sheep advocacy non-governmental organizations have also had an important effect on mountain sheep conservation and management. The most important non-governmental organization for funding mountain sheep conservation projects is the Wild Sheep Foundation (1974–present) and its state and provincial chapters and affiliates (Schultz et al. 1999). During the fiscal year 2014–2015 through the fiscal year 2023–2024 alone, the Wild Sheep Foundation raised and directed $63,182,000 for mountain sheep conservation, education, research and management programs, with most of the funding directed at state, provincial, and tribal and First Nation mountain sheep management programs (Wild Sheep Foundation 2024).

### 2.4.4 LOOKING TO THE FUTURE (1980–PRESENT)

Management and conservation of mountain sheep has rapidly advanced, but challenges remain. First and foremost, the wildlife profession continues to need dedicated biologists with passion for the resource. Natural history is the glue that holds the understanding of wildlife, including mountain sheep, together (Berger et al. 2024). This concept is not new and was recognized by Leopold (1933) as the field of wildlife management was developing. During the transition to professional wildlife biology, important individual characteristics were paramount: the ability to interact with others, a technical understanding of nature, forestry training camp education, and an aptitude for critical thinking. Those unable to accomplish the tasks of self-education tended to drop out or were replaced (Leopold 1933). Modern biological education programming is now multi-faceted, developing technical skills while also creating competencies in computer modeling, artificial intelligence, statistics, and life history information. As Leopold (1933) noted, there remains a fundamental need for a field biologist to have a deep enthusiasm for wildlife and its conservation, but he also stated that this alone is not enough as, "Mere enthusiasm, however, has failed to conserve game in the past, and is not likely to succeed better in the future" (Leopold 1933:413). So, wildlife biologists today must recognize the complexity of the environments and ecosystems they work within; being scientific investigators who demonstrate an ability to think independently, while at the same time actively engaging with, they must be able to regularly reach out to others to share local, traditional, cultural, ecological, and scientific knowledge. A lone wolf philosophy will not work in today's world any more than it did in Leopold's time.

In many ways, the crossroads biologists face today are similar to the issues of pioneering wildlife biologists. Things change and they are changing rapidly with big data, artificial intelligence, eDNA, machine learning, computer modeling, sophisticated data analysis, meta-analysis, remote sensing, climate change, and the paramount importance of understanding what stakeholders want from agencies, non-governmental organizations, universities, and others working in the arena. The field work, the glue that holds everything together, cannot be lost in the mix (Berger et al. 2024). Of course, biologists must continue to keep up with modern technology to ensure that management decisions are scientifically influenced, and they need to continue to be successful in obtaining funds so the solid research that is needed can continue. There are too many challenges facing mountain sheep for any reduction in effort.

Mountain sheep are among the most valued large mammals in the world because of their habitats, beauty and grace, popularity with the public, their role in the environment, and their ubiquitous ecological mysteries (Krausman 2017). We have learned a lot about successful and unsuccessful translocations of mountain sheep, interactions with domestic livestock and wildlife, the influence of climate change on mountain sheep, diseases and disease transmission, encroachment by humans into mountain sheep habitat, key factors related to habitat, predator-prey relationships, community dynamics, and an increasing human population, among other relationships. All of this highlights the need for future research on the response of mountain sheep to anthropogenic influences and large-scale and long-term studies that monitor the responses of increasing human growth to mountain sheep (Chapter 23), with collaboration among land-use planners, wildlife managers, and ecologists to ensure that pertinent research is integrated into regulations and policies (Krausman 2017). Leopold set the stage for scientific research in the profession nearly 100 years ago. Wildlife biologists have enhanced and continued to develop the profession ever since. Biologists, managers, administrators, and policy makers have risen to the challenge in the past as challenges arose and will continue to do so to ensure a healthy future for mountain sheep in North America (Chapter 24).

## 2.5 SUMMARY

Historical perspectives on population trends of mountain sheep in North America beginning in the late 1600s are based on the perceived numbers in remote and rugged wilderness areas across most of their range. Early historical accounts suggest robust populations of mountain sheep in Alaska, mountain ranges in western North America, and deserts in the United States and Mexico, at the arrival of early explorers and settlers. Those settlers' perceptions of wildlife and mountain sheep abundance were influenced through experiences from the Old World and were traditionally rooted in consumptive and commercial use. This differed from Indigenous Peoples' more holistic connections to mountain sheep that were formed through generations of spiritual, cultural, and consumptive use, but also through a long-term connection with mountain sheep populations and the landscapes and ecology of mountain sheep habitats. Although accounts are limited, Native Peoples probably did not negatively affect populations of mountain sheep to the extent of the consumptive activities of early Europeans. Still, in hearing stories and reading written records there appears to be a timeless, shared appreciation for mountain sheep in North America.

It is generally accepted that mountain sheep abundance before written records was influenced by environmental changes, in many of the same ways as it is today. Early explorers kept sporadic and incomplete records describing varying and sometimes conflicting assessments of mountain sheep abundance across the West, depending on year or location. As Western settlement progressed, mountain sheep populations declined dramatically, with the consensus that this was due to unregulated subsistence and market hunting, excessive livestock grazing, mining, recreation, and unfavorable weather patterns. While those factors all had influences that could be significant in localized geographic or temporal scales, we have come to understand that the then-unknown effects from domestic sheep pathogens likely played the single largest part in wildlife and mountain sheep population declines and extirpations. Historically, without the scientific understanding and technology that we have today, those early attempts to make sense of the losses logically gravitated toward more obvious causes like illegal hunting, because it made sense on a superficial level. The result is that mountain sheep were rare in most areas of the United States by the close of the 1800s and in parts of Canada by the early 1900s. This motivated management changes in some areas to make the landscape more conducive to the restoration of mountain sheep populations through the middle and later 1900s. Most bighorn sheep thrive in a variety of undisturbed montane habitats that are often fire-maintained ecosystems; thinhorns more typically are associated with climax alpine and northern grassland habitats. Fire suppression in either of these ecosystems can increase understory and shrubby vegetation growth, which can reduce available forage, visibility (a key factor in predator detection and avoidance), and generally detracts from optimal mountain sheep habitat. Land-use changes and increasingly effective harvest restrictions allowed bighorn sheep populations to rebound throughout the 1960s and peak in different areas in the late 1970s through the early 1980s, in part because of reintroduction and translocation efforts. By the mid-1990s and continuing into the early 2000s, population trends in bighorns had again shifted, with some jurisdictions experiencing declines in mountain sheep populations. Mountain sheep in the United States presently number less than a quarter million while most of the other big game species surpass a million. Mountain sheep conservation and management, however, have been a success. They were on the verge of extirpation in the western continental United States. Prominent naturalists predicted their demise, and indeed they were extirpated in several states;

however, today, some jurisdictions host large, stable, and increasing populations. Fortunately, widespread conservation efforts and the application of innovative management techniques assisted with increasing mountain sheep populations in the United States, Canada, and the deserts of Mexico. The stage has been set and everyone interested in wildlife should continue to build on the philosophy of Leopold (1933) and the many advances in the profession ever since.

## REFERENCES

Adorno, R., and P. Pautz, editors. 1999. *Álvar Núñez Cabeza de Vaca: His Account, His Life, and the Expedition of Panfilo de Narváez.* Three volumes. University of Nebraska Press, Lincoln, Nebraska, USA.

Allen, G. M. 1942. *Extinct and Vanishing Mammals of the Western Hemisphere with Marine Species of All the Oceans.* American Committee for International Wild Life Protection, New York, New York, USA.

Allen, J. A. 1912. Historical and nomenclatorial notes on North American sheep. *Bulletin of the American Museum of Natural History* 31:1–29.

Alvarez, T. 1976a. Background summary and results of the 1974 desert sheep hunting season. Pages 22–28 *in* T. Álvarez, editor. *Borrego Cimarrón. Baja California Temporada: 1974.* Subsecretaría Forestal y de la Fauna, México, D.F., México.

Alvarez, T. 1976b. Status of desert bighorns in Baja California. *Transactions of the Desert Bighorn Council* 20:18–21.

American Museum of Natural History. 1885. *Visitors' Guide to the Collection of Mammals in the American Museum of Natural History.* American Museum of Natural History, New York, New York, USA.

Anderson, R. M. 1919. Field studies of life histories of Canadian mammals. *Canadian Field Naturalist* 33:86–90.

Audubon, J. J., and J. Bachman. 1846–1854. *The Viviparous Quadrupeds of North America.* Three volumes. V. G. Audubon, New York, New York, USA.

Audubon, M. R., editor. 1897. *Audubon and His Journals with Zoological and Other Notes by Elliot Coues.* Charles Scribner's Sons, New York, New York, USA.

Ayers, L. A., C. D. Broadhead, and S. H. Anderson, editors. 1999. *A Bibliography to Rocky Mountain Bighorn Sheep Literature.* Wyoming Cooperative Fish and Wildlife Research Unit, Cheyenne, Wyoming, USA.

Bailey, V. 1932. Mammals of New Mexico. *North American Fauna* 53:1–412.

Bailey, V. 1936. The mammals and life zones of Oregon. *North American Fauna* 55:1–416.

Baillie-Grohman, W. A. 1882. *Camps in the Rockies.* Sampson Low, Marston, Searle, and Rivington, London, United Kingdom.

Banfield, A. W. F. 1974. *The Mammals of Canada.* University of Toronto Press, Toronto, Ontario, Canada.

Bannon, A. H. 1911. *A Hunter's Summer in Yukon Territory.* Privately printed, Columbus, Ohio, USA.

Barrows, P., and J. Holmes. 1990. *Colorado's Wildlife Story.* Colorado Division of Wildlife, Denver, Colorado, USA.

Bates, R. H. 1988. *Mountain Man: The Story of Belmore Browne: Hunter, Explorer, Artist, Naturalist and Preserver of Our Northern Wilderness.* Amwell Press, Clinton, New Jersey, USA.

Bawtree, A., and L. Zabek. 2011. Fur, gold, and settlement: the building blocks of range management in British Columbia. *Society for Range Management* 53:45–49.

Benedict, J. B. 1992. Footprints in the snow: high-altitude cultural ecology of the Colorado Front Range, U.S.A. *Arctic and Alpine Research* 24:1–16.

Benedict, J. B. 1996. *The Game Drives of the Rocky Mountain National Park*. Research Report 7. Center for Mountain Archaeology, Ward, Colorado, USA.

Berger, J., V. C. Bleich, and R. T. Bowyer. 2024. Must we lose our biological connection to nature to endure changing times? *Journal of Wildlife Management* 88:e22639.

Besser, T. E., and E. F. Cassirer. 2023. Epizootic bighorn sheep pneumonia caused by *Mycoplasma ovipneumoniae*. Pages 145–163 in D. A. Jessup and R. W. Radcliffe, editors. *Wildlife Disease and Health in Conservation*. Johns Hopkins University Press, Baltimore, Maryland, USA.

Biscotti, M. L. 2019. *American Sporting Periodicals: An Annotated Bibliography*. Rowan and Littlefield, Lanham, Maryland, USA.

Bishop, C. J., and M. W. Hubbard. 2018. The role of field research in state wildlife management. Pages 208–219 in T. J. Ryder, editor. *State Wildlife Management and Conservation*. Johns Hopkins University Press, Baltimore, Maryland, USA.

Blair, N. 1987. *The History of Wildlife Management in Wyoming*. Wyoming Game and Fish Department, Cheyenne, Wyoming, USA.

Bleich, V., and D. J. Thompson. 2018. State management of big game. Pages 78–95 in T. J. Ryder, editor. *State Wildlife Management and Conservation*. Johns Hopkins University Press, Baltimore, Maryland, USA.

Blood, D. A. 1961. An ecological study of California bighorn sheep Ovis canadensis californiana (Douglas) in southern British Columbia. Thesis, University of British Columbia, Vancouver, British Columbia, Canada.

Bolton, H. E. 1930. *Anza's California Expeditions*. Volume 4. University of California Press, Berkeley, California, USA.

Bowman, D. M., G. L. Perry, S. I. Higgins, C. N. Johnson, S. D. Fuhlendorf, and B. P. Murphy. 2016. Pyrodiversity is the coupling of biodiversity and fire regimes in food webs. *Philosophical Transactions of the Royal Society: Biological Sciences* 371:20150169.

Boyd, R. T., editor. 2022. *Indians, Fire, and the Land in the Pacific Northwest*. Second edition. Oregon State University Press, Corvallis, Oregon, USA.

Bradley, J. R. 1904. *Hunting Big Game in Northwest British Columbia*. Privately printed, Mail and Express Job Print, New York, New York, USA.

Breen, D. H. 2024. Ranching history. *The Canadian Encyclopedia. Historica Canada*. https://www.thecanadianencyclopedia.ca/en/article/ranching-history. Accessed 26 October 2024.

Brewer, C. E., and V. Bleich. 2023. Desert bighorn sheep. Pages 19–33 in K. M. Lehr and J. Schwab, editors. *Records of North American Big Game. Volume 2.* Fifteenth edition. Boone and Crockett Club, Missoula, Montana, USA.

Brewer, C. E., V. C. Bleich, J. A. Foster, T. Hosch-Hebdon, D. E. McWhirter, E. M. Rominger, M. W. Wagner, and B. P. Wiedmann. 2014. *Bighorn Sheep: Conservation Challenges and Management Strategies for the 21st Century*. Wild Sheep Working Group, Western Association of Fish and Wildlife Agencies, Cheyenne, Wyoming, USA.

Brooks, A. 1923. The Rocky Mountain sheep in British Columbia. *Canadian Field Naturalist* 37:23–25.

Brown, D. E. 1989. Early history. Pages 1–11 in R. M. Lee, editor. *The Desert Bighorn Sheep in Arizona*. Arizona Game and Fish Department, Tucson, Arizona, USA.

Brown, D. E. 2009. Shifting attitudes and the beginnings of wildlife restoration: the onset of laws and bureaucrats. Pages 263- 359 in D. E. Brown, editor. *Arizona Wildlife: The Territorial Years, 1863–1912*. Arizona Game and Fish Department, Phoenix, Arizona, USA.

Brown, D. E. 2012. *Bringing Back the Game: Arizona Wildlife Management, 1912–1962*. Arizona Game and Fish Department, Phoenix, Arizona, USA.

Brown, N. A., K. E. Ruckstuhl, S. Donelon, and C. Corbett. 2010. Changes in vigilance, grazing behaviour and spatial distribution of bighorn sheep due to cattle presence in the Royal Provincial Park, Alberta. *Agriculture, Ecosystems and Environment* 135:226–231.

Buechner, H. K. 1960. The bighorn sheep in the United States, its past, present, and future. *Wildlife Monographs* 4:1–174.

Bunch, T. D., C. Wu, Y. P. Zhang, and S. Wang. 2006. Phylogenetic analysis of snow sheep (*Ovis nivicola*) and closely related taxa. *Journal of Heredity* 97:21–30.

Bureau of Land Management. 2023. *Wild Horse and Burro Program: Highlights from Fiscal Year 2023*. Bureau of Land Management, Washington, D.C., USA.

Burroughs, R. E. 1961. *The Natural History of the Lewis and Clark Expedition*. Michigan State University Press, East Lansing, Michigan, USA.

Campos, E., J. Cuéllar, O. Salvador, E. García-Trejo, and F. Poreira. 2020. The genetic diversity and phylogeography of Mexican domestic sheep. *Small Ruminant Research* 187:106109.

Carmony, N. B., and D. E. Brown. 1993. *The Wilderness of the Southwest: Charles Sheldon's Quest for Desert Bighorn Sheep and Adventures with the Havasupai and Seri Indians*. University of Utah Press, Salt Lake City, Utah, USA.

Carson, B. 1941. *Man, the Greatest Enemy of Desert Bighorn Mountain Sheep: A History of a Game Animal Rapidly Disappearing from Its Native Habitat in the Mountains of West Texas*. Bulletin No. 21. Texas Game, Fish and Oyster Commission, Austin, Texas, USA.

Castellanos-Perez, E., M. Valencia-Castro, and J. J. Quiñones. 2002. Goats and the need for range management in Mexico. *Rangelands* 24:24–27.

Chambers, C. L., and R. S. Holthausen. 2000. Montane ecosystems used as rangelands. Pages 213–280 in R. Jemison and C. Raish, editors. *Livestock Management in the American Southwest: Ecology, Society, and Economics*. Elsevier, Amsterdam, Netherlands.

Cook, R. L., compiler. 1994. *A Historical Review of Reports, Field Notes and Correspondence on the Desert Bighorn Sheep in Texas*. Revised. Texas Parks and Wildlife, Austin, Texas, USA.

Couey, F., and A. Schallenberger. 1971. Bighorn sheep. Pages 97–105 in T. W. Mussehl and F. W. Howell, editors. *Game Management in Montana*. Montana Fish and Game Department, Helena, Montana, USA.

Couey, F. M. 1950. *Rocky Mountain Bighorn Sheep of Montana*. Bulletin No. 2. Montana Fish and Game Commission, Helena, Montana, USA.

Cowan, I. M. 1940. Distribution and variation in the native sheep of North America. *American Midland Naturalist* 24:505–580.

Crosby, A. W. 1972. *The Columbian Exchange: Biological and Cultural Consequences of 1492*. Greenwood Press, Westport, Connecticut, USA.

Crosby, H. W. 1984. *The Cave Paintings of Baja California: Discovering the Great Murals of an Unknown People.* Revised edition. Sunbelt Publications, Chula Vista, California, USA.

Cudahy, J. 1928. *Mañanaland. Adventures with Camera and Rifle Through California in Mexico.* Duffield and Company, New York, New York, USA.

Czech, K., and R. Valdez. 2016. *An Annotated Bibliography of Books Related to Hunting the Mountain Sheep and Goats of the World.* Safari Press, Long Beach, California, USA.

Damm, G. R., and N. Franco. 2014. *The CIC Caprinae Atlas of the World.* Two volumes. CIC International Council for Game and Wildlife Conservation. Budakeszi, Hungary in cooperation with Rowland Ward Publications, Johannesburg, Republic of South Africa.

Demarchi, R. A., and C. L. Hartwig. 2004. *Status of Thinhorn Sheep in British Columbia.* Wildlife Bulletin B-119. British Columbia Ministry of Water, Land and Air Protection, Biodiversity Branch, Victoria, Canada.

Demarchi, R. A., C. L. Hartwig, and D. A. Demarchi. 2000. *Status of the California Bighorn Sheep in British Columbia.* Wildlife Bulletin No. B-99. Wildlife Branch, British Columbia Ministry of the Environment, Lands and Parks, Victoria, Canada.

Dixon, J. S. 1938. Birds and mammals of Mount McKinley National Park, Alaska. *United States National Park Service, Fauna Series* 2:1–236.

Donahue, D. L. 1999. *The Western Range Revisited: Removing Livestock from Public Lands to Conserve Native Biodiversity.* University of Oklahoma Press, Norman, Oklahoma, USA.

Duda, M. D., M. F. Jones, and A. Criscione. 2010. *The Sportsman's Voice: Hunting and Fishing in America. Volume 1.* Responsive Management, Harrisonburg, Virginia, USA.

Dunmire, W. W. 2014. *New Mexico's Spanish Livestock Heritage: Four Centuries of Animals, Land, and People.* University of New Mexico Press, Albuquerque, USA.

Dunraven, W. T. 1876. *The Great Divide.* Chatto and Windus, London, United Kingdom.

Eaton-González, R. A. A. Guevara Carrizales, and J. Tapia-Mercado, coordinadores. 2017. *Estudios sobre el Borrego Cimarrón en el Noroeste de México.* Universidad Autónoma de Baja California, Mexicali, Baja California, México.

Emmet, W. T. 1901. *Good hunting! A Record of Certain Occurrences in British Columbia during the Summer of 1888.* Privately printed, New York, New York, USA.

Ffolliott, P. F., L. A. Cabrera, and C. M Guido. 1996. Use of fire in the future: benefits, concerns, constraints. Pages 177–222 *in* P. F. Ffollliot, et al., technical coordinators. *Effects of Fire on Madrean Province Ecosystems: A Symposium Proceedings.* U.S. Forest Service General Technical Report RM-GTR-289, Fort Collins, Colorado, USA.

Fleischner, T. L. 1994. Ecological costs of livestock grazing in western North America. *Conservation Biology* 8:629–644.

Fleischner, T. L. 2010. Livestock grazing and wildlife conservation in the American West: historical, policy and conservation biology perspectives. Pages 235–265 *in* J. T. du Toit, R. Kock and J. C. Deutsch, editors. *Wild Rangelands: Conserving Wildlife While Maintaining Livestock in Semi-Arid Ecosystems.* Wiley-Blackwell, Hoboken, New Jersey, USA.

Frison, G. C. 2004. *Survival by Hunting: Prehistoric Human Predation and Animal Prey.* University of California Press, Berkeley, California, USA.

Frison, G. C., C. A. Reher, and D. N. Walker. 1990. Prehistoric mountain sheep hunting in the central Rocky Mountains of North America. Pages 208–240 *in* L. B. Davis and B. O. K. Reeves, editors. *Hunters of the Recent Past.* Unwin Hyman, London, United Kingdom.

Game Department of British Columbia. 1957. *Eleventh Annual Game Convention Report of Proceedings.* Game Department of British Columbia, Victoria, Canada.

Gámez-Brunswick, C., and O. Rojas-Soto. 2020. New insights into palaeo-distributions based on Holocene rock art. *Journal of Biogeography* 47:2543–2553.

Garfinkel, A. P., D. A. Young, and R. F. M. Yoke II. 2010. Bighorn hunting, resource depression, and rock art in the Coso Range, eastern California: a computer simulation model. *Journal of Archaeological Science* 37:42–51.

Garfinkel, A. P., and D. R. Austin. 2011. Reproductive symbolism in Great Basin rock art: bighorn sheep hunting, fertility, and forager ideology. *Cambridge Archaeological Journal* 21:1–19.

Garrison, K. R., J. W. Cain III, E. M. Rominger, and E. Goldstein. 2016. Sympatric cattle grazing and desert bighorn sheep foraging. *Journal of Wildlife Management* 80:197–207.

Geist, V. 1993. *Mountain Sheep Country.* North Word Press, Minocqua, Wisconsin, USA.

Gilman, C. 2003. *Lewis and Clark: Across the Divide.* Smithsonian Institution, Washington, D.C., USA.

Goldstein, E., and E. Rominger. 2003. *Plan for the Recovery of Bighorn Sheep in New Mexico, 2003–2013.* New Mexico Department of Game and Fish, Santa Fe, USA.

Grant, C. 1980. The desert bighorn and aboriginal man. Pages 7–39 *in* G. Monson and L. Sumner, editors. *The Desert Bighorn: Its Life History, Ecology, and Management.* University of Arizona Press, Tucson, Arizona, USA.

Grant, C., J. W. Baird, and J. K. Pringle. 1968. *Rock Drawings of the Coso Range, Inyo County, California.* Maturango Museum, China Lake, California, USA.

Green, H. U. 1949. *The Bighorn Sheep of Banff National Park.* National Parks and Historic Sites Service, Development Services Branch, Ottawa, Ontario, Canada.

Grinnell, G. B. 1904. The mountain sheep and its range. Pages 270–348 *in* G. B. Grinnell, editor. *American Big Game and Its Haunts.* Forest and Stream Publishing, New York, New York, USA.

Gulliford, A. 2018. *The Wooly West: Colorado's Hidden History of Sheepscapes.* Texas A&M University Press, College Station, Texas, USA.

Gwich'in Elders. 1997. *Hành' Kak Geenjit Gwich'in Ginjik/ Gwich'in Words about the Land.* Gwich'in Renewable Resource Board, Inuvik, Northwest Territories, Canada.

Hammond, G. P., and D. A. Rey. 1953. *Don Juan de Oñate, Colonizer of New Mexico.* University of New Mexico Press, Albuquerque, New Mexico, USA.

Head, H. F., and R. D. Child. 1994. *Rangeland Ecology and Management.* Westview Press, Boulder, Colorado, USA.

Heller, M. 1997. *American Hunting and Fishing Books: An Annotated Bibliography of Books and Booklets on American Hunting and Fishing, 1800–1970.* Nimrod and Pescatore Press, Mesilla, New Mexico, USA.

Hockett, B. S., and E. Dillingham, with contributions by C. A. Shaw and M. O'Brien. 2023. *Large-Scale Traps of the Great Basin.* Texas A&M University Press, College Station, Texas, USA.

Honess, R. F., and N. M. Frost. 1942. *A Wyoming Bighorn Sheep Study*. Bulletin No. 2. Wyoming Game and Fish Department, Cheyenne, Wyoming, USA.

Hornaday, W. T. 1899. *Popular Official Guide to the New York Zoological Park, with Maps, Plans and Illustrations*. New York Zoological Society, New York, New York, USA.

Hornaday, W. T. 1904. *The American Natural History: A Foundation of Useful Knowledge of the Higher Animals of North America*. Charles Scribner's Sons, New York, New York, USA.

Hornaday, W. T. 1906. *Camp-Fires in the Canadian Rockies*. Charles Scribner's Sons, New York, New York, USA.

Hornaday, W. T. 1908. *Camp-Fires on Desert and Lava*. Charles Scribner's Sons, New York, New York, USA.

Hornaday, W. T. 1909. *Popular Guide to the New York Zoological Park*. Tenth edition. New York Zoological Society, New York, USA.

Husby, F. M., G. A. Mitchel, D. Grindle, and J. Hanscom, editors. 1998. Agroborealis 30:12, 22 (Spring). *Agricultural and Forestry Experiment Station*, University of Alaska, Fairbanks, USA. https://scholarworks.alaska.edu/handle/11122/1595. Accessed 27 October 2024.

Jackson, H. H. T. 1944. *Big-Game Resources of the United Staes, 1937–1942*. Research Report 8. United States Fish and Wildlife Service, Washington, D.C., USA.

Jacoby, K. 2001. *Crimes against Nature: Squatters, Poachers, Thieves, and the Hidden History of American Conservation*. University of California Press, Berkeley, California, USA.

Jones, F. L. 1950. A survey of the Sierra Nevada bighorn. *Sierra Club Bulletin* 35:29–76.

Jones, F. L. 1980. Competition. Pages 197–216 in G. Monson and L. Sumner, editors. *The Desert Bighorn: Its Life History, Ecology, and Management*. University of Arizona Press, Tucson, Arizona, USA.

Jones, K. G. 2004. *Trends in the U. S. Sheep Industry*. Agricultural Information Bulletin 787. United States Department of Agriculture, Economic Research Service, Washington, D.C., USA.

Kay, C. E. 1994. Aboriginal overkill: the role of Native Americans in structuring western ecosystems. *Human Nature* 5:359–398.

Kay, C. E. 2002a. Are ecosystems structured from the top-down or bottom-up? Pages 179–214 in C. E. Kay and R. L. Simmons, editors. *Wilderness and Political Ecology: Aboriginal Influences and the Original State of Nature*. University of Utah Press, Salt Lake City, Utah, USA.

Kay, C. E. 2002b. Afterword: false gods, ecological myths, and biological reality. Pages 238–259 in C. E. Kay and R. L. Simmons, editors. *Wilderness and Political Ecology: Aboriginal Influences and the Original State of Nature*. University of Utah Press, Salt Lake City, Utah, USA.

Krausman, P. R. 2017. *And Then There Were None: The Demise of Desert Bighorn Sheep in the Pusch Ridge Wilderness*. University of New Mexico Press, Albuquerque, New Mexico, USA.

Krausman, P. R., R. Valdez, and J. A. Bissonette. 1996. Bighorn sheep and livestock. Pages 237–243 in P. R. Krausman, editor. *Rangeland Wildlife*. Society for Range Management, Denver, Colorado, USA.

Krausman, P. R., V. C. Bleich, W. M. Block, D. E. Naugle, and M. C. Wallace. 2011. An assessment of rangeland activities on wildlife populations and habitats. Pages 253–290 in D. D. Briske, editor. *Conservation Benefits of Rangeland Practices*. United States Department of Agriculture, Natural Resource Conservation Service, Washington, D.C., USA.

LaBelle, J. M., and S. R. Pelton. 2013. Communal hunting along the continental divide of northern Colorado: results from the Olson Drive (5BL147), USA. *Quaternary International* 297:45–63.

Lee, R., and D. E. Mellink. 1996. Status of bighorn sheep in Mexico – 1995. *Desert Bighorn Council Transactions* 40:35–39.

Lemke, A. 2022. *The Architecture of Hunting: The Built Environment of Hunter-Gatherers and Its Impact on Mobility, Property, Leadership, and Labor*. Texas A&M University Press, College Station, Texas, USA.

Leopold, A. 1933. *Game Management*. Charles Scribner's Sons, New York, New York, USA.

Lewis, H. T. 1978. Traditional uses of fire by Indians in northern Alberta. *Current Anthropology* 19:401–402.

Lewis, H. T., and T. A. Ferguson. 1988. Yards, corridors, and mosaics: how to burn a boreal forest. *Human Ecology* 16:57–77.

Ligon, J. S. 1927. *Wildlife of New Mexico: Its Conservation and Management*. New Mexico Department of Game and Fish, Santa Fe, New Mexico, USA.

Lumholtz, K. S. 1912. *New Trails in Mexico: An Account of One Year's Exploration in Northwestern Sonora, Mexico and Southwestern Arizona, 1909–1910*. Charles Scribner's Sons, New York, New York, USA.

Lyman, R. L. 2004. Aboriginal overkill in the intermountain west of North America: Zooarchaeological tests and implications. *Human Nature* 15:169–208.

Mackie, R. J. 2000. History and management of large mammals of North America. Pages 282–320 in S. Demarais and P. R. Krausman, editors. *Ecology and Management of Large Mammals in North America*. Prentice Hall, Upper Saddle River, New Jersey, USA.

MacLachlan, I. 2006. *The Historical Development of Cattle Production in Canada*. University of Lethbridge, Lethbridge, Alberta, Canada. https://opus.uleth.ca/items/1cbd7ff1-1c55-493c-a9f0-fb90ff6c88d3. Accessed 26 October 2024.

Marshal, J., V. C. Bleich, and N. G. Andrew. 2008. Evidence for interspecific competition between feral ass *Equus asinus* and mountain sheep *Ovis canadensis* in a desert environment. *Wildlife Biology* 14:228–236.

Martin, P. S. 1967. Prehistoric overkill. Pages 75–120 in P. S. Martin and H. E. Wright Jr., editors. *Pleistocene Extinctions: The Search for a Cause*. Yale University Press, New Haven, Connecticut, USA.

Maschner, H., and O. K. Mason. 2013. The bow and arrow in northern North America. *Evolutionary Anthropology* 22:133–138.

McGillivray, D. 1803. Memorandum respecting the mountain ram of North-America. *Medical Repository* 6:238–240.

McKnight, T. 1964. Feral livestock in Anglo-America. *University of California Publications in Geography* 16:1–67.

Medellín, R. A., C. Manterola, M. Valdéz, D. G. Hewitt, D. Doan-Crider, and T. E. Fulbright. 2005. History, ecology, and conservation of the pronghorn antelope, bighorn sheep, and black bear in Mexico. Pages 387–404 in J.- L. E. Cartron, G. Ceballos, and R. S. Felger, editors. *Biodiversity, Ecosystems, and Conservation in Northern Mexico*. Oxford University Press, New York, New York, USA.

Mellado, M., A. Olvera, A. Quero, and G. Mendoza. 2005. Diets of prairie dogs, goats, and sheep on a desert rangeland. *Rangeland Ecology and Management* 58:373–379.

Mellink, E. 1993. The president spoke. Pages 201–220 in G. P. Nabhan, editor. *Counting Sheep: Twenty Ways of Seeing Desert Bighorn*. University of Arizona Press, Tucson, USA.

Mellink, E., and M. E. Riojas-López. 2020. Livestock and grassland interrelationships along five centuries of ranching the semiarid grasslands on the southern highlands of the Mexican Plateau. *Science in the Anthropocene* 8:20.

Miller, D. S., E. Hoberg, G. Wesier, K. Aune, M. Atkinson, and C. Kimberling. 2012. A review of hypothesized determinants associated with bighorn sheep (*Ovis canadensis*) die-offs. *Veterinary Medicine International* 2012:796527.

Monson, G. 1980. Distribution and abundance. Pages 40–51 *in* G. Monson and L. Sumner, editors. *The Desert Bighorn: Its Life History, Ecology, and Management.* University of Arizona Press, Tucson, USA.

Moulton, G. E., editor. 1987. *The Definitive Journals of Lewis and Clark. Volume 4. From Fort Mandan to Three Forks.* University of Nebraska Press, Lincoln, Nebraska, USA.

Muir, J. 1874. The wild sheep of California. *Overland Monthly* 12:358–363.

Muir, J. 1894. *The Mountains of California.* Century Company, New York, New York, USA.

Murie, A. 1944. The wolves of Mount McKinley. *United States National Park Service, Fauna Series No.* 5:1–238.

Murray, W. B., and A. Espinosa -T. 2006. The natural setting of bighorn petroglyphs in the eastern Sierra Madre (Nuevo Leon, Coahuila, Mexico). *American Indian Rock Art* 32:45–51.

Nagaoka, L. 2012. The overkill hypothesis and its impact on conservation biology. Pages 110–138 *in* S. Wolverton and R. L. Lyman, editors. *Conservation Biology and Applied Zooarchaeology.* University of Arizona Press, Tucson, Arizona, USA.

Nagaoka, L., T. Rick, and R. T. Wolverton. 2018. The overkill model and its impact on environmental research. *Ecology and Evolution* 8:9683–9696.

Nelson, E. W. 1921. Lower California and its natural resources. *Memoirs of the National Academy of Sciences* 16:1–194.

O'Connor, J. 1974. *Sheep and Sheep Hunting.* Winchester Press, New York, New York, USA.

Ogren, H. A. 1954. *A Population Study of Bighorn Sheep (Ovis canadensis canadensis) on Wildhorse Island.* Pittman-Robertson Federal Aid Project, Job No. V-A-W-1-R-13 and W-60-R-1. Montana Game and Fish Department, Helena, Montana, USA.

Organ, J. F., and R. E. McCabe. 2018. History of state wildlife management in the United States. Pages 1–23 *in* T. J. Ryder, editor. *State Wildlife Management and Conservation.* Johns Hopkins University Press, Baltimore, Maryland, USA.

Pafici, M., A. Cristiana, and M. Lumbierres. 2023. Drivers of habitat availability for terrestrial mammals: unravelling the role of livestock, land conversion and intrinsic traits in the past 50 years. *Global Change Biology* 29:6900–6911.

Palliser, J. 1853. *Solitary Rambles and Adventures of a Hunter in the Prairies.* J. Murray, Long, London, United Kingdom.

Parkman, F. 1849. *The California and Oregon Trail. Being Sketches of Prairie Rocky Mountain life.* George P. Putnam, New York, New York, USA.

Petersen, S. L., J. D. Scasta, K. A. Schoenecker, and J. D. Hennig. 2023. Feral equids. Pages 735–757 *in* L. B. McNew, D. K. Dahlgren, and J. L. Beck, editors. *Rangeland Wildlife Ecology and Conservation.* Springer Nature, Cham, Switzerland.

Peyton, J. 2011. Imbricated geographies of conservation and consumption in the Stikine Plateau. *Environment and History* 17:555–581.

Phillips, J. C. 1930. *American Game Mammals and Birds. A Catalogue of Books, 1582 to 1925. Sport, Natural History, and Conservation.* Houghton Mifflin, Boston, Massachusetts, USA.

Picton, H. D., and T. N. Lonner. 2008. *Montana's Wildlife Legacy: Decimation to Restoration.* Media Works, Bozeman, Montana, USA.

Pigoti, J. S., K. B. Springer, J. S. Honke, et al. 2023. Independent age estimates resolve the controversy of ancient human footprints at White Sands. *Science* 382:73–75.

Post, G., editor. 1971. *An Annotated Bibliography of the Mountain Sheep of North America.* Rachelwood Wildlife Research Preserve, New Florence, Pennsylvania, USA.

Rich, E. E., and A. M. Johnson, editors. 1955. *Black's Rocky Mountain Journal – 1924: A Journal of a Voyage from Rocky Mountain Portage in Peace River to the Sources of Finlays Branch and North West Ward in Summer 1824.* Hudson's Bay Record Society, London, United Kingdom.

Risenhoover, K. L., and J. A. Bailey. 1985. Foraging ecology of mountain sheep: implications for habitat management. *Journal of Wildlife Management* 49:797–804.

Roosevelt, K. 1920. *The Happy Hunting-Grounds.* Charles Scribner's Sons, New York, New York, USA.

Roosevelt, T. 1885. *Hunting Trips of a Ranchman.* G. P. Putnam's Sons, New York, New York, USA.

Roosevelt, T. 1888. *Ranch Life and the Hunting-Trail.* The Century Company, New York, New York, USA.

Roosevelt, T. 1893. *The Wilderness Hunter.* G. P. Putnam's Sons, New York, New York, USA.

Roosevelt, T. 1905. *Outdoor Pastimes of an American Hunter.* Charles Scribner's Sons, New York, New York, USA.

Rubin, E. S., D. Conrad, L. E. Harding, and B. M. Russo. 2024. Associations between a feral equid and the Sonoran Desert ecosystem. *Wildlife Monographs* 215:e1083.

Rush, W. M. 1927. Notes on diseases in wild game animals. *Journal of Mammalogy* 8:163–165.

Russell, O. 1921. *Journal of a Trapper: or Nine Years in the Rocky Mountains, 1834–1843.* Sims York, Boise, Idaho, USA.

Russo, J. P. 1955. *The Desert Bighorn Sheep in Arizona: A Research and Management Study.* Arizona Game and Fish Department, Phoenix, Arizona, USA.

Rutledge, L. 1989. *That Some May Follow: The History of Guide Outfitting in British Columbia.* Guide Outfitters Association of British Columbia, Mile House, Canada.

Seton, E. T. 1901. *Lives of the Hunted. Containing a True Account of the Doings of Five Quadrupeds and Three Birds, and in Elucidation of the Same, 200 Drawings.* Charles Scribner's Sons, New York, New York, USA.

Seton, E. T. 1927. *Lives of the Game Animals. Accounts of Those Land Animals in America, North of the Mexican Border, Which Are Considered "Game," Either Because They Have Held the Attention of the Sportsmen, or Received the Protection of the Law.* Volume 3. Hoofed mammals. Double, Doran and Company, Garden City, New York, USA.

Sandoval, A. V., R. Valdez, and A. Espinosa -T. 2019. Desert bighorn sheep in Mexico. Pages 350–365 *in* R. Valdez and J. A. Ortega- S., editors. *Wildlife Ecology and Management in Mexico.* Texas A&M University Press, College Station, Texas, USA.

Sands, J. L. 1967. Bighorn sheep. Pages 69–82 *in New Mexico Wildlife Management.* New Mexico Department of Game and Fish, Santa Fe, USA.

Sayre, N. F. 2023. A history of North American rangelands. Pages 49–71 in L. B. McNew, D. K. Dahlgren, and J. L. Beck, editors. *Rangeland Wildlife Ecology and Conservation*. Springer Nature, Cham, Switzerland.

Schneider, J. S., R. S. Begole, M. Jorgensen, E. S. Rubin, and L. L. Jee. 2014. Prehistoric bighorn sheep procurement tactics in the Colorado Desert: a hypothesis for a stone-feature complex in Yaqui Pass, Anza-Borrego Desert State Park, California. *Journal of California and Great Basin Anthropology* 34:181–210.

Schultz, R. A., D. A. Pedrotti, and S. C. Reneau, editors. 1999. *Putting Sheep on the Mountain: The Foundation for North American Wild Sheep: Twenty-Five Years Dedicated to Mountain Sheep, 1974–1999*. Foundation for North American Wild Sheep, Cody, Wyoming, USA.

Seegmiller, R. F., and R. D. Ohmart. 1981. Ecological relationships of feral burros and desert bighorn sheep. *Wildlife Monographs* 78:1–58.

Shaw, G. 1804. Ovis canadensis. *Naturalist's Miscellany* 15: Text (unpaginated) attached to Plate 610.

Sheldon, C. 1925. The big game of Chihuahua, Mexico. Pages 138–181 in G. B. Grinnell and C. Sheldon, editors. *Hunting and Conservation, the Book of the Boone and Crockett Club*. Yale University Press, New Haven, Connecticut, USA.

Sherwood, M. 1981. *Big Game in Alaska: A History of Wildlife and People*. Yale University Press, New Haven, Connecticut, USA.

Shichi, K., T. Goebel, M. Izuho, and K. Kashiwaya. 2023. Climate amelioration, abrupt vegetation recovery, and the dispersal of *Homo sapiens* in Baikal Siberia. *Science Advances* 9:eadi0189.

Slifer, D. 1998. *Signs of Life: Rock Art of the Upper Rio Grande*. Ancient City Press, Santa Fe, New Mexico, USA.

Smith, D. R. 1954. *The Bighorn Sheep in Idaho: Its Status, Life History and Management*. Wildlife Bulletin No. 1. Idaho Game and Fish Department, Boise, Idaho, USA.

Southesk, Earl of. 1875. *Saskatchewan and the Rocky Mountains: A Diary and Narrative of Travel, Sport, and Adventure during a Journey through the Hudson's Bay Company's Territories in 1859 and 1860*. Edmonston and Douglas, Edinburgh, United Kingdom.

Spencer, J. B., and A. A. MacMillan, editors. 1937. *Sheep Husbandry in Canada*. Eighth edition. Farmers' Bulletin 30. Department of Agriculture, Field Services, Live Stock Branch, Ottawa, Canada.

Sugden, L. G. 1961. *The California Bighorn in British Columbia, with Particular Reference to the Churn Creek Herd*. British Columbia Department of Recreation and Conservation, Victoria, Canada.

Surovell, T. A. 2008. Extinction of big game. *Encyclopedia of Archaeology* 2:1365–1374.

Tapia Landeros, A. 2008. *Homo-Ovis: El Borrrego Cimarrón En México*. Universidad Autónoma de Baja California, Mexicali, México.

Tarango, L. A., and P. R. Krausman. 1997. Desert bighorn sheep in Mexico. *Desert Bighorn Council Transactions* 41:1–7.

Taylor, D. E. 2016. *The Rise of the American Conservation Movement: Power, Privilege, and Environmental Protection*. Duke University Press, Durham, North Carolina, USA.

Taylor, W. P. 1924. The basic importance of life-history studies. *Journal of Mammalogy* 5:44–48.

Taylor, W. P. 1948. Outlines for study of mammalian ecology and life histories. *United States Fish and Wildlife Service, Wildlife Leaflet* 304:1–26.

Taylor, W. T. T., P. Librado, Chief Joseph American Horse, Carlton Shield Chief Gover, J. Arterberry, Antonia Loretta Afraid of Bear Cook, H. L. Heron, Robert Milo Yellow Hair, M. Gonzalez, B. Means, et al. 2023. Early dispersal of domestic horses into the Great Plains and northern Rockies. *Science* 379:1316–1323.

Thiessen, J. 2020. *Idaho Wildlife: History, Exploitation, Politics and Management*. Privately printed. Lewiston, Idaho, USA.

Tinker, B. 1978. *Mexican Wilderness and Wildlife*. University of Texas Press, Austin, Texas, USA.

Toweill, D. E., and V. Geist. 1999. *Return of Royalty: Mountain Sheep of North America*. Boone and Crockett Club and Foundation for North American Wild sheep, Missoula, Montana, USA.

Trefethen, J. B., editor. 1975. The Wild Sheep in Modern North America. Proceedings of the workshop on the management biology of North American Wild Sheep. Boone and Crockett Club, Missoula, Montana, USA.

Tsukamoto, G. 1987. Bighorn sheep: desert cliff-hanger. Pages 161–175 in H. Kallman, editor. *Restoring America's Wildlife, 1937–1987: The First 50 Years of the Federal Aid in Wildlife Restoration (Pittman-Roberston) Act*. United States Fish and Wildlife Service, Washington, D.C., USA.

Tubbs, S. A., and C. S. Jenkinson. 2003. *The Lewis and Clark Companion. An Encyclopedic Guide to the Voyage of Discovery*. Henry Holt, New York, New York, USA.

Valdez, R. 1988. *Wild Sheep and Wild Sheep Hunters of the New World*. Wild Sheep and Goat International, Mesilla, New Mexico, USA.

Valdez, R. 1997. Mexico. Pages 303–307 in D. M. Shackleton, editor. *Wild Sheep and Goats and Their Relatives: Status Survey and Conservation Action Plan for Caprinae*. International Union for Conservation of Nature and Natural Resources, Gland, Switzerland.

Valdez, R. 2011. Snow sheep (*Ovis nivicola*). Pages 737–738 in D. E. Wilson and R. A. Mittermeier, editors. *Handbook of the Mammals of the World*. Volume 2. *Hoofed Mammals*. Lynx Edicions, Barcelona, Spain.

Valdez, R. 2019. Historical and conservation perspectives of wildlife in Mexico. Pages 1–18 in R. Valdez and J. A. Ortega- S., editors. *Wildlife Ecology and Management in Mexico*. Texas A&M University Press, College Station, Texas, USA.

Valdez, R. 2021. A value worth preserving: cultural diversity shaped Aldo Leopold's life and philosophy. *The Wildlife Professional* 15:44–45.

Valdez, R., and P. R. Krausman. 1999. Description, distribution, and abundance of mountain sheep in North America. Pages 3–22 in R. Valdez and P. R. Krausman, editors. *Mountain Sheep of North America*. University of Arizona Press, Tucson, Arizona, USA.

Vandersommers, D. 2016. Narrating animals history from the crags: a turn of the century tale about mountain sheep. *Journal of American Studies* 51:751–777.

Vankat, J. L., and J. Major 1978. Vegetation changes in Sequoia National Park. *Journal of Biogeography* 5:377–402.

Venegas, M. 1757. *Noticia de la California y su Conquista hasta el Tiempo Presente. Volume* 1. Emprenta de la Viuda de Manuel Fernandez. Madrid, Spain.

Wagner, F. H. 1978. Livestock grazing and the livestock industry. Pages 121–145 in H. P. Brokaw, editor. *Wildlife and America. Contributions to an Understanding of American Wildlife and Its Conservation*. Council for Environmental Quality, Washington, D.C., USA.

Wakeling, B. F., R. Lee, D. Brown, R. Thompson, M. Tluczek, and M. Weisenberger. 2009. The restoration of desert bighorn sheep in the Southwest, 1951–2007: factors influencing success. *Desert Bighorn Council Transactions* 50:1–15.

Warner, T. J., editor. 2004. *The Dominguez-Escalante Journal; Their Expedition through Colorado, Utah, Arizona, and New Mexico in 1776.* University of Utah Press, Salt Lake City, Utah, USA.

Warren, L. S. 1997. *The Hunter's Game: Poachers and Conservationists in Twentieth Century America.* Yale University Press, New Haven, Connecticut, USA.

Welles, R. E., and F. B. Welles. 1961. The bighorn of Death Valley. *Fauna of the National Parks of the United States, Fauna Series No.* 6:1–242.

Whetten, N. L. 1948. *Rural Mexico.* University of Chicago Press, Chicago, Illinois, USA.

White, D. A., compiler. 2001. *Plains and Rockies, 1800–1865. One Hundred Twenty Proposed Additions to the Wagner-Camp and Becker Bibliography of Travel and Adventure in the American West.* Arthur H. Clark, Spokane, Washington, USA.

Whiting, J. C., V. C. Bleich, R. T. Bowyer, and C. W. Epps. 2023. Restoration of bighorn sheep: history, successes, and remaining conservation issues. *Frontiers in Ecology and Evolution* 11:1083350

Whitley, D. S. 2013. Archaeologists, Indians, and aspects of rock art research. *Time and Mind* 6:81–88.

Whitley, D. S. 2023. Ethnography, shamanism, and far western North American rock art. *Boletín del Museo Chileno de Arte Precolombino* 28:275–299.

Wildeman, G., and J. H. Brock. 2000. Grazing in the southwest: history of land use and grazing since 1540. Pages 1–25 *in* R. Jemison and C. Raish, editors. *Livestock Management in the American Southwest: Ecology, Society, and Economics.* Elsevier Science, Amsterdam, Netherlands.

Wild Sheep Foundation. 2024. *Wild Sheep Foundation: Conservation Impact.* Wild Sheep Foundation, Bozeman, Montana, USA.

Wild Sheep Initiative. 2024. *Wild Sheep Population Estimates, License Sales and Harvest 1990–2023, ver. 2024.* Western Association of Fish and Wildlife Agencies, Boise, Idaho, USA.

Wild Sheep Working Group. 2015. *Records of Wild Sheep Translocations-United States and Canada, 1922-Present.* Western Association of Fish and Wildlife Agencies, Cheyenne, Wyoming, USA.

Williams, G. W. 2002. Aboriginal use of fire: are there any "natural" plant communities? Pages 179–214 *in* C. E. Kay and R. T. Simmons, editors. *Wilderness and Political Ecology: Aboriginal Influences and the Original State of Nature.* University of Utah Press, Salt Lake City, Utah, USA.

Wilshire, H. G., J. E. Neilson, and R. W. Hazlett. 2008. *The American West at Risk: Science, Myths, and Politics of Land Abuse and Recovery.* Oxford University Press, New York, New York, USA.

Winship, G. P. 1896. The Coronado Expedition. Pages 329–613 *in The Fourteenth Annual Report of the Bureau of Ethnology. Part 1.* Smithsonian Institution, Washington, D.C., USA.

Wishart, W. 1975. Report on recommendations of the Rocky Mountain bighorn workshop group. Pages 165–207 *in* J. B. Trefethen, editor. *The Mountain Sheep in Modern North America. Proceedings of the Workshop on the Management Biology of North American Wild Sheep.* Boone and Crockett Club, Missoula, Montana, USA.

Witte, S. S., and M. V. Gallagher, editors. 2010. *The North American Journals of Prince Maximillian of Wied. Volume 2, April-September 1833.* University of Oklahoma Press, Norman, Oklahoma, USA.

Wright, G. M., J. S. Dixon, and B. H. Thompson. 1933. *A Preliminary Survey of Faunal Relationships in National Parks.* United States National Park Service, Washington, D.C., USA.

# 3 Range-Wide Status of Mountain Sheep

*Daryl W. Lutz, Froylán Hernández, and Darren L. Bruning*

## 3.1 INTRODUCTION

Mountain sheep (bighorn sheep [*Ovis canadensis*], thinhorn sheep [*Ovis dalli*]) are distributed throughout suitable landscapes in western North America (Figure 3.1). The number of mountain sheep was estimated to be 500,000 (Valdez 1988) to 4,000,000 (Seton 1929). Early American naturalist Ernest Thompson Seton estimated there may

have been as many as 2 million mountain sheep in the lower 48 states and another 2 million in Canada and Alaska in the 1800s. Regardless of the accuracy of these early accounts, mountain sheep populations declined sharply as European settlement spread into the western part of North America. The distribution of bighorn sheep also contracted considerably from the mid-1800s to the 2000s (Figure 3.2), and by the mid-1950s it was estimated there were fewer than

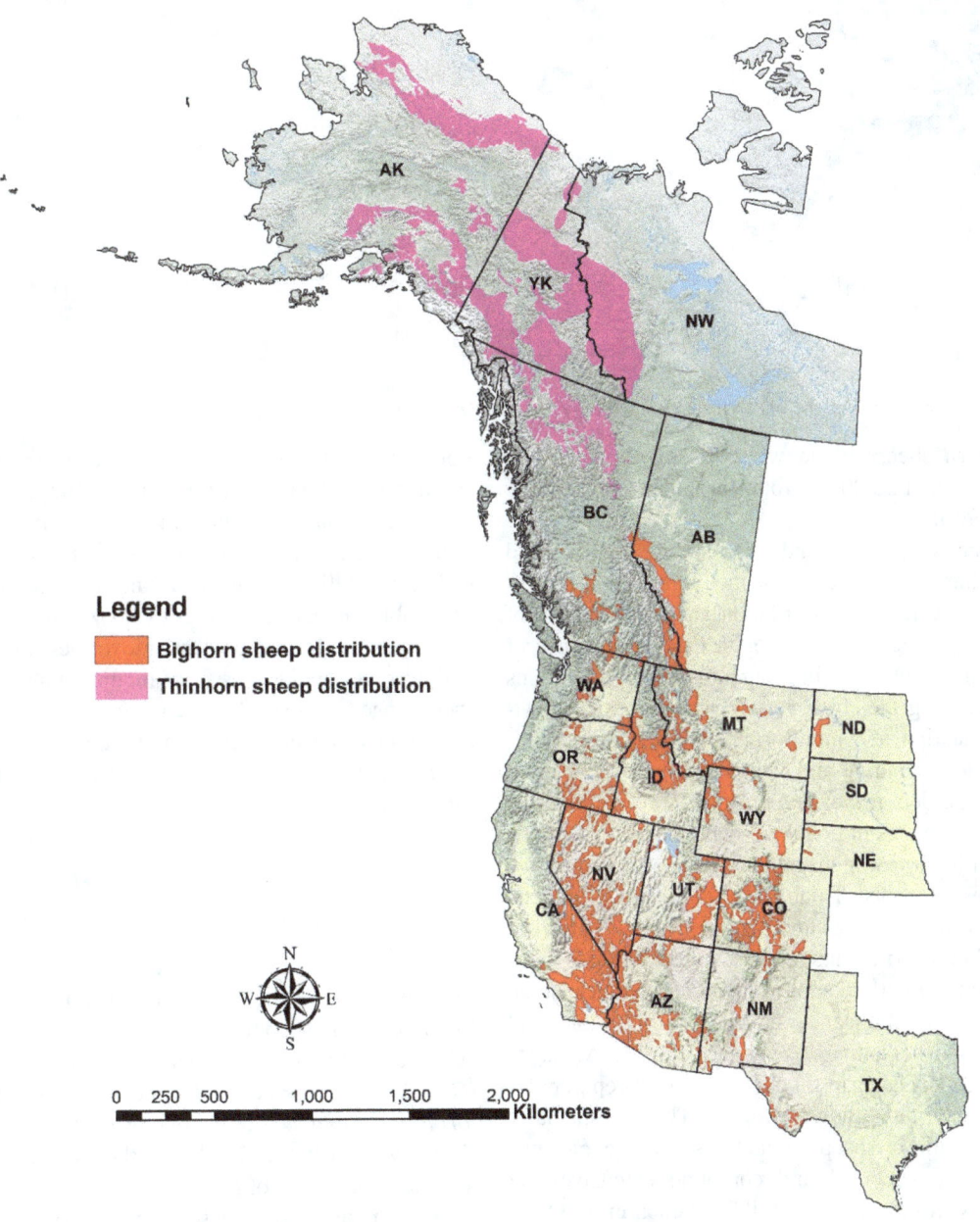

**Legend**

🟧 Bighorn sheep distribution
🟪 Thinhorn sheep distribution

**FIGURE 3.1**  Mountain sheep distribution in North America.

DOI: 10.1201/9781003518686-4

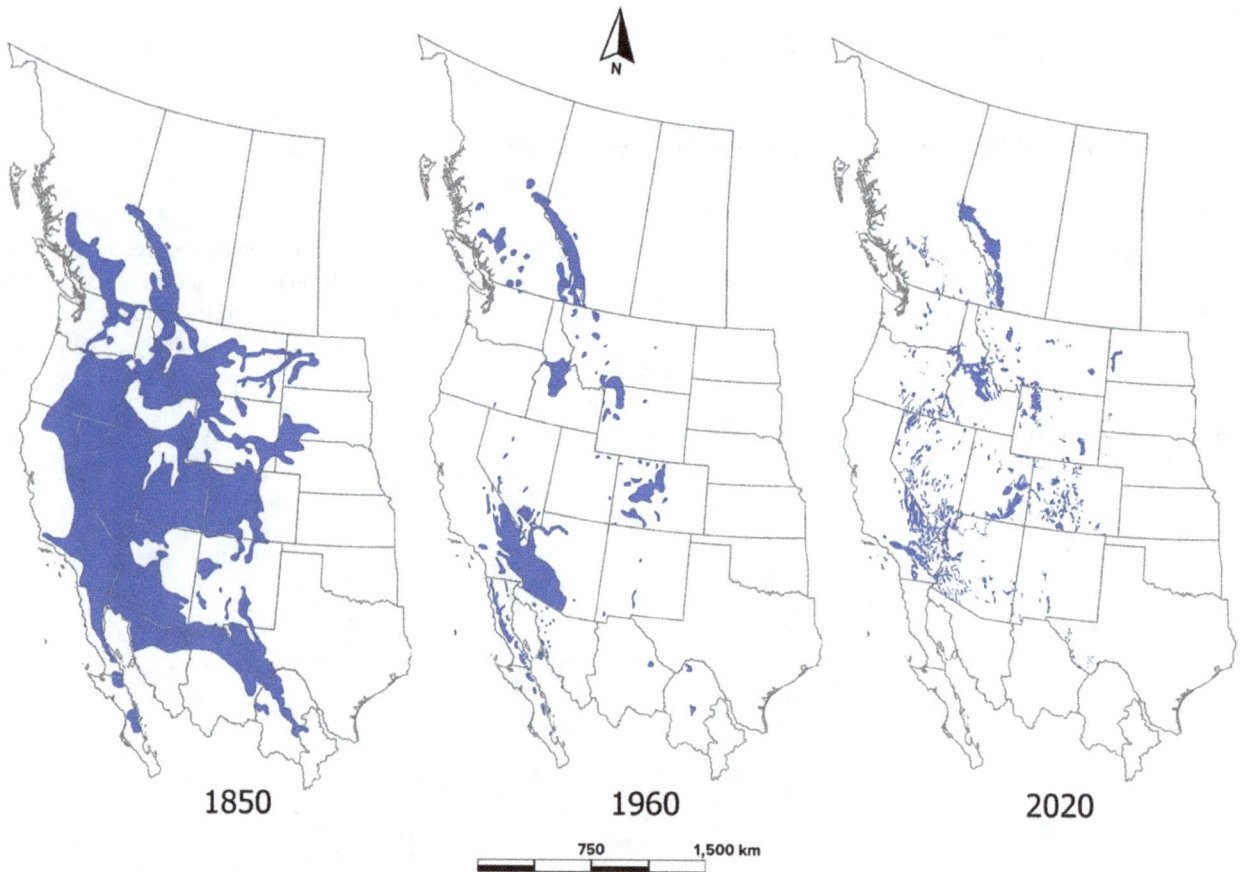

**FIGURE 3.2**   Distribution of the bighorn sheep in North America, 1850–2012.

17,000 bighorn sheep in the western United States and perhaps fewer than 25,000 in total throughout their range (Buechner 1960).

Knowledge of past abundance, status, trends, and population fluctuation patterns of mountain sheep is fundamental for the future conservation and management of these iconic species. Broad reasons for the importance of understanding the changes in mountain sheep populations include monitoring the effectiveness of management strategies, identification and monitoring of threats and influences, and understanding and the prevention of unwanted herd population declines or extirpation. More specifically, knowledge of changes in population abundance and trends can assist with harvest management, the status of reintroductions, the outcomes of habitat manipulations, and the effects of land-use practices and regulatory protections.

All jurisdictions (i.e., state and provincial wildlife management agencies) with mountain sheep have at least some population status, abundance or trend information, and data (Western Association of Fish and Wildlife Agencies [WAFWA] 2024). Data gaps about mountain sheep population change and status, however, exist. This is especially the case for thinhorn sheep populations where expansive, remote landscapes curtail and confound extensive and regular survey efforts (Simmons 1982, Veitch et al. 1998, Alaska Department of Fish and Game 2014, Wild Sheep

Foundation 2014, Jex et al. 2016). Regarding wildlife conservation in Mexico, including the management of mountain sheep, advancements in wildlife management are stifled by the constant shifting of federal agencies responsible for wildlife management and because of limited and unreliable financial resources (Valdez et al. 2009).

In this chapter, we present an overview of the historical and current population size of mountain sheep throughout their range in North America. We also describe the current trends and changes of mountain sheep populations, their distribution, and the value of continued monitoring throughout their range.

## 3.2   NUMBERS AND POPULATION STATUS OF MOUNTAIN SHEEP

Alaska supports the largest numbers of mountain sheep, followed by Canada, the contiguous United States, and Mexico. Dall's sheep (*O. dalli dalli*) far outnumber all other subspecies, followed in abundance by Rocky Mountain (*O. c. canadensis*) and California bighorn sheep (*O. c. californiana*) combined, desert bighorn, and Stone's sheep (*O. d. stonei*; WAFWA 2024). Presently, there is no range-wide abundance estimate of North American mountain sheep.

Each management jurisdiction employs different techniques to enumerate and estimate the status of mountain sheep populations within their political borders. Many jurisdictions do not attempt to estimate population size but rely on changes in trend based on varying metrics such as harvest, hunter effort, or trend counts. The WAFWA Wild Sheep Initiative compiles mountain sheep population status and abundance annually from these jurisdictional estimates (Tables 3.1–3.3; WAFWA 2024).

Rocky Mountain and California bighorn sheep populations are generally stable across North America. In Alberta, British Columbia, Canada; Colorado, Montana, Nebraska, Nevada, New Mexico, South Dakota, and Utah, United States, there has been some evidence of recent or longer-term downward trends (Figure 3.3). In contrast, mountain sheep populations in Arizona, North Dakota, Oregon, Washington, and Wyoming have been stable or increased. In many jurisdictions, population increases are attributed to recent reintroduction and augmentation efforts (WAFWA 2024). Rocky Mountain bighorn sheep populations are increasing in Idaho, whereas California bighorn sheep populations are declining. In British Columbia, California

## TABLE 3.1

**Estimated Population, License Sales, and Harvest of Rocky Mountain and California Bighorn Sheep in the United States and Canada, 2023**

|  | Estimated Population | Male Licenses | Male Harvest | Female Licenses | Female Harvest |
|---|---|---|---|---|---|
| Alberta | 6,000 | 2,482[a] | 199 | 173 | 62 |
| Arizona | 1,300 | 19 | 19 | n/a | n/a |
| British Columbia[b] | 4,250 | 179 | n/a | 7 | n/a |
| California | n/a | n/a | n/a | n/a | n/a |
| Colorado | 7,000 | 250 | 174 | 91 | 58 |
| Idaho | 3,400 | 96 | 78 | 0 | n/a |
| Montana | 4,748 | 120[c] | 132[d] | 77 | 31 |
| Nebraska | 215 | 2 | 2 | 0 | 0 |
| Nevada | 2,000 | 41 | 31 | 0 | 0 |
| New Mexico | 1,565 | 28 | 27 | 30 | 30 |
| North Dakota | 490 | 6 | 6 | 0 | n/a |
| Oregon | 5,500 | 79[e] | 71 | 22 | 8 |
| South Dakota | 335 | 11 | 11 | 0 | 0 |
| Utah | 1,646 | 63 | 60 | 5 | 4 |
| Washington | 1600 | 25 | 25 | 28 | 25 |
| Wyoming | 6,987 | 186 | 154 | 50 | 28 |

[a] Alberta; includes licenses allocated to outfitters, and general and special (limited entry hunt) licenses sold to Alberta residents.

[b] British Columbia; data listed under licenses includes only the numbers of available draws/permits issued in limited entry hunt zones.

[c] Limited entry hunt units only.

[d] Montana; includes limited and unlimited hunting district male harvest.

[e] Includes controlled hunts, one auction tag, and one raffle tag.

*n/a=not available.*

## TABLE 3.2

**Estimated Population, License Sales, and Harvest of Desert Bighorn Sheep in the United States and Canada, 2023**

|  | Estimated Population | Male Licenses | Male Harvest | Female Licenses | Female Harvest |
|---|---|---|---|---|---|
| Arizona | 5,600 | 137 | 131 | n/a | n/a |
| California | 4,500 | 26 | 20 | n/a | n/a |
| Colorado | 500 | 15 | 14 | 0 | 0 |
| Nevada | 7,100 | 240[a] | 193 | 90 | 44 |
| New Mexico | 1,168 | 29 | 29 | 0 | 0 |
| Texas | 530 | 4 | 2[b] | 0 | 0 |
| Utah | 3,273 | 85 | 75 | 0 | 0 |

[a] Total male tags including management male hunt for one-horned males.

[b] Two outstanding tags at time of print; hunt season ends 31 July.

n/a=not available

**TABLE 3.3**

**Estimated Population, License Sales, and Harvest of Thinhorn Sheep in the United States and Canada, 2023**

|  | Estimated Population | Male Licenses | Male Harvest | Female Licenses | Female Harvest |
|---|---|---|---|---|---|
| **Dall's sheep** |  |  |  |  |  |
| Alaska | Decline | 4,147 | 424 | 0 | 0 |
| British Columbia[a] | 700 | 46 | n/a | 0 | 0 |
| NW Territories | n/a | n/a | n/a | n/a | n/a |
| Yukon | 22,500 | 1,742 | 225 | 0 | 0 |
| **Stone's sheep** |  |  |  |  |  |
| British Columbia | 12,300 |  | n/a | 0 | 0 |
| Yukon | 100 |  | n/a | n/a | n/a |

[a] British Columbia; data listed under licenses includes only the numbers of available draws or permits issued in limited entry hunt zones.

n/a = not available.

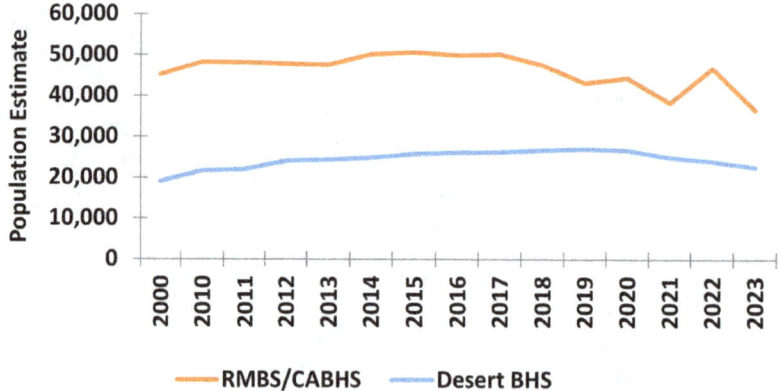

**FIGURE 3.3** Bighorn sheep (Rocky Mountain and California bighorn sheep – RMBS, CABS) and desert bighorn sheep population trend throughout North America, 2000–2023.

bighorn sheep have declined significantly over the past 5 years (WAFWA 2024).

Desert bighorn sheep have and continue to occupy the arid mountain ranges in California, Nevada, Utah, Arizona, Colorado, New Mexico, and Texas and the Mexican states of Baja California, Baja California Sur, Sonora, Chihuahua, Coahuila, and Nuevo Leon. Since 2000, their numbers have remained stable throughout their range within the United States (Figure 3.3). In certain jurisdictions, however, such as Nevada and Texas, desert bighorn sheep populations have declined since 2019. Nevada's population peaked in 2019 at 10,300 sheep but in 2023, they declined to an estimated 7,200 animals (WAFWA 2024). In Texas, the desert bighorn sheep population experienced a steady increase and remained disease-free from 1993 through 2019. In 2019, the first disease event associated with *Mycoplasma ovipneumoniae* (*M. ovi*) was detected in one population in Texas. Shortly after, several other disease events associated with *M. ovi* occurred and resulted in a decrease from around 1,500 to 600 animals (WAFWA 2024). In Colorado and California, desert bighorn sheep populations have remained stable at about 500 and 4,500 animals,

respectively (WAFWA 2024). In New Mexico, they have steadily increased from about 150 (E. Rominger, New Mexico Department of Game and Fish, unpublished data) to about 1,200 animals since 1990 (T. Batter, New Mexico Department of Game and Fish, unpublished data). Trend data for the Mexican jurisdictions is lacking, but there were an estimated 12,655 animals in the early 2010s (Sandoval et al. 2019) and in 2023 the population was estimated at 17,000 animals (Brewer and Bleich 2023; refer to chapter 17 for a detailed review of mountain sheep status in Mexico).

In Alaska, United States and Yukon, Northwest Territories, and British Columbia, Canada, thinhorn sheep populations occupy suitable landscapes distributed across historical range. In Alaska, trend data and anecdotal information point to an overall decline in abundance of Dall's sheep since roughly 2010, with a recent (2021–2024) sharp decline in many populations due to harsh winter conditions over multiple years (WAFWA 2024, Table 3.3). Historically, throughout Alaska's vast Dall's sheep range, there is high variability in population trends. Since 2010, even as some of Alaska's sheep populations declined, other populations have increased (Alaska Department of Fish and Game 2014).

In the Yukon, some Dall's sheep populations have declined 50%–60% since 2015 (Jex et al. 2025). Almost all populations of Dall's sheep in the Northwest Territories were declining near the start of the 21st century. Range-wide voluntary observations from outfitter clients have also indicated a similar and concurrent decline in sheep seen per hunter-day (Chan 2025). In British Columbia, Stone's and Dall's sheep numbers have declined since 2018 (WAFWA 2024, Figure 3.4), with localized variability of estimated declines ranging from 13% to 50% (Jex et al. 2025). Also, in British Columbia new genetic information) expanded the range of Dall's sheep to include areas previously identified as Stone's sheep range (Sim et al. 2019). The adjustment of the range of the two subspecies has resulted in revisions to population estimates for both Stone's and Dall's sheep in British Columbia (WAFWA 2024).

## 3.3 PRIMARY CAUSES FOR SHIFTING STATUS OF MOUNTAIN SHEEP

While many mountain sheep populations are currently stable, some are below management goals in many jurisdictions, due to a variety of factors of which disease, primarily pneumonia, harsh winter conditions, and climate change are perhaps the most important (Jex et al. 2016, Cassirer et al. 2017). Wildlife managers, veterinarians, and researchers have been seeking solutions to population decline with some success (WAFWA 2024).

*M. ovi,* a bacterial pathogen considered an initiating agent of pneumonia in mountain sheep, is of utmost concern to mountain sheep managers throughout North America. Today, there are few bighorn sheep populations that have avoided infection; many have experienced catastrophic all-age die-offs when exposed (Chapter 8). Recently, a strain of enzootic *M. ovi* not found in mountain sheep populations in the lower 48 states was detected in several wild ungulate populations in Alaska, including Dall's sheep (Lieske

et al. 2022). Exposure to novel pathogens could be catastrophic to thinhorn sheep populations in northern Canada and Alaska. Some populations of bighorn sheep, such as the northern badlands population in North Dakota, have demonstrated resilience after a die-off with population recovery (WAFWA 2024), while others have not been able to recover at all (e.g., Whiskey Mountain Herd, Wyoming; Wyoming Game and Fish Department 2019). Other populations have recovered in the short term only to be re-infected and once again experience a population decline (e.g., Yakima Canyon Herd, WA; Bernatowicz et al. 2017). Lamb survival and recruitment underpins population stagnation after an infection as lambs are readily exposed by pathogen carriers in the population and simply cannot overcome the disease and die (Cassirer et al. 2013, Smith et al. 2014). Separation of mountain sheep from domestic sheep and goats (WAFWA 2025) and barbary sheep or aoudads (*Ammotragus lervia*; WAFWA 2024) is paramount to minimize the transmission of this pathogen. Other diseases or pathogens of concern are discussed in Chapter 8.

Harsh winter conditions and climate change have also contributed to changes in mountain sheep populations and their dynamics (Lohuis et al. 2018, Van de Kerk et al. 2020). Winter severity, primarily snow depth, underpins many mountain sheep population declines (WAFWA 2024). For example, Dall's sheep in Alaska and the Yukon declined after extraordinarily harsh winters with record accumulated snow and icing conditions (Rattenbury et al. 2018, Nelson 2023, Yukon Department of Environment 2023, Wells 2024, 2025; Osburn 2025).

Climate change influences on high-elevation alpine and mountain ecosystems are occurring rapidly (Parmesan 2006, Gude 2022, Van de Kerk et al. 2020, White et al. 2025). In northern latitudes, there is concern that changing climate (Olsen et al. 2011) and the corresponding severe weather events coupled with decreasing habitat suitability due to increased "shrubification" (Dial et al. 2016) could

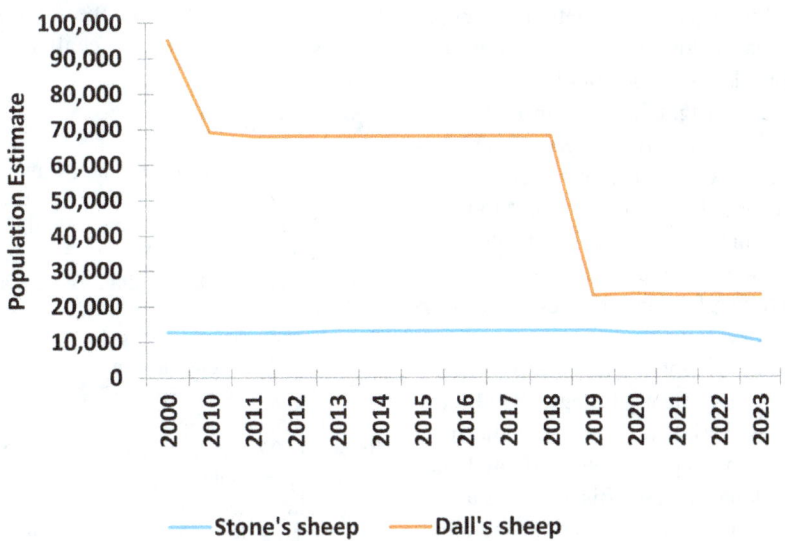

**FIGURE 3.4** Thinhorn sheep population trend in British Columbia and Yukon, Canada 2000–2023.

lead to population declines of Dall's sheep (Van de Kerk et al. 2020, Jex et al. 2025). For example, Jex et al. (2025) documented all-age mortalities, and reduced lamb survival and recruitment of Dall's sheep in Alaska, Yukon, Northwest Territories, and British Columbia because of the winters of 2019–2020, 2020–2021, and 2021–2022. Other weather-related events, such as severe multiyear droughts, are impacting mountain sheep populations in California and Nevada (WAFWA 2024).

Coupled with changes in climate and subsequent habitat suitability and availability, managers are concerned about increased human presence and activities including: backcountry skiing, military activities, hiking, biking, authorized and unauthorized off-road vehicle use, road and trail establishment, housing and resort development, energy and mineral development, and fire suppression (Krausman et al. 1995, Papouchis et al. 2001, Longshore et al. 2013, Jex et al. 2016, Sproat et al. 2020). Predation is another factor impacting mountain sheep populations. In New Mexico and Nevada, lamb recruitment in small, predation-depressed herds has been unable to overcome mortalities caused by mountain lion (*Felis (not Puma) concolor*) predation. Mountain sheep and feral horses overlap in resource use (Stoner et al. 2021) and the presence of horses has caused mountain sheep to avoid water sources (Ostermann-Kelm et al. 2008). In Nevada, for example, important mountain sheep habitat, including riparian areas and water sources, have been degraded by feral horses and burros (Nevada Department of Wildlife 2024).

## 3.4 MONITORING RANGE-WIDE STATUS OF MOUNTAIN SHEEP

Wildlife managers rely on current population status and trends to make appropriate decisions to guide management actions to achieve goals and objectives. Mountain sheep managers use population data to prescribe sustainable harvest, to respond to the values and general interest of publics, and to advise decisions about special designations for populations (e.g., species of greatest conservation need identified in state's wildlife action plans). Understanding historical population trends provides insight into the influences of population fluctuations. Additionally, historical and current population assessments are important in recognizing environmental change in mountain sheep habitat. Mountain sheep can also serve as an environmental indicator, and monitoring their range-wide status provides awareness of the health of biomes, and therefore, contributes to identifying potential threats to ecosystems.

Monitoring the status of mountain sheep populations can assist with management strategies and conservation actions and is essential to prioritizing the most suitable ranges or populations to preserve genetic and ecogeographical viability. Managers use this information to adapt to changes in abundance and distribution, including adjusting long-employed and long-accepted methods and strategies including harvest, habitat manipulations, and transplants. Considering the ever-increasing presence of disease, biologists may need to consider moving away from widespread, traditional trap and transplant operations that have been used to populate historically occupied mountain sheep ranges. Instead, it might become more prudent to focus management resources on those relatively healthy populations that exist in the locations with the most suitable environmental conditions to ensure the long-term viability of mountain sheep.

Because wildlife species are not bound to socio-political boundaries, timely population status reporting and information sharing is vital to the development and implementation of better management strategies contributing to successful wildlife management programs. Thereby, understanding the changes in mountain sheep populations contributes to the prevention of population declines and possible extirpation.

## 3.5 SUMMARY

It is clear the number and distribution of bighorn sheep has contracted considerably since the mid-1800s (Figure 3.2), and the status of mountain sheep populations varies throughout North America. Jurisdictions use different methods to estimate population size or trend of mountain sheep populations (Chapter 13) based on available resources and management needs and priorities. Regardless of survey methodology, data suggests we are in a period of diminished abundance for at least some mountain sheep populations. This reality and the reasons for this diminished abundance must continue to be documented, monitored, researched, and communicated to all who manage or otherwise care about mountain sheep to maintain and hopefully improve current population status.

It is imperative we continue to chronicle the past so better management decisions for mountain sheep are made today and in the future. This is especially important because the environment that mountain sheep exist in is constantly changing and the factors affecting population change become even more challenging.

## REFERENCES

Alaska Department of Fish and Game. 2014. *Trends in Alaska Sheep Populations, Hunting, and Harvests*. Division of Wildlife Conservation, Wildlife Management Report ADFG/DWC/WMR-2014-3, Juneau, Alaska, USA.

Bernatowicz, J., D. Bruning, E. F. Cassirer, R. B. Harris, K. Mansfield, and P. Wik. 2017. Management responses to pneumonia outbreaks in three Washington state bighorn herds: lessons learned and questions yet unanswered. *Biennial Symposium of the Northern Wild Sheep and Goat Council* 20:38–61.

Brewer, C. E., and V. C. Bleich. 2023. Desert bighorn sheep: species spotlight. Pages 19–33 *in* K. M. Lehr and J. Schwab, editors. *Records of North American Big Game*, 15th Edition, Volume II. Boone and Crockett Club, Missoula, Montana, USA.

Buechner, H. K. 1960. The bighorn sheep in the United States, its past, present, and future. *Wildlife Monograph* 4:3–174.

Cassirer, E. F., K. R. Manlove, E. S. Almberg, P. L. Kamath, M. Cox, P. Wolff, A. Roug, J. Shannon, R. Robinson, R. B. Harris, B. J. Gonzales, R. K. Plowright, P. J. Hudson, P. C. Cross, A. Dobson, and T. E. Besser. 2017. Pneumonia in bighorn sheep: risk and resilience. *Journal of Wildlife Management* 82:32–45.

Cassirer, E. F., R. K. Plowright, K. R. Manlove, P. C. Cross, A. P. Dobson, K. A. Potter, and P. J. Hudson. 2013. Spatio-temporal dynamics of pneumonia in bighorn sheep. *Journal of Animal Ecology* 82:518–528.

Chan, K. 2025. *Mackenzie Mountain Non-resident and Non-resident Alien Hunter Harvest Summary 2018–2021*. Environment and Natural Resources, Government of the Northwest Territories. Manuscript report, Canada.

Dial, R. J., T. Scott Smeltz, P. F. Sullivan, C. L. Rinas, K. Timm, J. E. Geck, S. Carl Tobin, T. S. Golden, and E. C. Berg. 2016. Shrubline but not treeline advance matches climate velocity in montane ecosystems of south-central Alaska. *Global Change Biology* 22:1841–1856.

Gude, J. A., N. J. DeCesare, K. M. Proffitt, S. N. Sells, R. A. Garrott, I. Rangwala, M. Biel, J. Coltrane, J. Cunningham, T. Fletcher, K. Loveless, R. Mowry, M. O'Reilly, R. Rauscher, and M. Thompson. 2022. Demographic uncertainty and disease risk influence climate-informed management of an alpine species. *Journal of Wildlife Management* 86:e22300.

Jex, B. A., J. B. Ayotte, V. C. Bleich, C. E. Brewer, D. L. Bruning, T. M. Hegel, N. C. Larter, R. A. Schwanke, H. M. Schwantje, and M. W. Wagner. 2016. *Thinhorn Sheep: Conservation Challenges and Management Strategies for the 21st Century*. Wild Sheep Working Group, Western Association of Fish and Wildlife Agencies, Boise, Idaho, USA.

Jex, B. A., K. Chan, M. Larivee, B. R. Wendling, and D. W. Lutz. 2025. *Observed Changes in Populations of Thinhorn Sheep in Western Canada and Alaska*. Wild Sheep Initiative, Western Association of Fish and Wildlife Agencies, Boise, Idaho, USA.

Krausman, P. R., W. W. Shaw, R. C. Etchberger, and L. K. Harris. 1995. The decline of bighorn sheep in the Santa Catalina Mountains, Arizona. Pages 245–250 *in* L. F. DeBano, P. F. Folliott, A. Ortega-Rubio, G. J. Gottfried, R. H. Hamre, and C. B. Edminster, technical coordinators. *Biodiversity and Management of the Madrean Archipelago: The Sky Islands of the Southwestern United States and Northeastern Mexico*. U. S. Department of Agriculture Forest Service, RM-GTR-264.

Lieske, C. L., D. R. Herndon, M. A. Highland, and K. B. Beckmen. 2022. Laboratory concordance study for the molecular detection of *Mycoplasma ovipneumoniae*. *Journal of Wildlife Disease* 58:257–268.

Lohuis, T., K. Smith, L. Metherell, and R. Dial. 2018. Dall's sheep population declines in Alaska's Chugach Range may be related to climate and weather patterns. *Biennial Symposium of the Northern Wild Sheep and Goat Council* 21:76.

Longshore, K., C. Lowrey, and D. B. Thompson. 2013. Detecting short-term responses to weekend recreation activity: desert bighorn avoidance of hiking trails. *Wildlife Society Bulletin* 37:698–706.

Nelson, M. 2023. *Dall Sheep Management Report and Plan, Game Management Units 20B, 20F, and 25C, White Mountains*. Alaska Department of Fish and Game, Species Management Report and Plan ADFG/DWC/SMRP-2023-4, Juneau, Alaska, USA.

Nevada Department of Wildlife. 2024. *Policy 67: Feral Horses and Burros. Report to the Nevada Board of Wildlife Commissioners*. Nevada Department of Wildlife, Reno, Nevada, USA.

Olsen, M. S., T. V. Callaghan, J. D. Reist, L. O. Reiersen, D. Dahl-Jensen, M. A. Granskog, B. Goodison, G. K. Hovelsrud, M. Johansson, R. Kallenborn, and J. Key. 2011. *The changing Arctic cryosphere and Likely Consequences: an overview*. Ambio 40:111–118.

Osburn, C. R. 2025. *Dall Sheep Management Report and Plan, Game Management Units 23 and 26A: Report Period 1 July 2016–30 June 2021, and Plan Period 1 July 2021–30 June 2026*. Alaska Department of Fish and Game, Species Management Report and Plan ADF&G/DWC/SMRP-2025-10, Juneau, Alaska, USA.

Ostermann-Kelm, S., E. R. Atwill, E. S. Rubin, M. C. Jorgensen, W. M. Boyce. 2008. Interactions between feral horses and desert bighorn sheep at water. *Journal of Mammalogy* 89:459–466.

Papouchis, C. M., F. J. Singer, and W. B. Sloan. 2001. Responses of desert bighorn sheep to increased human recreation. *Journal of Wildlife Management* 65:573–582.

Parmesan, C. 2006. Ecological and evolutionary responses to recent climate change. *Annual Review of Ecology, Evolution, and Systematics* 37:637–669.

Rattenbury, K. L., J. H. Schmidt, D. K. Swanson, B. L. Borg, B. A. Mangipane, and P. J. Sousanes. 2018. *Delayed spring onset drives declines in abundance and recruitment in a mountain ungulate*. Ecosphere 9:e02513.

Sandoval, A. V., R. Valdez, and A. Espinosa -T. 2019. Desert bighorn sheep in Mexico. Pages 350–365 *in* R. Valdez and J. A. Ortega-S., editors. *Wildlife Ecology and Management in Mexico*. Texas A&M University Press, College Station, Texas, USA.

Seton, E. T. 1929. The bighorn. Pages 519–573 *in* E. T. Seton, editor. *Lives of the Game Animals. Vol. 3. Part 2*. Doubleday, Doran Company, Garden City, New York, USA.

Sim, Z., C. S. Davis, B. A. Jex, T. Hegel, and D. W. Coltman. 2019. Management implications of highly resolved hierarchical population genetic structure in thinhorn sheep. *Conservation Genetics* 20:185–201.

Simmons, N. 1982. Seasonal ranges of Dall's sheep in the MacKenzie Mountains, Northwest Territories. *Arctic* 35:512–518.

Smith, J. B., J. A. Jenks, T. W. Grovenburg, and R. W. Klaver. 2014. Disease and predation: sorting out causes of a bighorn sheep (*Ovis canadensis*) decline. *PLoS One* 9:e88271.

Sproat, K. K., N. R. Martinez, T. S. Smith, W. B. Sloan, J. T. Flinders, J. W. Bates, J. G. Cresto, and V. C. Bleich. 2020. Desert bighorn sheep responses to human activity in southeastern Utah. *Wildlife Research* 47:16–24.

Stoner, D. C., M. T. Anderson, C. A. Schroeder, C. A. Bleke, and E. T. Thacker. 2021. Distribution of competition potential between native ungulates and free-roaming equids on western rangelands. *Journal of Wildlife Management* 85:1062–1073.

Valdez, R. 1988. *Wild Sheep and Wild Sheep Hunters of the New World*. Wild Sheep and Goat International, Mesilla, New Mexico, USA.

Valdez, R., J. Guzmán-Aranda, F. Abarca, L. Tarango-Arambula, F. Clemente-Sánchez. 2009. Wildlife conservation and management in Mexico. *Wildlife Society Bulletin* 34:270–282.

Van de Kerk, M., S. Arthur, M. Bertram, B. Borg, J. Herriges, J. Lawler, B. Mangipane, C. Lambert Koizumi, B. Wendling, L. Prugh. 2020. Environmental influences on Dall's sheep survival. *Journal of Wildlife Management* 84:1127–1138.

Veitch, A., E. Simmons, J. Adamczewski, and R. Popko. 1998. Status, harvest, and co-management of Dall's sheep in the Mackenzie Mountains, Northwest Territories. *Biennial Symposium of the Northern Wild Sheep and Goat Council* 11:134–153.

WAFWA (Western Association of Fish and Wildlife Agencies). 2025. *Recommendations for Domestic Sheep and Goat Management in Wild Sheep Habitat.* Western Association of Fish and Wildlife Agencies, Bosie, Idaho, USA.

WAFWA. 2024. *2024 Range-Wide Status of Bighorn and Thinhorn Sheep in North America.* Western Association of Fish Wildlife Agencies, Wild Sheep Initiative. Boise, Idaho, USA.

Wells, J. J. 2024. *Dall Sheep Management Report and Plan, Game Management Unit 20A: Report Period 1 July 2016–30 June 2021, and Plan Period 1 July 2021–30 June 2026.* Alaska Department of Fish and Game, Species Management Report and Plan ADFG/DWC/SMRP-2024-13, Juneau, Alaska, USA.

Wells, J. J. 2025. *Dall Sheep Management Report and Plan, Game Management Units 12, 13C, and 20D, Tok Management Area: Report Period 1 July 2016–30 June 2021, and Plan Period 1 July 2021–30 June 2026.* Alaska Department of Fish and Game, Species Management Report and Plan ADFG/DWC/SMRP-2025-11, Juneau, Alaska, USA.

White, K. S., B. Cadsand, S. D. Côté, T. Graves, S. Hamel, R. Harris, F. P. Hayes, E. Hood, K. Hurley, T. Jessen, B. A. Jex, E. Peitzsch, W. Sarmento, H. Schwantje, and J. Berger. 2025. Mountain sentinels in a changing world: review and conservation implications of weather and climate effects on mountain goats (*Oreamnos americanus*). *Global Ecology and Conservation* 57:e03364. https://doi.org/10.1016/j.gecco.2024.e03364.

Wild Sheep Foundation. 2014. Thinhorn Sheep Summit Synthesis and Summary. DOC27_2014–05–19ThinhornSheepSummitSummarySynthesis.pdf. Accessed 10 March 2025.

Wyoming Game and Fish Department. 2019. *Whiskey Mountain Bighorn Sheep Plan.* Wyoming Game and Fish Department, Cheyenne, Wyoming, USA.

Yukon Department of Environment. 2023. Kluane Region Sheep Population Surveys—2023. Kluane Region Sheep Population Surveys 2023 - Summary | Yukon.ca. Accessed 10 March 2025.

# 4 Physical Characteristics and Horn Development

*Kevin L. Monteith, Seth T. Rankins, Jaron T. Kolek, and Tayler N. LaSharr*

## 4.1 INTRODUCTION

Mountain sheep (bighorn [*Ovis canadensis*], thinhorn [*Ovis dalli*]), one of the most iconic and charismatic species of North America, epitomize physical adaptations for life ranging from the alpine to the desert and some of the most rugged and inaccessible terrain (Hornaday 1901, McCann 1956). Their physical attributes are adept for negotiating precipitous terrain and their affinity for such terrain underpins their use of habitat (McCann 1956, Geist 1971, Hansen 1980a). Topping off their distinguishing morphological features is a crown of spiraling horns that can exceed 12% of the body mass of an adult male (Geist 1966). Their weighty horns have captivated fascination of wildlife fanatics and hunters worldwide (Hornaday 1901). The factors that influence the size, conformation, and growth of mountain sheep may be of considerable interest to wildlife managers (Monteith et al. 2013). Both body and horn size can be important indicators of fitness and animal quality and, given their interest among wildlife enthusiasts, managers, and conservationists, have become a point of consideration in management.

## 4.2 BODY CHARACTERISTICS

Mountain sheep have evolved adaptations that allow them to persist in steep terrain and extreme environments (Geist 1971). Mountain sheep have specialized hooves, with rough pads and split toes that allow them to grip loose, rocky terrain and navigate rugged landscapes with ease. Both thinhorn and bighorn sheep are small to medium-sized bovids with short legs and a stocky build (Hansen 1980b, Krausman and Bowyer 2003). Rocky Mountain bighorn sheep tend to be slightly larger than thinhorn or desert sheep (Hansen 1980b, Krausman and Bowyer 2003), but both species are sexually dimorphic with adult males being 48%–88% heavier than mature females (Festa-Bianchet et al. 1996, Wilson and Ruff 1999, Polák and Frynta 2009, Table 4.1). Both female and male bighorn sheep attain more than 95% of their maximum skeletal growth by 2 or 3 years of age but continue to gain mass until approximately 6 years of age (Seip and Bunnell 1984, Nichols and Bunnell 1999). The internal organs of mountain sheep make up approximately 33% of their live mass (Hansen 1980b). During autumn, up to 28% of the live mass, or 41% of the ingesta-free body mass of mountain sheep

---

## TABLE 4.1

**Range in Reported Morphology Metrics of Body Size for Mountain Sheep Across Their Range in North America (Reported Ranges Often were Derived from Small Sample Sizes and May Not be Representative of All Populations)**

| Species | Subspecies | Sex | Newborn Mass (kg) | Adult Mass (kg) | Body Length (cm) | Height to Shoulder (cm) | Tail Length (cm) | Ear Length (cm) |
|---|---|---|---|---|---|---|---|---|
| Bighorn | Rocky Mountain | Female | 2.3–5.9 | 38–91 | 102–179 | No data | 8.0–12.1 | 7.0–12.7 |
| | | Male | 2.8–6.2 | 51–143 | 127–195 | No data | 9.5–12.7 | 6.3–12.7 |
| | Desert | Female | 2.5–4.2 | 34–87 | 117–189 | 78–86 | 7.7–13.0 | 7.0–11.4 |
| | | Male | 2.1–4.9 | 44–102 | 106–174 | 77–103 | 7.0–13.0 | 9.5–10.8 |
| | Sierra Nevada | Female | 3.5[a] | 37–72 | 114–167 | No data | No data | No data |
| | | Male | 3.4–3.7 | 47–110 | 135–179 | No data | No data | No data |
| Thinhorn | Dall's | Female | 3.2–4.1 | 46–64 | 138–162 | 76–84 | 7.0–9.9 | 8.0–9.0 |
| | | Male | 4.1[b] | 64.9–114 | 134–174 | 87–109 | 7.0–11.5 | 7.0–9.2 |
| | Stone's | Female | 2.8–3.7[c] | 44.5–61 | 132[b] | 70–95 | 7.7[b] | No data |
| | | Male | | 77[b] | 130–178 | No data | 7.5–10.9 | 6.2–8.5 |

[a] *Sources:* Cowan (1940), Hall (1946), Blood et al. (1970), Geist (1971), Blunt et al. (1977), Bunnell and Olsen (1976), Nichols (1978), Wishart (1978), Heimer (1979), Seip and Bunnell (1984), Shackleton (1985), Hass (1995), Krausman and Bowyer (2003), Parr et al. (2018), Parikh et al. (2025), unpublished data (California Department of Fish and Wildlife), unpublished data (K. Monteith).

[b] Values are from a single animal.

[c] Range is for five lambs of unknown sex.

---

DOI: 10.1201/9781003518686-5

can be fat stores, which are used to partially fuel survival in seasonal environments (Stephenson et al. 2020, Denryter et al. 2022). Thus, the body mass of an animal can fluctuate throughout the year, particularly for sheep that inhabit highly seasonal environments (Hansen 1980b, Festa-Bianchet et al. 1996, Smiley et al. 2022).

Mountain sheep in North America range in color from pure white in Dall's sheep (*O. dalli dalli*) to light-faced chocolate and slate body-colored in Stone's sheep (*O. dalli stonei*), to predominantly black in some desert bighorn sheep (*Ovis canadensis nelsoni*; Hornaday 1901, Loehr et al. 2008, Sim et al. 2019, Bleich 2017, 2024; Jex and Sim 2021, Figure 4.1). Bighorn sheep range from cream or nearly white (leucistic or isabelline coloration) to a chocolate brown with white highlights along the belly, insides of the legs, muzzle, and rump (Bleich 2017, 2024). These white highlights are characteristic of both species and are most striking in bighorn and Stone's sheep with darker coats. Some record keeping organizations (e.g., Boone and Crockett Club) have the opinion that thinhorn sheep with even one black hair anywhere other than the tail are Stone's (or Fannin's) sheep (Nesbitt et al. 2009); however, several authors have clarified the genetic assignment and lineage of the Fannin phenotype as most strongly linked to Dall's sheep origins (Worley et al. 2004, Sim et al. 2019, Jex and Sim 2021; Chapter 19 presents alternate data). At adulthood, the guard hairs of mountain sheep along their back is approximately 25 to 50 mm in length (Hansen 1980b, Shackleton 1985, Bowyer and Leslie 1992). Mountain sheep have straight guard hairs and the more prominent underfur that is crimped in a zigzag pattern (Hansen 1980b, Nicholls and Bunnell 1999). The crimped underfur is useful in distinguishing between hair of mountain sheep and mule deer (*Odocoileus hemionus*; which is always straight) in the field.

Major glands of mountain sheep are the lacrimal (or preorbital), inguinal, interdigital, and sweat glands (Hansen 1980b, Valdez and Krausman 1999). Species in the genus *Ovis* lack a subcaudal gland (Valdez and Krausman 1999, Krausman and Bowyer 2003). Although mountain sheep have functioning sweat glands, they likely are not important for heat dissipation, which primarily occurs through panting (Schmidt-Nielsen 1979, Hansen 1980b). Furthermore, mountain sheep have a very broad thermal neutral range (−20°C to +10°C for Rocky Mountain bighorn [*O. c. canadensis*]; Chapter 12), and rarely exhibit signs of thermal stress despite the temperature extremes (−40°C to more than 49°C) they experience (Chappel and Hudson 1978, Hansen 1980a, Krausman and Bowyer 2003). Interdigital and lacrimal glands in mountain sheep are used primarily for scent marking and intraspecific signaling (Geist 1971, Hansen 1980b, Shackleton 1985). Glands may serve more than one purpose, however, and little research has been done on the endocrine system of mountain sheep in comparison to domestic sheep. The exact purpose of the inguinal gland in mountain sheep is unknown, but there is speculation that it may play a role in mother-offspring relationships and stimulate suckling behavior in neonates (Alexandre-Pires et al. 2017).

Little is known about the acuity of taste, smell, or touch sensory systems in mountain sheep, although they are selective foragers, implying they possess the ability to sense food quality and palatability. The optic system of mountain sheep is that of a prey species. The eyeballs of mountain sheep are located on the sides of the skull, providing a wide field of view (>300°). The wide field of view provided by their laterally located eyeballs comes at a cost to depth perception though, explaining why bighorn sheep may walk into wire fences (Hansen 1980c). Furthermore, mountain sheep possess a visual streak, which is why the pupil of their eye appears to be horizontally flattened. Light passing through the eye is focused on the band of receptor cells increasing the peripheral vision of mountain sheep. Mountain sheep possess a tapetum lucidum, as evidenced by eyeshine. The tapetum lucidum acts as a mirror reflecting light back through the retina, doubling the chance that a rod or cone will be stimulated (Lesiuk and Braekevelt 1983). The presence of a tapetum lucidum in their eye suggests that mountain sheep possess reasonable night vision (Walls 1942, Lesiuk and Braekevelt 1983). Although bighorn sheep are primarily diurnal (Wagner 2000, Glass et al. 2022, Terry et al. 2022), nocturnal activity may not be uncommon (Longshore et al. 2009). Dall's sheep inhabit high latitudes that have limited daylight during the winter (Bowyer and Leslie 1992), and it seems likely that they have good night vision to live in these environments. Bighorn sheep have poorer hearing than humans across the range of frequencies that have been tested (clicks from 1 to 4 kHz; DeYoung et al. 1993).

Sheep belonging to the genus *Ovis*, including mountain sheep, typically have 32 permanent teeth (i 0/3, c 0/1, p3/3, m 3/3; Shackleton 1985, Krausman and Bowyer 2003). Mountain sheep have a dental pad on their upper palate in place of upper incisors. The canine teeth are incisiform, meaning that the canine teeth look like a fourth incisor, rather than a true canine. The premolars and molars are subhypsodont. The incisors (including the incisiform canines) are separated from the premolars and molars by a diastema, or gap, along the lower jaw. There is some variation in the number of teeth erupted from the gums in newborn lambs; however, by 1 month of age, lambs have their full complement of 20 deciduous, or milk teeth (Hansen and Deming 1980). Mountain sheep can be reliably aged up to 4 years of age based upon tooth replacement patterns of the incisors and incisiform canines (Krausman and Bowyer 2003). There is some interindividual variation in the exact timing of tooth replacement however; when estimating age of mountain sheep near their birthday confirming age by at least two methods (e.g., tooth replacement, counting horn annuli) may be desirable. Like deer (*Odocoileus* spp.), mountain sheep have a tricuspid deciduous premolar 4 and a bicuspid permanent premolar 4 (Hemming 1969). Age can also be estimated via counts of cementum annuli in the cross section of the root of a tooth (Laws 1952, Hemming 1969). Using cementum annuli has been considered the most accurate method to age old mountain sheep (Hemming

1969, Hoefs and König 1984). Nevertheless, estimating age by horn annuli may be more accurate, especially for males, than cementum annuli techniques (Wehausen et al. 2024). Although counting cementum annuli is highly accurate for determining age of animals in seasonal environments, in environments with more stochastic variation in resource scarcity (e.g., the Sonoran Desert), determining age with cementum annuli is less accurate (Marshal et al. 2002, Foley et al. 2022). Mountain sheep only grow permanent molars, and occasionally have supernumerary (i.e., extra) teeth like maxillary canines (Bradley and Allred 1966, Bunch et al. 1984, Jansen and Krausman 2006).

## 4.3 HORN CHARACTERISTICS

In North America, mountain sheep carry the heaviest weapons among extant ruminants relative to body size. These secondary sexual characteristics likely evolved primarily for male-male combat, including display and intimidation related to such interactions (Geist 1966, 1971; Clutton-Brock 1982, Goss 1983, Bubenik and Bubenik 1990). Despite the evolutionary link to male-male combat, mountain sheep of both sexes possess horns. Horns of females, however, are much smaller and shorter than those of males (Figure 4.1). Additionally, female horns are relatively thin and gently curved, whereas male horns are comparatively massive at the base, growing in a curled spiral that tapers at its end.

Though the principal components of mountain sheep horns are consistent across species, size and growth form vary substantially, especially between thinhorn (Dall's, Stone's) and bighorn (Rocky Mountain, desert, Sierra [*Ovis canadensis sierrae*]) sheep. The horns of thinhorn sheep are longer, thinner, and tend to flare outward, away from the face when compared with the more massive and compact spiral of bighorns. Within the bighorns, the horns of the Rocky Mountain subspecies form the tightest curl, those of desert bighorns usually are also curled close to the face but may flare widely outward depending on the individual, whereas Sierra bighorn have the most splayed horn shape with the least pronounced curl (Figure 4.1). Rocky Mountain

desert bighorn sheep

Rocky Mountain bighorn sheep

Sierra Nevada bighorn sheep

Stone's sheep (thinhorn)

Dall's sheep (thinhorn)

**FIGURE 4.1** Mountain sheep in North America range from small to medium-sized bovids that are well adapted to living in rugged landscapes. Mountain sheep range in color from pure white to brown or nearly black. All subspecies of mountain sheep in North America exhibit strong sexual dimorphism in horn and body size. Males have large, heavy horns that curl around their head, whereas females have short, small horns that are relatively thin and gently curved compared with males. Thinhorn and bighorn sheep have slightly different horn morphology, which is most evident in males. Thinhorn sheep typically have horns that are longer, thinner, and more flared compared with the more massive and compact spiral of the horns that bighorn sheep possess. (Illustrations by Ben Regan; used with permission.)

**TABLE 4.2**

**Reported Morphology Metrics of Horn Size for Rocky Mountain Bighorn Sheep, Sierra Nevada Bighorn Sheep, Desert Bighorn Sheep, Dall's Sheep, and Stone's Sheep, North America**

| | Sex | Maximum Basal Circumference (cm) | Maximum Tip-to-Tip (cm) | Maximum Length (cm) |
|---|---|---|---|---|
| Rocky Mountain Bighorn Sheep | Male | 41.9 | 65.4 | 126.4 |
| Sierra Nevada Bighorn Sheep | Male | 39 | – | 86 |
| | [Female] | [17] | – | [38] |
| Desert Bighorn Sheep | Male | 43.2 | 65.1 | 111.1 |
| Dall's Sheep | Male | 37.5 | 87.3 | 121.6 |
| Stone's Sheep | Male | 37.5 | 78.7 | 131.1 |

*Sources:* California Department of Fish and Wildlife (unpublished data), Boone and Crockett Club (unpublished data).

subspecies tend to possess the greatest potential in size of horns with other subspecies generally being smaller in mass (Table 4.2).

The development of horns likely begins early in gestation as the epidermis above the frontal bone thickens, vacuolates, and subsequently develops a dense network of nerve bundles (Nasoori 2020). At the time of birth, horn buds have not yet ossified, are independent of the bone of the skull, and typically are visible only as tufts of hair. Development proceeds rapidly as a pair of ossicones develop under the skin and fuse with the underlying frontal bone with hardened knobs protruding above the hair by 1 month of age (Hansen and Deming 1980). Ossicones continue to develop into bony cores that form the template for future horn while the epidermal covering of early ossicones becomes keratinized forming the hardened sheath that typically is referred to as horn. Given that true horns are permanent structures, horn size typically continues to increase over the lifetime of an animal, despite substantial declines in the rate of growth later in life (Bergeron et al. 2008). Brooming (i.e., the breakage of the lamb tip portion of the horn caused through collision and compaction of horn tips generally caused during male-male combat) is an exception to the continued increase in size with age, because it results in the loss of the distal portion of horns (Geist 1971, Shackleton and Hutton 1971, Bubenik and Bubenik 1990). Consequently, annual horn growth in later life may not be adequate to compensate for horn loss associated with brooming. Thinhorn sheep often carry their lamb tips through their lifetime, maintaining their shape through grooming and polishing behaviors, and are less apt to experience brooming in the conventional sense. This may be a function of mating and social rank behaviors more typically being displayed through posturing rather than physical conflict; horn breakage from falls and other mishaps does still occur.

Horn cores grow continuously throughout an animal's life and consist of a thin layer of compact bone filled with a dense network of thin bony plates, or trabecular bone, that serve as struts creating a porous or foam-like layer that surrounds a large sinus at the center of the core (Figure 4.2). Although struts of trabecular bone are common as mechanical support that is lightweight in spines, skulls, and the ends of long bones, the plate-like structure of these bones

in mountain sheep is rather unique. Typical trabecular bone struts in human femurs, for example, are rod-like and have thicknesses ranging from 0.09 to 0.018 mm while those of sheep horns are more plate-like with thicknesses ranging from 2.5 to 22 mm (Drake et al. 2016).

Each year a new layer of keratin or sheath develops over the underlying horn core forcing the horn annulus that was grown the previous year away from the skull. Thus, horn sheaths are grown one inside another like nested cones (Figure 4.2); except for the first or lamb horn, only a small portion of the horn growth of each year is visible. Additionally, this nested growth pattern creates a distinct break where growth stops in the early winter and begins again the following spring. Horn sheath thus allows individuals to be aged by counting the horn rings or annuli that develop each year (Wehausen et al. 2024). Annual horn growth additionally can provide insight into the status of an individual or population given that annual growth rings will reflect the energy and resources available to devote to horn growth in a given year (Loehr et al. 2010).

Everything from the hard keratinized sheath to the interfacial tissue to the bony core plays important, but distinct, roles in the head-to-head ramming that occurs during intraspecific battles for dominance in mountain sheep. Potential force in collisions can range as high as 3,390 joules in bighorn sheep (Kitchener 1988). Keratin-based materials are more resistant to fractures than mineralized bone and are thus better suited to resist the localized forces experienced at the impact site (Currey 1999, Drake et al. 2016). The interfacial tissue, made up of a complex network of collagen and keratin fibers, facilitates load transfer between the impacted horn and the bony core (Fuller and Donahue 2021). The bony core, largely a result of the presence of trabecular bone, provides flexibility and absorbs the majority of strain energy that follows initial impact (Drake et al. 2016). Additionally, the bulk of energy being transferred to the bony core provides an advantage as bone can be regrown if damaged, unlike the horn sheaths. Finally, the spiralized geometry of the horns further dissipates kinetic energy by loading and uploading as does a spring during impact and vibrating thereafter (Drake et al. 2016). Collectively, the keratinized sheath provides impact resistance, the interfacial tissue yields robust attachment and transfer of energy,

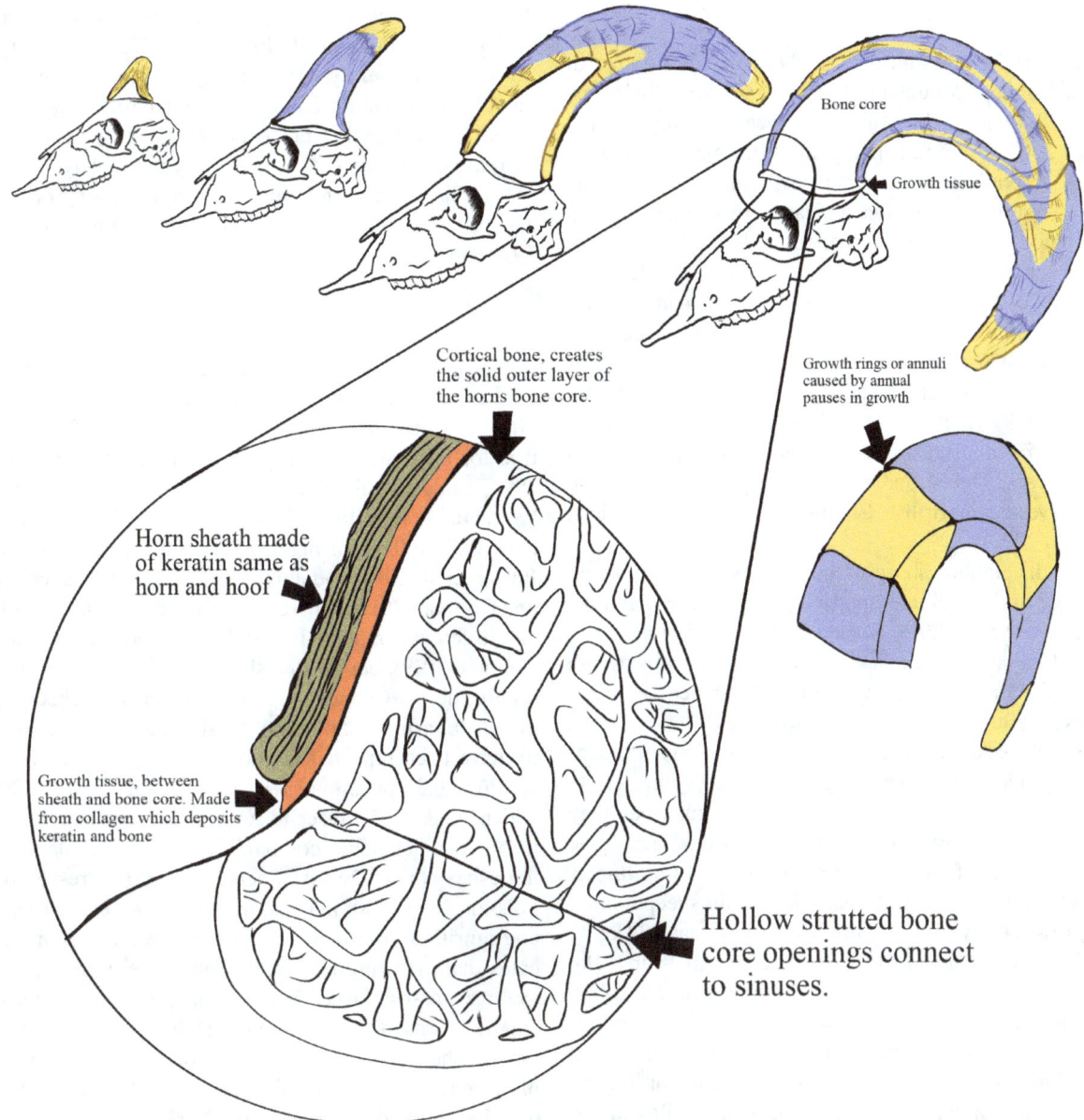

**FIGURE 4.2** Horns of North American mountain sheep grow continuously throughout their lives. Horns consist of a keratin sheath that covers growth tissue and the bony core of the horn. (Illustrations by Ben Regan; used with permission.)

while the bony core and spiralized geometry absorb and dissipate energy, protecting the brain cavity and allowing the individual to sustain incredible impact loads (Wheatley et al. 2023).

## 4.4 AGE, GENETICS, AND NUTRITION

Understanding the influencers of horn and body size in mountain sheep is often of interest to wildlife managers and the general public, as the phenotype of these animals represents substantial biological and sociological importance. Often, body size and horn size are closely linked, with animals that have large horns also having large bodies (Festa-Bianchet et al. 2004, Monteith et al. 2018). Body and

horn size can be indicators of animal quality, where animals with larger bodies or more impressive horns may be more reproductively successful (Coltman et al. 2002, Kruuk et al. 2002, Preston et al. 2003), more resistant to parasites (Ezenwa and Jolles 2008), and produce better or more sperm (Malo et al. 2005). Consequently, horn and body size can be honest indicators of fitness (Solberg and Saether 1993, Ditchkoff et al. 2001, Vanpé et al. 2007, Bonenfant et al. 2009). Additionally, across the range of mountain sheep, there is substantial interest in harvesting males with large and impressive horns (Chapter 16). There are three primary mechanisms that determine the size, shape, and morphology of horns and bodies in mountain sheep: age, genetics, and nutrition (Monteith et al. 2018).

### 4.4.1 AGE

Age has the largest influence on horn and body size in mountain sheep. Mountain sheep that survive their first year of life typically live 10 or more years (Krausman and Bowyer 2003). As a function of their life-history strategy, females generally have higher survival than males and the oldest sheep in a population typically are females even in the absence of hunting (McCullough 1999). In the wild, females can attain 14–15 years of age, whereas males typically live 10–11 years of age. Although rare, some mountain sheep may live for more than 20 years (DeCesare 2002). The oldest Dall's sheep recorded was a female of 19 years 10 months in Kluane National Park in the winter of 1988–1989 (Hoefs 1991). Rocky Mountain, California (*O. c. californiana*), Dall's and Stone's males as old as 16 (bighorns and Dall's) and 17 (Stone's) years of age have been harvested in British Columbia (Province of British Columbia 2023).

At birth, bighorn sheep of both sexes typically weigh 2.7–6.2 kg and males are approximately 10% larger than females (Parr et al. 2018, Karsch 2014, Parikh et al. 2025, Table 4.1). For thinhorn sheep, newborns typically weigh between 2.95 and 6.22 kg, with little differences between the sexes (Bunnell 1982, Nicholls and Bunnell 1999, Scotton 1997). Adult females in some populations of mountain sheep may not become pregnant in every year despite being sexually mature (Proffitt et al. 2021), but when they do, they are almost exclusively monotocous (i.e., give birth to a single offspring), and litter size is not an important factor influencing birth mass. Mountain sheep exhibit rapid growth early in life, with the ability to reach 14 kg by 3 weeks of age and can weigh anywhere from 18 to 40 kg at weaning in bighorn sheep (Festa-Bianchet et al. 2000); thinhorn sheep grow at a slower rate reaching approximately 30 kg by early winter in their first year (Bunnell and Olsen 1976). Growth is fastest in the first 2 years of life and slows with age; at 6 months of age, bighorn females can be 40% of their adult mass and males up to 30% of their adult mass. In the following year, females have attained up to 70% of their adult body mass and males up to 50% of their adult body mass (Festa-Bianchet et al. 2000).

The initiation of horn growth of mountain sheep begins during gestation (Nasoori 2020); horns of both sexes are visible shortly after birth, and horn growth continues throughout their life. In their first week of life, lambs of both sexes develop tufts of hair where their horns will grow and by 1 month of age their horns protrude slightly above their hair (Hansen and Deming 1980). Female bighorn sheep can grow horns that are 27 cm long by 2 years of age. In bighorn and thinhorn sheep, males reach 75% of asymptotic horn size by age 4 (Bunnell 1978, Jorgenson et al. 1998, Festa-Bianchet et al. 2004), and unlike other mountain ungulates (e.g., Alpine ibex [*Capra ibex*], chamois [*Rupicapra rupicapra*], mountain goats [*Oreamnos americanus*]) are unable to offset poor growth that occurred during their first few years of life

once they reach older age classes (i.e., compensatory growth; Jorgenson et al. 1998, Metcalfe and Monaghan 2001). Consequently, males that have small horns at the age of 4 will remain comparatively small for their entire lives (Festa-Bianchet et al. 2004). Given that horn size increases with age, old males will typically have the largest and heaviest horns unless severe brooming or breakage has occurred (Geist 1971, Shackleton and Hutton 1971, Bubenik and Bubenik 1990).

### 4.4.2 GENETICS

Size and configuration of horns in mountain sheep is, at least in part, heritable (Coltman et al. 2003, Poissant et al. 2008, Pigeon et al. 2016, Sim and Coltman 2019), as is also true for ungulate weapons more broadly (Williams et al. 1994, Lukefahr and Jacobson 1998, Kruuk et al. 2002). Estimated heritability ($h^2$) of horn length for males ranges from 0.37 to 0.69 for bighorn sheep (Coltman et al. 2003, Pigeon et al. 2016) and about 0.33 for thinhorn sheep (Sim and Coltman 2019), whereas estimated heritability of base circumference of horns is 0.36 for thinhorn sheep (Sim and Coltman 2019), but may be slightly higher in other species within the *Ovis* genus (e.g., $h^2 \sim 0.44$ in Soay sheep [*Ovis aries*]; Johnston et al. 2010). Similar patterns of heritability have been reported for horn volumes (i.e., $h^2$ of horn volume $\sim 0.36$ in thinhorn sheep; Sim and Coltman 2019), which appears to transcend sex with estimates of heritability for females being comparable to that of males in bighorn sheep ($h^2 = 0.24 \pm 0.09$ versus $0.32 \pm 0.12$, respectively; Poissant et al. 2008). Thus, all else being equal (i.e., age and nutrition), heritability of horn size and volume indicates that offspring of parents with large horns would on average possess larger horns than those of parents with smaller horns. There also exists some level of heritability among phenotypic traits such as body size in parents and offspring (Poissant et al. 2008). Heritability of horn characteristics and body size may hold implications for management programs, because management actions that target a heritable phenotypic trait (e.g., harvest of large males), especially one that is linked to fitness, can have implications for growth and other life-history traits (Poissant et al. 2008, Mysterud 2011).

Given horns and body size are heritable, selective pressures can act to influence the evolutionary trajectory of horn and body size. The fitness benefits that large horns and bodies confer for males (e.g., increased reproductive success) demonstrate the advantage that these phenotypic traits have for mountain sheep. Large horns and bodies can give males an advantage in fighting for access to females, resulting in selective pressure where males with more impressive horns have a greater chance of reproductive success, and thus are more likely to pass on their genes to future generations (Coltman et al. 2002, Preston et al. 2003). Although, alternative mating tactics like being sneaky can simultaneously weaken the expected relationship in weapon size and reproductive success (Hogg 1984). Perhaps more contested in the scientific literature, however, is the role humans play in imposing artificial selection that results in evolutionary change for mountain sheep (Pigeon et al. 2016, Festa-Bianchet and Mysterud 2018, Heffelfinger 2018). Humans can influence the genetic structure of a population whenever they alter the animals responsible for breeding and producing the next generation (Conover et al. 2009).

Across their range, mountain sheep are desirable game animals, with hunters often seeking males that have large, heavy, or impressive horns (Chapter 16; Damm 2008, Knox 2011, Messner 2011, Simon 2016, Heffelfinger 2018). Thus, the influence of harvest practices on the growth and size of horns has received considerable attention across academic, management, and public spheres (Festa-Bianchet et al. 2015, Pigeon et al. 2016, Boyce and Krausman 2018, Festa-Bianchet and Mysterud 2018, Heffelfinger 2018). When harvest is sufficiently selective (i.e., based on a morphological attribute) and intensive (i.e., most males removed that attain the morphological minimum for harvest), removal of individuals with large weaponry via hunting can affect selective pressures against rapid growth of large horns (Festa-Bianchet et al. 2000, Hard et al. 2006, Monteith et al. 2013, Pigeon et al. 2016, LaSharr et al. 2019). Nevertheless, consequences for horn size are not a foregone conclusion in the presence of selective harvest (LaSharr et al. 2019). Circumstances that lead to evolutionary changes require harvest practices that are highly selective and intensive (Heffelfinger 2018), a situation that is uncommon across much of mountain sheep range in North America today (Chapter 16).

Genetic contributions of males is not the only factor that influences the size and morphology of horns in a population; age, nutrition, maternal effects, cohort effects, contribution of genetics from females (which are not selectively removed based on horn size), and refugia all can limit the potential for evolutionary change that may occur through selective harvest practices (Heffelfinger 2018, Poisson et al. 2020). Nevertheless, detecting evolutionary changes can be difficult in long-lived species, like mountain sheep, because these changes occur slowly and over many generations (Mysterud and Bischof 2010, Coulson et al. 2018). Despite the many factors that can inhibit evolutionary change occurring via harvest, evolutionary changes have been empirically demonstrated in mountain sheep, with declines in horn length occurring in as few as five generations when harvest was exclusively based on a morphological criterion (i.e., 4/5 curl), was intense (Pigeon et al. 2016), and harvest intensity occurred in a regionalized geographic scale (Douhard et al. 2016). Nevertheless, across much of North America, hunting practices are more conservative (Chapter 16), and thus, harvest practices do not always result in detectable changes to horn size over time (LaSharr et al. 2019). Although genetics contributes to the observed phenotype, the full expression of genetic potential in horns and body size is contingent on individual fitness realized through environmental conditions, especially nutrition.

### 4.4.3 Nutrition

Age and genetics play a fundamental role in shaping body size, horn growth, and conformation; however, the nutrition animals are exposed to throughout their lives ultimately determines the ability of animals to reach, and express, their full genetic potential (Monteith et al. 2018). Mountain sheep inhabit environments that undergo often dramatic fluctuations in resource availability and quality seasonally and annually. As resources change throughout the year, animals must balance their energy allocation across reproduction, growth, maintenance, and ultimately survival (Varpe et al. 2009, Smiley et al. 2022). For mountain sheep, how animals allocate their energy throughout their lives determines, in large part, the size of both their bodies and horns. The influence of nutrition on body and horn size occurs primarily through two pathways: direct environmental effects and maternal effects. Both pathways can be further complicated by a variety of factors, including animal density or disease, and habitat quality.

Perhaps the clearest example of the cumulative effects of nutrition across the lifetime of a mountain sheep on horn growth was demonstrated via a comparison of horn growth in Dall's sheep between wild sheep in Kluane National Park, Yukon, Canada, and a captive herd that was established from the same animals but fed a high-quality diet (Hoefs and Nowlan 1997). Despite having the same genetic origins in that they originated from the same population, by 3 years old, captive males attained horns exceeding 60 cm in length, a length that required another year of growth for wild males (Figure 4.3). Further, by 7 years old, horn volume of captive males fed the high-quality diet was 57% greater than their wild counterparts that were subject to environmental conditions (Hoefs and Nowlan 1997).

Across six populations of Sierra bighorn sheep, the absence of hunting and monitoring associated with recovery efforts yielded the opportunity to evaluate population-level linkages between nutritional condition and size of males. Males aged 6–10 years across the six populations ranged in mean horn size from 194 to 220 cm (sum of horn lengths and basal circumferences), and in body mass during spring from 71 to 81 kg (Monteith et al. 2018, Figure 4.4). More than 80% of the variation in age-corrected horn size of males across the six populations could be explained by average nutritional condition of females in spring or autumn (Figure 4.5). Horn size of 7-year-old males was predicted to increase by 4 cm for every 1 percentage point increase in ingesta-free body fat of adult females in spring (Monteith et al. 2018). Nutrition, which is both maternally mediated and realized by the individual, has massive consequences for growth and size of mountain sheep.

*Direct Environment.* The environments that mountain sheep inhabit are extreme; across their range, mountain sheep have evolved to persist in areas with exceptionally harsh winters (Denryter et al. 2022), high alpine environments subject to marked seasonal variation in timing and

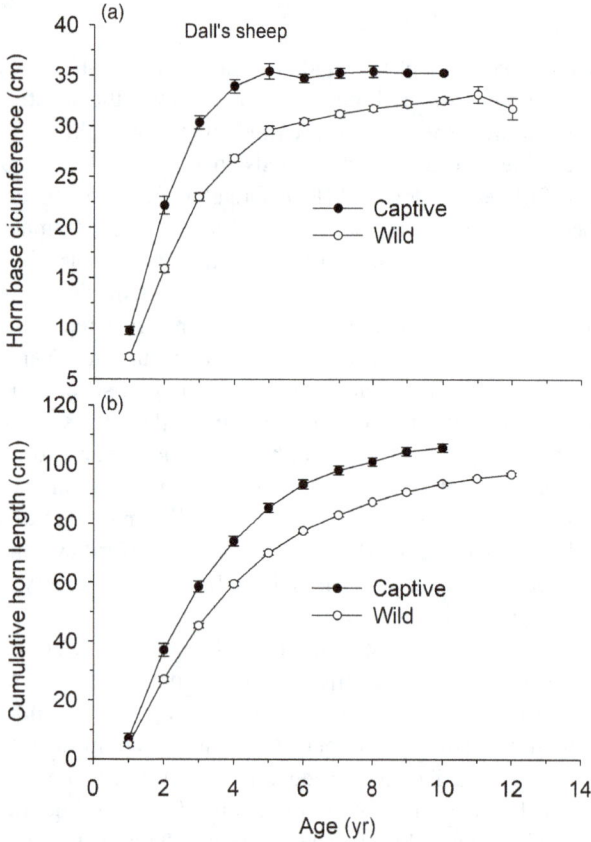

**FIGURE 4.3** Difference in horn length and horn circumference of wild (n=24) and captive (n=10) Dall's sheep from 1969 to 1992. The wild animals were at a high density during the study period in Kluane National Park, southwest Yukon, Canada, and the captive population originated from the park but were fed an alfalfa hay and pellet diet, providing excellent nutrition. (Data originally adapted from Hoefs and Nowlan 1997; adapted figure originally produced in Monteith et al. 2018; used with permission.)

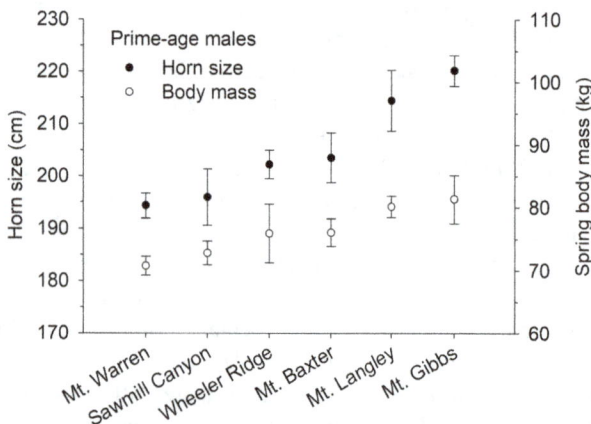

**FIGURE 4.4** Average horn size (cm±SE; n=65) and body mass (kg±SE; n=42) during spring for male Sierra Nevada bighorn sheep aged 6–10 years across six populations from 2002 to 2016 in Sierra Nevada, California, United States. Horn size was calculated as the combined length and basal circumference of both horns. (Figure originally produced in Monteith et al. 2018; used with permission.)

growth of forage (Smiley et al. 2022), and environments that are shaped by extreme drought (Gedir et al. 2020). For animals that inhabit landscapes with harsh and unpredictable conditions, individuals accrue somatic reserves (e.g., protein, fat) that can be used as capital to finance survival when resources in the landscape are limited or inaccessible. Thus, animals must carefully balance how they use energy throughout the year.

During the first few years of life, young animals do not incur the energetic costs of reproduction, and often put most of their energy beyond basal metabolism into growth of somatic tissue. Given the amount of body growth that occurs during the first few years of life for mountain sheep, the resources animals have access to before reaching asymptotic body size can affect their phenotype for their entire lives (Hamel et al. 2009). Once animals have reached independence and reproductive maturity, they must then balance allocation of resources to current reproduction and their survival. For males, individuals continue to balance energy allocation between body growth and horn growth until they achieve asymptomatic body size, which often does not occur until 6 years of age in bighorns (Festa-Bianchet et al. 1996); in thinhorn sheep males, asymptotic body size likely is attained at 7 years of age (Environment Yukon 2017). Horns are nonessential to survival for mountain sheep, and during the first few years of life, males may prioritize growth, maintenance, and acquisition of somatic tissue at the cost of horn growth when exposed to poor-quality or limited resources during the growing season (Jorgenson et al. 1998). Given that males can attain 75% of asymptotic horn size by age 4, the resources that animals have available to them during this period ultimately will determine the maximal size of their horns (Bunnell 1978, Jorgenson et al. 1998, Nichols and Bunnell 1999, Festa-Bianchet et al. 2004).

Interannual variation in resource availability during growth can accumulate to have additive effects on size of horns and be responsible for directional changes in harvested males over time (Hik and Carey 2000, LaSharr et al. 2019). Spring weather and its influence on primary productivity were a key determinant of interannual variation in horn growth for Dall's sheep (Bunnell 1978, Loehr et al. 2010, Karabatsos et al. 2024). The annulus of growth is a direct reflection of conditions experienced during that year (Bunnell 1978, Hik and Carey 2000), which is cumulative to horn size over the life of a male.

*Maternal Effects.* For long-lived ungulates that persist in unpredictable and variable environments, both fundamental attributes of mountain sheep, opportunities to reproduce multiple times over their lifetime are central to their life history and require securing survival over reproduction (Festa-Bianchet and Jorgenson 1998, Smiley et al. 2022). Dramatic seasonal fluctuations in resource availability can affect maternal conditions during gestation and early offspring development (Pigeon et al. 2021). Consequently, female mountain sheep pass the costs of reproduction to their offspring, which can pose lifelong consequences on growth

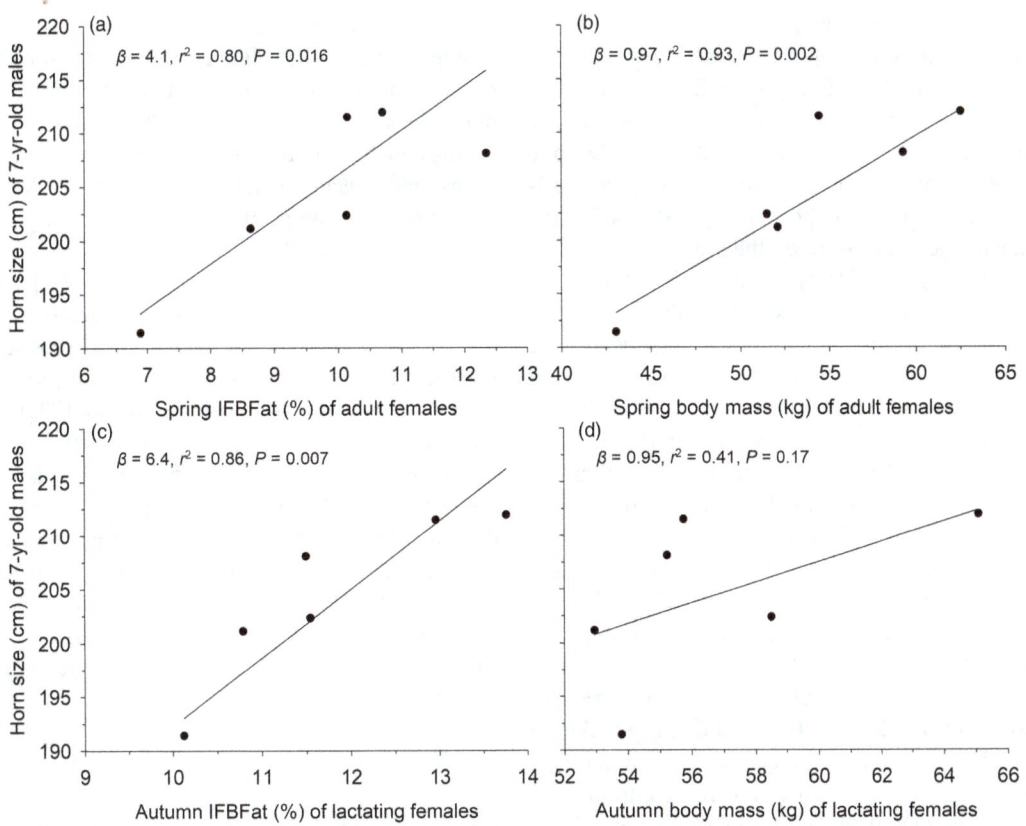

**FIGURE 4.5** Percent ingesta-free body fat (IFBFat) and body mass (kg) of adult (more than 2 years old) female Sierra Nevada bighorn sheep measured in spring ($n = 115$), and for lactating females measured in autumn ($n = 96$) across six populations in Sierra Nevada, California, United States, from 2002 to 2016. Body fat explained more than 80% of the variation in horn size of males in the population, demonstrating the importance of nutrition in producing males with large and heavy horns. (Figure originally produced in Monteith et al. 2018; used with permission.)

and adult size of their young (Martin and Festa-Bianchet 2010, Monteith et al. 2018, Festa-Bianchet et al. 2019, Pigeon et al. 2021). With poor maternal nutrition, maternal provisioning declines, growth of young can be suppressed (Festa-Bianchet and Jorgenson 1998, Festa-Bianchet et al. 2019), and mortality of young increases (Portier et al. 1998). For long-lived females, risk-sensitive allocation of resources favors allocation to maintenance and survival of mother (Festa-Bianchet and Jorgenson 1998, Smiley et al. 2022), thus projecting the nutritional limitation onto the offspring (Martin and Festa-Bianchet 2010, Festa-Bianchet et al. 2019). Though phenotype of the offspring depends on genes inherited from mother and father, the condition of the mother during gestation and early provisioning of offspring can lead to lifelong consequences on growth and, ultimately, size attained as an adult (Monteith et al. 2009). Notably, whether conditions improve later in life or not, a negative maternal effect can compromise growth through adulthood and, thus, nutritional signatures from gestation and early growth last for an animal's entire life (Monteith et al. 2018). Environmental signatures, likely borne out via the nutritional condition of the mother during the year a male was born (i.e., cohort effects), can be persistent through the life of a male (Bunnell 1978, Hik and Carey 2000, Hamel et al. 2009).

*Density.* Acknowledgment that density is limiting to populations has been met with reluctance, is fraught with misunderstanding (Bowyer et al. 2014), and yet, density dependence is central to the life history of long-lived herbivores like mountain sheep (McCullough 1999). As population density increases, the finite resources within an environment are divided among more individuals, leading to intensified competition for those resources (Chapter 10), which can directly influence the size of secondary sexual characteristics, like horns (Jorgenson et al. 1993, 1998; Festa-Bianchet et al. 1998), and the condition or body size of animals (Bowyer et al. 2014). Resources are finite and environments can only support so many animals; consequently, density and the ensuing consequences for animal nutrition are inextricably linked (Monteith et al. 2014, Chapter 7).

Some of the clearest empirical demonstrations of the role of density dependence in populations of wild ungulates come from work in mountain sheep (Jorgenson et al. 1993, 1998; Festa-Bianchet et al. 1998, Portier et al. 1998, Mitchell et al. 2015), especially with the associated nutritional limitation that holds short- and long-term consequences on body and horn growth. A tripling in population density for a population of bighorn sheep in Alberta, Canada, led to a 22% reduction in growth of young over summer compared with only a 9% reduction in mass gain over summer by adult females (Festa-Bianchet and Jorgenson 1998). The same rapid increases in the density of bighorn sheep was the primary contributor to marked declines in annual horn growth during the first 4 years of life (Jorgenson et al. 1993, Douhard et al. 2017). Despite the higher number of animals in the population, the availability and yield of legal bighorn males (i.e., 4/5 curl) remained unchanged, because of the density-dependent reductions in horn growth (Jorgenson et al. 1998). With elevated competition for resources at high density, a necessary trade-off in resource allocation to prioritization of somatic maintenance over the growth of secondary sexual characteristics leads to a reduction in growth of horns. Given the importance of energy allocation in the first few years of life to horn and body size, young animals that live in populations at high densities are likely to exhibit stunted growth (Jorgenson et al. 1998, Chapter 23).

*Disease.* Across their range, pneumonia is one of the most pervasive and pressing management concerns for mountain sheep (Chapter 8). Pneumonia can have devastating and long-term consequences for populations, and outbreaks can result in die-offs that many populations struggle to recover from (Cassirer et al. 2013, 2018). Though population crashes immediately following an exposure are a constant concern for managers, the insidious effects of chronic and long-term presence of pathogens are measurable and detrimental. Clearing infections of any kind comes at an energetic cost, and contending with pathogens associated with pneumonia incurs measurable nutritional costs in bighorn sheep (Smiley et al. 2024). Consistent with nutritional costs of pathogens and nutritional underpinning of horn growth, disease outbreaks can result in diminished horn growth (Martin et al. 2022). The presence of disease in populations of bighorn sheep reduced annual increments of horn growth by 12%–35%, culminating in a 3%–13% reduction in horn length of individuals that had been exposed to disease in their lifetime (Martin et al. 2022). Individuals that survive disease outbreaks may have compromised nutritional condition, reduced horn growth, and low reproductive success. The cascading effects of pneumonia and the influence on horn size and, by inference, body size can result in populations composed of smaller, less-fit individuals, which may have implications for the genetic diversity and resilience of populations through time.

## 4.5 IMPLICATIONS FOR MANAGEMENT

A common theme of management for mountain sheep is the consideration for the number, age structure, and size of males within populations (Wild Sheep Foundation Professional Biologist Meeting Attendees 2008, Environment Yukon 2017, LaSharr et al. 2019). Factors affecting growth of horns have consequently become common considerations in management of mountain sheep (Monteith et al. 2018). Although management focused on horn size often originates from public desires, the production of impressive males is only possible in populations that are harvested

sustainably and have access to good food (Traill et al. 2014, Monteith et al. 2018). Because horns are nonessential for survival, mountain sheep prioritize allocation in their body maintenance and survival over horn growth (Geist 1971, Festa-Bianchet et al. 2004). Therefore, growth of large horns only occurs when other energetic demands have been met, meaning growth is predicated on good nutrition (Monteith et al. 2018).

In North America, management of big game species, including mountain sheep, often consists of adjusting the number of animals on the landscape via harvest or altering the quality and quantity of forage to provide more or better-quality food and cover (Leopold 1933). For mountain sheep, however, disease is one of the most pressing management concerns for many populations across their range (Chapter 8) and can come with sizable energetic costs that can affect horn growth and limit nutritional condition (O'Brien et al. 2014, Martin et al. 2022, Smiley et al. 2024). Thus, the three primary management considerations for promoting rapid horn growth in populations of mountain sheep are harvest management, habitat manipulation, and disease management.

### 4.5.1 HARVEST

Harvest practices play a role in determining yield within populations experiencing density-dependent survival and reproduction (Errington 1934, McCullough 1979, Bowyer et al. 2014). For mountain sheep, harvest practices can also influence horn growth. As is common with most ungulates, the reproductive ecology of mountain sheep (i.e., polygyny) can allow for liberal harvest of bighorn males without negatively influencing population trajectories (Bowyer et al. 2014). Consequently, harvest of males is more common than females in North America (Chapter 16). Harvest practices can influence horn size of mountain sheep through three primary pathways: shifts in the age structure, genetic change, and by decreasing competition for food. First, given that horns continue to grow through the lifetime of an animal, increasing harvest of males can lower the overall age structure of the male population, reducing the number of old males that possess the largest horns. In this scenario, smaller horns do not necessarily indicate lower quality animals, just that males were harvested before they reached asymptotic horn size. Liberal harvest in populations where males exhibit rapid horn growth has greater implications for age structure than in populations with slower horn growth (Bonenfant et al. 2009, Schindler et al. 2017). Secondly, when harvest is sufficiently selective and intense, shifts in horn size also may occur through selective pressures against rapid horn growth (Pigeon et al. 2016, Festa-Bianchet and Mysterud 2018). In areas where harvest is relatively liberal, managers often impose morphological criterion (i.e., curl restrictions). Thus, age and overall horn size of harvested animals through time are important factors when determining how harvest is affecting mountain sheep populations (LaSharr et al. 2019).

When ungulate populations are approaching nutritional carrying capacity (i.e., the number of animals that the forage in an area will sustainably support) competition for nutritional resources depresses reproductive output, limiting population growth (Robbins 1973, McCullough 1979, DeYoung et al. 2000). When nutrition becomes limiting, horn growth slows given its nonessential functions (Geist 1971). Populations well below carrying capacity, therefore, will have the greatest rates of horn growth and ultimately reach the largest overall horn size (Geist 1971). Although the harvest of females is relatively rare in mountain sheep management, decreasing population numbers, primarily through the harvest of females, can increase the amount of forage available for each sheep (Jorgenson et al. 1993, 1998). The increased forage on a per animal basis allows males to achieve greater horn growth (Jorgenson et al. 1993, 1998). Although horn growth may be increased through population reduction, decreasing the population through female harvest will have little influence on horn growth when a population is not nutritionally limited.

### 4.5.2 Habitat Manipulation

Although abundance of mountain sheep is currently a fraction of historical numbers, habitat may still be limiting growth for some populations (Buechner 1960, Whiting et al. 2023a, Wagler et al. 2024, Chapter 19). Mechanisms influencing declines in range quality are numerous and can have compounding effects. Across their range, mountain sheep may contend with domestic grazing, fire suppression, habitat loss, and climate change. Understanding the primary causes of decline in habitat quality for a population is a fundamental first step in identifying appropriate management actions to improve growth in mountain sheep through habitat improvements.

Much of western North America has experienced or is experiencing overgrazing (Ditomaso 2000, Krausman et al. 2009). Overabundant livestock, exotic species, and wildlife can cause long-term changes in plant communities and degrade rangeland quality (Ditomaso 2000, Krausman et al. 2009). Long-term and historic overgrazing can increase the amount of shrubs, which are generally less palatable and digestible for grazers including mountain sheep (Chapter 7; Hobbs et al. 1983, Ditomaso 2000, Krausman et al. 2009). The increase in browse and decrease in forbs or graminoids available on rangelands in western North America has likely resulted in mountain sheep consuming lower quality diets that support survival but may be insufficient to enhance horn growth. While spread of disease often is the primary concern surrounding domestic livestock (O'Brien et al. 2014, Anderson et al. 2022), understanding the influence of domestic and feral animals on restructuring native plant communities (Ditomaso 2000, Krausman et al. 2009) should be of equal concern to managers, especially in areas where forage is limiting or where introduction and expansion of non-native, invasive plant species is a risk. Limiting grazing in and near wild sheep populations can

be an important tool for managers to bolster the nutritional resources available to mountain sheep.

Decades of fire suppression in western North America have increased forest cover in some areas (Rodman et al. 2019, Hagmann et al. 2021), which decreases the available habitat and nutritional resources for mountain sheep (Risenhoover and Bailey 1985, Werdel et al. 2023, Chapter 20). The use of fire, both prescribed and wildfire, is the most studied and recommended habitat management strategy for mountain sheep (Hobbs and Spowart 1984, Bleich et al. 2008, Greene et al. 2012, Sittler et al. 2019, Donovan et al. 2021). Wildfire often decreases woody cover and converts forest to early successional grassland habitat. The decrease in visual obstruction and concealment cover for ambush predators increases the amount of habitat available for mountain sheep (Karsch et al. 2016). Additionally, regrowth of graminoids following a fire often is more palatable and nutritious for mountain sheep in the short term (Hobbs and Spowart 1984, Ruckstuhl et al. 2000, Sittler et al 2019). Although wildfire often is associated with improved nutrition (Hobbs and Spowart 1984), and consequently increased horn growth for mountain sheep, large intense wildfires followed by drought conditions may decrease the available forage available to mountain sheep and result in poor nutritional condition or starvation (Clapp and Beck 2016). The response of mountain sheep to wildfire, therefore, is dependent on soil moisture and the availability of alternative forage while graminoids regrow after combustion (Clapp and Beck 2016, Donovan et al. 2021). Although prescribed fire may be used to prevent woody encroachment in existing grasslands and improve forage quality (with adequate soil moisture), prescribed fire is not usually intense enough to remove stands of trees. When managers wish to expand the amount of habitat available, as opposed to maintaining existing habitat, prescribed fire may need to be used in conjunction with other habitat management actions such as the use of mechanical thinning techniques (Smith et al. 1999, Chapter 20).

Other management strategies that have been used to improve habitat for mountain sheep include mechanical tree removal, grazing management, and herbicide application (Chapter 20). The mechanical removal of trees and understory increases sightlines and habitat visibility, and promotes the growth of graminoids, forages for mountain sheep (Smith et al. 1999, Dibb and Quinn 2006, Chapter 20). Although the benefits of mechanical tree removal for species such as bighorn sheep are as good as or better than wildfire, it is often difficult to implement mechanical tree removal in the steep terrain favored by mountain sheep (Smith et al. 1999, Dibb and Quinn 2006). Therefore, the practicality of using mechanical tree removal to promote increased nutritional resources that lead to increased horn growth in mountain sheep habitat is limited. When exotic plant species, such as cheatgrass (Bromus tectorum) or Russian thistle (Salsola tragus) outcompete more nutritious native forage plants, the application of herbicide may be warranted (Antill et al. 2012, Amundson 2022). In some scenarios grazing management targeted at reducing grazing pressure in areas with some exotic plant species may also reduce the density of unpalatable and poor-quality forage (Vavra et al. 2007). Removing competing herbivores, such as domestic sheep, aoudad (*Ammotragus lervia*), or feral horses, when they are in the same habitat as mountain sheep, also may increase nutritional resources available for mountain sheep and promote individual fitness and horn growth (Buechner 1960, Gallizioli 1977, Smith and Krausman 1987, Marshal et al. 2008, García et al. 2023, Whiting et al. 2023b).

### 4.5.3 Disease Management

Given the extensive energetic burden that pneumonia-related pathogens can incur on mountain sheep, pathogens can limit horn growth through nutritionally mediated pathways (Martin et al. 2022). Therefore, an additional concern for managing bighorn sheep with endemic pathogens is the nutritional burden of sub-clinical (i.e., when symptoms of disease are not readily observed) respiratory disease (Cassirer et al. 2018, Martin et al. 2022, Wagler et al. 2023, Smiley et al. 2024). Although not simple in practice, an intuitive recommendation for increasing the size and number of large-horned mountain sheep is to ensure mountain sheep populations remain free of respiratory pathogens (Martin et al. 2022). The primary method to limit and prevent the spread of respiratory pathogens to mountain sheep is to limit the intermingling of domestic sheep, goats, and mountain sheep (O'Brien et al. 2014, Anderson et al. 2022). After populations of mountain sheep have been exposed to respiratory pathogens there are few management strategies for treatment, and success often is limited. The most common strategy for eliminating respiratory pathogens from mountain sheep populations is test and cull (Chapter 8). Concerns with the energetic burden of respiratory disease limiting horn growth in mountain sheep are primarily a concern in bighorn sheep, because domestic livestock are rare throughout the range of thinhorn sheep (Heimer 2020, Chapter 8).

### 4.6 SUMMARY

There is no single strategy that will produce large-horned and large-bodied mountain sheep. The growth and size of horns and bodies in mountain sheep result from a complex interplay of factors, including age, genetics, and nutrition. Though all three pieces influence the size, conformation, and growth of horns and bodies in mountain sheep, nutrition, which can act through multiple pathways, often is the most influential piece underpinning the phenotypic expression of mountain sheep. Moreover, nutrition is one of the pieces that managers may be able to influence through conservation and management actions. Understanding the factors that influence body and horn size is central to the management of mountain sheep populations; thus, factors like harvest practices, disease management, and habitat management which interact to affect individual growth and population dynamics are important conservation considerations.

## REFERENCES

Alexandre-Pires, G., C. Martins, A. M. Galvão, M. Miranda, O. Silva, D. Ligeiro, T. Nunes, and G. Ferreira-Dias. 2017. Understanding the inguinal sinus in sheep (*Ovis aries*)—Morphology, secretion, and expression of progesterone, estrogens, and prolactin receptors. *International Journal of Molecular Sciences* 18:1516.

Amundson, R. 2022. Wildfire recovery to provide optimal habitat for bighorn sheep in the Douglas Creek bighorn sheep herd. *Biennial Symposium of the Northern Wild Sheep and Goat Council* 23:54.

Anderson, K., M. L. Cahn, T. R. Stephenson, A. P. Few, B. E. Hatfield, D. W. German, J. M. Weissman, and B. Croft. 2022. Cost distance models to predict contact between bighorn sheep and domestic sheep. *Wildlife Society Bulletin* 46:e1329.

Antill, T. M., M. A. Naeth, E. W. Bork, and A. L. Westhaver. 2012. Russian thistle (*Salsola tragus* L.) control on bighorn sheep winter ranges in Jasper National Park. *Natural Areas Journal* 32:391–397.

Bergeron, P., M. Festa-Bianchet, A. Von Hardenberg, and B. Bassano. 2008. Heterogeneity in male horn growth and longevity in a highly sexually dimorphic ungulate. *Oikos* 117:77–82.

Bleich, V. C. 2017. Abnormal coloration in bighorn sheep. *Bulletin of the Southern California Academy of Sciences* 16:54–59.

Bleich, V. C. 2024. Isabelline coloration: A heretofore unrecognized chromatic aberration in bighorn sheep. *California Fish and Wildlife Journal* 110:e8.

Bleich, V. C., H. E. Johnson, S. A. Holl, L. Konde, S. G. Torres, and P. R. Krausman. 2008. Fire history in a chaparral ecosystem: Implications for conservation of a native ungulate. *Rangeland Ecology and Management* 61:571–579.

Blood, D. A., D. R. Flook, and W. D. Wishart. 1970. Weights and growth of Rocky Mountain bighorn sheep in western Alberta. *Journal of Wildlife Management* 34:451–455.

Blunt, F. M., H. A. Dawson, and E. T. Thorne. 1977. Birth weights and gestation in a captive Rocky Mountain bighorn sheep. *Journal of Mammalogy* 58:106–106.

Bonenfant, C., F. Pelletier, M. Garel, and P. Bergeron. 2009. Age-dependent relationship between horn growth and survival in wild sheep. *Journal of Animal Ecology* 78:161–171.

Bowyer, R. T., and D. M. Leslie, Jr. 1992. Ovis dalli. *Mammalian Species* 393:1–7.

Bowyer, R. T., V. C. Bleich, K. M. Stewart, J. C. Whiting, and K. L. Monteith. 2014. Density dependence in ungulates: A review of causes, and concepts with some clarifications. *California Fish and Game* 100:550–572.

Boyce, M. S., and P. R. Krausman. 2018. Special section: Controversies in mountain sheep management. *Journal of Wildlife Management* 82:5–7.

Bradley, W. G., and L. G. Allred. 1966. A comparative study of dental anomalies in desert bighorn sheep. *Desert Bighorn Council Transactions* 10:78–85.

Bubenik, G. A., and A. B. Bubenik, editors. 1990. *Horns, Pronghorns, and Antlers: Evolution, Morphology, Physiology, and Social Significance.* Springer, New York, New York, USA.

Buechner, H. K. 1960. The bighorn sheep in the United States, its past, present, and future. *Wildlife Monographs* 4:1–174.

Bunch, T. D., M. Hoefs, and R. L. Glaze. 1984. Upper canines in Dall's sheep (*Ovis dalli dalli*). *Journal of Wildlife Diseases* 20:158–161.

Bunnell, F. L. 1978. Horn growth and population quality in Dall sheep. *Journal of Wildlife Management* 42:764–775.

Bunnell, F. L. 1982. The lambing period of mountain sheep: Synthesis, hypotheses, and tests. *Canadian Journal of Zoology* 60:1–14.

Bunnell, F. L., and N. A. Olsen. 1976. Weights and growth of Dall sheep in Kluane Park Reserve, Yukon Territory, Canada. *Canadian Field-Naturalist* 98:479–484.

Cassirer, E. F., K. R. Manlove, E. S. Almberg, P. L. Kamath, M. Cox, P. Wolff, A. Roug, J. Shannon, R. Robinson, R. B. Harris, B. J. Gonzales, R. K. Plowright, P. J. Hudson, P. C. Cross, A. Dobson, and T. E. Besser. 2018. Pneumonia in bighorn sheep: Risk and resilience. *Journal of Wildlife Management* 82:32–45.

Cassirer, E. F., R. K. Plowright, K. R. Manlove, P. C. Cross, A. P. Dobson, K. A. Potter, and P. J. Hudson. 2013. Spatio-temporal dynamics of pneumonia in bighorn sheep. *Journal of Animal Ecology* 82:518–528.

Chappel, R. W., and R. J. Hudson. 1978. Winter bioenergetics of Rocky Mountain bighorn sheep. *Canadian Journal of Zoology* 56:2388–2393.

Clapp, J. G., and J. L. Beck. 2016. Short-term impacts of fire-mediated habitat alterations on an isolated bighorn sheep population. *Fire Ecology* 12:80–98.

Clutton-Brock, T. H. 1982. The function of antlers. *Behaviour* 79:108–124.

Coltman, D. W., M. Festa-Bianchet, J. T. Jorgenson, and C. Strobeck. 2002. Age-dependent sexual selection in bighorn rams. *Proceedings of the Royal Society B: Biological Sciences* 269:165–172.

Coltman, D. W., P. O'Donoghue, J. T. Jorgenson, J. T. Hogg, C. Strobeck, and M. Festa-Bianchet. 2003. Undesirable evolutionary consequences of trophy hunting. *Nature* 426:655–658.

Conover, D. O., S. B. Munch, and S. A. Arnott. 2009. Reversal of evolutionary downsizing caused by selective harvest of large fish. *Proceedings of the Royal Society B: Biological Sciences* 276:2015–2020.

Coulson, T., S. Schindler, L. Traill, and B. E. Kendall. 2018. Predicting the evolutionary consequences of trophy hunting on a quantitative trait. *Journal of Wildlife Management* 82:46–56.

Cowan, I. M. 1940. Distribution and variation in the native sheep of North America. *American Midland Naturalist* 24:505–580.

Currey, J. D. 1999. The design of mineralised hard tissue for their mechanical functions. *The Journal of Experimental Biology* 202:3285–3294.

Damm, G. R. 2008. Recreational Trophy Hunting: "What do we know and what should we do?" Pages 5–11 *in* R. D., Baldus, G. R., Damm, and K. Wollscheid, editors. *Best Practices in Sustainable Hunting—A Guide to Best Practices from Around the World.* CIC—International Council for Game and Wildlife Conservation, Budakeszi, Hungary.

DeCesare, N. J. 2002. *Movement and Resource Selection of Recolonizing Bighorn Sheep in Western Montana.* Thesis, University of Montana, Missoula, USA.

Denryter, K., M. M. Conner, T. R. Stephenson, D. W. German, and K. L. Monteith. 2022. Survival of the fattest: How body fat and migration influence survival in highly seasonal environments. *Functional Ecology* 36:2569–2579.

DeYoung, D. W., P. R. Krausman, L. E. Weiland, and R. C. Etchberger. 1993. Baseline ABRs in mountain sheep and desert mule deer. *International Congress on Noise as a Public Health Problem* 6:251–254.

DeYoung, R. W., E. C. Hellgren, T. E. Fulbright, W. F. Robbins, and I. D. Humphreys. 2000. Modeling nutritional carrying capacity for translocated desert bighorn sheep in western Texas. *Restoration Ecology* 8:57–65.

Dibb, A. D., and M. S. Quinn. 2006. Response of bighorn sheep to restoration of winter range. *Biennial Symposium of the Northern Wild Sheep and Goat Council* 15:59–69.

Ditchkoff, S. S., R. L. Lochmiller, R. E. Masters, S. R. Hoofer, and R. A. Van Den Bussche. 2001. Major-histocompatibility-complex-associated variation in secondary sexual traits of white-tailed deer (*Odocoileus virginianus*): Evidence for good-genes advertisement. *Evolution* 55:616–625.

Ditomaso, J. M. 2000. Invasive weeds in rangelands: Species, impacts, and management. *Weed Science* 48:255–265.

Donovan, V. M., S. P. H. Dwinnell, J. L. Beck, C. P. Roberts, J. G. Clapp, G. S. Hiatt, K. L. Monteith, and D. Twidwell. 2021. Fire-driven landscape heterogeneity shapes habitat selection of bighorn sheep. *Journal of Mammalogy* 102:757–771.

Douhard, M., G. Pigeon, M. Festa-Bianchet, D. W. Coltman, S. Guillemette, and F. Pelletier. 2017. Environmental and evolutionary effects on horn growth of male bighorn sheep. *Oikos* 126:1031–1041.

Douhard, M., M. Festa-Bianchet, F. Pelletier, J.-M. Gaillard, and C. Bonenfant. 2016. Changes in horn size of Stone's sheep over four decades correlate with trophy hunting pressure. *Ecological Applications* 26:309–321.

Drake, A., T. L. Haut Donahue, M. Stansloski, K. Fox, B. B. Wheatley, and S. W. Donahue. 2016. Horn and horn core trabecular bone of bighorn sheep rams absorbs impact energy and reduces brain cavity accelerations during high impact ramming of the skull. *Acta Biomaterialia* 44:41–50.

Environment Yukon. 2017. *Science-Based Guidelines for Management of Thinhorn Sheep in Yukon.* Yukon Fish and Wildlife Branch Report MR-16-03. Whitehorse, Yukon, Canada.

Errington, P. L. 1934. Vulnerability of bob-white populations to predation. *Ecology* 15:110–127.

Ezenwa, V. O., and A. E. Jolles. 2008. Horns honestly advertise parasite infection in male and female African buffalo. *Animal Behaviour* 75:2013–2021.

Festa-Bianchet, M., and A. Mysterud. 2018. Hunting and evolution: Theory, evidence, and unknowns. *Journal of Mammalogy* 99:1281–1292.

Festa-Bianchet, M., D. W. Coltman, L. Turelli, and J. T. Jorgenson. 2004. Relative allocation to horn and body growth in bighorn rams varies with resource availability. *Behavioral Ecology* 15:305–312.

Festa-Bianchet, M., J.-M. Gaillard, and J. T. Jorgenson. 1998. Mass- and density-dependent reproductive success and reproductive costs in a capital breeder. *American Naturalist* 152:367–379.

Festa-Bianchet, M., and J. T. Jorgenson. 1998. Selfish mothers: Reproductive expenditure and resource availability in bighorn ewes. *Behavioral Ecology* 9:144–150.

Festa-Bianchet, M., J. T. Jorgenson, and D. Réale. 2000. Early development, adult mass, and reproductive success in bighorn sheep. *Behavioral Ecology* 11:633–639.

Festa-Bianchet, M., J. T. Jorgenson, W. J. King, K. G. Smith, and W. D. Wishart. 1996. The development of sexual dimorphism: Seasonal and lifetime mass changes in bighorn sheep. *Canadian Journal of Zoology* 74:330–342.

Festa-Bianchet, M., S. Schindler, and F. Pelletier. 2015. Record books do not capture population trends in horn length of bighorn sheep. *Wildlife Society Bulletin* 39:746–750.

Festa-Bianchet, M., S. D. Côté, S. Hamel, and F. Pelletier. 2019. Long-term studies of bighorn sheep and mountain goats reveal fitness costs of reproduction. *Journal of Animal Ecology* 88:1118–1133.

Foley, A. M., J. S. Lewis, O. Cortez, M. W. Hellickson, D. G. Hewitt, R. W. DeYoung, C. A. DeYoung, and M. J. Schnupp. 2022. Accuracies and biases of ageing white-tailed deer in semiarid environments. *Wildlife Research* 49:237–249.

Fuller, L. H., and S. W. Donahue. 2021. Material properties of bighorn sheep (*Ovis canadensis*) horncore bone with implications for energy absorption during impacts. *Journal of the Mechanical Behavior of Biomedical Materials* 114:104224.

Gallizioli, S. 1977. Overgrazing on desert bighorn ranges. *Transactions Desert Bighorn Council* 21:21–22.

García, A. R., F. N. González Saldívar, C. M. Cantú Ayala, and J. I. Uvalle Sauceda. 2023. Food competition between Barbary sheep (*Ammotragus lervia* Pallas, 1777) and Bighorn sheep (*Ovis canadensis* Shaw, 1804) in Coahuila State. *Revista Mexicana de Ciencias Forestales* 14:294–316.

Gedir, J. V., J. W. Cain, T. L. Swetnam, P. R. Krausman, and J. R. Morgart. 2020. Extreme drought and adaptive resource selection by a desert mammal. *Ecosphere* 11:e03175.

Geist, V. 1966. The evolution of hornlike organs. *Behaviour* 27:175–214.

Geist, V. 1971. *Mountain Sheep: A Study in Behavior and Evolution.* The University of Chicago Press, Chicago, Illinois, USA.

Glass, D. M., P. R. Prentice, A. D. Evans, and O. J. Schmitz. 2022. Local differences in maximum temperature determine water use among desert bighorn sheep populations. *Journal of Wildlife Management* 86:e22313.

Goss, R. J. 1983. *Deer Antlers: Regeneration, Function and Evolution.* Academic Press, Cambridge, United Kingdom.

Greene, L., M. Hebblewhite, and T. R. Stephenson. 2012. Short-term vegetation response to wildfire in the eastern Sierra Nevada: Implications for recovering an endangered ungulate. *Journal of Arid Environments* 87:118–128.

Hagmann, R. K., P. F. Hessburg, S. J. Prichard, N. A. Povak, P. M. Brown, P. Z. Ful, R. E. Keane, E. E. Knapp, J. M. Lydersen, K. L. Metlen, et al. 2021. Evidence for widespread changes in the structure, composition, and fire regimes of western North American forests. *Ecological Applications* 31:e02431.

Hall, E. R. 1946. *Mammals of Nevada.* University of California Press, Berkley, California, USA.

Hamel, S., J. M. Gaillard, M. Festa-Bianchet, and S. D. Côté. 2009. Individual quality, early-life conditions, and reproductive success in contrasted populations of large herbivores. *Ecology* 90:1981–1995.

Hansen, C. G. 1980a. Habitat. Pages 64–79 *in* G. Monson and L. Sumner, editors. *The Desert Bighorn: Its Life History, Ecology, and Management.* The University of Arizona Press, Tucson, Arizona, USA.

Hansen, C. G. 1980b. Physical characteristics. Pages 52–63 *in* G. Monson and L. Sumner, editors. *The Desert Bighorn: Its Life History, Ecology, and Management.* The University of Arizona Press, Tucson, Arizona, USA.

Hansen, C. G. 1980c. Senses and intelligence. Pages 113–123 *in* G. Monson and L. Sumner, editors. *The Desert Bighorn: Its Life History, Ecology, and Management.* The University of Arizona Press, Tucson, Arizona, USA.

Hansen, C. G., and O. V. Deming. 1980. Growth and development. Pages 152–171 *in* G. Monson and L. Sumner, editors. *The Desert Bighorn: Its Life History, Ecology, and Management.* University of Arizona Press, Tucson, Arizona, USA.

Hard, J. J., L. Scott Mills, and J. M. Peek. 2006. Genetic implications of reduced survival of male red deer *Cervus elaphus* under harvest. *Wildlife Biology* 12:427–441.

Hass, C. C. 1995. Gestation period and birth weights of desert bighorn sheep in relation to other Caprinae. *The Southwestern Naturalist* 40:139–147.

Heffelfinger, J. R. 2018. Inefficiency of evolutionarily relevant selection in ungulate trophy hunting. *Journal of Wildlife Management* 82:57–66.

Heimer, W. E. 1979. *Interior sheep studies.* Juneau, Alaska, USA.

Heimer, W. E. 2020. Management update and summary: Alaskan Dall's Sheep 2020. *Biennial Symposium of Northern Wild Sheep and Goat Council* 22:49–56.

Hemming, J. E. 1969. Cemental deposition, tooth succession, and horn development as criteria of age in Dall sheep. *Journal of Wildlife Management* 33:552–558.

Hik, D. S., and J. Carey. 2000. Cohort variation in horn growth of Dall sheep rams in the southwest Yukon, 1969–1999. *Biennial Symposium of Northern Wild Sheep and Goat Council* 12:88–100.

Hobbs, N. T., and R. A. Spowart. 1984. Effects of prescribed fire on nutrition of mountain sheep and mule deer during winter and spring. *Journal of Wildlife Management* 48:551–560.

Hobbs, N. T., D. L. Baker, and R. B. Gill. 1983. Comparative nutritional ecology of montane ungulates during winter. *Journal of Wildlife Management* 47:1–16.

Hoefs, M. 1991. A longevity record for Dall's Sheep, *Ovis dalli dalli*, Yukon Territory. *Canadian Field Naturalist* 105:397–398.

Hoefs, M., and R. König. 1984. Reliability of aging old Dall sheep ewes by the horn annulus technique. *Journal of Wildlife Management* 48:980–982.

Hoefs, M., and U. Nowlan. 1997. Comparison of horn growth in captive and free-ranging Dall's rams. *Journal of Wildlife Management* 61:1154–1160.

Hogg, J. T. 1984. Mating in bighorn sheep: Multiple creative male strategies. *Science* 255:526–529.

Hornaday, W. T. 1901. *Notes on the Mountain Sheep of North America, with a Description of a New Species.* New York Zoological Society, New York, New York, USA.

Jansen, B. D., and P. R. Krausman. 2006. Maxillary canines in bighorn sheep. *The Southwestern Naturalist* 51:434–436.

Jex, B. A., and Z. Sim. 2021. A rare phenotype of thinhorn sheep—the dark phased Dall's: New genetic analysis tools help re-map thinhorn sheep subspecies distributions in North America. *Caprinae News—Newsletter of the IUCN Caprinae Specialist Group* 1:14:16

Johnston, S. E., D. Beraldi, A. F. McRae, J. M. Pemberton, and J. Slate. 2010. Horn type and horn length genes map to the same chromosomal region in Soay sheep. *Heredity* 104:196–205.

Jorgenson, J. T., M. Festa-Bianchet, M. Lucherini, and W. D. Wishart. 1993. Effects of body size, population density, and maternal characteristics on age at first reproduction in bighorn ewes. *Canadian Journal of Zoology* 71:2509–2517.

Jorgenson, J. T., M. Festa-Bianchet, and W. D. Wishart. 1998. Effects of population density on horn development in bighorn rams. *Journal of Wildlife Management* 62:1011–1020.

Karabatsos, S., N. C. Larter, D. G. Allaire, K. Eykelboom, C. Estevo, M. Iravani, I. C. Barrio, and D. S. Hik. 2024. Dall's sheep horn growth and harvest management in the Mackenzie Mountains, Northwest Territories, Canada. *Journal of Wildlife Management* 88:e22536.

Karsch, R. C. 2014. Desert bighorn sheep adult female and lamb survival, cause-specific mortality, and parturient female habitat selection in the Peloncillo Mountains, New Mexico. Thesis, New Mexico State University, Las Cruces, New Mexico, USA.

Karsch, R. C., J. W. Cain, E. M. Rominger, and E. J. Goldstein. 2016. Desert bighorn sheep lambing habitat: Parturition, nursery, and predation sites. *Journal of Wildlife Management* 80:1069–1080.

Kitchener, A. 1988. An analysis of the forces of fighting of the blackbuck (*Antilope cervicapra*) and the bighorn sheep (*Ovis canadensis*) and the mechanical design of the horn of bovids. *Journal of Zoology* 214:1–20.

Knox, W. M. 2011. The antler religion. *Wildlife Society Bulletin* 35:45–48.

Krausman, P. R., D. E. Naugle, M. R. Frisina, R. Northrup, V. C. Bleich, W. M. Block, M. C. Wallace, and J. D. Wright. 2009. Livestock grazing, wildlife habitat, and rangeland values. *Rangelands* 31:15–19.

Krausman, P. R., and R. T. Bowyer. 2003. Mountain sheep (*Ovis canadensis* and *O. dalli*). Pages 1095–1115 *in* Feldhamer G. A., Thompson B. C., and Chapman J. A., editors. *Wild Mammals of North America: Biology, Management, and Conservation*. 2nd edition. John Hopkins University Press, Baltimore, Maryland, USA.

Kruuk, L. E. B., J. Slate, J. M. Pemberton, S. Brotherstone, F. Guinness, and T. Clutton-Brock. 2002. Antler size in red deer: Heritability and selection but no evolution. *Evolution* 56:1683–1695.

LaSharr, T. N., R. A. Long, J. R. Heffelfinger, V. C. Bleich, P. R. Krausman, R. T. Bowyer, J. M. Shannon, R. W. Klaver, C. E. Brewer, M. Cox, A. A. Holland, A. Hubbs, C. P. Lehman, J. D. Muir, B. Sterling, and K. L. Monteith. 2019. Hunting and mountain sheep: Do current harvest practices affect horn growth? *Evolutionary Applications* 12:1823–1836.

Laws, R. M. 1952. A new method of age determination for mammals. *Nature* 169:972–973.

Leopold, A. 1933. *Game Management*. Scribner's, New York, New York, USA.

Lesiuk, T. P., and C. R. Braekevelt. 1983. Fine structure of the canine tapetum lucidum. *Journal of Anatomy* 136:157–164.

Loehr, J., J. Carey, H. Ylönen, and J. Suhonen. 2008. Coat darkness is associated with social dominance and mating behaviour in a mountain sheep hybrid lineage. *Animal Behaviour* 76:1545–1553.

Loehr, J., J. Carey, R. B. O'hara, and D. S. Hik. 2010. The role of phenotypic plasticity in responses of hunted thinhorn sheep ram horn growth to changing climate conditions. *Journal of Evolutionary Biology* 23:783–790.

Longshore, K., C. Lowrey, M. Jeffress, and D. B. Thompson. 2009. Nocturnal movements of desert bighorn sheep in the Muddy Mountains, Nevada. *Desert Bighorn Council Transactions* 51:18–31.

Lukefahr, S. D., and H. A. Jacobson. 1998. Variance component analysis and heritability of antler traits in white-tailed deer. *Journal of Wildlife Management* 62:262–268.

Malo, A. F., J. J. Garde, A. J. Soler, A. J. García, M. Gomendio, and E. R. S. Roldan. 2005. Male fertility in natural populations of red deer is determined by sperm velocity and the proportion of normal spermatozoa. *Biology of Reproduction* 72:822–829.

Marshal, J. P., P. R. Krausman, V. C. Bleich, W. B. Ballard, and J. S. Mckeever. 2002. Rainfall, El Niño, and dynamics of mule deer in the Sonoran Desert, California. *Journal of Wildlife Management* 66:1283–1289.

Marshal, J. P., V. C. Bleich, and N. G. Andrew. 2008. Evidence for interspecific competition between feral ass *Equus asinus* and mountain sheep *Ovis canadensis* in a desert environment. *Wildlife Biology* 14:228–236.

Martin, A. M., J. T. Hogg, K. R. Manlove, T. N. LaSharr, J. M. Shannon, D. E. McWhirter, H. Miyasaki, K. L. Monteith, and P. C. Cross. 2022. Disease and secondary sexual traits: Effects of pneumonia on horn size of bighorn sheep. *Journal of Wildlife Management* 86:e22154.

Martin, J. G. A., and M. Festa-Bianchet. 2010. Bighorn ewes transfer the costs of reproduction to their lambs. *American Naturalist* 176:414–423.

McCann, L. J. 1956. Ecology of the mountain sheep. *The American Midland Naturalist* 56:297–324.

McCullough, D. R. 1979. *The George Reserve Deer Herd: Population Ecology of a K-Selected Species*. University of Michigan Press, Ann Arbor, Michigan, USA.

McCullough, D. R. 1999. Density dependence and life-history strategies of ungulates. *Journal of Mammalogy* 80:1130–1146.

Messner, T. C. 2011. White-tailed deer Management strategies and domestication processes. *Human Ecology* 39:165–178.

Metcalfe, N. B., and P. Monaghan. 2001. Compensation for a bad start: Grow now, pay later? *Trends in Ecology & Evolution* 16:254–260.

Mitchell, C. D., R. Chaney, K. Aho, J. G. Kie, and R. T. Bowyer. 2015. Population density of Dall's sheep in Alaska: Effects of predator harvest? *Mammal Research* 60:21–28.

Monteith, K. L., L. E. Schmitz, J. A. Jenks, J. A. Delger, and R. T. Bowyer. 2009. Growth of male white-tailed deer: Consequences of maternal effects. *Journal of Mammalogy* 90:651–660.

Monteith, K. L., R. A. Long, T. R. Stephenson, V. C. Bleich, R. T. Bowyer, and T. N. Lasharr. 2018. Horn size and nutrition in mountain sheep: Can ewe handle the truth? *Journal of Wildlife Management* 82:67–84.

Monteith, K. L., R. A. Long, V. C. Bleich, J. R. Heffelfinger, P. R. Krausman, and R. T. Bowyer. 2013. Effects of harvest, culture, and climate on trends in size of horn-like structures in trophy ungulates. *Wildlife Monographs* 183:1–28.

Monteith, K. L., V. C. Bleich, T. R. Stephenson, B. M. Pierce, M. M. Conner, J. G. Kie, and R. T. Bowyer. 2014. Life-history characteristics of mule deer: Effects of nutrition in a variable environment. *Wildlife Monographs* 186:1–62.

Mysterud, A. 2011. Selective harvesting of large mammals: How often does it result in directional selection? *Journal of Applied Ecology* 48:827–834.

Mysterud, A., and R. Bischof. 2010. Can compensatory culling offset undesirable evolutionary consequences of trophy hunting? *Journal of Animal Ecology* 79:148–160.

Nasoori, A. 2020. Formation, structure, and function of extra-skeletal bones in mammals. *Biological Reviews* 95:986–1019.

Nesbitt, W. H., P. L. Wright, E. L. Buckner, C. R. Byers, and J. Reneau. 2009. *Measuring and Scoring North American Big Game Trophies*. Third edition. Boone and Crockett Club, Missoula, Montana, USA.

Nichols, L. 1978. Dall's sheep. Pages 173–189 *in* J. L. Schmidt and D. L. Gilbert, editors. *Big Game of North America, Ecology and Management*. Stackpole Books, Harrisburg, Pennsylvania, USA.

Nichols, L., and F.L. Bunnell. 1999. Natural history of Thinhorn Sheep. Pp. 23–77 *in* R. Valdez and P. R. Krausman, eds. *Mountain Sheep of North America*. University of Arizona Press, Tucson, Arizona, USA.

O'Brien, J. M., C. S. O'Brien, C. McCarthy, and T. E. Carpenter. 2014. Incorporating foray behavior into models estimating contact risk between bighorn sheep and areas occupied by domestic sheep. *Wildlife Society Bulletin* 38:321–331.

Parikh, G. L., R. C. Karsch, J. W. Cain III, E. M. Rominger, and E. J. Goldstein. 2025. Neonate Morphometrics and Lambing Season Characteristics of Desert Bighorn Sheep. *Mammalia* 89:121–130.

Parr, B. L., J. B. Smith, J. A. Jenks, and D. J. Thompson. 2018. Population dynamics of a bighorn sheep (*Ovis canadensis*) herd in the southern Black Hills of South Dakota and Wyoming. *American Midland Naturalist* 179:1–14.

Pigeon, G., M. Festa-Bianchet, D. W. Coltman, and F. Pelletier. 2016. Intense selective hunting leads to artificial evolution in horn size. *Evolutionary Applications* 9:521–530.

Pigeon, G., J. Landes, M. Festa-Bianchet, and F. Pelletier. 2021. Do early-life conditions drive variation in senescence of female bighorn sheep? *Frontiers in Cell and Developmental Biology* 9:637692.

Poissant, J., A. J. Wilson, M. Festa-Bianchet, J. T. Hogg, and D. W. Coltman. 2008. Quantitative genetics and sex-specific selection on sexually dimorphic traits in bighorn sheep. *Proceedings of the Royal Society B: Biological Sciences* 275:623–628.

Poisson, Y., M. Festa-Bianchet, and F. Pelletier. 2020. Testing the importance of harvest refuges for phenotypic rescue of trophy-hunted populations. *Journal of Applied Ecology* 57:526–535.

Polák, J., and D. Frynta. 2009. Sexual size dimorphism in domestic goats, sheep, and their wild relatives. *Biological Journal of the Linnean Society* 98:872–883.

Portier, C., M. Festa-Bianchet, J. -M. Gaillard, J. T. Jorgenson, and N. G. Yoccoz. 1998. Effects of density and weather on survival of bighorn sheep lambs (*Ovis canadensis*). *Journal of Zoology* 245:271–278.

Preston, B. T., I. R. Stevenson, J. M. Pemberton, D. W. Coltman, and K. Wilson. 2003. Overt and covert competition in a promiscuous mammal: The importance of weaponry and testes size to male reproductive success. *Proceedings of the Royal Society B: Biological Sciences* 270:633–640.

Proffitt, K. M., A. B. Courtemanch, S. R. Dewey, B. Lowrey, D. E. McWhirter, K. Monteith, J. T. Paterson, J. Rotella, P. J. White, and R. A. Garrott. 2021. Regional variability in pregnancy and survival rates of Rocky Mountain bighorn sheep. *Ecosphere* 12:e03410.

Province of British Columbia. 2023. *Compulsory Inspection Data 1900 to 2022*. Wildlife Branch, Ministry of Water, Land and Resource Stewardship, Victoria, British Columbia, Canada.

Risenhoover, K. L., and J. A. Bailey. 1985. Foraging ecology of mountain sheep: Implications for habitat management. *Journal of Wildlife Management* 49:797–804.

Robbins, C. T. 1973. *The Biological Basis for the Determination of Carrying Capacity*. Cornell University, Ithaca, New York, USA.

Rodman, K. C., T. T. Veblen, S. Saraceni, and T. B. Chapman. 2019. Wildfire activity and land use drove 20th-century changes in forest cover in the Colorado front range. *Ecosphere* 10:e02594.

Ruckstuhl, K. E., M. Festa-Bianchet, and J. T. Jorgenson. 2000. Effects of prescribed grassland burns on forage availability, quality and bighorn sheep use. *Biennial Symposium of Northern Wild Sheep and Goat Council* 12:11–25.

Schindler, S., M. Festa-Bianchet, J. T. Hogg, and F. Pelletier. 2017. Hunting, age structure, and horn size distribution in bighorn sheep. *Journal of Wildlife Management* 81:792–799.

Schmidt-Nielsen, K. 1979. *Desert Animals: Physiological Problems of Heat and Water*. Dover Publications, Mineola, New York, USA.

Scotton, B. D. 1997. *Estimating Rates and Causes of Neonatal Lamb Mortality of Dall Sheep in the Central Alaska Range*. Alaska Department of Fish and Game, Juneau, Alaska, USA.

Seip, D. R., and F. L. Bunnell. 1984. Body weights and measurements of Stone's sheep. *Journal of Mammalogy* 65:513–514.

Shackleton, D. M. 1985. *Ovis canadensis*. *Mammalian Species* 320:1–9.

Shackleton, D. M., and D. A. Hutton. 1971. An analysis of the mechanism of brooming of mountain sheep horns. *Zeitschrift für Säugetierkunde* 36:342–350.Sim, Z., and D. W. Coltman. 2019. Heritability of horn size in thinhorn sheep. *Frontiers in Genetics* 10:959.

Simon, A. 2016. Against trophy hunting: A Marxian-Leopoldian critique. *Monthly Review* 68:17.

Sittler, K. L., K. L. Parker, and M. P. Gillingham. 2019. Vegetation and prescribed fire: Implications for stone's sheep and elk. *Journal of Wildlife Management* 83:393–409.

Smiley, R. A., B. L. Wagler, T. N. LaSharr, K. A. Denryter, T. R. Stephenson, A. B. Courtemanch, T. W. Mong, D. Lutz, D. McWhirter, D. Brimeyer, P. Hnilicka, B. Lowrey, and K. L. Monteith. 2022. Heterogeneity in risk-sensitive allocation of somatic reserves in a long-lived mammal. *Ecosphere* 13:e4161.

Smiley, R. A., B. L. Wagler, W. H. Edwards, J. Jennings-Gaines, K. Luukkonen, K. Robbins, M. Johnson, A. B. Courtemanch, T. W. Mong, D. Lutz, D. McWhirter, J. L. Malmberg, B. Lowrey, and K. L. Monteith. 2024. Infection-nutrition feedbacks: Fat supports pathogen clearance but pathogens reduce fat in a wild mammal. *Proceedings of the Royal Society B: Biological Sciences* 291:20240636.Smith, D. R., and P. R. Krausman. 1987. Diet of desert bighorn sheep in the Virgin Mountains, Arizona. *Desert Bighorn Council Transactions* 31:11–14.

Smith, T. S., P. J. Hardin, and J. T. Flinders. 1999. Response of bighorn sheep to clear-cut logging and prescribed burning. *Wildlife Society Bulletin* 27:840–845.

Solberg, E. J., and B. E. Saether. 1993. Fluctuating asymmetry in the antlers of moose (*Alces alces*): Does it signal male quality? *Proceedings of the Royal Society B: Biological Sciences* 254:251–255.

Stephenson, T. R., D. W. German, E. Frances Cassirer, D. P. Walsh, M. E. Blum, M. Cox, K. M. Stewart, and K. L. Monteith. 2020. Linking population performance to nutritional condition in an alpine ungulate. *Journal of Mammalogy* 101:1244–1256.

Terry, P. J., A. C. Alvidrez, and C. W. Black. 2022. Factors affecting bighorn sheep activity at water developments in southwestern Arizona. *Journal of Wildlife Management* 86:e22134.

Traill, L. W., S. Schindler, and T. Coulson. 2014. Demography, not inheritance, drives phenotypic change in hunted bighorn sheep. *Proceedings of the National Academy of Sciences of the United States of America* 111:13223–13228.

Valdez, R., and P. R. Krausman. 1999. Description, distribution, and abundance of mountain sheep in North America. Pages 3–22 in R. Valdez and P. R. Krausman, editors. *Mountain Sheep of North America*. The University of Arizona Press, Tucson, Arizona, USA.

Vanpé, C., J. M. Gaillard, P. Kjellander, A. Mysterud, P. Magnien, D. Delorme, G. V. Laere, F. Klein, O. Liberg, and A. J. M. Hewison. 2007. Antler size provides an honest signal of male phenotypic quality in roe deer. *The American Naturalist* 169:481–493.

Varpe, Ø., C. Jørgensen, G. A. Tarling, and Ø. Fiksen. 2009. The adaptive value of energy storage and capital breeding in seasonal environments. *Oikos* 118:363–370.

Vavra, M., C. G. Parks, and M. J. Wisdom. 2007. Biodiversity, exotic plant species, and herbivory: The good, the bad, and the ungulate. *Forest Ecology and Management* 246:66–72.

Wagler, B. L., R. A. Smiley, A. B. Courtemanch, D. Lutz, D. McWhirter, D. Brimeyer, P. Hnilicka, and K. L. Monteith. 2024. Disparate home range dynamics reflect nutritional inadequacies on summer range for a large herbivore. *Ecosphere* 15:e4864.

Wagler, B. L., R. A. Smiley, A. B. Courtemanch, D. Lutz, D. McWhirter, D. Brimeyer, P. Hnilicka, T. J. Robinson, and K. L. Monteith. 2023. Implications of forage quality for population recovery of bighorn sheep following a pneumonia epizootic. *Journal of Wildlife Management* 87:e22452.

Wagner, G. D. 2000. Diet selection, activity patterns and bioenergetics of bighorn ewes in central Idaho. Thesis, University of Idaho, Moscow, Idaho, USA.

Walls, G. L. 1942. *The Vertebrate Eye and Its Adaptive Radiation.* Hafner Publishing Company, New York, New York, USA.

Wehausen, J. D., C. J. O'Brien, and D. R. McCullough. 2024. Reliability of tooth cementum rings to age bighorn sheep: A blind test. *California Fish and Wildlife Journal* 110:e19.

Werdel, T. J., J. A. Jenks, J. T. Kanta, C. P. Lehman, and T. J. Frink. 2023. Resource selection and herbaceous biomass at foraging sites of translocated bighorn sheep. *Rangeland Ecology and Management* 87:141–149.

Wheatley, B. B., E. C. Gilmore, L. H. Fuller, A. M. Drake, and S. W. Donahue. 2023. How the geometry and mechanics of bighorn sheep horns mitigate the effects of impact and reduce the head injury criterion. *Bioinspiration and Biomimetics* 18:026005.

Whiting, J. C., V. C. Bleich, R. T. Bowyer, and C. W. Epps. 2023a. Restoration of bighorn sheep: History, successes, and remaining conservation issues. *Frontiers in Ecology and Evolution* 11:1–17.

Whiting, J. C., V. C. Bleich, R. T. Bowyer, K. Manlove, and K. White. 2023b. Bighorn sheep and mountain goats. Pages 759–790 in L. B. McNew, D. K. Dahlgren, and J. L. Beck, editors. *Rangeland Wildlife Ecology and Conservation.* Springer Nature, Berlin, Germany.

Wild Sheep Foundation Professional Biologist Meeting Attendees. 2008. Ram harvest strategies for Western States and Provinces. *Biennial Symposium of the Northern Wild Sheep and Goat Council* 16:92–98.

Williams, J. D., W. F. Krueger, and D. H. Harmel. 1994. Heritabilities for antler characteristics and body weight in yearling white-tailed deer. *Heredity* 73:78–83.

Wilson, D. E., and S. Ruff, editors. 1999. *The Smithsonian Book of North American Mammals.* Smithsonian Institution Press, Washington, D.C., USA.

Wishart, W. 1978. Bighorn sheep. Pages 161–171 *in* J. L. Schmidt and D. L. Gilbert, editors. *Big Game of North America: Ecology and Management.* Stackpole Books, Harrisburg, Pennsylvania, USA.

Worley, K., C. Strobeck, S. Arthur, J. Carey, H. Schwantje, A. Veitch, and D. W. Coltman. 2004. Population genetic structure of North American thinhorn sheep (*Ovis dalli*). *Molecular Ecology* 13:2545–2556.

# 5 Social Behavior and Reproductive Strategies

*Marco Festa-Bianchet and Fanie Pelletier*

## 5.1 INTRODUCTION

Mountain sheep are social ungulates, relying on learning and tradition to migrate between seasonal ranges (Jesmer et al. 2018). In North America, they occur in mountains from the Arctic to deserts in Mexico and the southwestern United States. Over this vast range, ecological conditions vary substantially as seasonality, predator and competitor guilds, physical terrain, vegetation composition and structure change. Human effects on sheep and their habitat also vary over their range, from pristine roadless habitats to populations using golf courses or coal mines. Consequently, aspects of mountain sheep behavior and reproductive tactics can vary among populations. For example, synchrony of births, dictated by seasonality and vegetation phenology, generally declines from north to south (Bunnell 1982). The degree of population synchrony in births reflects synchrony in timing of estrus (Hogg et al. 2017), which in turn may affect breeding sex ratio, competition among males, and mating tactics (Pelletier et al. 2006). In this chapter, we will first consider the social structure of mountain sheep, then examine their mating system, the determinants of reproductive success, and, finally, discuss the evolution of reproductive strategies. Not all aspects of behavior are equally well documented for populations facing contrasting ecological conditions. For many aspects of behavior, more literature is available for bighorn sheep in the Rocky Mountains than for thinhorn or desert sheep.

Several aspects of mountain sheep behavior appear to be adaptations that optimize their survival and reproductive success by avoiding cursorial predators such as wolves (*Canis lupus*) and coyotes (*Canis latrans*). Behaviors that improve predator avoidance include gregariousness, vigilance, preference for open habitat near escape terrain, and use of traditional trails to migrate or to move to safety when predators are sighted. Those behavioral tactics come at the cost of restricting foraging to safe areas, often with lower quantity and quality of forage, prioritizing safety from cursorial predators over forage intake. To optimize lifetime reproductive success, females adopt a conservative reproductive strategy and allocate resources to reproduction only after they have secured enough for their own maintenance (Festa-Bianchet and Jorgenson 1998). Therefore, the main factor limiting female reproduction is low-risk access to forage near escape terrain, even though there may be better forage further away, where the chances of surviving a predator attack are reduced. Use of escape terrain is an efficient tactic to reduce predation by coursing predators, but avoiding ambushing and stalking predators such as cougars (*Puma concolor*) requires a different behavioral tactic. Sheep cannot optimize avoidance of both types of predators. Similarly, wild Asiatic sheep (*Ovis* spp.) and Asiatic ibex (*Capra sibirica*) face a behavioral trade-off in avoiding wolves or snow leopards (*Panthera uncia*; Kachel et al. 2023). For males, the main factors limiting reproductive success are the number of breeding females and competition with other males (Martin et al. 2016). Male-male competition is at the origin of pronounced sexual dimorphism in body and horn size, riskier foraging strategy by males with less reliance on escape terrain, and alternative male mating tactics to secure the highest possible mating success by either outrunning or outfighting competitors. Those sexual differences in behavior and reproductive tactics have important implications for age-specific survival, population dynamics, and hunting management.

## 5.2 SOCIAL STRUCTURE

### 5.2.1 SEXUAL SEGREGATION

Sexual segregation outside the rut is a major characteristic of social structure in mountain sheep and it is often associated with differences in area and habitat use (Ruckstuhl and Neuhaus 2002). Male groups sometimes forage further from escape terrain than female groups, likely because the trade-offs between predator avoidance and reproductive success differ between the sexes. Males appear more willing than females to take risks to maximize growth, because large horn and body size determine the outcome of competition for estrous females. It is also likely that some predators such as coyotes, that are a threat to females and lambs, do not threaten adult males because of their much larger body size.

Young males remain in nursery groups with females, lambs, and yearlings, until 2–4 years of age, then gradually move to bachelor groups of adult males (Festa-Bianchet 1991). The move from nursery to bachelor groups is gradual, with males aged 2–3 years switching back and forth between the two types of groups (Festa-Bianchet 1991, Ruckstuhl and Festa-Bianchet 2001). When present in sufficient numbers, subadult males, aged 2 or 3 years, may form distinct groups (Ruckstuhl and Festa-Bianchet 2001). Formation of age-specific male groups has been observed also in Alpine ibex (*Capra ibex*), and may be related to either social preferences for animals of the same age or to

DOI: 10.1201/9781003518686-6

differences in body size that affect foraging time budgets. Young adult males are larger than adult females but smaller than mature males (Festa-Bianchet et al. 1996), and may forage more efficiently when in a group of similarly sized individuals (Ruckstuhl 1999), as smaller animals require more daily feeding time than larger individuals. Given that the optimal duration of grazing and ruminating bouts varies with body size, sheep may best synchronize time budgets when in a group of similar-sized animals (Ruckstuhl 1998). Synchronization of foraging and ruminating bouts is essential to maintain group cohesion. If a large male joined a nursery herd, it will fill its rumen faster than the adult females, and its optimal behavior would be to lay down and ruminate (Ruckstuhl 1998). If the rest of the group continues feeding, however, they will move away, and the male would be left out of the group. Sheep that are substantially smaller or larger than the mean of other members of the group may be forced to adopt suboptimal time budgets to remain within that group (Ruckstuhl and Neuhaus 2002).

### 5.2.2 GREGARIOUSNESS AND ASSOCIATION PATTERNS

The social structure of mountain sheep is characterized by fission-fusion dynamics, where the size and composition of groups change over time. As discussed above, individuals usually associate based on their sex and age (Figure 5.1). Gregariousness likely facilitates learning of seasonal ranges and movement corridors, but it also increases the risk of disease transmission (Manlove et al. 2014). The main advantage of gregariousness in mountain sheep, however, is to reduce predation risk through increased vigilance and dilution effects (Rieucau and Martin 2008). In larger groups, each individual can spend less time vigilant, yet the overall level of vigilance increases as more eyes are looking for predators. When a predator attacks, an individuals is less likely to be killed in a large than in a small group (Rieucau and Martin 2008). When local population size is small, mountain sheep may form mixed-sex groups simply because avoiding predators trumps synchronization of time budgets (Ruckstuhl and Festa-Bianchet 2001).

Although nearly all females remain in their natal group for life, after weaning they do not associate preferentially with their mother or other relatives. Most young adults co-exist with their mother until 4–5 years of age (Figure 5.2). Yet, in the Sheep River population females aged 1–3 years were not in the same group as their mother about 75% of the time. When population and group sizes were accounted for, the distribution of mothers and their adult daughters among groups was not different from random (Festa-Bianchet 1991). Strong philopatry but no preferential kin association was also reported for desert bighorn females (Boyce et al. 1998). At Ram Mountain, Canada, proximity collars, that register the number and duration of instances when two individuals come within 1.5 m of each other, revealed that some pairs of adult females associated more than expected if distribution within a group was random, although the number of proximity encounters was highly variable between individuals and years (Vander Wal et al. 2016). Network analyses of the yearly association patterns revealed variability in strength of association and social structure. Those associations, however, were not explained by similarities in age, reproductive status, or social dominance, and the effect of kin relationship was null or slightly negative (Vander Wal et al. 2016). While it is important for females to be in a group, the available evidence suggests there are no consistent patterns of association.

Sociality in mountain sheep has fitness consequences for both sexes, comparable to the effects of variation in population density or body mass (Vander Wal et al. 2015). In particular, females that have a higher social centrality, a measure of connectedness to other females in the population, have higher survival and are more likely to produce lambs. Individual females also appear to change their social behavior according to reproductive status, being more central in years when they are nursing a lamb (Vander Wal et al. 2015). Given that gregariousness is likely an adaptation to avoid predation, these results suggest that fine-scale behavior within a group may affect the direct and indirect effects of predation (Chapter 23).

**FIGURE 5.1** (a) Mixed-age male group and (b) nursery group led by a 14-year-old female. Sheep River, Alberta. (Photo by F. Pelletier, used with permission.)

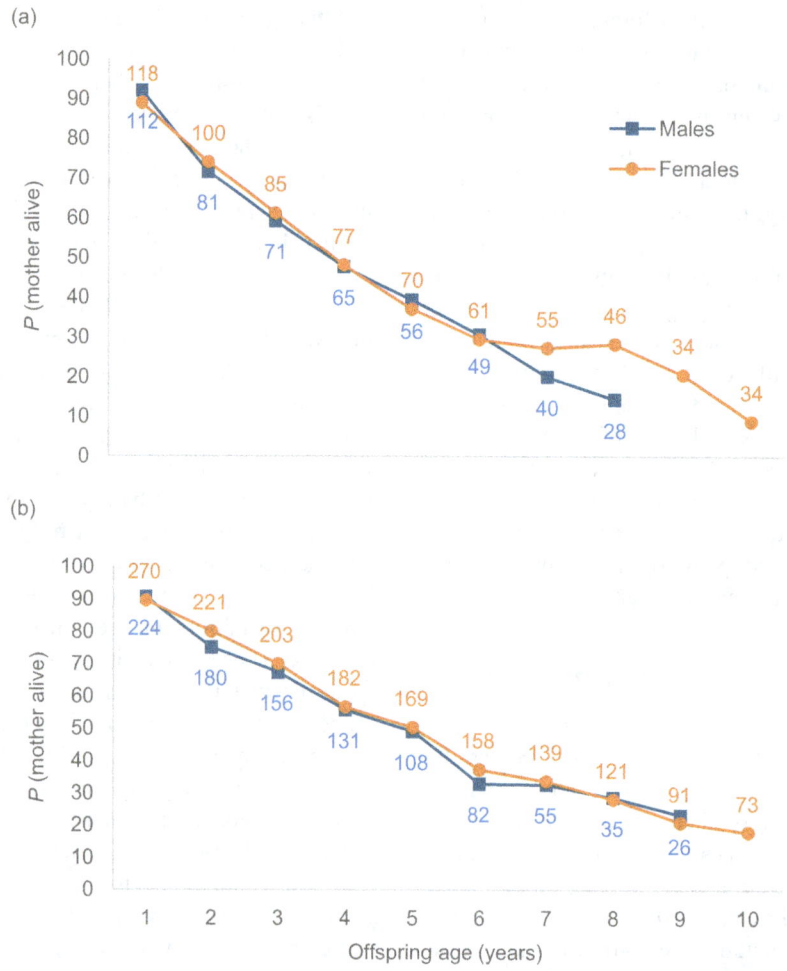

**FIGURE 5.2** Proportion of bighorn sheep males (■) and females (•) whose mothers were alive according to offspring age for (a) Sheep River (1981–2003) and (b) Ram Mountain (1979–2020). Numbers indicate sample sizes. Sheep whose mother was shot, poached, or collected are not included. (Data on panel A is updated from Festa-Bianchet (1991) while panel B illustrates unpublished data.)

### 5.2.3 Dominance Relationships

Dominance is a characteristic of two individuals, called a dyad. It is typically defined as asymmetry in aggression by one animal toward another, with a predictable outcome, where the dominant can displace the subordinate, often without physical contact (Drews 1993). Dominance is thought to be established through agonistic interactions, but, especially for males, it may also rely on visual signals such as obvious differences in body or horn size. Well-established dyadic dominance relationships can save time and energy from costly behaviors such as fights, establish order of access to resources, and reduce the risk of injury by avoiding physical interactions. The term dominance hierarchy is used in different ways across taxa and contexts (Tibbetts et al. 2022). In mountain sheep, it describes the linear hierarchy within a group resulting from dyadic relationships. A dominance hierarchy is linear when there are few or no circular relationships: if A is dominant to B and B is dominant to C, C will be subordinate to A. Linear hierarchies are at the basis of social interactions among sheep of both sexes (Pelletier and Festa-Bianchet 2006, Favre et al. 2008). The

presence and stability of social hierarchies suggest that, within a population, sheep recognize other individuals and are aware of their relative dominance status.

*Females.* Physical interactions among females mostly involve threats, displacements from resting sites, or horn butts to the body. Horn clashes also occur, often without a clear winner (Festa-Bianchet 1991). Agonistic interactions between females are less frequent than among males, and rarely escalate. It is not obvious what advantages females obtain from dominance, given that in most cases their food is widely distributed and not easily defended. Perhaps dominant females may displace subordinates from snow craters in winter as reported in caribou (*Rangifer tarandus*) (Barrette and Vandal 1986), or from some high-value foods such as broken cacti in deserts (Warrick and Krausman 1989). Dominance rank in females is mostly determined by age, possibly because dyadic relationships are established when one female is aged 2 and the other is a yearling, and 2-year-olds are larger than yearlings. As suggested for red deer (*Cervus elaphus*), ranks may then be maintained for life even if the younger female becomes larger, mostly because there are no obvious advantages from being dominant (Thouless and Guinness 1986). For example, at Sheep River, the younger female won only 7% of 301 interactions among known-age females (Festa-Bianchet 1991). In the Ram Mountain population, however, mass increased dominance rank for females aged 7 years and older, while horn size did not play a detectable role (Favre et al. 2008).

*Males.* Interactions among males occur year-round, but are most frequent during the pre-rut and rut. During the pre-rut (early October to mid-November in the Canadian Rockies), males form congregations (Festa-Bianchet 1986), and frequent social interactions establish or reinforce the social hierarchy. Males use several types of agonistic interactions (Figure 5.3). In rare cases, interactions escalate to combat. The frontal clashes of mature males are possibly the most iconic behavior of mountain sheep (Figure 5.3a), yet they are rare. They only occur when two males are similar in size and neither can easily establish dominance. These battles can go on for hours and are mostly ritualized, with males facing each other a few meters apart, rearing up on their hind legs and smashing horns with a sideways twist. Interactions that are more often observed are less impressive and include body butt (Figure 5.3b), eye rubbing (Figure 5.3c), and front kick (Figure 5.3d), the latter often accompanied by grunt-like vocalizations by the dominant male. During clashes, the horn tips can hit the horn base of the opponent, often breaking off in what is called brooming. Some mature males may have the first or even the first two horn increments completely broomed off, which can cause some difficulties in aging them. In most dyadic encounters, however, one male quickly establishes dominance through threats, mounting, or a single horn clash (Geist 1971).

Bighorn males form a linear hierarchy where the positive effect of body mass on social rank increases after they reach their asymptotic mass, usually between 6 and 8 years of age. Among males of the same age, about a quarter of dyads show a reversal of dominance status over time, where A is dominant to B 1 year, but B becomes dominant to A in a later year. The frequency of reversals drops rapidly with increasing differences in age: it is 10% for 1 year difference but only 1% for dyads with 2 years difference in age (Pelletier and Festa-Bianchet 2006). The outcome of repeated interactions between two males within the same year is highly predictable: during a 5-year study at Sheep River, Alberta, the male that won the first observed interaction won 93%–96% of subsequent interactions. The main determinants of social rank were age and mass (Figure 5.4). Age and mass in males are highly correlated until about 5 years of age. Age explained 77% of the variance in rank among males aged 1–5 years, but only a nonsignificant 12% for those aged 6 years and older (Pelletier and Festa-Bianchet 2006). In other words, up to age 5 the older male is nearly always dominant, but after age 6 factors such as body mass and aggressiveness come into play, so dominance is no longer mostly determined by age. Although horn size may also play a role in establishing dominance rank, it is correlated with age and mass, and we are unaware of studies that quantified the relative role of horn size on social dominance after accounting for age and body mass. Because larger males have larger horns, dominant males have larger horns than subordinates. Male dominance rank plays a very important fitness role, because the most dominant males can use the highly successful mating tactic of

(a) Frontal clash

(b) Body Butt

(c) Eye rub

(d) Front kick

**FIGURE 5.3** Social interactions among bighorn males at Sheep River, Alberta. (Photo by F. Pelletier; used with permission.)

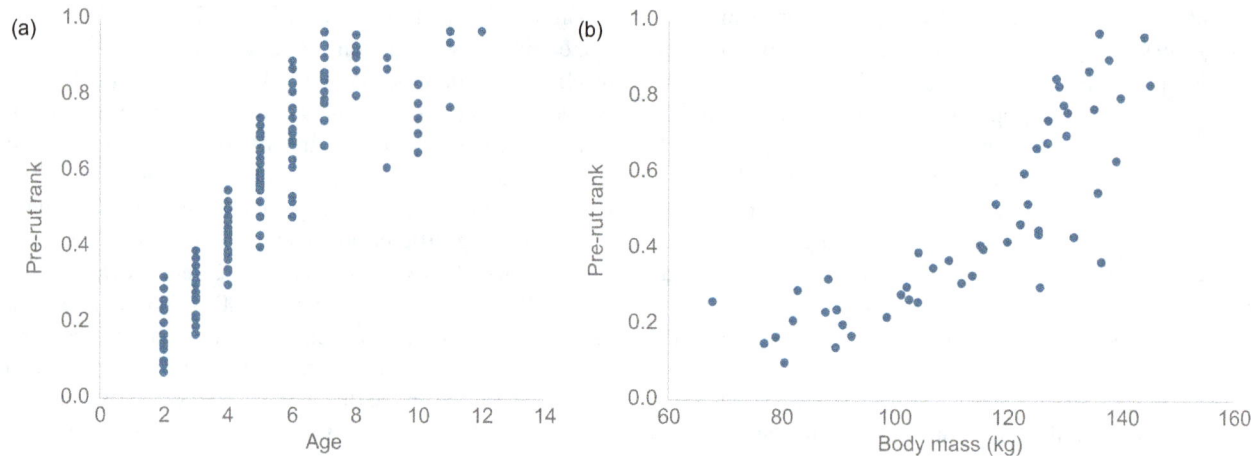

**FIGURE 5.4** (a) Age and social rank for bighorn males aged 2 years and older. A linear mixed model explained 86% of the variance (age: F1,64=558.54, $N$=119 observations from 53 individuals, $P$<0.001; age²: F1,64=61.02, $P$<0.001). (b) Body mass in early autumn and social rank for bighorn males aged 2 years and older. A linear mixed model explained 73% of variance (mass: F1,23=118.61, $N$=51 observations from 26 individuals, $P$<0.001; mass²: F1,23=7.85, $P$=0.01). (Reproduced from Pelletier and Festa-Bianchet (2006).)

tending single estrous females. Consequently, it can be very rewarding for a male to attain high dominance rank, and as males gain size and strength they may challenge closely ranked individuals to which they are subordinate. The social rank established during the pre-rut will determine female access and the mating tactics used by males for reproduction, as presented in the next section.

## 5.3   MATING SYSTEM AND MALE REPRODUCTIVE TACTICS

The mating system of mountain sheep involves female-defense serial polygyny by top-ranking males and alternative mating tactics by other males. Dominant males serially defend single estrous females, a tactic known as tending. Subordinate males attempt to separate the estrous female from the defending male and force a copulation, a tactic known as coursing. Less frequently, subordinates may attempt to prevent an estrous female from joining breeding groups, a tactic known as blocking (Hogg 1984). Copulations typically last less than 2 seconds. Estrous females are defended by dominant males for 1–2 days. Only males at the top of the social hierarchy can use the tending tactic, because typically there are very few females in estrus at the same time. The most dominant male in a group, known as the alpha male, has priority of access to estrous females. The second-ranked, or beta, male can adopt the tending tactic if there are at least two females in estrus at the same time, while the alpha male is tending another ewe. If the alpha male ends its tending bout and moves over, he will take over the tending role from the beta male, that will join the groups of coursing males. The third-ranked male in a breeding group can only adopt the tending tactic if there are three females in estrus at the same time, an uncommon event in most populations. In a large population with 50 breeding females, over a 15-day breeding season and assuming a 24-hour estrus, on average there should be

three estrous females/day. If a population is divided into multiple breeding groups in isolated patches of habitat, the highest-ranked male in each breeding group will adopt the tending tactic. Therefore, a male ranked number four or five in the hierarchy may have a tending opportunity simply because more dominant males may be in other isolated groups. When no estrous female is available within their group, however, the highest-ranked males frequently move among breeding groups looking for estrous females.

The main mating tactic used by subordinate males is coursing. Coursing males try to separate the estrous female from the tending male. A wild chase then ensues, with up to eight to ten males attempting to force copulations while the female is running, while pushing and butting each other (Figure 5.3b). It may take from a few minutes to over an hour before the tending male regains control of the female. This is a dangerous situation for both sexes, running over steep and often frozen terrain. Falls during these chases can result in death (Festa-Bianchet 1987). The female will typically attempt to rejoin the tending male, possibly because it can provide some safety against coursing males. When terrain features allow it, the female may climb a rock face and the tending male may then block access to her. Estrous females can mate with multiple males, including the tending and one or more coursing males, so that sperm competition plays a role in siring success. After the tending male regains control of the female following a coursing chase, he will typically copulate with her several times. These retaliatory copulations likely increase the chances that his sperm will fertilize her egg (Hogg 1988).

### 5.3.1   THE CHALLENGES OF MEASURING MALE REPRODUCTIVE SUCCESS

While mating behavior of mountain sheep, including different mating tactics and copulations, can be documented through observations (Hogg 1984, Pelletier et al. 2006), assessing male reproductive success requires genetic analyses and parentage assignment. That involves extensive tissue sampling of mothers, lambs, and candidate fathers, extraction of DNA and use of software that accounts for unsampled males, genetic diversity of different alleles, and other variables that may affect parentage assignment (Kalinowski et al. 2007). Even with intensive sampling and genetic analysis, the estimation of male reproductive success is much more difficult than that of female reproductive success. If the lamb of a marked female is marked and monitored, that female's reproductive success can be precisely measured. If the lamb dies at birth or before it is marked, observations of lactation and sightings of a lamb can be used to determine female reproductive status each year. On the other hand, while many paternities can be confidently assigned, others cannot. For example, in some years in the isolated population at Ram Mountain, even though all resident males were marked and genotyped, about a third of lambs could not be assigned a father. These lambs were fathered by unsampled males that were only present during the rut. Similarly, if a marked male leaves a study population for the rut, its reproductive success elsewhere cannot be documented. If a lamb dies before it is sampled, its father will remain unknown. Given these difficulties, it is not surprising that reliable data on male reproductive success exist for only a few species of ungulates. By reliable data we mean enough paternity assignments to compare different males and relate siring success to male age, physical traits, and population variables such as sex ratio or male age structure (Festa-Bianchet 2012). For most species, these data exist for just one population, so that their generality cannot be assessed. Rocky Mountain bighorn sheep are the only species with reliable data on male reproductive success for three populations: the National Bison Range in Montana, United States (Hogg and Forbes 1997), and two populations in Alberta, Canada: Sheep River (Hogg and Forbes 1997) and Ram Mountain (Martin et al.

2016). Results from these three populations reveal broadly similar patterns, presented below. We likely know more about reproductive success of bighorn males than for males of any other wild ungulate. There are, however, no data on male reproductive success for thinhorn sheep or for desert bighorns. The long reproductive season of desert bighorns (Rubin et al. 2000) likely increases the temporal and spatial unpredictability of estrous ewes, with unknown effects on male reproductive success.

### 5.3.2 Factors Affecting Male Reproductive Success

The main factor limiting male reproductive success is competition with other males. Therefore, body size and social dominance are the main determinants of siring success. The tending tactic is much more successful than the alternative tactics of coursing and blocking. The tending male has a 40%–50% chance of fathering the lamb of the defended female (Hogg and Forbes 1997, Coltman et al. 2002). Therefore, about half of paternities in a population are shared among males other than those at the very top of the dominance hierarchy. Age is a major determinant of reproductive success (Figure 5.5) likely because it is correlated with both size and dominance status (Figure 5.4). Older males tend to be dominant and larger, both in terms of body mass and horn size, so they are more likely to use the tending tactic and have high reproductive success. Because coursing males also achieve some copulations, however, males as young as 2 years, and very occasionally yearlings, also father some lambs (Martin et al. 2016). For males aged 2–4 years, dominance, horn size, and body mass play a secondary role in reproductive success, which may be mainly influenced by differences in behavior, particularly agility and speed, during coursing chases or willingness to take risks (Ritchot et al. 2021). Coursing chases can involve up to eight to ten males running after one female and jockeying

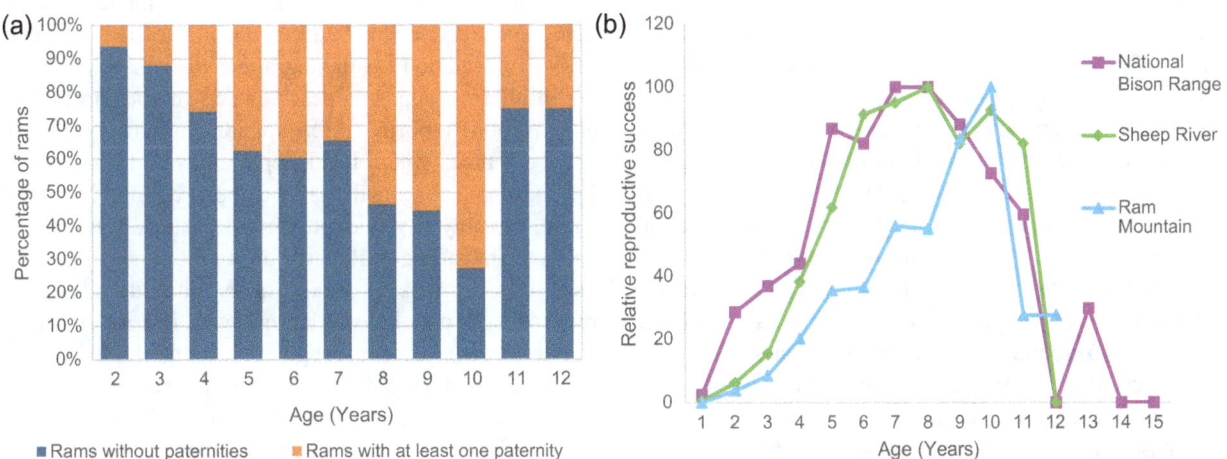

**FIGURE 5.5** Age-specific reproductive success of bighorn males. (a) Proportion of males aged 2–12 years at Ram Mountain that fathered at least one lamb, ruts of 1988–2022. (b) Relative age-specific reproductive success of bighorn males at Sheep River and Ram Mountain, Alberta, and National Bison Range, Montana. For each age, the graph shows the number of lambs fathered relative to the age class with the greatest reproductive success (8 years at Sheep River and Bison Range, 10 years at Ram Mountain). (Modified from Festa-Bianchet (2012), with updated data for Ram Mountain to include the ruts of 1988–2022.)

for position. During these chases, there is some pushing and shoving as males try to secure a position immediately behind the female, but few or no dominance interactions. That is because the female would be long gone if two males stopped to interact with each other. We suspect that luck plays a large role in the siring success of coursing males.

Mating skew, the proportion of paternities obtained by the most successful males, likely varies according to population characteristics such as size, sex ratio, and age structure. At Sheep River and at the National Bison Range, the top three males in the hierarchy fathered about 30%–40% of each lamb cohort (Hogg and Forbes 1997), while at Ram Mountain the most successful male fathered 20%–45% of lambs with assigned paternities over six cohorts (Coltman et al. 2002). There is some evidence of a decrease in reproductive success with advanced age (Figure 5.5), but the sample size of older males is small. Rams suffer high natural mortality at ages 6–10 (Loison et al. 1999), and both Sheep River and Ram Mountain were trophy-hunted, so that very few males survived past 10 years of age. For example, at Ram Mountain only eight males survived to 11 years. In all three populations, reproductive success appeared to decline for males aged 11 and older (Figure 5.5). Males aged 6–8 years appeared to have a lower relative siring success at Ram Mountain than at Sheep River or the National Bison Range (Figure 5.5), but the reasons for these possible age differences have not been investigated. Males aged 7–10 years had the highest reproductive success. For example, at Ram Mountain from 1988 to 2021, only 6% of 158 2-year-olds were known to sire at least one lamb. The proportion siring peaked at 73% for 10-year-olds then declined to 25% for males aged 11 or older. The greatest reproductive skew was in 2021, when 9 of 13 lambs were fathered by the same 7-year-old.

Horn size and body mass tend to be correlated, and no study has teased apart the relative contributions of horn and body size on male reproductive success. At Ram Mountain, however, the correlation with siring success was stronger for horn length than for body mass in 5 of 6 years (Coltman et al. 2002). If two males have the same mass but different horn sizes, the larger-horned one will likely be dominant. The mating system and the factors affecting individual reproductive success in males have important implications for harvest management (Chapter 23).

It has been suggested that large-horned males expend much energy during the rut and then suffer high winter mortality (Geist 1966). After observing marked males of known age, social rank, and, in many cases, mass, however, Pelletier (2005) found that although larger ones spent more time in rutting activities, highly successful tending males did not reduce the time they spent foraging and did not lose more mass over the rut compared to less successful coursing males. It is likely that in mountain sheep the main energetic cost of reproduction for males is in trying to reproduce, not in actually succeeding (Festa-Bianchet 2012). A coursing male that spends much time involved in chases and moving between breeding groups in deep snow may well expend more energy than a dominant male that spends much of its time resting or feeding while tending estrous females, and can easily take over the tending position when challenging a subordinate male.

Operational sex ratio and male age structure can be affected by trophy hunting, and there are suggestions that mating behavior may differ in hunted and unhunted populations, with greater harassment of females when most males are young adults (Singer and Zeigenfuss 2002). At Ram Mountain, in years with a greater proportion of males aged 2–4 years the reproductive success of males aged 3 years increased. Male age structure, however, had no effect on the siring success of males aged 2 or 4 years (Ritchot et al. 2021).

### 5.3.3 Dispersal and Breeding Migrations

Demographic dispersal, where a sheep is born in one population and moves permanently to another population where it will breed, is rare. Over 70 population-years of monitoring bighorn sheep at Ram Mountain and Sheep River, we detected 11 emigrants and 26 immigrants among 601 males aged ≥1 year. We detected two emigrants and nine immigrants among 606 females aged ≥1 year. We likely underestimate emigration as very few sheep were radiocollared, therefore some may have emigrated but we never saw them in other populations. We found out about some emigrant males only through the compulsory registration of hunter kills. Because over 95% of adult residents were marked in both populations in nearly all years, however, we likely detected all immigrants, which were unmarked when they arrived. It appears that less than 10% of males and less than 3% of females disperse. Therefore, mountain sheep populations are demographically isolated, and local density seems unlikely to lead to changes in dispersal: changes in population size are determined almost entirely by births and deaths (Turgeon et al. 2024). Perhaps dispersal is more frequent in special circumstances, for example, when fire, land management, or retreating glaciers create new habitat.

No information on demographic dispersal is available for thinhorn sheep. For desert sheep in California, Epps et al. (2004) reported three apparent natural recolonizations of ranges where local populations had gone extinct. Demographic dispersal may be more common in desert sheep, possibly as an adaptation to a naturally fragmented habitat. The strong philopatry of mountain sheep can hinder reintroductions: it usually takes many decades before reintroduced populations learn about historic migration routes between seasonal ranges (Jesmer et al. 2018).

While demographic dispersal appears uncommon, genetic dispersal through male breeding migrations is a key aspect of mountain sheep biology, relevant for population genetics, gene flow within metapopulations, male reproductive tactics, conservation and management. After the pre-rut congregation (Festa-Bianchet 1986), several males leave their native population, visit other female groups up to 80 km away looking for breeding opportunities, then return to their native population after the rut (Hogg 2000). These breeding migrations can involve over half the resident males (Lassis et al. 2022a). Males appear to undertake breeding migrations based on their perception of local mating opportunities: they are more likely to leave if middle-ranked, when aged 4–8 years, when their natal population includes many competing males and a male-biased sex ratio (Hogg 2000, Lassis et al. 2022a). Some males undertake exploratory movements in October-November, then return to their native population, and may leave then return multiple times before the rut. Presumably, during these excursions they assess the availability of females and competing males in nearby populations (Lassis et al. 2022a). A recent study of male movement in Nevada reported that apparent breeding migrations are more frequent in populations with shorter breeding seasons, but also found contrasting effects of sex ratio and the number of females on the probability of these forays (Ricci et al. 2024). The apparent contrast in frequency and causes of breeding migrations suggests that mating competition and determinants of siring success may differ among mountain sheep species, subspecies, and ecological conditions. Future studies of siring success in thinhorn sheep and desert bighorns may reveal important differences from the results presented here that are mostly based on three northern populations of Rocky Mountain bighorns.

Breeding migrations have several important implications for conservation of mountain sheep. First, if routes are blocked by human development, isolation could decrease genetic variability, as reported for desert bighorn populations isolated by major highways (Epps et al. 2005). Landscape genetic analyses in Alberta suggest that breeding migrations may provide gene flow up to 100 km, and that major river valleys are natural obstacles (Deakin et al. 2020). Second, on long-range movements, males can carry contagious diseases such as pneumonia, contracted mostly from domestic sheep, to multiple populations (Cassirer et al. 2013) so that one contact with domestic sheep can lead to pneumonia die-offs over a wide range. Third, during breeding migrations males can move between hunted and protected populations, exposing those from protected populations to risk of harvest.

Given that males appear more likely to leave their native population if they detect a high level of mating competition (Lassis et al. 2022a), it seems reasonable to speculate that more males will move from protected to hunted areas than vice versa. That is because in hunted areas, some potential competitors are shot before the rut. If breeding migrations take place after the hunting season, males exiting protected areas may provide some genetic rescue from intense selective hunting (Lassis et al. 2022b). If, however, breeding migrations overlap with the hunting season and trophy hunting pressure is intense, modeling suggests that selective hunting outside protected areas can lead to the evolution of smaller horns inside protected areas (Lassis et al. 2023). For example, in Alberta there is a pronounced peak in trophy male harvest in the last week of October,

just before the hunting season closes. Compared to males taken earlier in the hunting season, late-season males are shot about 20% closer to protected areas and have slightly larger horns after accounting for age, suggesting that many originate from populations inside protected areas with weaker selection against large horns (Pelletier et al. 2014). Breeding migrations in late October present a challenge to managers because the number of harvestable males in an area during the hunting season may differ from that estimated during winter population surveys. Closing the hunting season in mid-October could increase the opportunity for genetic rescue, and gene flow in general, from protected to hunted areas.

### 5.3.4 FEMALE CHOICE

Female sheep are smaller than males, and likely cannot avoid forced copulations. During coursing chases, females sometime appear to run in a circle and return to the tending male. Rather than actively choosing to mate with a dominant male, they may seek its protection against the injury risks involved in coursing chases. Females may also exert some mate choice by selecting where to be when in estrus, which can then force males to select specific sites (Hogg 1987). Given that females may copulate with multiple males, they have an opportunity for postcopulatory mate choice, as reported in other species (Calsbeek and Bonneaud 2008). There are, however, no studies on postcopulatory sperm selection in mountain sheep. Intriguingly, the sex ratio of lambs sired by highly successful males tends to be slightly male-biased, and females allocate relatively more maternal resources to sons if the father is a successful male and to daughters if the father is a relatively unsuccessful male (Douhard et al. 2016). Given that male lambs lead to greater fitness costs than female lambs (Bérubé et al. 1996), mothers in poor condition may be advantaged to avoid mating with highly successful males. Like many studies of potentially adaptive offspring sex ratio in mammals (Festa-Bianchet 1996), however, effect sizes are small and these expectations remain speculative. Finally, there is no evidence that sheep of either sex avoid mating with close relatives (Rioux-Paquette et al. 2010), raising inbreeding concerns for small populations.

## 5.4 FEMALE REPRODUCTIVE STRATEGIES AND PARTURITION

Female mountain sheep may reproduce once a year and typically produce only one lamb, although rare cases of twinning have been reported, especially in captivity (Hoefs 1978, Eccles and Shackleton 1979). In Rocky Mountain bighorn sheep, given that gestation time is fixed at about 173 days and can only vary by a few days (Hogg et al. 2017), the timing of breeding is the main determinant of the timing of parturition. Northern populations have very short parturition seasons. Typically, about 80% of lambs are born over a 3-week period (Rachlow and Bowyer 1991, Feder et al. 2008). For Stone's sheep, the median lambing date for 23 lambs was May 17 (range May 3 to June 14) (Enns et al. 2024). Parturition in desert bighorn sheep is much less seasonal, likely because the timing of vegetation growth is less predictable than in northern environments. Parturition in desert bighorn sheep also appears highly variable in timing and synchronicity among populations (Thompson and Turner 1982, Rubin et al. 2000). Given the weaker seasonality, it may be possible for desert bighorn females to vary the interbirth interval, for example, by conceiving a few months earlier following the loss of a lamb to predation or delaying subsequent conception by a few months if maternal allocation depletes resources. Flexibility in interbirth interval could affect population dynamics through its effects on yearly productivity, but to our knowledge it has not been investigated. Interbirth intervals are highly constrained in northern sheep because lambs born late suffer high mortality (Festa-Bianchet 1988, Feder et al. 2008). Lambs born after the green-up peak are lighter at weaning (Renaud et al. 2022). Primiparity can be at 2 to 5 years and varies with environmental conditions (Bunnell and Olsen 1981, Martin and Festa-Bianchet 2012). After primiparity, gestation rate in bighorn sheep tends to be above 90% (Proffitt et al. 2021), except for females in very poor condition (Smiley et al. 2024). Given that lamb survival to weaning is very variable (van de Kerk et al. 2020), gestation rate is a very unprecise measure of recruitment. For thinhorn sheep, gestation rate after primiparity may be more variable. One study in Alaska based on 133 female-years of monitoring over 5 years reported an average of 77% and a yearly range from 44% to 96% (Lohuis 2016). Generally, gestation rate starts to decline at 11–13 years of age, but it remains more than 50% until about 16 years of age (Martin and Festa-Bianchet 2011). Given that very few females survive past 16 years, because mortality increases substantially with age (Loison et al. 1999), reproductive senescence is unlikely to be important for population dynamics in most populations of mountain sheep.

### 5.4.1 THE CONSERVATIVE REPRODUCTIVE STRATEGY OF FEMALE MOUNTAIN SHEEP

Female mountain sheep allocate resources to their lambs only as long as that allocation does not compromise their survival. That is because lamb survival, independently of maternal allocation, is much lower on average and much more variable from year to year than the survival of mothers (Gaillard et al. 1998). Therefore, the best strategy for a female is to ensure her own survival so that she can reproduce again (Festa-Bianchet and Jorgenson 1998, Gaillard et al. 1998). Even at an advanced age, the residual reproductive value of mothers is likely higher than that of their lamb. Reproduction in female mountain sheep is costly: compared to nonreproductive females, or to females whose lamb dies soon after birth, females that wean a lamb give birth later

the following year (Renaud et al. 2019) and are less likely to wean a lamb the following year (Festa-Bianchet et al. 1998), particularly if they have weaned a son (Bérubé et al. 1996). Male lambs grow faster than female lambs and are therefore more energetically costly for their mothers. As conditions deteriorate, such as at high population density, lactating females allocate fewer resources to their lambs, so that summer mass gain by lambs declines substantially, while mass gain by lactating females is mostly independent of population density (Martin and Festa-Bianchet 2010). Lambs that are smaller at weaning have lower survival (Festa-Bianchet et al. 1997). We suggest that reduced maternal allocation, leading to greater lamb mortality, is the main mechanism of density dependence in bighorn sheep population dynamics.

Another manifestation of the conservative reproductive strategy of bighorn sheep is variability in age of primiparity, a vital rate that is generally strongly density-dependent in ungulates (Gaillard et al. 2000). In thinhorn sheep, resource availability affects age of primiparity, that can range from as young as 13 months in captivity with unlimited forage (Hoefs and Nowlan 1993) to 4 or 5 years in populations with limited resources (Bunnell and Olsen 1981). At Ram Mountain at low population density, heavy yearling female bighorn sheep were primiparous at age 2, but as density increased age of primiparity shifted to 3–4 years and was independent of mass as a yearling (Martin and Festa-Bianchet 2012). In contrast, age of primiparity in the Sheep River population was not affected by density (Festa-Bianchet et al. 1995). Primiparous females gain less mass in summer and lose more mass in winter than multiparous or nulliparous females. Although early primiparity led to greater lifetime reproductive success, females that were small as yearlings and delayed primiparity mostly caught up in mass to other females by age 6 (Martin and Festa-Bianchet 2012). Our long-term studies were able to assess the costs of reproduction for females because they partially accounted for environmental conditions and individual reproductive potential (Chapter 23). For example, the reduction in reproductive success in the year following the weaning of a lamb was greater at high population density and for lighter females than at low density or for heavy females (Festa-Bianchet et al. 1998). Some females have consistently higher weaning success than others (Hamel et al. 2009), and these individual differences can be partly explained by differences in mass as young adults. Heavier females have greater reproductive success, lower costs of reproduction, and greater longevity (Bérubé et al. 1999). Ideally, the costs of reproduction in the wild should be assessed by experimental manipulation, such as the use of contraceptives as was done for eastern grey kangaroos (*Macropus giganteus*) (Gélin et al. 2015). No such manipulations have been published for mountain sheep.

Female reproductive allocation in mountain sheep mostly varies with resource availability and individual reproductive potential. Additional evidence of the conservative reproductive strategy of bighorn females comes from studies of the indirect effects of predation and from how females allocate resources when faced with pathogens. During episodes of intense cougar predation, females react to the stress caused by predator attacks by reducing allocation to reproduction, presumably to increase their own chances of survival. At Ram Mountain, in years of intense cougar predation, fecundity dropped by about 14% and lamb weaning mass by about 8% (Cloutier et al. 2024). Given that smaller lambs are less likely to survive, Bourbeau-Lemieux et al. (2011) estimated that indirect effects may account for up to a third of the decrease in lamb survival in years of high cougar predation. Females infected by pathogens allocate resources to the immune system, leading to fat loss and decreased reproduction (Smiley et al. 2024). Females also delay primiparity after pneumonia epizootics, possibly to allocate more resources to maintenance and fighting pathogen infections (Turgeon et al. 2024)

## 5.4.2 CONSERVATION IMPLICATIONS

Behavior and mating strategies of mountain sheep have important implications for their conservation. The predictability of seasonal area use, open habitat, and gregariousness makes mountain sheep a sought-after species for ecotourism, particularly in protected areas where sheep are habituated to human observers. Behavioral adaptations to avoid cursorial predators and high vulnerability to ambush or stalking predators mean that the impact of predation on population dynamics depends on the type of predator. Avoidance of forest and preference for open habitats makes prescribed fires an important tool to counter the forest or shrub encroachment that is degrading the habitat of many alpine and northern populations. The conservative reproductive strategy of females means that populations generally grow slowly and may not show strong density dependence, affecting decisions about harvests of sex-age classes other than mature males. Finally, the combination of age-specific mating tactics, horn growth, and effects of horn size on dominance rank of males affects the potential for evolutionary shrinkage of horns under intense selective harvest.

## 5.5 SUMMARY

Mountain sheep are sexually segregated outside the rut and rely on habitat near escape terrain to avoid coursing predators. Both sexes are highly philopatric and permanent dispersal is rare. Adult females join their natal group but do not preferentially associate with kin. Female reproductive strategy is conservative. Females favor their own survival over allocation to their current lamb. When resources are scarce females delay primiparity from 2 to 4 years of age, and lactating females appear to produce less milk so that lamb mass gain decreases and lamb survival declines. Lambs born in the year after their mother had weaned a lamb have lower growth and survival. The mating system involves serial female-defense polygyny by highly dominant males that defend a single estrous female at a time. Subordinate males use alternative mating tactics such as coursing, attempting to separate the female from the defending male. The defending male has a 40%–50% probability of siring the lamb. Subordinate males using alternative mating tactics share the remaining paternities. The positive correlation of horn size and siring success is weak until about age 6 and strong for older males. The mating system and the age-specific relationship between horn size and mating success make mountain sheep highly susceptible to hunting-induced evolution of smaller horns if trophy harvests are intense. Male breeding migrations provide gene flow among populations and can be blocked by human development, leading to loss of genetic variation.

## REFERENCES

Barrette, C., and D. Vandal. 1986. Social rank, dominance, antler size, and access to food in snow-bound wild woodland caribou. *Behaviour* 97:118–146.

Bérubé, C., M. Festa-Bianchet, and J. T. Jorgenson. 1999. Individual differences, longevity, and reproductive senescence in bighorn ewes. *Ecology* 80:2555–2565.

Bérubé, C. H., M. Festa-Bianchet, and J. T. Jorgenson. 1996. Reproductive costs of sons and daughters in Rocky Mountain bighorn sheep. *Behavioral Ecology* 7:60–68.

Bourbeau-Lemieux, A., M. Festa-Bianchet, J. M. Gaillard, and F. Pelletier. 2011. Predator-driven component Allee effects in a wild ungulate. *Ecology Letters* 14:358–363.

Boyce, W. M., R. R. Ramey, T. C. Rodwell, E. S. Rubin, and R. S. Singer. 1998. Population subdivision among desert bighorn sheep (*Ovis canadensis*) ewes revealed by mitochondrial DNA analysis. *Molecular Ecology* 8:99–106.

Bunnell, F. L. 1982. The lambing period of mountain sheep: synthesis, hypothesis, and tests. *Canadian Journal of Zoology* 60:1–14.

Bunnell, F. L., and N. A. Olsen. 1981. Age-specific natality in Dall's sheep. *Journal of Mammalogy* 62:379–380.

Calsbeek, R., and C. Bonneaud. 2008. Postcopulatory fertilization bias as a form of cryptic sexual selection. *Evolution* 62:1137–1148.

Cassirer, E. F., R. K. Plowright, K. R. Manlove, P. C. Cross, A. P. Dobson, K. A. Potter, and P. J. Hudson. 2013. Spatio-temporal dynamics of pneumonia in bighorn sheep. *Journal of Animal Ecology* 82:518–528.

Cloutier, Z., M. Festa-Bianchet, and F. Pelletier. 2024. Direct and indirect effects of cougar predation on bighorn sheep fitness. *Ecology* 105:e4374.

Coltman, D. W., M. Festa-Bianchet, J. T. Jorgenson, and C. Strobeck. 2002. Age-dependent sexual selection in bighorn rams. *Proceedings of the Royal Society B* 269:165–172.

Deakin, S., J. C. Gorrell, J. Kneteman, D. S. Hik, R. M. Jobin, and D. W. Coltman. 2020. Spatial genetic structure of Rocky Mountain bighorn sheep (*Ovis canadensis canadensis*) at the northern limit of their native range. *Canadian Journal of Zoology* 98:317–330.

Douhard, M., M. Festa-Bianchet, D. W. Coltman, and F. Pelletier. 2016. Paternal reproductive success drives sex allocation in a wild mammal. *Evolution* 70:358–368.

Drews, C. 1993. The concept and definition of dominance in animal behaviour. *Behaviour* 125:283–313.

Eccles, T. R., and D. M. Shackleton. 1979. Recent records of twinning in North American mountain sheep. *Journal of Wildlife Management* 43:974–976.

Enns, G. E., B. A. Jex, and M. S. Boyce. 2024. Stone's sheep (*Ovis dalli stonei*) lambing and nursery habitat selection. *Canadian Journal of Zoology* 102:691–707.

Epps, C. W., D. R. McCullough, J. D. Wehausen, V. C. Bleich, and J. L. Rechel. 2004. Effects of climate change on population persistence of desert-dwelling mountain sheep in California. *Conservation Biology* 18:102–113.

Epps, C. W., P. J. Palsboll, J. D. Wehausen, G. K. Roderick, R. R. Ramey, and D. R. McCullough. 2005. Highways block gene flow and cause a rapid decline in genetic diversity of desert bighorn sheep. *Ecology Letters* 8:1029–1038.

Favre, M., J. G. A. Martin, and M. Festa-Bianchet. 2008. Determinants and life-history consequences of social dominance in bighorn ewes. *Animal Behaviour* 76:1373–1380.

Feder, C., J. G. A. Martin, M. Festa-Bianchet, C. Bérubé, and J. Jorgenson. 2008. Never too late? Consequences of late birthdate for mass and survival of bighorn lambs. *Oecologia* 156:773–781.

Festa-Bianchet, M. 1986. Site fidelity and seasonal range use by bighorn rams. *Canadian Journal of Zoology* 64:2126–2132.

Festa-Bianchet, M. 1987. Bighorn sheep, climbing accidents, and implications for mating strategy. *Mammalia* 51:618–620.

Festa-Bianchet, M. 1988. Birthdate and survival in bighorn lambs (*Ovis canadensis*). *Journal of Zoology* 214:653–661.

Festa-Bianchet, M. 1991. The social system of bighorn sheep: grouping patterns, kinship and female dominance rank. *Animal Behaviour* 42:71–82.

Festa-Bianchet, M. 1996. Offspring sex ratio studies of mammals: does publication depend upon the quality of the data or the direction of the results? *Écoscience* 3:42–44.

Festa-Bianchet, M. 2012. The cost of trying: weak interspecific correlations among life-history components in male ungulates. *Canadian Journal of Zoology* 90:1072–1085.

Festa-Bianchet, M., J.-M. Gaillard, and J. T. Jorgenson. 1998. Mass- and density-dependent reproductive success and reproductive costs in a capital breeder. *American Naturalist* 152:367–379.

Festa-Bianchet, M., and J. T. Jorgenson. 1998. Selfish mothers: reproductive expenditure and resource availability in bighorn ewes. *Behavioral Ecology* 9:144–150.

Festa-Bianchet, M., J. T. Jorgenson, C. H. Bérubé, C. Portier, and W. D. Wishart. 1997. Body mass and survival of bighorn sheep. *Canadian Journal of Zoology* 75:1372–1379.

Festa-Bianchet, M., J. T. Jorgenson, M. Lucherini, and W. D. Wishart. 1995. Life-history consequences of variation in age of primiparity in bighorn ewes. *Ecology* 76:871–881.

Festa-Bianchet, M., J. T. Jorgenson, W. J. King, K. G. Smith, and W. D. Wishart. 1996. The development of sexual dimorphism: seasonal and lifetime mass changes of bighorn sheep. *Canadian Journal of Zoology* 74:330–342.

Gaillard, J.-M., M. Festa-Bianchet, and N. G. Yoccoz. 1998. Population dynamics of large herbivores: variable recruitment with constant adult survival. *Trends in Ecology and Evolution* 13:58–63.

Gaillard, J.-M., M. Festa-Bianchet, N. G. Yoccoz, A. Loison, and C. Toïgo. 2000. Temporal variation in fitness components and population dynamics of large herbivores. *Annual Review of Ecology and Systematics* 31:367–393.

Geist, V. 1966. The evolutionary significance of mountain sheep horns. *Evolution* 20:558–566.

Geist, V. 1971. *Mountain sheep*. University of Chicago Press, Chicago.

Gélin, U., M. E. Wilson, G. Coulson, and M. Festa-Bianchet. 2015. Experimental manipulation of female reproduction demonstrates its fitness costs in kangaroos. *Journal of Animal Ecology* 84:239–248.

Hamel, S., S. D. Côté, J. M. Gaillard, and M. Festa-Bianchet. 2009. Individual variation in reproductive costs of reproduction: high-quality females always do better. *Journal of Animal Ecology* 78:143–151.

Hoefs, M. 1978. Twinning in Dall sheep. *Canadian Field-Naturalist* 92:292–293.

Hoefs, M., and U. Nowlan. 1993. Minimum breeding age of Dall sheep, *Ovis dalli dalli*, ewes. *Canadian Field-Naturalist* 107:241–243.

Hogg, J. T. 1984. Mating in bighorn sheep: multiple creative male strategies. *Science* 225:526–529.

Hogg, J. T. 1987. Intrasexual competition and mate choice in Rocky Mountain bighorn sheep. *Ethology* 75:119–144.

Hogg, J. T. 1988. Copulatory tactics in relation to sperm competition in Rocky Mountain bighorn sheep. *Behavioral Ecology and Sociobiology* 22:49–59.

Hogg, J. T. 2000. *Mating systems and conservation at large spatial scales*. Pages 214–252 in M. Apollonio, M. Festa-Bianchet, and D. Mainardi, editors. *Vertebrate Mating Systems*. World Scientific, Singapore.

Hogg, J. T., and S. H. Forbes. 1997. Mating in bighorn sheep: frequent male reproduction via a high-risk "unconventional" tactic. *Behavioral Ecology and Sociobiology* 41:33–48.

Hogg, J. T., S. J. Dunn, J. Poissant, F. Pelletier, and J. A. Byers. 2017. Capital vs. income-dependent optimal birth date in two North American ungulates. *Ecosphere* 84:e01766.

Jesmer, B. R., J. A. Merkle, J. R. Goheen, E. O. Aikens, J. L. Beck, A. B. Courtemanch, M. A. Hurley, D. E. McWhirter, H. M. Miyasaki, K. B. Monteith, and M. J. Kauffman. 2018. Is ungulate migration culturally transmitted? Evidence of social learning from translocated animals. *Science* 361:1023–1025.

Kachel, S. M., R. Bayarakcismith, Z. Kubanychbekov, R. Kulunbekov, T. M. McCarthy, B. V. Weckworth, and A. J. Wirsing. 2023. Ungulate spatiotemporal responses to contrasting predation risk from wolves and snow leopards. *Journal of Animal Ecology* 92:142–157.

Kalinowski, S. T., M. L. Taper, and T. C. Marshall. 2007. Revising how the computer program CERVUS accommodates genotyping error increases success in paternity assignment. *Molecular Ecology* 16:1099–1106.

Lassis, R., M. Festa-Bianchet, and F. Pelletier. 2022a. Breeding migrations by bighorn sheep males are driven by mating opportunities. *Ecology and Evolution* 12:e8692.

Lassis, R., M. Festa-Bianchet, and F. Pelletier. 2022b. Effects of hunting pressure and timing of harvest on bighorn sheep (*Ovis canadensis*) horn size. *Canadian Journal of Zoology* 100:507–516.

Lassis, R., M. Festa-Bianchet, J. Van de Walle, and F. Pelletier. 2023. Genetic rescue from protected areas is modulated by migration, hunting rate, and timing of harvest. *Evolutionary Applications* 16:1105–1118.

Lohuis, T. 2016. *Ewe Dall's Sheep Survival, Pregnancy and Parturition Rates, and Lamb Recruitment in GMU 14C, Chugach Mountains, AK*. Alaska Department of Fish and Game, Juneau, Alaska.

Loison, A., M. Festa-Bianchet, J.-M. Gaillard, J. T. Jorgenson, and J.-M. Jullien. 1999. Age-specific survival in five populations of ungulates: evidence of senescence. *Ecology* 80:2539–2554.

Manlove, K. R., E. F. Cassirer, P. C. Cross, R. K. Plowright, and P. J. Hudson. 2014. Costs and benefits of group living with disease: a case study of pneumonia in bighorn lambs (*Ovis canadensis*). *Proceedings of the Royal Society B* 281:20142331.

Martin, A. M., M. Festa-Bianchet, D. W. Coltman, and F. Pelletier. 2016. Demographic drivers of age-dependent sexual selection. *Journal of Evolutionary Biology* 29:1437–1446.

Martin, J. G. A., and M. Festa-Bianchet. 2010. Bighorn ewes transfer the costs of reproduction to their lambs. *American Naturalist* 176:414–423.

Martin, J. G. A., and M. Festa-Bianchet. 2011. Age-independent and age-dependent decreases in reproduction of females. *Ecology Letters* 14:576–581.

Martin, J. G. A., and M. Festa-Bianchet. 2012. Determinants and consequences of age of primiparity in bighorn ewes. *Oikos* 121:752–760.

Pelletier, F. 2005. Foraging time of rutting bighorn rams varies with individual behavior, not mating tactic. *Behavioral Ecology* 16:280–285.

Pelletier, F., J. T. Hogg, and M. Festa-Bianchet. 2006. Male mating effort in a polygynous ungulate. *Behavioral Ecology and Sociobiology* 60:645–654.

Pelletier, F., and M. Festa-Bianchet. 2006. Sexual selection and social rank in bighorn rams. *Animal Behaviour* 71:649–655.

Pelletier, F., M. Festa-Bianchet, J. T. Jorgenson, C. Feder, and A. Hubbs. 2014. Can phenotypic rescue from harvest refuges buffer wild sheep from selective hunting? *Ecology and Evolution* 4:3375–3382.

Proffitt, K. M., A. Courtemanch, S. Dewey, B. H. Lowrey, D. E. McWhirter, K. B. Monteith, J. T. Paterson, J. J. Rotella, P. J. White, and R. A. Garrott. 2021. Regional variability in pregnancy and survival rates of Rocky Mountain bighorn sheep. *Ecosphere* 12:e03410.

Rachlow, J. L., and R. T. Bowyer. 1991. Interannual variation in timing and synchrony of parturition in Dall's sheep. *Journal of Mammalogy* 72:487–492.

Renaud, L. A., G. Pigeon, M. Festa-Bianchet, and F. Pelletier. 2019. Phenotypic plasticity in bighorn sheep reproductive phenology: from individual to population. *Behavioral Ecology and Sociobiology* 73:50.

Renaud, L. A., M. Festa-Bianchet, and F. Pelletier. 2022. Testing the match–mismatch hypothesis in bighorn sheep in the context of climate change. *Global Change Biology* 28:21–32.

Ricci, L. E., M. Cox, and K. R. Manlove. 2024. Movement decisions driving metapopulation connectivity respond to social resources in a long-lived ungulate, bighorn sheep (*Ovis canadensis*). *Philosophical Transactions of the Royal Society B* 379:20220533.

Rieucau, G., and J. G. A. Martin. 2008. Many eyes or many ewes: vigilance tactics in female bighorn sheep *Ovis canadensis* vary according to reproductive status. *Oikos* 117:501–506.

Rioux-Paquette, E., M. Festa-Bianchet, and D. W. Coltman. 2010. No inbreeding avoidance in an isolated population of bighorn sheep. *Animal Behaviour* 80:865–871.

Ritchot, Y., M. Festa-Bianchet, D. W. Coltman, and F. Pelletier. 2021. Determinants and long-term costs of early reproduction in males of a long-lived polygynous mammal. *Ecology and Evolution* 11:6829–6845.

Rubin, E. S., W. M. Boyce, and V. C. Bleich. 2000. Reproductive strategies of desert bighorn sheep. *Journal of Mammalogy* 81:769–786.

Ruckstuhl, K. E. 1998. Foraging behaviour and sexual segregation in bighorn sheep. *Animal Behaviour* 56:99–106.

Ruckstuhl, K. E. 1999. To synchronise or not to synchronise: a dilemma for young bighorn males? *Behaviour* 136:805–818.

Ruckstuhl, K. E., and M. Festa-Bianchet. 2001. Group choice by subadult bighorn rams: trade-offs between foraging efficiency and predator avoidance. *Ethology* 107:161–172.

Ruckstuhl, K. E., and P. Neuhaus. 2002. Sexual segregation in ungulates: a comparative test of three hypotheses. *Biological Reviews* 77:77–96.

Singer, F. J., and L. C. Zeigenfuss. 2002. Influence of trophy hunting and horn size on mating behavior and survivorship of mountain sheep. *Journal of Mammalogy* 83:682–698.

Smiley, R. A., B. L. Wagler, W. H. Edwards, J. E. Jennings-Gaines, K. Luukkonen, K. Robbins, M. Johnson, A. Courtemanch, T. W. Mong, D. Lutz, D. E. McWhirter, J. L. Malmberg, B. H. Lowrey, and K. B. Monteith. 2024. Infection–nutrition feedbacks: fat supports pathogen clearance but pathogens reduce fat in a wild mammal. *Proceedings of the Royal Society B* 291:20240636.

Thompson, R. W., and J. C. Turner. 1982. Temporal geographic variation in the lambing season of bighorn sheep. *Canadian Journal of Zoology* 60:1781–1793.

Thouless, C. R., and F. E. Guinness. 1986. Conflict between red deer hinds: the winner always wins. *Animal Behaviour* 34:1166–1171.

Tibbetts, E. A., J. Pardo-Sanchez, and C. Weise. 2022. The establishment and maintenance of dominance hierarchies. *Philosophical Transactions of the Royal Society B* 377:20200450.

Turgeon, R., F. Pelletier, S. D. Côté, M. Festa-Bianchet, and S. Hamel. 2024. Sporadic events have a greater influence on the dynamics of small isolated populations than density-dependence and environmental conditions. *American Naturalist* 204:574–588.

van de Kerk, M., S. M. Arthur, M. Bertram, B. L. Borg, J. Herriges, J. Lawler, B. A. Mangipane, C. Lambert Koizumi, B. Wendling, and L. R. Prugh. 2020. Environmental influences on Dall's sheep survival. *Journal of Wildlife Management* 84:1127–1138.

Vander Wal, E., A. Gagné-Delorme, M. Festa-Bianchet, and F. Pelletier. 2016. Dyadic associations and individual sociality in bighorn ewes. *Behavioral Ecology* 27:560–566.

Vander Wal, E., M. Festa-Bianchet, D. Réale, D. W. Coltman, and F. Pelletier. 2015. Sex-based differences in the adaptive value of social behavior contrasted against morphology and environment. *Ecology* 96:631–641.

Warrick, G. D., and P. R. Krausman. 1989. Barrel cacti consumption by bighorn sheep. *Southwestern Naturalist* 34:483–486.

# 6 Application of Genetics to Taxonomy, Biology, Conservation, and Management

*Clinton W. Epps, Michael R. Buchalski, Joshua P. Jahner, and Zijian Sim*

## 6.1 INTRODUCTION

Genetic and genomic approaches have been used in many aspects of the conservation and management of mountain sheep (bighorn [*Ovis canadensis*], thinhorn [*Ovis dalli*]) in North America. We review the history of genetics as applied to wild populations of bighorn and thinhorn sheep, including the accelerating development of new techniques that enable asking and answering new questions, genetic and genomic efforts to describe species and subspecies divergence, application of population genetics to describing population history and assisting with management, direct management of genetic characteristics of populations, genomic approaches to understanding adaptive variation and consequences of inbreeding or low genetic diversity, and genetic applications for other taxa relevant to mountain sheep management. We conclude by describing developing and potential directions for genomic-based research on mountain sheep. Key terms are further defined in the glossary at the end of the chapter.

We briefly review the history of methods used to study genetically based questions. Worldwide, the study of species and populations has benefited from the rapidly evolving ability to evaluate genetic variation in mountain sheep populations. The study of genetics in mountain sheep of North America mirrors that progression (Epps et al. 2019). Early genetic work on wild sheep used protein electrophoresis to assess amino acid variation in proteins (allozymes), rather than direct evaluation of differences in DNA. That type of marker has relatively low variation in wild populations, as protein variants reflect genetic mutations that lead to expressed phenotypic changes, a relatively rare phenomenon. Sage and Wolff (1986) used allozymes to evaluate genetic diversity and evolutionary history of Dall's sheep (*O. d. dalli*). By the 1990s, allozymes were used to address questions such as whether heterozygosity (a measure of genetic diversity) of bighorn males correlated with horn size (Fitzsimmons et al. 1995) or whether genetic diversity of translocated bighorn populations differed sharply from their source population (Fitzsimmons et al. 1997). The first studies in mountain sheep to directly assess DNA variation investigated mitochondrial DNA (mtDNA). mtDNA is much more abundant than nuclear DNA (nDNA) in mammalian cells, and so early studies that relied on isolating DNA without amplification (Ramey II 1993, 1995; Luikart

and Allendorf 1996) were more feasible. Moreover, because mtDNA is normally inherited solely from the maternal line, it typically does not vary within an individual, whereas mammals possess two copies of the nuclear genome (i.e., the complete DNA assembly in the nucleus of the cell) that differ from each other due to maternal and paternal inheritance. Initial DNA-based studies of mountain sheep in North America used techniques such as restriction fragment length polymorphisms (RFLPs), which allowed coarse distinction of mitochondrial lineages using samples from individuals captured from wild populations (Ramey II 1993, 1995; Luikart and Allendorf 1996). Direct evaluation of DNA variation was facilitated by the adoption of the polymerase chain reaction (PCR; Mullis et al. 1986), which allows targeting and copying specific genetic regions of interest, allowing easier analysis of those regions. Early techniques for directly assessing genetic variation could not easily accommodate multiple versions of the same regions of DNA in the same sample, and the recombination that occurs in nDNA complicated phylogenetic analysis; thus, mtDNA became the marker of choice for exploring taxonomic relationships, evolutionary history, and population genetic variation. As DNA sequencing (i.e., directly identifying the order and identity of base pairs making up a region of DNA) became available and more affordable, researchers took advantage of the ability to directly assess nucleotide variation within short sections of mitochondrial genes or regions, thus gaining more resolution of differences among individuals and populations than offered by RFLPs. That permitted assessing population structure of mountain sheep at finer scales, such as evaluating differences among desert bighorn (*O. canadensis nelsoni*) female groups within the Peninsular Range of California (Boyce et al. 1999).

In the 1990s, a new class of genetic markers (i.e., microsatellites) offered the ability to assess variation of nDNA at levels allowing differentiation of individuals by genotype (Buchanan et al. 1993, Forbes et al. 1995). In addition to forensic applications (Taberlet et al. 1996), these markers offered opportunity for robust population genetic assessments of recent demographic history, population structure and levels of differentiation (Forbes and Hogg 1999, Worley et al. 2004, Epps et al. 2005), estimates of bottlenecks (Luikart et al. 1998, Ramey et al. 2000), migration rates, effective population size (Waples and England 2011),

DOI: 10.1201/9781003518686-7

heritability of fitness-related traits (Coltman et al. 2005), inbreeding depression and genetic rescue (i.e., reversing effects of inbreeding and low genetic diversity through gene flow with more diverse populations; Hogg et al. 2006, Johnson et al. 2011), and other parameters highly relevant for conservation and management of mountain sheep. Studies in this era also began to assess genetic variation directly associated with fitness for mountain sheep, such as at the major histocompatibility complex or in other genes linked to immune system function (Boyce et al. 1997, Luikart et al. 2008a). Non-invasive genetic techniques (Taberlet et al. 1996), particularly those using microsatellites, were increasingly employed after 2000 in mountain sheep research (Wehausen et al. 2004, Luikart et al. 2008b). Those methods allowed much more efficient sampling of mountain sheep on large landscapes, as capture was no longer required. Instead, fecal samples could be employed as a source of DNA, enabling landscape genetic studies of mountain sheep that evaluated how fragmentation and human activities shaped gene flow and connectivity (Epps et al. 2005, 2006, 2007) and estimates of population size (Schoenecker et al. 2015, Pfeiler et al. 2020).

Genomic approaches, distinguished from genetics largely based on the amount of genetic data obtained per individual and subsequent analytical opportunities, have seen increasing application in wildlife research (Allendorf et al. 2010). Genomic studies employ more markers or much more sequence than are typically used in analyses employing microsatellites, short sections of mtDNA, or small numbers of nuclear genes, enabling more thorough coverage across an animal's genome. Enabled by advances in sequencing technology, sometimes described as next-generation sequencing, genomic studies of mountain sheep emerged after 2010. Initially, such studies usually relied on single-nucleotide polymorphisms (SNPs), which are 1-base-pair sites that vary within a population or across the range of a species, and are assessed by simultaneous sequencing of many short DNA fragments through techniques such as double digest restriction-site associated DNA sequencing (ddRADSeq; Peterson et al. 2012), genotyping by sequencing (Parchman et al. 2012), or SNP arrays assessed by a standardized chip assay (Miller et al. 2011, Flesch et al. 2020). The SNP datasets may be used for many of the same questions as those addressed by microsatellite markers (above) but potentially offer more power to resolve relationships and estimate population genetic parameters precisely, depending on the number of SNPs assessed (usually 1,000s–10,000s). Changes in technology and continued decreases in sequencing costs continue to facilitate access to genomic data. For instance, a recently developed assay yields about 45,000 SNPs in Rocky Mountain bighorn (*O. c. canadensis*; Deakin and Coltman 2024). Whole-genome resequencing now offers the potential to sequence significant portions of individual genomes, albeit at low coverage (i.e., lower confidence due to less replication of sequence for any given region), for dozens or hundreds of individuals at reasonable cost (Kardos et al. 2015,

Li et al. 2020). Genomic work is greatly facilitated by reference genomes. A draft genome for a Rocky Mountain bighorn sheep is available (Miller et al. 2015a). The domestic sheep genome (Jiang et al. 2014, Davenport et al. 2022) is also frequently employed for reference in analyses, due to its high quality, extensive gene annotation, and close similarity to the genomes of mountain sheep in North America. Mountain sheep genetic research is strengthened by the close taxonomic relationship with a model domestic species (*Ovis aries*); thus, research in topics such as gene transcription (see Section 6.7) are typically well explored in the domestic congener before application to the wild species.

This rapid decrease in DNA sequencing costs and proliferation of new approaches opens opportunities to address more or different questions in evolution, conservation, and management than were possible in previous decades. These include methods to estimate inbreeding such as runs of homozygosity (ROH; Curik et al. 2014), historical demography such as estimates of broad changes in population size over recent millennia or centuries (Quinn et al. 2023), estimating genetic load (Mathur and DeWoody 2021), and testing for evidence of selection on genes or locations in genomes. We detail existing examples of those approaches as applied to mountain sheep subsequently in this review.

## 6.2 USING GENETICS TO ASSIST WITH TAXONOMY

The evolutionary relationships and taxonomic classifications among North American mountain sheep have been the subject of extensive scientific investigation and ongoing discussion. The fossil record for such species is poor because conditions for fossilization are not optimal in montane regions, and recurring glaciation has disrupted geologic strata overlying terrain historically inhabited by sheep. Therefore, DNA sequence-based estimates of lineage divergence often provide the only available means of reconstructing the evolutionary past. We synthesize the published genetic evidence supporting the divergence of North American mountain sheep species, starting with the closest living ancestor and tracing the taxonomic branches down to subspecies or ecotypes.

The evolutionary history of many taxa in North America is strongly influenced by the effects of Pleistocene glaciations, periods when ice covered large portions of the continent (Brunsfeld et al. 2001, Shafer et al. 2010). Studies have documented the effects of glacial cycles on the phylogeography of a diversity of mammalian (Galbreath et al. 2009), avian (Spellman et al. 2007), amphibian (Thompson and Russell 2005), fish (Reilly and Miller 2021), and plant (Golden and Bain 2000) species from northwestern North America. The capture of Earth's water as continental ice sheets brings about a corresponding lowering of sea level, exposing land bridges in shallow areas that connect areas previously separated by water. Of note is the Bering land bridge (i.e., Beringia), which connected the continent of Asia and North America in the area between the Chukotka and Seward Peninsulas in present-day Russia and the United States, respectively. The establishment of this land bridge is thought to have allowed for the migration of the predecessors of contemporary North American mountain sheep from ancestral Asia (Cowan 1940).

During times of glacial advance, species and populations are driven to ice-free refugia to escape the inhospitable conditions brought about by the ice sheets. Populations isolated in different glacial refugia often follow separate evolutionary trajectories due to genetic drift and local selective pressures. Two major ice sheets were known in North America during the Pleistocene, the western Cordilleran and eastern Laurentide glaciers (Shafer et al. 2010). During periods of glacial maxima, there were two major refugia in northwestern North America thought to have been occupied by mountain sheep: northern Beringia refugium and southern refugium (approximately south of the US-Canada border; Pielou 1991). This prolonged isolation would have been interspersed with periodic glacial retreat, creating ice-free corridors and allowing for gene flow to occur between previously isolated populations (Hewitt 1996, 2000). As for many other species on the continent, mountain sheep evolution is a complex interplay between colonization following sea level change, isolation in refugia, post-glacial range expansion and admixture, and other environmental factors.

### 6.2.1 PACHYCERIFORMS

The snow sheep (*Ovis nivicola*), indigenous to northeastern Siberia and the Russian Far East, represents the closest common ancestor to North American mountain sheep (Bunch et al. 2006, Rezaei et al. 2010, Chapter 1). Snow sheep and North American mountain sheep both belong to the Holarctic subgenus Pachyceros (Clark 1964, Geist 1971, Valdez 1982). Pachyceros diverged from other Eurasian *Ovis* sometime during the late Pliocene (Hiendleder et al. 2002, Bunch et al. 2006) to early Pleistocene (Rezaei et al. 2010), about 5.6–2.4 million years ago. Further, DNA analyses indicate snow sheep are genetically distinct from their North American counterparts, suggesting a long period of geographic isolation (Bunch et al. 2006, Rezaei et al. 2010, Dotsev et al. 2021). Mitochondrial cytochrome b gene sequences suggest Pachyceros split into two distinct lineages approximately 2.3–1.6 million years ago (Bunch et al. 2006, Rezaei et al. 2010, Dotsev et al. 2021), long before the dissolution of the Bering land bridge isolated North American mountain sheep from snow sheep. Rezaei et al. (2010) reported that snow sheep and North American mountain sheep form robust reciprocally monophyletic groups, after excluding a chimeric haplotype mischaracterized by Bunch et al. (2006), supporting the hypothesis of a single migration event of Asiatic sheep to North America across the Bering land bridge. This finding of monophyly among the Pachyceriforms was more recently supported by whole-genome sequencing data (Santos et al. 2021).

Despite their relatively recent shared ancestry, snow sheep and North American mountain sheep have different numbers of chromosomes ($2n=52$ and 54, respectively). This has been attributed to the fusion of acrocentric chromosomes (i.e., those for which the centromere is not central and is instead located near the end of the chromosome arm). Acrocentric chromosomes are common in mammalian taxa and their fusion has resulted in major karyotype changes within both mountain and domestic sheep (Appels et al. 1998). Bunch et al. (2006) demonstrated the ancestral *Ovis* karyotype had 60 chromosomes, which is still maintained in the genus *Capra*. The fusion of acrocentric chromosomes 5 and 11 resulted in the $2n=54$ karyotype common to all Pachyceros. The most recent karyotype exclusive to snow sheep arose from fusion of acrocentrics 9 and 19 (Nadler and Bunch 1977, Bunch 1978, Bunch et al. 2006). The $2n=52$ karyotype of snow sheep likely occurred after dissolution of the Bering land bridge, even though mtDNA divergence between snow sheep and mountain species of North American *Ovis* began much earlier.

Following divergence from snow sheep, the North American mountain sheep species began to diverge approximately 1.6–0.95 million years ago (Loehr et al. 2006, Rezaei et al. 2010, Humphreys and Barraclough 2014). mtDNA and microsatellite analyses have consistently identified genetic differentiation associated with the vicariance of these two species (Worley et al. 2004, Loehr et al. 2006, Rezaei et al. 2010) in separate ice-free refugia north and south of the Laurentide Ice Sheet, a feature that once covered most of modern-day Canada. Loehr et al. (2006) investigated the glacial refugia of mountain sheep by analyzing their mtDNA. They found evidence for two major refugia, one in eastern Beringia and another south of the Laurentide Ice Sheet.

#### 6.2.1.1 *Ovis dalli*

Subspecific relationships in thinhorn sheep are strongly influenced by cycles of isolation and admixture from repeated glaciations. Genetic data have provided insights regarding the influence of glacial cycles on the origins of the thinhorn sheep subspecies, Dall's sheep and Stone's sheep (*O. d. stonei*). Early work postulated that thinhorn sheep used only the northern Beringian refugium (Geist 1971). Isolation in the two major refugia can account for much of the present-day large-scale genetic variation among thinhorn and bighorn sheep. Multiple minor refugia have been proposed to have harbored wildlife during glacial maxima, including Haida Gwaii, Alexander Archipelago (Soltis et al. 1997), Vancouver Island (Brunsfeld et al. 2001), and northern British Columbia (Catto et al. 1996).

Loehr et al. (2006) first suggested a small population of thinhorn sheep survived periodically isolated in a minor refugium between the Laurentide and Cordilleran glaciers. The asynchronous movements of the two ice sheets resulted in an ice-free region in northeast British Columbia near the city of Fort St. John which supported plant and animal life (Catto et al. 1996). To test the two-refugia hypothesis, Loehr et al. (2006) examined the phylogeography and historical demography of thinhorn and bighorn sheep samples from areas representing the modern-day locations of Beringia, the southern refugium, the Catto et al. (1996) minor refugia, and areas recolonized after retreat of the Cordilleran and Laurentide ice sheets. The authors sequenced a 604 bp portion of the mtDNA control region in 223 thinhorn and bighorn sheep collected from the Brooks Range, Yukon-Charley Rivers National Preserve, Alaskan Range, Ogilvie Mountains, Mackenzie Mountains, Pelly Mountains, central and southwest Yukon, northern British Columbia, and Alberta. Their analyses supported the two-refugia scenario. First, haplotypes for both bighorn and thinhorn sheep sampled in British Columbia were found to be more similar to each other than to haplotypes from other populations from those respective species. A star-shaped haplotype network was also observed for both species of mountain sheep in British Columbia, which together with the divergent haplotypes provides evidence for a population bottleneck (Loehr et al. 2006, figure 3). Furthermore, they used molecular dating to estimate the split between Dall's sheep inhabiting present-day Beringia and Stone's sheep at 220 thousand years ago (kya), which predates the beginning of the last glacial period. The authors also dated a rapid population expansion in Stone's sheep about 7.3 kya, which coincides with the melting of ice sheets in the region around the minor refugium (Catto et al. 1996). Finally, phylogenetic analysis also revealed extensive paraphyly between Stone's sheep from British Columbia and bighorn sheep from Alberta (Loehr et al. 2006, figure 3), which is consistent with historical hybridization. The discovery of a smaller refugium is significant because it challenges previous understandings of species survival during glacial periods and the role past thinhorn and bighorn hybridization played in the formation of the Stone's sheep subspecies.

This finding of hybridization between Stone's sheep and bighorn sheep was later confirmed by the whole-genome sequencing results of Santos et al. (2021). Those authors used data from bighorn, Stone's, Dall's (Chen et al. 2021), and snow sheep (Upadhyay et al. 2020) and domestic goat to assess the phylogenomic history and patterns of ancient introgression among North American mountain sheep. Genealogical analysis indicated that Stone's sheep originated from a hybridization between Dall's and bighorn sheep, with a relatively greater contribution from the former (71%). Multiple genomic fragments containing genes in the melanogenesis pathway showed signs of introgression between the Stone's and bighorn sheep genomes. Santos et al. (2021) postulated this is the origin of the dark coats in Stone's sheep, while the lack of hybridization explains the light color in Dall's sheep.

Sim et al. (2016) attempted to replicate the results of Loehr et al. (2006) with about 10,000 nDNA SNP markers genotyped using the OvineHD SNP array (Kijas et al. 2014) in 55 samples collected across the species' distribution.

Phylogenetic analysis supported the two-refugia scenario by showing a basal, reciprocally monophyletic relationship between a clade consisting mainly of thinhorn sheep south of the Yukon-British Columbia, Canada, border and the rest of the species. This southern clade consists entirely of Stone's sheep and is geographically closest to the minor refugium compared to any other sampling group in the study. Bayesian clustering of SNP marker data also supports a north-south split consistent with results from phylogenetic inference. Thinhorn sheep occupying the traditional subspecies transition zone bounded by the Pelly Mountains in the Yukon to the north and the Cassiar and Atlin Mountains in British Columbia to the south and east are informally known as Fannin sheep (Sheldon 1911). Sim et al. (2016) reported that population to be consistently admixed, although both phylogenetic and clustering analyses indicated closer affinity to Dall's than to Stone's sheep. They suggested that admixture is a signature of interbreeding between populations with different refugial origins because of post-glacial range expansion. The finding that Fannin sheep are more closely related to Dall's sheep answers a taxonomic question that has long intrigued the thinhorn sheep community (Nichols 1978).

Thinhorn sheep subspecies have traditionally been sorted by color: white Dall's and dark Stone's. In the contact zone between Dall's and Stone's sheep, however, coat colors can range from nearly all white to almost completely dark due to extensive admixture between the two subspecies. Prior to genetic studies, the taxonomic status of Fannin sheep was unknown. Since Fannin sheep is not a formally recognized taxon, the convention is for thinhorn sheep with any recognizable dark pelage to be classified as Stone's sheep when subspecies must be formally declared (e.g., hunter registration, record keeping). Microsatellite (Worley et al. 2004) and SNP (Sim et al. 2019) marker data now show that most individuals in the transition zone between the two subspecies are more closely related to Dall's sheep than Stone's sheep, which sets up a direct conflict between the established and readily observable color-based scheme and genetic data. This has direct and indirect conservation implications. Foremost, the re-drawing of subspecies boundaries to reflect genetic data (Figure 6.1) will mean that Stone's sheep, which is already the less abundant of the two subspecies, now has an even smaller range and total population. The area being re-classified represents over 20% of the global Stone's sheep population (Demarchi and Hartwig 2004) including almost all Stone's sheep in the Yukon Territory. This information can be used in decision-making processes such as the designation of a limited-entry harvest zone to preserve the evolutionary legacy of dark-phased Dall's sheep. Jex and Sim (2021) suggested that these individuals can be considered a unique dark-phased phenotype of Dall's sheep, perhaps representing an uneasy compromise between the eye test and genetics. The current assignment of these dark-phased Dall's sheep continues to challenge harvest record keeping organizations (e.g., Boone & Crockett Club), which have

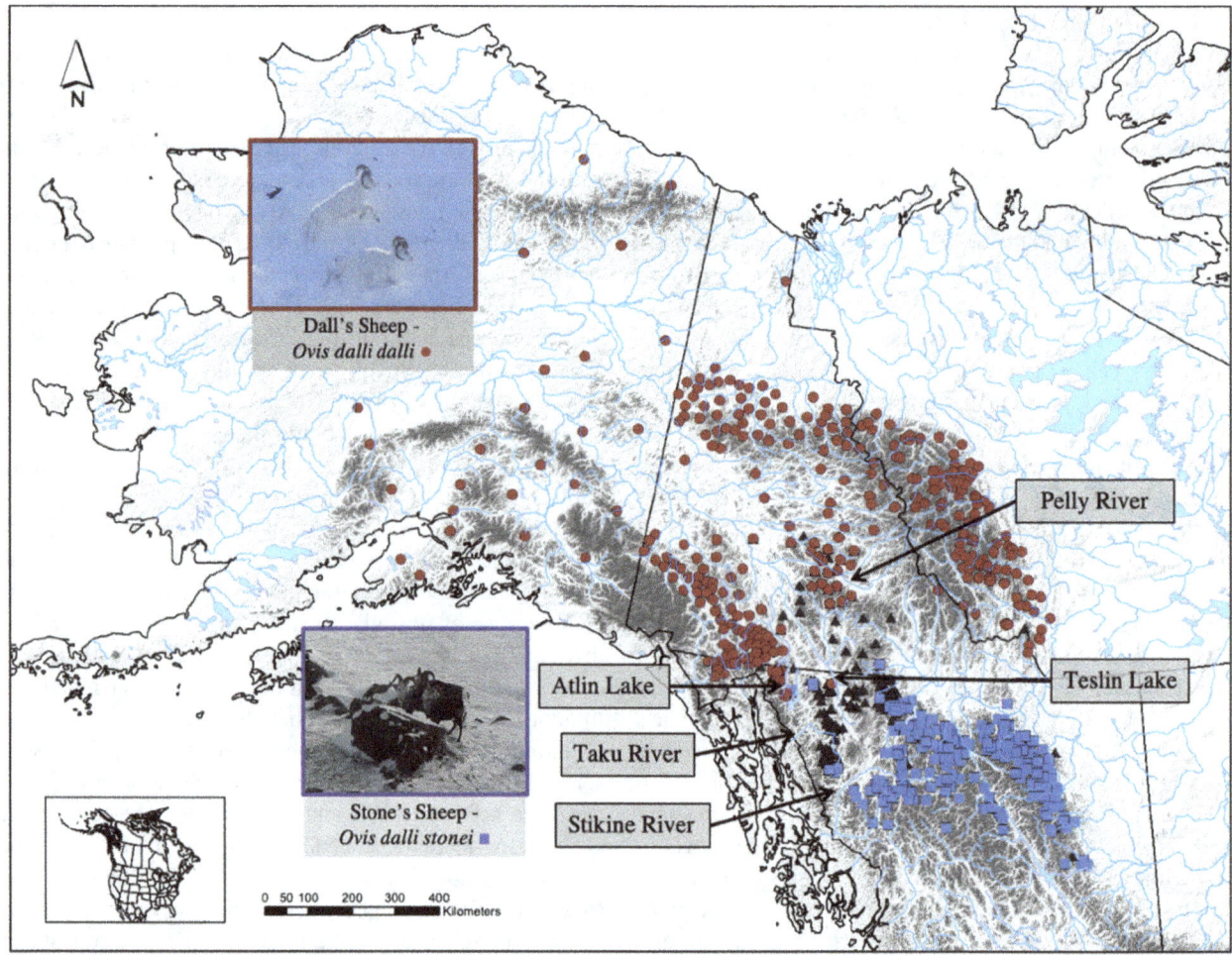

**FIGURE 6.1** Thinhorn sheep subspecies identity derived from Bayesian clustering analysis based on 2,820 males sampled (1996–2015) in northwestern North America, genotyped at 153 nuclear single-nucleotide polymorphisms, in Sim et al. (2019; filled circle-red=Dall's sheep; filled square-blue=Stone's sheep; filled triangle=admixed individuals). Representative pictures for mature Dall's and Stone's sheep males are shown (photo by Bill A. Jex; used with permission).

some entries of Stone's sheep harvested from ranges where genetic assignment would indicate the males are of more Dall's sheep origins.

### 6.2.1.2 *Ovis canadensis*

The evolutionary history of bighorn sheep and phylogenetic relationships among subspecies reflect the complex dynamics of glacial cycles and the associated habitat shifts within the southern refugium, phenomena that are not well understood. The best available data suggests multiple pocket refugia may have existed south of the Laurentide Ice Sheet. Fossil evidence indicates bighorn sheep continuously occupied at least two areas of the southern refugium during the late Pleistocene: one in the current Mojave Desert, established about 300 kya, prior to the Illinoian glaciation (Jefferson 1991), and another near Natural Trap Cave, Wyoming (Martin and Gilbert 1978, Wang 1988), established during the Sangamon interglacial (~100 kya).

This complex history is further obfuscated by human activity over the past 100 and more years. Once common throughout mountain ranges of the western United States

and southwestern Canada, numerous bighorn sheep populations were extirpated, primarily through the introduction of domestic livestock diseases (Buechner 1960, Jessup 1985, Onderka and Wishart 1988). Management actions reintroducing bighorn sheep to former range and augmentation of remnant populations to increase demographic numbers did not always use the appropriate genetic stock (Hurley et al. 2015, Wild Sheep Working Group 2015), with some translocations occurring over great distances (Wiedmann and Sargeant 2014). The result is a complex, continent-wide mosaic of bighorn sheep genetic diversity that continues to challenge wildlife genetic managers.

Genetic analyses have contributed significantly to the revision of bighorn sheep taxonomy since 1995. Initially, subspecific taxonomy (Figure 6.2) was based solely on comparisons of skull morphology (Cowan 1940). In the Rocky Mountains and Dakota Badlands two subspecies were described: Rocky Mountain bighorn sheep and the now-extinct Audubon's bighorn sheep (*O. c. auduboni*) which was later synonymized with Rocky Mountain sheep by Wehausen and Ramey II (2000). Cowan grouped animals from southwestern British

Columbia, the Cascade Range of Washington and Oregon, and remnant herds in the Sierra Nevada of California into California bighorn sheep (*O. c. californiana*). In the Southwest, Cowan described the distributions of four desert-adapted subspecies: Nelson (*O. canadensis nelsoni*), Mexican (*O. c. mexicana*), Peninsular (*O. c. cremnobates*), and Weem's bighorn sheep (*O. c. weemsi*).

Cowan's subspecies designations have influenced management decisions (e.g., source stock selection, protection status, trophy categorization) for decades and some reference texts continue to recognize many of these designations (Wilson and Reeder 2005). More modern morphometric studies have demonstrated some of Cowan's subspecies represent artifacts of small sample size and age-related size differences (Ramey II 1993, Wehausen and Ramey II 1993), and have elevated Sierra Nevada bighorn sheep (*O. c. sierrae*) as a unique taxon distinct from California bighorn (Wehausen et al. 2005, Chapter 1). Throughout these taxonomic revisions, genetics has played a significant role in clarifying the evolutionary relationships between bighorn sheep populations.

The best available genetic data suggest some subspecies designated by Cowan are highly divergent. Buchalski et al. (2016) used mtDNA control region sequences from 804 samples collected throughout the Southwest and Canada to create a maximum likelihood tree, identifying three highly divergent clades corresponding to desert, Rocky Mountain, and Sierra Nevada bighorn sheep. Deep phylogenetic divergence among geographically endemic clades such as these implies long-term isolation (Avise 2000) consistent with subspecies formation. The associated estimates of time to most recent common ancestor placed the divergence between Rocky Mountain and desert bighorn at 680±130 kya, and Sierra Nevada and desert bighorn at 640±120 kya. Therefore, bighorn sheep lineages began to diverge during the late Pleistocene, prior to the Illinoian glaciation. Loehr et al. (2006) estimated divergence between desert and Rocky Mountain bighorn as more recent (~380 kya), but still well before the Last Glacial Maximum (LGM). This deep divergence between desert and Rocky Mountain bighorn was also substantiated by Malaney et al. (2015) in a Bayesian phylogenetic analysis. Those authors sequenced a 1,181 bp segment of the mtDNA NADH dehydrogenase 5 (*ND5*) gene for 110 samples of desert, Rocky Mountain, and individuals considered California bighorn sheep. Currently, genetic and skull morphology data suggest desert, Rocky Mountain, and Sierra Nevada bighorn sheep are the only taxa among Cowan's original designations warranting subspecies status, although analyses of microsatellite (Barbosa et al. 2021) and SNP data (Jahner et al. 2025) found support for strong genetic distinction of the British Columbia portion of Cowan's California subspecies (populations now found primarily in BC, ID, OR, NV, UT, WA). The following sections detail phylogenetic relationships within each of these taxonomic groups consistent with the presence of subspecific structure (i.e., ecotypes).

*Rocky Mountain bighorn sheep.* Luikart and Allendorf (1996) sampled 288 individuals from 22 indigenous populations (herds) located throughout the geographic range of Rocky Mountain bighorn sheep and screened for mtDNA variation at RFLPs from seven enzymes. Their phylogeographic analysis revealed four geographically widespread haplotypes and no genetic discontinuities indicative of long-term population isolation. This finding of minimal phylogeographic population structure (i.e., lack of concordance between phylogeny and geography) suggests historical gene flow occurred on a regional scale in the absence of long-term biogeographic barriers. This finding was further supported by the phylogenetic analysis of Buchalski et al. (2016) who reported minimal statistical support for subclades within the greater Rocky Mountain bighorn clade. Luikart and Allendorf (1996) attributed the lack of phylogenetic structure to the relatively recent colonization of the Rocky Mountains, suggesting populations did not experience sufficient isolation (i.e., many thousands of generations) for mtDNA variation to accumulate. Gene flow at the regional scale is possible given the high potential for dispersal in bighorn sheep (i.e., either natal or recolonization). Luikart and Allendorf (1996) also reported lower haplotype diversity in Canadian herds, consistent with a founder effect resulting from post-glacial expansion northward (i.e., recolonization) following the end of the Wisconsin Ice Age 10–20 kya.

Further, Luikart and Allendorf (1996) reported haplotype frequencies varied significantly among local populations. The authors attributed this frequency heterogeneity to female philopatry and forest expansion following retreat of Pleistocene glaciers (10 kya), as proposed by Geist (1971). They suggest Rocky Mountain bighorn sheep experienced substantial local population differentiation for thousands of years prior to anthropogenic habitat fragmentation. Thus, the overall pattern for Rocky Mountain bighorn is one of phylogeographic continuity as indicated by widespread haplotypes, with localized population structure as indicated by differences in haplotype frequencies. Luikart and Allendorf (1996) suggest that this seemingly contradictory pattern of genetic variation is possible if recent population fragmentation subdivides a previously well-connected metapopulation (i.e., a group of populations linked by occasional movements; Hanski and Gilpin 1991) into largely isolated local populations where genetic drift is strong. Alternatively, this pattern is also possible if gene flow is high enough to prevent mtDNA sequence divergence, but too low to prevent differentiation in haplotype frequencies; mtDNA haplotypes require thousands of generations to accumulate substantial mutational differences, particularly when defined by low-resolution techniques such as RFLP, whereas differences in haplotype frequencies can develop in a few generations if genetic drift is strong.

*California bighorn sheep.* Appropriateness of the California bighorn sheep subspecies designation has been challenged by both morphometric and genetic studies. Wehausen and Ramey II (2000) examined geographic variation in skull and horn characters of bighorn sheep ranging from the Great Basin to southern British Columbia and Alberta. They applied a more appropriate statistical analysis than was available to Cowan (1940), concluding that native California bighorn sheep from British Columbia and Washington were indistinguishable from Rocky Mountain bighorn sheep, and therefore assigned extant and extinct native populations from that region to the *O. c. canadensis* subspecies. The interpretation of these results was that California and Rocky Mountain bighorn represent opposite ends of a single, diverse taxon exhibiting clinal variation across mountainous habitats spanning more than 2,000 km.

This decision was later supported by the mtDNA phylogenetic analysis of Malaney et al. (2015), who examined variation among 110 samples of Rocky Mountain, California, and desert bighorn sheep translocated into their presumed native range in Nevada. Those authors reported California and Rocky Mountain bighorn formed a well-supported polyphyletic clade distinct from desert bighorn. Such close evolutionary relationship was further demonstrated by Buchalski et al. (2016) who reported Rocky Mountain bighorn, primarily of Alberta and Colorado origin, formed a well-supported polyphyletic clade with California bighorn from Spences Bridge, British Columbia. Despite similarities among California and Rocky Mountain bighorn mtDNA, Malaney et al. (2015) reported a subdivision between the two groups at nDNA microsatellite loci that was comparable to their differentiation from desert bighorn. The authors acknowledge that such genetic differentiation in the translocated range could be due to founder effects that can accentuate divergence. Barbosa et al. (2021) likewise employed nDNA microsatellite loci to evaluate the ancestry of bighorn sheep in British Columbia, Oregon, Washington, and Idaho, in part to determine whether bighorn sheep managed as the California subspecies could be distinguished from those managed as the Rocky Mountain subspecies. Their study showed clear distinction between those lineages in that region (Barbosa et al. 2021), but questions about the geographic distribution of California bighorn remained due to the limited study area. Jahner et al. (2025) followed this up with a range-wide SNP dataset to reconstruct the evolutionary history of bighorn lineages using population and phylogenetic analyses. These analyses showed strong differentiation between California bighorn sheep in, or originally translocated from, western British Columbia, Canada, and all other subspecies, in line with the results of Barbosa et al. (2021). Thus, while not supporting Cowan's (1940) original definition of the California subspecies, which included extirpated bighorn presumed to inhabit the Cascades Mountain Range of Oregon and Washington, and populations in the Sierra Nevada of California that are now recognized as Sierra Nevada bighorn sheep (see below), Jahner et al. (2025) concluded that California bighorn originating from western British Columbia and likely northern Washington (Figure 6.2) represent a distinct evolutionary lineage that is as much or more differentiated as other currently recognized subspecies.

*Desert bighorn sheep.* Ramey II (1995) screened 244 bighorn sheep from indigenous remnant populations in the desert southwest for mtDNA variation at five RFLPs. He reported a single haplotype was most common across the region with all other haplotypes showing minimal sequence divergence from that main haplotype. The resulting haplotype network was star shaped, suggesting populations of desert bighorn sheep had not experienced long periods of geographic isolation. Further, the majority of haplotype variance occurred within rather than between putative desert subspecies. Because the geographic distribution of mtDNA haplotype variation was not concordant with Cowan's (1940) original range maps for desert-dwelling bighorn sheep subspecies (Figure 6.2), Ramey concluded that those subspecies should not be recognized. This finding supported previous morphometric studies demonstrating Cowan's desert subspecies designations were largely the result of small sample sizes and different age distributions among samples (Ramey II 1993, Wehausen and Ramey II 1993).

Ramey II (1995) further suggested that desert bighorn have continuously inhabited the desert southwest since they diverged from Rocky Mountain bighorn, but population fluctuations and local extinctions had limited their genetic variation, making those populations appear recently derived. He proposed that constriction of suitable habitat during the climatic fluctuations of the late Pleistocene may have caused cyclic reductions in local population sizes and local extinctions, in turn resulting in the loss of mtDNA diversity. Such changes in habitat availability are consistent with known climatic and vegetational shifts in the desert southwest during Pleistocene glacial cycles, as forest cover advanced to elevations 1,200 m lower than contemporary distributions (Van Devender et al. 1987). This would have isolated populations on patches of suitable alpine habitat, a scenario similar to that of Rocky Mountain bighorn sheep as proposed by Luikart and Allendorf (1996). Additionally, the extensive Pleistocene pluvial lakes that filled many areas between mountain ranges in the Mojave and Great Basin deserts would have further isolated many populations (Laity 2009, figure 5.7). Repeated founder events during the recolonization of ranges would have also reduced mtDNA diversity, further obfuscating the evolutionary history of desert bighorn.

Ramey II (1995) acknowledged the presence of genetic structure within desert bighorn in the form of variation in mtDNA haplotype frequencies among local populations (e.g., eastern Mojave Desert, Death Valley and Great Basin, United States, and Baja California, Mexico). This finding of intra-subspecific genetic structure was further substantiated by Buchalski et al. (2016). Those authors tested genetic support for the desert subspecies designations of Cowan (1940) using 804 samples collected from 58 locations across the desert southwest. The maximum likelihood phylogeny based on a 515 bp segment of the mtDNA control region represented desert bighorn as a highly polyphyletic clade showing no support for deep divergence (i.e., comparable to Rocky Mountain or Sierra Nevada) among Cowan's Nelson (*O. c. nelsoni*), Peninsular (*O. c. cremnobates*), or Mexican (*O. c. mexicana*) bighorn. Network analysis, however,

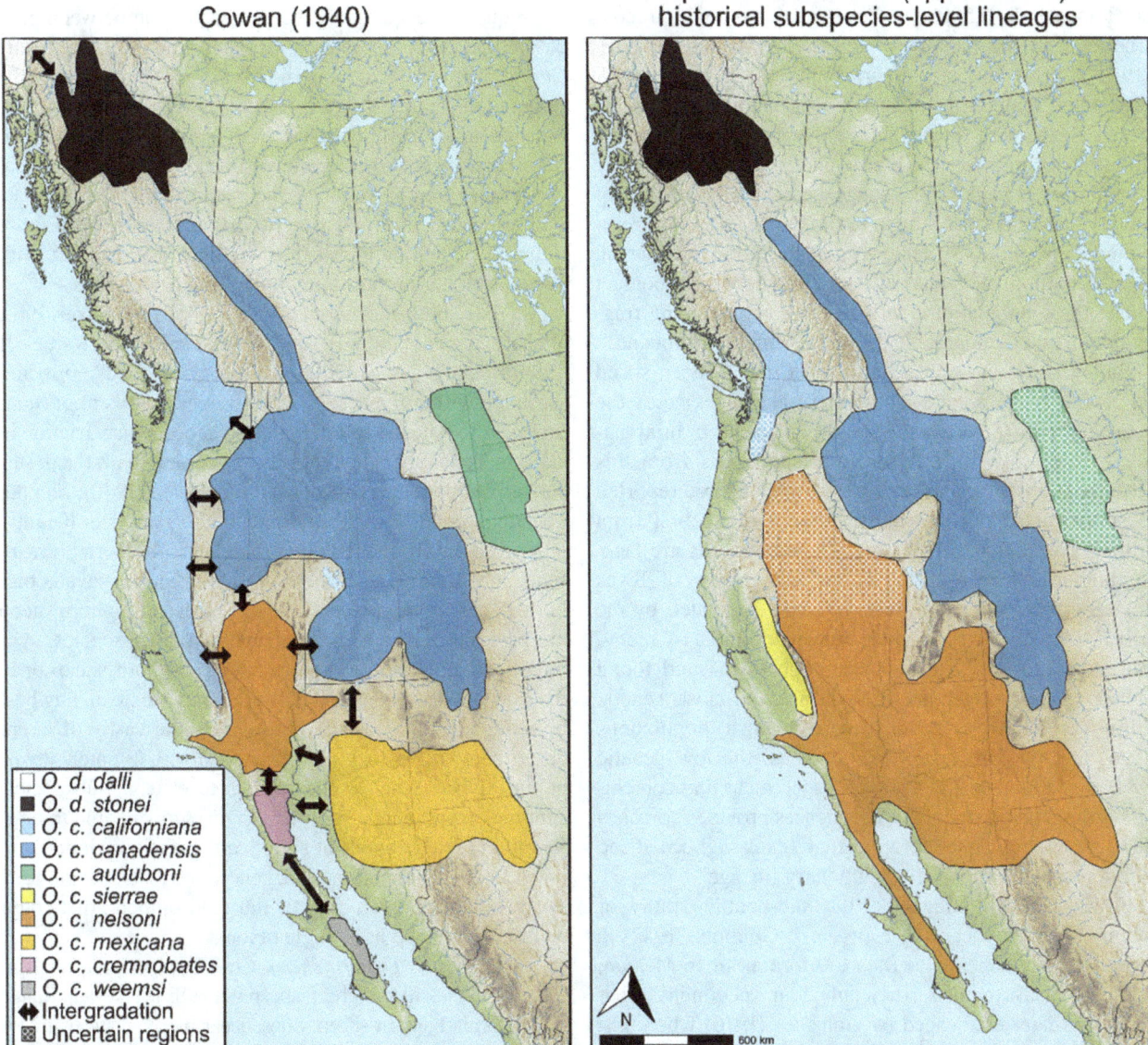

**FIGURE 6.2** Cowan (1940) made the first graphical attempt to describe the historical (pre-European settlement) geographic distributions of bighorn sheep lineages based on cranial morphology (left panel, recreated based on Cowan's map), which was used for planning many of the bighorn sheep translocations over the past century. Subsequent studies (reviewed in the Taxonomy section), based on both morphological and genetic datasets, have called those distributions into question. While acknowledging that subspecies distributions before the widespread extirpations of the 19th and 20th centuries may never be fully known, we offer a speculative revision (right panel) based on our current understanding of genetic variation in bighorn sheep; white circles accentuate geographic regions with high uncertainty. Audubon's bighorn (*O. c. auduboni*) has not been reevaluated genetically.

indicated minimal haplotype sharing among the three putative subspecies, and those geographic areas where sharing was observed (Buchalski et al. 2016, figure 4) largely coincided with zones of contact between subspecies identified by Cowan (1940). Those zones included areas of contact between Nelson's bighorn sheep (*O. c. nelsoni*) and the formerly recognized Mexican bighorn sheep (*O. c. mexicana*) in the northern Sonoran Desert within Arizona, United States, and between and the formerly recognized Peninsular bighorn (*O. c. cremnobates*) in the northern Peninsular Ranges in California, United States. Importantly, the network aged multiple endemic haplogroups within Peninsular

and Mexican bighorn that significantly predated the LGM, by 103–56 kya for Peninsular and 160–9 kya for Mexican. Interestingly, the mtDNA control region haplotype network of Buchalski et al. (2016) was sparse with many missing intermediate haplotypes. This finding is consistent with Ramey II's (1995) suggestion that historic desert bighorn demography was characterized by cycles of local extinction due to changes in habitat, followed by recolonization.

Coalescent simulations of effective splitting times, treating Nelson, Peninsular, and Mexican bighorn as populations, suggested relatively recent divergence. Mexican bighorn may have diverged as early as the late Pleistocene

(37–3 kya), with the two other populations separating in the Holocene (8–0.500 kya). Bayesian demographic reconstruction and the haplotype network by Buchalski et al. (2016) suggest modern desert bighorn reflect a small, fragmented remnant of a much larger historic population, one that expanded during the Sangamon interglacial followed by decline since the LGM. The estimated ages of the endemic desert haplogroups suggest potential occupancy of multiple desert pocket refugia in the greater southern refugium during the LGM. Following deglaciation, changes in the distribution of habitat allowed for secondary contact among these populations, with all three experiencing fragmentation and demographic decline during the Holocene.

Bailey (1912) described the now-extinct bighorn sheep indigenous to Texas and southeastern New Mexico as the Texas bighorn (*O. c. texianus*) based on unique cranial features. Cowan (1940) later reviewed specimens from this purported subspecies and using a sample of three reported no statistical support for morphological distinction from animals considered at that time. Though records are lacking, bighorn sheep from the Trans-Pecos region of Texas and eastern New Mexico were assumed extirpated by the early 1960s. Yet a recent study sequenced mtDNA recovered from pre-extirpation museum specimens and found bighorn sheep indigenous to Texas showed close genetic affinity to multiple extant desert bighorn populations range-wide (Wright et al. 2024a). Thus, the low genetic divergence found between museum specimens and contemporary desert bighorn mtDNA sequences provides corroborating genetic evidence refuting that Texas bighorn sheep once represented a distinct evolutionary lineage.

Cowan (1940) recognized the subspecific status of Weem's bighorn sheep (*O. c. weemsi*; Goldman 1937), a purported race endemic to Baja California Sur in Mexico, based on measurements from only four specimens. That status was later challenged by Gonzales (1976) who compared male skulls collected from six locations along a north-south transect of the Baja Peninsula, including the ranges of Weem's bighorn sheep and the adjacent Peninsular bighorn sheep as designated by Cowan (1940). Finding mostly non-significant differences among locations, Gonzales (1976) suspected Weem's bighorn sheep was invalid as he found no distinct discontinuity with Peninsular bighorn sheep, although he noted the material studied was inadequate for determining the taxonomic status of bighorn indigenous to the region. Ramey II (1993) later reported no morphological differences between Weem's and the adjacent Peninsular bighorn sheep, proposing all desert bighorn be synonymized. Subsequent genetic analyses have also found little support for the subspecific status of Weem's bighorn sheep. Using the only available mtDNA D-loop haplotype, Buchalski et al. (2016) demonstrated that even though this sequence fell within the desert bighorn clade, it was not highly divergent from other desert haplogroups. Interestingly, the sequence also was not sister to the geographically proximate Peninsular bighorn sheep, which is a parsimonious phylogeographic prediction.

Rodriguez-Rodriguez et al. (2015) used nuclear inter-simple sequence repeats (i.e., the genome region between microsatellites) to compare bighorn sheep from Baja California Sur to animals from Sonora, Mexico. They reported the populations in Baja California Sur were genetically distinct, but not highly differentiated from Sonora. The combined data suggest only minor intra-subspecific morphological differences among bighorn sheep indigenous to the Baja Peninsula and minimal genetic differentiation from those animals in mainland Mexico, findings inconsistent with Weem's bighorn representing a legitimate subspecies.

To date, both morphometric (Bradley and Baker 1967, Wehausen and Ramey II 1993) and genetic data (Ramey II 1995, Buchalski et al. 2016) have demonstrated significant variation within the desert bighorn lineage. It is important, however, to note the following: first, geographic variation in skull morphology is not entirely concordant with the desert subspecies boundaries proposed by Cowen (1940); second, genetic populations within desert bighorn are significantly less divergent from each other than desert bighorn are with other subspecies (i.e., Rocky Mountain, Sierra Nevada bighorn). These combined results suggest desert bighorn sheep are best referred to as a single, polytypic subspecies, synonymized as Nelson's bighorn sheep. Bighorn sheep once distinguished as Nelson, Peninsular, and Mexican may best be thought of as ecotypes considering the vastly different desert ecosystems they currently occupy. Ultimately, taxonomic divisions below the species level (e.g., subspecies, ecotype, evolutionary significant unit) are valuable to conservation if they represent significant genetic variation and potential local adaptation (i.e., genetic variants that confer a selective advantage in a particular environment, that could be lost if managed as a single taxon).

*Sierra Nevada bighorn sheep.* Cowan's (1940) morphological analysis classified bighorn sheep in the Sierra Nevada Range as California bighorn sheep. Subsequent morphological studies using larger sample sizes and more sophisticated methods have demonstrated the distinctness of the population (Wehausen and Ramey II 1993, 2000) leading to its classification as a subspecies (Wehausen et al. 2005). This classification is also fully supported by genetic analyses. Ramey II (1995) reported bighorn sheep from the Sierra Nevada had a single RFLP haplotype unique to the subspecies. Buchalski et al. (2016) reported a similar lack of genetic diversity, with only a single control region haplotype unique to the population. These findings emphasize the extreme population bottleneck Sierra Nevada bighorn sheep experienced during the 20th century (Wehausen 1980, U. S. Fish and Wildlife Service (USFWS) 2007) and suggest all remaining individuals are descendants of a single maternal lineage. Buchalski et al. (2016) also reported the Sierra Nevada haplotype was highly divergent from haplotypes for bighorn in British Columbia; those animals also were previously considered California bighorn sheep (Cowan 1940). The survival of a single mtDNA control region haplotype in Sierra Nevada bighorn limits the phylogenetic methods that can be applied to this subspecies, and whole mitochondrial or nuclear genome sequencing may be necessary to better understand its evolutionary history.

Sierra Nevada bighorn sheep appear to have a complex evolutionary history with the desert subspecies that is not fully understood. Although Buchalski et al. (2016) estimated the two lineages diverged $640 \pm 120$ kya, the authors also reported evidence of historic gene flow. Within the Sierra Nevada clade, they found a basal desert haplotype with time to most recent common ancestor dated at $150 \pm 60$ kya. This desert haplotype was found in individuals from the adjacent southern Great Basin and northern Mojave Desert. The authors hypothesized the two subspecies experienced secondary contact that predated the LGM. The geographic distribution of these two haplotypes is consistent with an eastern expansion of Sierra Nevada bighorn sheep into adjacent mountain ranges unsuitable for desert bighorn at that time. Secondary contact would have occurred during an interglacial period, as temperatures rose and forests and pluvial lakes receded, and desert bighorn expanded their distribution north. Further, hybridization between desert populations and Sierra Nevada eastern migrants would have occurred as habitats were shifting and desert-adapted genes were increasingly under selection; Jahner et al. (2025) reported that desert and Sierra Nevada bighorn were more similar than other subspecies based on population and phylogenetic analysis of 14,622–24,795 SNP loci. Despite the possibility of ancient gene flow, there is no evidence of contemporary gene flow between desert and Sierra Nevada bighorn based on microsatellite genotypes (Buchalski et al. 2016) or SNP data (Jahner et al. 2025).

## 6.3 USING GENETICS TO INFER POPULATION PROCESSES AND RECENT HISTORY

Genetic approaches have provided extensive insights on population history of mountain sheep, including connectivity, population bottlenecks, translocation history, colonization, dispersal, and animal movements. As evolving methods made non-invasive genetic sampling practical, and as agencies have honed their ability to capture large numbers of animals efficiently for global positioning system collaring, translocation, disease surveillance, and genetic sampling, landscape-scale genetic datasets now allow comprehensive estimation of those elements of population history.

### 6.3.1 GENETIC STRUCTURE

Genetic estimates of population structure (i.e., the degree of genetic differentiation of herds, populations, or other units) have been unusually informative for interpreting the population history of mountain sheep compared to other species, largely because of the small size and fragmented distribution of many mountain sheep populations. Because local populations rarely exceed a few hundred individuals and the polygynous mating system of mountain sheep, the effective population sizes of those populations are very small, making genetic drift (random changes in allele frequencies over time) very rapid, resulting in changes in genetic structure that can reflect very recent population history (Epps and

Keyghobadi 2015). Therefore, unlike in the much larger or more continuous populations such as those now typical for deer (*Odocoileus* spp.) in North America, genetic structure of bighorn sheep populations can change sharply even over a few generations. For example, Epps et al. (2005) detected strong decreases in gene flow, and thus dispersal, between desert bighorn populations separated by interstate highways for less than 50 years at the time of that study; more isolated populations had lower genetic diversity. When new interpopulation connections arise after natural recolonizations of empty habitat, translocations, behavioral shifts (e.g., road-crossing behaviors; Epps et al. 2018), or even human interventions (e.g., construction of highway overpasses; Chapter 20), detectable changes in population genetic structure can be observed over even shorter time frames. Epps et al. (2018) described clear instances of genetic shifts after only 12 years, during which bighorn sheep in one location apparently learned how to cross an interstate highway, and population pairwise $F_{ST}$ (a measure of genetic differentiation) declined from moderate (0.11) in 2000–2003 to low (0.04) in 2013–2015 (Figure 6.3). Within the largely native metapopulations of the Mojave Desert of California, genetic structure estimates have served to determine sources for natural recolonizations, track the spread of translocated animals or their descendants between populations, and thus have helped with expectations around disease spread or metapopulation dynamics (Epps et al. 2010, 2018; Dugovich et al. 2023).

Estimates of genetic structure have proven to be helpful tools for evaluating long-term consequences of bighorn restoration efforts through translocation, including unintended hybridization of subspecies or unanticipated linkages (Latch et al. 2005, Epps et al. 2010, Malaney et al. 2015, Gille et al. 2019, Jahner et al. 2019, Flesch et al. 2020, Wright et al. 2024b). Despite the wealth of information on animal movement provided by global positioning system collars in the last several decades, comprehensive population genetic evaluations, including landscape genetic studies (i.e., studies that test models of movement or gene flow over complex landscapes against observed genetic structure; Manel et al. 2003), can offer long-term, integrated pictures of linkages in complex landscapes (Epps et al. 2007, Creech et al. 2014, Creech et al. 2017). Such analyses are complementary to global positioning system-based movement studies (Dekelaita et al. 2023). While genetic studies may fail to detect ephemeral or exploratory contacts by individuals briefly transiting nearby populations, which may be important for disease transmission (Biek and Real 2010), they are sometimes better suited to detecting infrequent movements important for gene flow. Many populations can be sampled over genetically short time frames, and even minimal gene flow can leave detectable signatures that last for years (Epps et al. 2005, 2010).

For both bighorn and thinhorn sheep, genetic structure has proved to be complex and hierarchical. As for most alpine specialists, dispersal and gene flow in mountain sheep tend to be limited due to specific habitat requirements

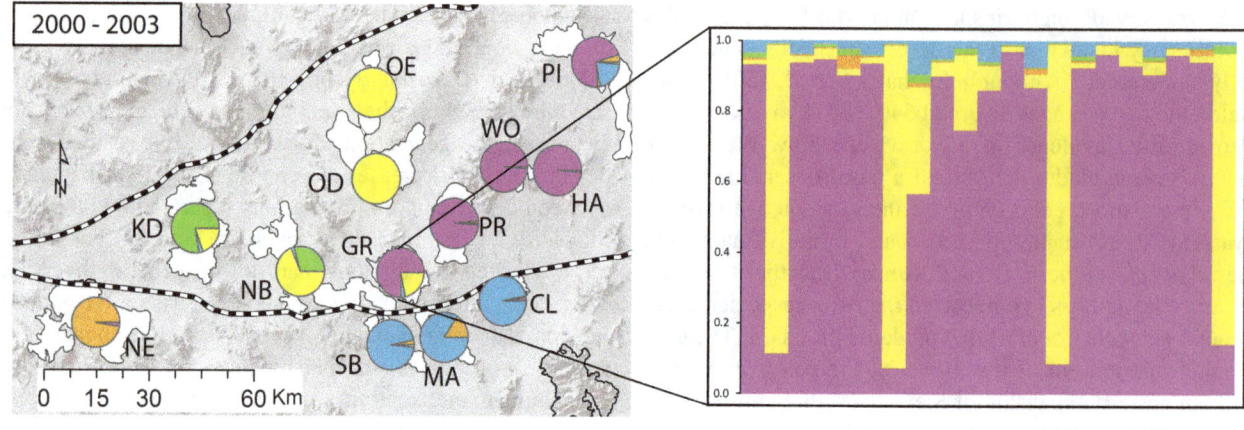

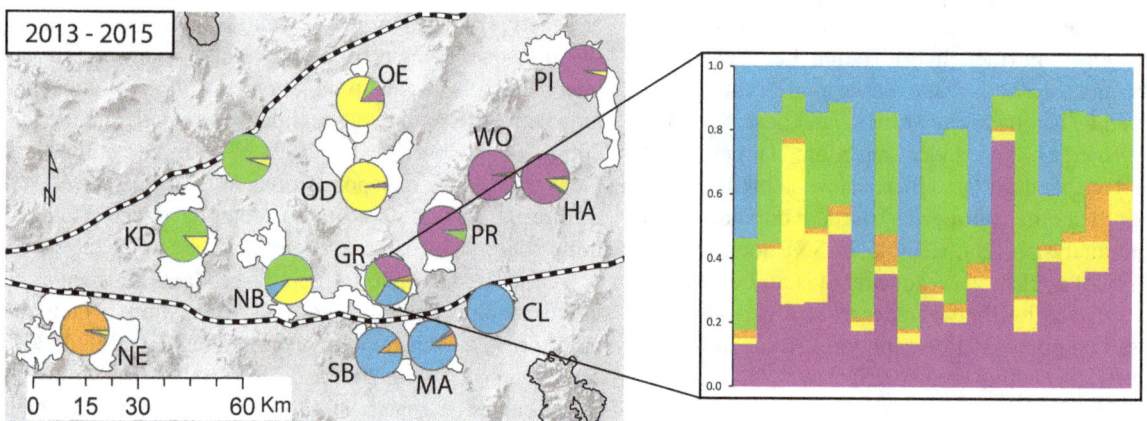

**FIGURE 6.3** Desert bighorn sheep populations (indicated by two-letter codes) genetically sampled at two time points (2000–2003 and 2013–2015, white polygons) in the Mojave Desert of California, with other nearby populations drawn with black outlines, and shaded topographic relief. The South Soda Mountains population, an apparent recent colonization, was sampled only in 2013–2015. Interstate highways are depicted with dashed lines. Average genetic assignments of individuals from desert bighorn sheep populations in 2000–2003 and 2013–2015 (*k*=5) from Program STRUCTURE, based on 16 microsatellite loci, are shown color-coded by proportional assignment to cluster by population (circles) and by individual (Granite Mountains [GR], where each vertical bar reflects an individual). In 2000–2003, no individuals bordering I-40 were assigned to populations on the opposite side, whereas in 2013–2015, five individuals in the Granite Mountains were at least 40% assigned to the populations south of I-40 (blue cluster). (Reproduced from Epps et al. (2018); used with permission.)

or the presence of landscape features that pose barriers to movements (e.g., very large water bodies, glaciers). Detectable and sometimes strong genetic differentiation can arise even when mountain ranges appear continuous due to populations being isolated in sky-islands, alpine areas separated by forested areas, or by water availability in desert systems. Mitochondrial (Loehr et al. 2006), microsatellite (Worley et al. 2004), and SNP (Sim et al. 2016, 2019) data indicate that the population genetic structure of thinhorn sheep is distributed in a hierarchical fashion due to an interplay between geographical distance, landscape features, and glacial history, among others. Species-wide, fine-scale mapping of population genetic structure was first conducted using microsatellites (Worley et al. 2004), then replicated with SNP markers (Sim et al. 2016, 2019). Both sets of markers indicate that population structure is influenced most strongly by glacial history at the species level

(as noted in the thinhorn sheep section). Unsurprisingly, genetic differentiation is greatest between populations of Dall's and Stone's sheep. Dall's and Stone's sheep separate along a northeast-southwest axis in the Teslin Lake-Taku River watersheds, with an admixture zone bounded by Atlin Lake and the Pelly River to the north, and the Stikine River to the south (Figure 6.1). Dall's sheep can be split into two larger population units along the Tintina Trench with one cluster inhabiting Brooks Range, Ogilvie Mountains, and Mackenzie Mountains (northern Dall's group), while the other occupies the Central Alaskan Range, Wrangell Range, and Coast Mountains (southern Dall's group). Genetic structure allows further partitioning of the northern Dall's group into three populations, including the Brooks Range, Ogilvie-Wernecke Mountains, and Mackenzie Mountains, while the southern Dall's group can be split into the Central Alaskan Range and Coast Mountains populations. Stone's sheep can be subdivided into three genetic clusters, including the Stikine-Skeena Mountains, Cassiar Mountains, and Rocky Mountains (Figure 6.4).

Beyond those distinctions, thinhorn sheep exhibit significant population genetic structure even within seemingly continuous mountain ranges. Roffler et al. (2014) conducted a spatial genetics study of Dall's sheep in the Wrangell-St. Elias National Park and Preserve. They reported genetic differentiation among populations inhabiting the park, influenced by timing of glacial retreat and landscape barriers. For instance, Dall's sheep occupying the Chisana and Nutzotin regions are separated from other populations by the St. Elias icefields to the south and Nabesna glacier to the west. Populations are also separated by the Wrangell icefield and the Chitina River valley. Notably, this study reported a lack of differences in dispersal between sexes, which is contrary to the male-dominant dispersal patterns found in most other species (Handley and Perrin 2007). The authors postulated that the lack of sex-biased dispersal may be due to patchy environments limiting suitable habitat, which favors philopatry for both sexes. Interestingly, Deakin et al. (2020) reported a similar lack of evidence for sex-biased dispersal in northern Rocky Mountain bighorn populations.

Bighorn sheep likewise exhibit strong hierarchical substructure even within subspecies (Buchalski et al. 2015). Within lineages, within regions, and even within metapopulations, moderate genetic structure (e.g., $F_{ST}$ values greater than 0.1) is observable between populations separated by only 15–30 km (Epps et al. 2005, Jahner et al. 2019), including in some systems with relatively continuous habitat such as Death Valley National Park in California (Epps et al. 2016), Grand Canyon National Park and associated areas in Arizona (Gille et al. 2019, Creech et al. 2020, Epps et al. 2024b), or the Peninsular Ranges of California (Boyce et al. 1999, Buchalski et al. 2015). Strongly isolated populations such as the San Gabriel Mountains and the White Mountains in California show differentiation at any level of genetic analysis (Epps et al. 2005, 2016; Jahner et al. 2025). These patterns likely result from climate shifts and resulting extinction-colonization dynamics, post-European contact extinctions and subsequent metapopulation dynamics, small census and effective population sizes, and strong natural fragmentation. The resulting strong, hierarchical genetic structure offers researchers an unusually high-resolution ability to track population processes with genetic tools (Epps et al. 2010, Olson et al. 2012, Jahner et al. 2019), while also complicating efforts to detect selection on genetic markers associated with environmental variation or even to clarify subspecies-level taxonomy.

## 6.3.2 Assignment Tests

Genetic assignment tests have provided insights on the movements of mountain sheep even to the individual level. Assignment tests typically estimate proportional ancestry of individuals to pre-defined populations or to genetic clusters, groups of individuals that are identified in the analysis to be genetically similar. Once such clusters are diagnosed, population metrics such as $F_{ST}$ can be used to describe relative differentiation of those clusters, and potential migrants can be identified based on their presence in one population but genetic assignment to another. Assignment tests have been widely applied in wildlife research, including for mountain sheep. Many studies have used them as a means of identifying genetic structure, particularly in cases where physical boundaries between populations (animals that interbreed freely) are unclear. For instance, population genetic structure within Peninsular desert bighorn was assessed in this manner (Buchalski et al. 2015). Because of the linear and relatively continuous nature of bighorn sheep habitat in this mountain range, extending from southern California into Mexico, the strength of gene flow or genetic structure was unknown; assignment tests indicated the presence of northern, central, and southern genetic clusters, albeit linked by stepwise gene flow. Epps et al. (2024b) used this approach to define populations of desert bighorn within Grand Canyon National Park; while this region appears to be a continuous habitat bisected by the Colorado River, they identified six genetically distinct clusters. This approach yielded insights on bighorn sheep interacting both along and occasionally across the Colorado River in two studies (Gille et al. 2019, Epps et al. 2024b).

Assignment tests are likewise often employed as a first step in describing and visualizing population ancestry in large study systems, particularly in regions where most or all populations were restored by translocation. Flesch et al. (2020) used assignment tests to group populations of bighorn sheep in Montana, Wyoming, Colorado, and Idaho by ancestry, reflecting both native population ancestry and translocation history; Malaney et al. (2015) and Jahner et al. (2019) accomplished similar analysis in Nevada, which holds translocated and native populations of three lineages or subspecies. Assignment tests have known complexity and pitfalls, however, when used to describe genetic structure: if sampling in a region is incomplete, sample sizes from some genetic clusters are highly unbalanced, or sampling is patchy across a continuous population exhibiting isolation by distance, assignment tests may suggest more structure than is truly present, or may erroneously assign individuals from under-sampled clusters to clusters that are well-represented in the data (Wang 2017, Perez et al. 2018). Where population structure is well defined by geography or previous study, however, and when genetic differentiation among populations is high, assignment tests offer the opportunity to reliably assign individuals or populations of unknown origin. This approach has helped diagnose changes in connectivity (Figure 6.3) and origins of stray bighorn detected in unexpected places (often important when assessing risk of spread of disease), and shed insight on metapopulation dynamics including dispersal and colonization (Epps et al. 2010, 2018, 2024a; Stowell et al. 2020).

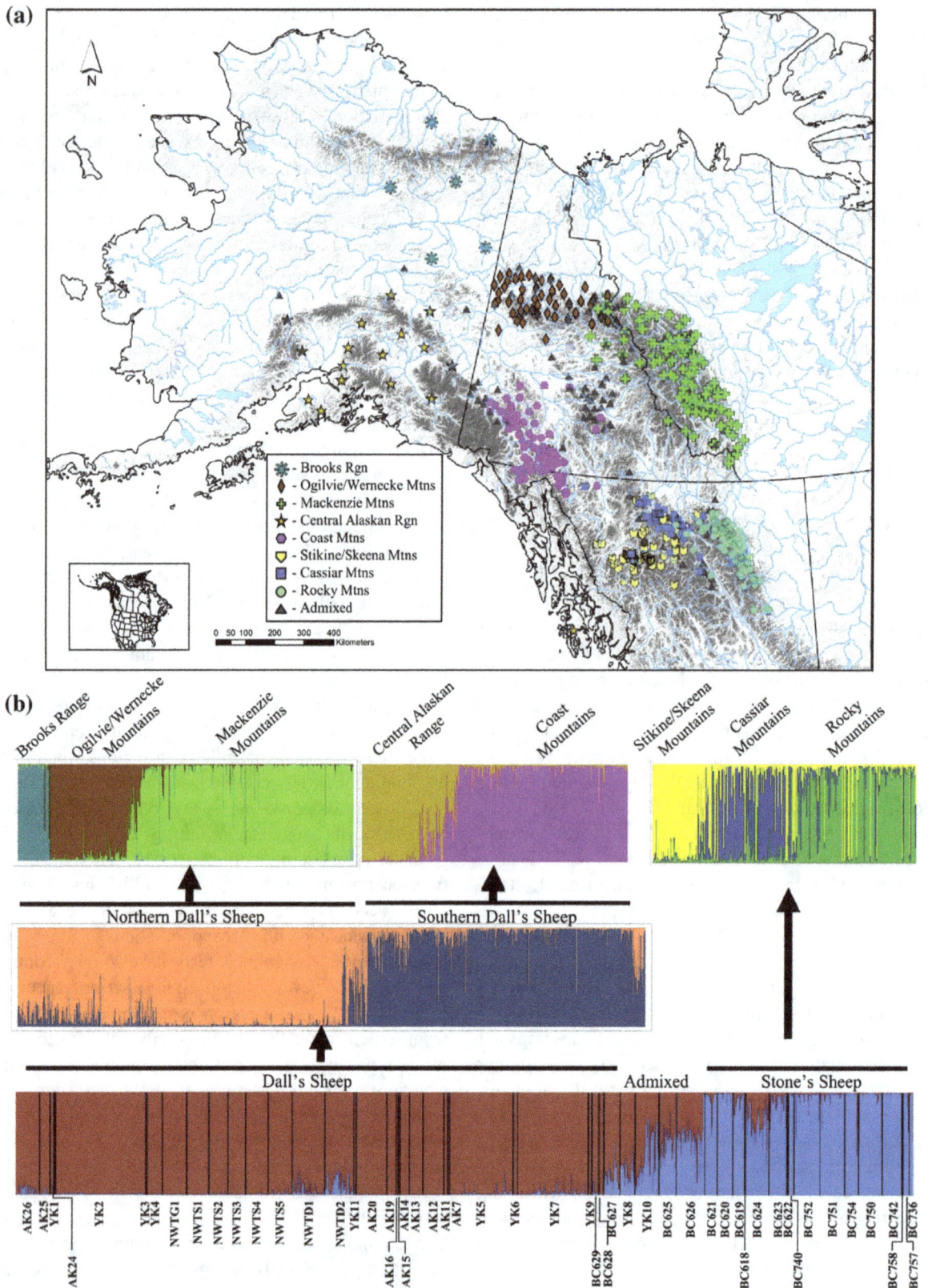

**FIGURE 6.4** (a) Hierarchical genetic cluster identities of thinhorn sheep sampled in Sim et al. (2019) and analysis based on 2,820 males sampled 1996–2015 in northwestern North America and genotyped at 153 nuclear single-nucleotide polymorphisms. (b) Relative genetic ancestry from each genetic cluster for individuals shown in (a) as indicated by Bayesian clustering analysis. The arrows indicate progression from species-wide to within-subspecies-level population structure. Sampling localities are below the species-wide bar plot and represent game management units in their respective jurisdictions (e.g., AK26=Alaska Game Management Unit 26; BC618=British Columbia Management Unit 6-18; YK1=Yukon Game Management Area 1; NWTD1=Northwest Territories Outfitter Management Area D/OT/01). (Reproduced from Sim et al. (2019); used with permission.)

### 6.3.3 INDIVIDUAL IDENTIFICATION AND POPULATION ESTIMATION

Non-invasive genetic studies also have used genetic recapture for insight into movement. Fecal samples collected over time or large landscapes can be evaluated for matching genotypes, including genotypes from invasively captured individuals; after accounting for genotyping error and variability of genetic markers involved, matching genotypes usually can be confidently assigned as originating from the same individual (Waits et al. 2001). Genetic recaptures offer direct evidence of movement and dispersal; Epps et al. (2018) detected one male bighorn on both sides of Interstate 40 through genotype recapture, and six bighorn were demonstrated to have crossed the Colorado River within a 16 km section of Grand Canyon National Park (Epps et al. 2024b). Where global position system telemetry is limited, and particularly where intensive non-invasive sampling is used for population estimation, genetic recapture can offer direct rather than inferential evidence of space use or rare events such as individuals crossing landscape features that usually act as dispersal barriers.

The ability to reliably identify individuals, even those that are closely related, from non-invasive genotypes appeared with the advent of microsatellite markers (Waits et al. 2001) but has also employed SNPs (Russello et al. 2015). Beyond genetic recapture, these techniques have enabled capture-recapture estimates of population size. Some mountain sheep populations are well suited to this, particularly those that are relatively small and attracted to localized resources, and this approach offers a useful alternative in systems where other approaches such as mark-resight with collared animals or aerial surveys are impractical due to regulatory environments or steep terrain (Epps et al. 2024b). Schoenecker et al. (2015) collected fecal pellets from a mineral lick in Rocky Mountain National Park, estimating a population of less than 100 individuals using mark-recapture analysis. Pfeiler et al. (2020) used fecal DNA samples from transects at water sources and on game trails to estimate the population size of desert bighorn in the Marble Mountains of California's Mojave Desert and compared that estimate to ground-based mark-resight of ear-tagged animals. They reported that the fecal DNA capture-recapture method was more cost-effective and more precise in this population of less than 200 individuals. Finally, Epps et al. (2024b) used fecal DNA collections over 5 years, with spatially explicit capture-recapture analysis, to estimate population size of bighorn sheep in Grand Canyon National Park, Arizona; that effort likewise yielded reasonable credible intervals despite being a much larger population (~550 individuals) with much greater spatial extent, spanning about 450 km of the Colorado River. This effort simultaneously yielded fine-scaled estimates of genetic structure and information on space use through genetic recaptures along and across the Colorado River. Fecal DNA-based population estimates are, however, only able to estimate the fraction of a population using the area sampled; in systems where sampling is focused on water or mineral licks, individuals using unmonitored resources will not be included (Schoenecker et al. 2015).

## 6.4 GENETIC MANAGEMENT

### 6.4.1 GENETIC DIVERSITY

Genetic diversity has been of keen interest to managers of mountain sheep because of the potential implications for long-term population viability (Frankham 1995) and adaptive capacity (i.e., the ability to respond to changes in environment due to standing genetic variation; Creech et al. 2020). Genetic diversity is one of the most reported summary statistics by conservation geneticists. Most genetic diversity metrics are based on an individual's heterozygosity, that is, the proportion of genetic markers (e.g., sequenced areas, microsatellites) with different alternative forms, or alleles. Studies of genetic diversity for the last several decades have largely relied on microsatellites or SNPs. Microsatellite datasets target areas of the genome where short DNA sequences, typically 2, 3, or 4 nucleotides long, are repeated many times. These short, repeated motifs are prone to mutation, so alleles take the form of different numbers of repeats (e.g., 14 vs. 20 repeats of ATC), expressed as different lengths of sequence. For SNP datasets, alleles are different nucleotides (A, C, G, T) at a single location in the genome. Levels of genetic diversity cannot easily be compared across datasets based on different genetic markers. Two different studies examining many of the same desert bighorn sheep populations in Nevada reported very different estimates of expected heterozygosity ($H_E$) for a microsatellite dataset (average $H_E = 0.677$; Malaney et al. 2015) and a SNP dataset ($H_E$ range $= 0.177$–$0.201$; Jahner et al. 2019). Thus, genetic diversity estimates are usually best considered an index that must be evaluated within a particular dataset, although inferences across similar classes of markers may be more comparable: for instance, most bighorn studies employing microsatellite loci use a mix of loci originally developed for domestic sheep or cattle, which allow for reasonable general comparisons.

Comparisons of genetic diversity across populations must consider that estimates are shaped by historical processes (e.g., glaciation, population bottlenecks) and by more contemporary factors (e.g., disease events, dispersal, management). For example, a population could have low diversity because it occurs on the periphery of the population's range where glaciation may have kept population sizes small (Deakin et al. 2020), or because it was recently founded via translocation using a small number of founders (Whittaker et al. 2004). Genetic diversity can be shaped strongly by ongoing and recent conditions. In the desert bighorn metapopulations of California's Mojave and Sonoran deserts, where population structure largely reflects geographically discrete mountain ranges separated by desert flats lacking escape terrain (Bleich et al. 1990), genetic diversity of markers not under strong selection has been shaped in part by habitat quality. Genetic diversity is positively correlated with mountain range elevation (Epps et al. 2006), which is also linked to population persistence (Epps et al. 2004) and measures of forage availability and diet quality (Creech 2016). The strongest influence on genetic diversity in that system, however, is connectivity (Epps et al. 2006) and habitat fragmentation (Epps et al. 2005). Thus, genetic diversity in such systems reflects recent metapopulation structure and stability: larger, more connected, and more stable populations in better habitats experience lower genetic drift and thus retain more diversity.

### 6.4.2 INBREEDING

Inbreeding estimates likewise have been of interest for managing mountain sheep. Some life history characteristics of mountain sheep seemingly increase the chances of inbreeding, including small population sizes, isolated populations, and polygynous mating with dominant males siring high proportions of offspring (Coltman et al. 2002, Hedrick and Wehausen 2014). A lack of inbreeding avoidance has been demonstrated in an isolated population of bighorn sheep (Rioux-Paquette et al. 2010), despite significantly lower over-winter survival of inbred female lambs (Rioux-Paquette 2011). Additionally, many mountain sheep populations have histories of bottlenecks (Ramey et al. 2000), where some past populations dwindled to a few dozen individuals, and frequent founder effects (essentially a bottleneck) stemming from widespread and sometimes sequential translocations or natural recolonizations involving small numbers of individuals (Hedrick et al. 2001, Olson 2013, Wild Sheep Working Group 2015). In some cases, although bottlenecks are suspected based on the small size of populations, gene flow among local populations obscures the expected signal (Driscoll et al. 2015). Population-level genetic diversity estimates are often reported, vary widely among populations, and are frequently interpreted as signals of potential inbreeding, which is suspected when heterozygosity estimates are very low (Gutierrez-Espeleta et al. 2000, Olson et al. 2012, Spaan et al. 2021). Mountain sheep studies often report the inbreeding coefficient, $F_{IS}$ (Wright et al. 2024b), but that statistic is best interpreted as a signal of non-random mating, such as that resulting from population structure, rather than actual inbreeding by close relatives. Genomic approaches have allowed better estimates of inbreeding in the sense of determining what proportions of both copies of an animal's genome are identical by descent from a recent ancestor.

Because inbred individuals have more areas of their genome where both copies are identical by descent, low individual-level genetic diversity estimates (heterozygosity) may indicate a history of inbreeding, although a variety of factors influence genetic diversity (Amos and Harwood 1998). Inbreeding, in the sense of mating between close relatives, causes decreases in heterozygosity and increases in homozygosity in individuals. This increase in homozygosity may increase expression of deleterious alleles, reducing survival and reproduction of inbred individuals relative to non-inbred

ones (Poirier et al. 2019). Inbreeding can be estimated from pedigrees, as in the Ram Mountain study population in Alberta, Canada (Pigeon et al. 2017, Rioux-Paquette et al. 2010, Rioux-Paquette 2011), but long pedigrees are not practical to estimate for most wild populations. Genomic measures of inbreeding can now be estimated by quantifying how much of an individual's genome exhibits ROH, regions where both copies of DNA are identical. Furthermore, the number, distribution, and length of ROH can be used to make inferences about the demographic history of the population; longer ROH can result from recent inbreeding events while smaller ROH can indicate inbreeding events in the more distant past, although selection and recombination rates also influence ROH (Shafer and Kardos 2025). That metric has been used to associate loci with fitness-related traits as well (Stoffel et al. 2021a). Applying this metric in wild populations requires that markers be mapped and at relatively high density, sequenced with high depth, and analyzed using model-based rather than rule-based approaches (Shafer and Kardos 2025). Other methods have also been used to estimate potential inbreeding. Flesch et al. (2022) used SNPs to estimate kinship in Rocky Mountain bighorn sheep populations in Colorado, Wyoming, and Montana, from which inbreeding in the subsequent generation can be inferred. They reported that translocation history and small population sizes correlated with higher degrees of kinship and thus higher potential inbreeding, and that dispersal between populations reduced average kinship most strongly. Wright et al. (2024b) also used SNPs to evaluate kinship in reintroduced desert bighorn populations in Texas as a signal of potential inbreeding and evaluate genetic diversity, concluding there is little evidence of inbreeding problems at this time in that system.

### 6.4.3 LONG-TERM GENETIC MONITORING

While an estimate of genetic information at a single time point is informative, significant additional insights can be gained through temporal monitoring. Localized extinction, colonization, migration, or hybridization between subspecies can influence genetic structure and diversity very quickly. Across the range of mountain sheep in North America, long-term or spatially extensive genetic research has increasingly been facilitated by efforts by state, provincial, and territorial agencies to archive tissue, blood, or horn shavings (although horn is less effective as a source of high-quality DNA) from hunter-harvested animals and animals captured for telemetry, disease surveillance, or translocation (Epps et al. 2019). Bighorn sheep have benefited from several long-term research programs with a genetic component (Chapter 23). Ram Mountain in Alberta, Canada, is the foremost example. Decades of research with regular capture and handling of most individuals in the population have provided a platform for integrating genetic and other data to address fundamental questions of ecology, evolution, and behavior of bighorn sheep, including how a variety of fitness measures have changed over time. Pedigrees informed by genetic data have been used to estimate heritability of traits such as body size and horn size in male and female bighorn, explore interactions between genetics, environment, and artificial selection from harvest, evaluate a lack of inbreeding avoidance, and link paternal phenotype with offspring traits and fitness (Coltman et al. 2003, Pelletier et al. 2007, Martin et al. 2014, Pigeon et al. 2016, 2017). The National Bison Range, a largely isolated population of reintroduced bighorn, has also yielded important insights; genetic estimates of inbreeding were correlated with fitness, showing clear consequences of long-term isolation of a typically sized mountain sheep population, and likewise the subsequent rescue of the population due to outbreeding with individuals from a different source (Hogg et al. 2006, Miller et al. 2012). Miller et al. (2015b) used both Ram Mountain and the National Bison Range populations to explore linkage (gametic) disequilibrium, a phenomenon that explores non-random association of alleles due to small population size, population structure, hybridization, or other factors, and to contrast levels between native and translocated (and thus bottlenecked) populations.

Other long-term research programs with a genetic monitoring component include Sierra Nevada bighorn sheep, where microsatellite and SNP data have been used to monitor translocations, movements among restored populations, paternity of lambs, and to guide translocations of males to improve genetic diversity within restored portions of the metapopulation where lower fitness of less genetically diverse individuals had been detected (Johnson et al. 2011, Flesch et al. 2020). Genetic research and monitoring of desert bighorn in California have established genetic datasets extending to nearly every bighorn population in the state (more than 70), with temporal extents up to 30 years in some locations. Temporal comparisons within those datasets have yielded insights on new colonization, changes in connectivity and barrier-crossing behaviors, and population expansions, while the spatial extent greatly strengthens identification of origins for foraying males or other individuals observed in unexpected locations (Epps et al. 2018, 2024a).

A different long-term genetic research program that is relevant for mountain sheep managers, despite being focused on domestic sheep, comes from the island of Hirta in the St. Kilda Archipelago off the northwestern coast of Scotland. The archipelago is home to a population of Soay sheep (*O. aries*) that was founded with approximately 100 individuals in 1932 (Clutton-Brock and Pemberton 2004). Researchers have taken genetic samples and recorded trait data (e.g., reproductive success, body size traits) for most Soay sheep individuals on the island since 1985. Accurate pedigrees based on genetic and later genomic datasets (Bérénos et al. 2014) have spurred important insights into the genetic factors that are associated with ecologically relevant traits, including lower first-year survival with higher frequency of long ROH (Stoffel et al. 2021b), diminished first-year reproductive success for males born in high-density years (Chapman et al. 2023), and reduced lifespans for individuals with shorter average telomere lengths (Froy et al. 2021). The

results from this system highlight what might be possible for wild systems where we currently know far less about genetic associations for fitness and adaptive traits.

### 6.4.4 MANAGING GENETIC DIVERSITY IN TRANSLOCATED OR RESTORED POPULATIONS

Over the past century, translocations have been extensively used by mountain sheep managers to reintroduce individuals to previously occupied habitats and augment existing population sizes (Wild Sheep Working Group 2015). The population declines and extirpations of the late 1800s and early 1900s did not have equal effects on all lineages and jurisdictions, so translocations have been more commonly employed in the United States, particularly for the California (derived from western British Columbia) and desert bighorn sheep subspecies. Several studies have characterized the genetic consequences of translocations in mountain sheep populations, and a few general trends have emerged. First, many translocated herds have reduced genetic diversity relative to source populations due to founder or bottleneck effects. For example, Hedrick and Wehausen (2014) reported a sharp decline in heterozygosity in the Pilares, Mexico, captive breeding population relative to the Tiburon Island herd from which most founders descended. Moreover, Olson (2013) demonstrated that the negative consequences of translocations for genetic diversity continue to worsen when populations are successively founded from translocated populations (sometimes called serial or dilution translocations). Using multiple source populations when establishing populations can help maintain genetic diversity better than single-source translocations (Olson 2013, Jahner et al. 2019). The Pilares, Mexico, population mentioned earlier, however, was also founded by several source populations and still had reduced diversity, suggesting that the relationship between the number of source herds and genetic diversity is likely dependent on multiple factors. In Oregon, genetic diversity of several single-source California bighorn herds was successfully supplemented by augmentation with bighorn from more diverse sources (Whittaker et al. 2004, Olson et al. 2012). Finally, as noted above, genetic rescue in the National Bison Range herd occurred after new individuals arrived and raised the population's genetic diversity, resulting in higher reproductive success (Hogg et al. 2006, Miller et al. 2012). Similarly, translocation of individuals from Cadomin, Alberta, to the population at Ram Mountain, intended to offset years of demographic decline, resulted in increased genetic diversity and higher survival of outbred offspring (Poirier et al. 2019). While no mountain sheep studies to date, including those mentioned above, have reported evidence for outbreeding depression (lowered fitness when mixing distantly related individuals due in part to adaptation of gene complexes to different environments), this potential complicating factor should still be considered when translocations involving individuals, particularly those from very different environments, may occur.

## 6.5 ADAPTIVE VARIATION, GENETIC BASIS OF TRAITS, AND ADAPTIVE CAPACITY

### 6.5.1 GENETIC BASIS OF TRAITS IN BIGHORN SHEEP

Many of the first studies identifying the genetic basis of bighorn traits are based on pedigrees and microsatellite data from long-term monitoring at Ram Mountain, Alberta (Reale et al. 1999, Reale and Festa-Bianchet 2000a,b; Coltman 2005, Coltman et al. 2005, Poissant et al. 2008, 2012; Wilson et al. 2005). These studies used quantitative genetics approaches to quantify the heritability of ecologically important traits (e.g., body mass, lifespan, reproductive success, and age-specific effects), characterize genetic correlations between traits, and identify quantitative trait loci (QTL), which are regions of the genome associated with variation in a trait measured over generations of known crossings. Pedigrees in wild populations require parentage analyses based on genetic markers. Pedigrees may also be used to assess inbreeding coefficients. In parentage analyses, lamb and maternal genotypes can be compared to genotypes of putative sires in a population, enabling determination of reproductive success for males, reproductive skew, and, in well-studied populations, offering the opportunity to link phenotypic data such as horn size to mating success (Coltman et al. 2002).

As genomic sequencing data has become more common, genome-wide association studies have been more widely employed than QTL mapping. For example, Kardos et al. (2015) quantified variation at 2.8 million SNPs to identify genomic regions associated with male horn size. Based on evidence of strong positive selection, the authors concluded that horn size in bighorn sheep is associated with variation in the *RXFP2* gene. That gene is found in a putative horn size QTL identified by Poissant et al. (2012) and has also been linked to horn size variation in Soay sheep (Johnston et al. 2011), though Miller et al. (2018) did not find an *RXFP2* association with horn morphology in genome-wide association studies of Ram Mountain individuals using the Ovine SNP chip.

Disease resistance is one of the most relevant adaptive traits for mountain sheep management, and a few studies have identified associations between disease prevalence and genetic variation. Dugovich et al. (2023) reported links between individual and population-level genetic diversity, immune response, and disease prevalence in a desert bighorn metapopulation in California. Luikart et al. (2008a) reported that parasitic lungworm abundance was higher in bighorn individuals in Rocky Mountain National Park with lower mean heterozygosity for seven microsatellite loci within disease-related genes, but this relationship disappeared when considering eight neutral loci. Similarly, for the Lostine herd in Oregon, persistent carriage of *Mycoplasma ovipneumoniae* was associated with lower heterozygosity in a microsatellite locus associated with immune function (Plowright et al. 2017). Martin et al. (2021) followed up with genome-wide association studies of *M. ovipneumoniae* carriage in the Lostine herd

using about 11,000 SNP loci and were able to explain around 55% of the phenotypic variance and identify several candidate loci, albeit with relatively few individuals ($n=52$). The development of generalizable and highly predictive models explaining the genetic basis of disease status holds promise for improving the management of mountain sheep populations, but this task will be challenging while individual sample sizes are small (Jahner et al. 2024).

There is much uncertainty surrounding the potential adaptive capacity and persistence of bighorn sheep populations in the face of future climate change. Landscape genetic approaches to estimate relative climate change vulnerability for 62 desert bighorn sheep herds found on United States National Park Service lands were undertaken by Creech et al. (2020). Based on a composite metric comprised of genetic diversity, genetic and geographic isolation, rate and direction of climate change, and elevation, populations in the Mojave Desert and southeastern Utah had the highest vulnerabilities, whereas remnant herds in Death Valley and the Grand Canyon had reduced risk. Even in vulnerable populations, low adaptive capacity could be mitigated if an adaptive allele is introduced via gene flow. Creech et al. (2017) used a landscape genetic model to evaluate the potential of this scenario in desert bighorn populations, finding faster spread of adaptive variation in landscapes with more homogeneous habitats and when adaptive alleles already exist in a population and do not have to be generated by mutation.

### 6.5.2 EVOLUTIONARY RESPONSE TO HARVEST

A quantitative genetic study of harvested and unharvested male bighorn sheep at Ram Mountain by Coltman et al. (2003) yielded another important finding: artificial selection imposed by heavy harvest of males based on a minimum horn size requirement affected horn size in that population. About 40% of legal-sized males were harvested each year, often before they had reproduced, and particularly in the case of males with the fastest-growing horns who were younger when they reached the horn size threshold for legal harvest and who had never contributed genetically to the population. Although environmental factors including density dependence likely also played a role in the 20-cm decline in horn length observed between 1972 and 2002, Coltman et al. (2003) determined that horn size in 2- to 4-year-old males was highly heritable, and that about 2.5 cm of the horn length decrease was attributable to a decline in the heritable component of horn quality. Although the conclusions of this study have been challenged (Boyce and Krausman 2018), later reanalysis of these data has supported the original conclusions (Morrissey et al. 2025, Pigeon et al. 2016). That level of artificial selection resulted from harvest regulations that allow very abundant tags and high harvest pressure, relying on minimum horn size restrictions (often of four-fifth curl) to limit the number of animals harvested. In other regions of North America, in contrast, harvest managers issue very few tags so harvest pressure remains low, restrict male harvest to full-curl individuals, or both.

### 6.5.3 Genetic Basis of Traits in Thinhorn Sheep

Relatively few studies have examined the genetic basis of traits in thinhorn sheep compared to bighorn sheep. The genetic basis of coat color has interested thinhorn sheep researchers since it plays a prominent role in subspecies identification. Loehr et al. (2008) examined the role of *MC1R* variants in coat color determination by sequencing the gene in 40 thinhorn sheep from across the species distribution. The *MC1R* gene is involved in the regulation of skin and hair color in many vertebrate species (Hofreiter and Schöneberg 2010). Loehr et al. (2008) reported no association between MC1R genotype and coat color. The outcomes of this study were likely affected by low sample size ($n = 40$), and results could have been confounded by strong population structure, since samples were collected over a large geographical area spanning over 1,200 km.

Worley et al. (2006) investigated three immune-associated genes, including major histocompatibility complex class II gene (DR molecule β-chain; DRB), interferon gamma (*IFNG*) gene, and natural resistance-associated macrophage polymorphism (*NRAMP*) gene, for signatures of selection. They collected data from genotyping microsatellite markers linked to the three genes and direct sequencing. Linked microsatellite data showed evidence of balancing selection in all three genes, but tests were non-significant after accounting for population structure. Translated coding sequence data was fixed for IFNG NRAMP. Sequence variation at the major histocompatibility complex DRB gene provided evidence for balancing selection through overdominance, where heterozygous individuals have higher fitness, based on an excess of non-synonymous substitutions and the presence of polymorphisms shared between thinhorn sheep and other ungulates examined (Worley et al. 2006).

Sim and Coltman. (2019) investigated heritability in horn size by genotyping 192 Dall's sheep from Yukon using the OvineHD SNP array. Horn length, base circumference, and horn volume data were obtained from compulsory inspections of harvested sheep. Heritability estimates for horn length, base circumference, and volume were 0.33 (SE = 0.02), 0.36 (SE = 0.03) and 0.36 (SE = 0.02), respectively, which are similar to estimates found in bighorn sheep (Miller et al. 2018). Genome-wide association analysis found two SNP loci to be suggestively associated with horn length located in the Glycosylphosphatidylinositol family receptor alpha-2 (*GFRA2*) and Fidgetin-like protein 2 (*FIGNL2*) genes. The authors cautioned that the associations were suggestive and saw no obvious link between the functions of those gene regions and horn size.

Roffler et al. (2016) assayed a set of candidate SNPs thought to be involved in immunological, metabolic, growth, and environmental signaling functions using microfluidic quantitative PCR-based SNP chips. The SNP discovery was performed using exon capture, which assesses sequence from coding genes in bighorn sheep and the domestic sheep genome, to identify regions of the thinhorn genome most likely to contain genes of adaptive significance. The authors identified 15–56 SNPs (depending on stringency of thresholds and analytical approach) to be potentially under selection according to $F_{ST}$ outlier analyses. Of the 15 most strongly supported markers for signatures of selection, nine were found to be under directional selection (favoring a particular allele), while six were thought to be under balancing selection (favoring diversity of alleles). These markers have roles in immune function, signaling for regulation, cellular processes, tumor suppression, respiratory health, reproduction, and olfactory receptors. All nine loci under directional selection were significantly associated with climatic variables, most notably annual mean temperature.

## 6.6 OTHER APPLICATIONS OF GENETICS

Genetic applications have strengthened other areas of mountain sheep research in recent decades. The adoption of PCR-based approaches for disease testing, particularly for respiratory pathogens, has allowed breakthroughs in diagnosing key respiratory pathogens, identifying sources, infection pathways, and patterns of infection, and consequences for population demography and management in mountain sheep that were not possible when culturing was the only means of detection. Since the importance of *M. ovipneumoniae* in the progression of respiratory pneumonia in bighorn was determined (Besser et al. 2008), screening for active infections during captures by swabbing nasal passages followed by quantitative PCR for pathogen detection and strain typing has elucidated the spread, extent, and strain types present in mountain sheep populations across multiple jurisdictions (Shirkey et al. 2021). Removal of sick or chronically infected individuals identified in that manner can improve population performance (Garwood et al. 2020, Wild Sheep Initiative 2023).

Other genetic-based techniques applied to mountain sheep include diet metabarcoding, microbiome analysis, and screening for parasites. For each, DNA is extracted from feces collected either during captures or after deposit by free-ranging individuals. Rather than targeting the DNA of the host, however, next-generation sequencing approaches are used to amplify DNA from bacteria present in the digestive tract (i.e., microbiome analysis; Couch et al. 2020), parasite DNA, the DNA of plants consumed by the animal (i.e., metabarcoding; Blum et al. 2023), or combinations of those elements (Nagati et al. 2024). In these analyses, sequences recovered from a sample are classified by screening them against public reference databases, although not all taxa are represented, which may challenge identification of diet items, parasites, or microbes to lower taxonomic levels. Gut microbes offer opportunities to explore relationships between nutrition, host population structure, disease, host genetics, and other factors that may be relevant to conservation and management (Couch and Epps 2022).

Molecular markers can be a powerful tool in the prosecution of wildlife crimes, making forensics a practical application of genetics to mountain sheep conservation. This is especially applicable to illegal hunting of mountain sheep since these species inhabit very remote areas where witnesses to the crime typically are not present. In such cases, the use of genetics may be the only way to link the suspect to the incident. The Genetic markers can be used in three broad ways: DNA matching, species identification, and population assignment. In DNA matching, DNA typing profiles (genotypes) from biological evidence recovered during an investigation can be compared to determine if they match. For example, microsatellite markers can be used to develop genotypes from blood found at an illegal kill site and antlers or horns seized from a suspect's residence, which can then be compared (Sim et al. 2021). Molecular tests can be used to identify the species origin of a biological material (Dawnay et al. 2007), as when mtDNA markers, such as cytochrome oxidase I, are used to determine the species identity of meat found in a suspect's freezer. Finally, population assignment tests using molecular markers can be used to estimate the origin of an individual, and, when combined with suitably extensive population allele frequency databases, can help investigators determine if a mountain sheep was harvested from a population that is protected from hunting (Manel et al. 2002).

## 6.7   NEW DIRECTIONS

Genetic methodologies have provided key insights into the biology, management, and conservation of North American mountain sheep. As sequencing technologies and analytical methods continue to advance, and as the associated costs for genome-scale datasets continue to decrease, novel avenues of research will become accessible to mountain sheep researchers. In this section, we review potential new directions.

### 6.7.1   Revising Taxonomic Classification and Historical Distribution of Subspecies

Much work remains to produce updated phylogenies of mountain sheep species, and below the species level, given the debates around past and present subspecies designations. Nuclear phylogenies based on genome-wide markers or whole-genome resequencing would best capture species history, whereas previous work based on mtDNA describes the history of those genes or genomic regions. A recent analysis (Jahner et al. 2025) used SNPs to evaluate relationships among desert, Sierra, Rocky Mountain, and bighorn originating in western British Columbia, previously considered part of California bighorn subspecies (Cowan 1940, Wehausen and Ramey 2000), and reported much deeper separation between those populations of California bighorn and Rocky Mountain bighorn populations than estimated by mtDNA analysis (Ramey II 1993, Wehausen and Ramey II 2000, Buchalski et al. 2016). Such analyses may be complicated by the likelihood of periodic gene flow among subspecies or even species (Santos et al. 2021), but will prove a key foundation for managing restored and native populations to maximize conservation of genomic variation associated with adaptation to

the wide variety of environments used by mountain sheep. Work based on larger portions of the mitochondrial genome (Davenport et al. 2018), rather than just the fast-evolving control region, which thus offers limited phylogenetic inference, could also be illuminating.

A related need is to update the range maps of the most divergent lineages or subspecies, to better guide subsequent restoration or management of existing populations (Figure 6.2). Cowan's (1940) taxonomy of subspecies in bighorn sheep was based on skull morphology; subsequent revisions by Ramey II (1993) and Wehausen and Ramey II (1993, 2000) used skull morphometrics and limited mtDNA sequence information. Later restoration relied largely on Cowan's maps, leading to potential ecological mismatch: for instance, California bighorn from British Columbia were used to restore most bighorn populations in Oregon, although arid habitat, connectivity to desert bighorn range, and some morphometric data suggest desert bighorn would have been more appropriate (Malaney et al. 2015, Wehausen and Ramey II 2000). Analysis of DNA from historical specimens from this region that predate extirpation, and other regions where original subspecies identities are unclear, could help guide decisions moving forward (Wright et al. 2024a). Specimens from museum collections, held in agency collections, recently discovered and still in remote field sites, or well-documented personal collections, can yield usable DNA. Such DNA is more difficult to isolate and is often heavily fragmented, but can yield sufficient sequence to link specimens, including bones or even well-preserved hair or feces, to lineages or subspecies defined by mtDNA or nDNA (Wilder et al. 2014). In some cases, museum specimens have allowed comparison of genetic diversity from historical (e.g., 19th-century) mammal populations to those of the present day (Rubidge et al. 2012).

### 6.7.2   Local Adaptation among Subspecies/Ecotypes

Managing mountain sheep populations, particularly in response to the widespread extirpations of the 19th and 20th centuries, has raised many questions about local adaptation (Wiedmann and Sargeant 2014, Bleich et al. 2018, Epps et al. 2018, Jahner et al. 2019). Subspecies are generally expected to reflect adaptation to different environments, such as the different temperature extremes experienced by desert and Sierra Nevada bighorn. Criteria for classifying mountain sheep subspecies have historically relied on morphometric data (Cowan 1940, Ramey II 1993, Wehausen and Ramey II 2000, Wehausen et al. 2005). Yet, local adaptation also can occur at levels below that of recognized subspecies, even at the population level (Homola et al. 2019, Farleigh and Jezkova 2023). Thus, finding evidence of local adaptation could help explain speciation among mountain sheep taxa and guide subsequent management or conservation actions. Moreover, human interventions and early and progressive management actions described by Hurley et al. (2015) may affect how local adaptations are represented in gene pools and expressed. Here, we explore three broad approaches to linking genetic data to local adaptation: evaluating gene expression differences, assessing genotype-environment association, and outlier analysis.

Until recently, mountain sheep genetic researchers have predominantly considered genetic variation at the level of changes in DNA sequence. Another level of genetic variation is that of gene expression, where researchers determine which genes are actively producing proteins within specific tissues or individuals. Such differences in gene expression are more dynamic concerning the immediate environment. Thus, comparisons of gene expression among species or populations can reveal the genetic architecture of speciation or local adaptation (Homola et al. 2019). Among individuals, differential gene expressions can reveal different physiological states, or, as expression often differs across tissues, indicate how organs respond to stressors or stimuli (Jia et al. 2022). Assessing gene expression requires capturing RNA, the molecule used by cells to copy a gene that will be expressed; that RNA is then carried to the protein-building organelles in the cell to initiate protein synthesis. The RNA is less stable than DNA, and gene expression is specific to tissue, time, individual, and place; thus, study design is paramount and the potential for confounding factors is high if not properly controlled. Broad assessment of gene transcription through a technique called RNA-seq (Mortazavi et al. 2008) can be used to generate a transcriptome; this provides a snapshot of relative levels of expression for all genes expressed within a given tissue. The various levels of transcription in immune genes can indicate that individuals are at different stages of a disease outbreak, or that some animals are chronic carriers of a pathogen, while others are not. Gene expression in many tissues can only be evaluated by very invasive procedures or sacrifice of the animal; thus, use in mountain sheep populations has been limited to expression of genes in blood cells (Bowen et al. 2022). Transcriptomics is now widely employed in domestic sheep to explore the genetic basis for controls of processes such as reproduction or metabolism, or effects of other factors such as nutrition on gene expression (An et al. 2024, Chen et al. 2024, Zhang et al. 2024). Currently, we know of no published research on mountain sheep in North America assessing transcriptomes, but Upadhyay et al. (2020) developed a snow sheep transcriptome as part of genome assembly for that species.

Gene expression can be assessed in a more targeted fashion using quantitative PCR: primers attach to specific sequences, such as RNA from genes of interest, and then use the rate at which copies are generated during PCR to assess the relative quantities of that sequence present in the original sample (Heid et al. 1996). Bowen et al. (2020, 2022) used this approach to develop a panel of gene transcripts associated with immune function in bighorn sheep and identified clear differences in transcription patterns across populations and between bighorn uninfected or chronically infected with the respiratory pathogen *M. ovipneumoniae*.

Genotype-environment association studies have a different goal. This work assesses whether DNA variation in a species, subspecies, or array of populations is associated with environmental conditions, such as precipitation, temperature, vegetation, or other habitat characteristics (Rellstab et al. 2015, Capblancq and Forester 2021). Loci that show allele frequencies strongly correlated with specific environmental conditions are then hypothesized to have been under selection for those conditions, although reliable inference from these patterns may be challenging (Lotterhos and Whitlock 2015, Hoban et al. 2016). An alternative approach, outlier analysis, scans for genetic markers that have significantly different patterns of differentiation than predicted under a model of neutral evolution (i.e., random processes versus natural selection; Hoban et al. 2016). Once biotic and abiotic factors associated with local adaptation are identified, a forecasting approach can be taken where predictions of how these factors change over time (e.g., from climate change models) can be used to highlight populations that are at increased risk of being maladapted to future conditions, thus exhibiting "genomic offsets" (Rellstab et al. 2021). As-yet unpublished efforts to apply genotype-environment association studies or outlier analysis to bighorn sheep have been complicated by the confounding of unique environmental factors with strong and hierarchical genetic structure, a very common outcome in mountain sheep as noted above, and translocations and loss of native bighorn sheep from large portions of their historical range.

### 6.7.3 New Insights for Genetic Management

An important need is to modernize our concept of inbreeding and inbreeding depression. As discussed above, $F_{IS}$, the traditional statistic reported, is unlikely to be predictive of inbreeding depression. The ROH statistic shows more promise, but as noted also reflects other factors (Shafer and Kardos 2025). Alternatively, genetic load, a measure of the number of deleterious alleles in an individual or population, is emerging as an important conservation genetic metric (Wilder et al. 2024). Overall genetic diversity may not always be predictive of conservation status, as some species or populations have existed at small population sizes with low diversity for many generations, while others have purged deleterious variants during population declines that led to genetic bottlenecks (Mathur and DeWoody 2021, Mathur et al. 2023). Thus, efforts to quantify how much of the standing genetic variation in populations and individuals are likely to be deleterious (i.e., their genetic load) are of increasing importance (Taylor et al. 2024). Genetic load can be estimated from genomic data by comparing variants among the target species and multiple outgroups, with the assumption that alleles private to the target species are recently derived versus ancestral (Quinn et al. 2024). The deleterious effects of derived alleles on protein function can then be predicted using gene annotations and a high-quality reference genome (Ng and Henikoff 2003, Cingolani et al. 2012), such as the domestic sheep genome (Davenport et al. 2022). Estimates of load then can be factored into breeding programs or translocation efforts (Wilder et al. 2024).

Epigenetics, or the study of how gene expression is modified for an existing DNA sequence, is another developing area of genetic research with applicability to wildlife. A key measure is assessing methylation, a modification of an individual's DNA that switches expression of a gene on or off. Methylation can be influenced by environmental conditions experienced by the individual or a parent and increases with age (De Paoli-Iseppi et al. 2017). Thus, one application with promise for mountain sheep, given the difficulty of reliably aging animals using horn rings or cementum analysis (Wehausen et al. 2024), would be aging of wild individuals based on an epigenetic clock developed by assessing methylation in known-age individuals, as has been achieved for taxa such as brown bears (*Ursus arctos*) using blood samples (Nakamura et al. 2023), mountain goats (*Oreamnos americanus*), and other taxa, with methylation also varying by sex (Czajka et al. 2024). Studies of other domestic or wild mammals report epigenetic markers of stress resulting from differences in handling or chemical exposures (Crossman et al. 2021, Narayan et al. 2022), suggesting other applications for epigenetic methods in the study of disease, response to human activity, nutrition, and other important factors influencing mountain sheep.

## 6.8   SUMMARY

Genetic research on mountain sheep has a rich past and a promising future. Their conservation status and high value as a game species, complicated evolutionary history, small population sizes, and close similarity to a well-researched model species (domestic sheep) have generated much creative research. Some studies likely have influenced management of other wildlife species, as they are commonly cited case studies of processes such as fragmentation, genetic rescue, or artificial selection from harvest (Coltman et al. 2003, Epps et al. 2005, Hogg et al. 2006). Strong collaborative relationships across state, provincial, and territorial agencies, academic institutions, and non-governmental organizations in North America have facilitated large-scale studies and effective research programs using genetic tools or addressing genetically related questions. Improvements in sample banking and archiving will aid future work (Epps et al. 2019). We urge continued attention to generating publicly archived genetic data, well-documented methods, and standardization of approaches to foster integration of genetic assessments with studies of disease, demography, nutrition, population structure, and harvest, to improve our fundamental understanding of species in the wild and how best to manage them.

Herein, we systematically reviewed genetic and genomic research on wild populations of bighorn and thinhorn sheep. We begin with a brief history and explanation of methodological approaches, supported with a glossary. We synthesize genetic evidence for differentiation of these species from Old World species, previously and currently recognized subspecies, and lower-level phylogenetic divisions. We present a map estimating revised historical ranges for bighorn sheep

subspecies. We then describe how genetics has informed estimation of population history, structure, connectivity, and demographics on ecological time scales, review long-term studies, address measures of genetic diversity and inbreeding, and how they have been and will be used to inform management of indigenous, restored, or translocated populations. We review research on adaptive variation and the genetic basis of traits and other genetics-based tools for study of mountain sheep including disease testing, gut microbiomes, and diet characterization. Finally, we consider recent developments or potential new directions including work on taxonomy, local adaptation, better measures of inbreeding depression, gene expression, and epigenetics.

## GLOSSARY OF TERMS

*Admixture* is the mixing of genetic ancestry between one population or lineage and another.

*Allele* is a potential version of a genetic marker; examples include different nucleotides (A, C, G, T), fragment lengths, or repeat lengths.

*Allozyme* is a molecular marker based on physical size differences of enzymes.

*Bottleneck* is a loss of genetic variation as a result of a reduction in population size, or founder effect (where only a small group of individuals establishes a new population).

*Chromosome* is a long stretch of DNA contained in the nucleus.

*Ecotype* is a distinct population or lineage that reflects localized adaptations (i.e., not occurring throughout the species) to particular habitats or environments.

*Effective population size* is the size of an ideal population experiencing the same amount of genetic drift as the real population; because real populations of most species exhibit non-random mating, uneven sex ratios, overlapping generations, and other factors that influence genetic drift, effective population sizes are typically much lower than the number of individuals alive in a population at any given time.

*Epigenetics* is the study of genetic factors (e.g., histone methylation) that do not involve physical changes to DNA.

$F_{IS}$ is often referred to as the inbreeding coefficient, this statistic is a measure of the reduction in heterozygosity in individuals relative to the subpopulation in which they are found; as such, it usually reflects non-random mating caused by unrecognized population substructure, and is not a good metric of inbreeding resulting from the mating of closely related individuals.

$F_{ST}$ is the fixation index, or a relative measure of differentiation among populations and lineages.

*Gene expression* is the cellular production of proteins or other functional products that are coded by particular DNA sequences.

*Gene transcription* is an early phase of gene expression, when ribonucleic acid (RNA) is produced by the cell from the DNA sequence of the gene that will be expressed; gene transcription studies rely on capturing RNA to determine what genes are being expressed in specific tissues at the time of sampling.

*Genetic diversity* is a metric similar to species diversity that measures the degree of heterozygosity in a population.

*Genetic drift* is a random shift in population allele frequencies across generations; this effect is much stronger in small populations.

*Genetic load* is genetic variants found in an individual or population that, if expressed, reduce fitness.

*Genetics* is the study of heritable variation derived from differences in DNA or RNA sequences in living organisms.

*Genome-wide association study* is an analysis that identifies regions of the genome that are associated with trait variation in a population.

*Genomics analyses* assess variation over a large proportion of the genome, ranging from thousands of markers to full genome sequences; this term is largely used as a contrast to studies based on a few dozen loci or sequence data from a single or small number of genes.

*Genotype* is an assessment of the alleles (variants) present in an individual at a genetic marker or across multiple genetic markers

*Haplotype* is one potential genetic sequence for a gene.

*Heterozygous* is a genetic state in which an individual has two different alleles at a given locus; *heterozygosity* is a measure of genetic diversity that reflects the proportion of markers that are heterozygous in an individual or a population.

*Homozygosity* is a genetic state in which an individual has two copies of the same allele at a given locus.

*Hybridization* is interbreeding between individuals from different lineages.

*Identical by descent* occurs when a particular genetic variant (allele) occurs in both genome copies of a diploid individual, or occurs in multiple individuals, due to inheritance from a common ancestor.

*Inbreeding depression* is a reduction in fitness caused by mating between closely related individuals.

*Isolation by distance* is a signature of genetic structure in which geographically close individuals or populations are more genetically similar than those that are more distant.

*Karyotype* is the number and physical makeup of chromosomes contained in the nucleus, which can differ between species.

*Linkage disequilibrium* is sometimes referred to as gametic disequilibrium, correlation of allele frequencies between loci or genomic regions resulting from loci that are physically close on the same chromosome, non-random mating due to population substructure, or other factors.

*Locus (plural loci)* is a genomic region showing sequence variation within or across individuals that serves as a genetic marker in population genetics and related applications.

*Major histocompatibility complex (MHC)* is a group of genes with known immune function that have commonly been investigated in conservation genetic studies.

*Microsatellite (e.g., simple sequence repeats; SSRs)* is a molecular marker that is made up of short tandem repeats of a DNA motif (e.g., ATCATCATC, representing three repeats of the motif ATC); individuals are scored for how many copies of the repeat they have in each copy of their genome.

*Migration rate* is an estimate of the rate of gene flow from one population (or lineage, in a phylogenetic context) to another in each generation.

*Mitochondrial DNA (mtDNA)* is a single-copy DNA found in the mitochondria of mammalian cells that has been used frequently to infer phylogenies of taxa at or below the level of species and to date genetic divergence times, and that is maternally inherited and does not undergo recombination as does nuclear DNA, thus, the entire mtDNA genome acts as a single genetic marker, which limits phylogenetic inference.

*Molecular dating* is a method of estimating the age of evolutionary events by comparing the accumulated changes in homologous DNA regions among different lineages.

*Monophyly* is a taxonomic grouping consisting of a set of lineages derived from a single common ancestor.

*Nuclear DNA* is the DNA found in the nucleus of multicellular organisms organized into several chromosomes; one copy of each chromosome is inherited from each parent.

*Outbreeding depression* is a reduction in fitness caused by gene flow or hybridization between highly divergent lineages.

*Paraphyly* is a taxonomic grouping consisting of some, but not all, lineages derived from a common ancestor.

*Pedigree* is a description of the relationships between members of a family through generations.

*Phylogenetics* is the study of how different evolutionary lineages diversified through time.

*Phylogeny* is a genetic hypothesis of how lineages diversified through time and how they are related to one another, often shown as a branching tree.

*Phylogeography* is the study of how geographic factors (mountain ranges, glaciers) have shaped the genetic structure of populations and lineages over time.

***Pleistocene glaciations*** is a period of repeated glacial cycles that occurred from about 2.6 million years ago to 11,000 years ago.

***Polyphyly (polyphyletic)*** occurs when taxa are treated as a group due to apparent similarity but have widely separate evolutionary origins; thus, the group does not include the most recent common ancestor to that group or all descendants of that ancestor.

***Population structure (genetic)*** is systematic differences in distribution of genetic variants among groups of organisms.

***Quantitative genetics*** is the study of how genetic factors affect trait variation among individuals in a population.

***Quantitative trait loci (QTL)*** are regions of the genome that affects trait variation among individuals in a population.

***Refugia (glacial)*** are areas with favorable conditions where organisms can survive during glacial periods.

***Restriction fragment length polymorphism (RFLP)*** is a method of assessing DNA sequence variation where restriction enzymes are used to cut the DNA at locations that match a particular base pair sequence for each enzyme; DNA sequence differences lead to different patterns of resulting fragments; fragments are visualized through electrophoresis, offering a way to evaluate certain types of sequence variation without directly sequencing the DNA; RFLPs provide a lower resolution picture of differences among sequences than can be achieved by direct sequencing.

***Ribonucleic acid (RNA)*** is a molecule that is part of the process of taking the information encoded in DNA and translating it into proteins.

***Runs of homozygosity (ROH)*** are long stretches of the genome that have no variation (homozygous) that can indicate a history of inbreeding.

***Selection*** is the process where survival of individuals is increased through the presence of a gene or trait that confers a fitness advantage; balancing selection maintains multiple variants of a gene in a population; directional selection favors a single variant in the population.

***Single-nucleotide polymorphism (SNP)*** is a molecular marker based on nucleotide variation (A, C, T, G) at a single location in the genome.

***SNP array*** is a type of DNA microarray that is used to detect SNPs.

***Whole-genome (re)sequencing*** is a method of determining all or nearly all of the DNA sequence in a genome.

## REFERENCES

Allendorf, F. W., P. A. Hohenlohe, and G. Luikart. 2010. Genomics and the future of conservation genetics. *Nature Reviews Genetics* 11:697–709.

Amos, W., and J. Harwood. 1998. Factors affecting levels of genetic diversity in natural populations. *Philosophical Transactions of the Royal Society of London Series B-Biological Sciences* 353:177–186.

An, L. X., Y. Y. Pan, M. J. Yuan, Z. H. Wen, L. Y. Qiao, W. W. Wang, J. H. Liu, B. J. Li, and W. Z. Liu. 2024. Full-length transcriptome and gene expression analysis of different *Ovis aries* adipose tissues reveals transcript variants involved in lipid biosynthesis. *Animals* 14:7.

Appels, R., R. Morris, B. S. Gill, and C. E. May. 1998. *Chromosome Biology*. Kluwer Academic, Boston, Massachusetts, USA.

Avise, J. C. 2000. *Phylogeography: The History and Formation of Species*. Harvard University Press, Cambridge, Massachusetts, USA.

Bailey, V. 1912. A new subspecies of mountain sheep from western Texas and northeastern New Mexico. *Proceedings of the Biological Society of Washington* 25:109–110.

Barbosa, S., K. R. Andrews, R. B. Harris, D. S. Gour, J. R. Adams, E. F. Cassirer, H. M. Miyasaki, H. M. Schwantje, and L. P. Waits. 2021. Genetic diversity and divergence among bighorn sheep from reintroduced herds in Washington and Idaho. *Journal of Wildlife Management* 85:1214–1231.

Bérénos, C., P. A. Ellis, J. G. Pilkington, and J. M. Pemberton. 2014. Estimating quantitative genetic parameters in wild populations: a comparison of pedigree and genomic approaches. *Molecular Ecology* 23:3434–3451.

Besser, T. E., E. F. Cassirer, K. A. Potter, J. VanderSchalie, A. Fischer, D. P. Knowles, D. R. Herndon, F. R. Rurangirwa, G. C. Weiser, and S. Srikumaran. 2008. Association of *Mycoplasma ovipneumoniae* infection with population-limiting respiratory disease in free-ranging Rocky Mountain bighorn sheep (*Ovis canadensis canadensis*). *Journal of Clinical Microbiology* 46:423–430.

Biek, R., and L. A. Real. 2010. The landscape genetics of infectious disease emergence and spread. *Molecular Ecology* 19:3515–3531.

Bleich, V. C., G. A. Sargeant, and B. P. Wiedmann. 2018. Ecotypic variation in population dynamics of reintroduced bighorn sheep. *Journal of Wildlife Management* 82:8–18.

Bleich, V. C., J. D. Wehausen, and S. A. Holl. 1990. Desert-dwelling mountain sheep: conservation implications of a naturally fragmented distribution. *Conservation Biology* 4:383–390.

Blum, M. E., K. M. Stewart, M. Cox, K. T. Shoemaker, J. R. Bennett, B. W. Sullivan, B. F. Wakeling, and V. C. Bleich. 2023. Variation in diet of desert bighorn sheep (*Ovis canadensis nelsoni*): tradeoffs associated with parturition. *Frontiers in Ecology and Evolution* 10:1071771.

Bowen, L., K. Longshore, P. Wolff, R. Klinger, M. Cox, S. Bullock, S. Waters, and A. K. Miles. 2020. Gene transcript profiling in desert bighorn sheep. *Wildlife Society Bulletin* 44:323–332.

Bowen, L., K. Manlove, A. Roug, S. Waters, N. LaHue, and P. Wolff. 2022. Using transcriptomics to predict and visualize disease status in bighorn sheep (*Ovis canadensis*). *Conservation Physiology* 10:coac046.

Boyce, M. S. and P. R. Krausman. 2018. Controversies in mountain sheep management. *Journal of Wildlife Management* 82:5–7.

Boyce, W. M., P. W. Hedrick, N. E. Muggli-Cockett, S. Kalinowski, M. C. T. Penedo, and R. R. Ramey. 1997. Genetic variation of major histocompatibility complex and microsatellite loci: a comparison in bighorn sheep. *Genetics* 145:421–433.

Boyce, W. M., R. R. Ramey, T. C. Rodwell, E. S. Rubin, and R. S. Singer. 1999. Population subdivision among desert bighorn sheep (*Ovis canadensis*) ewes revealed by mitochondrial DNA analysis. *Molecular Ecology* 8:99–106.

Bradley, W. G., and L. R. Baker. 1967. Range of variation in Nelson bighorn sheep from the Desert Game Range and its taxonomic significance. *Desert Bighorn Council Transactions* 11:114–140.

Brunsfeld, S., J. Sullivan, D. E. Soltis, and P. S. Soltis. 2001. Comparative phylogeography of northwestern North America: a synthesis. Pages 319–339 *in* J. Silvertown, and J. Antonovics, editors. *Integrating Ecology and Evolution in a Spatial Context*. Blackwell Science, London, United Kingdom.

Buchalski, M. R., A. Y. Navarro, W. M. Boyce, T. W. Vickers, M. W. Tobler, L. A. Nordstrom, J. A. Garcia, D. A. Gille, M. C. T. Penedo, O. A. Ryder, et al. 2015. Genetic population structure of Peninsular bighorn sheep (*Ovis canadensis nelsoni*) indicates substantial gene flow across US-Mexico border. *Biological Conservation* 184:218–228.

Buchalski, M. R., B. N. Sacks, D. A. Gille, M. C. T. Penedo, H. B. Ernest, S. A. Morrison, and W. M. Boyce. 2016. Phylogeographic and population genetic structure of bighorn sheep (*Ovis canadensis*) in North American deserts. *Journal of Mammalogy* 97:823–838.

Buchanan, F. C., R. P. Littlejohn, S. M. Galloway, and A. M. Crawford. 1993. Microsatellites and associated repetitive elements in the sheep genome. *Mammalian Genome* 4:258–264.

Buechner, H. K. 1960. The bighorn sheep in the United States: its past, present, and future. *Wildlife Monographs* 4:1–174.

Bunch, T. D. 1978. Fundamental karyotype in domestic and wild species of sheep: identity and ranking of autosomal acrocentrics involved in biarmed formations. *Journal of Heredity* 69:77–80.

Bunch, T. D., C. Wu, Y. P. Zhang, and S. Wang. 2006. Phylogenetic analysis of snow sheep (*Ovis nivicola*) and closely related taxa. *Journal of Heredity* 97:21–30.

Capblancq, T., and B. R. Forester. 2021. Redundancy analysis: a Swiss Army Knife for landscape genomics. *Methods in Ecology and Evolution* 12:2298–2309.

Catto, N., D. G. E. Liverman, P. T. Bobrowsky, and N. Rutter. 1996. Laurentide, Cordilleran, and Montane glaciation in the western Peace River Grande Prairie Region, Alberta and British Columbia, Canada. *Quaternary International* 32:21–32.

Chapman, E. G., J. G. Pilkington, and J. M. Pemberton. 2023. Correlates of early reproduction and apparent fitness consequences in male Soay sheep. *Ecology and Evolution* 13:e10058.

Chen, Y. L., R. Wang, R. Pang, Z. P. Sun, X. L. He, W. H. Tang, J. Y. Ou, H. M. Yi, X. Cheng, J. H. Chen, Y. et al. 2024. Transcriptome-based revelation of the effects of sleep deprivation on hepatic metabolic rhythms in Tibetan sheep (*Ovis aries*). *Animals* 14:3165.

Chen, Z. H., Y. X. Xu, X. L. Xie, D. F. Wang, D. Aguilar-Gómez, G. J. Liu, X. Li, A. Esmailizadeh, V. Rezaei, J. Kantanen, et al. 2021. Whole-genome sequence analysis unveils different origins of European and Asiatic mouflon and domestication-related genes in sheep. *Communications Biology* 4:1307.

Cingolani, P., A. Platts, L. L. Wang, M. Coon, T. Nguyen, L. Wang, S. J. Land, X. Y. Lu, and D. M. Ruden. 2012. A program for annotating and predicting the effects of single nucleotide polymorphisms, SnpEff: SNPs in the genome of *Drosophila melanogaster* strain w[1118]; *iso*-2; *iso*-3. *Fly* 6:80–92.

Clark, J. L. 1964. *The Great Arc of the Wild Sheep*. University of Oklahoma Press, Norman, Oklahoma, USA.

Clutton-Brock, T. H., and J. M. Pemberton, editors. 2004. *Soay Sheep: Dynamics and Selection in an Island Population*. Cambridge University Press, Cambridge, United Kingdom.

Coltman, D. W. 2005. Testing marker-based estimates of heritability in the wild. *Molecular Ecology* 14:2593–2599.

Coltman, D. W., M. Festa-Bianchet, J. T. Jorgenson, and C. Strobeck. 2002. Age-dependent sexual selection in bighorn rams. *Proceedings of the Royal Society B-Biological Sciences* 269:165–172.

Coltman, D. W., P. O'Donoghue, J. T. Hogg, and M. Festa-Bianchet. 2005. Selection and genetic (co)variance in bighorn sheep. *Evolution* 59:1372–1382.

Coltman, D. W., P. O'Donoghue, J. T. Jorgenson, J. T. Hogg, C. Strobeck, and M. Festa-Bianchet. 2003. Undesirable evolutionary consequences of trophy hunting. *Nature* 426:655–658.

Couch, C. E., H. K. Arnold, R. S. Crowhurst, A. E. Jolles, T. J. Sharpton, M. F. Witczak, C. W. Epps, and B. R. Beechler. 2020. Bighorn sheep gut microbiomes associate with genetic and spatial structure across a metapopulation. *Scientific Reports* 10:6582.

Couch, C. E., and C. W. Epps. 2022. Host, microbiome, and complex space: applying population and landscape genetic approaches to gut microbiome research in wild populations. *Journal of Heredity* 113:221–234.

Cowan, I. M. 1940. The distribution and variation in the native sheep of North America. *America Midland Naturalist* 24:505–580.

Creech, T. G. 2016. Predicting diet quality and genetic diversity of a desert-adapted ungulate with NDVI. *Journal of Arid Environments* 127:160–170.

Creech, T. G., C. W. Epps, E. L. Landguth, J. D. Wehausen, R. S. Crowhurst, B. Holton, and R. J. Monello. 2017. Simulating the spread of selection-driven genotypes using landscape resistance models for desert bighorn sheep. *Plos One* 12:e0176960.

Creech, T. G., C. W. Epps, R. J. Monello, and J. D. Wehausen. 2014. Using network theory to prioritize management in a desert bighorn sheep metapopulation. *Landscape Ecology* 29:605–619.

Creech, T. G., C. W. Epps, J. D. Wehausen, R. S. Crowhurst, J. R. Jaeger, K. Longshore, B. Holton, W. B. Sloan, and R. J. Monello. 2020. Genetic and environmental indicators of climate change vulnerability for desert bighorn sheep. *Frontiers in Ecology and Evolution* 8:279.

Crossman, C. A., L. G. Barrett-Lennard, and T. R. Frasier. 2021. An example of DNA methylation as a means to quantify stress in wildlife using killer whales. *Scientific Reports* 11:16822.

Curik, I., M. Ferencakovic, and J. Sölkner. 2014. Inbreeding and runs of homozygosity: a possible solution to an old problem. *Livestock Science* 166:26–34.

Czajka, N., J. M. Northrup, M. J. Jones, and A. B. A. Shafer. 2024. Epigenetic clocks, sex markers and age-class diagnostics in three harvested large mammals. *Molecular Ecology Resources* 24:e13956.

Davenport, K. M., D. M. Bickhart, K. Worley, S. C. Murali, M. Salavati, E. L. Clark, N. E. Cockett, M. P. Heaton, T. P. L. Smith, B. M. Murdoch, and B. D. Rosen. 2022. An improved ovine reference genome assembly to facilitate in-depth functional annotation of the sheep genome. *Gigascience* 11:giab096.

Davenport, K. M., M. Duan, S. S. Hunter, D. D. New, M. W. Fagnan, M. A. Highland, and B. M. Murdoch. 2018. Complete mitochondrial genome sequence of bighorn sheep. *Genome Announcements* 6:e00464–00418.

Dawnay, N., R. Ogden, R. McEwing, G. R. Carvalho, and R. S. Thorpe. 2007. Validation of the barcoding gene COI for use in forensic genetic species identification. *Forensic Science International* 173:1–6.

De Paoli-Iseppi, R., B. E. Deagle, C. R. McMahon, M. A. Hindell, J. L. Dickinson, and S. N. Jarman. 2017. Measuring animal age with DNA methylation: from humans to wild animals. *Frontiers in Genetics* 8:106.

Deakin, S., and D. W. Coltman. 2024. Development of a high-density sub-species-specific targeted SNP assay for Rocky Mountain bighorn sheep (*Ovis canadensis canadensis*). *Peerj* 12:e16946.

Deakin, S., J. C. Gorrell, J. Kneteman, D. S. Hik, R. M. Jobin, and D. W. Coltman. 2020. Spatial genetic structure of Rocky Mountain bighorn sheep (*Ovis canadensis canadensis*) at the northern limit of their native range. *Canadian Journal of Zoology* 98:317–330.

Dekelaita, D. J., C. W. Epps, D. W. German, J. G. Powers, B. J. Gonzales, R. K. Abella-Vu, N. W. Darby, D. L. Hughson, and K. M. Stewart. 2023. Animal movement and associated infectious disease risk in a metapopulation. *Royal Society Open Science* 10:220390.

Demarchi, R., and C. Hartwig. 2004. Status of thinhorn sheep in British Columbia. Ministry of Water, Land and Air Protection, Biodiversity Branch.

Dotsev, A. V., E. Kunz, V. R. Kharzinova, I. M. Okhlopkov, F. H. Lv, M. H. Li, A. N. Rodionov, A. V. Shakhin, T. P. Sipko, D. G. Medvedev, et al. 2021. Mitochondrial DNA analysis clarifies taxonomic status of the northernmost snow sheep (*Ovis nivicola*) population. *Life-Basel* 11:252.

Driscoll, C. C., J. G. Driscoll, C. Hazekamp, J. B. Mitton, and J. D. Wehausen. 2015. A tale of two markers: population genetics of Colorado Rocky Mountain bighorn sheep estimated from microsatellite and mitochondrial data. *Journal of Wildlife Management* 79:819–831.

Dugovich, B. S., B. R. Beechler, B. P. Dolan, R. S. Crowhurst, B. J. Gonzales, J. G. Powers, D. L. Hughson, R. K. Vu, C. W. Epps, and A. E. Jolles. 2023. Population connectivity patterns of genetic diversity, immune responses and exposure to infectious pneumonia in a metapopulation of desert bighorn sheep. *Journal of Animal Ecology* 92:1456–1469.

Epps, C. W., L. Bowen, M. R. Buchalski, E. F. Cassirer, D. Coltman, W. C. Conway, M. Cox, R. B. Harris, J. P. Jahner, M. D. Matocq, et al. 2019. Frequently-asked questions about wild sheep genetics and genomics. *Proceedings of the 2018 North American Wild Sheep and Goat Symposium* 2018:45–75.

Epps, C. W., M. R. Buchalski, R. C. Crowhurst, and C. M. Aiello. 2024a. Evaluation of foraying ram genetics relative to neighboring desert bighorn populations. Unpublished Report. Oregon State University, Corvallis, Oregon, USA.

Epps, C. W., R. S. Crowhurst, and B. S. Nickerson. 2018. Assessing changes in functional connectivity in a desert bighorn sheep metapopulation after two generations. *Molecular Ecology* 27:2334–2346.

Epps, C. W., P. B. Holton, R. J. Monello, R. S. Crowhurst, S. M. Gaulke, W. M. Janousek, T. G. Creech, and T. A. Graves. 2024b. Population and spatial dynamics of desert bighorn sheep in Grand Canyon during an outbreak of respiratory pneumonia. *Frontiers in Ecology and Evolution* 12:22.

Epps, C. W., and N. Keyghobadi. 2015. Landscape genetics in a changing world: disentangling historical and contemporary influences and inferring change. *Molecular Ecology* 24:6021–6040.

Epps, C. W., D. R. McCullough, J. D. Wehausen, V. C. Bleich, and J. L. Rechel. 2004. Effects of climate change on population persistence of desert-dwelling mountain sheep in California. *Conservation Biology* 18:102–113.

Epps, C. W., P. J. Palsboll, J. D. Wehausen, G. K. Roderick, and D. R. McCullough. 2006. Elevation and connectivity define genetic refugia for mountain sheep as climate warms. *Molecular Ecology* 15:4295–4302.

Epps, C. W., P. J. Palsboll, J. D. Wehausen, G. K. Roderick, R. R. Ramey II, and D. R. McCullough. 2005. Highways block gene flow and cause a rapid decline in genetic diversity of desert bighorn sheep. *Ecology Letters* 8:1029–1038.

Epps, C. W., J. D. Wehausen, V. C. Bleich, S. G. Torres, and J. S. Brashares. 2007. Optimizing dispersal and corridor models using landscape genetics. *Journal of Applied Ecology* 44:714–724.

Epps, C. W., J. D. Wehausen, P. J. Palsboll, and D. R. McCullough. 2010. Using genetic tools to track desert bighorn sheep colonizations. *Journal of Wildnlife Management* 74:522–531.

Epps, C. W., J. D. Wehausen, W. B. Sloan, S. Holt, T. G. Creech, R. S. Crowhurst, J. R. Jaeger, K. M. Longshore, and R. J. Monello. 2016. Fifty years after Welles and Welles: distribution and genetic structure of desert bighorn sheep in Death Valley National Park. Pages 70–91 *in* G. White, editor. *First Death Valley Natural History Conference Proceedings*. Death Valley Natural History Association, Death Valley, California, USA.

Farleigh, K., and T. Jezkova. 2023. Genetic signals of local adaptation in a desert rodent that occupies diverse climates and plant communities. *Landscape Ecology* 38:3269–3284.

Fitzsimmons, N. N., S. W. Buskirk, and M. H. Smith. 1995. Population history, genetic variability, and horn growth in bighorn sheep. *Conservation Biology* 9:314–323.

Fitzsimmons, N. N., S. W. Buskirk, and M. H. Smith. 1997. Genetic changes in reintroduced Rocky Mountain bighorn sheep populations. *Journal of Wildlife Management* 61:863–872.

Flesch, E., T. Graves, J. Thomson, K. Proffitt, and R. Garrott. 2022. Average kinship within bighorn sheep populations is associated with connectivity, augmentation, and bottlenecks. *Ecosphere* 13:e3972.

Flesch, E. P., T. A. Graves, J. M. Thomson, K. M. Proffitt, P. J. White, T. R. Stephenson, and R. A. Garrott. 2020. Evaluating wildlife translocations using genomics: a bighorn sheep case study. *Ecology and Evolution* 10:13687–13704.

Forbes, S. H., and J. T. Hogg. 1999. Assessing population structure at high levels of differentiation: microsatellite comparisons of bighorn sheep and large carnivores. *Animal Conservation* 2:223–233.

Forbes, S. H., J. T. Hogg, F. C. Buchanan, A. M. Crawford, and F. W. Allendorf. 1995. Microsatellite evolution in congeneric mammals: domestic and bighorn sheep. *Molecular Biology and Evolution* 12:1106–1113.

Frankham, R. 1995. Inbreeding and extinction: a threshold effect. *Conservation Biology* 9:792–799.

Froy, H., S. L. Underwood, J. Dorrens, L. A. Seeker, K. Watt, R. V. Wilbourn, J. G. Pilkington, L. Harrington, J. M. Pemberton, and D. H. Nussey. 2021. Heritable variation in telomere length predicts mortality in Soay sheep. *Proceedings of the National Academy of Sciences of the United States of America* 118:e2020563118.

Galbreath, K. E., D. J. Hafner, and K. R. Zamudio. 2009. When cold is better: climate-driven elevation shifts yield complex patterns of diversification and demography in alpine specialist (American pika, *Ochotona princeps*). *Evolution* 63:2848–2863.

Garwood, T. J., C. P. Lehman, D. P. Walsh, E. F. Cassirer, T. E. Besser, and J. A. Jenks. 2020. Removal of chronic *Mycoplasma ovipneumoniae* carrier ewes eliminates pneumonia in a bighorn sheep population. *Ecology and Evolution* 10:3491–3502.

Geist, V. 1971. *Mountain Sheep: A Study in Behavior and Evolution.* Univ. of Chicago Press, Chicago, Illinois, USA.

Gille, D. A., M. R. Buchalski, D. Conrad, E. S. Rubin, A. Munig, B. F. Wakeling, C. W. Epps, T. G. Creech, R. Crowhurst, B. Holton, et al. 2019. Genetic outcomes of translocation of bighorn sheep in Arizona. *Journal of Wildlife Management* 83:838–854.

Golden, J. L., and J. F. Bain. 2000. Phylogeographic patterns and high levels of chloroplast DNA diversity in four *Packera* (Asteraceae) species in southwestern Alberta. *Evolution* 54:1566–1579.

Goldman, E. A. 1937. A new mountain sheep from Lower California. *Proceedings of the Biological Society of Washington* 50:29–32.

Gonzales, P. D. 1976. Skull measurements of male desert bighorns from Baja California, Mexico. *Desert Bighorn Council Transactions* 6:43–52.

Gutierrez-Espeleta, G. A., S. T. Kalinowski, W. M. Boyce, and P. W. Hedrick. 2000. Genetic variation and population structure in desert bighorn sheep: implications for conservation. *Conservation Genetics* 1:3–15.

Handley, L. J. L., and N. Perrin. 2007. Advances in our understanding of mammalian sex-biased dispersal. *Molecular Ecology* 16:1559–1578.

Hanski, I., and M. E. Gilpin. 1991. Metapopulation dynamics: brief history and conceptual domain. *Biological Journal of the Linnean Society* 42:3–16.

Hedrick, P. W., G. A. Gutierrez-Espeleta, and R. N. Lee. 2001. Founder effect in an island population of bighorn sheep. *Molecular Ecology* 10:851–857.

Hedrick, P. W., and J. D. Wehausen. 2014. Desert bighorn sheep: changes in genetic variation over time and the impact of merging populations. *Journal of Fish and Wildlife Management* 5:3–13.

Heid, C. A., J. Stevens, K. J. Livak, and P. M. Williams. 1996. Real time quantitative PCR. *Genome Research* 6:986–994.

Hewitt, G. M. 1996. Some genetic consequences of ice ages, and their role in divergence and speciation. *Biological Journal of the Linnean Society* 58:247–276.

Hewitt, G. M. 2000. The genetic legacy of the Quaternary ice ages. *Nature* 405:907–913.

Hiendleder, S., B. Kaupe, R. Wassmuth, and A. Janke. 2002. Molecular analysis of wild and domestic sheep questions current nomenclature and provides evidence for domestication from two different subspecies. *Proceedings of the Royal Society B-Biological Sciences* 269:893–904.

Hoban, S., J. L. Kelley, K. E. Lotterhos, M. F. Antolin, G. Bradburd, D. B. Lowry, M. L. Poss, L. K. Reed, A. Storfer, and M. C. Whitlock. 2016. Finding the genomic basis of local adaptation: pitfalls, practical solutions, and future directions. *American Naturalist* 188:379–397.

Hofreiter, M., and T. Schöneberg. 2010. The genetic and evolutionary basis of colour variation in vertebrates. *Cellular and Molecular Life Sciences* 67:2591–2603.

Hogg, J. T., S. H. Forbes, B. M. Steele, and G. Luikart. 2006. Genetic rescue of an insular population of large mammals. *Proceedings of the Royal Society B-Biological Sciences* 273:1491–1500.

Homola, J. J., C. S. Loftin, K. M. Cammen, C. C. Helbing, I. Birol, T. F. Schultz, and M. T. Kinnison. 2019. Replicated landscape genomics identifies evidence of local adaptation to urbanization in wood frogs. *Journal of Heredity* 110:707–719.

Humphreys, A. M., and T. G. Barraclough. 2014. The evolutionary reality of higher taxa in mammals. *Proceedings of the Royal Society B-Biological Sciences* 281:1783.

Hurley, K., C. Brewer, and G. N. Thornton. 2015. The role of hunters in conservation, restoration, and management of North American wild sheep. *International Journal of Environmental Studies* 72:784–796.

Jahner, J. P., C. A. Buerkle, D. G. Gannon, E. M. Grames, S. E. McFarlane, A. Siefert, K. L. Bell, V. L. DeLeo, M. L. Forister, J. G. Harrison, et al. 2024. Interpretable and predictive models based on high-dimensional data in ecology and evolution. bioRxiv:2024.2003.2015.585297.

Jahner, J. P., M. D. Matocq, J. L. Malaney, M. Cox, P. Wolff, M. A. Gritts, and T. L. Parchman. 2019. The genetic legacy of 50 years of desert bighorn sheep translocations. *Evolutionary Applications* 12:198–213.

Jahner, J. P., T. L. Parchman, M. D. Matocq, M. Cox, R. S. Crowhurst, L. M. Galland, S. M. Burdo, M. Buchalski, J. M. Hallas, S. Barbosa, et al. 2025. Resolving the evolutionary history of bighorn sheep to inform future management: an answer to the California bighorn lineage question. bioRxiv doi: 10.1101/2025.03.13.643091.

Jefferson, G. T. 1991. Rancholabrean age vertebrates from the southeastern Mojave Desert, California. Pages 27–40 *in* R. E. Reynolds, editor. *Crossing the Borders: Quaternary Studies in Eastern California and Southwestern Nevada Mojave Desert.* Quaternary Research Center, San Bernardino County Museum Association, Redlands, California, USA.

Jessup, D. A. 1985. Diseases of domestic livestock which threaten bighorn sheep populations. *Desert Bighorn Council Transactions* 29:29–33.

Jex, B. A., and Z. Sim. 2021. A rare phenotype of thinhorn sheep - the dark phased Dall's: new genetic analysis tools help re-map thinhorn sheep subspecies distributions in North America. *Caprinae News – Newsletter of the IUCN Caprinae Specialist Group* 1:14–16.

Jia, B. Y., X. Wang, F. Q. Ma, X. Li, X. T. Han, L. L. Zhang, J. M. Li, N. C. Diao, K. Shi, C. X. Ge, et al. 2022. The combination of SMRT sequencing and Illumina sequencing highlights organ-specific and age-specific expression patterns of miRNAs in Sika Deer. *Frontiers in Veterinary Science* 9:1042445.

Jiang, Y., M. Xie, W. B. Chen, R. Talbot, J. F. Maddox, T. Faraut, C. H. Wu, D. M. Muzny, Y. X. Li, W. G. Zhang, et al. 2014. The sheep genome illuminates biology of the rumen and lipid metabolism. *Science* 344:1168–1173.

Johnson, H. E., L. S. Mills, J. D. Wehausen, T. R. Stephenson, and G. Luikart. 2011. Translating effects of inbreeding depression on component vital rates to overall population growth in endangered bighorn sheep. *Conservation Biology* 25:1240–1249.

Johnston, S. E., J. C. McEwan, N. K. Pickering, J. W. Kijas, D. Beraldi, J. G. Pilkington, J. M. Pemberton, and J. Slate. 2011. Genome-wide association mapping identifies the genetic basis of discrete and quantitative variation in sexual weaponry in a wild sheep population. *Molecular Ecology* 20:2555–2566.

Kardos, M., G. Luikart, R. Bunch, S. Dewey, W. Edwards, S. McWilliam, J. Stephenson, F. W. Allendorf, J. T. Hogg, and J. Kijas. 2015. Whole-genome resequencing uncovers molecular signatures of natural and sexual selection in wild bighorn sheep. *Molecular Ecology* 24:5616–5632.

Kijas, J. W., L. Porto-Neto, S. Dominik, A. Reverter, R. Bunch, R. McCulloch, B. J. Hayes, R. Brauning, J. McEwan, and C. Int Sheep Genomics. 2014. Linkage disequilibrium over short physical distances measured in sheep using a high-density SNP chip. *Animal Genetics* 45:754–757.

Laity, J. J. 2009. *Deserts and Desert Environments.* John Wiley & Sons, Hoboken, New Jersey, USA.

Latch, E. K., J. R. Heffelfinger, B. F. Wakeling, J. Hanna, D. Conrad, and O. E. Rhodes Jr. 2005. Genetic subspecies identification of a recently colonized bighorn sheep population in central Arizona. Pages 1–9 *in* J. W. Cain, and P. R. Krausman, editors. *Managing Wildlife in the Southwest.* Southwest Section of the Wildlife Society, Tucson, Arizona, USA.

Li, X., J. Yang, M. Shen, X. L. Xie, G. J. Liu, Y. X. Xu, F. H. Lv, H. Yang, Y. L. Yang, C. B. Liu, et al. 2020. Whole-genome resequencing of wild and domestic sheep identifies genes associated with morphological and agronomic traits. *Nature Communications* 11:2815.

Loehr, J., J. Carey, H. Ylönen, and J. Suhonen. 2008. Coat darkness is associated with social dominance and mating behaviour in a mountain sheep hybrid lineage. *Animal Behaviour* 76:1545–1553.

Loehr, J., K. Worley, A. Grapputo, J. Carey, A. Veitch, and D. W. Coltman. 2006. Evidence for cryptic glacial refugia from North American mountain sheep mitochondrial DNA. *Journal of Evolutionary Biology* 19:419–430.

Lotterhos, K. E., and M. C. Whitlock. 2015. The relative power of genome scans to detect local adaptation depends on sampling design and statistical method. *Molecular Ecology* 24:1031–1046.

Luikart, G., and F. W. Allendorf. 1996. Mitochondrial-DNA variation and genetic-population structure in Rocky Mountain bighorn sheep (*Ovis canadensis canadensis*). *Journal of Mammalogy* 77:109–123.

Luikart, G., K. Pilgrim, J. Visty, V. O. Ezenwa, and M. K. Schwartz. 2008a. Candidate gene microsatellite variation is associated with parasitism in wild bighorn sheep. *Biology Letters* 4:228–231.

Luikart, G., S. Zundel, D. Rioux, C. Miquel, K. A. Keating, J. T. Hogg, B. Steele, K. Foresman, and P. Taberlet. 2008b. Low genotyping error rates and noninvasive sampling in bighorn sheep. *Journal of Wildlife Management* 72:299–304.

Luikart, G., W. B. Sherwin, B. M. Steele, and F. W. Allendorf. 1998. Usefulness of molecular markers for detecting population bottlenecks via monitoring genetic change. *Molecular Ecology* 7:963–974.

Malaney, J. L., C. R. Feldman, M. Cox, P. Wolff, J. D. Wehausen, and M. D. Matocq. 2015. Translocated to the fringe: genetic and niche variation in bighorn sheep of the Great Basin and northern Mojave deserts. *Diversity and Distributions* 21:1063–1074.

Manel, S., M. K. Schwartz, G. Luikart, and P. Taberlet. 2003. Landscape genetics: combining landscape ecology and population genetics. *Trends in Ecology & Evolution* 18:189–197.

Manel, S., P. Berthier, and G. Luikart. 2002. Detecting wildlife poaching: identifying the origin of individuals with Bayesian assignment tests and multilocus genotypes. *Conservation Biology* 16:650–659.

Martin, A. M., E. F. Cassirer, L. P. Waits, R. K. Plowright, P. C. Cross, and K. R. Andrews. 2021. Genomic association with pathogen carriage in bighorn sheep (*Ovis canadensis*). *Ecology and Evolution* 11:2488–2502.

Martin, A. M., M. Festa-Bianchet, D. W. Coltman, and F. Pelletier. 2014. Sexually antagonistic association between paternal phenotype and offspring viability reinforces total selection on a sexually selected trait. *Biology Letters* 10:20140043.

Martin, L. D., and B. M. Gilbert. 1978. Excavations at Natural Trap Cave. *Transactions of the Nebraska Academy of Sciences and Affiliated Societies* VI:107–116.

Mathur, S., and J. A. DeWoody. 2021. Genetic load has potential in large populations but is realized in small inbred populations. *Evolutionary Applications* 14:1540–1557.

Mathur, S., J. M. Tomecek, L. A. Tarango-Arámbula, R. M. Perez, and J. A. DeWoody. 2023. An evolutionary perspective on genetic load in small, isolated populations as informed by whole genome resequencing and forward-time simulations. *Evolution* 77:690–704.

Miller, J. M., J. Poissant, J. T. Hogg, and D. W. Coltman. 2012. Genomic consequences of genetic rescue in an insular population of bighorn sheep (*Ovis canadensis*). *Molecular Ecology* 21:1583–1596.

Miller, J. M., J. Poissant, J. W. Kijas, D. W. Coltman, and C. Int Sheep Genomics. 2011. A genome-wide set of SNPs detects population substructure and long range linkage disequilibrium in wild sheep. *Molecular Ecology Resources* 11:314–322.

Miller, J. M., J. Poissant, R. M. Malenfant, J. T. Hogg, and D. W. Coltman. 2015b. Temporal dynamics of linkage disequilibrium in two populations of bighorn sheep. *Ecology and Evolution* 5:3401–3412.

Miller, J. M., M. Festa-Bianchet, and D. W. Coltman. 2018. Genomic analysis of morphometric traits in bighorn sheep using the Ovine Infinium® HD SNP BeadChip. *Peerj* 6:e4364.

Miller, J. M., S. S. Moore, P. Stothard, X. Liao, and D. W. Coltman. 2015a. Harnessing cross-species alignment to discover SNPs and generate a draft genome sequence of a bighorn sheep (*Ovis canadensis*). *BMC Genomics* 16:397.

Morrissey, R. B., R. M. Malenfant, F. Pelletier, D. W. Coltman, and M. Festa-Bianchet. 2025. The potential for rapid evolution in response to exploitation-induced selection. *Proceedings of the 2024 Northern Wild Sheep and Goat Council*: in press.

Mortazavi, A., B. A. Williams, K. McCue, L. Schaeffer, and B. Wold. 2008. Mapping and quantifying mammalian transcriptomes by RNA-Seq. *Nature Methods* 5:621–628.

Mullis, K., F. Faloona, S. Scharf, R. Saiki, G. Horn, and H. Erlich. 1986. Specific enzymatic amplification of DNA invitro: the polymerase chain-reaction. *Cold Spring Harbor Symposia on Quantitative Biology* 51:263–273.

Nadler, C. F., and T. D. Bunch. 1977. G-Band patterns of Siberian snow sheep (*Ovis nivicola*) and their relationship to chromosomal evolution in sheep. *Cytogenetics and Cell Genetics* 19:108–117.

Nagati, M., M. J. Bergeron, P. Gagne, A. Arsenault, A. Droit, P. Wilson, G. Pittoello, S. Kutz, M. Manseau, and C. Martineau. 2024. Exploring winter diet, gut microbiota and parasitism in caribou using multi-marker metabarcoding of fecal DNA. *Scientific Reports* 14:27960.

Nakamura, S., J. Yamazaki, N. Matsumoto, M. Inoue-Murayama, H. Y. Qi, M. Yamanaka, M. Nakanishi, Y. Yanagawa, M. Sashika, T. Tsubota, et al. 2023. Age estimation based on blood DNA methylation levels in brown bears. *Molecular Ecology Resources* 23:1211–1225.

Narayan, E., G. Sawyer, D. Fox, R. Smith, and A. Tilbrook. 2022. Interplay between stress and reproduction: novel epigenetic markers in response to shearing patterns in Australian merino sheep (*Ovis aries*). *Frontiers in Veterinary Science* 9:830450.

Ng, P. C., and S. Henikoff. 2003. SIFT: predicting amino acid changes that affect protein function. *Nucleic Acids Research* 31:3812–3814.

Nichols, L. 1978. Dall's sheep. Pages 172–189 *in* J. L. Gilbert, and D. L. Schmidt, editors. *Big Game of North America: Ecology and Management*. Stackpole Books, Harrisburg, Pennsylvania, USA.

Olson, Z. H. 2013. Translocation history and genetic diversity in reintroduced bighorn sheep. *Journal of Wildlife Management* 77:1553–1563.

Olson, Z. H., D. G. Whittaker, and O. E. Rhodes. 2012. Evaluation of experimental genetic management in reintroduced bighorn sheep. *Ecology and Evolution* 2:429–443.

Onderka, D. K., and W. D. Wishart. 1988. Experimental contact transmission of *Pasteurella haemolytica* from clinically normal domestic sheep causing pneumonia in Rocky Mountain bighorn sheep. *Journal of Wildlife Diseases* 24:663–667.

Parchman, T. L., Z. Gompert, J. Mudge, F. D. Schilkey, C. W. Benkman, and C. A. Buerkle. 2012. Genome-wide association genetics of an adaptive trait in lodgepole pine. *Molecular Ecology* 21:2991–3005.

Pelletier, F., D. Réale, D. Garant, D. W. Coltman, and M. Festa-Bianchet. 2007. Selection on heritable seasonal phenotypic plasticity of body mass. *Evolution* 61:1969–1979.

Perez, M. F., F. F. Franco, J. R. Bombonato, I. A. S. Bonatelli, G. Khan, M. Romeiro-Brito, A. C. Fegies, P. M. Ribeiro, G. A. R. Silva, and E. M. Moraes. 2018. Assessing population structure in the face of isolation by distance: are we neglecting the problem? *Diversity and Distributions* 24:1883–1889.

Peterson, B. K., J. N. Weber, E. H. Kay, H. S. Fisher, and H. E. Hoekstra. 2012. Double digest RADseq: an inexpensive method for *de novo* SNP discovery and genotyping in model and non-model species. *Plos One* 7:e37135.

Pfeiler, S. S., M. M. Conner, J. S. McKeever, T. R. Stephenson, D. W. German, R. S. Crowhurst, P. R. Prentice, and C. W. Epps. 2020. Costs and precision of fecal DNA mark-recapture versus traditional mark-resight. *Wildlife Society Bulletin* 44:531–542.

Pielou, E. 1991. *After the Ice Age: The Return of Life to Glaciated North America*. The University of Chicago Press, Chicago, Illinois, USA.

Pigeon, G., M. Festa-Bianchet, D. W. Coltman, and F. Pelletier. 2016. Intense selective hunting leads to artificial evolution in horn size. *Evolutionary Applications* 9:521–530.

Pigeon, G., T. H. G. Ezard, M. Festa-Bianchet, D. W. Coltman, and F. Pelletier. 2017. Fluctuating effects of genetic and plastic changes in body mass on population dynamics in a large herbivore. *Ecology* 98:2456–2467.

Plowright, R. K., K. R. Manlove, T. E. Besser, D. J. Paez, K. R. Andrews, P. E. Matthews, L. P. Waits, P. J. Hudson, and E. F. Cassirer. 2017. Age-specific infectious period shapes dynamics of pneumonia in bighorn sheep. *Ecology Letters* 20:1325–1336.

Poirier, M. A., D. W. Coltman, F. Pelletier, J. Jorgenson, and M. Festa-Bianchet. 2019. Genetic decline, restoration and rescue of an isolated ungulate population. *Evolutionary Applications* 12:1318–1328.

Poissant, J., C. S. Davis, R. M. Malenfant, J. T. Hogg, and D. W. Coltman. 2012. QTL mapping for sexually dimorphic fitness-related traits in wild bighorn sheep. *Heredity* 108:256–263.

Poissant, J., A. J. Wilson, M. Festa-Bianchet, J. T. Hogg, and D. W. Coltman. 2008. Quantitative genetics and sex-specific selection on sexually dimorphic traits in bighorn sheep. *Proceedings of the Royal Society B-Biological Sciences* 275:623–628.

Quinn, C. B., S. Preckler-Quisquater, M. R. Buchalski, and B. N. Sacks. 2024. Whole genomes inform genetic rescue strategy for montane red foxes in North America. *Molecular Biology and Evolution* 41:msae193.

Quinn, L., G. Garcia-Erill, C. Santander, A. Brüniche-Olsen, X. D. Liu, M. H. S. Sinding, M. P. Heaton, T. P. L. Smith, P. Pecnerová, L. D. Bertola, et al. 2023. Colonialism in South Africa leaves a lasting legacy of reduced genetic diversity in Cape buffalo. *Molecular Ecology* 32:1860–1874.

Ramey II, R. R. 1993. Evolutionary genetics and systematics of North American mountain sheep: implications for conservation. Dissertation, Cornell University, Ithaca, New York.

Ramey II, R. R. 1995. Mitochondrial DNA variation, population structure, and evolution of mountain sheep in the southwestern United States and Mexico. *Molecular Ecology* 4:429–439.

Ramey, R. R., G. Luikart, and F. J. Singer. 2000. Genetic bottlenecks resulting from restoration efforts: The case of bighorn sheep in Badlands National Park. *Restoration Ecology* 8:85–90.

Reale, D., and M. Festa-Bianchet. 2000a. Mass-dependent reproductive strategies in wild bighorn ewes: a quantitative genetic approach. *Journal of Evolutionary Biology* 13:679–688.

Reale, D., and M. Festa-Bianchet. 2000b. Quantitative genetics of life-history traits in a long-lived wild mammal. *Heredity* 85:593–603.

Reale, D., M. Festa-Bianchet, and J. T. Jorgenson. 1999. Heritability of body mass varies with age and season in wild bighorn sheep. *Heredity* 83:526–532.

Reilly, J. R., and J. M. Miller. 2021. A phylogeographic contact zone for Arctic grayling in Alberta, Canada. *North American Journal of Fisheries Management* 41:1619–1630.

Rellstab, C., B. Dauphin, and M. Exposito-Alonso. 2021. Prospects and limitations of genomic offset in conservation management. *Evolutionary Applications* 14:1202–1212.

Rellstab, C., F. Gugerli, A. J. Eckert, A. M. Hancock, and R. Holderegger. 2015. A practical guide to environmental association analysis in landscape genomics. *Molecular Ecology* 24:4348–4370.

Rezaei, H. R., S. Naderi, I. C. Chintauan-Marquier, P. Taberlet, A. T. Virk, H. R. Naghash, D. Rioux, M. Kaboli, and F. Pompanon. 2010. Evolution and taxonomy of the wild species of the genus *Ovis* (Mammalia, Artiodactyla, Bovidae). *Molecular Phylogenetics and Evolution* 54:315–326.

Rioux-Paquette, E. 2011. Sex-differential effects of inbreeding on overwinter survival, birth date and mass of bighorn lambs. *Journal of Evolutionary Biology* 24:121–131.

Rioux-Paquette, E., M. Festa-Bianchet, and D. W. Coltman. 2010. No inbreeding avoidance in an isolated population of bighorn sheep. *Animal Behaviour* 80:865–871.

Rodriguez-Rodriguez, M. A., J. Gasca-Pineda, R. A. Medellin, and L. E. Eguiarte. 2015. Analysis of genetic diversity of bighorn sheep (*Ovis canadensis*) from Mexican populations. *Journal of Mammalogy* 96:473–480.

Roffler, G. H., S. J. Amish, S. Smith, T. Cosart, M. Kardos, M. K. Schwartz, and G. Luikart. 2016. SNP discovery in candidate adaptive genes using exon capture in a free-ranging alpine ungulate. *Molecular Ecology Resources* 16:1147–1164.

Roffler, G. H., S. L. Talbot, G. Luikart, G. K. Sage, K. L. Pilgrim, L. G. Adams, and M. K. Schwartz. 2014. Lack of sex-biased dispersal promotes fine-scale genetic structure in alpine ungulates. *Conservation Genetics* 15:837–851.

Rubidge, E. M., J. L. Patton, M. Lim, A. C. Burton, J. S. Brashares, and C. Moritz. 2012. Climate-induced range contraction drives genetic erosion in an alpine mammal. *Nature Climate Change* 2:285–288.

Russello, M. A., M. D. Waterhouse, P. D. Etter, and E. A. Johnson. 2015. From promise to practice: pairing non-invasive sampling with genomics in conservation. *Peerj* 3:18.

Sage, R. D., and J. O. Wolff. 1986. Pleistocene glaciations, fluctuating ranges, and low genetic-variability in a large mammal (*Ovis dalli*). *Evolution* 40:1092–1095.

Santos, S. H. D., R. M. Peery, J. M. Miller, A. Dao, F. H. Lyu, X. Li, M. H. Li, and D. W. Coltman. 2021. Ancient hybridization patterns between bighorn and thinhorn sheep. *Molecular Ecology* 30:6273–6288.

Schoenecker, K. A., M. K. Watry, L. E. Ellison, M. K. Schwartz, and G. Luikart. 2015. Estimating bighorn sheep (*Ovis canadensis*) abundance using noninvasive sampling at a mineral lick within a national park wilderness area. *Western North American Naturalist* 75:181–191.

Shafer, A. B. A., C. I. Cullingham, S. D. Côté, and D. W. Coltman. 2010. Of glaciers and refugia: a decade of study sheds new light on the phylogeography of northwestern North America. *Molecular Ecology* 19:4589–4621.

Shafer, A. B. A., and M. Kardos. 2025. Runs of homozygosity and inferences in wild populations. *Molecular Ecology*: in press. doi:10.1111/mec.17641

Sheldon, C. 1911. *The Wilderness of the Upper Yukon; A Hunter's Explorations for Wild Sheep in Sub-Arctic Mountains*. T. Fisher Unwin, New York, New York, USA.

Sim, Z., C. S. Davis, B. A. Jex, T. Hegel, and D. W. Coltman. 2019. Management implications of highly resolved hierarchical population genetic structure in thinhorn sheep. *Conservation Genetics* 20:185–201.

Sim, Z., J. C. Hall, B. A. Jex, T. M. Hegel, and D. W. Coltman. 2016. Genome-wide set of SNPs reveals evidence for two glacial refugia and admixture from postglacial recolonization in an alpine ungulate. *Molecular Ecology* 25:3696–3705.

Sim, Z., L. Monderman, D. Hildebrand, T. Packer, and R. M. Jobin. 2021. Development and implementation of a STR based forensic typing system for moose (*Alces alces*). *Forensic Science International-Genetics* 53:102536.

Soltis, D. E., M. A. Gitzendanner, D. D. Strenge, and P. S. Soltis. 1997. Chloroplast DNA intraspecific phylogeography of plants from the Pacific Northwest of North America. *Plant Systematics and Evolution* 206:353–373.

Spaan, R. S., C. W. Epps, R. Crowhurst, D. Whittaker, M. Cox, and A. Duarte. 2021. Impact of *Mycoplasma ovipneumoniae* on juvenile bighorn sheep (*Ovis canadensis*) survival in the northern Basin and Range ecosystem. *Peerj* 9:e10710.

Spellman, G. M., B. Riddle, and J. Klicka. 2007. Phylogeography of the mountain chickadee (*Poecile gambeli*): diversification, introgression, and expansion in response to Quaternary climate change. *Molecular Ecology* 16:1055–1068.

Stoffel, M. A., S. E. Johnston, J. G. Pilkington, and J. M. Pemberton. 2021a. Genetic architecture and lifetime dynamics of inbreeding depression in a wild mammal. *Nature Communications* 12:2972.

Stoffel, M. A., S. E. Johnston, J. G. Pilkington, and J. M. Pemberton. 2021b. Mutation load decreases with haplotype age in wild Soay sheep. *Evolution Letters* 5:187–195.

Stowell, S. L. M., R. B. Gagne, D. McWhirter, W. Edwards, and H. B. Ernest. 2020. Bighorn sheep genetic structure in Wyoming reflects geography and management. *Journal of Wildlife Management* 84:1072–1090.

Taberlet, P., S. Griffin, B. Goossens, S. Questiau, V. Manceau, N. Escaravage, L. P. Waits, and J. Bouvet. 1996. Reliable genotyping of samples with very low DNA quantities using PCR. *Nucleic Acids Research* 24:3189–3194.

Taylor, R. S., M. Manseau, S. Keobouasone, P. Liu, G. Mastromonaco, K. Solmundson, A. Kelly, N. C. Larter, M. Gamberg, H. Schwantje, et al. 2024. High genetic load without purging in caribou, a diverse species at risk. *Current Biology* 34:1234–1236.

Thompson, M. D., and A. P. Russell. 2005. Glacial retreat and its influence on migration of mitochondrial genes in the long-toed salamander (*Ambystoma macrodactylum*) in Western North America. Pages 205–246 in A. M. T. Elewa, editor. *Migration of Organisms: Climate, Geography, Ecology*. Springer Berlin Heidelberg, New York, New York, USA.

Upadhyay, M., A. Hauser, E. Kunz, S. Krebs, H. Blum, A. Dotsev, I. Okhlopkov, V. Bagirov, G. Brem, N. Zinovieva, et al. 2020. The first draft genome assembly of snow sheep (*Ovis nivicola*). *Genome Biology and Evolution* 12:1330–1336.

USFWS. 2007. *Recovery Plan for Sierra Nevada Bighorn Sheep*. U. S. Fish and Wildlife Service, Sacramento.

Valdez, R. 1982. *The Wild Sheep of the World*. Wild Sheep and Goat International, Mesilla, New Mexico, USA.

Van Devender, T. R., R. S. Thompson, and J. L. Betancourt. 1987. Vegetation history of the deserts of southwestern North America: the nature and timing of the late Wisconsin-Holocene transition. Pages 323–352 in W. F. Ruddiman, and H. E. Wright Jr., editors. *North America and Adjacent Oceans during the Last Deglaciation*. Geological Society of America.

Waits, L. P., G. Luikart, and P. Taberlet. 2001. Estimating the probability of identity among genotypes in natural populations: cautions and guidelines. *Molecular Ecology* 10:249–256.

Wang, J. L. 2017. The computer program STRUCTURE for assigning individuals to populations: easy to use but easier to misuse. *Molecular Ecology Resources* 17:981–990.

Wang, X. 1988. Systematics and population ecology of late Pleistocene bighorn sheep (*Ovis canadensis*) of Natural Trap Cave, Wyoming. *Transactions of the Nebraska Academy of Sciences and Affiliated Societies* XVI:173–183.

Waples, R. S., and P. R. England. 2011. Estimating contemporary effective population size on the basis of linkage disequilibrium in the face of migration. *Genetics* 189:633–644.

Wehausen, J. D. 1980. Sierra Nevada bighorn sheep: history and population ecology. Dissertation, University of Michigan, Ann Arbor, Michigan, USA.

Wehausen, J. D., C. J. O'Brien, and D. R. McCullough. 2024. Reliability of tooth cementum rings to age bighorn sheep: a blind test. *California Fish and Wildlife Journal* 110:e19.

Wehausen, J. D., and R. R. Ramey II. 1993. A morphometric reevaluation of the Peninsular bighorn subspecies. *Desert Bighorn Council Transactions* 37:1–10.

Wehausen, J. D., and R. R. Ramey II. 2000. Cranial morphometric and evolutionary relationships in the northern range of *Ovis canadensis*. *Journal of Mammalogy* 81:145–161.

Wehausen, J. D., R. R. Ramey II, and C. W. Epps. 2004. Experiments in DNA extraction and PCR amplification from bighorn sheep feces: the importance of DNA extraction method. *Journal of Heredity* 95:503–509.

Wehausen, J. D., V. C. Bleich, and R. R. Ramey. 2005. Correct nomenclature for Sierra Nevada bighorn sheep. *California Fish and Game* 91:216–218.

Whittaker, D. G., S. D. Ostermann, and W. M. Boyce. 2004. Genetic variability of reintroduced California bighorn sheep in Oregon. *Journal of Wildlife Management* 68:850–859.

Wiedmann, B. P., and G. A. Sargeant. 2014. Ecotypic variation in recruitment of reintroduced bighorn sheep: implications for translocation. *Journal of Wildlife Management* 78:394–401.

Wilder, B. T., J. L. Betancourt, C. W. Epps, R. S. Crowhurst, J. I. Mead, and E. Ezcurra. 2014. Local extinction and unintentional rewilding of bighorn sheep (*Ovis canadensis*) on a desert island. *Plos One* 9:e91358.

Wilder, A. P., C. C. Steiner, S. Hendricks, B. C. Haller, C. Kim, M. L. Korody, and O. A. Ryder. 2024. Genetic load and viability of a future restored northern white rhino population. *Evolutionary Applications* 17:e13683.

Wild Sheep Initiative. 2023. *Pneumonia and Bighorn Sheep: Test and Remove.* Factsheet 7–14–2023. Wild Sheep Initiative, Western Association of Fish and Wildlife Agencies, Boise, Idaho, USA.

Wild Sheep Working Group. 2015. *Records of Wild Sheep Translocations: United States and Canada, 1922-Present.* Wild Sheep Working Group, Western Association of Fish and Wildlife Agencies, Boise, Idaho, USA.

Wilson, A. J., L. E. B. Kruuk, and D. W. Coltman. 2005. Ontogenetic patterns in heritable variation for body size: using random regression models in a wild ungulate population. *American Naturalist* 166:E177–E192.

Wilson, D. E., and D. M. Reeder. 2005. *Mammal Species of the World: A Taxonomic and Geographic Reference.* 3rd edition. John Hopkins University Press, Baltimore, Maryland, USA.

Worley, K., J. Carey, A. Veitch, and D. W. Coltman. 2006. Detecting the signature of selection on immune genes in highly structured populations of wild sheep (*Ovis dalli*). *Molecular Ecology* 15:623–637.

Worley, K., C. Strobeck, S. Arthur, J. Carey, H. Schwantje, A. Veitch, and D. W. Coltman. 2004. Population genetic structure of North American thinhorn sheep (*Ovis dalli*). *Molecular Ecology* 13:2545–2556.

Wright, E. A., J. D. Manthey, M. R. Buchalski, B. R. McKinney, D. A. Ray, C. D. Phillips, and R. D. Bradley. 2024b. Genomic affinity following restoration of a locally extirpated species: a case study of desert bighorn sheep in Texas. *Conservation Genetics* 25:1209–1230.

Wright, E. A., M. R. Buchalski, and R. D. Bradley. 2024a. Mitochondrial DNA indicates that extirpated *Ovis canadensis texianus* was a member of the desert bighorn sheep complex. *Occasional Papers, Museum of Texas Tech University* 390:1–29.

Zhang, M. S., X. J. An, C. Yuan, T. T. Guo, B. P. Xi, J. B. Liu, and Z. K. Lu. 2024. Integration analysis of transcriptome and metabolome revealed the potential mechanism of spermatogenesis in Tibetan sheep (*Ovis aries*) at extreme high altitude. *Genomics* 116:110949.

# 7 Nutritional Ecology of Mountain Sheep
## *From Bites to Populations*

*Kristin Denryter and Thomas R. Stephenson*

## 7.1 INTRODUCTION

Nutrition influences virtually every aspect of mountain sheep biology and ecology, from growth to survival and reproduction, to habitat use and seasonal movements, and interactions with and susceptibility to disease and predators (Cook 1990, Festa-Bianchet and Jorgenson 1998, Dekelaita et al. 2020, Van de Kerk et al. 2020, Denryter et al. 2024, Smiley et al. 2024). Because of the wide-reaching influences of nutrition, mountain sheep (bighorn [*Ovis canadensis*], thinhorn [*Ovis dalli*]) biologists are likely to encounter any number of nutritional problems that need their attention. Common nutritional problems include questions about starvation, competition (Chapter 10), nutritional carrying capacity, demographic rates (Chapter 15), population numbers (Chapter 3), or interactions with other potential limiting factors (i.e., something that restricts the performance of an individual or productivity of a population; Chapters 12 and 21). If nutrition is the main factor limiting demographic rates, biologists may want to institute habitat treatments, and to be effective, they would need to know what makes good nutrition for mountain sheep and where mountain sheep will access nutrition following habitat treatments. Biologists also would need to know what kind of treatment to employ to improve forage quality or availability for mountain sheep, such as a prescribed burn (Sittler et al. 2019). Just as importantly, if nutrition is not limiting, managers can explore other actions that may be more appropriate and cost-effective for sound conservation of mountain sheep (Chapter 20). While this reflects a pragmatic approach for managers, there are many biologists who study nutrition because we are fascinated thinking about how mountain sheep turn sparse landscapes of plants into muscle, fat, horns, and more sheep.

Relationships and interactions between mountain sheep nutrition and other variables can be complex and difficult to isolate, which is why rigorous studies are needed to evaluate the influence of nutrition on individuals and populations. As with other ungulates, rigorous evaluations of nutrition were rare historically, which may have contributed to perceptions that nutrition was relatively unimportant as a limiting factor or that winter represented the most important nutritional bottleneck of the year (Torbit et al. 1985). The importance of nutrition as a limiting factor, and especially summer nutrition, for mountain sheep has been increasingly recognized (Cook 1990, Denryter et al. 2022a, Smiley et al. 2022, Wagler et al. 2023). New, validated techniques are providing opportunities to evaluate nutritional status of individuals and populations (Stephenson et al. 2020) in ways that were never possible before, though some older techniques remain important.

In this chapter, we introduce fundamental concepts in nutrition and explore adaptations in the structure (morphology), function (digestive processes), and behavior (foraging, movement, habitat use) of mountain sheep that constrain nutritional interactions with their environment. We also discuss the consequences of nutrition to individual performance and population productivity. Throughout the chapter, we use individual performance to refer to the extent to which an animal synthesizes tissue through somatic growth, milk production, accretion of body reserves, or otherwise. We use population productivity to refer to the number of new animals the population generates through reproduction, survival, and recruitment. Lastly, we use nutritional condition to refer to the state of an animal's body components (Harder and Kirkpatrick 1994), as indicated by fat, protein, or both.

## 7.2 FUNDAMENTALS OF NUTRITION IN THE LIFE CYCLE OF MOUNTAIN SHEEP

Nutrition has fundamental roles in the life cycle of mountain sheep, with body mass and body fat influence various aspects of reproduction (Bunnell and Olsen 1981, Stephenson et al. 2020, Smiley et al. 2024) and effects of nutrition begin prior to conception. Fatter and heavier females typically breed and give birth earlier and produce larger offspring with better odds of survival than females in poorer nutritional condition (Bunnell 1980, 1982; Festa-Bianchet 1988a, Hogg et al. 1992, Rubin et al. 2000, Van de Kerk et al. 2020). A high plane of nutrition is associated with a lower age of first reproduction (Bunnell 1978, Nichols 1978, Jorgenson et al. 1993, Proffitt et al. 2021), which therefore has the potential to increase lifetime reproductive success and hence fitness. Bighorn sheep are capital breeders because they rely on stored reserves to partially meet the costs of reproduction (Festa-Bianchet et al. 1998, Stephenson et al. 2020, Smiley et al. 2022). For example, stored fat and protein support fetal growth (during gestation) and milk production in early lactation when food supplies are seasonally limited (Festa-Bianchet 1988a, Parker et al. 2009, Monteith et al. 2013). Nutritional deprivation during gestation is associated with increased incidence of spontaneous abortions, lower birth mass, delayed birth, and reduced offspring survival

DOI: 10.1201/9781003518686-8

(Verme and Ullrey 1984, Rachlow and Bowyer 1994, Feder et al. 2008). Undoubtedly, accretion and mobilization of body reserves represent important adaptations for reproduction in mountain sheep.

Growth of juvenile mountain sheep *ex utero* also is strongly tied to nutrition. Life history theory predicts that females should prioritize their survival and future opportunities for reproduction over current reproduction, known as risk-sensitive allocation (Smiley et al. 2022). Because females prioritize their survival over current reproductive investment, poor nutrition can be more costly to the lamb than the female (Festa-Bianchet and Jorgenson 1998). Reduced maternal investment manifests via a reduction in milk production in females and milk transfer to lambs (Berger 1979, Festa-Bianchet 1988b), which suppresses lamb growth (Festa-Bianchet and Jorgenson 1998, Martin and Festa-Bianchet 2010). Maternal nutritional condition during gestation and population density during lactation also can affect horn growth in males for years after birth (Hoefs and Nowlan 1997, Jorgenson et al. 1998, Monteith et al. 2018). Horn growth, however, also is strongly tied to forage quality (rather than quantity; Bunnell 1978, Seip 1983).

Nutrition is important for survival in mountain sheep. Lamb and adult survival is higher on ranges with better forage quality or greater forage biomass (Cook 1990, Dekelaita et al. 2020, Van de Kerk et al. 2020), but this can be variable (Bilodeau 2021). Through effects on forage, winter precipitation and frequency of freeze-thaw events are associated with reduced adult survival (Dekelaita et al. 2020, Van de Kerk et al. 2020), but winter weather does not always influence survival of mountain sheep (Jorgenson et al. 1997, Portier et al. 1998). The effect of winter weather may be moderated by a buffering effect of body fat acquired the previous summer or interactions between body fat and migratory tactic (Stephenson et al. 2020, Proffitt et al. 2021, Denryter et al. 2022a, Smiley et al. 2024).

Through effects on survival and reproduction, nutrition also affects recruitment and rates of population growth. In desert environments, increased winter or annual precipitation is associated with increased lamb recruitment and population growth (Wehausen et al. 1987, Bender and Weisenberger 2005), as is access to greater forage quality and quantity in more northern sheep (Wagler et al. 2023). The annual rate of population growth (Lambda) is a function of body fat levels in mountain sheep, which can be used to estimate animal-indicated nutritional carrying capacity (Stephenson et al. 2020; as opposed to estimating nutritional carrying capacity from the forage base).

### 7.2.1 Nutrition as a Limiting Factor

Nutrition operates as a bottom-up constraint to mountain sheep, bounding an upper limit of individual performance and population productivity. When nutrition is limiting individuals and populations experience reduced performance or productivity as compared to when adequate nutrition is available to meet metabolic needs.

Nutritional limitations can be density-dependent when population growth rates are a function of population density, or density-independent when population growth rate is not tied to population density. Competition with conspecifics for limited resources is implicated in density-dependent nutritional limitations and affects various demographic rates. Survival of adult female and yearling sheep can be density-dependent in northern and desert systems (Jorgenson et al. 1997, Dekelaita et al. 2020). Lamb birth rates, survival, and growth rates are suppressed via density-dependent nutritional limitations (Festa-Bianchet 1988a, Festa-Bianchet and Jorgenson 1998, Portier et al. 1998). High population density can delay age of first reproduction (Jorgenson et al. 1993) and reduce horn growth in males (Festa-Bianchet et al. 1993, 2004).

Density-independent nutritional limitations often are associated with weather, such as extreme snowfall or avalanches in northern systems or variable precipitation and primary productivity in desert systems (Bender and Weisenberger 2005, Conner et al. 2018). Forage quality that is below nutritional requirements for maintenance, growth, or other performance also represents a density-independent limitation (Cook 1990). Even forage quantity can be a density-independent limitation under some circumstances. For example, forage quantity can be limiting when: acquisition of forage is too slow to meet nutritional requirements because of high handling time, poor spatial distribution of bites leads to competition between searching for bites and acquisition, animals have to spend substantial time sorting desirable and undesirable plant parts, or snow limits access to forage (Seip 1983, Goodson et al. 1991a, Denryter 2017). Density-independent and density-dependent limitations also can interact to affect populations, including through effects on lamb survival (Portier et al. 1998).

Assessing nutritional limitations in mountain sheep can be accomplished in several ways. Combining information on the forage base, nutrient intake, and the nutritional requirements of mountain sheep collectively illustrates what attributes of the plant community may be limiting (Cook et al. 2016, Denryter 2017). Forage sampling can be intensive, intake rates are difficult to collect, and nutritional requirements are estimated (Chappel and Hudson 1980, Denryter et al. 2021, Denryter and White 2026) but not known empirically for free-ranging mountain sheep. Often a suitable alternative to range assessments and estimates of intake rate for determination of whether a nutritional limitation exists is to evaluate nutritional condition of individuals by evaluating their body fat (Stephenson et al. 2020). In particular, comparisons of body fat of lactating versus non-lactating females can be informative because when nutrition is limiting, lactating females accrete less body fat than non-lactating females (Cook et al. 2004). Body fat assessments do not provide information on what attributes of the plant community are limiting. Nonetheless, relations between body fat and population growth can provide insights to the influence of nutrition on population dynamics (Monteith et al. 2014, Stephenson et al. 2020).

## 7.2.2 Interactions between Nutrition and Other Limiting Factors

Nutrition-disease interactions are particularly important for mountain sheep afflicted with pneumonia or other infectious diseases. Poor nutrition may increase susceptibility to disease in mountain sheep, particularly lambs (Cook 1990). Susceptibility in adults, however, is less apparent (Cassirer et al. 2018), though disease can reduce the nutritional condition of infected females (Smiley et al. 2024). Good nutrition during summer can offset some costs of pneumonia infection in desert bighorn sheep (*Ovis canadensis* spp.; Dekelaita et al. 2020) and high levels of body fat are needed to support pathogen clearance (Smiley et al. 2024). Additionally, animals in good nutritional condition are better able to tolerate plant toxins than animals in poor nutritional condition (Lopez-Ortiz et al. 2004).

Nutrition can alter susceptibility to predation. Lambs may be more vulnerable when nutrition is poor (Cook 1990, Van de Kerk et al. 2020, Wagler et al. 2023). In other ungulate taxa, neonates are more vulnerable to predation when they have lower birth mass, are born after peak parturition, are born after severe winters, or are born to mothers in poor nutritional condition (Rognmo et al. 1983, Skogland 1990, Fairbanks 1993, Adams et al. 1995, Singer et al. 1997, Adams and Dale 1998a,b; Testa and Adams 1998, Keech et al. 2000, Côté and Festa-Bianchet 2001, Mech 2007, Carstensen et al. 2009). Even the grandmother's nutritional condition can have cascading effects on neonate survival (Mech et al. 1991). Adult ungulates in poor nutritional condition also can be more vulnerable to predation and other sources of mortality (Bender et al. 2008, Sand et al. 2012).

### 7.2.3 Nutrition and Spatial Ecology

The need to acquire nutrition influences spatial behaviors of mountain sheep including seasonal habitat use, movements, and distributions. Mountain sheep often have to balance nutritional demands with predation risk or other factors in selecting habitat (Festa-Bianchet 1988c, Aycrigg et al. 2021, Denryter et al. 2024), which can alter patterns of habitat use seasonally and under different foraging conditions. For example, when forage was sparse, female Dall's sheep (*Ovis dalli dalli*) favored access to forage over safety from predators, but not when forage was more abundant (Rachlow and Terry Bowyer 1998). When forage was limiting, lactating bighorn sheep increased home range size and traveled further from escape terrain to access forage, which can increase predation risk (Wagler et al. 2024).

Many mountain sheep migrate along elevational gradients, which allows them to access high-quality forage as it becomes available (Hebert 1973, Spitz 2015, Merkle et al. 2016, Courtemanch et al. 2017). Seasonal migrations also can be important for mountain sheep to access mineral licks, which allow them to meet mineral requirements that cannot be satisfied by plants alone (Ayotte et al. 2008, Enns 2021). As with habitat selection, nutritional condition of mountain sheep affects choice of migratory tactic, representing a balancing act among endogenous and exogenous resources, starvation risk, and predation risk (Figures 7.1 and 7.2; Denryter et al. 2022a, 2024). During winter, mountain sheep conserve energy by reducing activity, particularly in snowier habitats or as winter progresses and energy reserves are depleted (Simmons 1982, Denryter et al. 2021).

## 7.3 DIGESTIVE MORPHOLOGY, FOOD PROCESSING, AND FORAGING BEHAVIOR OF MOUNTAIN SHEEP

Mountain sheep are ruminants, which refers to morphological and physiological adaptations in the digestive tract that facilitate processing of a plant-based diet. The features that most distinguish ruminants from other herbivores are the four-chambered stomach, microbial foregut fermentation, and rumination (evidenced by cud chewing), which allow

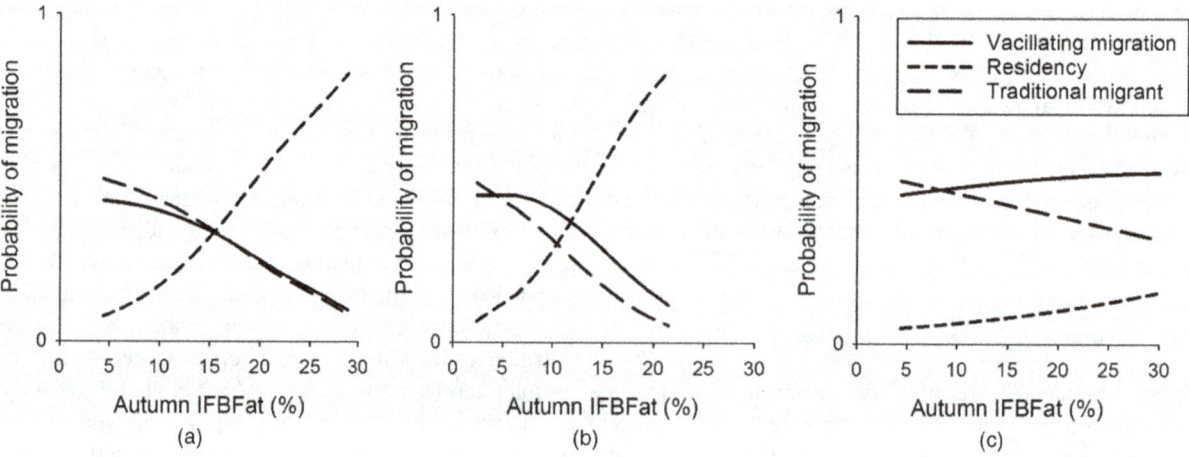

**FIGURE 7.1** Ingesta-free body fat (IFBFat) in autumn influenced choice of migratory tactic (traditional migrant, vacillating migrant, residency) in male Sierra Nevada bighorn sheep in California across three study areas from 2006 to 2019. Study areas were central (a), northern (b), and southern (c). Open Access; Denryter et al. (2024).

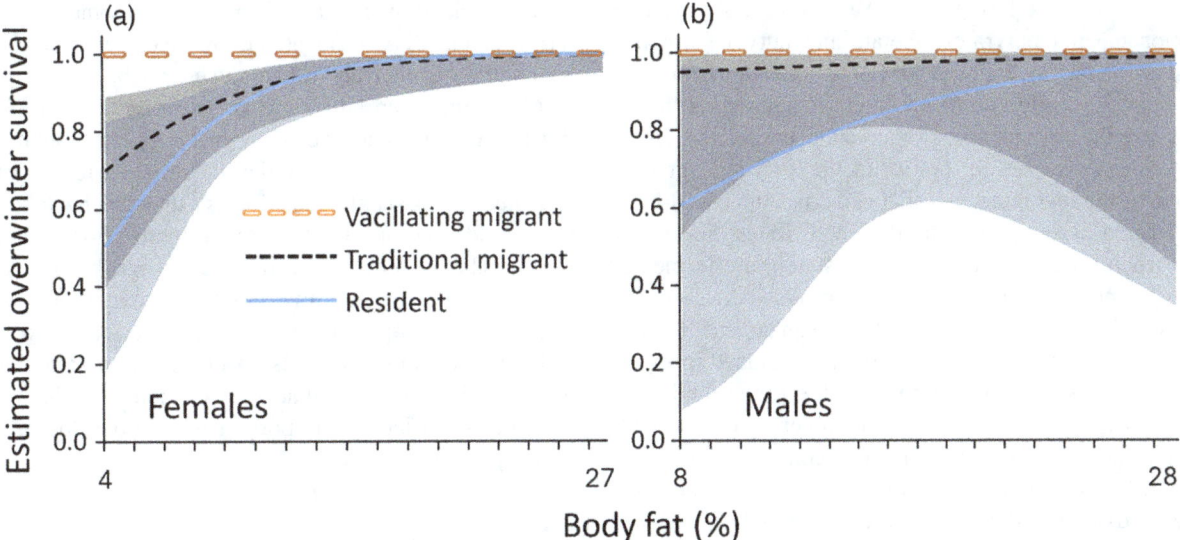

**FIGURE 7.2** Overwinter survival as a function of body fat for female (a) and male (b) Sierra Nevada bighorn sheep in California from 2006 to 2018 relative to three migratory tactics (traditional migrant, vacillating migrant, resident). Confidence intervals (95%) are shown in gray. Open Access; Denryter et al. (2022a).

ruminants to efficiently use fibers in the plant cell wall (primarily the carbohydrates cellulose and hemicellulose) as a source of energy (Van Soest 1982, Church 1993). Mammals lack enzymes needed to digest plant fibers, but ruminants have co-evolved a symbiosis with microbes that possess enzymes capable of digesting plant fibers. Microbes present in the rumen also synthesize many vitamins and synthesize protein from microbial protein and non-protein sources of nitrogen, such as urea, which ruminants can recycle to the rumen (Church 1993). The cost to ruminants of processing fibers is longer retention time in the digestive tract, which can sometimes be disadvantageous (see Section 7.3.4). Here, we provide a basic overview of ruminant digestive morphology and food processing, including implications to mountain sheep ecology, nutrition, and behavior.

### 7.3.1 STRUCTURE AND FUNCTION

Nutrient acquisition is initiated the moment a mountain sheep crops (tears) a bite of food; however, the chemical components of the food item must be transformed into products that are usable in the body. Digestion is the process by which ingested foods are transformed via physical, biological, and chemical processes in the digestive tract into products (e.g., volatile fatty acids [VFAs], amino acids, ammonia) that can be absorbed and used in the body for energy metabolism or assimilation as body tissue (e.g., fat, muscle). In mountain sheep, the digestive tract includes the mouth and associated structures (tongue, lips, lower incisors, dental pad, salivary glands, molars); the foregut, which includes the esophagus and four-chambered stomach (rumen, reticulum, omasum, abomasum), the midgut (small intestine); the hindgut (cecum, large intestine); and some accessory organs (pancreas, liver, gallbladder). Microbial

fermentation occurs in the foregut, prior to chemical digestion, and hence mountain sheep are foregut fermenters.

Prehensile structures of the mouth allow mountain sheep to select, grip, and sort bites of food. The dental pad is a unique feature of ruminants (and camelids), neither of which has upper incisors, and it is used in conjunction with the lower incisors to efficiently crop bites of food (Hofmann 1988). Ingested food is lubricated with saliva and the tongue moves food to the back of the mouth. Relatively large particles are swallowed and enter the esophagus initially because little chewing has occurred (most chewing occurs during rumination). Saliva facilitates transmission of food through the esophagus to the reticulum and rumen and acts as an important buffer to help counteract the effects of VFAs produced during digestion, aiding in maintenance of a neutral pH in the rumen. VFAs are the primary source of energy for ruminants as compared with glucose in monogastric animals including humans.

Digesta in the foregut flows freely between the first two chambers, the reticulum and rumen, which are contiguous and often considered together as the reticulo-rumen or rumino-reticulum. Both chambers function to store and delay food passage for enhanced digestion. The reticulum is a pouch-like structure, characterized by its honeycomb pattern of folds that sorts food particles by size. Small particles pass from the reticulum to the omasum, but large particles are returned to the rumen for further digestion and rumination. In rumination, a bolus of food (cud) is transmitted via the bi-directional esophagus to the mouth (regurgitation of ingesta) for additional chewing and reduction in particle size.

Food entering the reticulo-rumen undergoes extensive microbial fermentation, which is an essential symbiosis. Muscle contractions in the rumen move food to ensure contact with bacteria, protozoa, and fungi for fermentation.

Carbohydrates are fermented to VFAs (especially acetate, propionate, and butyrate). Acetate and butyrate are precursors for fat (including milk fat), whereas propionate is a precursor for glucose (from gluconeogenesis). Although VFAs are the primary energy source for ruminants, glucose is the energy source for brain function and fetal growth, as one of the few molecules that can cross the blood-brain barrier and placenta. Structural carbohydrates (fibers) and non-structural carbohydrates (sugar, starch) are fermented by different species of microbes and produce different end products (acetate from structural, propionate and butyrate from non-structural). Some proteins are degraded into peptides, amino acids, and ammonia in the rumen and other protein (bypass protein) escapes the rumen and is available for digestion and absorption in the abomasum and small intestine. Some lipids are hydrolyzed in the rumen, but most lipids bypass digestion in the rumen and are digested and absorbed through the small intestine.

Products of fermentation (e.g., VFAs, ammonia) in the rumen are absorbed across the rumen wall, which is enhanced by papillae that increase ruminal surface area. Rumen papillae respond to dietary changes, seasonally or otherwise, and can increase or decrease in number, size, and morphology (Church 1993, Zimmerman et al. 2006). VFAs and ammonia are routed through the bloodstream to various cells and organs for energy metabolism, urea recycling, and protein synthesis. Ammonia absorbed from the rumen is transferred to the liver and converted to urea. Urea and ammonia can be recycled to the rumen via the bloodstream or saliva and used by rumen microbes to generate microbial protein; excess urea is eliminated in urine (Cooper and Plum 1987, Owens and Zinn 1993, Alio et al. 2000). Urea recycling is an important adaptation for nitrogen conservation and survival on diets that are low in nitrogen (Owens and Zinn 1993, Barboza and Parker 2006, 2008; Barboza et al. 2009), including those experienced by mountain sheep at some times of year (Hansen 1996). Hence, urea recycling may be most relevant for desert sheep or for northern mountain sheep during winter.

Food particles small enough to leave the rumen move into the omasum and eventually through the rest of the digestive tract. The omasum acts as a pump to transmit foods from the reticulum to the abomasum and functions in absorption of water, VFAs, sodium, potassium, and some other nutrients. The abomasum is considered the true or glandular stomach of ruminants, with its enzymatic and hydrolytic digestion. The small intestine secretes various enzymes and receives bile and pancreatic enzymes required for digestion of lipids, sugars, and proteins. The small intestine also is an important site of absorption, owing to many villi, and is highly vascularized so nutrients absorbed here can be readily transported via the bloodstream to other parts of the body.

The cecum and large intestine are the final organs in the sequence of the ruminant digestive tract. The cecum occurs as a blind pouch where the small and large intestines meet. The cecum is a site of additional post-ruminal (hindgut) fermentation, yet it has a similar microbial population to the rumen. The large intestine primarily functions in absorbing water and electrolytes, waste formation, and some additional fermentation. VFAs fermented in the colon can be absorbed but contribute relatively little to the energy budget.

Mountain sheep lambs are not born with a fully functional ruminant digestive tract. In lambs, the esophageal groove, which is formed by the undeveloped reticulum, is activated during suckling and forms a tube that bypasses the rumen and omasum, shunting milk directly into the abomasum for chemical digestion. If the esophageal groove does not close, milk can enter the rumen, where its fermentation results in ruminal acidosis, which can be fatal. Ingestion of forage by lambs promotes development and maturation of the lamb rumen into that of a functional ruminant and butyrate is particularly important for the development of rumen papillae (Lyford 1993).

### 7.3.2 Feed Types and Diet

Ruminants have been organized into ecologically similar groups based on their diet and morphology of the digestive tract (Hofmann 1988). Concentrate selectors (sometimes considered browsers) consume diets low in fiber and high in cell solubles, primarily comprising deciduous shrubs and forbs. Bulk or roughage feeders (sometimes considered grazers) consume diets high in fiber, typically comprising grasses. Intermediate mixed feeders have evolved to live on diets comprising a mix of diet items, including forbs, leaves of deciduous shrubs, and a variety of graminoids. Historically, mountain sheep have been classified as grazers or bulk feeders, but they are highly selective foragers, meaning they choose diets discriminately from what is available, consuming specific forages or parts of forages (e.g., leaves, flowers) preferentially over others. Mountain sheep also have diverse diets that are not entirely consistent with the grazer classification. Empirical support for a universal, invariable grazer-browser classification scheme is generally lacking, but it remains a useful conceptualization for illustrating the spectrum of diets of wild ruminants (Gordon and Illius 1994, Robbins et al. 1995).

Mountain sheep forage across diverse landscapes with mosaics of plant communities that differ in nutritional value. Mountain sheep in the north consume a variety of graminoids (i.e., grasses, sedges, and rushes), forbs, and deciduous shrubs (Todd 1975). Diets vary seasonally, but grasses (bluegrass [*Poa* spp.], wild rye [*Elymus* spp.]) and sedges (*Carex* spp.) are important year-round for Stone's sheep (*Ovis dalli stonei*; Luckhurst 1973, Seip 1983), and Dall's sheep (Hansen 1996, Metherell 2023); during spring and autumn for Rocky Mountain bighorn sheep (Johnson 1980, Wagner and Peek 2006), and for Stone's sheep in winter (Luckhurst 1973). Forbs (e.g., locoweed [*Oxytropis* spp.], yarrow [*Achillea* sp.], sagewort [*Artemisia* spp.]) are important during summer and early autumn for Stone's sheep, Dall's sheep, and Rocky Mountain bighorn sheep (Johnson 1980, Seip 1983, Wagner and Peek 2006, Metherell 2023). Mushrooms and lichens also can be important seasonally (Seip 1983, Metherell 2023).

Deciduous shrubs (willow [*Salix* spp.] and blueberry [*Vaccinium* spp.]) are important during summer–autumn for Stone's and Dall's sheep and for Rocky Mountain bighorn sheep (Seip 1983, Wagner and Peek 2006, Metherell 2023). During winter, grasses are generally important with contributions of forbs and deciduous shrubs varying across landscapes (Blood 1961, Oldemeyer et al. 1971, Luckhurst 1973, Tilton and Willard 1981). Important grasses include wheatgrass (*Agropyron* spp.), bluegrass, fescue (*Festuca* spp.), needlegrass (*Stipa* spp.), bromegrass (*Bromus* spp.), and junegrass (*Koeleria* spp.; Blood 1961, Constan 1967, Tilton and Willard 1981, Keating et al. 1985). In some areas, rabbitbrush (*Chrysothamnus* spp.) and phlox (*Phlox* spp.) are important during winter (Oldemeyer et al. 1971, Keating et al. 1985).

Diets of desert bighorn sheep are diverse and include deciduous and evergreen shrubs, forbs, graminoids, and succulents (Brewer and Harveson 2007). Important shrubs include mountain mahogany (*Cercocarpus* spp.), buckthorn (*Ceanothus* spp.), acacia (*Acacia* spp.), sagebrush (*Artemisia* spp.), buckwheat (*Eriogonum* spp.), bitterbrush (*Purshia* spp.), ephedra (*Ephedra* spp.), saltbrush (*Atriplex* spp.), and currant (*Ribes* spp.; Seegmiller et al. 1990, Blum et al. 2023). Important forbs include sandwort (*Eremogone* spp.), buckwheat (*Eriogonum* spp.), and janusia (*Janusia* spp.; Miller and Gaud 1989, Seegmiller et al. 1990, Fulbright et al. 2001, Blum et al. 2023). Important graminoids include bromegrass, needlegrass, bluegrass, and muhly (*Muhlenbergia* spp.). Prickly pear (*Opuntia* spp.) is an important succulent (Seegmiller et al. 1990). Sierra Nevada bighorn (*O. c. sierrae*) sheep diets comprise species included in the diets of northern and desert taxa (Wehausen 1980, Greene et al. 2012). Water or nitrogen (protein) content of forages can influence forage selection in the desert (Cain et al. 2017), but not in more northern climates (Seip 1983).

What animals eat is an important part of their ecology, and diets can certainly be a good starting point for studies of nutrition. Diets, however, are only one piece of the overall nutritional picture. What really matters from a nutritional perspective is nutrient intake and how nutrient intake relates to the nutritional requirements of the individual. Robbins (1993: 1) warned that using diets "…to develop management schemes is all too often destined to failure because of the absolute need to understand the much broader nutritional interaction from an ecological perspective. Unfortunately, preoccupation with food habits has reduced our investigations of other equally important areas of nutrition…" Thus, rigorous studies of nutrition must necessarily go beyond diet composition.

### 7.3.3 Food Quality: Nutrients and Antinutritional Factors

*Nutrients.* Mountain sheep obtain nutrients from the foods they eat. Macronutrients are required in large amounts and include carbohydrates, proteins, lipids, and water.

Micronutrients are required in much lesser amounts and include vitamins and minerals.

Carbohydrates that are fermented to VFAs are the main source of energy for ruminants and are components of various other metabolites and nutrients, such as fatty acids and sugars for DNA and RNA. Carbohydrates are present in two forms in plant cells. Storage carbohydrates, which are in cytoplasm and comprise starch and various sugars (-saccharides) are highly digestible and therefore readily provide digestible energy. Structural carbohydrates (fibers) include cellulose, hemicellulose, and pectin, occur in the cell wall of the plant, and provide energy after being fermented by microbes into VFAs.

Proteins are the primary form of nitrogen in animals. They serve various roles in immunity, blood clotting, metabolism, and hormones, and can be an energy source if energy supplies are low. Other non-protein nitrogen sources are also important to ruminants including nucleic acids, ammonia, and urea.

Lipids include fats, oils, and other compounds, and are a concentrated source of energy. Plants contain few lipids but provide essential fatty acids to ruminants (Robbins 1993). Ruminants can synthesize most fatty acids from acetate derived from microbial fermentation of plant fibers.

Minerals are inorganic molecules with functions in cell structure, as catalysts, and in various physiological and metabolic processes. All minerals are micronutrients, being required in small quantities, but these are subdivided into macro- and trace minerals. Macrominerals are required in relatively large amounts in the diet (g/kg) and include calcium, phosphorus, and sodium. Trace minerals are required in relatively small amounts (mg, μg/kg) and include iron, zinc, manganese, copper, molybdenum, and selenium. Because mineral content of plants is variable, mountain sheep may exhibit geophagy (soil consumption) to acquire various minerals (Simmons 1982, Heimer 1988, Ayotte et al. 2008). Some minerals interact with one another to affect absorption and bioavailability, such as the effects of iron and zinc on copper absorption. In other cases, sparing effects can occur, such as between selenium and Vitamin E, meaning a slight deficiency in one may be compensated for by higher amounts of the other (Mehdi and Dufrasne 2016). Other minerals must be balanced within relatively narrow ratios (e.g., 2:1 ratio of Ca:P).

Calcium is important to bone health, density, and composition, and is used in cell signaling, blood clotting, nerve excitability, acid-base balance, muscle contraction, and enzyme activation (Carafoli 1991, Robbins 1993, Hurwitz 1996, Suttle 2010). Approximately 99% of body calcium is stored in the skeleton. Forages generally provide adequate calcium (Kincaid 1993, Suttle 2010), so it is rarely limiting.

Phosphorus is a component of teeth, bones, nucleic acids, cell membranes, and energy currencies (e.g., adenosine triphosphate). The rumen microbiome requires phosphorus for cellulose digestion and microbial protein synthesis (Bucholtz and Bergen 1973, Komisarczuk et al. 1987, Kincaid 1993). Deficiencies in phosphorus can occur in domestic ruminants grazing poor-quality ranges (typically <0.25% phosphorous; Goff 2009) and are the second most common deficiency in grazing ruminants (Kincaid 1993). Experiments with deer (*Odocoileus* spp.) demonstrated adaptive, compensatory responses to phosphorus deficiency (Grasman and Hellgren 1993).

Sodium is an important cation and electrolyte, with roles in maintaining osmolarity, body fluid volume, acid-base balance, body pH, transmission of nerve impulses, muscle contractions, and appetite (Robbins 1993, Suttle 2010). Sodium deficiencies are unlikely, but may occur where water leaches nutrients from soils (e.g., alpine and montane environments; Robbins 1993, Barboza et al. 2009).

Iron is an important trace mineral with a significant role in oxygen transport, energy metabolism, and as an antioxidant (Miller et al. 1993, Suttle 2010). Many forages, especially legumes, provide adequate levels of iron to meet nutritional requirements (Miller et al. 1993, Suttle 2010), and iron deficiencies are generally uncommon in free-ranging herbivores (van den Top 2005, Suttle 2010). Parasitic infections can cause iron-deficiency anemia in ruminants and young ruminants are more susceptible to such anemias, especially if they have a history of illness (van den Top 2005, Suttle 2010).

Zinc is a component of various enzymes involved in DNA, RNA, and protein synthesis; is a cofactor of many enzyme systems including various antioxidant systems; and is involved in immune function (Miller et al. 1993, Robbins 1993, Prasad et al. 2004). Zinc levels vary seasonally (Albl et al. 1977) and deficiencies have occurred in domestic ruminants, manifesting as retarded growth, hair loss or roughness, reduced appetite and weight loss (Miller 1970, Miller et al. 1993, Shankar and Prasad 1998).

Selenium has roles as an antioxidant, as part of the enzyme glutathione peroxidase (Arthur 2000, Margis et al. 2008), and in thyroid metabolism (Rowntree et al. 2004), fertility (Maiorino et al. 1999, Ahsan et al. 2014), and immunity (Petrie et al. 1989). Selenium deficiencies have been associated with nutritional muscular dystrophy (white muscle disease) in wild and domestic ruminants (Hebert and Cowan 1971, Godwin et al. 1974, Wilson and Grace 2001), which manifests as ill thrift, poor growth rates, stiff gaits, and animals being too weak to stand or nurse. Selenium deficiencies also can result in decreased fertility. Wild ruminants may be more adept at selenium conservation and have greater glutathione peroxidase activity than domestic ruminants (Robbins et al. 1985); increased glutathione peroxidase activity means wild animals may have greater antioxidant activity from a given level of selenium compared to domestic taxa. Hence, wild ruminants can persist on lower levels of selenium than domestics. Generally, even when selenium levels of wild ungulates have been low, animals have not displayed signs of deficiency, nor have population-level effects been noted (Harper 1984, Vikøren et al. 2011). Selenium has a narrow safety margin and is the essential trace element most likely to become toxic (Suttle 2010).

The primary role of manganese in the body is to activate various enzymes (Culotta et al. 2005, Studer et al. 2022), including those involved in intermediary metabolism of lipids and carbohydrates (Scrutton et al. 1972, Robinson 1998). Manganese deficiencies can manifest as low birth weight of lambs, skeletal deformities, or reproductive issues.

Copper is part of many enzymes including those involved in cellular respiration, central nervous system function, and immune function (Spears 2000, Spears and Weiss 2008). Copper deficiencies manifest via abnormalities in the hair coat, reduced growth, and lower reproductive rates, but reproductive effects may be secondary to poor nutritional condition rather than copper deficiency per se (Handeland et al. 2008).

Molybdenum is involved in various enzyme systems in animals (de Renzo et al. 1953, Gardner et al. 2003), but also is important for its role in copper deficiencies through the formation of insoluble copper thiomolybdate (Robbins 1993, Gardner et al. 2003). Deficiencies may manifest as neurologic symptoms and death (Sardesai 1993, Schwarz et al. 2009).

Vitamins are organic molecules that serve as coenzymes in metabolic processes. Mountain sheep are unlikely to experience vitamin deficiencies because rumen microbes can synthesize all B vitamins and Vitamin K. Mountain sheep can synthesize Vitamin C from glucose and Vitamin D following exposure to sunlight, and most green forages provide adequate quantities of precursors of Vitamin A and Vitamin E (Robbins 1993, Barboza et al. 2009). B vitamins have many roles, especially in metabolism and as neurotransmitters and enzymes. Vitamin C has roles in immunity, bone formation, and iron absorption. Vitamin A is important for vision, immunity, bone maintenance, reproduction, and more. Vitamin D is important to calcium-phosphorus homeostasis, which is important for bone health, immunity, transmission of nerve impulses, and more. Vitamin E is an important antioxidant that has sparing effects with selenium, and deficiencies of Vitamin E or selenium can contribute to capture myopathy in ungulates. Vitamin K is essential for blood clotting, calcium metabolism, and more.

*Antinutritional factors.*—Other components of mountain sheep foods, including plant secondary metabolites, are antinutritional, meaning they reduce the availability of other nutrients in the diet. Plant secondary metabolites are compounds that serve no primary metabolic function in plants, but often are involved in plant defenses, and include three predominant classes: phenolics, alkaloids, and terpenes.

Two phenolics with nutritional relevance to herbivores are condensed tannins and lignin. Tannins are present in some forbs and woody vegetation; they inhibit protein digestion by forming insoluble complexes with proteins. When tannin content of forage is high, digestible protein intake can be negative (Robbins et al. 1987a, McArt et al. 2009, Spalinger et al. 2010, Denryter et al. 2022b). Negative digestible protein intake occurs because animals lose more protein during digestion (i.e., from sloughing of endogenous protein from the digestive tract) than is replaced by the forage. Lignin provides rigidity to plant cell walls, as part of a matrix with structural carbohydrates; it is indigestible even by rumen microbes and inhibits digestion of carbohydrates in its matrix.

Alkaloids and terpenoids pose different challenges to mountain sheep than phenolics. Alkaloids can be neurotoxic and fatal to mammals and occur in species such as lupine (*Lupinus* spp.) and larkspur (*Delphinium* spp.; Pfister et al. 1996, Lopez-Ortiz et al. 2004). Consequently, mountain sheep usually avoid or consume only small amounts of alkaloid-containing forages. Terpenoids occur in plant essential oils and resins, particularly those found in sagebrush (*Artemisia* spp.) and conifers. They are high in gross energy content (Personius et al. 1987), but not very digestible (Striby et al. 1987), so little energy is available to the animal.

### 7.3.4 FOOD QUALITY: DIGESTIBILITY AND DIGESTIVE CONSTRAINTS

Food quality refers to the nutritional value of foods, which is a function of their nutrient content and digestibility. Food quality is important to mountain sheep and can be limiting even when biomass of available forage is high (Seip 1983, Cook 1990) because of constraints on digestion. Fiber is one of the main determinants of digestibility (Mertens 1993). Tannins and lignin reduce digestibility, with effects of tannins being most pronounced in protein digestion (Robbins et al. 1987a).

Digestive constraints arise because of adaptations that make the ruminant digestive tract so effective at digesting fiber. Because the rumen has a limited capacity, some food must exit the rumen before more food can enter it (Blaxter et al. 1961, Conrad et al. 1964). Foods that are less digestible will pass more slowly through the digestive tract than foods that are more digestible, thereby limiting intake until the rumen has capacity again. Because of these digestive constraints (i.e., gut fill, passage rates), ruminants cannot simply eat more food to meet nutritional requirements when forage quality is low. Ruminants can even starve to death with a rumen full of food because of digestive constraints (Doman and Rasmussen 1944, Pearson 1969, Gasaway and Coady 1974, Dean et al. 1975, Sanford et al. 1993).

### 7.3.5 HIGH-QUALITY FORAGE

High-quality forage refers to forage that can meet nutritional requirements for a specified level of individual performance (e.g., fattening, growth), productivity (i.e., reproduction), or both. Characteristics of high-quality forage during summer include high digestibility (i.e., low fiber and low concentrations of antinutritional factors), high energy content (≥12.1 kJ/g digestible energy), and high protein content (≥13% crude protein; Seip 1983, Hansen 1996). Seip (1983) estimated summer maintenance requirements could be met on forages as low as 48% digestibility, but

requirements for gestation, growth, and lactation would be higher (≥61%–68% digestibility). For desert bighorn sheep, high-quality forages are available during the rainy season, with quality declining through the dry season.

Plant phenology underpins seasonal cycles in forage quality and abundance. In northern environments, initiation of plant growth occurs during spring, the start of which varies across latitudinal gradients. During spring, forages are fast growing, highly digestible, and high in protein and energy. Plant quality peaks in spring–summer, begins to decline as plants senesce in summer–autumn, and is lowest in winter when deciduous foliage is unavailable. As plants age, they undergo senescence, which includes increasing deposition of lignin; however, plant toxins decrease as plants mature and senesce (Striby et al. 1987, Ralphs et al. 1988, Pfister et al. 1996).

## 7.4 NUTRITIONAL REQUIREMENTS

### 7.4.1 ENERGY AND PROTEIN REQUIREMENTS

Energy and protein are two of the most limiting nutritional currencies for ruminants (Verme and Ozoga 1980, Syrjälä-Qvist and Salonen 1983, Robbins 1993, Cook et al. 2016, Denryter et al. 2022b), even more so than minerals or vitamins, though water can be an important limiting nutrient for mountain sheep in arid environments (Cain et al. 2017). Energy is not a nutrient but a property of nutrients that powers all cellular functions, growth, reproduction, survival, and is required for protein use (Syrjälä-Qvist and Salonen 1983). Starvation occurs when animals do not have adequate energy to power these processes; in the most extreme cases, starvation leads to death, but in less severe cases, growth and reproduction may be reduced. Energy intake and efficiency of energy use, but not protein, generally limit animal production (Sachdeva et al. 1973, Abler et al. 1976, Verme and Ozoga 1980, Owens and Zinn 1993).

Protein is necessary for immune function, growth, reproduction, and can serve as a source of energy. Rumen microbes convert non-protein nitrogen to microbial protein, which can be digested in the abomasum and small intestine to provide ruminants with another source of protein besides dietary protein. Microbial protein can supply 100% of the protein requirements of non-lactating adults, but only about 75% of the elevated protein requirements of growing or lactating animals can be met by rumen microbial production (Owens and Zinn 1993, Wu 2017, Amaral-Philips 2020). Additionally, a minimum level of nitrogen, often as ammonia, is required for ruminal fermentation, and when diets are nitrogen-deficient, feed intake and efficiency of digestion decline (Van Soest 1982, Goodson et al. 1991b). Ruminants can mobilize body protein stores and recycle nitrogenous wastes (e.g., urea) to help meet protein requirements (Bryant and Robinson 1961, Ørskov and McDonald 1979, Barboza and Parker 2006, 2008; Firkins et al. 2007).

Requirements for energy and protein reflect the quantity of each needed to support a specific metabolic objective (e.g., growth, fattening, reproduction). The total requirement for an individual mountain sheep is the sum of its requirements for maintenance, activity, production, and reproduction. Once estimates of nutritional requirements are available, they provide a baseline against which energy, protein, or other intakes (e.g., minerals) can be compared to evaluate nutritional value of habitats or landscapes (Parker et al. 1999, Cook et al. 2016, Denryter et al. 2022b). Energy and protein requirements have been estimated for captive mountain sheep using indirect calorimetry (respirometry; Chappel and Hudson 1979, Dailey and Hobbs 1989). Expenditures have been combined in bioenergetic models to estimate energy requirements for free-ranging mountain sheep (Denryter et al. 2021).

### 7.4.2  Maintenance Requirements

Maintenance requirements are those required to sustain homeostasis while maintaining a constant (neither increasing or decreasing) body mass. Maintenance requirements are the first priority for nutrients in the body and make up a large proportion of the nutritional requirements for mountain sheep year-round (Denryter et al. 2021). Metabolic requirements are a primary component of maintenance requirements and scale allometrically with body size, such that larger-bodied individuals have greater absolute requirements but lower relative (mass-specific) requirements. In addition to metabolism, maintenance requirements for energy typically include costs of normal activities and thermoregulation (Robbins 1993). Maintenance protein requirements reflect protein lost from homeostatic activities in feces and urine, skin sloughing, and hair molt (Robbins 1993, National Research Council 2007, Denryter and White 2026). Many variables influence maintenance requirements including level of food intake, weather and environmental conditions, behavioral and physiological adaptations, sex, season, and body mass (Chappel and Hudson 1980, Cook et al. 1998, Signer et al. 2011, Denryter et al. 2021). Absolute (total) requirements for maintenance are typically greater in males than females, in reproductive than non-reproductive females, in juveniles than non-growing animals, and during summer than winter (Chappel and Hudson 1980).

### 7.4.3  Requirements for Locomotion, Activity, and Thermoregulation

Mountain sheep require energy to support locomotion and other activities, such as foraging, fleeing from predators, and migration; minimal levels of activity are included with maintenance requirements. Activity costs vary, but mountain sheep that migrate longer distances, move faster or across steeper terrain, or move through deeper snow will have higher energy requirements for activity (Parker et al. 1984, Luick and White 1986, Dailey and Hobbs 1989). Supplemental energy

expenditures for thermoregulation also increase substantively in winter at temperatures below −20°C in Rocky Mountain bighorn sheep (Chappel and Hudson 1979).

### 7.4.4  Production Requirements: Growth and Fattening

Productive costs include those for fattening and growth. Productive costs are highly variable depending on production goals and timelines. In seasonal environments, the greater an animal's deficits in body mass and body fat entering summer, the greater its protein and energy requirements will be prior to winter. When nutrition is limiting, expenditures on production may be reduced to compensate, exhibited by reductions in somatic or horn growth and fat accretion.

### 7.4.5  Requirements for Reproduction

Reproductive costs largely refer to costs of gestation and lactation, with lactation having the highest requirements for energy and protein (Oftedal 1985, Robbins 1993, National Research Council 2007), but they also can include costs of searching for and competing for mates in males. During gestation and lactation, females may rely on energy and protein reserves to supplement nutrient intake to support fetal growth and milk production (Oftedal 2000, Barboza and Parker 2008, Taillon et al. 2013, Weber et al. 2013).

### 7.4.6  Water Requirements

Water comprises most (50%–60%) of the body of adult mammals and is an important solvent with roles in metabolism, thermoregulation, milk production, growth, digestion, and excretion among others (Robinson 1957). Water balance of mountain sheep is constantly in flux as they lose water to their environment (e.g., urination, evaporative loss) and gain it from various sources. Mountain sheep acquire water by consuming free water that is not bound in other compounds such as water in a stream; from preformed water contained in food; or from metabolic water produced via metabolism. Mountain sheep acquiring more preformed water require less free water and in extreme cases may survive dry periods even without free water (Warrick and Krausman 1989). Use of free water increases during summer when desert plants have less moisture (Whiting et al. 2010, Terry et al. 2022) and can be affected by water quality (Meehan et al. 2021) and invasive wildlife species, such as feral horses (Ostermann-Kelm et al. 2008, Hall et al. 2018).

Where water is limiting for wild herbivores, generally so is food and hence it is important to evaluate food and water requirements and availabilities before concluding water is limiting (Robbins 1993). Provisioning additional water can help increase populations only if water is limiting and all other needs, including food, are met (Robbins 1993). Experimental removal of artificial water sources did not result in changes to desert bighorn sheep home range size or movement rates, nor did the presence of artificial water sources prevent mortalities during a drought (Cain et al. 2008). Mortalities instead were linked to inadequate forage quality and quantity (Cain et al. 2008).

### 7.4.7 MINERAL AND VITAMIN REQUIREMENTS

Requirements of mountain sheep and other wild ruminants for minerals and vitamins are poorly understood and therefore, requirements for domestic livestock (National Research Council 2007) often are used as crude guidelines. Wild ruminants may be better adapted to lower-mineral diets than domestic ruminants that have undergone artificial selection for traits required for high levels of production (Robbins et al. 1985). Hence, requirements for domestic ruminants should be applied cautiously to wild ruminants. Geophagy and migration are behavioral adaptations that allow mountain sheep to acquire minerals from inorganic sources (e.g., soil; Ayotte et al. 2008, Oster et al. 2018, Enns 2021).

### 7.4.8 FOOD INTAKE: THE FUNCTIONAL RESPONSE IN MOUNTAIN SHEEP

Rates of forage intake are a function of the size and quality of available bites, spatial distributions of available bites, and biomass of available food. The relationship between the rate of food intake and food density (biomass) is the functional response (Solomon 1949, Holling 1959). In sheep, the functional response is sometimes better explained by grass leaf length (or height) than by forage density (Seip 1983). Bite mass and biomass are typically the strongest predictors of intake rate when food quality is relatively high (Shipley and Spalinger 1992). Bite rate is inversely related to bite mass, and hence as bite rate increases, bite mass decreases. Increasing bite rate is a compensatory mechanism for small bite mass but cannot fully compensate (Cook et al. 2016, Denryter 2017). Further, there is not a strong relationship between bite rate and intake rate, which limits the utility of bite rate as an indicator of nutritional value (Cook et al. 2016, Denryter 2017).

Intake rates of mountain sheep have rarely been documented, but bite rates have been. Mountain sheep consume almost twice as many bites of food during summer (18,000) than winter (9,600) (Seip 1983, Hansen 1996). Mountain sheep take fewer bites during winter than summer because they spend less time foraging each day (4–8 hours foraging in winter versus 10–13 hours during summer; Seip 1983, Hansen 1996) and snow restricts access to forage (Goodson et al. 1991b). During periods of snow in late-winter-spring, mountain sheep increase foraging time and bite mass, and prioritize biomass intake, whereas during snow-free periods they prioritize intake relative to forage quality (Goodsons et al. 1991b).

Mountain sheep face important constraints to forage intake relative to their gregarious social structure and access to escape terrain. Foraging efficiency (i.e., the ratio of food or nutrient gain compared to time or energy spent foraging) of mountain sheep decreases as they move further from escape terrain, which can exacerbate food limitations (Risenhoover and Bailey 1985, Cook 1990). Large group size (>10) and increased visibility of habitats are associated with increased foraging efficiency and allow mountain sheep to increase access to forage and foraging efficiency (Risenhoover and Bailey 1985, Mooring et al. 2004, Rankins et al. 2024). Thus, group foraging possibly represents a behavioral adaptation that allows for use of habitats with greater predation risk (Risenhoover and Bailey 1985, Mooring et al. 2004, Rankins et al. 2024). Population density, however, influences competition for forage resources among individuals and can affect intake rates (Hobbs 2024).

## 7.5 TECHNIQUES FOR EVALUATING DIET COMPOSITION, FOOD QUALITY, AND FOOD QUANTITY

### 7.5.1 DIET COMPOSITION

Diet composition refers to the specific makeup and amounts of all foods consumed by an animal over a given period, often presented on a proportional basis. Diet composition differs from food selection in that diet composition does not provide information on dietary choices relative to availability of all the different foods. Estimating diet composition is difficult in free-ranging herbivores including mountain sheep. Methods for estimating diet composition are broadly categorized as direct, in which observers watch the animal eat, and indirect, in which diets are inferred from samples collected after eating has occurred, typically from samples of feces, rumen contents, tissue samples, or feeding site investigations (Shipley et al. 2020a). Each technique has advantages and disadvantages.

Direct observations often rely on the bite count technique (Wallmo et al. 1977), in which the number of bites an animal consumes of specific forages is recorded during a specified collection period and mass of those bites is estimated by clipping and drying simulated bites that approximate the size of bites consumed by the forager. Bite count techniques have been employed with spotting scopes or binoculars to study diets of free-ranging Dall's sheep (Metherell 2023), Stone's sheep (Seip 1983), and desert bighorn sheep (Goodson et al. 1991b). Challenges with this technique include estimating bite mass and data that are biased when the mouth of the focal sheep disappears from view. Only counting bites, without paired bite masses, is inadequate to estimate intake rates or diet composition because bite mass of individual plant species can vary up to 1,000-fold (Denryter 2017, Shipley et al. 2020a). The same inadequacy is present with a newer technique, the use of video collars to estimate diet composition. Video collars have not been validated for determining diet composition in mountain sheep and studies on other ungulates identified significant errors in diet composition as determined from video collars (K. Denryter, Alaska Department of Fish and Game, unpublished data).

Historically, fecal samples were collected and evaluated using microhistological techniques in which plant parts remaining in the feces were identified with the use of a microscope. The primary disadvantage of this technique is that less-digestible forages are overrepresented in the feces, relative to the proportion they comprise in the diet, and more digestible forages are underrepresented (Boertje et al. 1985). In some cases, digestibility correction factors have been developed, but these are not universally applicable (Russell et al. 1993). Recently, DNA metabarcoding has been applied to mountain sheep diets, but this method has not been validated and may face some of the same biases as microhistological techniques. For example in ruminants, nucleic acids are rapidly broken down in the rumen with high rates of disappearance and are further digested by pepsin in the abomasum in ruminants, affecting their presence in fecal samples (Smith and Mcallan 1970, McAllan and Smith 1973, Liu et al. 2015). Currently, the utility of DNA metabarcoding and video collars probably is limited to providing a list of species consumed by foragers.

## 7.5.2 Food Quality

Typically, wildlife biologists are interested in determining digestible energy and digestible protein content of forages available to mountain sheep because these are the two nutritional currencies most likely to be limiting for ruminants. To estimate digestible energy, measures of gross (total) energy and digestibility of the forage item are needed. Gross energy is a measure of the total energy content of a food item released during combustion in a bomb calorimeter, but not all this energy is available to the animal. Energy is lost to feces, urine, methane production, and heat production during digestion (i.e., the heat increment of feeding). Digestible energy reflects the difference between the gross (total) energy content of the forage and what is lost to feces. Digestible energy is the product of gross energy and digestibility. In some cases, biologists may desire to determine metabolizable energy content of forages, which accounts for energy losses to urine and methane production; digestible energy in ungulates generally is converted to metabolizable energy by multiplying digestible energy by 0.82.

Digestibility of forage is measured via *in vitro* dry matter digestibility (in an incubated test tube with rumen fluid and a buffer; Tilley and Terry 1963), sequential fiber analysis (Goering and Van Soest 1970), or *in vivo* (in a live animal for which all food intake and feces are measured, although this is rare). *In vitro* dry matter digestibility is affected by the source of rumen fluid, typically from domestic ruminants, and the diet the rumen donor was consuming (Shipley et al. 2020a), which may not reflect the microbial community in mountain sheep. Sequential fiber analysis allows for extraction and estimation of individual constituents of digestible contents of the plant cell (starch and sugars) versus less-digestible, structural components of the cell wall (cellulose, hemicellulose). Digestibility is determined by substituting the outputs of each sequential fiber analysis step into summative equations (Robbins et al. 1987a,b; Cook et al. 2022). Summative equations were developed using cervids and have not been confirmed in mountain sheep but can be used to approximate digestibility by mountain sheep. The advantage of sequential fiber analysis is that it is highly repeatable and not subject to biases of *in vitro* dry matter digestibility (Vogel et al. 1999, Shipley et al. 2020a, Cook et al. 2022).

Protein content of forages is estimated from nitrogen content of forages as determined in an elemental analyzer. Elemental analyzers use the Dumas method (rather than the Kjeldahl method). Nitrogen content of the sample is multiplied by 6.25 to estimate crude protein content because plant proteins average approximately 16% nitrogen (100/16=6.25). The effects of tannins on protein digestibility are accounted for by measuring the protein-precipitating capacity of the sample using bovine serum albumin procedures (Martin and Martin 1983, McArt et al. 2009). Digestible protein is estimated using summative equations for cervids (Robbins et al. 1987a). Metabolizable protein is difficult to estimate and is not often used for wild ruminants.

Other methods for estimating quality of forage include near-infrared reflectance spectroscopy, enzymatic digestion methods, and analysis of fecal nitrogen. Near-infrared reflectance spectroscopy requires thousands of samples for each forage (and forage part), for each unique growing environment, for calibration against wet chemistry methods (e.g., sequential fiber analysis). While near-infrared reflectance spectroscopy has been validated in some cases for livestock, it has not been validated for mountain sheep or other wild ruminants, which consume other forages in other growing environments (though there may be some overlap). Enzymatic digestion methods leave substantially more unexplained variation in digestibility than sequential fiber analyses. Fecal nitrogen is still commonly used to index forage quality, though it can be biased and an insensitive indicator of dietary nitrogen, particularly if some proteins are bound by tannins and not available to the animal (Robbins et al. 1987a). Even small differences in digestibility can scale up to have important multiplier effects on animal performance (White 1983), so choosing an accurate and precise method for determination of diet quality is of paramount importance.

To determine forage quality, forage samples should be collected according to study objectives, by how the animals consume the forages (e.g., separating by plant part [leaves, flowers, stems]), by plant functional group, by plant species, or by how biomass samples are collected if the goal is to pair quality with biomass. After collection, forage samples should be frozen or placed directly in a lyophilizer (freeze drier) to preserve and prepare them for sequential fiber or protein analysis. Not freezing samples risks altering plant chemistry and hence estimates of forage quality. Other methods of sample preservation may be required for plant secondary metabolites but may alter samples for digestible energy and protein analyses (Shipley et al. 2020a).

### 7.5.3 Food Quantity

Various methods are available to estimate food quantity and an important consideration is what the animals actually eat, in terms of species, plant part, and biomass (Shipley et al. 2020b). Intensive, destructive plant sampling generally provides the most accurate and precise measures of available food but can be time-consuming and expensive. Double-sampling techniques have gained increasing popularity for their rapid execution in the field, but require training and calibration for each observer (Greene et al. 2012, Monzingo et al. 2022). Methods that estimate metrics of the plant community that are not directly relevant to mountain sheep foraging behavior and selection, such as forage cover, may not reflect nutritional relationships. Thus, such indirect measures of forage should be used cautiously, particularly if study objectives relate to understanding animal-centric relationships between the forager and its environment (Searle et al. 2007, Cook et al. 2018).

## 7.6 FAT DYNAMICS

### 7.6.1 The Animal Indicator Concept

Mountain sheep are a product of the environment in which they live. Franzmann and Schwartz (1985) proposed the animal indicator concept and suggested that the physiological attributes of an animal could be used to determine the quality of their habitat. The body composition of an animal represents the integration of numerous extrinsic environmental variables, including the nutritive value of forage and weather variables that limit access to forage and the nutritional state of the animal (e.g., growing, lactating), which influence energy intake, expenditures, and requirements. Intrinsic factors such as reproductive success also have a strong influence on an animal's nutritional condition (Cook et al. 2004). Energy is the primary currency that promotes growth, reproduction, and survival. Consequently, the indicator that indexes habitat quality should reflect the availability of energy for metabolic processes. Fat and protein are the primary somatic reserves and function as a reservoir of stored energy. A combination of fat measured using ultrasonography and a body palpation score provides the most accurate and precise estimate of body fat in ungulates and has been validated across the full range of nutritional condition, including in mountain sheep (Cook et al. 2007, 2010, 2021; Stephenson et al. 2020).

### 7.6.2 Assessment of Body Fat Post Mortem and In Vivo

Estimation of body fat should rely on a method that has a linear relationship across the full range of nutritional condition. The range of body fat in bighorn sheep is between 0.5% and 33% (Stephenson et al. 2020). Adipose tissue is >90% fat and muscle tissue is only 20% protein; the high water content of muscle reduces the energy it provides.

Given that fat and protein provide 38 and 17 kJ/g of energy respectively, there is a stark difference in the energy provided by catabolism of adipose (>34.2 kJ/g) versus muscle (3.4 kJ/g) tissues; fat provides the most stored energy by far. Furthermore, adipose tissue may be almost entirely catabolized to release energy but only a third of muscle may be catabolized because it is needed for structural support and mobility (Cook et al. 2021).

There are three primary reserves of body fat in an animal and their order of catabolism varies. The three reserves are catabolized simultaneously but their order of complete mobilization differs. Subcutaneous fat disappears first, followed by abdominal fat, and marrow fat is fully mobilized last (Riney 1955). Understanding the pattern of catabolism is important in the selection and interpretation of methods for estimating body fat.

Post-mortem and *in vivo* methods exist that predict body fat with a high degree of accuracy and precision (Cook et al. 2007, 2010, 2021). Ultrasonography and body condition scoring via palpation can quickly predict body fat in live bighorn sheep during capture (Stephenson et al. 2020). The Kistner score provides the most accurate estimate of body fat in whole carcasses. If carcasses have been preyed upon and largely consumed, femur marrow fat can be a useful indicator of the degree of body fat depletion. Even though femur marrow fat is not metabolized until body fat declines below about 7%–8%, it provides a useful measure of whether an animal was close to starvation when it died. Often, any depletion of marrow fat is indicative of poor nutritional condition (Mech and Delgiudice 1985, Cook et al. 2001). The most useful and recommended indices of body fat were noted here; for a more complete review of the efficacy of measures of nutritional condition, see Cook et al. (2021) for an exhaustive review or Stephenson et al. (2020) for those validated for bighorn sheep.

### 7.6.3 Seasonal Dynamics in Fat Accretion and Catabolism

Mautz (1978) equated the fat cycle in deer to that observed in hibernating animals whereby ungulates accumulate fat reserves during the summer and catabolize those fat reserves during winter. The degree to which accretion and catabolism of fat reserves occurs in any single season is determined by available nutrition (Smiley et al. 2022) but also the state of the reserves that an animal carries into a season (Monteith et al. 2013), that is, they are state-dependent. Nutritional carryover permits storage of resources when they are abundant for use in later seasons when they are scarce (Harrison et al. 2011).

### 7.6.4 Differences between Males and Females

Adult bighorn sheep males are 41%–46% heavier than females (Massing et al. 2024). Females tend to incur their greatest energetic costs of the year during lactation, whereas males incur their highest energy expenditures during rut.

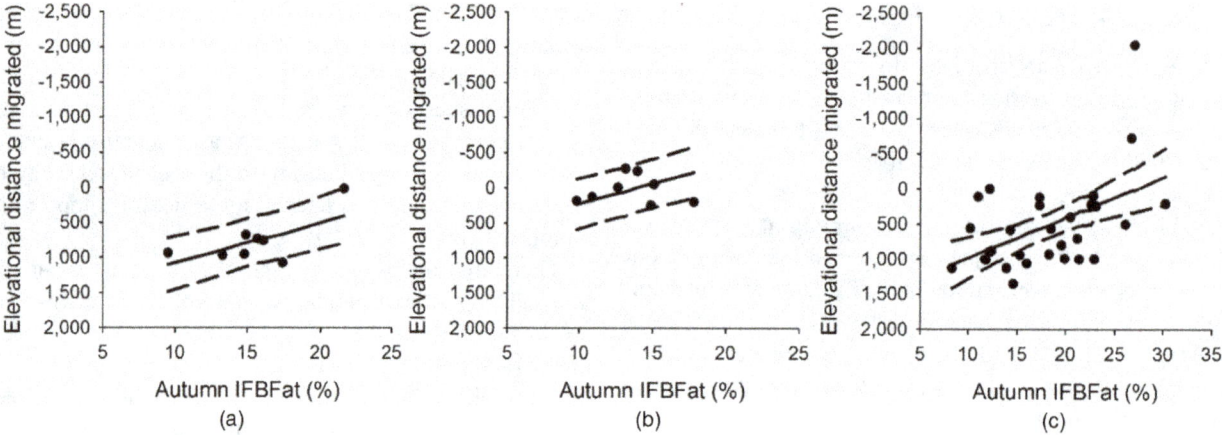

**FIGURE 7.3** Ingesta-free body fat (IFBFat, ±95% confidence interval) in autumn influenced how far in elevation migratory male Sierra Nevada bighorn sheep migrated during winter in California from 2006 to 2018 across three study areas: central (a), northern (b), and southern (c). Open Access; Denryter et al. (2024).

Male and female bighorn sheep exhibited a range of body fat in autumn, but males did not tend to exhibit fat levels as low as females (Denryter et al. 2022a). Some females may have exhibited lower body fat in autumn because of lactation costs and rutting males may not have exhibited the full cost of rut in October, when many animals were sampled. Males, being in their worst nutritional condition of the year following rut, adjust their migratory behaviors relative to body fat to balance risk of starvation versus predation over winter (Figure 7.3; Denryter et al. 2024).

The gastrocentric hypothesis proposed that males, with larger digestive capacity, may be able to persist on lower-quality diets than females, which could explain sexual segregation (Barboza and Bowyer 2000). Data needed to test the hypothesis empirically (digestion trials), however, are lacking and support from empirical studies is equivocal. Studies in desert bighorn sheep revealed no differences in diet composition between males and females and other studies attributed dietary differences between sexes to a function of females trading off nutrition to reduce predation risk to their lambs (Bleich et al. 1997, Brewer and Harveson 2007).

### 7.6.5   Fat Costs of Reproduction

Costs of reproduction may be viewed through their effects on maternal mass or on future reproduction (Festa-Bianchet et al. 1998) or in energetic terms (Stephenson et al. 2020). In Sierra Nevada bighorn sheep, lactating females had 6.6 percentage points less body fat in autumn than non-lactating females; consequently, lactating females entered winter with 130.5 MJ less stored energy than non-lactating females, equivalent to about 20 days' worth of energy requirements (Stephenson et al. 2020). The advantage of directly calculating reproductive costs using energy as a currency is that it permits calculations under an endless array of scenarios. For example, the effects of density, winter severity, and duration of lactation (including in successive years) may be calculated. Energetics modeling can be used to evaluate mechanisms associated with various combinations of physiological

states (e.g., nutritional state, lactation status) and environmental conditions (Denryter et al. 2021).

## 7.7   NUTRITIONAL LINKS TO POPULATION DYNAMICS

### 7.7.1   Links between Nutrition and Demography

Aside from immigration and emigration, survival and reproduction influence population change and both are linked to nutritional condition. The rate of population change in ungulates is most sensitive to adult female survival (Gaillard et al. 1998). Body fat >14% ensured >90% survival in adult female Sierra Nevada bighorn sheep that wintered in a harsh climate regardless of their migratory tactic (Stephenson et al. 2020, Denryter et al. 2022a, Figure 7.2a). Nutritional condition also influenced an array of reproductive parameters. Body fat measured in autumn predicted probability of pregnancy (Stephenson et al. 2020). Pregnancy rates in adult females were higher than 90% when body fat exceeded 7.7% in desert and Sierra Nevada bighorn sheep, but declined markedly at low levels of body fat (i.e., <6%), with about half of individuals experiencing a reproductive pause (Figure 7.4a; Heimer 1978, Cameron 1994, Stephenson et al. 2020). Body fat alone does not directly regulate fertility; rather body fat influences ovulation and fertility through effects on a female's total energy balance, which also includes intake from available forage (Bronson and Manning 1991). Body fat represents shorter term nutritional condition whereas body mass, particularly in spring, is more indicative of body size and the influence of nutrition before reaching adulthood. Body mass appears to dictate the age of primiparity in female bighorn sheep, indicating a threshold that is necessary to reproduce (Martin and Festa-Bianchet 2012). Growth of lambs is linked to nutrients provided through lactation and foraging, but in many populations, the high prevalence of respiratory disease complicates interpretation of the role of disease versus nutrition in assessing drivers of lamb survival (Smiley

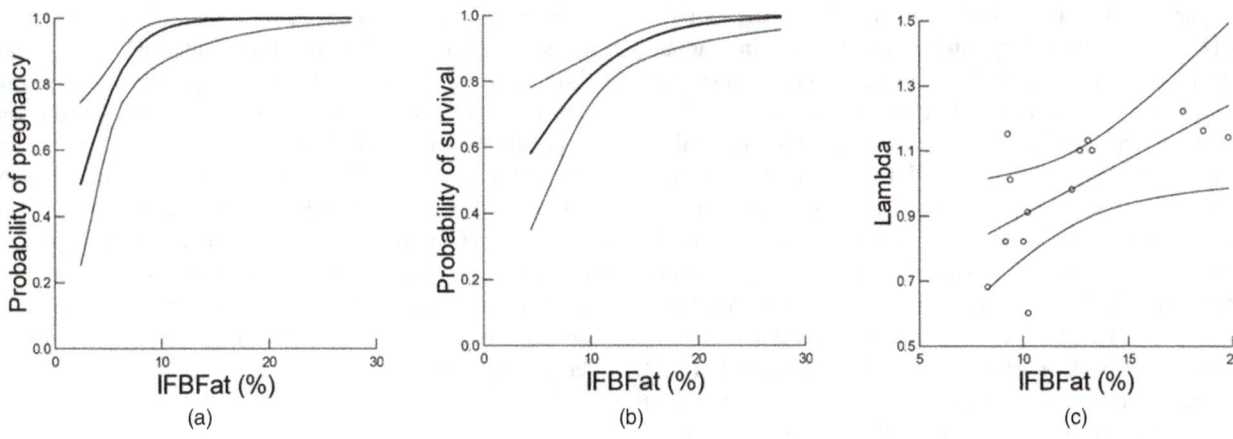

**FIGURE 7.4** Ingesta-free body fat (IFBFat, ±95% confidence interval) influenced probability of pregnancy (a), probability of adult survival (b), and lambda (c), in desert and Sierra Nevada bighorn sheep in Nevada and California from 2003 to 2018. Open Access; Stephenson et al. (2020).

et al. 2024). Nevertheless, lamb survival was positively related to maternal body fat (Bilodeau-Hussey et al. 2025) as was adult survival (Figure 7.4b; Stephenson et al. 2020).

The ultimate measure linking nutritional condition to population dynamics is the ability to predict population growth rate (Lambda, $\lambda$). Body fat of adult female mule deer measured in spring predicted $\lambda$ in California (Monteith et al. 2014). Similarly, autumn body fat of lactating Sierra Nevada bighorn sheep was related to $\lambda$ (Figure 7.4c; Stephenson et al. 2020).

Understanding the broad influence of nutrition on demography requires examining the effects under the full range of variation in nutritional condition. Earlier literature suggested that adult females were relatively invulnerable to severe winters (Jorgenson et al. 1997, Conner et al. 2018). Long-term studies that included extreme variation in weather now reveal the effects of severe weather on survival of bighorn sheep under differing migratory tactics (Denryter et al. 2022a). Unfortunately, heavy winters have a strongly detrimental effect on bighorn sheep survival (Conner et al. 2018, Denryter et al. 2022a), even though such winters may occur infrequently. Climate change may increase the incidence of extreme weather events that have the potential to negatively influence the nutritional energetics of bighorn populations.

### 7.7.2 Animal-Indicated Nutritional Carrying Capacity

Determining the proportion of the food supply that a population may use is a primary challenge in assessing nutritional carrying capacity. Rather than measuring the quantity and quality of forage in an occupied range (DeYoung et al. 2000), the nutritional condition of animals on a range can indicate the degree to which a population is able to assimilate the forage. Animal-indicated nutritional carrying capacity uses the mean nutritional condition of females in a population to assess the proximity of the population to carrying capacity (Monteith et al. 2014).

As populations approach nutritional carrying capacity, population growth slows to zero. Monteith et al. (2014) used the level of body fat associated with $\lambda = 1$ to signify the relative position of that population relative to its annual food supply. In addition to evaluating carrying capacity, nutritional condition (body fat) indicates the intrinsic capacity of habitat to meet nutritional requirements (Stephenson et al. 2020) or the degree to which animals can assimilate nutritional resources if Allee effects (e.g., density-dependent reductions in foraging efficiency due to decreased population size) reduce their optimal foraging (Rankins et al. 2024).

### 7.7.3 Consequences to Mortality: The Role of Nutrition in Understanding Predation

Predation is often the proximate cause of mortality when ungulate carcasses are investigated, but quantifying nutrition is essential for interpreting the ultimate causes of mortality. Bighorn sheep die of starvation when fat reserves decline below the level at which they can no longer be catabolized for energy (0.5%–1.0% body fat); these animals represent the doomed surplus (Errington 1946). At the population level, a combination of factors interacts to determine nutritional condition: the intrinsic capacity of the habitat to produce high-quality forage, weather (e.g., severe winters, drought), and animal density. In stochastic environments, the annual food supply can vary substantially from one year to the next. Consequently, the level of body fat (which indicates nutritional condition) exhibited by mountain sheep in a population can vary on an annual basis and in years when it is low more animals will be predisposed to starvation. Data on the nutritional status of a population is essential for interpretation of the underlying causes of mortality. Predation that is compensatory occurs when animals would die of another cause (e.g., starvation or doomed surplus) if they were not preyed upon; compensatory mortality replaces other causes of mortality (Errington 1956). Consequently, when predation is compensatory the consequences on population dynamics are minimal when compared to additive mortality. A mountain sheep population that is not limited by nutrition and exhibits individuals with high levels of body fat is more likely to experience predation that is additive to baseline mortality.

Large-scale field studies have attempted to manipulate either predation or nutrition to determine whether predation was compensatory. In some cases, predator removal did not improve mule deer survival (Bartmann et al. 1992) and predation was determined to be largely compensatory (Hurley et al. 2011). Enhanced nutrition improved body fat, increased survival, and reduced predation mortality of mule deer in Colorado, suggesting that predation was compensatory in the unsupplemented treatments (Bishop et al. 2009). Studies lacking a nutritional component have concluded that mountain lion (*Puma concolor*) predation was additive (Johnson et al. 2010, Gammons et al. 2021). A powerful approach to evaluate the influence of maternal nutritional condition on survival and recruitment of mule deer used the residuals of recruitment to represent the difference between potential recruitment (due to nutrition) and realized recruitment (accounting for effects of predation; Monteith et al. 2014). Whether predation was compensatory or additive was influenced by maternal nutritional condition and migratory behavior (Monteith et al. 2014). In mountain sheep, it is also imperative that disease be considered as a predisposing factor in mortality investigations (Smiley et al. 2024).

## 7.8 NUTRITION OF MOUNTAIN SHEEP: FROM PAST TO PRESENT

Mountain sheep, and particularly bighorn sheep, continue to recover from extensive population declines across their range. Aside from continuing to contend with disease from domestic sheep and goats, migration patterns were lost that undoubtedly have influenced the nutrition of mountain sheep (Lowrey et al. 2020, Berger et al. 2022). In addition, the small size of many populations of mountain sheep requires that we consider Allee effects and their influence on nutritional condition. Population density can alter behavior and risk; bighorn sheep moved farther from escape terrain into habitat with higher forage biomass as population density increased (Rankins et al. 2024).

Advancing our understanding of the nutritional ecology of mountain sheep will require a strong command of fundamentals provided in this chapter. New tools can be transformative, as exemplified by ultrasonography techniques, and can offer ways to address previously unanswerable questions. We also must approach new tools with an appropriate level of skepticism to ensure they accurately reflect the biology we aim to understand. Perseverance will be required as efforts to develop tools, such as nutritional landscapes, can prove daunting (Bilodeau-Hussey et al. 2025).

As the name implies, wildlife managers typically have a goal of managing wildlife. Unfortunately, we may not be able to manage all resources that would benefit a population of interest. Nevertheless, agencies can benefit from managing the expectations of their constituents and the broader public. If we have a deep knowledge of nutrition and how it constrains a mountain sheep population, that information can be shared with those who want us to act. If predation on mountain sheep is of concern, having nutritional data to aid in determining whether it is compensatory or additive is powerful. If a population of sheep resides in wilderness, there will likely be challenges in modifying habitat, but sharing the science on nutritional limitation can be used by mountain sheep advocates to influence change, even if it is over the long term.

## 7.9 SUMMARY

Nutrition influences virtually every aspect of mountain sheep biology. Nutritional demands of mountain sheep vary on an annual cycle, relative to reproductive status, and with changes in the environment. When nutrition is limiting, either through density-dependent or density-independent pathways, mountain sheep experience declines in productivity (e.g., pregnancy, calf survival, recruitment), performance (e.g., fattening, growth), and survival. The presence of nutritional limitations, however, does not negate the importance of other potential limiting factors; nutrition interacts with other factors, including disease and predation, to influence demographic rates of populations. Evaluating nutritional status of mountain sheep is best accomplished by directly evaluating their body fat levels via validated ultrasonography and manual palpation techniques; this information can be related to demographic rates to evaluate how nutrition influences population dynamics. Information on mountain sheep nutrition can provide critical background knowledge to determine whether a management action to address nutrition is warranted, likely to succeed, or both. Managing nutrition for mountain sheep across landscapes and jurisdictions can be incredibly challenging, but success is most achievable with a strong understanding of nutritional ecology of mountain sheep.

## REFERENCES

Abler, W. A., D. E. Buckland, R. L. Kirkpatrick, and P. F. Scanlon. 1976. Plasma progestins and puberty in fawns as influenced by energy and protein. *Journal of Wildlife Management* 40:442–446.

Adams, L. G., and B. W. Dale. 1998a. Reproductive performance of female Alaskan caribou. *Journal of Wildlife Management* 62:1184–1195.

Adams, L. G., and B. W. Dale. 1998b. Timing and synchrony of parturition in Alaskan caribou. *Journal of Mammalogy* 79:287–294.

Adams, L. G., F. J. Singer, and B. W. Dale. 1995. Caribou calf mortality in Denali National Park, Alaska. *Journal of Wildlife Management* 59:584–594.

Ahsan, U., Z. Kamran, I. Raza, S. Ahmad, W. Babar, M. H. Riaz, and Z. Iqbal. 2014. Role of selenium in male reproduction: A review. *Animal Reproduction Science* 146:55–62.

Albl, P., P. A. Boyazoglu, and J. P. Bezuidenhout. 1977. Observations on the mineral status of springbok (Antidorcas marsupialis Zimmerman) in south west Africa. *Madoqua* 10:79–83.

Alio, A., C. B. Theurer, O. Lozano, J. T. Huber, R. S. Swingle, A. Delgado-Elorduy, P. Cuneo, D. DeYoung, and K. E. Webb. 2000. Splanchnic nitrogen metabolism by growing beef steers fed diets containing sorghum grain flaked at different densities. *Journal of Animal Science* 78:1355–1363.

Amaral-Philips, D. M. 2020. Are your dairy cows getting the protein they need? University of Kentucky Cooperative Extension Service, Lexington, Kentucky, USA.

Arthur, J. R. 2000. The glutathione peroxidases. *Cellular and Molecular Life Sciences* 57:1825–1835.

Aycrigg, J. L., A. G. Wells, E. O. Garton, B. Magipane, G. E. Liston, L. R. Prugh, and J. L. Rachlow. 2021. Habitat selection by Dall's sheep is influenced by multiple factors including direct and indirect climate effects. *PLoS ONE* 16:1–19.

Ayotte, J. B., K. L. Parker, and M. P. Gillingham. 2008. Use of natural licks by four species of ungulates in northern British Columbia. *Journal of Mammalogy* 89:1041–1050.

Barboza, P. S., and R. T. Bowyer. 2000. Sexual segregation in dimorphic deer: a new gastrocentric hypothesis. *Journal of Mammalogy* 81:473–489.

Barboza, P. S., and K. L. Parker. 2006. Body protein stores and isotopic indicators of N balance in female reindeer (*Rangifer tarandus*) during winter. *Physiological and Biochemical Zoology* 79:628–644.

Barboza, P. S., and K. L. Parker. 2008. Allocating protein to reproduction in arctic reindeer and caribou. *Physiological and Biochemical Zoology* 81:835–855.

Barboza, P. S., K. L. Parker, and I. D. Hume. 2009. *Integrative Wildlife Nutrition*. 1st edition. Springer-Verlag, Heidelberg, Germany.

Bartmann, R. M., G. C. White, and L. H. Carpenter. 1992. Compensatory mortality in a Colorado mule deer population. *Wildlife Monographs* 121:3–39.

Bender, L. C., J. G. Cook, R. C. Cook, and P. B. Hall. 2008. Relations between nutritional condition and survival of North American elk (*Cervus elaphus*). *Wildlife Biology* 14:70–80.

Bender, L. C., and M. E. Weisenberger. 2005. Precipitation, density, and population dynamics of desert bighorn sheep on San Andres National Wildlife Refuge, New Mexico. *Wildlife Society Bulletin* 33:956–964.

Berger, D. J., D. W. German, C. John, R. Hart, T. R. Stephenson, and T. Avgar. 2022. Seeing is beleaving: perception drives seasonal migration in Sierra Nevada bighorn sheep (*Ovis canadensis sierrae*). *Frontiers in Ecology and Evolution* 10:742275.

Berger, J. 1979. Weaning conflict in desert and mountain bighorn sheep (*Ovis canadensis*): an ecological interpretation. *Zeitschrift für Tierpsychologie* 50:188–200.

Bilodeau-Hussey, N. M., K. S. Huggler, E. F. Cassirer, H. Miyasaki, M. A. Hurley, L. A. Shipley, and R. A. Long. 2025. Effects of maternal condition, disease status, and behavior on survival of juvenile bighorn sheep. *Journal of Wildlife Management* 89:e22721.

Bilodeau, N. M. 2021. Context-dependent effects of nutrition and dam behavior on neonatal survival in a long-lived herbivore. MS Thesis, College of Natural Resources, University of Idaho, Moscow, USA.

Bishop, C. J., G. C. White, D. J. Freddy, B. E. Watkins, and T. R. Stephenson. 2009. Effect of enhanced nutrition on mule deer population rate of change. *Wildlife Monographs* 172:1–28.

Blaxter, K. L., F. W. Wainman, and R. S. Wilson. 1961. The regulation of food intake by sheep. *Animal Production* 3:51–61.

Bleich, V. C., R. T. Bowyer, and J. D. Wehausen. 1997. Sexual segregation in mountain sheep: resources or predation? *Wildlife Monographs* 48:1248–1254.

Blood, D. A. 1961. An ecological study of California bighorn sheep *Ovis canadensis california* (Douglas) in southern British Columbia. Department of Zoology, University of British Columbia, Vancouver, British Columbia, Canada.

Blum, M. E., K. M. Stewart, M. Cox, K. T. Shoemaker, J. R. Bennet, B. W. Sullivan, B. F. Wakeling, and V. C. Bleich. 2023. Variation in diet of desert bighorn sheep around parturition: tradeoffs associated with parturition. *Frontiers in Ecology and Evolution* 10:1071771.

Boertje, R. D., J. L. Davis, and P. Valkenburg. 1985. Uses and limitations of fecal analyses in Rangifer studies. Pages 307–316 in T. C. Meredith and A. M. Martell, editors. *Proceedings of the Second North American Caribou Workshop*, Val Morin, Quebec, Canada.

Brewer, C. E., and L. A. Harveson. 2007. Diets of bighorn sheep in the Chihuahuan Desert, Texas. *Southwestern Naturalist* 52:97–103.

Bronson, F. H., and J. M. Manning. 1991. The energetic regulation of ovulation: a realistic role for body fat. *Biology of Reproduction* 44:945–950.

Bryant, M. P., and I. M. Robinson. 1961. Studies on the nitrogen requirements of some ruminal cellulolytic bacteria. *Applied Microbiology* 9:96–103.

Bucholtz, H. F., and W. G. Bergen. 1973. Microbial phospholipid synthesis as a marker for microbial protein synthesis in the rumen. *Applied Microbiology* 25:504–513.

Bunnell, F. L. 1978. Horn growth and population quality in Dall sheep. *Journal of Wildlife Management* 42:764–775.

Bunnell, F. L. 1980. Factors controlling lambing period of Dall's sheep. *Canadian Journal of Zoology* 58:1027–1031.

Bunnell, F. L. 1982. The lambing period of mountain sheep: synthesis, hypotheses, and tests. *Canadian Journal of Zoology* 60:1–14.

Bunnell, F. L., and N. A. Olsen. 1981. Age-specific natality in Dall's sheep. *Journal of Mammalogy* 62:379–380.

Cain, J. W., J. V Gedir, J. P. Marshal, P. R. Krausman, J. D. Allen, G. C. Duff, B. D. Jansen, and J. R. Morgart. 2017. Extreme precipitation variability, forage quality and large herbivore diet selection in arid environments. *Oikos* 126:1459–1471.

Cain, J. W., P. R. Krausman, J. R. Morgart, B. D. Jansen, and M. P. Pepper. 2008. Responses of desert bighorn sheep to removal of water sources. *Wildlife Monographs* 171:1–32.

Cameron, R. D. 1994. Reproductive pauses by female caribou. *Journal of Mammalogy* 75:10–13.

Carafoli, E. 1991. Calcium pump of the plasma membrane. *Physiological Reviews* 71:129–153.

Carstensen, M., G. D. Delgiudice, B. A. Sampson, and D. W. Kuehn. 2009. Survival, birth characteristics, and cause-specific mortality of white-tailed deer neonates. *Journal of Wildlife Management* 73:175–183.

Cassirer, E. F., K. R. Manlove, E. S. Almberg, P. L. Kamath, M. Cox, P. Wolff, A. Roug, J. Shannon, R. Robinson, R. B. Harris, B. J. Gonzales, R. K. Plowright, P. J. Hudson, P. C. Cross, A. Dobson, and T. E. Besser. 2018. Pneumonia in bighorn sheep: risk and resilience. *Journal of Wildlife Management* 82:32–45.

Chappel, R. W., and R. J. Hudson. 1979. Energy cost of feeding in Rocky Mountain bighorn sheep. *Acta Theriologica* 23:359–363.

Chappel, R. W., and R. J. Hudson. 1980. Prediction of energy expenditures by Rocky Mountain bighorn sheep. *Canadian Journal of Zoology* 58:1908–1912.

Church, D. C., editor. 1993. *The Ruminant Animal: Digestive Physiology and Nutrition*. Waveland Press, Long Grove, Illinois, USA.

Conner, M. M., T. R. Stephenson, D. W. German, K. L. Monteith, A. P. Few, and E. H. Bair. 2018. Survival analysis: informing recovery of Sierra Nevada bighorn sheep. *Journal of Wildlife Management* 82:1442–1458.

Conrad, H. R., A. D. Pratt, and J. W. Hibbs. 1964. Regulation of feed intake in dairy cows. I. Change in importance of physical and physiological factors with increasing digestibility. *Journal of Dairy Science* 47:54–62.

Constan, J. 1967. Food habits, range use and relationships of bighorn sheep to mule deer and elk in winter, Gallatin Canyon, Montana. Montana State University, Bozeman, Montana, USA.

Cook, J. G. 1990. Habitat, nutrition, and population ecology of two transplanted bighorn sheep populations in southcentral Wyoming. University of Wyoming, Laramie, Wyoming, USA.

Cook, J. G., B. K. Johnson, R. C. Cook, R. A. Riggs, L. D. Bryant, and L. L. Irwin. 2004. Effects of summer-autumn nutrition and parturition date on reproduction and survival of elk. Wildlife Monographs 155:1–61.

Cook, J. G., L. L. Irwin, L. D. Bryant, R. A. Riggs, and J. W. Thomas. 1998. Relations of forest cover and condition of elk: a test of the thermal cover hypothesis in summer and winter. *Wildlife Monographs* 141:3–61.

Cook, J. G., R. C. Cook, R. W. Davis, and L. L. Irwin. 2016. Nutritional ecology of elk during summer and autumn in the Pacific Northwest. *Wildlife Monographs* 195:1–81.

Cook, J. G., R. C. Cook, R. W. Davis, M. M. Rowland, R. M. Nielson, M. J. Wisdom, J. M. Hafer, and L. L. Irwin. 2018. Development and evaluation of a landscape nutrition model for elk in western Oregon and Washington. In: Modeling elk nutrition and habitat use in western Oregon and Washington. *Wildlife Monographs* 199:13–30.

Cook, R. C., J. A. Crouse, J. G. Cook, and T. R. Stephenson. 2021. Evaluating indices of nutritional condition for caribou (*Rangifer tarandus*): which are the most valuable and why? *Canadian Journal of Zoology* 99:596–613.

Cook, R. C., J. G. Cook, D. L. Murray, P. Zager, B. K. Johnson, and M. W. Gratson. 2001. Development of predictive models of nutritional condition for Rocky Mountain elk. *Journal of Wildlife Management* 65:973–987.

Cook, R. C., J. G. Cook, T. R. Stephenson, W. L. Myers, S. M. McCorquodale, D. J. Vales, L. L. Irwin, P. B. Hall, R. D. Spencer, S. L. Murphie, K. A. Schoenecker, and P. J. Miller. 2010. Revisions of rump fat and body scoring indices for deer, elk, and moose. *Journal of Wildlife Management* 74:880–896.

Cook, R. C., L. A. Shipley, J. G. Cook, M. J. Camp, D. S. Monzingo, S. L. Robatcek, S. L. Berry, I. T. Hull, W. L. Myers, K. Denryter, and R. A. Long. 2022. Sequential detergent fiber assay results used for nutritional ecology research: evidence of bias since 2012. *Wildlife Society Bulletin* 46:e1348.

Cook, R. C., T. R. Stephenson, W. L. Myers, J. G. Cook, and L. A. Shipley. 2007. Validating predictive models of nutritional condition for mule deer. *Journal of Wildlife Management* 71:1934–1943.

Cooper, A. J. L., and F. Plum. 1987. Biochemistry and physiology of brain ammonia. *Physiological Reviews* 67:440–519.

Côté, S. D., and M. Festa-Bianchet. 2001. Birthdate, mass and survival in mountain goat kids: effects of maternal characteristics and forage quality. *Oecologia* 127:230–238.

Courtemanch, A. B., M. J. Kauffman, S. Kilpatrick, and S. R. Dewey. 2017. Alternative foraging strategies enable a mountain ungulate to persist after migration loss. *Ecosphere* 8:1–16.

Culotta, V. C., M. Yang, and M. D. Hall. 2005. Manganese transport and trafficking: lessons learned from Saccharomyces cerevisiae. *Eukaryotic Cell* 4:1159–1165.

Dailey, T. V., and N. T. Hobbs. 1989. Travel in alpine terrain: energy expenditures for locomotion by mountain goats and bighorn sheep. *Canadian Journal of Zoology* 67:2368–2375.

Dean, R. E., M. D. Strickland, J. L. Newman, E. T. Thorne, and W. G. Hepworth. 1975. Reticulo-rumen characteristics of malnourished mule deer. *Journal of Wildlife Management* 39:601–604.

Dekelaita, D. J., C. W. Epps, K. M. Stewart, J. S. Sedinger, J. G. Powers, B. J. Gonzales, R. K. Abella-Vu, N. W. Darby, and D. L. Hughson. 2020. Survival of adult female bighorn sheep following a pneumonia epizootic. *Journal of Wildlife Management* 84:1268–1282.

Denryter, K. 2017. Foraging ecology of woodland caribou in boreal and montane ecosystems of northern British Columbia. Dissertation, University of Northern British Columbia, Prince George, Canada.

Denryter, K., M. M. Conner, T. R. Stephenson, D. W. German, and K. L. Monteith. 2022a. Survival of the fattest: how body fat and migration influence survival in highly seasonal environments. *Functional Ecology* 36:2569–2579.

Denryter, K., R. C. Cook, J. G. Cook, and K. L. Parker. 2022b. Animal-defined resources reveal nutritional inadequacies for woodland caribou during summer in northeastern British Columbia. *Journal of Wildlife Management* 86:e22161.

Denryter, K., D. W. German, T. R. Stephenson, and K. L. Monteith. 2021. State- and context-dependent applications of an energetics model in free-ranging bighorn sheep. *Ecological Modelling* 440:109349.

Denryter, K., and R. G. White. 2026. Nutritional requirements of North American ungulates. in J. G. Cook, E. H. Merrill, and R. C. Cook, editors. *Nutritional Ecology of North American Herbivores*. Johns Hopkins University Press, Baltimore, Maryland, USA.

Denryter, K., T. R. Stephenson, and K. L. Monteith. 2024. Migratory behaviours are risk-sensitive to physiological state in an elevational migrant. *Conservation Physiology* 12:coae029.

DeYoung, R. W., E. C. Hellgren, T. E. Fulbright, W. F. Robbins Jr., and I. D. Humphreys. 2000. Modeling nutritional carrying capacity for translocated desert bighorn sheep in western Texas. *Restoration Ecology* 8:57–65.

Doman, E. R., and D. I. Rasmussen. 1944. Supplemental winter feeding of mule deer in northern Utah. *Journal of Wildlife Management* 8:317–338.

Enns, G. E. 2021. Seasonal habitat use, habitat selection, and migratory behaviours of Stone's sheep (*Ovis dalli stonei*) in northern British Columbia, Canada. Department of Biological Sciences, University of Alberta, Edmonton, Alberta, Canada.

Errington, P. L. 1946. Predation and vertebrate populations. *American Naturalist* 21:144–177.

Errington, P. L. 1956. Factors limiting higher vertebrate populations. *Science* 124:304–307.

Fairbanks, W. S. 1993. Birthdate, birthweight, and survival in pronghorn fawns. *Journal of Mammalogy* 74:129–135.

Feder, C., J. G. A. Martin, M. Festa-Bianchet, C. Bérubé, and J. T. Jorgenson. 2008. Never too late? Consequences of late birthdate for mass and survival of bighorn lambs. *Oecologia* 156:773–781.

Festa-Bianchet, M. 1988a. Birthdate and survival in bighorn lambs (*Ovis canadensis*). *Journal of the Zoological Society of London* 214:653–661.

Festa-Bianchet, M. 1988b. Nursing behaviour of bighorn sheep: correlates of ewe age, parasitism, lamb age, birthdate and sex. *Animal Behaviour* 36:1445–1454.

Festa-Bianchet, M. 1988c. Seasonal range selection in bighorn sheep: conflicts between forage quality, forage quantity, and predator avoidance. *Oecologia* 75:580–586.

Festa-Bianchet, M., D. W. Coltman, L. Turelli, and J. T. Jorgenson. 2004. Relative allocation to horn and body growth in bighorn rams varies with resource availability. *Behavioral Ecology* 15:305–312.

Festa-Bianchet, M., J.-M. Gaillard, and J. T. Jorgenson. 1998. Mass- and density-dependent reproductive success and reproductive costs in a capital breeder. *The American Naturalist* 152:367–379.

Festa-Bianchet, M., and J. T. Jorgenson. 1998. Selfish mothers: reproductive expenditure and resource availability in bighorn ewes. *Behavioral Ecology* 9:144–150.

Festa-Bianchet, M., J. T. Jorgenson, W. J. King, K. G. Smith, and W. D. Wishart. 1993. The development of sexual dimorphism: seasonal and lifetime mass changes in bighorn sheep. *Canadian Journal of Zoology* 74:330–342.

Firkins, J. L., Z. Yu, and M. Morrison. 2007. Ruminal nitrogen metabolism: perspectives for integration of microbiology and nutrition for dairy. *Journal of Dairy Science* 90 (E Supp):E1–E16. Elsevier.

Franzmann, A. W., and C. C. Schwartz. 1985. Moose twinning rates: a possible population condition assessment. *Journal of Wildlife Management* 49:394–396.

Fulbright, T. E., W. Frank Robbins, E. C. Hellgren, R. W. DeYoung, and I. D. Humphreys. 2001. Lack of diet partitioning by sex in reintroduced desert bighorn sheep. *Journal of Arid Environments* 48:49–57.

Gaillard, J.-M., M. Festa-Bianchet, and N. G. Yoccoz. 1998. Population dynamics of large herbivores: variable recruitment with constant adult survival. *Trends in Ecology & Evolution* 13:58–63.

Gammons, D. J., J. L. Davis, D. W. German, K. Denryter, J. D. Wehausen, and T. R. Stephenson. 2021. Predation impedes recovery of Sierra Nevada bighorn sheep. *California Fish and Wildlife* 107:444–470.

Gardner, W. C., K. Broersma, J. D. Popp, Z. Mir, P. S. Mir, and W. T. Buckley. 2003. Copper and health status of cattle grazing high-molybdenum forage from a reclaimed mine tailing site. *Canadian Journal of Animal Science* 83:479–485.

Gasaway, W. C., and J. W. Coady. 1974. Review of energy requirements and rumen fermentation in moose and other ruminants. *Le Naturaliste Canadien* 101:227–262.

Godwin, K. O., R. E. Kuchel, and C. N. Fuss. 1974. Some biochemical features of white muscle disease in lambs, and the influences of selenium. *Australian Journal of Biological Sciences* 27:633–643.

Goering, H. K., and P. J. Van Soest. 1970. *Forage Fiber Analysis*. Agriculture Handbook No. 379. Agricultural Research Service, U.S. Dept. of Agriculture, Washington, D.C., USA.

Goff, J. P. 2009. Phosphorous deficiency. Pages 135–137 *in* D. E. Anderson and D. M. Rings, editors. *Food Animal Practice*. 5th edition. Saunders, Philadelphia, Pennsylvania, USA.

Goodson, N. J., D. R. Stevens, and J. A. Bailey. 1991a. Winter-spring foraging ecology and nutrition of bighorn sheep on montane ranges. *Journal of Wildlife Management* 55:422–433.

Goodson, N. J., D. R. Stevens, and J. A. Bailey. 1991b. Effect of snow on foraging ecology and nutrition of bighorn sheep. *Journal of Wildlife Management* 55:214–222.

Gordon, I. J., and A. W. Illius. 1994. The functional significance of the browser-grazer dichotomy in African ruminants. *Oecologia* 98:167–175.

Grasman, B. T., and E. C. Hellgren. 1993. Phosphorous nutrition in white-tailed deer: nutrient balance, physiological responses, and antler growth. *Ecology* 74:2279–2296.

Greene, L., M. Hebblewhite, and T. R. Stephenson. 2012. Short-term vegetation response to wildfire in the eastern Sierra Nevada: implications for recovering an endangered ungulate. *Journal of Arid Environments* 87:118–128.

Hall, L. K., R. T. Larsen, R. N. Knight, and B. R. McMillan. 2018. Feral horses influence both spatial and temporal patterns of water use by native ungulates in a semi-arid environment. *Ecosphere* 9:e02096.

Handeland, K., A. Bernhoft, and M. S. Aartun. 2008. Copper deficiency and effects of copper supplementation in a herd of red deer (*Cervus elaphus*). *Acta Veterinaria Scandinavica* 50:1999–2002.

Hansen, M. C. 1996. Foraging ecology of female Dall's sheep in the Brooks Range, Alaska. PhD Thesis, University of Alaska Fairbanks, Fairbanks, Alaska, USA.

Harder, J. D., and R. L. Kirkpatrick. 1994. Physiological methods in wildlife research. Pages 275–306 *in* T. A. Bookhout, editor. *Research and Management Techniques for Wildlife and Habitats*. 5th edition. The Wildlife Society, Bethesda, Maryland, USA.

Harper, W. L. 1984. Pregnancy rate and early lamb survival of California bighorn sheep (*Ovis canadensis californiana*, Douglas 1871) in the Ashnola Watershed, British Columbia. Department of Animal Science, University of British Columbia, Vancouver, British Columbia, Canada.

Harrison, X. A., J. D. Blount, R. Inger, D. R. Norris, and S. Bearhop. 2011. Carry-over effects as drivers of fitness differences in animals. *Journal of Animal Ecology* 80:4–18.

Hebert, D. M. 1973. Altitudinal migration as a factor in the nutrition of bighorn sheep. PhD Thesis, Department of Zoology, University of British Columbia, Vancouver, Canada.

Hebert, D. M., and I. M. Cowan. 1971. White muscle disease in the mountain goat. *Journal of Wildlife Management* 35:752–756.

Heimer, W. E. 1978. Alternate year reproduction in a low quality, declining Dall sheep population: management considerations. Pages 30–41 *in* H. Cleator, editor. *Proceedings of the Biennial Symposium of the Northern Wild Sheep and Goat Council*, Penticton, British Columbia, Canada.

Heimer, W. E. 1988. A magnesium-driven hypothesis of dall sheep mineral lick use: preliminary tests and management relevance. *Bienn. Symposium. North Wild Sheep and Goat Council* 6:269–279.

Hobbs, N. T. 2024. A general, resource-based explanation for density dependence in populations of large herbivores. *Ecological Monographs* 94:e1600.

Hoefs, M., and U. Nowlan. 1997. Comparison of horn growth in captive and free-ranging Dall's rams. *Journal of Wildlife Management* 61:1154–1160.

Hofmann, R. R. 1988. Anatomy of the gastro-intestinal tract. Pages 14–43 *in* D. C. Church, editor. *The Ruminant Animal: Digestive Physiology and Nutrition,* Prentice-Hall, Upper Saddle River, New Jersey, USA.

Hogg, J. T., C. C. Hass, and D. A. Jenni. 1992. Sex-biased maternal expenditure in Rocky Mountain bighorn sheep. *Behavioral Ecology and Sociobiology* 31:243–251.

Holling, C. S. 1959. Some characteristics of simple types of predation and parasitism. *Canadian Entomology* 91:385–398.

Hurley, M. A., J. W. Unsworth, P. Zager, M. Hebblewhite, E. O. Garton, D. M. Montgomery, J. R. Skalski, and C. L. Maycock. 2011. Demographic response of mule deer to experimental reduction of coyotes and mountain lions in southeastern Idaho. *Wildlife Monographs* 178:1–33.

Hurwitz, S. 1996. Homeostatic control of plasma calcium concentration. *Critical Reviews in Biochemistry and Molecular Biology* 31:41–100.

Johnson, B. K. 1980. Bighorn sheep food habits, forage preferences, and habitat selection in alpine and subalpine communities. Department of Fishery and Wildlife Biology, Colorado State University, Fort Collins, Colorado, USA.

Johnson, H. E., L. S. Mills, T. R. Stephenson, and J. D. Wehausen. 2010. Population-specific vital rate contributions influence management of an endangered ungulate. *Ecological Applications* 20:1753–1765.

Jorgenson, J. T., M. Festa-Bianchet, J.-M. Gaillard, and W. D. Wishart. 1997. Effects of age, sex, disease, and density on survival of bighorn sheep. *Ecology* 78:1019–1032.

Jorgenson, J. T., M. Festa-Bianchet, and M. Lucherini. 1993. Effects of body size, population density, and maternal characteristics on age at first reproduction in bighorn ewes. *Canadian Journal of Zoology* 71:2509–2517.

Jorgenson, J. T., M. Festa-Bianchet, and W. D. Wishart. 1998. Effects of population density on horn development in bighorn rams. *Journal of Wildlife Management* 62:1011–1020.

Keating, K. A., L. R. Irby, and W. F. Kasworm. 1985. Mountain sheep winter food habits in the Upper Yellowstone Valley. *Journal of Wildlife Management* 49:156–161.

Keech, M. A., R. T. Bowyer, J. M. Ver Hoef, R. D. Boertje, B. W. Dale, and T. R. Stephenson. 2000. Life-history consequences of maternal condition in Alaskan moose. *Journal of Wildlife Management* 64:450–462.

Van de Kerk, M., S. Arthur, M. Bertram, B. Borg, J. Herriges, J. Lawler, B. Mangipane, C. Lambert Koizumi, B. Wendling, and L. Prugh. 2020. Environmental influences on Dall's sheep survival. *Journal of Wildlife Management* 84:1127–1138.

Kincaid, R. 1993. Macro elements for ruminants. Pages 326–341 *in* D. C. Church, editor. *The Ruminant Animal, Digestive Physiology and Nutrition.* Waveland Press, Long Grove, Illinois, USA.

Komisarczuk, S., R. J. Merry, and A. B. McAllan. 1987. Effect of different levels of phosphorus on rumen microbial fermentation and synthesis determined using a continuous culture technique. *British Journal of Nutrition* 57:279–290.

Liu, Y., Y. Zhang, P. Dong, R. An, C. Xue, Y. Ge, L. Wei, and X. Liang. 2015. Digestion of nucleic acids starts in the stomach. *Nature Publishing Group* 5:11936.

Lopez-Ortiz, S., K. E. Panter, J. A. Pfister, and K. L. Launchbaugh. 2004. The effect of body condition on disposition of alkaloids from silvery lupine (*Lupinus argenteus* Pursh) in sheep. *Journal of Animal Science* 82:2798–2805.

Lowrey, B., D. E. Mcwhirter, K. M. Proffitt, K. L. Monteith, A. B. Courtemanch, P. J. White, J. T. Paterson, S. R. Dewey, and R. A. Garrott. 2020. Individual variation creates diverse migratory portfolios in native populations of a mountain ungulate. *Ecological Applications* 30:1–14.

Luckhurst, A. J. 1973. Stone sheep and their habitat in the northern rocky mountain foothills of British Columbia. Department of Plant Science, University of British Columbia, Vancouver, British Columbia, Canada.

Luick, B. R., and R. G. White. 1986. Oxygen consumption for locomotion by caribou calves. *Journal of Wildlife Management* 50:148–152.

Lyford, S. J. 1993. Growth and development of the ruminant digestive system. Pages 44–63 *in* D. C. Church, editor. *The Ruminant Animal: Digestive Physiology and Nutrition.* Waveland Press, Long Grove, Illinois, USA.

Maiorino, M., L. Flohe, A. Roveri, P. Steinert, J. B. Wissing, and F. Ursini. 1999. Selenium and reproduction. *BioFactors* 10:251–256.

Margis, R., C. Dunand, F. K. Teixeira, and M. Margis-Pinheiro. 2008. Glutathione peroxidase family: an evolutionary overview. *Federation of European Biochemical Societies Journal* 275:3959–3970.

Martin, J. G. A., and M. Festa-Bianchet. 2012. Determinants and consequences of age of primiparity in bighorn ewes. *Oikos* 121:752–760.

Martin, J. G. A., and M. Festa-Bianchet. 2010. Bighorn ewes transfer the costs of reproduction to their lambs. *The American Naturalist* 176:414–423.

Martin, J. S., and M. M. Martin. 1983. Tannin assays in ecological studies: precipitation of ribulose-1,5-bisphosphate carboxylase/oxygenase by tannic acid, quebracho, and oak foliage extracts. *Journal of Chemical Ecology* 9:285–294.

Massing, C. P., D. W. German, and T. R. Stephenson. 2024. Morphologic differences in bighorn populations of California. *Southwestern Naturalist* 67:263–273.

Mautz, W. W. 1978. Sledding on a bushy hillside: the fat cycle in deer. *Wildlife Society Bulletin* 6:88–90.

McAllan, A. B., and R. H. Smith. 1973. Degradation of nucleic acids in the rumen. *British Journal of Nutrition* 29:331–345.

McArt, S. H., D. E. Spalinger, W. B. Collins, E. R. Schoen, T. Stevenson, and M. Bucho. 2009. Summer dietary nitrogen availability as a potential bottom-up constraining on moose in south-central Alaska. *Ecology* 90:1400–1411.

Mech, L. D. 2007. Femur-marrow fat of white-tailed deer fawns killed by wolves. *Journal of Wildlife Management* 71:920–923.

Mech, L. D., and G. D. Delgiudice. 1985. Limitations of the marrow-fat technique as an indicator of body condition. *Wildlife Society Bulletin* 13:204–206.

Mech, L. D., M. E. Nelson, and R. E. Mcroberts. 1991. Effects of maternal and grandmaternal nutrition on deer mass and vulnerability to wolf predation. *Journal of Mammalogy* 72:146–151.

Meehan, M. A., G. Stokka, and M. Mostrom. 2021. Livestock water requirements. *NDSU Extension* 1763:4.

Mehdi, Y., and I. Dufrasne. 2016. Selenium in cattle: a review. *Molecules* 21:1–14.

Merkle, J. A., K. L. Monteith, E. O. Aikens, M. M. Hayes, K. R. Hersey, A. D. Middleton, B. A. Oates, H. Sawyer, B. M. Scurlock, and M. J. Kauffman. 2016. Large herbivores surf waves of green-up in spring. *Proceedings of the Royal Society of London B: Biological Sciences* 283:1–8.

Mertens, D. R. 1993. Kinetics of cell wall digestion and passage in ruminants. Pages 535–570 *in* H. G. Jung, D. R. Buxton, R. D. Hatfield, and J. Ralph, editors. *Forage Cell Wall Structure and Digestibility.* American Society of Agronomy, Crop Science Society of America, and Soil Science Society of America, Madison, Wisconsin, USA.

Metherell, L. 2023. Female Dall's sheep summer nutrition in the Chugach Mountains, Alaska. Thesis, University of Alaska Fairbanks, Fairbanks, Alaska, USA.

Miller, G. D., and W. S. Gaud. 1989. Composition and variability of desert bighorn sheep diets. *Journal of Wildlife Management* 53:597–606.

Miller, J. K., N. Ramsey, and F. C. Madsen. 1993. The trace elements. Pages 342–400 *in* D. C. Church, editor. *The Ruminant Animal, Digestive Physiology and Nutrition.* Waveland Press, Long Grove, Illinois, USA.

Miller, W. J. 1970. Zinc nutrition of cattle: a review. *Journal of Dairy Science* 53:1123–1135.

Monteith, K. L., V. C. Bleich, T. R. Stephenson, B. M. Pierce, M. M. Conner, J. G. Kie, and R. T. Bowyer. 2014. Life-history characteristics of mule deer: effects of nutrition in a variable environment. *Wildlife Monographs* 186:1–61.

Monteith, K. L., R. A. Long, T. R. Stephenson, V. C. Bleich, R. T. Bowyer, and T. N. Lasharr. 2018. Horn size and nutrition in mountain sheep: can ewe handle the truth? *Journal of Wildlife Management* 82:67–84.

Monteith, K. L., T. R. Stephenson, V. C. Bleich, M. M. Conner, B. M. Pierce, and R. T. Bowyer. 2013. Risk-sensitive allocation in seasonal dynamics of fat and protein reserves in a long-lived mammal. *Journal of Animal Ecology* 82:377–388.

Monzingo, D. S., J. G. Cook, L. A. Shipley, and R. C. Cook. 2022. Factors influencing predictions of understory vegetation biomass from visual cover estimates. *Wildlife Society Bulletin* 46:e1300.

Mooring, M. S., T. A. Fitzpatrick, T. T. Nishihira, and D. D. Reisig. 2004. Vigilance, predation risk, and the Allee effect in desert bighorn sheep. *Journal of Wildlife Management* 68:519–532.

National Research Council. 2007. *Nutrient Requirements of Small Ruminants: Sheep, Goats, Cervids, and New World Camelids.* National Academies Press, Washington, D.C., USA.

Nichols, L. 1978. Dall sheep reproduction. *Journal of Wildlife Management* 42:570–580.

Oftedal, O. T. 1985. Pregnancy and lactation. Pages 215–238 *in* R. J. Hudson and R. G. White, editors. *Bioenergetics of Wild Herbivores.* CRC Press, Boca Raton, Florida, USA.

Oftedal, O. T. 2000. Use of maternal reserves as a lactation strategy in large mammals. *Proceedings of the Nutrition Society* 59:99–106.

Oldemeyer, J. L., W. J. Barmore, and D. L. Gilbert. 1971. Winter ecology of bighorn sheep in Yellowstone National Park. *Journal of Wildlife Management* 35:257–269.

Ørskov, E.-R., and I. McDonald. 1979. The estimation of protein degradability in the rumen from incubation measurements weighted according to rate of passage. *The Journal of Agricultural Science* 92:499–503.

Oster, K. W., P. S. Barboza, D. D. Gustine, K. Joly, and R. D. Shively. 2018. Mineral constraints on arctic caribou (*Rangifer tarandus*): a spatial and phenological perspective. *Ecosphere* 9:e02160.

Ostermann-Kelm, S., E. R. Atwill, E. S. Rubin, M. C. Jorgensen, and W. M. Boyce. 2008. Interactions between feral horses and desert bighorn sheep at water. *Journal of Mammalogy* 89:459–466.

Owens, F. N., and R. Zinn. 1993. Protein metabolism of ruminant animals. Pages 227–249 *in* D. C. Church, editor. *The Ruminant Animal, Digestive Physiology and Nutrition.* Waveland Press, Long Grove, Illinois, USA.

Parker, K. L., P. S. Barboza, and M. P. Gillingham. 2009. Nutrition integrates environmental responses of ungulates. *Functional Ecology* 23:57–69.

Parker, K. L., M. P. Gillingham, T. A. Hanley, and C. T. Robbins. 1999. Energy and protein balance of free-ranging black-tailed deer in a natural forest environment. *Wildlife Monographs* 143:1–48.

Parker, K. L., C. T. Robbins, and T. A. Hanley. 1984. Energy expenditures for locomotion by mule deer and elk. *Journal of Wildlife Management* 48:474–488.

Pearson, H. A. 1969. Starvation in antelope with stomachs full of feed. Research Note RM 148.

Personius, T. L., C. L. Wambolt, J. R. Stephens, and R. G. Kelsey. 1987. Crude terpenoid influence on mule deer preference for sagebrush. *Journal of Range Management* 40:84.

Petrie, H. T., L. W. Klassen, P. S. Klassen, J. R. O'Dell, and H. D. Kay. 1989. Selenium and the immune response: 2. Enhancement of murine cytotoxic T-lymphocyte and natural killer cell cytotoxicity in vivo. *Journal of Leukocyte Biology* 45:215–220.

Pfister, J. A., G. D. Manners, D. R. Gardner, K. W. Price, and M. H. Ralphs. 1996. Influence of alkaloid concentration on acceptability of tall larkspur (*Delphinium* spp.) to cattle and sheep. *Journal of Chemical Ecology* 22:1147–1168.

Portier, C., M. Festa-Bianchet, J.-M. Gaillard, J. T. Jorgenson, and N. G. Yoccoz. 1998. Effects of density and weather on survival of bighorn sheep lambs (*Ovis canadensis*). *Journal of Zoology* 245:271–278.

Prasad, A. S., B. Bao, F. W. J. Beck, O. Kucuk, and F. H. Sarkar. 2004. Antioxidant effect of zinc in humans. *Free Radical Biology and Medicine* 37:1182–1190.

Proffitt, K. M., A. B. Courtemanch, S. R. Dewey, B. Lowrey, D. E. McWhirter, K. L. Monteith, J. T. Paterson, J. Rotella, P. J. White, and R. A. Garrott. 2021. Regional variability in pregnancy and survival rates of Rocky Mountain bighorn sheep. *Ecosphere* 12:e03410.

Rachlow, J. L., and R. T. Bowyer. 1994. Variability in maternal behavior by Dall's sheep: environmental tracking or adaptive strategy? *Journal of Mammalogy* 75:328–337.

Rachlow, J. L., and R. T. Bowyer. 1998. Habitat selection by Dall's sheep (*Ovis dalli*): maternal trade-offs. *Journal of Zoology* 245:457–465.

Ralphs, M. H., J. D. Olsen, J. A. Pfister, and G. D. Manners. 1988. Plant-animal interactions in larkspur poisoning in cattle. *Journal of Animal Science* 66:2334–2342.

Rankins, S. T., T. R. Stephenson, and K. L. Monteith. 2024. Sociality modulates nutritional carrying capacity of an endangered species. *Frontiers in Ecology and Evolution* 12:1417970.

de Renzo, E. C., E. Kaleita, P. Heytler, J. J. Oleson, B. Hutchings, and J. H. Williams. 1953. The nature of the xanthine oxidase factor. *Journal of the American Chemical Society* 75:753.

Riney, T. 1955. Evaluating condition of free ranging red deer (*Cervus elaphus*), with special reference to New Zealand. *New Zealand Journal of Science and Technology. B.* 36:429–463.

Risenhoover, K. L., and J. A. Bailey. 1985. Foraging ecology of mountain sheep: implications for habitat management. *Journal of Wildlife Management* 49:797–804.

Robbins, C. T. 1993. *Wildlife Feeding and Nutrition.* 2nd edition. Academic Press Inc., San Diego, California, USA.

Robbins, C. T., D. E. Spalinger, and W. Van Hoven. 1995. Adaptation of ruminants to browse and grass diets: are anatomical-based interpretations valid? *Oecologia* 103:208–213.

Robbins, C. T., S. M. Parish, and B. L. Robbins. 1985. Selenium and glutathione peroxidase activity in mountain goats. *Canadian Journal of Zoology* 63:1544–1547.

Robbins, C. T., S. Mole, A. E. Hagerman, and T. A. Hanley. 1987b. Role of tannins in defending plants against ruminants: reduction in dry matter digestion. *Ecology* 68:1606–1615.

Robbins, C. T., T. A. Hanley, A. E. Hagerman, O. Hjeljord, D. L. Baker, C. C. Schwartz, and W. W. Mautz. 1987a. Role of tannins in defending plants against ruminants: reduction in protein availability. *Ecology* 68:98–107.

Robinson, B. H. 1998. The role of manganese superoxide dismutase in health and disease. *Journal of Inherited Metabolic Disease* 21:598–603.

Robinson, J. R. 1957. Functions of water in the body. *Proceedings of the Nutrition Society* 16:108–112.

Rognmo, A., K. A. Markussen, E. Jacobsen, H. J. Grav, and A. S. Blix. 1983. Effects of improved nutrition in pregnant reindeer on milk quality, calf birth weight, growth, and mortality. *Rangifer* 3:10–18.

Rowntree, J. E., G. M. Hill, D. R. Hawkins, J. E. Link, M. J. Rincker, G. W. Bednar, and R. A. Kreft. 2004. Effect of Se on selenoprotein activity and thyroid hormone metabolism in beef and dairy cows and calves. *Journal of Animal Science* 82:2995–3005.

Rubin, E. S., W. M. Boyce, and V. C. Bleich. 2000. Reproductive strategies of desert bighorn sheep. *Journal of Mammalogy* 81:769–786.

Russell, D. E., A. M. Martell, and W. A. C. Nixon. 1993. Range ecology of the Porcupine caribou herd in Canada. *Rangifer Special Issue* 13:1–170.

Sachdeva, K. K., O. P. S. Sengar, S. N. Singh, and I. L. Lindahl. 1973. Studies on goats. I. Effect of plane of nutrition on the reproductive performance of does. *Journal of Agricultural Science* 80:375–379.

Sand, H., C. Wikenros, P. Ahlqvist, T. H. Strømseth, and P. Wabakken. 2012. Comparing body condition of moose (*Alces alces*) selected by wolves (*Canis lupus*) and human hunters: consequences for the extent of compensatory mortality. *Canadian Journal of Zoology* 90:403–412.

Sanford, S. E., G. K. A. Josephson, and A. J. Rehmtulla. 1993. Winter death syndrome (exposure and starvation) in deer. *Canadian Veterinary Journal* 34:443–443.

Sardesai, V. M. 1993. Molybdenum: an essential trace element. *Nutrition in Clinical Practice* 8:277–281.

Schwarz, G., R. R. Mendel, and M. W. Ribbe. 2009. Molybdenum cofactors, enzymes and pathways. *Nature* 460:839–847.

Scrutton, M. C., P. Griminger, and J. C. Wallace. 1972. Pyruvate carboxylase. *Journal of Biological Chemistry* 247:3305–3313.

Searle, K. R., N. T. Hobbs, and I. J. Gordon. 2007. It's the "foodscape", not the landscape: using foraging behavior to make functional assessments of landscape condition. *Israel Journal of Ecology and Evolution* 53:297–316.

Seegmiller, R. F., P. R. Krausman, W. H. Brown, and F. M. Whiting. 1990. Nutritional composition of desert bighorn sheep forage in the Harquahala Mountains, Arizona. *Desert Plants* 10:87–90.

Seip, D. R. 1983. Foraging ecology and nutrition of Stone's sheep. Department of Forestry, University of British Columbia, Vancouver, British Columbia, Canada.

Shankar, A. H., and A. S. Prasad. 1998. Zinc and immune function: the biological basis of altered resistance to infection. *American Journal of Clinical Nutrition* 68:447s–463s.

Shipley, L. A., R. C. Cook, and D. G. Hewitt. 2020a. Techniques in wildlife nutritional ecology. Pages 439–482 in N. Silvy, editor. *The Wildlife Techniques Manual.* Volume 1, *Research Methods.* Johns Hopkins University Press, Baltimore, Maryland, USA.

Shipley, L. A., and D. E. Spalinger. 1992. Mechanics of browsing in dense food patches: effects of plant and animal morphology on intake rate. *Canadian Journal of Zoology* 70:1743–1752.

Shipley, L. A., R. C. Cook, and D. G. Hewitt. 2020b. Techniques for wildlife nutritional ecology. Pages 439–482 in N. Silvy, editor. *The Wildlife Techniques Manual.* Volume 1, *Research Methods.* 8th edition. Johns Hopkins University Press, Baltimore, Maryland, USA.

Signer, C., T. Ruf, and W. Arnold. 2011. Hypometabolism and basking: the strategies of alpine ibex to endure harsh over-wintering conditions. *Functional Ecology* 25:537–547.

Simmons, N. M. 1982. Seasonal ranges of Dall's sheep, Mackenzie Mountains, Northwest Territories. *Arctic* 35:512–518.

Singer, F. J., A. Harting, K. K. Symonds, and M. B. Coughenour. 1997. Density dependence, compensation, and environment effects on elk calf mortality in Yellowstone National Park. *Journal of Wildlife Management* 61:12–25.

Sittler, K. L., K. L. Parker, and M. P. Gillingham. 2019. Vegetation and prescribed fire: implications for stone's sheep and elk. *Journal of Wildlife Management* 83:393–409.

Skogland, T. 1990. Density dependence in a fluctuating reindeer herd: maternal versus offspring effects. *Oecologia* 84:442–450.

Smiley, R. A., B. L. Wagler, T. N. Lasharr, K. Denryter, T. R. Stephenson, A. B. Courtemanch, T. W. Mong, D. Lutz, D. Mcwhirter, D. Brimeyer, P. Hnilicka, B. Lowrey, and K. L. Monteith. 2022. Heterogeneity in risk-sensitive allocation of somatic reserves in a long-lived mammal. *Ecosphere* 13:e4161.

Smiley, R. A., B. L. Wagler, W. H. Edwards, J. E. Jennings-Gaines, K. Luukkonen, K. Robbins, M. Johnson, A. B. Courtemanch, T. W. Mong, D. Lutz, D. McWhirter, J. L. Malmberg, B. Lowrey, and K. L. Monteith. 2024. Infection–nutrition feedbacks: fat supports pathogen clearance but pathogens reduce fat in a wild mammal. *Proceedings of the Royal Society B* 291:20240636.

Smith, R. H., and A. B. Mcallan. 1970. Nucleic acid metabolism in the ruminant 2. *Formation of microbial nucleic acids in the rumen in relation to the digestion of food nitrogen, and the fate of dietary nucleic acids. *British Journal of Nutrition* 24:545–556.

Solomon, M. E. 1949. The natural control of animal populations. *Journal of Animal Ecology* 18:1–35.

Spalinger, D. E., W. B. Collins, T. A. Hanley, N. E. Cassara, and A. M. Carnahan. 2010. The impact of tannins on protein, dry matter, and energy digestion in moose (*Alces alces*). *Canadian Journal of Zoology* 88:977–987.

Spears, J. W. 2000. Micronutrients and immune function in cattle. *Proceedings of the Nutrition Society* 59:587–594.

Spears, J. W., and W. P. Weiss. 2008. Role of anti-oxidant and trace element in health and immunity of transition dairy cows. *The Veterinary Journal* 176:70–76.

Spitz, D. B. 2015. Does migration matter? Cause and consequences of migratory behavior in Sierra Nevada bighorn sheep. Dissertation, University of Montana, Missoula, Montana, USA.

Stephenson, T. R., D. W. German, E. F. Cassirer, D. P. Walsh, M. E. Blum, M. Cox, K. M. Stewart, and K. L. Monteith. 2020. Linking population performance to nutritional condition in an alpine ungulate. *Journal of Mammalogy* 101:1244–1256.

Striby, K. D., C. L. Wambolt, R. G. Kelsey, and K. M. Havstad. 1987. Crude terpenoid influence on in vitro digestibility of sagebrush. *Journal of Range Management* 40:244.

Studer, J. M., W. P. Schweer, N. K. Gabler, and J. W. Ross. 2022. Functions of manganese in reproduction. *Animal Reproduction Science* 238:106924.

Suttle, N. F. 2010. *Mineral Nutrition of Livestock.* 4th edition. CABI, Wallingford, United Kingdom.

Syrjälä-Qvist, L., and J. Salonen. 1983. Effect of protein and energy supply on nitrogen utilization in reindeer. *Acta Zoologica Fennica* 175:53–55.

Taillon, J., P. S. Barboza, and S. D. Côté. 2013. Nitrogen allocation to offspring and milk production in a capital breeder. *Ecology* 94:1815–1827.

Terry, P. J., A. C. Alvidrez, and C. W. Black. 2022. Factors affecting bighorn sheep activity at water developments in southwestern Arizona. *Journal of Wildlife Management* 86:e22134.

Testa, J. W., and G. P. Adams. 1998. Body condition and adjustments to reproductive effort in female moose (*Alces alces*). *Journal of Mammalogy* 79:1345–1354.

Tilley, J. M. A., and R. A. Terry. 1963. A two-stage technique for the in vitro digestion of forage crops. *Grass and Forage Science* 18:104–111.

Tilton, M. E., and E. E. Willard. 1981. Winter food habits of mountain sheep in Montana. *Journal of Wildlife Management* 45:548–553.

Todd, J. W. 1975. Foods of Rocky Mountain bighorn sheep in southern Colorado. *Journal of Wildlife Management* 39:108–111.

Torbit, S. C., L. H. Carpenter, D. M. Swift, and A. W. Alldredge. 1985. Differential loss of fat and protein by mule deer during winter. *Journal of Wildlife Management* 49:80–85.

van den Top, A. M. 2005. Reviews on the mineral provision in ruminants (VIII): iron metabolism and requirements in ruminanta. CVN documentation report 40. Central Veevoederbureau, Lelystad, Netherlands.

Van Soest, P. J. 1982. *Nutritional Ecology of the Ruminant.* Cornell University Press, Ithaca, New York, USA.

Verme, L. J., and J. J. Ozoga. 1980. Influence of protein-energy intake on deer fawns in autumn. *Journal of Wildlife Management* 44:305–314.

Verme, L. J., and D. E. Ullrey. 1984. Physiology and Nutrition. Pages 91–118 *in* L. K. Halls, editor. *White-Tailed Deer: Ecology and Management.* Stackpole Books, Harrisburg, Pennsylvania, USA.

Vikøren, T., A. B. Kristoffersen, S. Lierhagen, and K. Handeland. 2011. A comparative study of hepatic trace element levels in wild moose, roe deer, and reindeer from Norway. *Journal of Wildlife Diseases* 47:661–672.

Vogel, K. P., J. F. Pedersen, S. D. Masterson, and J. J. Toy. 1999. Evaluation of a filter bag system for NDF, ADF, and IVDMD forage analysis. *Crop Science* 39:276–279.

Wagler, B. L., R. A. Smiley, A. B. Courtemanch, D. Lutz, D. McWhirter, D. Brimeyer, P. Hnilicka, and K. L. Monteith. 2024. Disparate home range dynamics reflect nutritional inadequacies on summer range for a large herbivore. *Ecosphere* 15:1–14.

Wagler, B. L., R. A. Smiley, A. B. Courtemanch, D. Lutz, D. McWhirter, D. Brimeyer, P. Hnilicka, T. J. Robinson, and K. L. Monteith. 2023. Implications of forage quality for population recovery of bighorn sheep following a pneumonia epizootic. *Journal of Wildlife Management* 87:e22452.

Wagner, G. D., and J. M. Peek. 2006. Bighorn sheep diet selection and forage quality in central Idaho. *Northwest Science* 80:246–258.

Wallmo, O. C., L. W. Carpenter, W. L. Regelin, R. B. Gill, and D. L. Baker. 1977. Evaluation of deer habitat on a nutritional basis. *Journal of Range Management* 30:122–126.

Warrick, G. D., and P. R. Krausman. 1989. Barrel cacti consumption by desert bighorn sheep. *Southwestern Naturalist* 34:483–486.

Weber, C., C. Hametner, A. Tuchscherer, B. Losand, E. Kanitz, W. Otten, S. P. Singh, R. M. Bruckmaier, F. Becker, W. Kanitz, and H. M. Hammon. 2013. Variation in fat mobilization during early lactation differently affects feed intake, body condition, and lipid and glucose metabolism in high-yielding dairy cows. *Journal of Dairy Science* 96:165–180.

Wehausen, J. D. 1980. Sierra Nevada bighorn sheep: history and population ecology. PhD Dissertation, University of Michigan, Ann Arbor, Michigan, USA.

Wehausen, J. D., V. C. Bleich, B. Blong, and T. L. Russi. 1987. Recruitment dynamics in a southern California mountain sheep population. *Journal of Wildlife Management* 51:86–98.

White, R. G. 1983. Foraging patterns and their multiplier effects on productivity of northern ungulates. *Oikos* 40:377–384.

Whiting, J. C., R. T. Bowyer, J. T. Flinders, V. C. Bleich, and J. G. Kie. 2010. Sexual segregation and use of water by bighorn sheep: implications for conservation. *Animal Conservation* 13:541–548.

Wilson, P. R., and N. D. Grace. 2001. A review of tissue reference values used to assess the trace element status of farmed red deer (*Cervus elaphus*). *New Zealand Veterinary Journal* 49:126–132.

Wu, G. 2017. *Principles of Animal Nutrition.* CRC Press, Boca Raton, Florida, USA.

Zimmerman, T. J., J. A. Jenks, and D. M. Leslie. 2006. Gastrointestinal morphology of female white-tailed and mule deer: effects of fire, reproduction, and feeding type. *Journal of Mammalogy* 87:598–605.

# 8 Diseases

*Michael W. Miller, Karen A. Fox, Margo J. Pybus, and Helen M. Schwantje*

## 8.1 INTRODUCTION

Parasites and other agents of disease are natural occurrences in North America's bighorn (*Ovis canadensis*) and thinhorn (*Ovis dalli*) sheep. Yet disease is, generally, rare. Throughout their lives, most mountain sheep provide suitable habitat for a myriad of organisms. These usually live in harmony (or détente) inside or on the outside of individual sheep with little or no consequence. But not always we hope the information in this chapter can provide a foundation of knowledge and principles for better understanding of diseases of mountain sheep and guiding appropriate and effective responses.

The effects of most recognized diseases on mountain sheep are limited to impairing individual or local (i.e., herd-level) productivity, survival, or resilience. A few diseases can have more widespread or lasting effects that limit population growth, persistence, and resilience and can complicate and limit options for management (Brewer et al. 2014, Jex et al. 2016). Diseases in their broadest context may have had more cumulative, range-wide consequences for bighorn sheep than for any other native ungulate species. Several apparently non-native (i.e., introduced) organisms played particularly pivotal roles in historical declines in Rocky Mountain (*O. c. canadensis*), desert (*O. c. nelsoni*), and California (*O. c. californiana*) bighorn sheep abundance and distribution (Hornaday 1901, Warren 1910, Grinnell 1928, Shillinger 1937, Spencer 1939, Honess and Frost 1942, McTaggart Cowan 1951, Buechner 1960). These and other factors (i.e., habitat suitability, fragmentation) continue to limit population resilience and abundance, hampering population restoration and impeding conservation efforts for the bighorn subspecies. North America's thinhorn (Dall's [*O. d. dalli*] and Stone's [*O. d. stonei*]) sheep populations are vulnerable to similar harm from introduced disease agents but have been spared from their effects to date, likely due to their remoteness and because such introductions either have not occurred or have been so limited in frequency, severity, or geographic distribution that they have gone undetected.

We provide an overview of an array of diseases that can affect mountain sheep in North America. We use an expansive definition of disease to include "any impairment that interferes with or modifies the performance of normal functions, including responses to environmental factors such as nutrition, toxicants, and climate; infectious [or otherwise transmissible] agents; inherent or congenital defects; or combinations of these factors" (Wobeser 1997:1, 2007:5). We focus on diseases of apparent significance,

acknowledging that not every disease ever seen in mountain sheep has been covered. The review of mountain sheep diseases by Bunch et al. (1999) provided a solid foundation on this subject, which we have built upon here. This chapter opens with elements of key principles and patterns characteristic of diseases in animal populations, with specific relevance and applications to mountain sheep ecology and management. This is followed by brief summaries of the apparent cause(s) and effects of some diseases that affect mountain sheep, with updated information and supporting references for readers seeking further details. Conservation and management implications of select diseases are included in context. Our intent is to provide those interested in the history, ecology, management, and conservation of North American mountain sheep with an accessible primer on native and introduced diseases reported to affect these species.

## 8.2 DISEASE ECOLOGY AND CONTROL: GENERAL PRINCIPLES AND PATTERNS

Understanding and, where warranted, attempting to control diseases that can affect mountain sheep or other free-ranging wildlife populations is complicated. Despite the variety and unique features of specific diseases of interest or concern, general epidemiological principles and patterns common to diseases of domestic animals and humans seem worth appreciating because they tend to apply in free-ranging animals. Some familiarity with these basics should help biologists and managers understand the likely influences of observed disease patterns on local scales, anticipate changes, and assess the need and options for a response.

### 8.2.1 THREE-PART DISHARMONY

Disease is an outcome. The widely embraced host-agent-environment model (Hanson 1969, Martin et al. 1987, Wobeser 2007) holds that diseases result from the interactions among the host or victim (mountain sheep, in this context), the disease agent (a pathogen, toxin, or other risk factor), and the surrounding environment (e.g., climate, weather, habitat conditions, forage availability, anthropogenic influences). The basic principles of disease investigation and control are similar for infectious and non-infectious causes, although in some contexts, the term risk factor, rather than causative agent, may be more appropriate in describing non-infectious diseases (Wobeser 2007). It is common practice, and often a pitfall, to narrowly focus on the specific host-agent interaction(s) that result in the

DOI: 10.1201/9781003518686-9

specific disease of interest, but natural systems are far more complicated (Hanson 1969, Wobeser 2007, Stephen 2022). Broader consideration of factors and interactions that can contribute or lead to a consequential number of cases (i.e., a disease outbreak) has led to a better understanding of disease ecology and may, in turn, better serve conservation and management.

## 8.2.2 All Parasites, Large and Small

Among the diseases having long-lasting adverse effects on mountain sheep abundance in North America, those arising from exposure to transmissible agents seem of greatest immediate concern. These agents, all technically parasites of one form or another, fall into two broad groups based on key ecological differences (Anderson and May 1979, Wobeser 2007). Viruses, bacteria, protozoa, and parasitic fungi are quite small, have short generation times, and multiply directly and rapidly in the host; in short, they are relatively r-selected parasites with limited habitat options. Helminths and arthropods have much longer generation times and multiply more slowly, often with part of their life cycles off their sheep hosts; they are relatively k-selected parasites with greater habitat flexibility.

These differences influence expected disease patterns. The microscopic agents (microparasites; Anderson and May 1979) tend to produce relatively brief infections (compared to their host's lifespan) and may induce some level of immunity to reinfection. Multiple species and genetic variants of species (strains) can infect the same host. The larger multicellular agents (macroparasites; Anderson and May 1979) tend to produce more persistent infections (if internal) or infestations (if external) and accommodate reinfection, and both the immune response and the damage (pathology) produced depend on the number of parasites harbored by the host. There are of course, exceptions, but understanding these broad differences can help managers consider whether control is appropriate and select a control strategy most likely to interrupt any related disease processes.

## 8.2.3 Coming to Terms with Disease as an Outcome of (Some) Host-Agent Interactions

As framed in the foregoing disease model, disease outcomes are a product of the interplay between the causative agent's actions and the host's responses in the context of environmental influences. The terms pathogen, pathogenicity, and virulence are often encountered in discussions of disease in mountain sheep. We have used these terms as follows:

- Pathogen. An agent (small or large) capable of causing host damage that can lead to disease; any damage to the host will be the result of agent's actions, host responses, or both;
- Pathogenicity. The capacity of an agent to cause damage that leads to disease in the host; and

- Virulence. The relative capacity to cause damage or severe disease.

In this chapter, the adjectives primary, opportunistic, commensal, and emergent are used sparingly about individual pathogens. These terms may lead to confusion or become a distraction in the context of mountain sheep diseases. Moreover, invoking these modifiers tends to oversimplify the complex relationships and processes that lead to observable disease (Casadevall and Pirofski 1999, 2001). The involvement of multiple hosts and pathogens, measurable variation of pathogenicity and virulence within some agents (strain variation), differing exposure or interaction histories across space and over time, and the potential for both host(s) and pathogen(s) to change independently or in response to one another or to environmental conditions over time can and often do influence the likelihood and severity of disease outcomes.

Severe disease reflects the most extreme adverse outcomes of host-agent interactions in a susceptible host (Casadevall and Pirofski 1999, 2001). A specific agent may or may not show pathogenicity consistently over time. Modifiers of virulence can originate from the agent (e.g., production of toxins, protective capsules, hiding within host cells, all features and substances collectively called virulence factors) and from the host (e.g., immune system responses, physiological status including nutritional or other stress responses, behavior), from the environment (weather events, climate, anthropogenic influences), and interactions among any or all of these. Viewing virulence as an ongoing damage-response dynamic rather than a static outcome (Casadevall and Pirofski 1999) makes sense in the context of most diseases of management consequence for mountain sheep.

## 8.2.4 About Time

The frequency of and variation in severity of disease occurrence over time has implications for host population dynamics and thus motivations for control responses (Martin et al. 1987). Diseases that emerge suddenly with a rapid increase in case numbers (epidemic or epizootic disease) garner attention from agencies, the public, and media, potentially creating interest in some form of management response. Diseases occurring with lower frequency and in more predictable patterns (endemic, enzootic disease) may cause less concern. Diseases that occur infrequently and with no discernable pattern (sporadic disease) can be difficult to detect in free-ranging situations and rarely rise to the level of a management concern.

Because natural selection generally favors (with notable exceptions) nonlethal agents, most host-agent relationships tend to move toward some form of coexistence over time, shifting from epidemic to endemic disease patterns and perhaps eventually to sporadic occurrence under otherwise stable environmental conditions (Martin et al. 1987). This shift may come at some cost in abundance or reproductive performance to both the host and the agent. Some agents (e.g., terrestrial rabies virus strains) seem less adapted toward reaching a commensal state than others and are maintained in natural systems by moving from host to host across relatively large geographic areas (Yekutiel 1980, Wobeser 2007).

It follows that sporadic or endemic disease patterns in most host species, including mountain sheep, are likely to reflect a longer association and coadaptation or coevolution with the causative agent(s); some of these agents may be described as native. Epidemic patterns may be more suggestive of a (relatively) recent introduction of a novel agent, genetic variant (i.e., strain), or virulence factor, or of some other perturbation or change in host or agent or environmental factors that gives rise to an imbalance favoring the disease agent. The apparent decline in occurrences of mange (a mite-induced skin disease; also see Section 8.14) in multiple bighorn ranges, which changed from severe epidemics in the late 1800s to sporadic cases more recently (Seton 1929, Buechner 1960, Boyce and Weisenberger 2005), likely reflects some degree of relative coadaptation following historical introductions of novel mites of the genus *Psoroptes*. In contrast, the severe mange outbreaks reported in British Columbia and Washington over the last decade are linked to a more recent introduction of a different novel *Psoroptes* spp. (Hering et al. 2021) and, as might be expected, resemble the patterns reported historically (Seton 1929, Brooks 1923).

When attempting to investigate, understand, and compare outbreaks of a particular disease, it is important to recognize that various agents and other relevant host and environmental factors may or may not be involved in a particular location at a point in time. These processes seem best viewed as likely to occur on a continuum, with the duration often uncertain. Depending on the agents involved and the herd history (both origin and health), adaptations and responses among the respective host-agent-environment relationships may add complexity to patterns of disease observed in the field.

### 8.2.5 An Unfortunate Reunion

A complicating but sometimes overlooked facet of host-agent and disease dynamics in mountain sheep is their close-yet-distant relationship to the domesticated ruminant species from Europe and elsewhere, which arrived on western North American ranges after the mid-1500s (Seton 1929, Shillinger 1937, Melville 1994). Mountain sheep and domestic sheep can interbreed and produce fertile but remarkably unappealing hybrid offspring, despite being different species with somewhat mismatched gestation periods. This underscores their physiological similarity and a shared biological suitability to host various infectious agents. Even so, mountain sheep species in North America diverged from their closest known Eurasian relatives about 1.6 million (M) years ago (ya; Rezaei et al. 2010). They likely carried a community of commensal parasites and encountered new agents that reshaped their innate host

responses to parasitism as their range expanded into North America and the thinhorn and bighorn species separated (~1 Mya; Rezaei et al. 2010). In Eurasia, meanwhile, sheep species and their complement of parasites evolved over about 2.5 M years and ultimately gave rise to the species that was domesticated by humans (~11,000 ya; Hiendleder et al. 2002, Rezaei et al. 2010). Domestication altered selective pressures on, and responses of, the hosts and their associated agents (Cao et al. 2021). It also provided new opportunities for novel pathogens to emerge and redistribute among domesticated sheep, goats, cattle, and other species. Given this background, the unfortunate consequences of introducing domestic ruminants and their divergently coadapted agents into mountain sheep and their habitats in recent centuries are consistent with a general pattern seen in the wake of reacquainting long-diverged lineages of various species, including humans (Shillinger 1937, Diamond 1997). Thus, we should not be surprised at the disease outcomes.

### 8.2.6 Control: Take It or Leave It

The choice to attempt control of any disease is based on anthropocentric concerns about the consequences for humans or the animals they care about (Martin et al. 1987, Wobeser 2007, Stephen 2022). The three broad objectives for managing a disease are: prevention (i.e., avoid introduction of disease agents of concern), control (i.e., reduce occurrences of an established disease to a biologically or economically justifiable or tolerable level), or eradication (i.e., eliminate the disease agent[s] of concern from a defined geographic area).

Of these, eradication requires the most extreme measures, is by far the most challenging, and offers the lowest prospects for success in most wildlife situations (Yekutiel 1980, Martin et al. 1987, Wobeser 2007). Regardless of the objective, expectations that a wildlife disease management campaign be simultaneously cheap, quick, and good are unlikely to be met. Desirability, feasibility, beneficiaries, costs and benefits, availability and viability of tools or approaches, goals and scope, and measures of effectiveness should all be considered in decision-making related to wildlife disease control (Wobeser 2007).

Commonalities in the principles and patterns covered thus far can be useful in the considerations for disease control. Control for most transmissible disease agents aims to depress their reproductive rate (Wobeser 2007) by disrupting one or more of the steps of the process required for transmission: the agent leaving one host or reservoir, being moved directly or indirectly, and gaining entry (e.g., via oral or respiratory uptake, direct or indirect contact, an insect bite, eating a tainted snail) to a naïve or susceptible host.

Numerous approaches have been applied in trying to manage animal diseases (Table 8.1; Martin et al. 1987, Wobeser 2007, Gillin 2022). Prospective control strategies can be developed using knowledge on how the disease agent exits and enters a host and its mode(s) of transmission or exposure to target the aspect(s) most vulnerable to intervention

## TABLE 8.1

**Disease Management Approaches**

| Disease Control Activity | Brief Description of Objective | Example(s) of Management Actions or Treatment Application in Mountain Sheep | Information Sources |
| --- | --- | --- | --- |
| Lethal removals | Eliminate infected individuals or carriers | Hunting and culling, applied selectively (test and remove) or non-selectively (depopulation); scales range from individuals to entire remnant herds | Garwood et al. (2020), Western Association of Fish and Wildlife Agencies (2021), and Wild Sheep Initiative (2023) |
| Restrict movements | Contain infected individuals or carriers (also termed quarantine) | Prevent or reduce movement of infected or exposed herds or regions (e.g., curtail translocations) | George et al. (2009), jurisdictional herd plans |
| Reduce contact | Use physical barriers or other measures to prevent transmission | Double-fencing or other means of physical and temporal separation to prevent contact between wild sheep and livestock | Wild Sheep Working Group (2012) |
| Chemical applications | Use disinfectants or therapeutics to combat agent | Field applications of antiparasitic drugs, antibiotics, mineral supplements | Schmidt et al. (1979), Lange et al. (1980), Boyce et al. (1992), and Sirochman et al. (2012) |
| Modify host resistance | Use vaccines or genetic selection to lower host susceptibility | Vaccines against microparasite(s) of concern | Cassirer et al. (2001) and George et al. (2008) |
| Environmental manipulations | Change environment to interrupt life cycle, transmission pathway, or agent persistence | Broadcast trace mineral applications, prescribed burns, dispersed water access | Proffitt et al. (2022) and Tsuchida et al. (2024) |
| Biological control | Use one biological agent against another | Screwworm (*Cochliomyia hominivorax*) control (indirectly benefited southern ranges) | Gutierrez et al. (2019) |
| Education | Teach humans about disease control | Outreach, information brochures, workshops on disease problems of concern, and ways to reduce risk of livestock-mountain sheep interaction | Wild Sheep Working Group (2012) and Wild Sheep Initiative (2024) |

Approaches common to disease control programs in domestic and wild animals have been applied or could be considered in attempts to manage disease in mountain sheep. Specific activities for controlling disease (adapted from Martin et al. 1987) are listed, along with published examples (where available), illustrating their applications in mountain sheep management.

(Martin et al. 1987, Wobeser 2007). Such approaches can apply in principle to either transmissible or non-transmissible agents. Several well-established strategies have been used or considered in attempts to manage disease in mountain sheep (Table 8.1). Regardless of the approaches adopted, additional aspects of disease control to consider include sustaining the effort for enough time to cause a measurable change (e.g., in disease frequency or distribution) and having a way to measure any change, along with the ability and willingness to modify management actions and perhaps goals to evolve with changing circumstances (Martin et al. 1987, Wobeser 2007, Gillin 2022, Stephen 2022).

## 8.3 DISEASES OF CONSERVATION AND MANAGEMENT CONSEQUENCE

Any disease may have consequences for the affected individual, its offspring, or its local conspecifics, but a subset of diseases, where and when they occur, have disproportionately large effects on conservation and management. Some of these diseases impair population stability and persistence, likely to hold abundance and population resilience below what otherwise might be supported by available forage and habitat. Other diseases may impinge on conservation or management options on local or larger

scales because of potential spillover to domestic livestock or humans. Although experience and an understanding of the effects of these diseases on free-ranging wild sheep are largely from observations in affected bighorn sheep populations, there is potential for equal or greater consequences in thinhorn sheep given their lack of prior exposure, population interconnectedness, and limited observability.

## 8.4 THE RESPIRATORY DISEASE COMPLEX

Respiratory disease is the predominant large-scale health problem in mountain sheep, likely having affected many of the herds and populations of the bighorn subspecies since the arrival of Europeans and their livestock in North America. Since at least early in the 20th century, outbreaks (often termed die-offs) have been periodically reported. Death loss (mortality) attributed to pneumonia occurs across all demographic groups, with losses in some outbreaks exceeding 50% of the affected herd (Rush 1927, McTaggart Cowan 1951, Buechner 1960, Bear and Jones 1973, Onderka and Wishart 1984, Cassirer et al. 1996). Outbreaks may be followed by successive years of little or no recruitment because of fatal respiratory disease in lambs under 6 months of age (Festa-Bianchet 1988, Cassirer and Sinclair 2007, George et al. 2008, Besser et al. 2012, Butler et al. 2013, Ryder et al. 1992, 1994). In some cases, successive years of respiratory disease-associated lamb mortalities may not be preceded by an appreciable all-age mortality event (Huwer 2015, Manlove et al. 2022). The overarching lesson is that disease outbreaks and effects are not always the same, and not always predictable.

The term respiratory disease covers a range of observable (i.e., clinical) field signs in mountain sheep (Spraker and Hibler 1982), including a wet or runny nose (i.e., nasal discharge), sneezing, head shaking, ear droop, coughing, wheezing or labored breathing, extended head and neck, and general malaise (i.e., depression, lack of energy or activity). The signs seen may vary, at least to some degree, with the stage at which illness is observed and other factors, including the extent of monitoring by agencies or the public. Some signs may not be unique to respiratory disease. For example, head shaking, sneezing, and coughing may occur with insect harassment and in dusty environments; mite and tick infestations can cause ear drooping. In its most severe forms, respiratory disease in mountain sheep presents as cases of sudden death, also termed acute or peracute mortality, with no warning or cause apparent from field observations. The foregoing signs appear in various combinations depending primarily on the agent(s) involved, the severity and duration of infection(s), and the individual host's age, prior exposure, physiological status, and immune system function. Similarly, the rate of occurrence can range from single or small groups of affected individuals to entire herds or herd complexes. Outbreaks can last for weeks, extending to months and years. There is no absolute seasonality to the outbreaks involving all demographic groups, but some tendency toward occurrences in autumn through late winter when social aggregations are largest. Outbreaks affecting primarily young lambs more consistently occur in the first several months after the appearance of the new lamb crop. Cumulative death losses in mountain sheep herds affected by respiratory disease range from negligible (Johnson et al. 2022) to catastrophic (Rush 1927, McTaggart Cowan 1951, Bear and Jones 1973, Onderka and Wishart 1984), with outbreaks of greater severity typically receiving the greatest management and public attention. Possible explanations vary, but some outbreaks do not appear to have an extended effect even without management intervention; some lead to endemic disease problems that may dampen population performance and facilitate subsequent epidemics; and some persist long enough to eventually result in long-term functional extirpation of the affected herd.

Like the variation in clinical signs of respiratory disease in mountain sheep, internal damage (i.e., pathology) also differs among tissues and among different individuals (Spraker and Hibler 1982). Nonetheless, when respiratory disease is the cause of death, a common thread for most fatalities is the presence of pneumonia. Classically, pneumonia in mountain sheep involves some form of bronchopneumonia, or inflammation of the bronchi (air passages to the lungs) and lungs. Due to a combination of normal lung anatomy and gravity, if normal protective mechanisms fail, the bronchi deliver inhaled material to the portions of the lung closest to the source (the head; also called cranial or anterior), and lowest in the chest (ventral). If that material contains a pathogen that stimulates an immune response leading to tissue damage, there is a consistent recognizable (macroscopic, or gross [not a subjective term in this context]) appearance of firm, red, maroon or purple colored lung tissue, in the most cranioventral lung lobe on the right side. Mucous and inflammatory exudates (fluid ooze) fill the affected airways and the sponge-like alveoli (gas exchange pockets). The pneumonic lung takes on a firm consistency that feels solid (consolidated), more like liver tissue, because there is no longer any air present, whereas healthy, pink lung tissue feels spongy, because it is air-filled. As disease progresses, larger portions of the lungs and other nearby tissues, such as the pleura (lining of the ribcage and covering of the lungs), are affected by the process until the sheep's immune system either fends off the infection or the individual sheep dies. The affected lung may become adhered or stuck to the lining of the ribs, to the heart sac, and to itself, and eventually, abscesses (pus pockets) may form. When the result is death, it usually occurs from a combination of respiratory compromise compounded by other effects of bacterial infection, such as spillover of bacterial toxins into the blood (e.g., endotoxemia). Some affected individuals who survive may suffer from long-standing (chronic) aftereffects of lung inflammation with severe scarring and respiratory compromise that can result in poor condition and a relative inability to move when chased, predisposing the individual to predation.

The progression of bronchopneumonia can be easily appreciated under a microscope. At the margin of healthy and diseased tissue, there is early lung damage and the arrival of inflammatory cells (white blood cells). Within the most cranioventral portions of the lung lie the more long-standing lesions. Here, the tissue is beyond repair and often has been colonized by bacteria suited to thriving in dead and dying tissue with little or no oxygen. Evidence of scarring and abscesses may eventually appear in chronic pneumonia cases.

The progressive nature of bronchopneumonia means that multiple agents can be found in pneumonic mountain sheep lungs at various stages of the infection. Some organisms may initiate and advance the infection while others play a more significant role in later tissue damage, but the cumulative effects from all stages of bronchopneumonia contribute to fatal outcomes. With laboratory examinations, the most easily identified or prominent bacteria, virus, or worm in a pneumonic lung sample, is easy to label as the most important, but that may not necessarily represent the disease stages occurring elsewhere in the lung or at earlier points in time or in other individuals in the herd. Most pneumonia cases in mountain sheep appear to be polymicrobial, defined as arising from contributions of multiple pathogens. Properly understanding a respiratory disease outbreak thus tends to involve a thorough investigation of multiple cases.

The cumulative findings from such investigations over time and across locations, combined with outcomes from controlled studies, have revealed that a suite of transmissible agents can contribute to the onset, severity, and duration of outbreaks and the resulting respiratory disease in mountain sheep. A number of these agents likely originated from domestic livestock, either immediately before the outbreak or at some point farther in the past (Seton 1929, Shillinger 1937, Onderka et al. 1988, Miller et al. 2013, Andrews et al. 2024). Introduced agents may arrive singly or in groups, depending on source(s) and circumstances. Some may be more readily transmitted than others between and within host species. Other historically native (endemic or commensal) parasites, host factors, and environmental factors also may be involved. It follows that various combinations of introduced and endemic agents and local circumstances may arise. The resulting interactions have been aptly described as a web of causation (Wobeser 2007).

The next several subsections explore the respective contributions of select pathogens and other factors to occurrences of pneumonia in mountain sheep. The inherent complexity of the respiratory disease syndrome presents challenges in prevention and control. Nonetheless, knowledge about interactions of hosts, agents, and their environment can help interpret observed disease patterns and help to develop long-term adaptive management strategies.

### 8.4.1 MYCOPLASMA OVIPNEUMONIAE AND OTHER MYCOPLASMAS

*Mycoplasma ovipneumoniae*, one species of relatively small and simple bacteria, has emerged as an important component of respiratory disease epidemics in mountain sheep. As suggested for other animal diseases caused by mycoplasmas (Markham and Noormohammadi 2005), its relatively new recognition in mountain sheep respiratory disease may be more attributable to advances in laboratory techniques for detection and differentiation than to a recent emergence of its involvement (Besser et al. 2012, Besser and Cassirer 2023).

*Mycoplasma ovipneumoniae* colonizes the outermost (mucosal) surfaces of the respiratory tract. Interaction with these mucosal surfaces and the resulting host immune responses to such interactions, which are themselves complex, can lead to respiratory tract damage, secondary infections, and disease. One common trait of disease-associated mycoplasmas is their ability to elude and persist in the face of their host's immune responses, thereby frustrating the host's normal mechanisms for clearing out invading pathogens and leading to chronic inflammation and sustained tissue damage (Simecka 2005). Variation in both disease-causing capacity (i.e., virulence) of mycoplasmas and in the immune system responses of the host species they infect may contribute to the variable outcomes of infection observed in mountain sheep (Besser et al. 2021, Johnson et al. 2022, Walsh et al. 2023, Andrews et al. 2024).

Multiple genetic variants (strains) of *M. ovipneumoniae* occur in mountain sheep (Kamath et al. 2019, Lieske et al. 2022, Andrews et al. 2024). The majority of those studied to date share relatively close genetic relatedness to the strains

that occur in domestic sheep or goats (Kamath et al. 2019, Andrews et al. 2024). A strain of *M. ovipneumoniae* identified in Alaskan Dall's sheep and caribou (*Rangifer tarandus*) has not been associated thus far with epidemic disease (Lieske et al. 2022, Andrews et al. 2024). Strain diversity, variable expression of potential virulence factors, and perhaps host anatomy may contribute to *M. ovipneumoniae*'s circumnavigation of host defenses and the variable severity of associated disease observed in the field (Besser et al. 2021, Johnson et al. 2022, Manlove et al. 2022, Walsh et al. 2023).

Because none of the agents involved in the mountain sheep respiratory disease complex tend to act alone in natural cases and outbreaks, it is difficult to attribute any particular clinical signs or pathologic findings to specific agents. Observations in natural and experimental settings, and comparative pathology from other species indicate that pathogenic strains of *M. ovipneumoniae* cause coughing and nasal discharge of varying severity and duration in mountain sheep. Infection of the middle ear can also occur, causing ear droop in lambs. Prolonged respiratory tract infection damages the lining and protective mechanisms within airways (cell-associated hairs or cilia that remove inhaled particles). Microscopic features (also seen in other mycoplasma-associated pneumonias of people and animals) include airways that are dilated, lined by reactive, mucus-producing cells, and surrounded by chronic inflammation (Besser et al. 2014, Besser and Cassirer 2023). In addition to direct effects, this damage can also facilitate invasion of lung tissue by other respiratory pathogens (e.g., Pasteurellaceae), thereby compounding disease severity and increasing the likelihood of death rather than recovery (Dassanayake et al. 2010a). Some surviving individuals do not completely clear the infection, instead becoming carriers that intermittently shed these bacteria and contribute to the recurrent respiratory disease outbreaks and related losses in young lambs that impair herd recovery after the initial outbreak event.

Respiratory mycoplasmas can be transmitted via aerosols and may not require prolonged contact or intensive interaction between carriers and susceptible hosts (Besser et al. 2014, Besser and Cassirer 2023, Walsh et al. 2023). Outbreaks involving *M. ovipneumoniae* often seem to arise from a recent introduction of novel strains into affected herds, but it can be difficult to determine the precise timing and mechanism of exposure and introductions in field settings. Genetic analyses suggest the ultimate sources of pathogenic *M. ovipneumoniae* strains affecting mountain sheep likely originate from domestic sheep or goats (Kamath et al. 2019, Besser and Cassirer 2023, Andrews et al. 2024). Mountain sheep can also serve as at least short- to medium-term reservoirs for *M. ovipneumoniae* (Manlove et al. 2022) as demonstrated by strains being maintained in herds for several years. Field data also suggest the possibility of local introductions or perpetuation via other ruminant species (e.g., mountain goats [*Oreamnos americanus*]; Wolff et al. 2019).

Control measures (Table 8.1) have been directed toward preventing new introductions from domestic livestock using physical and temporal separation and by eliminating chronic carriers within wild sheep herds through selective culling (Garwood et al. 2020, Western Association of Fish and Wildlife Agencies 2021). A more detailed review of *M. ovipneumoniae*-related respiratory disease in bighorn sheep was conducted by Besser and Cassirer (2023).

Other mycoplasma species (e.g., *M. arginini, M. bovirhinis*, unidentified *Mycoplasma* sp.) have also been reported in connection with some respiratory disease cases or outbreaks in mountain sheep. Cases of *M. bovis* infection have been confirmed in two bighorn sheep mortalities in British Columbia (C. Thacker, Wildlife Veterinarian, Province of British Columbia, personal communication); its capacity for causing severe disease in North American ungulates including bison (*Bison bison*; Register et al. 2019), pronghorn (*Antilocapra americana*; Malmberg et al. 2020), and mule deer (*Odocoileus hemionus*; Malmberg et al. 2025) suggests there may be some potential for concern.

## 8.4.2 Pasteurellaceae

Bacteria in the family Pasteurellaceae contribute to a gamut of animal diseases (Adlam and Rutter 1989, Miller 2001). Mountain sheep appear extremely susceptible to pasteurellosis, the generic term for disease caused by these bacteria. Pasteurellosis can take a variety of forms, ranging from a localized infection (abscess), to pneumonia to systemic or blood-borne infections (septicemia). Pneumonia is the most common form reported in mountain sheep, but systemic pasteurellosis (identified as hemorrhagic septicemia in older literature) also occurs.

A variety of Pasteurellaceae species and strains have been isolated from mountain sheep herds and their presence has been widely reported in respiratory disease investigations. This may be in part because Pasteurellaceae, unlike mycoplasmas, readily grow in the standard diagnostic culture medias available for decades (Post 1962, Adlam and Rutter 1989, Jaworski et al. 1998). Epidemics involving Pasteurellaceae have been reported from bighorn sheep for nearly a century (Potts 1937, Marsh 1938, McTaggart Cowan 1951, Post 1962, Miller 2001). It seems likely that introduced pathogens, including some Pasteurellaceae, have been affecting bighorn populations since the earliest time of settlement by Europeans and the introduction of their domestic livestock into western North America.

Multiple Pasteurellaceae species inhabit the upper respiratory tract (i.e., mouth, throat, nasal passages) in various combinations and ecological relationships. Most have adaptations (e.g., capsules and biofilms that are protective) that minimize sustained confrontation with their host, at least under favorable conditions (McDougald et al. 2012, Pillai et al. 2018). Some carry and exchange genetic traits expressed in response to adverse conditions, including elevated temperatures and other host responses (Fisher et al. 1999, McNeil et al. 2002, 2003; Kelley et al. 2007). These traits can also contribute to virulence. For example, numerous *Mannheimia* spp. and *Bibersteinia trehalosi* strains carry a gene (*lktA*) encoding a toxin (leukotoxin) that targets

and ruptures host leukocytes (white blood cells designed to attack bacteria). Leukotoxin-producing strains can cause severe pneumonia and systemic disease in mountain sheep. Consequently, *lktA* serves as one marker for identifying potentially harmful Pasteurellaceae. Other bacterial products including adhesins (for attachment), lipopolysaccharide (or endotoxin), fibrinogen binding protein (McNeil et al. 2002), and a novel protease (McNeil et al. 2003) also may contribute to an individual strain's potential virulence (Gioia et al. 2006, Khamesipour et al. 2014).

The extent of coadaptive or coevolutionary relationships between mountain sheep and the pathogenic Pasteurellaceae that are commonly carried in livestock seems uncertain. Some apparently commensal (and perhaps ancestral) *Bibersteinia* spp. commonly found in bighorns do not carry *lktA* (Miller et al. 2013). This seems consistent with observations that leukocytes from bighorn sheep are more vulnerable to leukotoxin than are those from domestic sheep (Silflow et al. 1989, 1991, 1993; Silflow and Foreyt 1994, Dassanayake et al. 2009). Pathogenic strains and strains modified to be less pathogenic in livestock (attenuated) are much more likely to cause severe disease in bighorn and thinhorn sheep than in domestic ruminants (Onderka et al. 1988, Foreyt et al. 1994, 1996). These observations suggest relatively recent introductions of pathogenic Pasteurellaceae and lack of coevolution in mountain sheep hosts as compared to domestic sheep and cattle, although coadaptation may have occurred to a degree in some locations over the last century or more.

*Bibersteinia* spp., *B. trehalosi*, *Pasteurella multocida*, and some *Mannheimia* spp. are common in today's mountain sheep populations. These Pasteurellaceae may or may not be truly native and some may represent past introductions that have persisted (Kelley et al. 2007, Miller and Wolfe 2011). Either way, their persistence as variably endemic and situationally pathogenic bacteria in mountain sheep seems to loosely resemble the relationships in domestic ruminants. Host responses to factors including trauma, stress, or intercurrent disease (e.g., mycoplasmosis, viral or lungworm infection) may trigger isolated cases and epidemics involving these endemic Pasteurellaceae strains (Thorne 1982b, Kraabel and Miller 1997, Bunch et al. 1999, Pillai et al. 2018). Some interspecies competition among bacteria likely occurs (Dassanayake et al. 2010b, Boukahil and Czuprynski 2018), but the presence of endemic Pasteurellaceae in a mountain sheep herd does not seem to prevent novel strain introductions from domestic or wild sources.

*Mannheimia haemolytica* is also considered an upper respiratory tract commensal in domestic sheep and cattle that can cause disease under circumstances including viral infection, stress, or other immune or respiratory compromise (Rice et al. 2007). Whether a similar commensal relationship occurs in mountain sheep remains unclear. Some *M. haemolytica* strains seem to be absent from healthy bighorn herds (Miller et al. 2013). Apparent introductions from livestock have been linked to epidemics (George et al. 2008, Wolfe et al. 2010, Miller et al. 2013), but observed patterns suggest that *M. haemolytica* may eventually disappear or be displaced by other Pasteurellaceae after introduction rather than becoming endemic in at least some locations (Dassanayake et al. 2010b).

Interpreting how Pasteurellaceae contribute to pneumonia outbreaks is complicated by their diversity and seeming ubiquity, and limitations of detection and identification (Miller et al. 2013, MacGlover 2023). Newer diagnostic tools (Conrad et al. 2020, Fox et al. 2023, MacGlover 2023, Kayal et al. 2024) could help clarify associations between observed disease patterns and various Pasteurellaceae species and strains, like approaches being used with mycoplasmas. Even without advanced diagnostics, the overall appearance of the pneumonia-affected lung (fibrin on lung surfaces, hemorrhage, abscessation in later stages) can be regarded as suspicious of *Pasteurellacae* involvement. Under the microscope, necrosis (tissue death) and exudation (fluid oozing) may manifest as fibrin at the surfaces of the lungs, suppuration (fluid containing dead cells and bacteria) within airways, or discrete abscesses (pus pockets). Other bacteria including *Fusobacterium necrophorum* and *Trueperella pyogenes* may contribute, particularly to abscessation. A microscopic hallmark of pasteurellosis is evidence of oat cells (i.e., damaged leukocytes that resemble oat seeds; Onderka et al. 1988, Gilmour and Gilmour 1989) in pneumonia cases, indicating leukotoxin involvement.

Pasteurellaceae can be transmitted via direct contact and, to a degree, via aerosol; lambs become infected from their dams within a few days of birth. These bacteria can persist outside a host for several days under favorable conditions. The potential for indirect transmission via contaminated water or feed could facilitate spillovers from other host species (e.g., cattle; Wolfe et al. 2010). Stressors or other agents, when present, could play a role in precipitating shedding (Pillai et al. 2018) of the agent in carriers, but may not be necessary to spark or sustain epidemics in a naïve mountain sheep herd. Recovery from infections (or vaccination) can provide some degree of immunity to reinfection and blunt the effects of subsequent exposure (Miller et al. 1991, Kraabel et al. 1998).

Domestic sheep, goats, and cattle are likely the most common sources of novel strains of Pasteurellaceae, although other ruminant species including mountain sheep could carry these under some circumstances. Introductions of novel Pasteurellaceae strains or species to wild sheep from contact with domestic livestock or other wild sheep populations, apparently, have contributed to disease outbreaks despite the presence of endemic Pasteurellaceae. Control measures (Table 8.1) have been attempted with varied effectiveness; of these, preventing new introductions from domestic sheep, cattle, goats, and other livestock or from wild sheep translocations seems most practicable. Population management and habitat enhancement also help minimize the frequency of pasteurellosis outbreaks arising from endemic Pasteurellaceae.

### 8.4.3 LUNGWORMS

The most significant lungworms in mountain sheep are *Protostrongylus stilesi* and *P. rushi*. These two species have common occurrence and wide distribution across the range of mountain sheep and can have marked health and population effects. In addition, infections of other lungworms such as *Muellerius capillaris* and *Parelaphostrongylus odocoilei*, are reported but have not been associated with significant, population-level health effects in mountain sheep.

The epidemiology and pathology of protostrongylosis in mountain sheep are well-described (Forrester 1971, Hibler et al. 1982). The complexity of the *Protostrongylus* spp. life cycle suggests a long-standing and very coevolved host-parasite relationship. Among other adaptations, these lungworms use a variety of small terrestrial snail species to support their development, environmental persistence, and transmission to a new host (Forrester 1971, Hibler et al. 1982). Moreover, *P. stilesi* developed the ability for immature worms in a pregnant host to cross the placenta into the late-gestation fetus, thereby circumventing the female's potent immune system by proceeding with maturation and reproduction in an immunologically naïve host (i.e., her lamb). Lambs are then born with the infection.

Generally, infections by native lungworms do not appear to compromise the overall health of individual mountain sheep. This is consistent with an endemic, well-established host-parasite relationship. In nearly every wild mountain sheep outside desert ranges, healthy or diseased, *Protostrongylus* spp. lungworms reside in nodules in parenchyma (spongy tissue) of the dorsocaudal (toward the spine and away from the head) portions of the lung (*P. stilesi*) or lie freely within major bronchi and bronchioles (*P. rushi*). Production of larvae is tolerable at a low level; inflammation in the nodules or association with eggs and larvae in other lung tissue is characteristic of a chronic, sustained but mild irritation with sometimes significant scarring in older animals. When mountain sheep behavior or physiological factors or environmental circumstances result in excessive numbers of lungworms, the health of infected individuals can be compromised (Jenkins et al. 2007).

The outcomes of lungworm infection can be different in young, immunologically naïve lambs. In some cases, lung damage in neonatal lambs can be pronounced, with secondary bacterial infection so concentrated around the larvae that pneumonia at this age is at least in part associated with this situation. Greater numbers of lungworms may open the door for more extensive lung damage in lambs. Reducing density of animals across their range or prescribed burning, can help reduce the potential for lungworm transmission associated with high-density hot-spot areas by limiting infections in intermediate hosts (snails); both have been used as management strategies. The value of prophylactic lungworm treatment using anthelmintics varies (Miller et al. 2000). Challenges include effective delivery of anthelmintics for each individual and the potential development of anthelmintic resistance.

In contrast to the sporadic nature of lungworm-associated pneumonia in the northern portion of mountain sheep ranges (Jenkins et al. 2007), the mountain sheep-lungworm relationship has become more complicated in southern ranges where non-native pathogens also occur. The immune responses mountain sheep have honed to suppress lungworm development and reproduction seem less well adapted to clearing other, relatively novel respiratory pathogens like mycoplasmas and Pasteurellaceae from lung tissue. The immune process and response to native lungworms may accelerate the tissue damage caused by these invading bacteria. For example, an intense inflammatory reaction naturally designed to combat lungworm eggs and first-stage larvae (Hibler et al. 1982, Bunch et al. 1999) may be maladaptive (counter-productive) against bacteria that evolved to take advantage of such host responses. This could partially explain the relatively severe outcomes of respiratory disease in mountain sheep as compared to domestic sheep when infected with the same bacterial agents. Similarly, the timing of respiratory disease outbreaks in young lambs may reflect the combined effects of maturation and the onset of reproductive activity of transplacentally acquired lungworms over the first several weeks after birth. In bighorn herds where *P. stilesi* is endemic and lambs die with pneumonia, the timing of disease tends to coincide with the first release of eggs and larval hatching after transplacentally acquired worms mature into adult worms in the lungs (~4–8 weeks after birth; Hibler et al. 1982). Varying agent involvement also could explain the apparent but inconsistent improvements in lamb survival after application of herd-level antiparasitic treatments designed to reduce lungworms in free-ranging herds. Moreover, if outcomes similar to those observed in experiments involving rabbits, a respiratory bacterium, and gastrointestinal helminths (Pathak et al. 2012, Nguyen et al. 2022) occur in mountain sheep, then the mere presence of lungworms and other naturally occurring helminth or arthropod parasites may render the mountain sheep's bacterial clearance mechanisms less effective. In this way, endemic lungworm infections could have profound effects on susceptibility to infection of the lower respiratory tract, agent shedding, and development of carrier states for other commensal or novel respiratory pathogens of mountain sheep.

## 8.4.4  RESPIRATORY VIRUSES

Viruses also can be associated with respiratory disease in mountain sheep, including outbreaks involving Parainfluenza 3 virus and respiratory syncytial virus. Serosurveys (screening blood serum for antibodies) suggest exposure to both viruses are relatively common in some mountain sheep populations, but reliability of the respective laboratory tests for mountain sheep and the importance of this exposure are rarely understood. Both viruses are also very common in domestic livestock. Infections in the absence of bacteria tend to be relatively mild and self-limiting, with substantial rates of apparent illness (morbidity) but relatively little direct mortality compared to

outcomes produced by other respiratory pathogens affecting mountain sheep (Dassanayake et al. 2013). Although the viruses currently in circulation do not seem directly responsible for large-scale losses, contemporary respiratory virus infections can set the stage for more severe disease resulting from concurrent infection with other pathogens. Such increased risk may come from direct damage to the host's respiratory system or from distracting or exhausting the host immune responses, allowing other agents the opportunity for deeper penetration of the respiratory tract. The ubiquity of these viruses and their distribution in a variety of ruminant host species add to the variability and complexity of respiratory disease outbreaks and their effects on mountain sheep populations. Although bacterial agents are often the focus of mountain sheep respiratory disease investigations, the apparent lack of viral activity in dead animals could reflect the hit-and-run nature of some viral infections.

## 8.4.5  PARANASAL SINUS TUMORS AND CHRONIC SINUSITIS (INCLUDING NASAL BOT INFESTATIONS)

*Paranasal sinus tumors.* These tumors are a relatively newly recognized complication of the respiratory disease complex in mountain sheep. Thus far, paranasal sinus tumors are known only in bighorn sheep, and only in some locations. The term tumor is a bit of a misnomer, as only a very small subset of these growths has evidence of neoplastic transformation (i.e., cancer). Using the definition of tumor to mean any space-occupying mass or swelling, paranasal sinus tumors are identified by expansion of the sinus linings in the air-filled spaces of the skull, often filling those spaces with soft tissue. These changes to the sinus lining elicit a response in the underlying bone, with most paranasal sinus tumors causing a bony reaction ranging from bone destruction to bone remodeling, to sometimes solid bone filling the sinus spaces.

While paranasal sinus tumors can cause extensive damage to local tissues and bone structures, perhaps the most significant effect on bighorn sheep health is through alteration of normal anatomy and obstruction of outflow, preventing clearance of bacteria and associated exudate from the upper respiratory tract. In this way, even if an infected sheep survives bronchopneumonia in the lower respiratory tract (lungs), the pathogens involved may be retained in tumor-affected portions of the sinuses (part of the upper respiratory tract) for the remainder of the animal's life. The prevalence of sinus tumors in affected herds is difficult to determine because cases can only be diagnosed in carcasses (postmortem), but prevalence in some herds is estimated at 25% or more. In at least one case all seven individuals in a remnant herd were affected (Fox et al. 2011).

The cause of paranasal sinus tumors in bighorn sheep remains under investigation. Although not a direct source of large-scale losses, as mentioned above, the presence of paranasal sinus tumors may facilitate long-term persistence of other upper respiratory system pathogens (e.g., mycoplasmas, Pasteurellaceae) either by mechanically disrupting the sinus lining or by impinging on local immune and clearance responses. In this way, the syndrome may play a role in prolonging the adverse effects of other introduced or endemic agents in a herd, for example, contributing annually to sustained losses of lambs (Fox et al. 2015). Until more is known about the cause(s) and role(s) of paranasal sinus tumors in mountain sheep health, managers should screen herds when possible, particularly all suspicious cases, and avoid translocations from herds where this syndrome has become established.

*Chronic sinusitis.* Sinusitis is a similar but apparently distinct syndrome affecting the paranasal sinuses and surrounding tissues of mountain sheep. Although the cause has not been determined with certainty, infestations with sheep nasal botfly (*Oestrus ovis*) larvae and subsequent bacterial infections arising secondary to entrapment of fly larvae in the paranasal sinuses have been suggested as a mechanism (Bunch et al. 1999). Domestic sheep and goats serve as the common hosts for botfly larvae and life cycle maintenance, but bighorn sheep and other species, including deer (*Odocoileus* spp.), can become infested. Chronic sinusitis is reported primarily in desert bighorn sheep. Substantial proportions of adult desert bighorns show evidence of chronic sinusitis in some ranges, perhaps because environmental conditions accommodate prolonged or repeated exposure to adult botflies and larval deposition. Severe cases can show skull and nasal sinus changes, become debilitating, and lead to death. Whether nasal bot infestations could serve to facilitate infection or chronic carriage of other pathogens that contribute to pneumonia, as hypothesized for nasal sinus tumors, has not yet been investigated. Options for controlling sheep nasal botflies in free-ranging settings seem limited; however, regular administration of oral avermectins aided in controlling infestations in captive Rocky Mountain bighorns held in a northern Colorado research facility.

### 8.4.6 Host Factors

Divergent host-agent coevolutionary paths could explain observed differences in lung defense mechanisms between bighorn and domestic sheep (Cao et al. 2021), including intensity of reaction to lungworm and perhaps other helminth or arthropod parasites, and differential vulnerability of certain immune system cells to some bacterial toxins. These host attributes, in combination with genetically controlled traits of the various agents involved, may help explain the overall susceptibility of mountain sheep (e.g., as compared to domestic sheep). As noted above, bighorn and thinhorn sheep leukocytes are more readily damaged by leukotoxin than those of domestic sheep; this and other differences in immune responses may contribute to an increased vulnerability to some forms of pasteurellosis and other infections. Even so, recovery from prior exposure to an agent or agent strain may provide some immunological resistance to subsequent challenges (Miller et al. 1991,

Walsh et al. 2023). The social behavior of mountain sheep, for example, the gathering, mixing, and high density created by females and young lambs in nursery groups, most likely facilitates agent transmission or low-level pathogen exposure resulting in production of antibodies (seroconversion) but not necessarily disease. Dispersals, exploratory forays, and seasonal social realignments such as the gathering of all sexes and ages during fall rut also create opportunities for introduction and mixing of novel agents or agent strains. Moreover, the physical similarities between mountain sheep and domestic sheep and attraction during breeding and foraying or straying, also likely contribute to interspecies attraction and the potential for novel pathogen introductions and transmission.

### 8.4.7 Environmental Factors

Severe weather events, dust, crowding around limited or artificial resources, interactions with domestic livestock, or other environmental stressors, including competition for forage or habitat, poor nutrition, and trace mineral deficiencies, have been linked to respiratory disease in both captive and free-ranging bighorns. Stress-induced effects on the host immune system are believed to be part of the underlying triggering mechanism (Thorne 1982b, Spraker et al. 1984, Kraabel and Miller 1997, Bunch et al. 1999, Beldomenico and Begon 2015, Romeo et al. 2020). Environmental conditions, both natural and human-modified, can foster persistence of various agents in the environment or contribute to facilitating transmission among hosts. The ongoing changes in climate conditions in mountain sheep ranges may also affect the occurrence of respiratory disease in as yet undescribed ways.

### 8.4.8 Thoughts Moving Forward

Our understanding of respiratory disease in mountain sheep has improved steadily over the century since pneumonia was first described in Rocky Mountain bighorns. The complex interactions and contributions of various agents, host, and environmental factors in epidemics and associated large-scale mortality events seem more fully appreciated. The possible combinations of known agents and circumstances yield a dizzying array of potential outcomes. It seems doubtful that introductions are always limited to a single pathogen, adding to the tangled web of disease effects.

Moving forward, the development of comprehensive management approaches for mitigating the effects of respiratory disease in mountain sheep will benefit from an appreciation for its complexity and recognizing current uncertainties and the likelihood of changes over time. From a comprehensive management standpoint, the foregoing interrelationships and complexities underscore the merits of addressing respiratory disease in mountain sheep as a multifaceted and dynamic complex rather than a collection of unique or competing entities (e.g., verminous pneumonia vs. pasteurellosis vs. mycoplasmosis).

The cornerstones of a truly comprehensive approach to management will therefore include elements for stemming the introduction of new (to a particular herd or locale) pathogen species or strains, and for mitigating the effects of those endemic agents that contribute to epidemics. Fixing a stand-alone problem is unlikely to moderate a problem that has a suite of potential components.

## 8.5 HEMORRHAGIC DISEASES

For many years hemorrhagic disease in wild ruminants referred to a syndrome caused by two related orbiviruses: bluetongue virus (BTV) and epizootic hemorrhagic disease virus (EHDV). Although wild sheep are susceptible to both, BTV is the virus most identified in cases of hemorrhagic disease in mountain and domestic sheep. More recently, another viral disease (adenoviral hemorrhagic disease) was added to the hemorrhagic disease category in North America. The virus associated with adenoviral hemorrhagic disease (*deer atadenovirus A*) produces a hemorrhagic disease indistinguishable from EHDV and BTV in mule (*Odocoileus hemionus*) and blacktailed deer (Columbian blacktailed deer [*O. h. columbianus*], Sitka deer [*O. h. sitkensis*]). The diseases have similar pathologic processes, but with otherwise marked differences in the virus, epidemiology, and host range, including the apparent lack of sheep susceptibility (Imus et al. 2019). A comprehensive review of these diseases in North American ecosystems is provided by Ruder and Christensen (2023). The following information applies to BTV and EHDV, but in mountain sheep the most common culprit tends to be BTV.

Both BTV and EHDV are transmitted indirectly by biting midges of the genus *Culicoides*. These viruses occur worldwide in tropical and temperate habitats that support their insect vectors. At least 27 variant strains (called serotypes) of BTV and seven serotypes of EHDV are documented. A subset of these serotypes is considered endemic in North America. Exotic serotypes do occur and should be considered as a reason for increased hemorrhagic disease-associated mortality in mountain sheep.

The geographic distribution of reported BTV and EHDV infections in native mountain sheep has thus far been limited to the southern portions of their North American range (i.e., ranges of the bighorn subspecies), with latitude, altitude, and local climate conditions limiting the insect vectors and exposure or transmission. Because the other domestic and wild ruminant hosts that can serve as reservoirs for these viruses tend to be more common and abundant than mountain sheep in locations where outbreaks occur, it seems likely that the majority of cases and outbreaks in mountain sheep result, at least initially, from spillover of concurrent outbreaks involving cattle or deer (*Odocoileus* spp.) or perhaps other affected domestic or wild ruminant hosts. Locally, the spatial distribution of cases and association with other susceptible species mortalities, often

corresponds to habitats that support the insect vectors. Outbreaks tend to be seasonal, extending from late summer to autumn, and self-limiting across most of the bighorn sheep habitats. These orbiviruses and their insect vectors appear to circulate year-round with more endemic stability in white-tailed deer (*O. virginianus*) and perhaps livestock in some of the extreme southern ranges that bighorns occupy (Ruder et al. 2015). These regional epidemiological patterns likely influence the occurrence of hemorrhagic disease in mountain sheep but patterns may change with changing climate and weather patterns that favor survival of the insect vectors (Purse et al. 2005, Stallknecht et al. 2015, Navarro Mamani et al. 2023).

Large-scale, regional hemorrhagic disease outbreaks tend to be associated with a single serotype of one or the other virus each year, while subsequent outbreaks may involve a different virus or serotype. Exposed mountain sheep that survive infection may carry some level of immunological resistance or immunity to the serotype(s) involved in preceding outbreaks (Ruder et al. 2015). Local outbreaks may be difficult to detect in free-ranging settings and typically rely on reports of sick animals or carcasses, often from outdoor recreationists. Both viruses and multiple serotypes can circulate in areas with endemic stability. Control of hemorrhagic disease in free-ranging mountain sheep seems limited by practicality to a few preventive measures (e.g., avoiding habitat modifications that may attract reservoir hosts or vectors, or using caution in translocating mountain sheep into areas where hemorrhagic disease exposure is likely). In captive bighorns, preventive measures including modification of holding and husbandry practices, vaccination, and application of permethrin-laced ear tags or collars have been somewhat successful in reducing annual losses to hemorrhagic disease.

Mountain sheep affected by hemorrhagic disease are often found dead in groups in the late summer through to autumn, in good body condition. If seen alive, infected animals show signs including breathing difficulties and depression. Because the disease course can be rapid, sudden, or acute, death may be the first indication of an outbreak. Carcasses will have blue-gray (congested) mucous membranes and often foam in the mouth or nostrils, with heavy, wet lungs (pulmonary edema) and tissue hemorrhages. Laboratory testing is necessary to confirm the presence and identify of the causative virus. The best tissue to collect is spleen, and even somewhat decomposed tissue can be frozen and may provide a diagnosis. Identifying the viral strain (serotype) can help explain epidemiological patterns. Small, localized outbreaks may go undetected in the field. Outbreaks and their effects may appear like, and be mistaken for respiratory disease die-offs, but the blue-gray membranes accompanied by foam in the mouth and nose indicate pulmonary edema; other lung changes are different from those of bronchopneumonia previously described. Regardless, routine samples should be taken and submitted to laboratories to confirm the diagnosis.

## 8.6 INFECTIOUS KERATOCONJUNCTIVITIS

An infectious disease syndrome of the eyes, infectious keratoconjunctivitis (sometimes called pinkeye), can occur in individual mountain sheep or outbreak situations (Meagher et al. 1992, Jansen et al. 2006). This disease also occurs in domestic ruminants and other wild ruminants in North America and Europe (López-Olvera et al. 2024). Inflammation is limited to the internal lining of the eyelids (conjunctiva) and subsequently the cornea (clear surface of the eye), resulting from a bacterial infection of the tissues. Clinical signs include excessive tearing and a thick discharge, cloudiness, and eventually ulcers (perforations) of the cornea. In extreme cases permanent scarring of the cornea and rupture of the eyeball may occur. The disease course can last for weeks before resolving, or the debilitating effects of impaired vision or blindness can lead to accidents or increased risk of predation.

As with respiratory disease, multiple agents can be associated with infectious keratoconjunctivitis. An outbreak in desert bighorns involved a combination of a *Mycoplasma* spp. (closely related to *M. conjunctivae*) and *Branhamella* spp. (Jansen et al. 2006); *Moraxella bovis*, *Chlamydia* spp., and additional agents were found in other infectious keratoconjunctivitis outbreaks (Thorne 1982a, Meagher et al. 1992). Transmission likely involves some contact with infected animals, but mechanical transmission of the pathogen(s) via flies may contribute to local spread. The sporadic pattern of outbreaks in mountain sheep suggests spillover of the causative agents from domestic or possibly other wild ruminant species. The outbreaks described thus far were self-limiting, with a range of outcomes in affected herds (Meagher et al. 1992, Jansen et al. 2007). Management practices directed toward preventing other pathogen introductions from domestic sheep, goats, and cattle seem likely to also lower the risk of infectious keratoconjunctivitis.

## 8.7 CONTAGIOUS ECTHYMA

Contagious ecthyma, also called sore mouth, scabby mouth, or orf, is a viral disease of wild and domestic ruminants affecting the skin and other surface (epithelial) tissues at or near specific locations on the body (Bunch et al. 1999, Robinson and Kerr 2001). The disease occurs in bighorn and thinhorn sheep, and domestic sheep and goats, and a variety of other species, including mountain goats. Outbreaks have occurred in locations with no known exposure to domestic livestock. Contagious ecthyma involves an environmentally resistant parapoxvirus that can persist for decades in scabs that drop off an infected individual and likely other contaminated material. Infections lead to raised growth on skin surfaces, particularly the lips, muzzle, mouth, udder, and other lightly haired areas. Epithelial damage and inflammation results in the development of crusty scabs laden with virus particles. In many cases, the disease is self-limiting, but secondary bacterial infection of lesions can lead to prolonged and more extensive disease. Infections typically are not fatal in adult sheep but may lead to losses in condition in nursing lambs, either from impaired suckling or reluctance of females to allow nursing when lesions develop on the udder. Individuals who recover from contagious ecthyma will likely carry at least temporary immunological resistance to reinfection.

Outbreaks of contagious ecthyma in mountain sheep tend to be associated with concentrations of animals, either from seasonal grouping or congregation at natural mineral licks or human-placed artificial mineral attractant sites, or salted roads. Whether specific circumstances trigger an outbreak is unclear. In some locations, outbreaks recur periodically in patterns likely arising from a combination of climate conditions and environmental contamination of preferred habitat use areas, recruitment of sufficient numbers of susceptible individuals into the herd, irritation or abrasion of tissues (i.e., from licking abrasive materials like salt or salted pavement), and eventually exposure to virus that persisted since the last outbreak. Because the host range of the causative parapoxvirus includes humans, advisories and personal protective measures are warranted when handling wildlife during outbreaks and during hunting seasons when wildlife harvesters can come into direct contact with infected tissue or scabs containing the virus.

## 8.8 BOVINE VIRAL DIARRHEA

The pestiviruses that cause bovine viral diarrhea (BVD; common in cattle) and border disease (common in domestic sheep) appear to be livestock pathogens with the potential to infect North American mountain sheep. The bovine viral diarrhea virus (BVDV) can infect various domestic and nondomestic ruminants including bighorn sheep, bison, and mountain goats (Passler and Walz 2010, Wolff et al. 2016, Fox et al. 2019, Hause et al. 2021). In cattle, most BVDV infections are transient. The resulting disease can differ in severity and may include gastrointestinal and respiratory involvement, reproductive loss, and immune system suppression that predisposes the individual to other illnesses. Infected cattle can shed the virus transiently or persistently depending on the circumstances surrounding their original exposure and infection (Thurmond 2005). Although contact with domestic cattle seems the likely origin and most common source of introduction to free-ranging ruminants, including mountain sheep, this virus apparently can be maintained and may be locally endemic in some North American wildlife populations (Wolff et al. 2016).

Serosurveys reveal exposure of Rocky Mountain bighorns to BVDV in multiple locations. In contrast, border disease virus exposure or infections have not been reported in mountain sheep, although susceptibility seems likely; however, specific testing for this virus is rarely performed. Reports of BVDV-associated disease in mountain sheep are limited despite ample evidence of exposure. Nonetheless, two examples illustrate the potential implications for management. In Nevada, a bighorn herd experiencing a respiratory disease outbreak attributed to *M. ovipneumoniae* and Pasteurellaceae infections also had evidence of pervasive, recent exposure to one BVD variant, BVDV1. Immunosuppression resulting from the viral infection was regarded as a potential predisposing factor for the respiratory disease epidemic (Wolff et al. 2016). In Colorado, captive bighorns contracted the gastrointestinal form of BVD from a vaccine contaminated with BVDV1, with over half of the vaccinated animals becoming ill and nearly half of those dying (Fox et al. 2019). The former example reemphasizes the potential complexity of respiratory disease in mountain sheep; the latter underscores the vulnerability of mountain sheep to livestock pathogens. Both serve to illustrate the risk of disregarding domestic cattle and their endemic pathogens in developing management approaches for preventing disease introduction into mountain sheep ranges.

## 8.9 MYCOBACTERIOSIS

Diseases caused by bacteria in the genus *Mycobacterium* can affect mountain sheep, but they are now relatively rare and with limited geographic distribution. Where they do occur, mycobacterial diseases do not appear to hamper population performance. Most cases of mycobacteriosis (any disease caused by *Mycobacterium* spp.) in mountain sheep are considered sporadic and likely represent spillovers from natural environmental sources of these bacteria. The presence of two mycobacterioses in mountain sheep, if and where present, can limit management flexibility because of potential spillover to domestic ruminants or humans.

### 8.9.1 PARATUBERCULOSIS

Paratuberculosis, also called Johne's disease, is a chronic intestinal disease caused by *Mycobacterium avium paratuberculosis*. It can occur in multiple domestic and wild ruminants and other mammalian species, including both mountain sheep species (Williams 2001, Garde et al. 2005). Although probably not widespread in mountain sheep, paratuberculosis has become endemic in Rocky Mountain bighorn herds in at least two locations (Colorado, United States, and Alberta, Canada), likely after spillover from domestic livestock at some point in the past. The *Mycobacterium avium paratuberculosis* isolated from the affected Alberta herd was the Type II strain associated with cattle (Forde et al. 2012). The endemic focus in Colorado has persisted since at least the 1970s, and also involves mountain goats (Williams et al. 1979).

Mountain sheep affected by paratuberculosis may appear healthy initially but ultimately develop intermittent to profuse diarrhea and emaciation in late stages of infection (Williams 2001). Terminally affected animals are emaciated and have diarrhea, with thickening of the small intestine wall and mesenteric lymph node enlargement. These field and necropsy observations accompanied by microscopic evidence of acid-fast bacteria in affected tissues are suggestive of paratuberculosis, which can be confirmed by demonstrating *Mycobacterium avium paratuberculosis*

via culture or molecular diagnostics (Forde et al. 2012). Paratuberculosis is spread by the fecal-oral route, with bacteria contaminating the environment. This disease does not appear to limit the abundance of affected mountain sheep in locations where it occurs, but concentrating infected herds via feeding, baiting, or other artificial attractants will facilitate direct and indirect transmission, leading to greater prevalence and associated mortality. Nonetheless, the presence of this disease limits management options for mountain sheep because of concern for spillovers to livestock or transfer to other locations via translocation or emigration of infected animals.

### 8.9.2 BOVINE TUBERCULOSIS

Tuberculosis is the broad term for diseases caused by multiple species of bacteria collectively grouped in the *Mycobacterium tuberculosis* complex (Miller 2023). Of these, bovine tuberculosis, caused by infection with *Mycobacterium bovis,* occurs most widely in domestic and nondomestic ungulates in North America and is of greatest potential concern for mountain sheep. Based on the wide range of reported hosts, including domestic sheep and goats, mountain sheep are likely susceptible to bovine tuberculosis. In nondomestic ungulates, most infections are detected postmortem, with suspicion arising from the presence of tubercules, circumscribed yellowish masses with a center ranging in consistency from liquefied to firm, surrounded by a layered and sometimes gritty tissue capsule, in lungs or other organs (Miller 2023). The recent review by Miller (2023) provides a more comprehensive discussion of tuberculosis in wild ungulates. Eradication and control campaigns in North America have reduced the occurrence of bovine tuberculosis in domestic livestock and thus the potential for outbreaks on ranges shared with mountain sheep. Nonetheless, given the potential for human infection from handling infected tissues, if suspected tubercules are noticed in a mountain sheep carcass then immediate precautions should be taken to protect against human infection. Considering the economic, public health, and regulatory importance of tuberculosis and related diseases, carcasses of mountain sheep showing pathology suggestive of bovine tuberculosis should be submitted for laboratory testing to screen for (and hopefully rule out) infection with *M. bovis* or other species within the *Mycobacterium tuberculosis* complex.

## 8.10 TOXOPLASMOSIS

*Toxoplasma gondii* is a microscopic protozoal parasite that primarily lives in wild or domestic cats (the definitive host), but that can infect a wide range of intermediate host species including humans. Intermediate hosts are infected through ingestion of oocysts that are shed in felid feces. The parasite matures to a dormant tissue cyst phase in the intermediate host, and the life cycle is complete when the tissue cysts are ingested by predatory felids. Intermediate hosts do not typically show signs of disease (toxoplasmosis), although dormant cysts can be activated during pregnancy and subsequently transferred to a fetus, usually resulting in death of the fetus. *Toxoplasma gondii* has been associated with fetal and neonatal deaths in many species (Dubey and Odening 2001, VanWormer and Shapiro 2023) including bighorn sheep (Baszler et al. 2000, Fisk et al. 2023). Serum antibody titers suggesting substantial (more than 40%) rates of exposure to *T. gondii* have been reported in herds of bighorn (Fisk et al. 2023) and Dall's sheep (Thacker 2020). Whether toxoplasmosis has population-limiting effects in bighorn or thinhorn sheep remains to be determined. Nonetheless, reproductive losses from toxoplasmosis in a herd could complicate or confound assessments of neonatal losses to recurrent respiratory disease or other causes, in lambs. Human health risks associated with harvesting and consuming infected animals also bear consideration, especially in heavily exposed herds.

## 8.11  OTHER HELMINTH, ARTHROPOD PARASITES, AND ASSOCIATED DISEASES

Other helminth and arthropod parasites also may be associated with disease in mountain sheep. Most of these are commensal parasites and cases of disease may reflect some imbalance in individual or local conditions that contribute to individual cases or outbreaks.

## 8.12  HELMINTHS

Parasitic worms (helminths) include a wide range of animals that spend part of their lives in the shared internal environment of other living organisms, as endoparasites. Though readily seen, helminth species in mountain sheep are of limited conservation or management concern. They generally exist as a natural component of mountain sheep in a natural balance with each host species and individual. This balance can be tipped toward increased helminth occurrence or abundance when individual sheep experience stress triggered or facilitated by nutritional, immunologic, physiologic, biologic, and cognitive stressors (Beldomenico and Begon 2015, Romeo et al. 2020) or when populations reach high densities. Their presence could influence responses to concurrent infections and may influence other demographic factors (Aleuy et al. 2018, 2020), but in doing so also may help buffer against host overabundance and its environmental consequences.

Major groups of helminths include nematodes (roundworms), trematodes (flatworms), and cestodes (tapeworms). These groups share similarities in body form and life history strategies but are not particularly close relatives. They are soft-bodied, often cylindrical, animals in the adult stage and produce eggs that leave the initial host, and as their larval stages develop, they move through various biotic and abiotic environments on their way back into individuals of the original host species. Information provided herein is limited to those helminths that potentially occur in wild sheep.

### 8.12.1  NEMATODES (ROUNDWORMS)

These long slender hydrostatic round helminths live in various internal organs and tissues. The most visible nematodes in mountain sheep include species that live in the lungs (see *Lungworms* in Section 8.4.3). Somewhat less visible but far more diverse and numerous, a relatively large suite of gastrointestinal nematodes (various genera and species) share space throughout the digestive system, most often in the abomasum and small intestines. They gain nutrients and shelter from the sheep and are associated with little actual damage to the gut tissues. Most coevolved with mountain sheep and are a relatively benign commensal community of coexisting species present in all mountain sheep populations, and in most individuals. Nonetheless, imbalances in host-parasite-environment relationships can have consequences. For example, the intensity of infection with *Marshallagia marshalli,* the most prevalent and abundant of nine gastrointestinal nematode species in Dall's sheep from the Mackenzie Mountains, was negatively associated with body condition and pregnancy (Aleuy et al. 2018).

The helminth populations within each individual arise from seasonal reproductive efforts tailored to support sufficient and appropriate transmission to sustain each helminth species, often tied to physiological and behavioral differences within the host across different annual seasons, such as gestation or rut (Rijal et al. 2024). Life history strategies include tolerance to heat, drought, and high-altitude conditions for the survival of life stages outside the host (Rijal et al. 2024). Changing climate could contribute to or amplify imbalances that might lead to adverse effects in mountain sheep (Aleuy et al. 2018). Additional information regarding gastrointestinal helminths is summarized in Hoberg et al. (2001). Of note, in the absence of further reports, a finding of carotid worm (*Elaeophora schneideri*) in two bighorns found dead or in poor body condition in New Mexico (Boyce et al. 1999) appears to have been of limited significance.

### 8.12.2  TREMATODES (FLATWORMS)

Among the suite of animals that live in mountain sheep, trematodes (small flat helminths) are conspicuously absent. All adult trematodes live a parasitic lifestyle and have complicated life history strategies involving sexual and asexual reproduction and one or more intermediate hosts, often aquatic. Given suitable conditions, the relationships between host, environment, and parasite can have health repercussions for individuals or local populations.

One such example involves the giant liver fluke (*Fascioloides magna*). This large trematode normally infects wild cervids, primarily white-tailed deer and elk (*Cervus canadensis*), and domestic livestock and occurs in some areas that overlap bighorn ranges (Pybus 2001). Transmission is focused around wetlands with emergent or submergent vegetation that sustain the development of infective larvae. The life cycle includes serial passage from adult flukes in deer, larval stages in aquatic snails, encysted larvae on vegetation, and ingestion by foraging herbivores. Changes in water access for a group of bighorn sheep in southeastern British Columbia resulted in sheep deaths attributed to severe liver damage from giant liver fluke infections. The affected bighorns grazed a local golf course with water sources from wetlands that contained fluke larvae (Mathieu et al. 2022). Recognizing the underlying changes in habitat and resource use that facilitated exposure helped managers identify preventative actions tailored to effectively reduce or eliminate risk associated with further giant liver fluke transmission.

### 8.12.3 CESTODES (TAPEWORMS)

Cestodes are long flat ribbon-like helminths that consist of a connected series (stobila) of identical components (proglottids) attached to a head (scolex, often with a holdfast structure with hooks or suckers) and a short neck from which new proglottids arise. Adult tapeworms live in the intestines of their host where, lacking a digestive system, they absorb nutrients through their specialized outer cuticle. Each proglottid contains complete male and female reproductive systems, which mature over time and through self- or cross-fertilization give rise to mature ova. The oldest proglottids at the distal end of the strobila are filled with eggs and detach and pass from the host among the fecal material. Few cestodes occur in mountain sheep and those that do rarely result in conservation or management concern. One species, *Wyominia tetoni*, present in some bighorn populations (Uhazy and Holmes 1971, Colwell et al. 1975, Kistner et al. 1977) is of novel interest because the adults live in the bile ducts of the liver, an unusual extra-intestinal location for a tapeworm. Although sometimes numerous, they do not appear to create health issues.

### 8.13 ECTOPARASITES

Ectoparasites specialize in living on the outside of other animals. In broad categories, these include: insects such as flies and fly larvae, lice, fleas, and keds (louse flies); as well as non-insects such as ticks and mites, which are classified as arachnids. For the most part, mountain sheep do not offer suitable habitats for ectoparasites, or the behavior and social dynamics of the two species do not support sustainable populations of animals like flies, fleas, or keds. Reports of the sheep bot fly (*Oestrus ovis*) on wild sheep in Montana and Wyoming (Capelle 1966) and elsewhere may be yet another example of a shared parasite that originated in domestic sheep, but the effects on bighorns are relatively limited. Ticks (*Dermacentor* spp., *Otobius* spp.) occur on mountain sheep in some areas. Heavy infestations may reflect spillovers from other locally abundant mountain ungulates, like elk, moose (*Alces alces*), and mule deer. For example, several small migratory subpopulations of Stone's sheep in northeastern British Columbia have recurring episodes of heavy winter tick (*D. albipictus*)

infestation where their winter ranges overlap with Rocky Mountain elk (*C. c. nelsoni*) and moose. Tick infestations in these sheep resulted in hair loss on the neck and trunk (in some cases extensive hair loss was observed) caused by overgrooming or skin irritation and are associated with poor body condition and exposure, and blood loss and anemia if tick burdens are high (Wood et al. 2010, Thacker 2020), but field study did not identify major health effects on the population. Although there are no reports to date of native ticks being associated with large-scale conservation or management concerns for mountain sheep, introduced (exotic) tick species could have more consequential effects. Moreover, changing climates may promote tick survival and density, which could affect individual mountain sheep fitness where higher tick infestations result. In desert climates with diminished native tick survival and densities, behavioral grooming by individuals may still be observed, perhaps reflecting an ancient evolution of desert bighorns in a tick-infested environment (Mooring et al. 2006).

### 8.14 MANGE

Mange, called scab in older literature and scabies when it involves humans, is a general term for any skin disease caused by infestations of parasitic mites. Mites associated with mange in wildlife differ from those associated with scabies in most human infections (Bornstein et al. 2001, Pence and Ueckermann 2002). The most common species of mange mite (*Sarcoptes scabiei*) has a worldwide distribution and burrows into the skin of more than 100 species of wild and domestic animals, but has not been reported in wild sheep.

### 8.14.1 PSOROPTIC MANGE

The mites responsible for mange can cause serious outbreaks and debilitating effects in individual bighorns (Seton 1929, McTaggart Cowan 1951, Buechner 1960, Lange et al. 1980, Lange 1982, Bunch et al. 1999, Boyce and Weisenberger 2005). Non-burrowing mites in the genus *Psoroptes* are the primary contemporary cause of mange in mountain sheep. Most infestations with *Psoroptes* spp. are superficial and limited to the ears of mountain sheep. Psoroptic mites, primarily *Psoroptes ovis*, have been detected in multiple wild sheep populations across western North America (Boyce et al. 1991). Outbreaks associated with more recent emergence of another species (*P. cuniculi*) in neighboring British Columbia and Washington bighorn herds are suspected to have originated as a spillover from a domestic rabbit source to bighorns held in captivity (Hering et al. 2021). The severity of disease and outbreaks reported with this contemporary novel mite introduction strongly resemble the patterns described historically when bighorns were first exposed to then-novel *Psoroptes* spp. mites introduced by livestock (Hornaday 1901, Warren 1910, Brooks 1923, Seton 1929, McTaggart Cowan 1951, Buechner 1960).

Psoroptic mites do not cause direct damage by burrowing through skin tissues. Instead, they feed on the skin surface cells. These mites have remarkable reproductive capacity and often occur in dense colonies in or around the ears of infested sheep. The mite feeding activities and the host behaviors of scratching, rubbing, and biting, to try to remove mites, activate strong immune responses in the colonized skin surfaces and promote secondary bacterial infections. In severe cases, crusty sheets of damaged hair, skin, and debris can plug the ears, and more extreme infestations can extend significant skin damage and sloughing of the skin to the head, neck, and body. When sufficiently severe, *Psoroptes* spp. infestations can result in decreased appetite, weight loss, blood loss (anemia), and emaciation. Mortality can come from debilitation, dehydration, exposure, or indirectly by increasing vulnerability to predators.

Mange in bighorns has an interesting history. Before European colonization, local indigenous peoples had no apparent concept of mange on mountain sheep (Beuchner 1960). The earliest reported outbreaks coincided with the appearance of settlers and livestock on western landscapes in the United States (Hornaday 1901, Seton 1929, Shillinger 1937) and likely occurred in western Canada (Brooks 1923, McTaggart Cowan 1951, Wishart 1958) in conjunction with significant bighorn population declines. This suggests the mites on bighorns perhaps originated with domestic sheep circa 1800s or earlier. Historical accounts and recent records suggest these mites now occur as locally endemic parasites in some locations throughout much of the bighorn species range, but not (yet) in thinhorn ranges. *Psoroptes* spp. mites have been ostensibly eradicated from domestic sheep in Canada since the 1920s and in the United States since the 1970s. Occasional cases of psoroptic mange are still found in bighorn subspecies, at least in some jurisdictions. In a well-documented situation, the combined effects of the mites, drought, and predation apparently eliminated mange mites from the San Andres Mountains (Boyce and Weisenberger 2005). The frequency of large-scale epidemics has diminished substantially over the last century or more, with no significant reported outbreaks associated with now-endemic mites since the 1990s. It is perhaps not a coincidence that mange was largely eradicated from domestic livestock in the United States and Canada by the late 1900s.

Sporadic cases of mange, like other parasitic diseases, can indicate underlying health concerns in an individual or population that facilitate opportunistic infections. As with many disease or parasite situations, the occurrence of multiple cases in a relatively small geographic scale likely merits field investigation and possibly management intervention. A limited trial using treated collars or ear tags suggested some cumulative effect over time in reducing the skin damage scores in a small free-ranging sheep population (Bleich et al. 2015). Avermectins (a class of antiparasitic drugs; Boyce et al. 1992, Foreyt 1993) and other newly developed acaricides (e.g., fluralaner) are longer-lasting, and certain forms may be more effective in reducing or eliminating *Psoroptes* spp. in mountain sheep (Hering et al. 2021). Delivery of therapeutics to effectively control mange in free-ranging sheep presents challenges, but ongoing research in British Columbia in collaboration with methods being developed in the United Kingdom (Burgess and Stubbings 2023) may provide promise. Given the uncertainty surrounding the circumstances that lead to epidemics and the difficulty in treating the parasite in free-ranging populations, preventing introductions of *Psoroptes* spp. into naïve thinhorn and bighorn sheep herds seems prudent.

## 8.15 OTHER VIRAL, BACTERIAL, PROTOZOAL, AND FUNGAL DISEASES

Several diseases associated with endemic or introduced viruses, bacteria, protozoa, and fungi are either rarely encountered in mountain sheep, never reported in mountain sheep but suspected to occur, or are so recently identified that mountain sheep susceptibility or management consequences are currently unknown. Although these diseases are not likely to pose a threat to mountain sheep populations in the near term, they bear mention herein. It is important to consider rare or unlikely disease entities, particularly in the context of unusual cases. For example, rabies virus is unlikely to infect mountain sheep but like all mammals, sheep are susceptible. Similarly, mountain sheep are likely susceptible to other viral infections with a predilection for infecting brain tissues (e.g., encephalitides such as eastern, western, and Venezuelan equine encephalitis) that occur in North America and may become more common, especially as insect vector distributions expand due to factors such as climate change.

Other diseases are important to consider due to implications for human health, especially if not quickly diagnosed after the person is exposed. For example, the sudden death of multiple mountain sheep in a location may indicate the aftermath of a lightning strike or avalanche, but could also indicate an anthrax outbreak; the latter being far more consequential in terms of herd implications and potential risk to other species, including humans and livestock.

Finally, some diseases may exist in mountain sheep but are currently under-recognized. For example, a virus (herpesvirus) identified in bighorn sheep resembles those causing malignant catarrhal fever (Slater et al. 2017, Cunha et al. 2019). Although a few cases of disease attributed to malignant catarrhal fever-associated virus infections have been reported in mountain sheep (Cunha et al. 2019, Thacker 2020), the potential for consequential outbreaks or losses remains unknown. Serological surveys have shown widespread exposure of both bighorn and thinhorn sheep to malignant catarrhal fever-associated viruses with no reported clinical signs (Zarnke et al. 2002, Thacker 2020).

Wise stewardship, management, and conservation of mountain sheep should consider the value in following standardized herd health monitoring approaches that could help detect new or (re)emerging agents with the potential to infect mountain sheep, considering the historical and contemporary effects of novel pathogens on their survival and abundance.

### 8.15.1 Incidental Diseases

Several diseases may be incidentally encountered in individual mountain sheep with minimal implications to herd performance, but could have real or perceived implications for hunters. For example, bacteria found normally in the environment can result in abscesses under the skin, within the muscles, under horn sheaths and hooves, or within internal organs of individual sheep because of wounds or damage to normal skin or mucosal barriers. Abscesses within internal organs, including the liver and lungs indicate bacterial spread through the blood, and ingestion of the meat of such animals is not recommended. Abscesses isolated to a single location under the skin or in the muscle are likely due to local penetration of the skin, and in these cases the affected portion can be discarded. Incidental abscesses can be differentiated from mycobacterial tubercles (by appearance and odor), but it is important to remember that the disease descriptions in this chapter are meant to help the reader appreciate the wide range of diseases that can affect mountain sheep, and not to serve as a guide for diagnoses. Examination of cases of concern by experienced wildlife professionals using appropriate protocols suitable for each situation and augmented by targeted diagnostic testing can best inform subsequent hunter, public, or management actions.

## 8.16   MALNUTRITION

Malnutrition or starvation in mountain sheep is most likely to be diagnosed during winter and early spring when forage is limited or of low quality; however, it may reflect a process that started the previous summer if forage was of poor quality or quantity during that season. Starving animals are usually thin and have minimal observable internal fat stores. Fat stored under the skin, within the tissues (omental and mesenteric) that surround the internal organs (viscera), especially around the kidneys, within the coronary groove of the heart, and the bone marrow are key indicators used to assess general fat stores and body condition. In addition, the eyes may appear sunken due to dehydration and loss of fat backing the eye sockets in emaciated animals. The femoral or other large bone marrow can be examined by breaking the bone along its long axis. Marrow or other fat depots with a strawberry-jam color and gelatinous consistency indicate serous atrophy and strongly suggest starvation.

If forage is limited, the rumen volume may decrease. When only low-quality forage is available the rumen may appear full, even in a starving animal. The colon may be empty or may be full of feces, especially when diet quality is low. The liver may decrease in size with sharp edges due to loss of energy stores (glycogen depletion). Small lambs will be more susceptible to winter malnutrition if late autumn energy stores are inadequate. Older aged animals are at greater risk for malnutrition, in part due to tooth

wear leading to inefficient chewing of forage, but tooth wear and abnormalities from trauma, including damage caused by poor quality forage, can affect any age of animal. In these cases, the rumen contains long-fiber-length forage that is largely undigestible. Starved animals may have other signs of poor condition, including heavy parasitism or other opportunistic infections. Deficiencies or imbalances in trace minerals, including copper and selenium, have been particularly documented in wild sheep populations, and specific micronutrient impairments may contribute to poor health in certain individuals or herds (McKinney et al. 2006, Thacker 2020, Tsuchida et al. 2024).

Multiple collaring studies in bighorn sheep revealed starvation in lambs during the neonatal period (Smith et al. 2014, Lohuis 2015, Grigg et al. 2017, Parr et al. 2018). In at least one of these studies, starvation was attributed to abandonment by the female after lambs were collared (Grigg et al. 2017). The main necropsy findings in a starving neonatal lamb are dehydration, obvious thinness, and a stomach that is either empty or contains dirt or plant material. During the neonatal period, lambs stay close to females and nurse frequently. A neonatal lamb with an empty stomach in the absence of other findings is particularly suspicious for abandonment. Even lambs that died from pneumonia tend to have clotted milk in their stomach if the female is still in attendance. The maternal behaviors of mountain sheep differ from those of cervids or other ungulates that hide and leave their young for short periods. These ungulate species may be more likely to return to a capture site to care for a collared neonate, whereas mountain sheep may be more likely to abandon lambs after separation during the disturbance of capture.

## 8.17   ACIDOSIS

Another form of malnutrition in mountain sheep is related to artificial feeding of an inappropriate diet, especially excess carbohydrate feeds, including corn. This is a common outcome of supplementing wildlife diets inappropriately and is not specific to mountain sheep. The ensuing disease is often referred to as rumen acidosis because sudden changes in the carbohydrate content of feed cause a shift in the rumen to a more acidic pH. In severe cases, the animal may die quickly from the immediate associated metabolic disturbances. In more prolonged cases, the animals may die from bacterial infections that occur after bacteria enter the bloodstream through the damaged rumen mucosa, often with severe diarrhea. Shifts in the species composition of rumen microfauna may also cause associated damage to the brain due to loss of thiamine producing bacteria (termed polioencephalomalacia). Finally, animals that survive acidosis may struggle with long-term digestive issues due to changes in or loss of normal gut fauna or may suffer from other related problems such as hoof abnormalities from laminitis (damage to the hoof tissue), also known as founder. Attracting mountain sheep to artificial feed sources can also affect the health of the larger population, as congregation can enhance the spread of respiratory pathogens, helminth and arthropod parasites, or other infectious organisms, and create localized overuse of the range. These potential outcomes should be considered whenever a feeding or baiting operation is proposed or considered. Such activities may create unforeseen effects and harm for individuals, but also for local herds and populations.

## 8.18  TOXINS

Mountain sheep are presumably susceptible to the same toxins as domestic sheep, goats, and other ruminants. Specific examples of intoxications include ingestion of toxic plants including oleander (*Nerium* spp.) and prune (*Prunus* spp.) species by reintroduced bighorn sheep (*O. c. nelsoni*), thought to be associated with urbanization of bighorn habitat (Ostermann et al. 2001). Similarly, human activity was associated with mortality events in bighorn sheep following ingestion of nitrates at a coal mining site and in Stone's sheep exposed to arsenic from the remnant contaminants at a gold mine (Poole et al. 2020). Contaminated water sources can also be a source of sudden or large-scale mortality. As an example, contamination of a water source (i.e., guzzler) with drowned animals led to an outbreak of botulism in a herd of desert bighorn sheep (Swift et al. 2000).

Toxins and toxic effects can be particularly difficult to detect or diagnose in wild populations and may be under-recognized in mountain sheep. For example, manifestations including congenital defects are unlikely to be recognized if affected neonates are efficiently removed by predators. Sublethal toxic effects may be even more difficult to appreciate among other contributing factors in an affected herd.

## 8.19  CAPTURE-ASSOCIATED MORTALITY

Capture-associated mortality refers to severe and immediate trauma that kills or necessitates euthanasia of the captured animal, or it can refer to more subtle aftermath and physiological complications that happen during or after the capture or handling. For example, the syndrome of capture myopathy (Bunch et al. 1999) is a progression of clinical signs and tissue damage that can occur suddenly (peracutely) or downstream after muscle damage from exertion and restraint (acutely, subacutely, or chronically, depending on the timing) when muscle function occurs in a low oxygen environment. Death may occur at any point during the progression of changes, including peracute cardiac arrest caused by release of excess potassium from damaged muscle cells into the circulation, chronic kidney damage arising from myoglobin release days to weeks after the capture, and damaged muscle that heals as scar tissue and later ruptures weeks after the capture event when the affected animal is chased again (e.g., by a predator). For this reason, capture myopathy and other capture-related complications are difficult to diagnose following capture due to factors such as predation, scavenging, and other obstacles to recovering carcasses for diagnostics. Therefore, death within 2 weeks of capture is typically attributed to capture-related causes in the absence of a thorough diagnostic investigation.

## 8.20  SUMMARY

Mountain sheep provide suitable habitat for a wide range of other, smaller organisms that can live in harmony with or upon them. Disease is generally an outlier condition. The effects of most diseases are limited to impairing individual or local (i.e., herd-level) productivity, survival, or resilience. Nonetheless, a few can have more widespread or lasting effects that limit population growth or persistence, or management options. Diseases in their broadest context, particularly those involving non-native (i.e., introduced) agents, have had cumulative, range-wide consequences for bighorn sheep. Some continue to undermine conservation of bighorn subspecies and threaten similar harm to native thinhorn subspecies. This chapter overviews an array of diseases that can affect mountain sheep in North America, emphasizing those of significance to mountain sheep or their populations. We provide brief summaries of the apparent cause(s) and effects, including conservation and management implications where relevant. We intend to provide those interested in the history, ecology, management, and conservation of North American mountain sheep with an accessible primer on native and introduced disease agents that can affect these species.

## REFERENCES

Adlam, C., and J. M. Rutter, editors. 1989. *Pasteurella and Pasteurellosis*. Academic Press, Inc., San Diego, California, USA.

Aleuy, O. A., K. Ruckstuhl, E. P. Hoberg, A. Veitch, N. Simmons, and S. J. Kutz. 2018. Diversity of gastrointestinal helminths in Dall's sheep and the negative association of the abomasal nematode, *Marshallagia marshalli*, with fitness indicators. *PLoS ONE* 13:e0192825.

Aleuy, O.A., E. Serrano, K.E. Ruckstuhl, E. P. Hoberg, and S. Kutz. 2020. Parasite intensity drives fetal development and sex allocation in a wild ungulate. *Scientific Reports* 10:e15626.

Anderson, R. M., and R. M. May. 1979. Population biology of infectious disease: Part 1. *Nature* 280:361–367.

Andrews, K. R., T. E. Besser, T. Stalder, E. M. Top, K. N. Baker, et al. 2024. Comparative genomic analysis identifies potential adaptive variation in *Mycoplasma ovipneumoniae*. *Microbial Genomics* 10:e001279.

Baszler, T. V., J. P. Dubey, C. V. Löhr, and W. J. Foreyt. 2000. Toxoplasmic encephalitis in a free-ranging Rocky Mountain bighorn sheep from Washington. *Journal of Wildlife Diseases* 36:752–754.

Bear, G. D., and C. W. Jones. 1973. History and distribution of bighorn sheep in Colorado. Colorado Division of Wildlife Special Report Number 66.

Beldomenico, P. M., and M. Begon. 2015. Stress-host-parasite interactions: a vicious triangle? *Revista FAVE (Sección Ciencias veterinarias)* 14:6–19.

Besser, T. E., E. F. Cassirer, K. A. Potter, K. Lahmers, J. L. Oaks, S. Shanthalingam, S. Srikumaran, and W. J. Foreyt. 2014. Epizootic pneumonia of bighorn sheep following experimental exposure to *Mycoplasma ovipneumoniae. PLoS ONE* 9:e110039.

Besser, T. E., and E. F. Cassirer. 2023. Epizootic bighorn sheep pneumonia caused by *Mycoplasma ovipneumoniae.* Pages 145–163 in D. A. Jessup and R. W. Radcliffe, editors. *Wildlife Disease and Health in Conservation.* Johns Hopkins University Press, Baltimore, Maryland, USA.

Besser T. E., E. F. Cassirer, A. Lisk, D. Nelson, K. R. Manlove, P. C. Cross, and J. T. Hogg. 2021. Natural history of a bighorn sheep pneumonia epizootic: source of infection, course of disease, and pathogen clearance. *Ecology and Evolution* 11:14366–14382.

Besser, T. E., M. A. Highland, K. Baker, E. F. Cassirer, N. J. Anderson, et al. 2012. Causes of pneumonia epizootics among bighorn sheep, Western United States, 2008–2010. *Emerging Infectious Diseases* 18:406–414.

Bleich, V. C., C. K. Johnson, S. G. Torres, J. H. Davis, J. M. Ramsey, J. T. Villepique, and B. J. Gonzales. 2015. *Psoroptes* infestation and treatment in an isolated population of bighorn sheep (*Ovis canadensis*). *Journal of Zoo and Wildlife Medicine* 46:491–497.

Bornstein, S., T. Mörner, and W. M. Samuel. 2001. *Sarcoptes scabei* and sarcoptic mange. Pages 107–119 in W. M. Samuel, M. J. Pybus, and A. A. Kocan, editors. *Parasitic Diseases of Wild Mammals.* 2nd edition. Iowa State University Press, Ames, Iowa, USA.

Boukahil, I., and C. J. Czuprynski. 2018. Mutual antagonism between *Mannheimia haemolytica* and *Pasteurella multocida* when forming a biofilm on bovine bronchial epithelial cells in vitro. *Veterinary Microbiology* 216:218–222.

Boyce, W., A. Fisher, H. Provencio, E. Rominger, J. Thilsted, and M. Ahlm. 1999 Elaeophorosis in bighorn sheep in New Mexico. *Journal of Wildlife Diseases* 35:786–789.

Boyce, W. M., and M. E. Weisenberger. 2005. The rise and fall of psoroptic scabies in bighorn sheep in the San Andres Mountains, New Mexico. *Journal of Wildlife Diseases* 41:525–531.

Boyce, W. M., D. A. Jessup, and R. K. Clark. 1991 Serodiagnostic antibody responses to *Psoroptes* sp. infestations in bighorn sheep. *Journal of Wildlife Diseases* 27:10–15.

Boyce, W. M., J. A. Miller, D. A. Jessup, and R. K. Clark. 1992. Use of ivermectin implants for the treatment of psoroptic scabies in free-ranging bighorn sheep. *Journal of Zoo and Wildlife Medicine* 23:211–213.

Brewer C. E., V. C. Bleich, J. A. Foster, T. Hosch-Hebdon, D. E. McWhirter, E. M. Rominger, M. W. Wagner, and B. P. Wiedmann. 2014. *Bighorn Sheep: Conservation Challenges and Management Strategies for the 21st Century.* Wild Sheep Working Group, Western Association of Fish and Wildlife Agencies, Cheyenne, Wyoming, USA.

Brooks, A. 1923. The Rocky Mountain sheep (Ovis canadensis) in British Columbia. *The Canadian Field-Naturalist* 37:23–25.

Buechner, H. K. 1960. The bighorn sheep in the United States, its past, present, and future. *Wildlife Monographs* 4:1–174.

Bunch, T. D., W. M. Boyce, C. P. Hibler, W. R. Lance, T. R. Spraker, and E. S. Williams. 1999. Diseases of North American wild sheep. Pages 209–237 in R. Valdez and P. R. Krausman, editors. *Mountain Sheep of North America.* The University of Arizona Press, Tucson, Arizona, USA.

Burgess, S., and L. Stubbings. 2023. Diagnosis and treatment of sheep scab. *In Practice* 45:477–484.

Butler, C. J., R. A. Garrott, and J. J. Rotella. 2013. *Correlates of Recruitment In Montana Bighorn Sheep Populations.* Fish and Wildlife Management and Ecology Department, Montana State University, Bozeman, Montana, USA.

Cao, Y.-H., S.-S. Xu, M. Shen, Z.-H. Chen, L. Gao, et al. 2021. Historical introgression from wild relatives enhanced climatic adaptation and resistance to pneumonia in sheep. *Molecular Biology and Evolution* 38:838–855.

Capelle, K. J. 1966. The occurrence of *Oestrus ovis* L. (Diptera: Oestridae) in the bighorn sheep from Wyoming and Montana. *The Journal of Parasitology* 52:618–621.

Casadevall, A., and L. Pirofski. 1999. Host-pathogen interactions: redefining the basic concepts of virulence and pathogenicity. *Infection and Immunity* 67:3703–3713.

Casadevall, A., and L. Pirofski. 2001. Host-pathogen interactions: the attributes of virulence. *The Journal of Infectious Diseases* 184:337–344.

Cassirer, E. F., and A. R. E. Sinclair. 2007. Dynamics of pneumonia in a bighorn sheep metapopulation. *Journal of Wildlife Management* 71:1080–1088.

Cassirer, E. F., K. M. Rudolph, P. Fowler, V. L. Coggins, D. L. Hunter, and M. W. Miller. 2001. Evaluation of ewe vaccination as a tool for increasing bighorn lamb survival following pasteurellosis epizootics. *Journal of Wildlife Diseases* 37:49–57.

Cassirer, E. F., L. E. Oldenburg, V. L. Coggins, P. Fowler, K. Rudolph, D. L. Hunter, and W. J. Foreyt. 1996. Overview and preliminary analysis of a bighorn sheep dieoff, Hells Canyon 1995–96. *Biennial Symposium of the Northern Wild Sheep and Goat Council* 10:78–86.

Colwell, D. A., J. S. Dunlap, and R. L. Johnson. 1975. *Wyominia tetoni* (Cestoda: Thysanosomatinae) from bighorn sheep in Washington. *Journal of Wildlife Diseases* 11:193–194.

Conrad, C. C., R. K. Daher, K. Stanford, K. K. Amoako, M. Boissinot, et al. 2020. A sensitive and accurate recombinase polymerase amplification assay for detection of the primary bacterial pathogens causing bovine respiratory disease. *Frontiers in Veterinary Science* 7:e208.

Cunha, C. W., O. M. Slater, B. Macbeth, P. J. Duignan, A. Warren, M. A. Highland, and H. Li. 2019. Domestic sheep and bighorn sheep carry distinct gammaherpesviruses belonging to the genus *Macavirus. Virus Research* 272:e197729.

Dassanayake, R. P., D. R. Call, A. A. Sawant, N. C. Casavant, G. C. Weiser, D. P. Knowles, and S. Srikumaran. 2010b. *Bibersteinia trehalosi* inhibits growth of *Mannheimia haemolytica* by a proximity-dependent mechanism. *Applied Environmental Microbiology* 76:1008–1013.

Dassanayake, R. P., S. Shanthalingam, C. N. Herndon, P. K. Lawrence, E. F. Cassirer, K. A. Potter, W. J. Foreyt, K. D. Clinkenbeard, and S. Srikumaran. 2009. *Mannheimia haemolytica* serotype A1 exhibits differential pathogenicity in two related species, *Ovis canadensis* and *Ovis aries. Veterinary Microbiology* 133:366–371.

Dassanayake, R. P., S. Shanthalingam, C. N. Herndon, R. Subramaniam, P. K. Lawrence, J. Bavananthasivam, E. F. Cassirer, G. J. Haldorson, W. J. Foreyt, F. R. Rurangirwa, and D. P. Knowles. 2010a. *Mycoplasma ovipneumoniae* can predispose bighorn sheep to fatal *Mannheimia haemolytica* pneumonia. *Veterinary Microbiology* 145:354–359.

Dassanayake, R. P., S. Shanthalingam, R. Subramaniam, C. N. Herndon, J. Bavananthasivam, G. J. Haldorson, W. J. Foreyt, J. F. Evermann, L. M. Herrmann-Hoesing, D. P. Knowles, and S. Srikumaran. 2013. Role of *Bibersteinia trehalosi*, respiratory syncytial virus, and parainfluenza-3 virus in bighorn sheep pneumonia. *Veterinary Microbiology* 162:166–172.

Diamond, J. 1997. *Guns, Germs, and Steel: The Fates of Human Societies.* W. W. Norton and Company, New York, New York, USA.

Dubey, J. P., and K. Odening. 2001. Toxoplasmosis and related infections. Pages 478–519 *in* W. M. Samuel, M. J. Pybus, and A. A. Kocan, editors. *Parasitic Diseases of Wild Mammals.* 2nd edition. Iowa State University Press, Ames, Iowa, USA.

Festa-Bianchet, M. 1988. A pneumonia epizootic in bighorn sheep, with comments on preventive management. *Biennial Symposium of the Northern Wild Sheep and Goat Council* 6:66–76.

Fisher, M. A., G. C. Weiser, D. L. Hunter, and A. C. S. Ward. 1999. Use of a polymerase chain reaction method to detect the leukotoxin gene *lktA* in biogroup and biovariant isolates of *Pasteurella haemolytica* and *P trehalosi. American Journal of Veterinary Research* 60:1402–1406.

Fisk, E. A., E. F. Cassirer, K. S. Huggler, A. P. Pessier, L. A. White, J. D. Ramsay, E. W. Goldsmith, H. R. Drankhan, R. M. Wolking, K. R. Manlove, T. Nordeen, J. T. Hogg, and K. R. Taylor. 2023. Abortion and neonatal mortality due to *Toxoplasma gondii* in bighorn sheep (*Ovis canadensis*). *Journal of Wildlife Diseases* 59:37–48.

Forde, T., S. Kutz, J. De Buck, A. Warren, K. Ruckstuhl, M. Pybus, and K. Orsel. 2012. Occurrence, diagnosis, and strain typing of *Mycobacterium avium* subspecies *paratuberculosis* infection in Rocky Mountain bighorn sheep (*Ovis canadensis canadensis*) in southwestern Alberta. *Journal of Wildlife Diseases* 48:1–11.

Foreyt, W. J. 1993. Efficacy of in-feed formulation ivermectin against *Psoroptes* sp. in bighorn sheep. *Journal of Wildlife Diseases* 29:85–89.

Foreyt, W. J., K. P. Snipes, and J. E. Lagerquist. 1996. Susceptibility of Dall sheep (*Ovis dalli dalli*) to pneumonia caused by *Pasteurella haemolytica. Journal of Wildlife Diseases* 32:586–593.

Foreyt, W. J., K. P. Snipes, and R. W. Kasten. 1994. Fatal pneumonia following inoculation of healthy bighorn sheep with *Pasteurella haemolytica* from healthy domestic sheep. *Journal of Wildlife Diseases* 30:137–145.

Forrester, D. J. 1971. Bighorn sheep lungworm pneumonia complex. Pages 158–173 *in* J. W. Davis and R. C. Anderson, editors. *Parasitic Diseases of Wild Mammals.* Iowa State University Press, Ames, Iowa, USA.

Fox, K. A., C. A. W. MacGlover, K. A. Blecha, and M. D. Stenglein. 2023. Assessing shared respiratory pathogens between domestic (*Ovis aries*) and bighorn (*Ovis canadensis*) sheep; methods for multiplex PCR, amplicon sequencing, and bioinformatics to characterize respiratory flora. *PLoS ONE* 18:e0293062.

Fox, K. A., J. H. Kopanke, J. S. Lee, L. L. Wolfe, K. L. Pabilonia, and C. E. Mayo. 2019. Bovine viral diarrhea in captive Rocky Mountain bighorn sheep associated with administration of a contaminated modified-live bluetongue virus vaccine. *Journal of Veterinary Diagnostic Investigation* 31:107–112.

Fox, K. A., N. M. Rouse, K. P. Huyvaert, K. A. Griffin, H. J. Killion, J. Jennings-Gaines, W. H. Edwards, S. L. Quackenbush, and M. W. Miller. 2015. Bighorn sheep (*Ovis canadensis*) sinus tumors are associated with coinfections by potentially pathogenic bacteria in the upper respiratory tract. *Journal of Wildlife Diseases* 51:19–27.

Fox, K. A., S. K. Wootton, S. L. Quackenbush, L. L. Wolfe, I. K. Levan, M. W. Miller, and T. R. Spraker. 2011. Paranasal sinus masses of Rocky Mountain bighorn sheep (*Ovis canadensis canadensis*). *Veterinary Pathology* 48:706–712.

Garde, E., S. Kutz, H. Schwantje, A. Veitch, E. Jenkins, and B. Elkin. 2005. Examining the risk of disease transmission between wild Dall's sheep and mountain goats, and introduced domestic sheep, goats, and llamas in the Northwest Territories. Northwest Territories Agricultural Policy Framework, Environment and Natural Resources, Government of the Northwest Territories, Canada.

Garwood, T. J., C. P. Lehman, D. P. Walsh, E. F. Cassirer, T. E. Besser, and J. A. Jenks. 2020. Removal of chronic *Mycoplasma ovipneumoniae* carrier ewes eliminates pneumonia in a bighorn sheep population. *Ecology and Evolution* 10:3491–3502.

George, J. L., D. J. Martin, P. M. Lukacs, and M. W. Miller. 2008. Epidemic pasteurellosis in a bighorn sheep population coinciding with the appearance of a domestic sheep. *Journal of Wildlife Diseases* 44:388–403.

George, J. L., R. Kahn, M. W. Miller, and B. Watkins. 2009. *Colorado Bighorn Sheep Management Plan 2009–2019.* Special Report Number 81. Colorado Division of Wildlife, Department of Natural Resources, Denver, Colorado, USA.

Gillin, C. M. 2022. Health protection and promotion for disease management in free-ranging wildlife populations. Pages 113–125 *in* C. Stephen, editor. *Wildlife Population Health.* Springer Nature Switzerland AG, Cham, Switzerland.

Gilmour, N. J. L., and J. S. Gilmour. 1989. Pasteurellosis of sheep. Pages 223–262 *in* C. Adlam and J. M. Rutter, editors. *Pasteurella and Pasteurellosis.* Academic Press, Inc., San Diego, California, USA.

Gioia, J., X. Qin, H. Jiang, K. Clinkenbeard, R. Lo, et al. 2006. The genome sequence of *Mannheimia haemolytica* A1: insights into virulence, natural competence, and Pasteurellaceae phylogeny. *Journal of Bacteriology* 188:7257–7266.

Grigg, J. L., L. L. Wolfe, K. A. Fox, H. J. Killion, J. Jennings-Gaines, M. W. Miller, and B. P. Dreher. 2017. Assessing timing and causes of neonatal lamb losses in a bighorn sheep *Ovis canadensis canadensis* herd via use of vaginal implant transmitters. *Journal of Wildlife Diseases* 53:596–601.

Grinnell, G. B. 1928. Mountain sheep. *Journal of Mammalogy* 9:1–9.

Gutierrez, A. P., L. Ponti, and P. A. Arias. 2019. Deconstructing the eradication of new world screwworm in North America: retrospective analysis and climate warming effects. *Medical and Veterinary Entomology* 33:282–295.

Hanson, R. P. 1969. Koch is dead. *Bulletin of the Wildlife Disease Association* 5:150–156.

Hause, B. M., A. Pillatzki, T. Clement, T. Bragg, J. Ridpath, and C. C. L. Chase. 2021. Persistent infection of American bison (*Bison bison*) with bovine viral diarrhea virus and bosavirus. *Veterinary Microbiology* 252:e108949.

Hering, A. M., N. B. Chilton, T. Epp, H. M. Schwantje, F. Cassirer, A. Walker, C. McLean, P. R. Thampy, E. Hanak, P. Wolff, M. Drew, K. D. Bardsley, and M. Woodbury. 2021. Traceback of the *Psoroptes* outbreak in British Columbian bighorn sheep (*Ovis canadensis*). *International Journal for Parasitology: Parasites and Wildlife* 14:273–279.

Hibler, C. P., T. R. Spraker, and E. T. Thorne. 1982. Protostrongylosis in bighorn sheep. Pages 208–213 in E. T. Thorne, N. Kingston, W. R. Jolley, and R. C. Bergstrom, editors. 1982. *Diseases of Wildlife in Wyoming*. 2nd edition. Wyoming Game and Fish Department, Cheyenne, Wyoming, USA.

Hiendleder, S., B. Kaupe, R. Wassmuth, and A. Janke. 2002. Molecular analysis of wild and domestic sheep questions current nomenclature and provides evidence for domestication from two different subspecies. *Proceedings of the Royal Society of London. Series B: Biological Sciences* 269:893–904.

Hoberg, E. P., A. A. Kocan, and L. G. Rickard. 2001. Gastrointestinal strongyles in wild ruminants. Pages 193–221 *in* W. M. Samuel, M. J. Pybus, and A. A. Kocan, editors. *Parasitic Diseases of Wild Mammals*. 2nd edition. Iowa State University Press, Ames, Iowa, USA.

Honess, R. F., and N. M. Frost. 1942. *A Wyoming Bighorn Sheep Study*. Bulletin Number 1. Wyoming Game and Fish Department, Cheyenne, Wyoming, USA.

Hornaday, W. T. 1901. *Notes on the Mountain Sheep of North America, with a Description of a New Species*. Appendix. Fifth Annual Report, New York Zoological Society, New York, New York, USA.

Huwer, S. L. 2015. *Population Estimation, Survival Estimation and Range Delineation for the Georgetown Bighorn Sheep Herd: Final Report*. Technical Publication Number 46, Colorado Parks and Wildlife, Denver, Colorado, USA.

Imus, J. K., H. D. Lehmkuhl, and L. W. Woods. 2019. Resistance of colostrum-deprived domestic lambs to infection with deer adenovirus. *Journal of Veterinary Diagnostic Investigation* 31:78–82.

Jansen, B. D., J. R. Heffelfinger, T. H. Noon, P. R. Krausman, and J. C. deVos, Jr. 2006. Infectious keratoconjunctivitis in bighorn sheep, Silver Bell Mountains, Arizona, USA. *Journal of Wildlife Diseases* 42:407–411.

Jansen, B. D., P. R. Krausman, J. R. Heffelfinger, T. H. Noon, and J. C. Devos, Jr. 2007. Population dynamics and behavior of bighorn sheep with infectious keratoconjunctivitis. *Journal of Wildlife Management* 71:571–575.

Jaworski, M. D., D. L. Hunter, and A. C. S. Ward. 1998. Biovariants of isolates of *Pasteurella* from domestic and wild ruminants. *Journal of Veterinary Diagnostic Investigation* 10:49–55.

Jenkins, E. J., A. M. Veitch, S. J. Kutz, T. K. Bollinger, J. M. Chirino-Trejo, B. T. Elkin, K. H. West, E. P. Hoberg, and L. Polley. 2007. Protostrongylid parasites and pneumonia in captive and wild thinhorn sheep (*Ovis dalli*). *Journal of Wildlife Diseases* 43:189–205.

Jex, B. A., J. B. Ayotte, V. C. Bleich, C. E. Brewer, D. L. Bruning, T. M. Hegel, N. C. Larter, R. A. Schwanke, H. M. Schwantje, and M. W. Wagner. 2016. *Thinhorn Sheep: Conservation Challenges and Management Strategies for the 21st Century*. Wild Sheep Working Group, Western Association of Fish and Wildlife Agencies, Boise, Idaho, USA.

Johnson, B. M., J. Stroud-Settles, A. Roug, and K. R. Manlove. 2022. Disease ecology of a low-virulence *Mycoplasma ovipneumoniae* strain in a free-ranging desert bighorn sheep population. *Animals* 12:1029.

Kamath, P. L., K. Manlove, E. F. Cassirer, P. C. Cross, and T. E. Besser. 2019. Genetic structure of *Mycoplasma ovipneumoniae* informs pathogen spillover dynamics between domestic and wild *Caprinae* in the western United States. *Scientific Reports* 9:15318.

Kayal, A., N. Nahar, L. Barker, T. Tran, M. Williams, P. J. Blackall, C. Turni, and L. Omaleki. 2024. Molecular identification and characterisation of *Mannheimia haemolytica*. *Veterinary Microbiology* 288:109930.

Kelley, S. T., E. F. Cassirer, G. C. Weiser, and S. Safaee. 2007. Phylogenetic diversity of Pasteurellaceae and horizontal gene transfer of leukotoxin in wild and domestic sheep. *Infection, Genetics and Evolution* 7:13–23.

Khamesipour, F., H. Momtaz, and M. Azhdary Mamoreh. 2014. Occurrence of virulence factors and antimicrobial resistance in *Pasteurella multocida* strains isolated from slaughter cattle in Iran. *Frontiers in Microbiology* 5:e536.

Kistner, T. P., S. M. Matlock, D. Wyse, and G. E. Masons. 1977. Helminth parasites of bighorn sheep in Oregon. *Journal of Wildlife Diseases* 13:125–130.

Kraabel, B. J., and M. W. Miller. 1997. Effect of simulated stress on susceptibility of bighorn sheep neutrophils to *Pasteurella haemolytica* leukotoxin. *Journal of Wildlife Diseases* 33:558–566.

Kraabel, B. J., M. W. Miller, J. A. Conlon, and H. J. McNeil. 1998. Evaluation of a multivalent *Pasteurella haemolytica* vaccine in bighorn sheep: protection from experimental challenge. *Journal of Wildlife Diseases* 34:325–333.

Lange, R. E. 1982. Psoroptic scabies. Pages 244–247 in E. T. Thorne, N. Kingston, W. R. Jolley, and R. C. Bergstrom, editors. *Diseases of Wildlife in Wyoming*. 2nd edition. Wyoming Game and Fish Department, Cheyenne, WY.

Lange, R. E., A. V. Sandoval, and W. P. Meleney. 1980. Psoroptic scabies in bighorn sheep (*Ovis canadensis mexicana*) in New Mexico. *Journal of Wildlife Diseases* 16:77–82.

Lieske, C. L., R. Gerlach, M. Francis, and K. B. Beckmen. 2022. Multilocus sequence typing of *Mycoplasma ovipneumoniae* detected in Dall's sheep (*Ovis dalli dalli*) and caribou (*Rangifer tarandus grantii*) in Alaska, USA. *Journal of Wildlife Diseases* 58:625–630.

Lohuis, T. 2015. Ewe Dall's sheep survival, pregnancy and parturition rates, and lamb recruitment in GMU 13D, Chugach Mountains, Alaska. Federal Aid in Wildlife Restoration Project AKW-4. Alaska Department of Fish and Game, Juneau.

López-Olvera, J.R., E. Ramírez, C. Martínez-Carrasco, and J. E. Granados. 2024. Wildlife–livestock host community maintains simultaneous epidemiologic cycles of *Mycoplasma conjunctivae* in a mountain ecosystem. *Veterinary Sciences* 11:217.

MacGlover, C. A. 2023. Characterization of respiratory pathogens associated with bighorn sheep pneumonia; bacterial genotypes, biotypes, and microbiome. Dissertation, University of Wyoming, Laramie, Wyoming, USA.

Malmberg, J. L., D. O'Toole, T. Creekmore, E. Peckham, H. Killion, et al. 2020. *Mycoplasma bovis* infections in free-ranging pronghorn, Wyoming, USA. *Emerging Infectious Diseases* 26:2807–2814.

Malmberg, J. L., J. Alder, H. Killion, D. Buttke, K. M. Pepin, and G. Wittemyer. 2025. Cross–species transmission at the wildlife–livestock interface: a case study of epidemiological inference from mule deer GPS collar data. *Ecology and Evolution* 15:e71182.

Manlove, K. R., A. Roug, K. Sinclair, L. E. Ricci, K. R. Hersey, et al. 2022. Bighorn sheep show similar in-host responses to the same pathogen strain in two contrasting environments. *Ecology and Evolution* 12:e9109.

Markham, P. F., and A. H. Noormohammadi. 2005. Diagnosis of mycoplasmosis in animals. Pages 355–382 *in* A. Blanchard and G. Browning, editors. *Mycoplasmas Molecular Biology Pathogenicity and Strategies for Control*. Horizon Bioscience, Wymondham, Norfolk, United Kingdom.

Marsh, H. 1938. Pneumonia in Rocky Mountain bighorn sheep. *Journal of Mammalogy* 19:214–219.

Martin, S. W., A. H. Meek, and P. Willeberg. 1987. *Veterinary Epidemiology: Principles and Methods.* 1st edition. Iowa State University Press, Ames, Iowa, USA.

Mathieu, A., C. Thacker, I. Teske, E. Jenkins, B. Wagner, B. Macbeth, S. Raverty, and M. Pybus. 2022. *Fascioloides magna* in free-ranging Rocky Mountain bighorn sheep (*Ovis canadensis*). *Journal of Wildlife Diseases* 58:592–598.

McDougald, D., S. Rice, N. Barraud, P. D. Steinberg, and S. Kjelleberg. 2012. Should we stay or should we go: mechanisms and ecological consequences for biofilm dispersal. *Nature Reviews Microbiology* 10:39–50.

Mckinney, T., T. W. Smith, and J.C., Devos, Jr. 2006. Evaluation of factors potentially influencing a desert bighorn sheep population. *Wildlife Monographs* 164:1–36.

McNeil, H. J., P. E. Shewen, R. Y. C. Lo, J. A. Conlon, and M. W. Miller. 2002. *Mannheimia haemolytica* serotype 1 and *Pasteurella trehalosi* serotype 10 culture supernatants contain fibrinogen-binding proteins. *Veterinary Immunology and Immunopathology* 90:107–110.

McNeil, H. J., P. E. Shewen, R. Y. C. Lo, J. A. Conlon, and M. W. Miller. 2003. Novel protease produced by a *Pasteurella trehalosi* serotype 10 isolate from a pneumonic bighorn sheep: characteristics and potential relevance to protection. *Veterinary Microbiology* 93:145–152.

McTaggart Cowan, I. 1951. The diseases and parasites of big game mammals of western Canada. Pages 37–64 in Province of British Columbia, Ministry of Recreation and Conservation, Fish and Wildlife Branch *Report of Proceedings of the Fifth Annual Game Convention.* Don McDiarmid, Printer, Victoria, British Columbia, Canada.

Meagher, M., W. J. Quinn, and L. Stackhouse. 1992. Chlamydial-caused infectious keratoconjunctivitis in bighorn sheep of Yellowstone National Park. *Journal of Wildlife Diseases* 28:171–176.

Melville, E. G. K. 1994. *A Plague of Sheep: Environmental Consequences of the Conquest of Mexico.* Cambridge University Press, Cambridge, United Kingdom.

Miller, M. A. 2023. Tuberculosis in free-ranging wild animals. Pages 389–406 in D. A. Jessup and R. W. Radcliffe, editors. *Wildlife Disease and Health in Conservation.* Johns Hopkins University Press, Baltimore, Maryland, USA.

Miller, M. W. 2001. Pasteurellosis. Pages 330–339 in E. S. Williams and I. K. Barker, editors. *Infectious Diseases of Wild Mammals.* 3rd edition. Iowa State University Press, Ames, Iowa, USA.

Miller, M. W., and L. L. Wolfe. 2011. Pasteurellaceae from Colorado bighorn sheep herds. *Journal of Wildlife Diseases* 47:800–804.

Miller, M. W., J. E. Vayhinger, D. C. Bowden, S. Roush, T. Verry, A. Torres, and V. Jurgens. 2000. Drug treatment for lungworm in bighorn sheep: reevaluation of a 20-year-old management prescription. *Journal of Wildlife Management* 64:505–512.

Miller, M. W., N. T. Hobbs, and E. S. Williams. 1991. Spontaneous pasteurellosis in captive Rocky Mountain bighorn sheep (*Ovis canadensis canadensis*): clinical, laboratory, and epizootiological observations. *Journal of Wildlife Diseases* 27:534–542.

Miller, M.W., B. M. Hause, H. J. Killion, K. A. Fox, W. H. Edwards, and L. L. Wolfe. 2013. Phylogenetic and epidemiologic relationships among Pasteurellaceae from Colorado bighorn sheep herds. *Journal of Wildlife Diseases* 49:653–660.

Mooring, M. S., B. L. Hart, T. A. Fitzpatrick, D. D. Reisig, T. T. Nishihira, I. C. Fraser, and J. E. Benjamin. 2006. Grooming in desert bighorn sheep (*Ovis canadensis mexicana*) and the ghost of parasites past. *Behavioral Ecology* 17:364–371.

Navarro Mamani, D. A., H. Ramos Huere, R. Vera Buendia, M. Rojas, W. A. Chunga, E. Valdez Gutierrez, W. Vergara Abarca, H. Rivera Gerónimo, and M. Altamiranda-Saavedra. 2023. Would climate change influence the potential distribution and ecological niche of bluetongue virus and its main vector in Peru? *Viruses* 15:e892.

Nguyen, N. T. D., A. K. Pathak, and I. M Cattadori. 2022. Gastrointestinal helminths increase *Bordetella bronchiseptica* shedding and host variation in supershedding. *eLife* 11:e70347.

Onderka, D. K., and W. D. Wishart. 1984. A major bighorn sheep dieoff from pneumonia in southern Alberta. *Biennial Symposium of the Northern Wild Sheep and Goat Council* 4:356–363.

Onderka, D. K., S. A. Rawluk, and W. D. Wishart. 1988. Susceptibility of Rocky Mountain bighorn sheep and domestic sheep to pneumonia induced by bighorn and domestic livestock strains of *Pasteurella haemolytica*. *Canadian Journal of Veterinary Research* 52:439–444.

Ostermann, S. D., J. R. Deforge, and W. D. Edge. 2001. Captive breeding and reintroduction evaluation criteria: a case study of peninsular bighorn sheep. *Conservation Biology* 15:749–760.

Parr, B. L., J. B. Smith, and J. A. Jenks. 2018. Population dynamics of a bighorn sheep (*Ovis canadensis*) herd in the southern Black Hills of South Dakota and Wyoming. *The American Midland Naturalist* 179:1–14.

Passler, T., and P. H. Walz. 2010. Bovine viral diarrhea virus infections in heterologous species. *Animal Health Research Reviews* 11:191–205.

Pathak, A. K., C. Pelensky, B. Boag, and I. M. Cattadori. 2012. Immuno-epidemiology of chronic bacterial and helminth co-infections: observations from the field and evidence from the laboratory. *International Journal for Parasitology* 42:647–655.

Pence, D. B. and E. Ueckermann. 2002. Sarcoptic mange in wildlife. *Revue Scientifique et technique-Office international des Epizooties* 21:385–398.

Pillai, D. K., E. Cha, and D. Mosier. 2018. Role of the stress-associated chemicals norepinephrine, epinephrine and substance P in dispersal of *Mannheimia haemolytica* from biofilms. *Veterinary Microbiology* 215:11–17.

Poole, K., I. Teske, K. Podrasky, J. Berdusco, C. Conroy, R. MacDonald, R. Davies, H. Schwantje, E. Chow, C. van Rensen, and T. Ayele. 2020. Bighorn sheep cumulative effects assessment report Elk Valley, Kootenay-Boundary Region, Government of British Columbia. Cumulative Effects Management Framework. Cranbrook, British Columbia, Canada. Unpublished Report.

Post, G. 1962. Pasteurellosis of Rocky Mountain bighorn (*Ovis canadensis canadensis*). *Wildlife Disease* 23:1–14.

Potts, M. K. 1937. Hemorrhagic septicemia in the bighorn of Rocky Mountain National Park. *Journal of Mammalogy* 18:105–106.

Proffitt, K., V. Boccadori, D. Walsh, and H. Manninen. 2022. Highland bighorn sheep population management evaluation annual interim report, Dec 2022. Montana Fish, Wildlife & Parks, Bozeman, Montana, USA. Unpublished Report. https://fwp.mt.gov/binaries/

content/assets/fwp/conservation/bighorn-sheep/2---an-nual-pr-report-2022---highland-mtn-bighorn-project.pdf. Accessed 17 February 2025.

Purse, B. V., P. S. Mellor, D. J. Rogers, A. R. Samuel, P. P. Mertens, and M. Baylis. 2005. Climate change and the recent emergence of bluetongue in Europe. *Nature Reviews Microbiology* 3:171–181.

Pybus, M. J. 2001. Liver flukes. Pages 121–149 *in* W. M. Samuel, M. J. Pybus, and A. A. Kocan, editors. *Parasitic Diseases of Wild Mammals*. 2nd edition. Iowa State University Press, Ames, Iowa, USA.

Register, K. B., M. D. Jelinski, M. Waldner, W. D. Boatwright, T. K. Anderson, et al. 2019. Comparison of multilocus sequence types found among North American isolates of *Mycoplasma bovis* from cattle, bison, and deer, 2007–2017. *Journal of Veterinary Diagnostic Investigation* 31:899–904.

Rezaei, H. R., S. Naderi, I. C. Chintauan-Marquier, P. Taberlet, A. T. Virk, H. R. Naghash, D. Rioux, M. Kaboli, and F. Pompanon. 2010. Evolution and taxonomy of the wild species of the genus *Ovis* (Mammalia, Artiodactyla, Bovidae). *Molecular Phylogenetics and Evolution* 54:315–326.

Rice, J. A., L. Carrasco-Medina, D. C. Hodgins, and P. E. Shewen. 2007. *Mannheimia haemolytica* and bovine respiratory disease. *Animal Health Research Reviews* 8:117–128.

Rijal, S., P. Neuhaus, J. Thorley, N. Caulkett, S. Kutz, and K.E. Ruckstuhl. 2024. Patterns of gastrointestinal parasite infections in bighorn sheep, *Ovis canadensis*, with respect to host sex and seasonality. *International Journal for Parasitology: Parasites and Wildlife* 24:100950.

Robinson, A. J., and P. J. Kerr. 2001. Parapoxvirus infections. Pages 179–201 in E. S. Williams and I. K. Barker, editors. *Infectious Diseases of Wild Mammals*. 3rd edition. Iowa State University Press, Ames, Iowa, USA.

Romeo, C., L. A. Wauters, F. Santicchia, B. Dantzer, R. Palme, A. Martinoli, and N. Ferrari. 2020. Complex relationships between physiological stress and endoparasite infections in natural populations. *Current Zoology* 66:449–457.

Ruder, M. G., and S. A. Christensen. 2023. Bluetongue, epizootic hemorrhagic disease, and cervid adenoviral hemorrhagic disease. Pages 180–197 in D. A. Jessup and R. W. Radcliffe, editors. *Wildlife Disease and Health in Conservation*. Johns Hopkins University Press, Baltimore, Maryland, USA.

Ruder, M. G., T. J. Lysyk, D. E. Stallknecht, L. D. Foil, D. J. Johnson, C. C. Chase, D. A. Dargatz, and E. P. J. Gibbs. 2015. Transmission and epidemiology of bluetongue and epizootic hemorrhagic disease in North America: current perspectives, research gaps, and future directions. *Vector-Borne and Zoonotic Diseases* 15:348–363.

Rush, W. M. 1927. Notes on diseases in wild game animals. *Journal of Mammalogy* 8:163–165.

Ryder, T. J., E. S. Williams, and S. L. Anderson. 1994. Residual effects of pneumonia on the bighorn sheep of Whiskey Mountain, Wyoming. *Proceedings of the Biennial Symposium of the Northern Wild Sheep and Goat Council* 9:15–19.

Ryder, T. J., E. S. Williams, K. W. Mills, K. H. Bowles, and E. T. Thorne. 1992. Effect of pneumonia on population size and lamb recruitment in Whiskey Mountain bighorn sheep. *Proceedings of the Biennial Symposium of the Northern Wild Sheep and Goat Council* 8:136–146.

Schmidt, R. L., C. P. Hibler, T. R. Spraker, and W. H. Rutherford. 1979. An evaluation of drug treatment for lungworm in bighorn sheep. *Journal of Wildlife Management* 43:461–467.

Seton, E. T. 1929. The bighorn. Pages 515–573 *in Lives of Game Animals. An Account of Those Land Animals in America, North of the Mexican Border, Which Are Considered "Game," Either Because They Have Held the Attention of Sportsmen, or Received the Protection of Law. Volume* III—*Part II. Hoofed Animals*. Doubleday, Doran & Company, Inc., Garden City, New York, USA.

Shillinger, J. E. 1937. Disease relationship of domestic stock and wildlife. Pages 298–302 in *Transactions of the Second North American Wildlife Conference*. American Wildlife Institute, Washington, D.C., USA.

Silflow, R. M., and W. J. Foreyt. 1994. Susceptibility of phagocytes from elk, deer, bighorn sheep, and domestic sheep to *Pasteurella haemolytica* cytotoxins. *Journal of Wildlife Diseases* 30:529–535.

Silflow, R. M., W. J. Foreyt, and R. W. Leid. 1993. *Pasteurella haemolytica* cytotoxin dependent killing of neutrophils from bighorn and domestic sheep. *Journal of Wildlife Diseases* 29:30–35.

Silflow, R. M., W. J. Foreyt, S. M. Taylor, W. W. Laegried, and H. D. Liggitt. 1991. Comparison of arachidonate metabolism by alveolar macrophages from bighorn and domestic sheep. *Inflammation* 15:43–54.

Silflow, R. M., W. J. Foreyt, S. M. Taylor, W. W. Laegried, H. D. Liggitt, and R. W. Leid. 1989. Comparison of pulmonary defense mechanisms in Rocky Mountain bighorn sheep (*Ovis canadensis canadensis*) and domestic sheep. *Journal of Wildlife Diseases* 25:514–520.

Simecka, J. W. 2005. Immune responses following mycoplasma infection. Pages 485–534 *in* A. Blanchard and G. Browning, editors. *Mycoplasmas Molecular Biology Pathogenicity and Strategies for Control*. Horizon Bioscience, Wymondham, Norfolk, United Kingdom.

Sirochman, M. A., K. J. Woodruff, J. L. Grigg, D. P. Walsh, K. P. Huyvaert, M. W. Miller, and L. L. Wolfe. 2012. Evaluation of management treatments intended to increase lamb recruitment in a bighorn sheep herd. *Journal of Wildlife Diseases* 48:781–784.

Slater, O. M., J. Peters-Kennedy, M. Lejeune, D. Gummer, B. Macbeth, A. Warren, T. Joseph, H. Li, C. W. Cunha, and P. J. Duignan. 2017. Sheep-associated malignant catarrhal fever–like skin disease in a free-ranging bighorn sheep (*Ovis canadensis*), Alberta, Canada. *Journal of Wildlife Diseases* 53:153–158.

Smith, J. B., J. A. Jenks, T. W. Grovenburg, and R. W. Klaver. 2014. Disease and predation: sorting out causes of a bighorn sheep (*Ovis canadensis*) decline. *PLoS ONE* 9:e88271.

Spencer, C. C. 1939. Notes on the life history of Rocky Mountain bighorn sheep (*Ovis canadensis canadensis*) on the Pike National Forest with particular reference to breeding habits and parasites present. Unpublished report 2–6–1939, United States Department of Agriculture Forest Service, Washington, D.C., USA.

Spraker, T. R., and C. P. Hibler. 1982. An overview of the clinical signs, gross and histological lesions of the pneumonia complex of bighorn sheep. *Biennial Symposium of the Northern Wild sheep and Goat Council* 3:163–172.

Spraker, T. R., C. P. Hibler, G. G. Schoonveld, and W. S. Adney. 1984. Pathologic changes and microorganisms found in bighorn sheep during a stress-related die-off. *Journal of Wildlife Diseases* 20:319–327.

Stallknecht, D. E., A. B. Allison, A. W. Park, J. E. Phillips, V. H. Goekjian, V. F. Nettles, and J. R. Fischer. 2015. Apparent increase of reported hemorrhagic disease in the midwestern and northeastern USA. *Journal of Wildlife Diseases* 51:348–361.

Stephen, C. 2022. Causation in wildlife population health. Pages 65–76 in C. Stephen, editor. *Wildlife Population Health*. Springer Nature Switzerland AG, Cham, Switzerland.

Swift, P. K., J. D. Wehausen, H. B. Ernest, R. S. Singer, A. M. Pauli, H. Kinde, T. E. Rocke, and V. C. Bleich. 2000. Desert bighorn sheep mortality due to presumptive Type C Botulism in California. *Journal of Wildlife Diseases* 36:184–189.

Thacker, C. 2020. Health surveillance of thinhorn sheep (*Ovis dalli*) herds in British Columbia and Alaska. Thesis, University of Calgary, Calgary, Alberta, Canada.

Thorne, E. T. 1982a. Infectious keratoconjunctivitis (*Moraxella*). Pages 81–84 in E. T. Thorne, N. Kingston, W. R. Jolley, and R. C. Bergstrom, editors. 1982. *Diseases of Wildlife in Wyoming*. 2nd edition. Wyoming Game and Fish Department, Cheyenne, Wyoming, USA.

Thorne, E. T. 1982b. Pasteurellosis. Pages 72–81 in E. T. Thorne, N. Kingston, W. R. Jolley, and R. C. Bergstrom, editors. *Diseases of Wildlife in Wyoming*. 2nd edition. Wyoming Game and Fish Department, Cheyenne, Wyoming, USA.

Thurmond, M. C. 2005. Virus transmission. Pages 91–104 *in* S. M. Goyal and J. F. Ridpath, editors. *Bovine Viral Diarrhea Virus: Diagnosis, Management and Control*. Blackwell Publishing, Ames, Iowa, USA.

Tsuchida, D. Y., M. F. Gentzkow, R. S. Spaan, J. Burco, C. E. Couch, J. M. Spaan, C. W. Epps, and B. R. Beechler. 2024. Bighorn sheep (*Ovis canadensis*) with higher whole blood selenium levels have improved survival and altered immune responses. *Journal of Wildlife Diseases* 60:721–726.

Uhazy, L. S., and J. C. Holmes. 1971. Helminths of the Rocky Mountain bighorn sheep in western Canada. *Canadian Journal of Zoology* 49:507–512.

VanWormer, E., and K. Shapiro. 2023. Toxoplasmosis and one health. Pages 407–418 in D. A. Jessup and R. W. Radcliffe, editors. *Wildlife Disease and Health in Conservation*. Johns Hopkins University Press, Baltimore, Maryland, USA.

Walsh, D. P., B. L. Felts, E. F. Cassirer, T. E. Besser, and J. A. Jenks. 2023. Host vs. pathogen evolutionary arms race: effects of exposure history on individual response to a genetically diverse pathogen. *Frontiers in Ecology and Evolution* 10:1039234.

Warren, E. R. 1910A. The mountain sheep. Pages 9–12 *in* E. R. Warren, editor. *The Mammals of Colorado: An Account of the Several Species Found within the Boundaries of the State, together with a Record of Their Habits and of Their Distribution*. G. P. Putnam's Sons, The Knickerbocker Press, New York and London.

Western Association of Fish and Wildlife Agencies. 2021. Test and remove for wild sheep: a user guide. Working document June 2021. Western Association of Fish and Wildlife Agencies, Boise, Idaho, USA.

Wild Sheep Initiative. 2023. Pneumonia and bighorn sheep: test and remove. Fact Sheet (7–14–2023). Western Association of Fish and Wildlife Agencies, Boise, Idaho, USA.

Wild Sheep Initiative. 2024. 2024 range-wide status of bighorn and thinhorn sheep in North America (ver. 07082024). Western Association of Fish and Wildlife Agencies, Boise, Idaho, USA.

Wild Sheep Working Group. 2012. Recommendations for domestic sheep and goat management in wild sheep habitat. Western Association of Fish and Wildlife Agencies, Boise, Idaho, USA.

Williams, E. S. 2001. Paratuberculosis and other mycobacterial diseases. Pages 361–371 in E. S. Williams and I. K. Barker, editors. Infectious Diseases of Wild Mammals. 3rd edition. Iowa State University Press, Ames, Iowa, USA.

Williams, E. S., T. R. Spraker, and G. G. Schoonveld. 1979. Paratuberculosis (Johne's disease) in bighorn sheep and a Rocky Mountain goat in Colorado. *Journal of Wildlife Diseases* 15:221–227.

Wishart, W. D. 1958. The bighorn sheep of the Sheep River valley. Thesis, The University of Alberta, Edmonton, Canada.

Wobeser, G. A. 1997. *Diseases of Wild Waterfowl*. 2nd edition. Plenum Press, New York, New York, USA.

Wobeser, G. A. 2007. *Diseases in Wild Animals: Investigation and Management*. 2nd edition. Springer-Verlag, Berlin, Heidelberg, Germany.

Wolfe, L. L., B. Diamond, T. R. Spraker, M. A. Sirochman, D. P. Walsh, C. M. Machin, D. J. Bade, and M. W. Miller. 2010. A bighorn sheep die-off in southern Colorado involving a Pasteurellaceae strain that may have originated from syntopic cattle. *Journal of Wildlife Diseases* 46:1262–1268.

Wolff, P. L., C. Schroeder, M. McAdoo, M. Cox, D. D. Nelson, J. F. Evermann, and J. F. Ridpath. 2016. Evidence of bovine viral diarrhea virus infection in three species of sympatric wild ungulates in Nevada: life history strategies may maintain endemic infections in wild populations. *Frontiers in Microbiology* 7:e292.

Wolff, P. L., J. A. Blanchong, D. D. Nelson, P. J. Plummer, C. McAdoo, M. Cox, T. E. Besser, J. Muñoz-Gutiérrez, and C. A. Anderson. 2019. Detection of *Mycoplasma ovipneumoniae* in pneumonic mountain goat (*Oreamnos americanus*) kids. *Journal of Wildlife Diseases* 55:206–212.

Wood, M.D., B.A. Culling, D.E. Culling, and H.M. Schwantje. 2010. Ecology and health of Stone's sheep (*Ovis dalli stonei*) in the Dunlevy/Schooler area, north-eastern British Columbia. Report No. 342. Peace/Williston Fish and Wildlife Compensation Program, Prince George, British Columbia, Canada.

Yekutiel, P. 1980. Eradication of infectious diseases: a critical study. *in* Contributions to Epidemiology and Biostatistics, Volume 2. S. Karger, Basel, Switzerland.

Zarnke, R. L., H. Li, and T. B. Crawford. 2002. Serum antibody prevalence of malignant catarrhal fever viruses in seven wildlife species from Alaska. *Journal of Wildlife Diseases* 38:500–504.

# 9 Relationships among Mountain Sheep and Their Predators

*Brian F. Wakeling, Vernon C. Bleich, Marcus E. Blum, and Justin A. Dellinger*

## 9.1 INTRODUCTION

The effect of predators on, and their relationship with, prey populations has been debated among managers involved in the conservation of North American artiodactyls in general and mountain sheep (bighorn [*Ovis canadensis*], thinhorn [*Ovis dalli*]) in particular. At the extremes of the debate, one theory suggests that predators regulate populations of mountain sheep through a top-down influence; the opposing theory suggests that habitat regulates mountain sheep population abundance through a bottom-up influence (Hunter and Price 1992, Power 1992). Most authors advocate for some mix of these two influences (Pierce et al. 2012, Monteith et al. 2014), but the relative degree of the effects from the two extremes remains poorly understood. Differences in the life-history characteristics among prey species may explain some differences between top-down and bottom-up regulation (Bowyer et al. 2005, 2014), but challenges remain in determining precise cause and effect even within mountain sheep (Rominger 2018). Understanding the reasons for the limited information supporting the differing theories may be as influential as the rationale for the theories that fuel the debate.

In the simplest of situations, a single prey species may be supported by habitat available in a limited geographic area, and that taxon may be preyed on by a single predator without the complicating influence of disease or anthropogenic habitat perturbations. The historical coexistence of moose (*Alces alces*) and wolves (*Canis lupus*) on Isle Royale, Michigan, may serve as an example (McLaren and Peterson 1994, Nelson et al. 2011). Using annual counts of predator and prey, a mathematical relationship between the numbers of prey and predators can be developed to explain the variation between populations of the two species. That relationship is nonlinear, and nonlinear mathematical relationships have multiple solutions. In short, even the simplest relationships may not be that simple (Nelson et al. 2011, Rodriguez Curras et al. 2024). These relationships are compounded in systems supporting multiple prey and predator species by the rates of predation, whether mortalities are additive or compensatory, and the age structure of the prey population of interest (Roemer et al. 2002, Bryant and Page 2005, Mills 2007), as has been demonstrated in some populations of mountain sheep (Rominger et al. 2004, Johnson et al. 2013, Rominger 2018).

Many theories have been developed to explain predator-prey relationships, and many have been incorporated as textbook examples to illustrate and support various explanations about the relationships among predators and prey (Ballard and Van Ballenberghe 1997). These textbook examples help contemporary biologists understand the underpinnings of these complex relationships, yet the data used to support some of these examples fail to hold up. The classic predator-prey relationship between lynx (*Lynx canadensis*) and snowshoe hares (*Lepus americanu*) was developed using harvest and fur sales data from two different landscapes: lynx pelts from across Canada and snowshoe hare pelts from the area surrounding Hudson Bay (Smith 1966, 1980). The differing geographies in which data were collected, however, do not allow for adequate causal inference (Hall 1988). Even the frequently proffered example of the boom and bust of the Kaibab mule deer (*Odocoileus hemionus*) herd falters when the source and bias of the various estimates of the deer population are examined closely. Methods for developing population estimates were not standardized nor entirely objective, yet this formed the basis for comparisons among years of predator removal (Hall 1988). The lack of empirical data to support such relationships limits our understanding of the relationships between prey and their predators and may simply reflect a relationship an observer was expecting to see (Hall 1988).

Most predator and prey relationships are complicated in that multiple predators often pursue multiple prey species. Perhaps the reason our contemporary understanding of many relationships among predator and prey species remains limited is because of the effect of European settlement of the North American continent (Jensen et al. 2023). By the time settlers reached western North America, the effect of humans was already evident on the eastern portion of the continent. Accurate understanding of predator-prey relationships requires study populations that approximate a natural state in assemblages and abundance. Yet by the time the study of wildlife began, the passenger pigeon (*Ectopistes migratorius*), the Heath hen (*Tympanuchus cupido cupido*), and the Labrador duck (*Camptorhynchus labradorius*) already were extinct or were well on the way due to unregulated forestry practices and market hunting (Spann 1937). By 1900, most species of North American ungulates, including pronghorn (*Antilocapra americana*), mule deer, white-tailed deer (*Odocoileus virginianus*), bison (*Bison bison*), and elk (*Cervus canadensis*), were reduced in numbers by orders of magnitude below those estimated by the Lewis and Clark expedition 100 years earlier (Brown and Warnum 2009).

DOI: 10.1201/9781003518686-10

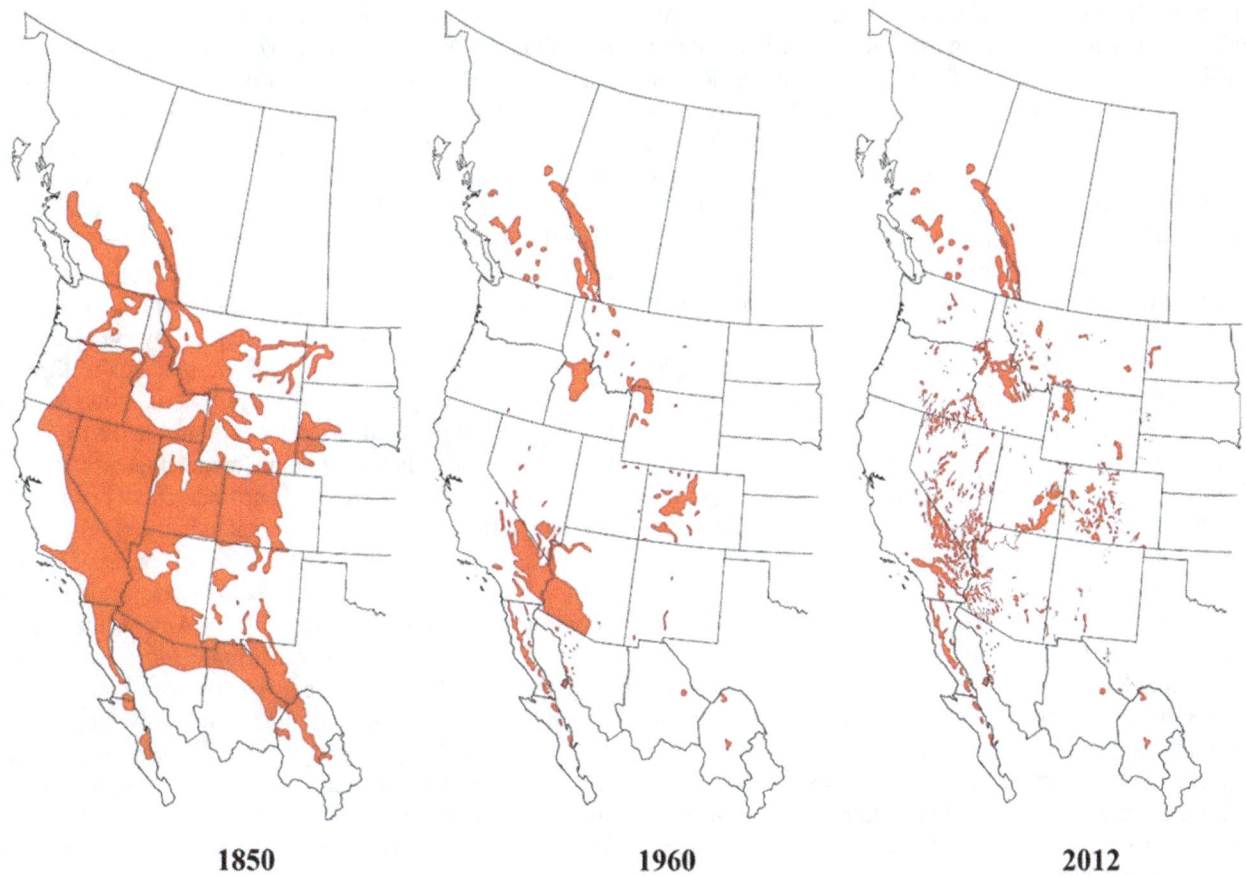

1850          1960          2012

**FIGURE 9.1**  Distribution of mountain sheep in North America in 1850, 1960, and 2012. (Adapted from Brewer et al. (2014).)

Along with the reduction in prey, predators were killed with religious fervor, and government agencies pursued them with indiscriminate toxicants and poisons into the 1960s within the United States (Brown and Warnum 2009). This practice continued even longer in Canada. Mountain sheep, perhaps because they inhabited some of the more remote and rugged areas on the continent, seemed to maintain historical numbers longer than did other North American artiodactyls. Nevertheless, most populations of these specialized mountain ungulates reached their low point in size and distribution by about 1960 (Figure 9.1).

The challenge of learning about and formulating an understanding of predator-prey relationships after their populations have been influenced by interactions with humans is remarkable. When describing this challenge, Smith (2016:29) wrote, "Unfortunately, the study of these complex relations would not emerge as a distinct field until about the same time as atomic physics did, in the 1890s—very, very late in history. Meanwhile, the process of obscuring whatever had been present in the oldest national parks was well under way before they could be studied. In fact, it was going on before the parks were even created...." As a result, restoration had to precede an understanding of historical relationships. Unfortunately, the study of wildlife occurred with altered assemblages and abundances of both predators and prey. This approach fostered partial understanding and frequent misunderstandings.

To further cloud our understanding, historical distributions of wildlife are based on observations recorded by individuals of that earlier era (Jensen et al. 2023). Those explorers and naturalists lacked the survey methodologies commonplace today and the objectivity that peer referees force upon researchers. Sightings and observations were recorded, at times because the observation was novel or unexpected, and in some cases contemporary biologists have shaded in blank areas of range maps to include the cumulative placement of dots on maps (Shaw 1993). As an example, the estimated distribution of mountain sheep in Nevada during 1850 included every square mile of the state (Figure 9.1). Historical ranges may not have been fully occupied, and abundance probably varied dramatically from area to area. This incomplete understanding led some to conclude that mountain sheep addressed predation from some predators, like mountain lions (*Puma concolor*), by adopting allopatric ranges (Shaw 1993). Today, mountain lions and mountain sheep share sympatric ranges, and mountain lion numbers influence mountain sheep abundance directly (McKinney et al. 2006, Gammons et al. 2021) or indirectly (Wehausen 1996).

A more contemporary example of the way that human activities have influenced predation on mountain sheep revolves around political decisions that eliminated the commercial trapping of bobcats (*Lynx rufus*; California Code of Regulations 1998, National Park Service 2002) in areas where the take of those furbearers had occurred uninterrupted for many decades (Bleich and Pauli 1999). An increasing presence of feral or free-ranging dogs in mountain sheep habitat potentially resulted in heretofore unrecognized consequences (Young et al. 2011). Changes in predator assemblages sometimes change the degree and influence of predation on prey populations. Due to the rapidity of such changes, researchers have insufficient time to replicate studies and improve understanding.

The return of native predators to historical ranges influences and further compounds our understanding of predator-prey relationships (Ripple et al. 2024). For example, the return of wolves to multiple areas of the continent is relatively recent and, as with anthropogenic changes, alters predator assemblages and the degree and influence of predation on prey populations. And because these changes are recent, replicating studies are not possible. As a result, our understanding remains incomplete but continues to evolve (Rominger 2018).

Our objective in this chapter is to describe the current understanding of relationships between mountain sheep and their predators, the gaps in our knowledge, and to posit some suggestions as to where our knowledge might be improved. Existing hypotheses in need of testing may be a useful way of identifying future research needs (Wakeling et al. 2009, Rominger 2018), and the testing of those hypotheses may foster improved theoretical understanding of relationships. Weinberg (1993:204) cautioned researchers, however, that, "…in order for a theory to be regarded as satisfactory it not only must agree with the results of experiments that have been done but also must make predictions that are at least plausible for experiments that in principle could be done." Hypotheses and theories should not be cavalier.

## 9.2   PREDATORS OF MOUNTAIN SHEEP

All taxa of Old-World wild sheep are well adapted to existing with the risk of predation, and North American mountain sheep are no exception. The ancestors of Dall's sheep (*Ovis dalli* spp.) and bighorn sheep (*Ovis canadensis* spp.) arose in Asia during the Villafranchian, some 2.5 million years ago and arrived in North America after crossing the Bering Land Bridge (Geist 1971, Valdez and Krausman 1999). Those specialized mountain ungulates, and the numerous species of predators that arrived similarly, then dispersed throughout much of western North America, and became widespread among the rugged, mountainous regions of the continent from northern Alaska to central Mexico, where they have persisted since they arrived in the New World. The guild of predators with which North American mountain sheep evolved has changed substantially over the millennia, with fewer species of large carnivores extant today (Seersholm et al. 2020).

Understanding the effects of predators on mountain sheep is rarely simple. Interactions between predators capable of killing bighorn or thinhorn sheep are frequently reported in the literature, but these interactions do not always result

in successful kills by the predator (Simmons 1964, Woolf and O'Shea 1968, Hornocker 1969, Shank 1977, Kelly 1980, Weaver and Mensch 1970, Bleich 1996, 1999; Goodson and Stevens 1994, Sawyer and Lindzey 2002). Further, previous authors have described potential predators of bighorn sheep simply due to range sympatry or observations of interspecific encounters, rather than documenting predators confirmed to have killed mountain sheep (Kelly 1980, Sawyer and Lindzey 2002, Nichols and Bunnell 2000). Moreover, the outcomes of predator encounters involving mountain sheep may be a function of circumstances involving habitat attributes, behavioral responses of the predator to threats from the intended prey, behavioral responses of the prey to the presence of the predator, prior exposure to the threat of predation, or other heretofore unrecognized factors.

Most students of mountain sheep understand that predation is a daily part of the life and death battle (Figure 9.2). Most also understand that these habitat specialists evolved with a suite of mammalian predators capable of killing young or adult individuals (Table 9.1). Among the extant taxa are the gray wolf, coyote (*Canis latrans*), bobcat, lynx, mountain lion, black bear (*Ursus americanus*), grizzly bear (*Ursus arctos horribilis*), and wolverine (*Gulo gulo*); additionally, the golden eagle (*Aquila chrysaetos*) is a capable and efficient aerial predator of young mountain sheep (Sawyer and Lindzey 2002). Jaguar (*Panthera onca*), ocelot (*Leopardus pardalis*), and gray fox (*Urocyon cinereoargenteus*) likely were, or remain, lesser threats due, in part, to fewer opportunities to encounter mountain sheep because of differences in habitat requirements or, in the case of the gray fox, its small size (Kelly 1980). Recently, free-roaming or feral domestic dogs have come to the forefront as predators of bighorn sheep, particularly in remote, sparsely inhabited areas occupied by desert bighorn sheep (*O. c. nelsoni*; Figures 9.3 and 9.4) and, in some cases, could have population-level effects (Holl 2004, Anderson 2008).

Among thinhorn sheep, the effect of predation is variable and relates to a combination of factors that include predator species, alternate prey abundance, and season. The wolf is most often identified as a main source of mortality caused by predation, but often species such as golden eagles and coyotes emerge as more frequent predators in specific areas (Barichello et al. 1989, Scotton 1998, Environment Yukon 2017). Wolf predation is more influential during winter when they have a mobility advantage. Predation by wolves and coyotes has been recorded from northern Alaska to northern British Columbia, but wolverine, lynx, grizzly bears, and black bears also are suspected of killing thinhorn sheep. The frequency with which mountain lion predation on thinhorn sheep occurs is unknown and is currently limited by their largely allopatric distribution. Should mountain lion populations continue to expand further north into thinhorn sheep range, the arrival of an additional predator using a novel hunting technique will influence specific populations at least until learned avoidance strategies develop (Demarchi and Hartwig 2004).

**FIGURE 9.2** Bobcats seldom have been confirmed as predators of bighorn sheep. Still, such events may be more common than realized because these felids are apt to occupy habitat used primarily by female bighorn sheep more so than coyotes, especially during periods of sexual segregation (Bleich et al. 1997). (Old Dad Mountain, California, USA, circa 2000. Photograph © G. C. Kerr; used with permission.)

Whether mortality is compensatory or additive is an important consideration; when deaths are compensatory, one source of mortality replaces another; when mortality is additive, the effects are cumulative (Bowyer et al. 2014, 2020). For example, following a harsh winter and the loss of a large proportion of the population of thinhorn sheep, predation by some smaller predators could be additive and effectively delay, or even preclude recovery of a population that had been reduced. In situations in which resource availability regulates the size or density of a population, losses to predators may be largely compensatory (Bowyer et al. 2020), with mortalities resulting from predation compensating for other potential sources of mortality, such as that associated with inadequate availability of nutrients (Bowyer et al. 2020). The influence of predation on thinhorn sheep population dynamics is linked to prey population size and trend, carrying capacity of the range, and the presence of sympatric predators (Jex et al. 2016).

## 9.3 FACTORS INFLUENCING PREDATION RATES

### 9.3.1 NUTRITIONAL INFLUENCES ON PREDATION

Mountain sheep are adapted to mountainous regions with precipitous terrain and open habitats that allow them to detect and evade predators. Yet not all habitats used by mountain sheep are of equal quality from nutritional or predator avoidance aspects. Precipitous terrain and steep slopes are commonly associated with predator evasion strategies of mountain sheep (Bleich et al. 1997, Blum et al.

2023a). Areas with abundant escape terrain, however, may be lacking in nutritional resources needed for these species to meet their metabolic requirements, especially immediately following parturition for females (Blum et al. 2023b) when lactation creates increased energy and protein needs (Holt et al. 1992; Parker et al. 2009; Enns et al. 2023, 2024). Forage patches proximate to escape terrain may provide sheep with nutrients that are adequate in abundance and quality, while simultaneously reducing predation risk. Nonetheless, high-quality forage and escape cover may not occur on the same portion of the landscape, so mountain sheep commonly make tradeoffs that allow individuals to meet their metabolic needs while also increasing individual or offspring survival. Mountain sheep often select lower quality habitat when females have offspring present, and males are more likely to use areas with higher quality forage and greater predation risk than those areas used by females (Berger 1991; Bleich et al. 1997; Courtemanch 2014; Blum et al. 2023a,b; Enns et al. 2023, 2024). Trading better-quality forage for enhanced safety has direct survival benefits for offspring and adults; these tradeoffs, however, may not be sustainable. Female mountain sheep with young may eventually need to use areas of greater nutrient availability but with higher predation risk to meet their nutritional needs (Blum et al. 2023a,b; Demarchi and Hartwig 2004).

These tradeoffs are not restricted to the use of habitat, as variation in behavior also allows individuals to increase nutrient intake while simultaneously reducing predation risk (Blum et al. 2023b). For instance, mountain sheep group size increases as animals increase distance to escape terrain (Rachlow and Bowyer 1998). Further, increased

**TABLE 9.1**

**Predators Confirmed to Have Killed Thinhorn Sheep or Bighorn Sheep, Location(s) of Occurrence, Source(s) of Documentation, and Area(s) of Probable Sympatry between Predator Species and North American Mountain Sheep**

| Species of Mountain Sheep | Confirmed Predation by | Location and Source | Additional Areas of Probable Sympatry if Mountain Sheep Are Present on the Landscape |
|---|---|---|---|
| Thinhorn | Wolf | AK[a,b,c,d], YT[e], BC[f] | NT (Paquet and Carbyn 2003) |
| | Wolverine | AK[d,g] | YT, NT, BC (Copeland and Whitman 2003) |
| | Coyote | AK[d,g,h], YT[i] | NT, BC (Bekoff and Gese 2003) |
| | Lynx | AK[j,k] | YT, NT, BC (Anderson and Lovallo 2003) |
| | Black Bear | AK[g] | YT, NT, BC (Pelton 2003) |
| | Golden Eagle | AK[d], YT[l] | NT, BC (CLO 2024) |
| | Grizzly Bear | AK[d] | YT, NT, BC (Schwartz et al. 2003) |
| | Mountain Lion | Unconfirmed | BC, AK, YT (Pierce and Bleich 2003) |
| Bighorn | Wolf | AB[m,n] | BC, YT, NT, MT, WY, ID, NM, AZ, CA, CO (Paquet and Carbyn 2003) |
| | Coyote | BC[o,p], WY[q], AB[r], CA[s], NV[t] | WA, OR, ID, MT, ND, SD, NE, UT, CO, NE, AZ, NM, TX, BCN, BCS, CHH, COA, DUR, NLE, SON (Bekoff and Gese 2003) |
| | Gray Fox | AZ[u,v] | CA, NV, UT, AZ, NM, TX, NE, SD, WY, BCN, BCS, CHH, COA, DUR, NLE, SON (Cypher 2003) |
| | Mountain Lion | AZ[w], NM[x], CA[y,z,aa,bb,cc,s,dd], MT[ee], AB[ff], BC[gg,hh] | WA, ID, ND, SD, NE, WY, NV, CO, TX, OR, BCN, BCS, CHH, COA, DUR, NLE, SON (Pierce and Bleich 2003) |
| | Bobcat | SD[ii], CA[ji],AZ[kk], NV[kk] | BC, WA, OR, ID, MT, ND, SD, WY, NE, NV, NM, TX, BCN, BCS, CHH, COA, DUR, NLE, SON (Anderson and Lovallo 2003) |
| | Lynx | Unconfirmed | BC, AB, ID, MT, WY, UT, WY, CO (Anderson and Lovallo 2003) |
| | Wolverine | Unconfirmed | AB, WA, MT, OR, ID, CA, CO (Copeland and Whitman 2003) |
| | Golden Eagle | ID[ll], CA[mm], NM[nn] | All additional jurisdictions inhabited by bighorn sheep (CLO 2024) |
| | Feral Dog | CA[oo,pp] | Plausible throughout all additional jurisdictions inhabited by bighorn sheep |
| | Black Bear | Unconfirmed | BC, AB, WA, ID, WY, MT, WY, UT, CO, CA, NM, TX, SON, CHH, COA, DUR, NL (Pelton 2003, Villalobos et al. 2019) |
| | Grizzly Bear | Unconfirmed | MT, WY, ID, BC, AB (Schwartz et al. 2003) |

This compilation, which should not be considered all-inclusive, includes the Canadian, American, and Mexican jurisdictions inhabited by mountain sheep. Two-letter postal codes denote location(s) of observations or probable sympatry between the specific predator and mountain sheep in the United States or Canada, three-letter postal codes denote location(s) in Mexico, and footnotes identify the citation(s) or other source(s) of documentation.

[a] Murie (1944); [b]Heimer and Smith (1972); [c]Heimer and Stephenson (1982); [d]Scotton (1998); [e]Hoefs et al. (1986); [f]Child et al. (1978); [g]Nichols and Bunnell (1999); [h]Prugh (2004); [i]Hoefs and Cowan (1979); [j]Stephenson et al. (1991); [k]Sheldon (1930); [l]Nette et al. (1984); [m]Anderson (2008); [n]Gibson (2008); [o]Dekker (1986); [p]Ashcroft (1986); [q]Thorne et al. (1979); [r]Shank (1977); [s]Jorgensen and Turner (1975); [t]McQuivey (1975); [u]Nichol (1937); [v]Nichol (1940); [w]Cunningham et al. (1999); [x]Rominger et al. (2004); [y]Hayes et al. (2000); [z]Wehausen (1996); [aa]Bleich et al. (1997); [bb]Schaefer et al. (2000); [cc]Cronemiller (1948); [dd]Gammons et al. (2021); [ee]Williams et al. (1995); [ff]Ross et al. (1997); [gg]British Columbia Ministry of Forests, Lands, Natural Resource Operations and Rural Development (2021); [hh]Harrison (1990); [ii]Parr et al. (2014); [ji]G. C. Kerr, Society for the Conservation of Bighorn Sheep (figure 2); [kk]Elliott (1961); [ll]Bleich et al. (2004); [mm]Ober (1931); [nn]Kennedy (1948); [oo]Anderson (2008); [pp]Holl (2004).

group size was related to lower rates of vigilance, resulting in increased forage intake rates (Berger 1978).

The extent and length of time that such tradeoffs occur likely depends on the somatic reserves accumulated during periods of abundant, high-quality forage, which has been linked to enhanced population performance (Stephenson et al. 2020). Greater fat reserves will allow females to spend more time in escape terrain while still providing their lambs with nutrients necessary for growth. Females in greater nutritional condition give birth to larger offspring with increased growth rates and fitness benefits for the female than those in lower condition (Monteith

et al. 2009, 2014). Increased vigilance and efforts to evade predators can reduce mass gain among females and offspring before weaning (Cloutier et al. 2024), and female body mass may be lower during the following breeding season. Chapter 7 addresses nutrition in more detail.

## 9.3.2 Disease Influences Susceptibility to Predation

Diseases influence the ability of mountain sheep to respond to various environmental factors, including predation. These effects are challenging to quantify in the natural environment because we rarely see the underlying conditions

**FIGURE 9.3** Feral dogs harassing two male bighorn sheep in the San Bernardino Mountains, California, December 2007. (Photograph © J, J, Restoration Service; used with permission.)

**FIGURE 9.4** Remains of a male lamb killed by a free-ranging domestic dog (inset), San Gabriel Mountains, California, December 2007. (Photographs courtesy of J. T. Villepique; used with permission.)

that lead to predation; we can, however, make informed assumptions about the indirect effects disease may have on predation based on our understanding of influences from parasites and disease.

Pathogens and parasites often weaken infected animals, which may reduce an individual's ability to effectively avoid predation (Joly and Messier 2004, Nava et al. 2023). Some diseases can impair cognitive function and motor coordination, increasing an individual's vulnerability to predation. For example, ataxia resulting from parasites (e.g., *Parelaphostrongylus tenuis* and *Parelaphostrongylus odocoilei*) may compromise movement and awareness

(Pybus et al. 1996, Jenkins et al. 2005), making affected individuals more susceptible to predators. Moreover, impaired cognitive function may reduce vigilance in mountain sheep, diminishing their ability to detect approaching predators. Such behavioral alterations can further increase predation risk, particularly if affected individuals become isolated from the herd, thereby losing the benefits associated with group vigilance. Affected individuals may select more risky habitats (e.g., lower elevations and further from escape terrain). Maladies (e.g., *Mycoplasma ovipneumoniae* and sinusitis) and parasites (e.g., *Psoroptes* mites) may also compromise senses, reducing an individual's ability to detect and respond to approaching predators.

Risk of predation may increase stress levels for some ungulates (Dulude-de Broin et al. 2019, Chitwood et al. 2022). These changes may result in decreased immune responses, potentially increasing susceptibility to various diseases in mountain sheep (Wagler et al. 2023, Smiley et al. 2024). Predator avoidance activities may increase susceptibility to disease and influence the likelihood of die-offs (Miller et al. 2012). In the presence of increased predator abundance, mountain sheep may use habitats with lower quality forage to reduce predation risk, a behavioral change that can influence body condition (Wehausen 1996). Unsuitable vegetative conditions, such as those resulting from drought, may limit habitat selection choices and the ability of mountain sheep to mitigate predation risk (Villepique et al. 2015, Huggler 2024). If the selection of less risky habitat further limits nutrient intake, mountain sheep may need to prioritize nutrient acquisition rather than selecting habitat with lower predation risk (Stearns 1989); a healthy, effective immune system is costly from an energetic standpoint (Lochmiller and Deerenberg 2000, Hasselquist and Nilsson 2012). Chapter 8 addresses disease in detail.

### 9.3.3  Weather Influences Predation Risk

In addition to disease, weather patterns (e.g., altitudinal, geographic, or vacillating) can affect the timing of seasonal migrations and influence susceptibility to predation (Sawyer and Lindzey 2002, Enns et al. 2023, Chauveau et al. 2025). Migratory mountain sheep can experience increased spatial overlap or move closer to other, more abundant ungulates (e.g., mule deer, elk, mountain goat [*Oreamnos americanus*], moose, caribou [*Rangifer tarandus*]) during winter (Johnson et al. 2013, Enns et al. 2023). These migratory herds may be more vulnerable to predation due to competitive interactions (Jex et al. 2016) or they may become secondary prey for mountain lions that generally prey on more abundant mule deer (Ross et al. 1997, Hayes et al. 2000, Sawyer and Lindzey 2002). Years with prolonged snowpack at higher elevations may result in higher rates of predation for mountain sheep during seasonal movements (Johnson et al. 2013). Migration and seasonal movements may increase predation risk from mountain lions for some desert bighorn sheep, such as the endangered Sierra Nevada bighorn sheep (*O. c. sierrae*; Gammons et al. 2021). For thinhorn sheep, wind-hardened snowpacks can increase the amount of terrain accessible by some predators, such as wolves, while reducing the effectiveness of what would have been escape terrain if snow were absent. This may increase winter predation risk, particularly to female and lamb groups (Koizumi and Derocher 2019). Predation from wolves and coyotes is greater for Dall's sheep during periods of deep snow (Burles and Hoefs 1984), indicating that the effects of weather on winter predation are not restricted to migratory mountain sheep.

### 9.3.4  Anthropogenic Disturbance Influences Habitat Use and Predation

Human presence and activity can influence predator effects on mountain sheep in either a positive or a negative way. At times, mountain sheep may choose to occupy habitat close to humans or even loud and disruptive mining activity because of forage and water availability, and because predators may not be prevalent in these areas (Wakeling et al. 2007). Mountain sheep may select areas such as parks or subdivisions for similar reasons, but humans may not always be tolerant of their presence (Chapter 22). The benefits to mountain sheep in terms of forage, water, and predator avoidance must outweigh the risks associated with highways, harassment, and disturbance before the use of disturbed habitats becomes prevalent.

The negative influence of anthropogenic disturbance on predation is often subtle and typically is manifest in reductions in habitat suitability or ability to detect and evade predators. Negative effects on mountain sheep may result from increased human presence, activity, or disturbance in or near preferred foraging sites or seasonally important areas (e.g., lambing sites, migratory stop-over sites, mineral licks; Brushett et al. 2023, Freeman and Maclean 2021, Enns et al. 2023). Increases in human activity in or near mountain sheep habitat can first lead to changes in patterns of movement between seasonally important areas (Courtemanch 2014, Freeman and Maclean 2021), and in activity budgets resulting from increased vigilance, which can reduce time for other activities such as foraging (Sproat et al. 2019). Decreased foraging can result in a decline in body condition and cause individuals to seek areas with higher quality forage (Brushett et al. 2023). Areas supporting higher quality forage, however, might contain less escape terrain, which increases vulnerability to predation and influences survival rates (Feist 1997).

Mountain sheep may respond to human activity like they respond to increased predation risk. Increased human presence in sensitive habitat areas can affect habitat selection (Papouchis et al. 2001), behavior, and reproductive success by mountain sheep (Weidmann and Bleich 2014, Chapter 21). Increased recreational hiking displaced female mountain sheep and led to the eventual abandonment of historical lambing sites (Papouchis et al. 2001, Wiedmann and Bleich 2014). Altered use of lambing sites resulted in decreased recruitment following the increase in human activity, and when compared with nearby herds with less human activity. Reduced site fidelity, especially concerning lambing sites, can increase predation risk for mountain sheep (Brushett et al. 2023) and contribute to changes in recruitment and population trend. Increases in activity close to areas important to mountain sheep seasonal ranges may affect foraging strategies, habitat use, and demographics (Papouchis et al. 2001, Bleich and Wiedmann 2014, Sproat et al. 2019, Chapter 21).

## 9.4 POPULATION LIMITATION AND REGULATION BY PREDATION

Any mortality factor that reduces rate of population growth is a limiting factor. This definition can include density-dependent and density-independent factors. Regulating factors, by contrast, refer to any density-dependent processes that ultimately keep populations at or near the carrying capacity of the area that the population occupies. Thus, regulating factors are a subset of limiting factors, characterized by negative-feedback mechanisms, that depress population growth as animal abundance increases (Messier 1991, Ballard et al. 2001). Yet, interactions between regulating and limiting factors may complicate the interpretation of effects in some situations (Mitchell et al. 2015). Individual animal tolerances and behaviors affected by a complex array of tradeoffs that an individual animal makes in real time can manifest in a risk-reward, tolerance-based behavior (Frid and Dill 2002, Lowrey and Longshore 2017), further complicating the interpretation of both limiting and regulating factors.

Predation may influence prey species directly or indirectly. Direct influences occur through numerical changes because of mortality, whereas indirect effects are generally the result of a behaviorally mediated change. An example of a direct effect is mountain lion predation on desert bighorn sheep, whereas an indirect effect may be the reduction in habitat suitability resulting from brush encroachment due to fire suppression, which may limit the ability of mountain sheep to detect a predator.

Mountain sheep predominantly occupy habitat that limits the hunting efficacy of most predators (Shaw 1993, Sawyer and Lindzey 2002). Historically, mountain sheep occurred in steep, mountainous terrain and low densities of shrubby vegetation, whereas mountain lions generally occupied brushy vegetative types with low visibility, greater abundance of prey, and features that facilitate ambush predation. Consequently, mountain lions may have had a reduced ability to limit mountain sheep before European colonization of North America. Because mountain sheep select a specific suite of terrain and vegetative features that differ from other native ungulates (Valdez and Krausman 1999), predation may have had a reduced role in their population dynamics until anthropogenic disturbances changed predator assemblages, prey assemblages, and habitat suitability. For example, Dall's and Stone's sheep (*O. d. stonei*) still occupy most of their historical range, which remain largely unaffected by anthropogenic disturbance (Nichols 1975, Jex et al. 2016) with little detectable effect from predation

(Gasaway et al. 1983). The effects of human activities at local scales have resulted in measurable declines in some populations (Demarchi and Hartwig 2004, Jex et al. 2016, 2025) even within these relatively natural settings.

The decline of mountain sheep populations into the 1960s created many smaller, isolated populations. Small, isolated populations of mountain sheep are more likely to be limited by predation because they consist of few individuals and the likelihood of immigration is reduced with isolation (Sawyer and Lindzey 2002). The expansion of mule deer in some areas may have influenced the distribution and abundance of mountain lions by creating a stable prey base. Changes in the composition of predator communities, such as the elimination of wolves, may have allowed mountain lions and coyotes to become more prevalent in certain areas. Moreover, anthropogenic features (e.g., roadways, canals) may have isolated some populations, primarily within the arid ecoregions supporting desert bighorn sheep, affecting access to traditionally used seasonal ranges, rates of recolonization, or opportunities for genetic interchange. Northern thinhorn sheep ranges may be vulnerable in the future. Under these conditions, contemporary predators may become a primary factor limiting population growth, and even population persistence, of some mountain sheep populations (Bailey 2000, Logan and Sweanor 2001).

As with other large mammals, sources of mortality for mountain sheep, such as predation, are more likely to be additive and exert population-level effects if adult females increasingly succumb to that source of mortality (Bleich et al. 1997, Hayes et al. 2000, Eberhardt 2002). Moreover, high predation rates for extended periods on lambs and juveniles can influence population trajectories (Barichello et al. 1989, Koizumi and Derocher 2019). Yet, predation on non-reproductive segments of a population, such as lambs and males, is more likely to be compensatory than that on females of reproductive age (Conner et al. 2018, Proffitt et al. 2021). Generally, most predators of mountain sheep do not kill sufficiently large numbers of individuals within a population to have a population-level effect (Sawyer and Lindzey 2002). Coyotes may be a major source of mortality for lambs in some areas (Thorne et al. 1979, Fairaizl 1980, Bleich 1996, Prugh and Arthur 2015). Golden eagles may have substantial effects on lamb and yearling numbers in some years or within some geographic areas. Mountain lions seem to be the one predator of mountain sheep that can effectively prey on all sex and age classes (Logan and Sweanor 2001, Rominger 2018). Mountain lions that have learned to specialize on mountain sheep can cause the extirpation of isolated populations (Festa-Bianchet et al. 2006), and predation by those secretive felids may limit more populations of mountain sheep than any other source of predation.

Predation by mountain lions is most prevalent in areas where bighorn sheep and more abundant ungulates, such as mule deer, are sympatric (Johnson et al. 2013). Further, bighorn sheep do not generally occur at densities sufficient to sustain a population of mountain lions if they are the sole species of large prey (Schaefer et al. 2000). If abundant ungulate populations that serve as primary prey decline, increased effects from mountain lion predation on mountain sheep may become evident (Leopold and Krausman 1986, Ross et al. 1997, Rominger and Weisenberger 2000, Logan and Sweanor 2001). As mountain lion populations expand into thinhorn sheep range, mountain lion predation effects are likely to be similar to that observed within bighorn sheep range.

Predation may influence mountain sheep populations in less direct ways than simply reducing numbers. Stress-induced breeding suppression in female mountain goats resulted from increased predation risk (Dulude-de Broin et al. 2019), and it is possible that mountain sheep could respond similarly. The effort required to avoid mountain lion predation may reduce foraging time and influence forage selection (Blum et al. 2023a,b), either of which may affect individual fitness and the potential for population growth (Cloutier et al. 2024). Years with increased mountain lion predation were associated with noticeable declines in lamb production, and both lambs and females gained less mass when compared to years with little or no mountain lion predation.

Predation seems to be most influential in small, isolated populations in the desert Southwest (Conner et al. 2018, Rominger 2018, Logan and Sweanor 2001, Clark et al. 2021). Large, interconnected, and widely distributed populations of mountain sheep, such as those in the interior Rocky Mountains or Alaska, seem to experience limited population-level effects from predation (Sawyer and Lindzey 2002). Small populations of mountain sheep generally are more susceptible to stochastic events (e.g., disease, harsh weather) that, when combined with predation, can increase the likelihood of local extirpations and enhance the need for management intervention (Johnson et al. 2013, Conner et al. 2018, Clark et al. 2021).

## 9.5 MANAGING PREDATION TO BENEFIT MOUNTAIN SHEEP

Actions to attenuate the influence predation has on prey species can take many forms, and we refer to this strategy as predation management. For instance, limited lethal removal of coyotes to benefit pronghorn populations frequently is employed with measurable benefits to population growth (Smith et al. 1986, Phillips and White 2003). Another largely theoretical option includes the introduction of wolves, which may displace coyotes through competition, intraguild predation, or both, and improve recruitment (Berger et al. 2008, Berger and Conner 2008). Thus, the lethal removal of a primary competitor or the introduction of a competing predator each can be considered forms of predation management, as can management to improve habitat suitability. Evaluating the benefits of any investment in predation management should consider more than the cost of management actions and the revenues generated from hunting licenses and consider less tangible benefits such as wildlife viewing or the cost to restore extirpated populations (Wakeling et al. 2015).

Predation management is generally effective when predation is a limiting factor, and predator removal efforts must be targeted in geographic scope, sustained for 3–5 years, and reduce predator populations by at least 70% (Ballard et al. 2001). For mountain sheep, predation management is most effective in areas where populations are small, limited in distribution, isolated from others, newly introduced populations, or when specialized predator behaviors have been identified. Efforts to reduce predation through lethal removal of wolves were ineffective in an interconnected population of Dall's sheep in Alaska, in part because they were a minor prey item, whereas moose and caribou were influenced to a greater extent by wolf predation (Gasaway et al. 1983). In contrast, removal of mountain lions through hunter harvest in Arizona yielded increases in desert bighorn sheep population indices despite drought conditions (McKinney et al. 2006). Desert bighorn sheep populations are often fewer in number and more likely to be isolated from each other when compared to most other populations of mountain sheep (Krausman and Leopold 1986, Etchberger et al. 1989), although exceptions exist among Rocky Mountain bighorn sheep in the southwestern United States (Rominger et al. 2004) and at the northern extent of their range in British Columbia and Alberta (Wild Sheep Initiative 2023). When mountain sheep populations number less than 200, the likelihood that predation may be a limiting factor increases (Ross et al. 1997). Predation management activities may be more effective frequently in the southwestern United States due to the isolated nature of mountain sheep populations and because mountain lions are common predators. The efficacy of predation management efforts within small populations may be enhanced because efforts can be more focused and targeted (Ballard et al. 2001). Habitat fragmentation, habitat change, and changes to predator and prey communities may increase the likelihood that intervention via predation management will become more widespread in the future (Brewer et al. 2014).

The largest population of desert bighorn sheep that was limited by mountain lions occupied the Kofa National Wildlife Refuge, Arizona. Before 2000, no resident mountain lions had been detected (Germaine et al. 2000). Mountain lions expanded into the area, and between 2000 and 2006, desert bighorn sheep declined from a population of about 800 to 400 (Harris et al. 2009). By 2009, at least 11 mountain lions occurred on the Kofa National Wildlife Refuge (Naidu 2009), and the addition of mountain lion predation alone potentially explained the decline in bighorn sheep numbers (Harris et al. 2009). The lethal removal of almost 70% of the known mountain lions resulted in a substantial increase in the bighorn sheep population (Brewer et al. 2013), and an increase in desert bighorn sheep to more than 1,100 animals in 2024 corresponds with the continued removal of mountain lions from this range (E. Butler, Arizona Game and Fish Department, unpublished data).

Mountain lion predatory behavior differs from other North American carnivores capable of killing bighorn sheep because they generally approach prey to within striking distance using stealth; as a result, the ratio of successful attempts compared with total attempts is substantially higher than for most other predators (Hornocker 1970). Because mountain lions are not coursing predators that take primarily weaker animals, they often take as many deer and elk in good condition as in poor condition (Hornocker 1970), and the same may apply to mountain sheep. Most evidence suggests that mountain lions occupy mountain sheep habitat when mule deer occupy sympatric ranges and are the primary prey (Schaefer et al. 2000, Johnson et al. 2013), but domestic livestock may influence mountain lion populations or distribution (Rominger et al. 2004). Mountain lions seem to prey on mountain sheep less when mule deer are abundant (Logan and Sweanor 2001) and increase predation on alternate prey when mule deer numbers decline (Leopold and Krausman 1986). A decline in mule deer abundance may increase mountain lion predation on mountain sheep and lead to subsequent declines in mountain sheep populations (Rominger and Weisenberger 2000, Holl and Bleich 2010). Reductions of mountain lion numbers are the only predation management action to influence mountain sheep abundance (McKinney et al. 2006).

Habitat management is often discussed as a factor that might influence predation on mountain sheep. Like the removal of mountain lions, little evidence suggests this has been effective in areas other than the southwestern United States. Desert bighorn sheep depend on visibility associated with sparsely vegetated habitat to detect potential predators at greater distances, and fire suppression may hurt habitat suitability to a greater extent than would exist under a natural fire regime (Etchberger et al. 1989). As a result, the use of prescribed fire, or carefully managed wildfires, may be an effective means of reducing predation rates on mountain sheep, particularly in the chaparral-dominated ecosystems of the sky islands of Arizona or the coastal mountain ranges of southwestern California (Bleich and Holl 1982; Holl and Bleich 1983, 2010; Etchberger et al. 1989; Holl et al. 2004, 2012). When habitat enhancement efforts are undertaken, sympatric ungulates may also increase in numbers. Under these conditions, the potential exists for increased interference competition among ungulates (Sittler et al. 2015, 2019; Sittler 2019).

## 9.6 INVESTIGATIONS TO EXPLORE

Much remains to be learned about the demographic consequences of interactions between predators and prey, but the design of future experiments and adaptive management approaches should focus on specific questions that remain unanswered. Is there a specific set of circumstances (e.g., mountain sheep population size, relation to carrying capacity, proportional reduction of predators) at which reducing predator numbers or density will result in an increase in mountain sheep numbers? Is there a set of habitat conditions within which predators do not limit mountain sheep populations, and can managers create those conditions using prescribed fire, silvicultural treatments, or other vegetative management techniques? Are there anthropogenic

influences that change regulating factors into limiting factors for mountain sheep, or are there thresholds within which mountain sheep still may thrive despite the presence of anthropogenic influences? Are there optimum assemblages or densities of multiple predators that would mitigate overall predation rates on mountain sheep? Are there optimum densities or assemblages of large prey species that could mitigate predation rates on mountain sheep or among all prey species? Obtaining answers to each of these questions, and more, will enhance our understanding of the predator-prey relationships of mountain sheep to improve their conservation and management.

## 9.7 SUMMARY

Our understanding of the effects of predators on prey continues to develop. Part of the reason that these relationships are so difficult to understand and explain is that many ecological communities have been highly altered when compared to the original situations in which these relationships developed, but the restoration of fully functioning ecosystems is a substantial, if not impossible, challenge (Bleich 2020, 2021, 2022). Understanding predator-prey relationships is difficult under the best of conditions. Yet when we try to learn how they work after altering the assemblages of prey, predators, and habitats, it is even more difficult. If we were able to restore the conditions that formerly existed, we may learn how predators influence one another (Moll et al. 2021) and how predation influences multiple prey (Sawyer and Lindzey 2002). This is, however, unlikely to occur.

An additional challenge is the complexity of the relationships among the multiple prey and predator species that occupy the diverse landscapes inhabited by mountain sheep. Although many predators kill mountain sheep, mountain lions may be the greatest limiting influence. This is due in part to the differences in how they, as ambush predators, hunt when compared with coursing or opportunistic predators. Within small and isolated populations of desert bighorn sheep, these secretive felids are apt to have the greatest influence on population dynamics.

For thinhorn sheep, abundance of multiple predator species, interactions with each other, and interactions with season (e.g., snow accumulations in winter) have successive and compounding effects. Thinhorn sheep are influenced by wolves and potentially wolverines in winter months when the functionality of escape terrain is reduced by wind-hardened snow accumulations. Predation management thus far has been ineffective at increasing thinhorn sheep numbers. Additionally, avian predators may have a more substantial effect on thinhorn sheep, although bighorn sheep also are subject to predation by golden eagles. Attempts to attenuate the influence of predators on mountain sheep through predation management, including habitat alteration, have been effective only for small, isolated populations of mountain sheep, primarily in xeric areas where limited numbers of alternative prey exists. Predation management should include more than lethal removal of some predators and must consider the abundance of other prey, the presence and interspecific relationships of additional carnivores, and habitat management options.

## REFERENCES

Anderson, E. M., and M. J. Lovallo. 2003. Bobcat and lynx. Pages 758–786 *in* G. A. Feldhamer, B. C. Thompson, and J. A. Chapman, editors. *Wild mammals of North America*. Second edition. Johns Hopkins University Press, Baltimore, Maryland, USA.

Anderson, T. 2008. Rural West going to the dogs: feral and free-roaming canines wreak havoc on wildlife and livestock. High Country News May 26, 2008. https://www.hcn.org/issues/issue-371/rural-west-going-to-the-dogs. Accessed 2 August 2024.

Bailey, J. A. 2000. Management of wild sheep in North America. *Proceedings of the North American Wild Sheep Conference* 2:355–371.

Ballard, W. B., D. Lutz, T. W. Keegan, L. H. Carpenter, and J. C. deVos, Jr. 2001. Deer-predator relationships: a review of recent North American studies with an emphasis on mule and black-tailed deer. *Wildlife Society Bulletin* 29:99–115.

Ballard, W. B., and V. Van Ballenberghe. 1997. Predator-prey relationships. Pages 247–273 *in* A. W. Franzmann and C. C. Schwartz, editors. *Ecology and Management of the North American Moose*. Smithsonian Institution, Washington, D.C., USA.

Barichello, N., J. Carey, R. Sumanik, R. Hayes, and A. Baer. 1989. The effects of wolf predation on Dall sheep populations in the southwest Yukon. Yukon Department of Renewable Resources, Whitehorse, Yukon, Canada.

Bekoff, M. C., and E. M. Gese. 2003. Coyote. Pages 467–481 *in* G. A. Feldhamer, B. C. Thompson, and J. A. Chapman, editors. *Wild mammals of North America*. Second edition. Johns Hopkins University Press, Baltimore, Maryland, USA.

Berger, J. 1978. Group size, foraging, and antipredator ploys: an analysis of bighorn sheep decisions. *Behavioral Ecology and Sociobiology* 4:91–99.

Berger, J. 1991. Pregnancy incentives, predation constraints and habitat shifts: experimental and field evidence for wild bighorn sheep. *Animal Behaviour* 41:61–64.

Berger, K. M., E. M. Gese, and J. Berger. 2008. Indirect effects and trophic cascades: a test involving wolves, coyotes, and pronghorn. *Ecology* 89:818–828.

Berger, K. M., and M. M. Conner. 2008. Recolonizing wolves and mesopredator suppression of coyotes: impacts on pronghorn population dynamics. *Ecological Applications* 18:599–612.

Bleich, V. C. 1996. Interactions between coyotes (*Canis latrans*) and mountain sheep (*Ovis canadensis*). *Southwestern Naturalist* 41:81–82.

Bleich, V. C. 1999. Mountain sheep and coyotes: patterns of predator evasion in a mountain ungulate. *Journal of Mammalogy* 80:283–289.

Bleich, V. C. 2020. Mountain lions, minimum viable populations, and intact ecosystems: a cautionary note. *Ecological Applications* 30:e01990.

Bleich, V. C. 2021. Fire on the mountain—run boys, run! *California Fish and Wildlife Journal* 107:33–43.

Bleich, V. C. 2022. Feral horses, feral asses, and professional politicians: broodings from a beleaguered biologist. *Human-Wildlife Interactions* 16:337–342.

Bleich, V. C., and S. A. Holl. 1982. Management of chaparral habitat for mule deer and mountain sheep in southern California. Pages 247–254 in C. E. Conrad and W. C. Oechel Technical Coordinators. *Proceedings of the symposium on the dynamics and management of Mediterranean-type ecosystems*. USDA Forest Service, General Technical Report PSW-58.

Bleich, V. C., and A. M. Pauli. 1999. Distribution and intensity of hunting and trapping activity in the East Mojave National Scenic Area, California. *California Fish and Game* 85:148–160.

Bleich, V. C., E. F. Cassirer, L. E. Oldenburg, V. L. Coggins, and D. L. Hunter. 2004. Predation by a golden eagle, *Aquila chrysaetos*, on a juvenile mountain sheep, *Ovis canadensis*. *California Fish and Game* 90:91–93.

Bleich, V. C., R. T. Bowyer, and J. D. Wehausen. 1997. Sexual segregation in mountain sheep: resources or predation? *Wildlife Monographs* 134:1–50.

Blum, M. E., K. M. Stewart, K. T. Shoemaker, M. Cox, B. F. Wakeling, T. E. Dilts, J. R. Bennett, and V. C. Bleich. 2023a. Changes in selection of resources with reproductive state in a montane ungulate. *Movement Ecology* 11:20.

Blum, M. E., K. M. Stewart, M. Cox, K. T. Shoemaker, J. R. Bennett, B. W. Sullivan, B. F. Wakeling, and V. C. Bleich. 2023b. Variation in diet of desert bighorn sheep (*Ovis canadensis nelsoni*): tradeoffs associated with parturition. *Frontiers in Ecology and Evolution* 10:1071771.

Bowyer, R. T., D. K. Person, and B. M. Pierce. 2005. Detecting top-down versus bottom-up regulation of ungulates by large carnivores: implications for conservation of biodiversity. Pages 342–361 in J. C. Ray, K. H. Redford, R. S. Steneck, and J. Berger, editors. Large Carnivores and Conservation of Biodiversity. Island Press, Covelo, California, USA.

Bowyer, R. T., K. M. Stewart, V. C. Bleich, J. C. Whiting, K. L. Monteith, M. E. Blum, and T. N. LaSharr. 2020. Metrics of harvest for ungulate populations: misconceptions, lurking variables, and prudent management. *Alces* 56:15–38.

Bowyer, R. T., V. C. Bleich, K. M. Stewart, J. C. Whiting, and K. L. Monteith. 2014. Density dependence in ungulates: a review of causes, and concepts with some clarifications. *California Fish and Game* 100:550–572.

Brewer, C., R. S. Henry, E. J. Goldstein, J. D. Wehausen, and E. M. Rominger. 2013. Strategies for managing mountain lion and desert bighorn sheep interactions. *Desert Bighorn Council Transactions* 52:1–15.

Brewer, C. E., V. C. Bleich, J. A. Foster, T. Hosch-Hebdon, D. E. McWhirter, E. M. Rominger, M. W. Wagner, and B. P. Wiedmann. 2014. *Bighorn Sheep: Conservation Challenges and Management Strategies for the 21st Century*. Western Association of Fish and Wildlife Agencies, Cheyenne, Wyoming, USA.

British Columbia Ministry of Forests, Lands, Natural Resource Operations and Rural Development. 2021. Kootenay Region bighorn sheep management plan. British Columbia Ministry of Forests, Lands, Natural Resource Operations and Rural Development, British Columbia, Cranbrook, Canada.

Brown, R. D., and L. H. Warnum. 2009. *A Brief History of Wildlife Conservation and Research in North America*. Boone and Crockett Club, Missoula, Montana, USA.

Brushett, A., J. Whittington, B. Macbeth, and J. M. Fryxell. 2023. Changes in movement, habitat use, and response to human disturbance accompany parturition events in bighorn sheep (*Ovis canadensis*). *Movement Ecology* 11:36.

Bryant, A. A., and R. E. Page. 2005. Timing and causes of mortality in the endangered Vancouver Island marmot (*Marmota vancouverensis*). *Canadian Journal of Zoology* 83:674–682.

Burles, D. W., and M. Hoefs. 1984. Winter mortality of Dall sheep, *Ovis dalli dalli*, in Kluane National Park, Yukon. *Canadian Field Naturalist* 37:324–326.

California Code of Regulations. 1998. Title 14, Subdivision 2—game, furbearers, nongame, and depredators. California State Legislature, Sacramento, California, USA.

Chauveau, V., M. Garel, C. Toigo, P. Anderwald, M. Apollonil, B. Bassano, M. Beurier, M. Bouche, A. Brambilla, F. Brivio, Y. Bunz, F. Cagnacci, M. Canut, J. Cavailhes, K. Champly, F. Filli, A. Frey-Roos, G. Gressmann, S. Grignolio, I. Herfindal, F. Jurgeit, L. Martinelli, L. Maurino, R. Papet, E. Petit, M. Ramanzin, P. Semenzato, E. Vannard, A. Coulon, A. Loison, and P. Marchand. 2025. It's time to go: drivers and plasticity of migration phenology in a short-distance migratory ungulate. *Journal of Animal Ecology* 94:1204-1220. https://doi.org/10.1111/1365-2656.70031

Child, K. N., K. K. Fujino, and M. W. Warren. 1978. A gray wolf (*Canis lupus columbianus*) and Stone sheep (*Ovis dalli stonei*) fatal predator-prey encounter. *Canadian Field Naturalist* 92:399–401.

Chitwood, M. C., C. Baruzzi, and M. A. Lashley. 2022. "Ecology of fear" in ungulates: opportunities for improving conservation. *Ecology and Evolution* 12:e8657.

Clark, T. J., J. S. Horne, M. Hebblewhite, and A. D. Luis. 2021. Stochastic predation exposes prey to predator pits and local extinction. *Oikos* 130:300–309.

Cloutier, Z., M. Festa-Bianchet, and F. Pelletier. 2024. Direct and indirect effects of cougar predation on bighorn sheep fitness. *Ecology* 105:e4374.

Conner, M. M., T. R. Stephenson, D. W. German, K. L. Monteith, A. P. Few, and E. H. Bair. 2018. Survival analysis: informing recovery of Sierra Nevada big-horn sheep. *Journal of Wildlife Management* 82:1442–1458.

Copeland, J. P., and J. S. Whitman. Wolverine. Pages 672–682 *in* G. A. Feldhamer, B. C. Thompson, and J. A. Chapman, editors. *Wild Mammals of North America*. Second edition. Johns Hopkins University Press, Baltimore, Maryland, USA.

Cornell Lab of Ornithology. 2024. Golden eagle range map. Cornell Lab of Ornithology, Ithaca, New York. https://www.allaboutbirds.org/guide/Golden_Eagle/maps-range. Accessed 5 August 2024.

Courtemanch, A. B. 2014. Seasonal habitat selection and impacts of backcountry recreation on a formerly migratory bighorn sheep population in northwest Wyoming. Thesis, University of Wyoming, Laramie, Wyoming, USA.

Cronemiller, F. P. 1948. Mountain lion preys on bighorn. *Journal of Mammalogy* 29:68.

Cunningham, S. C., C. R. Gustavson, and W. B. Ballard. 1999. Diet selection of mountain lions in southeastern Arizona. *Journal of Range Management* 52:202–207.

Cypher, B. L. 2003. Foxes. Pages 511–546 *in* G. A. Feldhamer, B. C. Thompson, and J. A. Chapman, editors. *Wild Mammals of North America*. Second edition. Johns Hopkins University Press, Baltimore, Maryland, USA.

Demarchi, R. A., and C. L. Hartwig. 2004. Status of thinhorn sheep in British Columbia. Wildlife Bulletin B-119. *Ministry of Water, Land and Air Protection*, Victoria, British Columbia, Canada.

Dulude-de Broin, F., S. Hamel, G. F. Mastromonaco, and S. D. Côté. 2019. Predation risk and mountain goat reproduction: evidence for stress-induced breeding suppression in a wild ungulate. *Functional Ecology* 34: 1003–1014.

Eberhardt, L. L. 2002. A paradigm for population analysis of long-lived vertebrates. *Ecology* 83:2841–2854.

Elliott, H. N. 1961. Bobcats and bighorn sheep. *Desert Bighorn Council Transactions* 5:38–41.

Enns, G. E., B. A. Jex, and M. S. Boyce. 2023. Diverse migration patterns and seasonal habitat use of Stone's sheep (*Ovis dalli stonei*). *Peer Journal* 11:e15215.

Enns, G. E., B. A. Jex, and M. S. Boyce. 2024. Stone's sheep (*Ovis dalli stonei*) lambing and nursery habitat selection. *Canadian Journal of Zoology* 102:691–707.

Environment Yukon. 2017. Science-based guidelines for management of thinhorn sheep in Yukon. Yukon Fish and Wildlife Branch Report MR-16-03. Environment Yukon, Whitehorse, Yukon, Canada.

Etchberger, R. C., P. R. Krausman, and R. Mazaika. 1989. Mountain sheep habitat characteristics in the Pusch Ridge Wilderness, Arizona. *Journal of Wildlife Management* 53:902–907.

Fairaizl, S. D. 1980. Population characteristics of transplanted California bighorn sheep in western North Dakota. Northern Wild Sheep and Goat Council Proceedings 2:70–87.

Feist, J. J. 1997. Bighorn sheep (*Ovis canadensis*) ecology and demography in the North Dakota badlands. Thesis, University of North Dakota, Grand Forks, USA.

Festa-Bianchet, M., T. Coulson, J.-M. Gaillard, J. T. Hogg, and F. Pelletier. 2006. Stochastic predation events and population persistence in bighorn sheep. *Proceedings of the Royal Society B* 273:1537–1543.

Freeman, S. D. and N. Maclean. 2021. Habitat use and movement of the Dome Mountain Stone's sheep. Tahltan Guide Outfitters Association, Dease Lake, British Columbia, Canada.

Frid, A., and L. M. Dill. 2002. Human-caused disturbance stimuli as a form of predation risk. *Conservation Ecology* 6:11.

Gammons, D. J., J. L. Davis, D. W. German, K. Denryter, J. D. Wehausen, and T. R. Stephenson. 2021. Predation impedes recovery of Sierra Nevada bighorn sheep. *California Fish and Wildlife Special CESA Issue*:444–470.

Gasaway, W. C., R. O. Stephenson, J. L. Davis, P. E. K. Sheperd, and O. E. Burris. 1983. Interrelationships of wolves, prey, and man in interior Alaska. *Wildlife Monographs* 84:1–50.

Geist, V. 1971. Mountain sheep: a study in behavior and evolution. University of Chicago Press, Chicago, Illinois, USA.

Germaine, S. S., K. D. Bristow, and L. A. Haynes. 2000. Distribution and population status of mountain lion in southwestern Arizona. *Southwestern Naturalist* 45:333–338.

Gibson, M. 2008. Wolf attack on big horn sheep - Kananaskis. https://www.youtube.com/watch?v=Gjo94EOgoC4. Accessed 1 August 2024.

Goodson, N. J., and D. R. Stevens. 1994. Cooperative defense by female bighorn sheep. *Northwestern Science* 75:76–77.

Hall, C. A. S. 1988. An assessment of several of the historically most influential theoretical models used in ecology and of the data provided in their support. *Ecological Modelling* 43:5–31.

Harris, G., L. Smythe, and R. Henry. 2009. Predation by mountain lions is capable of causing desert bighorn sheep population decline at Kofa National Wildlife Refuge, Arizona. *Desert Bighorn Council Transactions* 50:40–53.

Harrison, S. 1990. Cougar predation on bighorn sheep in the Junction Wildlife Management Area, British Columbia. Dissertation, University of British Columbia, Vancouver, BC, Canada.

Hasselquist, D., and J.-A. Nilsson. 2012. Physiological mechanisms mediating costs of immune responses: what can we learn from studies of birds? *Animal Behaviour* 83:1303–1312.

Hayes, C. L., E. S. Rubin, M. C. Jorgensen, and W. M. Boyce. 2000. Mountain lion predation of bighorn sheep in the Peninsular Ranges, California. *Journal of Wildlife Management* 64:954–959.

Heimer, W. E., and A. C. Smith. 1972. Appendix V. Pages 1–3 *in* L. Nichols and W. Heimer, editors. *Sheep Report. Federal Aid in Wildlife Restoration Annual Project Report 13*. Alaska Department of Fish and Game, Juneau, Alaska, USA.

Heimer, W. E., and R. O. Stephenson. 1982. Responses to Dall sheep populations of wolf control in interior Alaska. *Northern Wild Sheep and Goat Council Proceedings* 3:320–329.

Hoefs, M., and I. M. Cowan. 1979. Ecological investigation of a population of Dall sheep (*Ovis dalli dalli* Nelson). *Syesis* 12(supplement 1): 1–81.

Hoefs, M., H. Hoefs, and D. Burles. 1986. Observations on Dall sheep, *Ovis dalli dalli*, and grey wolf, *Canis lupus pambasilens*, in Kluane Lake area, Yukon. *Canadian Field-Naturalist* 100:78–84.

Holl, S. A. 2004. Implementation strategy to restore the San Gabriel Mountains bighorn sheep population. Los Angeles County Fish and Game Commission, U.S. Forest Service, and California Department of Fish and Game. Los Angeles Fish and Game Commission, Los Angeles, California, USA.

Holl, S. A., and V. C. Bleich. 2009. Reconstructing the San Gabriel Mountains bighorn sheep population. *California Fish and Game* 95:77–87.

Holl, S. A., and V. C. Bleich. 1983. San Gabriel mountain sheep: biological and management considerations. USDA Forest Service, San Bernardino National Forest, San Bernardino, California, USA.

Holl, S. A., and V. C. Bleich. 2010. Responses of large mammals to fire and rain in the San Gabriel Mountains, California. *Northern Wild Sheep and Goat Council Proceedings* 17:139–156.

Holl, S. A., V. C. Bleich, B. W. Callenberger, and B. Bahro. 2012. Simulated effects of two fire regimes on bighorn sheep: the San Gabriel Mountains, California, USA. *Fire Ecology* 8(3):88–103.

Holl, S. A., V. C. Bleich, and S. G. Torres. 2004. Population dynamics of bighorn sheep in the San Gabriel Mountains, California, 1967–2002. *Wildlife Society Bulletin* 32:412–426.

Holt, B. S., W. H. Miller, and B. F. Wakeling. 1992. Composition and quality of mountain sheep diets in the Superstition Mountains, Arizona. *Desert Bighorn Council Transactions* 36:36–40.

Hornocker, M. G. 1969. Defensive behavior in female bighorn sheep. *Journal of Mammalogy* 50:128.

Hornocker, M. G. 1970. An analysis of mountain lion predation upon mule deer and elk in the Idaho Primitive Area. *Wildlife Monographs* 21:3–39.

Huggler, K. S. 2024. Feast or famine: interactive effects of life-history, nutrition, and behavior on performance of a long-lived herbivore. Dissertation, University of Wyoming, Laramie, Wyoming, USA.

Hunter, M. D., and P. W. Price. 1992. Playing chutes and ladders: heterogeneity and the relative roles of bottom-up and top-down forces in natural communities. *Ecology* 73:724–732.

Jenkins, E. J., E. P. Hoberg, and L. Polley. 2005. Development and pathogenesis of *Parelaphostrongylus odocoilei* (Nematoda: Protostrongylidae) in experimentally infected thinhorn sheep (*Ovis dalli*). *Journal of Wildlife Diseases* 41:669–682.

Jensen, W. F., V. C. Bleich, and D. G. Whittaker. 2023. Historical trends in black-tailed deer, mule deer, and their habitats. Pages 25–42 *in* J. R. Heffelfinger and P. R. Krausman, editors. *Ecology and Management of Black-Tailed and Mule Deer of North America*. CRC Press, Boca Raton, Florida, USA.

Jex, B. A., K. Chan, M. Larivee, B. R. Wendling, and D. W. Lutz. 2025. Observed changes in populations of thinhorn sheep in western Canada and Alaska. Wild Sheep Initiative, Western Association of Fish and Wildlife Agencies, Boise, Idaho, USA.

Jex, B. A., J. B. Ayotte, V. C. Bleich, C. E. Brewer, D. L. Bruning, T. M. Hegel, N. C. Larter, R. A. Schwanke, H. M. Schwantje, and M. W. Wagner. 2016. *Thinhorn sheep: conservation challenges and management strategies for the 21st Century*. Wild Sheep Working Group, Western Association of Fish and Wildlife Agencies, Boise, Idaho, USA.

Johnson, H. E., M. Hebblewhite, T. R. Stephenson, D. W. German, B. M. Pierce, and V. C. Bleich. 2013. Evaluating apparent competition in limiting the recovery of an endangered ungulate. *Oecologia* 171:295–307.

Joly, D. O., and F. Messier. 2004. The distribution of *Echinococcus granulosus* in moose: evidence for parasite-induced vulnerability to predation by wolves? *Oecologia* 140:586–590.

Jorgensen, M. C., and R. E. Turner. 1975. Desert bighorn of the Anza-Borrego Desert State Park. *Desert Bighorn Council Transactions* 19:51–53.

Kelly, W. 1980. Predator relationships. Pages 186–196 *in* G. Monson and L. Sumner, editors. *The Desert Bighorn:* Its Life History, Ecology, and Management. University of Arizona Press, Tucson, Arizona, USA.

Kennedy, C. A. 1948. Golden eagle kills bighorn lamb. *Journal of Mammalogy* 29:68–69.

Koizumi, L. C., and A. E. Derocher. 2019. Predation risk and space use of a declining Dall sheep (*Ovis dalli dalli*) population. *PLoS ONE* 14(4):e0215519.

Krausman, P. R., and B. D. Leopold. 1986. Habitat components for desert bighorn sheep in the Harquahala Mountains, Arizona. *Journal of Wildlife Management* 50:504–508

Leopold, B. D., and P. R. Krausman. 1986. Diets of 3 predators in Big Bend National Park, Texas. *Journal of Wildlife Management* 50:290–295.

Lochmiller, R. L., and C. Deerenberg. 2000. Trade-offs in evolutionary immunology: just what is the cost of immunity? *Oikos* 88:87–98.

Logan, K., and L. Sweanor. 2001. *Desert Puma: Evolutionary Ecology and Conservation of an Enduring Carnivore*. Island Press, Washington, D.C., USA.

Lowrey, C., and K. M. Longshore. 2017. Tolerance to disturbance regulated by attractiveness of resources: a case study of desert bighorn sheep within the River Mountains, Nevada, USA. *Western North American Naturalist* 77:82–98.

McKinney, T., T. W. Smith, and J. C. deVos, Jr. 2006. Evaluation of factors potentially influencing a desert bighorn sheep population. *Wildlife Monographs* 164:1–36.

McLaren, B. E., and R. O. Peterson. 1994. Wolves, moose, and tree rings on Isle Royale. *Science* 266:1555–1558.

McQuivey, R. P. 1975. Bighorn research in Nevada. *Desert Bighorn Council Transactions* 19:15–18.

Messier, F. 1991. The significance of limiting and regulating factors on the demography of moose and white-tailed deer. *Journal of Animal Ecology* 60:377–393.

Miller, D. S., E. P. Hoberg, G. C. Weiser, K. Aune, M. Atkinson, and C. Kimberling. 2012. A review of hypothesized determinants associated with bighorn sheep (*Ovis canadensis*) die-offs. *Veterinary Medicine International* 4:796527.

Mills, L. S. 2007. *Conservation of Wildlife Populations: Demography, Genetics and Management.* Blackwell Publishers, Malden, Massachusetts, USA.

Mitchell, C. D., R. Chaney, K. Aho, J. G. Kie, and R. T. Bowyer. 2015. Population density of Dall's sheep in Alaska: effects of predator harvest? *Mammal Research* 60:21–28.

Moll, R. J., P. J. Jackson, B. F. Wakeling, C. W. Lackey, J. P. Beckmann, J. J. Millspaugh, and R. A. Montgomery. 2021. An apex carnivore's life history mediates a predator cascade. *Oecologia* 196:223–234.

Monteith, K. L., L. E. Schmitz, J. A. Jenks, J. A. Delger, and R. T. Bowyer. 2009. Growth of male white-tailed deer: consequences of maternal effects. *Journal of Mammalogy* 90:651–660.

Monteith, K. L., V. C. Bleich, T. R. Stephenson, B. M. Pierce, M. M. Conner, J. G. Kie, and R. T. Bowyer. 2014. Life-history characteristics of mule deer: effects of nutrition in a variable environment. *Wildlife Monographs* 186:1–56.

Naidu, A. 2009. Genetic analysis of mountain lion (*Puma concolor*) feces from Kofa National Wildlife Refuge, Arizona. MS thesis, University of Arizona, Tucson, Arizona, USA.

National Park Service. 2002. *Mojave National Preserve General Management Plan.* Mojave National Preserve, Barstow, California, USA.

Nava, M., L. Corlatti, N. Formenti, T. Trogu, L. Pedrotti, A. Gugiatti, P. Lanfranchi, C. Luzzago, and N. Ferrari. 2023. Parasite-mediated manipulation? *Toxoplasma gondii* infection increases risk behaviour towards culling in red deer. *Biology Letters* 19:20230292.

Nelson, M. P., J. A. Vucetich, R. O. Peterson, and L. M. Vucetich. 2011. The Isle Royale wolf–moose project (1958-present) and the wonder of long-term ecological research. *Endeavour* 35:31–39.

Nette, T., D. Burles, and M. Hoefs. 1984. Observations of golden eagle (*Aquila chrysaetos*) predation on Dall sheep (*Ovis dalli dalli*) lambs. *Canadian Field-Naturalist* 98:252–254.

Nichol, A. A. 1937. Desert bighorn sheep study—1937. Arizona Game and Fish Department Files, Phoenix, Arizona, USA.

Nichol, A. A. 1940. The desert bighorn sheep in Arizona. Arizona Game and Fish Department Files, Phoenix, Arizona, USA.

Nichols, L. 1975. Report from Alaska. Pages 8–13 *in* J. B. Trefethan, editor. *The Wild Sheep in Modern North America.* The Boone and Crockett Club and Winchester Press, New York, New York, USA.

Nichols, L., and F. L. Bunnell. 1999. Natural history of thinhorn sheep. Pages 23–77 *in* R. Valdez and P. R. Krausman, editors. *Mountain Sheep of North America.* University of Arizona Press, Tucson, Arizona, USA.

Ober, E. H. 1931. The mountain sheep in California. *California Fish and Game* 17:27–39.

Papouchis, C. M., F. J. Singer, and W. B. Sloan. 2001. Responses of desert bighorn sheep to increased human recreation. *Journal Wildlife Management* 65:573–582.

Paquet, P. C., and L. N. Carbyn. 2003. Gray wolf. Pages 482–510 *in* G. A. Feldhamer, B. C. Thompson, and J. A. Chapman, editors. *Wild Mammals of North America.* Second edition. Johns Hopkins University Press, Baltimore, Maryland, USA.

Parker, K. L., P. S. Barboza, and M. P. Gillingham. 2009. Nutrition integrates environmental responses of ungulates. *Functional Ecology* 23:57–69.

Parr, B. L., J. Kanta, J. Sandrini, D. J. Thompson, and J. A. Jenks. 2014. Bobcat predation on a bighorn lamb in the western Black Hills of South Dakota. *Prairie Naturalist* 46:41–43.

Pelton, M. R. 2003. Black bear. Pages 547–555 *in* G. A. Feldhamer, B. C. Thompson, and J. A. Chapman, editors. *Wild Mammals of North America.* Second edition. Johns Hopkins University Press, Baltimore, Maryland, USA.

Phillips, G. E., and G. C. White. 2003. Pronghorn population response to coyote control: modeling and management. *Wildlife Society Bulletin* 31:1162–1175.

Pierce, B. M., and V. C. Bleich. 2003. Mountain lion. Pages 744–757 *in* G. A. Feldhamer, B. C. Thompson, and J. A. Chapman, editors. *Wild Mammals of North America.* Second edition. Johns Hopkins University Press, Baltimore, Maryland, USA.

Pierce, B. M., V. C. Bleich, K. L. Monteith, and R. T. Bowyer. 2012. Top-down versus bottom-up forcing: evidence from mountain lions and mule deer. *Journal of Mammalogy* 93:977–988.

Power, M. E. 1992. Top-down and bottom-up forces in food webs: do plants have primacy? *Ecology* 73:733–746.

Proffitt, K. M., A. B. Courtemanch, S. R. Dewey, B. Lowrey, D. E. McWhirter, K. L. Monteith, J. T. Paterson, J. Rotella, P. J. White, and R. A. Garrott. 2021. Regional variability in pregnancy and survival rates of Rocky Mountain bighorn sheep. *Ecosphere* 12:e03410.

Prugh, L. 2004. Foraging ecology of coyotes in the Alaska Range. Dissertation, University of British Columbia, Vancouver, Canada.

Prugh, L., and S. Arthur. 2015. Optimal predator management for mountain sheep conservation depends on the strength of mesopredator release. *Oikos* 124:1241–1250.

Pybus, M. J., S. Groom, and W. M. Samuel. 1996. Meningeal worm in experimentally-infected bighorn and domestic sheep. *Journal of Wildlife Diseases* 32:614–618.

Rachlow, J. L., and R. T. Bowyer. 1998. Habitat selection by Dall's sheep (*Ovis dalli*): maternal trade-offs. *Journal of Zoology* 245:457–465.

Ripple, W. J., C. Wolf, R. L. Beschta, A. D. Craig, Z. S. Curcija, E. J. Lundgren, L. C. Satterfield, S. T. Woodrich, and A. J. Wirsing. 2024. A shifting ecological baseline after wolf extirpation. *BioScience:* biae034. https://doi.org/10.1093/biosci/biae034

Rodriguez Curras, M., M. C. Romanski, and J. N. Pauli. 2024. The pulsed effects of reintroducing wolves on the carnivore community of Isle Royale. *Frontiers in Ecology and the Environment* 22:e2750. https://doi.org/10.1002/fee.2750

Roemer, G. W., C. J. Donlan, and F. Courchamp. 2002. Golden eagles, feral pigs, and insular carnivores: how exotic species turn native predators into prey. *Proceedings of the National Academy of Sciences of the USA* 99:791–796.

Rominger, E. M. 2018. The Gordian Knot of mountain lion predation and bighorn sheep. *Journal of Wildlife Management* 82:19–31.

Rominger, E. M., H. A. Whitlaw, D. L. Weybright, W. C. Dunn, and W. B. Ballard. 2004. The influence of mountain lion predation on bighorn sheep translocations. *Journal of Wildlife Management* 68:993–999.

Rominger, E. M., and M. E. Weisenberger. 2000. Biological extinction and a test of the "conspicuous individual hypothesis" in the San Andres Mountains, New Mexico. *Transactions of the North American Wild Sheep Conference* 2:293–307.

Ross, P. I., M. G. Jalkotzy, and M. Festa-Bianchet. 1997. Cougar predation on bighorn sheep in southwestern Alberta during winter. *Canadian Journal of Zoology* 74:771–775.

Sawyer, H., and F. Lindzey. 2002. A review of predation on bighorn sheep (*Ovis canadensis*). Wyoming Cooperative Fish and Wildlife Research Unit, Laramie, Wyoming, USA.

Schaefer, R. J., S. G. Torres, and V. C. Bleich. 2000. Survivorship and cause-specific mortality in sympatric populations of mountain sheep and mule deer. *California Fish and Game* 86:127–135.

Schwartz, C. C., S. D. Miller, and M. A. Haroldson. 2003. Grizzly bear. Pages 556–586 *in* G. A. Feldhamer, B. C. Thompson, and J. A. Chapman, editors. *Wild Mammals of North America*. Second edition. Johns Hopkins University Press, Baltimore, Maryland, USA.

Scotton, B. D. 1998. Timing and causes of neonatal Dall sheep mortality in the central Alaska Range. Thesis, University of Montana, Missoula, Montana, USA.

Seersholm, F. V., D. J. Werndly, A. Grealy, T. Johnson, E. M. K. Early, E. L. Lundelius, Jr., B. Winsborough, G. E. Farr, R. Toomey, A. J. Hansen, B. Shapiro, M. R. Waters, G. McDonald, A. Linderholm, T. W. Stafford Jr., and M. Bunce. 2020. Rapid range shifts and megafaunal extinctions associated with late Pleistocene climate change. *Nature Communications*. https://doi.org/10.1038/s41467-020-16502-3

Shank, C. C. 1977. Cooperative defense by bighorn sheep. *Journal of Mammalogy* 58:243–244.

Shaw, H. G. 1993. Only prey. Pages 221–241 *in* G. B. Nabhan, editor. *Counting Sheep: 20 Ways of Seeing Desert Bighorn*. University of Arizona Press, Tucson, Arizona, USA.

Sheldon, C. 1930. *The Wilderness of Denali*. Charles Scribner's Sons, New York, New York, USA.

Simmons, N. M. 1964. A desert bighorn sheep study: part two. *Desert Bighorn Council Transactions* 8:103–112.

Sittler, K. L. 2019. Stone's sheep and elk forage characteristics – 7-years post fire. Wildlife Infometrics Report No. 663. Wildlife Infometrics Inc., Mackenzie, British Columbia, Canada.

Sittler, K. L., K. L. Parker, and M. P. Gillingham. 2015. Resource separation by mountain ungulates on a landscape modified by fire. *Journal of Wildlife Management* 79:591–604.

Sittler, K. L., K. L. Parker, and M. P. Gillingham. 2019. Vegetation and prescribed fire: implications for Stone's sheep and elk. *Journal of Wildlife Management* 83:393–409.

Smiley, R. A., B. L. Wagler, W. H. Edwards, J. E. Jennings-Gaines, K. Luukkonen, K. Robbins, M. Johnson, A. B. Courtemanch, T. W. Mong, D. Lutz, D. McWhirter, J. L. Malmberg, B. Lowrey, and K. L. Monteith. 2024. Infection–nutrition feedbacks: fat supports pathogen clearance but pathogens reduce fat in a wild mammal. *Proceedings of the Royal Society B* 291:20240636.

Smith, J. F. 2016. *Engineering Eden*. Crown Publishing, New York, New York, USA.

Smith, R. H., D. J. Neff, and N. G. Woolsey. 1986. Pronghorn response to coyote control: a benefit:cost analysis. *Wildlife Society Bulletin* 14:226–231.

Smith, R. L. 1966. *Ecology and Field Biology*. Harper & Row, New York, New York, USA.

Smith, R. L. 1980. *Ecology and Field Biology*. Third edition. Harper & Row, New York, New York, USA.

Spann, M. C. 1937. The effect of the white man on the bird life of America. *Bios* 8:210–219.

Sproat, K. K., N. R. Martinez, T. S. Smith, W. B. Sloan, J. T. Flinders, J. W. Bates, J. G. Cresto, and V. C. Bleich. 2019. Desert bighorn sheep responses to human activity in south-eastern Utah. *Wildlife Research* 47:16–24.

Stearns, S. C. 1989. Trade-offs in life-history evolution. *Functional Ecology* 3:259–268.

Stephenson, R. O., D. V. Grangaard, and J. Burch. 1991. Lynx, *Felis lynx*, predation on red foxes, *Vulpes vulpes*, caribou, *Rangifer tarandus*, and Dall sheep, *Ovis dalli*, in Alaska. *Canadian Field-Naturalist* 2:255–262.

Stephenson, T. R., D. W. German, E. F. Cassirer, D. P. Walsh, M. E. Blum, M. Cox, K. M. Stewart, and K. L. Monteith. 2020. Linking population performance to nutritional condition in an alpine ungulate. *Journal of Mammalogy* 101:1244–1256.

Thorne, E. T., G. Butler, T. Varcalli, K. Becker, and S. Hayden-Wing. 1979. The status, mortality, and response to management of the bighorn sheep of Whiskey Mountain. Wildlife Technical Report 7. Wyoming Game and Fish Department, Cheyenne, Wyoming, USA.

Valdez, R., and P. R. Krausman. 1999. Description, distribution, and abundance of mountain sheep in North America. Pages 3–22 *in* R. Valdez and P. R. Krausman, editors. *Mountain Sheep of North America*. University of Arizona Press, Tucson, Arizona, USA.

Villalobos, J. A. D., S. G. Isern, M. A. G. Aranda, and R. V. Rincon. 2019. Black bear in Mexico. Pages 206–221 *in* R. Valdez and J. A. Ortega-S., editors. *Wildlife Ecology and Management in Mexico*. Texas A&M University Press, College Station, Texas, USA.

Villepique, J. T., B. M. Pierce, V. C. Bleich, A. Andic, and R. T. Bowyer. 2015. Resource selection by an endangered ungulate: a test of predator-induced range abandonment. *Advances in Ecology* 2015:357080. https://doi.org/10.1155/2015/357080

Wagler, B. L., R. A. Smiley, A. B. Courtemanch, D. Lutz, D. McWhirter, D. Brimeyer, P. Hnilicka, T. J. Robinson, and K. L. Monteith. 2023. Implications of forage quality for population recovery of bighorn sheep following a pneumonia epizootic. *Journal of Wildlife Management* 87:e22452.

Wakeling, B. F., H. S. Najar, and J. C. O'Dell. 2007. Mortality of bighorn sheep along U. S. Highway 191 in Arizona. *Desert Bighorn Council Transactions* 49:18–22.

Wakeling, B. F., R. L. Day, and A. A. Munig. 2015. The efficacy and economics of limited lethal removal of coyotes to benefit pronghorn in Arizona. *Proceedings of the Biennial Pronghorn Workshop* 26:66–87.

Wakeling, B. F., R. Lee, D. Brown, R. Thompson, M. Tluczek, and M. Weisenberger. 2009. The restoration of desert bighorn sheep in the Southwest, 1951–2007: factors influencing success. *Desert Bighorn Council Transactions* 50:1–17.

Weaver, R. A., and J. L. Mensch. 1970. Observed interaction between bighorn sheep and reported predator species. *California Fish and Game* 56:206–207.

Wehausen, J. D. 1996. Effects of mountain lion predation on bighorn sheep in the Sierra Nevada and Granite Mountains of California. *Wildlife Society Bulletin* 24:471–479.

Weinberg, S. 1993. *Dreams of a Final Theory: The Scientist's Search for the Ultimate Laws of Nature*. Random House, Inc., New York, New York, USA.

Wiedmann, B. P., and V. C. Bleich. 2014. Demographic responses of bighorn sheep to recreational activities: a trial of a trail. *Wildlife Society Bulletin* 38:773–782.

Wild Sheep Initiative. 2023. Range-wide status report of bighorn and thinhorn sheep in North America. Wild Sheep Initiative, Western Association of Fish and Wildlife Agencies, Boise, Idaho, USA.

Williams, J. S., J. J. McCarthy, and H. D. Picton. 1995. Cougar habitat use and food habits on the Montana Rocky Mountain Front. *Intermountain Journal of Sciences* 1:16–28.

Woolf, A., and T. O'Shea. 1968. Two bighorn sheep coyote encounters. *Journal of Mammalogy* 49:770.

Young, J. K., K. A. Olson, R. P. Reading, S. Amgalanbaatar, and J. Berger. 2011. Is wildlife going to the dogs? Impacts of feral and free-roaming dogs on wildlife populations. *BioScience* 61:125–132.

# 10 Competition among Mountain Sheep and Other Ungulates

*Jericho C. Whiting, Vernon C. Bleich, and Tom Smith*

## 10.1 INTRODUCTION

Competition for food, water, cover, and space with native, feral, and domestic ungulates plays an important role in habitat use and the population persistence of species of North American mountain sheep (Buechner 1960, Douglas and Leslie 1999). Usually, competition among these ungulates is related inversely to forage, water, and space use, and competition can increase if availability of these resources is reduced by common use. Competition among bighorn sheep (*Ovis canadensis*) and native, feral, or domestic ungulates was recognized early as a potential detriment to bighorn sheep, especially during times of deep snow or drought (Mills 1937, Davis and Taylor 1939, Jones 1955, Villa 1959, Weaver 1959). Feral equids can have serious effects on the availability of forage, water, and cover accessible to Rocky Mountain (*O. c. canadensis*) and, especially, desert bighorn sheep (*O. c. nelsoni, O. c. mexicana*; McKnight 1958, Seegmiller and Simpson 1979, Seegmiller and Ohmart 1981). Similar concerns over competition with native, feral, or domestic ungulates have been reported for Dall's (*O. dalli*) and Stone's sheep (*O. d. stonei*; Hoefs and Brink 1978, Bleich 2009, Jung et al. 2015a,b, Jex et al. 2016).

Interspecific relationships, including interspecific competition, are dynamic processes and cannot be adequately investigated in a short period of time or at a single location (Akenson 1992); in part, this is because habitat selection and behavior of potential competitors in areas of sympatry may involve a combination of interspecific interactions and environmental factors. Those differences, even if due to pre-existing preferences in use of habitat, could also reflect the outcome of historical competition or be a consequence of current competition (Schoener 1974, Hansson 1995, Marshal et al. 2012).

Now and in the future, climate change and anthropogenic pressures will continue to modify niches and, thus, competitive interactions among species (Sax et al. 2013, Brewer et al. 2014, Gaynor et al. 2018, Berger et al. 2020). Proper conservation of mountain sheep populations and their habitats will require biologists to understand the historical competition between mountain sheep and other ungulates and to recognize and quantify how climate change, contemporary ecological conditions, and anthropogenic pressures will alter competitive interactions in the future. Herein, we discuss the potential for competition for food, water, cover, and space among mountain sheep and native, feral, and domestic ungulates, and the ways in which competition can affect mountain sheep habitat and populations.

## 10.2 OVERVIEW OF CONCEPTS

### 10.2.1 THE ROLE OF NICHE IN COMPETITION AND GRADIENTS OF COMPETITIVE INTENSITY

A species niche refers to the total adaptation of a particular species to the environment it occupies (Grinnell 1917, Feldhamer et al. 2020). The fundamental niche is the full range of resources and environmental conditions that a species can use or occupy indefinitely without the influence of constraining factors (e.g., predation, competition), whereas the realized niche is a smaller range of resources or environmental conditions that a species can use or occupy due to the presence of constraining factors (Hutchinson 1957, Morrison 2013, Feldhamer et al. 2020). Niches occupied by animals are multidimensional and complex (Hutchinson 1957, Sax et al. 2013, Feldhamer et al. 2020), and niche axes can vary across diel activity, seasons, and sex (Hutchinson 1957, Bleich et al. 1997, Stewart et al. 2002, Bowyer 2004, Bowyer 2022, Cox et al. 2023). As species interact in the environment, competition may ensue if they share limited resources (Birch 1957). Interactions among species in the same environment can affect the relative abundance of populations, influence population persistence, and affect the phenotypes of animals, thereby contributing to natural selection among coexisting taxa (Birch 1957, Seegmiller and Simpson 1979, Feldhamer et al. 2020).

The competitive exclusion principle is important for understanding niche relationships and various forms of competition (Feldhamer et al. 2020). This principle asserts that two species cannot coexist on the same limited resource indefinitely; the species with the competitive advantage will displace the other species (Birch 1957, Hardin 1960). If the realized niche of a species becomes small enough, local extirpation can occur for that species (Feldhamer et al. 2020). In nature, however, niches can overlap without competitive exclusion occurring (Birch 1957, Cole 1960). Moreover, some animals may be most competitive at the center of their niche space and gradually less so near the periphery of their niche space (Hutchinson 1957, Sax et al. 2013, Bowyer et al. 2023).

## 10.2.2   THE ROLE OF NUMBERS IN RESOURCE COMPETITION

The number of individuals involved in competition plays an important role in determining the intensity of competition between or among species. Higher population densities can lead to increased competition, as individuals are more likely to encounter one another and vie for the same resources (Stewart et al. 2002). Additionally, one species may reduce the abundance of a second species in a particular habitat by directly depleting resources, interfering with the second species' ability to obtain those resources, or by causing the second species to expend energy obtained from those resources in aggressive encounters (Schoener 1974). Additionally, resource availability decreases as the number of competing species increases. Resource scarcity, in turn, increases competition as individuals compete more intensely for the same resources (Festa-Bianchet et al. 2019, Berger et al. 2022).

Resource abundance is affected when competition occurs; behavioral changes may also result, manifesting as increased aggression or territoriality, which can lead to increased energy expenditure and a higher risk of injury (Berger et al. 2022). Another potential response by one species to competition with another is to abandon areas that are used (Ferretti and Mori 2020). For example, as mule deer (*Odocoileus hemionus*) and pronghorn (*Antilocapra americana*) interacted with bighorn sheep, those bovids were displaced, possibly settling for poorer grazing and security (Ferretti and Mori 2020). Finally, as the number of competitors increases, intense competition can lead to evolutionary changes, such as the development of adaptations that enhance competitive ability. Such changes can lead to niche differentiation, where competing species evolve to exploit different resources or habitats, thereby reducing direct competition (Finke and Snyder 2008).

## 10.2.3   THE ROLE OF SEASONALITY AND RESOURCE AVAILABILITY

Seasonality and resource availability play important roles in shaping interactions and competition among species, including mountain sheep and native, feral, and domestic ungulates. Seasonality influences resource fluctuations in the abundance of forage and water, and behavioral adaptations, such as migration. The responses to these seasonal fluctuations vary; in some cases, when plant growth is abundant and provides ample forage, there is an increase in diet overlap between species (Marshal et al. 2012). In other situations, when forage availability is reduced on a seasonal basis, decreases in diet overlap among herbivores have been reported (Schwartz and Ellis 1981, Baldi et al. 2004, Liu and Jiang 2004), indicating that as forage becomes scarce, large herbivores focus more on the vegetation each is best adapted to use (Rosenzweig 1981, Schoener 1982).

Whatever the situation, forage is most apt to be abundant during the growing season, usually spring and summer for mountain ungulates, and the potential for resource competition is lessened when compared with other seasons characterized by lesser and poorer forage and water resources (Murray and Illius 2000). In contrast, winter typically brings a scarcity of forage at high elevations, forcing large herbivores to rely on stored fat reserves or migrate to areas with better resources (Shimadzu et al. 2013). Many populations of mountain sheep migrate seasonally (Geist 1971, DeCesare and Pletscher 2006, Jesmer et al. 2018, Enns et al. 2023), and if they are translocated to unfamiliar areas the loss of knowledge of seasonal ranges can threaten persistence of the population (Stevens and Goodson 1993, Robinson et al. 2019).

Seasonal changes also lead to behavioral adaptations in mountain sheep and their competitors, all of which exhibit behaviors that have evolved to cope with these changes. For instance, during summer, mountain sheep often migrate to higher elevations to access newly emerging vegetation (Spitz et al. 2018). But in winter, heavy snow often forces mountain sheep and other ungulates to lower elevations where snow cover is less severe (Shimadzu et al. 2013).

## 10.2.4   RESOURCE PARTITIONING

Resource partitioning refers simply to differences in resource use between species, regardless of the origins of the differences (Walter 1991). As such, resource partitioning plays an important role in allowing species with similar niches, such as mountain sheep, elk (*Cervus canadensis*), and mountain goats (*Oreamnos americanus*) to access important habitats and preferred forage while minimizing conflict. Resource partitioning does not wholly eliminate interspecies competition and conflict but lessens its effect as competing species use the same resources at different times, or shift focus to resources not favored by another species. Resource partitioning between potential competitors can be confirmed by comparing the habitats selected, the temporal use of those same habitats, differences in dietary preferences, and spatial segregation. In the following paragraphs, we will discuss each of these forms of resource partitioning as they relate to mountain sheep and their competitors.

*Species-specific habitat use.* Mountain sheep prefer steep, rocky terrain where they can escape predators (Geist 1971, Krausman and Bowyer 2003, Chapter 19). Within these areas, however, mountain sheep move about to optimize forage quality and quantity (Reed 2001), access water (Whiting et al. 2009, 2010), avoid predation (Festa-Bianchet 1988), and to meet other needs (e.g., rutting, lambing; Whiting et al. 2012, Robinson et al. 2020); whereas, elk, for example, tend to favor open meadows and forest edges for grazing (Collins and Urness 1983), providing some spatial separation from bighorn sheep, though range overlaps exist. Elk are more likely than mountain sheep to occur at low elevations during winter when snow cover is shallow (Stewart et al. 2002). Another example are mountain goats, which are well-adapted to rugged, alpine environments and most often inhabit areas with cliffs and rocky ridges that are steeper than those used by mountain sheep (Whiting et al. 2023b). Moreover, goats generally occupy high elevations, even above the tree line (Taylor et al. 2006, Whiting et al. 2023b).

*Temporal habitat use.* Resource partitioning occurs when competing species access the same resources at different times of day. For instance, mountain sheep primarily feed during the day (Krausman et al. 1999, Shackleton et al. 1999), whereas elk feed more frequently at night (Griffin and Silliman 2011). Alternately, mountain goats typically feed during the early morning and late afternoon to evening and are most active during these cooler parts of the day, especially in warmer seasons, to avoid overheating (Côté and Festa-Bianchet 2003). This temporal partitioning of mountain sheep and their competitors helps to minimize competition and conflict.

*Shifts in dietary preference.* Another way in which resource partitioning may be achieved is through differences in dietary preferences between competing species. For example, mountain sheep primarily consume grasses, sedges, and forbs (Krausman et al. 1999, Shackleton et al. 1999), whereas elk are more generalists, consuming a wider variety of vegetation, including grasses, shrubs, and tree bark (Collins and Urness 1983). These dietary differences help minimize competition between these species. Mountain goats primarily graze on grasses, sedges, and forbs in alpine meadows in summer. As winter approaches and these food sources become scarce, their diet shifts to browsing shrubs, conifers, and lichens (Hibbs 1967).

*Spatial segregation of habitats.* Resource partitioning can be achieved through the spatial separation of habitats used by potentially competing species (Figure 10.1). By occupying different zones within the same ecosystem, direct competition for resources and the probability of agonistic interactions are minimized. For instance, mountain sheep generally use higher, steeper areas than elk, which prefer lower, more open areas (Irwin and Peek 1983, Smith et al. 1991). Mountain goat habitat is characterized by high alpine meadows, vertical ledges, and slide rocks, areas that are not often frequented by either elk or mountain sheep (Laundré 1994).

In summary, resource partitioning enables species to coexist by reducing direct competition and allowing them to exploit different ecological niches. This partitioning is essential for maintaining ecosystem stability. The probability of competition among mountain sheep and native, feral, or domestic ungulates varies, and opportunities for competition differ depending on whether thinhorn sheep or bighorn sheep are the subject of concern or interest. The potential for interspecific competition is dependent on sympatry between the species involved and is likely exacerbated by location, habitat fragility, primary productivity, and species persistence (Jex et al. 2016). Habitat niches occupied by a suite of native, feral, or domestic ungulates that represent potential competitors to mountain sheep range from wide-open, productive grasslands characterized by high visibility and forgiving terrain to the sometimes sparsely vegetated, rugged mountainous terrain and steep cliffs that mountain sheep inhabit (Figure 10.1).

## 10.2.5 TYPES OF COMPETITION

Interference competition, exploitative competition, and apparent competition are all mechanisms by which species interact and influence each other, but they differ in how these effects occur. All three mechanisms can influence species populations and community structure, shaping ecosystem dynamics in different ways.

*Interference competition (direct, behavior based).* Interference competition can occur when one species impedes another, or excludes another from accessing resources. For example, mountain goats have been observed agonistically interacting with mountain sheep at mineral licks (Reed 2001). Similarly, mule deer and pronghorn can displace bighorn (Ferretti and Mori 2020). Those investigators concluded that mountain sheep were also socially intolerant of elk and therefore avoided grazing close to them. The larger bodies of elk and feral horses likely intimidate smaller grazers, such as mountain sheep, resulting in competitive displacement or exclusion (Berger 1985).

*Exploitative competition (indirect, resource based).* Exploitative competition occurs when species use the same limited resource. The use of a resource by one species subsequently reduces its availability for the other species (Krausman and Bowyer 2003, Pianka 2011). For instance, domestic horses, cattle, and sheep may reduce resources (e.g., water, forage, space) to the point of excluding bighorn sheep from preferred ranges (Buechner 1960, Bissonette and Steinkamp 1996, Ostermann-Kelm et al. 2008). Range over-exploitation by domestic livestock has removed available forage for bighorn sheep, has caused shifts in forage composition, and has had an additional negative effect on mountain sheep and other native ungulates (Eldridge et al. 2016).

*Apparent competition (indirect, predator based).* Apparent competition is an indirect interaction where two species appear to compete but do not actually share resources. This occurs when two or more prey species indirectly affect each other through a shared predator (Holt and Lawton 1993,1994). An increase in one prey species can boost the predator population, leading to increased predation on the other, perhaps secondary, prey species (Johnson et al. 2013, Jex et al. 2016, Rominger 2018). This phenomenon may explain why mountain lion (*Puma concolor*) predation on bighorn sheep has become an increasing concern, in part due to mule deer range expansion, fire suppression, reductions in the number of bighorns in herds, and a decline in mountain lion removals in recent years (Rominger et al. 2004).

As climate change continues and anthropogenic pressures increase (e.g., habitat loss, mineral extraction, poaching, recreation, introduction of invasive plants and animal species; Larsen et al. 2012, Bowyer et al. 2019, Bates et al. 2021, Whiting et al. 2023a), these problems will affect niches occupied by native species and potentially affect population persistence and evolutionary processes (Sax et al. 2013, Gaynor et al. 2018, Cox et al. 2023). Such niche alterations will increase the degree of competition among ungulates (Johnson et al. 2013, Rominger 2018, Berger

**FIGURE 10.1**   General habitat-niche breadths of selected North American native, feral, or domestic ungulates relative to that of big-horn sheep in North America. The probability of competitive interactions between mountain sheep and other ungulates is reduced substantially by the propensity of mountain sheep to be limited by the availability of rugged terrain. (Adapted from Bowyer et al. (2023).)

et al. 2020), especially in high-latitude, high-altitude environments, such as those occupied by many mountain sheep populations (Figure 10.2; Berger et al. 2022).

## 10.3   COMPETITION WITH MULE DEER, ELK, AND PRONGHORN

Native species that could compete with mountain sheep include mule deer, elk, and pronghorn; however, the potential for competition to occur differs depending on whether thinhorn sheep or bighorn sheep are the taxon of interest. The distribution of mule deer has expanded northward in recent years (Heffelfinger and Latch 2023, Jensen et al. 2023), but the probability of competition between thinhorn sheep and these cervids is less than between mule deer and bighorn sheep. In part, this is due to the limited distribution of mule deer in the high latitudes occupied by thinhorn sheep.

The potential for competition between mule deer and big-horn sheep is comparatively greater than that for mule deer and thinhorn sheep, in large part because the geographic distributions of mule deer and bighorn sheep overlap extensively, but often these species differ in habitat preferences. Nonetheless, habitat use overlaps where bighorn sheep and

mule deer use the same winter ranges (Schallenberger 1965, Wehausen 1996, Johnson et al. 2013), coastal chaparral vegetation on a year-round basis (Holl and Bleich 1983), and particularly in deserts where both species make extensive use of slopes or bajadas (Russo 1956, Jones 1980, O'Farrill et al. 2019). The common use of these areas, even on a seasonal basis, increases the potential for competitive interactions (Bowyer et al. 2023); however, opinions vary regarding the degree to which dietary similarities are likely to lead to forage competition between bighorn sheep and mule deer.

Despite overlapping geographic distributions across various ecosystems, subtle differences in the life-history strategies of mule deer and bighorn sheep may reduce the likelihood of competition for forage. Bighorn sheep depend largely on their visual acuity and the openness of vegetation to detect and evade predators and generally occupy more rugged and open terrain than mule deer. For bighorn sheep, the predator evasion strategy may be affected more by the presence of, or proximity to, vegetative cover (Bleich 1999). Thus, small-scale separation between these species may limit the potential for resource competition to occur. In mesic habitats occupied by mule deer and bighorn sheep on a year-round basis, there is an inverse relationship between forage availability and density of cover (Figure 10.3) and its

**FIGURE 10.2** Bighorn sheep and thinhorn sheep are sometimes sympatric with mountain goats and share common resources, which can result in niche overlap and aggressive interactions among the species. (Figure adapted from Olsen (2022).)

relative value among areas with differing fire histories (Holl and Bleich 2010, Holl et al. 2012). When combined with differences in predator evasion strategies, the differing values may further reduce the potential for resource competition.

Following extensive extirpations, elk have increased in numbers and geographic distribution (Wisdom and Thomas 1996) because of concerted efforts to restore this species to its historical range (Lyon and Thomas 1987).

Earlier investigators reported elk and bighorn sheep cohabiting overlapping winter ranges in Wyoming and Montana, United States, and Alberta, Canada (Murie 1940, Flook 1964, Constan 1972), indicating that competition between those species was possible if shared winter ranges became overpopulated by either of those native species (Lyon and Ward 1982). Nevertheless, differential use of habitat and forage by elk and bighorn sheep on winter ranges occupied by

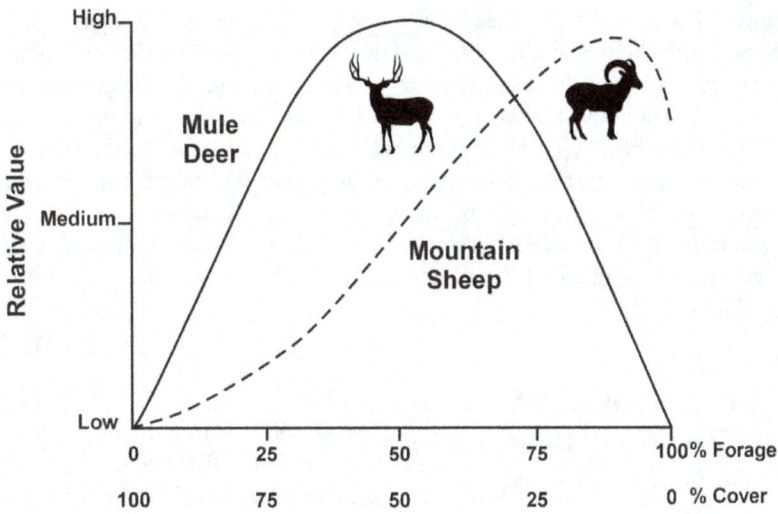

**FIGURE 10.3** In the fire-dependent chaparral habitat occupied by bighorn sheep and mule deer in the Transverse Ranges of southern California, the relative values of forage availability and vegetation density shift with the age (i.e., time since fire) of the vegetation. Moreover, there is an inverse relationship between the proportion of habitat that provides value as cover or forage for these two species. Areas of chaparral in which vegetation density obscures visibility are of low value to mountain sheep; however, even areas with very little vegetative cover shortly after a fire are of more nutritional value for those mountain ungulates than is older, mature vegetation. A similar pattern is observed for mule deer, but the relative value of chaparral ranges as a function of the proportion of cover and forage is skewed to the right for bighorn sheep when compared with mule deer. Holl et al. (2012) emphasized the conservation value of small, patchy fires in chaparral-dominated systems, in part because they minimize the potential for resource competition. (Figure adapted from Bleich and Holl (1982).)

both likely reduces the potential for resource competition (Schallenberger 1965, McCullough and Schneegas 1966, McCullough 1969).

Given the potential for resource competition on winter ranges occupied by bighorn sheep and elk, a similar outcome might be expected on ranges occupied by thinhorn sheep (i.e., Stone's sheep) that have been colonized by elk. Sittler et al. (2015), however, suggested that habitat overlap between those taxa is likely minimal, in large part because of habitat partitioning by landscape features, including vegetation category, slope, ruggedness, and time since fire. Nevertheless, concern has arisen regarding the potential for interference competition and apparent competition to occur due to elk colonizing winter habitat used by Stone's sheep (Jex et al. 2016, Sittler et al. 2015, 2019). Elk are regarded widely as adaptive foragers (Houston 1982, Christianson and Creel 2007) and are potential competitors with other native ungulates (Johnson et al. 2000, Stewart et al. 2002) across much of their distribution. Currently, Stone's sheep and elk have minimized the potential for resource competition by partitioning winter range, with Stone's sheep being habitat specialists that strongly select for steep slopes and high elevations with access to escape terrain as a predator evasion strategy (Walker et al. 2006, 2007; Sittler et al. 2015).

To date, we are unaware of situations where pronghorn have substantially competed with either thinhorn sheep or bighorn sheep (Ferretti and Mori 2020), although the geographic distributions of bighorn sheep and pronghorn overlap extensively. Pronghorn occupy grasslands, open-shrub communities, sagebrush (*Artemisia tridentata*) grasslands, and desert scrub but generally use areas with low-growing vegetation and high visibility that are flat in terms of topographic relief, when compared with mountain sheep that select rugged, open slopes having high visibility (Bleich et al 1997), and habitat overlap between those species is minimal (McCullough 1980, Yoakum 2004). Nonetheless, various levels of dietary overlap have been reported between pronghorn and bighorn sheep across several study areas (McCullough 1980, Hansen et al. 2001), and these species may use the same water sources in some areas. Differences in habitat use, however, reduce the likelihood of those species competing for forage (Figure 10.1).

## 10.4 COMPETITION WITH MOUNTAIN AND NORTHERN UNGULATES (MOUNTAIN GOAT, CARIBOU, AND BISON)

The niches of mountain goats, caribou (*Rangifer tarandus*), and bison (*Bison bison*) overlap with those of mountain sheep, resulting in potential resource competition (Jex et al. 2016). Differences in the forage niches, especially during winter, however, may reduce, if not eliminate, competition between these large herbivores (Adams et al. 1982).

Mountain goats and mountain sheep often compete due to extensive niche overlap. Mountain goats have an edge in the competition; however, their habitat is a subset of

that used by mountain sheep, with the latter using rolling meadows near cliffs and vertical ledge habitats that are not extensively used by mountain goats (Feldhamer et al. 2003). Mountain goats won aggressive encounters with bighorn sheep 98% of the time (Berger et al. 2022) and can displace the latter from mineral sources or foraging areas. This conflict is lessened seasonally, however, when bighorn sheep vacate high-elevation alpine areas and migrate to lower elevations as snow accumulates, while mountain goats are more likely to remain behind in steep, rocky habitats. This pattern is less apt to occur between mountain goats and thinhorn sheep because thinhorn sheep remain high on windswept plateaus adjacent to escape terrain, and encounters with mountain goats are more apt to occur when sheep forage at high elevations.

Caribou and thinhorn sheep exhibit some dietary overlap; however, caribou primarily feed on arboreal lichens and other vegetation (Adams et al. 1982, Webber et al. 2022), whereas mountain sheep graze on grasses, forbs, and shrubs (Buechner 1960, Krausman et al. 1999, Shackleton et al. 1999). Competition with caribou for food resources can be moderate but currently is not as intense as between mountain sheep and mountain goats, because caribou generally do not overlap in habitat use with mountain sheep (Geist 1971, Webber et al. 2022). A changing climate, however, may alter this pattern if caribou increase use of snow-free areas at high elevations.

Bison and thinhorn sheep also share some dietary preferences, particularly in high-elevation areas where forage is generally least abundant (Jung et al. 2015a). Thus, there is potential for food competition between reintroduced bison and thinhorn sheep during winter, when forage availability is most limited; however, this competition is generally not a problem during the remainder of the year (Jung et al. 2015a). Although bison had been introduced to the area in which the investigations occurred, these species may have co-evolved along niche axes such that spatial or temporal segregation is minimized, and bison are merely filling a niche that had been vacant since they were locally extirpated 350 years ago (Jung et al. 2015b).

## 10.5 COMPETITION WITH EXOTIC OR FERAL UNGULATES

The presence of free-ranging populations of non-indigenous ungulates in native habitats has important implications for the maintenance of biodiversity (Spear and Chown 2009) and can also be socially and economically valuable (Forsyth et al. 2009). When established on the landscape, these non-native species can have ecosystem-level effects through hybridization, resource competition or apparent competition, the introduction of diseases, alterations to vegetation structure and composition, predation, facilitation, trophic cascades, and changes to soil system function (Spear and Chown 2009). Such non-indigenous species are often referred to as exotic, invasive, alien, or feral. Hereafter, we refer to these non-indigenous species simply as exotic or feral and consider those that have become integral components of the various ecosystems inhabited by mountain sheep. Exotic species that may be competitors with mountain sheep originated largely in Africa or Asia. In North America, these non-native ungulates occupy habitat niches similar to those they occupy naturally. Whether released intentionally or escaped from captivity, they have become established as self-perpetuating populations on private and public lands in many areas occupied by mountain sheep, and are best described as animals that have gone feral following domestication or are descended from other stock and are now established on the landscape (McKnight 1961).

Among the exotic species that are of concern from the standpoint of competition with mountain sheep are the aoudad or Barbary sheep (*Ammotragus lervia*), that are native to northern Africa, the Persian ibex (*Capra aegagrus aegagrus*), which is native to the Middle East, and the gemsbok (*Oryx gazella*), which is native to southern Africa. Aoudads are of greatest concern and currently inhabit areas formerly occupied by desert bighorn sheep, are expanding into areas occupied by desert bighorn sheep, or are currently sympatric with desert bighorn sheep (Krausman et al. 1999). There is less concern about gemsbok competing with desert bighorn sheep due to the differences in habitat preferences between these two species (Saiz 1975).

Aoudads were introduced to portions of southwestern North America, initially for zoological exhibitions and later to provide sporting opportunities (Cassinello 2015, Wright et al. 2022). Recently, genetic analyses indicated that multiple sources of origin are represented among free-ranging populations of this species (Wright et al. 2022). These exotic ungulates were released in areas occupied by desert bighorn sheep to fill an ecological niche purportedly vacated following the extinction of some large herbivore during the Pleistocene (Seegmiller and Simpson 1979).

Initially, it was thought that aoudads would not compete with desert bighorn sheep because they were expected to occupy differing habitat niches (Seegmiller and Simpson 1979). Aoudads have, however, dispersed widely and across long distances, and are now free-ranging throughout a large area in which they are sympatric with desert bighorn sheep in New Mexico and western Texas. The ability of aoudads to disperse widely, reproduce rapidly, and their dietary preferences like those of desert bighorn sheep have subsequently raised concerns that this species may have negative effects on native bighorn sheep (Seegmiller and Simpson 1979, Simpson and Krysl 1981, Brewer et al. 2014, Brewer and Bleich 2023). This potentiality was raised by Barrett (1967), nearly 60 years ago, who noted that if aoudads and desert bighorn sheep were sympatric and both species preferred the same habitat, resource competition would be inevitable, and desert bighorn sheep would be at a disadvantage.

Aoudads are capable of producing twins and triplets when resources are adequate; they give birth during all seasons (DeArment 1971, Gray and Simpson 1980, Habibi 1982), and they may give birth more than once each year

(Blunt 1963). Moreover, they are behaviorally dominant over and compete with desert bighorn sheep for forage and water (Brewer et al. 2014). Previous investigators (Barrett 1967, Seegmiller and Simpson 1969) hypothesized that aoudads would outcompete bighorn sheep in areas of sympatry, and such appears to be occurring in western Texas where aoudads reportedly number more than 30,000, and desert bighorn sheep numbered about 1,500 until recently (Wright et al. 2022) but have declined to about 500 (Chapter 3).

The development of sympatry and, thus, increased potential for contact between aoudads and native bighorn sheep persists, in part is a result of opportunities taken by landowners to diversify sources of income (Dvorak 1980, Demarais et al. 1990, Radke 2025). In California, approximately 2,000 free-ranging aoudads are on private land near the central coast (California Wildlife Habitat Relationships Program 2016, Mori et al. 2017). These exotic ungulates are classified as nongame animals and may be hunted year-round under the regulatory authority of the California Fish and Game Commission (Cornman 2024). Their presence has important economic benefits for landowners. In another example likely founded on a potential economic benefit, a proposal to release aoudads on private lands proximate to habitat occupied by Rocky Mountain bighorn sheep was recently denied by the North Dakota Department of Agriculture. Given the potential monetary returns associated with such ventures, however, similar proposals are apt to continue.

The closest location thought to have been occupied by bighorn sheep in California (Buechner 1960) is the La Panza Range, and the most proximate area currently occupied by bighorn sheep is San Rafael Peak (Bleich et al. 2019), both of which are approximately 75 and 200 km, respectively, from the area currently inhabited by aoudads. The potential for contact between aoudads and bighorn sheep, however, was not a consideration when bighorn sheep were translocated to San Rafael Peak in 1985 (Bleich et al. 2019). Given the dispersal abilities of the aoudad (Barrett 1967, Simpson et al. 1978, Dickinson and Simpson 1980) and the proximity of the aoudad population to habitat historically occupied by bighorn sheep, additional translocations of bighorn sheep closer to the area occupied by aoudads are unlikely as long as the aoudad population remains extant.

The aoudad is a highly adaptable herbivore that has evolved under conditions perhaps even more harsh than those of the desert bighorn sheep. In addition to concerns about resource competition between aoudads and bighorn sheep, especially on a seasonal basis or in a changing environment (Etchart 2021, Parikh et al. 2025), there is also concern over the potential for disease transmission to native bighorn sheep. *Mycoplasma ovipneumoniae*, a pathogen of substantial concern that predisposes bighorn sheep to polymicrobial respiratory disease (Besser et al. 2013, Cassirer et al. 2013, 2018), has been detected in aoudads and bighorn sheep in Texas (Wright et al. 2024), and the prion genotype that confers average susceptibility to diseases such as scrapie (Goldmann 2008) has been detected in several aoudads

(Wright et al. 2022). These recent findings may represent previously unrecognized challenges to the conservation of bighorn sheep and could be even more significant and concerning than the competitive advantages of aoudads described previously.

Feral ungulates that are sympatric with mountain sheep and have been or are of the greatest concern from the standpoint of resource competition are goats, horses, and burros (Figure 10.4). Goats were introduced to the New World by Spanish explorers and missionaries hundreds of years ago (Browne 1869, Wagoner 1952, Holechek et al. 1989). These ungulates are well known for the ability to exploit steep, rocky terrain and use lower-quality forage than mountain sheep, and have been recognized as potential competitors with bighorn sheep (Kelly 1960, Van den Akker 1960). That concern generally was resolved by 1978 (Jones 1980), but goats continue to be viewed as potential competitors with desert bighorn sheep in the southwestern United States and northern Mexico (Bleich 2015).

In California alone, past or contemporary evidence of feral or free-ranging goats has been reported in or proximate to 21 mountain ranges occupied by desert bighorn sheep (Bleich 2015). The use of domestic goats for brush control, escapees from domestic goat operations, or pack goats that have escaped from recreational trekkers may have been the founders of feral populations in the past; however, the continued presence of domestic goats on private land proximate to habitat occupied by bighorn sheep remains a concern. Goats have been implicated in disease episodes in multiple western states (Gross 1960, Coggins 2002, Jansen et al. 2006, Wiedmann and Hosek 2013, Wiedmann et al. 2024). Competition aside, the potential for pathogen transmission from captive, free-ranging, or feral goats remains (United States Fish and Wildlife Service 2000, 2007; Western Association of Fish and Wildlife Agencies 2024), and is addressed in detail in Chapter 8.

During the 1500s, Spanish colonists also introduced the progenitors of feral horses and burros currently inhabiting many arid regions of western North America (Olsen 2016).

Population growth among feral equids has influenced ecosystem function, habitat quality for native herbivores, and prompted public concern about the sustainability of western rangelands, where feral equids, particularly feral burros, are serious competitors of bighorn sheep (Rubin et al. 2024).

In 1971, feral equids in the United States were protected by the Wild Free-Roaming Horses and Burros Act (Act; Douglas and Leslie 1996). The Act was well-intentioned and established guidelines for maintaining the distributions of feral horses and burros that existed at that time. The Act also mandated that allowable population densities in those areas be established by the agencies best suited to do so; however, subsequent legislation has had a significant influence on the Act as originally written. Public pressure and subsequent political expediency collectively have handicapped the United States Bureau of Land Management and the Forest Service, the agencies responsible for implementing the Act, to effectively curb the proliferation of feral equids and their long-term effects on public rangelands (National Research Council 2013, Beever et al. 2018, Schoenecker et al. 2021, Hennig et al. 2023). As a result, the influence of feral equids on native species will continue unabated, if not exacerbated, as populations of these feral animals continue to increase (Bleich 2022).

Feral horses and burros are not ruminants, but rather hind-gut fermenters (Van Soest 1994). Both species can persist on less digestible forage than native mountain sheep (Janis 1976). These equids have a digestive system that is specialized to consume large quantities of low-quality forage when compared with ruminants (Schoenecker et al. 2016), which generally consume forage of greater digestibility (Marshal et al. 2008). When compared with bighorn sheep, the ability of feral equids to persist on coarse, albeit more abundant, vegetation is especially notable during years of poor forage production in deserts occupied by bighorn sheep (Marshal et al. 2008). When available, feral horses and burros incorporate higher-quality forage, such as that selected by sympatric ruminants, into their diets; however, they also focus on lower-quality forage that may be unsuitable for ruminants during periods of forage

FIGURE 10.4    (a) Feral burro and desert bighorn sheep are sympatric in many parts of the Mojave, Sonoran, and Chihuahuan deserts of North America, where competition for water can be severe. Whipple Mountains, California, September 1970. (Photograph by Dan Smith, California Department of Fish and Game.) (b) A domestic sheep and a desert bighorn sheep together in typical bighorn sheep habitat, Anza-Borrego Desert State Park, California, July 2024. (Photograph by Janene Colby, California Department of Wildlife.) (c) Domestic livestock and desert bighorn sheep both use limited resources on federal grazing allotments throughout the Great Basin, Mojave, Sonoran, and Chihuahuan deserts of western North America. Ord Mountain, California, July 2009. (Photograph © Carlos Gallinger, Society for the Conservation of Bighorn Sheep; used with permission.)

scarcity (Marshal et al. 2008). This plasticity in dietary preference allows an advantage over bighorn sheep occupying arid, highly variable environments typical of most deserts (Marshal et al. 2012).

Feral burros are potential competitors with desert bighorn sheep for limited resources, such as water and forage, due to overlap in habitat use by the two species (Halloran 1949, Sumner 1959, Weaver 1959, St. John 1965, Hansen and Martin 1973, Weaver 1973, Seegmiller and Ohmart 1981). The question of competition, whether for forage or for water in the form of interference or resource competition, surfaced as a primary concern many years ago (Halloran 1949, Sumner 1959, Welles 1960, Welles and Welles 1961), and conflicting opinions linger.

Seegmiller and Ohmart (1981) concluded that exploitation competition between these species was occurring or could be imminent in areas of sympatry and commented that desert bighorn sheep are too limited in distribution and numbers to accept the risks of coexistence with feral burros. Those authors recommended removing feral burros from areas of sympatry with desert bighorn sheep and from additional areas that have the potential for reintroducing desert bighorn sheep. As a result, methods of precluding accessibility to water sources have been developed and installed to prevent use by feral equids yet can still be accessed by native ungulates (Halloran and Deming 1958, Andrew et al. 1997, Bleich et al. 2020), and management direction is to continue to do so (Brewer et al. 2014). In the absence of persistent and intensive management, however, competition for forage will remain a concern and damage to vegetation will continue unabated. Moreover, it is not possible to preclude access to every water source used by feral equids and bighorn sheep.

Despite early concerns about resource competition between feral burros and bighorn sheep (Halloran 1949), investigators have begun to provide stronger evidence that resource competition is a problem. The differences in the digestive efficiencies of ruminants and equids may have resulted in estimates of dietary overlap being biased downward. In part, this is because there are difficulties associated with manipulating populations of large mammals for experimental purposes in a wild setting (Wagner 1983, Stewart et al. 2002, Ostermann-Kelm et al. 2008), and experimental removals followed by the long-term absence of feral equids will be necessary to provide further insight into the question of how those non-native herbivores affect native mountain sheep (Marshal et al. 2012, Iacono et al. 2024). In the absence of a demographic consequence to mountain sheep, however, the question will continue to linger (Wagner 1983).

The question of the relationship between feral horses and mountain sheep similarly is complex, and published opinions vary from the possibility that overt facilitation is of direct benefit to bighorn sheep and other native species (Berger 1986, Coates and Schemnitz 1994) to the conclusion that competition for water by feral horses has affected desert bighorn sheep populations (Halloran and Deming 1958). Behavioral displacement of bighorn sheep by horses has been clearly demonstrated, as evidenced by

bighorn sheep moving to water sources where horses were absent (Ostermann-Kelm 2008). Schoenecker et al. (2004) reported no influence of feral horses or their presence on the demography of bighorn sheep in a study in Montana and Wyoming; those investigators concluded, however, that areas with horses had significantly lower plant productivity, suggesting competition for resources existed, and that the distribution of bighorn sheep may be affected by horses. Additional investigators have expressed concern that the potential for competition, whether resource-based or via interference, does have an effect. Despite the clear spatial overlap in the habitats used by desert bighorn sheep and horses in their study area, Ostermann-Kelm et al. (2008) reported a 76% reduction in the number of bighorn sheep groups that came to water when just one horse was present in the area. Although those authors reported no evidence of aggressive interactions between the species and concluded that evidence of direct competition was not apparent, their results provided evidence of indirect interference competition (Diamond 1978, Loft et al. 1991).

Elsewhere, Berger (1977) reported that desert bighorn sheep were sympatric with feral horses and feral burros, but there was no overlap between the two feral congeners; this was attributed to the use of similar habitats by desert bighorn sheep and feral burros, but not between bighorn sheep and feral horses. Additionally, other investigators described considerable spatial and habitat separation between horses and Rocky Mountain bighorn sheep during all seasons; however, neither species appeared to use all available habitat (Gudorf et al. 1996, Kissell 1996), which may have precluded opportunities for resource competition to arise. It was forewarned by Hoefs and Brink (1978), though, that removal of forage by feral horses sympatric with Dall's sheep in Kluane National Park had the potential to decrease forage availability in those areas and be followed by a decline of the sheep population due to overgrazing by sympatric herbivores, and such is postulated for Rocky Mountain bighorn sheep in several Canadian national parks (Stelfox 1974).

Differences between mountain sheep and feral equids, particularly in terms of habitat preferences, anatomical adaptations, foraging strategies, and behavior, complicate the debate regarding competition between these species. What is clear, however, is that the long-term, continuing, and overabundant presence of feral equids has had profound ecological effects on sensitive arid or semi-arid ecosystems (Beever 2003, Ostermann-Kelm et al. 2009, Davies et al. 2014, Hall et al. 2016, Davies and Boyd 2019, Rubin et al. 2024) occupied by native ungulates, including bighorn sheep. When combined with the ineffective management of feral equids since the passage of the Act (Bleich 2022), the failure to consider the ecological and evolutionary relationships or adaptations of non-native perissodactyls in the context of sympatry with native species (Hennig et al. 2023) has compounded concerns. When further combined with an overabundance of politically motivated enthusiasm for feral equids (Bleich 2022), the resulting synergy has led to widespread ecological change.

The effects of feral equids on bighorn sheep, primarily through resource competition for water, were recognized early by Halloran (1949), who warned that feral burros competing with bighorn sheep for water must be either eliminated or reduced to very low numbers. It is ironic, though, that those who were among the earliest to question whether feral burros affected bighorn sheep also proclaimed, "For the sake of the entire biota, the burros must be controlled and their numbers kept down…" (Welles and Welles 1961:178). The management of feral burros and feral horses, however, continues to be hampered by societal views regarding the status of equid populations and is neutered by a lack of management options (Hennig et al. 2023). In the absence of corrective legislation, the potential for competition between mountain sheep and feral equids will continue to increase and habitat quality will continue to be degraded. Moreover, the well-being of other native species and those feral perissodactyls will remain threatened (Bleich 2022).

## 10.6 COMPETITION WITH DOMESTIC UNGULATES (CATTLE, SHEEP, AND GOATS)

For desert bighorn sheep, overgrazing by livestock on rangelands occupied by both can result in large-scale changes in the composition and density of plant species used by those native ungulates and lead to a reduction in carrying capacity for bighorn sheep (Figures 10.4 and 10.5; Buechner 1960, Krausman et al. 1999, Garrison et al. 2016). Domestic sheep and goats have historically been the most prevalent competitor of desert bighorn sheep, due to the similarity in feeding habits, forage preferences, and, in the case of domestic goats, use of similar rugged terrain (Davis and Taylor 1939, Villa 1959, Weaver 1972, Gallizioli 1977, Krausman et al. 1999). The presence of livestock also affects the distribution and abundance of water sources used by desert bighorn sheep (Halloran and Deming 1958, McCutchen 1981, Brewer et al. 2014). Potential competition between desert bighorn sheep and livestock can also pose an obstacle to identifying suitable locations when reintroducing desert

bighorn sheep populations (Krausman et al. 1999, Singer et al. 2000a,b).

Among Rocky Mountain bighorn sheep, competition for forage and spatial interactions with livestock can occur seasonally (Buechner 1960, Risenhoover et al. 1988). Rocky Mountain bighorn sheep and cattle often eat grass-dominated diets (Shackleton et al. 1999), and dietary overlap may occur (Smith et al. 1991, Brewer et al. 2014). This overlap can be especially frequent during drought or other times of reduced forage abundance (Buechner 1960, Chaikina and Ruckstuhl 2006, Whiting et al. 2023b). For example, cattle grazing can cause changes in the quality and quantity of forage available to wildlife, decreasing the availability of forage and digestible energy, especially during winter. Long-term grazing could also alter vegetation composition (Figure 10.5; Brown and Ruckstuhl 2012). For bighorn sheep, forage bite rates decreased and vigilance increased when cattle were nearby (Brown et al. 2010). Moreover, spatial competition between livestock and Rocky Mountain bighorn sheep can occur (Risenhoover et al. 1988). For example, Rocky Mountain bighorn sheep avoided areas with cattle and decreased use when cattle were present (Bissonette and Steinkamp 1996). Potential competition between Rocky Mountain bighorn sheep and livestock can also be an obstacle when identifying otherwise suitable locations for the reintroduction of those native ungulates (Smith et al. 1991, Singer et al. 2000b).

Among thinhorn sheep, little is known about competition with domestic ungulates. Such competition could increase in the future and become problematic, especially with landscape-level alterations by humans, climate change, or following habitat enhancement projects (e.g., prescribed fire management practices) in thinhorn habitat (Bleich 2009, Jex et al. 2016, Berger et al. 2022). For example, competition could develop if livestock take advantage of early seral habitats in areas occupied by thinhorn sheep following fire (Jex et al. 2016), and biologists have recommended that future research address knowledge gaps regarding competition between thinhorn sheep and livestock after implementation of habitat enhancements (Bleich 2009, Jex et al. 2016).

**FIGURE 10.5** Domestic cattle have been grazed in bighorn sheep habitat across much of western North America and historically have been viewed as potential competitors with bighorn sheep and other native ruminants. Cattle can remove substantial amounts of vegetation, or water, upon which native species may be dependent. (a) Note the abundance of forage at this spring, and (b) the presence of multiple cattle the next day. (c) Within 5 days, available forage was reduced and cattle dung was visible. Newberry Mountains, California, August 2009. (Photographs © Carlos Gallinger, Society for the Conservation of Bighorn Sheep; used with permission.)

## 10.7 SUMMARY

Competition for food, water, cover, or space with native, feral, and domestic ungulates plays an important role in determining habitat use and the persistence of mountain sheep populations. Usually, competition among these ungulates is related inversely to forage or water availability, and competition can increase if availability of those resources is reduced by common use. In the future, as climate change continues and anthropogenic pressures intensify, competition among mountain sheep and native, feral, and domestic ungulates is expected to increase. Future research should focus on how climate change and anthropogenic pressures alter the niches of these ungulates and, subsequently, the competition among them. The increased use of advanced technology (Chapter 13) will substantially aid in understanding how these species may or may not compete temporally and spatially (Tomkiewicz et al. 2010). Proper conservation of mountain sheep populations and their habitats will require biologists to understand and mitigate competition among mountain sheep and native, feral, and domestic ungulates.

## REFERENCES

Adams, L., K. Risenhoover, and J. Bailey. 1982. Ecological relationships of mountain goats and Rocky Mountain bighorn sheep. *Northern Wild Sheep and Goat Council Proceedings* 3:9–22.

Akenson, H. A. 1992. Spatial relationships and behavior of bighorn sheep sharing a winter range with mule deer and elk in central Idaho. Thesis, University of Idaho, Moscow, Idaho, USA.

Andrew, N. G., L. M. Lesicka, and V. C. Bleich. 1997. An improved fence design to protect water sources for native ungulates. *Wildlife Society Bulletin* 25:823–825.

Baldi, R., A. Pelliza-Sbriller, D. Elston, and S. Albon. 2004. High potential for competition between guanacos and sheep in Patagonia. *Journal of Wildlife Management* 68:924–938.

Barrett, R. 1967. Some comparisons between the Barbary sheep and the desert bighorn. *Desert Bighorn Council Transactions* 11:16–26.

Bates, S. B., J. C. Whiting, and R. T. Larsen. 2021. Comparison of effects of shed antler hunting and helicopter surveys on ungulate movements and space use. *Journal of Wildlife Management* 85:437–448.

Beever, E. 2003. Management implications of the ecology of free-roaming horses in semi-arid ecosystems of the western United States. *Wildlife Society Bulletin* 31:887–895.

Beever, E. A., L. Huntsinger, and S. L. Petersen. 2018. Conservation challenges emerging from free-roaming horse management: a vexing social-ecological mismatch. *Biological Conservation* 226:321–328.

Berger, J. 1977. Sympatric and allopatric relationships among desert bighorn sheep and feral equids in Grand Canyon. *Southwestern Naturalist* 22:540–543.

Berger, J. 1985. Interspecific interactions and dominance among wild Great Basin ungulates. *Journal of Mammalogy* 66:571–573.

Berger, J. 1986. *Wild Horses of the Great Basin: Social Competition and Population Size.* University of Chicago Press, Chicago, Ilinois, USA.

Berger, J., M. Biel, and F. P. Hayes. 2022. Species conflict at Earth's edges–contests, climate, and coveted resources. *Frontiers in Ecology and Evolution* 10:991714.

Berger, J., T. Wangchuk, C. Briceno, A. Vila, and J. E. Lambert. 2020. Disassembled food webs and messy projections: modern ungulate communities in the face of unabating human population growth. *Frontiers in Ecology and Evolution* 8:128.

Besser, T. E., E. F. Cassirer, M. A. Highland, P. Wolff, A. Justice-Allen, K. Mansfield, M. A. Davis, and W. Foreyt. 2013. Bighorn sheep pneumonia: sorting out the cause of a polymicrobial disease. *Preventive Veterinary Medicine* 108:85–93.

Birch, L. C. 1957. The meanings of competition. *American Naturalist* 91:5–18.

Bissonette, J. A., and M. J. Steinkamp. 1996. Bighorn sheep response to ephemeral habitat fragmentation by cattle. *Great Basin Naturalist* 56:319–325.

Bleich, V. C. 1999. Mountain sheep and coyotes: patterns of predator evasion in a mountain ungulate. *Journal of Mammalogy* 80:283–289.

Bleich, V. C. 2009. Perceived threats to wild sheep: levels of concordance among states, provinces, and territories. *Desert Bighorn Council Transactions* 50:32–39.

Bleich, V. C. 2015. Records of goats and other caprines in or near bighorn sheep habitat in California. *Desert Bighorn Council Transactions* 53:24–33.

Bleich, V. C. 2022. Feral horses, feral asses, and professional politicians: broodings from a beleaguered biologist. *Human-Wildlife Interactions* 16:1–6.

Bleich, V. C., M. E. Blum, K. T. Shoemaker, D. Sustaita, and S. A. Holl. 2019. Habitat selection by bighorn sheep in a mesic ecosystem: the San Rafael Mountains, California, USA. *California Fish and Game* 105:205–224.

Bleich, V. C., M. W. Oehler Sr, and J. G. Kie. 2020. Managing rangelands for wildlife. Pages 126–148 in N. J. Silvy, editor. *Techniques for Wildlife Investigations and Management.* Eighth edition. Volume II. Johns Hopkins University Press, Baltimore, Maryland, USA.

Bleich, V. C., R. T. Bowyer, and J. D. Wehausen. 1997. Sexual segregation in mountain sheep: resources or predation? *Wildlife Monographs* 134:1–50.

Bleich, V. C., and S. A. Holl. 1982. Management of chaparral habitat for mule deer and mountain sheep in southern California. USDA Forest Service, General Technical Report PSW-58:247–254, Berkeley, California, USA.

Blunt, F. 1963. Barbary sheep: not in Wyoming, we hope. *Wyoming Wildlife* 27:28–31.

Bowyer, R. T. 2004. Sexual segregation in ruminants: definitions, hypotheses, and implications for conservation and management. *Journal of Mammalogy* 85:1039–1052.

Bowyer, R. T. 2022. *Sexual Segregation in Ungulates: Ecology, Behavior, and Conservation.* Johns Hopkins University Press, Baltimore, Maryland, USA.

Bowyer, R. T., K. M. Stewart, J. W. Cain III, and B. R. McMillan. 2023. Competition with other ungulates. Pages 141–157 in J. R. Heffelfinger and P. R. Krausman, editors. *Ecology and Management of Black-Tailed and Mule Deer of North America.* CRC Press, Boca Raton, Florida, USA.

Bowyer, R. T., M. S. Boyce, J. R. Goheen, and J. L. Rachlow. 2019. Conservation of the world's mammals: status, protected areas, community efforts, and hunting. *Journal of Mammalogy* 100:923–941.

Brewer, C., V. C. Bleich, J. Foster, T. Hosch-Hebdon, D. McWhirter, E. Rominger, M. Wagner, and B. Wiedmann. 2014. Bighorn sheep: conservation challenges and management strategies for the 21st century. Wild Sheep Working Group, Western Association of Fish and Wildlife Agencies, Cheyenne, Wyoming, USA.

Brewer, C. E., and V. C. Bleich. 2023. Trophy spotlight: desert bighorn sheep. Pages 18–33 in K. M. Lehr and J. Schwab, editors. *Records of North American big game*. Boone and Crockett Club, Missoula, Montana, USA.

Brown, N. A., and K. E. Ruckstuhl. 2012. Differences in vegetation and native ungulate use between exclosures and cattle grazed plots in Sheep River Provincial Park, Alberta. *Journal of Ecosystem & Ecography* 2:1–9.

Brown, N. A., K. E. Ruckstuhl, S. Donelon, and C. Corbett. 2010. Changes in vigilance, grazing behaviour and spatial distribution of bighorn sheep due to cattle presence in Sheep River Provincial Park, Alberta. *Agriculture, Ecosystems & Environment* 135:226–231.

Browne, J. R. 1869. *Resources of the Pacific Slope: A Sketch of the Settlement and Exploration of Lower California*. D. Appleton and Company, New York, New York, USA.

Buechner, H. K. 1960. The bighorn sheep in the United States, its past, present, and future. *Wildlife Monographs* 4:1–174.

California Wildlife Habitat Relationships Program. 2016. Complete list of amphibian, reptile, bird and mammal species in California. California Department of Fish and Wildlife, Sacramento, California, USA. https://wildlife.ca.gov/Data/CWHR. Accessed 26 March 2025.

Cassinello, J. 2015. Datasheet report for *Ammotragus lervia* (aoudad). Endangered Species Compendium. https://www.cabi-digitallibrary.org/doi/full/10.1079/cabicompendium.94507. Accessed 26 March 2025.

Cassirer, E. F., K. R. Manlove, E. S. Almberg, P. L. Kamath, M. Cox, P. Wolff, A. Roug, J. Shannon, R. Robinson, and R. B. Harris. 2018. Pneumonia in bighorn sheep: risk and resilience. *Journal of Wildlife Management* 82:32–45.

Cassirer, E. F., R. K. Plowright, K. R. Manlove, P. C. Cross, A. P. Dobson, K. A. Potter, and P. J. Hudson. 2013. Spatio-temporal dynamics of pneumonia in bighorn sheep. *Journal of Animal Ecology* 82:518–528.

Chaikina, N. A., and K. E. Ruckstuhl. 2006. The effect of cattle grazing on native ungulates: the good, the bad, and the ugly. *Rangelands* 28:8–14.

Christianson, D. A., and S. Creel. 2007. A review of environmental factors affecting elk winter diets. *Journal of Wildlife Management* 71:164–176.

Coates, K. P., and S. D. Schemnitz. 1994. Habitat use and behavior of male mountain sheep in foraging associations with wild horses. *Great Basin Naturalist* 54:86–90.

Coggins, V. 2002. Rocky Mountain bighorn sheep/domestic sheep and domestic goat interactions: a management prospective. *Northern Wild Sheep and Goat Council Proceedings* 13:165–174.

Cole, L. C. 1960. Competitive exclusion. *Science* 132:348–349.

Collins, W. B., and P. J. Urness. 1983. Feeding behavior and habitat selection of mule deer and elk on northern Utah summer range. *Journal of Wildlife Management* 47:646–663.

Constan, K. J. 1972. Winter foods and range use of three species of ungulates. *Journal of Wildlife Management* 36:1068–1076.

Cornman, A. 2024. Committee staff summary for January 15, 2024 WRC. Staff summary California Fish and Game Commission, Sacramento, California, USA. Available at https://wildlife.ca.gov/Search-Results?q=aoudad#gsc.tab=0&gsc.q=aoudad&gsc.page=1. Accessed 12 March 2025.

Côté, S. D. and M. Festa-Bianchet. 2003. Mountain goat. Pages 1061–1075 in G. A. Feldhamer, B. C. Thompson, and J. A. Chapman. *Wild Mammals of North America: Biology, Management, and Conservation*. Johns Hopkins University Press, Baltimore, Maryland, USA.

Cox, D. T., A. S. Gardner, and K. J. Gaston. 2023. Diel niche variation in mammalian declines in the Anthropocene. *Scientific Reports* 13:1031.

Davies, K. W., and C. S. Boyd. 2019. Ecological effects of free-roaming horses in North American rangelands. *BioScience* 69:558–565.

Davies, K. W., G. Collins, and C. S. Boyd. 2014. Effects of feral free-roaming horses on semi-arid rangeland ecosystems: an example from the sagebrush steppe. *Ecosphere* 5:1–14.

Davis, W. B., and W. P. Taylor. 1939. The bighorn sheep of Texas. *Journal of Mammalogy* 20:440–455.

DeArment, R. 1971. Reaction and adaptability of introduced aoudad sheep. Final Report, Federal Aid in Wildlife Restoration Project W-45-R-21. Texas Parks and Wildlife Department, Austin, Texas, USA.

DeCesare, N. J., and D. H. Pletscher. 2006. Movements, connectivity, and resource selection of Rocky Mountain bighorn sheep. *Journal of Mammalogy* 87:531–538.

Demarais, S., D. A. Osborn, and J. J. Jackley. 1990. Exotic big game: a controversial resource. *Rangelands* 12:121–125.

Diamond, J. M. 1978. Niche shifts and the rediscovery of interspecific competition. *American Scientist* 66:322–331.

Dickinson, T. G., and C. D. Simpson. 1980. Dispersal and establishment of Barbary sheep in southeast New Mexico. Pages 33–45 in C. D. Simpson, editor. *Proceedings of the Symposium on Ecology and Management of Barbary Sheep*. Texas Tech University, Lubbock, Texas, USA.

Douglas, C. L., and D. M. Leslie. 1996. Feral animals on rangelands. Pages 281–292 in P. R. Krausman, editor. *Rangeland Wildlife*. Society for Range Management, Denver, Colorado, USA.

Douglas, C. L., and D. M. Leslie. 1999. Management of bighorn sheep. Pages 238–262 in R. Valdez and P. R. Krausman, editors. *Mountain Sheep of North America*. University of Arizona Press, Tucson, Arizona, USA.

Dvorak, D. F. 1980. A brief history and status of aoudad sheep in Palo Duro Canyon, Texas. Pages 23–24 in C. D. Simpson, editor. *Proceedings of the Symposium on Ecology and Management of Barbary Sheep*. Texas Tech University, Lubbock, Texas, USA.

Eldridge, D. J., A. G. Poore, M. Ruiz-Colmenero, M. Letnic, and S. Soliveres. 2016. Ecosystem structure, function, and composition in rangelands are negatively affected by livestock grazing. *Ecological Applications* 26:1273–1283.

Etchart, J. L. 2021. Evaluating water use and seasonal ranges of desert bighorn sheep and aoudad in the Sierra Vieja Mountains, Texas. Thesis, Sul Ross State University, Alpine, Texas, USA.

Enns, G. E., B. A. Jex, and M. S. Boyce. 2023. Diverse migration patterns and seasonal habitat use of Stone's sheep (*Ovis dalli stonei*). *PeerJ* 11:e15215.

Feldhamer, G. A., B. C. Thompson, and J. A. Chapman. 2003. *Wild Mammals of North America: Biology, Management, and Conservation*. John Hopkins University Press, Baltimore, Maryland, USA.

Feldhamer, G. A., J. F. Merritt, C. Krajewski, J. L. Rachlow, and K. M. Stewart. 2020. *Mammalogy: Adaptation, Diversity, Ecology*. Johns Hopkins University Press, Baltimore, Maryland, USA.

Ferretti, F. and E. Mori. 2020. Displacement interference between wild ungulate species: does it occur? *Ethology Ecology & Evolution* 32:2–15.

Festa-Bianchet, M. 1988. Seasonal range selection in bighorn sheep: conflicts between forage quality, forage quantity, and predator avoidance. *Oecologia* 75:580–586.

Festa-Bianchet, M., S. D. Côté, S. Hamel, and F. Pelletier. 2019. Long-term studies of bighorn sheep and mountain goats reveal fitness costs of reproduction. *Journal of Animal Ecology* 88:1118–1133.

Finke, D. L., and W. E. Snyder. 2008. Niche partitioning increases resource exploitation by diverse communities. *Science* 321:1488–1490.

Flook, D. R. 1964. Range relationships of some ungulates native to Banff and Jasper national parks, Alberta. Pages 119–228 in D. J. Crisp, editor. *Grazing in Terrestrial and Marine Environments*. Blackwell, Oxford, United Kingdom.

Forsyth, D. M., J. P. Parkes, A. P. Woolnough, G. Pickles, M. Collins, and I. Gordon. 2009. Environmental and economic factors determine the number of feral goats commercially harvested in Western Australia. *Journal of Applied Ecology* 46:101–109.

Gallizioli, S. 1977. Overgrazing on desert bighorn ranges. *Desert Bighorn Council Transactions* 21:21–23.

Garrison, K. R., J. W. Cain III, E. M. Rominger, and E. J. Goldstein. 2016. Sympatric cattle grazing and desert bighorn sheep foraging. *Journal of Wildlife Management* 80:197–207.

Gaynor, K. M., C. E. Hojnowski, N. H. Carter, and J. S. Brashares. 2018. The influence of human disturbance on wildlife nocturnality. *Science* 360:1232–1235.

Geist, V. 1971. *Mountain Sheep: A Study in Behavior and Evolution*. University of Chicago Press, Chicago, Illinois, USA.

Goldmann, W. 2008. PrP genetics in ruminant transmissible spongiform encephalopathies. *Veterinary Research* 39:1–14.

Gray, G. G., and C. D. Simpson. 1980. *Ammotragus lervia*. *Mammalian Species* 44:1–7.

Griffin, J., and B. Silliman. 2011. Resource partitioning and why it matters. *Nature Education Knowledge* 3:49.

Grinnell, J. 1917. The niche-relationships of the California thrasher. *Auk* 34:427–433.

Gross, J. 1960. History, present, and future status of the desert bighorn sheep (*Ovis canadensis mexicana*) in the Guadalupe Mountains of southeastern New Mexico and northwestern Texas. *Desert Bighorn Council Transactions* 4:66–71.

Gudorf, M., P. Y. Sweanor, F. J. Singer, A. Blankenship, V. C. Bleich, T. Easterly, J. Emmerich, C. Eustace, L. Irby, D. Jaynes, B. Jellison, R. Kissel, J. Lindsay, J. Parks, T. Peters, K. Reid, S. Stewart, and T. Voss. 1996. Bighorn sheep habitat assessment of the Greater Bighorn Canyon National Recreation Area. National Park Service and National Biological Service, Lovell, Wyoming, USA.

Habibi, K. 1982. Some aspects of population dynamics of aoudad in the Hondo Valley, New Mexico. *Desert Bighorn Council Transactions* 26:12–15.

Hall, L. K., R. T. Larsen, M. D. Westover, C. C. Day, R. N. Knight, and B. R. McMillan. 2016. Influence of exotic horses on the use of water by communities of native wildlife in a semi-arid environment. *Journal of Arid Environments* 127:100–105.

Halloran, A. F. 1949. Desert bighorn management. *Transactions of the North American Wildlife Conference* 14:527–537.

Halloran, A. F., and O. V. Deming. 1958. Water development for desert bighorn sheep. *Journal of Wildlife Management* 22:1–9.

Hansen, M., J. Yoakum, W. Pyle, and R. Anthony. 2001. New strategies for pronghorn food habit studies. *Biennial Pronghorn Antelope Workshop Proceedings* 19:71–94.

Hansen, R. M., and P. S. Martin. 1973. Ungulate diets in the lower Grand Canyon. *Journal of Range Management* 26:380–381.

Hansson, S. 1995. Effects of exploitative food competition on food niche dynamics: a simulation analysis. *Ecological Modelling* 77:167–187.

Hardin, G. 1960. The competitive exclusion principle: an idea that took a century to be born has implications in ecology, economics, and genetics. *Science* 131:1292–1297.

Heffelfinger, J. R., and E. K. Latch. 2023. Origin, classification, and distribution. Pages 3–24 in J. R. Heffelfinger and P. R. Krausman, editors. *Ecology and Management of Black-Tailed and Mule Deer of North America*. CRC Press, Boca Raton, Florida, USA.

Hennig, J. D., C. J. Duchardt, S. Esmaeili, S. D. Fuhlendorf, J. L. Beck, T. I. Francisco, and J. D. Scasta. 2023. A crossroads in the rearview mirror: the state of United States feral equid management in 2023. *BioScience* 73:404–407.

Hibbs, L. D. 1967. Food habits of the mountain goat in Colorado. *Journal of Mammalogy* 48:242–248.

Hoefs, M., and V. C. Brink. 1978. Forage production and utilization of a Dall sheep winter range, southwest Yukon Territory. *Northern Wild Sheep and Goat Council Proceedings* 1:87–105.

Holechek, J. L., R. D. Pieper, and C. H. Herbel. 1989. *Range Management Principles and Practices*. Prentice-Hall, Englewood Cliffs, New Jersey, USA.

Holl, S. A., and V. C. Bleich. 1983. San Gabriel mountain sheep: biological and management considerations. USDA Forest Service, San Bernardino National Forest, San Bernardino, California, USA.

Holl, S. A., and V. C. Bleich. 2010. Responses of large mammals to fire and rain in the San Gabriel Mountains, California. *Northern Wild Sheep and Goat Council Proceedings* 17:139–156.

Holl, S. A., V. C. Bleich, B. W. Callenberger, and B. Bahro. 2012. Simulated effects of two fire regimes on bighorn sheep: the San Gabriel Mountains, California, USA. *Fire Ecology* 8:88–103.

Holt, R. D., and J. H. Lawton. 1993. Apparent competition and enemy-free space in insect host-parasitoid communities. *American Naturalist* 142:623–645.

Holt, R. D. and J. H. Lawton. 1994. The ecological consequences of shared natural enemies. *Annual Review of Ecology and Systematics* 25:495–520.

Holt, R. D. and M. B. Bonsall. 2017. Apparent competition. *Annual Review of Ecology, Evolution, and Systematics* 48:447–471.

Houston, D. G. 1982. *The Northern Yellowstone Elk: Ecology and Management*. MacMillan Publishing Company, New York, New York, USA.

Hutchinson, G. E. 1957. Concluding remarks. *Cold Springs Harbor Symposium on Quantitative Biology* 22:415–427.

Iacono, P. C., K. A. Schoenecker, K. R. Manlove, P. J. Jackson, and D. C. Stoner. 2024. Evaluating mountain lion diet before and after a removal of feral horses in a semiarid environment. *Ecosphere* 15:e4919.

Irwin, L. L., and J. M. Peek. 1983. Elk habitat use relative to forest succession in Idaho. *Journal of Wildlife Management* 47:664–672.

Janis, C. 1976. The evolutionary strategy of the Equidae and the origins of rumen and cecal digestion. *Evolution* 30:757–774.

Jansen, B. D., J. R. Heffelfinger, T. H. Noon, P. R. Krausman, and J. C. deVos Jr. 2006. Infectious keratoconjunctivitis in bighorn sheep, Silver Bell Mountains, Arizona, USA. *Journal of Wildlife Diseases* 42:407–411.

Jensen, W. F., V. C. Bleich, and D. G. Whittaker. 2023. Historical trends in black-tailed deer, mule deer, and their habitats. Pages 25–42 *in* J. R. Heffelfinger and P. R. Krausman, editors. *Ecology and Management of Black-Tailed and Mule Deer of North America*. CRC Press, Boca Raton, Florida, USA.

Jesmer, B. R., J. A. Merkle, J. R. Goheen, E. O. Aikens, J. L. Beck, A. B. Courtemanch, M. A. Hurley, D. E. McWhirter, H. M. Miyasaki, K. L. Monteith, and M. J. Kauffman. 2018. Is ungulate migration culturally transmitted? Evidence of social learning from translocated animals. *Science* 361:1023–1025.

Jex, B. A., J. B. Ayotte, V. C. Bleich, C. E. Brewer, D. L. Bruning, T. M. Hegel, N. C. Larter, R. A. Schwanke, H. M. Schwantje, and M. W. Wagner. 2016. Thinhorn sheep: conservation challenges and management strategies for the 21st century. Wild Sheep Working Group, Western Association of Fish and Wildlife Agencies, Cheyenne, Wyoming, USA.

Johnson, B. K., J. W. Kern, M. J. Wisdom, S. L. Findholt, and J. G. Kie. 2000. Resource selection and spatial separation of mule deer and elk during spring. *Journal of Wildlife Management* 64:685–697.

Johnson, H. E., M. Hebblewhite, T. R. Stephenson, D. W. German, B. M. Pierce, and V. C. Bleich. 2013. Evaluating apparent competition in limiting the recovery of an endangered ungulate. *Oecologia* 171:295–307.

Jones, F. L. 1955. Bighorn management problems in California. *Annual Conference of the Western Association of State Game and Fish Commissioners Proceedings* 35:177–181.

Jones, F. L. 1980. Competition. Pages 197–216 *in* G. Monson and L. Sumner, editors. *The Desert Bighorn: Its Life History, Ecology, and Management*. University of Arizona Press, Tucson, USA.

Jung, T. S., S. A. Stotyn, and S. M. Czetwertynski. 2015a. Dietary overlap and potential competition in a dynamic ungulate community in northwestern Canada. *Journal of Wildlife Management* 79:1277–1285.

Jung, T. S., T. M. Hegel, S. A. Stotyn, and S. M. Czetwertynski. 2015b. Co-occurrence of reintroduced and resident ungulates on a shared winter range in northwestern Canada. *Ecoscience* 22:7–16.

Kelly, W. E. 1960. Bighorn sheep management recommendations for the state of Arizona. *Desert Bighorn Council Transactions* 4:41–44.

Kissell, R. E., Jr. 1996. Population dynamics, food habits, seasonal habitat use, and spatial relationships of bighorn sheep, mule deer, and feral horses in the Pryor Mountains. Dissertation, Montana State University, Bozeman, USA.

Krausman, P. R., A. V. Sandoval, and R. C. Etchberger. 1999. Natural history of desert bighorn sheep. Pages 139–191 *in* R. Valdez and P. R. Krausman, editors. *Mountain sheep of North America*. University of Arizona Press, Tucson, USA.

Krausman, P. R., and R. T. Bowyer. 2003. Mountain sheep (*Ovis canadensis* and *O. dalli*). Pages 1095–1115 *in* G. A. Feldhamer, B. C. Thompson, and J. A. Chapman, editors. *Wild Mammals of North America: Biology, Management, and Conservation*. John Hopkins University Press, Baltimore, Maryland, USA.

Larsen, R. T., J. A. Bissonette, J. T. Flinders, and J. C. Whiting. 2012. Framework for understanding the influences of wildlife water developments in the western United States. *California Fish and Game* 98:148–163.

Laundré, J. W. 1994. Resource overlap between mountain goats and bighorn sheep. *Great Basin Naturalist* 54:114–121.

Liu, B., and Z. Jiang. 2004. Dietary overlap between Przewalski's gazelle and domestic sheep in the Qinghai Lake region and implications for rangeland management. *Journal of Wildlife Management* 68:241–246.

Loft, E. R., J. W. Menke, and J. G. Kie. 1991. Habitat shifts by mule deer: the influence of cattle grazing. *Journal of Wildlife Management* 55:16–26.

Lyon, L., and J. Thomas. 1987. Elk: Rocky Mountain majesty. Pages 145–159 *in* H. Kallman, editor. *Restoring America's Wildlife 1937–1987: The First 50 Years of the Federal Aid in Wildlife Restoration (Pittman-Robertson) Act*. U.S. Fish and Wildlife Service, Washington, D.C., USA.

Lyon, L. J., and A. L. Ward. 1982. Elk and land management. Pages 443–478 *in* J. W. Thomas and D. E. Toweill, editors. *Elk of North America: Ecology and Management*. Stackpole Books, Harrisburg, Pennsylvania, USA.

Marshal, J. P., V. C. Bleich, and N. G. Andrew. 2008. Evidence for interspecific competition between feral ass *Equus asinus* and mountain sheep *Ovis canadensis* in a desert environment. *Wildlife Biology* 14:228–236.

Marshal, J. P., V. C. Bleich, P. R. Krausman, M.-L. Reed, and A. Neibergs. 2012. Overlap in diet and habitat between the mule deer (*Odocoileus hemionus*) and feral ass (*Equus asinus*) in the Sonoran Desert. *Southwestern Naturalist* 57:16–25.

McCullough, D. R. 1969. The tule elk: its history, behavior and ecology. *University of California Publications in Zoology* 88:1–209.

McCullough, D. R., and E. R. Schneegas. 1966. Winter observations on the Sierra Nevada bighorn sheep. *California Fish and Game* 52:68–84.

McCullough, Y. B. 1980. Niche separation of seven North American ungulates on the National Bison Range, Montana. Dissertation, University of Michigan, Ann Arbor, Michigan, USA.

McCutchen, H. E. 1981. Desert bighorn zoogeography and adaptation in relation to historic land use. *Wildlife Society Bulletin* 9:171–179.

McKnight, T. L. 1958. The feral burro in the United States: distribution and problems. *Journal of Wildlife Management* 22:163–179.

McKnight, T. L. 1961. A survey of feral livestock in California. *Yearbook of the Association of Pacific Coast Geographers* 23:28–42.

Mills, H. B. 1937. A preliminary study of the bighorn of Yellowstone National Park. *Journal of Mammalogy* 18:205–212.

Mori, E., G. Mazza, L. Saggiomo, A. Sommese, and B. Esattore. 2017. Strangers coming from the Sahara: an update on the worldwide distribution, potential impacts and conservation of alien aoudad. *Annales Zoologici Fennici* 54:373–386.

Morrison, M. L. 2013. Wildlife restoration. Pages 246–261 *in* P. R. Krausman and J. W. Cain, editors. *Wildlife Management and Conservation: Contemporary Principles and Practices*. John Hopkins University Press, Baltimore, Maryland, USA.

Murie, A. 1940. *Ecology of the Coyote in the Yellowstone. Fauna of the National Parks of the United States 4*. U.S. Government Printing Office, Washington, D.C., USA.

Murray, M. G., and A. W. Illius. 2000. Vegetation modification and resource competition in grazing ungulates. *Oikos* 89:501–508.

National Research Council. 2013. *Using Science to Improve the BLM Wild Horse and Burro Program: A Way Forward*. The National Academies Press, Washington, D.C., USA.

O'Farrill, G., R. A. Medellín, S. F. Matter, and G. N. Cameron. 2019. Habitat use and diet of desert bighorn sheep (*Ovis canadensis mexicana*) and endemic mule deer (*Odocoileus hemionus nelsonii*) on Tiburón Island, Mexico. *Southwestern Naturalist* 63:225–234.

Olsen, N. 2022. Mountain goats and bighorn sheep battle in climate crisis, new study shows. Available at https://source.colostate.edu/mountain-goats-and-bighorn-sheep-battle-in-climate-crisis-new-study-shows/. Accessed 9 January 2025.

Olsen, S. L. 2016. The roles of humans in horse distribution through time. Pages 105–120 *in* J. I. Ransom and P. Kaczensky, editors. *Wild Equids: Ecology, Management, and Conservation.* Johns Hopkins University Press, Baltimore, Maryland, USA.

Ostermann-Kelm, S., E. R. Atwill, E. S. Rubin, M. C. Jorgensen, and W. M. Boyce. 2008. Interactions between feral horses and desert bighorn sheep at water. *Journal of Mammalogy* 89:459–466.

Ostermann-Kelm, S. D., E. A. Atwill, E. S. Rubin, L. E. Hendrickson, and W. M. Boyce. 2009. Impacts of feral horses on a desert environment. *BMC Ecology* 9:1–10.

Parikh, G. L., J. L. Etchart, R. O'Shaughnessy, L. A. Harveson, and J. W. Cain. 2025. Feeding habits of sympatric aoudad *(Ammotragus lervia)* and desert bighorn sheep (*Ovis canadensis mexicana*) in West Texas. *Journal of Wildlife Management* e70008. https://doi.org/10.1002/jwmg.70008

Pianka, E. R. 2011. *Evolutionary Ecology*. Seventh edition. eBook. https://www.google.com/books/edition/Evolutionary_Ecology/giFL5bonGhQC?hl=en&gbpv=1&printsec=frontc over. Accessed 26 March 2025.

Radke, M. 2025. Rugged vulnerability. *Wildlife Professional* 19:18–26.

Reed, D. F. 2001. A conceptual interference competition model for introduced mountain goats. *Journal of Wildlife Management* 65:125–128.

Risenhoover, K. L., J. A. Bailey, and L. A. Wakelyn. 1988. Assessing the Rocky Mountain bighorn sheep management problem. *Wildlife Society Bulletin* 16:346–352.

Robinson, R. W., J. C. Whiting, J. M. Shannon, D. D. Olson, J. T. Flinders, T. S. Smith, and R. T. Bowyer. 2019. Habitat use and social mixing between groups of resident and augmented bighorn sheep. *Scientific Reports* 9:14984.

Robinson, R. W., T. S. Smith, J. C. Whiting, R. T. Larsen, and J. M. Shannon. 2020. Determining timing of births and habitat selection to identify lambing period habitat for bighorn sheep. *Frontiers in Ecology and Evolution* 8:97. doi:10.3389/fevo.2020.00097

Rominger, E. M. 2018. The Gordian Knot of mountain lion predation and bighorn sheep. *Journal of Wildlife Management* 82:19–31.

Rominger, E. M., H. A. Whitlaw, D. L. Weybright, W. C. Dunn, and W. B. Ballard. 2004. The influence of mountain lion predation on bighorn sheep translocations. *Journal of Wildlife Management* 68:993–999.

Rosenzweig, M. L. 1981. A theory of habitat selection. *Ecology* 62:327–335.

Rubin, E. S., D. Conrad, L. E. Harding, and B. M. Russo. 2024. Associations between a feral equid and the Sonoran Desert ecosystem. *Wildlife Monographs* 215:e1083.

Russo, J. 1956. The desert bighorn sheep in Arizona: a research and management study. *Arizona Game and Fish Department Wildlife Bulletin* 1:1–153.

Saiz, R. B. 1975. Ecology and behavior of the gemsbok at White Sands Missile Range, New Mexico. Thesis, Colorado State University, Fort Collins, Colorado, USA.

Sax, D. F., R. Early, and J. Bellemare. 2013. Niche syndromes, species extinction risks, and management under climate change. *Trends in Ecology & Evolution* 28:517–523.

Schallenberger, A. 1965. Big game forage competition in the Sun River Canyon. Thesis, Montana State University, Bozeman, Montana, USA.

Schoenecker, K. A., S. R. B. King, and T. A. Messmer. 2021. The wildlife professional's duty in achieving science-based sustainable management of free-roaming equids. *Journal of Wildlife Management* 85:1057–1061.

Schoenecker, K. A., S. R. B. King, M. K. Nordquist, D. Nandintsetseg, and Q. Cao. 2016. Habitat and diet of equids. Pages 41–57 *in* J. I. Ransom and P. Kaczensky, editors. *Wild Equids: Ecology, Management, and Conservation.* Johns Hopkins University Press, Baltimore, Maryland, USA.

Schoenecker, K. A., F. J. Singer, K. A. Grams, and J. E. Roelle. 2004. Bighorn sheep (*Ovis canadensis*) survivorship and habitat studies in Bighorn Canyon National Recreation Area and surrounding lands, Wyoming and Montana, 2000–2003. Pages 3–33 in K. A. Schoenecker, editor. *Bighorn Sheep Habitat Studies, Population Dynamics, and Population Modeling in Bighorn Canyon National Recreation Area, 2000–2003.* Open File Report 2004-1337. U.S. Geological Survey, Ft. Collins, Colorado, USA.

Schoener, T. W. 1974. Resource partitioning in ecological communities: research on how similar species divide resources helps reveal the natural regulation of species diversity. *Science* 185:27–39.

Schoener, T. W. 1982. The controversy over interspecific competition. *American Scientist* 70:586–595.

Schwartz, C. C., and J. E. Ellis. 1981. Feeding ecology and niche separation in some native and domestic ungulates on the shortgrass prairie. *Journal of Applied Ecology* 18:343–353.

Seegmiller, R. F., and R. D. Ohmart. 1981. Ecological relationships of feral burros and desert bighorn sheep. *Wildlife Monographs* 78:1–58.

Seegmiller, R. F., and C. D. Simpson. 1979. The Barbary sheep: some conceptual implications of competition with desert bighorn. *Desert Bighorn Council Transactions* 23:47–49.

Shackleton, D. M., C. C. Shank, and B. M. Wikeem. 1999. Natural history of Rocky Mountain and California bighorn sheep. Pages 78–138 *in* R. Valdez and P. R. Krausman, editors. *Mountain Sheep of North America.* University of Arizona Press, Tucson, Arizona, USA.

Shimadzu, H., M. Dornelas, P. A. Henderson, and A. E. Magurran. 2013. Diversity is maintained by seasonal variation in species abundance. *BMC Biology* 11:1–9.

Simpson, C. D., and L. J. Krysl. 1981. Status and distribution of Barbary sheep in the southwest United States. *Desert Bighorn Council Transactions* 25:9–15.

Simpson, C. D., L. J. Krysl, D. B. Hampy, and G. G. Gray. 1978. The Barbary sheep: a threat to desert bighorn survival. *Desert Bighorn Council Transactions* 22:26–31.

Singer, F. J., V. C. Bleich, and M. A. Gudorf. 2000a. Restoration of bighorn sheep metapopulations in and near western National Parks. *Restoration Ecology* 8:14–24.

Singer, F. J., C. M. Papouchis, and K. K. Symonds. 2000b. Translocations as a tool for restoring populations of bighorn sheep. *Restoration Ecology* 8:6–13.

Sittler, K. L., K. L. Parker, and M. P. Gillingham. 2015. Resource separation by mountain ungulates on a landscape modified by fire. *Journal of Wildlife Management* 79:591–604.

Sittler, K. L., K. L. Parker, and M. l. P. Gillingham. 2019. Vegetation and prescribed fire: implications for Stone's sheep and elk. *Journal of Wildlife Management* 83:393–409.

Smith, T. S., J. T. Flinders, and D. S. Winn. 1991. A habitat evaluation procedure for Rocky Mountain bighorn sheep in the Intermountain West. *Great Basin Naturalist* 51:205–225.

Spear, D., and S. L. Chown. 2009. Non-indigenous ungulates as a threat to biodiversity. *Journal of Zoology* 279:1–17.

Spitz, D. B., M. Hebblewhite, T. R. Stephenson, and D. W. German. 2018. How plastic is migratory behavior? Quantifying elevational movement in a partially migratory alpine ungulate, the Sierra Nevada bighorn sheep (*Ovis canadensis sierrae*). *Canadian Journal of Zoology* 96:1385–1394.

St. John, K. P. J. 1965. Competition between desert bighorn sheep and feral burros for forage in Death Valley National Monument. *Desert Bighorn Council Transactions* 9:89–92.

Stelfox, J. G. 1974. Range ecology of bighorn sheep in relation to self-regulation theories. *Northern Wild Sheep Council Proceedings* 3:67–76.

Stevens, D. R., and N. J. Goodson. 1993. Assessing effects of removals for transplanting on a high-elevation bighorn sheep population. *Conservation Biology* 7:908–915.

Stewart, K. M., R. T. Bowyer, J. G. Kie, N. J. Cimon, and B. K. Johnson. 2002. Temporospatial distributions of elk, mule deer, and cattle: resource partitioning and competitive displacement. *Journal of Mammalogy* 83:229–244.

Sumner, L. 1959. Effects of wild burros on bighorn in Death Valley National Monument. *Desert Bighorn Council Transactions* 3:4–8.

Taylor, S., W. Wall, and Y. Kulis. 2006. Habitat selection by mountain goats in south coastal British Columbia. *Northern Wild Sheep and Goat Council Proceedings* 15:141–157.

Tomkiewicz, S. M., M. R. Fuller, J. G. Kie, and K. K. Bates. 2010. Global positioning system and associated technologies in animal behaviour and ecological research. *Philosophical Transactions of the Royal Society B: Biological Sciences* 365:2163–2176.

United States Fish and Wildlife Service. 2000. *Recovery Plan for Bighorn Sheep in the Peninsular Ranges, California*. U.S. Fish and Wildlife Service, Portland, Oregon, USA.

United States Fish and Wildlife Service. 2007. *Recovery Plan for the Sierra Nevada Bighorn Sheep*. U.S. Fish and Wildlife Service, Sacramento, California, USA.

Van den Akker, J. B. 1960. Human encroachment on bighorn habitat. *Desert Bighorn Council Transactions* 4:38–40.

Van Soest, P. 1994. *Nutritional Ecology of the Ruminant*. Cornell University Press, Ithaca, New York, USA.

Villa-R., B. 1959. Brief notes on the present status and distribution of bighorn sheep in Mexico. *Desert Bighorn Council Transactions* 3:77–79.

Wagner, F. H. 1983. Status of wild horse and burro management on public rangelands. *North American Wildlife and Natural Resources Conference Transactions* 48:116–133.

Wagoner, J. J. 1952. History of the cattle industry in southern Arizona, 1540–1940. *University of Arizona Social Sciences Bulletin* 20:1–132.

Walker, A. B., K. L. Parker, and M. P. Gillingham. 2006. Behaviour, habitat associations, and intrasexual differences of female Stone's sheep. *Canadian Journal of Zoology* 84:1187–1201.

Walker, A. B., K. L. Parker, M. P. Gillingham, D. D. Gustine, and R. J. Lay. 2007. Habitat selection by female Stone's sheep in relation to vegetation, topography, and risk of predation. *Ecoscience* 14:55–70.

Walter, G. H. 1991.What is resource partitioning? *Journal of Theoretical Biology* 150:137–143.

Weaver, R. A. 1959. Effects of burros on desert water supplies. *Desert Bighorn Council Transactions* 3:1–3.

Weaver, R. A. 1972. Conclusions of the bighorn investigation in California. *Desert Bighorn Council Transactions* 16:56–65.

Weaver, R. A. 1973. Burro versus bighorn. *Desert Bighorn Council Transactions* 17:90–94.

Webber, Q. M., K. M. Ferraro, J. G. Hendrix, and E. Vander Wal. 2022. What do caribou eat? A review of the literature on caribou diet. *Canadian Journal of Zoology* 100:197–207.

Wehausen, J. D. 1996. Effects of mountain lion predation on bighorn sheep in the Sierra Nevada and Granite Mountains of California. *Wildlife Society Bulletin* 24:471–479.

Welles, R. E. 1960. Progress report on current Death Valley burro survey. *Desert Bighorn Council Transactions* 4:85–87.

Welles, R. E., and F. B. Welles. 1961. *The Bighorn of Death Valley. Fauna of the National Parks of the United States 6*. U.S. Government Printing Office, Washington, D.C., USA.

Western Association of Fish and Wildlife Agencies. 2024. *Recommendations for Domestic Sheep and Goat Management in Wild Sheep Habitat*. Wild Sheep Initiative, Western Association of Fish and Wildlife Agencies, Boise, Idaho, USA.

Whiting, J. C., D. D. Olson, J. M. Shannon, R. T. Bowyer, R. W. Klaver, and J. T. Flinders. 2012. Timing and synchrony of births in bighorn sheep: implications for reintroduction and conservation. *Wildlife Research* 39:565–572.

Whiting J. C., R. T. Bowyer, and J. T. Flinders. 2009. Diel use of water by reintroduced bighorn sheep. *Western North American Naturalist* 69:407–412.

Whiting, J. C., R. T. Bowyer, J. T. Flinders, V. C. Bleich, and J. G. Kie. 2010. Sexual segregation and use of water by bighorn sheep: implications for conservation. *Animal Conservation* 13:541–548.

Whiting, J. C., V. C. Bleich, R. T. Bowyer, and C. W. Epps. 2023a. Restoration of bighorn sheep: history, successes, and remaining conservation issues. *Frontiers in Ecology and Evolution* 11:1083350.

Whiting, J. C., V. C. Bleich, R. T. Bowyer, K. Manlove, and K. White. 2023b. Bighorn sheep and mountain goats. Pages 759–790 in L. B. McNew, D. K. Dahlgren, and J. L. Beck, editors. *Rangeland Wildlife Ecology and Conservation*. Springer International Publishing, Cham, Switzerland.

Wiedmann, B. P., and B. M. Hosek. 2013. North Dakota bighorn sheep management plan (2013–2023). Federal Aid in Wildlife Restoration Project Report A-213. North Dakota Game and Fish Department, Bismarck, North Dakota, USA.

Wiedmann, B. P., V. C. Bleich, and C. E. Penner. 2024. North Dakota bighorn sheep management plan (2024–2034). Federal Aid in Wildlife Restoration Project Report A-290. North Dakota Game and Fish Department, Bismarck, North Dakota, USA.

Wisdom, M. J., and J. W. Thomas. 1996. Elk. Pages 157–181 in P. R. Krausman, editor. *Rangeland Wildlife*. Society for Range Management, Denver, Colorado, USA.

Wright, E. A., G. G. Brugette, K. F. Buckert, F. Hernández, J. H. Reed, S. R. Wyckoff, J. C. Taylor, K. R. Manlove, C. D. Phillips, and R. D. Bradley. 2024. Multi-locus sequence typing indicates multiple strains of *Mycoplasma* in desert bighorn sheep and aoudad in Texas. *Journal of Wildlife Management* 88:e22594.

Wright, E. A., R. C. Wiedmeier, E. K. Roberts, D. R. Pipkin, F. Hernández, J. P. Bayouth, W. C. Conway, and R. D. Bradley. 2022. Distinct mtDNA lineages in free-ranging *Ammotragus* (aoudad) from the United States indicate multiple introductions from northern Africa. *Ecology and Evolution* 12:e8849.

Yoakum, J. D. 2004. Relationships with other herbivores. Pages 501–538 in B. W. O'Gara and J. D. Yoakum, editors. *Pronghorn Ecology and Management*. University of Colorado Press, Boulder, Colorado, USA.

# 11 Seasonal Migrations and Other Movements

*Blake Lowrey*

## 11.1 INTRODUCTION

Migration has evolved across taxa as an important behavioral response that allows animals to persist in seasonal environments. Seasonal changes in abiotic conditions, such as weather, or biotic conditions, such as forage, competition, or susceptibility to disease or predation, can result in varying conditions over the course of a year. Migration provides a behavioral mechanism whereby animals can move between seasonal ranges to exploit spatially and temporally variable resources or to avoid areas that are unfavorable due to predation or other threats (Dingle and Drake 2007). Foundationally, the act of migrating should enhance an individual's opportunities to survive through providing access to high-quality resources and safer environments (Fryxell and Sinclair 1988). Migratory behaviors span a continuum from range residency to long-distance migrations (Cagnacci et al. 2011, Barker et al. 2018), and are key components in promoting sustainable and abundant ungulate populations (Bolger et al. 2008). In this chapter, migration is defined broadly as an individual behavior in which animals make repeated seasonal movements between mostly nonoverlapping ranges (Dingle and Drake 2007). Migrations are round-trip movements, usually leaving a winter range in spring, traveling to a summer range, and returning to winter range in fall. Nonmigratory individuals that remain in a single annual range for the duration of the year are referred to as residents. Additionally, other nonmigratory movements such as forays, dispersals, or visits to specific resources such as mineral licks and water can help animals access important resources or aid in range expansion and are important components of the mountain sheep (bighorn [*Ovis canadensis]*, thinhorn [*Ovis dalli*]) movement portfolio.

The advent of global positioning system (GPS) collars in wildlife research provided an important step forward in documenting and detailing animal movement. By collecting a continuous stream of GPS locations from collared animals, for the first time, biologists could see clear migratory routes by connecting GPS locations and discern other attributes such as migration timing, duration, and distance, or the use of stopovers along a migratory route (Craighead et al. 1972, Sawyer and Kauffman 2011, Courtemanch et al. 2017). While mountain sheep are generally considered migratory across their broad distributions, there is a high degree of variability between and among mountain sheep species and populations, the details of which have become clearer though

the continued deployment of GPS collars as a management tool (Geist 1971, Hengeveld and Cubberley 2012, Lowrey et al. 2021, Proffitt et al. 2021). As a result, much of the current understanding of mountain sheep migration is relatively new and has not been synthesized in a single source. For example, since the last compiled volume on mountain sheep (Valdez and Krausman 1999), biologists aiding the recovery of the Sierra Nevada bighorn sheep (*O. c. sierrae*) in California have deployed GPS collars on over 300 animals across 15 subpopulations. Further, Wyoming Game and Fish Department and other collaborators have deployed nearly 550 GPS collars on Rocky Mountain bighorn sheep (*Ovis canadensis canadensis*) across Wyoming. Additional smaller scale collaring efforts have also been conducted for desert bighorn sheep (*O. c. nelsoni*) and thinhorn species (Dall's bighorn sheep [*O. dalli*], Stone's bighorn sheep [*O. d. stonei*]). The deployment of thousands of GPS collars across the range of mountain sheep over the last 30 years has greatly enhanced our understanding of the many facets of their spatial ecology. Using integrated approaches, these efforts have contributed valuable insights to mountain sheep spatial ecology (including migration), disease, population dynamics, habitat, and predation, among other topics. Building from historical records and studies, and more recent research efforts, this chapter provides a comprehensive overview of mountain sheep migration and other movements across their broad distribution in western North America (Chapter 1).

## 11.2 MIGRATORY BEHAVIORS OF MOUNTAIN SHEEP

Mountain sheep are generally considered migratory with individuals moving between discrete seasonal ranges, yet there is a high degree of variation between populations, and in many cases, variation among individuals in a single population (Geist 1971, Hurley 1985, Festa-Bianchet 1986a). While populations can be fully resident (DeVoe et al. 2020) or fully migratory (Lula et al. 2020), most mountain sheep populations are partially migratory and contain resident and migrant individuals, which overlap on a single range for a portion of the year (Chapman et al. 2011). Within the migrant component, differences in timing, duration, geographic distance, and varied elevational gradients can result in diverse patterns among migrant individuals. While some individuals may undertake relatively short migrations that

DOI: 10.1201/9781003518686-12

largely traverse elevational gradients, other individuals traverse broad geographic distances over complex landscapes. Further, resident behaviors can occur at low- and high-elevation areas whereby individuals remain within high-mountain environments or low-elevation canyon, badland, or prairie break landscapes for the annual period.

Migratory behaviors for mountain sheep are not species-specific. Rather, there are examples of many different behaviors across the broad range of mountain sheep species. Given the diversity of movements, discretizing the continuum of migratory behavior into separate classifications, each with a clean definition, is a difficult task. While the binary delineation between resident and migrant behaviors is relatively straightforward based on individual seasonal distributions, there is no standard approach to further classify migrants into more specific behaviors (e.g., elevational, geographic migrations). Commonly, migratory classifications follow a general theme the uses landscape complexity and migration distance to delineate migratory behaviors (Hurley 1985). Herein, mountain sheep migrations are described in three broad categories: elevational migrations and related variants, geographic migrations, and annual residents. These classifications are not discrete, yet provide a general framework for discussing migratory movements in mountain sheep.

## 11.2.1 ELEVATIONAL MIGRATIONS

Elevational (i.e., altitudinal) migrations largely occur along an elevation gradient whereby animals wintering at low elevations move to high-elevation summer ranges in the spring and then return to low elevations in fall (Figure 11.1). The migrations are relatively direct yet can span a range of elevation gradients and traverse varying geographic distances. On average, mountain sheep elevational migrations are relatively short and gain 500–1,500 m of elevation over a 5–15 km migration that occurs over 1–2 weeks (Geist 1971, Seip and Bunnell 1984, Demarchi and Hartwig 2004, Courtemanch et al. 2017, Lowrey et al. 2020). Spring migrations are a gradual process with animals commonly aligning their movements with the spring green-up and moving up in elevation during May and June, although inclement weather or snow can cause animals to return to winter range for short periods (Geist 1971, Seip 1983, Festa-Bianchet 1988). The onset of spring migration is commonly influenced by the emergence of green vegetation, which is influenced by other correlated factors such as temperature and snow (Hoefs 1976). Fall migrations often occur during October and November and align with the first significant snow events at high elevations, causing animals to move to lower elevations to prolong access to summer forage (Hoefs 1976, Lowrey et al. 2020). Elevational migrations and the strong association with spring green-up are behavioral adaptations to the seasonality of mountain environments. Consequently, these migrations are more common for Rocky Mountain bighorn sheep and thinhorn sheep inhabiting mountain environments. The movements

of bighorn sheep in arid environments such as canyons or prairie breaks are less aligned with spring and fall seasonality and are more related to water availability or exploratory movements between isolated habitat patches (see Section 11.6).

Elevational migrations are an expression of the forage maturation hypothesis whereby herbivores seek out and consume vegetation at an intermediate phenological state with maximum crude protein and digestibility to maximize energy intake (Fryxell 1991). In mountain environments, migrants can track the phenological wave of peak greenness in spring and move up in elevation to continually intake high-quality forage at the leading edge of spring green-up (Pettorelli et al. 2007, Merkle et al. 2016, Aikens et al. 2017). By migrating, elevational migrants effectively extend their access to the best available forage over spring and summer. Through experimental trials comparing body mass of individuals fed diets that mimicked forage conditions on year-round low-elevation ranges (i.e., the control) and diets with higher crude protein in spring and summer to simulate migration, researchers showed a benefit of migration through increased mass gain for bighorn sheep fed the simulated migratory diets (Hebert 1973). More recent work with wild populations has shown similar benefits in other ungulates (Middleton et al. 2018). Relatedly, by prolonging access to forage, migrants shorten the winter period of mass loss and reduce their reliance on fat reserves. As a result, at lower body mass levels, migrants can retain a higher over-winter survival when compared to residents, partially decoupling survival with over-winter fat reserves (Denryter et al. 2022). As the high-mountain environments become snow-covered in the late fall and early winter, the forage benefit is lessened, and animals move back to low-elevation winter ranges, which generally experience milder winter conditions. For males, the rut additionally influences timing of the fall migration as males will seek out rutting areas during the breeding season (see Section 11.4; Demarchi and Hartwig 2004).

While elevational migrations have traditionally been described as seasonal round-trip movements between low-elevation winter ranges and high-elevation summer ranges, more recent work has expanded the breadth of elevational migrations to include short-term and repeated movements between seasonal ranges that are separated by an elevational gradient. For example, abbreviated migrations describe the short-term spring or fall movement of high-elevation residents to lower elevation foraging areas for brief periods before returning to high-elevation ranges for the duration of summer or winter. Abbreviated migrations (i.e., reverse migrations) are not committed movements between two distinct ranges used for the duration of a season, but instead, function as abbreviated movements between high-elevation annual ranges and lower elevation foraging areas (Courtemanch et al. 2017). Although the term abbreviated migration is relatively new and was used to describe the migrations of Rocky Mountain bighorn sheep in the Teton Range of Wyoming (Courtemanch

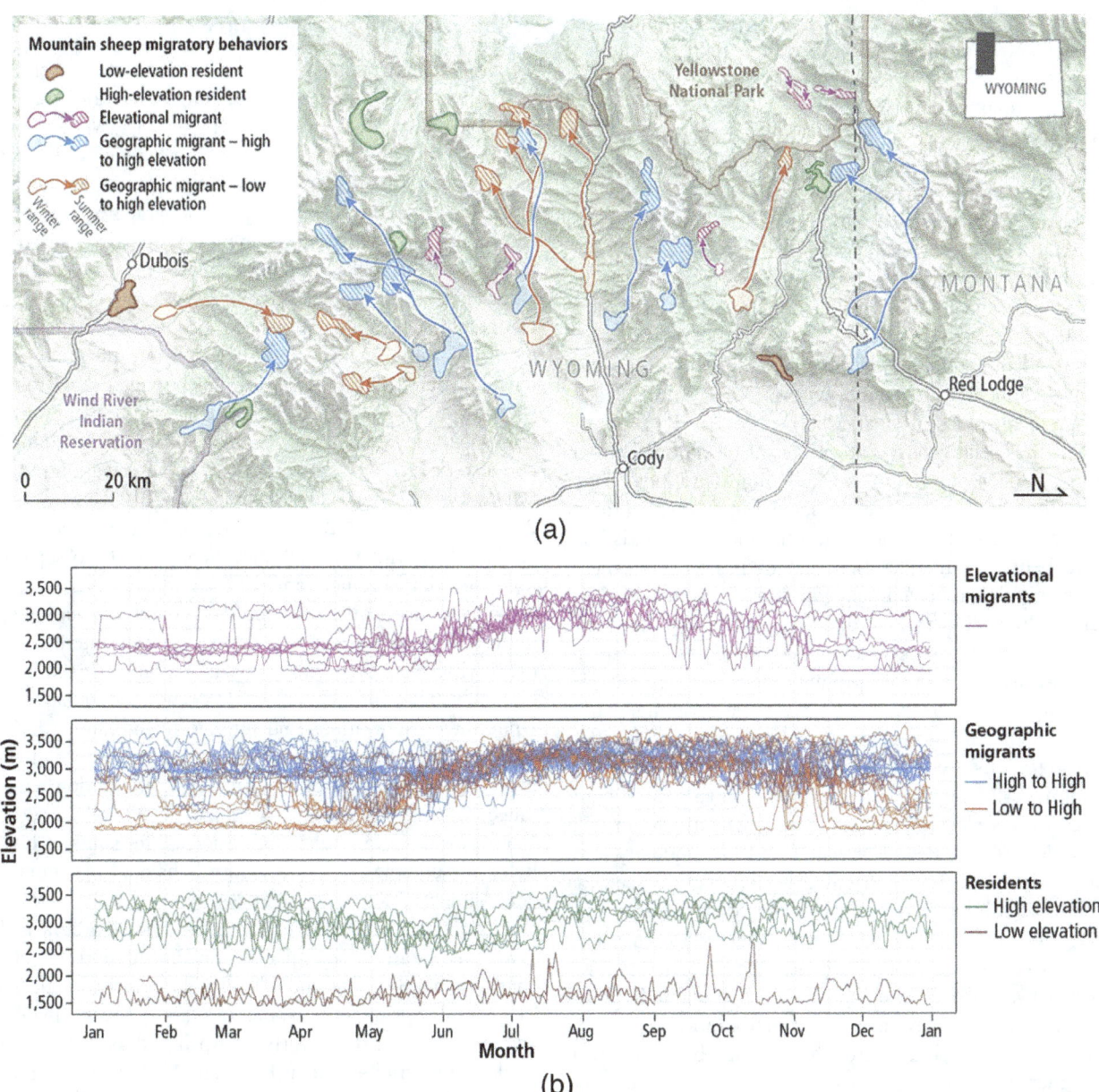

**FIGURE 11.1** Migratory behaviors of mountain sheep as illustrated by Rocky Mountain bighorn sheep in the Absaroka Range of western Wyoming. (a) Generalized seasonal ranges and migration routes for elevational migrants (purple); geographic migrants moving between low-elevation winter ranges and high-elevation summer ranges (orange) and between high-elevation winter ranges and high-elevation summer ranges (blue); and annual residents at low (brown) and high (green) elevations. Each of the generalized migrations and seasonal ranges represents seasonal movements from an individual that was monitored with a global positioning collar. The selected individuals show a sample of the diversity of movements in this region. (b) Corresponding elevation profiles for the same individuals. Note the abbreviated migrations for the high-elevation residents represented by the drop in elevation in May and the general increase in elevation in June and July. In tandem, the two figures help to depict the differences between the migratory behaviors concerning geographic distance (a) and the elevations used seasonally (b).

et al. 2017), descriptions of the behavior have been noted in early observational studies for thinhorn sheep. For example, Seip and Bunnell (1984), observed Stone's sheep moving down in elevation 1,000–1,500 m in late April to forage on lower elevation subalpine areas before moving back up to high-elevation annual ranges. Similar movements have been described for other thinhorn populations across their ranges (Demarchi and Hartwig 2004, Walker et al. 2007,

Aycrigg et al. 2021). By migrating down to lower elevations in spring, animals can access newly emergent vegetation 1–1.5 months before the green-up arrives in alpine areas. Similar movements occur in fall, whereby animals respond to the first significant snow in the high peaks by moving to mid-elevations to prolong access to snow-free forage areas. Once snow begins to accumulate at mid-elevations, however, animals ascend back to their high-elevation winter ranges, which are wind-blown and generally remain free of deep snow for the duration of winter (Seip and Bunnell 1984, Courtemanch et al. 2017, Enns et al. 2023). In general, abbreviated migrations occur between April and early June in the spring and in early October during fall. Animals cover a range of elevational and geographic distances, which are influenced by local geography and weather. Abbreviated migrations in the Teton Range are relatively gradual and traverse gradients of 500 m over 10 km (Courtemanch et al. 2017). Abbreviated migrations for Stone's sheep in British Columbia traversed 1,000–1,500 m of elevation over similar geographic distances (Seip and Bunnell 1984), although some populations have been documented descending as little as 150 m to access forage in spring (Enns et al. 2023).

Another relatively recent variant of elevational migrations has been referred to as vacillating migrations. These migrations are similar to abbreviated migrations, but rather than being a single spring or fall migration, animals vacillate between high-elevation and mid- or low-elevation areas, making repeated short-duration movements, often during winter (Denryter et al. 2021). Vacillating migrations are a flexible behavior and enable individuals to respond to real-time cues that may prompt movement, for example, to escape predation or pursue favorable foraging conditions. In the Sierra Nevada of California where vacillating migrations are best described, individuals that winter at high elevations have reduced predation risk when compared to low-elevation winter ranges, which overlap with mule deer (*Odocoileus hemionus*), and consequently have higher predation rates attributed to apparent competition as mediated by a shared predator, the cougar (*Puma concolor*; Johnson et al. 2013). While the high-elevation ranges may provide reduced predation risk, they are also characterized by relatively poor forage. As a result, individuals wintering at high elevations occasionally move to low and mid-elevations to forage throughout winter, but then return to higher elevations in response to local environmental conditions (John et al. 2024). The decision to undergo a vacillating migration is modulated by balancing the risk-reward tradeoff between predation and forage. In a recent study, fatter individuals had a decreased propensity to migrate and were able to stay resident in the relatively low-risk high-elevation winter ranges while thinner animals had an increased migratory propensity and also migrated further distances to access better forage on account of their lower fat reserves and decreased body condition during winter (Denryter et al. 2024). The flexibility of vacillating migrations allows individuals to exploit similar resource dynamics as long-distance geographic migrants, but do so within a semi-local scale and change ranges in response to real-time cues (John et al. 2024). While vacillating migrations have predominantly focused on elevational movements, similar back-and-forth geographic movements between seasonal ranges have been noted for Stone's sheep, which can make repeat movements over 10 km between summer and winter ranges (Enns et al. 2023).

## 11.2.2  Geographic Migrations

Geographic migrations occur over varying distances but are generally longer than elevational migrations and traverse complex landscapes (Hurley 1985, Figure 11.1). While elevational migrations classically ascend a simple elevation gradient, geographic migrations can traverse multiple drainage systems across broad landscapes. Geographic migrations can be further subset into different classifications based on distance thresholds, for example, short-, mid-, or long-distance migrations (Lowrey et al. 2020, Enns et al. 2023). Geographic migrations often have an elevational component whereby animals are moving between low-elevation winter ranges and high-elevation summer ranges. As a result, short-distance geographic migrations can be similar to traditional elevational migrations. Geographic migrations can also occur between two distinct high-elevation ranges (Hurley 1985, Hengeveld and Cubberley 2012, Lowrey et al. 2019, Enns et al. 2023, Figure 11.1). In these scenarios, animals that winter at high elevations migrate to separate high-elevation summer ranges. These movements seem to exploit season-specific resources and conditions but remain in the alpine environment throughout the year. Interestingly, the high-elevation winter ranges may also function as high-elevation summer ranges for other individuals that winter at low elevations and migrate to the alpine during spring. The swapping of seasonal ranges whereby individuals migrate as far as 50 km from a high-elevation winter range that is subsequently occupied by other individuals (both elevational migrants and residents) for summer highlights the complexity of migratory behaviors in mountain environments (Lowrey et al. 2020).

Geographic migration distances and dates are variable across mountain sheep species. Specific departure dates change annually as individuals respond to local conditions. In general, spring geographic migrations occur in May and June but can be as early as April and as late as July, while fall migrations occur between October to December (Hurley 1985, Hengeveld and Cubberley 2012, Lowrey et al. 2019, Lula et al. 2020, Enns et al. 2023). Spring and fall migrations occur over a 1- to 2-week period on average, but can range from a few days to 2 months. Lengths of geographic migrations range from 5 to over 50 km with averages between 20 and 30 km commonly reported across populations of mountain sheep (Hurley 1985, Lowrey et al. 2020, Enns et al. 2023). Similar distances have been documented for desert bighorn sheep moving between isolated ranges (see Section 11.6.2).

The degree to which mountain sheep surf the green wave during long-distance migrations is not well documented. While phenology likely plays an important role in the timing of spring geographic migrations and the prevalence of migratory behaviors (Aikens et al. 2020a), the migration routes themselves do not likely track a single phenological wave from winter to summer range. Rather, geographic migrations traverse multiple slope aspects across varying elevational gradients with asynchronous phenological patterns. While individuals may opportunistically forage on a south-facing aspect with early new growth, subsequent ridge crossings would expose migrants to north-facing slopes still covered in snow. Relative to other ungulate species where green wave surfing is well documented, mountain sheep migrations are relatively short and occur over a narrow time window. For example, long-distance mule deer migrations in the Red Desert Basin of southwest Wyoming, on average, cover 186 km over 72 days (Ortega et al. 2023). In these migrations, animals closely track spring phenology for over 2 months compared to an average of 2 weeks for mountain sheep. While green-wave surfing along the relatively short-distance and duration migrations of mountain sheep likely has nutritional benefits, the seasonal occupation of highly productive summer ranges as a result of the migration likely provides the fundamental benefit of migration for mountain sheep. Moreover, given the observation that one individual's high-elevation winter range will be subsequently inhabited during summer by an elevational migrant moving to high-elevation summer ranges raises the question: why migrate in the first place? Migrations in mountain sheep mirror the complexity of their mountain environment, and the development of individual migratory behaviors represents the evolution of a complex process in which current-day phenological patterns are only a single component.

### 11.2.3 ANNUAL RESIDENTS

Resident individuals inhabit a single range throughout the year (Figure 11.1). While general range shifts or expanded distributions in summer are common for residents, these movements are characterized by short distances or repeated back-and-forth movements that result in largely overlapping seasonal use without distinct seasonal ranges. Resident behavior can occur at both low and high elevations. High-elevation residents seek out rugged areas that remain relatively snow-free during winter due to prevailing winds (Seip and Bunnell 1984, Walker et al. 2007, Poole et al. 2016, Courtemanch et al. 2017, Bowyer et al. 2000). While this behavior is noted for bighorn sheep, it is more commonly described for thinhorn sheep (Bowyer et al. 2000, Demarchi and Hartwig 2004). Low-elevation residents have similar habitat requirements, but at low elevations, which generally receive less snow. Low-elevation residents further minimize exposure to snow by selecting for aspects that receive increased solar radiation and steep slopes that continually shed snow (DeCesare and Pletscher 2006, DeVoe et al. 2020, Lowrey et al. 2021).

Resident individuals can occur with other animals that migrate, as in partially migratory populations (Chapman et al. 2011). For most partially migratory taxa, the period of overlap between migrants and residents occurs on the range of scarcity where resources are less abundant (Chapman et al. 2011). For ungulates inhabiting temperate climates, overlap between migrants and residents generally occurs at low-elevation areas where residents are

year-round inhabitants and overlap with migrants, which use the low-lying areas during winter. Given the low- and high-elevation resident behavior in mountain sheep, however, overlap between residents and migrants also occurs on high-elevation ranges during summer when migrants move up from low-elevation areas to join high-elevation residents. Additionally, high-elevation residents may overlap with abbreviated or vacillating migrants that temporarily move down to mid- and low elevations to access forage while the residents remain at high elevations (Seip and Bunnell 1984, Walker et al. 2007).

Partially migratory populations are common in mountain environments where individuals need to balance the tradeoffs of migrating or remaining resident on a single annual range and may switch their behavior from year to year (see Section 11.5). In contrast to the prevalence of partially migratory populations in mountainous regions, prairie breaks or canyon and desert landscapes are commonly characterized by fully resident populations, mostly at mid- to low elevations (Hoefs 1974, Van Dyke 1978, DeVoe et al. 2020, Locke et al. 2005, Hoglander et al. 2015). In these landscapes, phenological patterns are more heterogeneous and do not follow elevational trends in spring green-up (DeVoe et al. 2020). The asynchronous nature of phenology in non-mountainous regions can modulate and dampen the realized benefits of migration and, in turn, lessen the number of migrant individuals (Aikens et al. 2017). Moreover, other limiting factors such as water scarcity more strongly influence bighorn sheep in arid regions (Cunningham and Hanna 1992, Cain et al. 2007, Gedir et al. 2020). Consequently, fully resident populations are more common in non-mountainous or arid environments inhabited by desert bighorn sheep or some populations of Rocky Mountain bighorn sheep in canyon or prairie break landscapes (Leslie and Douglas 1979, Risenhoover and Bailey 1985, DeVoe et al. 2020).

## 11.3 LEARNING TO MIGRATE AND MIGRATORY DIVERSITY

Our understanding of how mountain sheep learn to migrate has been aided by the varied management histories of mountain sheep and the widespread use of translocations as a management tool. Translocations have played an important role in expanding the distribution of mountain sheep into historical ranges or augmenting existing populations (Buechner 1960, Singer et al. 2000a, Wild Sheep Working Group 2015). As of 2015, nearly 1,500 restoration efforts resulted in the translocation of more than 21,500 bighorn sheep in North America, with lesser amounts for thinhorn species (Brewer et al. 2014, Wild Sheep Working Group 2015). These management actions have resulted in a varied distribution of native, augmented, and restored populations across the range of bighorn sheep, and to a lesser degree, thinhorn sheep. In this context, native populations are those that have remained extant on the landscape for generations and with sufficient abundance that has not warranted augmentation through translocation. On the other end of the spectrum, restored populations are within historical ranges but reestablished through translocations after population extirpation. Augmented populations are those that retained a native component that was later bolstered through translocations because of concerns over long-term persistence and low abundance (Lowrey et al. 2020). By characterizing the migration patterns across the management histories, recent work has provided insights into the mechanisms through which migrations evolve and persist.

Working with multiple populations with varied management histories, researchers reported that migratory propensity increased with the time since translocations occurred and was most common in native populations that had remained extant on the landscape for generations (Jesmer et al. 2018). While recently translocated populations were initially nonmigratory, after multiple decades animals gained knowledge of the surrounding landscape, resulting in an increased propensity to migrate and foraging that more closely tracked spring phenology. As individuals gain knowledge, likely through exploration, information about migration routes is then maintained through social learning and cultural transmission within social groups (Jesmer et al. 2018).

Relatedly, looking more broadly across the Rocky Mountain west, Lowrey et al. (2020) compared the diversity of migratory movements among individuals within restored, augmented, and native populations by comparing the elevational difference and geographic distance between core summer and winter ranges. Restored and augmented populations had limited migratory diversity and largely consisted of residents or elevational migrants. In contrast, native populations had a diversity of migratory behaviors, including residents at high and low elevations, elevational migrants, and short-, mid-, and long-distance geographic migrants. Within the context of socially learned and culturally transmitted migratory behaviors in ungulates, native populations represent a long evolutionary history on the landscape. When population knowledge is eliminated or greatly reduced, as in restored or augmented populations, the result is a reduction in migratory propensity (Jesmer et al. 2018) and a loss of migratory diversity, inclusive of long-distance migrations (Lowrey et al. 2020). From a conservation standpoint, the loss of native populations of mountain sheep will result in the loss of generations of knowledge regarding the spatiotemporal dynamics of forage and other drivers of migration (Jesmer et al. 2018). Consequently, used habitat will likely be left unoccupied because of the loss of memory and the migratory routes that access important seasonal ranges (Merkle et al. 2019).

The migratory behaviors of native mountain sheep reflect the complexity of their mountain environment and exploit multiple landscape patterns across varied spatial scales. As in other taxa (Schindler et al. 2010, Gilroy et al. 2016), within-population variation in migratory behaviors and spatial patterns may buffer populations from the effects of

interannual variation in environmental conditions and other perturbations that negatively affect the fitness of one migratory behavior versus another. For example, winter conditions will be experienced differently across a population due to the diversity of migratory behaviors and the resulting broad spatial spread of seasonal ranges. While high-elevation residents may experience harsh winter conditions with reduced survival in one year, elevational migrants at low elevations may experience more favorable winter conditions with increased survival. While the population dynamics of any single migratory behavior can be inherently volatile, the complementary dynamics of the aggregate portfolio at the population level can dampen variability and result in more stable demographic performance over time (Schindler et al. 2010, Lowrey et al. 2020), akin to the buffering effects of diversified financial portfolios (Markowitz 1952). While the idea that migratory behaviors and seasonal ranges can be a complex mixture of different social groups is not new (Geist 1971, Festa-Bianchet 1986a, Demarchi and Hartwig 2004), the details of the spatial and temporal dynamics concerning different migratory behaviors have only become clear in recent years with large-scale collaring efforts.

## 11.4   RUTTING AND LAMBING INFLUENCE ON MIGRATION

The relationship between migration and rutting or lambing is not well studied in ungulates or mountain sheep specifically. Nonetheless, the difference in reproductive needs between the sexes likely results in varying migratory patterns that may be expressed in spring or fall. For example, Rocky Mountain bighorn sheep males may leave winter ranges later in spring to remain in larger winter range congregations, which allow males to assess dominance among conspecifics in advance of the rut and engage in experimental antagonistic interactions (Festa-Bianchet 1986b). Consequently, male spring migrations may be less strongly linked with nutrition as they balance additional social and reproductive needs. In fall, male migrations are often more varied with respect to timing and routes as they make broad movements between multiple rutting areas (Demarchi and Hartwig 2004). For example, early field studies with detailed observations of known individuals commonly describe the disappearance of males during the rut and the sudden appearance of new males into the population, which would again disappear after the rut (Geist 1971, Festa-Bianchet 1986b). Additionally, when tracked with GPS collars, males tend to have much longer movements than females during the rut and are 3.5 times more likely to move between groups than females (Borg et al. 2016). In some areas, about half of the marked males would leave the study area to rut elsewhere (Lassis et al. 2022). These rut-based movements can occur in tandem with seasonal migrations, and while there is a general change in distribution resulting from external seasonal pressures, the movements between seasonal ranges for prime age breeding males can be strongly influenced by intrinsic reproductive needs (Festa-Bianchet 1986b).

For females, the relationship between migration and parturition may play an important role in lamb survival. Interestingly, from the limited observations that have been reported, multiple strategies have been observed in the same population. For example, in spring, pregnant females have been observed leaving winter ranges early in favor of high-elevation lambing ranges (Festa-Bianchet 1988). These early migrants leave winter ranges during peak productivity at low elevations and are often disconnected from local phenology, requiring that individuals subsist on relatively poor over-winter forage at the high-elevation lambing areas for 2–4 weeks until the green-wave advances upward. Other females in the same population remained on low-elevation winter ranges for lambing and then migrate to high elevations within 2 weeks of giving birth. Similar patterns of variation in female migration timing with respect to parturition have been noted for bighorn sheep in western Wyoming (K. Monteith, University of Wyoming, unpublished data). These scenarios likely represent tradeoffs for females balancing optimal forage intake with predator avoidance. While the high-elevation lambing areas were generally predator-free and provided lower risk during the lambing period, the reduced risk came with the cost of delayed access to optimal forage and a mismatch with spring green-up, which had not yet reached high-elevation ranges (Festa-Bianchet 1988).

## 11.5 MIGRATORY PLASTICITY AND SWITCHING

While most ungulates, mountain sheep included, are partially migratory (Chapman et al. 2011), questions remain on the factors that influence the maintenance of partial migration. Migratory behaviors may be fixed at the individual level. In this scenario, individuals do not change behaviors from year to year and instead function as separate groups of animals with fixed migratory behaviors, yet overlap on seasonal ranges. This paradigm is often underlying research that contrasts habitat characteristics of the resident and migrant population components with the assumption that behaviors are static for each individual. An alternative explanation is that partial migration is maintained as a conditional strategy whereby individual migratory behavior is plastic, with individuals changing between resident and migrant behaviors across years in response to internal or external cues (Spitz et al. 2018). The extent to which migration is plastic has important implications for population-level management. For example, resident and migrant population components experience different habitat conditions throughout the year, which can influence survival and reproduction. If migratory behaviors are fixed, the two population segments function independently of one another and management actions would be most effective if they targeted individuals with a specific migratory behavior. In contrast, if individual

behaviors are conditional and change from year to year, the population-level vital rates are dependent on the relative proportion of migrants versus residents and the switching of behaviors across years (Spitz et al. 2018). While migrants can have a higher survival compared to residents for a given season, this does not necessarily result in high fitness, which is contingent on lifetime reproductive success and may involve multiple behaviors over the lifespan of an individual (Denryter et al. 2022).

While the plasticity of migratory behavior is not well understood across mountain sheep species, limited research suggests that partial migration is maintained by a conditional strategy where animals switch behaviors across years (Spitz et al. 2018, 2020). Female Sierra Nevada bighorn sheep in an abbreviated migration system have a switching rate of 25%, indicating that on average, an individual changes between resident and migrant behaviors every 4 years (Sptiz et al. 2018). Males in the same system seemed to be more plastic with a switching rate of 54% (i.e., switching behaviors 60 out of the 111 opportunities; Denryter et al. 2024). On average, 9% of the Rocky Mountain bighorn individuals that were monitored for at least 2 years in western Wyoming switched between strategies (Lowrey et al. 2020). Switching rates were less common in restored and augmented populations (means of 6.25% and 15.2%, respectively) than in native populations where 18% of the collared animals switched behaviors among years. Across the multiple studies, switching rates were higher between the resident and migrant binary classifications than between the various migratory behaviors (i.e., vacillating, elevational, geographic). On average, the switching rates for mountain sheep are higher than for other regional ungulates (Spitz et al. 2018).

The determinants of switching are difficult to study due to the large datasets required to address intrinsic and extrinsic factors that may influence switching. In addition to long-term monitoring of single individuals, individual-based metrics such as body fat or pregnancy status can influence migratory propensity. Researchers working to restore the Sierra Nevada bighorn sheep have long-term monitoring of individuals and additionally collect autumn body condition. In this system the propensity to migrate was stronger for animals with lower body fat, representing a draw from high-elevation ranges to lower elevations, through vacillating and abbreviated migrations to seek out better forage conditions (Denryter et al. 2024). Here, animals balance tradeoffs between reduced predation risk at high elevations with better foraging conditions at lower elevations. Therefore, the decision to migrate is dependent on body fat levels and risk-reward tradeoffs with staying resident at high-elevation areas with low predation risk or migrating to low elevations to access better forage, but doing so under greater risk of predation (Spitz et al. 2020, Denryter et al. 2022, 2024). Because the potential ranges an individual may migrate to are often out of sight, the decision to migrate is also related to an individual's perception of current conditions based on memory of past experiences (Berger et al. 2022).

## 11.6 NONMIGRATORY MOVEMENTS OF MOUNTAIN SHEEP

In this chapter, mountain sheep migration is defined as repeated seasonal movements between mostly non-overlapping ranges that occur within the spring and fall. However, other nonmigratory movements are important components of the mountain sheep movement portfolio. In particular, the use of mineral licks, water sources, and exploratory-oriented movements have important implications for mountain sheep life histories.

### 11.6.1 MINERAL LICKS

Localized concentrations of minerals (i.e., mineral licks) are readily used by ungulates for mineral supplementation (Figure 11.2). Mineral licks are naturally occurring on the landscape, but can also be anthropogenic in origin such

as salt and minerals on roads (Hengeveld and Cubberley 2012). Trace minerals, such as sodium, calcium, magnesium, copper, and selenium, are in relatively high concentrations at mineral licks and are essential in the physiology of many animals (Dormaar and Walker 1996, Ayotte et al. 2006, Strand et al. 2025). Wild ungulates generally obtain essential trace minerals through their herbaceous diets but can augment trace mineral deficiencies through geophagia (i.e., the behavior of eating soil) at natural or anthropogenic lick sites (Robbins 1983). Most studies indicate that mineral licks provide an important source of sodium, which occurs at levels below the required intake in early spring forage (Dormaar and Walker 1996, Ayotte et al. 2006). Additionally, carbonates obtained at licks are thought to help stabilize the rumen pH in spring when animals transition from highly fibrous winter diets to lush spring plant growth (Ayotte et al. 2006). Calcium can support pregnancy and lactation in females (Ayotte et al. 2006). Selenium, copper, and zinc are additional trace minerals that support immune function, reproduction, and muscle and bone growth, yet are often lacking in mountain sheep diets (Ayotte et al.

**FIGURE 11.2** Example dry mineral lick used by Rocky Mountain bighorn sheep in western Wyoming (a). Many mineral licks are approached through a web of trails, which can be visible in remote imagery (b). Where mineral licks are used by multiple species, they may facilitate interspecific interactions with implications for disease transfer and competition as seen in Grand Teton National Park where a single lick was visited by a native bighorn sheep lamb and a non-native mountain goat with just over a minute between the two photos (c and d). (Photos by B. Garrott (a) and Grand Teton National Park personnel (c, d); used with permission (b) from Google Earth.)

2006, Anderson 2017). As a result, mountain sheep often seek out mineral licks to augment mineral deficiencies from regional forage (Bertram et al. 2018).

Across the range of mountain sheep, mineral licks can influence the distribution of animals, the timing and route of seasonal migrations, and provide access to important minerals during key times of the year (Seip 1983, Pitzman 1970). When mineral licks are close to seasonal ranges and consistently used, they can serve as anchor points for core use areas or the delineation of a specific mineral lick range (Geist 1971, Hengeveld and Cubberley 2012). When mineral licks are located outside of a seasonal range, individuals make numerous repeated visits, often retracing seasonal migrations for short-term mineral lick use (Seip and Bunnell 1984, Heimer 1973, Garrott et al. 2021). These movements are commonly observed despite the associated travel costs, increased stress, or high levels of juvenile mortality, further emphasizing the importance of mineral licks to mountain sheep (Rice 2010, Watts and Schemnitz 1985). Notably, different groups of desert bighorn sheep from a single population in New Mexico traveled 4 km over relatively flat and exposed terrain to a known mineral lick in the adjacent hills (Watts 1979). In all, the sheep visited the lick 18 times in 21 months, during which half of the ten lambs observed in the study disappeared while making the traverse or while in the adjacent hills. The lamb mortality associated with the mineral lick visitation was noted as the primary factor limiting population expansion (Watts 1979).

Mineral licks are most commonly used during spring and summer, coinciding with the changes from winter to spring diets and the onset of lactation in females. Nonetheless, across the range of mountain sheep, there are reports of consistent mineral lick use throughout the snow-free periods. Low-elevation licks are often used first, likely coinciding with early green-up in these areas (Walker et al. 2007). In some areas, lick usage is highest during the spring migration when animals are moving from winter to summer ranges (Heimer 1973). In this way, mineral licks can function as important stopover areas (Sawyer and Kauffman 2011, Hengeveld and Cubberley 2012). In desert areas, mineral lick use can coincide with the flush of new plant growth brought on by summer rains (Watts and Schemnitz 1985, Holl and Bleich 1987).

Time spent at the lick and actual time spent licking are variable and likely influenced by travel distances (Ayotte et al. 2008). For example, individuals that travel longer distances to mineral licks are likely more inclined to remain at the site for longer periods and bedding between licking bouts. In contrast, animals with a mineral lick within their seasonal range are likely more inclined to move away from the site after a single licking bout but return more frequently (Ayotte et al. 2008). Heimer (1973) reported that Dall's sheep spent an average of 75 minutes at the mineral lick site, during which an average of 93% of the time animals were actively licking. Lactating females are most often observed using mineral licks, likely a result of the demands of spring lactation (Heimer 1973). Mineral licks

are approached by strong directional movements and often characterized by a network of approach trails, which can be seen from landscape imagery and used to evaluate the presence of mineral licks (Figure 11.2; Seip 1983, Garrott et al. 2021). Travel distances to mineral licks vary depending on proximity, but movements of 15–20 km are common (Heimer 1973, Arthur et al. 2014, Bertram et al. 2018).

From a management perspective, mineral licks provide important opportunities to gain insights into mountain sheep populations. Mineral licks have served as focal points to observe animals and relocate known individuals (Geist 1971). When the fidelity to specific mineral licks is strong, lick associations have been used to define herd groupings for research (Walker et al. 2007). More recent efforts have used remote cameras positioned at mineral licks to help estimate population abundance (Butler and Dewey 2018). Mineral licks can also serve as focal points to document and characterize within and among species interactions with important implications for spatiotemporal disease dynamics (Plummer et al. 2018, Strand et al. 2025). For example, in Grand Teton National Park where non-native mountain goats (*Oreamnos americanus*) overlap with native bighorn sheep (Lowrey et al. 2018b), cameras at mineral licks have helped to characterize the extent to which the two species interreact at point-specific areas with important implications for disease transfer (Butler and Dewey 2018, Lowrey et al. 2018a, Figure 11.2). It is also suspected that mineral licks can benefit genetic diversity through enabling individuals, especially young males, to move from one social group to another (Heimer 1973, Hengeveld and Cubberley 2012).

Similar to learning migratory routes through generations of exploration and cultural knowledge (Jesmer et al. 2018), the locations of mineral licks are likely retained in individual memory (Merkle et al. 2019). In the case of reintroduced mountain sheep that are translocated into novel landscapes, animals may be physiologically compromised until they discover licks in their new environment. In addition to the many threats and challenges that influence translocation success (Singer et al. 2000a, Rominger et al. 2004), the time it takes for animals to learn mineral lick locations in a novel environment may additionally influence the survival of translocated animals. Mineral supplication has historically been included in bighorn sheep management, and may also aid translocation success (Coggins 2006, Cox 2022, Proffitt et al. 2022). While many of the mineral licks used by mountain sheep are naturally occurring in backcountry settings, visitation to anthropogenic licks such as roads is not uncommon. Anthropogenetic sources have lethal toxicities and air bore dust that can negatively influence sheep and expose animals to wildlife-vehicle collisions (Hengeveld and Cubberley 2012). To discourage sheep from accessing road-side salts, artificial salt sources have been added to remote locations for some populations, but the effectiveness of these measures has not been sufficiently addressed (Hengeveld and Cubberley 2012).

## 11.6.2 Movements in Relation to Water Sources

Water is a vital, yet limited resource for mountain sheep in arid landscapes and has important implications for seasonal movements and distributions (McKee et al. 2015, Gedir et al. 2020), particularly for desert bighorn sheep in the southwest United States (McQuivey 1975, Longshore et al. 2009). Guzzlers and other artificial water sources have been installed in arid landscapes since the 1940s to bolster regional sheep through augmenting natural water sources (Leslie and Douglas 1979). Throughout arid environments, desert and bighorn sheep generally stay close (i.e., within 2–4 km) to perennial water sources in summer (Locke et al. 2005, Cain et al. 2007, Waddell et al. 2007, Dolan 2006), although periodic rains can enable movements further from permanent sources (Leslie and Douglas 1979). When water is scarce, desert bighorn sheep will tolerate additional risk to access the limited resource. For example, water sources that are relatively far from preferred escape terrain are commonly used during summer despite the increased risk (Gedir et al. 2020). Similarly, when traveling to water sources in summer, animals tolerate higher traffic volumes and associated road crossings compared to winter tolerance thresholds when water is at less of a premium (Cunningham and deVos 1992, Gagnon et al. 2014). During cooler months, animals are less restricted to water sources and use more distal areas for longer durations, commonly occupying escape terrain for 2–3 days before returning to known water sources with quick and direct movements (Cunningham and Hanna 1992).

## 11.6.3 Exploratory Movements

Mountain sheep undertake long-distance exploratory movements beyond their seasonal home ranges that are not associated with seasonal migration or mineral lick visits, yet have important implications for disease spread and genetic diversity (Geist 1971, Bleich et al. 1990, O'Brien et al. 2014, Berger et al. 2022). In contrast to spring and fall migrations, which generally follow known routes with predictable date ranges, exploratory movements can occur at unpredictable times of the year and often result in long wanderings in which an individual may interact with domestic sheep or other native mountain sheep outside of their traditional social groupings (O'Brien et al. 2014). From a genetic perspective, exploratory movements, particularly during the rut, can help to maintain genetic diversity for populations that may be relatively isolated (Bleich et al. 1996, Epps et al. 2007). Within the context of disease transmission in mountain sheep, however, managers often strive to minimize risk through limiting external inter- and intra-specific social interactions (Cassirer et al. 2017). Consequently, exploratory movements provide a unique challenge to managers trying to mitigate risk due to the increased potential of transmitting novel pathogens within and among species (Lowrey et al. 2018b, Dekelaita et al. 2023).

In mountain sheep, exploratory movements are most commonly described as forays, which reference short-term, round-trip movements outside of an individual's established home range (Singer et al. 2001, O'Brien et al. 2014). The exploratory nature of forays, as defined by Singer et al. (2001), aligns with other related terminology, for example, extra-home range (DeCesare and Pletscher 2006) or intermountain (Dekelaita et al. 2023) movements. Forays are generally not season-specific or described relative to a specific purpose (i.e., breeding), but instead relate to unpredictable and individual-based movements (Marcisz 2021) that can occur year-round in males and females (DeCesare and Pletscher 2006, Dekelaita et al. 2023). Forays can also occur post-release for translocated animals, which can set an indiscriminate course away from the release site for up to 4 months post-release (Werdel et al. 2021).

Recent work has additionally used the term foray to describe breeding migrations, particularly for desert bighorn sheep in the southwest United States (Ricci et al. 2024). In desert systems, male breeding migrations have an exploratory nature and entail movements between isolated mountain habitats with unknown breeding opportunities (Bleich et al. 1990, Krausman et al. 1996). In this context, the term foray is used to describe breeding-specific movements in which males leave their traditional seasonal ranges to seek out breeding opportunities during the rut (Geist 1971, Festa-Bianchet 1986b, Hogg 2000).

Dispersals are additional exploratory movements that occur outside of the traditional spring and fall migration periods. While forays are generally described as round-trip movements (Singer et al. 2001), dispersals are one way, resulting in the establishment of a new home range in distant areas (Singer et al. 2000b). Incomplete monitoring of individuals and management actions focused on disease risk, such as capturing or killing individuals observed outside of their known distribution, can complicate the classification of a dispersal movement. When observed in areas outside of known distributions, individuals are often dispatched as a precautionary measure against disease transmission. In many instances, whether or not an individual completed (or would have completed) a round-trip movement is unknown.

Exploratory movement distances vary across studies, but average between 20 and 30 km (DeCesare and Pletscher 2006, O'Brien et al. 2014, Borg et al. 2016, Dekelaita et al. 2023). Within the context of disease transmission, however, distance outliers are particularly important and have been documented as far as 50–60 km for both sexes (O'Brien et al. 2014, Werdel et al. 2021). Given the difficulty in documenting these movements and the scant reporting in published studies, much longer travel distances likely occur. Moreover, while there are reports of unmarked sheep far beyond known distributions, in many instances it is not possible to assess travel distances without GPS data or other markings.

## 11.7 MANAGING MIGRATION: CURRENT THREATS AND RESEARCH OPPORTUNITIES

Given their affinity for rugged landscapes, mountain sheep often occupy public lands with some regulatory measures to help maintain and conserve the natural landscape character. While some populations are within publicly managed lands annually, others, particularly on low-elevation winter ranges, are closer to anthropogenic infrastructure and face varying degrees of direct or indirect human influences. As with other taxa, roads and other linear infrastructure, such as fences, can increase direct mortality through wildlife-vehicle collisions or have negative effects through impeding seasonal migrations or movements to mineral licks and water sources (Cunningham and Hanna 1992, Gagnon et al. 2014). While the prioritization of road crossing structures or fence removal and modification projects is often influenced by other ungulate species, targeted crossing structures and other management actions that support landscape connectivity can be effective management tools to facilitate bighorn sheep migration and movement (Bleich et al. 1990, Gagnon et al. 2022).

Measures that facilitate habitat connectivity are particularly important for mountain sheep within a metapopulation structure, such as in many desert environments where animals inhabit isolated and rugged mountain areas that are increasingly separated by a matrix of human development (Krausman and Leopold 1986). In these systems, population persistence at the metapopulation level is dependent on the dispersal of individuals between isolated mountain ranges, which can increase local abundance and enhance genetic diversity (McQuivey 1975, Krausman et al. 1996, Epps at al. 2005). Under current and predicted warming trends, continued reductions in the already limited water availability will increase population-level vulnerability to local extirpations through reduced forage quality and other stressors, further emphasizing the importance of connectivity between isolated habitat patches (Epps et al. 2004). Although some habitat islands may not support permanent populations, they can provide important stepping stones and enhance the connection between larger habitat cores (Bleich et al. 1990). Within the metapopulation structure, maintaining movement between isolated habitat patches in the face of continued human development and climate change is important for the long-term persistence of mountain sheep inhabiting these areas (Krausman et al. 1996, Epps et al. 2006).

In addition to aiding landscape connectivity, road crossing projects are often constructed in tandem with additional fencing to exclude animals from the roadway. While wildlife-vehicle collisions are generally low for mountain sheep when compared to other species (Paul et al. 2023), some populations can be particularly influenced, especially resident populations that regularly cross roadways or access roads and other travel corridors (i.e., train tracks) for anthropogenic mineral supplementation, access to water,

or snow-free travel routes in winter. For example, in the resident Thompson Falls population of western Montana, there were 385 documented mortalities from vehicles or trains between 1985 and 2009 (Montana Fish, Wildlife and Parks 2010). Wildlife-vehicle collisions on two separate 1.6-km-long sections accounted for 86% of the collision mortalities for this small population, which had average counts of 250 animals during the same period (Montana Fish, Wildlife and Parks 2010). Given the small abundance estimates of many remnant mountain sheep populations (Buechner 1960), minimizing wildlife-vehicle collisions through installing road crossing structures, fencing to exclude animals from the roadway, or flashing signs to alert drivers of high congestion areas are important measures to limit the additive mortality associated with vehicles.

The continued human expansion into mountain sheep core habitat areas, particularly during winter, is an additional threat with implications for migration and movement. For example, present-day Rocky Mountain bighorn sheep in Grand Teton National Park consist mostly of high-elevation residents that undergo abbreviated elevational migrations to gain early access to lower elevation forage in spring (Courtemanch et al. 2017). Historical migration patterns, however, contained elevational migrations between the current year-round ranges and low-elevation winter ranges in the surrounding valley bottoms (Whitfield 1983). The remnant bighorn sheep population is thought to have persisted by adopting a high-elevation resident strategy (Courtemanch et al. 2017). Yet, within the context of the migratory diversity observed in other native populations throughout the region, an alternative explanation is that both strategies existed historically, and that the migrant component either changed behaviors or was unable to persist with the loss of low-elevation winter ranges caused by anthropogenic development (Lowrey et al. 2020). Moreover, the current high-elevation residents are experiencing increased pressure from human recreation as backcountry use continues to grow in popularity (Courtemanch 2014). In the Teton Range and elsewhere, bighorn sheep have been observed avoiding areas of high human use in summer and winter, resulting in decreased habitat availability and tradeoffs concerning nutrition and predation (Courtemanch 2014, Kolek 2024). Resource extraction activities can also influence seasonal movements and distributions of mountain sheep (Poole et al. 2016). With the human development of low-elevation winter ranges and the subsequent loss of cultural knowledge of alternative migration routes and winter habitats, remnant populations have limited options when facing additional pressures from human recreation, climate change, and other disturbances.

Although the extent is variable, high latitudes and mountain environments preferred by many mountain sheep are some of the most influenced landscapes from global climate change and can experience warmer winter temperatures, earlier spring snow melt, and higher summer temperatures (Aycrigg et al. 2021, Hostetler et al. 2021, Intergovernmental Panel on Climate Change 2014,

Chapter 21). These stressors can disrupt the spatiotemporal dynamics of spring phenology with cascading effects on migratory ungulates (Post et al. 2008, John et al. 2024). For example, drought can shorten the duration of spring green-up and reduce the sequential nature of the green wave, thus reducing the benefit of tracking spring phenology (Epps et al. 2004, Aikens et al. 2020b). Additionally, the observed increase in rain-on-snow events can impede animal mobility and access to forage with negative effects on survival (Berger et al. 2018, Bertram et al. 2018, Pan et al. 2018). While mountain sheep can buffer themselves by moving to more favorable conditions through migration, mismatches can occur when animal migrations or habitat use occurs at sub-optimal times. Sierra Nevada bighorn sheep undergo vacillating migrations to higher elevations, which are generally cooler and provide thermal refugia from warmer spring temperatures. Nonetheless, by seeking cooler temperatures at higher elevations where snowmelt is still occurring, sheep can get ahead of the green wave. In other words, higher spring temperatures force animals to high elevations while optimal forage is still at mid- to low elevations (John et al. 2024). Importantly, females restricted to high-elevation lambing areas could experience phenological mismatch between the timing of lambing and high-quality forage availability at lower elevations that may influence long-term population growth given the reliance of fat reserves for successful lambing and over-winter survival (Stephenson et al. 2020, John et al. 2024). In systems where animals are using migration to modulate the risk-reward balance, the changing phenological dynamics from climate change may alter current patterns with implications for nutrition, predation, lamb survival, or the proportion of migratory individuals from year to year (Stephenson et al. 2020, Aikens et al. 2020b). Understanding the influence of these changing dynamics on mountain sheep migration and life histories remains an important research and conservation priority.

While the depth of knowledge on mountain sheep migration continues to grow, there are still many understudied aspects that have important implications for mountain sheep management and conservation. Respiratory disease is arguably the biggest management challenge facing bighorn sheep (Butler et al. 2017, Cassirer et al. 2017). Additional information integrating disease and movements, including traditional migrations and other nonmigratory movements (i.e., mineral lick visitation, forays), can help inform management actions related to disease transmission. For mountain sheep, the predominant management practice is to limit contacts between conspecifics outside of traditional seasonal ranges. This is especially important with domestic sheep, which can introduce novel pathogens into native mountain sheep populations, resulting in all age die-offs (Cassirer et al. 2017, Chapter 8). These movements may also increase the risk of transmitting novel strains from one native population to another. While studies from other migratory taxa suggest that migration can provide an escape from disease (Altizer et al. 2011), this has not been seen in mountain sheep, which remain highly susceptible to disease related all age die-offs and poor recruitment (Cassirer et al. 2017). Within the context of forays in mountain sheep, there is some evidence that animals infected with *M. ovipneumoniae* are less likely to undertake these long-distance movements (Dekelaita et al. 2023). In a species where inter- and intra-specific interactions are a primary management concern, more information relating disease transmission dynamics with migration and other movements has important implications in understanding and predicting disease transmission and spread.

Regardless of the stressor, whether it be human pressures, climate change, or disease, managing for diversity in migratory and nonmigratory movements can help to enhance population resilience in the face of environmental change (Anderson et al. 2014). Populations with diverse migratory portfolios have more options to adapt if certain behaviors become less beneficial in the future. Similarly, working to preserve the migratory landscape and the existing linkages between seasonal ranges can help to maintain the migratory process through preserving a range of conditions that promote migration (Lawler et al. 2015). Moreover, harvest and other management tools can help to maintain existing diversity through targeted efforts that do not disproportionately affect individuals within a single migratory component, for example, ensuring that hunting does not result in the over-harvest of individuals with a single migratory behavior within a population. Relatedly, translocations focused on expanding the distribution of existing populations have had positive results in Montana (Lula et al. 2020). Translocating individuals with specific migratory behaviors into their preferred habitats can help to diversify the migratory portfolios of extant populations while also minimizing concerns regarding disease transmission (Lowrey et al. 2020). By moving high-elevation residents into unoccupied high-elevation habitats, managers can provide new migratory behaviors to existing populations, while minimizing potential interactions with domestic sheep, which are more common at mid- and low elevations. The continued collection of GPS data from across the range of mountain sheep will undoubtedly continue to further our understanding of the complexities of migration in mountain sheep and aid management and conservation efforts into the future.

## 11.8 SUMMARY

Migration is common in mountain sheep and serves as an important life history trait that is critical for population persistence through enhancing foraging opportunities, minimizing predation, and providing access to more secure habitats. Other movements, such as visits to mineral licks or water sources, forays, breeding-related movements, or dispersals, provide access to essential resources and habitats. Across the range of mountain sheep, migrations and other seasonal movements define a complex movement portfolio that relates to and supports all aspects of mountain sheep ecology. Migratory behaviors in mountain sheep are not species-specific. Bighorn and thinhorn sheep have examples of migratory behaviors that span the continuum between annual residents and long-distance migrants. In general, migratory behaviors can be characterized as: low- or high-elevation residents; elevational migrations and other variants (i.e., abbreviated and vacillating migrations), which are relatively short and largely span an elevation gradient; and geographic migrations, which traverse broad and complex landscapes and can connect low-elevation winter ranges with high-elevation summer ranges or high-elevation winter ranges with high-elevation summer ranges. Native populations that have been extant on the landscape without notable human intervention have greater migratory propensity and more diverse migratory movements due to the maintenance of migration through cultural learning and social transmission. Restored and augmented populations, where the population-level knowledge of migration has been lost or greatly reduced, are largely nonmigratory, although translocations show some ability to restore short-distance elevational migrations. Seasonal spring and fall migrations are less common in desert bighorn sheep or bighorn sheep living in canyon or prairie break landscapes, although these populations rely on forays, dispersals, and other broad movements for population persistence. Mineral lick visitation is critical and common across the range of mountain sheep. Protecting existing migration and movement routes and minimizing human encroachment and disturbances are important management and conservation priorities. Managing for migratory diversity can help sustain migratory behavior in the face of climate change and other anthropogenic pressures, which can limit landscape connectivity between seasonal ranges or isolated habitats and alter the spatiotemporal dynamics of regional phenology with cascading effects to other biotic and abiotic interactions.

## REFERENCES

Aikens, E. O., A. Mysterud, J. A. Merkle, F. Cagnacci, I. M. Rivrud, M. Hebblewhite, M. A. Hurley, W. Peters, S. Bergen, J. De Groeve, S. P. H. Dwinnell, B. Gehr, M. Heurich, A. J. Mark Hewison, A. Jarnemo, P. Kjellander, M. Kröschel, A. Licoppe, J. D. C. Linnell, E. H. Merrill, A. D. Middleton, N. Morellet, L. Neufeld, A. C. Ortega, K. L. Parker, L. Pedrotti, K. M. Proffitt, S. Saïd, H. Sawyer, B. M. Scurlock, J. Signer, P. Stent, P. Šustr, T. Szkorupa, K. L. Monteith, and M. J. Kauffman. 2020a. Wave-like patterns of plant phenology determine ungulate movement tactics. *Current Biology* 30:3444–3449.e4.

Aikens, E. O., K. L. Monteith, J. A. Merkle, S. P. H. Dwinnell, G. L. Fralick and M. J. Kauffman. 2020b. Drought reshuffles plant phenology and reduces the foraging benefit of green-wave surfing for a migratory ungulate. *Global Change Biology* 26:4215–4225.

Aikens, E. O., M. J. Kauffman, J. A. Merkle, S. P. H. Dwinnell, G. L. Fralick, and K. L. Monteith. 2017. The greenscape shapes surfing of resource waves in a large migratory herbivore. *Ecology Letters* 20:741–750.

Altizer, S., R. Bartel, and B. A. Han. 2011. Animal migration and infectious disease risk. *Science* 331:296–302.

Anderson, G. 2017. Is selenium deficiency a major issue in the Whiskey Mountain bighorn sheep herd? Wyoming Game and Fish Department, Cheyenne, Wyoming, USA.

Anderson, S. C., J. W. Moore, M. M. McClure, N. K. Dulvy, and A. B. Cooper. 2014. Portfolio conservation of metapopulations under climate change. *Ecological Applications* 25:559–572.

Arthur, S. M., T. Craig, M. Cebrian, and P. Barboza. 2014. Demographics and spatial ecology of Dall sheep in the central Brooks Range. Project number 6.15. Alaska Department of Game and Fish, Juneau, Alaska, USA.

Aycrigg, J. L., A. G. Wells, E. O. Garton, B. Magipane, G. E. Liston, L. R. Prugh, and J. L. Rachlow. 2021. Habitat selection by Dall's sheep is influenced by multiple factors including direct and indirect climate effects. *PLoS ONE* 16(3):e0248763.

Ayotte, J. B., K. L. Parker, J. M. Arocena, and M. P. Gillingham. 2006. Chemical composition of lick soils: functions of soil ingestion by four ungulate species. *Journal of Mammalogy* 87:878–888.

Ayotte, J. B., K. L. Parker, and M. P. Gillingham. 2008. Use of natural licks by four species of ungulates in northern British Columbia. *Journal of Mammalogy* 89:1041–1050.

Barker, K. J., M. S. Mitchell, K. M. Proffitt, and J. D. DeVoe. 2018. Land management alters traditional nutritional benefits of migration for elk. *The Journal of Wildlife Management* 83:167–174.

Berger, D. J., D. W. German, C. John, R. Hart, T. R. Stephenson and T. Avgar. 2022. Seeing is be-leaving: perception informs migratory decisions in Sierra Nevada bighorn sheep (*Ovis canadensis sierrae*). *Frontiers in Ecology and Evolution* 10:742275.

Berger, J., C. Hartway, A. Gruzdev, and M. Johnson. 2018. Climate degradation and extreme icing events constrain life in cold-adapted mammals. *Scientific Reports* 8:1156.

Bertram, M. R., J. Herriges, C. T. Seaton, J. Lawler, K. Beckmen, and S. Dufford. 2018. Distribution, movements, and survival of Dall's Sheep (*Ovis dalli dalli*) in the White Mountains, Alaska. Refuge report 2018-002. U.S. Fish and Wildlife Service. Fairbanks, Alaska, USA.

Bleich, V. C., J. D. Wehausen, and S. A. Holl. 1990. Desert-dwelling mountain sheep: conservation implications of a naturally fragmented distribution. *Conservation Biology* 4:383–390.

Bleich, V. C., J. D. Wehausen, R. R. Ramey II, and J. L. Rechel. 1996. Metapopulation theory and mountain sheep: implications for conservation. Pages 353–373 *in* D. R. McCullough, editor. *Metapopulations and Wildlife Conservation*. Island Press, Covelo, California, USA.

Bolger, D. T., W. D. Newmark, T. A. Morrison, and D. F. Doak. 2008. The need for integrative approaches to understand and conserve migratory ungulates. *Ecology Letters* 11:63–77.

Borg, N. J., M. S. Mitchell, P. M. Lukacs, C. M. Mack, L. P. Waits, and P. R. Krausman. 2016. Behavioral connectivity among bighorn sheep suggests potential for disease spread. *The Journal of Wildlife Management* 81:38–45.

Bowyer, R. T., D. M. Leslie, Jr., and J. L. Rachlow. 2000. Dall's and Stone's sheep. Pages 491–516 *in* S. Demarais and P. R. Krausman, editors. *Ecology and Management of Large Mammals in North America*. Prentice Hall, Upper Saddle River, New Jersey, USA.

Brewer, C. E., Bleich, Vernon C., J. A. Foster, T. Hosch-Hebdon, D. E. McWhirter, E. M. Rominger, M. W. Wagner, and B. P. Wiedmann. 2014. Bighorn sheep: conservation challenges and management strategies for the 21st century. Wild Sheep Working Group, Western Association of Fish and Wildlife Agencies, Cheyenne, Wyoming, USA.

Buechner, H. K. 1960. The bighorn sheep in the United States, its past, present, and future. *Wildlife Monographs* 4:3–174.

Butler, C. J. and S. Dewey. 2018. Monitoring Teton Range bighorn sheep using trail-cameras. Report. Grand Teton National Park, Moose, Wyoming, USA.

Butler, C. J., W. H. Edwards, J. E. Jennings-Gaines, H. J. Killion, M. E. Wood, D. E. McWhirter, J. T. Paterson, K. M. Proffitt, E. S. Almberg, P. J. White, J. J. Rotella, and R. A. Garrott. 2017. Assessing respiratory pathogen communities in bighorn sheep populations: sampling realities, challenges, and improvements. *PLoS ONE* 12:e0180689.

Cagnacci, F., S. Focardi, M. Heurich, A. Stache, A. J. M. Hewison, N. Morellet, P. Kjellander, J. D. C. Linnell, A. Mysterud, M. Neteler, L. Delucchi, F. Ossi, and F. Urbano. 2011. Partial migration in roe deer: migratory and resident tactics are end points of a behavioral gradient determined by ecological factors. *Oikos* 120:1790–1802.

Cain, J. W., III, P. R. Krausman, J. R. Morgart, B. D. Jansen, and M. P. Pepper. 2007. Responses of desert bighorn sheep to removal of water sources. *Wildlife Monographs* 171:1–32.

Cassirer, E. F., K. R. Manlove, E. S. Almberg, P. L. Kamath, M. Cox, P. Wolff, A. Roug, J. Shannon, R. Robinson, R. B. Harris, B. J. Gonzales, R. K. Plowright, P. J. Hudson, P. C. Cross, A. Dobson, and T. E. Besser. 2017. Pneumonia in bighorn sheep: risk and resilience. *Journal of Wildlife Management* 82:32–45.

Chapman, B. B., C. Brönmark, J.-Å. Nilsson, and L. A. Hansson. 2011. The ecology and evolution of partial migration. *Oikos* 120:1764–1775.

Coggins, V. L. 2006. Selenium supplementation, parasite treatment, and management of bighorn sheep at Lostine River, Oregon. *Biennial Symposium North Wild Sheep Goat Council.* 15:98–106.

Courtemanch, A. B. 2014. Seasonal habitat selection and impacts of backcountry recreation on a formerly migratory bighorn sheep population in northwest Wyoming, USA. Thesis, University of Wyoming.

Courtemanch, A. B., M. J. Kauffman, S. Kilpatrick, and S. R. Dewey. 2017. Alternative foraging strategies enable a mountain ungulate to persist after migration loss. *Ecosphere* 8:1–16.

Cox, M. 2022. Effects of mineral supplements on California bighorn sheep in Northern Nevada. *Biennial Symposium North Wild Sheep and Goat Counci* 15:107–120.

Craighead, F., J. Craighead, C. Cote, and H. K. Buechner. 1972. Satellite and ground radiotracking of elk. NASA, Washington Animal Orientation and Navigation.

Cunningham, S. C., and J. C. deVos. 1992. Mortality of mountain bighorn sheep in the Black Canyon area of northwestern Arizona. *Desert Bighorn Council Transactions* 36:27–29.

Cunningham, S. C., and L. Hanna. 1992. Movements and habitat use of desert bighorn in the Black Canyon area. Arizona Game and Fish Department, Phoenix, Arizona, USA.

DeCesare, N. J., and D. H. Pletscher. 2006. Movements, connectivity, and resource selection of Rocky Mountain bighorn sheep. *Journal of Mammalogy* 87:531–538.

Dekelaita, D. J., C. W. Epps, D. W. German, J. G. Powers, B. J. Gonzales, R. K. Abella-Vu, N. W. Darby, D. L. Hughson, and K. M. Stewart. 2023. Animal movement and associated infectious disease risk in a metapopulation. *Royal Society Open Science* 10:220390.

Demarchi, R., and C. Hartwig. 2004. Status of thinhorn Sheep in British Columbia. British Columbia Ministry of Water, Land and Air Protection Biodiversity Branch. Wildlife Bulletin No. B-119. Victoria, Canada.

Denryter, K., M. M. Conner, T. R. Stephenson, D. W. German, and K. L. Monteith. 2022. Survival of the fattest: how body fat and migration influence survival in highly seasonal environments. *Functional Ecology* 36:2569–2579.

Denryter, K., T. R. Stephenson, and K. L. Monteith. 2021. Broadening the migratory portfolio of altitudinal migrants. *Ecology* 102:e03321. https://doi.org/10.1002/ecy.3321

Denryter, K., T. R. Stephenson, and K. L. Monteith. 2024. Migratory behaviours are risk-sensitive to physiological state in an elevational migrant. *Conservation Physiology* 12. https://doi.org/10.1093/conphys/coae029

DeVoe, J. D., B. Lowrey, K. M. Proffitt, and R. A. Garrott. 2020. Restoration potential of bighorn sheep in a prairie region. *The Journal of Wildlife Management* 84:1256–1267.

Dingle, H., and V. A. Drake. 2007. What is migration? *BioScience* 57:113–121.

Dolan, B. F. 2006. Water developments and desert bighorn sheep: implications for conservation. *Wildlife Society Bulletin* 34:642–646.

Dormaar, J. F., and B. D. Walker. 1996. Elemental content of animal licks along the eastern slopes of the Rocky Mountains in southern Alberta, Canada. *Canadian Journal of Soil Science* 76:509–512.

Enns, G. E., B. A. Jex, and M. S. Boyce. 2023. Diverse migration patterns and seasonal habitat use of Stone's sheep (*Ovis dalli stonei*). *PeerJ* 11:e15215. https://doi.org/10.7717/peerj.15215

Epps, C. W., D. R. McCullough, J. D. Wehausen, V. C. Bleich, and J. L. Rechel. 2004. Effects of climate change on population persistence of desert-dwelling mountain sheep in California. *Conservation Biology* 18:102–113.

Epps, C. W., P. J. Palsbøll, J. D. Wehausen, G. K. Roderick, and D. R. McCullough. 2006. Elevation and connectivity define genetic refugia for mountain sheep as climate warms. *Molecular Ecology* 15:4295–4302.

Epps, C. W., P. J. Palsbøll, J. D. Wehausen, G. K. Roderick, R. R. Ramey, and D. R. McCullough. 2005. Highways block gene flow and cause a rapid decline in genetic diversity of desert bighorn sheep. *Ecology Letters* 8:1029–1038.

Epps, C.W., J. D. Wehausen, V. C. Bleich, S. G. Torres, and J. S. Brashares. 2007. Optimizing dispersal and corridor models using landscape genetics. *Journal of Applied Ecology* 44:714–724.

Festa-Bianchet, M. 1986a. Seasonal dispersion of overlapping mountain sheep ewe groups. *Journal of Wildlife Management* 50:325–330.

Festa-Bianchet, M. 1986b. Site fidelity and seasonal range use by bighorn rams. *Canadian Journal of Zoology* 64:2126–2132.

Festa-Bianchet, M. 1988. Seasonal range selection in bighorn sheep: conflicts between forage quality, forage quantity, and predator avoidance. *Oecologia* 75:580–586.

Fryxell, J. M. 1991. Forage quality and aggregation by large herbivores. *The American Naturalist* 138:478–498.

Fryxell, J. M., and A. R. E. Sinclair. 1988. Causes and consequences of migration by large herbivores. *Trends in Ecology & Evolution* 3:237–241.

Gagnon, J. W., K. S. Ogren, S. C. Sprague, S. R. Boe, and R. E. Schweinsburg. 2022. Mitigating bighorn sheep–vehicle collisions and habitat fragmentation with overpasses and adaptive mitigation. *Human-Wildlife Interactions* 16:353–372.

Gagnon, J. W., N. L. Dodd, S. C. Sprague, C. Loberger, R. Nelson, S. Boe and R. E. Schweinsburg. 2014. Evaluation of measures to promote desert bighorn sheep highway permeability: US Route 93 (No. FHWA-AZ-14–677). Department of Transportation. Research Center, Arizona, USA.

Garrott, R., K. Proffitt, J. Rotella, E. Flesch, E. Lula, C. Butler, B. Lowrey, J. T. Paterson, J. DeVoe, and E. Grusing. 2021. Bighorn sheep ecology: an integrated science project to support restoration and conservation. Final Report for Federal Aid in Wildlife Restoration Grant #W-159-R. Montana Fish, Wildlife and Parks, Helena, Montana, USA.

Gedir, J. V., J. W. Cain, III, T. L. Swetnam, P. R. Krausman, and J. R. Morgart. 2020. Extreme drought and adaptive resource selection by a desert mammal. *Ecosphere* 11:e03175.

Geist, V. 1971. *Mountain Sheep: A Study in Behavior and Evolution*. University of Chicago Press, Chicago, Illinois, USA.

Gilroy, J. J., J. A. Gill, S. H. M. Butchart, V. R. Jones, and A. M. A. Franco. 2016. Migratory diversity predicts population declines in birds. *Ecology Letters* 19:308–317.

Hebert, D. M. 1973. Altitudinal migration as a factor in the nutrition of bighorn sheep. Dissertation, The University of British Columbia, Vancouver, Canada.

Heimer, D. W. E. 1973. Dall sheep movements and mineral lick use. Final Report, Alaska Department of Fish and Game. Juneau, Alaska, USA.

Hengeveld, P. E. and J. C. Cubberley. 2012. Sulpher/8 Mile Stone's sheep project: research summary and management considerations. Synergy Applied Ecology, Mackenzie BC, Canada.

Hoefs, M. 1974. Food selection of Dall's sheep (*Ovis dalli dalli*). V. Geist, and F. Walther, editors. *The Behaviour of Ungulates and Its Relation to Management*. Volume 2. International Union for Conservation of Nature 24:758-786.

Hoefs, M. 1976. Ecological investigation of Dall sheep (*Ovis dalli dalli, nelson*) and their habitat on Sheep Mountain, Kluane National Park, Yukon Territory, Canada. Dissertation, The University of British Columbia, Vancouver, Canada.

Hogg, J. T. 2000. Mating systems and conservation at large spatial scales. Pages 214–252 in M. Apollonio, M. Festa-Bianchet, and D. Mainardi, editors. *Vertebrate Mating Systems*. World Scientific Publishing Company, Singapore.

Hoglander, C., B. G. Dickson, S. S. Rosenstock, and J. L. Anderson. 2015. Landscape models of space use by desert bighorn sheep in the Sonoran Desert of southwestern Arizona. *Journal of Wildlife Management* 79:77–91.

Holl, S. A., and V. C. Bleich. 1987. Mineral lick use by mountain sheep in the San Gabriel Mountains, California. *The Journal of Wildlife Management* 51:383. The papers of an International Symposium held at the University of Calgary, Alberta, Canada 2–5 November 1971.

Hostetler, S., C. Whitlock, B. Shuman, D. Liefert, C. Drimal, and S. Bischke. 2021. Greater Yellowstone climate assessment: past, present, and future climate change in greater Yellowstone watersheds. Montana State University, Institute on Ecosystems, Bozeman, Montana, USA. https://doi.org/10.15788/GYCA2021

Hurley, K. 1985. The Trout Peak bighorn sheep herd, northwest Wyoming. Thesis, University of Wyoming, Laramie, Wyoming, USA.

Intergovernmental Panel on Climate Change. 2014. Climate Change Synthesis report. Contribution of Working Groups I, II and III to the Fifth Assessment Report of the Intergovernmental Panel on Climate Change.

Jesmer, B. R., J. A. Merkle, J. R. Goheen, E. O. Aikens, J. L. Beck, A. B. Courtemanch, M. A. Hurley, D. E. McWhirter, H. M. Miyasaki, K. L. Monteith, and M. J. Kauffman. 2018. Is ungulate migration culturally transmitted? Evidence of social learning from translocated animals. *Science* 361:1023–1025.

John, C., T. Avgar, K. Rittger, J. A. Smith, L. W. Stephenson, T. R. Stephenson, and E. Post. 2024. Pursuit and escape drive fine-scale movement variation during migration in a temperate alpine ungulate. *Scientific Reports* 14:15068.

Johnson, H. E., M. Hebblewhite, T. R. Stephenson, D. W. German, B. M. Pierce, and V. Bleich. 2013. Evaluating apparent competition in limiting the recovery of an endangered ungulate. *Oecologia* 171:295–307.

Kolek, J. T. 2024. Recreation in sheep country: assessing the response of an endangered ungulate to novel disturbance. Thesis, University of Wyoming, Laramie, Wyoming, USA.

Krausman, P. R., R. C. Etchberger, and R. M. Lee. 1996. Persistence of mountain sheep populations in Arizona. *Southwestern Naturalist* 41:399–402.

Krausman, P. R., and B. D. Leopold. 1986. The importance of small populations of desert bighorn sheep. *Transactions of the North American Wildlife and Natural Resources Conference* 51:52–61.

Lassis, R., M. Festa-Bianchet and F. Pelletier. 2022. Breeding migrations by bighorn sheep males are driven by mating opportunities. *Ecology and Evolution* 12:e8692.

Lawler, J. J., D. D. Ackerly, C. M. Albano, M. G. Anderson, S. Z. Dobrowski, J. L. Gill, N. E. Heller, R. L. Pressey, E. W. Sanderson, and S. B. Weiss. 2015. The theory behind, and the challenges of, conserving nature's stage in a time of rapid change. *Conservation Biology* 29:618–629.

Leslie, D. M., and C. L. Douglas. 1979. Desert bighorn sheep of the River Mountains, Nevada. *Wildlife Monographs* 66:3–56.

Locke S. L., C. E. Brewer and L. A. Harveson. 2005. Habitat use and movements of desert bighorn sheep in west Texas. *Desert Bighorn Council Transactions* 48:1–11.

Longshore, K. M., C. Lowrey, and D. B. Thompson. 2009. Compensating for diminishing natural water: predicting the impacts of water development on summer habitat of desert bighorn sheep. *Journal of Arid Environments* 73:280–286.

Lowrey, B., C. J. Butler, W. H. Edwards, M. E. Wood, S. R. Dewey, G. L. Fralick, J. Jennings-Gaines, H. Killion, D. E. McWhirter, H. M. Miyasaki, S. T. Stewart, K. S. White, P. J. White, and R. A. Garrott. 2018a. A survey of bacterial respiratory pathogens in native and introduced mountain goats (*Oreamnos americanus*). *Journal of Wildlife Diseases* 54:852–858.

Lowrey, B., D. E. McWhirter, K. M. Proffitt, K. L. Monteith, A. B. Courtemanch, P. J. White, J. T. Paterson, S. R. Dewey, and R. A. Garrott. 2020. Individual variation creates diverse migratory portfolios in native populations of a mountain ungulate. *Ecological Applications*. https://doi.org/10.1002/eap.2106

Lowrey, B., J. D. DeVoe, K. M. Proffitt, and R. A. Garrott. 2021. Behavior-specific habitat models as a tool to inform ungulate restoration. *Ecosphere* 12(8):e03687.

Lowrey, B., K. M. Proffitt, D. E. McWhirter, P. J. White, A. B. Courtemanch, S. R. Dewey, H. M. Miyasaki, K. L. Monteith, J. S. Mao, J. L. Grigg, C. J. Butler, E. S. Lula, and R. A. Garrott. 2019. Characterizing population and individual migration patterns among native and restored bighorn sheep (*Ovis canadensis*). *Ecology and Evolution*. https://doi.org/10.1002/ece3.5435

Lowrey, B., R. A. Garrott, D. E. McWhirter, P. J. White, N. J. DeCesare, and S. T. Stewart. 2018b. Niche similarities among introduced and native mountain ungulates. *Ecological Applications* 28:1131–1142.

Lula, E. S., B. Lowrey, K. M. Proffitt, A. R. Litt, J. A. Cunningham, C. J. Butler, and R. A. Garrott. 2020. Is habitat constraining bighorn sheep restoration? A case study. *Journal of Wildlife Management* 84:588–600.

Marcisz, E. 2021. Ranges and forays of bighorn sheep (*Ovis canadensis*) in the Thompson region of British Columbia. Thesis, Thompson Rivers University, Kamloops, British Columbia, Canada.

Markowitz, H. 1952. Portfolio selection. *The Journal of Finance* 7:77–91.

McKee, C. J., K. M. Stewart, J. S. Sedinger, A. P. Bush, N. W. Darby, D. L. Hughson, and V. C. Bleich. 2015. Spatial distributions and resource selection by mule deer in an arid environment: responses to provision of water. *Journal of Arid Environments* 122:76–84.

McQuivey, R. P. 1975. The status and trend of desert bighorn sheep in Nevada. Special report 75-1. Nevada Department of Fish and Game, Reno, Nevada.

Merkle, J. A., K. L. Monteith, E. O. Aikens, M. M. Hayes, K. R. Hersey, A. D. Middleton, B. A. Oates, H. Sawyer, B. M. Scurlock, and M. J. Kauffman. 2016. Large herbivores surf waves of green-up during spring. *Proceedings of the Royal Society B: Biological Sciences* 283:20160456.

Merkle, J. A., H. Sawyer, K. L. Monteith, S. P. H. Dwinnell, G. L. Fralick, and M. J. Kauffman. 2019. Spatial memory shapes migration and its benefits: evidence from a large herbivore. *Ecology Letters* 22:1797–1805.

Middleton, A. D., J. A. Merkle, D. E. McWhirter, J. G. Cook, R. C. Cook, P. J. White, and M. J. Kauffman. 2018. Green-wave surfing increases fat gain in a migratory ungulate. *Oikos* 127:1060–1068.

Montana Fish, Wildlife and Parks. 2010. Montana bighorn sheep conservation strategy. Wildlife Division. Helena, Montana, USA.

O'brien, J. M., C. S. O'Brien, C. McCarthy, and T. E. Carpenter. 2014. Incorporating foray behavior into models estimating contact risk between bighorn sheep and areas occupied by domestic sheep. *Wildlife Society Bulletin* 38:321–331.

Ortega, A. C., E. O. Aikens, J. A. Merkle, K. L. Monteith, and M. J. Kauffman. 2023. Migrating mule deer compensate en route for phenological mismatches. *Nature Communications* 14:2008.

Pan, C. G., P. B. Kirchner, J. S. Kimball, Y. Kim, and J. Du. 2018. Rain-on-snow events in Alaska, their frequency and distribution from satellite observations. *Environmental Research Letters* 13:075004.

Paul, K., J. Faselt, M. Bell, M. P. Huijser, D. Theobald, A. Keeley, and R. Ament. 2023. West-wide study to identify important highway locations for wildlife crossings. Center for Large Landscape Conservation, Western Transportation Institute – Montana State University, Bozeman, Montana, USA. https://doi.org/10.53847/QVYS3181

Pettorelli, N., F. Pelletier, A. V. Hardenberg, M. Festa-Bianchet, and S. D. Côte. 2007. Early onset of vegetation growth vs. rapid green-up: impacts on Juvenile Mountain ungulates. *Ecology* 88:381–390.

Pitzman, M. S. 1970. Birth behavior and lamb survival in mountain sheep in Alaska. M.S. Thesis, University of Alaska.

Plummer, I. H., C. J. Johnson, A. R. Chesney, J. A. Pedersen and M. D. Samuel. 2018. Mineral licks as environmental reservoirs of chronic wasting disease prions. *PLoS ONE* 13:e0196745. https://doi.org/10.1371/journal.pone.0196745

Poole, K. G., R. Serrouya, I. E. Teske, and K. Podrasky. 2016. Rocky Mountain bighorn sheep (*Ovis canadensis canadensis*) winter habitat selection and seasonal movements in an area of active coal mining. *Canadian Journal of Zoology* 94:733–745.

Post, E., C. Pedersen, C. C. Wilmers, and M. C. Forchhammer. 2008. Warming, plant phenology and the spatial dimension of trophic mismatch for large herbivores. *Proceedings of the Royal Society B: Biological Sciences* 275:2005–2013.

Proffitt, K. M., A. B. Courtemanch, S. R. Dewey, B. Lowrey, D. E. McWhirter, K. L. Monteith, J. T. Paterson, J. Rotella, P. J. White, and R. A. Garrott. 2021. Regional variability in pregnancy and survival rates of Rocky Mountain bighorn sheep. *Ecosphere* 3:e03410.

Proffitt, K. M., V. Boccadori, D. Walsh, and H. Manninen. 2022. Highland bighorn sheep population management evaluation annual interim report. Montana Fish, Wildlife & Parks.

Ricci, L. E., M. Cox, and K. R. Manlove. 2024. Movement decisions driving metapopulation connectivity respond to social resources in a long-lived ungulate, bighorn sheep (*Ovis canadensis*). *Philosophical Transactions of the Royal Society B: Biological Sciences* 379:20220533. https://doi.org/10.1098/rstb.2022.0533

Rice, C. G. 2010. Mineral lick visitation by mountain goats, *Oreamnos Americanus*. *Canadian Field Naturalist* 124:225–237.

Risenhoover K. L., and J. A. Bailey. 1985. Foraging ecology of mountain sheep: implications for habitat management. *Journal of Wildlife Management* 49:797–804.

Robbins, C. T. 1983. *Wildlife Feeding and Nutrition*. Academic Press Inc, New York, New York, USA.

Rominger, E. M., H. A. Whitlaw, D. L. Weybright, W. C. Dunn, and W. B. Ballard. 2004. The influence on mountain lion predation on bighorn sheep translocations. *Journal of Wildlife Management* 68:993–999.

Sawyer, H., and M. J. Kauffman. 2011. Stopover ecology of a migratory ungulate. *Journal of Animal Ecology* 80:1078–1087.

Schindler, D. E., R. Hilborn, B. Chasco, C. P. Boatright, T. P. Quinn, L. A. Rogers, and M. S. Webster. 2010. Population diversity and the portfolio effect in an exploited species. *Nature* 465:609–612.

Seip, D. R. 1983. Foraging ecology and nutrition of Stone's sheep. Dissertation, The University of British Columbia, Vancouver, Canada.

Seip, D. R., and F. L. Bunnell. 1984. Foraging behaviour and food habits of Stone's sheep. *Canadian Journal of Zoology* 63:1638–1646.

Singer, F. J., M. E. Moses, S. Bellew, and W. Sloan. 2000b. Correlates to colonizations of new patches by translocated populations of bighorn sheep. *Restoration Ecology* 8:66–74.

Singer, F. J., L. C. Zeigenfuss, and L. Spicer. 2001. Role of patch size, disease, and movement in rapid extinction of bighorn sheep. *Conservation Biology* 15:1347–1354.

Singer, F. J., C. M. Papouchis, and K. K. Symonds. 2000a. Translocations as a tool for restoring populations of bighorn sheep. *Restoration Ecology* 8:6–13.

Spitz, D. B., M. Hebblewhite, and T. R. Stephenson. 2020. Habitat predicts local prevalence of migratory behavior in an alpine ungulate. *Journal of Animal Ecology* 89:1032–1044.

Spitz, D. B., M. Hebblewhite, T. R. Stephenson, and D. W. German. 2018. How plastic is migratory behavior? quantifying elevational movement in a partially migratory alpine ungulate, the Sierra Nevada bighorn sheep (*Ovis canadensis sierrae*). *Canadian Journal of Zoology* 96:1385–1395.

Stephenson, T. R., D. W. German, E. F. Cassirer, D. P. Walsh, M. E. Blum, M. Cox, K. M. Stewart, and K. L. Monteith. 2020. Linking population performance to nutritional condition in an alpine ungulate. *Journal of Mammalogy* 101:1244–1256.

Strand, L. T., V. Martinsen, K. S. Utaaker, M. Davey, G. R. Rauset, O. Strand, J. Aiken, A. Kuznetsova and B. Ytrehus. 2025. Soil characteristics at artificial salt licks and their potential impacts on occurrence of chronic wasting disease. *Ecosphere* 16(3):e70221.

Valdez, R., and P. R. Krausman, editors. 1999. *Mountain Sheep of North America*. University of Arizona Press, Tucson, Arizona, USA.

Van Dyke, W. A. 1978. Population characteristics and habitat utilization of bighorn sheep, Steens Mountain, Oregon. M.S. Thesis, Oregon State University, Corvallis, Oregon, USA.

Waddell, R. B., C. S. O'Brien, and S. S. Rosenstock. 2007. Bighorn sheep use of a developed water in southwestern Arizona. *Desert Bighorn Council Transactions* 49:8–17.

Walker, A. B. D., K. L. Parker, M. P. Gillingham, D. D. Gustine, and R. J. Lay. 2007. Habitat selection by female Stone's sheep in relation to vegetation, topography, and risk of predation. *Ecoscience* 14:55–70.

Watts, T. J. 1979. Status of the Big Hatchet desert sheep population, New Mexico. *Desert Bighorn Council Transactions* 22:92–94.

Watts, T. J., and S. D. Schemnitz. 1985. Mineral lick use and movement in a remnant desert bighorn sheep. *The Journal of Wildlife Management* 49:994–996.

Werdel, T. J., J. A. Jenks, J. T. Kanta, C. P. Lehman, and T. J. Frink. 2021. Space use and movement patterns of translocated bighorn sheep. *Mammalian Biology* 101:329–344.

Whitfield, M. B. 1983. Bighorn sheep history, distributions and habitat relationships in the Teton Mountain Range, Wyoming. Thesis, Idaho State University, Pocatello, Idaho, USA.

Wild Sheep Working Group. 2015. Records of wild sheep translocations United States and Canada 1922 to Present. Western Association of Fish and Wildlife Agencies, Cheyenne, Wyoming, USA.

# 12 Weather, Water, and Precipitation

*James W. Cain III and Madelon Van de Kerk*

## 12.1 INTRODUCTION

Wildlife agencies have increasingly engaged in wildlife habitat enhancement efforts by actively manipulating vegetative communities or local hydrology. Research continues to reveal methods of habitat manipulation that improve habitat use and quality for mountain sheep (bighorn sheep [*Ovis canadensis*], thinhorn sheep [*Ovis dalli*]). In some areas increasing shrubs, graminoids, and forbs may be an underlying theme in the manager's quest for increasing habitat quality, and in others, reducing vegetation cover may be the objective. The relative importance of providing drinking water sources varies by geographic region and season. Where mountain sheep occur in arid regions, they drink regularly and will often undertake regular movements to water sources. Wildlife managers have increased the availability of drinking water through modification of natural water sources and by construction of wildlife water developments (e.g., water storage, guzzlers). When considering water management options, it is worth considering the goals, costs (including annual maintenance and subsequent replacement), and anticipated benefits to mountain sheep from any water project and ensuring wildlife water developments are located and constructed properly to maximize their effectiveness as a permanent feature once built.

Nutritional condition of mountain sheep can influence survival, reproduction, and recruitment. Climatic conditions affect forage availability and nutritional quality, hence many of the changes in demographic rates of mountain sheep species are linked to climatic conditions, which can enhance survival and juvenile recruitment or reduce them. Furthermore, extreme weather events, such as droughts, extreme snowfall, or rain-on-snow events, can also be associated with acute decreases in survival of adults and lambs. This chapter will highlight the state of our knowledge on the influence of weather and precipitation and on habitat improvement options for mountain sheep with a focus on water sources.

## 12.2 WATER

Water is critical for physiological processes, including joint lubrication, digestion, thermoregulation, nutrient transport, and waste excretion (Robinson 1957). Animals obtain water from food, metabolic water, and free water in the environment (e.g., lakes, streams, rain, snow). Preformed water content in forage varies by plant species, plant parts, and season, ranging from less than 10% in senescent grasses, to 30%–60% in woody browse, and more than 70% in graminoids, forbs, and other succulent plants (cacti; Cain et al. 2008a, 2017) during the growing season. Metabolic water is produced during the oxidation of fats, proteins, and carbohydrates with the oxidation of 1 g of fat, protein, and carbohydrate producing 1.07, 0.4, and 0.56 g of water, respectively; some water from oxidation of protein is lost due to excretion of nitrogenous waste (Barboza et al. 2009).

### 12.2.1 OVERVIEW OF WATER BALANCE AND THERMOREGULATION

Mountain sheep inhabit environments with extreme seasonal climatic conditions ranging from arid environments with high summer temperatures and limited water availability, to northern regions with extreme cold and snowy winter conditions (Chapter 19). These environmental conditions can pose challenges for maintaining temperature and water balance for mountain sheep. Mountain sheep can use behavioral mechanisms to avoid physiologically challenging conditions, or they can employ morphological and physiological adaptations to cope with challenges in maintaining temperature and water balance under extreme environmental conditions (Louw and Seely 1982, Louw 1984). Simultaneous use of multiple adaptations can enhance the ability of animals to cope with challenges of maintaining water and temperature balance (Chappel and Hudson 1978, Nichols and Bunnell 1999, Cain et al. 2006, Kihwele et al. 2020). Although there are comparatively few ecophysiological studies on mountain sheep, mechanisms described for other ungulates of similar body size and occupying areas with similar climatic conditions as outlined in the following sections may provide insight into how mountain sheep cope with temperature and water stress.

### 12.2.2 WATER BALANCE AND TURNOVER

To support essential body functions, animals must maintain water balance within physiologically acceptable limits. Body water occurs in one of four metabolic compartments including fluid within cells (i.e., intracellular); interstitial fluid in body tissues; extracellular fluid in the rumen, digestive tract, and excretory system; and blood plasma. The distribution of water within the body reflects the animal's hydration status and can vary by species according to their evolutionary history (Louw 1993). For example, species with the capacity to occupy environments with limited drinking water maintain blood plasma volume and prevent circulatory failure by moving fluid from other

DOI: 10.1201/9781003518686-13

compartments when they become severely dehydrated (Silanikove 1994, Barboza et al. 2009).

The rate of movement of water through the body tissues is the water turnover rate. Water turnover rates also vary among species and are lower in animals adapted to arid environments with limited availability of drinking water (Nagy and Peterson 1988, Grenot 1992). Water turnover rate increases with increasing ambient temperature in normally hydrated animals (Longhurst et al. 1970) and when metabolic rate increases because of an individual's metabolism or activity level (Denryter et al. 2021). When animals are dehydrated, water turnover decreases regardless of ambient temperature (Maloiy 1973). Water turnover rates are also related to age, sex, and time of day. Finally, reproductive status of females can affect turnover rates, with rates being 40%–50% higher in pregnant and/or lactating females (Maloiy et al. 1979, Cain et al. 2006).

## 12.2.3  WATER LOSS

Animals can lose body water via feces, urine, and cutaneous and pulmonary evaporation. Females also lose water due to lactation. Excessive water loss can pose challenges for maintaining water balance and this may affect the efficiency of an individual's metabolism. Mountain sheep occupying mesic environments with abundant sources of free water can easily compensate for water losses. Loss of body water, however, can pose a serious physiological challenge for mountain sheep occupying arid environments. Mountain sheep in hot, dry environments rely on adaptations to minimize water loss, improve their tolerance to substantial amounts of water loss, or both (McNab 2002).

## 12.2.4  THERMOREGULATION

Like all mammals, mountain sheep need to maintain body temperature within a specific range across widely varying climatic conditions. The range of body temperatures that can be maintained without incurring significant energetic cost (i.e., thermal neutral zone), and upper and lower critical temperatures, varies across species of mountain sheep and is related to their adaptations to the climatic conditions of the environments in which they evolved. Body temperature is a product of the environmental conditions and heat produced via metabolic pathways internal to the animal (Porter and Gates 1969). Heat exchange between the body and the environment is affected by thermal radiation, air temperature, humidity, wind speed, and vapor pressure deficit (difference between amount of moisture in the air and the amount of moisture when saturated). The rate and direction of heat exchange depends on the temperature gradient between the animal and its environment. Animals lose heat to the environment non-evaporatively via radiation, convection, and conduction when their body temperature is higher than the ambient temperature (Porter and Gates 1969). Conversely, when body temperature is lower than the ambient temperature, animals gain heat from the environment,

increasing total heat load. Heat exchange between an animal and the environment can be affected by physiological, behavioral, and morphological mechanisms.

## 12.2.5  PHYSIOLOGICAL ADAPTATIONS, THERMOREGULATION, AND WATER BALANCE

Mountain sheep shed excess body heat using evaporative heat loss during periods when ambient temperature exceeds body temperature. Evaporative heat loss is accomplished passively via diffusion of water through the skin and respiratory tract, or actively by sweating and panting. Although mountain sheep have sweat glands, most evaporative heat loss in sheep occurs via panting. Evaporative cooling, whether cutaneous or respiratory, is an efficient way to lose excess body heat when free water sources are widely available, but there can be substantial costs when drinking water is limited.

The primary challenge of using evaporative cooling to maintain suitable body temperatures in arid environments is the loss of body water. Arid-adapted ungulates often have physiological adaptations to minimize water loss while maintaining temperature balance. For instance, some desert-adapted species use facultative evaporative cooling. When hydrated, the rate of sweating and panting increases with body temperature, but when dehydrated the rate of sweating decreases and animals maintain lower respiratory rates to minimize water loss (Maloiy 1970, Taylor 1970b, Finch and Robertshaw 1979, Parker and Robbins 1984, Baker 1989, Ostrowski et al. 2006). Also, the body temperature trigger point to initiate sweating or panting is often higher when animals are dehydrated (Schmidt-Nielsen et al. 1957, Taylor 1969, 1970a; Maloiy and Hopcraft 1971, Finch 1972).

Rocky Mountain (*O. canadensis canadensis*), California bighorn (*O. c. californiana*), and thinhorn sheep (*O. dalli*) are well adapted to cold temperatures during winter. The lower critical temperature of Rocky Mountain sheep during winter is −20°C when they have their winter coat (Chappel and Hudson 1978). During summer, temperatures are generally mild and free water is widely available on most ranges, limiting the need for evaporative cooling in northern species of mountain sheep (Shackleton et al. 1999, Nichols and Bunnell 1999).

Water loss can be minimized by reductions in fecal moisture content and urine volume, and increasing urine concentration (Maloiy et al. 1979, Kihwele et al. 2020). The capacity to reduce water losses in feces and urine is generally higher in species adapted to water-limited environments. Fecal moisture content in arid-adapted ungulates ranges from 40% to 50%, while that of species adapted to mesic environments is often >70% (Maloiy et al. 1979, Woodall and Skinner 1993). Fecal moisture content can be reduced further when dehydrated, with some species, including desert bighorn sheep (*O. c. nelsoni*), reducing fecal moisture content by 17%–50% (Maloiy and Hopcraft 1971, Turner 1973). Similarly, thinhorn sheep likely can reduce fecal moisture content during winter when surface water is limited and forages are senescent, restricting water intake. In comparisons of normally hydrated versus dehydrated ungulates, reductions in urine volume greater than 75% and increases in urine concentration more than two times have been documented, primarily in desert-adapted species (Maloiy and Hopcraft 1971, Ali et al. 1982, Mohamed et al. 1988). Desert bighorn sheep have highly efficient kidneys, with a relative medullary thickness nearly twice that of domestic sheep, enabling production of highly concentrated urine (i.e., 3,900 mOsm; Horst and Langworthy 1971, Turner 1973).

Metabolic heat contributes to the total heat load of mountain sheep. In a hot environment, a reduction in metabolic rate can decrease metabolic heat gain and reduce the need for use of evaporative cooling. Species that occupy arid climates have lower metabolic rates after accounting for allometric relationships; this lower metabolic rate is associated with lower rates of water turnover in desert ungulates (Maloiy et al. 1979). Although data are not specifically available for desert bighorn, other desert ungulates have reduced metabolic rates when dehydrated (Ostrowski et al. 2006).

The metabolic rate of northern species of mountain sheep fluctuates seasonally, with lower metabolic rate observed during winter, due to reduced forage intake (Chappel and Hudson 1980). Rocky Mountain and California bighorn, and thinhorn sheep use metabolic heat production to aid in maintaining body temperatures during winter. Chappel and Hudson (1978) reported that the lower critical temperature during winter for Rocky Mountain bighorn sheep was −20°C, but that metabolic rate increased when temperatures dropped to −30°C.

Dehydration-induced hyperthermia in desert ungulates can reduce evaporative cooling and loss of body water by allowing body temperature to rise during the day, then passively lose heat at night when the thermal gradient becomes favorable for non-evaporative heat loss (Mitchell et al. 2002). Animals frequently use countercurrent heat exchange, where heat is transferred from warm arterial blood headed toward the brain to cool venous blood heading back to the heart, to maintain brain temperatures below body temperature (Langman et al. 1979). This mechanism increases the temperature at which animals begin to thermoregulate via evaporative cooling, thus conserving water (Mitchell et al. 2002, Fuller et al. 2005). Similarly, nasal heat exchange has been documented in cold-adapted ungulates where exhaled air temperature was much cooler than body temperature, thus conserving body heat (Langman 1985).

Regional heterothermy, a process where different parts of an organism's body are at different temperatures, can also influence the direction and rate of heat exchange with the environment. For mountain sheep inhabiting hot environments, vasodilation of blood vessels near the skin increases blood flow, allowing more heat to be lost to the environment. Mountain sheep in cold environments use countercurrent heat exchange in the legs and vasoconstriction of blood vessels near the skin to conserve body heat. Fatty acid composition in the marrow of leg bones has a higher proportion of unsaturated fatty acids distally, allowing for more fluid fat storage in cold temperatures and more efficient heat exchange (West and Shaw 1975, Turner 1979).

### 12.2.6 BEHAVIORAL REGULATION OF BODY TEMPERATURE AND WATER BALANCE

The water content of forage (i.e., preformed water) contributes to total water intake and if forage water content is high relative to water demands, some ungulates may be able to survive extended periods without access to free water (Taylor 1968, 1969; Gedir et al. 2016). Graminoids (e.g., grasses) generally have lower water content than forbs or browse, particularly once they begin to senesce. Mountain sheep with diets composed primarily of grasses can be more dependent on free water than those with a higher proportion of forbs and browse in their diets (Maloiy 1973, Kay 1997). Rocky Mountain bighorn sheep and thinhorn sheep, however, generally occupy areas where forage moisture content remains high throughout the growing season. In desert ranges, herbaceous forages (i.e., grasses and forbs) with high moisture content are available for a short time following wet periods but quickly dry out, whereas woody browse provides more consistent levels of forage water content except during periods when drought deciduous species lose their leaves (Cain et al. 2017). Succulent plants (e.g., cacti) have a very high water content and consumption by desert bighorn increases during hot, dry periods (Warrick and Krausman 1989, Cain et al. 2017); however, the nutritional quality of cacti is generally low (Cain et al. 2017). Adaptive foraging decisions by desert bighorn sheep may reduce or eliminate the deficit between water requirements and preformed water intake under some conditions (Gedir et al. 2016). For Rocky Mountain bighorn sheep and thinhorn sheep, forage moisture content is typically lowest during winter when forage species are dormant, while forage moisture content for desert bighorn sheep is generally lowest during summer months before the start of summer rains.

Animals can influence their heat loads by adjusting the duration, timing, and location of daily activities. During hot periods, desert-adapted animals spend less time being active or shift activities to cooler periods of the day (Hetem et al. 2012a,b). For example, desert bighorn sheep shift activity periods, foraging, and travel to the cooler crepuscular periods and at night during hot summer months (Simmons 1969, Miller et al. 1984, Alderman et al. 1989). In cold northern climates, thinhorn sheep are more active during winter daylight periods and during periods with favorable weather, improving their ability to thermoregulate.

Mountain sheep exploit microclimates to aid in thermoregulation, with species in hot environments seeking cooler microclimates and those in cold environments seeking bedding areas on southerly aspects with higher solar/heat loading areas (Chapter 19). The use of cooler microclimates by desert bighorn sheep reduces environmental heat loads and helps maintain a temperature gradient that facilitates

(a)																		(b)

**FIGURE 12.1**    Desert bighorn sheep in the Sierra Pinta Mountains, Cabeza Prieta National Wildlife Refuge, Arizona, and Dall's sheep in the Northwest Territories use caves as thermal cover. (Photos by Robert Henry (a) and Mike Schroeder (b); used with permission.)

non-evaporative heat loss, while the use of warmer microclimates by northern bighorn and thinhorn sheep reduces energy expenditure and increases access to forage during winter (Nichols and Bunnell 1999). During inactive periods, thinhorn sheep and desert bighorn sheep commonly use caves (Figure 12.1), vegetation, and other environmental features as a thermal refuge during midday in summer or they use landscape features that allow them to forage in cooler areas (Cain et al. 2008b, Gedir et al. 2020). Bedding down on cool substrates can increase conductive heat loss, just as bedding down on dark-colored shale and scree on south-facing slopes can improve heat absorption. Thinhorn sheep preferences for rugged terrain in all seasons provides for security and escape terrain, but they also offer solar loading and wind-swept foraging areas in winter, and cooling winds that help with thermoregulation while also reducing biting fly numbers in summer (Chapter 19; Hengeveld and Cubberley 2011, Enns et al. 2023).

Mountain sheep can use their body orientation to affect heat exchange with the environment. During hot periods, active animals in areas that lack shade can orient their body position parallel to the sun, reducing surface area exposed to solar radiation, thus minimizing environmental heat load. Animals can increase convective heat loss by adjusting body position relative to wind direction to increase non-evaporative and evaporative heat loss. In cold environments, animals select southerly aspects in rugged terrain with favorable microclimates, orienting their body to maximize heat gain from solar radiation and reduce convective heat loss from the wind.

### 12.2.7    MORPHOLOGICAL ADAPTATIONS FOR THERMOREGULATION AND WATER BALANCE

Body size and shape, pelage, and horns are morphological characteristics that can influence regional heterothermy, and the dynamics of thermoregulation and water balance of

mountain sheep living in extreme environments. The surface-area-to-volume ratio affects heat exchange dynamics between animals and their environment. Larger animals, with lower surface area:volume ratios, exchange heat with the environment more slowly, while small species, with larger surface area:volume ratios, have higher rates of heat loss and gain to and from the environment (Phillips and Heath 1995). Similarly, body and appendage shape influences surface area:volume ratios and affects heat transfer rates between the animal and the environment, with long, thin appendages minimizing radiant heat gain and maximizing convective heat loss. Desert-adapted species characteristically have longer, thinner appendages with higher surface area:volume ratios, thereby facilitating heat loss compared to similar species inhabiting cooler environments, which typically need to retain body heat (Phillips and Heath 1995). Desert bighorn sheep have lower body mass and have longer, thinner appendages than mountain sheep from colder climates (Figure 12.2; McCutchen 1981).

Heat transfer between mountain sheep and their environment can be affected by pelage and horn characteristics. Thin pelage provides little insulation but facilitates heat loss. Further, sparse pelage on ventral surfaces enhances evaporative and non-evaporative heat loss. Conversely, thick pelage increases insulation but impedes evaporative and convective cooling (Hofmeyr 1985). Desert-adapted ungulates typically have glossy, light-colored pelage, which reflects more radiation than does dark-colored pelage. Dark-colored pelage in bighorns and Stone's sheep (*O. d. stonei*) may absorb more solar radiation and acquire more radiative heat loads than light-colored pelage; however, due to the structure of the hair, much of the radiation initially remains at the surface of the pelage, reducing the amount of heat radiation that reaches the skin (Walsberg 1983). The ability of the hair to transport heat or block heat is ultimately affected by the coat and hair structure and how

(a)                                      (b)                                      (c)

**FIGURE 12.2**   Photos depicting body size differences between Dall's sheep (a), desert bighorn, and Rocky Mountain bighorn sheep. (Photos by Emily Mesner (a), Rod Covin (b), and Diane Renkin (c); used with permission.)

an animal manipulates the orientation of the outer guard hairs and insulating under-fur (e.g., flattened or raised). The color, or lack thereof, in species such as mountain goats (*Oreamnos americanus*) and Dall's sheep (*O. dalli*), and similar to polar bears (*Ursus maritimus*), can benefit solar heat gain by the guard hairs acting as light guides, transporting solar radiation to the skin (Walsberg 1983); in these species, the skin is often darker colored or with black pigments in some areas. This moves the radiative heat loading from the surface of the hair coat to the surface of the skin, improving solar heat acquisition. Dall's sheep adaptations in the hair's structure and their ability to manipulate the orientation of their guard hairs allow for greater insulating effect and more heat capture in winter. In summer, the light pelage attracts less surficial solar loading, and when combined with the animal's ability to flatten their hair coat, surface solar reflectance is increased and the animal's core remains cooler.

The vascularized horns of bovids may contribute to thermoregulation with species from warm environments having larger horn cores and thinner keratin sheaths, facilitating heat loss, than species occupying cooler environments (Picard et al. 1999). This pattern is also observed in mountain sheep specifically, with thinhorn sheep having the smallest horn cores and thickest keratin sheaths while desert bighorn sheep have larger horn cores and thinner keratin sheaths; Rocky Mountain bighorn sheep are intermediate for both horn core size and sheath thickness (Picard et al. 1996, Hoefs 2000).

## 12.3   WATER REQUIREMENTS OF MOUNTAIN SHEEP

The water requirements of mountain sheep vary depending on environmental heat load, diet, and morphological, behavioral, and physiological adaptations for maintaining temperature and water balance (Cain et al. 2006). Estimating the relative contributions of free, preformed, and metabolic water toward maintenance of water balance can be particularly challenging because accurate assessments should be made from free-ranging animals exposed to variation in climatic and forage conditions, which determines water losses and the amount of preformed and metabolic water,

respectively. This challenge is responsible for the lack of data on water requirements for many species, including most species of mountain sheep.

Desert bighorn sheep may be able to meet water requirements during some periods of the year with preformed and metabolic water, provided they make adaptive foraging decisions (Gedir et al. 2016). When unable to meet their water requirements from these sources, they need to have access to free water to maintain water balance. Estimates of water requirements are not available for most species of mountain sheep, with most studies on water requirements limited to desert bighorn. Desert bighorn sheep require about 4% of their body mass in water per day (Turner 1973). Water use by desert bighorn is related to forage quality, affecting production of metabolic water, preformed water content in forage and seasonal water demands (Turner 1973, Krausman et al. 1985, Warrick and Krausman 1989). Desert bighorn consume 2–4 L of water per visit to water sources (Knudsen 1963, Hailey 1967) and visit water every 2–3 days (Bradley 1963, Knudsen 1963). Unsurprisingly, desert bighorn sheep increase visitation to water sources during periods with high temperatures and low precipitation (Glass et al. 2022, Terry et al. 2022). Depending on forage conditions, some of their water requirements can be met by preformed and metabolic water. Data on the contribution of preformed water to overall water requirements are not available but likely vary seasonally, depending on forage moisture content (Turner 1973, Gedir et al. 2016). For example, Krausman et al. (1985) continuously monitored two female desert bighorn sheep and reported that they did not visit water sources for more than 10 days but during this period, both females were regularly observed eating barrel cactus (*Ferocactus* spp.) which typically has moisture content greater than 75% (Seegmiller et al. 1990, Cain et al. 2017). While succulents may be a good source of preformed water, they typically have low nutritional value with less than 5% protein content (Hughes and Smith 1990, Cain et al. 2017).

The availability of perennial water sources is often used to define habitat quality for desert bighorn (Longshore et al. 2009, Bleich et al. 2010). The distribution of desert bighorn sheep is often restricted to within 4 km of perennial water sources during hot, dry periods (Wilson 1971, Turner et al.

2004, Cain et al. 2008a, Gedir et al. 2020). Females with lambs have higher water requirements due to lactation, thus they remain closer to water sources than males or females without lambs (Bleich et al. 1997). Some desert bighorn sheep populations, however, occupy ranges lacking perennial water sources (Watts 1979, Krausman et al. 1985, Alderman et al. 1989).

There have been few experimental examinations of the influence of perennial water sources on mountain sheep distribution, movements, behavior, or demography. Cain et al. (2008a) evaluated the responses of desert bighorn sheep to the removal of water sources in southern Arizona, including assessments of diet selection, home range size, movement, survival, and recruitment. Water removal did not result in the predicted changes in the response variables, although there were some confounding effects between the treatment and climatic conditions. During severe drought, the presence of water sources did not prevent high adult mortality rates, indicating that forage was more limiting than water; anthropogenic sources of surface water do not compensate for malnourishment caused by a lack of forage during severe or extreme droughts. Rainfall increased during the period when water removal treatment occurred, which likely dampened any potential treatment effects because of improved forage conditions, higher forage moisture content, and increased availability of ephemeral sources of water. Water may not be limiting for desert bighorn during periods of above-average precipitation. Thus, additional experimental studies are needed to determine whether and under what conditions water developments have population-level influences. During periods when nutritional content of forage with lower moisture content is at or above maintenance levels, the presence of perennial water sources may help desert sheep avoid making a trade-off between selecting forage with high moisture content at the expense of nutritional quality as reported for Sonoran pronghorn (*Antilocapra americana sonoriensis*; Hervert et al. 2005). In addition, questions remain regarding how changing climatic conditions will affect forage moisture content, water requirements, and thermoregulatory challenges of desert bighorn, which may shift the role of water provisioning for management and conservation of mountain sheep populations. Experimental studies on the role of water sources have not been conducted for Rocky Mountain and California bighorn, or thinhorn sheep.

## 12.4 WATER MANAGEMENT FOR MOUNTAIN SHEEP

Water developments have been used in areas where natural resources were scarce or unreliable, to mitigate the loss or accessibility of naturally occurring water sources, habitat loss, and fragmentation from human developments and infrastructure (Rosenstock et al. 1999, Krausman et al. 2006, Simpson et al. 2011). In addition, water developments have been built to expand the distribution of animals (Weaver

1973, Bleich 2008). Water is assumed to limit the distribution and productivity of mountain sheep in arid regions; thus, water developments have become integral to the management of desert bighorn and some, Rocky Mountain bighorn sheep populations (Simpson et al. 2011). Water developments can play a role in conservation and management of desert bighorn metapopulations by increasing use of otherwise suitable ranges that lack reliable water sources, increasing the chance that dispersing animals encounter water in areas with other habitat components and improving the chances that immigrants encounter resident conspecifics (Bleich 2009).

Water developments for mountain sheep are typically precipitation catchments, which harvest rain or snow on a collection surface and store the water for future use by wildlife. Precipitation catchments are most common in areas with limited subsurface water availability and in arid and semi-arid areas with lower precipitation but where storms often occur as brief, intense thunderstorms that generate substantial runoff. Common precipitation catchments include guzzlers, modified tinajas, adits, retention dams, and sand tanks (Figures 12.3–12.5). In areas with subsurface water, wells and windmills, horizontal wells, and developed springs and seeps may be viable options for providing water for mountain sheep and other wildlife. In addition, water developments for livestock may be used by mountain sheep, particularly when they occur near escape terrain. Additional information on water catchments for mountain sheep is available in Bleich et al. (2020), and references therein.

## 12.5 POTENTIAL BENEFITS OF WATER MANAGEMENT FOR MOUNTAIN SHEEP

Provisioning of water sources for mountain sheep is a common practice by wildlife management agencies, particularly for desert bighorn sheep in arid regions; however, water sources are also maintained for some Rocky Mountain and California bighorn populations occupying areas with limited availability of naturally occurring sources of free water and in ranges where nearby ground-water pumping for residential and agricultural uses is reducing flow of springs and seeps (Whiting et al. 2009). Free water is generally not limited in ranges occupied by thinhorn sheep (Nichols and Bunnell 1999). The goals of water development include increasing abundance, increasing available habitat (thereby increasing population size and distribution), improving habitat quality, and mitigating the effects of drought (Rosenstock et al. 1999, Simpson et al. 2011, Whiting et al. 2012a).

The abundance of some populations has been associated with the availability of water sources, with populations having higher abundance in areas with perennial water sources. Some population increases in desert bighorn sheep were associated with construction of additional water sources (Leslie and Douglas 1979), and population declines (Douglas 1988) and mortalities have been attributed to loss

**FIGURE 12.3**  Halfway Tank, man-made water tank with a sand tank and diversion gabion in the Cabeza Prieta Mountains, Cabeza Prieta National Wildlife Refuge, Arizona. (Photo by J. W. Cain III.)

**FIGURE 12.4**  Heart Tank, a modified tinaja located in the Sierra Pinta, Cabeza Prieta National Wildlife Refuge, Arizona. (Photo by J. W. Cain III.)

of water sources (Allen 1980). The ability of desert bighorn sheep populations to expand distributions into some geographies where water is a limiting factor in range distribution has been addressed through habitat interventions such as the construction of water guzzlers (e.g., Chupadera Mountains, New Mexico Wild Sheep Foundation). Some researchers of desert bighorn sheep, however, did not report an influence of water sources on population size (Krausman and Leopold 1986, Krausman and Etchberger 1995). Ranges only seasonally used by desert bighorn sheep began to be used all year after construction of water sources (Leslie and Douglas 1979). It is unclear whether some of the changes in sheep density around water sources following construction are related to an increase in abundance or simply a shift in the

**FIGURE 12.5**   Buck Peak Tank, an adit located in the Cabeza Prieta Mountains, Cabeza Prieta National Wildlife Refuge, Arizona. (Photo by J. W. Cain III.)

distribution of animals (Weaver 1973, Bleich 2008, Bleich et al. 2010, Cain et al. 2022), but in either outcome, the provisioning of water does seem to have changed sheep distributions. Data on the relationships between water availability and abundance of Rocky Mountain and California bighorn, and thinhorn sheep are lacking in the literature.

For populations displaying wide annual fluctuations in survival and reproductive success, addition of water sources may improve habitat quality, provided that the variation in survival and reproductive success is related to a lack of reliable water sources. In cases where forage (or other resources) is limited, the construction of water sources is unlikely to improve overall habitat quality or result in improved reproductive success or juvenile recruitment. That said, increasing temperatures and aridity associated with changing climatic conditions, however, may alter the availability of natural water sources, increasing the importance of water developments in maintaining mountain sheep populations. For example, desert bighorn sheep populations in the Mojave Desert that occupy higher-elevation desert mountain ranges with more precipitation and the presence of natural springs were less likely to decrease than those occupying lower-elevation mountain ranges with limited sources of perennial water (Epps et al. 2004).

## 12.6   POTENTIAL NEGATIVE EFFECTS OF WATER DEVELOPMENTS

The potential adverse effects of constructing water developments for mountain sheep and other wildlife include disease transmission, poor water quality, higher levels of predation or competition, direct mortality, and altered plant community structure and composition (Rosenstock et al. 1999, Simpson et al. 2011, Whiting et al. 2012a). Water developments may contribute to wildlife disease dynamics (Benavides et al. 2012, Titcomb et al. 2021). Hematophagous gnats in the genus *Culicoides* are vectors of bluetongue and epizootic hemorrhagic disease, both of which affect mountain sheep populations. Although some have suggested that water developments may serve as habitat for larval stages of *Culicoides*, Rosenstock et al. (2004) reported that *several* species of *Culicoides*, including the known vector (*C. sonorensis*), were common at both wildlife waters and in sand at control sites and suggested that modified natural water sources typically lack the appropriate substrate material and nutrient loads that provide the best conditions for larval *Culicoides*. There has been one documented occurrence of at least 45 desert bighorn sheep mortalities after consuming water contaminated with the bacterium *Clostridium botulinum* caused by the entrapment, drowning, and decomposition of bighorn in a storage tank (Swift et al. 2000). Subsequently, Bleich (2003) reported that it is not uncommon for both natural water sources and anthropogenic water developments to have conditions favorable to produce botulinum toxins. Concentration of animals at limited water sources may also increase direct pathogen transfer; however, this would not be unique to anthropogenic water developments because interactions between animals should be similar at naturally occurring water sources.

Poor water quality can also be a concern for wildlife water developments, and some authors suggest that water quality may influence the health of wildlife (Kubly 1990, Broyles 1995). Potential water quality issues of concern include pH, heavy metals, toxic chemicals, dissolved solids, and toxins resulting from blue-green algae blooms. Examinations of water quality in water developments in the Southwest suggested that water quality parameters were generally within limits established for livestock (Rosenstock et al. 2005, Bleich et al. 2006); some sites had higher pH, alkalinity, and fluoride but levels were not high enough to be of concern for wildlife health (Rosenstock et al. 2005, Bleich et al. 2006). Toxic algal blooms are rare in wildlife water developments; however, the death of more than 100 elk (*Cervus canadensis*) using an earthen livestock tank in northern New Mexico in 2013 was attributed to neurotoxin produced by the cyanobacteria *Anabaena* (New Mexico Department of Game and Fish 2013).

Others have suggested that providing water developments can result in more consistent predator occupancy of arid regions, resulting in increased predation on ungulates (Broyles 1995, Harris et al. 2020). Although the concentration of predation events on ungulates near water sources has been widely reported in African savanna systems (de Boer et al. 2010, Davidson et al. 2013), there has been relatively little research on the role of water sources in exacerbating predation on ungulates in North America (Rosenstock et al. 1999, Simpson et al. 2011). Visitation rates to water sources in New Mexico by bobcat (*Lynx rufus*), coyote (*Canis latrans*), and mountain lion (*Puma concolor*) were associated with higher temperature and prey visitation rates (Harris et al. 2015), and long-term monitoring with remote cameras documented interactions between predators and some ungulates (J. Erz, U.S. Fish and Wildlife Service, unpublished data), but documentation of predation events at water sources was rare. A study of mountain lion kill sites in relation to water sources in five study areas in the Chihuahuan and Sonoran deserts reported that only 1.2% ($n=20$) of 1,556 ungulate lion kills were within 100 m of a wildlife water source, and only 3% ($n=39$) were within 250 m. Desert bighorn sheep kills comprised 9% ($n=145$) of the ungulate kills and were present at only three of the study areas. When comparing mountain lion kill sites of desert bighorn to random locations within mountain lion home ranges, the probability of a site being a desert bighorn sheep kill site increased closer to man-made water sources and was highest during the hot-dry season (Prude 2020). Only 11% of bighorn kills were within 500 m of a water source, four kills were within 250 m of water, and only one kill was less than 100 m from water; the mean distance from desert bighorn sheep killed by lions to water sources was 2,467 m. Therefore, the effect of man-made water sources on the distribution of kill sites was likely more strongly associated with the influence of water sources on desert bighorn distribution rather than the use of water sources as stalking areas by mountain lions (Prude 2020).

Increased exploitative and interference competition (Chapter 21) could occur if water developments increase overlap between mountain sheep and other ungulates with similar habitat and dietary needs. For example, water sources constructed along the base of mountains occupied by desert bighorn sheep or in less rugged areas near escape terrain used by Rocky Mountain bighorn sheep could increase competition for forage with mule deer (*Odocoileus hemionus*) and elk. When construction of water developments results in range expansion of non-native ungulates, mountain sheep can face increased competition for forage or water (Bleich et al. 2010, Marshal et al. 2012). For example, feral horse and burro can affect mountain sheep by reducing availability of preferred forage species or by antagonistic interactions at water sources, preventing use by mountain sheep (Weaver 1959, Berger 1985, Ostermann-Kelm et al. 2008). Competition for forage can be particularly exacerbated during droughts with reduced forage production (Marshal et al. 2012). The potential for competition can be reduced if water developments are placed to enhance spatial separation between mountain sheep and their potential competitors; water sources may be fenced using cable or pipe fencing that sheep can jump over, but excludes feral equids. Desert bighorn sheep avoid water sources in the presence of feral horses and feral burros (Dunn and Douglas 1982, Ostermann-Kelm et al. 2008). Similarly, desert bighorn sheep in west Texas were less likely to use wildlife waters in areas with high activity of non-native Barbary sheep (*Ammotragus lervia*; Foster et al. 2005).

Wildlife can occasionally become entrapped and drown in wildlife water sources that have not been equipped with exit ramps or other self-rescue structures. Most documented mortalities are small mammals, birds, and reptiles (Andrew et al. 2001, Rosenstock et al. 2004), but there are some documented instances of drowning by mule deer, coyote, and desert bighorn sheep (Rosenstock et al. 2004). Most modern designs for wildlife water sources include escape ramps and many older drinkers have been retrofitted to minimize drowning of wildlife (Simpson et al. 2011, Bleich et al. 2020). Overall, the number of documented drowning instances does not indicate that wildlife waters are having population-level effects due to drowning.

Persistent congregation of mountain sheep and other herbivores around point resources, such as wildlife waters, can result in changes in plant community structure and composition, known as piosphere effects. Researchers have documented piosphere effects of increased forage use, decreases in forage availability, and changes in plant community structure immediately around the water sources (Thrash 1998, Parker and Witkowski 1999). The effects of high rates of herbivory and plant trampling near water sources can also increase invasive species and soil nutrient concentrations (Tolsma et al. 1987, Brooks et al. 2006, Jawuoro et al. 2017). Piospheres in the southwestern United States are typically observed in areas associated with domestic livestock and feral equids (Welles and Welles 1961, Andrew 1988, Davies and Boyd 2019). Because mountain sheep often occur at low densities, detrimental influences to vegetation around water sources are less likely to occur unless water sources are accessible to other ungulate species. During drought periods, desert bighorn sheep selected habitat farther from water sources, indicating that palatable forages may have been depleted close to water developments (Gedir et al. 2020).

## 12.7 EFFECTS OF WEATHER ON DEMOGRAPHY

### 12.7.1 WEATHER VARIABLES

Weather conditions play an important role in shaping survival, reproduction, and population dynamics of mountain sheep. As changes to climates and weather cycles occur, understanding these weather variables, variation therein, and effects thereof is important for understanding how they influence mountain sheep demography (Chapter 21).

*Temperature.* Mountain sheep behavior, physiology, and demographics are all affected by temperature. Changes in temperature can affect metabolic rates, forage availability, the timing of reproductive cycles, and patterns of movement. High temperatures can be particularly challenging for desert bighorn sheep, having to survive in extreme conditions, where daytime temperatures can exceed 40°C (Cain et al. 2008b). In this heat, they face risks of dehydration, and heat stress, which can lead to higher mortality rates, especially in prolonged heat waves (Epps et al. 2004, Bender and Weisenberger 2005). In contrast, Rocky Mountain and California bighorn, and thinhorn sheep live in cooler, high-altitude environments, where in winter months they are exposed to sub-zero temperatures, severe storm events, and localized heavy snowfall. These cause increased energy expenditure to maintain body heat and can result in frostbite and hypothermia, particularly in lambs (Festa-Bianchet 1988a). Thinhorn sheep will significantly reduce their movements during extreme and prolonged cold events and winter storms, often remaining in the same spot for periods longer than 6–8 hours, likely in an effort to reduce energy expenditure and maintain body temperature (Bill A. Jex, Wildlife Branch, British Columbia, unpublished data).

*Precipitation.* Availability and access to food and water are directly affected by precipitation patterns. In arid environments, desert bighorn sheep rely on sparse rainfall events to sustain the vegetation they forage on and availability of drinking water (Bender and Weisenberger 2005, Cain et al. 2017). Forage quantity and quality are determined by the timing and amount of rainfall, influencing the body condition and reproductive success of sheep (Douglas and Leslie 1986, Wehausen et al. 1987, Hik and Carey 2000, Overstreet et al. 2014, Cain et al. 2017). In mountainous regions, glaciers and winter snowpacks provide a steady water supply essential for plant growth and hydration needs of Rocky Mountain and thinhorn sheep by slowly melting throughout spring and summer. Snow that arrives too early in the season, stays too late, or accumulates in excess can negatively affect alpine ungulate populations (Cosgrove et al. 2021) by increasing energy expenditure for moving through deep snow and blocking access to autumn, winter,

and spring forage (Parker et al. 1984, Robinson and Merrill 2012, Cosgrove et al. 2021).

*Seasonal variation.* Both temperature and precipitation fluctuate seasonally, affecting localized resource availability, survival, and breeding cycles. Climate patterns, ocean temperatures, and currents can have dramatic year-to-year effects on mountain sheep populations across their range (Jex et al. 2025, Chapter 21). In temperate climates, increasing temperatures and snowmelt in spring bring a surge of vegetation growth, offering abundant forage for lactating females and newborn lambs (Festa-Bianchet 1988a). Later in summer, temperatures can soar, and water sources may become scarce or dry up, resulting in declines in forage quality. For desert bighorn sheep, seasonal dynamics are less predictable, the availability of resources instead being closely related to sporadic and episodic rainfall events (Cain et al. 2017).

*Extreme weather events.* Sporadic but extreme events such as droughts, heavy snowfall, and storms can have immediate and severe effects on mountain sheep populations. These events can exacerbate existing stressors and cause new ones, resulting in significant demographic changes. Droughts can affect desert bighorn sheep, leading to decreased forage quality and quantity and limiting water availability (Cain et al. 2008a, 2017). Prolonged droughts can lead to poor body condition, decreased reproductive success, and higher mortality rates (Epps et al. 2004, Bender and Weisenberger 2005, Cain et al. 2008a). For Rocky Mountain and thinhorn sheep, heavy snowfalls, rain-on-snow events, and prolonged snow periods can limit access to forage and impede movement, increasing energy expenditure and vulnerability to predation (Festa-Bianchet 1988a, Van de Kerk et al. 2020). Cold temperatures combined with wind can also directly reduce lamb survival and fitness (Festa-Bianchet 1988a, Cosgrove et al. 2021, Jex et al. 2025).

### 12.7.2 EFFECT OF WEATHER ON DEMOGRAPHY OF ROCKY MOUNTAIN AND SIERRA NEVADA BIGHORN SHEEP

Bighorn sheep populations must navigate steep and rocky terrains and endure a wide range of climate conditions, including harsh winters and variable summers, subsisting on sub-maintenance diets for much of the year (Stephenson et al. 2020). The survival, reproduction, and overall population dynamics are therefore closely intertwined with local weather patterns, causing them to be particularly vulnerable to the effects of climate change and variability.

*Survival.* Although survival of Rocky Mountain bighorn sheep is affected by a complex combination of many factors, weather plays a pivotal role. These sheep are well adapted to cold climates, but challenges posed by harsh weather have significant implications for survival, particularly in winter, which is the most critical season for Rocky Mountain and Sierra Nevada bighorn sheep (Poole

2013, Douhard et al. 2018, Stephenson et al. 2020, Proffitt et al. 2021). Winter conditions in habitat of northern bighorn sheep are characterized by low temperatures, heavy snowfall, and extended periods of snow cover, the severity of which directly influences the availability of forage and energy expenditure necessary for survival (Douglas and Leslie 1986, Festa-Bianchet et al. 1997, Portier et al. 1998, Hamel et al. 2009a, Proffitt et al. 2021).

When snow accumulates to significant depths, sheep often must move to lower elevations and south-facing slopes with less snow cover to access patches of vegetation (Festa-Bianchet 1988a, Berger 1991) or to higher elevations where wind-swept habitats provide better access to forage (Poole 2013). But even in these more favorable microhabitats, there can be substantial energetic costs associated with foraging, because snow not only covers forage but also can become compacted and drifted, requiring more energy to move through it (Festa-Bianchet 1988b). The resulting calorie deficit from increased costs of locomotion means that sheep often must rely on fat reserves accumulated in the growing season to survive winter (Festa-Bianchet et al. 1997, Feder et al. 2008, Stephenson et al. 2020). When these nutritional reserves are insufficient in particularly harsh winters, higher mortality rates result, especially in the most vulnerable cohorts such as older sheep and lambs (Festa-Bianchet et al. 1997, Jorgenson et al. 1997).

Several researchers have reported a strong correlation between winter severity and mortality rates of Rocky Mountain bighorn sheep. For example, researchers demonstrated that Rocky Mountain bighorn sheep in the Greater Yellowstone Ecosystem experienced significant population declines during winters with prolonged cold periods and heavy snowfall (White et al. 2008). These conditions resulted in a substantial increase in mortality rates, particularly among lambs, which lacked the fat reserves and body size required to survive harsh winters. Calorie deficits during winter not only affect mortality rates directly but also have indirect effects by increasing disease susceptibility (Jorgenson et al. 1997, Singer et al. 2001). Sheep that are weakened by nutritional stress are more vulnerable to respiratory infections, parasites, and other health issues. This increased susceptibility can cause additional mortality during particularly harsh winters, resulting in even stronger population declines (Singer et al. 2001). Sheep populations living in areas with higher habitat quality and who are in better body condition appear to have a higher probability of clearing pathogens (Smiley et al. 2024) and exhibit faster recovery from disease events (Wagler et al. 2023). From an energetic perspective, an animal could survive for up to 23 days on the amount of fat lost to high levels of infection (Smiley et al. 2024).

The beginning of a new growing season at the transition from winter to spring marks another important period for Rocky Mountain bighorn sheep. The timing of snowmelt and the start of the growing season influence when the sheep can begin recovery from the nutritional stress of winter and start regaining their fat reserves (Festa-Bianchet

1988a, Douhard et al. 2018). Bighorn sheep can improve their overall body condition quickly during early, warm, and wet springs, which generally result in an abundance of high-quality forage (Jorgenson et al. 1993, Festa-Bianchet et al. 1997, Pelletier and Festa-Bianchet 2004). In contrast, during springs with late snowmelt, the delayed growth of new forage can extend the period of nutritional stress, which can be particularly detrimental for lactating females, which require high nutritional intake during late gestation and to produce milk for their lambs (Festa-Bianchet 1989, Bérubé et al. 1996, Ruckstuhl and Festa-Bianchet 1998, Festa-Bianchet et al. 2019). Both females and lambs may suffer from malnutrition in the absence of sufficient quality forage, leading to higher mortality rates for both life stages (Festa-Bianchet et al. 1997, Pelletier and Festa-Bianchet 2004, Feder et al. 2008). Chronically poor maternal condition will manifest in reduced productivity and impaired population growth and resilience and can instigate life-long effects on the size and growth of offspring (Monteith et al. 2018). Weather conditions in summer affect survival of Rocky Mountain and Sierra Nevada bighorn sheep, through the effect of precipitation on plant growth. Survival of Sierra Nevada bighorn sheep, however, is related to the length of the growing season the previous year, as indexed by the date of peak normalized difference vegetation index (Connor et al. 2018). When temperatures are high and precipitation is low, both forage quality and quantity are reduced (Festa-Bianchet 1988a,b; Hamel et al. 2009b). In those conditions, sheep may struggle to recover from winter, be unable to maintain adequate body condition, be more vulnerable to disease and predation, and be unable to build up enough fat reserves for the following winter. The spread of parasites, such as lungworm, can also be exacerbated by warm, dry conditions, further compromising the health of sheep populations (Festa-Bianchet 1989, Pelletier et al. 2005).

*Reproduction.* The timing of northern bighorn sheep reproduction is highly synchronized with the seasons, with the birth of lambs closely aligned with the availability of forage (Festa-Bianchet 1988b, Feder et al. 2008). Rocky Mountain and Sierra Nevada bighorn sheep typically have their lambs in late spring or early summer, when snow has melted and given way to new vegetation growth. The alignment of lambing with this time of year is important, as females need access to this new high-quality forage for milk production to support the rapid growth and survival of their lambs (Ruckstuhl and Festa-Bianchet 1998, Stephenson et al. 2020). This new forage has elevated concentrations of potassium which consequentially causes herbivores such as bighorn sheep to lose trace minerals and electrolytes (e.g., sodium, magnesium), making mineral lick visitation necessary to maintain nutritional equilibriums in the face of the demands of lactation and regaining body fat lost during the winter months (Ayotte et al. 2006). In years with late snowmelt, access to nutritional resources (e.g., preferred spring forage, mineral licks) can be affected so lambs born before adequate forage is available may be at higher risk of increased mortality from starvation due in large part to poor maternal condition (Feder et al. 2008, Grusing et al. 2020). On the other hand, if lambing occurs too late in the season, there may not be enough high-quality forage available to support milk production, also negatively affecting lamb survival (Festa-Bianchet 1988b, Feder et al. 2008). Weather conditions can also affect the timing of lambing, with the previous winter conditions affecting maternal body condition. Females that are in poor condition due to harsh winter conditions can give birth later and produce smaller, weaker lambs that are less likely to survive (Ruckstuhl and Festa-Bianchet 1998, Hamel et al. 2009a, Monteith et al. 2018). They are also more prone to miscarriages and stillbirths and are more likely to abandon their lambs, particularly in years where forage is scarce and the energetic cost of rearing lambs is high (Berger 1991, Festa-Bianchet et al. 1994). Finally, the timing and synchronizations of lambing are important for predator avoidance. A synchronized pulse in birth of lambs may allow more lambs to survive long enough to become more mobile, enhancing their ability to evade predators. Lambs born before or after the main birth pulse may have lower survival rates. Additionally, lambs born too early are more vulnerable to predation if they are too small or weak to keep up with their mothers, whereas lambs born too late might not gain weight enough to survive their first winter (Festa-Bianchet 1988b, Feder et al. 2008).

Compared to adult survival, lamb survival of Rocky Mountain bighorn sheep tends to be low, highly variable, and associated with weather conditions, and thus is a key factor in population dynamics (Gaillard et al. 1998). In years with favorable conditions, such as warm springs and moist summers with abundant forage, lamb survival is usually high (Portier et al. 1998). In years with late snowmelt, dry springs and summers, wet-cold weather conditions during lambing, or early frosts and arrival of snow, lamb survival can be substantially decreased. In these years, late and overall decreased availability of high-quality forage can lead to malnutrition, increasing lamb vulnerability to disease and predation (Portier et al. 1998). Additionally, snow accumulation can force lambs and females into more exposed areas and slow down movement, further increasing vulnerability to predation (Festa-Bianchet 1988a). In years with poor forage availability females may already be in weakened condition and therefore be less able to defend lambs from predators (Berger 1991, Festa-Bianchet et al. 2006, Cloutier et al. 2024). Loss of lambs to predation can significantly decrease recruitment, particularly influencing small populations, where large changes in recruitment can have dramatic long-term consequences, particularly if adult mortality is also high (White et al. 2008, Proffitt et al. 2021). Reproduction is covered in more detail in Chapter 5.

*Population dynamics.* With vital rates of Rocky Mountain and Sierra Nevada bighorn sheep so closely related to weather patterns, fluctuations and trends in population size and structure are shaped by short-term weather events and long-term climate variability. In years with favorable weather patterns, high survival rates and successful reproduction tend to result in a growing population (Berger 1990, Festa-Bianchet et al. 2019, Stephenson et al. 2020). For example, in years with mild winters and an abundance of high-quality forage in spring and summer, both adult survival and lamb recruitment are typically high, resulting in population growth (Festa-Bianchet 1991). In years with adverse weather conditions, such as harsh winters, late springs, and droughts, populations can experience substantial declines due to decreased survival and reproduction. Periods of prolonged, harsh winters can cause high mortality rates for both lambs and adults, potentially leading to population crashes and, when accumulating over time, cause long-term reductions in population size (White et al. 2008). Long-term studies of bighorn populations have demonstrated a cyclical pattern in population dynamics, with periods of slow growth alternated by periods of sharp declines related to adverse weather (Festa-Bianchet et al. 2019). Such dynamics were, for example, documented in the Greater Yellowstone Ecosystem, where research noted populations expanding during favorable weather conditions and contracting during periods of harsh winters or drought, highlighting the role of weather as a key influence of Rocky Mountain bighorn sheep population dynamics (White et al. 2008). These documented fluctuations in population size can have long-term implications for population structure and genetic heterogeneity. Periods of population declines can lead to a decrease in genetic diversity, particularly in small, isolated populations (Poirier et al. 2018). This decrease can be compounded by habitat loss and fragmentation and reduce a population's ability to adapt to changing environmental conditions such as climate change and other anthropogenic factors (Rioux-Paquette et al. 2010). In the face of climate change, maintaining habitat connectivity, to preserve the ability of a population to shift its range in response to changing weather patterns, is crucial to maintain genetic diversity and population resilience (Hogg et al. 2006).

### 12.7.3 EFFECT OF WEATHER ON DEMOGRAPHY OF THINHORN SHEEP

Thinhorn sheep are native to mountainous, alpine areas of northwestern North America, inhabiting steep, rugged, remote terrain that is characterized by alpine meadows, rocky outcrops, and snow-covered ridges (Nichols and Bunnell 1999, Aycrigg et al. 2021, Chapter 19). Like Rocky Mountain and Sierra Nevada bighorn sheep, thinhorn sheep populations are sensitive to weather conditions and resulting changes in food, water, and shelter availability. The high-altitude, inhospitable environments that thinhorn sheep occupy are characterized by short, cool summers, and long, harsh winters, and weather patterns can fluctuate dramatically (Aycrigg et al. 2021). Although thinhorn sheep have evolved physiological and behavioral adaptations to these conditions, increasing variability in weather due to changing climate poses significant challenges (Jex et al. 2016, Chapter 21).

*Survival.* As with many alpine species, survival of thinhorn sheep is influenced by environmental conditions, particularly winter severity and associated forage availability. Survival during winter largely depends on the energy reserves that sheep accumulated during the previous autumn and summer combined with winter forage availability, which becomes limited as snow accumulates in the high mountains. Thinhorn sheep rely on sites where snow is shallow or absent, such as wind-exposed ridges, and steep, south-facing slopes, where they can graze on any remaining vegetation (Aycrigg et al. 2021). These areas may become inaccessible when snowfall is unusually heavy or persistent, forcing sheep to spend additional energy moving through deep snow in search of forage (Mahoney et al. 2018, Sivy et al. 2018, Cosgrove et al. 2021). This can result in malnutrition when energy reserves are already low (Nichols and Bunnell 1999). Another challenge of winter conditions is icing, which occurs when rain falls on snow and subsequently freezes, or when temperature fluctuations cause freeze-thaw-refreeze cycles (Bertram et al. 2018, Van de Kerk et al. 2020) or temperature inversions. These events result in a thick, icy layer or crust over vegetation, which, unlike snow, is almost impossible for sheep to break through to access forage. Icing events are a significant cause of mortality for thinhorn sheep populations; entire herds can die from starvation in winters where icing persists for extended periods, resulting in catastrophic population declines (Toweill and Geist 1999). Alaska Fish and Game reported a loss in the western Brooks Range in winter 2013–2014 of an estimated 80% of a Dall's sheep population as a result of starvation and diminished access to escape terrain to avoid predators, resulting from ground-fast ice (D. Brunning, Alaska Department of Fish and Game, personal communication). Therefore, increasing frequency in icing events due to climate change is a significant threat to thinhorn sheep population stability (Intergovernmental Panel on Climate Change 2013, Van de Kerk et al. 2020, Chapter 21).

The influence of severe winter conditions particularly affects lambs and older sheep that are less capable of enduring prolonged periods of nutritional stress (Van de Kerk et al. 2020). Older individuals may have worn teeth or other deteriorating physical traits, limiting their ability to forage in winter and to build up enough fat reserves in summer and autumn. Lambs, born in late spring, may also lack sufficient body mass and fat reserves to survive long, harsh winters (Bunnell 1980, Scotton 1997). In general, hunting success of the primary predators of thinhorn sheep, including wolverine (*Gulo gulo*), wolves (*Canis lupus*), coyotes, and lynx (*Lynx canadensis*), tends to increase during harsh winters because weaker, malnourished individuals are easier to catch (Prugh and Arthur 2015). Additionally, deep snowpack limits sheep mobility and ability to evade predators, and the search for food leaves them more exposed to predators (Mahoney et al. 2018, Cosgrove et al. 2021). The mortality risk is compounded by increased susceptibility

of malnourished sheep to bacterial and parasitic infections (Downs et al. 2018, Chapter 8). Lungworm, lumpy jaw, and gastrointestinal helminths in particular affect thinhorn sheep, reducing their fitness and overall condition (Valdez and Krausman 1999, Hoefs and Bunch 2001, Aleuy et al. 2018, Thacker 2020, Chapter 8).

Although winter severity influences thinhorn sheep survival, summer and autumn conditions also affect survival, particularly through their effect on the sheep's ability to build up enough fat reserves to survive the winter (Van de Kerk et al. 2020, Cosgrove et al. 2021). The availability of high-quality forage during the short growing season is vital to survive the long, cold winter months. If summer forage quality or quantity declines due to drought, sheep may not gain enough mass to survive the following winter or successfully rear a lamb. Another challenge of hot, dry summers includes harassment from biting insects such as mosquitoes (*Aedes, Anopheles, Culex, Culiseta* spp.), black flies (*Simulium vittatum*), and bot flies (*Hypoderma tarandi, Cephenemyia trompe*). Although thinhorn sheep often move to higher, windier elevations to avoid these insects, this increased movement depletes energy and reduces the time spent foraging, limiting their ability to accumulate fat reserves (Scotton 1997). Sheep can also be forced to travel long distances to find drinking water in dry years, as water sources become scarce. This not only increases their energy expenditure, but competition for limited water sources also increases intraspecific conflict, especially among males during the mating season (Loehr et al. 2010). In general, survival challenges for thinhorn sheep are likely to intensify as climate change increasingly affects weather patterns, posing a serious threat to their long-term population viability (Chapter 21).

*Reproduction.* Reproductive outcomes of thinhorn sheep populations are heavily influenced by weather conditions during key phases of the reproductive cycle, including rut, gestation, lambing, and early lamb survival. During the rut, in late autumn or early winter, males expend significant energy competing for access to females, leaving them in weakened condition once they enter winter (Geist 1971). The success of rut is affected by the quality and quantity of forage in the previous summer and autumn. If adults of both sexes enter mating season in good body condition and optimal health, the males are more capable of competing for mates, and the females are more likely to conceive (Scotton 1997). After dry summers with limited forage availability, males are less competitive during rut and females can experience lower conception rates or even delay conception, resulting in lower overall reproductive success. Additionally, females that do conceive may carry smaller, weaker fetuses due to poor maternal condition (Downs et al. 2018).

Timing of rut is closely linked to environmental cues such as day length and ensures that, after a gestation of approximately 175 days, thinhorn lambs are born when conditions are most favorable. Lambs are usually born in late May or early June (Enns et al. 2024), coinciding with the availability of new vegetation emerging in spring.

Females depend on this high-quality forage to support the high energy demands of pregnancy and lactation, and its availability immediately following birth is essential for maternal recovery and lamb survival (Geist 1971). While high-quality spring forage is important, it typically has increased concentrations of trace minerals (e.g., potassium), and through the process of digestion, herbivores lose electrolytes, like sodium and magnesium. To compensate for digestive losses, visitation to mineral licks is necessary (Ayotte et al. 2006). Peak use of mineral licks coincides with the demands of lactation and regaining body fat lost during winter (Grusing et al. 2020), so if mineral licks are not locally available within seasonal home ranges, mountain sheep may make short-term migrations to visit natural or anthropogenic mineral sources (e.g., roadside salt residues).

Success of lambing is further tied to the health and condition of females during early gestation (Downs et al. 2018, Cosgrove et al. 2021). Females that enter winter in poor condition due to inadequate summer forage are less likely to carry a pregnancy to term. In harsh winters, nutritional stress can lead to spontaneous abortions or stillbirths, and females that do give birth may deliver small, weaker lambs with lower survival rates. Variation in weather patterns can also influence gestation outcomes. In springs with late snowmelt, delaying the start of the growing season, females may give birth when forage is still scarce (Van de Kerk et al. 2018). Conversely, warm winters and early snowmelt can cause vegetation to grow earlier than usual. These mismatches between the times of lambing and optimal forage conditions can compromise the ability of females to nurse their lambs, consequently increasing lamb mortality rates.

Thinhorn sheep are most vulnerable during the period immediately following lambing, because lambs are particularly susceptible to predation, malnutrition, and exposure. To reduce the likelihood of predation, lambing occurs at high elevation on steep, inaccessible cliffs (Aycrigg et al. 2021, Enns et al. 2024). Lambs, however, are born with only a thin layer of fat and limited ability to regulate their body temperature, making them vulnerable to cold and wet weather conditions. The risk of hypothermia for lambs is highest during cool, wet springs, particularly if harsh weather coincides with their birth, or if mothers are unable to find adequate shelter. Lamb mortality is higher in years with late snowmelt or unusually cold or wet weather during the lambing season (Van de Kerk et al. 2018). For example, not a single lamb was observed during surveys in a region of the Brooks Range, Alaska, after the exceptionally late spring of 2013 (Rattenbury et al. 2016, Jex et al. 2025). Lamb mortality is also higher in years where unusual weather patterns result in more variability and reduced synchronization of lambing; however, Rachlow and Bowyer (1991) reported a shift in the parturition period for Dall's sheep associated with a late spring snowstorm but did not report a difference in mortality rates. Lambs born late in the season may not have enough time to gain enough energy reserves to survive winter (Rachlow and Bowyer 1991, Scotton 1997).

Long-term reproductive success of thinhorn sheep populations are variable and closely tied to variability in weather patterns. Prolonged periods of severe winters or poor summer forage can result in multiple consecutive years of low lamb recruitment and subsequent population declines (Jex et al. 2025). In the same conditions that negatively affect reproductive success, mortality rates are high, compounding the detrimental effect on the population. As climate change is expected to increase the weather variability and the frequency of extreme weather events, matching the timing of lambing to the time of optimal forage will become increasingly harder, limiting a population's ability to recover from years of low recruitment (Chapter 21).

*Population dynamics.* The dynamics of thinhorn sheep populations are shaped by the effects of environmental factors on survival and reproduction (Van de Kerk et al. 2018, 2020). Uniquely adapted to life in harsh, high-altitude environments, thinhorn sheep are also particularly sensitive to fluctuations in predator pressure, habitat availability, and climate conditions. Variability in weather patterns plays a particularly large role in influencing population dynamics and long-term viability of thinhorn sheep populations. Prolonged periods of favorable weather, resulting in high-quality forage and consequently higher survival rates and reproductive success, trigger gradual population growth. These periods are often followed by sharp declines in population size due to harsh weather events. The population declines tend to be influenced by a reduction in lamb survival, which is particularly sensitive to fluctuations in weather. Lamb mortality can be exceptionally high in years with heavy snowfall and prolonged winters, as the extended period of nutritional stress causes females to give birth to smaller lambs and be unable to support lactation. Severe winters also reduce the number of females that successfully conceive, and the number of females that survive to reproduce the next year. Therefore, the cumulative effects of several harsh winters in a row can lead to rapid population declines.

Because of their sensitivity to weather conditions, climate change is expected to have a profound influence on thinhorn sheep populations; their environments are also changing most rapidly in response to changing climatic conditions (Van de Kerk et al. 2018, Cosgrove et al. 2021, Chapter 21). The expected increase in frequency and severity of extreme weather events trends toward warmer winters, cooler summers, and generally wetter annual conditions (Bevington et al. 2019) may cause populations to decline beyond their potential for recovery. The increased frequency of heavy snowfall and icing events from warm, variable weather during winter poses a threat to survival and reproductive success, while summer droughts reduce forage availability and cause sheep to go into the winter with suboptimal fat reserves. Extirpation risk may be particularly high for small populations in isolated habitats, where movement of individuals between populations is difficult, reducing gene flow and limiting potential to migrate or shift to more suitable environments. The combined effects of climate change, habitat fragmentation, and loss of genetic diversity pose a serious threat to long-term viability of thinhorn sheep populations. Conservation efforts aimed at maintaining and restoring habitat connectivity are essential for the ability of sheep to adapt to changing environmental conditions (Chapter 21).

### 12.7.4  Effect of Weather on Demography of Desert Bighorn Sheep

*Survival.* Desert bighorn sheep are typically not exposed to winter weather severe enough to influence survival of healthy adult sheep. Some ranges, however, do experience short periods of cold, wet, or snowy conditions, which may overlap with the lambing period of some populations of desert bighorn sheep and could influence lamb survival. Some examples of winter weather reducing lamb survival include reports by Hansen (1960) that cold winter weather resulted in lower lamb survival in the Desert Game Range (Desert National Wildlife Refuge), Nevada, and Douglas (2001) who reported a negative relationship between precipitation during April and lamb survival in Canyonlands National Park, Utah. Similarly, lamb:female ratios in the San Gabriel Mountains, California, were inversely related to precipitation during the lambing period (Holl and Bleich 1983). Other researchers, however, have reported a positive influence of winter precipitation during late gestation and immediately post-parturition on lamb survival (Douglas and Leslie 1986, Wehausen et al. 1987, Douglas 2001, Wehausen 2005, Longshore et al. 2016). Therefore, instances of abnormally cold, wet weather immediately post-parturition reducing lamb survival appear to be the exception.

Drought conditions can strongly affect the survival of all age classes of desert bighorn sheep. Lambs and younger animals are generally more sensitive to drought-induced declines in forage nutritional quality, commonly leading to reduced recruitment following droughts. For example, McKinney et al. (2001) reported that drought-related decreases in winter precipitation were related to declines in recruitment rates of desert bighorn sheep in Arizona. Cain et al. (2008a) reported increases in yearling:female ratios with increasing summer precipitation. Extended drought, however, can lead to malnutrition of adults, resulting in lower rates of adult survival. For example, during a 4-year study in southwestern Arizona spanning climatic conditions ranging from extreme drought to above-average precipitation, survival of adult females was positively associated with precipitation (Cain et al. 2008a). Adult survival increased with increasing annual precipitation in the current and previous year. Survival rates were lowest during extreme drought, declining to 33%–65% during the hot summer (Cain et al. 2008a). Drought, particularly during winter and summer, has also been related to population declines in desert bighorn sheep populations (Monson 1960, McKinney et al. 2001, 2006).

*Reproduction.* Timing of rut in desert bighorn sheep is typically longer than in mountain sheep from cooler temperate regions, and peaks range from July to September (Turner and Hansen 1980), a period that generally coincides with summer rains in much of the southwestern United States. Lambing season is directly linked to the timing of the rut. Lambing season for desert bighorn sheep does not have a synchronized peak in parturition and lambing season can be protracted over several months (Lenarz 1979, Bunnell 1982, Rubin et al. 2000, Robinson et al. 2020, Parikh et al. 2024). Lambing season varies across desert bighorn populations but generally occurs from January through March or April (Turner and Hansen 1980, Parikh et al. 2024). From 10% to 35% of lambs in some desert ranges are born outside of what would be considered the normal lambing season (Turner and Hansen 1980). Timing of parturition can also vary across years within populations (Parikh et al. 2024). In southern Arizona, lambs were born in every month except October (Witham 1983).

Lamb parturition synchrony in temperate environments coincides with warmer spring climatic conditions and peak forage availability, maximizing the growth of neonates and enhancing survival through their first winter (Rutberg 1987, Festa-Bianchet 1988b, Rubin et al. 2000). Lambs born early could face harsh early spring weather conditions and those born late may not have enough time for growth and accumulating nutritional reserves to survive winter (Bunnell 1982, Feder et al. 2008). Parturition synchrony can also result in lowered risk of predation per individual (Testa 2002). Synchrony of births may be associated with higher survival in Rocky Mountain bighorn sheep (Whiting et al. 2012b) and other species (Singer et al 1997, Cook et al. 2004, Gregg et al. 2001). Low primary productivity of desert environments combined with high spatiotemporal variation in precipitation result in the lack of a predictable peak in resource availability. Climatic conditions during late winter and early spring are generally mild, with good forage conditions associated with winter rains (Rubin et al. 2000, Cain et al. 2017). The less predictable phenology of forage quality and availability in desert systems likely favors a protracted lambing season in desert bighorn (Leslie and Douglas 1979, Thompson and Turner 1982). Further, interannual variation in the timing of lambing could be associated with the influence of the previous year's reproductive effort and forage conditions on maternal nutritional condition (Overstreet et al. 2014, Proffitt et al. 2021, but see Rubin et al. 2000).

Higher birth mass has been demonstrated to contribute to higher survival rates in numerous ungulate species (Thorne et al. 1976, Fairbanks 1993, Cook et al. 2004, Heffelfinger et al. 2018). Neonate birth mass is strongly influenced by maternal body condition, with young born to females in good nutritional condition typically having higher birth mass and growth rates post-parturition. Forage availability and nutritional quality are linked to maternal body condition of desert bighorn sheep (Overstreet et al. 2014) and forage dynamics in desert bighorn ranges are largely determined by precipitation patterns (Noy-Meir 1973, Krausman et al. 1989, Marshal et al. 2005, Cain et al. 2017), thus climatic conditions largely determine lamb survival and population performance of desert bighorn sheep. For example, autumn lamb:female ratios from desert bighorn sheep populations are often positively correlated with winter rainfall during late gestation (Douglas and Leslie 1986, Wehausen et al. 1987, Douglas 2001, McKinney et al. 2001, Wehausen 2005, Longshore et al. 2016). Therefore, rainfall links forage conditions during late gestation and early post-parturition with juvenile recruitment, likely through their effects on fetal growth rates, neonate birth mass, and milk quality provided by the female. Furthermore, precipitation during autumn of the previous year during early to mid-gestation has also been linked to lamb survival, likely through the effects of forage on maternal body condition (Douglas and Leslie 1986, Wehausen et al. 1987, Krausman et al. 1999). In addition, in southern Utah, temperatures in May during lambing and winter precipitation 2 years prior were associated with lamb survival in the current year (Douglas 2001). Conversely, too much precipitation at the wrong time of year can be associated with lower lamb survival. Hansen (1960) reported that cloudy and cold winter weather was related to higher incidence of pneumonia and internal parasites, resulting in lower lamb survival on the Desert Game Range. Douglas (2001) reported a negative relationship between precipitation during April and lamb survival in Canyonlands National Park, Utah, and Holl and Bleich (1983) reported a similarly negative relationship between lamb:female ratios and precipitation during lambing, which they attributed to weather-related mortality of lambs. In New Mexico, lamb:female ratios from 2000 to 2022 for desert bighorn sheep had a quadratic relationship with precipitation during the pre-parturition period (3 months prior to start of current lambing season); lamb:female ratios were highest at intermediate levels of precipitation (Padilla et al. 2024). Furthermore, lamb:female ratios declined with increasing mean daily temperature prior to parturition (Padilla et al. 2024).

*Population dynamics.* Population dynamics of desert bighorn sheep are primarily influenced by food supply; however, predation, disease, and other factors can keep desert bighorn numbers below what could be supported by available forage. Predation losses can be particularly limiting for translocated populations (Kamler et al. 2002, Rominger et al. 2004) small populations (Wehausen 1996); however, predation has also been implicated in the decline of large populations of desert bighorn sheep (Holl et al. 2004).

Survival of adult females can have the greatest influence on growth rates of large ungulate populations. Population modeling efforts for desert bighorn sheep are consistent with adult female survival having the highest elasticities (largest relative contribution of demographic rate to population growth rate; de Kroon et al. 1986, Rubin et al. 2002, Harris et al. 2022). When survival of adult females is high, however, with low annual variation, realized changes

in population growth are influenced by juvenile survival, which is typically low and more variable (Gaillard et al. 1998, 2000). High variation in lamb survival (Cain et al. 2019) and recruitment rates (Douglas and Leslie 1986, Douglas 2001, Harris et al. 2022) has been commonly reported for desert bighorn sheep populations. Under conditions without disease outbreaks or excessive predation, desert bighorn sheep population dynamics are largely influenced by lamb survival and recruitment.

High spatial and temporal variability in rainfall in desert systems influences forage dynamics, yet changes in forage conditions occur over a shorter period than can be tracked by changes in population vital rates (i.e., survival, recruitment; Marshal et al. 2009). A weak temporal link between changes in forage resources and population abundance would reduce the role of density dependence in desert bighorn sheep populations inhabiting ranges with high spatial and temporal variation in rainfall. Marshal et al. (2009) evaluated changes in abundance of 20 desert bighorn populations in the Sonoran Desert across Arizona and southeastern California in relation to sheep density and rainfall to evaluate extrinsic and intrinsic factors related to population growth. The study populations spanned a gradient in mean annual precipitation. Variation in annual rainfall across ranges occupied by the study populations was inversely related to total annual rainfall such that populations with the lowest mean annual rainfall also experienced the highest annual variation in rainfall. Populations with growth rates showing density-dependent effects (i.e., effect of abundance the previous year on current year's population growth) had lower variation in annual precipitation than populations where density-dependent effects were not detected. Population showing a strong effect of rainfall had lower mean annual precipitation than populations lacking a rainfall effect (Marshal et al. 2009). Similarly, Bender and Weisenberger (2005) modeled population abundance as a function of rainfall and previous population size and similarly failed to find a strong density-dependent influence on population growth and lamb:female ratios were related to previous precipitation and largely unrelated to population size in the previous year (but see Rominger et al. 2008, Bender and Weisenberger 2009). Conversely, Wehausen et al. (1987) suggested that density-dependent effects lowered recruitment of desert bighorn in the Santa Rosa Mountains, California.

## 12.8 SUMMARY

Ranges occupied by mountain sheep have extreme seasonal climatic conditions ranging from high temperatures and limited surface water in desert environments, to northern areas with extreme cold and snowy winter conditions. These environmental conditions can pose challenges for maintaining temperature and water balance for mountain sheep. Mountain sheep use behavioral, physiological, and morphological adaptations to maintain temperature and water balance. Mountain sheep meet water requirements from free water in the environment, preformed water in their forage, and via metabolic water produced during the oxidation of fats, carbohydrates, and proteins. When preformed water and metabolic water are not sufficient to meet water demands, mountain sheep need access to sources of drinking water. Natural perennial sources of drinking water are limited in desert environments; thus, wildlife managers have constructed and maintained sources of water for desert bighorn sheep and some populations of Rocky Mountain bighorn sheep occupying more arid ranges. Climatic conditions also drive survival, reproduction, and population dynamics of mountain sheep. Extremely high summer temperatures in desert environments can lead to dehydration and heat stress, potentially resulting in higher mortality rates. In contrast, northern species of mountain sheep (Rocky Mountain bighorn and thinhorn sheep) must contend with extreme winter conditions and are exposed to freezing temperatures and heavy snowfall, which can increase energy expenditure to maintain body heat. Increased snow depth can increase costs of locomotion and impede access to forage. These conditions can be particularly challenging for young of the year. Extreme weather events, such as droughts, heavy snowfall, and storms, can have immediate and severe effects on mountain sheep populations. These events can exacerbate existing stressors and cause new ones, resulting in significant demographic changes. Availability and quality of forage is directly affected by precipitation patterns. Forage quantity and quality are determined by the timing and amount of rainfall, which influences body condition and reproductive success of mountain sheep. Droughts can affect desert bighorn sheep leading to decreased forage quality and quantity and limiting water availability, and prolonged drought can reduce lamb survival, recruitment, and adult survival. Heavy winter and spring snow can restrict access to forage for northern bighorn and thinhorn sheep, affecting survival and recruitment rates.

## REFERENCES

Alderman, J. A., P. R. Krausman, and B. D. Leopold. 1989. Diel activity of female desert bighorn sheep in western Arizona. *Journal of Wildlife Management* 53:264–271.

Aleuy, O. A., K. Ruckstuhl, E. P. Hoberg, A. Veitch, N. Simmons, and S. J. Kutz. 2018. Diversity of gastrointestinal helminths in Dall's sheep and the negative association of the abomasal nematode, *Marshallagia marshalli*, with fitness indicators. *PLoS ONE* 13:e0192825.

Ali, K. E., H. M. Mousa, and J. D. Hume. 1982. Total body water and water economy in camels, desert goats and desert sheep during water restriction and deprivation. *Iugoslavica Physiologica et Pharmacologica Acta* 18:229–236.

Allen, R. W. 1980. Natural mortality and debility. Pages 172–185 in G. Monson and L. Sumner, editors. *The Desert Bighorn.* University of Arizona Press, Tucson, Arizona, USA.

Andrew, M. H. 1988. Grazing impact in relation to livestock watering points. *Trends in Ecology and Evolution* 12:336–339.

Andrew, N. G., V. C. Bleich, A. D. Morrison, L. M. Lesicka, and P. J. Cooley. 2001. Wildlife mortalities associated with artificial water sources. *Wildlife Society Bulletin* 29:175–280.

Aycrigg, J. L., A. G. Wells, E. O. Garton, B. Magipane, G. E. Liston, L. R. Prugh, and J. L. Rachlow. 2021. Habitat selection by Dall's sheep is influenced by multiple factors including direct and indirect climate effects. *PLoS ONE* 16:e0248763.

Ayotte, J. B., K. L. Parker, J. M. Arocena, and M. P. Gillingham. 2006. Chemical composition of lick soils: functions of soil ingestion by four ungulate species. *Journal of Mammalogy* 87:878–888.

Baker, M. A. 1989. Effects of dehydration and rehydration on thermoregulatory sweating in goats. *Journal of Physiology* 417:421–435.

Barboza, P. S., K. L. Parker, and I. D. Hume. 2009. Integrative Wildlife Nutrition. Springer-Verlag, Berlin and Heidelberg, Germany.

Benavides, J., P. D. Walsh, L. A. Meyers, M. Raymond, and D. Caillaud. 2012. Transmission of infectious diseases en route to habitat hotspots. *PLoS ONE* 7(2):e31290.

Bender, L. C., and M. E. Weisenberger. 2005. Precipitation, density, and population dynamics of desert bighorn sheep on San Andres National Wildlife Refuge, New Mexico. *Wildlife Society Bulletin* 33:956–964.

Bender, L. C., and M. E. Weisenberger. 2009. Criticisms biologically unwarranted and analytically irrelevant: reply to Rominger et al. *Journal of Wildlife Management* 73:806–810.

Berger, J. 1985. Interspecific interactions and dominance among wild Great Basin ungulates. *Journal of Mammalogy* 66:571–573.

Berger, J. 1990. Persistence of different-sized populations: an empirical assessment of rapid extinctions in bighorn sheep. *Conservation Biology* 4:91–98.

Berger, J. 1991. Pregnancy incentives, predation constraints and habitat shifts: experimental and field evidence for wild bighorn sheep. *Animal Behaviour* 51:65–77.

Bertram, M. R., J. Herriges, C. T. Seaton, J. Lawler, K. Beckmen, and S. Dufford, S. 2018. Distribution, movements, and survival of Dall's sheep (*Ovis dalli dalli*) in the White Mountains, Alaska. Refuge report 2018-002. U.S. Fish and Wildlife Service. Fairbanks.

Bérubé, C. H., M. Festa-Bianchet, and J. T. Jorgenson. 1996. Reproductive costs of sons and daughters in Rocky Mountain bighorn sheep. *Behavioral Ecology* 7:60–68.

Bevington, A. R., H. E. Gleason, V. N. Foord, W. C. Floyd, and H. P. Griesbauer. 2019. Regional influence of ocean–atmosphere teleconnections on the timing and duration of MODIS-derived snow cover in British Columbia, Canada. *Cryosphere* 10:2693–2712.

Bleich, V. C. 2003. The potential for botulism in desert-dwelling mountain sheep. *Desert Bighorn Council Transactions* 47:2–8.

Bleich, V. C. 2008. Reprovisioning wildlife water developments: considerations for determining priorities to transport water. ESCAPE Technical Report 2008-01. Eastern Sierra Center for Applied Population Ecology, Bismarck, North Dakota, USA.

Bleich, V. C. 2009. Factors to consider when reprovisioning water developments used by mountain sheep. *California Fish and Game* 95:153–159.

Bleich, V. C., J. P. Marshal, and N. G. Andrew. 2010. Habitat use by a desert ungulate: predicting effects of water availability on mountain sheep. *Journal of Arid Environments* 74:638–645.

Bleich, V. C., M. W. Oehler, and J. G. Kie. 2020. Managing range-lands for wildlife. Pages 126–148 in N. J. Silvy, editor. *The Wildlife Management Techniques Manual. Volume 2.* Johns Hopkins University Press, Baltimore, Maryland, USA.

Bleich, V. C., N. G. Andrew, M. J. Martin, G. P. Mulcahy, A. M. Pauli, and S. S. Rosenstock. 2006. Quality of water available to wildlife in desert environments: comparisons among anthropogenic and natural sources in southeastern California. *Wildlife Society Bulletin* 34:625–630.

Bleich, V. C., R. T. Bowyer, and J. D. Wehausen. 1997. Sexual segregation in mountain sheep: resources or predation? *Wildlife Monographs* 134:3–50.

Bradley, W. G. 1963. Water metabolism in desert mammals with special reference to desert bighorn sheep. *Desert Bighorn Council Transactions* 7:26–39.

Brooks, M. L., J. R. Matchett, and K. H. Berry. 2006. Effects of livestock water sites on alien and native plants in the Mojave Desert, USA. *Journal of Arid Environments* 67(supplement):125–147.

Broyles, B. 1995. Desert wildlife water developments: questioning use in the Southwest. *Wildlife Society Bulletin* 23:663–675.

Bunnell, F. L. 1980. Factors controlling lambing period of Dall's sheep. *Canadian Journal of Zoology* 58:1027–1031

Bunnell, F. L. 1982. The lambing period of mountain sheep: synthesis, hypotheses, and tests. *Canadian Journal of Zoology* 60: 1–14.

Cain, J. W. III, J. V. Gedir, J. P. Marshal, P. R. Krausman, J. D. Allen, G. C. Duff, B. D. Jansen, and J. R. Morgart. 2017. Extreme precipitation variability, forage quality and large herbivore diet selection in arid environments. *Oikos* 126:1459–1471.

Cain, J. W. III, B. D. Jansen, R. R. Wilson, and P. R. Krausman. 2008b. Potential thermoregulatory advantages of shade use by desert bighorn sheep. *Journal of Arid Environments* 72:1518–1525.

Cain, J. W. III, P. R. Krausman, J. R. Morgart, B. D. Jansen, and M. P. Pepper. 2008a. Responses of desert bighorn sheep to removal of water sources. *Wildlife Monographs* 171. https://doi.org/10.2193/2007-209

Cain, J. W. III, P. R. Krausman, S. S. Rosenstock, and J. C. Turner. 2006. Mechanisms of thermoregulation and water balance in desert ungulates. *Wildlife Society Bulletin* 34:570–581.

Cain, J. W. III, P. R. Krausman, S. T. Boyle, and S. S. Rosenstock. 2022. Water and other welfare factors. Pages 239–267 in P. R. Krausman and J. W. Cain III, editors. *Wildlife Management: Contemporary Principles and Practices,* 2nd Edition. The Wildlife Society and Johns Hopkins University Press, Baltimore, Maryland, USA.

Cain, J. W. III, R. C. Karsch, E. J. Goldstein, E. M. Rominger, and W. R. Gould. 2019. Survival and cause-specific mortality of desert bighorn sheep lambs. *Journal of Wildlife Management* 83:251–259.

Chappel, R. W., and R. J. Hudson. 1978. Winter bioenergetics of Rocky Mountain bighorn sheep. *Canadian Journal of Zoology* 56:2388–2393.

Chappel, R. W., and R. J. Hudson. 1980. Prediction of energy expenditures by Rocky Mountain bighorn sheep. *Canadian Journal of Zoology* 58:1908–1912.

Cloutier, Z., M. Festa-Bianchet, and F. Pelletier. 2024. Direct and indirect effects of cougar predation on bighorn sheep fitness. *Ecology* 105:e4374.

Connor, M. M., T. R. Stephenson, D. W. German, K. L. Monteith, A. P. Few, and E. H. Blair. 2018. Survival analysis: informing recovery of Sierra Nevada bighorn sheep. *Journal of Wildlife Management* 82:1442–1458.

Cook, J. G., B. K. Johnson, R. C. Cook, R. A. Riggs, T. Delcurto, L. D. Bryant, and L. L. Irwin. 2004. Effects of summer-autumn nutrition and parturition date on reproduction and survival of elk. *Wildlife Monograph* 155:1–61.

Cosgrove, C. L., J. Wells, A. W. Nolin, J. Putera, and L. R. Prugh, 2021. Seasonal influence of snow conditions on Dall's sheep productivity in Wrangell-St Elias National Park and Preserve. *PLoS ONE,* 16:e0244787.

Davidson, Z., M. Valeix, F. Van Kesteren, A. J. Loveridge, J. E. Hunt, F. Murindagomo, and D. W. Macdonald. 2013. Seasonal diet and prey preference of the African lion in a waterhole-driven semi-arid savanna. *PLoS ONE* 8:e0055182.

Davies, K. W., and C. S Boyd. 2019. Ecological effects of free-roaming horses in North American rangelands. *BioScience* 69:558–565.

de Boer, W. F., M. J. Vis, H. J. De Knegt, C. Rowles, E. M. Kohi, F. Van Langevelde, and S. E. Van Wieren. 2010. Spatial distribution of lion kills determined by the water dependency of prey species. *Journal of Mammalogy* 91:1280–1286.

de Kroon, H., A. Plaisier, J. M. van Groenendaal, and H. Caswell 1986. Elasticity: the relative contribution of demographic parameters to population growth rate. *Ecology* 67:1427–1431.

Denryter, K., D. W. German, T. R. Stephenson, and K. L. Monteith. 2021 State- and context-dependent applications of an energetics model in free-ranging bighorn sheep. Ecological Modelling Volume 440.

Douglas, C. L. 1988. Decline of desert bighorn sheep in the Black Mountains of Death Valley. *Desert Bighorn Council Transactions* 25:36–38.

Douglas, C. L. 2001. Weather, disease, and bighorn lamb survival during 23 years in Canyonlands National Park. *Wildlife Society Bulletin* 29:297–305.

Douglas, C. L., and D. M. Leslie, Jr. 1986. Influence of weather and density on lamb survival of desert mountain sheep. *Journal of Wildlife Management* 50:153–156.

Douhard, M., S. Guillemette, M. Festa-Bianchet, and F. Pelletier. 2018. Drivers and demographic consequences of seasonal mass changes in an alpine ungulate. *Ecology* 99:724–734.

Downs, C. J., B. V. Boan, T. D. Lohuis, and K. M. Stewart. 2018. Investigating relationships between reproduction, immune defenses, and cortisol in Dall sheep. *Frontiers in Immunology* 9:105.

Dunn, W. C., and C. L. Douglas. 1982. Interactions between desert bighorn sheep and feral burros at spring areas in Death Valley. *Desert Bighorn Council Transactions* 26:87–96.

Enns, G. E., B. A. Jex, and M. S. Boyce. 2023. Diverse migration patterns and seasonal habitat use of Stone's sheep (*Ovis dalli stonei*). *PeerJ* 11:e15215.

Enns, G. E., B. A. Jex, and M. S. Boyce. 2024. Stone's sheep (*Ovis dalli stonei*) lambing and nursery habitat selection. *Canadian Journal of Zoology* 102:691–707.

Epps, C. W., D. R. McCullough, J. D. Wehausen, V. C. Bleich, and J. L. Rechel. 2004. Effects of climate change on population persistence of desert-dwelling mountain sheep in California. *Conservation Biology* 18:102–113.

Fairbanks, W. S. 1993. Birthdate, birthweight, and survival in pronghorn fawns. *Journal of Mammalogy* 74:129–135.

Feder, C., J. G. A. Martin, M. Festa-Bianchet, C. Bérubé, and J. Jorgenson. 2008. Never too late? Consequences of late birthdate for mass and survival of bighorn lambs. *Oecologia* 156:773–781.

Festa-Bianchet, M. 1988b. Birthdate and survival in bighorn lambs (*Ovis canadensis*). *Journal of Zoology* 214:653–661.

Festa-Bianchet, M. 1988a. Seasonal range selection in bighorn sheep: conflicts between forage quality, forage quantity, and predator avoidance. *Oecologia* 75:580–586.

Festa-Bianchet, M. 1989. Individual differences, parasites, and the costs of reproduction for bighorn ewes (*Ovis canadensis*). *Journal of Animal Ecology* 58:785–795.

Festa-Bianchet, M. 1991. The social system of bighorn sheep: grouping patterns, kinship and female dominance rank. *Animal Behaviour* 42:71–82.

Festa-Bianchet, M., S. D. Côté, S. Hamel, and F. Pelletier. 2019. Long-term studies of bighorn sheep and mountain goats reveal fitness costs of reproduction. *Journal of Animal Ecology* 88:1118–1133.

Festa-Bianchet, M., J. T. Jorgenson, C. H. Berube, C. Portier, and W. D. Wishart. 1997. Body mass and survival of bighorn sheep. *Canadian Journal of Zoology* 75:1372–1379.

Festa-Bianchet, M., J. T. Jorgenson, and W. D. Wishart. 1994. Early weaning in bighorn sheep, *Ovis canadensis* affects growth of males but not of females. *Behavioral Ecology* 5:21–27.

Festa-Bianchet, M., T. Coulson, J. M. Gaillard, J. T. Hogg, and F. Pelletier. 2006. Stochastic predation events and population persistence in bighorn sheep. *Proceedings of the Royal Society B: Biological Sciences* 273:1537–1543.

Finch, V. A. 1972. Thermoregulation and heat balance of the East African eland and hartebeest. *American Journal of Physiology* 222:1374–1379.

Finch, V. A., and D. Robertshaw. 1979. Effect of dehydration on thermoregulation in eland and hartebeest. *American Journal of Physiology* 6:R192–196.

Foster, J. A., L. A. Harveson, and M. T. Pittman. 2005. Use of guzzlers by bighorn sheep in the Chihuahuan Desert. *Desert Bighorn Council Transactions* 48:12–22.

Fuller, A., P. R. Kamerman, S. K. Maloney, A. Matthee, G. Mitchell, and D. Mitchell. 2005. A year in the thermal life of a free-ranging herd of springbok *Antidorcas marsupialis*. *Journal of Experimental Biology* 208:2855–2864.

Gaillard, J. M., M. Festa-Bianchet, and N. G. Yoccoz. 1998. Population dynamics of large herbivores: variable recruitment with constant adult survival. *Trends in Ecology and Evolution* 13:58–63.

Gaillard, J. M., M. Festa-Bianchet, N. G. Yoccoz, A. Loison, and C. Toïgo. 2000. Temporal variation in fitness components and population dynamics in large herbivores. *Annual Review of Ecology, Evolution and Systematics* 31:367–393.

Gedir, J. V., J. W. Cain III, P. R. Krausman, J. D. Allen, G. C. Duff, and J. R. Morgart. 2016. Potential foraging decisions by a desert ungulate to balance water and nutrient intake in a water-stressed environment. *PLoS ONE* 11:e0148795.

Gedir, J. V., J. W. Cain III, T. L. Swetnam, P. R. Krausman, and J. R. Morgart. 2020. Extreme drought and adaptive resource selection by a desert mammal. *Ecosphere* 11:e03175.

Geist, V. 1971. *Mountain Sheep: A Study in Behavior and Evolution*. University of Chicago Press, Chicago/London.

Glass, D. M., P. R. Prentice, A. D. Evans, and O. J. Schmitz. 2022. Local differences in maximum temperature determine water use among desert bighorn sheep populations. *Journal of Wildlife Management* 86:e22313.

Gregg, M. A., M. Bray, K. M. Kilbride, and M. R. Dunbar. 2001. Birth synchrony and survival of pronghorn fawns. *Journal of Wildlife Management* 65:19–24.

Grenot, C. J. 1992. Ecophysiological characteristics of large herbivorous mammals in arid Africa and the Middle East. *Journal of Arid Environments* 23:125–155.

Grusing, E. C., B. Lowrey, J. DeVoe, & R. A. Garrott. 2020. Evaluating characteristics of mineral licks used by two mountain ungulates. *22nd Biennial Northern Wild Sheep and Goat Council Symposium, Alberta (virtual)*. Alberta, Canada.

Hailey, T. L. 1967. Reproduction and water utilization of Texas transplanted desert bighorn sheep. *Desert Bighorn Council Transactions* 11:53–58.

Hamel, S., S. D. Côté, J. M. Gaillard, and M. Festa-Bianchet. 2009a. Individual variation in reproductive costs of reproduction: high-quality females always do better. *Journal of Animal Ecology* 78:143–151.

Hamel, S., M. Garel, M. Festa-Bianchet, J. M. Gaillard, and S. D. Côté. 2009b. Spring Normalized Difference Vegetation Index (NDVI) predicts annual variation in timing of peak faecal crude protein in mountain ungulates. *Journal of Applied Ecology* 46:582–589.

Hansen, C. G. 1960. Lamb survival on the Desert Game Range. *Desert Bighorn Council Transactions* 4:60–61.

Harris, G., J. G. Sanderson, J. Erz, S. E. Lehnen, and M. J. Butler. 2015. Weather and prey predict mammals' visitation to water. *PLoS ONE* 10:e0141355.

Harris, G. M., D. R. Stewart, D. Brown, L. Johnson, J. Sanderson, A. Alvidrez, T. Waddell, and R. Thompson. 2020. Year-round water management for desert bighorn sheep corresponds with visits by predators not bighorn sheep. *PLoS ONE* 15:e0241131.

Harris, G. M., M. J. Butler, D. R. Stewart, and J. W. Cain III. 2022. The abundance and persistence of Caprinae populations. *Scientific Reports* 12:13807.

Heffelfinger, L. J., K. M. Stewart, A. P. Bush, J. S. Sedinger, N. W. Darby, and V. C. Bleich. 2018. Timing of precipitation in an arid environment: Effects on population performance of a large herbivore. *Ecology and Evolution* 8:3354–3366.

Hengeveld, P. E., and J. C. Cubberley, editors. 2011. Stone's sheep population dynamics and habitat use in the Sulphur/8 Mile oil and gas pre-tenure plan area, northern British Columbia, 2005–2010. Synergy Applied Ecology, Mackenzie, British Columbia, Canada.

Hervert, J. J., J. L. Bright, R. S. Henry, L. A. Piest, and M. T. Brown. 2005. Home-range and habitat-use patterns of Sonoran pronghorn in Arizona. *Wildlife Society Bulletin* 33:8–15.

Hetem, R. S., W. M. Strauss, L. G. Fick, S. K. Maloney, L. C. R. Meyer, M. Shobrak, A. Fuller, and D. Mitchell. 2012a. Activity re-assignment and microclimate selection of free-living Arabian oryx: responses that could minimize the effects of climate change on homeostasis? *Zoology* 115:411–416.

Hetem, R. S., W. M. Strauss, L. G. Fick, S. K. Maloney, L. C. R. Meyer, M. Shobrak, A. Fuller, and D. Mitchell. 2012b. Does size matter? comparison of body temperature and activity of free-living Arabian oryx (*Oryx leucoryx*) and the smaller Arabian sand gazelle (*Gazella subgutturosa marica*) in the Saudi desert. *Journal of Comparative Physiology B* 182:437–449.

Hik, D. S., and J. Carey. 2000. Cohort variation in horn growth of Dall sheep rams in the southwest Yukon, 1969–1999. *Biennial Symposium of the Northern Wild Sheep and Goat Council* 12:88–100.

Hoefs, M. 2000. The thermoregulatory potential of Ovis horn cores. *Canadian Journal of Zoology* 78:1419–1426.

Hoefs, M., and T. D. Bunch. 2001. Lumpy jaw in wild sheep and its evolutionary implications. *Journal of Wildlife Diseases* 37:39–48.

Hofmeyr, M. D. 1985. Thermal properties of the pelages of selected African ungulates. *South African Journal of Zoology* 20:179–189.

Hogg, J., F. Pelletier, and K. Ruckstuhl. 2006. Breeding migration, gene flow and management for connectivity in bighorn sheep. Biennial Symposium of the Northern Wild Sheep and Goat Council, 15. Whitefish, Montana, USA.

Holl, S. A., and V. C. Bleich. 1983. San Gabriel mountain sheep: biological and management considerations. Administrative report, San Bernardino National Forest, San Bernardino, California, USA.

Holl, S. A., V. C. Bleich, and S. G. Torres. 2004. Population dynamics of bighorn sheep in the San Gabriel Mountains, California, 1967–2002. *Wildlife Society Bulletin* 32:412–426.

Horst, R., and M. Langworthy. 1971. Observations on the kidney of the desert bighorn sheep. *Anatomical Record* 2:343.

Hughes, K. S., and N. S. Smith. 1990. Sonoran pronghorn use of habitat in southwest Arizona. Final Report 14–16–009–1564 RWO #6. Arizona Cooperative Fish and Wildlife Research Unit, Tucson, Arizona, USA.

Intergovernmental Panel on Climate Change. 2013. Climate change 2013: the physical science basis. Contribution of working group I to the fifth assessment report of the intergovernmental panel on climate change.

Jawuoro, S. O., O. K. Koech, G. N., Karuku, and J. S. Mbau. 2017. Plant species composition and diversity depending on piospheres and seasonality in the southern rangelands of Kenya. *Ecological Processes* 6:16.

Jex, B. A., K. Chan, M. Larivee, B. R. Wendling, and D. W. Lutz. 2025. Observed changes in populations of thinhorn sheep in western Canada and Alaska. Sheep Initiative, Western Association of Fish and Wildlife Agencies, Boise, Idaho, USA.

Jex, B. A., J. B. Ayotte, V. C. Bleich, C. E. Brewer, D. L. Bruning, T. M. Hegel, N. C. Larter, R. A. Schwanke, H. M. Schwantje, and M. W. Wagner. 2016. Thinhorn sheep: conservation challenges and management strategies for the 21st century. Wild Sheep Working Group. Western Association of Fish and Wildlife Agencies, Boise, Idaho, USA.

Jorgenson, J. T., M. Festa-Bianchet, and J. M. Gaillard. 1997. Effects of age, sex, disease, and density on survival of bighorn sheep. *Ecology* 78:1019–1032.

Jorgenson, J. T., M. Festa-Bianchet, M. Lucherini, and W. D. Wishart. 1993. Effects of body size, population density, and maternal characteristics on age at first reproduction in bighorn ewes. *Canadian Journal of Zoology* 71:2509–2517.

Kamler, J. F., R. M. Lee, J. C. deVos Jr, W. B. Ballard, and H. A. Whitlaw. 2002. Survival and cougar predation of translocated bighorn sheep in Arizona. *Journal of Wildlife Management* 66:1267–1272.

Kay, R. N. B. 1997. Responses of African livestock and wild herbivores to drought. *Journal of Arid Environments* 37:683–694.

Kihwele, E. S., V. Mchomvu, N. Owen-Smith, R. S. Hetem, M. C. Hutchinson, A. B. Potter, H. Olff, and M. P. Veldhuis. 2020. Quantifying water requirements of African ungulates through a combination of functional traits. *Ecological Monographs* 90:e01404.

Krausman, P. R., A. V. Sandoval, and R. C. Etchberger. 1999. Natural history of desert bighorn sheep. Pages 139–191 in R. Valdez and P. R. Krausman, editors. *Mountain sheep of North America*. University of Arizona Press, Tucson, Arizona, USA.

Krausman, P. R., and B. D. Leopold. 1986. Habitat components for desert bighorn sheep in the Harquahala Mountains, Arizona. *Journal of Wildlife Management* 50:504–508.

Krausman, P. R., B. D. Leopold, R. F. Seegmiller, and S. G. Torres. 1989. Relationships between desert bighorn sheep and habitat in western Arizona. *Wildlife Monographs* 102:3–66.

Krausman, P. R., S. S. Rosenstock, and J. W. Cain III. 2006. Developed waters for wildlife: science, perception, values, and controversy. *Wildlife Society Bulletin* 34:563–569.

Krausman, P. R., S. Torres, L. L. Ordway, J. J. Hervert, and M. Brown. 1985. Diel activity of ewes in the Little Harquahala Mountains, Arizona. *Desert Bighorn Council Transactions* 29:24–26.

Krausman, P. R., and R. C. Etchberger. 1995. Response of desert ungulates to a water project in Arizona. *Journal of Wildlife Management* 59:292–300.

Knudsen, M. F. 1963. A summer waterhole study at Carrizo Spring, Santa Rosa Mountains of Southern California. *Desert Bighorn Council Transactions* 7:185–192.

Kubly, D. M. 1990. Limnological features of desert mountain rock pools. Pages 103–120 in G. K. Tsukamoto and S. J. Stiver, editors. *Proceedings of the Wildlife Water Development Symposium*. Nevada Chapter, The Wildlife Society, U.S. Department of the Interior, Bureau of Land Management, Washington, DC, and Nevada Department of Wildlife, Reno, USA.

Langman, V. A. 1985. Nasal heat exchange in a northern ungulate, the reindeer (Rangifer tarandus). *Respiration Physiology* 59:279–287.

Langman, V. A., G. M. O. Maloiy, K. Schmidt-Nielsen, and R. C. Schroter. 1979. Nasal heat exchange in the giraffe and other large mammals. *Respiration Physiology* 37:325–333.

Lenarz, M. S. 1979. Social structure and reproductive strategy in desert bighorn sheep (*Ovis canadensis mexicana*). *Journal of Mammalogy* 60:671–678.

Leslie, D. M., Jr., and C. L. Douglas. 1979. Desert bighorn sheep of the River Mountains, Nevada. *Wildlife Monographs* 66:1–56.

Loehr, J., J. Carey, R. B. O'Hara, and D. S. Hik. 2010. The role of phenotypic plasticity in responses of hunted thinhorn sheep ram horn growth to changing climate conditions. *Journal of Evolutionary Biology* 23:783–790.

Longhurst, W. M., N. F. Baker, G. E. Connolly, and R. A. Fisk. 1970. Total body water and water turnover in sheep and deer. *American Journal of Veterinary Research* 31:673–677.

Longshore, K., C. Lowrey, and P. Cummings. 2016. Foraging at the wildland-urban interface decouples weather as a driver of recruitment for desert bighorn sheep. *Wildlife Society Bulletin* 40:494–499.

Longshore, K. M., C. Lowrey, and D. B. Thompson. 2009. Compensating for diminishing natural water: predicting the impacts of water developments on summer habitat of desert bighorn sheep. *Journal of Arid Environments* 73:280–286.

Louw, G. N. 1984. Water deprivation in herbivores under arid conditions. Pages 106–126 in F. M. C. Gilchrist and R. I. Mackie, editors. *Herbivore Nutrition in the Subtropics and Tropics*. Science Press, Craighall, South Africa.

Louw, G. N. 1993. *Physiological Animal Ecology*. Longman Scientific and Technical, Burnt Mill, United Kingdom.

Louw, G. N., and M. Seely. 1982. *Ecology of Desert Organisms*. Longman Group, Burnt Mill, United Kingdom.

Mahoney, P. J., G. E. Liston, E. LaPoint, E. Gurarie, B. Mangipane, A. G. Wells, T. J. Brinkman, J. U. H. Eitel, M. Hebblewhite, A. W. Nolin, N. Boelman, and L. R. Prugh. 2018. Navigating snowscapes: scale-dependent responses of mountain sheep to snowpack properties. *Ecological Applications* 28:1715–1729.

Maloiy, G. M. O. 1970. Water economy of the Somali donkey. *American Journal of Physiology* 219:1522–1527.

Maloiy, G. M. O. 1973. Water metabolism of East African ruminants in arid and semi-arid regions. *Journal of Animal Breeding and Genetics* 90:219–228.

Maloiy, G. M. O., and D. Hopcraft. 1971. Thermoregulation and water relations of two East African antelopes: the hartebeest and impala. *Comparative Biochemistry and Physiology A* 38:525–534.

Maloiy, G. M. O., W. V. Macfarlane, and A. Shkolnik. 1979. Mammalian herbivores. Pages 185–209 in G. M. O. Maloiy, editor. *Comparative Physiology of Osmoregulation in Animals*. Volume 2. Academic, London, United Kingdom.

Marshal, J. P., J. W. Cain III, V. C. Bleich, and S. S. Rosenstock. 2009. Intrinsic and extrinsic sources of variation in the dynamics of large herbivore populations. *Canadian Journal of Zoology* 87:103–111.

Marshal, J. P., P. R. Krausman, and V. C. Bleich. 2005. Dynamics of mule deer forage in the Sonoran Desert. *Journal of Arid Environments* 60:593–609.

Marshal, J. P., V. C. Bleich, P. R. Krausman, M. Reed, and A. Neibergs. 2012. Overlap in diet and habitat between the mule deer (*Odocoileus hemionus*) and feral ass (*Equus asinus*) in the Sonoran Desert. *Southwestern Naturalist* 51:16–25.

McCutchen, H. E. 1981. Desert bighorn zoogeography and adaptation in relation to historic land use. *Wildlife Society Bulletin* 9:171–179.

McKinney, T., T. W. Smith, and J. C. deVos, Jr. 2006. Evaluation of factors potentially influencing a desert bighorn sheep population. *Wildlife Monographs* 164:1–36.

McKinney, T., T. W. Smith, and J. D. Hanna. 2001. Precipitation and bighorn sheep in the Mazatzal Mountains, Arizona. *Southwestern Naturalist* 46:345–353.

McNab, B. K. 2002. *The Physiological Ecology of Vertebrates*. Cornell University Press, Ithaca, New York, USA.

Miller, G. D., M. H. Cochran, and E. L. Smith. 1984. Nighttime activity of desert bighorn sheep. *Desert Bighorn Council Transactions* 28:23–25.

Mitchell, D., S. K. Maloney, C. Jessen, H. P. Laburn, P. R. Kamerman, G. Mitchell, and A. Fuller. 2002. Adaptive heterothermy and selective brain cooling in arid-zone mammals. *Comparative Biochemistry and Physiology B* 131:571–585.

Mohamed, S. M., B. H. Ali, and T. Hassan. 1988. Some effects of water deprivation on dorcas gazelle (*Gazella dorcas*) in the Sudan. *Comparative Biochemistry and Physiology* 90A:225–228.

Monson, G. 1960. Effects of climate on desert bighorn numbers. *Desert Bighorn Council Transactions* 4:12–14.

Monteith, K. L., R. A. Long, T. R. Stephenson, V. C. Bleich, R. T. Bowyer, and T. N. LaSharr. 2018. Horn size and nutrition in mountain sheep: can ewe handle the truth? *Journal of Wildlife Management* 82:67–84.

Nagy, K. A., and C. C. Peterson. 1988. *Scaling of Water Flux Rates in Animals*. University of California Press, Berkeley, California, USA.

New Mexico Department of Game and Fish. 2013. Toxic algae cause of 100 elk deaths in northeastern NM. News Release. New Mexico Department of Game and Fish. https://www. wildlife.state.nm.us/legacy/publications/press_releases/ documents/2013/102213elkfinding.html. Date accessed 30 December 2024.

Nichols, L., and F. L. Bunnell. 1999. Natural history of thinhorn sheep. Pages 23–77 in R. Valdez and P. R. Krausman, editors. *Mountain Sheep of North America*. University of Arizona Press, Tucson, Arizona, USA.

Noy-Meir, I. 1973. Desert ecosystems: environment and producers. *Annual Review of Ecology and Systematics* 4:25–51.

Ostermann-Kelm, S., E. R. Atwill, E. S. Rubin, M. C. Jorgensen, and W. M. Boyce. 2008. Interactions between feral horses and desert bighorn sheep at water. *Journal of Mammalogy* 89:459–466.

Ostrowski, S., J. B. Williams, P. Mésochina, and H. Sauerwein. 2006. Physiological acclimation of a desert antelope, Arabian oryx (*Oryx leucoryx*), to long-term food and water restriction. *Journal of Comparative Physiology B* 176:191–201.

Overstreet, M., C. A. Caldwell, and J. W. Cain III. 2014. Adult survival, apparent lamb survival, and body condition of desert bighorn sheep in relation to habitat and precipitation on the Kofa National Wildlife Refuge, Arizona. USFWS and USGS Cooperator Science Series 109–2014.

Padilla, C. J., C. Q. Ruhl, J. W. Cain III, and M. E. Gompper. 2024. Effects of *Mycoplasma ovipneumoniae*, abundance, and environmental conditions on bighorn sheep lamb:ewe ratios and adult survival in New Mexico. *Ecosphere* 15:e70095.

Parikh, G. L., R. C. Karsch, J. W. Cain III, E. M. Rominger, and E. J. Goldstein. 2024. Neonate morphometrics and lambing season characteristics of desert bighorn sheep. Mammalia 2024–0074.

Parker, A. H., and E. T. F. Witkowski. 1999. Long-term impacts of abundant perennial water provision for game on herbaceous vegetation in a semi-arid African savannah woodland. *Journal of Arid Environments* 41:309–321.

Parker, K. L., and C. T. Robbins. 1984. Thermoregulation in mule deer and elk. *Canadian Journal of Zoology* 62:1409–1422.

Parker, K. L., C. T. Robbins, and T. A. Hanley. 1984. Energy expenditures for locomotion by mule deer and elk. *Journal of Wildlife Management* 48:474–488.

Pelletier, F., K. A. Page, T. Ostiguy, and M. Festa-Bianchet. 2005. Fecal counts of lungworm larvae and reproductive effort in bighorn sheep, *Ovis canadensis*. *Oikos* 110:473–480.

Pelletier, F., and M. Festa-Bianchet. 2004. Effects of body mass, age, dominance and parasite load on foraging time of bighorn rams, *Ovis canadensis*. *Behavioral Ecology and Sociobiology* 56:546–551.

Phillips, P. K., and J. E. Heath. 1995. Dependency of surface temperature regulation on body size in terrestrial mammals. *Journal of Thermal Biology* 20:281–289.

Picard, K. 1996. The cost of horniness: heat loss may counter sexual selection for large horns in temperate bovids. *Ecoscience* 3:280–284.

Picard, K., D. W. Thomas, M. Festa-Bianchet, F. Belleville, and A. Laneville. 1999. Differences in the thermal conductance of tropical and temperate bovid horns. *Ecoscience* 6:148–158.

Poirier, M. A., D. W. Coltman, F. Pelletier, J. T. Jorgenson, and M. Festa-Bianchet. 2018. Genetic decline, restoration and rescue of an isolated ungulate population. *Evolutionary Applications* 12:1318–1328.

Poole, K. G. 2013. Habitat use, seasonal movements, and population dynamics of bighorn sheep in the Elk Valley. Aurora Wildlife Research, Nelson, British Columbia, Canada.

Porter, W. P., and D. M. Gates. 1969. Thermodynamic equilibria of animals with environment. *Ecological Monographs* 39:227–244.

Portier, C., M. Festa-Bianchet, J. M. Gaillard, J. T. Jorgenson, and N. G. Yoccoz. 1998. Effects of density and weather on survival of bighorn sheep lambs (*Ovis canadensis*). *Journal of Zoology* 245:271–278.

Proffitt, K. M., A. B. Courtemanch, S. R. Dewey, B. Lowrey, D. E. McWhirter, K. L. Monteith, J. T. Paterson, J. Rotella, P. J. White, and R. A. Garrott. 2021. Regional variability in pregnancy and survival rates of Rocky Mountain bighorn sheep. *Ecosphere* 12:e03410.

Prude, C. H. 2020. Influence of habitat heterogeneity and water sources on kill site locations and puma prey composition. MS Thesis, New Mexico State University, Las Cruces, New Mexico, USA.

Prugh, L. R., and S. M. Arthur. 2015. Optimal predator management for mountain sheep conservation depends on the strength of mesopredator release. *Oikos* 124:1241–1250.

Rachlow, J. L. and R. T. Bowyer. 1991. Interannual variation in timing and synchrony of parturition in Dall's sheep. *Journal of Mammalogy* 72:487–492.

Rattenbury, K., J. Schmidt, L. Phillips, S. Arthur, B. Borg, J. Burch, K. Joly, J. Lawler, B. Mangipane, and J. Putera. 2016. Recent trends in Dall's sheep populations in Alaska's National Parks and Preserves. Alaska Dall Sheep Working Group Meeting, February 20–21, 2016, Anchorage, Alaska, USA.

Rioux-Paquette, E., M. Festa-Bianchet, and D. W. Coltman. 2010. No inbreeding avoidance in an isolated population of bighorn sheep. *Animal Behaviour* 80:865–871.

Robinson, B. G., and E. H. Merrill. 2012. The influence of snow on the functional response of grazing ungulates. *Oikos* 121:28–34.

Robinson, J. R. 1957. Functions of water in the body. *Proceedings of the Nutritional Society* 16:108–112.

Robinson, R. W., T. S. Smith, J. C. Whiting, R. T. Larsen, and J. M. Shannon. 2020. Determining timing of births and habitat selection to identify lambing period habitat for bighorn sheep. *Frontiers in Ecology and Evolution* 8: 97.

Rominger, E. M., H. W. Whitlaw, D. L. Weybright, W. C. Dunn, and W. B. Ballard. 2004. The influence of mountain lion predation on bighorn sheep translocations. *Journal of Wildlife Management* 68:993–999.

Rominger, E. M., E. J. Goldstein, and M. A. Evans. 2008. Biological and statistical errors make inferences circumspect: response to Bender and Weisenberger. *Journal of Wildlife Management* 72:580–582.

Rosenstock, S. S., M. J. Rabe, C. S. O'Brien, and R. B. Waddell. 2004. Studies of wildlife water developments in southwestern Arizona: wildlife use, water quality, wildlife diseases, wildlife mortalities, and influences in native pollinators. Technical Guidance Bulletin No. 8. Arizona Game and Fish Department, Phoenix, Arizona, USA.

Rosenstock, S. S., V. C. Bleich, M. J. Rabe, and C. Reggiardo. 2005. Water quality and wildlife water sources in the Sonoran Desert, United States. *Rangeland Ecology and Management* 58:623–627.

Rosenstock, S. S., W. B. Ballard, and J. C. deVos. 1999. Benefits and impacts of wildlife water developments. *Journal of Range Management* 52:302–311.

Rubin, E. S., W. M. Boyce, and V. C. Bleich. 2000. Reproductive strategies of desert bighorn sheep. *Journal of Mammalogy* 81:769–786.

Rubin, E. S., W. M. Boyce, and E. P. Caswell-Chen. 2002. Modeling demographic processes in an endangered population of bighorn sheep. *Journal of Wildlife Management* 66:796–810.

Ruckstuhl, K. E., and M. Festa-Bianchet. 1998. Do reproductive status and lamb gender affect the foraging behavior of bighorn ewes? *Ethology* 104:941–954.

Rutberg, A. T. 1987. Adaptive hypotheses of birth synchrony in ruminants: an interspecific test. *American Naturalist* 130:692–710.

Scotton, B. D. 1997. Estimating rates and causes of neonatal lamb mortality of Dall sheep in the Central Alaska Range. Alaska Department of Game and Fish Report

Schmidt-Nielsen, K., B. Schmidt-Nielsen, S. Jarnum, and T. R. Houpt. 1957. Body temperature of the camel and its relation to water economy. *American Journal of Physiology* 188:103–112.

Seegmiller, R. F., P. R. Krausman, W. H. Brown, and F. M. Whiting. 1990. Nutritional composition of desert bighorn sheep forage in the Harquahala Mountains, Arizona. *Desert Plants* 10:87–90.

Shackleton, D. M., C. C. Shank, and B. M. Wikeem. 1999. Natural history of Rocky Mountain and California bighorn sheep. Pages 78–138 in R. Valdez and P. R. Krausman, editors. *Mountain Sheep of North America.* University of Arizona Press, Tucson, Arizona, USA.

Silanikove, N. 1994. The struggle to maintain hydration and osmoregulation in animals experiencing severe dehydration and rapid rehydration: the story of ruminants. *Experimental Physiology* 79:281–300.

Simmons, N. M. 1969. The social organization, behavior and environment of the desert bighorn sheep on the Cabeza Prieta Game Range, Arizona. Dissertation, University of Arizona, Tucson, Arizona, USA.

Simpson, N. O., K. M. Stewart, and V. C. Bleich. 2011. What have we learned about water developments for wildlife? not enough! *California Fish and Game* 97:190–209.

Singer, F. J., A. Harting, K. K. Symonds, and M. B. Coughenour. 1997. Density dependence, compensation, and environmental effects on elk calf mortality in Yellowstone National Park. *Journal of Wildlife Management* 61:12–25.

Singer, F. J., L. C. Zeigenfuss, and L. Spicer. 2001. Role of patch size, disease, and movement in rapid extinction of bighorn sheep. *Conservation Biology* 15:1347–1354.

Sivy, K. J., A. W. Nolin, C. L. Cosgrove, and L. R. Prugh. 2018. Critical snow density threshold for Dall's sheep (*Ovis dalli dalli*). *Canadian Journal of Zoology* 96:1170–1177.

Smiley, R. A., B. L. Wagler, W. H. Edwards, J. Jennings-Gaines, K. Luukkonen, K. Robbins, M. Johnson, A. B. Courtenmanch, T. W. Mong, D. Lutz, D. McWhirter, J. L. Malmberg, B. Lowrey, and K. L. Monteith. 2024. Infection–nutrition feedbacks: fat supports pathogen clearance but pathogens reduce fat in a wild mammal. *Proceedings of the Royal Society B* 291:20240636.

Stephenson, T. R., D. W. German, F. E. Cassirer, D. P. Walsh, M. E. Blum, M. Cox, K. M. Stewart, and K. L. Monteith. 2020. Linking population performance to nutritional condition in an alpine ungulate. *Journal of Mammalogy* 101:1244–1256.

Swift, P. K., J. D. Wehausen, H. B. Ernest, R. S. Singer, A. M. Pauli, H. Kinde, T. E. Rocke, and V. C. Bleich. 2000. Desert bighorn sheep mortality due to presumptive type C botulism in California. *Journal of Wildlife Diseases* 36:184–189.

Taylor, C. R. 1968. Hygroscopic food: a source of water for desert antelopes. *Nature* 219:181–182.

Taylor, C. R. 1969. Metabolism, respiratory changes and water balance of an antelope, the eland. *American Journal of Physiology* 217:317–320.

Taylor, C. R. 1970a. Dehydration and heat: effects on temperature regulation of East African ungulates. *American Journal Physiology* 219:1136–1139.

Taylor, C. R. 1970b. Strategies of temperature regulation: effects on evaporation in East African ungulates. *American Journal Physiology* 219:1131–1135.

Terry, P. J., A. C Alvidrez, and C. W. Black. 2022. Factors affecting bighorn sheep activity at water developments in southwestern Arizona. *Journal of Wildlife Management* 86:e22134.

Testa, J. W. 2002. Does predation on neonates inherently select for earlier births? *Journal of Mammalogy* 83:699–706.

Titcomb, G., J. N. Mantas, J. Hulke, I. Rodriguez, D. Branch, and H. Young. 2021. Water sources aggregate parasites with increasing effects in more arid conditions. *Nature Communications* 12:art no. 7066.

Thacker, C. 2020. Health surveillance of thinhorn sheep (*Ovis dalli*) herds in British Columbia and Alaska. M.Sc. Thesis, University of Calgary, Calgary, Alberta, Canada.

Thorne, E. T., R. E. Dean, and W. G. Hepworth. 1976. Nutrition during gestation in relation to successful reproduction in elk. *Journal of Wildlife Management* 40:330–335.

Thompson, R. W., and J. C. Turner. 1982. Temporal geographic variation in the lambing season of bighorn sheep. *Canadian Journal of Zoology* 60:1781–1793.

Thrash, I. 1998. Impact of water provision on herbaceous vegetation in Kruger National Park, South Africa. *Journal of Arid Environments* 38:437–450.

Tolsma, D. J., W. H. O. Ernst, and R. A. Verwey. 1987. Nutrients in soil and vegetation around two artificial water points in eastern Botswana. *Journal of Applied Ecology* 24:991–100.

Toweill, D., and V. Geist. 1999. Return of royalty: wild sheep of North America. Boone and Crockett Club and Foundation for North American Wild Sheep, Missoula, Montana, USA.

Turner, J. C. Jr. 1973. Water, energy, and electrolyte balance in the desert bighorn sheep, Ovis canadensis. Ph.D. Dissertation, University of California, Riverside, California, USA.

Turner, J. C. Jr. 1979. Adaptive strategies of selective fatty acid deposition in the bone marrow of desert bighorn sheep. *Comparative Biochemistry and Physiology Part A: Physiology* 62:599–604.

Turner, J. C. Jr., C. L. Douglas, C. R. Hallum, P. R. Krausman, and R. R. Ramey. 2004. Determination of critical habitat for the endangered Nelson's bighorn sheep in southern California. *Wildlife Society Bulletin* 32:427–448.

Turner, J. C. and C. G. Hansen. 1980. Reproduction. Pages 145–151 in G. Monson and L. Sumner, editors. *The Desert Bighorn, Its Life History, Ecology, and Management.* University of Arizona Press, Tucson, Arizona, USA.

Valdez, R., and P. R. Krausman. 1999. *Mountain Sheep of North America.* University of Arizona Press, Tucson, Arizona, USA.

Van de Kerk, M., S. Arthur, M. Bertram, B. Borg, J. Herriges, J. Lawler, B. Mangipane, C. L. Lambert, B. Wendling, and L. Prugh. 2020. Environmental influences on Dall's sheep survival. *Journal of Wildlife Management* 84:1127–1138.

Van de Kerk, M., D. Verbyla, A. W. Nolin, K. J. Sivy, and L. R. Prugh. 2018. Range-wide variation in the effect of spring snow phenology on Dall sheep population dynamics. *Environmental Research Letters* 13:075008.

Wagler, B. L., R. A. Smiley, A. B. Courtemanch, D. Lutz, D. McWhirter, D. Brimeyer, P. Hnilicka, T. J. Robinson, and K. L. Monteith. 2023. Implications of forage quality for population recovery of bighorn sheep following a pneumonia epizootic. *Journal of Wildlife Management* 87:e22452.

Walsberg, G. E. 1983. Coat color and solar heat gain in animals. *BioScience* 33:88–91.

Warrick, G. D., and P. R. Krausman. 1989. Barrel cactus consumption by desert bighorn sheep. *Southwestern Naturalist* 34:483–486.

Watts, T. J. 1979. Status of the Big Hatchet desert sheep population, New Mexico. *Desert Bighorn Council Transactions* 23:92–94.

Weaver, R. A. 1959. Effects of wild burros on desert water supplies. *Desert Bighorn Council Transactions* 3:1–3.

Weaver, R. A. 1973. California's bighorn management plan. *Desert Bighorn Council Transactions* 17:22–42.

Wehausen, J. D. 2005. Nutrient predictability, birthing seasons, and lamb recruitment for desert bighorn sheep. Pages 37–50 in *Symposium Proceedings for the Sweeney Granite Mountains Desert Research Center 1978–2003: A Quarter Century of Research and Teaching.* University of California Natural Reserve Program.

Wehausen, J. D. 1996. Effects of mountain lion predation on bighorn sheep in the Sierra Nevada and Granite Mountains of California. *Wildlife Society Bulletin* 24:471–479.

Wehausen, J. D., V. C. Bleich, B. Blong, and T. L. Russi. 1987. Recruitment dynamics in a southern California mountain sheep population. *Journal of Wildlife Management* 51:86–98.

Welles, R. E., and F. B. Welles. 1961. The bighorn of Death Valley. Fauna Series No. 6. US National Park Service, Washington, D.C., USA.

West, G. C., and D. L. Shaw. 1975. Fatty acid composition of Dall sheep bone marrow. *Comparative Biochemistry and Physiology, B* 50:599–601

White, P., T. Lemke, D. Tyers, and J. Fuller. 2008. Initial effects of reintroduced wolves Canis lupus on bighorn sheep Ovis canadensis dynamics in Yellowstone National Park. *Wildlife Biology* 14:138–146.

Whiting, J. C., D. D. Olson, J. M. Shannon, R. T. Bowyer, R. W. Klaver, and J. T Flinders. 2012b. Timing and synchrony of births in bighorn sheep: implications for reintroduction and conservation. *Wildlife Research* 39:565–572.

Whiting, J. C., R. T. Bowyer, and J. T. Flinders. 2009. Diel use of water by reintroduced bighorn sheep. *Western North American Naturalist* 69:407–412.

Whiting, J. C., V. C. Bleich, R. T. Bowyer, and R. T. Larsen. 2012a. Water availability and bighorn sheep: life-history characteristics and persistence of populations. *Advances in Environmental Research* 21:131–163.

Wilson, L. O. 1971. The effect of free water on desert bighorn home range. *Desert Bighorn Council Transactions* 15:82–89.

Witham, J. H. 1983. Desert bighorn sheep in southwestern Arizona (reproduction, survival, telemetry). Dissertation, Colorado State University, Fort Collins, Colorado, USA.

Woodall, P. F., and J. D. Skinner. 1993. Dimensions of the intestine, diet, and faecal water loss in some African antelope. *Journal of Zoology* 229:457–471.

# Section 2

## Population Management

# 13 Mountain Sheep Survey Methods

*Vernon C. Bleich, Daryl W. Lutz, Marcus E. Blum,*
*Danielle M. Glass, Steven G. Torres, and Jericho C. Whiting*

## 13.1 INTRODUCTION

Among the most basic questions asked by managers or researchers is, how many are there? Although fundamental to many of the decisions made regarding conservation or management actions (Rabe et al. 2002, Panaccio et al. 2024), the answer(s) to that question can be elusive. Nevertheless, "No form of wildlife management … is possible without reliable information on the numbers, population dynamics, and movements of the animals concerned" (Norton-Griffiths 1978:1). Information on total numbers, population structure, and distribution is important to conservation and management of wildlife, but obtaining reliable results can be especially challenging when working with mountain ungulates (Harris 1994). Methods that have been used for decades continue to provide meaningful data, but new information on technological advances now appears in professional literature frequently. For example, at least 50 professional journals currently specialize in conservation, ecology, or similar disciplines (Landhuis 2016). Further, new statistical methods continually have enhanced the reliability of estimating population size or structure, methodological advances have resulted in many novel methods of obtaining or manipulating information, and the evolution of technology continues at a seemingly increasing rate. This productivity and the number of journals currently publishing information of this type make it virtually impossible to keep abreast of advances in methods, applications, or results that currently appear on a regular basis (Berger et al. 2024).

Population estimates and demographic assessments are dependent upon information obtained from a variety of methods, the majority of which we discuss in this chapter. Such estimates can potentially be used to track fluctuations in abundance across years and could be helpful when monitoring populations before, during, and after habitat alterations, unanticipated meteorological events, or disease outbreaks (Taylor et al. 2021). Nevertheless, we here distinguish methods employed to obtain information from the statistical or mathematical manipulations that are dependent on that information. In short, there is a distinct separation between the methods used to obtain raw data (e.g., evidence of presence, or observations) and the inferences derived from that information. In this chapter we are concerned with the methods used to obtain that information, which then is manipulated mathematically or statistically to produce estimates of numbers or densities and then used in conservation or management decisions (Bart et al. 2004).

Investigators are reminded that in the absence of clearly defined objectives, the selected methodology, even if considered to be the best in the world, will yield questionable results. Thus, it is important that investigators consider the purpose, management goals, the measure of precision, and population boundaries or survey closure prior to implementing any survey methodology, whether the data will be used to derive population estimates, minimum numbers, demographic characteristics, or other information (Table 13.1). Methods available include distance sampling, census counts, mark-resight surveys, and index-level approaches, all of which can be implemented by personnel using ground or aerial methods, or by gathering information indirectly, among which are the use of pellet counts, genetic identification (i.e., DNA methods) of individuals, remote cameras, or other rapidly evolving technologies.

In planning population assessments and the most efficacious method of obtaining the desired information, careful research and consideration of how the data will be applied are essential. Indeed, some methods involve considerable risk to study animals or survey personnel or raise concerns about animal welfare; some are costly when compared to others that would also provide comparable information and would be adequate for the intended purpose; and, in some situations, careful thought may obviate the need for the desired information. Thus, clearly defined objectives including risks, costs, necessity, and use of the planned work warrant careful scrutiny, and all surveys must be well designed and have clear objectives.

Knowledge of the natural history of mountain sheep (bighorn [*Ovis canadensis*], thinhorn [*Ovis dalli*]) is essential, and a familiarity with past assessments and monitoring efforts are foundational to any survey effort. For example, sexual segregation results in temporal and spatial differences in seasonal groupings of males and females, and thereby has implications for population surveys, habitat management, research outcomes, and, ultimately, for conservation (Rubin and Bleich 2005, Gissi et al. 2024); if ignored, sexual segregation and aggregation have the potential to yield biased information obtained when any method of sampling is employed. Thus, this natural history phenomenon can have a potentially profound influence on survey results if those surveys are not conducted when the sexes are most apt to be aggregated (Bleich et al. 1997, Rubin and Bleich 2005, Bowyer 2022). Given that necessary assumptions are met by investigators, these considerations will optimize the efficiency of any survey and monitoring efforts.

DOI: 10.1201/9781003518686-15

**TABLE 13.1**

**Methods Commonly Used to Conduct Surveys of Mountain Sheep Include a Variety of Ground Surveys, the Use of Remote Cameras, Aerial Surveys, and Genetic Methods**

| Survey Method Being Considered | | | |
|---|---|---|---|
| **Ground or Hiking Surveys** | **Camera Surveys** | **Aerial Surveys** | **DNA Methods** |
| Population exists in a limited area that can be covered by available personnel in a reasonable amount of time | Population occupies habitat that is hot, arid, and includes a limited number of water sources | Population occurs over such an extensive area that it is impractical to survey by hiking | Population is readily accessible and occurs in areas wherein feces remain suitable to yield viable DNA |
| | OR | + | + |
| | Habitat contains other locations or chokepoints where sheep congregate, such as mineral licks | Population does not contain water sources, other chokepoints, or is too large to make camera surveys efficacious | Population's habitat contains known areas of high sheep use where fecal samples may be found |
| | OR | + | + |
| | Population is large, or dense enough to capture bighorn images using random grid sampling methodologies | Cost of the aircraft is not an obstacle | There is no necessity to know the population's age structure |
| | | OR | |
| | | Personnel hours available to survey are limited | |
| | | + | |
| | | Aerial surveys are the most efficient expenditure of time per unit return on data | |

Terrain, season, accessibility, cost, and survey objectives all influence the method to be used. It is important that investigators consider the purpose, management goals, the measure of precision, and population boundaries or survey closure prior to implementing any of these methods, and the context in which the results will be used.

Protocols that generate demographic estimates with a low coefficient of variation are the gold standard, but these often are not achievable, and sometimes not needed. Realistically, however, estimates with such precision are often not within the budget or long-term sustainability of some resource management agencies. Other agencies, however, have developed standardized protocols or guidelines to assist with survey design and implementation; examples include the Province of British Columbia, and the States of Nebraska and North Dakota (Resources Inventory Committee 1998, Nordeen et al. 2019, Wiedmann et al. 2024). Guidelines and protocols such as these ensure consistency in methodologies and enhance the likelihood that survey results and the metrics derived or estimated therefrom are directly comparable over extended periods of time.

## 13.2  A NOVEL INTERPRETATION OF ANCIENT ROCK ART, AND ITS HISTORICAL IMPLICATIONS

Perhaps the earliest records of North American mountain sheep in any given area are those shared through oral histories and prehistoric artwork created by Native Americans. Often in the form of petroglyphs (images created by removing part of a rock surface by incising, picking, carving, or abrading; Figure 13.1) or pictographs (images painted onto rock surfaces; Figure 13.2), numerous hypotheses to explain the meanings of these ancient works have been proffered. Within the geographic distribution of wild sheep, those ancient drawings likely indicated a close connection among hunters and the animals (Grant 1968, 1980), between the making of sheep drawings and the hunting of the animal, and it has been suggested that such images accurately portrayed aspects of the behavior and ecology of wild sheep (Matheny et al. 1997). These sites often are situated along, "… well-used game trails in direct association with watering holes (natural tanks) in the steep walled canyon bottoms. Some glyphs are at gorge entrances next to hunting blinds. The largest groupings of rock drawings are [associated with] natural sheep 'traps' (cul-de-sacs and hunting enclosures) where game could be driven past hidden hunters [and] are also found on stony promontories astride saddles between drainages. Smaller concentrations are located near springs" (Garfinkel 2006:224). Springs and natural rock tanks, often referred to as tinajas, that may hold up to 100,000 L on a seasonal basis (Bleich et al. 2020), represent sources of surface water for bighorn sheep in arid ecosystems, and in some areas tinajas provide the only surface water available during summer (Bleich et al. 2020), a resource upon which desert bighorn sheep are dependent (Turner 1973).

**FIGURE 13.1** The intended meanings of petroglyphs created by Native Americans are uncertain, and potential interpretations are abundant and remain open to debate. Is it possible that at least one purpose was to convey information about the relative abundance of mountain sheep and, as in this image, mule deer in a specific geographic area? If so, such images would be among the earliest attempts to convey index-level population information on a species inextricably linked to the survival and culture of those ancient peoples. (Coso Range, Inyo Co., California, October 2019. Photograph © Pat Woods and used with permission.)

**FIGURE 13.2** Pictographs differ from petroglyphs in that they are not scraped, carved, chiseled, or otherwise incised into rock, but rather are painted on the surface. As with petroglyphs, the meanings of pictographs remain open to interpretation. (Ayres Rock, Inyo County, California, March 2015. Photograph © Pat Woods; used with permission.)

Despite a diversity of published opinions posited to explain concentrations of petroglyphs and pictographs (Matheny et al. 1997, Garfinkel 2006, Garfinkel and Rogers 2016), none of which we are aware considering the possibility that they represent some index to the relative abundance of the species depicted, or otherwise indicated the locations at which a particular species was most apt to be encountered, or where hunters were most apt to be successful. Thus, we wonder if the 93 rock panels described by Garfinkel and Austin (2011) depicting multiple bighorn sheep when compared to those 29 depicting singular animals, or if the 95 panels illustrating ≤5 bighorn sheep when compared to the 27 containing from 6 to 50 individuals, were indicative of the relative abundance of bighorn sheep in different parts of the Coso Range, California. Further, there is evidence linking the rate of rock-art production with the rise and fall of human predation on bighorn sheep (Gilreath and Hildebrandt 2008, Garfinkel et al. 2010). Is it possible that the frequency of bighorn sheep compared to mule deer (*Odocoileus hemionus*), both of which occupy the Coso Range, in these panels provided information on the relative abundance of those species? Is it possible that the frequency of bighorn sheep when compared to elk (*Cervus canadensis*) among the glyphs in Nine Mile Canyon, Utah, provided information on the relative abundance of those two species at the time?

We are neither anthropologists nor archaeologists, but as wildlife biologists we speculate that the concentrations of petroglyphs or pictographs associated with various locations represent some measure of the relative abundances of animals on a geographic basis. We suggest these Native Peoples provided information on the commonness or rarity of the animals in question. Mule deer and bighorn sheep are sympatric in the Coso Range (Longhurst et al. 1952, Eerkens et al. 2007, Abella et al. 2011), but the abundance of sheep glyphs when compared to those of mule deer may be more than a curiosity. Similarly, the abundance of sheep glyphs when compared to those of elk in Nine Mile Canyon, where both species occur in sympatry (Matheny et al. 1997), or near Moab, Grand County, Utah, where elk and bighorn sheep (Figure 13.3) occupy areas proximate to each other, also may be more than curiosities. In short, they

may have represented efforts to describe the probability of encountering bighorn sheep, mule deer, or elk at or near specific locations, or were an indication of the relative abundance of a particular species in a specific geographic area. Informal or historical records must be viewed with caution and subjected to intense scrutiny before being accepted as factual or even meaningful (Diaz and Corti 2024) but can also enhance our comprehension of interactions among humans, animals, and the past environment (Groves et al. 2022). In that context, these pictographs and petroglyphs may represent early attempts to apply index-level information to the status of populations of mountain sheep, mule deer, or elk. As noted by Gamez-Brunswick and Rojas-Soto (2020:2550), rock art can "In some ways … be … a valuable contribution to our knowledge of prehistoric animal populations…".

## 13.3 CONTEMPORARY SURVEY METHODS

Methods used over millennia, and the results of which have been conveyed through traditional and oral histories and might have been used to index populations of large mammals, have been used for decades by contemporary investigators to provide information applicable to the conservation or management of mountain sheep. Among the most thorough summaries of the methods used to gather, analyze, and interpret information on the sizes of wildlife populations were those of Caughley (1977) and Lancia et al. (1994), but techniques and statistical methods have changed substantially over the past 50 years (Pierce et al. 2020). New technologies, statistical methods, modeling applications, and computer power have had an influence on the methods used to estimate population size, distribution, or

**FIGURE 13.3** Petroglyph portraying an abundance of what appear to be bighorn sheep and a single elk, suggesting that bighorn sheep were much more common or, at the least, were encountered much more frequently than were elk by the Native Americans that historically occupied this region. (Moab, Grand County, Utah, October 2024. Photograph © Pat Woods; used with permission.)

structure (Buckland et al. 1993, White 2005, Wang et al. 2024), and enhancements of statistical and technological methods continue at an increasing rate (Harris et al. 2024).

Readers of this chapter should recognize that survey methods are not synonymous with the statistical derivation of metrics or parameters based on survey results. The former involves methods associated with data acquisition, and the latter with the analyses of those data. In the previous section, we described what may have been the earliest method of assigning index-level information to the abundance of North American mountain sheep. Much of the contemporary literature deals with newly developed technologies that yield data from remote locations, methods of modeling that automatically generate results from incoming data that are subject to interpretation by individuals that may or may not be familiar with the species they purportedly are investigating (Krausman 2020, 2023), or with techniques that presumably will provide even further insight into the complexities of the natural world than currently is the situation. Below, we review more recent methods used to survey populations, or to otherwise obtain demographic information, and other potentially valuable, albeit ancillary, data. Each method includes a discussion of its historical use, its use or limitations as a survey technique, and where or when it is best used. We close the chapter with an overview and a brief discussion of emerging methods that are under development and may be applicable to mountain sheep surveys in the future.

Mountain sheep have been important subjects in the development of survey methods, associated technological advances, and resultant conclusions dating back to the work of Seton (1927), who extrapolated his observations of bighorn sheep in Glacier National Park, Montana, to the continent of North America. He wrote, "In 1916, I spent some weeks camping in the Park, and after using all sources of information, calculated that the 2,500 square miles of the Park contained not less than 1,500 Sheep (*sic*). In the official Report (*sic*) this estimate of mine was raised to 2,000, or about 1 Sheep (*sic*) to the square mile of this possible territory. The Park is far from ideal range. It contains very little true Sheep (*sic*) pasture of the kind that they sought a hundred years ago. One can travel about a whole month, as I proved in 1916, and never see a Sheep (*sic*). In the Big Horn Basin, judging by Lewis and Clark's accounts, and those of later travellers (*sic*), Sheep (*sic*) may have been ten times as numerous as the above Park estimate. Offsetting this, however, is the sparse population of the arid desert regions. This, of course, does not include the Slim-horned Sheep of the north, whose numbers must be at least as great" (Seton 1927:534–535).

From these numbers and densities, Seton (1927) used simple extrapolation to conclude the number of bighorn sheep that occupied North America about 100 years prior, as noted by his reference to Lewis and Clark, was between 1.5 and 2.0 million individuals, and as many thinhorn sheep, "… whose numbers may be at least as great" (Seton 1927:535). Although these extrapolations have been

questioned (Demarchi 1977, Wishart 1978, Valdez 1988, Hurley 2021) they remain entrenched in the contemporary literature and continue to be cited in the context of historical numbers. It also paved the way for the realization that additional methods, both old and new and described in this chapter, were essential to understanding the numbers of mountain sheep on a local scale. There remain many benefits to developing standardized methodologies applicable to mountain sheep, and other ungulates, to better understand and compare results among geographic areas (Mason et al. 2006). Efforts to do so have occurred within jurisdictions (Bleich 1979, Hervert et al. 1998, Wehausen and Bleich 2007), but there appears to be no such movement toward continental, species-specific, or habitat standards for conducting surveys of mountain sheep.

### 13.3.1 INDIRECT EVIDENCE

A curiosity about natural history and its implications for conservation was emphasized early on by Leopold (1933), but the value and importance of spending time in the field appear to have been devalued in recent years (Bleich and Oehler 2000, Bleich 2018, Berger et al. 2024). Nevertheless, a historical account of efforts to verify the presence of or to quantify the numbers of bighorn sheep in a given area began at about the time the last petroglyphs or pictographs of that species were being created. The earliest, albeit indirect, descriptions of mountain sheep and their distributions were provided by missionaries, naturalists, or explorers that previously had not encountered either species. Among those were Francisco Maria Piccolo, a Spanish padre who in 1702 made reference to and then proffered a description of desert bighorn sheep (*O. c. spp.*). From 1800 to 1843 Duncan McGillvray collected several Rocky Mountain bighorn sheep (*O. c. canadensis*) from near the Bow River in Alberta, Canada; and members of the Lewis and Clark expedition, John James Audubon, and Prince Maximillian Wied-Nuwied verified the presence of bighorn sheep in western North America (Valdez 1982, Valdez and Krausman 1999). Near this same time in 1824 Samuel Black recorded his observations of thinhorn sheep in the Finlay River region of British Columbia (Rich and Johnson 1955, cited by Demarchi and Hartwig 2004), which created awareness of these new species of sheep to Europeans (Chapter 2). Several decades later, additional descriptions of thinhorn sheep (*O. dalli spp.*) from the upper Yukon River in Alaska, and Stone's sheep (*O. d. stonei*) from near the Stikine River in northern British Columbia, Canada, became formally known to science (Valdez 1982, Valdez and Krausman 1999).

Following these first- or second-hand references to the existence of North American mountain sheep, a substantial reliance on evidence of the presence of those specialized ungulates and other wildlife developed, whether or not the animals themselves were observed (Coonen 1951, Jones 1980, Liebenberg 1990), and those methods have evolved into the fields of investigation termed presence-absence, and occupancy, detection (MacKenzie 2005a, Kery and Schmidt 2008, Steerweg et al. 2017). MacKenzie (2005a:849) reminded readers, however, that presence-absence investigations might better be referred to as "presence-not detected" methods because a failure to detect does not equate to species absence. The result has been the evolution of an entire field of endeavor concerned with the development of statistical techniques and correction factors that can be applied to issues associated with imperfect rates of detection (MacKenzie 2005b, MacKenzie et al. 2009, 2018; Panaccio et al. 2024).

Whether or not an area is occupied on a seasonal or year-round basis is a separate question, but it also suffers from the same possibility of incomplete detection. If identified by a knowledgeable investigator familiar with the available evidence, the presence of tracks, beds, fecal material, trails, or even hair indicates that a mountain sheep was present in the area of interest at one time or another. In areas of potential sympatry between mountain sheep and mule deer, however, investigators must be aware that only recently have economical and reliable laboratory methods been developed to accurately distinguish between species. It was not until Wehausen (2024) demonstrated an efficacious method based on genetic characteristics that such came into existence. Readers are cautioned, however, that females are more philopatric than males (Bleich et al. 1990, Rubin and Bleich 2005), and given the vagility of male mountain sheep and their tendency to range widely, investigators should not infer permanent occupancy if only feces (pellets) deposited by male mountain sheep are detected; thus, it is essential that species and sex be ascertained before drawing conclusions (Wehausen 2023).

Wildlife biologists also should not depend on indirect methods such as presence-absence data, occupancy information, pellet counts, or even more advanced methods such as DNA techniques in cases when age structure of the population (Belant et al. 2024) or phenotype (i.e., horn curl; Balsara 2024) are important considerations, as when setting harvest regulations. Most of the indirect methods described here are a relatively inexpensive means of confirming use of an area by mountain sheep, documenting colonization events, and distinguishing use of an area by those species even in the absence of visual confirmation. Nevertheless, when properly designed and implemented, sampling schemes involving genetic analyses can be a useful method of estimating the abundance of mountain sheep or other wildlife in predefined geographic areas (Poole et al. 2011, Pfeiler et al. 2020, Epps et al. 2024), as described elsewhere in this chapter.

### 13.3.2  Sign Reading

If not actually observed in an area, the current or past presence of mountain sheep sometimes is discernible by evidence left behind by those ungulates. Such evidence may be in the form of pellets, tracks, trailing, beds, urine stains or scent posts, evidence of foraging, or physical remains

of dead animals (Jones 1980). Further, "None of the differences between modern science and the art of tracking require a fundamentally new way of thinking" (Liebenberg 1990:165).

Recently, Tomecek and Evans (2020) emphasized that the presence of tracks and signs was the first information that prehistoric humans learned and made use of to understand the wildlife around them. It also is noteworthy that for the first time in its long history of publication, the current edition of *The Wildlife Techniques Manual* (Silvy 2020) has included a chapter on sign reading and interpretation. In part this is explained by a tendency to overlook the presence of wildlife because sophisticated methods and advanced technology have replaced the very field skills that prior managers and researchers had been taught in natural history courses, or that many others had learned while growing up. Because, " … fewer students of wildlife programs have information from a childhood spent outdoors, it [became] increasingly important to include a chapter on these [sign reading] skills" (Tomecek and Evans 2020:220).

Numerous authors have facilitated interest in and the development of sign reading skills through publications (Brunner 1909, Murie 1954, Liebenberg 1990, Elbroch 2003). The use of information using those skills has not yet been lost, and sign reading has played important and meaningful roles in rare species management, predator management, game species management, nuisance wildlife management, urban wildlife management (Tomecek and Evans 2024), and seasonal habitat identification and management. Although not foolproof (Grigione et al. 1999), and sometimes combined with more modern techniques such as the use of remote cameras (Long et al. 2003, Goward 2004), it is important that traditional skills and associated methods be retained (Tomecek and Evans 2024). This is especially desirable among those working in the remote and rugged areas characteristic of the habitats most often associated with mountain sheep.

### 13.3.3  Pellet Counts

Pellet counts are labor-intensive and inefficient in terms of the return on time invested, but they have been used for decades by wildlife managers for a variety of purposes (Flinders and Crawford 1977, MacCracken and Van Ballenberghe 1987). Data obtained from pellet counts can yield meaningful information on trends, numbers, responses to habitat treatments, and variations in density associated with distribution of ungulate populations (Neff 1968, Simmons and Hansen 1980), and when pellet counts are properly designed the data are amendable to statistical analysis (Neff 1968). Because the method is so dependent on adequate staffing, funding, accessibility of the survey areas, and because there have been many improvements in technologies applicable to assessing ungulate populations, as will be described elsewhere in this chapter, pellet group counts are seldomly used today to survey mountain sheep.

In situations where investigators are interested in the use of a specific resource, for example, natural sources of surface water or wildlife water developments, and manpower is sufficient to allow scheduled visits to those sites, investigators can obtain valuable information on the relative use of those resources by mountain sheep, information on seasonal use, and the sex of animals using the resource if pellets are collected and subjected to appropriate analyses. In areas where mountain goats (*Oreamnos americanus*), or mule deer and mountain sheep are sympatric, confirmation that the pellets were deposited by mountain sheep, however, is essential (Wehausen 2023). Pellet counts have been used to obtain information on responses of bighorn sheep to environmental perturbations, using a before-after-control-impact design (Eberhardt 1976, Conner et al. 2015). For example, pellet group counts on permanent plots were conducted biannually to obtain information on trends and seasonal use by mountain sheep occupying areas proximate to recent anthropogenic water developments, although plots had not been established to serve as randomly selected controls (Bleich 1979).

### 13.3.4 DNA Applications

Identification of individuals comprising a population has been possible for nearly 30 years (Taberlet et al. 1996, 1997, 1999) and has application to the conservation and management of mountain sheep. With the relative ease of collecting suitable samples of scat without the requirement of exposing an animal to stressful capture conditions, the use of fecal DNA in wildlife research has grown exponentially over the recent past (Waits and Paetkau 2005). Biologists use DNA to estimate population size (Lounsberry et al. 2015, Schoenecker et al. 2015, Pfeiler et al. 2020, Stewart et al. 2023, Epps et al. 2024), understand metapopulation structure (Epps et al. 2005, 2010, 2018, Chapter 6), detect uncommon or cryptic species (Hatfield et al. 2021, Peralta et al. 2022, Wehausen 2023), identify individual animals (Ernest et al. 2002), confirm historical distributions (Wilder et al. 2014), and understand animal diets (Ando et al. 2020). Estimates of population size using fecal DNA analyses are often more precise (i.e., has a lower coefficient of variation) and less expensive than traditional ground or aerial surveys (Pfeiler et al. 2020, Stewart et al. 2023) when populations are distributed over small ranges when compared to large study areas.

Estimating population size using fecal DNA entails four basic steps: undertaking multiple rounds of fecal sample collections from the field, extracting the DNA from the samples, identifying individual animals using the resulting genetic information, and producing a population estimate using capture-recapture modeling. Samples can be obtained immediately after defecation (Schoenecker et al. 2015) but can be collected days to months (Epps et al. 2024, Pfeiler et al. 2020, Wehausen 2023) after deposition. Samples collected immediately or in frozen condition typically yield a higher success rate of DNA extraction, but the

rate can remain high on older samples if environmental conditions allow (Bach et al. 2022) and when appropriate post-collection storage procedures are used.

Fecal samples can be collected using several sampling methods, among which are randomly distributed searches, opportunistic collections, or targeted approaches. For example, samples can be collected along randomly placed transects (Stewart et al. 2023) from which information on population density or total numbers in a given area can be derived, or near water sources (Pfeiler et al. 2020), mineral licks (Schoenecker et al. 2015), or other point locations where mountain sheep congregate (Epps et al. 2024) and the results used by investigators to develop information on the sex and number of animals at localized sites. They also can be collected opportunistically throughout the suspected range of a population or in specific geographic areas to investigate the presence or absence of mountain sheep (Epps et al. 2024, Wehausen 2024). The use of the samples for estimating populations, however, is affected by the demographic objective, research design, and sample collection strategy (Table 13.1). Readers are reminded that multiple sampling periods within a short time period are required for traditional capture-recapture modeling; nonetheless, spatial capture-recapture modeling allows for the use of more opportunistic sampling (Epps et al. 2024).

Following acquisition, pellets are dried and stored in paper bags (Pfeiler et al. 2020, Epps et al. 2024), stored in ethanol (Schoenecker et al. 2015, Stewart et al. 2023), or swabbed in the field and the swab placed in a buffered solution (Bach et al. 2022). For the paper-bag and ethanol methodologies, DNA is extracted using a razor blade from the outermost layer of pellets where the digested material came in contact with the epithelial cells lining the individual's intestine (Wehausen et al. 2004). Swabbing the pellets while in the field was reported to be the most efficient method of obtaining DNA from ungulate fecal material (Bach et al. 2022). When using either the paper-bag or ethanol methodology, a shorter time between sample collection and DNA extraction also yields higher extraction rates (Brinkman et al. 2010). Precipitation and other moisture in the environment decrease the success of DNA extraction (Brinkman et al. 2010, Poole et al. 2011, Stewart et al. 2023), explaining why multiple investigators have used fecal DNA to estimate populations of bighorn sheep (Schoenecker et al. 2015, Pfeiler et al. 2020, Wehausen 2023, Epps et al. 2024). Although some populations of thinhorn sheep occupy arid and dry, albeit cold, areas, and pellets may remain frozen for extended periods following deposition, investigators have not yet applied capture-recapture methods to Dall's sheep based on the results of fecal DNA collections. In part, this likely is a function of the remote locations occupied by thinhorn sheep, costs involved in accessing those locations, and perhaps time constraints on personnel.

Bighorn sheep are individually genotyped using polymerase chain reaction (PCR) on microsatellite loci and a marker for sex identification (Schoenecker et al. 2015, Pfeiler et al. 2020, Wehausen 2023, Epps et al. 2024).

The number of microsatellite loci required to identify individuals depends on the allelic variability at each locus and the general order of magnitude of the population size (Epps et al. 2018, Pfeiler et al. 2020). Single nucleotide polymorphisms have not currently been used for fecal population estimates of mountain sheep, but may be used in the future as methods and technologies evolve. Capture-recapture modeling can incorporate genotyping error, allowing researchers to produce population estimates based on their level of confidence in their genetic samples (McKelvey and Schwartz 2004, Lukacs and Burnham 2005). Population estimation efforts should not use fecal DNA when wildlife managers also wish to know the animal age (Table 13.1). In many instances, however, methods using fecal DNA can provide a precise and relatively inexpensive means of confirming use of mountain sheep in an area, documenting colonization events, determining sex of the individuals detected, distinguishing between different taxa in the absence of visual confirmation, and estimating population size when sampling effort is adequate.

Environmental DNA (eDNA) has become an increasingly popular non-invasive method to detect the presence of various species (Bohmann et al. 2014, Harper et al. 2019). Researchers collect environmental samples such as soil and water to determine the presence of a species in a particular area by analyzing those samples for DNA from the species they hope to detect. While this method is appropriate for confirming the presence of mountain sheep, it is unlikely to be a useful technique for population surveys or deriving population estimates. As an example, desert bighorn sheep commonly congregate at water sources, which would provide an ideal source of eDNA. The concentration of genetic material from various individuals congregating at the water source, however, will hamper the investigator's ability to discern unique individuals from the sample, or the number of replicated visits by individuals, both of which are requisite to develop population estimates. While this technique is not useful for deriving population estimates, it likely has other applications, such as detection of use by a previously identified individual animal (or animals) or the presence of a specific pathogen within a mountain range or at a particular location.

### 13.3.5 Direct Evidence

Despite the reliance on the sophisticated genetic techniques described above, and aerial techniques or remote cameras, which will be discussed later in this chapter, ground-based survey methods have a number of advantages, and can be conducted by individuals on the ground or, under the right conditions, from wheeled vehicles (Norton-Griffiths 1978, Simmons and Hansen 1980), the latter of which seldom are encountered in areas occupied by mountain sheep. For example, investigators can obtain information, albeit subjective, on the degree of use and amount of time mountain sheep spend in a particular area, age and sex ratios, habitat utilization, and patterns of distribution or movement

(Simmons and Hansen 1980). Some of the methods that have proven valuable in conducting ground-based surveys will be considered archaic by many recent graduates, but those long-term studies (Chapter 23) have yielded substantial amounts of population-specific information on the ecology of mountain sheep and otherwise have contributed in meaningful ways to the conservation of those specialized ungulates. Despite this, few monitoring programs are based on a truly random spatial sample (Rabe et al. 2002), and convenience sampling commonly is employed (Anderson 2001). There are situations in which convenience sampling can yield useful information, but the lack of randomization in the survey design affects the bias and the power to detect differences (Bornstein et al. 2013) and precludes the ability to calculate sampling error from the results.

Readers are referred to Found and Patterson (2020) for an extremely useful review of methods used to conduct ungulate surveys, including ground-based and aerial techniques. In this section, we discuss a number of these methods and provide examples of how data obtained from those efforts have contributed to the conservation and management of mountain sheep. Because of the remote areas and difficult environmental conditions that characterize habitats occupied by thinhorn sheep, the majority of these techniques have been applied to bighorn sheep. Among these ground-based but labor-intensive methods are the aforementioned methods based on pellet group surveys (Simmons and Hansen 1980), hiking or transect surveys whereby investigators travel either on standardized routes or on routes that have been established randomly and record observations of mountain sheep and other biological information (Holl and Bleich 1983), and counts at waterholes, mineral licks, or other locations at which bighorn or thinhorn sheep congregate on a seasonal basis that have yielded data (Wehausen et al. 1987, Rubin et al. 1998). Some ground-based methods, which sometimes include the use of wheeled vehicles, used to survey populations of Rocky Mountain bighorn sheep, Sierra Nevada bighorn sheep (*O. c. sierrae*), or desert bighorn sheep occupying chaparral-dominated habitat in southern California (Bleich and Holl 1982, Holl and Bleich 1983) are best carried out on the winter ranges upon which a large proportion of the animals comprising a population are present for extended periods; examples include the White Mountains, Sierra Nevada, and San Gabriel Mountains of California and the Whiskey Basin of Wyoming. These methods are discussed in the following paragraphs.

### 13.3.6 Hiking Surveys

Records of mountain sheep seen by observers on the ground have provided useful information over many decades. For example, observations of bighorn sheep recorded by the Lewis and Clark expedition in 1805 provided the first records of that species in western North Dakota and eastern Montana (Wiedmann et al. 2024). Albeit not intended to quantify information on the numbers of bighorn sheep, such ground-based efforts played an important role in defining the distribution, habitat, and habitat use of North America's mountain sheep. As emphasized by Simmons and Hansen (1980), surveys that put boots on the ground in any given area can yield substantial amounts of information on the distribution, density, and patterns of use in an area; habitat associations; seasonal presence or use of a particular area; and demographic information, such as group size(s), lamb:female (bighorn sheep) or lamb:female-like (thinhorn sheep) and male:female ratios.

The value of information obtained from these labor-intensive and time-consuming efforts is emphasized by investigators that worked primarily with desert bighorn sheep. For example, Jones et al. (1953) provided the first detailed information on the distribution of desert bighorn sheep in the Santa Rosa Mountains, Riverside County, California. Goodman (1963) later emphasized the value of hiking surveys in specific canyons of that mountain range and taking the time to search for sign or observe carefully for bighorn sheep, while simultaneously acknowledging the manpower requirements to do so effectively. Among well-known examples of this method is that of Geist (1971), who investigated the ecology of Stone's sheep in northern British Columbia for many years, and that reported by Weaver and his colleagues, who completed the first comprehensive effort to ascertain the status of bighorn sheep in California during 1968–1972 (Bleich and Weaver 2007). Other examples include the systematic surveys of investigators that began many years ago and continued for multiple decades in the Sierra Nevada (Wehausen 1980, U.S. Fish and Wildlife Service 2007) or several Mojave Desert mountain ranges or Great Basin mountain ranges (Wehausen 1983, 1986, 1992, 2005, 2024; Schroeder 2004). Similar ground efforts (Holl and Bleich 1983, 2009a,b; Holl et al. 2004) in the San Gabriel Mountains, California, yielded important information for that population of bighorn sheep. Additionally, ground-based surveys, whether hiking or motorized, are used in Wyoming, Nebraska, and other jurisdictions to provide demographic information on multiple winter ranges. Moreover, hiking surveys, albeit augmented with the advantage of telemetry data, are used exclusively to obtain demographic information and information on animal health in some jurisdictions, whereas the placement and continued monitoring of telemetry devices facilitate the observation of bighorn sheep social groups by ground personnel, and decrease disturbance associated with aerial surveys. The hiking method is used to detect issues with herd health, develop demographic profiles, perform censuses, and to assist the regulatory process with harvest recommendations. Prominent examples of jurisdictions employing this methodology include Nebraska and North Dakota (Nordeen et al. 2019, Wiedmann et al. 2024).

Results obtained from repeated hiking surveys can be combined with data obtained from aerial surveys and, as sample sizes accumulate, estimates of demographic parameters can be derived; this method has been termed multiple direct sampling (Wehausen 1990, Jaeger et al. 1991).

Independent simultaneous double-counts of the same geographic areas also can be used to derive similar estimates (Magnusson et al. 1978). A derivation of this method initially used to estimate the population of bighorn sheep occupying the San Gabriel Mountains, Los Angeles, and San Bernardino counties, California (DeForge 1980), was refined by Holl and Bleich (1983), and has been in use for decades in efforts to derive trends and population estimates of bighorn sheep occupying that range (Holl et al. 2004, Bleich et al. 2008, Holl and Bleich 2009a,b; Daly and Dierkes 2024). Additionally, modern methods incorporate information using multiple variables, build models from multiple sources of information, and create reasonable models from older approaches and with fewer parameters (White 2005). For example, Johnson et al. (2010) used sophisticated statistical methods to combine information obtained independently by investigators using piecemeal results of limited sample size obtained from multiple methods including hiking surveys, intermittent data collection, inconsistent methodologies, and only a subset of important parameters or covariates for conservation decisions affecting Sierra Nevada bighorn sheep, a state and federally endangered taxon (Wehausen et al. 2005, U.S. Fish and Wildlife Service 2007).

### 13.3.7 Waterhole Surveys

Bighorn sheep are dependent upon the availability of surface water during summer (Turner 1973), and counts conducted at watering sites during hot, arid conditions were among the earliest methods used by managers to obtain demographic information on that species. In most cases those early efforts were neither systematic nor standardized, and little was known about the visitation rates of individual animals using the waterholes, the influences of changing weather conditions, or statistical methods that could be useful in refining the results of those surveys (Simmons and Hansen 1980). Waterhole counts became more refined with time, and historically were conducted at natural sources of surface water upon which bighorn sheep were dependent. More recently, such counts have been employed to ascertain the use of anthropogenic sources of water developed specifically for the benefit of bighorn sheep, or to develop information on the relative level of use by bighorn sheep at natural and artificial water sources (Bleich et al. 1982).

There are a number of shortcomings associated with data obtained from this method. Among these are (1) manpower limitations (i.e., at least one person must be present at each location); (2) coverage of every location apt to be used by bighorn sheep in a predefined area must be certain; (3) the weather must be suitable (i.e., hot and dry) throughout the count; (4) there is no way of knowing how long to extend each count; (5) one cannot be certain of the rate at which bighorn sheep visit water; and (6) in the absence of marked, or otherwise identifiable animals, there is no certainty with respect to the rate at which individual animals are repeat visitors (Simmons and Hansen 1980). In addition to these issues, male and female bighorn sheep segregate by sex for much of the year, and males occur significantly further from water sources than do females on a year-round basis (Bleich et al. 1997), and this phenomenon must be a consideration in survey design (Bleich et al. 1997, Rubin and Bleich 2005, Bowyer 2022).

Despite the aforementioned shortcomings, waterhole counts have been conducted for many decades in some locations, and have continued to be important sources of information in the absence of, or as augmentations to, other methods. Two well-known long-term efforts have yielded important demographic information on bighorn sheep occupying the peninsular ranges of southern California. In one example, data collected during waterhole counts conducted over 30 years clarified the factors affecting the population dynamics of bighorn sheep occupying the Santa Rosa Mountains, California (Wehausen et al. 1987). In another example, information obtained from waterhole counts conducted over a period of 26 years shed new insight on the metapopulation structure and population dynamics of bighorn sheep occupying several mountain ranges in Anza-Borrego Desert State Park, California (Rubin et al. 1998). Standardized methods, competent personnel, complete coverage of all water sources in a given area, and cooperative weather all enhance the value and use of data obtained from waterhole surveys.

Remote photography as a reliable means of recording bighorn sheep visits to surface water sources during hot, dry periods has made the manpower requirements for such counts much less of an issue, provides reliable information, and produces results comparable to those obtained from other types of ground surveys or aerial surveys (Perry et al. 2010, Glass et al. 2022). Further, battery life has been extended substantially by incorporation of solar panels into remote camera battery systems (Wehausen 2005). Similar to the use of DNA techniques if accurate sex ratios are a consideration, as when setting harvest objectives, caution is advised because of the different rates at which male and female bighorn sheep visit specific water sources (Bleich et al. 1997), the effects of sexual segregation on use of water sources (Whiting et al. 2010, 2012), and the seasonal differences in water requirements of males and females.

### 13.3.8   Mineral Lick Surveys

As is the case with counts of bighorn sheep using waterholes, which provide a resource upon which desert bighorn sheep are dependent with few exceptions (Krausman et al. 1985), mountain sheep occurring at more northern latitudes appear to be dependent on mineral licks (Honess and Frost 1942, Packard 1946, Smith 1954, Geist 1971, Heimer 1973, 2022; Skipworth 1974). Counts at these point sources of minerals have been important sources of demographic information in northern environments inhabited by Rocky Mountain bighorn sheep, Stone's sheep, or Dall's sheep that use these sites (Heimer 1973, 2022; Dormaar and Walker 1996, Ayotte et al. 2008, Schoenecker et al. 2015).

Desert bighorn sheep appear not to be as dependent on mineral licks as mountain sheep at more northern latitudes (Watts and Schemnitz 1985, Holl and Bleich 1987), but in some areas use point sources that provide concentrations of trace minerals or macro-minerals (Hamilton et al. 1982, Watts and Schemnitz 1985). In many areas desert bighorn sheep may be able to obtain adequate amounts of minerals from vegetation consumed (Kennedy 1957, Mahon 1969, Watts and Schemnitz 1985), but the concentration of each mineral in the same forage species varies across geographic areas (Fox et al. 2000, Bleich et al. 2017) and the use of point sources of minerals likely varies as a function of mineral concentrations in forage plants. Despite this variation, a variety of methods have been used to obtain information on populations that are dependent on mineral licks or use specific sites with regularity, including on-site observations, remote photography, pellet counts, and fecal DNA techniques.

### 13.3.9   Winter Range Surveys

Winter range surveys can be labor-intensive, but have proven to be an effective method of obtaining long-term data useful in assessing the status of well-studied populations. Additionally, aerial surveys of bighorn sheep occupying well-defined winter ranges have proven useful in understanding long-term population trends; when combined with additional information obtained from ground observations, it has been extremely useful in collection of demographic data over extended periods (Becker et al. 1978, Wehausen 1983, Holl et al. 2004, Wyoming Game and Fish Department 2004, U.S. Fish and Wildlife Service 2007, Johnson et al. 2010).

### 13.3.10   Remote Photography

Automated cameras, referred to early on as camera-sets, have been used by wildlife researchers for nearly a century to investigate cryptic species, record animal behavior, investigate animal abundance, and for enforcing wildlife laws and regulations (Young and Goldman 1946, Kucera and Barrett 1993, Pierce et al. 1998, Hossain et al. 2016). The early camera-sets relied on mechanical linkages that activated the camera when the subject was within the frame of the photograph to be captured and triggered the shutter (Gregory 1938, Young and Goldman 1946). Cameras have been used to gather information on mountain sheep since the early 1960s and, not surprisingly, were placed strategically at locations used on a regular basis by those mountain ungulates (Knudsen 1963, Helvie 1972; Constantino 1973, 1974). Time-lapse cameras later were fitted with photoelectric cells that cut power to the camera at night to conserve batteries and film (Davis and Bleich 1980, Montalbano et al. 1985, Jaeger et al. 1991). Shortly thereafter, investigators added solar panels to recharge batteries, and this modification allowed cameras in remote locations to function until the film was exhausted. These devices were efficient and were used for many years to investigate desert bighorn sheep population ecology in many areas of the Mojave Desert for extended periods (Wehausen 2005).

Motion-sensor cameras that record images electronically have since become readily available and are being used extensively by investigators to monitor mountain sheep behavior, resource use, or demography across the many regions and diverse habitats occupied by these mountain ungulates (Whiting et al. 2010, Dertien et al. 2017, Glass et al. 2022, Enns et al. 2023). In colloquial terminology, time-lapse cameras and motion-sensor cameras are referred to as trail cameras, remote cameras, and use of the term camera traps has become common in the ecological literature (Parker et al. 2020).

During the 1970s, investigators used cameras to document behavior and movement of desert bighorn sheep in Arizona and Nevada (Leslie and Douglas 1980, Campbell and Remington 1981). In the 1980s and 1990s, investigators used cameras to investigate timing of sexual segregation in areas occupied by desert bighorn sheep in California (Bleich et al. 1997). Cameras have been used more recently to document use of water sources and investigate sexual segregation in Rocky Mountain bighorn sheep in Utah, and to assess habitat use by thinhorn sheep in Alaska and British Columbia (Whiting et al. 2009a,b, 2010; Dertien et al. 2017, Enns et al. 2023).

Across those decades and habitats, remote cameras have also been instrumental in increasing our understanding of population parameters, behaviors, and patterns of habitat use by mountain sheep. Advances in battery life and the use of high-capacity memory cards allow multiple sampling periods, increasing the precision of estimates derived using this technology (Dekelaita et al. 2020). Cameras set at water sources have provided information on the abundance of males, females, and young (Jaeger et al. 1991, Wehausen 2005, Harris et al. 2020a, Taylor et al. 2021); for marked or otherwise identifiable animals, individual female-lamb relationships also have been identified (Enns et al. 2023). Information can be generated from group compositions derived from individual images (Taylor et al. 2021), or from sequences of photos (Dekelaita et al. 2020), and data derived from the use of remote cameras can be used to calculate estimates using mark-resight techniques (Taylor et al. 2021), or distance sampling when marked animals are not present in the population of interest. These devices have also been used to document the use of artificial water sources (i.e., wildlife water developments) and natural sources of surface water (Davis and Bleich 1980, Whiting et al. 2010, Rich et al. 2019, Terry et al. 2022), and to investigate competition and its effect on bighorn sheep (Ostermann-Kelm et al. 2008).

Remote cameras have been deployed to investigate movement patterns of bighorn sheep across wildlife crossing structures (Gagnon et al. 2022, Aiello et al. 2024) and to determine the extent of, and shifts in, range use by bighorn sheep in response to human disturbance (Campbell and Remington 1979, Leslie and Douglas 1980). Others have

employed remote cameras to ascertain use of specific habitats by competitors of bighorn sheep, including feral equids, or record the presence or relative abundance of predators of bighorn sheep (Bleich et al. 1997, Ostermann-Kelm et al. 2008, Harris et al. 2020b). Additionally, researchers have employed this method to investigate habitat use by Dall's sheep, Stone's sheep, and bighorn sheep across a variety of seasons and study areas (Whiting et al. 2010, Dertien et al. 2017, Enns et al. 2023), and Goward (2022) used this technology to merge indigenous understanding with contemporary methods.

The use of remote cameras in mountain sheep surveys has expanded substantially because of technological advancements that have made cameras smaller, more durable, reliable, and affordable (Wearn and Glover-Kapfer 2019, Parker et al. 2020, Taylor et al. 2021). Despite the initial purchase costs, sampling with cameras is often less expensive and intrusive than other methods (Taylor et al. 2021), even though some types of camera lighting (e.g., infrared flashes) do appear to be visible to wild sheep. Still, these devices can gather and store large quantities of images (Tabak et al. 2019, Taylor et al. 2020), or deliver images via cellular phone networks directly to the researcher or their computer that can store and archive large amounts of photographic data. Storing, sorting, and manually classifying millions of images, however, are challenging and time consuming tasks (Fegraus et al. 2011, Yu et al. 2013), and financial or temporal constraints often make it unrealistic for researchers to analyze every image collected by remote cameras (Yu et al. 2013, Willi et al. 2019, Taylor et al. 2020, Gadot et al. 2024). Currently, computer-automated systems are advancing rapidly, facilitating efficient storage, sorting, classification, and analysis of images (Tabak et al. 2019, Vargas-Felipe et al. 2021). These systems can save substantial time and money (Taylor et al. 2020, Vargas-Felipe et al. 2021) but do currently have limitations associated with the accuracy of species assignments, enumeration of individuals, and group-size interpretations.

Proper remote camera equipment, placement, and protocols are necessary for species detection, interpretation of results, and repeatability of studies (Rovero et al. 2013, Meek et al. 2014, Resources Information Standards Committee 2019), especially when participating in large-scale surveys (Shamon et al. 2024). When remote cameras are used to collect data on mountain sheep, consideration should be given to the number of cameras used, spatial extent of sampling, camera brand and model, settings (e.g., sensitivity), delay between consecutive triggers, distance from camera to sampling area, height and orientation at which the camera is set, and other considerations specific to survey objectives (Burton et al. 2015). Further, managers or researchers should encourage the use of common terminology, describe standard data storage and analytical processes (Harris et al. 2010, Burton et al. 2015), and enhance guidelines for standardized sampling (Rovero et al. 2013). For example, studies of bighorn sheep have elucidated the number of days needed to sample with cameras at water sources to estimate the abundance of Rocky Mountain and desert bighorn sheep (Perry et al. 2010, Taylor et al. 2020), and the number of survey days necessary to obtain the desired information may depend on local meteorological conditions (Glass et al. 2022). Thus, it is essential that proper camera equipment and protocols be implemented when using remote cameras to conduct surveys of mountain sheep.

The future is bright for using remote cameras in mountain sheep research. Consideration should be given, however, to collaborating and following standardized sampling methods, which will help to link remote camera data to ecological processes across species, populations, and study areas (Burton et al. 2015, Shamon et al. 2024). Such collaborations will allow for stronger inferences and better conservation monitoring (Burton et al. 2015, Shamon et al. 2024). Additionally, estimates of abundance and distribution derived from remote camera data across years can supplement aerial survey data and help determine when a population of bighorn sheep is sufficiently large to hunt or to withstand removal of individuals for translocation (Stevens and Goodson 1993, Taylor et al. 2021). Such estimates can potentially be used to track fluctuations in abundance across years and could be helpful when monitoring populations before, during, and after habitat alterations, unanticipated meteorological events, or disease outbreaks (Taylor et al. 2021). Future investigators also can use geographic coordinates obtained from collared animals to identify the season(s) or locations at which to deploy remote cameras to maximize the efficacy of this methodology (Gagnon et al. 2022, Glass et al. 2022, Aiello et al. 2024). Finally, as remote cameras in bighorn sheep research are often set at water sources, long-term monitoring of water sources, even those not currently used by bighorn sheep, can provide important insight into how water use changes before and after large wildfires, habitat alterations, and in a changing climate (Epps et al. 2006, Whiting et al. 2010, Larsen et al. 2012, Rich et al. 2019).

## 13.3.11 Aerial Surveys

While technology continues to develop and improve, having a good biologist in the ship to count and classify mountain sheep is invaluable and, in the words of Brewer, "You seem to find them right where you think you will" (Anderson 2020:125); experience matters. Certainly, camera traps, infrared photography, mathematical models, and the other techniques to monitor mountain sheep discussed in this chapter are valuable, but firsthand knowledge based on observations made by experienced field staff and biologists is irreplaceable. These observations provide context to information on minimum direct counts of the mountain sheep observed that contribute to estimates of population size and trend, sex and age ratios (lambs:females:rams; males:females), information on relative abundance (i.e., animals seen per flight hour), number of males categorized by age-class, minimum number known alive, and information on animal distribution and habitat use (Geist 1968, Douglas-Hamilton 1996, White

2005, Litoroh et al. 2010). The use of aircraft to count mountain sheep dates prior to World War II (Sumner 1948), and thousands of hours have been expended conducting countless aerial surveys, beginning with fixed-wing aircraft and then shifting primarily to helicopters starting in the 1960s (Woodhead 1969, Bleich 1983).

Fixed-wing aircraft have been used for many years to survey thinhorn sheep populations (Sumner 1948, Nichols 1970, Simmons and Robertson 1970, Simmons et al. 1984), but have not proven to be an effective method of observing and classifying bighorn sheep (Russo 1956, Nichols 1970, Simmons and Hansen 1980, Irby 1994, Strickland et al. 1994). It is particularly interesting that the observations Nichols (1970:25) made as he assessed fixed-wing aerial survey effectiveness for Dall's sheep: "Because they are white in color and inhabit almost exclusively the treeless alpine, Dall sheep (*sic*) are one of the best suited of all species to aerial census". Yet, Nichols found that this assumption was a trap and that these animals could not be counted with a reliable degree of accuracy.

The ability to observe wildlife from the air has limitations, particularly from a fixed-wing aircraft (Bleich et al. 2001). While the use of fixed-wing aircraft is limited for mountain sheep counts, their use in the development of sightability and other population indices continues. They are used to determine the number of marked (collared) animals in the sampling area prior to helicopter-based surveys enabling calculation of sightability correction factors and provide an estimate of the probability of observing or missing animals (Whitten 1996, Taylor et al. 2002).

Helicopter surveys enable observers to find and count mountain sheep in the remote and rugged habitats they typically occupy, and this technique is used by nearly all jurisdictions inhabited by mountain sheep for surveys that include total or minimum counts, distance sampling, or animal classification. Managers have been refining the use of helicopters to collect mountain sheep demographic data for years (Pollock and Kendall 1987, Bodie et al. 1990, McDonald et al. 1990, Bodie and Oldenburg 1994, Blum et al. 2023) and it is understood that the use of helicopters disturbs mountain sheep dramatically (Krausman and Hervert 1983, Bleich et al. 1990, 1994; Stockwell et al. 1991, Frid 2003). In particular, the noise and presence of the aircraft flying in pursuit of sheep cause mountain sheep to move considerable distances or into avalanche-prone areas, leaving optimal habitat and potentially the sampling units (e.g., count blocks), each of which violates survey assumptions and confounds survey efforts and analyses (Bleich et al. 1994). Moreover, the responses of individual mountain sheep are often shaped by prior experience, so it cannot explicitly be known, and has been deemed the Bo-Peep effect because investigators cannot be certain of the locations of animals that have responded to the influence of the aircraft either before or after being encountered during the survey flight (Bleich et al. 1990).

Classification surveys of mountain sheep conducted from the air, whether from a helicopter or fixed-wing aircraft, can be less accurate than observations made from the ground when classifying age and sex of the animals (Wyoming Game and Fish Department 2004), particularly when those observations are not recorded using high-resolution photography that can be referred to during post-survey data quality processes. Mountain sheep observed from the ground often can be viewed for longer periods of time, enabling observers to better and more accurately count and classify animals according to age and sex categories. In the end, however, experience gained through many hours of observing animals, whether in the air or on the ground, is an important factor (Bleich 1982, 1998; Smith 1988). Despite the advantages of using experienced observers, photographs taken from the aircraft to verify group compositions can be used to resolve uncertainties associated with age or sex classifications and is recommended (Resources Information Standards Committee 2013).

During a typical helicopter classification survey, observers record the total numbers of males, females, and young-of-the-year. Male and female yearling sheep can be difficult to identify from adult females, but observers should attempt to do so; in some jurisdictions, however, ratios of young per female are expressed as young per adult-like females or young per female-like sheep, as is the case for thinhorn sheep, which acknowledges that difficulty (Bleich 1998, Schmidt and Rattenbury 2018, van de Kerk et al. 2018). Whether thinhorn or bighorn sheep, males ≥2 years of age are often classified into four age classes based on horn morphology (Geist 1968), but horn growth occurs at different rates among areas and is, at least in part, a function of environmental factors such as nutrient availability (Shackleton et al. 1999, Hik and Carey 2000, LaSharr et al. 2019). Surveys are typically done when animals are most concentrated, perhaps when on winter range or at the peak of sexual aggregation when mountain sheep are in rut. For example, bighorn sheep surveys in the Rocky Mountains of the western United States typically are done in late November through mid-December after hunting seasons, and when mountain sheep are concentrated on winter ranges and males are rutting. Further north in southern British Columbia and Alberta aerial surveys for California (*O. c. californiana*) and Rocky Mountain bighorn sheep typically are flown during January through mid-March (Resources Information Standards Committee 2002, Stepnisky et al. 2012), when sheep are concentrated on winter ranges (both high- and low-elevation ranges), because movements to winter ranges are dependent on early winter snows (Poole et al. 2016) and often are not completed by mid-December.

Thinhorn sheep surveys also are flown primarily during winter, in part because it is often easier to find individuals and tracks since sheep are more concentrated at higher elevations, due to the nature of snow accumulations in their wind-swept winter habitats. Late winter surveying also supports collection of information on recruitment of young-of-the-year, and enhanced encounter rates and classification of males that are more difficult to locate during summer and fall months. Further, winter surveys help identify patterns of use within important seasonal habitats

(e.g., winter bedding, foraging, movement as verified by track densities), habitats that may expose the herd or individuals to risks (e.g., highway crossings, lower elevation interface areas), and pursuits that can discourage use of certain areas by mountain sheep such as the use of snowmobiles, back-country skiing, or other recreational activities (Courtemanch 2014, Wiedmann and Bleich 2014, Sproat et al. 2020, Bates et al. 2021, Evans 2024, Kolek 2024). Moreover, winter is considered the limiting season in the north for mountain sheep and most other ungulates; thus, winter surveys can provide an understanding of the severity of the winter and provide insights into population change (Bill A. Jex, British Columbia Ministry of Water, Land and Resource Stewardship, personal communication).

At the southerly latitudes occupied by most desert bighorn sheep, however, longer growing seasons that are accompanied by increased rates of horn growth, an earlier onset of sexual aggregation, and peak of rut (Rubin et al. 2000) dictate that earlier surveys are appropriate. Investigators should, however, avoid conducting aerial surveys during the early neonatal period given the responses of lambs to human disturbance and their vulnerability to injury or death when fleeing from the helicopter (Wehausen and Bleich 2007, Brushett et al. 2023, Blum et al. 2024b). Failure to account for the periods of sexual segregation and aggregation can lead to misleading interpretations of sex ratios (Rubin and Bleich 2005), which often are used to estimate population size and in setting harvest recommendations or removals for translocations (Lonner et al. 2024). Moreover, loss of lambs, whether detected or not, may influence survey results and subsequent interpretations or analyses.

In some herds or circumstances, mountain sheep are classified most effectively using a combination of helicopter and ground surveys (Becker et al. 1978, Holl and Bleich 1983, Irby 1994, Wyoming Game and Fish Department 2004). Some herds are distributed in remote habitats that are difficult or impossible to access, but in habitats that are easily accessible on the ground. In areas where observers have good vehicular access to large concentrations of mountain sheep wintering at low elevations, combined ground and aerial surveys may complement one another and add to the sample size needed to estimate population metrics (i.e., sex and age ratios, or minimum population size). Nevertheless, large numbers of mountain sheep also winter at scattered high-elevation sites that are accessible only by air. In such circumstances, ground surveys and flights should be conducted as concurrently as possible to avoid duplicate counts of groups that may move between high- and low-elevation winter habitats. Data from both surveys can be separated or combined to estimate sex and age ratios and, if conducted simultaneously, methods are available to incorporate the results of both surveys (and others) into a single population estimate (Magnusson et al. 1978, Johnson et al. 2010).

Mountain sheep population trends can be tracked through time if a consistent protocol or sampling design is followed, such as using pre-determined survey units or count blocks, and the same amount of effort per survey unit (Hervert et al. 1998, Wehausen and Bleich 2007). Specifically, the surveys must take place at the same time of year, under similar conditions, must cover the same area, and the survey intensity must be the same. A trend count represents the number of animals observed under a given set of conditions; it is not a total count but, more realistically, is a minimum count. Further, demographic characteristics of the population (i.e., sex and age structure) are important considerations when evaluating population trends. For example, lamb recruitment may be high relative to the number of female bighorn sheep observed, but if recruitment is not replacing mortality, the population could be in decline. Climatic conditions and observer biases (detection rates) change from year to year and, as a result, varying proportions of animals are missed between surveys. Therefore, trend analyses should be based on data from years in which environmental conditions (i.e., snow cover, cloud cover, temperature, wind speeds) are reasonably comparable, survey intensity is the same, and observers are equally experienced in the art of seeing and classifying mountain sheep on the landscape. Trend counts provide corroborating data to verify and refine population estimates.

Many wildlife management agencies have adopted sightability surveys (Bodie and Oldenburg 1994, Bodie et al. 1995, Taylor et al. 2002, Cassirer and Zager 2004, Udevitz et al. 2006) or simple mark-resight surveys (Perry et al. 2010) for mountain sheep. Other agencies have adopted sightability correction factors using double-survey methods (Magnusson et al. 1978, Graham and Bell 1989) or modifications thereof. Such surveys are calibrated by aerial observers and then are used to estimate mountain sheep population size and a measure of error around the estimate. Recently, Blum et al. (2024a:1) described a "Bayesian integrated data model" to leverage information from telemetry data, helicopter surveys, and habitat characteristics to estimate abundance of bighorn sheep while accounting for availability and perception probability (i.e., detection given availability).

The data collected using aircraft, particularly helicopters, is essential to the management of mountain sheep and other populations of large mammals (Bender et al. 2003, Choquenot et al. 2008, Dyal et al. 2022). Helicopters are a useful platform to conduct mountain sheep surveys, though they do cause disturbance to the animals that must be accounted for when designing sampling surveys and conducting data analyses. Additionally, when funding program constraints dictate the use of inferior machines or unskilled pilots, they expose personnel to the risk of injury or death that, sadly, has occurred repeatedly during our respective careers (Bleich 1982, Sasse 2003, Baker et al. 2010). Regardless, and when properly equipped, helicopters provide good visibility for the observer(s) and when the machine is matched to the type of flying conditions in which the survey will occur, they have the power and performance capability to conduct mountain sheep surveys in the rugged and high-elevation terrain mountain sheep

typically inhabit. Certainly, anytime managers or researchers take to the air to conduct wildlife monitoring or data collection, there is a measure of risk. Nevertheless, they are widely used and among the most efficacious survey methods available.

## 13.4   WHAT'S ON THE HORIZON?

Drones, which are referred to widely as unmanned aerial vehicles, have become increasingly popular and effective tools for conducting ungulate population surveys for some species. This relatively new technology offers advantages that appeal to wildlife biologists, including increased precision and accuracy (McMahon et al. 2020, 2021; Zabel et al. 2023, Blum et al. 2024a, Pfeffer et al. 2024), cost-effectiveness (Linchant et al. 2015), reduced disturbance when compared to that associated with a helicopter survey (Hodgson and Koh 2016, McMahon et al. 2021), safety (Jones et al. 2006, Watts et al. 2012), and enhanced post-processing capabilities (Lee et al. 2021). To the best of our knowledge, there are no published studies on the efficacy of this technology for surveying mountain sheep, but studies involving other ungulates are increasing in Africa, Asia, and elsewhere in North America but in very different terrain when compared to that occupied by mountain ungulates (Witczuk et al. 2017, Guo et al. 2018, Linchant et al. 2018, Wang et al. 2019, Yang et al. 2019, vanVuuren et al. 2023, Blum et al. 2024a). Many of these investigators reported results that have important implications for mountain sheep population surveys, but there is a glaring difference between the terrain in which these investigations have been conducted when compared to the rugged topography occupied by thinhorn and bighorn sheep. The use of drones as a method to obtain information for use in deriving population estimates is a complex and in-depth topic that cannot be fully covered in a subsection of a single chapter; therefore, we recommend those interested in the application of this technology consult the peer-reviewed literature, and reference texts such as Wich and Koh (2018), or forthcoming reviews (Blum et al. 2025). For the time being, the application of unmanned aerial vehicles to the repertoire of techniques available to conduct mountain sheep surveys remains unproven in its use.

In addition to drones, there are a variety of new technologies being deployed that potentially will increase the efficacy of mountain sheep surveys. Various types of artificial intelligence, especially machine learning algorithms, are being used to process and analyze large datasets while increasing accuracy of target detection (Torney et al. 2019, Eikelboom et al. 2019). Camera and drone surveys are known for their ability to produce large datasets that can be especially time-consuming to process. Convolutional neural networks are a commonly used machine learning algorithm that researchers can train to detect and classify various species in images from camera traps and drone surveys (Villa et al. 2017, Gray et al. 2018, Torney et al. 2019, Islam et al. 2023). While commonly used for red, green, blue images, these algorithms can also be used to differentiate between heat signatures across the landscape, expanding their use to thermal imagery (Lyu et al. 2024). The use of this technology increases the feasibility of surveys that accumulate very large datasets for wildlife agencies that may not have the capacity to analyze such volumes of information, while also providing accuracy similar to that derived from expert counts (Torney et al. 2019). And, as we move forward, artificial intelligence may become more user-friendly and provide increased accuracy, which will likely have major implications for mountain sheep surveys.

In recent years there has been an increase in the use of camera- or video-collars in ungulate research (Lavelle et al. 2012, Brockman et al. 2017, Vuillaume et al. 2021). These global positioning system collars are typically used to assess behavior, survival, movement, and diet, but they may be useful for population estimates, or in developing understanding of reproductive rates (e.g., lamb:female), or lamb survival when marked animals are repeatedly photographed. Biologists may be able to use capture-recapture techniques to assess mountain sheep population sizes with this technology, although serious challenges lie ahead regarding variable detection, rates of contact outside of an animal's herd, and cost. Given the increase in popularity of these collars, they may provide a new method of surveying populations soon, albeit with substantial challenges and model assumptions yet to be resolved.

Very high-resolution satellite imagery (i.e., <1-m resolution) has also become an increasingly popular method for deriving estimates of wildlife populations and the availability and frequency of the captured images has been significantly improved with many areas on 12–24 hour cycles. In combination with artificial intelligence, this technology has proven effective for some ungulate species (Xue et al. 2017, Wang et al. 2019, Duporge et al. 2021). Detection of animals using this imagery, however, is reliant upon the openness of the landscape, body size of the target species, contrasting color of the target animal and its habitat, image resolution, and cloud cover (LaRue et al. 2017, Delplanque et al. 2024). Further, cost may be a limiting factor for some applications; Wang et al. (2019) reported that the price of 0.5-m imagery ranged from $14.00 to $27.50 US$/km$^2$, which would represent a cost of up to $8,250 for a study area of 300 km$^2$. The aforementioned constraints may provide substantial challenges for mountain sheep population surveys using satellite imagery. While mountain sheep typically prefer more open habitats, these rugged landscapes offer many nooks and crannies that hide sheep, and some individuals use caves to avoid increased temperatures. Additionally, these mountain ungulates remain cryptic throughout the year, and bighorn sheep may not provide enough contrast with the surrounding landscape to be consistently identified by artificial intelligence or human experts analyzing satellite imagery. Dall's sheep and Stone's sheep also are cryptic across the landscape throughout the year regardless of whether the ground is completely snow-covered, or the snow is distributed in patches across the landscape. Costs

for this imagery may currently be a limiting factor, but an increase in the number of satellites and greater availability of high-resolution sensors specifically for wildlife surveys likely will reduce costs in the future.

## 13.5 CONSIDERATIONS FOR ESTABLISHING POPULATION MONITORING FRAMEWORKS

In planning population assessments, it is important to consider monitoring purpose, management goals, needed precision, and population boundaries or survey closure. Of course, knowledge of the natural history of mountain sheep and a familiarity with past assessments and monitoring efforts are foundational to any such project. These considerations will optimize the efficiency of any survey and monitoring effort. Protocols that generate demographic estimates with a low coefficient of variation are the gold standard, but these often are not achieved and sometimes not needed. Realistically, estimates with such precision are often not within the budget or long-term sustainability within resource management agencies. This dilemma was recognized and described by Walters (1986), and has inspired the implementation of adaptive management harvest strategies to allow the use of harvest or dead-side data, population responses, or other feedbackto enhance the applicability of population assessments.

Walters (1986) challenged traditional approaches of uncertainty in the management or conservation of renewable natural resources, and argued that scientific understanding will best be achieved through experiences gained while treating management or conservation actions as an ongoing, adaptive, and experimental process, rather than the traditional approaches of applying current theory or basic research. Not only has an adaptive management approach to population surveys been widespread among the various methods that have evolved to conduct surveys or establish population numbers, but adaptive management science is particularly relevant to mountain sheep conservation especially when evaluating removals for translocations or assessing impacts of disease events. As advocated by McNab (1983), such events represent manipulations; moreover, they are among those most apt to affect population size or persistence, and disease events are of great concern among managers of North American mountain sheep (Bleich 2009).

Despite the diversity of survey methodologies, they all share a common need to have clearly documented sampling protocols, complete data sheets, and staff proficiency. Standardization of any procedure is an essential requirement to ensure that any estimates are repeatable, consistent, and comparable. When resources are limited, surveys can generate initial data representing the distribution of animals and relative abundance to assist with future efforts. Such pioneering work may result only in index-level count data without variance estimates, but by following simple principles of standardization of procedures, they can fulfill some of the desired management goals. Whether or not these procedures are carried out by direct ground or aerial observations, photographic techniques, genetic methods, or techniques still in developmental stages, all include defining survey boundaries as emphasized by Norton-Griffiths (1978) and documenting survey effort that was expended in terms of kilometers walked, hours flown, days of direct or indirect observation, and specific locations within the area of interest. This level of survey can only establish the most basic form of relative comparability within, or among, areas and over periods of time; but if conducted annually and with the additional support of independent survey approaches, it can provide sufficient insight into the status and trajectory of individual populations for most management or conservation purposes, as demonstrated repeatedly (Wehausen 1983, 1986, 1992, 2005, 2024; Wehausen et al. 1987, Rubin et al. 1998, Holl et al. 2004, Holl and Bleich 2009a,b, Holl et al. 2012).

Considering population assessments over broader regions is both a responsibility and challenge for state, provincial, and territorial resource agencies. In light of managing a system of independent populations as metapopulations (more appropriately referred to as metapopulation fragments in California [Epps et al. 2005, Abella et al. 2011, Prentice et al. 2019, Vu et al. 2021]), Torres et al. (1994) demonstrated the utility of placing those individual populations into robust size classes to characterize the distribution of those categories within each metapopulation fragment. This method has been used in California continuously since its inception (Torres et al. 1994, 1996; Epps et al. 2003, 2005; Abella et al. 2011, Prentice et al. 2019, Vu et al. 2021) and serves as an additional starting point and dynamic method of comparisons to help assess future changes in what now represent fragments of what once was a large metapopulation, yet provide insight for changes and priorities for conservation and management decisions. A similar approach has been adopted elsewhere (Wiedmann et al. 2024), and the realities of funding likely will result in an expansion of this approach.

## 13.6  SUMMARY

North American mountain sheep have played prominent roles in the ecology and culture of human beings for millennia. Dating to the Holocene, the hunters and gatherers that occupied the continent long after the progenitors of those specialized ungulates arrived from Siberia were among the earliest peoples to depend on wild sheep as sources of sustenance, possibly for inspiration, and perhaps were the first to develop an index-level approach to the abundance of mountain sheep. As the exploration of North America progressed, the presence of wild sheep became of increasing interest, leading to descriptions of multiple taxa, estimates of numbers, and the evolution of methods of conducting surveys to better determine their abundance. From the rock art of the earliest North Americans to the advent of aerial surveys, and the later technological achievements such as thermal imaging, progress in survey methods has been constant, while the objective of understanding how many sheep there are continues unabated. This chapter explores the historical aspects of that progression, provides insight into the advantages and disadvantages of the various methods used and currently in use, and in the end raises questions about the level of precision necessary to conserve and properly ensure the persistence of these iconic mountain ungulates in a constantly changing world.

## REFERENCES

Abella, R., V. C. Bleich, R. A. Botta, B. J. Gonzales, T. R. Stephenson, S. G. Torres, and J. D. Wehausen. 2011. Status of bighorn sheep in California—2010. *Desert Bighorn Council Transactions* 51:54–68.

Aiello, C., N. Galloway, K. Fratella, P. Prentice, N. Darby, D. Hughson, and C. Epps. 2024. Highway underpasses offer little fragmentation relief for desert bighorn sheep near Mojave National Preserve, California. *California Fish and Wildlife Journal* 110:e5.

Anderson, D. R. 2001. The need to get the basics right in wildlife field studies. *Wildlife Society Bulletin* 29:1294–1297.

Anderson, R. M. 2020. *Rimrock Man: The Early Life of W. B. Carson*. E. and L. Davis Publishers, Cotulla, Texas, USA.

Ando, H., H. Mukai, T. Komura, T. Dewi, M. Ando, and Y. Isagi. 2020. Methodological trends and perspectives of animal dietary studies by noninvasive fecal DNA metabarcoding. *Environmental DNA* 2:391–406.

Ayotte, J. B., K. L. Parker, and M. P. Gillingham. 2008. Use of natural licks by four species of ungulates in northern British Columbia. *Journal of Mammalogy* 89:1041–1050.

Bach, B. H., A. B. Quigley, K. M. Gaynor, A. McInturff, K. L. Charles, J. Dorcy, and J. S. Brashares. 2022. Identifying individual ungulates from fecal DNA: a comparison of field collection methods to maximize efficiency, ease, and success. *Mammalian Biology* 102:863–874.

Baker, J., V. Bleich, D. Casady, T. Evans, E. Kleinfelter, B. Lloyd, C. Ogata, R. Mayfield, R. Mohr, T. Palmisano, B. Payne, R. Schaefer, T. Stephenson, C. Stowers, S. Torres, and M. Yaremko. 2010. Helicopter accident report and recommendations to the Director for future operations—accident date: January 5, 2010. Wildlife Branch Administrative Report, California Department of Fish and Game, Sacramento, California, USA.

Balsara, M. 2024. The evolutionary response to hunting-induced selection on horn characteristics in Yukon thinhorn sheep (*Ovis dalli*). Thesis, Western University, London, Ontario, Canada.

Bart, J., S. Droege, P. Geissler, B. Peterjohn, and C. J. Ralph. 2004. Density estimation in wildlife surveys. *Wildlife Society Bulletin* 32:1242–1247.

Bates, S. B., J. C. Whiting, and R. T. Larson. 2021. Comparison of effects of shed antler hunting and helicopter surveys on ungulate movements and space use. *Journal of Wildlife Management* 85:437–448.

Becker, K., T. Varcalli, E. T. Thorne, and G. B. Butler. 1978. Seasonal distribution patterns of Whiskey Mountain bighorn sheep. *Northern Wild Sheep and Goat Council Proceedings* 1:1–16.

Belant, J. L., K. Denker, and K. F. Kellner. 2024. Age-based scoring as a complementary approach to sustainable trophy hunting. *BioScience* 74:737–739.

Bender, L. C., W. L. Myers, and W. R. Gould. 2003. Comparison of helicopter and ground surveys for North American elk *Cervus elaphus* and mule deer *Odocoileus hemionus* population composition. *Wildlife Biology* 9:199–205.

Berger, J., V. C. Bleich, and R. T. Bowyer. 2024. Must we lose our biological connection to nature to endure changing times? *Journal of Wildlife Management* 88:e22639.

Bleich, V. C. 1979. Development and operations. Annual Progress Report, Federal Aid in Wildlife Restoration Project W-26-D-34. California Department of Fish and Game, Sacramento, California, USA.

Bleich, V. C. 1982. An illustrated guide to aging the lambs of mountain sheep. *Desert Bighorn Council Transactions* 26:59–62.

Bleich, V. C. 1983. Comments on helicopter use by wildlife agencies. *Wildlife Society Bulletin* 11:304–306.

Bleich, V. C. 1998. Importance of observer experience in determining age and sex of mountain sheep. *Wildlife Society Bulletin* 26:877–880.

Bleich, V. C. 2009. Perceived threats to mountain sheep: levels of concordance among states, provinces, and territories. *Desert Bighorn Council Transactions* 50:32–39.

Bleich, V. C. 2018. Maintaining momentum for conservation: bighorn sheep as an example. *Wildlife Society Bulletin* 42:540–546.

Bleich, V. C., C. S. Y. Chun, R. W. Anthes, T. E. Evans, and J. K. Fischer. 2001. Visibility bias and development of a sightability model for tule elk. *Alces* 37:315–327.

Bleich, V. C., H. E. Johnson, S. A. Holl, L. Konde, S. G. Torres, and P. R. Krausman. 2008. Fire history in a chaparral ecosystem: implications for conservation of a native ungulate. *Rangeland Ecology and Management* 61:571–579.

Bleich, V. C., L. J. Coombes, and G. W. Sudmeier. 1982. Volunteers and wildlife habitat management: twelve years together. *Transactions of the Western Section of the Wildlife Society* 18:64–68.

Bleich, V. C., and M. W. Oehler. 2000. Wildlife education in the United States: thoughts from agency biologists. *Wildlife Society Bulletin* 28:542–545.

Bleich, V. C., and R. A. Weaver. 2007. Status of mountain sheep in California: comparisons between 1957 and 2007. *Desert Bighorn Council Transactions* 49:55–67.

Bleich, V. C., R. T. Bowyer, A. M. Pauli, M. C. Nicholson, and R. W. Anthes. 1994. Mountain sheep *Ovis canadensis* and helicopter surveys: ramifications for the conservation of large mammals. *Biological Conservation* 70:1–7.

Bleich, V. C., R. T. Bowyer, A. M. Pauli, R. L. Vernoy, and R. W. Anthes. 1990. Responses of mountain sheep to aerial sampling using helicopters. *California Fish and Game* 76:197–204.

Bleich, V. C., R. T. Bowyer, and J. D. Wehausen. 1997. Sexual segregation in mountain sheep: resources or predation? *Wildlife Monographs* 134:1–50.

Bleich, V. C., and S. A. Holl. 1982. Management of chaparral habitat for mule deer and mountain sheep in southern California. Pages 247–254 *in* C. E. Conrad and W. C. Oechel, Technical Coordinators. *Proceedings of the Symposium on the Dynamics and Management of Mediterranean-type Ecosystems*. USDA Forest Service, General Technical Report PSW-58.

Blum, J., A. M. Foley, R. W. DeYoung, D. G. Hewitt, J. Baumgardt, M. W. Hellickson, and H. L. Perotto-Baldivieso. 2024a. Evaluation of drone surveys for ungulates in southwestern rangelands. *Wildlife Society Bulletin* 48:e1515.

Blum, M. E., C. P. Lehman, J. T. Hogg, T. D. Nordeen, M. Cox, H. M. Miyasaki, E. F. Cassirer, and K. L. Monteith. 2024b. Capture and handling of neonates. Pages 8.1–8.20 *in* V. C. Bleich and D. W. Lutz, editors. *Wild Sheep Capture and Handling Guidelines*. Second edition. Western Association of Fish and Wildlife Agencies, Boise, Idaho, USA.

Blum, M. E., F. E. Buderman, J. R. Bennett, K. M. Stewart, M. Cox, and P. J. Williams. 2023. Comparing contemporary models to traditional indices to estimate abundance of desert bighorn sheep. *Journal of Wildlife Management* 88:e22517.

Blum, M. E., J. Blum, and V. C. Bleich. 2025. Drone technology and its utility for mountain sheep surveys: a brief review. *California Fish and Wildlife Journal*: In Press.

Bodie, W. L., E. O. Garton, E. R. Taylor, and M. McCoy. 1995. A sightability model for bighorn sheep in canyon habitats. *Journal of Wildlife Management* 59:832–840.

Bodie, W. L., E. R. Taylor, M. McCoy, and D. Toweill. 1990. Status and distribution of California bighorn sheep in Idaho. *Northern Wild Sheep and Goat Council Proceedings* 7:12–16.

Bodie, W. L. and L. E. Oldenburg. 1994. A standardized technique for helicopter surveys of bighorn sheep. *Northern Wild Sheep and Goat Council Proceedings* 9:65–68.

Bohmann, K., A. Evans, M. T. P. Gilbert, G. R. Carvalho, S. Creer, M. Knapp, D. W. Yu, and M. de Bruyn. 2014. Environmental DNA for wildlife biology and biodiversity monitoring. *Trends in Ecology and Evolution* 29:358–367.

Bornstein, M. H., J. Jager, and D. L. Putnick. 2013. Sampling in developmental science: situations, shortcomings, solutions, and standards. *Developmental Review* 33:357–370.

Bowyer, R. T. 2022. *Sexual Segregation in Ungulates: Ecology, Conservation, and Behavior*. Johns Hopkins University Press, Baltimore, Maryland, USA.

Brinkman, T. J., M. K. Schwartz, D. K. Person, K. L. Pilgrim, and K. J. Hundertmark. 2010. Effects of time and rainfall on PCR success using DNA extracted from deer fecal pellets. *Conservation Genetics* 11:1547–1552.

Brockman, C. J., W. B. Collins, J. M. Welker, D. E. Spalinger, and B. W. Dale. 2017. Determining kill rates of ungulate calves by brown bears using neck-mounted cameras. *Wildlife Society Bulletin* 41:88–97.

Brunner, J. 1909. *Tracks and Tracking*. Outing Publishing Company, New York, New York, USA.

Brushett, A., J. Whittington, B. Macbeth, and J. M. Fryxell. 2023. Changes in movement, habitat use, and response to human disturbance accompany parturition events in bighorn sheep (*Ovis canadensis*). *Movement Ecology* 11(1):36.

Buckland, S. T., D. R. Anderson, K. P. Burnham, and J. L. Laake. 1993. *Distance Sampling: Estimating Abundance of Biological Populations*. Chapman and Hall, New York, New York, USA.

Burton, A. C., E. Neilson, D. Moreira, A. Ladle, R. Steenweg, J. T. Fisher, E. Bayne, and S. Boutin. 2015. Wildlife camera trapping: a review and recommendations for linking surveys to ecological processes. *Journal of Applied Ecology* 52:675–685.

Campbell, B. H., and R. Remington. 1979. Bighorn use of artificial water sources in the Buckskin Mountains, Arizona. *Desert Bighorn Council Transactions* 23:50–56.

Campbell, B. H., and R. Remington. 1981. Influence of construction activities on water-use patterns of desert bighorn sheep. *Wildlife Society Bulletin* 9:63–65.

Cassirer, E. F., and P. Zager. 2004. A sightability model for helicopter surveys of bighorn sheep in Hells Canyon. *Northern Wild Sheep and Goat Council Proceedings* 14:191.

Caughley, G. 1977. *Analysis of Vertebrate Populations.* John Wiley & Sons, New York, New York, USA.

Choquenot, D., N. Bolton, and D. Woods. 2008. Evaluating helicopter-based surveys for estimating densities of Himalayan thar. *Wildlife Research* 35:358–364.

Constantino, G. M. 1973. Time-lapse photography census of bighorns at the Desert National Wildlife Range. *Desert Bighorn Council Transactions* 17:59–72.

Constantino, G. M. 1974. Additional time-lapse photography field techniques. *Desert Bighorn Council Transactions* 18:29–30.

Conner, M. M., W. C. Saunders, N. Bouwes, and C. Jordan. 2015. Evaluating impacts using a BACI design, ratios, and a Bayesian approach with a focus on restoration. *Environmental Monitoring and Assessment* 188(10):555.

Coonen, L. P. 1951. The prehistoric roots of biology. *Scientific Monthly* 73:154–165.

Courtemanch, Alyson A. B. 2014. Seasonal habitat selection and impacts of backcountry recreation on a formerly migratory bighorn sheep population in northwest Wyoming, USA. Thesis, University of Wyoming, Laramie, WY, USA.

Daly, T., and D. Dierkes. 2024. Volunteers needed for bighorn sheep survey. https://wildlife.ca.gov/News/Archive/volunteers-needed-for-bighorn-sheep-survey. Accessed 14 October 2024.

Davis, J. H., and V. C. Bleich. 1980. Time-lapse photography: a new focus on wildlife. *Outdoor California* 41(4):7–9.

DeForge, J. R. 1980. Ecology, behavior, and population dynamics of desert bighorn sheep, *Ovis canadensis nelsoni*, in the San Gabriel Mountains of California. Thesis, California State Polytechnic University, Pomona, California, USA.

Dekelaita, D. J., C. W. Epps, K. M. Stewart, J. S. Sedinger, J. G. Powers, B. J. Gonzales, R. K. Abella-Vu, N. W. Darby, and D. L. Hughson. 2020. Survival of adult female bighorn sheep following a pneumonia epizootic. *Journal of Wildlife Management* 84:1268–1282.

Delplanque, A., J. Théau, S. Foucher, G. Serati, S. Durand, and P. Lejeune. 2024. Wildlife detection, counting and survey using satellite imagery: are we there yet? *GIScience and Remote Sensing* 61. https://doi.org/10.1080/15481603.2024.2348863

Demarchi, R. A. 1977. Canada's mountain sheep: their present status and future prospects. Pages 46–50 *in* T. Mosquin and C. Suchal, editors. *Canada's Threatened Species and Habitats.* Canadian Wildlife Federation and World Wildlife Fund, Ottawa, Ontario, Canada.

Demarchi, R. A., and C. L. Hartwig. 2004. *Status of Thinhorn Sheep in British Columbia.* Wildlife Bulletin B-119. British Columbia Ministry of Water, Land and Air Protection, Biodiversity Branch, Victoria, British Columbia, Canada.

Dertien, J. S., P. F. Doherty Jr, C. F. Bagley, J. A. Haddix, A. R. Brinkman, and E. S. Neipert. 2017. Evaluating Dall's sheep habitat use via camera traps. *Journal of Wildlife Management* 81:1457–1467.

Diaz, N. I., and P. Corti. 2024. Ensuring the quality of historical data for wildlife conservation: a methodological framework. *Perspectives in Ecology and Conservation* 22(4). https://doi.org/10.1016/j.pecon.2024.12.003

Dormaar, J. F., and B. D. Walker. 1996. Elemental content of animal licks along the eastern slopes of the Rocky Mountains in southern Alberta, Canada. *Canadian Journal of Soil Science* 76:509–512.

Douglas-Hamilton, I. 1996. Counting elephants from the air: total counts. Pages 28–37 *in* K. Kagwana, editor. *Studying Elephants.* African Wildlife Foundation, Nairobi, Kenya.

Duporge, I., O. Isupova, S. Reece, D. W. Macdonald, and T. Wang. 2021. Using very-high-resolution imagery and deep learning to detect and count African elephants in heterogeneous landscapes. *Remote Sensing in Ecology and Conservation* 7:369–381.

Dyal, J. R., K. V. Miller, M. J. Cherry, and G. J. D'Angelo. 2022. White-tailed deer movement in response to helicopter surveys. *Wildlife Society Bulletin* 46:e1383.

Eberhardt, L. L. 1976. Quantitative ecology and impact assessment. *Journal of Environmental Management* 4:27–70.

Eerkens, J. W., J. Rosenthal, D. C. Young, and J. King. 2007. Early Holocene landscape archaeology in the Coso Basin, northwestern Mojave Desert, California. *North American Archaeologist* 28:87–112.

Eikelboom, J. A. J., J. Wind, E. van de Ven, L. M. Kenana, B. Schroder, H. J. de Knegt, F. van Langevelde, and H. H. T. Prins. 2019. Improving the precision and accuracy of animal population estimates with aerial image detection. *Methods in Ecology and Evolution* 10:1875–1887.

Elbroch, M. 2003. *Mammal Tracks and Sign: A Guide to North American Species.* Stackpole Books, Harrisburg, Pennsylvania, USA.

Enns, G. E., B. A. Jex, and M. S. Boyce. 2023. Stone's sheep (*Ovis dalli stonei*) lambing and nursery habitat selection. *Canadian Journal of Zoology* 102:691–707.

Epps, C. W., P. B. Holton, R. J. Monello, R. S. Crowhurst, S. M. Gaulke, W. M. Janousek, T. G. Creech, and T. A. Graves. 2024. Population and spatial dynamics of desert bighorn sheep in Grand Canyon during an outbreak of respiratory pneumonia. *Frontiers in Ecology and Evolution* 12:1377214.

Epps, C. W., P. J. Palsboll, J. D. Wehausen, G. K. Roderick, and D. R. McCullough. 2006. Elevation and connectivity define genetic refugia for mountain sheep as climate warms. *Molecular Ecology* 15:4295–4302.

Epps, C. W., P. J. Palsboll, J. D. Wehausen, G. K. Roderick, R. R. Ramey II, and D. R. McCullough. 2005. Highways block gene flow and cause rapid decline in genetic diversity of desert bighorn sheep. *Ecologyl Letters* 8:1029–1038.

Epps, C. W., R. S. Crowhurst, and B. S. Nickerson. 2018. Assessing changes in functional connectivity in a desert bighorn sheep metapopulation after two generations. *Molecular Ecology* 27:2334–2346.

Epps, C. W., V. C. Bleich, J. D. Wehausen, and S. G. Torres. 2003. Status of bighorn sheep in California. *Desert Bighorn Council Transactions* 47:20–35.

Ernest, H. B., E. S. Rubin, and W. M. Boyce. 2002. Fecal DNA analysis and risk assessment of mountain lion predation of bighorn sheep. *Journal of Wildlife Management* 66:75–85.

Evans, A. D. 2024. The influence of recreational disturbance on desert bighorn sheep behavior and stress in western Colorado. Thesis, New Mexico State University, Las Cruces, New Mexico, USA.

Fegraus, E. H., K. Lin, J. A. Ahumada, C. Baru, S. Chandra, and C. Youn. 2011. Data acquisition and management software for camera trap data: a case study from the TEAM Network. *Ecological Informatics* 6:345–353.

Flinders, F. T., and J. A. Crawford. 1977. Composition and degradation of jackrabbit and cottontail fecal pellets. *Journal of Range Management* 30:217–220.

Found, R., and B. R. Patterson. 2020. Assessing ungulate populations in temperate North America. *Canadian Wildlife Biology and Management* 9:21–42.

Fox, L. M., P. R. Krausman, M. L. Morrison, and T. H. Noon. 2000. Mineral content of Sonoran pronghorn forage. *California Fish and Game* 86:159–174.

Frid, A. 2003. Dall's sheep responses to overflights by helicopter and fixed-wing aircraft. *Biological Conservation* 110:387–399.

Gadot, T., S. Istrate, H. Kim, D. Morris, S. Beery, T. Birch, and J. Ahumada. 2024. To crop or not to crop: comparing whole-image and cropped classification on a large dataset of camera trap images. *IET Computer Vision* 2024:1–16. https://doi.org/10.1049/cvi2.12318

Gagnon, J. W., C. D. Loberger, K. S. Ogren, S. C. Sprague, S. R. Boe, and R. E. Schweinsburg. 2022. Mitigating bighorn sheep-vehicle collisions and habitat fragmentation with overpasses and adaptive mitigation. *Human-Wildlife Interactions* 16:353–372.

Gamez-Brunswick, C., and O. Rojas-Soto. 2020. New insights into palaeo-distributions based on Holocene rock art. *Journal of Biogeography* 47:2543–2553.

Garfinkel, A. P. 2006. Paradigm shifts, rock art studies, and the "Coso Sheep Cult" of eastern California. *North American Archaeologist* 27:203–244.

Garfinkel, A. P., and A. K. Rogers (compilers). 2016. Appendix D. Pages 56–72 *in* A. K. Rogers and R. L. Kaldenberg. *The Euro-American Discovery of the Coso Petroglyphs, Inyo County,* California, USA. Maturango Museum, Ridgecrest, California, USA.

Garfinkel, A. P., and D. R. Austin. 2011. Reproductive symbolism in Great Basin rock art: bighorn sheep hunting, fertility and forager ideology. *Cambridge Archaeological Journal* 21:453–471.

Garfinkel, A. P., D. A. Young, and R. M. Yohe II. 2010. Bighorn hunting, resource depression, and rock art in the Coso Range of eastern California: a computer simulation model. *Journal of Archaeological Science* 37:42–51.

Geist, V. 1968. On interrelation of external appearance, social behavior and social structure of mountain sheep. *Zeitschrift für Tierpsychologie* 25:199–215

Geist, V. 1971. *Mountain Sheep: A Study in Behavior and Evolution.* University of Chicago Press, Chicago, Illinois, USA.

Gilreath, A. J., and W. R. Hildebrandt. 2008. Coso rock art within its archaeological context. *Journal of California and Great Basin Anthropology* 28:1–22.

Gissi, E., R. T. Bowyer, and V. C. Bleich. 2024. Sex-based differences affect conservation. *Science* 384:1309–1310.

Glass, D. M., P. R. Prentice, A. D. Evans, and O. J. Schmitz. 2022. Local differences in maximum temperature determine water use among desert bighorn sheep populations. *Journal of Wildlife Management* 86:e22313.

Goodman, J. D. 1963. A report on the first mid-winter and spring bighorn sheep census in the Santa Rosa Mountains, 1962–1963. *Desert Bighorn Council Transactions* 7:135–138.

Goward, S. L. 2022. I spy through a camera's eye: divii in the Gwich'in Settlement Area. *Arctic* 75:510–515.

Goward, S. L. 2004. Using remote camera traps to monitor population demographics and community ecology of divii (Dall's sheep): part of a community-based monitoring program in the Northern Richardson Mountains, NT. Thesis, University of Victoria, Victoria, British Columbia, Canada.

Graham, A., and R. Bell. 1989. Investigating observer bias in aerial survey by simultaneous double-counts. *Journal of Wildlife Management* 53:1009–1016.

Grant, C. 1968. Desert bighorn rock drawings of the Coso Range, Inyo County, California. *Desert Bighorn Council Transactions* 12:40–49.

Grant, C. 1980. The desert bighorn and aboriginal man. Pages 7–39 *in* G. Monson and L. Sumner, editors. *The Desert Bighorn: Its Life History, Ecology, and Management.* University of Arizona Press, Tucson, Arizona, USA.

Gray, P. C., A. B. Fleishman, D. J. Klein, M. W. McKown, V. S. Bézy, K. J. Lohmann, and D. W. Johnston. 2018. A convolutional neural network for detecting sea turtles in drone imagery. *Methods in Ecology and Evolution* 10:345–355.

Gregory, T. 1938. Lion in the Carmens. *Chicago Naturalist* 1(3):70–81; (4):110–120.

Grigione, M. M., P. Burman, V. C. Bleich, and B. M. Pierce. 1999. Identifying individual mountain lions (*Felis concolor*) by their tracks: refinement of an innovative technique. *Biological Conservation* 88:25–32.

Groves, P., D. H. Mann, and M. L. Kunz. 2022. Prehistoric perspectives can help interpret the present: 14000 years of moose (*Alces alces*) in the Western Arctic. *Canadian Journal of Zoology* 100:732–746.

Guo, X., Q. Shao, Y. Li, Y. Wang, D. Wang, J. Liu, J. Fan, and F. Yang. 2018. Application of UAV remote sensing for a population census of large wild herbivores: taking the headwater region of the Yellow River as an example. *Remote Sensing* 10:1041.

Hamilton, K. M., S. A. Holl, and C. L. Douglas. 1982. An evaluation of the effects of recreational activity on bighorn sheep in the San Gabriel Mountains, California. *Desert Bighorn Council Transactions* 26:50–55.

Harper, L. R., L. L. Handley, A. I. Carpenter, M. Ghazali, C. D. Muri, C. J. Macgregor, T. W. Logan, A. Law, T. Breithaupt, D. S. Read, A. D. McDevitt, and B. Hänfling. 2019. Environmental DNA (eDNA) metabarcoding of pond water as a tool to survey conservation and management priority mammals. *Biological Conservation* 238:108225.

Harris, G., R. Thompson, J. L. Childs, and J. G. Sanderson. 2010. Automatic storage and analysis of camera trap data. *Bulletin of the Ecological Society of America* 91:352–360.

Harris, G. M., D. R. Stewart, D. Brown, L. Johnson, J. Sanderson, A. Alvidrez, T. Waddell, and R. Thompson. 2020b. Year-round water management for desert bighorn sheep corresponds with visits by predators not bighorn sheep. *PLoS ONE* 15:e0241131.

Harris, G. M., D. R. Stewart, M. J. Butler, E. M. Rominger, C. Q. Ruhl, D. T. McDonald, and P. M. Schmidt. 2024. N-Mixture models with camera trap imagery produce accurate abundance estimates of ungulates. *Scientific Reports* 14:31421. https://doi.org/10.1038/s41598-024-83011-4

Harris, G. M., M. J. Butler, D. R. Stewart, E. M. Rominger, and C. Q. Ruhl. 2020a. Accurate population estimation of Caprinae using camera traps and distance sampling. *Scientific Reports* 10:17729.

Harris, R. B. 1994. Dealing with uncertainty in counts of mountain ungulates. *International Snow Leopard Symposium Proceedings* 7:105–111.

Hatfield, B. E., J. M. Runcie, E. A. Siemion, C. B. Quinn, and T. R. Stephenson. 2021. New detections extend the known range of the state-threatened Sierra Nevada red fox. *California Fish and Wildlife Journal* 107 (Special CESA Issue):438–443.

Heimer, W. E. 1973. Dall sheep movements and mineral lick use. Final Report, Federal Aid in Wildlife Restoration Projects W-17-R-2, W-17-R-3, W-17-R-4, and W-17-R-5. Alaska Department of Fish and Game, Fairbanks, Alaska, USA.

Heimer, W. E. 2022. *Dall Sheep Management in Alaska from Pleistocene to Present*. Privately Published, Fairbanks, Alaska, USA.

Helvie, J. B. 1972. Census of desert bighorn sheep with time-lapse photography. *Desert Bighorn Council Transactions* 16:3–8.

Hervert, J. J., R. S. Henry, M. T. Brown, and R. L. Kearns. 1998. Sighting rates of bighorn sheep during helicopter surveys on the Kofa National Wildlife Refuge, Arizona. *Desert Bighorn Council Transactions* 42:11–26.

Hik, D. S., and J. Carey. 2000. Cohort variation in horn growth of Dall (*sic*) sheep rams in southwest Yukon, 1969–99. *Northern Wild Sheep and Goat Council Proceedings* 12:88–100.

Hodgson, J. C., and L. P. Koh. 2016. Best practice for minimising unmanned aerial vehicle disturbance to wildlife in biological field research. *Current Biology* 26:R404–R405.

Holl, S. A., and V. C. Bleich. 1983. San Gabriel mountain sheep: biological and management considerations. USDA Forest Service, San Bernardino National Forest, San Bernardino, California, USA.

Holl, S. A., and V. C. Bleich. 1987. Mineral lick use by mountain sheep in the San Gabriel Mountains, California. *Journal of Wildlife Management* 51:381–383.

Holl, S. A., and V. C. Bleich. 2009a. Responses of large mammals to fire and rain in the San Gabriel Mountains, California. *Northern Wild Sheep and Goat Council Proceedings* 17:139–156.

Holl, S. A., and V. C. Bleich. 2009b. Reconstructing the San Gabriel Mountains bighorn sheep population. *California Fish and Game* 95:77–87.

Holl, S. A., V. C. Bleich, B. W. Callenberger, and B. Bahro. 2012. Simulated effects of two fire regimes on bighorn sheep: the San Gabriel Mountains, California, USA. *Fire Ecology* 8(3):88–103.

Holl, S. A., V. C. Bleich, and S. G. Torres. 2004. Population dynamics of bighorn sheep in the San Gabriel Mountains, California, 1967–2002. *Wildlife Society Bulletin* 32:412–426.

Honess, R. F., and K. B. Frost. 1942. A Wyoming bighorn sheep study. Wyoming Game and Fish Department Report 1. Wyoming Game and Fish Department, Cheyenne, Wyoming, USA.

Hossain, A. N. M., A. Barlow, C. G. Barlow, A. J. Lynam, S. Chakma, and T. Savini. 2016. Assessing the efficacy of camera trapping as a tool for increasing detection rates of wildlife crime in tropical protected areas. *Biological Conservation* 201:314–319.

Hurley, K. 2021. It's all about the numbers! *Wild Sheep* 9(3):20.

Irby, L. R. 1994. Utility of summer fixed-wing aerial surveys in predicting lamb:ewe ratios observed on winter range. *Northern Wild Sheep and Goat Council Proceedings* 9:51–55.

Islam, S. B., D. Valles, T. J. Hibbits, W. A. Ryberg, D. K. Walkup, and M. R. J. Forstner. 2023. Animal species recognition with deep convolutional neural networks from ecological camera trap images. *Animals* 13:1526.

Jaeger, J. R., J. D. Wehausen, and V. C. Bleich. 1991. Evaluation of time-lapse photography to estimate population parameters. *Desert Bighorn Council Transactions* 35:5–8.

Johnson, H. E., L. S. Mills, J. D. Wehausen, and T. R. Stephenson. 2010. Combining ground count, telemetry, and mark-resight data to infer population dynamics in an endangered species. *Journal of Applied Ecology* 47:1083–1093.

Jones, F. L. 1980. Sign reading and field identification. Pages 236–259 *in* G. Monson and L. Sumner, editors. *The Desert Bighorn: Its Life History, Ecology, and Management*. University of Arizona Press, Tucson, Arizona, USA.

Jones, F. L., G. Flittner, and R. Gard. 1953. Report on a survey of bighorn sheep in the Santa Rosa Mountains, Riverside County. *California Fish and Game* 43:179–191.

Jones, G. P., L. G. Pearlstine, and H. F. Percival. 2006. An assessment of small unmanned aerial vehicles for wildlife research. *Wildlife Society Bulletin* 34:750–758.

Kennedy, C. E. 1957. Discussion. *Desert Bighorn Council Transactions* 1:50.

Kery, M. and B. Schmidt. 2008. Imperfect detection and its consequences for monitoring for conservation. *Community Ecology* 9:207–216.

Knudsen, M. F. 1963. A summer waterhole study at Carrizo spring, Santa Rosa Mountains of southern California. *Desert Bighorn Council Transactions* 7:185–192.

Kolek, J. T. 2024. Recreation in sheep country: assessing the response of an endangered ungulate to novel disturbance. Thesis, University of Wyoming, Laramie, Wyoming, USA.

Krausman, P. R. 2020. Important considerations when using models. *Journal of Wildlife Management* 84:1221–1223.

Krausman, P. R. 2023. Managing artificial intelligence. *Journal of Wildlife Management* 87:e22492.

Krausman, P. R., and J. J. Hervert. 1983. Mountain sheep responses to aerial surveys. *Wildlife Society Bulletin* 11:372–375.

Krausman, P. R., S. G. Torres, L. L. Ordway, J. J. Hervert, and M. Brown. 1985. Diel activity of ewes in the Little Harquahala Mountains Arizona. *Desert Bighorn Council Transactions* 29:24–26.

Kucera, T. E., and R. A. Barrett. 1993. The Trailmaster* camera system for detecting wildlife. *Wildlife Society Bulletin* 21:505–508.

Lancia, R. A., J. D. Nichols, and K. H. Pollock. 1994. Estimating the number of animals in wildlife populations. Pages 215–253 in T. A. Bookhout, editor. *Research and Management Techniques for Wildlife and Habitats*. Fifth edition. The Wildlife Society, Bethesda, Maryland, USA.

Landhuis, E. 2016. Scientific literature: information overload. *Nature* 535:457–458.

Larsen, R. T., J. A. Bissonette, J. T. Flinders, and J. C. Whiting. 2012. Framework for understanding the influences of wildlife water developments in the western United States. *California Fish and Game* 98:148–163.

LaRue, M. A., S. Stapleton, and M. Anderson. 2017. Feasibility of using high-resolution satellite imagery to assess vertebrate wildlife populations. *Conservation Methods* 31:213–220.

LaSharr, T. N., R. A. Long, J. R. Heffelfinger, V. C. Bleich, P. R. Krausman, R. T. Bowyer, J. M. Shannon, R. W. Klaver, C. E. Brewer, M. Cox, A. A. Holland, A. Hubbs, C. P. Lehman,

J. D. Muir, B. Sterling, and K. L. Monteith. 2019. Hunting and mountain sheep: do current harvest practices affect horn growth? *Evolutionary Applications* 12:1823–1836.

Lavelle, M. J., S. E. Hygnstrom, A. M. Hildreth, T. A. Campbell, D. B. Long, D. G. Hewitt, J. Beringer, and K. C. VerCauteren. 2012. Utility of improvised video-camera collars for collecting contact data from white-tailed deer: possibilities in disease transmission studies. *Wildlife Society Bulletin* 36:828–834.

Lee, S., Y. Song, and S.-H. Kil. 2021. Feasibility analyses of real-time detection of wildlife using UAV-derived thermal and RGB images. *Remote Sensing* 13:2169.

Leopold, A. 1933. *Game Management*. Charles Scribner's Sons, New York, New York, USA.

Leslie, D. M., and C. L. Douglas. 1980. Human disturbance at water sources of desert bighorn sheep. *Wildlife Society Bulletin* 8:284–290.

Liebenberg, L. 1990. *The Art of Tracking: The Origin of Science*. David Philip Publishers Ltd., Claremont, South Africa.

Linchant, J., S. Lhoest, S. Quevauvillers, P. Lejeune, C. Vermeulen, J. S. Ngabinzeke, B. L. Belanganayi, W. Delvingt, and P. Bouche. 2018. UAS imagery reveals new survey opportunities for counting hippos. *PLoS ONE* 13:e0206413.

Linchant, J., J. Lisein, J. Semeki, P. Lejeune, and C. Vermeulen. 2015. Are unmanned aircraft systems (UASs) the future of wildlife monitoring? A review of accomplishments and challenges. *Mammal Review* 45:239–252.

Litoroh, L., F. W. Ihwagi, R. Mayienda, J. Bernard, and I. Douglas-Hamilton. 2010. Total aerial count of elephants in Laikipia-Samburu Ecosystem in November 2008. Kenya Wildlife Service, Nairobi, Kenya.

Long, E. S., D. M. Fecske, R. A. Sweitzer, J. A. Jenks, B. M. Pierce, and V. C. Bleich. 2003. Efficacy of photographic scent stations to detect mountain lions. *Western North American Naturalist* 63:529–532.

Longhurst, W. M., A. S. Leopold, and R. F Dasmann. 1952. A survey of California deer herds, their ranges and management problems. *Game Bulletin* 6:1–136.

Lonner, B. N., T. R. Stephenson, and M. Cox. 2024. Translocation, release, and monitoring of wild sheep. Pages 9.1–9.33 *in* V. C. Bleich and D. W. Lutz, editors. *Wild Sheep Capture and Handling Guidelines*. Second edition. Western Association of Fish and Wildlife Agencies, Boise, Idaho, USA.

Lounsberry, Z. T., T. D. Forrester, M. T. Olegario, J. L. Brazeal, H. U. Wittmer, and B. N. Sacks. 2015. Estimating sex-specific abundance in fawning areas of a high-density Columbian black-tailed deer population using fecal DNA. *Journal of Wildlife Management* 79:39–49.

Lukacs, P. M., and K. P. Burnham. 2005. Estimating population size from DNA-based closed capture-recapture data incorporating genotyping error. *Journal of Wildlife Management* 69:396–403.

Lyu, H., F. Qiu, L. An, D. Stow, R. Lewison, and E. Bohnett. 2024. Deer survey from drone thermal imagery using enhanced faster R_CNN based on ResNets and FPN. *Ecological Informatics* 79:102383.

MacCracken, J. G., and V. Van Ballenberghe. 1987. Age- and sex-related differences in fecal pellet dimensions of moose. *Journal of Wildlife Management* 51:360–364.

MacKenzie, D. I. 2005a. What are the issues with presence-absence data for wildlife managers? *Journal of Wildlife Management* 69:849–860.

MacKenzie, D. I. 2005b. Was it there? Dealing with imperfect detection for species presence/absence data. *Australian and New Zealand Journal of Statistics* 47:65–74.

MacKenzie, D. I., J. D. Nichols, M. A. Royle, K. H. Pollock, L. L. Bailey, and J. E. Hines. 2018. *Occupancy Estimation and Modeling: Inferring Patterns and Dynamics of Species Occurrence*. Second edition. Academic Press, London, United Kingdom.

MacKenzie, D. I., J. D. Nichols, M. E. Seamans, and R. J. Gutierrez. 2009. Modeling species occurrence dynamics with multiple states and imperfect detection. *Ecology* 90:823–835.

Magnusson, W. E., G. J. Caughley, and G. C. Grigg. 1978. A double-survey estimate of population size from incomplete counts. *Journal of Wildlife Management* 42:174–176.

Mahon, C. L. 1969. Mineral deficiencies in desert bighorns and domestic livestock in San Juan County. *Desert Bighorn Council Transactions* 13:27–32.

Mason, R., L. H. Carpenter, M. Cox, J. C. deVos, J. Fairchild, D. J. Freddy, J. R. Heffelfinger, R. H. Kahn, S. M. McCorquodale, D. F. Pac, D. Summers, G. C. White, and B. K. Williams. 2006. A case for standardized ungulate surveys and data management in the western United States. *Wildlife Society Bulletin* 34:1238–1242.

Matheny, R. T., T. S. Smith, and D. G. Matheny. 1997. Animal ethology reflected in the rock art of Nine Mile Canyon, Utah. *Journal of California and Great Basin Anthropology* 19:70–103.

McDonald, L. L., H. B. Harvey, F. J. Mauer, and A. W. Brackney. 1990. Design of aerial surveys for Dall (*sic*) sheep in the Arctic National Wildlife Refuge, Alaska. *Northern Wild Sheep and Goat Council Proceedings* 7:176–193.

McKelvey, K. S., and M. K. Schwartz. 2004. Genetic errors associated with population estimation using non-invasive molecular tagging: problems and new solutions. *Journal of Wildlife Management* 68:439–448.

McMahon, M. C., M. A. Ditmer, E. J. Isaac, S. A. Moore, and J. D. Forester. 2021. Evaluating unmanned aerial systems for the detection and monitoring of moose in northeastern Minnesota. *Wildlife Society Bulletin* 45:312–324.

McMahon, M. C., M. A. Ditmer, and J. D. Forester. 2020. Comparing unmanned aerial systems with conventional methodology for surveying a wild white-tailed deer population. *Wildlife Research* 49:54–65.

McNab, J. 1983. Wildlife management as scientific experimentation. *Wildlife Society Bulletin* 11:397–401.

Meek, P., G. Ballard, A. Claridge, R. Kays, K. Moseby, T. O'Brien, A. O'Connell, J. Sanderson, D. Swann, and M. Tobler. 2014. Recommended guiding principles for reporting on camera trapping research. *Biodiversity and Conservation* 23:2321–2343.

Montalbano, F., III, P. W. Glanz, M. W. Olinde, and L. S. Perrin. 1985. A solar-powered time-lapse camera to record wildlife activity. *Wildlife Society Bulletin* 13:178–182.

Murie, O. J. 1954. *A Field Guide to Animal Tracks*. First edition. Houghton Mifflin Company, Boston, Massachusetts, USA.

Neff, D. J. 1968. The pellet-group count technique for big game trend, census, and distribution: a review. *Journal of Wildlife Management* 32:597–614.

Nichols, L. 1970. Aerial inventory and classification of Dall (*sic*) sheep in Alaska. *Northern Wild Sheep Council Transactions* 1:25–33.

Nordeen, T., W. Inselman, and A. Hardin. 2019. *Nebraska Bighorn Sheep Management Plan*. Nebraska Game and Parks Commission, Lincoln, Nebraska, USA.

Norton-Griffiths, M. 1978. *Counting Animals*. Second edition. Serengeti Ecological Monitoring Programme of the African Wildlife Leadership Foundation, Nairobi, Kenya.

Ostermann-Kelm, S., E. R. Atwill, E. S. Rubin, M. C. Jorgensen, and W. M. Boyce. 2008. Interactions between feral horses and desert bighorn sheep at water. *Journal of Mammalogy* 89:459–466.

Packard, F. M. 1946. An ecological study of the bighorn sheep in Rocky Mountain National Park, Colorado. *Journal of Mammalogy* 27:3–28.

Panaccio, M., A. Bambrilla, B. Bassano, T. Smith, and A. von Hardenberg. 2024. A new double observer-based census framework to improve abundance estimations in mountain ungulates and other gregarious species with a reduced effort. *Ecological Solutions and Evidence* 5:e12405. https://doi.org/10.1002/2688-8319.12405

Parker, I. D., R. R. Lopez, and S. L. Locke. 2020. Use of remote cameras in wildlife ecology. Pages 325–333 in N. J. Silvy, editor. *The wildlife Techniques Manual*. Eighth edition. Volume 1. Johns Hopkins University Press, Baltimore, Maryland, USA.

Peralta, D., T. Vaz-Freire, C. Ferreira, T. Mendes, A. Mira, S. Santos, P. C. Alves, X. Lambin, P. Beja, J. Paupério, and R. Pita. 2022. From species detection to population size indexing: the use of sign surveys for monitoring a rare and otherwise elusive small mammal. *European Journal of Wildlife Research* 69:9.

Perry, T. W., T. Newman, and K. M. Thibault. 2010. Evaluation of methods to estimate size of a population of desert bighorn sheep (*Ovis canadensis mexicana*) in New Mexico. *Southwestern Naturalist* 55:517–524.

Pfeffer, D. G., J. A. Foster, and J. C. Kinsey. 2024. Using unmanned aerial vehicles equipped with thermal cameras to survey a known population of white-tailed deer. *Journal of Fish and Wildlife Management* 15:283–288.

Pfeiler, S. S., M. M. Conner, J. S. McKeever, T. R. Stephenson, D. W. German, R. S. Crowhurst, P. R. Prentice, and C. W. Epps. 2020. Costs and precision of fecal DNA mark–recapture versus traditional mark–resight. *Wildlife Society Bulletin* 44:531–542.

Pierce, B. L., R. R. Lopez, and N. J. Silvy. 2020. Estimating animal abundance. Pages 297–324 in N. J. Silvy, editor. *The Wildlife Techniques Manual*. Eighth edition. Volume 1. Johns Hopkins University Press, Baltimore, Maryland, USA.

Pierce, B. M., V. C. Bleich, C. L.-B. Chetkiewicz, and J. D. Wehausen. 1998. Timing of feeding bouts of mountain lions. *Journal of Mammalogy* 79:222–226.

Pollock, K. H., and W. L. Kendall. 1987. Visibility bias in aerial surveys: a review of estimation procedures. *Journal of Wildlife Management* 51:502–510.

Poole, K. G., D. M. Reynolds, G. Mowat, and D. Paetkau. 2011. Estimating mountain goat abundance using DNA from fecal pellets. *Journal of Wildlife Management* 75:1527–1534.

Poole, K. G., R. Serrouya, I. E. Teske, and K. Podrasky. 2016. Rocky Mountain bighorn sheep (*Ovis canadensis canadensis*) winter habitat selection and seasonal movements in an area of active coal mining. *Canadian Journal of Zoology* 94:733–745.

Prentice, P. R., J. Colby, L. E. Greene, C. P. Massing, and T. R. Stephenson. 2019. Status of bighorn sheep in California, 2019. *Desert Bighorn Council Transactions* 55:70–84.

Rabe, M. J., S. S. Rosenstock, and J. C. deVos, Jr. 2002. Review of big game survey methods used by wildlife agencies of the western United States. *Wildlife Society Bulletin* 30:36–52.

Rich, E. E., and A. M. Johnson. 1955. Black's Rocky Mountain journal — 1924: a journal of a voyage from Rocky Mountain Portage in Peace River to the sources of Finlays Branch and North West Ward in summer 1824. Hudson's Bay Record Society, London, United Kingdom.

Rich, L. N., S. R. Beissinger, J. S. Brashares, and B. J. Furnas. 2019. Artificial water catchments influence wildlife distribution in the Mojave Desert. *Journal of Wildlife Management* 83:855–865.

Resources Information Standards Committee. 2002. *Aerial-based inventory methods for selected ungulates: bison, mountain goat, mountain sheep, moose, elk, deer and caribou*. Standards for components of British Columbia's biodiversity 32. British Columbia Ministry of Forests, Lands, and Natural Resource Operations and Ministry of Environment. Victoria, British Columbia, Canada.

Resources Information Standards Committee. 2013. *Aerial-based inventory methods for selected ungulates: mule deer*. Standards for components of British Columbia's biodiversity 32. Addendum. British Columbia Ministry of Forests, Lands, and Natural Resource Operations and Ministry of Environment. Victoria, British Columbia, Canada.

Resources Information Standards Committee. 2019. *Wildlife camera metadata protocol: standards for components of British Columbia's biodiversity 44*. British Columbia Ministry of Environment and Climate Change Strategy and British Columbia Ministry of Forests, Lands, Natural Resource Operations and Rural Development. Victoria, British Columbia, Canada.

Resources Inventory Committee. 1998. *Species inventory fundamentals. Standards for British Columbia's biodiversity 1*. British Columbia Ministry of the Environment, Lands and Parks. Victoria, British Columbia, Canada.

Rovero, F., F. Zimmermann, D. Berzi, and P. Meek. 2013. Which camera trap type and how many do I need? A review of camera features and study designs for a range of wildlife research applications. *Hystrix* 24:148–156.

Rubin, E. S., and V. C. Bleich. 2005. Sexual segregation: a necessary consideration in wildlife conservation. Pages 379–391 in K. E. Ruckstuhl and P. Neuhaus, editors. *Sexual Segregation in Vertebrates: Ecology of the Two Sexes*. Cambridge University Press, Cambridge, United Kingdom.

Rubin, E. S., W. M. Boyce, and V. C. Bleich. 2000. Reproductive strategies of desert bighorn sheep. *Journal of Mammalogy* 81:769–786.

Rubin, E. S., W. M. Boyce, M. C. Jorgensen, S. G. Torres, C. L. Hayes, C. S. O'Brien, and D. A. Jessup. 1998. Distribution and abundance of bighorn sheep in the peninsular ranges, California. *Wildlife Society Bulletin* 26:539–551.

Russo, J. P. 1956. The desert bighorn sheep in Arizona. *Arizona Game and Fish Department Wildlife Bulletin* 1:1–153.

Sasse, D. B. 2003. Job-related mortality of wildlife workers in the United States, 1937–2000. *Wildlife Society Bulletin* 31:1000–1003.

Schmidt, J. H., and K. L. Rattenbury. 2018. An open-population distance sampling framework for assessing population dynamics in group-dwelling species. *Methods in Ecology and Evolution* 9:936–945.

Schoenecker, K. A., M. K. Watry, L. E. Ellison, M. K. Schwartz, and G. Luikart. 2015. Estimating bighorn sheep (*Ovis canadensis*) abundance using noninvasive sampling at a mineral lick within a national park wilderness area. *Western North American Naturalist* 75:181–191.

Schroeder, C. A. 2004. Bighorn sheep in the White Mountains, California: 2004 survey results. Administrative Report, California Department of Fish and Game, Bishop, California, USA.

Seton, E. T. 1927. *Lives of Game Animals.* Volume III. Doubleday, Doran and Company, Garden City, New York. 1953, Reprint. Charles T. Branford Company, Boston, Massachusetts, USA.

Shackleton, D. R., C. C. Shank, and B. M. Wikeem. 1999. Natural history of Rocky Mountain and California bighorn sheep. Pages 78–138 *in* R. Valdez and P. R. Krausman, editors. *Mountain sheep of North America.* University of Arizona Press, Tucson, Arizona, USA.

Shamon, H., R. Maor, M. V. Cove, R. Kays, J. Adley, P. D. Alexander, D. N. Allen, M. L. Allen, C. L. Appel, and E. Barr. 2024. SNAPSHOT USA 2021: A third coordinated national camera trap survey of the United States. *Ecology* 105:e4318.

Silvy, N. J., editor. 2020. *The Wildlife Techniques Manual.* Eighth edition. Volume 1. Johns Hopkins University Press, Baltimore, Maryland, USA.

Simmons, N. M., and C. G. Hansen. 1980. Population survey methods. Pages 260–272 *in* G. Monson and L. Sumner, editors. *The Desert Bighorn: Its Life History, Ecology, and Management.* University of Arizona Press, Tucson, Arizona, USA.

Simmons, N. M., and J. R. Robertson. 1970. Progress and problems—marking and counting Dall sheep in the Mackenzie Mountains, Northwest Territories. *Northern Wild Sheep Council Proceedings* 1:5–19.

Simmons, N. M., M. B. Bayer, and L. O. Sinkey. 1984. Demography of Dall's sheep in the Mackenzie Mountains, Northwest Territories. *Journal of Wildlife Management* 48:156–162.

Skipworth, J. P. 1974. Ingestion of grit by bighorn sheep. *Journal of Wildlife Management* 38:880–883.

Smith, B. L. 1988. Simulated field test of age and sex classification criteria for mountain goats. *Northern Wild Sheep and Goat Council Proceedings* 6:204–209.

Smith, D. R. 1954. *The Bighorn Sheep in Idaho.* Wildlife Bulletin 1. Idaho Department of Fish and Game, Boise, Idaho, USA.

Sproat, K. K., N. R. Martinez, T. S. Smith, W. B. Sloan, J. T. Flinders, J. W. Bates, J. G. Cresto, and V. C. Bleich. 2020. Desert bighorn sheep responses to human activity in southeastern Utah. *Wildlife Research* 47:16–24.

Steerweg, R., M. Hebblewhite, J. Whittington, P. Lukacs, and K. McKelvey. 2017. Sampling scales define occupancy and underlying occupancy-abundance relationships in animals. *Ecology* 99:172–183.

Stepnisky, D., C. Thiessen, and R. Stavne. 2012. Wildlife management unit 445 bighorn sheep. Pages 29–33 *in* M. Ranger and S. Webb, editors. *Delegated Big Game Surveys, 2009–2010 Survey Season.* Data Report D-2011-001, Alberta Conservation Association, Sherwood Park, Alberta, Canada.

Stevens, D. R., and N. J. Goodson. 1993. Assessing effects of removals for transplanting on a high-elevation bighorn sheep population. *Conservation Biology* 7:908–915.

Stewart, A. M., M. M. Conner, J. S. McKeever, A. Ellsworth, R. S. Crowhurst, C. W. Epps, and T. R. Stephenson. 2023. Comparing fecal DNA capture-recapture to mark-resight for estimating abundance of mule deer on winter ranges. *Journal of Wildlife Management* 87:e22350.

Stockwell, C. A., G. C. Bateman, and J. Berger. 1991. Conflicts in national parks: a case study of helicopters and bighorn sheep time budgets at the Grand Canyon. Biological Conservation 56:317–328.

Strickland, D., L. L. McDonald, J. Kern, T. Spraker, and A. Loranger. 1994. Analysis of 1992 Dall (*sic*) sheep and mountain goat survey data, Kenai National Wildlife Refuge. *Northern Wild Sheep and Goat Council Proceedings* 9:35–42.

Sumner, L. 1948. An air census of Dall (*sic*) sheep in Mount McKinley National Park. *Journal of Wildlife Management* 12:302–304.

Tabak, M. A., M. S. Norouzzadeh, D. W. Wolfson, S. J. Sweeney, K. C. Vercauteren, N. P. Snow, J. M. Halseth, P. A. Di Salvo, J. S. Lewis, and M. D. White. 2019. Machine learning to classify animal species in camera trap images: applications in ecology. *Methods in Ecology and Evolution* 10:585–590.

Taberlet, P., J. J. Camarra, S. Griffin, E. Uhres, O. Hanotte, L. P. Waits, C. Dubois-Paganon, T. Burke, and J. Bouvet. 1997. Noninvasive genetic tracking of the endangered Pyrenean brown bear population. *Molecular Ecology* 6:869–876.

Taberlet, P., L. P. Waits, and G. Luikart. 1999. Noninvasive genetic sampling: look before you leap. *Trends in Ecology and Evolution* 14:323–327.

Taberlet, P., S. Griffin, B. Goossens, S. Questiau, V. Manceau, N. Escaravage, L. P. Waits, and J. Bouvet. 1996. Reliable genotyping of samples with very low DNA quantities using PCR. *Nucleic Acids Research* 24:3189–3194.

Taylor, E., D. E. Toweill, and W. A. Van Dyke. 2002. Validation of a helicopter sightability model for bighorn sheep. *Northern Wild Sheep and Goat Council Proceedings* 13:40–48.

Taylor, J. C., S. B. Bates, J. C. Whiting, B. R. McMillan, and R. T. Larsen. 2020. Optimising deployment time of remote cameras to estimate abundance of female bighorn sheep. *Wildlife Research* 48:127–133.

Taylor, J. C., S. B. Bates, J. C. Whiting, B. R. McMillan, and R. T. Larsen. 2021. Using camera traps to estimate ungulate abundance: a comparison of mark-resight methods. *Remote Sensing in Ecology and Conservation* 8:32–44.

Terry, P. J., A. C. Alvidrez, and C. W. Black. 2022. Factors affecting bighorn sheep activity at water developments in southwestern Arizona. *Journal of Wildlife Management* 86:e22134.

Tomecek, J. M., and J. Evans. 2024. Identification of animals from field sign. Pages 222–235 *in* N. J. Silvy, editor. *The Wildlife Techniques Manual.* Eighth edition, Volume 1. Johns Hopkins University Press, Baltimore, Maryland, USA.

Torney, C. J., D. J. Lloyd-Jones, M. Chevallier, D. C. Moyer, H. T. Maliti, M. Mwita, E. M. Kohi, and G. C. Hopcraft. 2019. A comparison of deep learning and citizen science techniques for counting wildlife in aerial survey images. *Methods in Ecology and Evolution* 10:779–787.

Torres, S. G., V. C. Bleich, and J. D. Wehausen. 1994. Status of bighorn sheep in California, 1993. *Desert Bighorn Council Transactions* 38:17–28.

Torres, S. G., V. C. Bleich, and J. D. Wehausen. 1996. Status of bighorn sheep in California, 1995. *Desert Bighorn Council Transactions* 40:27–34.

Turner, J. C. 1973. Water, energy and electrolyte balance in the desert bighorn sheep, *Ovis canadensis.* Dissertation, University of California, Riverside, California, USA.

Udevitz, M. S., B. S. Shults, L. G. Adams, and C. Kleckner. 2006. Evaluation of aerial survey methods for Dall's sheep. *Wildlife Society Bulletin* 34:732–740.

U.S. Fish and Wildlife Service. 2007. *Recovery Plan for the Sierra Nevada Bighorn Sheep.* U.S. Fish and Wildlife Service, Sacramento, California, USA.

Valdez, R. 1982. *The Wild Sheep of the World.* Wild Sheep and Goat International, Mesilla, New Mexico, USA.

Valdez, R. 1988. *Wild Sheep and Wild Sheep Hunters of the New World*. Wild Sheep and Goat International, Mesilla, New Mexico, USA.

Valdez, R., and P. R. Krausman. 1999. Description, distribution, and abundance of mountain sheep in North America. Pages 3–22 *in* R. Valdez and P. R. Krausman, editors. *Mountain Sheep of North America*. University of Arizona Press, Tucson, Arizona, USA.

van de Kerk, M., D. Verbyla, A. W. Nolin, K. J. Sivy, and L. R. Prugh. 2018. Range-wide variation in the effect of spring snow phenology on Dall (*sic*) sheep population dynamics. *Environmental Research Letters* 13:075008.

vanVuuren, M., R. vanVuuren, L. M. Silverberg, J. Manning, K. Pacifici, W. Dorgeloh, and J. Campbell. 2023. Ungulate responses and habituation to unmanned aerial vehicles in Africa's savanna. *PLoS ONE* 18:e0288975.

Vargas-Felipe, M., L. Pellegrin, A. A. Guevara-Carrizales, A. P. Lopez-Monroy, H. J. Escalante, and J. A. Gonzalez-Fraga. 2021. Desert bighorn sheep (*Ovis canadensis*) recognition from camera traps based on learned features. *Ecological Informatics* 64:101328.

Villa, A. G., A. Salazar, and F. Vargas. 2017. Towards automatic wild animal monitoring: identification of animal species in camera trap images using very deep convolutional neural networks. *Ecological Informatics* 41:24–32.

Vu, R. K., R. Ianniello, J. Colby, E. Schaeffer, J. N. Sanchez, J. T. Villepique, L. E. Greene, and T. R. Stephenson. 2021. Status of bighorn sheep in California, 2021. *Desert Bighorn Council Transactions* 56:87–95.

Vuillaume, B., J. H. Richard, and S. D. Côté. 2021. Using camera collars to study survival of migratory caribou calves. *Wildlife Society Bulletin* 45:325–332.

Waits, L. P., and D. Paetkau. 2005. Noninvasive genetic sampling tools for wildlife biologists: a review of applications and recommendations for accurate data collection. *Journal of Wildlife Management* 69:1419–1433.

Walters, C. J. 1986. *Adaptive Management of Renewable Resources*. MacMillan Publishing Company, New York, New York, USA.

Wang, D., Q. Shao, and H. Yue. 2019. Surveying wild animals from satellites, manned aircraft and unmanned aerial systems (UASs): a review. *Remote Sensing* 11:1308.

Wang, L., C. Diao, and Y. Lu. 2024. The role of remote sensing in species distribution models: a review. *International Journal of Remote Sensing* 46:661–685.

Watts, A. C., V. G. Ambrosia, and E. A. Hinkley. 2012. Unmanned aircraft systems in remote sensing and scientific research: classification and considerations of use. *Remote Sensing* 4:1671–1692.

Watts, T. J., and S. D. Schemnitz. 1985. Mineral lick use and movement in a remnant desert bighorn sheep population. *Journal of Wildlife Management* 49:994–996.

Wearn, O. R., and P. Glover-Kapfer. 2019. Snap happy: camera traps are an effective sampling tool when compared with alternative methods. *Royal Society Open Science* 6:181748.

Wehausen, J. D. 1980. Sierra Nevada bighorn sheep: history and population ecology. Dissertation, University of Michigan, Ann Arbor, Michigan, USA.

Wehausen, J. D. 1983. White Mountain bighorn sheep: an analysis of management alternatives. Administrative Report, Inyo National Forest, Bishop, California, USA.

Wehausen, J. D. 1986. Bighorn sheep in the White Mountains: past and recent history. Pages 180–182 *in* C. A. Hall and D. J. Young, editors. *Natural History of the White-Inyo Range, Eastern California and Western Nevada and High-Altitude Physiology*. University of California White Mountain Research Station, Bishop, California, USA.

Wehausen, J. D. 1990. *Cattle impacts on mountain sheep in the Mojave Desert. Contract Report III*. California Department of Fish and Game, Bishop, CA.

Wehausen, J. D. 1992. *Demographic Studies on Mountain Sheep in the Mojave Desert: Report IV*. Interagency Agreement FG 1411. California Department of Fish and Game, Sacramento, California, USA.

Wehausen, J. D. 2005. Nutrient predictability, birthing seasons, and lamb recruitment for desert bighorn sheep. Pages 37–50 *in* J. Goerrissen and J. M. Andre, editors. *Sweeney Granite Mountains Desert Research Center 1978–2003: A Quarter Century of Research and Teaching*. University of California Natural Reserve Program, Riverside, California, USA.

Wehausen, J. D. 2023. A simple genetic method to distinguish mule deer and bighorn sheep fecal pellets and its application to detecting bighorn sheep colonization events in California. *California Fish and Wildlife Journal* 109:e18.

Wehausen, J. D. 2024. Status of bighorn sheep herds that utilize Yosemite National Park 2019–2024. Final report to Yosemite National Park, Interagency Cooperative Agreement P19AC01192. Sierra Nevada Bighorn Sheep Foundation, Bishop, California, USA.

Wehausen, J. D., and V. C. Bleich. 2007. Influence of aerial search time on survey results. *Desert Bighorn Council Transactions* 49:23–29.

Wehausen, J. D., V. C. Bleich, B. Blong, and T. L. Russi. 1987. Recruitment dynamics in a southern California mountain sheep population. *Journal of Wildlife Management* 51:86–98.

Wehausen, J. D., V. C. Bleich, and R. R. Ramey II. 2005. Correct nomenclature for Sierra Nevada bighorn sheep. *California Fish and Game* 91:216–218.

Wehausen, J. D., R. R. Ramey, and C. W. Epps. 2004. Experiments in DNA extraction and PCR amplification from bighorn sheep feces: the importance of DNA extraction method. *Journal of Heredity* 95:503–509.

White, G. C. 2005. Correcting wildlife counts using detection probabilities. *Wildlife Research* 32:211–216.

Whiting, J. C., R. T. Bowyer, and J. T. Flinders. 2009a. Annual use of water sources by reintroduced Rocky Mountain bighorn sheep *Ovis canadensis canadensis*: effects of season and drought. *Acta Theriologica* 54:127–136.

Whiting, J. C., R. T. Bowyer, and J. T. Flinders. 2009b. Diel use of water by reintroduced bighorn sheep. *Western North American Naturalist* 69:407–412.

Whiting, J. C., R. T. Bowyer, J. T. Flinders, V. C. Bleich, and J. G. Kie. 2010. Sexual segregation and use of water by bighorn sheep: implications for conservation. *Animal Conservation* 13:541–548.

Whiting, J. C., V. C. Bleich, R. T. Bowyer, and R. T. Larsen. 2012. Water availability and bighorn sheep: life-history characteristics and persistence of populations. Pages 127–158 *in* J. A. Daniels, editor. *Advances in Environmental Research, Volume* 21. Nova Science Publishers, New York, New York, USA.

Whitten, K. R. 1996. Estimating population size and composition of Dall (*sic*) sheep in Alaska: assessment of previously used methods and experimental implementation of new techniques. Progress Report, Federal Aid in Wildlife Restoration Act Project W-24-3. Alaska Department of Fish and Game, Juneau, Alaska, USA.

Wich, S. A., and L. P. Koh. 2018. *Conservation Drones: Mapping and Monitoring Biodiversity.* Oxford University Press, Oxford.

Wiedmann, B. P., and V. C. Bleich. 2014. Responses of bighorn sheep to recreational activities: a trial of a trail. *Wildlife Society Bulletin* 38:773–782.

Wiedmann, B. P., V. C. Bleich, and C. E. Penner. 2024. North Dakota bighorn sheep management plan (2024–2034). Federal Aid in Wildlife Restoration Project Report A-290. North Dakota Game and Fish Department, Bismarck, North Dakota, USA.

Wilder, B. T., J. L. Betancourt, C. W. Epps, R. S. Crowhurst, J. I. Mead, and E. Ezcurra. 2014. Local extinction and unintentional rewilding of bighorn sheep (*Ovis canadensis*) on a desert island. *PLoS ONE* 9:e91358.

Willi, M., R. T. Pitman, A. W. Cardoso, C. Locke, A. Swanson, A. Boyer, M. Veldthuis, and L. Fortson. 2019. Identifying animal species in camera trap images using deep learning and citizen science. *Methods in Ecology and Evolution* 10:80–91.

Wishart, W. 1978. Bighorn sheep. Pages 161–171 *in* J. L. Schmidt and D. L. Gilbert, editors. *Big Game of North America: Ecology and Management.* Stackpole Books, Harrisburg, PA.

Witczuk, J., S. Pagacz, A. Zmarz, and M. Cypel. 2017. Exploring the feasibility of unmanned aerial vehicles and thermal imaging for ungulate surveys in forests: preliminary results. *International Journal of Remote Sensing* 39:5504–5521.

Woodhead, D. J. 1969. The use of helicopters in wildlife management. *East African Agricultural and Forestry Journal* 34(supplement 1):105–107.

Wyoming Game and Fish Department. 2004. *Handbook of Biological Techniques.* Wyoming Game and Fish Department, Cheyenne, Wyoming, USA.

Xue, Y., T. Wang, and A. K. Skidmore. 2017. Automatic counting of large mammals from very high resolution panchromatic satellite imagery. *Remote Sensing* 9:878.

Yu, X., J. Wang, R. Kays, P. A. Jansen, T. Wang, and T. Huang. 2013. Automated identification of animal species in camera trap images. *Journal on Image and Video Processing* 2013:52.

Yang, F., Q. Shao, and Z. Jiang. 2019. A population census of large herbivores based on UAV and its effects on grazing pressure in the Yellow-River-Source National Park, China. *International Journal of Environmental Research and Public Health* 16:4402.

Young, S. P., and E. A. Goldman. 1946. *The Puma: Mysterious American Cat.* The American Wildlife Institute, Washington, DC, and the Monumental Printing Co., Baltimore, Maryland, USA.

Zabel, F., M. A. Findlay, and P. J. C. White. 2023. Assessment of the accuracy of counting large ungulate species (red deer *Cervus elaphus*) with UAV-mounted thermal infrared cameras during night flights. *Wildlife Biology* 2023:e01071.

# 14 Capture and Translocation of Mountain Sheep

*Daryl W. Lutz and Vernon C. Bleich*

## 14.1 INTRODUCTION

Mountain sheep (bighorn [*Ovis canadensis*], thinhorn [*Ovis dalli*]) populations have declined since the 1900s with a subsequent reduction in their geographical distributions. Declines are attributed primarily to habitat degradation and loss coupled with changing climates, interspecific competition, disease, and unregulated hunting (Buechner 1960, Epps et al. 2004, Pfeifer et al. 2010, Rattenbury et al. 2018, Aycrigg et al. 2021). As a result, management has been directed at elimination or mitigation of anthropogenic factors and biologists are learning more about disturbance dynamics that influence populations and sheep use of habitats. Understanding disturbance dynamics and the responses of mountain sheep populations is important because many populations are small and prone to extirpation or are remote and rarely assessed (Whiting et al. 2023a). Progress has been made, and numerous management efforts have sought to reduce or eliminate anthropogenic influences on these specialized ungulates (e.g., removal of livestock from mountain sheep ranges, minimizing effects of human activities), but challenges remain (Bleich 2009a, Brewer et al. 2014, Jex et al. 2016, Whiting et al. 2023a,b; Bleich and Lutz 2024).

The capture of wildlife has its roots in providing nutrients for human consumption, whether for the immediate use as a source of protein, or for trapping wild animals for domestication (Drew 2020). Native Americans occupying western North America constructed traps for numerous species of artiodactyls, including mountain sheep (Muir 1901, Egan 1917, Anell 1969, Grant 1980, Figures 14.1–14.4). Those winged enclosures were designed to hold and concentrate multiple animals in a small area; mountain sheep that were pushed into the enclosures by hunters likely were dispatched immediately (Grant 1980). Hockett and Dillingham (2023) have described these structures and methods in detail.

During contemporary times in North America, the capture of wildlife was conducted primarily for the conservation or restoration of species with restricted numbers or distributions. Mountain sheep have been at the forefront of such efforts

**FIGURE 14.1** Bighorn sheep are depicted in this ancient pictograph traveling along a fence line or a driveline possibly constructed to direct them into a trap or other area from which they could not escape near Black Point, Lincoln County, Nevada (Nissen 1982). (Figure adapted from Hockett and Dillingham (2023).)

DOI: 10.1201/9781003518686-16

**FIGURE 14.2** Pictograph of a natural bighorn sheep trap that may have consisted of the top of a mesa and a narrow canyon or gully, or wings to direct the sheep onto the mesa, from which there was no way to escape because the sheep could not negotiate the steep sides of the mesa and were easily slaughtered. Alternatively, this pictograph may depict a corral that had been constructed to hold sheep after hunters had pushed them through the wings of the trap near Moab, Grand County, Utah. (Figure adapted from Hockett and Dillingham (2023).)

**FIGURE 14.3** The Tukudika or Sheep Eater Indians constructed traps such as this one to capture and kill mountain sheep. The sheep were herded into the trap using wings constructed of trees and other material to funnel them to the trap where they were killed for food and other uses such as tool making. (Photo by Lynn Stewart; used with permission.)

(Krausman 2000, Krausman and Bleich 2013, Hurley 2021), but thinhorn sheep have been translocated to a far lesser extent (Jex et al. 2016). The methods employed have evolved in terms of utility, efficacy, and safety; moreover, recently there have been improvements in capture and handling techniques that are used by mountain sheep managers and researchers in the conservation of bighorn sheep and thinhorn sheep (Bleich and Lutz 2024). In this chapter, we explore techniques and considerations germane to the establishment, maintenance, or augmentation of mountain sheep populations.

**FIGURE 14.4** Pit blind used by the Tukudika or Sheep Eater Indians to conceal them to either herd mountain sheep in to a trap or to ambush with bow and arrow. (Photo by Lynn Stewart; used with permission.)

## 14.2 CAPTURE METHODS FOR MOUNTAIN SHEEP

The earliest recorded captures of wild sheep for purposes of conservation occurred in California during the first few years of the 20th century when an entrepreneur used specially modified leg hold traps to provide specimens for exhibition at zoological gardens or for agricultural experiments (Bleich et al. 2023). Limited use of that method continued until the mid-1940s (Aldous 1958), but many additional methods have been used to capture mountain sheep. Among these have been leg snares, corral traps, rocket-nets, drop-nets, drive-nets that depended on ground crews or aircraft to move animals into them, chemical immobilization from the ground or aircraft, and the use of net-guns fired from an aircraft (Aldous 1958, Jessup et al. 1982, 2014; Coggins 1999). Other techniques, including the use of a net-gun from the ground, throw-nets, or hand-capture, also have been rarely used (Lacey 1973, Heimer et al. 1980, Scotton 1998, Silvy et al. 2020a).

Capture and handling of wild sheep for any purpose requires extensive planning, is labor-intensive, and costly (Bleich 1990). General recommendations for the handling of wild animals have been developed by professional societies (Gannon et al. 2007, Sikes et al. 2011, 2016). Some wildlife agencies have developed guidelines or recommendations for

the capture, handling, sampling, treatment, or translocation of wildlife in general (Jessup and Clark 1980, 1982; Jessup et al. 1986, Wildlife Investigations Lab 2018), and other agencies eventually developed guidelines specifically for the capture and translocation of wild sheep (Remington and Fuller 1989, George et al. 2008, Wyoming Game and Fish Department 2022). Such information was specific to states and limited to that species. With the urging of wild sheep biologists from throughout North America, Foster (2005) developed a set of guidelines to address that shortcoming; those recommendations were in use for nearly 20 years. The Western Association of Fish and Wildlife Agencies' Wild Sheep Initiative updated and expanded those recommendations in 2024 (Bleich and Lutz 2024). That document consists of separate contributions prepared by experienced practitioners, and addresses planning and logistics (Bleich and Lutz 2024, Torland et al. 2024), health and veterinary care (Justice-Allen et al. 2024), capture methods (Rominger et al. 2024, Bleich et al. 2024b, Nordeen et al. 2024, Blum et al. 2024), marking (Hernández et al. 2024), and transport and translocation (Lonner et al. 2024) of mountain sheep.

Physiological effects, survival rates, and efficacy of several capture methods have been evaluated over an extended period (Kock et al. 1987a–c; Jessup et al. 1988, Wagler et al. 2022). Details on the use of drive-nets to capture wild sheep are available (Foster 2005) but this technique has not been used widely recently. Contemporary information and guidance on the use of chemical immobilization, the drop-net, and net-gun, which account for most wild sheep capture methods in use today, are provided by Bleich and Lutz (2024).

### 14.2.1 CHEMICAL IMMOBILIZATION

Chemical immobilization is an appropriate technique for capturing wild sheep if circumstances allow the investigator to approach within the effective range of the shooter and projector and performance is not anticipated to be affected by the topography or wind (Figure 14.5). The remote delivery of drugs in darts can be an effective tool to selectively capture, mark, collar, or sample wild sheep in several field situations. Care must be taken, however, to avoid accidentally striking non-target animals or immobilizing individuals in unsafe terrain. When done correctly, and with the advancement of new drugs and technology, chemical immobilization is a safe and reliable method of capturing wild sheep (Nordeen et al. 2024).

Wildlife professionals considering chemical immobilization must recognize that given the terrain conditions and unique responses to aerial pursuit, immobilization of mountain sheep in field settings can be difficult; wildlife professionals interested in this technique are referred to Woodbury (1996), Jorgensen et al. (1990), Kreeger (1999), and Nordeen et al. (2024) for information on capture protocols and human safety. The drugs and drug combinations listed in Nordeen et al. (2024) are commonly used or frequently recommended for safe chemical immobilization,

FIGURE 14.5   Whether from the air or the ground, patience is required when using chemical immobilization. Shots taken with a dart at close range are most effective to be certain the drug is placed in a large muscle usually on the hind quarter. Darts fired from excessive range or under inappropriate ambient conditions likely will result in a miss, or injuries or mortalities to the target animal. (Photograph provided by Todd Nordeen; used with permission.)

but that is not an exhaustive list. Drugs used to immobilize wildlife are federally administered, so their availability varies by jurisdiction. Furthermore, their use in mountain sheep capture is considered off-label, and veterinary oversight is required to prescribe, dispense, or report adverse reactions (Justice-Allen et al. 2024, Nordeen et al. 2024).

## 14.2.2   DROP-NET

Drop-netting (Figures 14.6–14.8) is an effective method of capturing mountain sheep and has been in widespread use across their range (Erickson 1970, Spraker 1977, Schmidt et al. 1978). This technique involves luring wild sheep under a net by habituating them to bait, usually consisting of alfalfa, fermented apple pulp, water, or salt (Rutherford and Schmidt 1973, Schmidt et al. 1978, Kock et al. 1987c, Heimer et al. 1980, Fisher and Humphreys 1999, Jessup et al. 2014). Drop-nets have been effective in capturing wild sheep in North America from elevations near sea level to those approaching 4,000 m (Fisher and Humphreys 1999, Jessup et al. 2014, Bleich et al. 2019), and can be used under most weather conditions. Drop-nets are effective in targeting specific social groups, facilitating whole-herd medical treatments, or for translocation. Drop-net captures require detailed planning because they involve large numbers of personnel (Bleich 1990). Two to three handlers are needed for each mountain sheep anticipated to be captured and additional personnel are needed to collect samples, collar or mark animals, and to record data. For example, if the anticipated number of mountain sheep to be captured is 20–30 sheep 80–100 personnel will be required. Like other

methods, a decision to use the drop-net should be considered carefully if animals are in the third trimester of gestation or when neonates are present (Bleich et al. 2024a). Overall, drop-nets have an excellent safety record with a mortality rate of <2% resulting from capture myopathy and <1% due to accidental deaths resulting from strangulation, broken bones, or other terminal injuries (Kock et al. 1987a).

Drop-nets remain an effective method for capturing wild sheep by wildlife managers in most jurisdictions, and many volunteers frequently can participate, thereby reducing personnel costs while providing opportunities for meaningful participation by interested stakeholders (Bleich 1990, Bleich et al. 2024). The cost of $474/animal captured and translocated using the drop-net approximated the cost per animal translocated but captured with a net-gun when agency biologists served as capture personnel (i.e., gunners and muggers) and only the helicopter and pilot were contracted and assuming the agency biologists' salaries were paid regardless of the method. For example, from 1983 to 1989 helicopter contract costs (US $) for net-gun captures averaged $479/animal translocated when agency biologists served as shooters and a helicopter was necessary to transport animals from the capture location. During the same period, the average cost per animal translocated that was captured in a drop-net was $493 when a helicopter was necessary to transport animals from the capture location (Bleich 1990). Those costs are a function of the animals translocated and the distance transported by the aircraft (Bleich 1990), and under identical circumstances but adjusted for inflation, costs per animal captured by drop-net would be $1,250, and by net-gun would be $1,330 (2023), respectively  in today's

**FIGURE 14.6**   Drop-nets require adequate space as they are generally very large depending on the number of mountain sheep to be captured. This drop-net is ready to be deployed on mountain bighorn sheep on Torrey Rim in Wyoming. (Photo by Wyoming Game and Fish; used with permission.)

**FIGURE 14.7**   Mountain sheep captured using a drop-net in Montana. Getting mountain sheep accustomed to the net, and then getting them under it, requires the use of bait such as apple pumice and alfalfa often for several days and perhaps a few weeks. (Photo by Richard (Ric) Horst; used with permission.)

dollars. Currently, translocation costs for animals captured with a net-gun likely are far greater than can be expected than those captured for translocation with a drop-net because many, if not most, agencies now contract capture services for the aircraft and capture crews and rates vary on a contract by contract basis; some agencies, however, continue to train and support in-house expertise (Rominger et al. 2024).

### 14.2.3   NET-GUN

Aerial net-gunning (Figures 14.9 and 14.10) is a method of capturing large mammals in which animals are restrained via a net deployed from a helicopter (Krausman et al. 1985). The net-gun was developed in New Zealand to capture free-ranging ungulates (Barrett et al. 1982, Krausman et al. 1985, Kock et al. 1987c). Wildlife biologists and health professionals must understand the dangers inherent in

**FIGURE 14.8**  Once the drop-net is released, mountain sheep get tangled in the net trying to escape. Having adequate personnel to handle each sheep under the net is essential to minimize injury to the sheep trying to escape the net. (Photo by Richard (Ric) Horst; used with permission.)

**FIGURE 14.9**  Net-gunning wild sheep requires experienced personnel especially the pilot, gunner, and mugger. This work is not for the faint of heart and can be dangerous, but it is an effective technique to capture mountain sheep. Net-gunning during cold weather is often preferred as denser air provides better and safer flying conditions especially at higher elevations as shown here during a capture of Stone's Sheep in British Columbia. (Photo by Bill A. Jex; used with permission.)

helicopter net-gun work (Rominger et al. 2024). Helicopter net-gunning is used widely because it allows for efficient, representative, and capture of targeted animals across a wide variety of rugged and remote landscapes. Although net-gunning from a helicopter often has lower risk of direct mortality (ranging from 0.2% to 2.4%; Rominger et al. 2024) than other methods (Firchow et al. 1986, Kock et al. 1987c, Webb et al. 2008), there remains the potential for lethal and non-lethal injuries to occur (Jessup et al. 1988, Northrup et al. 2014, Van de Kerk et al. 2020). This capture method continues to be the technique of choice considering that hundreds or thousands of mountain sheep have been captured

**FIGURE 14.10** Net-gunning mountain sheep occurs in a variety of vegetation types and climates. Net-gunning this desert bighorn sheep while in what appears less severe terrain can be hampered by warmer temperatures and thinner air. This young male is being pushed through a shallow, non-rocky saddle by Pilot Don Landells and the gunner, Rick Clark, just prior to the shot. As a testament to the dangers involved in this type of work and out of respect for their contributions to conservation, we note that Don and BLM wildlife biologist Jim Bicket perished in a tragic accident during a bighorn sheep survey on 6 October 1986. (Photograph by Mike Kock, Mojave Desert, San Bernardino Co., California, circa 1985, and used with permission.)

using it. But risk is a function of time spent aloft (Bleich et al. 2001) and, "[If] one is not airborne, one cannot fall out of the sky" (Bleich et al. 2005:334). Thus, we emphasize that this work is hazardous and that aircraft accidents are the leading cause of death for biologists, technicians, pilots, gunners, and muggers conducting wildlife survey or capture work (Bleich 1983, Sasse 2003).

## 14.3 TRANSLOCATION OF MOUNTAIN SHEEP

The terms introduction, reintroduction, and augmentation have been adapted from long-standing documents (International Union for Conservation of Nature 1987) and a recent review (Hale and Koprowski 2018) in an effort to standardize terminologies associated with translocation events. These terms are the intended or unintended movement of an organism out of its native range, the intentional movement of an organism into native range from which it has been extirpated, and movement of members of a species to augment the number of individuals remaining in an original habitat, respectively (Hale and Koprowski 2018). All three of these strategies have been used in the conservation of mountain sheep, but managers must consider several aspects of mountain sheep movements and resulting opportunities, or challenges, that these events create.

Translocation is a management tool to reestablish a population in unoccupied, suitable habitats, or to augment existing populations. Translocations of mountain sheep have been done for a variety of reasons since the early 1900s (Bleich et al. 2023) and many have been successful,

but others have not (Krausman 2000). From 1922 to 2006, at least 1,460 separate projects resulted in the translocation of ≥21,000 mountain sheep in the United States and Canada (Wild Sheep Working Group [WSWG] 2015) and translocations have become a more regular occurrence in Mexico because of the ecological and economic advantages associated with such efforts (Janke et al. 2019, Brewer and McEnroe 2020, Bleich et al. 2021).

Whether establishing a new population or augmenting an existing population, there is risk of transmission of pathogens from the source herd to the target population and even to nearby populations of mountain sheep. In Nevada, newly established populations have created a situation whereby mountain sheep are infecting neighboring mountain sheep populations with pathogens that cause respiratory diseases (Manlove 2021). Managers need to consider and assess this risk to ensure extant populations are not compromised by pathogens introduced because of translocations.

Several factors are important considerations when translocating mountain sheep (Wilson et al. 1975, Hansen et al. 1980, Leslie and Douglas 1999, Foster 2005, Lonner et al. 2024). Today, translocations, while still being used, are becoming more complex with the myriad of diseases affecting mountain sheep. Managers considering a translocation must be certain the transmission of pathogens such as *Mycoplasma ovipneumoniae* does not occur because of this restoration strategy. Chapter 8 provides additional information on diseases and pathogens.

Translocations are appropriate when adequate habitats exist that are similar to those the source herd occupied and help to ensure mountain sheep survive in their new environment. Careful and thorough assessments of habitats, disease status, and the age and sex structure of the source herd need to be conducted prior to any translocation effort and vetted through all affected entities including land-management agencies and the public (Lonner et al. 2024). The number of sheep to release at one time is subject to numerous variables including capture and transportation logistics and management objectives in the short and long term. Generally, the establishment of a mountain sheep population is considered successful, and limited hunter harvest (Chapter 16) may be appropriate when a population is on an upward trend, or stable and at least 100 mountain sheep occur in the population.

Managers must recognize that habitat not suitable for mountain sheep, high rates of predation, disease, or a combination of these and other factors should be addressed prior to implementing translocations for the purposes of introduction, reintroduction, or augmentation. In some instances, the implications of public access to translocated populations and associated habitats should be considered when choosing translocation sites (Lonner et al. 2024).

For a translocation to be successful, mountain sheep habitat must be sufficient in quantity and quality to meet the needs of the population on a year-round basis. Prior to any translocation, an on-site evaluation by mountain sheep habitat experts is necessary; geographic information system and spatial analysis techniques may be helpful to evaluate many

attributes associated with habitat suitability (Table 14.1). Coordination among individuals conducting such evaluations will help ensure important considerations are not overlooked and maximize the probability of a successful management action. When adequate data exist, use of models to help evaluate the degree to which proposed release locations provide appropriate habitat for mountain sheep is recommended (Wagler and Monteith 2021). Following a translocation, sophisticated modeling techniques and field observations can be used to estimate and assess whether habitat selection or distribution of the translocated animals occurred as predicted. Chapter 19 addresses the habitat of mountain sheep in detail.

Ultimately, managers and other proponents of translocation projects must consider habitat attributes in the context of other factors, including jurisdictional management goals and objectives, societal expectations, and political interests, among others. The quality of available habitat must be at the forefront of these considerations to maximize wild sheep population performance. Further, monitoring these habitats post-release may provide insights into management needs, potential enhancement options, or future conservation priorities, thus helping refine future modeling and interpretations of habitat associations.

Attributes of mountain sheep habitat including minimum area requirements (Table 14.1) must be considered because guiding parameters will vary among locations

## TABLE 14.1

**Physical Habitat Attributes, Definitions, and Minimum Area Requirements to be Evaluated for Proposed Wild Sheep Translocation Sites; These Parameters and Definitions of Suitable Habitat, Winter Range, Lambing Habitat, Summer Range, and Escape Terrain were Gleaned from the Literature for General Guidance and Do Not Fully Represent All Circumstances**

| Habitat Attribute | Definition | Minimum Area Requirement |
|---|---|---|
| Suitable physical habitat | Land areas with slopes ≥60% (escape terrain) and the contiguous land within 300 m; land within 1,000 m if escape terrain is proximate on two sides, and lacking dense vegetation, human developments, or man-made or natural barriers. | Bighorn sheep; 137 km$^2$; >320 km$^2$ Thinhorn sheep; 84 mi$^2$ (218 km$^2$; ewes); 147 mi$^2$ (381 km$^2$; rams) |
| Winter range | Land areas defined as suitable habitat with southern (SE, S, SW) aspects (bighorns) and west/southwest aspects (thinhorns); and <10 in (25 cm) winter snowpack. | >4% of suitable habitat (bighorn and thinhorn) |
| Lambing habitat | All suitable habitat in ≥2-ha patches with mid- to upper-slopes ≥60%; and within 1,000 m of water with southern, eastern, or western aspects. | Steep, rugged sites (~65% for thinhorns) with adequate solar radiation and level birth sites near escape terrain (<400 m). Approximately 10.3 km$^2$ for bighorns |
| Summer Range | Suitable higher elevation habitat minus slopes >60% with sufficient availability of perennial grasses and forbs; and <16 km; bighorns or 10–14 km; thinhorns from other seasonal ranges. | 32 mi$^2$ (83 km$^2$; bighorns) Thinhorn sheep: 88 km$^2$; females; 97 km$^2$; males |
| Escape terrain | Areas with slopes ≥60% plus the contiguous land within 150 m with slopes of 32%–60%. | 23 km$^2$ (bighorns) and general availability of escape terrain proximal to daily movements |
| Property ownership | Property ownership to be mapped as private, USFS, BLM, National Park Service, state land, or other jurisdictional options. | >75% of all habitats on lands that allow habitat improvements, public access, and other management actions are preferred. |

*Sources:* Rachlow and Bowyer (1991), Smith et al. (1991), Johnson and Swift (2000), Singer et al. (2000), Zeigenfuss et al. (2000), McKinney et al. (2003), Hengeveld and Cubberley (2011), Lowrey et al. (2020), Robinson et al. (2020), Enns et al. (2023). Details of habitat features are covered in Chapter 19.

and the ecological requirements of the mountain sheep to be translocated (Lonner et al. 2024). In addition to these metrics, the availability of migration corridors or routes (Epps et al. 2007) and other strategies necessary to provide access to suitable seasonal habitats must be considered. In addition to attributes of physical habitat, other variables of importance include proximity to domestic sheep or goats, proximity to other populations of mountain sheep, predators and the potential to manage them, forage quality and quantity, seasonal range requirements and opportunities for migration, water availability, anthropogenic disturbances, and availability of mineral licks (Lonner et al. 2024).

### 14.3.1  Source Stock for Translocation

In western North America, most translocations have involved mountain sheep being moved to what was considered by wildlife managers to be suitable habitat, albeit with little consideration of ecological differences between areas occupied by source populations and the intended release areas. Recently, the similarities between habitat occupied by a source population (i.e., source stock) and the location to which mountain sheep are translocated has become recognized as an important consideration (Whiting et al. 2012, Wiedmann and Sargeant 2014, Bleich et al. 2018, 2021), as have the genetic consequences of translocations (Whittaker et al. 2004, Jahner et al. 2019, Barbosa et al. 2021). Translocations must be carefully planned and then evaluated to maximize efficient use and survivorship of relocated animals and continued and careful monitoring of the effects of removals on the source population(s) is essential.

Source stock simply refers to an established population of mountain sheep (i.e., source population) to be used to introduce or reintroduce mountain sheep into vacant habitat, or to augment an existing, but separate, wild herd (i.e., recipient population). Similar to habitat evaluations, a number of variables regarding the source stock must be assessed including disease history and current status, demographic characteristics and structure (i.e., age, sex), behavioral and ecological characteristics of the source herd and release sites, and genetics (Whiting et al. 2012, 2023a; Brewer et al. 2014, Lonner et al. 2024). Moreover, managers also face continued uncertainty about the appropriateness of allowing mixing of mountain sheep from different evolutionary lineages, or continued questions about the most appropriate source of mountain sheep to be used in restoration efforts (Epps et al. 2019).

### 14.3.2  Strategies for Releasing Translocated Animals

Release sites should provide translocated animals with the best opportunity to survive under a variety of anticipated and unanticipated conditions. Managers need to be aware translocated animals often flee after being released; the release site should be in an area free of hazards and near enough to water so they will be able to locate it easily.

Releasing animals into areas that may support high densities of predators or dense cover that will decrease opportunities for released animals to detect or evade predators should be avoided (Moore 1958, Rominger et al. 2004). In environments with a harsh winter climate, it is best to avoid translocations when snowpack at the release site is above average or is deep enough to impede animal movements (Lonner et al. 2024). Translocations implemented during years of below average snowpack will may help ensure animals are able to acquire adequate forage, and it will allow animals to move about without the energetic demands of travel through deep snow (Fancy and White 1987, Sivy et al. 2018).

A hard release involves releasing sheep as soon as possible following transport to limit holding time to a minimum and to reduce stress; this is the most common method currently in use (Figure 14.11). Ideally, mountain sheep should not be held for more than one night. A soft release is one in which sheep are held for a period in some structure or facility (e.g., pen, corral, transport boxes; Figures 14.12 and 14.13) to allow the animals to acclimate, calm down, or imprint on the area to which they have been moved. Survival rates and indices of group cohesion have been evaluated between hard and soft releases but with variable results, in part because of a lack of consistency among holding times and facility design (Berbach 1987, Thompson et al. 2001, Jessup et al. 2014, Bleich et al. 2019, Daily et al. 2022).

Soft releases generally are used infrequently (Lonner et al. 2024). When aerial transport involves numerous flights to the release site, holding animals together in a pen (Bleich et al. 2019) or inside the transport boxes in which they were moved until all have been delivered will allow all animals to be released simultaneously rather than in multiple small groups. Releasing animals as a group may or may not help with group cohesion (Thompson et al. 2001) or decrease the potential for them to split up and move into unsuitable areas. When released upon arrival, members of every previously released group continue to scatter and move further from the point of release with each arrival of the helicopter, which can result in fragmentation of social groups (e.g., each crate load; Thompson et al. 2001, Jessup et al. 2014, Bleich et al. 2019).

Regardless of release type, animals should be released ≥2 hours before total darkness so they have time to familiarize themselves with their immediate surroundings prior to nightfall. Holding animals overnight in a trailer is preferable to releasing them at night, and releasing animals during the day provides managers an opportunity to observe animals as they disperse. If inclement weather conditions at the release are severe enough to cause concern, releasing the sheep near areas where they can find protection from the elements is recommended (Lonner et al. 2024).

### 14.3.3  Post-release Monitoring

A well-planned and organized monitoring strategy must be part of every translocation project. This planning includes

**FIGURE 14.11**   A "hard release" of mountain sheep in the Badlands of western North Dakota. Hard releases are the most common technique used today. Because most translocated sheep are fitted with telemetry collars, biologists are able to monitor survival, distribution, and habitat use and have found this release technique works well. (Photograph courtesy of North Dakota Game and Fish Department.)

**FIGURE 14.12**   To minimize scattering that occurs when bighorn sheep are released immediately upon aerial transport to the intended release site, animals can be held in transport crates until all animals have arrived at the release location, and then released simultaneously, but the number of transport boxes must be adequate to ensure all animals are on-site prior to being released. Special acknowledgment to Brian Wakeling of the Arizona Game and Fish Department for loaning the crates to the California Department of Fish and Game. Eagle Crags, China Lake Naval Weapons Center, San Bernardino Co., California, October 2006. (Photograph by Debra Hamilton; used with permission.)

FIGURE 14.13   A temporary holding pen into which translocated animals are placed after they have been moved to a release site by helicopter transport can be used to hold animals in a single group until all translocated animals have arrived on-site. After several hours, the animals can be released as a group to help prevent continued dispersal from the release site when animals are released immediately upon arrival. San Rafael Peak, Ventura Co., California, December 1985. (Photograph © by B. Moose Peterson; used with permission.)

routine visits to the release site and surrounding area to monitor sheep distribution and habitat use, demographic monitoring, mortality investigations, coordination and communication with public or private landowners, provision of updates to interested stakeholders, and predator management if necessary. Together, this information will allow managers to better understand the initial and long-term success (or failure) of the translocation and use the data to plan for future translocations (Lonner et al. 2024).

Sheep translocated to a new area will wander and explore their new surroundings for escape terrain, resources, routes to evade predators, habitats, and preferred lambing locations. Nonetheless, there should be no tolerance for individuals or groups venturing into or through areas occupied by domestic sheep or goats on public or private land. Plans to respond to or manage such forays must be defined and agreed upon in advance, and a commitment to conduct those actions (i.e., lethal or non-lethal removal) must be in place prior to translocations.

Post-translocation data will help define the efficacy of an initial habitat evaluation or habitat model and offer options for future management in the area, which may include additional translocations. Such data will also help the manager evaluate more specific use of habitat by sheep and potentially highlight previously unknown or unidentified issues that could affect sheep distributions (e.g., a new hiking or motorcycle trail adjacent to potential natal range) and support new management directives to mitigate those negative effects.

### 14.3.4   Predation

Predation by a host of carnivores has long been recognized as an important factor influencing the behavior and, potentially, the population dynamics of mountain sheep (Chapter 9). For example, most early discussions of mountain lion (*Puma concolor*) predation were anecdotal accounts of documented events (Blaisdell 1961, Kelly 1980), and data to address potential population-level effects of such predation were lacking until recently (Rominger 2018, Bleich et al. 2021). Nonetheless, some populations of mountain sheep have been extirpated because of predation and little skepticism remains regarding the role of predation as a factor limiting the success of translocations.

The number of animals involved in a translocation project generally is small (WSWG 2015), and translocated individuals can be especially vulnerable to predation because of the lack of familiarity with their new surroundings (Rominger et al. 2004) or vulnerabilities to stochastic events (Lande 1993). Thus, predator management may be needed prior to implementing translocations of mountain sheep; this is especially the case for sheep originating from source populations that never have been, or no longer are, susceptible to predation risk and, hence, may be especially vulnerable to predation in their new surroundings. Therefore, predator removal must be a consideration (Moore 1958). If predation losses are a concern, preemptive predator removal should be considered to help minimize mortality among the translocated animals (Whiting et al. 2023a). If predation becomes an issue after sheep are released and is a primary cause of mortality, options need to be in place to remove the offending individual(s). Ideally, any such effort is short-term and implemented to help establish the sheep population, but long-term management also could be necessary (Gammons et al. 2021).

Although predator removal has been successful in other jurisdictions, political considerations recently have superseded management needs and hindered efforts (Gammons et al. 2021) to recover the federally endangered Sierra Nevada bighorn sheep (*O. c. sierrae*). When managed appropriately mountain lion populations are resilient and recover rapidly (Dellinger and Torres 2020), but objections to the removal of lions for any reason are frequent and often gain substantial media attention (Vera 2020). Despite such opposition and the political influence of an uninformed public, elected politicians, and political appointees, the importance of predator management to translocation success has been emphasized repeatedly (Lonner et al. 2024).

## 14.4 CASE STUDIES

Herein, we describe five case histories of translocations and present details associated with the outcome of each. We consider factors that played prominent roles in the outcomes of these efforts, in which we provide succinct descriptions of earlier efforts to reestablish mountain sheep in Wyoming, New Mexico, California, and Nevada. These case histories are centered on the importance of selection of appropriate translocation stock; the compounding effect of predation on the effort to establish, reestablish, or augment populations; effects of disease outbreaks on translocation success; transmission of pathogens resulting from translocations; and the methods of releasing mountain sheep.

### 14.4.1 SOURCE STOCK MATTERS

Mountain sheep from the Whiskey Mountain Herd in Wyoming have been used as source stock for translocations throughout the West (WSWG 2015). Some efforts were successful and others were not (Lutz 2024). In Wyoming, most attempts to establish mountain sheep in low-elevation, xeric canyons or mountain ranges have been largely unsuccessful. Those unsuccessful attempts were likely the result of mismatching the source stock to their new environment and habitats. Two examples are described: the Devil's Canyon and Ferris-Seminoe herds. Subsequent translocations with mountain sheep from areas with habitat similar to those used by these two herds were later successful and demonstrate the importance of using source stock that is appropriately adapted to the environment and habitats where they are to be placed.

The Devils Canyon Herd was established in 1973 with the initial translocation of 39 sheep from the Whiskey Mountain Herd. This reintroduction effort was unsuccessful as recruitment and population growth was stagnant for the next 30 years. The later lambing period of migratory sheep from the high elevations of the Wind River Range

(Whiskey Mountain) was incongruous to the plant phenology of the xeric Devil's Canyon ecosystem. The solution to this problem was to match the landscape with non-migratory and early lambing (April or May instead of June) mountain sheep. This was first attempted in December of 2004 with a translocation of 20 sheep from the Deschutes River breaks of Oregon and again in January of 2006 with another 20 sheep from the Missouri River Breaks of Montana. The total population in this herd was estimated at 50–90 mountain sheep following these translocations. By 2009 this herd had grown to at least 130 mountain sheep based on an aerial count. Lamb ratios held between 32 and 59 lambs:100 females following the translocations (2008–2022). This herd continued to grow through 2018 when a pneumonia-related die-off occurred. Up to that time, attempts to curb population growth and keep the Devil's Canyon population near the objective of 175 sheep began in 2014 with captures of mountain sheep to be translocated into the Ferris-Seminoe Herd. Between 2014 and 2018 approximately 120 sheep were removed from Devil's Canyon for population augmentation efforts into the Ferris-Seminoe Herd (WSWG 2015).

Translocation of mountain sheep to the Ferris and Seminoe mountains is the most recent successful attempt to reestablish a mountain sheep population in Wyoming (Lutz 2024). This effort has been ongoing and early attempts included more than a half-dozen translocations ranging from seven to 100 animals and extending from the late 1940s through the early 1980s. Nearly all translocated sheep came from the Whiskey Mountain Herd. Mountain sheep translocated to both the Ferris and the Seminoe segments failed to thrive. Poor lamb survival in both sub-herds largely was because of lambs being born in June, well after the flush of spring vegetation in those low-elevation, xeric mountain ranges. Most females had only low protein forage available during hot summer months that likely left them unable to produce enough milk to foster lambs. Later and successful attempts began in 2009 and continued through 2018 and included 166 mountain sheep from the Diablo Mountains of Oregon and the Devil's Canyon herd in Wyoming. Source populations for these animals were non-migratory, adapted to low-elevation habitats, and typically gave birth in late April or early May. These non-migratory, early lambing mountain sheep have survived and reproduced in the Ferris, Seminoe, and Bennett mountain ranges, and the population is currently estimated to be near 400 animals (Umphlett 2023). Starting in 2024, managers are using harvest of males, females, and lambs to decrease and maintain the Ferris-Seminoe population at the management goal of an average count of 300 sheep in mid-winter (Umphlett 2023).

Mountain sheep adapted to migrations and lambing later in June were not suitable source stock for either of these herds, whereas the non-migratory, early lambing mountain sheep from Oregon and Montana and then Devil's Canyon appear to have been. Consequently, matching the source population of mountain sheep to their new environment was a key consideration as advocated by Wehausen (1991) and Ramey (1993, 1995), is consistent with results reported by others (Wiedmann and Sargeant 2014, Bleich et al. 2018), and has been widely recommended by managers (Brewer et al. 2014, Whiting et al. 2023a, Lonner et al. 2024).

### 14.4.2 Predation Has Consequences

Mountain sheep in New Mexico were translocated to two areas in attempts to reestablish populations of those ungulates on historical range. Thirty-two Rocky Mountain bighorn sheep (*Ovis canadensis canadensis*) and 31 desert bighorn sheep (*O. c.* spp.) an endangered taxon in New Mexico (Rominger et al. 2006) were monitored intensively over ≥7 years in the Wheeler Peak Wilderness Area and Sierra Ladron, respectively (Rominger et al. 2004). Density of mountain lions was greater in Sierra Ladron than in Wheeler Peak Wilderness Area (where no mountain sheep were predated by mountain lions). "Annual adult survival was higher ($P<0.005$) in the Wheeler Peak population (0.955) than in the Sierra Ladron population (0.784)", that "Annual lamb:ewe ratios also were higher ($P<0.001$) in the Wheeler Peak population (66.7 vs. 29.8)", and that "Mean annual exponential growth rate (*r*) in the Wheeler Peak population was 0.25 compared to –0.01 for the Sierra Ladron population" (Rominger et al. 2004:993). These findings are consistent with those of others, who concluded that intensive removal of mountain lions may be required to minimize the risk of extirpation of small populations of mountain sheep (Fisher et al. 1999, Ernest et al. 2002).

Virtually all translocations of mountain sheep involve a small number of animals, and the loss of these founders to any cause can be problematic. The conclusion that, "No short-term alternatives to high mountain lion predation—other than control—may be available" (Rominger et al. 2004:998) is especially applicable when the persistence of translocated populations is a consideration and predation rates are high.

### 14.4.3 Some Challenges Cannot Be Foreseen

The state of California lagged far behind the efforts of other governmental jurisdictions with respect to translocating mountain sheep to reestablish them on historically occupied ranges (Bleich et al. 2021). California did not become involved in the active management of mountain sheep until nearly 50 years following the initiation of translocations in other states or provinces (WSWG 2015). From 1922 to 2006, at least 1,460 separate projects resulted in the translocation of ≥21,000 mountain sheep in the United States and Canada, but only 477 mountain sheep had been translocated in 43 management efforts in California (WSWG 2015); as of December 2021, those numbers totaled 577 and 68, respectively. Translocations have become a more regular occurrence in Mexico because of the ecological and economic advantages associated with such efforts (Janke et al. 2019, Brewer and McEnroe 2020, Bleich et al. 2021, Brewer and Bleich 2023). We present a summary of events leading to the first translocation of mountain sheep

in California (Bleich et al. 2021). This undertaking did not have the advantages of knowledge of the benefits of using ecologically similar translocation stock or the realization that livestock disease could play an important role in the failure of the state's initial venture into the restoration of these specialized ungulates to historically occupied habitat.

Nearly 60 years ago, a report prepared by Leopold et al. (1963) for the United States Department of the Interior mentioned the possibility of restoring mountain sheep to historically occupied habitat at Lava Beds National Monument (LBNM) in northeastern California, but planning for that action was well underway prior to implementation (Blaisdell 1971). The Leopold Report also emphasized the need to restore mountain sheep to the southern Sierra Nevada. At that time, the historical range of mountain sheep from the southern Sierra Nevada to northeastern California was considered by Cowan (1940) to have been occupied by California bighorn sheep (*O. c. californiana*), and that taxon was listed as rare by the California Fish and Game Commission shortly after passage of the California Endangered Species Act (Leach et al. 1974). That listing likely stimulated interest in cooperative efforts (Blaisdell 1972, Keay et al. 1987, Bleich et al. 1991) to reestablish mountain sheep of that subspecies in the Warner Mountains, Modoc County, and the Sierra Nevada of Inyo and Mono counties to areas from which they had been extirpated.

The initial translocation occurred in 1971 when ten mountain sheep caught near Williams Lake, British Columbia, Canada, were trucked to California and released in a 500-ha enclosure at LBNM to serve as breeding stock intended to produce offspring for release. Ultimately, however, attempts to reestablish mountain sheep at LBNM and in the Warner Mountains, Modoc County, were unsuccessful because of a series of unanticipated mortality events.

An early issue involved the poaching of two adult males within the enclosure, and the replacement of one of those with a male from the Sheldon Reserve in Nevada (Blaisdell 1974), but additional deaths attributed to bluetongue virus (Blaisdell 1975) and losses of lambs to contagious ecthyma followed (Blaisdell 1982). Following an increase in that captive population to 42 individuals, which was thought capable of providing stock for translocation, an effort in 1980 to reestablish mountain sheep in the Warner Mountains went awry, and six of ten animals died during the operation (Blaisdell 1982, Jessup et al. 2014). One month later, ten mountain sheep were captured in the southern Sierra Nevada and translocated to the Warner Mountains (Camilleri and Thayer 1982) to augment the four surviving individuals from LBNM that had been released there previously (Sleznick 1980). Despite the presence of a fence, all the mountain sheep remaining at LBNM died later in 1980 from respiratory disease contracted from domestic sheep that were grazed on adjacent National Forest lands (Blaisdell 1982, Foreyt and Jessup 1982, Weaver and Clark 1988, Jessup et al. 2014). Following a considerable increase of mountain sheep in the Warner Mountains, that entire population ultimately succumbed to respiratory disease in

1988 following contact with domestic sheep (Weaver and Clark 1988). No additional translocations have been implemented in northeastern California.

### 14.4.4 RESTORATION CAN FACILITATE PATHOGEN TRANSFER

One of the largest and most aggressive mountain sheep restoration efforts was a huge success, until it failed. Nevada included mountain sheep habitat on more mountain ranges than any other state, and most of them historically were inhabited by mountain sheep. Within the Basin and Range Geographic Province, those rugged areas were proximate to each other with narrow valleys separating long, linear mountains. Indeed, Dutton (1886:116) had described the region as being, "… composed of many short, abrupt ranges or ridges, looking upon the map like an army of caterpillars crawling northward". In the southern portion of the state, opportunities for movement by mountain sheep between Mojave Desert mountain ranges were similar, with no real barriers to intermountain movements by those specialized ungulates.

Historically, between 30,000 and 50,000 mountain sheep are thought to have occupied Nevada (Nevada Division of Wildlife 2001). By the 1960s, however, <2,500 mountain sheep remained, primarily due to the consequences of pathogens contracted from domestic sheep, which numbered >2.5 million at their peak. At that time, however, pathogens contracted from domestic sheep had not been recognized for their role in the catastrophic extirpation of mountain sheep across the northern two-thirds of the state. Additionally, the propensity for mountain sheep to move between or among mountainous areas that form metapopulations (Schwartz et al. 1986, Bleich et al. 1990a, 1996) was not realized. Metapopulations are a series of spatially disjunct populations distributed across suitable patches of habitat but with restricted movement among those patches; metapopulation persistence depends on the combined dynamics of localized extinctions and colonizations of suitable habitat and the rate of colonization must exceed the extinction rate among the local populations comprising the metapopulation (McCullough 1996). In the absence of greater attention to the issue of pathogen transfer and metapopulation structure, a massive collaborative mountain sheep restoration program was launched, but with few rules: do not reintroduce mountain sheep directly overlapping domestic sheep on public land; and try to reestablish populations in areas where they occurred historically.

In 1968, the first experimental mountain sheep reintroductions were conducted (Broadbent 1969). By 2000, over 2,100 mountain sheep had been translocated into 59 separate mountain ranges, creating at least ten metapopulations that were comprised of neighboring herds. Through 2024, the total of translocated mountain sheep in Nevada had risen to 3,400 animals. The peak statewide mountain sheep population in modern times was reached in 2019 at 12,500

(Cox 2021, 2024) and it was distributed among 90 separate mountain sheep herds that formed several metapopulations. Major restoration goals for mountain sheep were achieved statewide because of this huge effort. But underlying this foundational success of mountain sheep restoration were nearby sources of virulent pathogens that were ignition sources and ready to spread rapidly by mountain sheep moving between or among occupied mountain ranges.

From 2014 to 2019, statewide surveillance efforts revealed that nearly one-half of the herds had been exposed to *Mycoplasma ovinpneumoniae* (*M. ovi*), and exposure was equally distributed among remnant and reintroduced herds. Further work revealed that some herds along the California border had been exposed to the same strain for years, if not decades, earlier. How severe herd declines were following initial exposure is unclear, but during the last decade some performed so well that female hunts were instituted to reduce populations. Unfortunately, this resulted in a false sense of security about the effects of respiratory disease facilitated by *M. ovi*. The Fairview strain epizootic, which began in 2007 and continues to the present, is chronicled in the following paragraphs.

The Fairview-Slate-Sand Springs mountain sheep population was re-established in 1996 with a translocation of 31 animals from the Pancake Range (WSWG 2015). By 2006 the herd had exceeded 300 adults and in October that year 21 mountain sheep were captured and relocated to Utah to help restore their populations. Beginning in 2007 the population experienced an all-age die-off during which 25% of adults died in 2007 and 2008 and was followed by 1 year of poor lamb recruitment. Banked samples later tested positive for a novel strain of *M. ovi* (i.e., the Fairview strain). This population proceeded to recover and grew to more than 500 adults by 2019, potentially increasing the probability of intermountain movements simply because there are more animals in large populations than in small populations (Bleich 2009a). During disease surveillance efforts in 2016, one yearling male of 15 mountain sheep sampled tested positive for *M. ovi*, but no action was taken to remove that individual.

In 2018 within the metapopulation created north and south of the Fairview herd, with ≤16 km separating adjacent mountain ranges, the mountain sheep population in the Clan Alpine Mountains experienced a die-off during which 40% of the adults perished and there were 4 years of poor lamb survival. In 2019, mountain sheep occupying the Stillwater Range, located across a narrow valley from the Clan Alpine Mountains, suffered a die-off during which 33% of the adults perished and was followed by 4 years of high lamb mortality. The same year, the Gabbs Valley Range to the south suffered a similar die-off, with a 50% loss of adults and 4 years of depressed lamb survival. During 2020, the population occupying the Tobin Range north of the Stillwater Range experienced the loss of 60% of the adults, but with only 1 year of poor lamb recruitment. South of the Gabbs Valley Range, the Monte Cristo Range experienced a die-off of 60% of the adults and was followed by 4 years of poor lamb recruitment.

This example represents just one metapopulation that succumbed to a cascade of disease outbreaks over 3 years and resulted in the loss of over 1,300 adults and likely a similar number of what would have been recruited lambs. Perhaps more importantly, this event involved mostly reintroduced herds. Are remnant herds somehow afforded more resilience? Possibly, but not in Nevada where remnant desert bighorn metapopulations in recent times have experienced equal, if not more debilitating and persistent pneumonia epizootics in the heart of the Mojave Desert. All of these affected populations are >240 km from the nearest public lands that are grazed by domestic sheep.

The application of metapopulation biology to mountain sheep originated with the work of Schwartz et al. (1986), and it has since become an important consideration in the conservation of mountain sheep (Brewer et al. 2014, Jex et al. 2016). Building mountain sheep metapopulations, as they existed historically, was a goal of Nevada's restoration program. The potential of pathogen transfer because of translocations or intermountain movements previously had been recognized (Dobson and May 1986, Simberloff and Cox 1987) but, unfortunately, knowledge of the dynamics of *M. ovi* infection within metapopulations was lacking at the time the restoration program was initiated. Even if only a fraction of the mountain ranges were repopulated with mountain sheep with ≥80 km between the closest herds, pathogen transmission may still have occurred, albeit taking more years to spread in concert with a domestic sheep source of pathogens.

### 14.4.5 THE QUANDARY OF RELEASE METHODS

As discussed earlier, mountain sheep generally have been released in one of two ways following translocation: a hard release directly from a transport vehicle, or a soft release that involves holding the sheep in a specially constructed facility for a time following their arrival at the release site. Biologists initially (Wilson et al. 1973, 1975) advocated the use of soft releases when translocating mountain sheep but that recommendation subsequently was revised (Wilson et al. 1982). As of 1980, soft releases had been used in Texas, California, North Dakota, Arizona, Idaho, Nevada, and New Mexico (Hansen et al. 1980, Wilson et al. 1982) with those except Nevada and Arizona intending to maintain broodstock within the enclosures and produce offspring for translocation to other areas; additional soft releases likely have occurred in other jurisdictions. Results of some of these efforts were closely monitored (Broadbent 1969, Elenowitz 1982, Berbach 1987, Morgart et al. 1992, Thompson et al. 2001, Bleich et al. 2019, Daily et al. 2022); however, on only three occasions have investigators compared the results of soft and hard releases to evaluate the efficacy of those methods (Morgart et al. 1992, Thompson et al. 2001, Daily et al. 2022).

In California, three soft releases were planned to restore mountain sheep to the Whipple Mountains; the first of these occurred in 1983, which were to be followed by additional soft releases in 1984 and 1985 (Weaver 1984). The use of multiple large holding pens was planned to encourage site fidelity among animals released into different parts of the mountain range with the intent of acclimating the founding populations to their new surroundings and to create separate female demes. Although the translocations were successful, the potential benefits ascribed to soft releases likely were not achieved because of unanticipated issues (Berbach 1987). Nutritional problems associated with overuse of forage within the enclosure(s) affected survival of translocated animals (Berbach 1987, Jessup et al. 1989) and resulted in the decision to use a hard release in 1985. Despite these setbacks, mountain sheep in the Whipple Mountains were estimated to number between 51 and 100 in 1993 (Torres et al. 1994), and the population is thought to consist of 25–50 adults over the long term (Epps et al. 2003, Prentice et al. 2019, Vu et al. 2021).

Three additional soft releases have occurred in California and were intended to reduce the scattering of mountain sheep after they were flown by helicopter to remote locations and released immediately upon arrival of each load (Weaver 1984, Jessup et al. 2014). During translocations to the Eagle Crags and within the San Gabriel Mountains, during which mountain sheep were released immediately as they arrived via helicopter, the sheep, which are sensitive to the sound generated by helicopters (Bleich et al. 1990b, 1994), dispersed further and further from the release site. In response, use of a portable holding pen was initiated at two release sites accessible only by helicopter so that all animals could be released simultaneously, and animals were retained in transport boxes for several hours following transport before being released simultaneously in a second translocation to the Eagle Crags (Figure 14.12). Research has revealed no difference in the grouping behavior of mountain sheep released from the enclosure (soft release) and those released simultaneously from a vehicle (hard release), but survival among animals released from temporary pens was lower than of those released from the vehicle, possibly the result of the extended stressor(s) associated with repeated arrivals of the helicopter (Thompson et al. 2001, Jessup et al. 2014).

Morgart et al. (1992), Thompson et al. (2001), and Daily et al. (2022) reported that soft releases did not improve survival of translocated animals. California's experiences in the Whipple Mountains clearly were negative, and there was no benefit to survival in the releases that were studied in California, Arizona, or Texas. Moreover, there was but a short-term benefit in terms of distances moved following release in Arizona (Morgart et al. 1992) or group cohesion in either California or in Texas (Thompson et al. 2001, Daily et al. 2022). As a result, investigators concluded there was no benefit to a soft release when compared with a hard release.

There remains a considerable discrepancy regarding the efficacy of conditioning through the use of soft releases on the behavior or survival of translocated animals (Tetzlaff et al. 2020). In part this conundrum exists because translocations generally have not been conducted as experiments (Seddon et al. 2007, Bannister et al. 2020) and that shortcoming has prevented many assumptions about the benefits or detriments of release techniques from being challenged (Berbach 1987, Tetzlaff et al. 2019). Despite the support of some (deVos et al. 1981, deVos 1982, Dodd 1983, Cunningham et al. 1989), the question of the potential benefits or detriments of both methods remains open to further investigation (Tetzlaff et al. 2020) and the efficacy of soft releases remains unconfirmed. As a consequence, the type of release employed may best be determined on a case-by-case basis.

## 14.5  SUMMARY

The capture of mountain sheep has happened since before modern settlement in North America. Native Americans used mountain sheep as a source of protein and tools, such as horn bows. Today, capture of mountain sheep is conducted as a matter of science and management. The capture and translocation of mountain sheep have been used as a management tool to better understand their ecology and to restore them throughout their historic range.

Several techniques have been developed to capture mountain sheep. Today, chemical immobilization, drop-netting, and helicopter net-gunning are the most common and effective methods. The technique chosen depends on capture goals and purpose, the environment captures occur in, budget, and personnel experience and preference. Each technique has advantages and disadvantages, and consideration must be given to personnel safety and minimization of injury or mortality to the mountain sheep captured (Bleich and Lutz 2024).

The translocation of mountain sheep is an important management tool used to restore mountain sheep to historically occupied habitats or to augment existing populations. "Use of translocations for restoring wild sheep began in 1922 with the capture of 20 bighorns in Alberta, Canada and subsequent release of 12 animals in Montana and 8 animals at Custer State Park, South Dakota. Since then, at least 1,460 separate projects have resulted in the translocation of more than 21,500 bighorn sheep in the United States and Canada combined. Nevada leads all jurisdictions with more than 4,000 wild sheep translocated through 174 projects. More than 1,300 wild sheep have been translocated across international boundaries from Canada and Mexico into 14 United States jurisdictions. The roots of today's bighorn populations in Nebraska, North Dakota, South Dakota and Texas can be traced entirely to translocations" (Wild Sheep Working Group 2015:3).

It has been estimated only 41% of mountain sheep translocations are considered successful (Singer et al. 2000). To maximize success, several factors need to be assessed when considering conducting a translocation including: habitat and the environment of the source herd and that of the location they are to be introduced, impact of predation, and the risk of foraying mountain sheep and disease transmission.

Biologists have learned a lot after conducting over 1,400 translocation projects. Perhaps the greatest lessons learned are: matching the source population of mountain sheep to habitats they are familiar with and most adapted to is essential; predation of naïve mountain sheep in a new environment can be substantial and must be addressed; translocations of mountain sheep, in the face of ever-increasing prevalence of pathogens that cause disease in mountain sheep, must be considered when and where translocations occur. Biologists must accept that more mountain sheep may not always be better, especially if mountain sheep are transmitting pathogens from one population to another; and there is much yet to learn with regard to capture and translocation of mountain sheep. For example, are soft or hard releases more effective? Does the use of predictive models to help evaluate the degree to which proposed release locations provide appropriate habitat for wild sheep helpful and increase project success? Is the use of landscape analyses, such as "circuitscape analysis", helpful in understanding how mountain sheep, particularly foraying sheep, change the risk of disease transmission?

Aspects of the capture and translocation of mountain sheep have been discussed by other authors in earlier publications, and many of those have appeared in the various editions of The Wildlife Society Techniques Manual, which has been published and updated over many decades. The eighth (Silvy 2020) includes chapters specific to the capture and handling of wild animals (Silvy et al. 2020a), chemical immobilization (Drew 2020), wildlife health and disease surveillance (Peterson and Ferro 2020), marking techniques (Silvy et al. 2020b), and radiotelemetry and remote monitoring (Silvy and Catanach 2020). Other material contained in Silvy (2020) and not addressed specifically in the most recent edition of the capture and handling guidelines (Bleich and Lutz 2024), but that remain germane to the capture, handling, or translocation of wildlife, include ethics (Peterson et al. 2020), communications and outreach (Jacobson et al. 2020), and adaptive management (Organ et al. 2020). At some point, one or more of these topics is likely to require additional attention but may or may not directly affect the outcome of capture, handling, or translocation operations (Bleich et al. 2024).

Additional summaries and reviews also are available. Some of those summaries are dated, but represent historical accounts of the use of immobilizing agents (Woodbury 1996, Kreeger 1999); the evolution and application of a variety of capture methods (Clover 1954, Beasom et al. 1980, Autenreith et al. 1981, Barrett et al. 1982, Krausman et al. 1985, Jessup et al. 2014) or release techniques (Thompson et al. 2001, Daily et al. 2022); or impacts of helicopters on wildlife during capture work (Autenreith et al. 1981, Barrett et al. 1982, Jessup 1982, Bleich 1983, Bleich et al. 1990b, 1994; Jessup et al. 2014); effects, use of, and safety of marks, collars, and telemetry (Bleich et al. 1990c, White and Garrott 1990, Millspaugh and Marzluff 2001, Krausman et al. 2004); and other aspects associated with the conservation of wild sheep. We encourage readers to further explore that history while keeping pace with ongoing advances and developments, and to apply information as appropriate and in the best interest of conservation, because mountain sheep deserve no less.

## REFERENCES

Aldous, C. M. 1958. Trapping and tagging of bighorn sheep. *Desert Bighorn Council Transactions* 2:36–39.

Anell, B. 1969. Running down and driving of game in North America. *Studia Ethnographica Upsaliensia* 30:1–129.

Autenreith, R. E., G. L. Copeland, and T. D. Reynolds. 1981. Capturing pronghorn using a helicopter and etorphine hydrochloride. *Wildlife Society Bulletin* 9:314–319.

Aycrigg, J. L., E. O. Garton, B. Magipane, G. E. Liston, L. R. Prugh, and J. L. Rachlow. 2021. Habitat selection by Dall's sheep is influenced by multiple factors including direct and indirect climate effects. *PLoS ONE* 16(3):e0248763.

Bannister, H. L., R. Brandle, S. Delean, D. C. Paton, and K. E. Moseby. 2020. Supportive release techniques provide no reintroduction benefit when efficacy and uptake is low. *Oryx* 54:206–214.

Barbosa, S., K. R. Andrews, R. B. Harris, D. S. Gour, J. R. Adams, E. F. Cassirer, H. M. Miyasaki, H. M. Schwantje, and L. P. Waits. 2021. Genetic diversity and divergence among bighorn sheep from reintroduced herds in Washington and Idaho. *Journal of Wildlife Management* 85:1214–1231.

Barrett, M. W., J. W. Nolan, and L. D. Roy. 1982. Evaluation of a hand-held net-gun to capture large mammals. *Wildlife Society Bulletin* 10:108–114.

Beasom, S. L., W. Evans, and L. Temple. 1980. The drive net for capturing western big game. *Journal of Wildlife Management* 44:478–480.

Berbach, M. W. 1987. The behavior, nutrition, and ecology of reintroduced desert mountain sheep in the Whipple Mountains, San Bernardino County, CA. Thesis, California State Polytechnic University, Pomona, California, USA.

Blaisdell, J. A. 1961. Bighorn-cougar relationships. *Desert Bighorn Council Transactions* 5:42–46.

Blaisdell, J. A. 1971. Progress report on selected National Park Service bighorn projects. *Desert Bighorn Council Transactions* 15:90–93.

Blaisdell, J. A. 1972. Progress report: Lava Beds bighorn reestablishment. *Desert Bighorn Council Transactions* 16:84–87.

Blaisdell, J. A. 1974. Lava Beds California bighorn: was 1973 typical? *Desert Bighorn Council Transactions* 18:46–47.

Blaisdell, J. A. 1975. Progress report: the Lava Beds reestablishment program. *Desert Bighorn Council Transactions* 19:36–37.

Blaisdell, J. A. 1982. Lava Beds wrap up: what did we learn? *Desert Bighorn Council Transactions* 26:32–33.

Bleich, V. C. 1983. Comments on helicopter use by wildlife agencies. *Wildlife Society Bulletin* 11:304–306.

Bleich, V. C. 1990. Costs of translocating mountain sheep. Pages 67–75 *in* P. R. Krausman and N. S. Smith, editors. *Managing Wildlife in the Southwest*. Arizona Chapter of The Wildlife Society, Phoenix, Arizona, USA.

Bleich, V. C. 2009a. Perceived threats to mountain sheep: levels of concordance among states, provinces, and territories. *Desert Bighorn Council Transactions* 50:32–39.

Bleich, V. C. 2009b. Factors to consider when reprovisioning water developments used by mountain sheep. *California Fish and Game* 95:153–159.

Bleich, V. C., C. D. Hargis, J. A. Keay, and J. D. Wehausen. 1991. Interagency coordination and the restoration of wildlife populations. Pages 277–284 *in* J. Edelbrock and S. Carpenter, editors. *Natural Areas and Yosemite: Prospects for the Future*. U.S. National Park Service, Denver Service Center, Denver, Colorado, USA.

Bleich, V. C., C. S. Y. Chun, R. W. Anthes, T. E. Evans, and J. K. Fischer. 2001. Visibility bias and development of a sightability model for tule elk. *Alces* 37:315–327.

Bleich, V. C., and D. W. Lutz (editors). 2024. *Wild Sheep Capture and Handling Guidelines*. Second edition. Western Association of Fish and Wildlife Agencies, Boise, Idaho, USA.

Bleich, V. C., G. A. Sargeant, and B. P. Wiedmann. 2018. Ecotypic variation in population dynamics of reintroduced bighorn sheep: implications for management. *Journal of Wildlife Management* 82:8–18.

Bleich, V. C., G. W. Sudmeier, M. W. Wolter, and R. A. Weaver. 2023. The first translocation of bighorn sheep in California: a critical review of the activities of William Franklin Frakes. *Human-Wildlife Interactions* 17:406–433.

Bleich, V. C., J. D. Wehausen, and S. A. Holl. 1990a. Desert-dwelling mountain sheep: conservation implications of a naturally fragmented distribution. *Conservation Biology* 4:383–390.

Bleich, V. C., J. D. Wehausen, J. A. Keay, J. G. Stahmann, and M. W. Berbach. 1990c. Radiotelemetry collars and mountain sheep: a cautionary note. *Desert Bighorn Council Transactions* 34:6–8.

Bleich, V. C., J. D. Wehausen, R. R. Ramey II, and J. L. Rechel. 1996. Metapopulation theory and mountain sheep: implications for conservation. Pages 353–373 *in* D. R. McCullough, editor. *Metapopulations and Wildlife Conservation*. Island Press, Covelo, California, USA.

Bleich, V. C., J. D. Wehausen, S. G. Torres, K. Anderson, and T. R. Stephenson. 2021. Fifty years of bighorn sheep translocations: details from California (1971–2020). *Desert Bighorn Council Transactions* 56:1–32.

Bleich, V. C., J. T. Villepique, T. R. Stephenson, B. M. Pierce, and G. M. Kutliyev. 2005. Efficacy of aerial telemetry as an aid to capture specific individuals: a comparison of two techniques. *Wildlife Society Bulletin* 33:332–336.

Bleich, V. C., M. E. Blum, K. T. Shoemaker, D. Sustaita, and S. A. Holl. 2019. Habitat selection by bighorn sheep in a mesic ecosystem: the San Rafael Mountains, California, USA. *California Fish and Game* 105:205–224.

Bleich, V. C., M. Fischer, R. A. Garrott, and R. Langley. 2024b. Capturing wild sheep with a drop-net. Pages 6.1–6.13 *in* V. C. Bleich and D. W. Lutz, editors. *Wild Sheep Capture and Handling Guidelines*. Second edition. Western Association of Fish and Wildlife Agencies, Boise, Idaho, USA.

Bleich, V. C., R. T. Bowyer, A. M. Pauli, M. C. Nicholson, and R. W. Anthes. 1994. Responses of mountain sheep *Ovis canadensis* to helicopter surveys: ramifications for the conservation of large mammals. *Biological Conservation* 45:1–7.

Bleich, V. C., R. T. Bowyer, A. M. Pauli, R. L. Vernoy, and R. W. Anthes. 1990b. Responses of mountain sheep to aerial sampling using helicopters. *California Fish and Game* 76:197–204.

Blum, M. E., C. P. Lehman, J. T. Hogg, T. D. Nordeen, M. Cox, H. M. Miyasaki, E. F. Cassirer, and K. L. Monteith. 2024. Capture and handling of neonates. Pages 8.1–8.20 *in* V. C. Bleich and D. W. Lutz, editors. *Wild Sheep Capture and Handling Guidelines.* Second edition. Western Association of Fish and Wildlife Agencies, Boise, Idaho, USA.

Brewer, C. E., and V. C. Bleich. 2023. Trophy spotlight: desert bighorn sheep. Pages 18–33 *in* K. M. Lehr and J. Schwab, editors. *Records of North American Big Game.* Fifteenth edition. Volume II. Boone and Crockett Club, Missoula, Montana, USA.

Brewer, C. E., V. C. Bleich, J. A. Foster, T. Hosch-Hebdon, D. E. McWhirter, E. M. Rominger, M. W. Wagner, and B. P. Wiedmann. 2014. *Bighorn Sheep: Conservation Challenges and Management Strategies for the 21st Century.* Western Association of Fish and Wildlife Agencies, Cheyenne, Wyoming, USA.

Brewer, C., and A. McEnroe. 2020. Justice league: WSF partner Mexico Council's stunning achievements. *Wild Sheep* Fall, Volume 8–Issue 3:46–56.

Broadbent, R. V. 1969. Nevada's 1968 transplant disappointment. *Desert Bighorn Council Transactions* 13:43–47.

Buechner, H. K. 1960. The bighorn sheep in the United States: its past, present and future. *Wildlife Monographs* 4:1–174.

Camilleri, E. P., and D. Thayer. 1982. Status of California bighorn in the South Warner Wilderness of California. *Desert Bighorn Council Transactions* 26:116–118.

Clover, M. R. 1954. A portable deer trap and catch-net. *California Fish and Game* 40:367–373.

Coggins, V. L. 1999. Oregon's corral type bighorn trap. *North American Wild Sheep Conference Proceedings* 2:249–251.

Cowan, I. McT. 1940. Distribution and variation in the native sheep of North America. *American Midland Naturalist* 24:505–580.

Cox, M. 2021. Status of bighorn sheep in Nevada, 2019–2020. *Desert Bighorn Council Transactions* 56:98–104.

Cox, M. 2024. In-depth statistical sheep information: Nevada. *Wild Sheep* 12(2):18–21.

Cunningham, S., N. Dodd, and R. Olding. 1989. Arizona's bighorn sheep reintroduction program. Pages 203–239 *in* R. M. Lee, editor. *The Desert Bighorn Sheep in Arizona.* Arizona Game and Fish Department, Phoenix, Arizona, USA.

Daily, T. S., C. E. Gonzalez, L. A. Harveson, W. C. Conway, and F. Hernandez. 2022. Comparing survival and cause-specific mortality of different translocation release methods for desert bighorn sheep. *Western North American Naturalist* 82:94–106.

Dellinger, J. A., and S. G. Torres. 2020. A retrospective look at mountain lion populations in California (1906–2018). *California Fish and Wildlife* 106:66–85.

de Vos, J. C. 1982. Preliminary report on four free releases of desert bighorn in Arizona. *Desert Bighorn Council Transactions* 26:111–112.

de Vos, J. C., W. Ough, D. Taylor, R. Miller, S. Wilcox, and R. Remington. 1981. Evaluation of a desert bighorn release. *Desert Bighorn Council Transactions* 25:29–30.

Daily, T. S., C. E. Gonzalez, L. A. Harveson, W. C. Conway, and F. Hernandez. 2022. Comparing survival and cause-specific mortality of different translocation release methods for desert bighorn sheep. *Western North American Naturalist* 82:94–106.

Dobson, A. P., and R. M. May. 1986. Disease and conservation. Pages 345–365 in M. E. Soule, editor. *Conservation Biology.* Sinauer Associates, Sunderland, Massachusetts, USA.

Dodd, N. L. 1983. Ideas and recommendations for maximizing desert bighorn sheep transplant efforts. *Desert Bighorn Council Transactions* 27:12–16.

Douglas, C. L., and D. M. Leslie. 1999. Management of bighorn sheep. Pages 239–269 *in* R. Valdez and P. R. Krausman, editors. *Mountain Sheep of North America.* University of Arizona Press, Tucson, Arizona, USA.

Drew, M. L. 2020. Chemical immobilization of wildlife. Pages 107–125 *in* N. J. Silvy, editor. *The Wildlife Techniques Manual.* Eighth edition. Volume I. Johns Hopkins University Press, Baltimore, Maryland, USA.

Dutton, C. E. 1886. Mount Taylor and Zuni Plateau. Pages 113–204 *in Report of the Secretary of the Interior Being Part of the Message and Documents Communicated to the Two Houses of Congress at the Beginning of the First Session of the Forty-ninth Congress.* Volume 3. Report of the Director of the United States Geological Survey. U.S. Government Printing Office, Washington, D.C., USA.

Egan, H. R. 1917. *Pioneering the West, 1846–1878.* Howard R. Egan Estate, Richmond, Utah, USA.

Elenowitz, A. 1982. Preliminary results of a desert bighorn transplant in the Peloncillo Mountains, New Mexico. *Desert Bighorn Council Transactions* 26:8–11.

Enns, G. E., B. A. Jex, and M. S. Boyce. 2023. Diverse migration patterns and seasonal habitat use of Stone's sheep (*Ovis dalli stonei*). *Peer Journal* 11:e15215.

Epps, C. W., D. R. McCullough, J. D. Wehausen, V. C. Bleich, and J. L. Rechel. 2004. Effects of climate change on population persistence of desert-dwelling mountain sheep in California. *Conservation Biology* 18:102–113.

Epps, C. W., E. Bowen, M. R. Buchalski, F. Cassirer, D. Coltman, W. C. Conway, M. Cox, R. Harris, J. P. Jahner, M. Matocq, T. L. Parchman, C. D. Phillips, H. Schwantje, Z. Sim, L. Waits, J. D. Wehausen, and the Wild Sheep Genomics Working Group. 2019. Frequently-asked questions about wild sheep genetics and genomics. *Northern Wild Sheep and Goat Council Proceedings* 21:45–75.

Epps, C. W., J. D. Wehausen, V. C. Bleich, S. G. Torres, and J. S. Brashares. 2007. Optimizing dispersal and corridor models using landscape genetics. *Journal of Applied Ecology* 44:714–724.

Epps, C. W., V. C. Bleich, J. D. Wehausen, and S. G. Torres. 2003. Status of bighorn sheep in California. *Desert Bighorn Council Transactions* 47:20–35.

Erickson, J. A. 1970. Use of drop net and collars in study of Dall sheep. *Northern Wild Sheep Council Proceedings* 1:20–21.

Ernest, H. B., E. S. Rubin, and W. M. Boyce. 2002. Tracking mountain lion predation of desert bighorn sheep with fecal DNA. *Journal of Wildlife Management* 66:75–85.

Fancy, S. G., and White, R. G. 1987. Energy expenditures for locomotion by barren-ground caribou. *Canadian Journal of Zoology* 65:122–128.

Firchow, K. M., M. R. Vaughan, and W. R. Mytton. 1986. Evaluation of the hand-held net gun for capturing pronghorns. *Journal of Wildlife Management* 50:320–322.

Fisher, A., and D. Humphreys. 1999. Methods for improving bighorn capture success. *North American Wild Sheep Conference Proceedings* 2:243–247.

Fisher, A., E. R. Rominger, P. Miller, and O. Byers. 1999. Population and habitat viability assessment workshop for the desert bighorn sheep of New Mexico (*Ovis canadensis*). Final report. IUCN/SSC Conservation Breeding Specialist Group, Apple Valley, Minnesota, USA.

Foreyt, W. J., and D. A. Jessup. 1982. Fatal pneumonia of bighorn sheep following association with domestic sheep. *Journal of Wildlife Diseases* 18:163–168.

Foster, C. L. 2005. Wild sheep capture guidelines. *Northern Wild Sheep and Goat Council Proceedings* 14:211–282.

Gammons, D. J., J. L. Davis, D. W. German, K. Denryter, J. D. Wehausen, and T. R. Stephenson. 2021. Predation impedes recovery of Sierra Nevada bighorn sheep. *California Fish and Wildlife* Special CESA Issue 107:444–470.

Gannon, W. L., and the Animal Care and Use Committee of the American Society of Mammalogists. 2007. Guidelines of the American Society of Mammalogists for the use of wild mammals in research. *Journal of Mammalogy* 88:809–823.

George, J., L. Wolfe, and M. Miller. 2008. *Bighorn Sheep Capture and Translocation Guidelines*. Colorado Division of Wildlife, Denver, Colorado, USA.

Grant, C. 1980. The desert bighorn and aboriginal man. Pages 7–39 *in* G. Monson and L. Sumner, editors. *The Desert Bighorn: Its Life History, Ecology, and Management*. University of Arizona Press, Tucson, Arizona, USA.

Hale, S. L., and J. L. Koprowski. 2018. Ecosystem-level effects of a keystone species reintroduction: a literature review. *Restoration Ecology* 26:439–445.

Hansen, C. G., T. L. Hailey, and G. I Day. 1980. Capturing, handling, and transplanting. Pages 273–287 *in* G. Monson and L. Sumner, editors. *The Desert Bighorn: Its Life History, Ecology, and Management*. University of Arizona Press, Tucson, Arizona, USA.

Heimer, W. E., S. D. DuBois, and D. G. Kelleyhouse. 1980. A comparison of rocket netting with other methods of capturing Dall sheep. *Northern Wild Sheep and Goat Council Proceedings* 2:601–614.

Hengeveld, P. E., and J. C. Cubberley, editors. 2011. *Stone's Sheep Population Dynamics and Habitat Use in the Sulphur/8 Mile Oil and Gas Pre-tenure Plan Area, Northern British Columbia, 2005–2010*. Synergy Applied Ecology, Mackenzie, British Columbia, Canada.

Hernández, F., M. Cox, and B. Jex. 2024. Collaring and marking wild sheep. Pages 4.1–4.17 *in* V. C. Bleich and D. W. Lutz, editors. *Wild Sheep Capture and Handling Guidelines*. Second edition. Western Association of Fish and Wildlife Agencies, Boise, Idaho, USA.

Hockett, B., and E. Dillingham. 2023. *Large-Scale Traps of the Great Basin*. Texas A&M University Press, College Station, Texas, USA.

Hurley, K. 2021. It's all about the numbers! *Wild Sheep* 9(3):20.

International Union for the Conservation of Nature. 1987. *Position Statement on Translocations of Living Organisms: Introductions, Reintroductions, and Re-stocking*. International Union for the Conservation of Nature and Natural Resources, Gland, Switzerland.

Jacobson, S. K., H. O. Brown, and B. S. Lowe. 2020. Communications and outreach. Pages 59–81 *in* N. J. Silvy, editor. *The Wildlife Techniques Manual*. Eighth edition. Volume II. Johns Hopkins University Press, Baltimore, Maryland, USA.

Jahner, J. P., M. D. Matocq, J. L. Malaney, M. Cox, P. Wolff, M. A. Gritts, and T. L. Parchman. 2019. The genetic legacy of 50 years of desert bighorn sheep translocations. *Evolutionary Applications* 12:198–213.

Janke, T. S., L. A. Harveson, and F. Hernandez. 2019. Binational movements of desert bighorn sheep between Texas, USA and Chihuahua, Mexico. *Desert Bighorn Council Transactions* 55:21–31.

Jessup, D. A. 1982. The use of the helicopter in the capture of free roaming wildlife. Pages 289–303 *in* L. Neilsen, J. C. Haigh, and M. E. Fowler, editors. *Chemical Immobilization of North American Wildlife*. Wisconsin Humane Society, Milwaukee, Wisconsin, USA.

Jessup, D. A., N. Kock, and M. Berbach. 1989. Coccidioidomycosis in a desert bighorn sheep (*Ovis canadensis nelsoni*) from California. *Journal of Zoo and Wildlife Medicine* 20:471–473.

Jessup, D. A., R. Mohr, and B. Feldman. 1982. A comparison of four methods of capturing bighorn. *Desert Bighorn Council Transactions* 26:21–25.

Jessup, D. A., R. K. Clark, R. A. Weaver, and M. D. Kock. 1988. Safety and cost effectiveness of netgun capture of desert bighorn sheep. *Journal of Zoo Animal Medicine* 19:208–213.

Jessup, D. A., S. R. deJesus, W. E. Clark, and V. C. Bleich. 2014. Evolution of ungulate capture techniques in California. *California Fish and Game* 100:491–526.

Jessup, D. A., W. E. Clark, and M. E. Fowler. 1986. Wildlife Restraint Handbook. Third edition. California Department of Fish and Game, Rancho Cordova, California, USA.

Jessup, D. A., and W. E. Clark. 1980. *Wildlife Restraint Handbook*. California Department of Fish and Game, Rancho Cordova, California, USA.

Jessup, D. A., and W. E. Clark. 1982. Wildlife Restraint Handbook. Second edition. California Department of Fish and Game, Rancho Cordova, California, USA.

Jex, B. A., J. B. Ayotte, V. C. Bleich, C. E. Brewer, D. L. Bruning, T. M. Hegel, N. C. Larter, R. A. Schwanke, H. M. Schwantje, and M. W. Wagner. 2016. *Thinhorn Sheep: Conservation Challenges and Management Strategies for the 21st Century*. Western Association of Fish and Wildlife Agencies, Boise, Idaho, USA.

Johnson, T. L., and D. M. Swift. 2000. A test of a habitat evaluation procedure for Rocky Mountain bighorn sheep. *Restoration Ecology* 8(4S):47–56.

Jorgensen, J., J. Samson, and M. Festa-Bianchet. 1990. Field immobilization of bighorn sheep with xylazine hydrochloride and antagonism with idazoxan. *Journal of Wildlife Diseases* 25:522–527.

Justice-Allen, A., B. Munk, E. Lantz, J. Burco, H. Schwantje, P. Wolff, and N. LaHue. 2024. Health and veterinary care of wild sheep. Pages 3.1–3.33 *in* V. C. Bleich and D. W. Lutz, editors. *Wild Sheep Capture and Handling Guidelines*. Second edition. Western Association of Fish and Wildlife Agencies, Boise, Idaho, USA.

Keay, J. A., J. D. Wehausen, C.D. Hargis, R. A. Weaver, and T. E. Blankinship. 1987. Mountain sheep reintroduction in the central Sierra: a cooperative effort. *Western Section of the Wildlife Society Transactions* 23:60–64.

Kelly, W. E. 1980. Predator relationships. Pages 186-196 *in* G. Monson and L. Sumner, editors. *The Desert Bighorn: Its Life History, Ecology, and Management.* University of Arizona Press, Tucson, Arizona, USA.

Kock, M. D., D. A. Jessup, R. K. Clark, and C. E. Franti. 1987b. Effects of capture on biological parameters in free ranging bighorn sheep: evaluation of dropnet, drivenet, chemical immobilization, and the netgun. *Journal of Wildlife Diseases* 23:641–651.

Kock, M. D., D. A. Jessup, R. K. Clark, C. E. Franti, and R. A. Weaver. 1987c. Capture methods in five subspecies of bighorn sheep: evaluation of dropnet, drivenet, chemical immobilization, and the netgun. *Journal of Wildlife Diseases* 23:634–640.

Kock, M. D., R. K. Clark, C. E. Franti, D. A. Jessup, and J. D. Wehausen. 1987a. Effects of capture on biological parameters in free ranging bighorn sheep: evaluation of normal, stressed and mortality outcomes and documentation of post-capture survival. *Journal of Wildlife Diseases* 23:652–662.

Krausman, P. R. 2000. An introduction to the restoration of bighorn sheep. *Restoration Ecology* 8(4S):3–5.

Krausman, P. R., J. J. Hervert, and L. L. Ordway. 1985. Capturing deer and mountain sheep with a net-gun. *Wildlife Society Bulletin* 13:71–73.

Krausman, P. R., and V. C. Bleich. 2013. Conservation and management of ungulates in North America. *International Journal of Environmental Studies* 70:372–382.

Krausman, P. R., V. C. Bleich, J. W. Cain III, T. R. Stephenson, D. W. DeYoung, P. W. McGrath, P. K. Swift, B. M. Pierce, and B. D. Jansen. 2004. Neck lesions in ungulates from collars incorporating satellite technology. *Wildlife Society Bulletin* 32:987–991.

Kreeger, T. J. 1999. *Handbook of Wildlife Chemical Immobilization.* Wildlife Pharmaceuticals, Inc., Fort Collins, Colorado, USA.

Lacey, E. N. 1973. Capture and care of four species of mountain sheep. *Desert Bighorn Council Transactions* 17:73–80.

Lande, R. 1993. Risks of population extinction from demographic and environmental stochasticity and random catastrophes. *American Naturalist* 142:911–927.

Leach, H. R., J. M. Brode, and S. J. Nicola. 1974. *At the Crossroads, a Report on California's Endangered and Rare Fish and Wildlife.* California Department of Fish and Game, Sacramento, California, USA.

Leopold, A. S., S. A. Cain, C. M. Cottam, I. N. Gabrielson, and T. L. Kimball. 1963. *Wildlife Management in the National Parks.* Report of the Advisory Board on Wildlife Management appointed by Secretary of the Interior Udall. U.S. Department of the Interior, Washington, D.C., USA. https://www.nps.gov/parkhistory/online books/leopold/leopold.htm. Accessed 29 Dec 2021.

Lonner, B. N., T. R. Stephenson, and M. Cox. 2024. Translocation, release, and monitoring of wild sheep. Pages 9.1–9.33 *in* V. C. Bleich and D. W. Lutz, editors. *Wild Sheep Capture and Handling Guidelines.* Second edition. Western Association of Fish and Wildlife Agencies, Boise, Idaho, USA.

Lowrey, B., D. E. McWhirter, K. M. Proffitt, K. L. Monteith, A. B. Courtemanch, P. J. White., J. T. Paterson, S. R. Dewey, and R. A. Garrott. 2020. Individual variation creates diverse migratory portfolios in native populations of a mountain ungulate. *Ecological Applications* 30:e2106.

Lutz, D. 2024. In-depth statistical sheep information: Wyoming. *Wild Sheep* 12(2):23–24.

Manlove, K. R. 2021. Nevada Department of Wildlife Subgrant Award #SG19-04: Bighorn disease ecology and herd response Annual Report - 2020–2021. Utah State University, Logan, Utah.

McCullough, D. R. 1996. Introduction. Pages 1–10 *in* D. R. McCullough, editor. *Metapopulations and Wildlife Conservation.* Island Press, Covelo, California, USA.

McKinney, T., S. R. Boe, and J. C. deVos. 2003. GIS-based evaluation of escape terrain and desert bighorn sheep populations in Arizona. *Wildlife Society Bulletin* 31:1229–1236.

Millspaugh, J. J., and J. M. Marzluff (editors). 2001. *Radio Tracking and Animal Populations.* Academic Press, San Diego, California, USA.

Moore, T. D. 1958. Transplanting and observations of transplanted bighorn sheep. *Desert Bighorn Council Transactions* 2:43–46.

Morgart, J. R., D. R. Smith, and P. R. Krausman. 1992. An evaluation of two methods used to translocate desert bighorn sheep. *Transactions of the International Union of Game Biologists* 18:147–152.

Muir, J. 1901. *The Mountains of California.* The Century Company, New York, New York, USA.

Nevada Division of Wildlife. 2001. *Nevada Division of Wildlife's Bighorn Sheep Management Plan.* Nevada Division of Wildlife, Reno, Nevada, USA.

Nissen, K. 1982. Images from the past: an analysis of six western Great Basin petroglyph sites. Dissertation, University of California, Berkeley, California, USA.

Nordeen, T. D., K. Beckmen, J. Broadhurst, A. B. Courtemanch, M. Cox, E. Kring, C. P. Lehman, C. Procter, and K. S. White. 2024. Capturing wild sheep using chemical immobilization. Pages 7.1–7.15 *in* V. C. Bleich and D. W. Lutz, editors. *Wild Sheep Capture and Handling Guidelines.* Second edition. Western Association of Fish and Wildlife Agencies, Boise, Idaho, USA.

Northrup, J. M., C. R. Anderson, and G. Wittemyer. 2014. Effects of helicopter capture and handling on movement behavior of mule deer. *Journal of Wildlife Management* 78:731–738.

Organ, J. F., D. J. Decker, S. J. Riley, J. E. McDonald Jr., and S. P. Mahoney. 2020. Adaptive management in wildlife conservation. Pages 93–106 *in* N. J. Silvy, editor. *The Wildlife Techniques Manual.* Eighth edition. Volume II. Johns Hopkins University Press, Baltimore, Maryland, USA.

Peterson, M. J., and P. J. Ferro. 2020. Wildlife health and disease surveillance, investigation, and management. Pages 174–199 in N. J. Silvy, editor. *The Wildlife Techniques Manual.* Eighth edition. Volume II. Johns Hopkins University Press, Baltimore, Maryland, USA.

Peterson, M. J., M. N. Peterson, T. R. Peterson, and E. Von Essen. 2020. Ethics in wildlife science and conservation. Pages 12–38 *in* N. J. Silvy, editor. Eighth edition. Volume II. Johns Hopkins University Press, Baltimore, Maryland, USA.

Pfeifer, E., J. Ruhlman, B. Middleton, D. Dye, and A. Acosta. 2010. Initial results from a study of climatic changes and the effect on wild sheep habitat in selected study areas of Alaska. U.S. Geological Survey Open-File Report 2010-1135:1–68.

Prentice, P. R., J. Colby, L. E. Greene, C. P. Massing, and T. R. Stephenson. 2019. Status of bighorn sheep in California, 2019. *Desert Bighorn Council Transactions* 55:70–84.

Rachlow, J. L., and R. T. Bowyer. 1991. Interannual variation in timing and synchrony of parturition in Dall's sheep. *Journal of Mammalogy* 72: 487-492.

Ramey, R. R. II. 1993. Evolutionary genetics and systematics of North American mountain sheep: implications for conservation. Dissertation, Cornell University, Ithaca, New York, USA.

Ramey, R. R. II. 1995. Mitochondrial DNA variation, population structure, and evolution of mountain sheep in the southwestern United States and Mexico. *Molecular Ecology* 4:429–443.

Rattenbury, K. L., J. H. Schmidt, D. K. Swanson, B. L. Borg, B. A. Mangipane, and P. J. Sousanes. 2018. Delayed spring onset drives declines in abundance and recruitment in a mountain ungulate. *Ecosphere* 9(11):e02513.

Remington, R., and A. Fuller. 1989. Capture and transplant techniques. Pages 195–197 *in* R. M. Lee, editor. *Desert Bighorn Sheep in Arizona*. Arizona Game and Fish Department, Phoenix, Arizona, USA.

Robinson, R. W., T. S. Smith, J. C. Whiting, R. T. Larsen, and J. M. Shannon. 2020. Determining timing of births and habitat selection to identify lambing period habitat for bighorn sheep. *Frontiers in Ecology and Evolution* 8:97.

Rominger, E. M. 2018. The Gordian Knot of mountain lion predation and bighorn sheep. *Journal of Wildlife Management* 82:19–31.

Rominger, E. M., H. A. Whitlaw, D. L. Weybright, W. C. Dunn, and W. B. Ballard. 2004. The influence of mountain lion predation on bighorn sheep translocations. *Journal of Wildlife Management* 68:993–999.

Rominger, E. M., M. Cox, A. Munig, B. A. Jex, T. Lohuis, T. R. Stephenson, C. Procter, G. Brennan, W. Livingston, and M. Schlegel. 2024. Capturing wild sheep with a net-gun. Pages 5.1–5.21 *in* V. C. Bleich and D. W. Lutz, editors. *Wild Sheep Capture and Handling Guidelines*. Second edition. Western Association of Fish and Wildlife Agencies, Boise, Idaho, USA.

Rominger, E. M., V. C. Bleich, and E. J. Goldstein. 2006. Bighorn sheep, mountain lions, and the ethics of conservation. *Conservation Biology* 20:1341.

Rutherford, W. H., and R. L. Schmidt. 1973. *Techniques for Supplementing Diet, Attracting and Baiting Bighorn Sheep*. Game Information Leaflet 95. Colorado Division of Wildlife, Denver, Colorado, USA.

Sasse, D. B. 2003. Job-related mortality of wildlife workers in the United States, 1937–2000. *Wildlife Society Bulletin* 31:1015–1020.

Schmidt, R. L., W. H. Rutherford, and F. M. Doderstam. 1978. Colorado bighorn capture techniques. *Wildlife Society Bulletin* 6:159–163.

Schwartz, O. A., V. C. Bleich, and S. A. Holl. 1986. Genetics and the conservation of mountain sheep *Ovis canadensis nelsoni*. *Biological Conservation* 37:179–190.

Scotton, B. D. 1998. Timing and causes of neonatal Dall sheep mortality in the central Alaska Range. Thesis, University of Montana, Missoula, Montana, USA.

Seddon, P. J., D. P. Armstrong, and R. F. Maloney. 2007. Developing the science of reintroduction biology. *Conservation Biology* 21:303–312.

Sikes, R. S., and the Animal Care and Use Committee of the American Society of Mammalogists. 2011. Guidelines of the American Society of Mammalogists for the use of wild mammals in research. *Journal of Mammalogy* 92:235–253.

Sikes, R. S., and the Animal Care and Use Committee of the American Society of Mammalogists. 2016. Guidelines of the American Society of Mammalogists for the use of wild mammals in research and education. *Journal of Mammalogy* 97:663–688.

Silvy, N. J. 2020. *The Wildlife Techniques Manual*. Eighth edition. Johns Hopkins University Press, Baltimore, Maryland, USA.

Silvy, N. J., R. R. Lopez, and M. J. Peterson. 2020b. Techniques for marking wildlife. Pages 236–261 *in* N. J. Silvy, editor. The wildlife techniques manual. Eighth edition. Volume I. Johns Hopkins University Press, Baltimore, Maryland, USA.

Silvy, N. J., R. R. Lopez, and T. A. Catanach. 2020a. Capturing and handling wild animals. Pages 62–106 *in* N. J. Silvy, editor. *The Wildlife Techniques Manual*. Eighth edition. Volume I. Johns Hopkins University Press, Baltimore, Maryland, USA.

Silvy, N. J., and T. A. Catanach. 2020. Radio telemetry, remote monitoring, and data analyses. Pages 262–296 *in* N. J. Silvy, editor. *The Wildlife Techniques Manual*. Eighth edition. Volume I. Johns Hopkins University Press, Baltimore, Maryland, USA.

Simberloff, D., and J. Cox. 1987. Consequences and costs of conservation corridors. *Conservation Biology* 1:63–71.

Singer, F. J., C. M. Papouchis, and K. K. Symonds. 2000. Translocations as a tool for restoring populations of bighorn sheep. *Restoration Ecology* 8:6–13.

Singer, F. J., V. C. Bleich, and M. A. Gudorf. 2000. Restoration of bighorn sheep metapopulations in and near western national parks. *Restoration Ecology* 8(4S):14–24.

Sivy, K. J., A. W. Nolin, C. L. Cosgrove, and L. R. Prugh. 2018. Critical snow density threshold for Dall's sheep (*Ovis dalli dalli*). *Canadian Journal of Zoology* 96:1170–1177.

Sleznick, J. 1980. Lava Beds bighorn sheep transplant to South Warner Mountains, Modoc National Forest. *Desert Bighorn Council Transactions* 24:62.

Smith, T. S., J. T. Flinders, and D. S. Winn. 1991. A habitat evaluation procedure for Rocky Mountain bighorn sheep in the inter-mountain west. *Great Basin Naturalist* 51:205–225.

Spraker, T. R. 1977. Capture myopathy of Rocky Mountain bighorn sheep. *Desert Bighorn Council Transactions* 21:14–16.

Tetzlaff, S. J., J. H. Sperry, B. A. Kingsbury, and B. A. DeGregorio. 2019. Captive-rearing duration may be more important than environmental enrichment for enhancing turtle head-starting success. *Global Ecology and Conservation* 20:e00797.

Tetzlaff, S. J., J. H. Sperry, and B. A. DeGregorio. 2020. Effects of antipredator training, environmental enrichment, and soft release on wildlife translocations: a review and meta-analysis. *Biological Conservation* 236:324–331.

Thompson, J. R., V. C. Bleich, S. G. Torres, and G. P. Mulcahy. 2001. Translocation techniques for mountain sheep: does the method matter? *Southwestern Naturalist* 46:87–93.

Torland, S., E. Partee, and D. G. Whittaker. 2024. Base camp operations. Pages 2.1–2.15 *in* V. C. Bleich and D. W. Lutz, editors. *Wild sheep capture and handling guidelines*. Second edition. Western Association of Fish and Wildlife Agencies, Boise, Idaho, USA.

Torres, S. G., V. C. Bleich, and J. D. Wehausen. 1994. Status of bighorn sheep in California, 1993. *Desert Bighorn Council Transactions* 38:17–28.

Umphlett, A. 2023. 2023 - Job Completion Report. Pages 77–79 *in Lander_JCR_2023*. Wyoming Game and Fish Department, Cheyenne, Wyoming, USA.

Van de Kerk, M., B. R. McMillan, K. R. Hersey, A. Roug, and R. T. Larsen. 2020. Effect of net-gun capture on survival of mule deer. *Journal of Wildlife Management* 84:813–820.

Vera, A. 2020. First California mountain lion killed under state's new depredation law. CNN website. Available from: https://www.cnn.com/2020/02/11/us/california-depredation-law-mountain-lion-killed/index.html. Accessed 16 Sept 2024.

Vu, R. K., R. Ianniello, J. Colby, E. Schaeffer, J. N. Sanchez, J. T. Villepique, L. E. Greene, and T. R. Stephenson. 2021. Status of bighorn sheep in California, 2021. *Desert Bighorn Council Transactions* 56:87–95.

Wagler, B. L. and K. L. Monteith. 2021. An assessment of bighorn sheep in the Sweetwater Rocks: habitat suitability and risk of contact with domestic sheep: Research Brief. Haub School of Environment and Natural Resources, University of Wyoming, Laramie, WY and Wyoming Cooperative Fish and Wildlife Research Unit, University of Wyoming, Laramie, Wyoming, USA.

Wagler, B. L., R. A. Smiley, A. B. Courtemanch, G. Anderson, D. W. Lutz, D. E. McWhirter, D. G. Brimeyer, P. Hnilicka, C. P. Massing, D. W. German, T. R. Stephenson, and K. L. Monteith. 2022. Effects of helicopter net-gunning on survival of bighorn sheep. *Journal of Wildlife Management* 86:e22181.

Weaver, R. A. 1984. Status of bighorn sheep in California. *Desert Bighorn Council Transactions* 28:48.

Weaver, R. A., and R. K. Clark. 1988. Status of bighorn sheep in California, 1987. *Desert Bighorn Council Transactions* 32:20.

Webb, S. L., J. S. Lewis, D. G. Hewitt, M. W. Hellickson, and F. C. Bryant. 2008. Assessing the helicopter and net gun as a capture technique for white-tailed deer. *Journal of Wildlife Management* 72:310–314.

Wehausen, J. D. 1991. Some potentially adaptive characters of mountain sheep populations in the Owens Valley region. Pages 124–135 in C. A. Hall Jr., V. Doyle-Jones, and B. Widawski, editors. *White Mountain Research Station Symposium 3*. White Mountain Research Station, Bishop, California, USA.

White, G. C., and R. A. Garrott. 1990. *Analysis of Wildlife Radio-Tracking Data*. Academic Press, San Diego, California, USA.

Whiting, J. C., D. D. Olson, J. M. Shannon, R. T. Bowyer, R. W. Klaver, and J. T. Flinders. 2012. Timing and synchrony of births in bighorn sheep: implications for reintroduction and conservation. *Wildlife Research* 39:565–572.

Whiting, J. C., V. C. Bleich, R. T. Bowyer, and C. W. Epps. 2023a. Restoration of bighorn sheep: history, successes, and remaining conservation issues. *Frontiers in Ecology and Evolution* 11:1083350.

Whiting, J. C., V. C. Bleich, R. T. Bowyer, K. Manlove, and K. White. 2023b. Bighorn sheep and mountain goats. Pages 759–790 in L. B. McNew, D. K. Dahlgren, and J. L. Beck, editors. *Rangeland Wildlife and Conservation*. Springer Nature Switzerland AG, Cham, Switzerland.

Whittaker, D. G., S. D. Ostermann, and W. M. Boyce. 2004. Genetic variability of reintroduced California bighorn sheep in Oregon. *Journal of Wildlife Management* 68:850–859.

Wiedmann, B. P., and G. A. Sargeant. 2014. Ecotypic variation in recruitment of reintroduced bighorn sheep: implications for translocation. *Journal of Wildlife Management* 78:397–401.

Wild Sheep Working Group. 2015. *Records of Wild Sheep Translocations: United States and Canada, 1922–Present*. Western Association of Fish and Wildlife Agencies, Cheyenne, Wyoming, USA.

Wildlife Investigations Lab. 2018. *Wildlife Restraint Handbook*. California Department of Fish and Wildlife, Rancho Cordova, California, USA.

Wilson, L. O. and C. L. Douglas. 1982. Revised procedures for capturing and re-establishing desert bighorn. *Desert Bighorn Council Transactions* 26:1–7.

Wilson, L. O., J. Day, J. Helvie, G. Gates, T. L. Hailey, and G. K. Tsukamoto. 1973. Guidelines for capturing and re-establishing desert bighorns. *Desert Bighorn Council Transactions* 17:137–154.

Wilson, L. O., J. Day, J. Helvie, G. Gates, T. L. Hailey, and G. K. Tsukamoto. 1975. Guidelines for establishing and capturing desert bighorn. Pages 269–295 in J. B. Trefethen, editor. *The Wild Sheep in Modern North America*. Winchester Press, New York, New York, USA.

Woodbury, M. 1996. *The Chemical Immobilization of Wildlife: Course Manual*. Canadian Association of Zoo and Wildlife Veterinarians, Saskatoon, Saskatchewan, Canada.

WSWG (Wild Sheep Working Group). 2015. *Records of Wild Sheep Translocations: United States and Canada, 1922–Present*. Western Association of Fish and Wildlife Agencies, Cheyenne, Wyoming, USA.

Wyoming Game and Fish Department. 2022. *Bighorn Sheep Capture and Handling Guidelines*. Wyoming Game and Fish Department, Cheyenne, Wyoming, USA.

Zeigenfuss, L. C., F. J. Singer, and M. A. Gudorf. 2000. Test of a modified habitat suitability model for bighorn sheep. *Restoration Ecology* 8(4S):38–46.

# 15 Modeling Population Dynamics of Mountain Sheep

*Mark S. Boyce and Evelyn H. Merrill*

## 15.1 INTRODUCTION

Fundamental to the application of modeling for North American mountain sheep (bighorn [*Ovis canadensis*], thinhorn [*Ovis dalli*]) populations is to focus on objectives (i.e., what is the right question?). Sometimes we simply wish to use models as book-keeping tools to organize and examine data, or to explore data for patterns (Williams et al. 2001). This might include demographic models that combine vital rates of survival and reproduction, where the question focuses on whether estimated vital rates suggest that a population is increasing or decreasing over a short time horizon (Burnham and Anderson 2004, Paterson et al. 2021).

Alternatively, the approach might involve ecological models with varying degrees of complexity, including interactions with other species and trophic levels (Maynard Smith 1974, Caughley and Lawton 1981, Tredennick et al. 2021). For wildlife managers demographic models might be combined with ecological models such as predator-prey or plant-herbivore models to address alternative harvesting strategies (Boyce et al. 2012b) or used for population viability analysis of an endangered population (Rubin et al. 2002). Identifying the best alternative models will depend on management objectives and how well the data support the model. Our objective is to review approaches for modeling the population dynamics of mountain sheep populations to identify applications that can assist management.

## 15.2 DEMOGRAPHY

Demographic models are also known as structured population models, book-keeping models that track the number of individuals in each age or stage class. Demographic models are composed of vital rates, including age- or stage-specific survival and recruitment probabilities (Figure 15.1). These structured population models can be full age-structured models, such as the Leslie Matrix, which requires a projection matrix for mountain sheep that includes approximately 14 age classes and age-specific survival and recruitment estimates (few mountain sheep live past 14 years). Seldom do we have access to sufficient data to parameterize such a large matrix with age-specific vital rates for both males and females. Because sample sizes would invariably be small for older ages, sampling error would be substantial (Ludwig 1999). Because data collected during monitoring typically are limited to the number of lambs, number of females, and number of males, a more reasonable structure

is a stage-based projection matrix times a column vector of the number of individuals in each stage class at time $t$:

$$\begin{bmatrix} 0 & F_{female} & 0 \\ 0.5S_{\female lamb} & S_{female} & 0 \\ 0.5S_{\male lamb} & 0 & S_{male} \end{bmatrix} \times \begin{bmatrix} n_{lamb} \\ n_{female} \\ n_{male} \end{bmatrix} \quad (15.1)$$

where $F_{female}$ is the fecundity of females averaged over all non-lamb females, $S_{female}$ is the survival of females, and $S_{male}$ is the survival of males, $S_{\female lamb}$ is the survival of female lambs, and $S_{\male lamb}$ is the survival of male lambs that survive to the male stage class. Sex ratio at birth is usually assumed to be 1:1 (Paterson et al. 2021). For the column vector, $n_{lamb}$ is the number of lambs, $n_{female}$ is the number of females, and $n_{male}$ is the count of males in the population (Boyce et al. 2012a). The product of these two matrices is a column vector of the number of individuals in each stage at time $t+1$. The total population is $N_t = n_{lamb} + n_{female} + n_{male}$. If classification data exist for yearlings, an additional stage class could be inserted for subadults yielding a $4 \times 4$ projection matrix. Regardless of the domain of the matrix, the dominant eigenvalue of $\lambda$, the constant projection matrix is the finite growth factor when the population achieves a stable age distribution (Caswell 2001).

A projection matrix describes geometric population growth as becoming infinitely large when the geometric mean $\lambda^* > 1$ or declining toward extinction if $\lambda^* < 1$; therefore, this model is applicable for only a short time. Over longer periods, vital rates can be a density-dependent function of abundance, and the population will exhibit logistic-like dynamics, converging on a carrying capacity where the geometric mean $\lambda^* = 1$. Similarly, models can incorporate variability in vital rates, causing the population projection to fluctuate accordingly (Boyce et al. 2012a). Sources of variation in the dynamics of different mountain sheep populations include a variety of external mechanisms including seasonality and changing habitat conditions (Smith et al. 1999), predation (Knopff and Boyce 2007), stochasticity in weather, and disease (Johnson et al. 2010, Renaud et al. 2019, Lula et al. 2020, Paterson et al. 2021, Turgeon et al. 2024). Furthermore, population growth and space use can differ for reintroduced populations compared to those on native ranges, due to behavioral factors (MacCallum 1989, Singer et al. 2000, Bleich et al. 2018, Jesmer et al. 2018).

Variation has been observed in vital rates among populations and over time (Paterson et al. 2021, Turgeon et al. 2024).

DOI: 10.1201/9781003518686-17

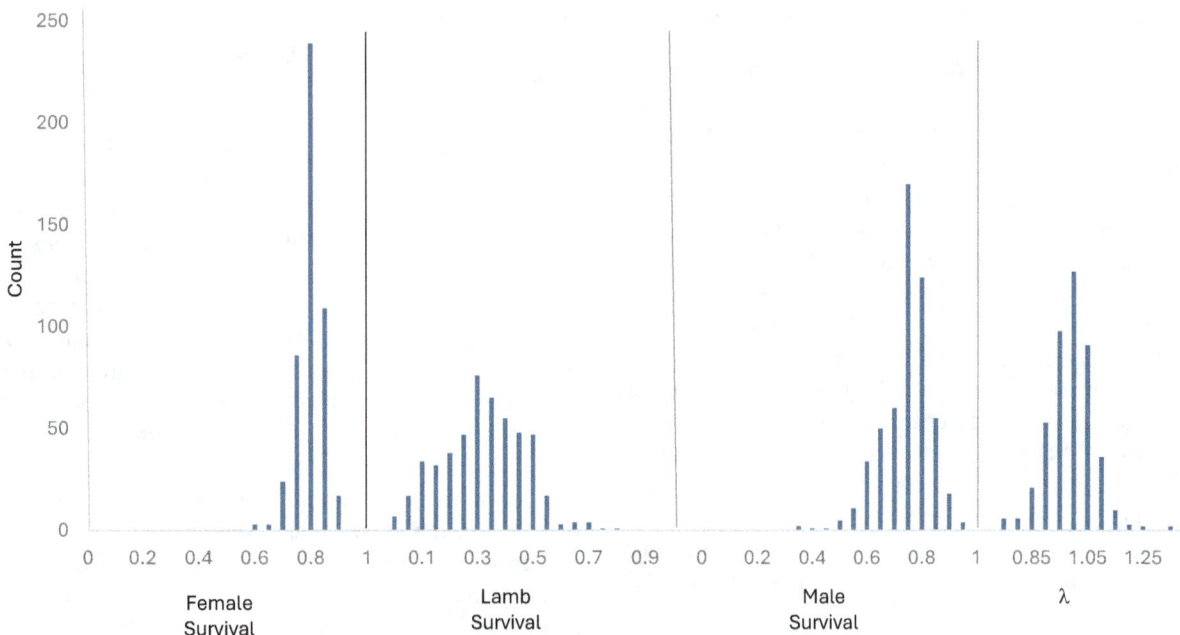

**FIGURE 15.1** Distribution of annual vital rates and population growth (λ) for 17 bighorn sheep populations in Montana and Wyoming, USA, 1983–2018. (From Paterson et al. (2021), open access.)

For long-lived species such as mountain sheep, highest sensitivity and elasticity typically exist for adult female survival (i.e., a change in adult survival will have an effect on population growth; Gaillard et al. 1998, Rubin et al. 2002), yet adult survival may not vary as much, indicating that changes in population abundance are more influenced by recruitment or juvenile survival even though sensitivity is less (Gaillard et al. 1998). Furthermore, the nature of the covariation in vital rates among stages due to external factors is key because when vital rates are correlated (i.e., a good year is a good year for all age and sex stages) this covariation influence is larger than the influence of any particular vital rate (Coulson et al. 2005). For example, winter severity is a major influence on mountain sheep populations, which varies substantially from year to year (Poole et al. 2016). Increased variation in vital rates also has the consequence of reducing long-term population growth rates (Boyce et al. 2006) and population size (Boyce and Daley 1980), thereby increasing the probability of local extirpation.

Structured populations also can undergo substantial short-term fluctuations simply because of shifts in age structure resulting in transient dynamics (Koons et al. 2005). Such transient population fluctuations will be masked in a collapsed stage-structured model (e.g., Eq. 15.1). In a study of 17 bighorn populations in Montana and Wyoming, USA, age structure deviations had no discernable consequences to fluctuations in population size and were trivial relative to the influence of survival and recruitment (Paterson et al. 2021).

Dispersal and large-scale movements can influence genetic viability and can expose bighorns to disease risk (DeCesare and Pletcher 2006, Cassirer et al. 2018). Males usually have much greater movements than females and propensity for movement can be a consequence of the history of the population (Jesmer et al. 2018). Translocated herds tend to move much less, apparently as they gradually become familiar with their new areas.

State and provincial wildlife management agencies have utilized a range of software packages for demographic simulations and setting harvest quotas. The most basic is software requiring inputs of age-specific vital rates and estimates of population composition. These include Populus, POP50, and POPR (Alstad 2001), and more complex structures such as in program ALCES (Stelfox et al. 2005). A fundamental problem with applications of each of these is the ability to secure sufficient data to obtain reliable estimates of vital rates and data on population structure, ideally recorded over several years to obtain estimates of process variance (Shenk et al. 1998, White 2000). Naturally, any attempt to model population fluctuations will require estimates of the temporal variability in vital rates. As a result, these matrix models can be useful for exploring the outcomes of projections when used iteratively with stochastic variation; however, caution is required for deterministic approaches. When data are limited, estimates of vital rates have been facilitated by using integrated population models that combine information from multiple sources using a Bayesian algorithm to improve the accuracy and precision of population projections (Schaub et al. 2007). Furthermore, for software to be useful for modeling mountain sheep populations, it needs to have the flexibility to incorporate density dependence (Mendelssohn 1976, Jorgenson et al. 1993, Bonenfant et al. 2009), plant-herbivore interactions (Caughley and Lawton 1981), and environmental influences (Botkin 1990, Nagy-Reis et al. 2021).

## 15.3   PLANT-HERBIVORE INTERACTIONS

All mountain sheep populations are influenced by access to vegetation, and single-population models for ungulates, ignoring the interaction with plants, are naïve (Caughley 1976). The herbivore's interaction with vegetation has substantial consequences for population dynamics because it sets the equilibrium level and stochasticity in vegetative growth can influence the variation in vital rates. Yet only a few population models for mountain sheep establish a link between the sheep population and vegetation (Monteith et al. 2018, Paterson et al. 2021, Beale et al. 2023).

Plant-herbivore systems can be broadly classified into two categories: non-interactive and interactive (Caughley and Lawton 1981). If non-interactive, the herbivore population does not affect the food resource or its rate of renewal. For example, granivorous birds are dependent upon seed production, but granivory may not necessarily influence the plant's population. In contrast, foraging by mountain sheep on herbaceous plants can have a direct effect on plant density or growth, which in turn may affect the nutrition of mountain sheep. Such interactive systems have been modeled by Caughley and Lawton (1981) as a pair of differential equations that had been studied mathematically by Kolmogorov (1936). Specifically, Caughley's interactive model included density dependence for the plants, but not for the herbivore, assuming density effects were manifested through the herbivore's effect on the vegetation. He used the following Ivlev-type functional and numerical responses:

$$dV/dt = bV(1 - V/K) - f_{max} N[1 - \exp(-a_1 N)] \qquad (15.2)$$

$$dV/dt = c_2 N[1 - \exp(-a_1 V)] - \delta N \qquad (15.3)$$

where $V$ is vegetation abundance or biomass, $b$ is the potential growth rate for the vegetation, $K$ is the carrying capacity for the vegetation, $f_{max}$ is the maximum rate at which mountain sheep can consume vegetation, $a_1$ is grazing efficiency when vegetation is sparse, $c_2$ is the rate at which the decline of an herbivore population is ameliorated by high vegetation abundance, $a_2$ is ability of sheep to increase when vegetation is sparse, and $\delta$ is the inherent death rate of mountain sheep when there is no vegetation.

The combination of density dependence in plants relative to a potential abundance of vegetation ($K$) and the functional response by the herbivore gives this model a fascinating range of dynamical behaviors. For a small $K$, when plants are sparse, the system predicts vegetation and herbivore populations that will approach equilibrium population sizes slowly after a disturbance. As we increase $K$ we find that the herbivore population tends to overshoot its equilibrium, but then will oscillate back toward a long-term average population size, with the magnitude of the oscillations tied to the increase in $K$ (Caughley and Lawton 1981). These dynamics are a common pattern observed in ungulate populations where there is an initial period of explosive herbivore growth, often called an eruption, which leads to high population numbers followed by forage depletion or shift in forage composition affecting the quality of the plant food resources resulting in a population crash, but ultimately recovering and oscillating toward some long-term average population size (Caughley 1976). Over 100 examples of such eruptive behavior in deer (*Odocoileus* spp.) populations were reported in North America during 1900–1945 (Leopold et al. 1947). In bighorn sheep, evidence of similar eruptive population patterns exists when they are released onto new ranges (MacCallum 1989). The existence of a plant refuge, whether herbivory-related (e.g., plant architecture changes to limit accessibility to new growth; Skarpe and Hester 2010) or herbivore behavior (i.e., avoiding areas or seeking out conspecifics due to predator risk; Rankins et al. 2024) can have a stabilizing effect. When combined with environmental perturbations resulting from fluctuations in weather (Bender and Weisenberger 2005), vegetation growth (Atkin et al. 1996), or predation (Turgeon et al. 2024), trophic-level interactions can lead to perpetual population fluctuations of varying magnitudes (Botkin 1990).

Parameterizing a Caughley-like model is challenging due to the numerous variables to be estimated and the seasonal ranges of mountain sheep (Jesmer et al. 2018, Enns et al. 2023). Vegetation growth and the functional responses of the herbivores can vary with the sheep's numerical response, reflecting the amount of time spent on seasonal ranges. Different environmental conditions will moderate whether herbivory reduces or stimulates vegetation renewal (i.e., $b$ term in Eq. 15.2; McNaughton et al. 1989), which can influence stability and the population equilibrium (Crawley 1983). Further, in many systems, it is the quality of the vegetation rather than the abundance that influences the herbivore's numeric response and given the variation in the seasonal quality of different plants, it would call for modeling individual or groups of plants simultaneously (Hobbs 2024). As with most ecological systems, there are seemingly an infinite number of complications that one might encounter in modeling plant-herbivore systems. Some of these could present serious difficulties for parameter estimation for a detailed modeling effort but might not affect the qualitative behavior of the system. For example, Caughley (1982) reported consequences of plant-herbivore interactions where the herbivore fed on several species of plants were similar to that for the single plant-species system such that combining several species of plants can be justified in the plant-herbivore model (Eqs. 15.2 and 15.3).

An alternative approach to parameterizing plant-herbivore models is to directly link forage conditions to animal performance (Gaillard et al. 2010). This includes making vital rates of different population segments to be a function of forage (Paterson et al. 2021, Proffitt et al. 2021), or indirectly via a nutrition index (e.g., ingesta-free body fat [IFBF]; Stephenson et al. 2020, Smiley et al. 2022, or fecal nitrogen; Blanchard et al. 2003) and including these variables in a population model. In mountain sheep, summer range is characterized by forages in a mixture of high-elevation meadows with reduced canopy cover, whereas winter ranges are typically at lower elevation areas on south aspects or high-elevation windswept alpine and subalpine ridges where snow depths are reduced (Demarchi 1986, Lula et al. 2020, Werdel et al. 2023). In all cases, access to forages can be constrained to areas nearby rugged and steep terrain to avoid predators (Rankins et al. 2024), proximity to water sources in dry areas (Turner et al. 2004, Bleich et al. 2019), and avoidance of human activity (Papouchis et al. 2001, Courtemanch 2014, Sproat et al. 2020). Although general patterns of range use exist in mountain sheep, nutritional resources on these ranges are highly variable in space and time (Proffitt et al. 2021). Even within the same population, variation in forage resources of similar areas like meadows that are within a few hundred meters of each other can be similar or dissimilar to meadows hundreds of kilometers apart (Klinger et al. 2024). This highlights the significant influence of factors in montane environments, such as snowpack, topography, soils, and prior herbivory, underscoring the importance of using local information when making predictions (Atkin et al. 1996).

Deriving functions of vital rates for varying population segments (e.g., sex, age class, reproductive state) from forage conditions requires considering the appropriate forage metric, and the relevant spatial and time frame that reflects an influence on animal performance. Common approaches for quantifying forage conditions include indirect measures of remote sensing, which include metrics such as the normalized difference vegetation index (NDVI) and associated metrics, like the instantaneous rate of green-up (Pettorelli et al. 2011). Indirect approaches are advantageous to use over broad areas (Barton et al. 2023) but provide little biological meaning for comparison or direction for vegetation improvements, especially because changes in green plant biomass and active growth are confounded. At a minimum, empirically linking ground-based measures to indirect measures is key, even if challenging (Klinger et al. 2024). In contrast, foodscapes estimate forage over broad-scale areas by extrapolating vegetation sampled in ground plots along environmental gradients or within mapped vegetation classes and can be used to reflect a variety of metrics, including total herbaceous biomass (Hebblewhite et al. 2008), acceptable biomass (Cook et al. 2016), or suitable biomass (i.e., biomass meeting specific nutritional requirements; Bilodeau-Hussey et al. 2025). These metrics are biologically more interpretable for a plant-herbivore system, but obtaining the data can be demanding, especially when forages are annually and seasonally dynamic (Walker 2023).

Because measuring intake, or even bite size that determines intake (Shipley 2007), can rarely be measured in free-ranging animals, by default, typically relationships assume plant consumption is related to animal use of an area (i.e., exposure to forage conditions). The nature of the relationship may be improved in relation to animal-based metrics, such as intake rate derived from tame animals feeding on the vegetation present (Denryter et al. 2022). Choosing the appropriate spatial and temporal scale of

exposure is important. For example, winter often has been considered the bottleneck for herbivore populations in montane environments (Coltrane and Barboza 2010, Kautz et al. 2020), but mountain sheep rely on body reserves accumulated during summer to survive winter and support reproduction (Proffitt et al. 2021, Smiley et al. 2022). Exploring lag times in modeling nutritional relationships among individuals may be important due to seasonal and annual carryover effects (Norris 2005, Harrison et al. 2011). When individual animals are the sample unit, units of exposure range from various buffered telemetry points, movement paths, and seasonal ranges (Huggler 2024, Bilodeau-Hussey et al. 2025). Population-level relationships can emerge when individuals in a population are exposed to variable forage conditions or across populations where population vital rates are linked to the full population home range (Portier et al. 1998, Stephenson et al. 2020, Paterson et al. 2021, Proffitt et al. 2021, Bilodeau-Hussey et al. 2025). In fact, examining the distributional properties (e.g., skewness, kurtosis) of performance metrics within and among populations at different densities can provide insight into density-dependent effects and population resistance to environmental stochasticity.

To date, several published relationships have been established between vital rates and forage conditions on bighorn sheep ranges, with notable variation among areas. In modeling vital rates of 19 bighorn sheep populations in Montana and Wyoming over 7 years, growing conditions on summer range were positively related to the probability of pregnancy (i.e., early-season NDVI) and juvenile winter survival (i.e., cumulative NDVI), where the evidence for a positive association between body fat in winter and pregnancy was inconclusive (Proffitt et al. 2021). When nutrient availability was indexed by extrapolating estimates of suitable biomass (*sensu* Hanley 2012) across sheep range, no direct forage-based relationship to juvenile survival was observed (Bilodeau-Hussey et al. 2025). Instead, IFBF of females in late winter had a strong positive effect on summer survival of juveniles. The mixed results might not be surprising given the high spatial and temporal variability in conditions across montane environments.

At the same time, it is evident that the signals emerging from these studies and others with large herbivores (Monteith et al. 2014, Lukacs et al. 2018, Merems et al. 2020) point to the necessity of modeling the interaction of mountain sheep with their seasonal forages in some form to capture and understand mountain sheep population dynamics due to spatial and annual variation within and across populations (Paterson et al. 2021). In the true sense of the Caughley and Lawton (1981) plant-herbivore model, density-dependent effects of the herbivore on its forage base can be reflected in the statistical interaction between forage metrics or their remotely sensed proxies and mountain sheep density (McKinney et al. 2001, Portier et al. 2005, Bender and Weisenberger 2005), but also might require approaches that link sheep density directly to their effects on nutritional resources as described by Hobbs (2024).

Local or regional models of vital rates as a function of forage conditions can be used in deterministic or stochastic projection matrix models with time-varying inputs on vegetation conditions (Boyce et al. 2012a), or parameters used as priors or even solved for in Bayesian models including integrated population models. The models can be used at varying levels to explore the contributions of vegetation management and environmental stochasticity to vital rates in the face of other perturbations, such as predation and disease.

## 15.4  CAUSE-SPECIFIC MORTALITY

Predation also contributes to population fluctuations, both directly and indirectly, by influencing distribution (Carpenter et al. 2014) and nutrition (Monteith et al. 2016, Smiley et al. 2024). For example, to avoid predation mountain sheep seldom venture far from escape terrain (MacCallum and Geist 1992, Beale and Boyce 2020, Rankins et al. 2024), which can limit sheep's access to forage overall or to efficiently exploiting spatial and temporal variability in forage conditions (Wang et al. 2006, 2009). Predation of bighorn sheep by cougars (*Puma concolor*) in Alberta, Canada, has been characterized as a stochastic event (Festa-Bianchet et al. 2006), acting as a limited perturbation (Turgeon et al. 2024). Desert bighorn sheep (*O. c. nelsoni*) can be targeted by cougars, sometimes limiting local populations (Rominger 2018). Released from predator control, cougar populations have increased in parts of western North America causing reduced desert bighorn populations, which might warrant renewed predator control (Rominger 2018).

Canid predator control at three populations in Alaska resulted in short-term increases in Dall's sheep (*O. dalli dalli*) abundance; however, due to habitat limitations, sheep populations decreased relative to a control population without canid removals (Mitchell et al. 2015). These results from Alaska suggest a Hydra effect, whereby predation increased the abundance of Dall's sheep by mitigating its influence on vegetation (Abrams 2009). Predator-prey models predict that mortality imposed on predator and prey results in an increase in the number of prey along with a decrease in abundance of predators, a prediction known as the Volterra principle (Boyce et al. 2025). When predators are having a significant effect, hunting mountain sheep and predators will result in an increase in mountain sheep abundance while reducing large carnivores that are often viewed as being in conflict with the interests of hunters.

Disease has become a significant limitation to many bighorn populations, particularly from *Mycoplasma ovipneumoniae* (*M. ovi*), which has severe consequences for mountain sheep (Cassirer et al. 2018). Disease limits distribution in areas where the distribution of mountain sheep overlaps with that of domestic sheep and goats, a process that has been modeled using a point of contact model (Carpenter et al. 2014, O'Brien et al. 2014, Wilson-Henjum et al. 2025). Variation in environment

over time influences the patterns of space use by bighorns and thereby the risk of transmission (Wilson-Henjum et al. 2025). *M. ovi* can have long-lasting effects because females can harbor latent infections that result in the death of their lambs. Presumably this can suppress population growth for extended periods, although recent analysis suggests that recovery is possible even for populations that have encountered *M. ovi* in the past (Paterson et al. 2021, Proffitt et al. 2021, Wagler et al. 2023).

Disease can be incorporated into a structured population model by making vital rates a function of disease prevalence and the animal's condition (Smiley et al. 2024). Also, disease can be considered a stochastic event (as for predation, Festa-Bianchet et al. 2006) whereby a random number generator is used to simulate future exposure and risk. Force of infection, exposure, condition, and recovery are all significant influences that vary depending on the disease. Chapter 8 provides an expanded discussion on disease.

## 15.5  HARVEST MODELS

Sustainable harvests are facilitated by density dependence in survival and reproduction (Mendelssohn 1976), as harvest can lead to increased survival (Boyce et al. 1999) and population size (Abrams 2009). Density-dependent effects on vital rates for mountain sheep are usually strongest for recruitment or juvenile survival (Douglas and Leslie 1986, Eberhardt 2002, Simard et al. 2020). In the case of mountain sheep, which can exist in small, isolated populations, density-dependent effects can be dwarfed by perturbations caused by drought, winter severity, predation, or disease events (McKinney et al. 2001, Bender and Weisenberger 2005, Turgeon et al. 2024) that can even result in local extirpation. These interactions make it challenging to estimate the coefficients for a density-dependent function, although estimating parameters for a density-dependent effect can be facilitated by integrated population models to extract the density-dependent signal that can improve estimates for harvest models (Abadi et al. 2012, Lewis et al. 2024). Alternatively, modeling harvests in a plant-herbivore model do not require density-dependent function *per se* because density dependence emerges by nutritional limitation (Caughley 1976).

The principles of harvest modeling are typically presented as single-species models that incorporate an underlying density-dependent model, discounted by harvest (Boyce et al. 2012b). A simple example with proportional harvest:

$$dN/dt = rN(1 - N/K) - hN \qquad (15.4)$$

where $h$ is the harvest rate from a population of $N$ animals. In this example with logistic growth, maximum sustainable yield would occur when $N$ is maintained at $K/2$. In practice, implementation of such a proportional harvest model would require knowledge of population size. This can be a limitation because precise estimation of population size can be expensive and challenging to achieve in mountainous terrain.

Additionally, we caution that maximizing sustained yield is a perilous target, as inevitable stochastic variation in the environment is likely to result in overharvesting (Boyce et al. 2012b).

A more practical and widely used practice in sheep management is a quota model, where a limited number of hunting licenses are issued. Adjusted for hunter success, the quota ($Q$) is removed annually by hunting:

$$dN/dt = rN(1 - N/K) - Q \qquad (15.5)$$

This basic structure underpins a structured population model that affords hunting based on quotas for males and non-trophy sheep (Varley and Boyce 2006).

All mountain sheep populations in North America occur in highly seasonal and variable environments. The seasonal timing of hunting seasons can influence long-term yields (Xu et al. 2005). Similarly, stochastic variation caused by climate is likely to reduce optimal harvests because variation in vital rates reduces long-term growth rates (Boyce et al. 2006). Climate change is expected to alter plant-herbivore dynamics in mountain sheep (e.g., due to severe winters, droughts, and floods; Boyce et al. 2006, 2012b). Perhaps the most powerful approach in our toolkit is adaptive management, where we first make predictions based on a harvest model, then manipulate the harvest to elicit a population response, followed by monitoring to ultimately evaluate the model's predictions (Walters 1986, Nagy-Reis et al. 2021). This is a never-ending iterative process whereby we adjust to management and to changing environments, ultimately improving our models to achieve improved predictive capability (Varley and Boyce 2006, Boyce et al. 2012b).

Potential yields for licensed sheep hunters might be reduced by aboriginal harvests (Lynch 2006) and predation by wolves (*Canis lupus*) and cougars. We might assume non-selective aboriginal harvests averaging some fraction of the population size. Predation can be modeled using a Holling Type III (sigmoidal) functional response (Varley and Boyce 2006) with predator densities and diets estimated based on recent studies of known predators of sheep.

## 15.6   WHAT'S A MANAGER TO DO?

We have reviewed a set of approaches to population modeling, none of which are likely to meet the needs of a wildlife biologist managing a population of mountain sheep. We have canvassed management biologists in most western states and provinces to learn that few use harvest models to establish quotas. Where models are used, they typically employ projection matrices that combine estimates of vital rates to anticipate changes in population size. Because these are exponential growth models, they must be used judiciously, as they cannot make reliable predictions beyond a short time horizon and are not suitable for harvest modeling (Mendelssohn 1976).

For small populations at risk of local extinction, population viability analysis can provide managers with insights on possible management alternatives (Rubin et al. 2002). These are of heuristic value but cannot be expected to yield reliable predictions of population dynamics, as small populations, by definition, are limited by the available data (Boyce 1993, Ludwig 1999). But the modeling exercise reveals data needs and encourages attention to population processes limiting abundance.

Of greatest interest to most wildlife managers will be exploring the behavior of a harvesting model (Boyce et al. 2012b). Throughout their distribution in North America, mountain sheep are managed as a trophy species, with harvests focused on mature males only (Boyce and Krausman 2018, Monteith et al. 2018), in contrast to a sustainable yield target for meat for most other ungulate taxa (Mahoney and Geist 2019). Few if any females are hunted in most jurisdictions (Western Association of Fish and Wildlife Agencies 2024), and therefore populations often are at the carrying capacity of the range, with the small number of males removed (typically only 2%–3% of the population) compensated by non-trophy sheep (McCullough 1990).

Consequently, population fluctuations for most mountain sheep are largely constrained by plant-herbivore interactions (Caughley and Lawton 1981), and harvesting has been modeled within a plant-herbivore framework (Caughley 1976); however, we are unaware of any state or provincial agency that uses such a model structure. Given the negative public attitudes toward trophy hunting (Brennan et al. 2019), we believe that hunting for mountain sheep would be viewed more favorably if the species were managed for sustainable yields, similar to other ungulates in North America.

Again, the most plausible guidelines for a manager would be to implement an adaptive management procedure, where a harvesting model serves as the framework for making predictions about the outcome of management interventions. This is then evaluated with data from monitoring to form a basis for revisiting the model and adjusting the model parameters or structure accordingly (Walters 1986).

## 15.7   SUMMARY

Mountain sheep populations are distributed in space and time, shaped by vital rates of births, deaths, immigration, and emigration. How we characterize mountain sheep populations depends on the questions and the potential application of the model results. Also, our ability to produce reliable models depends on having sufficient data; models are constrained by the available data (Nagy-Reis et al. 2021). Models for mountain sheep are used to synthesize data on vital rates using demographic models (i.e., structured population models; Paterson et al. 2021). Underlying models for demography, such as exponential growth models, are inadequate for modeling sustainable yield (Mendelsohn 1976). Yet, vital rates can be modeled as functions of population density, animal condition, vegetation, and environmental variables, allowing such complexity to be built into the model (Varley and Boyce 2006, Boyce et al. 2012a). The interaction with vegetation can replace the need for density dependence but would require detailed plant-herbivore interaction that is seldom available. Nevertheless, broad-scale characterization of habitats using NDVI (Hurley et al. 2017, Barton et al. 2023, Enns et al. 2024) or food-scape models can be combined with density dependence in vital rates. In many jurisdictions mountain sheep are managed strictly for trophies, but to do so requires sound population management that cannot be achieved by harvesting only old males (Monteith et al. 2018).

## REFERENCES

Abadi, F., O. Gimenez, H. Jakober, W. Stauber, R. Arlettaz, and M. Schaub. 2012. Estimating the strength of density dependence in the presence of observation errors using integrated population models. *Ecological Modelling* 242:1–9.

Abrams, P. A. 2009. When does greater mortality increase population size? The long history and diverse mechanisms underlying the hydra effect. *Ecology Letters* 12:462–474.

Alstad, D. 2001. *Basic Populus Models of Ecology*, Prentice Hall, Hoboken, New Jersey, USA.

Atkin, O. K., B. Botman, and H. Lambers. 1996. The causes of inherently slow growth in alpine plants: an analysis based on the underlying carbon economies of alpine and lowland *Poa* species. *Functional Ecology* 10:698–707.

Barton, O., J. R. Healey, S. Cordes, A. J. Davies, and G. Shannon. 2023. Predicting the spatial expansion of an animal population with presence-only data. *Ecology and Evolution* 13:e10778.

Beale, M. M., K. Knopff, C. C. Small, and D. C. McLean. 2023. Carrying capacity and cumulative effects management: a case study using bighorn sheep. *Conservation Science and Practice* 2023:e13027.

Beale, M. M., and M. S. Boyce. 2020. Ungulate habitat selection on reclaimed coal mines in west-central Alberta. *Restoration Ecology* 28:828–840.

Bender, L. C., and M. E. Weisenberger. 2005. Precipitation, density, and population dynamics of desert bighorn sheep on San Andres National Wildlife Refuge, New Mexico. *Wildlife Society Bulletin* 33:956–964.

Bilodeau-Hussey, N. M., K. S. Huggler, E. F. Cassirer, H. Miyasaki, M. A. Hurley, L. A. Shipley, and R. A. Long. 2025. Effects of maternal condition, disease status, and behavior on survival of juvenile bighorn sheep. *Journal of Wildlife Management* 2025:e22721.

Blanchard, P., M. Festa-Bianchet, J.-M. Gaillard, and J. T. Jorgenson. 2003. A test of long-term fecal nitrogen monitoring to evaluate nutritional status in bighorn sheep. *Journal Wildlife Management* 67:477–484.

Bleich, V. C., G. Sargeant, and B. P. Wiedmann. 2018. Ecotypic variation in population dynamics of reintroduced bighorn sheep: implications for management. *Journal of Wildlife Management* 82:8–18.

Bleich, V. C., M. E. Blum, K. T. Shoemaker, D. Sustaita, and S. A. Holl. 2019. Habitat selection by bighorn sheep in a mesic ecosystem: the San Rafael Mountains, California, USA. *California Fish Game* 105:205–224.

Bonenfant, C., J. M. Gaillard, T. Coulson, M. Festa-Bianchet, A. Loison, M. Garel, L. E. Loe, P. Blanchard, N. Pettorelli, N. Owen-Smith, J. Du Toit, and P. Duncan. 2009. Empirical evidence of density dependence in populations of large herbivores. *Advances in Ecological Research* 41:313–357.

Botkin, D. B. 1990. *Discordant Harmonies: A New Ecology for the Twenty-First Century*. Oxford University Press, New York, New York, USA.

Boyce, M. S. 1993. Population viability analysis: adaptive management for threatened and endangered species. *Transactions of the North American Wildlife and Natural Resources Conference* 58:520–527.

Boyce, M. S., A. R. E. Sinclair, and G. C. White. 1999. Seasonal compensation of predation and harvesting. *Oikos* 87:419–426.

Boyce, M. S., C. A. E. Carpentier, and J. D. C. Linnell. 2025. Coexisting with large carnivores based on the Volterra Principle. *Conservation Biology* 39(1):e14448.

Boyce, M. S., C. V. Haridas, and C. Lee. 2006. Demography in an increasingly variable world. *Trends in Ecology and Evolution* 21:141–148.

Boyce, M. S., and D. J. Daley. 1980. Population tracking of fluctuating environments and natural selection for tracking ability. *American Naturalist* 115:480–491.

Boyce, M. S., K. H. Knopff, J. Northrup, J. Pitt, and L. S. Vors. 2012b. Harvest models and climate change. Pages 293–306 in J. Brodie, E. Post, and D. Doak, editors. *Conserving Wildlife Populations in a Changing Climate*. University of Chicago Press, Chicago, Illinois, USA.

Boyce, M. S., P. Baxter, and H. Possingham. 2012a. Managing moose harvests by the seat of your pants. *Theoretical Population Biology* 82:340–347.

Boyce, M. S., and P. Krausman. 2018. Controversies in mountain sheep management. *Journal of Wildlife Management* 82:5–7.

Brennan, L. A., D. G. Hewitt, and S. P. Mahoney. 2019. Social, economic, and ecological challenges to the North American model of wildlife conservation. Pages 130–147 in S. P. Mahoney, and V. Geist, editors. *The North American Model of Wildlife Conservation*. Johns Hopkins University Press, Baltimore, Maryland, USA.

Burnham, K. P., and D. R. Anderson. 2004. Multimodel inference: understanding AIC and BIC in model selection. *Sociological Methods and Research* 33:261–304.

Carpenter, T. E., V. L. Coggins, C. McCarthy, C. S. O'Brien, J. M. O'Brien, and T. J. Schommer. 2014. A spatial risk assessment of bighorn sheep extirpation by grazing domestic sheep on public lands. *Preventive Veterinary Medicine* 114:3–10.

Cassirer, E. F., K. R. Manlove, E. S. Almberg, P. L. Kamath, M. Cox, P. Wolff, A. Roug, J. Shannon, R. Robinson, R. B. Harris, B. J. Gonzales, R. K. Plowright, P. J. Hudson, P. C. Cross, A. Dobson, and T. E. Besser. 2018. Pneumonia in bighorn sheep: risk and resilience. *Journal of Wildlife Management* 82:32–45.

Caswell, H. 2001. *Matrix Population Models: Construction, Analysis, and Interpretation*. Second edition. Sinauer Associates, Sunderland, Massachusetts, USA.

Caughley, G. 1976. Wildlife management and the dynamics of ungulate populations. *Applied Biology* 1:183–246.

Caughley, G. 1982. Vegetation complexity and the dynamics of modelled grazing systems. *Oecologia* 54:309–312.

Caughley, G., and J. H. Lawton. 1981. Plant-herbivore systems. Pages 132–166 in R. M. May, editor. *Theoretical Ecology: Principles and Applications*. Second edition. Sinauer Associates Inc., Publishers, Sunderland, Massachusetts, USA.

Coltrane, J. A., and P. S. Barboza. 2010. Winter as a nutritional bottleneck for North American porcupines (*Erethizon dorsatum*). *Journal Comparative Physiology B* 180:905–918.

Cook, J. G., R. W. Davis, and L. L. Irwin. 2016. Nutritional ecology of elk during summer and autumn in the Pacific Northwest. *Wildlife Monographs* 195:1–81.

Coulson, T., J.-M. Gaillard, and M. Festa-Bianchet. 2005. Decomposing the variation in population growth into contributions from multiple demographic rates. *Journal Animal Ecology* 74:789–801.

Courtemanch, A. B. 2014. Seasonal habitat selection and impacts of backcountry recreation on a formerly migratory bighorn sheep population in northwest Wyoming, USA. Thesis, Department of Zoology and Physiology, University of Wyoming, Laramie, Wyoming, USA.

Crawley, M. J. 1983. *Herbivory: The Dynamics of Animal-Plant Interactions*. Studies in Ecology, Volume 10. University of California Press, Berkeley, California, USA.

DeCesare, N. J., and D. H. Pletcher. 2006. Movements, connectivity, and resource selection of Rocky Mountain bighorn sheep. *Journal of Mammalogy* 87:531–538.

Demarchi, R. 1986. Biophysical resources of the East Kootenay Area: wildlife. British Columbia Ministry of the Environment Lands and Parks, Victoria. *British Columbia Technical Report* 22:1–134.

Denryter, K., R. C. Cook, J. C. Cook, and K. L. Parker. 2022. Animal-defined resources reveal nutritional inadequacies for woodland caribou during summer-autumn. *Journal Wildlife Management* 86:1–32.

Douglas, C. L., and D. M. Leslie. 1986. Influence of weather and density on lamb survival of desert mountain sheep. *Journal of Wildlife Management* 50:153–156.

Eberhardt, L. L. 2002. A paradigm for population analysis of long-lived vertebrates. *Ecology* 83:2841–2854.

Enns, G., B. A. Jex, and M. S. Boyce. 2023. Diverse migration patterns and seasonal habitat use of Stone's sheep (*Ovis dalli stonei*). *Peer J* 11:e15215.

Enns, G., B. A. Jex, and M. S. Boyce. 2024. Stone's sheep (*Ovis dalli stonei*) lambing and nursery habitat selection. *Canadian Journal of Zoology* 102:69–707.

Festa-Bianchet, M., T. Coulson, J.-M. Gaillard, J. T. Hogg, and F. Pelletier. 2006. Stochastic predation events and population persistence in bighorn sheep. *Proceedings of the Royal Society*, B 273:1537–1543.

Gaillard, J.-M., M. Festa-Bianchet, and N. G. Yoccoz. 1998. Population dynamics of large herbivores: variable recruitment with constant adult survival. *Trends in Ecology & Evolution* 13:58–63.

Gaillard, J.-M., M. Hebblewhite, A. Loison, M. Fuller, R. Powell, M. Basille, and B. Van Moorter. 2010. Habitat–performance relationships: finding the right metric at a given spatial scale. *Philosophical Transactions of the Royal Society B: Biological Sciences* 365: 2255–2265.

Hanley, T. A., D. E. Spalinger, K. J. Mock, O. L. Weaver, and G. M. Harris. 2012. Forage resource evaluation system for habitat – deer: an interactive deer habitat model. U.S. Forest Service General Technical Report PNW-GTR-858, Portland, Oregon, USA. doi.org/10.2737/PNW-GTR-858

Harrison, X. A., J. D. Bloun, R. Inger, D. R. Norris, and S. Bearhop. 2011. Carry-over effects as drivers of fitness differences in animals. *Journal Animal Ecology* 80:4–18.

Hebblewhite, M., E. Merrill, and G. McDermid. 2008. A multi-scale test of the forage maturation hypothesis in a partially migratory ungulate population. *Ecological Monographs* 78:41–311.

Hobbs, N. T. 2024. A general, resource-based explanation for density dependence in populations of large herbivores. *Ecological Monographs* 2024:e1600.

Huggler, K. S. 2024. Feast or famine: interactive effects of life-history, nutrition, and behavior on performance of a long-lived herbivore. Dissertation, University of Idaho, Moscow, Idaho, USA.

Hurley, M., M. Hebblewhite, P. M. Lukacs, J. J. Nowak, J.-M. Gaillard, and C. Bonenfant. 2017. Regional-scale models for predicting overwinter survival of juvenile ungulates. *Journal of Wildlife Management* 81:364–378.

Jesmer, B. R., J. A. Merkle, J. R. Goheen, E. O. Aikens, J. L. Beck, A. B. Courtemanch, M. A. Hurley, D. E. McWhirter, H. M. Miyasaki, K. L. Monteith, and M. J. Kauffman. 2018. Is ungulate migration culturally transmitted? Evidence of social learning from translocated animals. *Science* 361:1023–1025.

Johnson, H. E., L. S. Mills, T. R. Stephenson, and J. D. Wehausen. 2010. Population-specific vital rate contributions influence management of an endangered ungulate. *Ecological Applications* 20:1753–1765.

Jorgenson, J. T., M. Festa-Bianchet, and W. D. Wishart. 1993. Harvesting bighorn ewes: consequences for population size and trophy ram production. *Journal of Wildlife Management* 57:429–435.

Kautz, T. M., J. L. Belant, D. E. Beyer Jr., B. K. Strickland, and J. F. Duquette. 2020. Influence of body mass and environmental conditions on winter mortality risk of a northern ungulate: evidence for a late-winter survival bottleneck. *Ecology and Evolution* 10:1666–1677.

Klinger R, T. Stephenson, J. Letchinge, L. Stephenson, and S. Jacobs. 2024. The noise is the signal: spatio-temporal variability of production and productivity in high elevation meadows in the Sierra Nevada mountain range of North America. *Frontiers in Ecology and Evolution* 11:1184918.

Knopff, K., and M. S. Boyce. 2007. Prey specialization by individual cougars (*Puma concolor*) in multi-prey systems. *Transactions of the North American Wildlife and Natural Resources Conference* 72:194–210.

Kolmogorov, A. N. 1936. Sulla teoria di Volterra della lotta per l'esistenza. *Giornale dell'Instituto Italiano degli Attuari* 7:74–80.

Koons, D. N., J. B. Grand, B. Zinner, and R. F. Rockwell. 2005. Transient population dynamics: relations to life history and initial population state. *Ecological Modelling* 185:283–297.

Leopold, A., L. K. Sowls, and D. L. Spencer. 1947. A survey of over-populated deer ranges in the United States. *Journal of Wildlife Management* 11:162–177.

Lewis, W. B., C. R. Nater, J. A. Rectenwald, D. C. Sisson, and J. A. Martin. 2024. Use of integrated population models for assessing density-dependence and juvenile survival in Northern Bobwhites (*Colinus virginianus*). *PeerJ* 12:e18625.

Ludwig, D. 1999. Is it meaningless to estimate the probability of extinction? *Ecology* 80:298–310.

Lukacs, P. M., M.S. Mitchell, M. Hebblewhite, B. K. Johnson, H. Johnson, M. Kauffman, M K. M. Proffitt, P. Zager, J. Brodie, K. Hersey, A. A Holland, M. Hurley, S. McCorquodale, A. Middleton, M. Nordhagen, J. J. Nowak, D. P. Walsh, and P.J. White. 2018. Factors influencing elk recruitment across ecotypes in the Western United States. *Journal Wildlife Management* 82:698–710.

Lula, E. S., B. Lowrey, K. M. Proffitt, A. R. Litt, J. A. Cunningham, C. J. Butler, and R. A. Garrott. 2020. Is habitat constraining bighorn sheep restoration? A case study. *Journal of Wildlife Management* 84:588–600.

Lynch, G. M. 2006. Does First Nation's hunting impact moose productivity in Alberta? *Alces* 41:25–32.

MacCallum, B. 1989. Seasonal and spatial distribution of big-horn sheep at an open pit coal mine in the Alberta foothills. *Proceedings America Society of Mining and Reclamation* 1989:141–150.

MacCallum, N. B., and V. Geist. 1992. Mountain restoration: soil and surface wildlife habitat. *GeoJournal* 27:23–46.

Mahoney, S., and V. Geist. 2019. *The North American Model of Wildlife Conservation.* Johns Hopkins University Press, Baltimore, Maryland, USA.

Maynard Smith, J. 1974. *Models in Ecology.* Cambridge University Press, Cambridge, United, Kingdom.

McCullough, D. R. 1990. Detecting density dependence: filtering the baby from the bathwater. *Transactions of the North American Wildlife and Natural Resources Conference* 55:535–543.

McKinney, T., T. W. Smith, and J. D. Hanna. 2001. Precipitation and desert bighorn sheep in the Mazatzal Mountains, Arizona. *Southwestern Naturalist* 46:345–353.

McNaughton, S. J., M. Oesterheld, D. A. Frank, and K. J. Williams. 1989. Ecosystem-level patterns of primary productivity and herbivory in terrestrial habitats. *Nature* 341:142–144.

Mendelssohn, R. 1976. Optimization problems associated with a Leslie matrix. *American Naturalist* 110:339–349.

Merems, J. L., L. A. Shipley, T. Levi, J. Ruprecht, D. A. Clark, M. J. Wisdom, N. J. Jackson, K. M. Stewart, and R. A. Long. 2020. Nutritional-landscape models link habitat use to condition of mule deer (*Odocoileus hemionus*). *Frontiers in Ecology and Evolution* 8(98):1–13.

Mitchell, C. D., R. Chaney, K. Aho, J. G. Kie, and R. T. Bowyer. 2015. Population density of Dall's sheep in Alaska: effects of predator harvest? *Mammal Research* 60:21–28.

Monteith, K., R. Long, T. Stephenson, V. C. Bleich, R. T. Bowyer, and T. LaSharr. 2018. Horn size and nutrition in mountain sheep: can ewe handle the truth? *Journal of Wildlife Management* 82:67–84.

Monteith, K. L., T. R. Stephenson, V. C. Bleich, M. M. Conner, B. M. Pierce, and R. T. Bowyer. 2016. Risk-sensitive allocation in seasonal dynamics of fat and protein reserves in a long-lived mammal. *Journal of Animal Ecology* 82:377–388.

Monteith, K. L., V. C. Bleich, T. R. Stephenson, B. M. Pierce, M. M. Conner, J. G. Kie, and R. T. Bowyer. 2014. Life-history characteristics of mule deer: effects of nutrition in a variable environment. *Wildlife Monographs* 186:1–62.

Nagy-Reis, M., J. Reimer, M. A. Lewis, W. F. Jensen, and M. S. Boyce. 2021. Aligning population models with data: adaptive management for big game harvests. *Global Ecology and Conservation* 26:e1501.

Norris, D. R. 2005. Carry-over effects and habitat quality in migratory populations. *Oikos* 109:178–186.

O'Brien, J. M., T. E. Carpenter, C. S. O'Brien, and C. McCarthy. 2014. Incorporating foray behavior into models estimating contact risk between bighorn sheep and areas occupied by domestic sheep. *Wildlife Society Bulletin* 38:321–331.

Papouchis, C. M., F. J. Singer, and W. B. Sloan. 2001. Responses of desert bighorn sheep to increased human recreation. *Journal Wildlife Management* 65:573–582.

Paterson, J. T., K. Proffitt, J. Rotella, D. McWhirter, and R. Garrott. 2021. Drivers of variation in the population dynamics of bighorn sheep. *Ecosphere* 12(7):e03679.

Pettorelli, N., S. J. Ryan, T. Mueller, and N. Bunnefeld. 2011. The normalized difference vegetation index (NDVI): unforeseen successes in animal ecology. *Climate Research* 46:15–27.

Poole, K. G., R. Serrouya, I. E. Teske, and K. Podrasky. 2016. Rocky Mountain bighorn sheep (*Ovis canadensis canadensis*) winter habitat selection and seasonal movements in an area of active coal mining. *Canadian Journal of Zoology* 94:733–745.

Portier, C., M. Festa-Bianchet, J.-M. Gaillard, J. T. Jorgenson, and N. G. Yoccoz. 1998. Effects of density and weather on survival of bighorn sheep lambs (*Ovis canadensis*). *Journal of Zoology, London* 245:271–278.

Proffitt, K. M., A. B. Courtemanch, S. R. Dewey, B. Lowrey, D. E. McWhirter, K. L. Monteith, J. T. Paterson, J. Rotella, P. J. White, and R. A. Garrott. 2021. Regional variability in pregnancy and survival rates of Rocky Mountain bighorn sheep. *Ecosphere* 12:e03410.

Rankins, S. T., T. R. Stephenson, and K. L. Monteith. 2024. Sociality modulates nutritional carrying capacity of an endangered species. *Frontiers Ecology Evolution* 12:1417970.

Renaud, L. A., G. Pigeon, M. Festa-Bianchet, and F. Pelletier. 2019. Phenotypic plasticity in bighorn sheep reproductive phenology: from individual to population. *Behavioral Ecology and Sociobiology* 73:50. doi.1007/s00265-019-2656-1

Rominger, E. M. 2018. The Gordian knot of mountain lion predation and bighorn sheep. *Journal of Wildlife Management* 82:19–31.

Rubin, E. S., W. M. Boyce, and E. R. Caswell-Chen. 2002. Modeling demographic processes in an endangered population of bighorn sheep. *Journal of Wildlife Management* 66:796–810.

Schaub, M., O. Gimenez, A. Sierro, and R. El Arlettaz. 2007. Use of integrated modeling to enhance estimates of population dynamics obtained from limited data. *Conservation Biology* 21:945–955.

Shenk, T. M., G. C. White, and K. P. Burnham. 1998. Sampling-variance effects on detecting density dependence from temporal trends in natural populations. *Ecological Monographs* 68:445–463.

Shipley, L. A. 2007. The influence of bite size on foraging at larger spatial and temporal scales by mammalian herbivores. *Oikos* 116:1964–1974.

Simard, M. A., T. Coulson, A. Gingras, and S. D. Côté. 2010. Influence of density and climate on population dynamics of a large herbivore under harsh environmental conditions. *Journal of Wildlife Management* 74:1671–1685.

Singer, F. J., E. Williams, M. W. Miller, and L. C. Zeigenfuss. 2000. Population growth, fecundity, and survivorship in recovering populations of bighorn sheep. *Restoration Ecology* 8:75–84.

Skarpe, C., and A. J. Hester. 2010. Plant traits, browsing and grazing herbivores, and vegetation dynamics. Pages 217–261 in I. J. Gordon and H. H. T. Prins, editors. *The Ecology of Browsing and Grazing.* Ecological Studies, 195. Springer, Berlin, Heidelberg, Germany.

Smiley, R. A., B. L. Wagler, T. N. LaSharr, K. A. Denryter, T. R. Stephenson, A. B. Courtemanch, T. W. Mong, D. Lutz, D. McWhirter, D. Brimeyer, P. Hnilicka, B. Lowrey, and K. L. Monteith. 2022. Heterogeneity in risk-sensitive allocation of somatic reserves in a long-lived mammal. *Ecosphere* 13:1–15.

Smiley, R. A., B. L. Wagler, W. H. Edwards, J. Jennings-Gaines, K. Luukkonen, K. Robbins, M. Johnson, A. B. Courtemanch, T. W. Mong, D. Lutz, D. McWhirter, J. L. Malmberg, B. Lowrey, and K. L. Monteith. 2024. Infection-nutrition feedbacks: fat supports pathogen clearance but pathogens reduce fat in a wild mammal. *Proceedings of the Royal Society B-Biological Science* 291:20240636.

Smith, T. S., P. J. Hardin, and J. T. Flinders. 1999. Response of bighorn sheep to clear-cut logging and prescribed burning. *Wildlife Society Bulletin* 27:840–845.

Sproat, K. K., N. R. Martinez, T. S. Smith, W. B. Sloan, J. T. Flinders, J. W. Bates, J. G. Cresto, and V. C. Bleich. 2020. Desert bighorn sheep responses to human activity in south-eastern Utah. *Wildlife Research* 47:16–24.

Stelfox, B. 2005. ALCES: Landscape and land use. www. ALCES.ca

Stephenson, T. R., D. W. German, E. F. Cassirer, D. P. Walsh, M. E. Blum, M. Cox, K. M. Stewart, and K. L. Monteith. 2020. Linking population performance to nutritional condition in an alpine ungulate. *Journal of Mammalogy* 101:1244–1256.

Tredennick, A. T., G. Hooker, S. P. Ellner, and P. B. Adler. 2021. A practical guide to selecting models for exploration, inference, and prediction in ecology. *Ecology* 102:e03336.

Turgeon, R., F. Pelletier, S. D. Côté, M. Festa-Bianchet, and S. Hamel. 2024. Sporadic events have a greater influence on the dynamics of small, isolated populations than density dependence and environmental conditions. *American Naturalist* 204:574–588.

Turner, J. C., C. L. Douglas, C. R. Hallum, and P. R. Krausman. 2004. Determination of critical habitat for the endangered Nelson's bighorn sheep in southern California. *Wildlife Society Bulletin* 32:427–448.

Varley, N., and M. S. Boyce. 2006. Adaptive management for reintroductions: updating a wolf recovery model for Yellowstone National Park. *Ecological Modelling* 193:315–339.

Wagler, B. L., R. A. Smiley, A. B. Courtemanch, D. Lutz, D. McWhirter, D. Brimeyer, P. Hnilicka, T. J. Robinson, and K. L. Monteith. 2023. Implications of forage quality for population recovery of bighorn sheep following a pneumonia epizootic. *Journal of Wildlife Management* 87:e22452.

Walker, P. D. 2023. Summer spatial ecology of woodland caribou across northern Ontario. Dissertation, University of Alberta, Edmonton, Canada.

Walters, C. J. 1986. *Adaptive Management of Renewable Resources*. Collier Macmillan Publishers, London, United Kingdom.

Wang, G., N. T. Hobbs, R. B. Boone, A. W. Illius, I. J. Gordon, J. E. Gross, and K. L. Hamlin. 2006. Spatial and temporal variability modify density dependence in populations of large herbivores. *Ecology* 87:95–102.

Wang, G., N. T. Hobbs, S. Twombly, R. B. Boone, A. W. Illius, I. J. Gordon, and J. E. Gross. 2009. Density dependence in northern ungulates: interactions with predation and resources. *Population Ecology* 51:123–132.

Werdel, T. J., J. A. Jenks, J. T. Kanta, C. P. Lehman, and T. J. Frink. 2023. Resource selection and herbaceous biomass at foraging sites of translocated bighorn sheep. *Rangeland Ecology and Management* 87:141–149.

Western Association of Fish and Wildlife Agencies. 2024. Range-wide status of bighorn and thinhorn sheep in North America. Wild Sheep Initiative, Western Association of Fish and Wildlife Agencies (ver. 07082024). Boise, Idaho, USA.

White, G. C. 2000. Population viability analysis: data requirements and essential analyses. Pages 288–331 *in* L. Boitani and T. K. Fuller, editors. *Research Techniques in Animal Ecology*. Columbia University Press, Ithaca, New York, USA.

Williams, B. K., J. D. Nichols, and M. J. Conroy. 2001. *Analysis and Management of Animal Populations*. Academic Press, New York, New York, USA.

Wilson-Henjum, G., L. Ricci, D. C. Stoner, K. Longshore, and K. R. Manlove. 2025. Desert bighorn sheep home range and disease transmission risk responses to temporally dynamic environmental variation. *Journal of Wildlife Management* 2025:e22715.

Xu, C., M. S. Boyce, and D. J. Daley. 2005. Harvesting in seasonal environments. *Journal of Mathematical Biology* 50:663–682.

# 16 Harvest Management of Mountain Sheep in North America

*Chadwick Lehman, Todd J. Brinkman, and Brad R. Wendling*

## 16.1 INTRODUCTION

Mountain sheep (bighorn sheep [*Ovis canadensis*], thinhorn sheep [*Ovis dalli*]) symbolize wildness as they are renowned for their ability to negotiate rugged terrain and thrive in some of the most remote and inaccessible areas of North America. The uniqueness of mountain sheep behavior, distribution, and habitat use contributes to their high value (e.g., ecological, economic, cultural) as big-game species, providing relatively rare hunting opportunities. These unique characteristics require distinctive harvest management strategies. North American mountain sheep include Dall's (*O. dalli dalli*), Stone's (*O. dalli stonei*), and bighorn sheep (*O. canadensis*) (Valdez and Krausman 1999, Chapter 1). Thinhorn (Dall's and Stone's sheep) and bighorn are distinctively different phenotypically and subpopulations of mountain sheep have been geographically clustered based on genetics (Sim et al. 2019). Current sheep populations are much lower in most areas as compared to historic levels. Sheep interest groups have collectively agreed that efforts should be implemented to restore mountain sheep populations across their native range to address conservation concerns and create more hunting opportunities. We provide a history and timeline of harvest management for mountain sheep, going back to early regulations just after population declines and extirpation of bighorns, and the market hunting and unregulated overharvest of thinhorns. The chapter then focuses on current harvest management, the defining regulations, and considerations for future harvest management approaches. The chapter separates bighorns and thinhorns because of different causes of population declines, management and regulation strategies, and timelines.

## 16.2 BIGHORNS

Before their decimation in the late 1800s and early 1900s, bighorns were widely distributed across the prairies and mountain ranges of Canada to northern Mexico and from the Pacific coast eastward to the western areas of the Dakotas, Nebraska, and Texas. Mountain sheep numbers in the contiguous United States were roughly 2 million before their decline (Seton 1929). Following Western settlement, bighorn numbers declined precipitously and were extirpated from a significant portion of their historic range. A reference in South Dakota described how quickly bighorns were decimated across their range. Bighorn sheep were nearly eliminated from the Black Hills by about 1887 until 1899

when the last one was killed (Seton 1929). Today, there are fewer than 65,000 bighorn in North America (Wild Sheep Initiative 2023). Diseases and competition associated with domestic livestock, human encroachment, and unregulated market hunting were all connected to the decline of bighorn sheep in the latter half of the 19th century (Buechner 1960, Wishart 1978, Chapters 2, 3, and 8).

In the early 1900s, concerned sportsmen and conservationists saw the plight of bighorns and established sanctuaries and parks in an attempt to slow population declines. Reintroduction programs and regulated harvests revitalized populations from extirpation and are a conservation success story. This success story only occurred through the dedication and collaborative forces of wildlife and land management agencies, conservation organizations, hunters, Indigenous groups, and many other interested contributors (Chapter 2). Nonetheless, bighorn sheep occupy only a small portion of their Indigenous range and across North America current population estimates are far less than their historical abundance. The conservation of mountain sheep requires foresight and an astute understanding of how harvest influences mountain sheep in their remaining occupied range. With the excessive pressures of disease from contact with domestic livestock (sheep, goats) and loss of habitat due to human encroachment, we can afford little error when it comes to making sustainable harvest management decisions related to mountain sheep.

Recent attention associated with the effects of harvest on game species over several decades has spurred interest among scientists and wildlife managers in understanding the effects of harvest (Festa-Bianchet 2017, Monteith et al. 2018, Heffelfinger 2018). Bighorn sheep are unique among large ungulates in North America as they are almost exclusively managed as a trophy species for their large horns (Monteith et al. 2013, 2018; LaSharr et al. 2019). The expectation of mature males being available by bighorn hunters and the promotion of fund-raising tags, which emphasize large-horned bighorn sheep, furthers the management strategy of trophy management versus other potential strategies and is somewhat controversial (Damm 2008, Festa-Bianchet 2012, Simon 2016). Despite this, funding generated through the auctioning of harvest permits comprises a significant proportion of agency mountain sheep program funding (Wild Sheep Foundation 2024).

Bighorn jurisdictions typically employ conservative harvest practices to maintain larger horn size and these

DOI: 10.1201/9781003518686-18

strategies generally differ from the management of other species of big game in North America (LaSharr et al. 2019). Most management strategies include male-only harvest regulations that limit harvest rate to less than 4% of the bighorn population (Jorgenson et al. 1993). Harvest of large-horned males has led to a debate over the influence of removing these animals from the population. A case study indicated males from one population in Alberta, Canada, were genotypically smaller than five generations previous (Coltman et al. 2003, Pigeon et al. 2016). Regulations in this study area initially allowed for a higher rate of harvest of fast-growing males who were shot at a young age before they could breed (Coltman et al. 2003, Pigeon et al. 2016), and regulations were later changed to reduce the harvest rate and restrict harvest to full-curl males (in 1997), to prevent younger males from being harvested before they could reproduce (Festa-Bianchet et al. 2006). Another study evaluated trophy harvest scoring records (e.g., Boone and Crockett scoring system) over a large geographic area in North America encompassing 35 years and they reported the majority of hunt areas had stable or increasing horn size in mountain sheep (LaSharr et al. 2019); however, this same study also noted that age-specific horn size declined in 44% of hunt areas where harvest was regulated solely by morphological criteria. This secondary finding supports the notion that harvest practices that are simultaneously selective and intensive might lead to changes in horn growth (LaSharr et al. 2019). Hunt areas that had regulations for increased opportunity such as Alberta most likely saw a reduction in horn size related to harvest intensity and potential genetic selection against larger horned young males (Coltman et al. 2003, Festa-Bianchet et al. 2006, Pigeon et al. 2016, LaSharr et al. 2019). Across a range of hunt areas and harvest strategies, horn size can be related to age, environmental factors, harvest intensity, and demographics (Coltman et al. 2003, Pigeon et al. 2016, Monteith et al. 2018, LaSharr et al. 2019).

## 16.3 THINHORNS

During late 1800s and early 1900s, market hunting and unregulated overharvest to support mining and prospecting activities occurred across thinhorn range and caused significant declines in population numbers in localized areas (Demarchi and Hartwig 2004). The establishment of some national parks (e.g., Denali National Park) was partially motivated to protect thinhorn sheep from excessive harvest (Rawson 2001). After market hunting was banned and regulations were established in the early 1900s, populations were thought to have recovered (Murie 1944). Still, the horn curl aspects of harvest regulations continued to evolve due to conservation and population concerns, and today male harvest is predominantly restricted to mature and full-curl males, across all jurisdictions (Demarchi 1978, Taras and Manning 2017). Additional parks, protected areas, and special management zones have also been established in more recent times by some jurisdictions to again protect thinhorn sheep from excessive harvest pressure (e.g., Todagin South Slopes Park and Wildlife Management Area, British Columbia in 2001). Unlike bighorns, there is limited evidence of a long-term effect on population numbers and distribution from past unregulated hunting.

Since harvest was regulated, the causes of past and recent decline are likely related to winter events and climate-related changes in habitat (Murie 1944, Rattenbury et al. 2018, Van de Kerk et al. 2020), rather than from overharvest. Harvest strategies for thinhorn sheep have mainly been directed at sustaining healthy population structure (demography, age composition), and attempting to reverse population declines while preserving hunting opportunity. As indirect effect of the conservative strategy, and similar to bighorns, harvest has mainly focused on mature males with regulations that limit harvest to older individuals with larger horns (e.g., full-curl). The focus on mature males has been misconstrued as trophy management. These management strategies that limit harvest to a certain degree of horn curl and a general-open-season or over-the-counter license are intended to balance conservation and hunting opportunity (Demarchi and Hartwig 2004, Alaska Department of Fish and Game 2024). Similar to conservative harvest strategies for bighorns, highly accessible areas or where populations of thinhorns may be vulnerable to high levels of hunter harvest, most jurisdictions have at least one area or zone that is managed through a restricted lottery or draw, harvest permit structure.

## 16.4 BIGHORN SHEEP

### 16.4.1 HARVEST HISTORY AND EARLY REGULATIONS

The decline of North American bighorn sheep was associated with the expansion of people across the North American West and the disappearance of the frontier (Turner 1935, Buechner 1960, Chapter 2). Each state or jurisdiction documented the decline of bighorns in the late 1800s and early 1900s, and although historical information is sparse, it was clear in most jurisdictions that conservation measures were needed to try and stem the demise of bighorn sheep across the West. Harvest of bighorn sheep as with most large ungulates was unregulated through the late 1800s, and in South Dakota bighorn sheep were extirpated in 1899 (Seton 1929). A similar trend occurred in Nebraska, Washington, and Wyoming as market hunting and introductions of disease from livestock resulted in diminished populations and the extirpation of bighorn sheep in some regions, by the early 1900s (Cowan 1940, Honess and Frost 1942, Nebraska Game and Parks Commission 2019, Washington Department of Fish and Wildlife 2015, Chapters 2 and 8). Oregon implemented protections for bighorns from harvest starting in 1911 but that did not prevent the extirpation of bighorns in the 1940s caused by over-exploitation and domestic livestock diseases (Oregon Department of Fish and Wildlife 2003).

Some jurisdictions were more proactive in responding to the decline of bighorns; Montana, Nevada, and California were ahead of other western states in regulating bighorn sheep hunting. In Montana, starting in 1872, legislators passed its first hunting regulation by limiting the season length; further, in 1895, Montana limited licenses to eight bighorn sheep and ultimately closed bighorn hunting altogether in 1915 (Montana Department of Fish, Wildlife and Parks 2010). California passed laws in 1878 to protect bighorn sheep from hunting (United States Fish and Wildlife Service 2007). Nevada enacted a law for bighorn sheep regulating season length starting in 1861 and closed bighorn hunting in 1901 (Nevada Division of Wildlife 2001).

Bighorn sheep were not extirpated in Colorado, Idaho, Nevada, New Mexico, and Utah, or in the province of Alberta, but populations were low or severely limited by the early 1900s (Alberta Fish and Wildlife Services 1993, Nevada Division of Wildlife 2001, Colorado Parks and Wildlife 2009, Idaho Fish and Game 2010, New Mexico Department of Game and Fish 2003, Utah Division of Wildlife Resources 2008). Records from these jurisdictions indicated that protection of bighorns from extirpation was concomitant with remote rugged terrain. Remnant populations were in wilderness areas and such isolation prevented domestic sheep and goats from getting close to mountain sheep, and humans were provided only limited access for subsistence or market hunting.

After bighorns were extirpated from a large portion of their historical range restoration programs and conservation-focused regulations eventually brought bighorn populations back to sustainable levels, providing hunting opportunity (Buechner 1960, Tefethen 1975, Hoefs 1985). This rebound in some populations led to harvest regulations, which were conservative in the 1960s and 1970s, and a review of various hunting regulations across North America focused harvest management on herd growth while providing a limited harvest on surplus males (Demarchi 1978). Early in the management process of the 1960s and 1970s most jurisdictions imposed a horn-size regulation with various definitions of minimum dimensions of horn length (Demarchi 1978). For example, from 1937 to 1985 bighorn sheep harvest in Wyoming was regulated by a limited permit system and harvest was restricted to at least three-fourths curl horns (Emmerich 1986). Other horn curl dimensions used by various agencies to regulate size of males harvested included one-half, four-fifths, or full (Rominger 2008, Figure 16.1).

A 2007 survey that included 20 North American jurisdictions in which hunting of bighorn sheep was permitted indicated most used limited entry draws, although over-the-counter and general-open-season licenses, also known as unlimited license hunting, occurs in much of Alberta, parts of British Columbia, and Montana (Rominger 2008). Further, most jurisdictions have implemented any male with no horn curl or age restriction in at least some of their hunt units. Exceptions include California, Colorado, portions of British Columbia, and Alberta, where one-half,

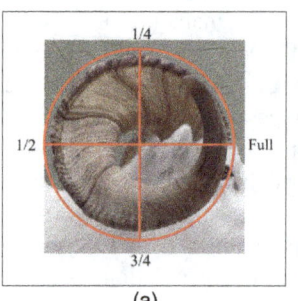 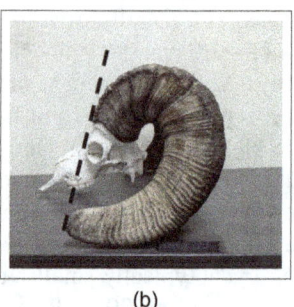

(a)                                  (b)

**FIGURE 16.1** A general guide for hunters to judge the difference in horn dimensions related to regulations for (a) one-half, three-fourths, and full-curl (Alaska Department of Fish and Game 2016). The province of Alberta defines (b) a four-fifths curl male as a straight line drawn from the most anterior point of the base of the horn to the tip of the horn extending beyond the anterior edge of the eye when viewed in profile (Alberta Fish and Wildlife Services 1993). Definitions of legal horn size vary by jurisdiction, and hunters should study the regulations before heading to the field.

three-fourths, four-fifths, full-curl, or mature male restrictions are in place (Figure 16.1).

Liberal hunting seasons in Montana from the late 1800s into the early 1900s resulted in the closure of bighorn hunting from 1915 to 1953. Hunting was reinitiated in 1953 with a limited draw of 30 permits, for three-fourths curl males, and permits were issued in three different areas (McCarthy 1986). Since 1956, Montana provided a diverse range of hunting options, including unlimited (over-the-counter or general-open-season licenses) hunting areas in at least three different areas in the state. Unlimited hunt zones are an active effort to offer a unique hunt opportunity to the public in the contiguous United States. In addition to other demographic data collected, quotas in these areas are based on numbers of three-fourths curl males observed in wintering areas. Although these units were unlimited in opportunity, they are protected from overharvest by having a threshold or maximum harvest quota set, limiting the number of animals that could be harvested in a given season. Successful hunters must report kills to the department within 48 hours and seasons are closed once the harvest quota is met. Rugged terrain and difficult access in these areas severely limit hunter success rates, enabling these unique hunts to provide maximized hunting opportunity without affecting the bighorn population (McCarthy 1986). Montana has also at times provided any-male and either-sex seasons. Montana Fish Wildlife and Parks have also implemented a one-half curl restriction in some units which is one-half curl or less, to remove younger males, three-fourths curl or better males, and female seasons depending on unit objectives (McCarthy 1986). Half curl seasons were only used for a few years in the 1980s and are no longer in place.

Limited hunting of males occurred in Oregon since 1965 and the primary harvest regulation through this period has been three-fourths curl or larger males (Van Dyke 1990). An any-male rule was put into place in 1991. Comparison of scores and age for harvested three-fourths curl or larger

(1965–1990) and any-male (1991–2004) showed slight increases in average age and average Boone and Crockett score (Oregon Division of Fish and Wildlife, unpublished data).

In Utah, legal hunting for desert bighorns resumed in 1967; and was primarily limited to a mature male until 1985. A mature male was defined as either 7 years of age or at least 144 Boone and Crockett points. Compliance with the age requirement was determined via horn annuli counts and scoring was assessed using a Boone and Crockett score or horn measurement method of size (Figure 16.2). This created situations where hunters misjudged animals and were given tickets for illegally harvesting non-trophy males and this stimulated a change in regulations to an any-male regulation (Smith and Flinders 1990). Utah does provide female seasons depending on unit objectives.

The primary approach for managing bighorn sheep limited harvest to <4% of the bighorn population (Jorgenson et al. 1993). An exception to the limited harvest idea occurred in few units in Montana as mentioned earlier, but also at Ram Mountain in Alberta, Canada (Coltman et al. 2003). In Alberta residents could harvest males with an unlimited harvest rate, cap or quota, but legal males had to be four-fifths curl or larger. Male bighorns complete about 80% of horn growth during the first 5 years of life and under four-fifths curl regulation, some males can reach legal size as early as 3 years of age (Jorgenson et al. 1998). Horn size appears to play little or no role in mating success until a bighorn male reaches 6–8 years old, and as a dominant male 8 years or older can tend estrous females (Festa-Bianchet et al. 2006). Over 30 years of unlimited harvest of four-fifths curl males from the Ram Mountain population unit, the average breeding value for horn and body size genetically correlated and declined due to harvest pressure (Coltman et al. 2003). Males, both phenotypically and genotypically, were smaller

than five generations previous. Fast-growing males were shot at a young age before they could breed at a higher rate, restricting the genetic contributions from young males who had fast-growing horns. Regulations changed to a full-curl in a small number of units in 1997 to try and allow those large-horned, younger males the opportunity to breed at least one time before being harvested (Festa-Bianchet et al. 2006). Today, however, most units in Alberta have remained at the four-fifths curl or larger restriction, but harvest pressure is managed through season length and, in some areas, limited opportunities through draws.

Agencies now have some experience of seeing how bighorn populations responded to earlier harvest strategies and observe population fluctuations over 40 years or more. With bighorn sheep populations fluctuating greatly due to habitat changes, disease epidemics, and recovery through reintroductions and other recovery programs, it was necessary for agencies to provide some guidelines on what triggers the opening or closing of hunting seasons. The minimum population size to hunt bighorns varied from 50 to 100 in most cases, but some jurisdictions opened very restricted harvest seasons with 25 animals available (Rominger 2008). Some agencies also consider lamb:female ratios, male:female ratios, male age structure, and public access as key components of what triggers opening (or maintaining) a hunting season.

Harvest of bighorn sheep (including all subspecies) remained relatively low (<1,300 annually) from 1990 through 2000 (Table 16.1; Figure 16.3; Chapter 3). Harvest of males has certainly been conservative, with data indicating the harvest of males/100 bighorn sheep ranges from 1.3 to 3.5 males/100 sheep (Rominger 2008). The highest ratios were from Montana and Wyoming at 3.5 males/100 bighorn sheep and lowest were 1.3 rams/100 bighorn sheep for Texas, and 1.5 rams/100 bighorn sheep for Arizona. With an estimated 72,000 bighorn sheep in the United States and

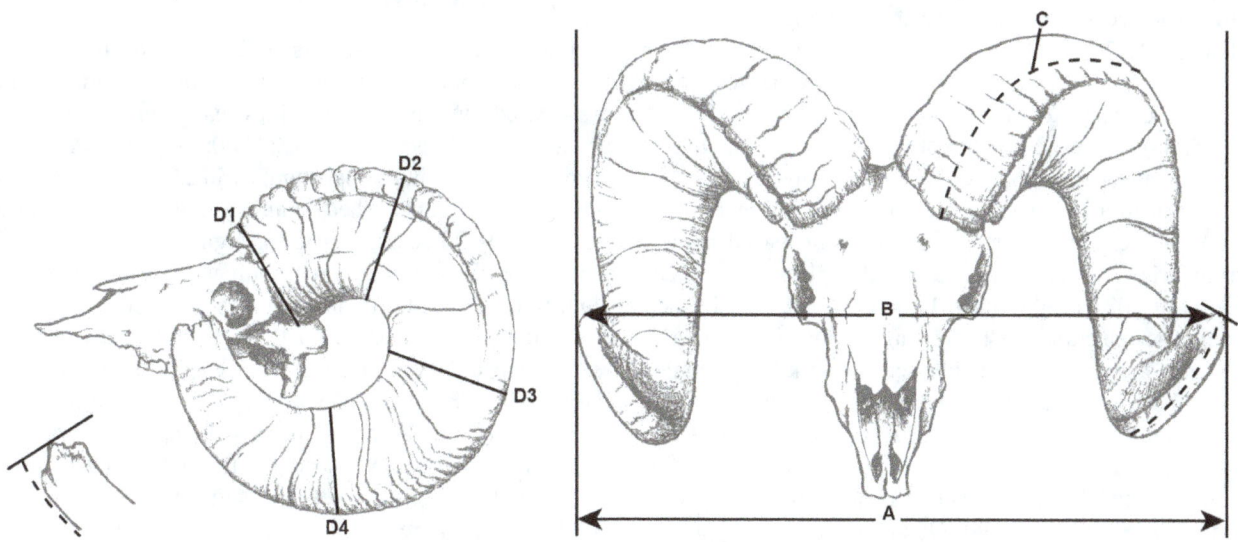

FIGURE 16.2 Measurement system used by the Boone and Crockett Record Book Program (Buckner and Reneau 2009). The full score was calculated as the cumulative sum of C for both horn lengths and all D circumference measurements for both horns.

TABLE 16.1

**Number of Licenses Issued and Harvest of Bighorn, Dall's, and Stone's Sheep in North America, 1990–2023**

| | Bighorn Sheep[a] | | | | Dall's Sheep[b] | | | | Stone's Sheep[b] | | | |
|---|---|---|---|---|---|---|---|---|---|---|---|---|
| Year | Licenses Issued | Rams Harvested | Ewes Harvested | Total Harvested | Licenses Issued | Rams Harvested | Ewes Harvested | Total Harvested | Licenses Issued | Rams Harvested | Ewes Harvested | Total Harvested |
| 1990 | 4,281 | 1,276 | 0 | 1,276 | 9,367 | 1,882 | 0 | 1,882 | 4,250 | 466 | 0 | 466 |
| 2000 | 3,298 | 1,223 | 0 | 1,223 | 8,285 | 1,124 | 0 | 1,124 | 1,910 | 280 | 0 | 280 |
| 2010 | 3,869 | 1,509 | 0 | 1,509 | 7,862 | 1,151 | 0 | 1,151 | 2,517 | 333 | 0 | 333 |
| 2011 | 3,705 | 1,365 | 0 | 1,365 | 7,916 | 1,194 | 0 | 1,194 | 2,501 | 330 | 0 | 330 |
| 2012 | 4,348 | 1,425 | 139 | 1,564 | 8,081 | 1,151 | 1 | 1,152 | 2,657 | 334 | 0 | 334 |
| 2013 | 4,195 | 1,396 | 145 | 1,541 | 8,147 | 1,140 | 2 | 1,142 | 2,692 | 339 | 0 | 339 |
| 2014 | 4,422 | 1,382 | 230 | 1,612 | 7,686 | 1,118 | 4 | 1,122 | 2,731 | 323 | 0 | 323 |
| 2015 | 4,714 | 1,469 | 318 | 1,787 | 7,506 | 1,110 | 7 | 1,117 | 2,815 | 298 | 1 | 299 |
| 2016 | 4,673 | 1,484 | 371 | 1,855 | 7,800 | 1,217 | 2 | 1,219 | 2,890 | 334 | 0 | 334 |
| 2017 | 4,524 | 1,505 | 331 | 1,836 | 7,504 | 1,230 | 4 | 1,234 | 2,846 | 330 | 1 | 331 |
| 2018 | 4,523 | 1,507 | 393 | 1,900 | 7,307 | 1,180 | 7 | 1,187 | 3,090 | 334 | 0 | 334 |
| 2019 | 4,584 | 1,523 | 374 | 1,897 | 7,977 | 1,278 | 2 | 1,280 | 3,164 | 290 | 0 | 290 |
| 2020 | 4,930 | 1,526 | 350 | 1,876 | 7,850 | 835 | 0 | 835 | 3,299 | 276 | 1 | 277 |
| 2021 | 3,922 | 1,503 | 388 | 1,891 | 7,334 | 861 | 0 | 861 | 3,419 | 330 | 0 | 330 |
| 2022 | 3,923 | 1,420 | 296 | 1,716 | 5,976 | 651 | 0 | 651 | 2,832 | 0 | 0 | 0 |
| 2023 | 3,924 | 1,381 | 291 | 1,672 | 5,643 | 565 | 0 | 565 | 102 | 0 | 0 | 0 |

[a] Bighorn sheep include all subspecies.

[b] Licenses issued doesn't accurately represent the number of people that actually hunted sheep because many resident hunters acquire over-the-counter licenses for all big-game species at the beginning of the season even when they may not intend to actively hunt sheep.

Canada in 2007 and approximately 1,310 males harvested, the average harvest ratio would be 1.8 males/100 bighorn sheep (Rominger 2008).

Male age at harvest was also closely monitored, and the percent of males that were 8 years old or older at harvest ranged from 30% to 73%, with a mean of 51% (Rominger 2008). The number of males harvested annually ranged from one to two in Nebraska to roughly 200 in Wyoming. Hunter harvest success ranged from 44% to 100% with a mean of 85%, and the majority of jurisdictions had harvest success rates ≥90%. In unlimited entry hunts in Alberta and Montana success rates were 7.5% in Alberta and 6.5% in Montana. In 2005, 43% of hunter numbers were from four unlimited entry hunts; however, 6% of the harvest came from these units (Rominger 2008).

Harvest of female bighorn sheep was rarely utilized as a management tool prior to 2007, with only six jurisdictions indicating they had female harvests (Rominger 2008); female harvest management as a tool has increased slightly as ten jurisdictions indicate they could implement this practice as of 2023 (Wild Sheep Initiative 2023, unpublished data). The reluctance of harvesting females is potentially related to agencies facing public scrutiny as killing females might be viewed as an unacceptable practice by their constituency (Monteith et al. 2018). Nonetheless, some agencies have and continue to provide female hunts with historical numbers of permits issued annually ranging from 95 to 374 (Rominger 2008). On an annual basis typically fewer than 300 females were harvested; Alberta harvested ~100 females per year, Colorado ~40, and Montana ~125. In Colorado primarily herds >100 bighorns were eligible

to be permitted for female hunting. Harvest success rates for female hunts is much lower than for males and hunter harvest success for females was 44% in Alberta, 39% in Colorado, and 75% in Montana. None of the jurisdictions harvested >10% of the estimated number of females within the hunted populations and the highest proportion of harvest (harvest rate) from within the hunted populations was 8.3% in Montana, 3.1% in Alberta, and 2.2% in Colorado (Rominger 2008).

### 16.4.2 CURRENT HARVEST MANAGEMENT

Bighorn sheep license holders in 2022 in North America totaled just under 4,000 (Table 16.1). Limited permits and draw-based hunting opportunities have not diminished demand as there was just over 243,000 applicants for bighorn sheep draw-hunt opportunities in 2020 (Wild Sheep Initiative 2023, unpublished data). Also, in 2020, there were just over 4,900 licenses issued for bighorns and outside of a couple unlimited entry opportunities, only 2% of sheep hunters are successful in drawing these coveted tags. Using the available data from 2010 through 2020, jurisdictions report an increase in demand or license applications increasing from 111,388 to 243,136, or roughly a twofold increase in 10 years (Wild Sheep Initiative 2023, unpublished data).

Agencies made some headway through recovery programs and bighorn harvest peaked in 2018 at 1,900 (Table 16.1; Figure 16.2; Wild Sheep Initiative 2023). Typically, greater than 40% of the harvest comes from the province of Alberta and the states of Colorado, Montana, and Wyoming, where each of these jurisdictions typically

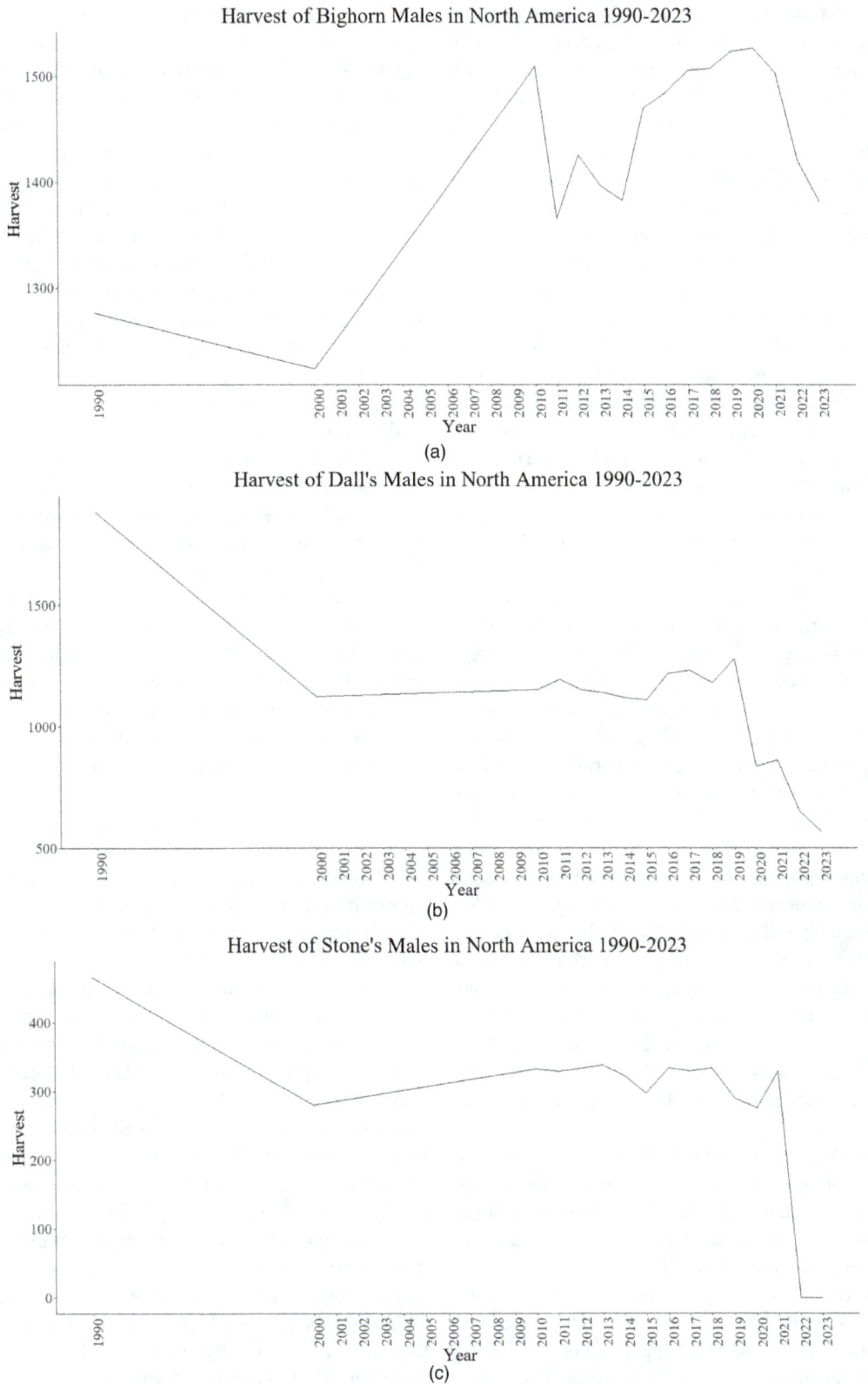

**FIGURE 16.3** The number of bighorn (a), Dall's (b), and Stone's (c) males harvested in North America from 1990 to 2023.

harvest over 100 bighorn sheep annually. Jurisdictions that have a moderate harvest (50–100 bighorn) include the province of British Columbia, and the states of Idaho, Nevada, New Mexico, Oregon, and Utah. The remaining jurisdictions typically harvest fewer than 50 bighorn sheep annually. Most of the bighorn harvest is typically the Rocky Mountain subspecies (65%–70%), but states such as Nevada, California, Colorado, Arizona, Texas, Utah, New Mexico, and Mexico provide some harvest for desert bighorn subspecies (30%–35%).

The preponderance of harvest is males; females make up less than 20% of the harvest in most years (Table 16.1). The value of harvesting females, given the role of density dependence in shaping herd performance in a population (Monteith et al. 2014, Bowyer et al. 2014, Monteith et al. 2018), may be overlooked, particularly when stakeholders and constituents are unsupportive. Horn growth may also be increased through improved nutritional condition because of having fewer females consuming limited resources (Jorgenson et al. 1993, Festa-Bianchet et al. 1998). Less competition with lowered density may also reduce vulnerability to disease and foster stable population growth in variable environments (Monteith et al. 2018). Population size may also influence disease dynamics in bighorn sheep populations with increased exposure to pneumonia-associated pathogens (Paolini et al. 2023). When herd densities are near asymptotes a large proportion of bighorn sheep populations (near 90%) experience pneumonia outbreaks resulting in population declines, and density-dependent mechanisms may play a role in such declines (Monello et al. 2001, Paolini et al. 2023).

A survey of 13 jurisdictions that allocate hunting licenses for bighorn sheep was conducted in 2018 (Wild Sheep Working Group 2018). All jurisdictions conduct bighorn population surveys to obtain population estimates and associated data on male:female ratios, lamb:female ratios, and total numbers of bighorn including males, females, and lambs. The population surveys can generally take place from August through April, but most of the surveys occur during fall or winter (62%). Three agencies (23%) use a sightability model with marked animals to estimate abundance on at least a portion of their herds, but most agencies use a minimum-count approach (77%) and provide an index of the population (Powell and Gale 2015). Bighorn population surveys provide the foundation from which agencies develop licensed hunting recommendations. Jurisdictions vary to a defined minimum number of animals needed to open and maintain a hunting season. Some agencies did not require a minimum number (five of 13) whereas eight agencies required a count of at least 50 animals before males could be harvested.

Once a hunting season is initiated there are primary guidelines for establishing license allocations and agency guidelines vary considerably across jurisdictions in North America. Only one jurisdiction-based license numbers on overall population abundance and licenses were allocated to 2.5% of the overall population size. The remaining jurisdictions based license allocations on the number of males in the population. Male age was not considered a guideline in six of 13 cases (46%), however, male class (mature males Class III or larger; Geist 1968) was considered in six of 13 cases. Where total males were used as the guideline, jurisdictions allocated licenses at 8%–10% of the total males counted; when mature males were used as a guideline, license allocations varied from 10% to 40% of the number of mature males counted (Wild Sheep Working Group 2018).

Once a bighorn is harvested there are several ways agencies check-in animals and collect harvest data. Most agencies (15 of 16) require a mandatory check-in to visually inspect and officially register the male by inserting a metal plug into the horn(s) of the animal harvested verifying its legal harvest status. Utah currently does not require an in-person check or a plug for a harvested bighorn, but has in the past; hunters must now complete a mandatory online harvest report within a prescribed time. Just over half of the agencies collect harvest data from hunters by using an online form through the agency website (nine of 16). Electronic tagging or E-tagging is gaining popularity due to its functionality of collecting data in real time, and some agencies (four of 16) collect this data through a cell phone application where the animal is tagged digitally, and the application is also used to collect harvest information. Three agencies still collect harvest data physically at a check-in site.

Typically, most bighorn licenses are allocated via limited draws, and in rare cases, exist for unlimited hunting opportunities. There are some additional special license opportunities for hunters through an auction tag or special raffle, conducted by non-profits to raise money for conservation (Hurley et al. 2015). In 2024 most jurisdictions (82%) provided auction or raffle opportunities, which varied from one to eight licenses being available per state; such licenses generated just under $6,300,000 for bighorn sheep conservation (Wild Sheep Foundation 2024).

### 16.4.3   FUTURE HARVEST MANAGEMENT

Bighorn sheep managers often reflect on what could be for opportunities as it relates to the hunting of mountain sheep if their populations were not limited by disease and range loss (Chapters 8, 19, and 21). One could only imagine that if bighorn sheep populations were not hampered in population growth and expansion, the realm of possibilities for harvest management. Considering historical estimates of bighorn sheep in the contiguous United States approaching 2 million, as suggested by Seton (1929), it is not inconceivable to think that bighorn harvests could total more than their current population estimates. Using a conservative 3% harvest rate on 2 million bighorn sheep equates to a potential harvest of 60,000 sheep, if bighorns were not limited. This potential is a remarkable number for us to ponder, and it motivates us as resource professionals to continue to try and develop solutions to overcome the current limitations for population growth. Unfortunately, the current reality is that there are fewer than 65,000 bighorns alive in North America (Wild Sheep Initiative 2023).

Past harvest management has been incredibly conservative, and most strategies were based on harvesting older mature males, at a rate of less than 4% of the bighorn population (Jorgenson et al. 1993, Hengeveld and Festa-Bianchet 2011, Wild Sheep Working Group 2018, Anderson 2020). Stakeholders have debated the merits of this approach, suggesting it causes the species to be almost exclusively managed from a hunting perspective as a trophy species across most of its range (Monteith et al. 2018). Certainly, this is a noble strategy for providing big-game animals with majestic ornaments for the bighorn sheep hunter (LaSharr et al. 2019), but is this the wisest course of action? Harvesting older mature males alone may not be the best strategy for keeping populations stable and helping to prevent erratic population fluctuations.

A conundrum might be that agencies and their stakeholders are afraid to harvest females as any surplus animals could be a possible source for future reintroductions of bighorns into other territories, states, or provinces, or in their state following a pneumonia die-off. In 2023, 44% of game agencies did not harvest females in their jurisdictions (Wild Sheep Initiative 2023). Healthy bighorn populations can grow at rates well above lambda values of 1.0, creating potential issues with range condition, dispersal, disease transmission, and decreased nutritional condition and horn growth (Jorgenson et al. 1993, Festa-Bianchet et al. 1998, Monteith et al. 2018, Paolini et al. 2023). Population growth rates vary depending upon several factors but harvesting females at a rate of 12% can provide a starting point for managers lacking sufficient population growth rate data (Jorgenson et al. 1993). The benefits of harvesting females may include reducing the risk of spillover of disease from domestic livestock such as domestic sheep or goats (Monello et al. 2001, Paolini et al. 2023). The harvest of females may also reduce density, improving nutritional condition of bighorns, which, in turn, may improve horn growth in males, particularly where there is limited opportunity to improve forage quality through habitat management (Jorgenson et al. 1993, Festa-Bianchet et al. 1998, Monteith et al. 2018). Finally, as we look to the future, managers should consider weighing the potential value in retaining females as a source for translocation stock against increasing female harvest opportunities for hunters, which help maintain populations below carrying capacity (Jorgenson et al. 1993, Monteith et al. 2018).

## 16.5 THINHORN SHEEP

### 16.5.1 Harvest History and Early Regulations

Indigenous groups have hunted thinhorn sheep for millennia in Alaska and Canada (Chapter 2). During the late 1800s and very early 1900s, unrestricted subsistence and market hunting likely resulted in excessive localized overharvest. Initial regulations, season lengths, and bag limits were established in the early 1900s (Demarchi and Hartwig 2004). Subsequent sheep population declines triggered the requirement for hunting licenses in the 1920s. Through time, season lengths and bag limits were reduced to further limit overharvest. In the early 1950s, the first horn curl requirements (≥3/4 curl) were gradually implemented in Alaska, and by the mid-1960s were standard throughout all thinhorn jurisdictions (Demarchi 1978). Horn-size requirements continued to become increasingly restrictive in many

areas, moving to a minimum of seven-eighth curl throughout Alaska, British Columbia, and Yukon Territories through the 1970s (Demarchi 1978), and by the 1980s and 1990s, harvest was further restricted to sheep with full-curl or greater horn lengths, or those males aged 8 years old or older; Alaska included options for harvesting males with both horns broken (horn tips missing) and this persists today in most areas. The Northwest Territories currently uses a three-fourths curl regulation, although most males harvested would be considered full-curl (Karabatsos et al. 2023). The legal definition of full-curl is slightly different across Alaska, British Columbia, and the Yukon Territory. The intent of the full-curl regulation was to create a conservative approach that ensured most males would have breeding opportunity before being vulnerable to harvest, protecting both age structure in the male cohort and males in general from overharvest. Males achieving full-curl status are a small component of the overall population and, predominantly by age 9 or 10 in some areas, begin to enter the senescence stage along with having a greater vulnerability to natural mortality (Murie 1944, Hoefs and Bayer 1983). When this occurs, harvest is more likely to be compensatory.

During the three-fourths curl requirement in Alaska, some managers speculated that most legal males were removed from some populations each year as a result of hunter harvest (Heimer and Watson 1984), or at the very least the three-fourths curl harvest requirement was removing breeders from the population before their successful participation in the rut, leading to population viability concerns. Survivorship curves constructed by personnel in the Central Alaska Range suggested that sheep in the heavily hunted area began to experience accelerated mortality at 3 years of age as compared to the adjacent unhunted Denali National Park, where accelerated mortality began at 8 years of age (Heimer and Watson 1984). Researchers examining female reproductive frequency in the heavily hunted area that had minimum curl change throughout their study from three-fourths to seven-eighth curl, versus an adjacent lightly hunted full-curl minimum harvest area, and concluded an earlier age of first reproduction in the heavily hunted area until the more restrictive seven-eighth curl hunting regulation was implemented (Heimer and Watson 1986). At that time, researchers speculated that the dominance-related mortality hypothesis (Geist 1968, 1971) applied to thinhorn sheep. This hypothesis suggested that thinhorn sheep have a structured social hierarchy that ensures an orderly rut where the largest and most mature males are the dominant breeders. Because of the physical demands of the rut on the breeders, the mortality rates of primary breeders are accelerated. Creating the full-curl requirement was thought to leave larger males (three-fourths to less than full-curl) on the landscape, thereby reestablishing the social hierarchy to optimize breeding and pregnancy rates.

The dominance-related mortality hypothesis has limited supporting evidence. In an early attempt to test the hypothesis, Singer and Zeigenfuss (2002) observed increased courtship activity by young Dall's sheep males in a heavily hunted area that had an absence of older males compared to an unhunted area nearby. In separate studies, Murphy et al. (1990) and Singer et al. (1991) did not detect differences in overwinter survival of males or females (as assessed by aerial observation) in heavily hunted and unhunted areas. Hoefs and Barichello (1984) reported that removal of up to 80% of mature males did not reduce life expectancy within the hunted population during the years of their study. More contemporary research indicates that 3- to 7-year-old males in both heavily or lightly hunted areas have similar mortality rates to older age cohorts (T. Lohuis, Alaska Department of Fish and Game, unpublished data). Also, horn morphometric research (B. R. Wendling, Alaska Department of Fish and Game, unpublished data) reported that on a statewide basis for 2016–2024, approximately 55%–64% of the males harvested each year were legally available for harvest at least once prior to their year of harvest, and after attaining 360° of curl. With this sample coming from harvested sheep, it should not be considered a random sample of the population; however, if hunters are more likely to take the first legal sheep seen, the sample may be representative of the population. This is a conservative estimate because the approximation only accounts for full-curl males. Males with broken or broomed horns, those that reach 8 years of age before achieving full-curl, and those full-curl individuals succumbing to natural mortality after the hunting season would also be legal; these individuals were not added to the estimate of legal males evading harvest. Therefore, current data suggest that at broad geographic scales, approximately half of legal males are not harvested each year. Although early research in Alaska argued that the dominance-related mortality hypothesis provided the biological justification of a full-curl requirement (Heimer and Watson 1986), this argument lacks evidence-based support. A reexamination of this argument by Whitten (2001) concluded that harvest never removed all mature males, harvest of ≥3/4 curl males did not affect survivorship of different age cohorts, and reduced productivity of females was correlated with weather and not lack of abundance of males three-fourths curl or larger. The full-curl or 8-year-old or greater requirement has remained intact, not because of the dominance-related mortality hypothesis, but because it errs on the side of conservation, and likely prevents overharvest. Also, because of the vast spatial distribution of sheep across thinhorn range and limited opportunities to get accurate estimates of population density on an annual basis, thinhorn sheep managers cannot often implement fine-scale harvest strategies. Ultimately, the full-curl or 8-year-old harvest management strategy should not result in suboptimal breeding opportunities or overharvest, but at the cost of not maximizing offtake.

The full-curl regulation also aligns with the interests of thinhorn hunters and wild sheep conservation organizations. Most sheep hunters (72%) are satisfied or very satisfied with the full-curl regulation (Brinkman 2014). When surveyed about the level of importance of multiple factors that influence hunter satisfaction, having an opportunity to harvest a full-curl male ranked among the top factors for resident hunters in Alaska (Brinkman 2014). Similarly, for non-resident hunters, 92% of survey respondents reported that having an opportunity to harvest a full-curl male was somewhat or very important (Brinkman 2014). The opportunity to harvest any thinhorn male (e.g., less than full-curl) received mixed support, with roughly half of residents and non-residents split between this opportunity being important and unimportant.

### 16.5.2 Current Harvest Management

The passage of the Alaska National Interest Lands Conservation Act (ANILCA) in 1980 and subsequent amendments to ANILCA had an influence on the regulatory structure of hunting sheep. First, ANILCA created Conservation System Units (e.g., national parks, wildlife refuges) that placed about 60% of the lands in the state into federal ownership for conservation purposes. This reduces hunting opportunities in those areas when comparing data between pre- and post-ANILCA as additional lands were closed to hunting after ANILCA's passage. Second, amendments to ANILCA created a dual management structure in Alaska by separating general hunting from subsistence hunting on federal lands. Subsistence is defined in state and federal laws as the customary and traditional uses of wild resources for food, clothing, fuel, transportation, construction, art, crafts, sharing, and customary trade. On federal lands, only federally qualified rural residents are provided a subsistence priority when it is necessary to protect the continued viability of the population or continue subsistence uses. This priority limits the harvest opportunities for non-federally qualified hunters.

The Alaska Board of Game establishes harvest regulations in Alaska, whereas the Federal Subsistence Board establishes restrictions on federal land based on qualification status. State regulations are developed through a public proposal process and by population and harvest data collected by state and federal agencies.

Current harvest regulations are a mix of draw, registration, and general harvest permits. Most thinhorn hunting opportunities are general-open-season hunts, a hunt that is open to an unlimited number of hunters (resident and non-resident). Draw-hunts are implemented for multiple reasons, including: allowing for archery-only hunts, providing opportunities to hunt in uncrowded and aesthetically pleasing conditions, and in some cases, harvesting under strategies other than full-curl (e.g., any-male hunts). Draw-hunts to manage for trophy opportunity (large full-curl males) exist in one area of Alaska (i.e., Tok Management Area). Registration hunts (i.e., unlimited number of hunters with seasons being closed when harvest goals are met) are less common (<1% of harvest) and implemented to allow emergency closures on short notice in the event a conservation need arises. In limited areas and on federal lands where federal regulations apply, any sheep (male, female) may be harvested by federally qualified rural residents for subsistence purposes. Harvest statistics are often limited for federally managed hunts because of lack of reporting requirements for federally qualified hunters. Nearly all harvest information is derived from the state harvest reporting system, and this information is thought to represent the majority (>90%) of annual sheep harvest statewide.

Season lengths are relatively consistent, starting in early August and ending in late September. For federally qualified users, on select federal lands, seasons may be extended until the end of April. Youth hunts are also available in some areas, allowing people less than 18 years old to hunt 10 days in advance of the normal hunting season start date. State harvest is limited to one sheep per year per person in nearly all areas and harvest is generally skewed toward the beginning of the season.

Hunters use a variety of ways to access hunting areas, with the most common being automobile, off-road vehicles, and aircraft. Roughly 60% of sheep are harvested by hunters using aircraft to access the hunting area. Some areas are designated as non-motorized hunt areas, meaning that hunters are prohibited from using motorized vehicles within those hunting areas. However, most federal lands, including wilderness, allow the use of fixed wing aircraft for access as a consequence of ANILCA. Concerning the use of aircraft, it is against the law to hunt sheep until 3 AM the day following the day a hunter has flown in on a non-regularly scheduled commercial aircraft. It is also prohibited to use aircraft to locate or direct hunters to sheep, while the hunting season is open. In terms of other special regulations, hunters must salvage all edible meat; non-residents must be personally accompanied in the field by a licensed guide or an Alaskan resident relative at least 19 years old within the second degree of kinship (e.g., parent, sibling, parent or sibling-in-law, step-parent, step-sibling). There is no limit to the number of commercial operators on state land, but federal lands have a limit on the number of tenured guides in designated areas (i.e., concession system).

Between 1962 and 2023 the average number of sheep harvested each season was 985, with a range of 336–1,454 (Table 16.1). Harvest was between 700 and 800 for much of the 2000s but has declined substantially beginning in 2020, with 629 males taken in that year. Initial declines in 2020 were followed by further decreases in statewide harvest, with 487 males taken under ANILCA in 2021, 424 in 2022, and 336 reported to date during the 2023 season. Comparing current harvest to the long-term average is confounded because early harvest happened before ANILCA, when more areas (e.g., national parks) were open to hunting; some areas have gone from general-open-season harvest to draw-hunt, thus limiting opportunity; and hunter participation has decreased (Wendling, unpublished data). Between 1980 and 2000 there was an average of 3,097 sheep hunters per year, and between 2001 and 2022 there was an average of 2,489 hunters per year. Harvest success rates were largely stable for both resident and non-resident hunters until the recent observed population declines. Annual weather conditions can affect survival (especially lamb)

and recruitment contributing to interannual variation in sheep harvest because of reductions in the availability of harvestable males 7–10 years after those events. For example, Cosgrove et al. (2021) identified a relationship between harsh weather, temperatures, snow depths, and decreased lamb production; these conditions were observed in consecutive winters between fall 2010 and spring 2015. Eight to 10 years after (2021–2023) saw additional challenging winter conditions for sheep; the total male harvests in 2021–2023 hunting years were the lowest harvest numbers on record for the state of Alaska. Weather conditions during the sheep hunting season may also have immediate effects on harvest. The presence of rain, wind, fog, and snow conditions affects hunter accessibility by aircraft and sheep sightability while hunting, likely influencing annual harvest success in the year in which they occur (Leorna et al. 2020).

Current harvest regulations for thinhorn sheep in British Columbia predominantly use a general-open-season, full-curl, or 8 year old and older harvest strategy, with season dates 1 August to 15 October. Resident hunter-only limited entry or draw-hunts are also used for Dall's sheep hunting and in some provincial parks (Halladay and Demarchi 1996). In the late 1980s harvest quotas were established for licensed guide outfitters and draw-hunt areas. To determine the quota allocation two approaches were employed, specifically: a population inventory model and a harvest age structure model. Both models incorporate data collected through harvest surveys and information obtained at the time of compulsory inspections and horn pinning. The population inventory model uses a combination of demographic and population inventory data to calculate a population estimate, to which a harvest rate is applied, and an annual allowable harvest is generated and then split between resident and non-resident hunter groups (Ayotte et al. 2022). In the absence of that type of data, the harvest age structure is employed. This approach uses data collected at the time of compulsory inspection, identifying the proportion of males aged less than 8; 8; and greater than 8 years in the harvest, and assesses those against established harvest conservation thresholds. A suggested metric for measuring sustainable harvest was to strive for 75% of the harvest to be comprised of males 8 years old or older (Demarchi and Hartwig 2004). Harvest allocations are determined in 5-year increments and can be adjusted based on the proportional representation of sheep that are ≥8 years old and older (Ayotte et al. 2022). Regardless of the model used, the objective is to have a harvest rate of less than 3% of the population (Ayotte et al. 2022).

Harvest regulations in the Northwest Territories have not changed since 1965, and restrict harvest to a male with three-fourths curl horn or larger during a 15 July to 31 October hunting season (Karabatsos et al. 2023). If hunters access an area via an aircraft (fixed-wing or helicopter), hunting is prohibited for 12 hours after landing (Government of Northwest Territories 2025). Resident and subsistence harvest is low at fewer than 20 sheep per year, with most of the harvest (~200 males; Larter et al. 2018) taken by non-residents or non-resident-aliens who are required to be accompanied by a guide (Jex et al. 2016). Eight hunting zones were established in 1971 and have remained unchanged (Latour and Maclean 1994). There is no established quota on the number of sheep that may be harvested; however, overall harvest rates are low and reported to range between 0.8% and 1.6% of the population (Veitch et al. 1998).

In 2019; Environment Yukon established science-based guidelines for the management of thinhorn sheep (Environment Yukon 2019). In the Yukon, hunters are restricted to the harvest of a full-curl male or a male that is at least 8 years old. Most hunting zones are open as general-open-seasons with dates generally between 1 August and 31 October, or 1 August and 15 September (Yukon Wildlife Act; https://laws.yukon.ca/cms/images/LEGISLATION/PRINCIPAL/2002/2002-0229/2002-0229.pdf. Accessed 31 Mar 2025).

Within subzones, if hunting pressure of harvest is determined to be too high, a draw-hunt may be implemented (Jex et al. 2016). Approximately 60% of the annual harvest are by non-residents (Jung et al. 2023). Non-resident sheep hunters must be accompanied by a registered guide. Aircraft, excluding helicopters, may be used to transport hunters; however, spotting of sheep or communicating information of sheep observed through the air is prohibited for 48 hours prior to hunting (Yukon Wildlife Act). Current guidelines strive to maintain a harvest rate under 4% of the non-lamb population and discourages the implementation of any female hunt (Environment Yukon 2019). In the absence of population survey data, the 5-year average of the proportion of the harvested males over 8 years old is used as a measure of sustainability (Environment Yukon 2019). Further management actions may be initiated if that average proportion is less than 60% (Environment Yukon 2019).

For the majority of sheep hunts across all jurisdictions, the skulls (with horns naturally attached) must be presented in person to established locations for compulsory inspection (sometime called sealing), measurement, and the installation of an aluminum horn pin that is uniquely numbered and thus creates a permanent marking set into the horn, within 30 days or lesser time if designated. In Northwest Territories, mandatory horn pins are only required for harvested sheep that are exported from the jurisdiction. In Alaska, sheep harvested under federal regulations, federally qualified users are not required to have a mandatory horn pin. Hunters must also complete a harvest report that includes details on hunt location, success, effort, mode of access, and whether a commercial service (e.g., guide, commercial transport) was used. Harvest data are compiled on an annual basis and aggregated to provide information on population and hunter demographics and to guide future harvest regulations. Aside from the opportunity to assess if the harvested male complies with the legal definition of full-curl (Figure 16.1), the compulsory inspection or sealing process provides the opportunity for biologists to gather data on age and horn morphometrics. Additionally, biological samples such as DNA and nasal swabs are sometimes collected to provide genetic and health (i.e., disease) information. These data provide large samples, both temporal and spatial, that allowed researchers to address questions regarding the age distribution of harvest, potential effects of selective harvest, assessment of population health, and broad scale evolutionary questions (Douhard et al. 2016, Lieske et al. 2022, Sim et al. 2019, Chapter 6).

Traditionally, northern Indigenous peoples in North America placed a high value on mountain sheep and used the entire animal, including the meat, hide, sinews, bones, and horns (Chapter 2). Today, thinhorn sheep are culturally and economically important to diverse sectors of Indigenous and non-Indigenous conservation stakeholders; they remain a highly prized game species, generally pursued by a relatively small proportion (as compared to other big-game species) but passionate group of hunters. Hunting thinhorn sheep is both a physically and mentally challenging activity, because of limited access to hunting areas, steep and rugged terrain, and inclement and unpredictable weather conditions. Even so, inhabitants of northern communities remain closely linked to wildlife, sustaining themselves on the resources of the land. The human dimensions of today's sheep hunting are different than those that resulted from a direct subsistence and cultural connection to the mountain sheep resource, and this new relationship remains a relatively understudied topic (Chapter 18).

During the 2000s and 2010s, the Alaska Board of Game received an increasing number of proposals from the public (mainly consumptive users) to alter the management and regulation of sheep, and this conservation-motivated interest can also be witnessed in other jurisdictions such as the Yukon and British Columbia. In summary, these proposals have reported a decline in the quality of sheep hunts because of declining sheep numbers and increased crowding and competition among user groups (Brinkman 2014). To solve this problem, most Alaskan proposals have suggested changes in season dates, permit allocation, and harvest limits; efforts in other jurisdictions have some similarities but have also focused on habitat interventions that include enhancement and predator management actions. Based on a survey of sheep hunters and commercial operators in Alaska, 74% of resident hunters and 84% of commercial operators agreed or strongly agreed that crowding was a problem (Brinkman 2014). When asked about the importance of multiple factors that affect the quality of the hunt, survey respondents indicated that level of crowding and the opportunity for seclusion from other hunters were most important, beyond the importance of being able to see or harvest sheep (Brinkman 2014). This suggests that sheep hunter satisfaction is strongly influenced by a remote and secluded experience. User group competition was also explored in British Columbia, but the results of that 2007 work disproved the competition hypothesis and concluded that the perception of competition held by resident hunters was likely more driven by the nature of sheep populations, hunter selection of habitat, and accessibility into those portions of the landscape (Addison 2007).

Thinhorn sheep are also considered an important economic resource to both government mountain sheep management programs but also commercial interests. In addition to revenue generated from licenses and tags, commercial guiding services and rural and remote communities benefit economically from sheep hunting. The current price of commercially guided Dall's sheep hunt in Alaska ranges from $25,000 to $45,000 per hunter and a Stone's sheep hunt in British Columbia can sell between $75,000 and $105,000. Unguided commercial transport by aircraft can range from $1,000 to $10,000 per hunter. The appeal of a thinhorn sheep hunt is illustrated each year when special permits are auctioned or raffled on behalf of agencies and government jurisdictions, which can generate several hundred thousand dollars per permit (Wild Sheep Foundation 2024).

### 16.5.3 FUTURE HARVEST MANAGEMENT

A minimum curl harvest strategy, whether three-fourths or full-curl, has been the pillar of thinhorn sheep management for more than a half a century. A misconception of this harvest strategy is that the species is being managed for mainly trophy value. The current and future harvest strategy aims at addressing declining population numbers and sustaining population resilience. Thinhorns occupy some of the last true and intact natural areas on the continent. Ecosystems that function as multi-predator and multi-prey environments in the harshest of winter conditions will require adaptive management as the climate rapidly changes (Jex et al. 2016, Chapter 21). In the face of changing climates in northern latitudes and an increase in the frequency of stochastic and severe weather events, future research on thinhorn populations and harvest will need to assess the duration and influence (negative vs. positive) of changing snow conditions and shifting plant phenology in northern sheep habitat (Van de Kerk et al. 2018). The vast spatial scale of these landscapes makes regular and complete population enumeration unachievable in consideration of agency capacity, current techniques, and financial constraints, for all areas open for harvest. The lack of population and reliable trend data, coupled with a challenge to understanding of how bottom-up or top-down environmental processes drive sheep population dynamics, particularly under rapid global climate change, leaves a minimum curl harvest strategy as the most practical option for conservative management. The byproduct of this strategy is that it is conservative by nature and may not maximize harvest opportunity, but it does provide an opportunity to a harvest of mature male that is considered by some as a trophy.

Future management of thinhorn sheep will continue to have high cultural, recreational, social, and economic importance. Balancing population and conservation need, hunt quality, and harvest opportunity will be an ongoing topic of discussion among sheep researchers, managers, Indigenous groups (e.g., First Nation, Alaska Native), and conservation-based organizations. Proposals submitted to the Alaska Board of Game by hunters and commercial operators suggest exploring the re-establishment of state guiding concessions to distribute hunting pressure and reduce conflict (Brinkman 2014). In the other thinhorn sheep jurisdictions this approach to non-resident harvest administration is already in place. Increasing research capacity to get higher resolution population data will support sustainable harvest management approaches, foster location-specific harvest strategies, and address issues related to hunter satisfaction.

## 16.6 SUMMARY

Bighorn and thinhorn sheep harvest management use different strategies to each achieve localized and/or regional goals and objectives. After bighorns were extirpated from a large portion of their historical range conservation-based regulations and efforts eventually recovered bighorn populations back to sustainable levels in many areas, thereby providing quality hunting opportunities. This recovery was a conservation success story and for many years conservative male-harvest regulations have been implemented by most jurisdictions. As managers face challenges associated with disease epizootics and habitat loss, it is incumbent upon resource professionals to develop innovative harvest strategies that improve our ability to manage, and at minimum, sustain this iconic species into the future. Future challenges for thinhorns currently appear to be influenced largely by climate-related changes in ecosystem conditions. The current general-open-season harvest strategy focused on mature males is a conservative approach that will likely continue except where harvest pressure results in conservation concerns; as observed in some locales the need to implement a draw-hunt has led to reductions in open-season hunt opportunities. Alternative harvest strategies will require more research on abiotic and biotic factors affecting productivity and recruitment. Public interests will continue to be researched and monitored to assist with allocation of hunting opportunity, and to assess approval of new regulatory actions. Because thinhorn population trajectories appear to be primarily influenced by factors outside of management control (e.g., weather), focusing on the resilience of population dynamics in an uncertain future will help with adaptive and innovative harvest management options.

## REFERENCES

Addison, C. 2007. An analysis of competition between resident and non-resident hunters of Stone's sheep (*Ovis dalli stonei*). Fish and Wildlife Branch, Ministry of Environment, Victoria, British Columbia, Canada.

Alaska Department of Fish and Game. 2016. Dall Sheep. Guide to judging sheep horns under the full-curl regulation. Alaska Department of Fish and Game, Division of Wildlife, Juneau, Alaska, USA.

Alaska Department of Fish and Game. 2024. 2024–2025 Alaska hunting regulations. Alaska Department of Fish and Game, Division of Wildlife, Juneau, Alaska, USA.

Alberta Fish and Wildlife Services. 1993. Management plan for bighorn sheep in Alberta. Environmental Protection, Fish and Wildlife Services, Edmonton, Alberta, Canada.

Anderson, N. 2020. Size-at-age of Alberta's bighorn sheep (*Ovis canadensis*). *Proceedings of the Biennial Symposium of the Northern Wild Sheep and Goat Council* 22:40–48.

Ayotte, J., B. A. Jex, M. Anderson, M. Bridger, K. Kriss, and A. Walker. 2022. Stewardship framework for thinhorn sheep management teams (draft). Prepared for the Fish and Wildlife Branch, British Columbia Ministry of Forests, Victoria, British Columbia, Canada.

Bowyer, R. T., V. C. Bleich, K. M. Stewart, J. C. Whiting, and K. L. Monteith. 2014. Density dependence in ungulates: a review of causes, and concepts with some clarifications. *California Fish and Game* 100:550–572.

Brinkman, T. 2014. Alaska sheep hunter survey: resident sheep hunter responses. Alaska Department of Fish & Game. https//:www.adfg.alaska.gov/static/home/library/pdfs/wildlife/mgt_rpts/14_sheep_hunter_survey_report.pdf. Accessed 20 Oct 2019.

Buckner, E. L., and J. Reneau. 2009. *Measuring and Scoring North American Big Game Trophies*. Third Edition. Boone and Crockett Club, Missoula, Montana, USA.

Buechner, H. K. 1960. The bighorn sheep in the United States, its past, present and future. *Wildlife Monographs* 4:1–174.

Colorado Parks and Wildlife. 2009. *Colorado Bighorn Sheep Management Plan 2009–2019*. Special Report Number 81. Colorado Division of Wildlife, Denver, Colorado, USA.

Coltman, D. W., P. O'Donoghue, J. T. Jorgenson, J. T. Hogg., C. Strobeck, and M. Festa-Bianchet. 2003. Undesirable evolutionary consequences of trophy hunting. *Nature* 426:655–658.

Cosgrove, C. L., J. Wells, A. W. Nolin, J. Putera, and L. R. Prugh. 2021. Seasonal influence of snow conditions on Dall's sheep productivity in Wrangell-St Elias National Park and Preserve. *PLoS ONE* 16:e0244787.

Cowan, I. M. 1940. Distribution and variation in the native sheep of North America. *American Midland Naturalist* 24:505–580.

Damm, G. R. 2008. Recreational trophy hunting: "What do we know and what should we do?" Pages 5–11 *in* R. D. Baldus, G. R. Damm, and K.-U. Wollscheid, editors. *Best Practices in Sustainable Hunting—A Guide to Best Practices from Around the World*. CIC–International Council for Game and Wildlife Conservation, Budakeszi, Hungary.

Demarchi, R. A. 1978. Evolution of mountain sheep horn curl regulations in British Columbia. Northern Wild Sheep and Goat Council. *Proceedings of the Biennial Symposium of the Northern Wild Sheep and Goat Council* 1:17–29.

Demarchi, R. A., and C. L. Hartwig. 2004. Status of thinhorn sheep in British Columbia. Wildlife Bulletin No. B-119. British Columbia Ministry of Water, Land and Air Protection, Biodiversity Branch, Victoria, British Columbia, Canada.

Douhard, M., M. Festa-Bianchet, F. Pelletier, J. M. Gaillard, and C. Bonenfant. 2016. Changes in horn size of Stone's sheep over four decades correlate with trophy hunting pressure. *Ecological Applications* 26:309–321.

Emmerich, J. M. 1986. Bighorn sheep harvest strategies in Wyoming. *Proceedings of the Biennial Symposium of the Northern Wild Sheep and Goat Council* 5:8–11.

Environment Yukon. 2019. Science-based guidelines for management of thinhorn sheep in Yukon (MR19-01). Environment Yukon, Whitehorse, Yukon, Canada.

Festa-Bianchet, M. 2012. Rarity, willingness to pay and conservation. *Animal Conservation* 15:12–13.

Festa-Bianchet, M. 2017. When does selective hunting select, how can we tell, and what should we do about it? *Mammal Review* 47:6–81.

Festa-Bianchet, M., D. Coltman, J. T. Hogg, and J. Jorgenson. 2006. Age-related horn growth, mating tactics, and vulnerability to harvest: why horn curl limits may select for small horns in bighorn sheep. *Proceedings of the Biennial Symposium of the Northern Wild Sheep and Goat Council* 15:42–49.

Festa-Bianchet, M., J. M. Gaillard, and J. T. Jorgenson. 1998. Mass- and density-dependent reproductive success and reproductive costs in a capital breeder. *American Naturalist* 152:367–379.

Geist, V. 1968. On the interrelation of external appearance, social behavior, and social structure of mountain sheep. *Ethology* 25:199–215.

Geist, V. 1971. *Mountain Sheep: A Study in Behavior and Evolution*. University of Chicago Press, Chicago, Illinois, USA.

Government of Northwest Territories. 2025. Northwest Territories summary of hunting and trapping regulations July 1, 2024–June 30, 2025. Wildlife Management Division Environment and Climate Change, Yellowknife, Northwest Territories, Canada.

Halladay, D. R., and R. A. Demarchi. 1996. Wildlife harvest strategy: improving British Columbia's wildlife harvest regulations. Wildlife Branch, British Columbia Ministry of Environment, Victoria, British Columbia, Canada.

Heffelfinger, J. R. 2018. Inefficiency of evolutionarily-relevant selection in ungulate trophy hunting. *Journal of Wildlife Management* 82:57–66.

Heimer, W. E., and S. M. Watson. 1984. Excess ram mortality in a heavily hunted Dall sheep population. *Proceedings of the Biennial Symposium of the Northern Wild Sheep and Goat Council* 4:425–432.

Heimer, W. E., and S. M. Watson. 1986. Comparative dynamics of dissimilar Dall sheep populations. Alaska Department of Fish and Game, Division of Game, Juneau, Alaska, USA.

Hengeveld, P. E., and M. Festa-Bianchet. 2011. Harvest regulations and artificial selection on horn size in male bighorn sheep. *Journal of Wildlife Management* 75:189–197.

Hoefs, M., editor. 1985. *Wild Sheep: Distribution, Abundance, Management and Conservation of the Sheep of the World and Closely Related Mountain Ungulates*. Northern Wild Sheep and Goat Council, Collins Books, Seattle, Washington, USA.

Hoefs, M., and M. Barrichello. 1984. Comparison between hunted and an unhunted Dall sheep population: a preliminary assessment of the impact of hunting. *Proceedings of the Biennial Symposium of the Northern Wild Sheep and Goat Council* 4:433–466.

Hoefs, M., and M. Bayer. 1983. Demographic characteristics of an unhunted Dall sheep (*Ovis dalli dalli*) population in southwest Yukon, Canada. *Canadian Journal of Zoology* 61:1346–1357.

Honess, R. F., and N. M. Frost. 1942. *A Wyoming Bighorn Sheep Study*. Wyoming Game and Fish Department Bulletin Number 1. Wyoming Game and Fish Department, Cheyenne, Wyoming, USA.

Hurley, K., C. Brewer, and G. N. Thornton. 2015. The role of hunters in conservation, restoration, and management of North American wild sheep. *International Journal of Environmental Studies* 72:784–796.

Idaho Fish and Game. 2010. Idaho bighorn sheep management plan. Idaho Fish and Game Department, Boise, Idaho, USA.

Jex, B. A., J. B. Ayotte, V. C. Bleich, C. E. Brewer, D. L. Bruning, T. M. Hegel, N. C. Larter, R. A. Schwanke, H. M. Schwantje, and M. W. Wagner. 2016. Thinhorn sheep: conservation challenges and management strategies for the 21st century. Wild Sheep Working Group, Western Association of Fish and Wildlife Agencies, Boise, Idaho, USA.

Jorgenson, J. T., M. Festa-Bianchet, and W. D. Wishart. 1993. Harvesting bighorn ewes: consequences for population size and trophy ram production. *Journal of Wildlife Management* 57:429–435.

Jorgenson, J. T., M. Festa-Bianchet, and W. D. Wishart. 1998. Effects of population density on horn development in bighorn rams. *Journal of Wildlife Management* 62:1011–1020.

Jung, T. S., C. Pinard, É. Bélanger, and C. Wall. 2023. Results of the 2022 Yukon sheep and goat hunter effort survey. Yukon Fish and Wildlife Branch Report SR-23-15, Environment Yukon, Whitehorse, Yukon, Canada.

Karabatsos, S., N. C. Larter, D. G. Allaire, K. Eykelboom, C. Estevo, M. Iravani, I. C. Barrio, and D. S. Hik. 2023. Dall's sheep horn growth and harvest management in the Mackenzie Mountains, Northwest Territories, Canada. *Journal of Wildlife Management* 88:e22536.

Larter, N. C., D. G. Allaire, and R. Mulders. 2018. Mackenzie Mountain non-resident and non-resident alien hunter harvest summary 2017. Environment and Natural Resources, Government of the Northwest Territories, Yellowknife, Northwest Territories, Canada.

LaSharr, T. N., R. A. Long, J. R. Heffelfinger, V. C. Bleich, P. R. Krausman, R. T. Bowyer, J. M. Shannon, R. W. Klaver, C. E. Brewer, M. Cox, A. A. Holland, A. Hubbs, C. P. Lehman, J. D. Muir, B. Sterling, and K. L. Monteith. 2019. Hunting and mountain sheep: do current harvest practices affect horn growth? *Evolutionary Applications* 12:1823–1836.

Latour, P. B., and N. Maclean. 1994. An analysis of data returned by outfitted hunters from the Mackenzie Mountains, NWT, 1979–1990. Department of Renewable Resources, Government of the Northwest Territories, Yellowknife, Northwest Territories, Canada.

Leorna, S., T. Brinkman, J. McIntyre, B. Wendling, and L. Prugh. 2020. Association between weather and Dall's sheep *Ovis dalli dalli* harvest success in Alaska. *Wildlife Biology* 2020(2):1–9.

Lieske, C. L., R. Gerlach, M. Francis, and K. B. Beckmen. 2022. Multilocus sequence typing of *Mycoplasma ovipneumoniae* detected in Dall's sheep (*Ovis dalli dalli*) and caribou (*Rangifer tarandus grantii*) in Alaska, USA. *Journal of Wildlife Disease* 58:625–630.

McCarthy, J. J. 1986. Bighorn sheep seasons in Montana, 1872–1985. *Proceedings of the Biennial Symposium of the Northern Wild Sheep and Goat Council* 5:14–23.

Monello, R. J., D. L. Murray, and E. F. Cassirer. 2001. Ecological correlates of pneumonia epizootics in bighorn sheep herds. *Canadian Journal of Zoology* 79:1423–1432.

Montana Department of Fish, Wildlife, and Parks. 2010. Montana bighorn sheep conservation strategy: 2010. Montana Department of Fish, Wildlife, and Parks, Helena, Montana, USA.

Monteith, K. L., R. A. Long, T. R. Stephenson, V. C. Bleich, R. T. Bowyer, and T. N. LaSharr. 2018. Horn size and nutrition in mountain sheep: can ewe handle the truth? *Journal of Wildlife Management* 82:67–84.

Monteith, K. L., R. A. Long, V. C. Bleich, J. R. Heffelfinger, P. R. Krausman, and R. T. Bowyer. 2013. Effects of harvest, culture, and climate on trends in size of horn-like structures in trophy ungulates. *Wildlife Monographs* 183:1–28.

Monteith, K. L., V. C. Bleich, T. R. Stephenson, B. M. Pierce, M. M. Conner, J. G. Kie, and R. T. Bowyer. 2014. Life-history characteristics of mule deer: effects of nutrition in a variable environment. *Wildlife Monographs* 186:1–62.

Murie, A. 1944. The wolves of Mount McKinley. *National Park Service Faunal Series* 5:1–238.

Murphy, E. C., F. J. Singer, and L. Nichols. 1990. Effects of hunting on survival and productivity of Dall sheep. *Journal of Wildlife Management* 54:284–290.

Nebraska Game and Parks Commission. 2019. Nebraska bighorn sheep management plan. Nebraska Game and Parks Commission, Lincoln, Nebraska, USA.

Nevada Division of Wildlife. 2001. Nevada bighorn sheep management plan. Nevada Division of Wildlife, Reno, Nevada, USA.

New Mexico Department of Game and Fish. 2003. Plan for recovery of desert bighorn sheep in New Mexico 2003–2013. New Mexico Department of Game and Fish, Santa Fe, New Mexico, USA.

Oregon Department of Fish and Wildlife. 2003. Oregon's bighorn sheep and Rocky Mountain goat management plan: December 2003. Oregon Department of Fish and Wildlife, Salem, Oregon, USA.

Paolini, K. E., M. K. Schwartz, M. M. Friggens, S. A. Cushman, J. K. Weckworth, and J. D. Holbrook. 2023. A review of population and landscape level dynamics associated with pneumonia outbreaks in bighorn sheep with implications for land management. *Conservation Science and Practice* 5:e12956.

Pigeon, G., M. Festa-Bianchet, D. W. Coltman, and F. Pelletier. 2016. Intense selective hunting leads to artificial evolution in horn size. *Evolutionary Applications* 9:521–530.

Powell, L. A., and G. A. Gale. 2015. *Estimation of Parameters for Animal Populations.* Caught Napping Publications, Lincoln, Nebraska, USA.

Rattenbury, K. L., J. H. Schmidt, D. K. Swanson, B. L. Borg, B. A. Mangipane, and P. J. Sousanes. 2018. Delayed spring onset drives declines in abundance and recruitment in a mountain ungulate. *Ecosphere* 9:e02513.

Rawson, T. 2001. *Changing Tracks: Predators and Politics in Mt. McKinley National Park.* University of Alaska Press, Fairbanks, Alaska, USA.

Rominger, E. 2008. Ram harvest strategies for western states and provinces. *Proceedings of the Biennial Symposium of the Northern Wild Sheep and Goat Council* 16:92–98.

Seton, E. T. 1929. The bighorn. Pages 519–573 *in* E. T. Seton, editor. *Lives of the Game Animals.* Volume 3 Part 2 Hoofed Animals. Doubleday, Doran Co., Garden City, New York, New York, USA.

Sim, Z., C. S. Davis, B. A. Jex, T. Hegel, and D. W. Coltman. 2019. Management implications of highly resolved hierarchical population genetic structure in thinhorn sheep. *Conservation Genetics* 20:185–201.

Simon, A. 2016. Against trophy hunting: a Marxian-Leopoldian critique. *Monthly Review* 68:17–31.

Singer, F. J., and L. C. Zeigenfuss. 2002. Influence of trophy hunting and horn size on mating behavior and survivorship of mountain sheep. *Journal of Mammalogy* 83:682–698.

Singer, F. J., E. C. Murphy, B. A. Cooper, and K. K. Laing. 1991. Activity in a hunted and an unhunted herd of Dall sheep. *Applied Animal Behaviour Science* 29:185–193.

Smith, T. S., and J. T. Flinders. 1990. Mountain sheep harvest regulations in Utah 1990. *Proceedings of the Biennial Symposium of the Northern Wild Sheep and Goat Council* 7:256–258.

Taras, M., and E. Manning. 2017. Dall's sheep news. Winter 2017 [newsletter]. Alaska Department of Fish and Game, Division of Wildlife Conservation, Juneau, Alaska, USA.

Tefethen, J. B. 1975. *The Wild Sheep in Modern North America.* Winchester Press, New York, New York, USA.

Turner, F. J. 1935. *The Frontier in American history.* Henry Holt and Sons, New York, New York, USA.

United States Fish and Wildlife Service. 2007. Recovery Plan for the Sierra Nevada bighorn sheep. California/Nevada Operations Office, Sacramento, California, USA.

Utah Division of Wildlife Resources. 2008. Utah bighorn sheep statewide management plan. Department of Natural Resources, Salt Lake City, Utah, USA.

Valdez, R., and P. R. Krausman, Editors. 1999. *Mountain Sheep of North America.* The University of Arizona Press, Tucson, Arizona, USA.

Van Dyke, W. 1990. Bighorn sheep harvest regulations in Oregon: management considerations. *Proceedings of the Biennial Symposium of the Northern Wild Sheep and Goat Council* 7:252–255.

Van de Kerk, M., D. Verbyla, A. W. Nolin, K. J. Sivy, and L. R. Prugh. 2018. Range-wide variation in the effect of spring snow phenology on Dall sheep population dynamics. *Environmental Research Letters* 13:075008.

Van de Kerk, M., S. Arthur, M. Bertram, B. Borg, J. Herriges, J. Lawler, B. Mangipane, C. L. Koizumi, B .Wendling, and L. Prugh. 2020. Environmental influences on Dall's sheep survival. *The Journal of Wildlife Management* 84:1127–1138.

Veitch, A. E., E. Simmons, J. Adamczewski, and R. Popko. 1998. Status, harvest, and co-management of Dall's sheep in the Mackenzie Mountains, Northwest Territories. *Proceedings of the Biennial Symposium of the Northern Wild Sheep and Goat Council* 11:134–153.

Washington Department of Fish and Wildlife. 2015. 2015–2021 Game management plan. Department of Fish and Wildlife, Olympia, Washington, USA.

Whitten, K. R. 2001. Effects of horn-curl regulations on demography of Dall's sheep: a critical review. *Alces* 37:483–495.

Wild Sheep Foundation. 2024. Wild Sheep Foundation conservation impact. Wild Sheep Foundation, Bozeman, Montana, USA.

Wild Sheep Initiative. 2023. Wild sheep westwide management metrics. Unpublished data. Western Association of Fish and Wildlife Agencies, Boise, Idaho, USA.

Wild Sheep Working Group. 2018. Wild sheep westwide management metrics. Unpublished data. Western Association of Fish and Wildlife Agencies, Boise, Idaho, USA.

Wishart, W. 1978. Bighorn sheep. Pages 161–171 *in* J. L. Schmidt and D. L. Gilbert, editors. *Big Game of North America.* Stackpole, Harrisburg, Pennsylvania, USA.

# 17 Mountain Sheep Management in Mexico

*Raul Valdez, Raymond M. Lee, Hugo Sotelo Gallardo, and Juan Manuel Segundo Galán*

## 17.1 INTRODUCTION

Bighorn sheep (*Ovis canadensis* spp.) in Mexico occur in five northern states: Coahuila, Chihuahua, Sonora, and Baja California, and Baja California Sur in the Baja California Peninsula; these states border the United States except for Baja California Sur. Three desert bighorn (*Ovis canadensis*) subspecies (based on the classification of Cowan; 1940) occur in Mexico: Mexican bighorn sheep (*Ovis c. mexicana*), peninsular bighorn sheep (*O. c. cremnobates*), and Weems' bighorn sheep (*O. c. weemsi*; Chapters 1 and 6). Mexican bighorn sheep occurred until the early 20th century in the northern Chihuahuan Desert in the states of Chihuahua and Coahuila where they were extirpated but have been reestablished principally in captive populations. Mexican bighorn sheep were not extirpated in the northwestern Sonoran Desert in the state of Sonora, where they occur in free-ranging and captive populations. Peninsular bighorn sheep occupy the northern region of the Baja California Peninsula and Weems' bighorn sheep occur in the southern portion; the boundary between the two subspecies has not been determined. Mountain sheep were not extirpated in the Baja California Peninsular and occur in free-ranging populations. The distribution of Rocky Mountain bighorn sheep (*O. c. canadensis*) did not historically extend into Mexico (Chapter 1). The desert bighorn sheep in Mexico is a priority species and is listed as a species of special protection (Diario Oficial de la Federación 2010), which permits hunting under strict measures and in the Convention on International Trade in Endangered Species of Wild Fauna and Flora (CITES) in Category II (CITES 2025a). In the early 1900s, in areas where they had not been extirpated, they survived in small ($n \sim 50$) isolated populations. These populations were imperiled by illegal hunting, habitat degradation due to land conversion, overstocked rangelands, the presence of feral animals that reduced availability of forage and watering sites, probable disease transmission from domestic goats, habitat fragmentation and anthropogenic movement barriers, invasive plants and tall vegetation encroachment, and loss of genetic diversity. Because of the unregulated exploitation of mountain sheep and their decreasing numbers, desert bighorn sheep hunting was prohibited in Mexico from 1922 to 1963, which did not prevent illegal hunting as law enforcement activities struggled to achieve effective compliance. Wildlife populations continued to be inadequately protected until the 1990s, and in some cases continue to have minimal protection, as in the state of Baja California where hunting is prohibited.

Two distinct North American mountain sheep (bighorn, thinhorn [*O. dalli*]) management schemes developed: top-down management in Canada and the United States in which management authority is vested in federal, state, and provincial agencies; and bottom-up management in Mexico where management authority of captive and free-ranging populations has been transferred to private landowners and communal entities (e.g., ejidos, comunidades). Sport hunting enterprises in Mexico have been the incentive for developing sustainable, economically viable wildlife management programs because of their high profit potential, and they do not require the extensive infrastructure of other forms of tourism. The mountain sheep populations in Mexico are the result of a management strategy shift from concentrating solely on free-ranging populations to equally emphasizing captive-breeding populations. This incentivized management strategy has increased the distribution and connectivity of bighorn sheep in Mexico. Translocations and population augmentations also have been important management tools; disease management has been a minor concern. Mexico has the most abundant population of desert bighorn sheep and Sonora is the jurisdiction with the highest number of bighorn sheep (11,600) in North America. There are an estimated 19,346 mountain sheep in Mexico of which 12,400 are free-ranging and 6,946 are in captive-breeding populations. Mexico has the most successful bighorn sheep captive-breeding program in the world and this is largely due to Mexico taking advantage of the vast big game hunting commercial and marketing industry to attract hunters, incentivizing desert bighorn sheep management through direct financial returns to private landowners and communal land managers. The high demand and exorbitant prices hunters are willing to pay for permits to hunt desert bighorn sheep are responsible for motivating the current trophy hunting management program; producing trophy males is the major emphasis of the bighorn sheep management program in Mexico. Mexico is currently the only country in the world that prioritizes a combined bighorn sheep captive-breeding and free-ranging population management strategy, an extremely successful management policy. Assessments to guide decision-making and certification to demonstrate professionalism should be incorporated in the desert bighorn sheep management program in Mexico. Our objective is to present an overview of the desert bighorn sheep management program in Mexico.

DOI: 10.1201/9781003518686-19

## 17.2   DISTRIBUTION OF DESERT BIGHORN SHEEP IN MEXICO

Bighorn sheep occurred in the northern Chihuahuan and northwestern Sonoran deserts of Mexico, a region encompassing the Mexican states of Coahuila, Chihuahua, and Sonora, and the states of Baja California, and Baja California Sur in the Baja California Peninsula. In this vast area, populations were distributed along the extensive interconnected mountain systems and adjoining hills and rocky areas (Leopold 1959, Valdez 1997, Tarango and Krausman 1997, Sandoval et al. 2019). Populations in Mexico were sympatric with populations in the United States in Texas and New Mexico; populations in northeastern Baja California still are sympatric with populations in southeastern California in the United States (Buchalski et al. 2015, Romero-Figueroa et al. 2017, 2024; Ruiz-Mondragón et al. 2018). They were extirpated in the states of Chihuahua and Coahuila but have been reestablished principally as captive populations in those states (Heffelfinger and Marquez-Muñoz 2005, Espinosa-T. et al. 2006, Sandoval et al. 2019, Figure 17.1). Bighorn sheep populations in the northwestern Sonoran Desert, in the state of Sonora and the Baja California Peninsula, were not extirpated but survived in small ($n \sim 50$) isolated populations and by the 1990s these populations had begun to stabilize and increase in some areas (Toweill and Geist 1999). The Mexican bighorn sheep was the most widespread subspecies in Mexico and was widely distributed in the northern Chihuahuan Desert in the states of Chihuahua and Coahuila. Bighorn sheep populations have been reestablished in Coahuila and Chihuahua where they had been extirpated, several in captive populations and one free-ranging population in each state (Figure 17.1). There is limited potential for connectivity between the northern and southern population clusters in Sonora (Figure 17.2). The peninsular bighorn occurs in the eastern versant of the northern Baja California Peninsula in the state of Baja California, from the United States-Mexico border, west of the Colorado River, to approximately the southern boundary of the state of Baja California (Figure 17.3). Weems' bighorn occupies the southern portion of the Baja California Peninsula, approximately from the Baja California-Baja California Sur boundary, south to near the city of La Paz which is the extent of the distribution of bighorn sheep in Baja California Sur (Figure 17.4). Detailed studies of subspecies boundaries are needed to determine if there are detectable differences in bighorn populations on the Baja California Peninsula and to assess the validity of subspecies designations (Chapter 1).

In Sonora, most populations of free-ranging bighorn sheep occupy disjunct areas where they are isolated in montane and hilly habitats, having low degrees of connectivity in northern areas but high connectivity in southern population core areas (Figure 17.2). In the southern area, the high connectivity among core areas is the result of relatively short movement corridors, where 35% of potential corridors were less than 10 km long and 67% of corridors

were less than 25 km long, distances within which desert bighorn sheep have been documented to traverse (Bleich et al. 1996, Segundo Galán et al. 2024). In the northern population area, characterized by lower connectivity and longer distances between core areas, only 20% of potential corridors were less than 10 km, 30% were less than 25 km, and potential corridors averaged 41 km. There is a large connectivity gap between northern and southern areas; the shortest potential corridor is 107 km and the four potential corridors average 131 km. Also, potential corridors may be impeded by agricultural development, paved roads with heavy traffic, fences, and other infrastructures (Segundo Galán et al. 2024). Enhancing connectivity involves prioritizing the reestablishment of extirpated populations in high-quality habitat corridors near core areas, increasing the likelihood of dispersing desert bighorn sheep encountering other populations (Bleich et al. 1996, Palfrey et al. 2022). Minimizing isolation, facilitating movement, and restoring extirpated populations is important to ensure the long-term viability of desert bighorn sheep in Mexico.

Comanaged, community-governed, and privately protected areas are increasingly being used as tools to increase protected area coverage and connectivity, complementing existing protected area networks (Bingham et al. 2021, Palfrey et al. 2022, Bezaury-Creel 2024, Fitzsimons and Mitchell 2024). Wildlife management units, officially titled Units for the Conservation, Management, and Sustainable Use of Wildlife (Unidades para la Conservación, Manejo y Aprovechamiento Sustentable de la Vida Silvestre, or their Mexican acronym UMAs), are important for increasing connectivity of free-ranging animals and act as protected areas for bighorn sheep populations in Mexico. Privately protected areas are increasingly being recognized for contributing to wildlife conservation (Fitzsimons and Mitchell 2024). Movement and metapopulation corridors can be facilitated by increasing the number of private and communal UMAs in mountain sheep habitats in Mexico. It is important to determine the habitat components that will improve landscape permeability for desert bighorn sheep (Hilty et al. 2019). The extent of corridor connectivity in the Baja California Peninsula has not been studied. The Sierra de la Rumorosa, the northernmost bighorn sheep-inhabited range in the state of Baja California, extends to the United States-Mexico border. This area provides an opportunity for a binational effort to facilitate bighorn sheep connectivity between the two countries (Romero-Figueroa et al. 2017, 2024) and to develop a joint research program (Buchalski et al. 2015, Fitzsimons and Mitchell 2024).

## 17.3   WILDLIFE CONSERVATION AND MANAGEMENT POLICY IN MEXICO

There are three major land tenure types in Mexico: federal, private, and communal lands. The principal communal lands are ejidos and comunidades. Comunidades are Indigenous communal landholdings, and ejidos are communal properties distributed to groups of individuals, but land ownership

**FIGURE 17.1** Historical distribution of extirpated desert bighorn sheep populations and areas of translocated free-ranging populations in Chihuahua and Coahuila, Mexico. (Adapted from Heffelfinger and Marquez-Muñoz (2005) and Espinosa-T. et al. (2006); used with permission. Cartography by Ramiro Velázquez Rincón.)

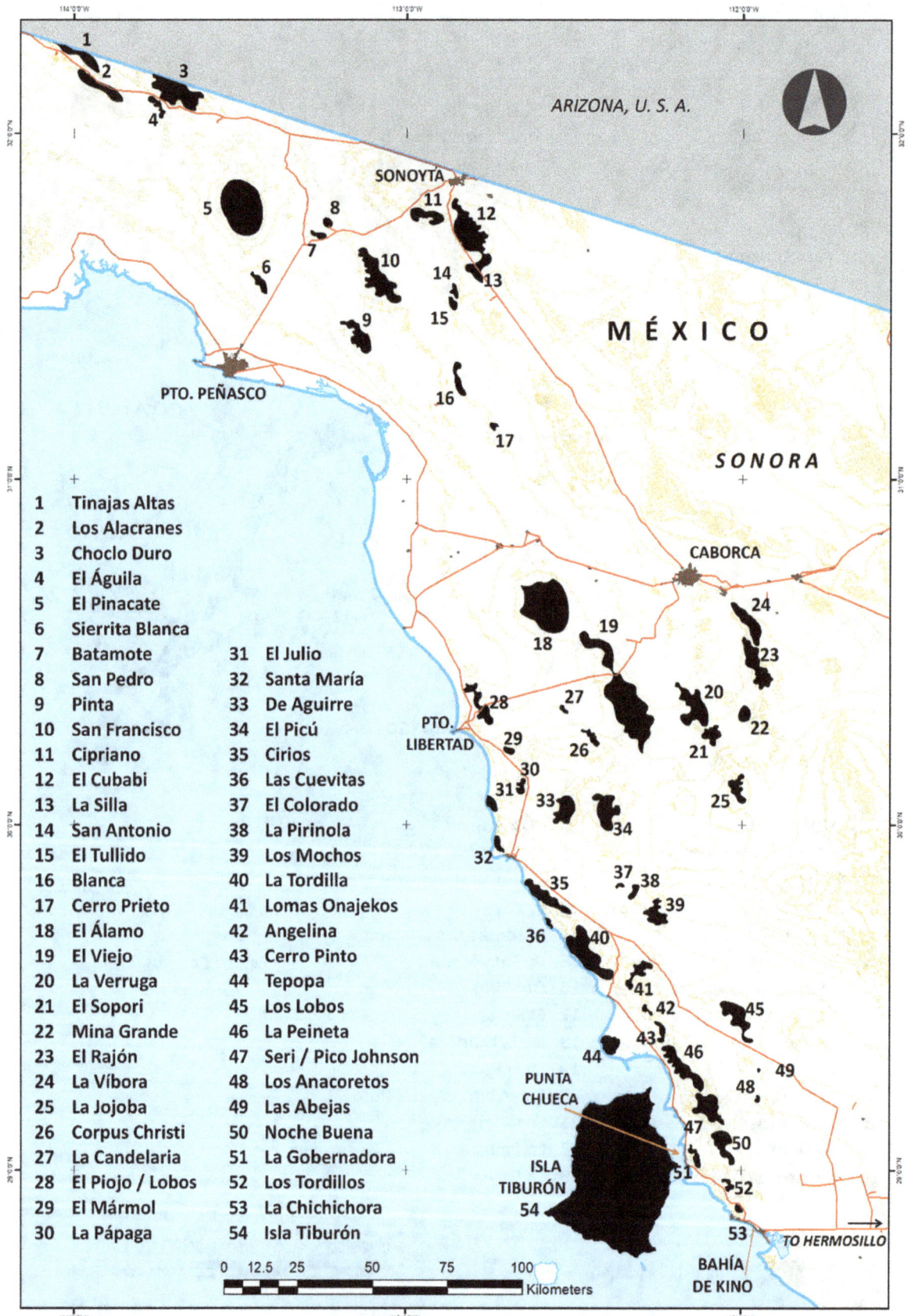

| 1 | Tinajas Altas | | |
|---|---|---|---|
| 2 | Los Alacranes | | |
| 3 | Choclo Duro | | |
| 4 | El Águila | | |
| 5 | El Pinacate | | |
| 6 | Sierrita Blanca | | |
| 7 | Batamote | 31 | El Julio |
| 8 | San Pedro | 32 | Santa María |
| 9 | Pinta | 33 | De Aguirre |
| 10 | San Francisco | 34 | El Picú |
| 11 | Cipriano | 35 | Cirios |
| 12 | El Cubabi | 36 | Las Cuevitas |
| 13 | La Silla | 37 | El Colorado |
| 14 | San Antonio | 38 | La Pirinola |
| 15 | El Tullido | 39 | Los Mochos |
| 16 | Blanca | 40 | La Tordilla |
| 17 | Cerro Prieto | 41 | Lomas Onajekos |
| 18 | El Álamo | 42 | Angelina |
| 19 | El Viejo | 43 | Cerro Pinto |
| 20 | La Verruga | 44 | Tepopa |
| 21 | El Sopori | 45 | Los Lobos |
| 22 | Mina Grande | 46 | La Peineta |
| 23 | El Rajón | 47 | Seri / Pico Johnson |
| 24 | La Víbora | 48 | Los Anacoretos |
| 25 | La Jojoba | 49 | Las Abejas |
| 26 | Corpus Christi | 50 | Noche Buena |
| 27 | La Candelaria | 51 | La Gobernadora |
| 28 | El Piojo / Lobos | 52 | Los Tordillos |
| 29 | El Mármol | 53 | La Chichichora |
| 30 | La Pápaga | 54 | Isla Tiburón |

**FIGURE 17.2**   Distribution of free-ranging desert bighorn sheep in the state of Sonora, Mexico. (Map courtesy of Juan Manuel Segundo Galán; used with permission. Cartography by Ramiro Velázquez Rincón.)

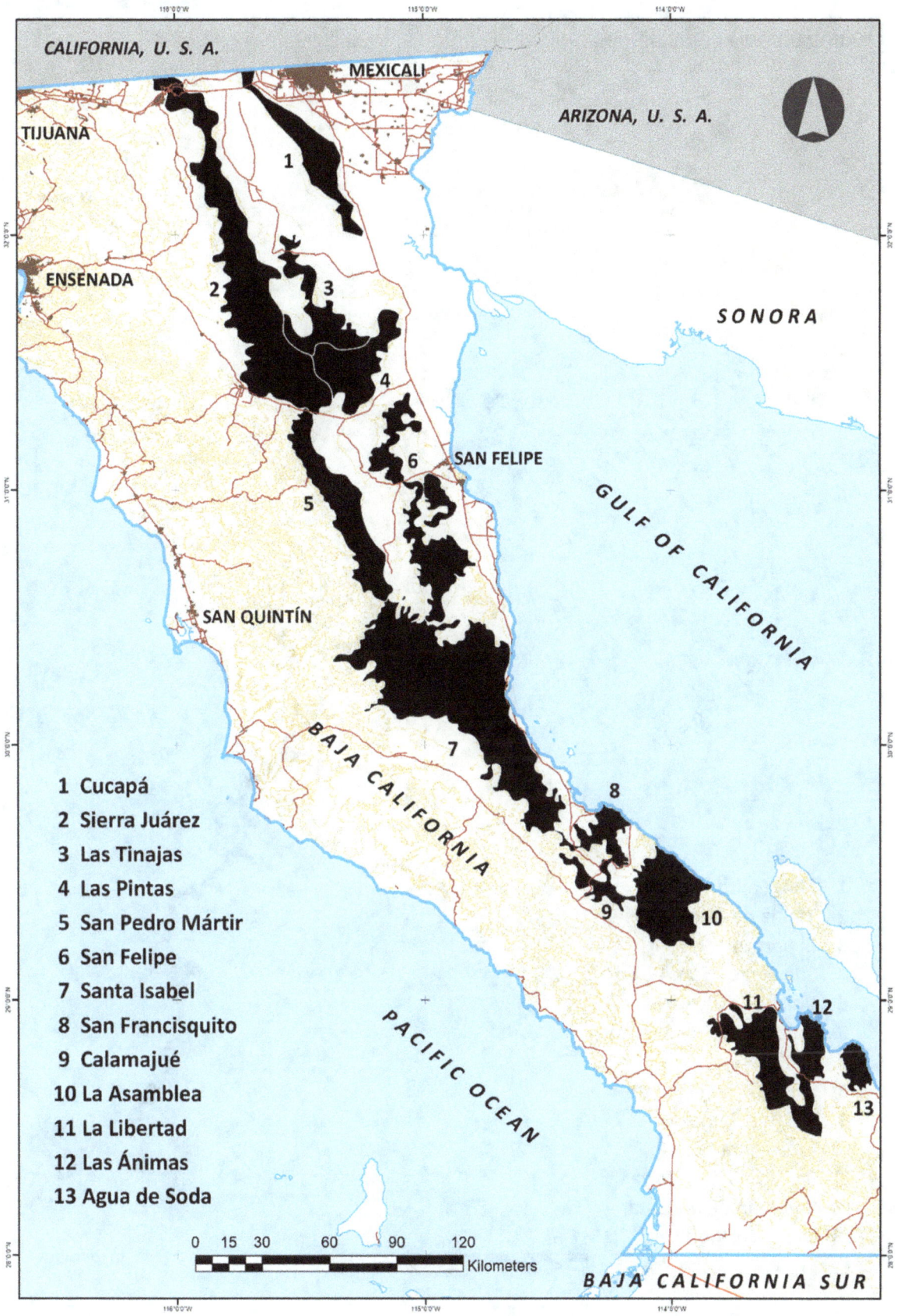

1 Cucapá
2 Sierra Juárez
3 Las Tinajas
4 Las Pintas
5 San Pedro Mártir
6 San Felipe
7 Santa Isabel
8 San Francisquito
9 Calamajué
10 La Asamblea
11 La Libertad
12 Las Ánimas
13 Agua de Soda

**FIGURE 17.3** Distribution of free-ranging desert bighorn sheep in the state of Baja California, Mexico. (Map courtesy of Enrique de Jesus Ruiz-Mondragón; used with permission. Cartography by Ramiro Velázquez Rincón.)

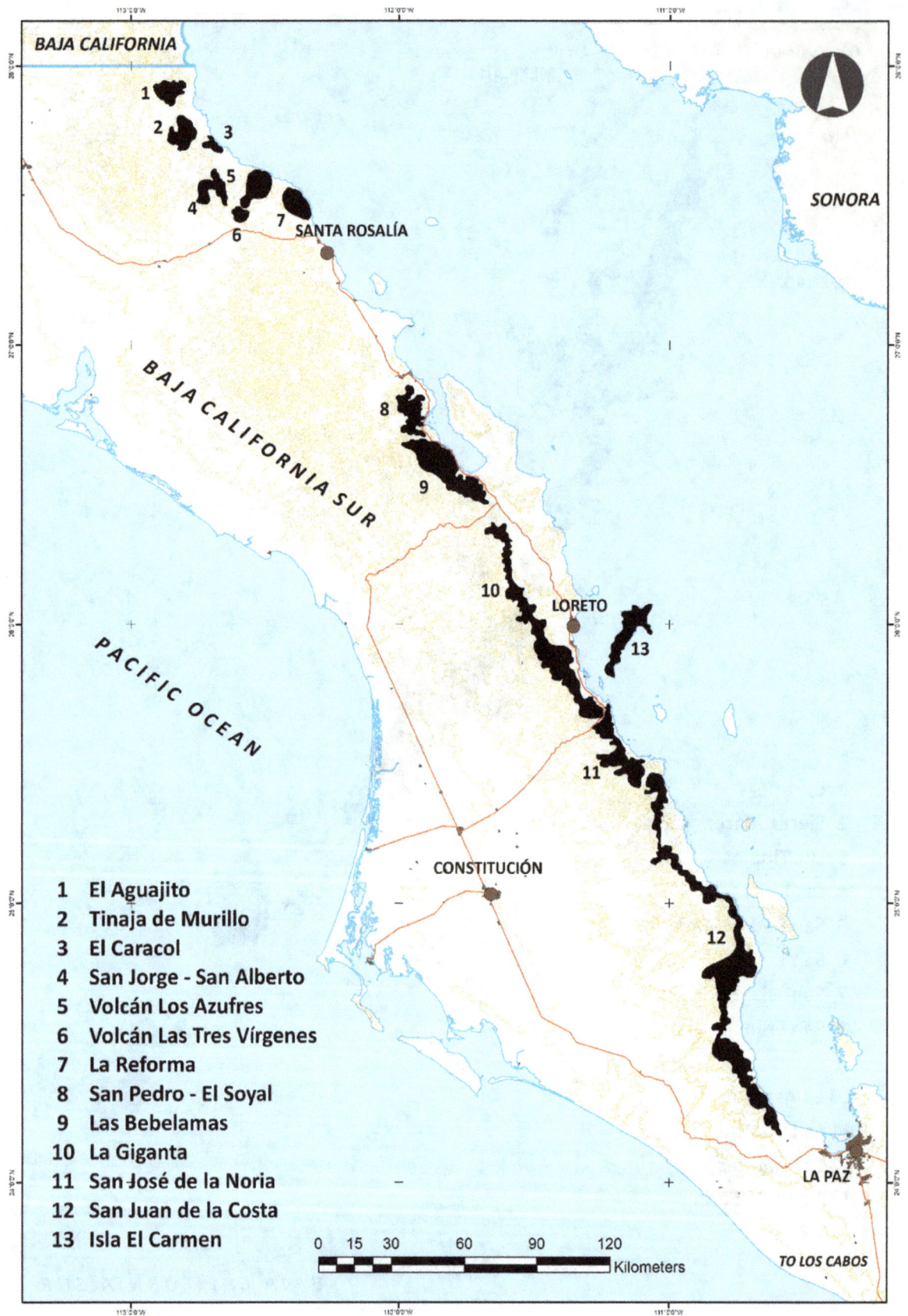

**FIGURE 17.4**  Distribution of free-ranging desert bighorn sheep in the state of Baja California Sur, Mexico. (Map courtesy of Juan Manuel Segundo Galán; used with permission. Cartography by Ramiro Velázquez Rincón.)

resides with the community, with collective administration and management of these lands and their resources, including wildlife and forests. Most of these lands are farmed. Initially, communal properties could not be sold and remained the property of the federal government, but regulatory changes in 1991 permitted their sale. Ejidos are an important form of land tenure in Mexico. There are 29,534 ejidos and 2,359 communal lands, including communidades, encompassing 84.5 (53%) million ha and 17.4 (19%) million ha, respectively, of rural lands of Mexico, and occupied by over 6 million communal landowners and their families (Morrret-Sánchez and Cosío-Ruiz 2017). Rural private property encompasses 79 (40%) million ha. To retain the right to continue to use them, communal property owners must manage resources productively and sustainably. Wildlife had been particularly neglected because it was not considered an economically viable resource (Valdez et al. 2006, Valdez 2019).

Mexico's current wildlife conservation and management policy, initiated in 1997, is incorporated in the Program of Wildlife Conservation and Productive Diversification of the Rural Sector (Programa de Conservación de la Vida Silvestre y Diversificación Productiva en el Sector Rural). The major objective of this program is to integrate environmental, economic, social, and legal strategies to address wildlife needs while promoting broader societal participation and creating realistic economic incentives. This program promotes participatory governance and conservation opportunities by involving key stakeholders in management decisions. It serves a complementary function for the conservation and sustainable use of biodiversity by creating opportunities for income, employment and foreign exchange in rural areas, incorporating beneficial opportunities provided by the wildlife resource (Instituto Nacional de Ecología 2000). This program has two strategies: the conservation and recovery of wildlife priority species, and the creation of a system of wildlife management units that emphasize the conservation, management, and sustainable use of wildlife. Priority species can be plants or animals and include those that are threatened or endangered, umbrella and charismatic species, and those that possess a cultural or economic value. Priority ungulate species include the desert bighorn sheep and pronghorn (*Antilocapra americana*; Diario Oficial de la Federación 2010, Instituto Nacional de Ecología 2000, Valdez et al. 2006, Ortega-Argueta et al. 2016, Gallardo 2018, González-Rebeles Islas et al. 2019b, Sandoval et al. 2019).

Wildlife management units also are an integral part of this program and an important legal component for the conservation and sustainable use of biodiversity. An UMA is a tract of federal, communal, or private land on which the sustainable use of flora and fauna is permitted. A total of 9,867 UMAs were registered by 2017 of which private lands represented 76.5%, ejidos 21.5%, and Indigenous land 2% (Romero-de-Diego et al. 2021, Secretaría de Medio Ambiente y Recursos Naturales [SEMARNAT] 2020). Effective wildlife conservation efforts in Mexico were not initiated until the 1990s with the formation of UMAs. The basic concept of UMAs was to create economic incentives by facilitating the integration of wildlife management programs in livestock, forestry, wildlife enterprises, and agricultural ventures. Wildlife uses (including plants) within UMAs are broadly interpreted to include research, environmental education, game farms and ranches, parks, and commercialization of wildlife by-products, which can be marketed through regulated laws. The UMAs are classified as extensive or intensive use areas. Extensive units are those in which wildlife is free-ranging such as game ranches. Intensive units are those in which wildlife or plants are raised under controlled management schemes, including botanical and zoological parks and wild animal breeding programs, such as the bighorn sheep captive-breeding projects (Instituto Nacional de Ecología 2000, González-Rebeles Islas et al. 2019a,b). Unidades para la Conservación, Manejo y Aprovechamiento Sustentable de la Vida Silvestre can vary in size depending on management objectives and economic viability. All UMAs must be registered with the federal Office of Wildlife (Dirección General de Vida Silvestre) and include a management plan. The management plan must incorporate census and monitoring methods, species-specific use criteria, harvest verification, and illegal harvest preventive measures. Secretaría de Medio Ambiente y Recursos Naturales is responsible for setting the number of permits issued each year based on helicopter or other scientifically reliable population surveys for free-ranging populations, which establish the population and number of animals of harvestable age (Valdez et al. 2006, Valdez 2019, Gallardo 2018, González-Rebeles Islas et al. 2019a,b). A major wildlife management change occurred when the administration of federal wildlife conservation regulations was transferred to some states through special agreements at the request of the states. At present, only the northern states of Baja California, Chihuahua, Coahuila, Nuevo León, Sonora, and Tamaulipas have participated in these agreements. Those states have assumed the regulatory functions of authorizing, registering, and supervising UMAs and their management, establishing standards for optimal management of captive wildlife, issuing permits for sport hunting, establishing hunting regulations, and promoting wildlife cultural traditions (González-Rebeles Islas et al. 2019b, SEMARNAT 2024b).

This paradigm shift in national wildlife conservation policy, initiated in 1997, was concordant with decentralized governance and community participatory approaches in managing biodiversity and natural resources (Valdez et al. 2006, Valdez 2019, Servín et al. 2018, González-Rebeles Islas et al. 2019a,b; Balachander et al. 2022). The empowerment of the rural populace, a people-oriented conservation policy, enabled the self-governance of rural communities to directly manage wildlife, a national policy that has been effective because of the ability of once disenfranchised peoples to now profit from the wildlife resource. Bighorn sheep management programs in Mexico were initiated by private landowners who had a tradition of livestock ranching, were well educated, had greater financial resources, and were willing to incorporate native wildlife populations into innovative multispecies sport hunting systems (Villarreal González 2013, Romero-de-Diego et al. 2021).

## 17.4 BIGHORN SHEEP HABITATS AND RELATED ISSUES IN MEXICO

Desert bighorn sheep in Mexico were widespread in the Chihuahuan and Sonoran deserts within the Southwest Deserts Ecoregion (Chapter 1). These deserts incorporated isolated or interconnected mountain ranges characterized by cliffs, steep-sloped canyons, and lower elevation undulating, rugged, hilly terrain associated with high temperatures and unreliable rainfall patterns that limit forage production and the availability of permanent watering sources (Hansen 1980, Álvarez-Cárdenas et al. 2009, Ruiz-Mondragón et al. 2018, 2025a; Segundo Galán et al. 2025). The expansive habitats in these regions provided movement corridors, but the harsh environmental conditions and anthropogenic land use made them prone to habitat fragmentation. The Chihuahuan Desert is the largest desert (518,000 km²) in Mexico, followed in area by the Sonoran Desert (275,000 km²). Most of the Chihuahuan Desert is in the state of Chihuahua, Mexico's largest state. Bighorn sheep populations in Sonora and the Baja California Peninsula occur in the Sonoran Desert. About 35% of Mexico's land area consists of desert, and these desert plant communities comprise the largest of Mexico's vegetation types (Challenger and Soberón 2008, Abd El-Ghani et al. 2017, Challenger 2019). Freezing temperatures occur during winter in the Chihuahuan Desert but rarely occur in the Sonora Desert; daily summer temperatures in both deserts can exceed 45°C (Laity 2008). Elevations in desert bighorn sheep habitats in Sonoran range from sea level along the Gulf of California to 2,625 m in inland ranges. Most precipitation falls in July (~86 mm) and least in May (~3.5 mm), averaging 450 mm annually. Most of the landholdings in habitats with bighorn sheep populations are privately owned in Coahuila, Chihuahua, and Sonora, and communally owned (ejidos) in the Baja California Peninsula (Ruiz-Mondragón 2014, Ruiz-Mondragón et al. 2023, 2025a). Private landowners are usually economically independent and can invest more in wildlife management programs than community members.

### 17.4.1 NATIVE UNGULATES, LARGE CARNIVORES, AND EXOTIC UNGULATES IN BIGHORN SHEEP HABITAT IN MEXICO

Wild artiodactyls, besides mountain sheep, in the Chihuahuan and Sonoran deserts include mule deer (*Odocoileus hemionus*), white-tailed deer (*O. virginianus*) (absent in the Baja California Peninsula except for the extreme northeastern corner), and collared peccary (*Pecari tajacu*). All ungulate species are sport hunted and combined species hunting opportunities are common. Large mammalian carnivores that are potential predators of adult mountain sheep include the mountain lion or puma (*Puma concolor*), bobcat (*Lynx rufus*), and coyote (*Canis latrans*; Hall 1981, Ceballos 2014). The wolf (*Canis lupus*), which

formerly occupied the entire region, has been extirpated (Brown 1983). Of the avifauna of Sonora, only the golden eagle (*Aquila chrysaetos*) is a potential predator. Puma predation was the primary factor limiting population growth in a reestablished desert bighorn sheep population in Coahuila (Sotelo Gallardo et al. 2018a,b); however, other studies of carnivore-mountain sheep interactions have not been conducted in Mexico. The aoudad (*Ammotragus lervia*) has been introduced to portions of Mexico and is the only invasive wild ungulate occurring in bighorn sheep habitats there, principally in Coahuila and Chihuahua (Valdez 2011, Sandoval et al. 2019). Aoudad populations have the potential to displace, out-reproduce, and outcompete desert bighorns for habitat and forage; they also have the potential to transmit domestic sheep and goat diseases to bighorn sheep (Sandoval et al. 2019, Wright et al. 2024, Chapter 2). Domestic ungulates in existing or potential mountain sheep habitats include domestic goats, cattle, donkeys, and horses; goats and donkeys occur in feral populations. Livestock ranching is a major economic activity in northern Mexico. Exotic grass species are widespread (Ortega-S. et al. 2019). Domestic goats are extensively grazed on the desert rangelands in communal grazing lands with high stocking rates, negatively affecting forage availability and carrying capacity in bighorn sheep habitats (Toweill and Geist 1999, Sandoval et al. 2019, Chapter 2). Goats are primarily browsers but with an opportunistic grazing strategy and are adapted to graze in rough, steep terrain; they are well-suited for grazing in desert bighorn sheep habitats (Mellado et al. 2005). Coahuila has the highest population of domestic goats (366,500), followed by Chihuahua (91,700) and Baja California Sur (71,400) (Instituto Nacional de Estadística y Geográfica 2023). More than 70% of the land area in northwestern Mexico, including areas formerly inhabited by desert bighorn sheep, is used for cattle ranching, most of which is exported to the United States (SEMARNAT 2023, Servicio de Información Alimentaria y Pesqueria 2023). These extensive cow-calf livestock management systems operate year-round with minimal husbandry practices. Given the potential for interspecific competition and disease transmission between feral and domestic animals and wildlife, a higher priority on research related to these topics should occur.

### 17.4.2 HISTORICAL BACKGROUND OF BIGHORN SHEEP HABITATS IN MEXICO

Desert regions in Mexico have been inhabited by Indigenous Tribal Communities for thousands of years before the European presence (Radding 1997, Tapia Landeros 2007, Mellink and Rioja-López 2020, Chapter 2). Livestock enterprises and mining industries have been present since the 1500s, however, until the early 19th century, only isolated areas with scattered ranches existed and in areas such as the Sonoran Desert, bighorn sheep habitat remained relatively undisturbed through the 1930s. Sheldon (Carmony and

Brown 1994), while on a journey to the Seri Tribal areas in December 1920 in northwestern Sonora, traveled (about 160 km) on horseback most of the way from Hermosillo to the Sea of Cortez, and recorded only scattered small ranches, some with horses but without cattle and few inhabitants. The area of the Native Seri Tribal (Comcáac) Homeland, where Sheldon hunted desert bighorn sheep and mule deer on the Pico Seri/Johnson area, and mule deer on Tiburon Island, had neither cattle, horses, donkeys, nor mules for transportation or for use as work animals, which continues to the present. Roosevelt (1920) made similar observations regarding low human populations during desert bighorn sheep hunting trips in Sonora in the early 1900s; North (1910) described even more desolate conditions in the Baja California Peninsula, where he hunted desert bighorn sheep in 1906 as did Cudahy (1928) in northern Baja California. Tinker (1978), an American appointed by the Mexican government in the early 1900s to enforce game laws in Sonora and Baja California, had similar experiences. O'Connor, shooting editor for an outdoor magazine, hunted bighorn sheep in northwestern Sonora from the Arizona border to the Sea of Cortez between 1934 and 1947, describing bighorn sheep populations as being widespread but in minimal numbers (O'Connor 1974, Anderson 2002).

### 17.4.3 Mountain Sheep Habitats in National Protected Areas and UMAs in Mexico

A protected area is a "geographically defined area which is designated or regulated and managed to achieve specific conservation objectives" (Convention on Biological Diversity 1993:2). National protected areas are the main biodiversity conservation instrument of Mexico's environmental stewardship programming, conserving natural areas that are representative of the biomes and genetic diversity of native species (Comisión Nacional de Áreas Naturales Protegidas 2018, 2024; González-Rebeles Islas et al. 2019b). Effective wildlife management in federally protected areas in Mexico suffers from high administrative and maintenance costs, with limited federal financial resources then being available for conservation programs, including law enforcement (Figueroa and Sánchez-Cordero 2008) and wildlife research (Valdez et al. 2006, Valdez 2019). There are 232 national protected areas in Mexico classified in six categories: biosphere reserves, flora and fauna protected areas, national parks, natural resource protected areas, sanctuaries, and natural monuments (Comisión Nacional de Áreas Naturales Protegidas 2024). These comprise 10.7% of the national land area. Because national protected areas encompass habitats with diverse fauna, wildlife conservation often is a priority (Watson et al. 2014, González-Rebeles Islas et al. 2019b, Wilson-Holt et al. 2022). One of the main differences in federal wildlife management policy between Canada, the United States, and Mexico is that in Mexico, hunting in biosphere reserves and national parks can be allowed, albeit limited and in specified areas. The

same policy applies to natural resource extraction, including mining and livestock grazing (Wakild 2015, Sandoval et al. 2019). Six national protected areas provide protected habitats for bighorn sheep in Mexico: El Pinacate and Gran Desierto de Altar Biosphere Reserve, Tiburon Island Seri Indigenous Community Nature Reserve in northwestern Sonora, Valle de los Cirios Flora and Fauna Protected Area, Sierra de San Pedro Mártir National Park in the state of Baja California, and the El Vizcaino Biosphere Reserve and Loreto II and Nopoló national parks in Baja California Sur. Biosphere reserves are a global network of protected areas of natural ecosystems promoted by United Nations Educational, Scientific and Cultural Organization's Man and the Biosphere Program (Secretariat of the Man and the Biosphere Program 2022). These reserves consist of a core area that conserve ecosystems and their functionality, and where exploitation activities that alter ecosystems are strictly prohibited. Within a surrounding buffer zone, human communities that inhabited the area at the time of the biosphere reserves' creation are permitted to engage in sustainable use of natural resources (Chester 2006). Biosphere reserves are established in biogeographic areas where ecosystems have not been altered by anthropogenic land use, and they are inhabited by emblematic or endangered species (United Nations Educational, Scientific and Cultural Organization 2024). The El Pinacate and Gran Desierto de Altar Biosphere Reserve has an area of 714,556 ha. Bighorn sheep inhabit a rugged, extinct volcanic region that constitutes 30% (213,227 ha) of the area with elevations ranging from 200 to 1,190 m. A major concern in the reserve is the depleted bighorn sheep population due to illegal hunting (Secretaría de Desarrollo Social 1993). Tiburon Island, in Sonora (Figure 17.2), is a federally designated natural protected reserve managed by the Seri (Comcáac) Indigenous Tribe for trophy hunting of bighorn sheep and mule deer, the only such designated protected area in Mexico. The Valle de los Cirios Flora and Fauna Protected Area (667,227 ha) in the state of Baja California has the largest area of bighorn sheep protected habitat in Mexico. About 25% of the area of distribution and 38% of the mountain sheep population in the state are in a protected area, principally in the Cirios reserve (Romero-Figueroa et al. 2024, Ruiz-Mondragón et al. 2025a). The Cirios reserve southern boundary borders the Vizcaino Biosphere Reserve and encompasses the southernmost population of mountain sheep in Baja California. The University of Baja California Mountain Sheep Sanctuary, a joint bighorn sheep conservation effort of the University Foundation, Autonomous University of Baja California, and the National Commission of Protected Areas, is in the Cirios protected area. It covers 30,000 ha in the Sierra de la Asamblea and has a population of 100 bighorn sheep (Figure 17.3). The university is responsible for managing the mountain sheep population, which consists of monitoring their numbers, surveillance, habitat improvement projects, research and integrating the local community in a co-management and environmental educational program (Ruiz-Mondragón 2024, Fundación Universidad Autónoma Baja California 2025). The Vizcaino Biosphere Reserve in Baja California Sur is the largest protected area in Mexico, covering 2,546,790 ha, of which sheep habitat includes 167,000 ha. The eight mountain ranges comprising the Tres Virgenes Mountain Complex in the Vizcaino Biosphere Reserve has the highest population (540) of mountain sheep in the Baja California Peninsula. Mountain sheep hunting within the Vizcaino Biosphere Reserve is permitted in the Ejido Bonfil with the objective of conserving wildlife and their habitats and alleviating poverty. The smaller mountain sheep habitats in the San Pedro Mártir National Park (900 ha) with a population of 200 in the state of Baja California, and the Loreto II (6,200 ha) and Nopoló (2,077 ha) national parks in La Giganta Mountain Range of Baja California Sur also provide mountain sheep protected habitats.

Examples of specific mountain sheep habitats include the Seri/Pico Johnson area in Sonora, Sierra Santa Isabel Range in northern Baja California, and the Sierra del Mechudo in the southern Baja California Peninsula. These habitats are in communal properties. The Sierra Seri/Pico Johnson is situated in the central coastal zone of Sonora, approximately 40 km north of Bahia Kino (Figure 17.2), has a maximum elevation of 1,060 m and with an average summer and winter rainfall of 194 and 45 mm, respectively. Average summer temperatures are 33°C and 15°C in winter (González-Saldívar et al. 2011, Gastelum-Mendoza et al. 2021). The Sierra Santa Isabel occupies the central eastern region of Baja California (2,072 km²), with an estimated population of 300–400 bighorn sheep (Romero-Figueroa et al. 2024, de Ruiz-Mondragón et al. 2025b, Figure 17.2). The maximum ambient temperature in the region is 45°C with average precipitation along the coastal region generally being less than 50 mm; in the central and northern region precipitation varies between 100 and 150 mm (Roberts and Ezcurra 2012). The steep terrain of the Sierra del Mechudo is located about 50 km north of La Paz in the southern end of the Sierra de la Giganta, near the Gulf of California (Álvarez-Cárdenas et al. 2009, Figure 17.3). Elevations here vary between 200 and 800 m, with annual temperatures ranging from 22°C to 35°C; summer temperatures can reach 45°C with an average of 28°C and extremely scarce rainfall that peaks in September. The vegetation of deserts in Mexico is detailed in Shreve and Wiggins (1964), Wiggins (1980), Brown (1982), Turner et al. (2005), González-Abraham et al. (2010), Rebman and Roberts (2012), and León de la Luz et al. (2018).

## 17.5 STATUS OF BIGHORN SHEEP POPULATIONS IN MEXICO

The present status of bighorn sheep in Mexico owes much of its success to translocations. Desert bighorns were reestablished in northeastern Chihuahua in 2000, within their historical range, on the Rancho La Guarida from free-ranging desert bighorn sheep captured in Sonora; several additional translocations were made to augment the herd (Uranga-Thomas and Valdez 2011). Additional captive-breeding populations were established in Chihuahua from animals from Rancho La Guarida and Sonora. There were 575 captive bighorn sheep in Chihuahua by 2015 of which 300 were in La Guarida and 275 in three other captive-breeding facilities (Sandoval et al. 2019). Bighorn sheep hunting was reopened in Chihuahua in 2009 on the La Guarida Ranch where a state hunting permit was issued for a male, which was successfully hunted, after a closed hunting season of more than 50 years. The Rancho Pilares UMA in Coahuila, owned by CEMEX, a cement producing company, encompasses 12,000 ha with 200 desert bighorn sheep in a 5,000 ha enclosure. The founding population consisted of 48 bighorn sheep (37 females and 11 males) from Sonora of which ten originated from two captive populations, 30 from a wild population (Cirios Mountains) in Sonora, and 18 from Tiburon Island (Durán Álvarez 2013). Mountain sheep originating from the CEMEX captive herd were released to establish a free-ranging herd in Coahuila (Sotelo Gallardo et al. 2018a,b; Sandoval et al. 2019). These initial captive management efforts yielded population increases. The Carmen Island (Figure 17.2) population originated from two translocations of 26 individuals of Weems' desert bighorn sheep from mainland communal landholdings; 40 individuals from this island population have been used to augment those mainland populations (Jiménez Lezama and Cavazos 2008). Perhaps the most successful transplant has been that on Tiburon Island, Sonora, where, after an initial release of 20 desert bighorn sheep, the population has increased to over 1,000 animals (Table 17.1; Sandoval et al. 2019).

There are about 19,350 mountain sheep in Mexico of which 12,400 are free-ranging and 6,950 are in captive-breeding populations (Table 17.1). In the state of Sonora, there are an estimated 11,600 mountain sheep in 52 UMAs of which 7,400 are free ranging; this includes 1,100 on Tiburon Island and 4,200 in captive-breeding herds (Segundo Galán et al. 2024). In the state of Baja California, there are an estimated 1,700 mountain sheep (Romero-Figueroa et al. 2024) in 13 mountain ranges and none in captive-breeding programs. In the state of Baja California Sur, there are 2,100 desert bighorn sheep in two disjunct populations, one on the mainland with an estimated 1,400 animals and the other population on Carmen Island with 700 desert bighorn sheep (S. D. Jiménez Lezama, Organización Vida Silvestre, personal communication). There are no captive-breeding programs in Baja California Sur. There are 3,800 desert bighorn sheep in the Baja California Peninsula. In Chihuahua, there are 2,100 desert bighorn sheep in eight separate herds, of which 600 are in a free-ranging population. In Coahuila, there are an estimated 1,825 desert bighorn sheep of which 600 are free-ranging. In the state of Nuevo Leon, there are 21 bighorn sheep of which 19 are in a captive-breeding

## TABLE 17.1

## Desert Bighorn Sheep Status and State Population Estimates in Mexico

| States | Population | Source |
|---|---|---|
| Sonora | | |
|    Captivity | 4,200 | Segundo Galán et al. (2024) |
|    Free-ranging | 6,300 | Segundo Galán et al. (2024) |
|    Tiburon Island | 1,100 | Segundo Galán et al. (2024), Secretaría de Agricultura, Ganadería, Recursos Hidráulicos, Pesca y Acuacultura (2024) |
| **Subtotal Sonora** | **11,600** | |
| Baja California | 1,700 | Romero Figueroa et al. (2024) |
| Baja California Sur | 1,400 | Segundo Galán et al. (2023) |
|    Carmen Island | 700 | S. D. Jiménez Lezama, personal communication |
| **Subtotal Baja Peninsula** | **3,800** | |
| Chihuahua | | |
|    Captivity | 1,500 | Secretaría de Desarrollo Urbano y Ecología (2024) |
|    Free-ranging | 600 | Jose Antonio Vallina, Rancho La Guarida, personal communication |
| Coahuila | | |
|    Captivity | 1,225 | Secretaría de Medio Ambiente (2024) |
|    Free-ranging | 600 | E. R. Woodyard, La Palmosa, personal communication |
| Nuevo León | | |
|    Captivity | 21 | Parques y Vida Silvestre (2024) |
| **Subtotal Chihuahua, Coahuila and Nuevo León** | **3,946** | |
| **TOTAL** | **19,346** | |

facility and two in a zoological park. There are 111 bighorn sheep hunting management units in Mexico encompassing 2,780,400 ha. Bighorn sheep in Mexico occur in every state where they have historically been documented. There may be bighorn sheep populations in private ranches that have not been recorded by state wildlife agencies because some owners prefer to keep that information private. In summary, the population trend of free-ranging populations of bighorn sheep is stable in Baja California and Baja California Sur and increasing in Sonora; captive-breeding populations in Chihuahua, Coahuila, and Sonora also are increasing in number.

Bighorn sheep populations in Mexico, based on published estimates, have increased particularly between 2012 and 2022 for free-ranging populations, and between 2017 and 2022 in captive-breeding populations. This is principally due to the captive-breeding program and the population increases of the Carmen and Tiburon Island populations. In 2013, there were an estimated 12,655 bighorn sheep in Mexico (Sandoval et al. 2019). There were an estimated 14,000 (Festa-Bianchet 2020) to 17,000 (Brewer and Bleich 2023) reported between 2019 and 2023. The state population estimates for 2019 and 2023 were provided by the Wild Sheep Foundation Mexico Council, which is comprised of private landowners that manage sport-hunted desert bighorn sheep populations in Mexico and whose main aim is to conserve and sustainably manage these populations on their properties. The Council's 2019 and 2023 population estimates do not differentiate between captive or free-ranging populations, lack specific sources of estimated population numbers, and include a population estimate of

1,000 mountain sheep in Nuevo Leon. Hence the Council's estimates cannot be directly compared with state and academic-sourced estimates.

### 17.5.1 Numbers of Desert Bighorn Sheep Management Units and Area Encompassed in Individual States in Mexico

In 2024, there were 111 registered wildlife management units (UMAs) authorized for free-ranging and captive-breeding desert bighorn programs in six states: two in Nuevo Leon (Parques y Vida Silvestre 2024), eight in Chihuahua (Secretaría de Desarrollo Urbano y Ecología 2024), 12 in Baja California Sur (SEMARNAT 2024a), 16 in Coahuila (Secretaría de Medio Ambiente 2024), 21 in Baja California (Secretaría de Medio Ambiente y Desarrollo Sustentable 2024), and 52 in Sonora (Segundo-Galán et al. 2024, Table 17.2). These UMAs encompass 2,780,286 ha and range in size from 1 ha in Sonora to 454,000 ha in Baja California Sur. Additional information is available for UMAs in Sonora. Of the 52 UMAs in Sonora, 41 are privately owned and cover an area of 420,545 ha, ten are ejidos with an area of 180,100 ha, plus the Tiburon Island Seri Indigenous Communal Property with an area of 120,756 ha. The 40 UMAs with bighorn sheep in captive enclosures cover an area of 36,110 ha. The smallest UMA covers 1 ha with 15 sheep in the enclosure and the largest covers 5,813 ha and has 229 sheep. The UMA with the largest captive desert bighorn sheep population ($n=530$) covers 1,650 ha, the UMA with the second largest population ($n=445$) covers 275 ha, the third largest population ($n=443$)

**TABLE 17.2**

**Desert Bighorn Sheep Management Units (UMAs) and Area Encompassed in Individual States in Mexico**

| State | Number of Management Units (UMAs) | Area (ha) | References |
|---|---|---|---|
| Nuevo León | 2 | 1,500 | Parques y Vida Silvestre (2024) |
| Chihuahua | 8 | 53,267 | Secretaría de Desarrollo Urbano y Ecología (2024) |
| Baja California Sur | 12 | 1,790,150 | Secretaría de Medio Ambiente y Recursos Naturales (2024a) |
| Coahuila | 16 | 102,519 | Secretaría de Medio Ambiente (2025) |
| Baja California | 21 | 111,450 | Secretaría de Medio Ambiente y Desarrollo Sustentable (2024) |
| Sonora | 52 | 721,400 | Segundo-Galán et al. (2024) |
| Total | 111 | 2,780,286 | |

is in an enclosure of 365 ha, and the fourth largest population ($n=426$) is in an enclosure of 365 ha. The combined area of the Sonoran UMAs encompasses 721,400 ha of which 84% of the UMAs are within the distribution of desert bighorn sheep populations equivalent to 88% of the distribution of the species. The UMAs in Sonora are distributed in 46 of the 54 mountain ranges, which have been identified as having bighorn sheep populations (Segundo Galán, wildlife consultant, personal communication). Wildlife management units have been pivotal in the recovery and increase of bighorn sheep populations and in expanding their distributions in Mexico (CITES 2025b).

## 17.6 BIGHORN SHEEP TROPHY HUNTING MANAGEMENT PROGRAM IN MEXICO

### 17.6.1 HUNTING PERMITS ISSUED IN MEXICO, 2000–2022

Approximately 3,899 hunting permits were issued in Mexico for the 2000–2022 hunting seasons (SEMARNAT 2024b, Secretaría de Agricultura, Ganadería, Recursos Hidráulicos, Pesca y Acuacultura 2025, Figure 17.5). The number of permits issued can be related to the five stages in the mountain sheep recovery program in Mexico. (1) During the hunting seasons between 2000 and 2005, less than 100 permits were issued per year. This low number of permits is because hunting permits were issued only for Baja California Sur and Sonora, which had reduced trophy male populations. (2) During the 2006–2010 hunting seasons, the number of permits increased slightly; these stabilized around 130, which reflects the increased bighorn sheep populations in Baja California Sur (estimated population then of 1,227; SEMARNAT 2023) and in Sonora (estimated population then of 2,500; Secretaría de Agricultura, Ganadería, Recursos Hidráulicos, Pesca y Acuacultura 2011). (3) The first legal hunting permits in Chihuahua and Coahuila were issued in 2009 and 2010, respectively, after the reestablishment of bighorn sheep in captive populations and after more than 50 years of being closed to hunting (Sandoval et al. 2014, 2019). The added availability of

hunting permits in these two states further increased the availability of hunting permits, which rose from 161 to 184 during the 2011–2013 hunting seasons. (4) During the 2014–2019 hunting seasons, the increased demand for and added availability of captive-bred trophy males enabled an increase of hunting permits from 200 to 230 (Figure 17.5). (5) During the 2020–2022 bighorn sheep hunting seasons, the number of hunting permits issued exceeded 295 in congruence with the then 130 existing bighorn sheep hunting UMAs, stable free-ranging desert bighorn sheep populations in Baja California Sur and increasing in Sonora, and greatly increased captive-breeding populations in Sonora, Chihuahua, and Coahuila, all of which provided expanded hunting opportunities.

### 17.6.2 CAPTIVE AND FREE-RANGING DESERT BIGHORN TROPHIES EXPORTED FROM MEXICO TO OTHER COUNTRIES

Mexico became a signatory to CITES in 1991. Desert bighorn sheep in Mexico are listed in CITES Appendix II (CITES 2025a). Bighorn sheep hunted in Mexico by foreign hunters require a CITES export permit, which is recorded in the CITES Trade Database (CITES 2025c). Of the 3,553 certified trophy males exported from Mexico during 2000–2022 to 26 foreign countries, 2,624 were hunted in free-ranging (Figure 17.6) and 929 in captive-breeding populations (Figure 17.7). Of these, 3,235 were exported to the United States, 223 to Canada, 23 to Spain, and 18 to Russia. Of the other 22 countries for which data is available, the number of trophies exported did not exceed ten desert bighorn sheep (CITES 2025c; Figure 17.8). Beginning in 2016, there is a decrease in the number of exported free-ranging males (Figure 17.6) but compensated by the increase of males sourced from captive-breeding populations (Figure 17.7). The steep decline in captive trophy males exported in 2019 (163 males) and in 2020 (108 males; Figure 17.7), and a similar decline in free-ranging males in 2019 and 2020 (Figure 17.6), is probably due to lower number of hunters during the Covid pandemic.

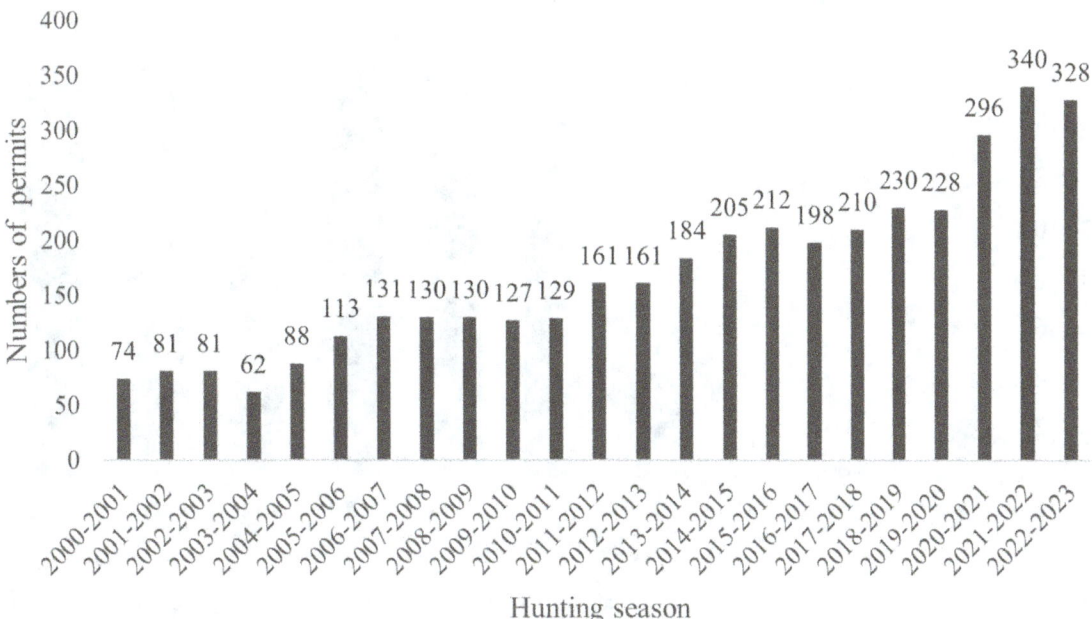

**FIGURE 17.5** Desert bighorn sheep hunting permits issued per year in Mexico during the 2000–2001 to 2022–2023 hunting seasons [Secretaría de Medio Ambiente y Recursos Naturales (2024b), Secretaría de Agricultura, Ganadería, Recursos Hidráulicos, Pesca y Acuacultura (2025)].

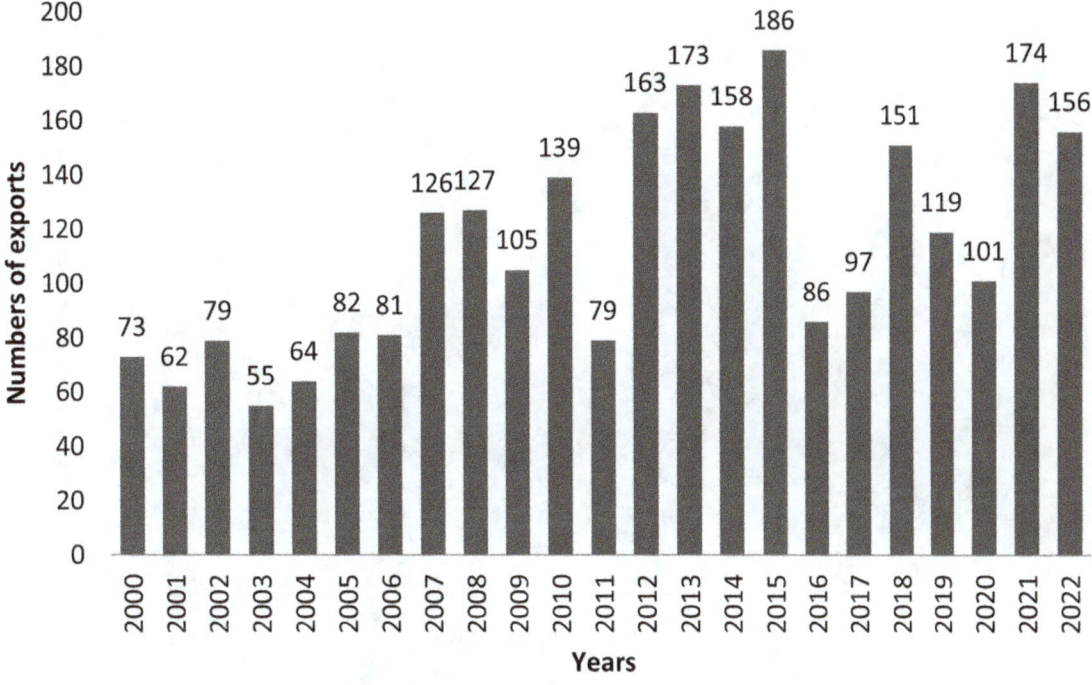

**FIGURE 17.6** Free-ranging desert bighorn trophy males exported per year from Mexico during the 2000–2001 to 2022–2023 hunting seasons (Convention on International Trade in Endangered Species of Wild Fauna and Flora 2025c).

## 17.7 ECONOMIC AND CONSERVATION ADVANTAGES OF A SUSTAINABLE BIGHORN SHEEP MANAGEMENT PROGRAM IN MEXICO

Trophy hunting, a category of legal hunting, refers to the payment of a fee by a foreign or resident hunter for a hunting experience, usually guided, for one or more individuals of a particular species with specific desired characteristics (International Union for the Conservation of Nature and Natural Resources 2016, Bichel and Hart 2023). The massive horns of a mature (+5 years of age) male are the desired characteristic of a mountain sheep trophy. Horn length and volume are the principal criteria in assessing trophy quality (Reneau et al. 2016). Mountain sheep trophy hunting is sex and age-biased, and expensive. In several jurisdictions

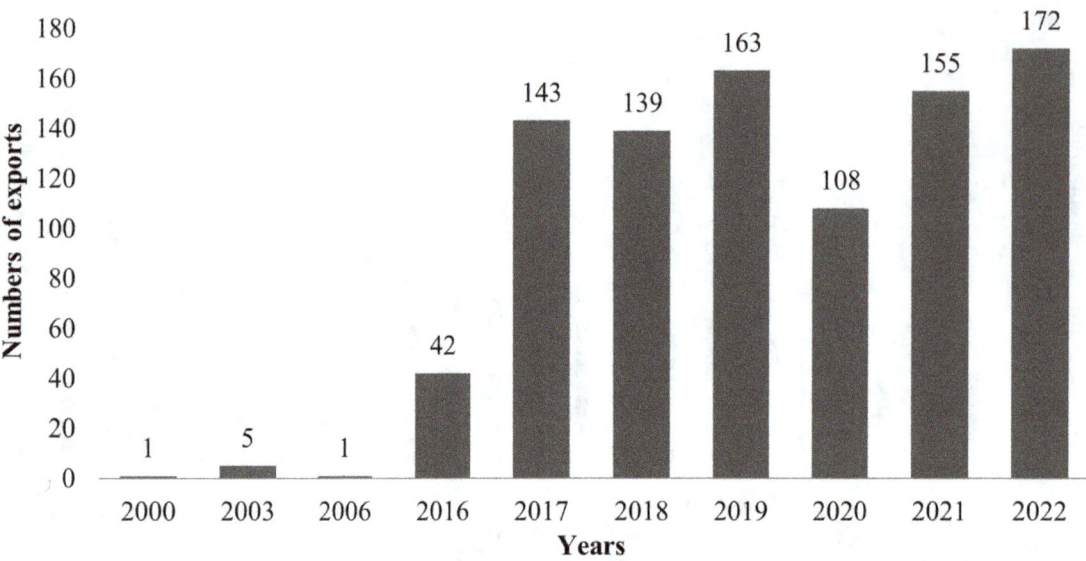

**FIGURE 17.7**  Captive-bred desert bighorn sheep trophy males exported per year from Mexico during the 2000–2001 to 2022–2023 hunting seasons (Convention on International Trade in Endangered Species of Wild Fauna and Flora 2025c).

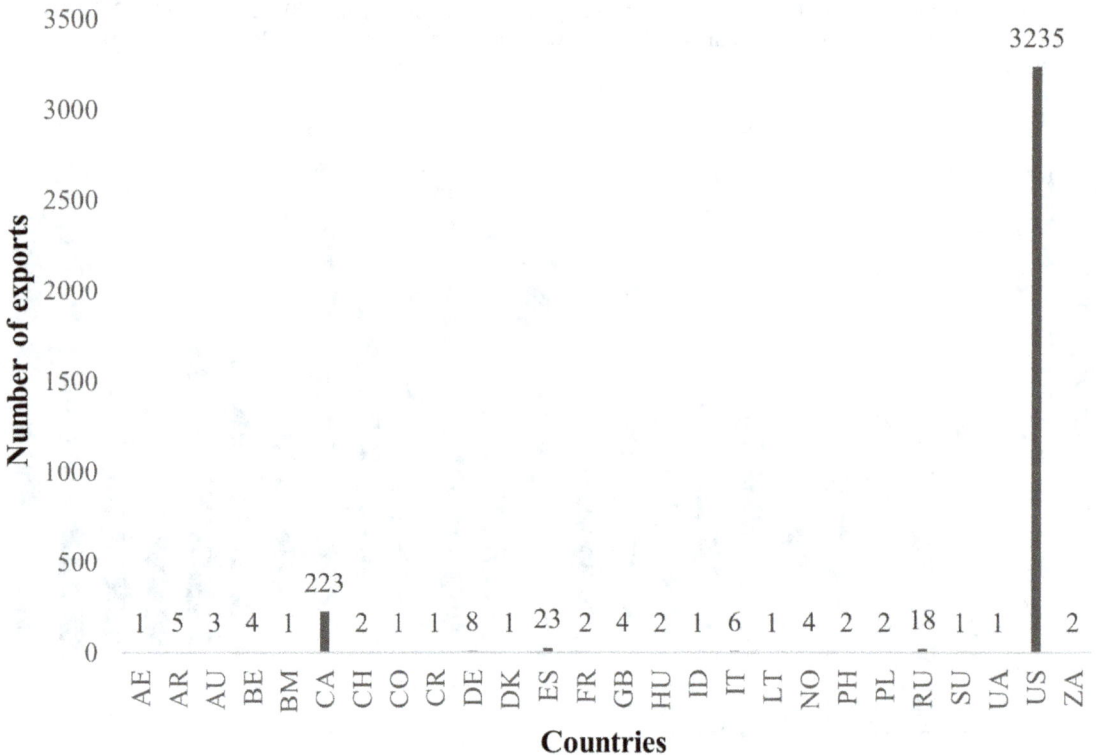

**FIGURE 17.8**  Desert bighorn sheep trophy males exported from Mexico to individual foreign countries, 2000–2022 (Convention on International Trade in Endangered Species of Wild Fauna and Flora 2025c). United Arab Emirates (AE), Argentina (AR), Australia (AU), Belgium (BE), Bermuda (BM), Canada (CA), Switzerland (CH), Colombia (CO), Costa Rica (CR), Germany (DE), Denmark (DK), Spain (ES), France (FR), United Kingdom (GB), Hungary (HU), Indonesia (ID), Italy (IT), Lithuania (LT), Norway (NO), Philippines (PH), Poland (PL), Russia (RU), Sudan (SU), Ukraine (UA), United States (US), South Africa (ZA).

in North America, one of the major objectives of mountain sheep management is to produce trophy quality males. Sustainable, adaptively managed trophy hunting program considerations should include being administered and regulated by a legitimate body such as a government, community-based organization, or non-government organization (NGO); ensuring harvest management occurs only in viable populations with a harvestable surplus that is sustainably managed through application of a low harvest rate for the hunted species; maintaining populations with an adaptive gene pool; discouraging the introduction of exotic species; creating a favorable environment for investment in the sustainable use of hunted species; and where generated funds are invested in management activities including population monitoring, law enforcement, and research (International Union for the Conservation of Nature and Natural Resources 2012). Recreational hunting, which includes trophy hunting, can promote biodiversity conservation (Wall and Child 2009, Hoffmann et al. 2010, Palazy et al. 2011, Bowyer et al. 2019, Stronza et al. 2019), but mismanagement of trophy-hunted populations can have detrimental effects (Pitman et al. 2016, Muposhi et al. 2017).

Ejido, communal, and private landholdings that provide trophy sport hunting have been pivotal in the recovery, continued increase, and expanded distribution of bighorn sheep populations in Mexico (SEMARNAT 2015, CITES 2025b). These successes have been due primarily to the ecological, economic, and social incentives and the enactment of hunting regulations and laws that benefit land managers and wildlife populations (Valdez et al. 2006, Valdez 2019, Ortega-S. et al. 2019, Sandoval et al. 2019). Large herbivores occurring outside of protected areas generate limited economic returns and are thus particularly vulnerable to habitat loss and fragmentation, population declines and extirpation due to overexploitation, loss of habitat, and competition with livestock: they require greater conservation efforts (Hoffmann et al. 2010, Ripple et al. 2015, Atwood et al. 2020, Torres-Romero et al. 2020). The UMAs have been criticized because they are focused on increasing game species populations and their habitats at the expense of nongame species (Gallina-Tessaro et al. 2009); however, nongame species have also benefited (e.g., passerine birds; Villarreal González 2006, 2009; Ramírez and Mondragón 2010, Ortega-S. et al. 2019). Globally, sport hunting has been a major economic incentive for nature conservation as exemplified in the increase in African ungulates, elephants, and large carnivore populations in eastern and southern African nations, because of the economic incentives derived from the international demand for big game trophy hunting. Trophy hunting has the potential to be the most important generator of revenue regionally in Mexico and in many other developing nations (Di Minin et al. 2021). The resulting ecological benefits to wildlife populations, habitat protection, nature conservation, economic benefits to private, community, and government entities, and socioeconomic improvements in rural human communities in Africa have been documented (Child 2000, Child et al. 2012, Cooney et al. 2012, Crosmary et al. 2014, 2015; Buckley and Mossaz 2015, Holechek and Valdez 2018, Naidoo et al. 2016, Taylor et al. 2021, Denner et al. 2024).

Wildlife conservation in Mexico has had a similar wildlife-economic-incentive history, especially about the sustainable trophy hunting management programs associated with wild ungulates (Morret-Sánchez and Cosío 2017, Lee 2008, Cooney et al. 2019, Valdez 2019, González-Rebeles Islas et al. 2019a,b; Ortega-S. et al. 2019, Sandoval et al. 2019, Ruiz-Mondragón et al. 2023, 2025a). Game ranching enterprises of captive and free-ranging ungulate populations are widespread in the United States and Canada (White 2000,

Peek et al. 2002, Anderson et al. 2017). Organized game ranching in Mexico was initiated in 1987 with the formation of the National Association of Diversified Cattle Ranchers (Asociación Nacional de Ganaderos Diversificados), who integrated white-tailed deer (*Odocoileus virginianus*) sport hunting management programs in livestock enterprises (Villarreal González 2006, 2013; González-Rebeles Islas et al. 2019a). These profitable wildlife ranching enterprises promote the legal use of wildlife and, in the process of instituting conservation and stewardship management practices, increase biodiversity and ecological integrity, but some game ranches include exotic ungulates (Teer 2003, Mungall 2024). The commercial sale of live bighorn sheep to landowners in Mexico to establish herds and augment populations has also been profitable. Bighorn sheep populations in Mexico have been successfully reestablished in Coahuila, Chihuahua, and populations augmented in Sonora through the purchase of bighorn sheep by private landowners. Hunting has been legally reinitiated in all states except for Baja California, where hunting is prohibited because of socio-political issues. The major reason for the prohibition of hunting in that state was that bighorn sheep are the mascot of the University of Baja California, which created a negative attitude toward desert bighorn sheep hunting among students and the public. The hunting program was also tainted by accusations of corruption in issuing bighorn sheep hunting licenses. And in response, the governor closed bighorn sheep hunting indefinitely in 1990, largely due to political pressure (Mellink 1993, Tapia Landeros 2007); mule deer hunting in Baja California, however, continues to be permitted. Removing the value associated with wildlife by prohibiting hunting could result in land use practices that do not promote habitat and wildlife conservation, especially in ecologically fragile arid lands. This threatens the persistence of wildlife in unprotected areas. Conservation actions directed at hunted species often create umbrella effects that benefit many other species, and in the example of sustainable hunting opportunities, the creation of economic benefits from conserving natural habitats in private and communal landholdings is inherently tied to sustainable, native wildlife populations (Hoffmann et al. 2010).

Mexico is in an especially favorable position to attract international hunters and benefit from economic investment because of its abundant wildlife populations and proximity to the United States and Canada. Dedicated mountain sheep hunters become absorbed in achieving an *Ovis* grand slam of mountain sheep, which entails collecting the four principal varieties of North America mountain sheep: Dall's (*Ovis dalli*), Stone's (*O. d. stonei*), Rocky Mountain and desert bighorn sheep. Desert bighorn sheep are the second least abundant mountain sheep taxon in North America; Stone's sheep are the least abundant (Chapter 1). Mexico has the most abundant populations of desert mountain sheep and Sonora is the jurisdiction with the highest population (11,600) of desert bighorn sheep in North America. The state of Nevada is second with 10,100 desert bighorn sheep (Brewer and Bleich 2023). The relative scarcity of adult desert bighorn sheep in the United States cannot supply the demand for hunting permits for trophy males. Abundant desert bighorn sheep populations in Mexico offer an accessible alternative hunting opportunity, especially for American and Canadian mountain sheep hunters. Mexico has taken advantage of the vast commercial and marketing hunting industry to attract hunters, and this is one of the reasons for its successful and highly profitable mountain sheep hunting program.

Mountain sheep are desirable big game hunting trophies for which sheep hunters are willing to pay high prices for the opportunity to hunt them (Lee 2011, Collins 2025, Chapter 2). Mexico's captive-breeding and free-ranging desert bighorn sheep populations are of excellent quality, with abundant high-scoring trophy males. Because males from captive populations are excluded from competitive trophy record consideration, there is a premium for free-roaming trophy animals, and as a consequence, these command higher prices. Of the 30 highest-scoring free-ranging desert bighorn sheep trophies listed in the Boone and Crockett Club's (B&C's) record book, 15 are from Mexico. Of those, 11 are from the Baja California Peninsula and three are from Sonora; the locality of one trophy male from Mexico is not recorded. The highest-ranking or world's record desert bighorn sheep is that of a male skull and horns found in the Sierra San Pedro Mártir in northern Baja California (Lehr and Schwab 2023). Mountain sheep hunters have differentiated themselves socially from other big game hunters and consider mountain sheep the ultimate big game trophy. Mountain sheep hunters, industry professionals, government biologists, and affiliated hunting organizations have an annual convention sponsored by the Wild Sheep Foundation (Chapter 2), where they socialize, participate in auction hunts to raise money for mountain sheep conservation projects, purchase hunting and outdoor-related merchandise, and arrange hunts through big game hunting outfitters (person or company that offers guided hunting services) and booking agents. The sale of hunting tags provides substantial financial resources for the ejidos and comunidades, which are economically strapped rural populations. These auctions provide the opportunity for philanthropists to direct funding into wildlife conservation efforts in various jurisdictions across Mexico, Canada, and the United States. Bighorn sheep hunting permits in Mexico sell from $35,000 to more than $100,000. In 2025, an auctioned bighorn hunt conservation permit in New Mexico sold $1.3 million, the highest amount ever paid for a state-sponsored mountain sheep conservation permit (Collins 2025).

Prices for desert bighorn sheep hunts in Mexico are variable; those for free-ranging males are considerably higher priced than those from captive-reared or high-fenced populations. We obtained mountain sheep male hunt prices for private ranch and Tiburon Island hunts from internet advertisements for free-ranging male hunts to illustrate the variety of pricing in the several Mexican states.

These pricing examples are based on a single ranch in each state: Chihuahua, $65,000; Coahuila, $85,000; Sonora, pricing on this ranch is based on B&C horn-scoring system (Reneau et al. 2016), where a score of 179 B&C points=$60,000; 180–184 B&C score=$70,000; 185–189 B&C score=$80,000, and higher price for a 190 B&C point trophy. In Baja California Sur, a hunt will cost $60,000 in the municipalities of Loreto and La Paz, with a hunt zone area of 485,000 ha. The average price for a bighorn sheep hunting permit in Baja California Sur is $48,000 (SEMARNAT 2023). On Tiburon Island, desert bighorn sheep hunting permits range from $90,000 to more than $100,000. These hunts are the most expensive because of the high probability of obtaining a high-scoring trophy male (≥180 B&C points). In comparison, a private ranch offering captive-reared big game hunts based on native ungulates is located within an hour's drive from the city of Chihuahua. The ranch has commingled desert bighorn sheep and mule deer populations in a 3,642 ha high-fenced enclosure. The desert bighorn sheep herd was initiated in 2010 with six bighorn sheep and subsequently augmented with ten animals. The herd has increased to 150 bighorn sheep. There are four male permits available per year and they sell for $39,500 each (A. C. E. Villalobos, Santa Rosa Ranch, personal communication; www.santarosahuntingranch.com. Accessed 20 Jan 2024). The free-range hunts are highly valued when compared to hunts for high-fenced bighorn sheep or other free-ranging North American horned or antlered big game species. For example, desert mule deer hunts in Mexico sell for up to $35,000, and those of whitetail for $7,500. Some of these funds are equitably distributed in rural communities directly or directed in support of social programs, enforcement of hunting regulations, and promote enhancement of habitat, such as providing watering sources for free-ranging populations. Desert bighorn sheep sport hunters have demonstrated a responsibility and willingness to insist that investments into promoting the conservation of Mexico's bighorn sheep occur. The recovery and continued population growth and expanded distribution of mountain sheep populations in Mexico are partly a result of those conservation investments and efforts.

## 17.8 CAPTIVE-BREEDING AND FREE-RANGING DESERT BIGHORN SHEEP MANAGEMENT IN MEXICO

### 17.8.1 Differences between Bighorn Sheep Management Programs in Mexico, the United States, and Canada

There exists a major difference in wildlife conservation efforts between Mexico, Canada, and the United States (Lee 2011). Conservation efforts in the United States have been primarily government funded; in Canada, responsibility for funding conservation programs is often shared, with contributions from federal, provincial, territorial, and non-government agencies. In contrast, in Mexico, most of the wildlife management efforts, including those of desert bighorn sheep, have been privately funded. Private land-owners have a financially vested interest in actively managing wildlife resources and are more motivated than state or federal agencies, which have only limited funding. Also, private landowners in Mexico's northern states, which are geographically close to the United States, took advantage of the enormous American hunting interest, developing profitable sport hunting enterprises. Northern Mexico has traditionally been a livestock-producing region. Mexican cattle growers in northern Mexico have adopted the management of multispecies, wildlife-livestock enterprises, like those first established in the United States (Villarreal González 2013, Ortega-S. et al. 2019, Mungall 2024).

There is also a major management difference in managing desert bighorn sheep populations between countries in North America. Mountain sheep in the United States and Canada are managed by state, provincial, territorial, and federal regulations, with policies that limit management options, especially regarding captive breeding and wildlife management. There has been a trenchant unwillingness of desert bighorn sheep management agencies outside of Mexico to break with traditional management philosophies especially regarding devolving management authority to private landowners and enterprises. There have been several state mountain sheep captive management programs in the United States, but these have been of relatively short duration (Cook 1994, Chapter 2). There has only been one non-governmental (but management authority resides with the state wildlife agency) organization mountain sheep captive management program in the United States, that of the Bighorn Institute in California (Rivkin and Burbank 2019) and several private ranches with free-ranging populations (e.g., the Turner Ranch in southern New Mexico, United States; Krausman et al. 2001, and several private ranches in Texas; Brewer and Bleich 2023). Mexico has successfully adopted a bighorn sheep management policy of devolving bighorn sheep management authority to private and communal entities with the inclusion of managing captive-breeding and free-ranging desert bighorn sheep populations (Sandoval et al. 2014, 2019).

### 17.8.2 BIGHORN SHEEP CAPTIVE BREEDING

Primary management actions for desert bighorn sheep include providing life requisites such as adequate food, water, and space (Figure 17.9). With captive herds, these items can be externally provided, but captive herds can quickly outgrow their limited habitat. In free-ranging populations, these resources must be managed to ensure their availability. Managers must also consider the effects of predation, potential competition for food and water in habitats with both wild and domestic animals, and transmission of domestic disease. Diseases, particularly respiratory diseases, have adversely influenced many mountain sheep populations. In free-ranging populations, the disease vector has been primarily from domestic sheep or goats (Chapter 8). In captive populations, stress and poor sanitation can also lead to diseases, and the high densities

**FIGURE 17.9** A herd of desert bighorn females and males shows the high quality of trophy males and healthy females in captive facilities in Sonora, Mexico. (Photo by Juan Manuel Segundo Galán; used with permission.)

can speed up their transmission. On the other hand, captive herds can be more closely monitored and treated for diseases or injuries under controlled conditions (Muths and McCallum 2016). An advantage of captive breeding is that population recovery goals can be attained in a shorter time frame, as is the case in Mexico. This is due to captive herds generally having better access to higher-quality foods and much lower mortality rates. Captive bighorn sheep can attain their maximum body and horn growth, and reproductive genetic potential, due to the high nutritional plane possible in captive-reared populations (Figure 17.9). Additional benefits of captive breeding include enhanced reproduction and recruitment, increased survivorship, control of disease exposure, and selective breeding. Reasons for these benefits of captive breeding can be attributed to the exclusion of predators, optimized nutrition, protection from extreme environmental conditions and catastrophes, consistent water availability, enhanced veterinary care through quick access to sick and injured animals, allowing administration of medications and other treatments: essentially effective animal husbandry (Ernest 2001). Establishing a disease awareness program has proven essential (Peterson and Ferro 2020).

Of particular concern to wildlife managers are the genetic limitations presented in smaller and isolated populations (sometimes referred to as founder effects and genetic drift;

**FIGURE 17.10**  Six free-ranging desert bighorn sheep males in a Sonoran Desert vegetation background. (Photo by Jonas Guinn; used with permission.)

Chapter 6). As populations lose genetic diversity (heterozygosity) and genetic adaptations for a specific environment (phenotypic expression), the need for these populations to be adaptable in response to a changing climate (phenotypic plasticity) becomes increasingly important if they are to thrive in those changing environments (Biebach et al. 2016). The recovery plans for the endangered peninsular pronghorn (*Antilocapra americana peninsularis*) population specifically attempts to preserve genetic heterozygosity and aspects of the plan could apply to managing desert bighorn sheep populations (Klimova et al. 2022). Many small, reintroduced bighorn sheep populations in Mexico originate primarily from the Tiburon Island population, which because of its low genetic diversity requires managers to consider genetic heterozygosity that could deleteriously effect mating and reproductive outcomes when planning options for augmenting or establishing desert bighorn sheep populations (Durán Alvarez 2013, Gasca-Pineda et al. 2013, Hedrick and Wehausen 2014, Biebach et al. 2016). Sound genetics management is an important aspect in promoting trophy horn quality (Figure 17.10; Chapters 4 and 16).

### 17.8.3  CHALLENGES OF THE BIGHORN SHEEP MANAGEMENT PROGRAM IN MEXICO

The desert bighorn sheep management program in Mexico lacks much of the basic information necessary to ensure effective conservation planning, and what is known is shared minimally among managers. Consequently, it is not possible to develop a free-ranging or captive-breeding management plan for bighorn sheep populations in Mexico or to evaluate the effectiveness of the program(s). Management programs should be transparent, but we were not able to obtain demographic information from captive-breeding populations, and only a minimum of information was available for free-ranging populations. This is especially the situation relative to the lack of female biological data. Management plans require detailed studies of population dynamics, productivity, survivorship, habitat use, connectivity and metapopulation analysis, and genetic and disease monitoring information (Biebach et al. 2016, Gitzen et al. 2016, Hayward and Slotow 2016, Muths and McCallum 2016, Johnson and Dinsmore 2020, Oyler-McCance et al. 2020), all of which are presently lacking. Further, there remains an absence of harvest management and quantitative information on wildlife and habitat ecology and management, and particularly among populations confined to island ecosystems that have, or will, exceeded ecological carrying capacity (*K*) for mountain sheep. The potential consequences of the introduction of desert bighorn sheep to Tiburon Island, which harbors several unique populations of vertebrates, and the potential consequence(s) of an overshoot of *K* for those endemic taxa should have been addressed before that management action. The demand for these hunting opportunities continues, but harvest management information to validate any concern that trophy quality has not, or is not, in decline remains unknown because long-term records of horn size and body measurements are unavailable. On the positive side, helicopter surveys continue to provide information necessary to quantify population size, age structure, and sex ratios for some free-ranging populations. Additionally, there have been substantial improvements to habitat conservation in Sonora that have centered on the exclusion of domestic or feral livestock, removal of fences that impede movements, adding water developments, and effective law enforcement in some bighorn sheep management areas, all of which should be applied in bighorn sheep habitats throughout Mexico. We contend that the conservation of desert bighorn sheep in Mexico will benefit substantially from the development of a nationwide management plan that has its foundation in a strong connection between researchers and managers (Bleich and Thompson 2018, Brennan et al. 2020), and recognizes the importance of private, ejido, and communal landholders, all of which have contributed substantially to the conservation of desert bighorn sheep in Mexico (Sandoval et al. 2019, Brewer and Bleich 2023) and whose contributions will be essential in the future. Bighorn sheep in Mexico have the potential of raising funds for conservation and education and enhancing environmental awareness, and can be aligned with local ecological, conservation, and social contexts (McGowan et al. 2020). Thus, formal efforts to encourage and support community-based participation in conservation programs should be expanded and technical advice provided on all aspects of implementation (Romero-de-Diego et al. 2021, Ruiz-Mondragón et al. 2023, 2025a,b).

One of the most evident shortcomings is the lack of collaboration and coordination in the bighorn sheep management program. If the mountain sheep conservation program is to be successful at the landscape level, desert bighorn sheep managers in Mexico need to apply a joint conservation effort within and among Mexican states, and internationally with the United States and potentially Canada. There is much biological and management knowledge based on scientific wildlife research in North America but it has not, as yet, been applied widely to the bighorn sheep program in Mexico. Conservation actions are frequently focused on a single species, but likely is most effective and often better achieved through the establishment and conservation of viable wildlife populations in their natural habitats with concomitant benefits to overall biodiversity that in some regions may require expanded networks of private and communal protected areas (Dunham et al. 2016, Hilty et al. 2019, Segundo Galán et al. 2024). Bighorn sheep management programs in Mexico have failed to take advantage of research afforded by captive-breeding populations. These populations provide the opportunity to conduct essential disease, reproductive, growth, nutritional, physiological, social behavior, and genetic studies under controlled conditions, the results of which can be applied to free-ranging populations. A multidisciplinary and multi-institutional science-based research program that links academia, governments, non-governmental organizations, and landholders should be a priority. Communal or private landholders in Mexico can gain economic benefits by advocating a biodiversity conservation framework, albeit with a focus on the sustainable management of large wild desert herbivores that also create beneficial outcomes for associated native animals and plants of the Sonoran and Chihuahuan deserts (Latiy 2008, Sandoval et al. 2019, Whitford and Duval 2020).

**FIGURE 17.11** A healthy desert bighorn female on Carmen Island, Baja California Sur, Mexico. Females are the nucleus for establishing high-quality mountain sheep populations. (Photo courtesy of Organización Vida Silvestre, A.C.; used with permission.)

The combined long-term wildlife conservation and management investments did enable the successful recovery of once widespread, extirpated, and decreasing bighorn sheep populations in Mexico. The biodiversity of desert ecosystems can equally benefit under the umbrella of mountain sheep conservation (Figure 17.11). Landholders in some areas of Mexico have implemented land management actions to mitigate or reverse the negative effects of land use mismanagement practices. The continued success of the desert bighorn sheep conservation program lies in the development of cooperative plans or programs that emphasize adaptive management as a collaborative means of attaining sustainable population goals and objectives. A primary component of any such program should incorporate assessments to guide strategic decision-making and certification to demonstrate professionalism, but with the ultimate goal being the recovery of desert ecosystems. Such an achievement will, in turn, be of benefit to biodiversity and the restoration of ecological integrity and natural processes. Incorporating assessment and certification in the desert bighorn sheep management program in Mexico will enable the formulation of a transparent, cohesive vision and a strategy for sustainable wildlife conservation.

## 17.9   SUMMARY

There is a significantly greater economic value attached to bighorn sheep in Mexico than any other big game mammal. Bighorn sheep are the highest-priced, sport-hunted big game mammal in Mexico and are in high demand internationally by big game hunters who are willing to pay exorbitant prices for hunting permits. This is in large part due to the rarity of desert bighorn sheep hunting opportunities globally. Mexico has the highest population of desert bighorn sheep in North America. Because it is a native desert-adapted large herbivore, desert bighorn sheep can be considered as a potential flagship species for arid mountainous ecosystems within its distribution in northern Mexico and the Baja California Peninsula. Bighorn sheep are a charismatic and economically valuable species that promote a behavioral awareness of the ecological and conservation values of wildlife and inspire the development of a financially incentivized social responsibility to manage wildlife resources and their habitats sustainably. A change in environmental awareness among community members within protected and surrounding areas involves obviating social ineffectiveness and creating changes in the values, attitudes, and behaviors that are vital to successful communal natural resources management programs that generate sustainable economic returns to those communities. Captive-bred bighorn sheep have been used in restoring naturalized bighorn sheep populations in Chihuahua and Coahuila, where they had been extirpated. The number of bighorn sheep has increased because of the establishment of additional wild populations, augmentation of existing populations, and successful captive breeding. Mexico is the only country in North America which prioritizes a combined mountain sheep captive-breeding and free-ranging population management strategy, a management policy that has had great success.

## REFERENCES

Abd El-Ghani, M. M., F. M. Huerta-Martínez, L. Hongyan, and R. Qureshi. editors. 2017. The deserts of Mexico. Pages 1–7 *in Plant Responses to Hyperarid Desert Environments*. Springer, Cham, Switzerland.

Álvarez-Cárdenas, C., P. Gallina-Tessaro, S. Díaz-Castro, I. Guerro-Cárdenas, A. Castellanos-Vera, and Z. Mesa-Zavalo. 2009. Evaluación de elementos estructurales del hábitat del borrego cimarrón en la Sierra del Mechudo, Baja California Sur, México. *Tropical Conservation Science* 2:189–203. [In Spanish]

Anderson, D. P., J. L. Outlaw, M. L. Earle, and J. W. Richardson. 2017. Economic impact of United States deer breeding and hunting operations. Research Report 17-4. Agricultural Food Policy Center, Texas A&M University, College Station, Texas, USA.

Anderson, J. 2002. *Jack O'Connor: The Legendary Life of America's Greatest Gunwriter*. Safari Press, Long Beach, California, USA.

Atwood, T. B., S. A. Valentine, E. Hammill, D. J. McCauley, E. M. P. Madin, K. H. Beard, and W. E. Pearse. 2020. Herbivores at the highest risk of extinction among mammals, birds, and reptiles. *Science Advances* 6:eabb8458.

Balachander, G., M. W. A. Halmy, B. Parlee, D. Biggs, et al. 2022. The drivers of the sustainable use of wild species. Pages 451–731 *in* J. M. Fromentin, M. R. Emery, J. Donaldson, M. C. Danner, A. Hallosserie, and D. Kieling, editors. *Thematic Assessment Report on the Sustainable Use of Wild Species of the Intergovernmental Science-Policy Platform on Biodiversity and Ecosystem Services*. IPBES Secretariat, Bonn, Germany.

Bezaury-Creel, J. E. 2024. Privately protected areas in Mexico: a 2012–2023 update. *Frontiers in Conservation Science* 4:1304771.

Bichel, N., and A. Hart. 2023. *Trophy Hunting*. Springer Nature, Republic of Singapore.

Biebach, I., D. M. Leigh, K. Sluzek, and L. F. Keller. 2016. Genetic issues in reintroduction. Pages 149–183 *in* D. S. Jachowski, J. J. Millspaugh, P. L. Angermeier, and R. Slotow, editors. *Reintroduction of Fish and Wildlife Populations*. University of California Press, Oakland, California, USA.

Bingham, H., J. A. Fitzsimons, B. A. Mitchell, K. H. Redford, and S. Stolton. 2021. Privately protected areas: missing pieces of the global conservation puzzle. *Frontiers in Conservation Science* 2:748127.

Bleich, V. C., and D. J. Thompson. 2018. State management of big game. Pages 75–95 *in* T. J. Ryder, editor. *State Wildlife Management and Conservation*. Johns Hopkins University Press, Baltimore, Maryland, USA.

Bleich, V. C., R. R. Ramey II, and J. L. Rechel. 1996. Metapopulation theory and mountain sheep: implications for conservation. Pages 353–373 *in* D. M. McCullough, editor. *Metapopulations and Wildlife Conservation*. Island Press, Washington, D.C., USA.

Bowyer, R. T., M. S. Boyce, J. R. Goheen, and J. I. Rachlow. 2019. Conservation of the world's mammals: status, protected areas, community efforts, and hunting. *Journal of Mammalogy* 100:923–941.

Brennan, L. A., S. J. Demaso, J. P. Sands, and M. J. Schnupp. 2020. Strengthening connections between research and management. Pages 1–11 in N. J. Silvy, editor. *The Wildlife Techniques Manual: Management.* Volume 2. Eighth edition. Johns Hopkins University Press, Baltimore, Maryland, USA.

Brewer, C. E., and V. C. Bleich. 2023. Trophy spotlight: desert bighorn sheep. Pages 18–33 *in* K. M. Lehr and J. Schwab, editors. *Records of North American Big Game.* Volume 2. Fifteenth edition. Boone and Crockett Club, Missoula, Montana, USA.

Brown, D. E., editor. 1982. Biotic communities of the American Southwest: United States and Mexico. *Desert Plants* 4:4–342.

Brown, D. E., editor. 1983. *The Wolf in the Southwest: The Making of an Endangered Species.* University of Arizona Press, Tucson, Arizona, USA.

Buchalski, M. R., A. Y. Navarro, W. M. Boyce, et al. 2015. Genetic structure of peninsular bighorn sheep (*Ovis canadensis nelsoni*) indicates substantial gene flow across the U.S.-Mexico border. *Biological Conservation* 184:218–228.

Buckley, R., and A. Mossaz. 2015. Hunting tourism and animal conservation. *Animal Conservation* 18:133–135.

Carmony, N. B., and D. E. Brown, editors. 1994. *The Wilderness of the Southwest: Charles Sheldon's Quest for Desert Bighorn Sheep and Adventures with the Havasupai and Seri Indians.* University of Utah Press, Salt Lake City, Utah, USA.

Ceballos, G., editor. 2014. *Mammals of Mexico.* Johns Hopkins University Press, Baltimore, Maryland, USA.

Challenger, A. 2019. Natural and human landscapes in Mexico. Pages 19–47 *in* R. Valdez and J. A. Ortega-S., editors. *Wildlife Conservation and Management in Mexico.* Texas A&M University Press, College Station, Texas, USA.

Challenger, A., and J. Soberón. 2008. Los ecosistemas terrestres. Pages 87–108 *in* J. Soberón, R. González, and I. J. March, compilers. *Capital Natural de México.* Volume 1. Conocimiento actual de la biodiversidad. CONABIO, México, D.F., México. [In Spanish]

Chester, C. C. 2006. *Conservation Across Borders: Biodiversity in an Interdependent World.* Island Press, Washington, D.C., USA.

Child, B. 2000. Making wildlife pay: converting wildlife's comparative advantage into real incentives for having wildlife in African savannas, case studies from Zimbabwe and Zambia. Pages 335–387 *in* H. H. T. Prins, J. G. Grootenhuis, and T. T. Dolan, editors. *Wildlife Conservation by Sustainable Use.* Kluwer Academic Publishers, Boston, Maryland, USA.

Child, B. A., J. Musengezi, G. D. Parent, and G. F. T. Child. 2012. The economics and institutional economics of wildlife on private land in Africa. *Pastoralism* 2:18.

Collins, D. 2025. $1.3 million bighorn tag sets record for most expensive auction tag ever. outdoorlife.com/hunting/million-dollar-sheep-tag. Accessed 20 Mar 2025.

Comisión Nacional de Áreas Naturales Protegidas. 2018. *100 años de Conservación en México: Áreas Naturales Protegidas de México.* SEMARNAT-CONANP, Ciudad de México, México. [In Spanish]

Comisión Nacional de Áreas Naturales Protegidas. 2024. Áreas naturales protegidas. https://www.gob.mx/conanp/documentos/areas-naturales-protegidas-278226?state=published. Accessed 5 Feb 2025. [In Spanish]

Convention on Biological Diversity. 1993. Secretariat of the Convention on Biological Diversity, Montreal, Quebec, Canada.

Convention on International Trade in Endangered Species of Wild Fauna and Flora (CITES). 2025a. Mexico. https://cites.org/esp/parties/country-profiles/mx. Accessed 14 Feb 2025.

Convention on International Trade in Endangered Species of Wild Fauna and Flora (CITES). 2025b. Mexico: Biannual report 2021–2023. CITES, Geneva, Switzerland. https://cites.org/eng/parties/country-profiles/mx/reports. Accessed 14 Feb 2025.

Convention on International Trade in Endangered Species of Wild Fauna and Flora (CITES). 2025c. Trade Database. https://trade.cites.org/. Accessed 14 Feb 2025.

Cook, R. L., compiler. 1994. *A Historical Review of Reports, Field Notes and Correspondence on the Desert Bighorn Sheep in Texas.* Revised. Texas Parks and Wildlife, Austin, Texas, USA.

Cooney, R., C. Freese, H. Dublin, D. Roe, et al. 2012. The baby and the bathwater: trophy hunting, conservation and rural livelihoods. *Unasylva* 68:3–16.

Cooney, R., P. M. Reidl, and L. G. Muñoz Lacy. 2019. *Community-Based Trophy Hunting of Bighorn Sheep in Mexico.* CITES Secretariat, Geneva, Switzerland.

Cowan, I. M. 1940. Distribution and variation in the native sheep of North America. *American Midland Naturalist* 24:505–580.

Crosmary, W.-G., S. D. Côté, and H. Fritz. 2014. Does trophy hunting matter to long-term population trends in African herbivores of different dietary guilds? *Animal Conservation* 18:117–130.

Crosmary, W.-G., S. D. Côté, and H. Fritz. 2015. The assessment of the role of trophy hunting in wildlife conservation. *Animal Conservation* 18:136–137.

Cudahy, J. 1928. *Mañanaland: Adventuring with Camera and Rifle Through California and Mexico.* Duffield and Company, New York, New York, USA.

Denner, C., H. S. Clements, M. F. Child, and A. De Vos. 2024. The diverse socioeconomic contributions of wildlife ranching. *Conservation Science and Practice* 6:e13166.

Di Minin, E., H. S. Clements, R. A. Correia, et al. 2021. Consequences of recreational hunting for biodiversity conservation and livelihoods. *One Earth* 4:238–251.

Diario Oficial de la Federación. 2010. Segunda sección, Secretaría de Medio Ambiente y Recursos Naturales. Norma Oficial Mexicana NOM-059-SEMARNAT-2010, Protección ambiental-Especies nativas de México de flora y fauna silvestres-Categorías de riesgo y especificaciones para su inclusión, exclusión o cambio-Lista de especies en riesgo. SEMARNAT, México D.F., México. [In Spanish]

Dunham, J. B., R. White, C. S. Allen, B. G. Marcot, and D. Shively. 2016. The reintroduction landscape: finding success at the intersection of ecological, social, and institutional dimensions. Pages 79–103 in D. S. Jachowski, J. J. Millspaugh, P. L. Angermeier, and R. Slotow, editors. *Reintroduction of Fish and Wildlife Populations.* University of California Press, Oakland, California, USA.

Durán Álvarez, C. 2013. *Genética de Poblaciónes del Borrego Cimarrón (Ovis canadensis) en Pilares, Coahuila, México. Su Manejo y Conservación en Cautiverio.* Thesis, Universidad Nacional Autónoma de México, Ciudad de México, México. [In Spanish]

Ernest, H. B. 2001. *Captive Breeding Contingency Plan: A Guide for Captive Breeding of Sierra Nevada Bighorn Sheep.* Report for Interagency Agreement # P9980059 between California Department of Fish and Game and Wildlife Health Center, School of Veterinary Medicine, University of California, Davis, California, USA.

Espinosa-T., A., A. V. Sandoval, and A. J. Contreras-B. 2006. Historical distribution of desert bighorn sheep (*Ovis canadensis mexicana*) in Coahuila, Mexico. *Southwestern Naturalist* 51:282–288.

Festa-Bianchet, M. 2020. *Ovis canadensis*. The IUCN Red List of Threatened Species 2020: e.T15735A22146699. https://dx.doi.org/10.2305/IUCN.UK.2020-2.RLTS. T15735A22146699.en. Accessed 12 Apr 2025.

Figueroa, F., and V. Sánchez-Cordero. 2008. Effectiveness of natural protected areas to prevent land use and land cover change in Mexican biosphere reserves. *Biodiversity Conservation* 17:3223–3240.

Fitzsimons, J. A., and B. A. Mitchell. 2024. Research priorities for protected areas. *Frontiers in Conservation Science* 5:1340887.

Fundación Universidad Autónoma de Baja California. 2025. Santuario Cimarrón. https://santuariocimarron.com/. Accessed 15 Aug 2025. [In Spanish]

Gallardo, J. G. 2018. Wildlife policy and laws in Mexico. Pages 343–355 *in* B. D. Leopold, W. B. Kessler, and J. L. Cummins, editors. *North American Wildlife Policy and Law*. Boone and Crockett Club, Missoula, Montana, USA.

Gallina-Tessaro, S. A., A. H. Hernández-Huerta, C. A. Delfín-Alonso, and A. González-Gallina. 2009. Unidades para la conservación, manejo y aprovechamiento sustentable de la vida silvestre en México (UMA). Retos para su correcto funcionamiento. *Investigación Ambiental* 1:143–152. [In Spanish]

Gasca-Pineda J., I. Cassaigne, R. A. Alonso, and L. E. Eguiarte. 2013. Effective population size, genetic variation, and their relevance for conservation: the bighorn sheep in Tiburon Island and comparisons with managed artiodactyls. *PLoS ONE* 8:e78120.

Gastelum-Mendoza, F. I., L. A. Tarango-Arámbula, G. Olmos-Oropeza, et al. 2021. Diet and sexual segregation of bighorn sheep (*Ovis canadensis mexicana* Merriam) in Sonora. *Agro Productividad* 14:31–39.

Gitzen, R. A., B. J. Keller, M. A. Miller, et al. 2016. Effective and purposeful monitoring of species reintroductions. Pages 283–317 *in* D. S. Jachowski, J. J. Millspaugh, P. L. Angermeier, and R. Slotow, editors. *Reintroduction of Fish and Wildlife Populations*. University of California Press, Oakland, California, USA.

González-Abraham, C., P. P. Garcillán, E. Ezcurra, y Grupo de Trabajo de Ecorregiones. 2010. Ecorregiones de la península de Baja California: una síntesis. *Boletín de la Sociedad Botánica de México* 87:69–82. [In Spanish]

González-Rebeles Islas, C., M. M. Méndez Méndez, and R. Valdez. 2019a. Wildlife as a public and economic resource in Mexico. Pages 48–62 *in* R. Valdez and J. A. Ortega-S., editors. *Wildlife Ecology and Management in Mexico*. Texas A&M University Press, College Station, Texas, USA.

González-Rebeles Islas, C., M. M. Méndez Méndez, and R. Valdez. 2019b. Evolution of wildlife laws and policy in Mexico. Pages 367–377 *in* R. Valdez and J. A. Ortega-S., editors. *Wildlife Ecology and Management in Mexico*. Texas A&M University Press, College Station, Texas, USA.

González-Saldívar, F., L. A. Tarango Arámbula, C. Cantú Ayala, J. Uvalle Sauceda, J. Marmolejo Moncivais, and C. A. Ríos Saldaña. 2011. Estudio poblacional y de distribución del borrego cimarrón (*Ovis canadensis mexicana* Merriam 1901) en Sonora. *Revista Mexicana de Ciencias Forestales* 2:63–76. [In Spanish]

Hall, E. R. 1981. *The Mammals of North America*. Second edition. John Wiley & Sons, New York, New York, USA.

Hansen, C. 1980. Habitat. Pages 64–78 *in* G. Monson and L. Sumner, editors. *The Desert Bighorn: Its Life History, Ecology and Management*. University of Arizona Press, Tucson, Arizona, USA.

Hayward, M. W., and R. Slotow. 2016. Management of reintroduced wildlife populations. Pages 319–340 *in* D. S. Jachowski, J. J. Millspaugh, P. L. Angermeier, and R. Slotow, editors. *Reintroduction of Fish and Wildlife Populations*. University of California Press, Oakland, California, USA.

Hedrick, P. W., and J. D. Wehausen. 2014. Desert bighorn sheep: changes in genetic variation over time and the impact of merging populations. *Journal of Fish and Wildlife Management* 5:3–13.

Heffelfinger, J. R., and E. Marquez-Muñoz. 2005. Historical occurrence and distribution of desert bighorn sheep in Chihuahua, Mexico. *Desert Bighorn Council Transactions* 48:28–38.

Hilty, J. A., A. T. H. Keeley, W. Z. Lidicker Jr., and A. M. Merenlender. 2019. *Corridor Ecology: Linking Landscapes for Biodiversity Conservation and Climate Adaptation*. Second edition. Island Press, Washington, D.C., USA.

Hoffmann, M., C. Hilton-Taylor, A. Angulo, et al. 2010. The impact of conservation on the status of the world's vertebrates. *Science* 330:1503–1509.

Holechek, J. L., and R. Valdez. 2018. Wildlife conservation on the rangelands of eastern and southern Africa. *Rangeland Ecology and Management* 71:245–258.

Instituto Nacional de Ecología. 2000. *Programa de Manejo de la Reserva de la Biosfera El Vizcaíno*. Instituto Nacional de Ecología, México, D.F., México. [In Spanish]

Instituto Nacional de Estadística y Geográfica. 2023. Resultados definitivos del censo agropecuario 2022 en el estado de Coahuila. Comunicado de prensa núm. 354/23 a 05 de junio de 2023. [In Spanish]

International Union for the Conservation of Nature and Natural Resources (IUCN). 2012. IUCN Species Survival Commission guiding principles on trophy hunting as a tool for creating conservation incentives. Version 1.0. IUCN, Gland, Switzerland.

International Union for the Conservation of Nature and Natural Resources (IUCN). 2016. IUCN Briefing Paper: informing decisions on trophy hunting. https://www.iucn.org/sites/dev/files/iucn_sept_briefing_paper_informingdecisionstrophy-hunting. Accessed 6 Jan 2024.

Jiménez Lezama, S., and M. C. Hernández Cavazos. 2008. Programa de conservación del borrego cimarrón (*Ovis canadensis weemsi*) en isla El Carmen, BCS. Pages 135–154 *in* O. V. Espina Barros, F. J. Franco Guerra, J. Hernández Hernández, and S. Romero Castañón, editors. *Conservación y Manejo de Fauna Cinegética de México*. Volume 1. Benemérita Universidad Autónoma de Puebla, Puebla, México. [In Spanish]

Johnson, D. H., and S. J. Dinsmore. 2020. Population analysis in wildlife ecology. Pages 334–364 *in* N. J. Silvy, editor. *The Wildlife Techniques Manual: Research*. Volume 1. Eighth edition. Johns Hopkins University Press, Baltimore, Maryland, USA.

Klimova, A., J. Gutiérrez-Rivera, V. Sánchez-Sotomayor, et al. 2022. The genetic consequences of captive breeding, environmental change and human exploitation in the endangered peninsular pronghorn. *Science Reports* 12:11253.

Krausman, P. R., P. Bangs, K. Kunkel, M. K. Phillips, P. Parsons, and E. Rominger. 2001. Mountain sheep restoration through private/public partnership. Pages 231–242 *in* D. S. Maehr, R. F. Noss, and J. L. Larkin, editors. *Large Mammals Restoration: Ecological and Sociological Challenges in the 21st Century*. Island Press, Washington, D.C., USA.

Laity, J. 2008. *Deserts and Desert Environments*. Wiley-Blackwell, Oxford, United Kingdom.

Lee, R. M. 2008. Hunting as a tool for wildlife conservation: the case of sheep hunting in Mexico. Pages 53-58 *in* R. D. Aldus, G. R. Damm, and K. Wollscheid, editors. *Best Practices in Sustainable Hunting. CIC Technical Services Publication Number 1*. International Council for Game and Wildlife Conservation, Budakeszi, Hungary.

Lee, R. M. 2011. Economic aspects of and the market for desert bighorn sheep. *Desert Bighorn Council Transactions* 51:46–49.

Lehr, K. M, and J. Schwab, editors. 2023. *Records of North American big game. A Book of the Boone and Crockett Club Containing Tabulations of Outstanding North American Big Game Trophies, Compiled from Data in the Club's Big Game Records Archives*. Volume 2. Fifteenth edition. Boone and Crockett Club, Missoula, Montana, USA.

León de la Luz, J. L., J. P. Redman, T. R. Van Devender, et al. 2018. El conocimiento florístico actual del Noroeste de México: desarrollo, recuento y análisis del endemismo. *Botanical Sciences* 96:555–568. [In Spanish]

Leopold, A. S. 1959. *The Wildlife of Mexico: The Game Birds and Mammals*. University of California Press, Berkeley, California, USA.

McGowan, J., L. J. Beaumont, R. J. Smith, et al. 2020. Conservation prioritization can resolve the flagship species conundrum. *Nature Communications* 11:994.

Mellado, M., A. Olvera, A. Quero, and G. Mendoza. 2005. Diets of prairie dogs, goats, and sheep on a desert rangeland. *Rangeland Ecology and Management* 58:373–379.

Mellink, E. 1993. The president spoke. Pages 201–220 *in* G. P. Nabhan, editor. *Counting Sheep: 20 Ways of Seeing Desert Bighorn*. University of Arizona Press, Tucson, Arizona, USA.

Mellink, E., and M. E. Riojas-López. 2020. Livestock and grassland interrelationships along five centuries of ranching the semiarid grasslands on the southern highlands of the Mexican Plateau. *Elementa: Science in the Anthropocene* 8:20.

Morret-Sánchez, J. C., and C. Cosío-Ruiz. 2017. Panorama de los ejidos y comunidades agrarias en México. *Agricultura, Sociedad y Desarrollo* 14:125–152. [In Spanish]

Mungall, E. C. 2024. *Exotic Animal Guide: Nonnative Hoofed Mammals in the United States*. Second edition. Texas A&M University Press, College Station, Texas, USA.

Muposhi, V. K., E. Gandiwa, S. M. Makuza, and P. Bartels. 2017. Ecological, physiological, genetic trade-offs and socio-economic implication of trophy hinting as a conservation tool: a narrative review. *Journal of Animal and Plant Sciences* 27:1–14.

Muths, E., and H. McCallum. 2016. Why you cannot ignore disease when you reintroduce animals. Pages 217–243 *in* D. S. Jachowski, J. J. Millspaugh, P. L. Angermeier, and R. Slotow, editors. *Reintroduction of Fish and Wildlife Populations*. University of California Press, Oakland, California, USA.

Naidoo, R., L. C. Weaver, R. W. Diggle, et al. 2016. Complementary benefits of tourism and hunting to communal conservancies in Namibia. *Conservation Biology* 30:628–638.

North, A. W. 1910. *Camp and Camino in Lower California: A Record of the Adventures of the Author While Exploring Peninsular California, Mexico*. Baker and Taylor, New York, New York, USA.

O'Connor, J. 1974. *Sheep and Sheep Hunting*. Winchester Press, New York, New York, USA.

Ortega-Argueta, A., A. A. González-Zamora, and A. Contreras-Hernández. 2016. A framework and indicators for evaluating policies for conservation and development: the case of wildlife management units in Mexico. *Environmental Science and Policy* 63:91–100.

Ortega-S., J. A., J. G. Villarreal González, S. Mandujano, et al. 2019. Wildlife conservation management: challenges and strategies in Mexico. Pages 378–390 *in* R. Valdez and J. A. Ortega-S., editors. *Wildlife Conservation and Management in Mexico*. Texas A&M University Press, College Station, Texas, USA.

Oyler-McCance, S. J., E. K. Latch, and P. L. Leberg. 2020. Conservation genetics and molecular ecology in wildlife management. Pages 554–580 in N. J. Silvy, editor. *The Wildlife Techniques Manual: Research*. Volume 1. Eighth edition. Johns Hopkins University Press, Baltimore, Maryland, USA.

Palazy, L., C. Bonenfant, J. M. Gaillard, and F. Courchamp. 2011. Rarity, trophy hunting, and ungulates. *Animal Conservation* 2011:1–8.

Palfrey, R., J. A. Oldekop, and G. Holmes. 2022. Privately protected areas increase global protected area coverage and connectivity. *Nature Ecology and Evolution* 6:1–8.

Parques y Vida Silvestre. 2024. Oficio Número PFPVS/117/2024. Gobierno del Estado de Nuevo León, Monterrey, México. [In Spanish]

Peek, J. M., K. T. Schmidt, M. J. Dorrance, and B. L. Smith. 2002. Supplemental feeding and farming of elk. Pages 617–647 in D. E. Toweill and J. W. Thomas, editors. *North American elk: Ecology and Management*. Smithsonian Institution, Washington, D.C., USA.

Peterson, M. J., and P. J. Ferro. 2020. Wildlife health and disease surveillance, investigation, and management. Pages 174–199 in N. J. Silvy, editor. *The Wildlife Techniques Manual: Research*. Volume 1. Eighth edition. Johns Hopkins University Press, Baltimore, Maryland, USA.

Pitman, R. T., J. Fattebert, S. T. Williams, et al. 2016. The conservation costs of game ranching. *Conservation Letters* 10:1–11.

Radding, C. 1997. *Wandering Peoples: Colonialism, Ethnic Spaces, and Ecological Frontiers in Northwestern Mexico, 1700–1850*. Duke University Press, Durham, North Carolina, USA.

Ramírez, F., and M. E. Mondragón. 2010. Conservación y aprovechamiento sustentable: La Cuenca Palo Blanco. Pages 98–99 *in* J. Carabias, J. Sarukhán, J. de la Masa, and C. Galindo Leal, editors. *Patrimonio Natural de México: Cien Casos de éxito*. CONABIO, México, D.F., México. [In Spanish]

Rebman, J. P., and N. C. Roberts. 2012. *Baja California Plant Field Guide*. Third edition. San Diego Natural History Museum, San Diego, California, USA.

Reneau, J., J. Spring, and C. Lacey. 2016. *How to Score North American Big Game*. Fifth edition. Boone and Crockett Club-Pope and Young Club, Missoula, Montana, USA.

Ripple, W. J., T. M. Newsome, C. Wolf, et al. 2015. Collapse of the world's largest herbivores. *Sciences Advances* 1:e1400103.

Rivkin, M., with R. Burbank. 2019. *Sheep on the Brink: A History of the Bighorn Institute*. Bighorn Institute, Palm Desert, California, and Silverfish Press, Rancho Mirage, California, USA.

Roberts, N., and E. Ezcurra. 2012. Climate. Pages 1–11 in J. P. Rebman and N. Roberts. *Baja California Plant Field Guide*. San Diego Natural History Museum and Sunbelt Publications, Chula Vista, California, USA.

Romero-de-Diego, C., A. Dean, A. Jagadish, B. Witt, M. B. Mascia, and M. Mills. 2021. Drivers of adoption and spread of wildlife management initiatives in Mexico. *Conservation Practice* 3:438.

Romero-Figueroa, G., E. de J. Ruiz-Mondragón, E. Shahriary, C. Yee-Romero, A. A. Guevara-Carrizales, R. Paredes-Montesinos, J. M. Corrales-Sauceda, I. Guerrero-Cárdenas, and R. Valdez. 2024. Population and conservation status of bighorn sheep in the state of Baja California, Mexico. *Animals* 14:504.

Romero-Figueroa, G., R. Texis Texis, E. A. Lozano-Cavazos, V. Ortiz-Avila, M. A. Garcia-Aranda, and O. Garcia Jurado. 2017. Descripción y modelación de hábitat potencial del borrego cimarrón (*Ovis canadensis*) en la Rumerosa, Baja California. Pages 120–142 in G. Ruiz-Campos and G. Laniz Garcia, compilers. *Estudios sobre el Borrego Cimarrón en el Noroeste de México*. Universidad Autónoma de Baja California, Mexicali, México. [In Spanish]

Roosevelt, K. 1920. *The Happy Hunting Grounds*. Charles Scribner's Sons, New York, New York, USA.

Ruiz-Mondragón, E. de J. 2014. Estado actual de la población del borrego cimarrón (*Ovis canadensis weemsi*) en la uma Ejido La Purísima, Baja California Sur, México. Thesis, Universidad Nacional Autónoma de México, Estado de México, México. [In Spanish]

Ruiz-Mondragón, E. de J., F. I. Gastelum-Mendoza, G. Romero-Figueroa, et al. 2025a. Delimitation of regional management units for desert bighorn sheep in Baja California. An application of the potential species distribution model. *Therya* 16:249–258.

Ruiz-Mondragón, E. de J. 2024. Modelo de manejo del borrego cimarrón, basado en el monitoreo comunitario participativo y la modelación de nicho ecológico, para Baja California, México. Dissertation, Universidad Autónoma de Baja California, Ensenada, México. [In Spanish]

Ruiz-Mondragón, E. de J., F. I. Gastelum-Mendoza, G. Romero-Figueroa, et al. 2025b. Use of trail cameras to estimate abundance and population structure of bighorn sheep in Baja California, Mexico. *Agrociencia* 59:2–15.

Ruiz-Mondragón, E. de J., G. Romero-Figueroa, M. A. García-Aranda, E. A. Lozano-Gavazos, and R. Valdez. 2018. Potential distribution model of *Ovis canadensis* in northern Baja California, Mexico. *Therya* 9:219–226.

Ruiz-Mondragón, E. de J., G. Romero-Figueroa, R. Paredes-Montesinos, et al. 2023. Community-based workshops to involve rural communities in wildlife management, case studies: bighorn sheep in Baja California, Mexico. *Animals* 13:3171.

Sandoval, A. V., R. Valdez, and A. Espinosa-T. 2014. El borrego cimarrón en México. Pages 489–518 in R. Valdez and J. A. Ortega-S., editors. *Ecología y Manejo de Fauna Silvestre en México*. Editorial del Colegio de Postgraduados, Montecillo, Texcoco, México. [In Spanish]

Sandoval, A. V., R. Valdez, and A. Espinosa-T. 2019. Desert bighorn sheep in Mexico. Pages 350–365 in R. Valdez. and J. A. Ortega-S., editors. *Wildlife Conservation and Management in Mexico*. Texas A&M University Press, College Station, Texas, USA.

Secretaría de Agricultura, Ganadería, Recursos Hidráulicos, Pesca y Acuacultura. 2011. Proyecto para la reintroducción y repoblación del borrego cimarrón (*Ovis canadensis mexicana*) en el Estado de Sonora, Hermosillo, México. https://www.hunting.sonora.gob.mx/convenios/proyectorepoblacion.pdf. Accessed 14 Feb 2025. [In Spanish]

Secretaría de Agricultura, Ganadería, Recursos Hidráulicos, Pesca y Acuacultura. 2024. Oficio Número 12–06/UT-057. Gobierno del Estado de Sonora, Hermosillo, México. [In Spanish]

Secretaría de Agricultura, Ganadería, Recursos Hidráulicos, Pesca y Acuacultura. 2025. Cintillos expedidos por temporada de caza. Gobierno del Estado de Sonora, Hermosillo, México. https://www.hunting.sonora.gob.mx/principal.php?op=16. Accessed 14 Feb 2025. [In Spanish]

Secretaría de Desarrollo Social. 1993. Decreto por el que se declara área natural protegida con el carácter de Reserva de la Biosfera, la región conocida como El Pinacate y Gran Desierto de Altar, ubicada en los municipios de Plutarco Elías Calles, Puerto Peñasco y San Luis Río Colorado, Son. Diario Oficial jueves 10 de junio de 1993. Secretaría de Desarrollo Social, Ciudad de México, México. [In Spanish]

Secretaría de Desarrollo Urbano y Ecología. 2024. Oficio Número UT-197/2024. Gobierno del Estado de Chihuahua, Chihuahua, México. [In Spanish]

Secretaría de Medio Ambiente. 2024. Oficio Número SMA-VS-01/0088–2024. Gobierno del Estado de Coahuila, Saltillo, México. [In Spanish]

Secretaría de Medio Ambiente. 2025. Relación de unidades de manejo para la conservación y aprovechamiento de la vida silvestre en el Estado de Coahuila 2023–2024. Gobierno del Estado de Coahuila, Saltillo, México. https://sma.gob.mx/vsunidades-de-manejo/. Accessed 14 Feb 2025. [In Spanish]

Secretaría de Medio Ambiente y Desarrollo Sustentable. 2024. SMDS/DRNA/VS/ENS/3660/2024. Gobierno del Estado de Baja California, Ensenada, México. [In Spanish]

Secretaría de Medio Ambiente y Recursos Naturales (SEMARNAT). 2020. Registro de Unidades de Manejo Extensivas para la Conservación de la Vida Silvestre. https://datos.gob.mx/busca/dataset/registros-de-unidades-demanejo-para-la-conservacion-de-la-vida-silvestre-uma. Accessed 20 Oct 2024. [In Spanish]

Secretaría de Medio Ambiente y Recursos Naturales (SEMARNAT). 2023. *Plan de Manejo Tipo para la Conservación y el Aprovechamiento Sustentable en vida libre de Borrego Cimarrón (Ovis canadensis) en Baja California Sur, México*. DGVS-SEMARNAT, Ciudad de México, México. [In Spanish]

Secretaría de Medio Ambiente y Recursos Naturales (SEMARNAT). 2024a. Oficio Número SEMARNAT/UCVSDHT/UT/1391/2024. Gobierno de México, Ciudad de México, México. [In Spanish]

Secretaría de Medio Ambiente y Recursos Naturales (SEMARNAT). 2024b. Oficio Número SEMARNAT/UVCSDHT/UT/1467/2024. Gobierno de México, Ciudad de México, México. [In Spanish]

Secretaría del Medio Ambiente y Recursos Naturales (SEMARNAT). 2015. Modificación del Anexo Normativo III, Lista de especies en riesgo de la Norma Oficial Mexicana NOM059-SEMARNAT-2010, protección ambiental-especies nativas de México de flora y fauna silvestres-categorías de riesgo y especificaciones para su inclusión, exclusión o cambio – lista de especies en riesgo. – Diario Oficial de la Federación. https://www.dof.gob.mx/nota_detalle.php?codigo=5578808&fecha=14/11/2019#gsc.tab=0. Accessed 20 Oct 2024. [In Spanish].

Secretariat of the Man and the Biosphere Program. 2022. *Technical Guidelines for Biosphere Reserves*. UNESCO, Paris, France.

Segundo Galán, J. 2023. Monitoreo aéreo de la población silvestre del borrego cimarrón (*Ovis canadensis*) en las sierras de Baja California Sur. Reporte Final. A. F. Flotillas, Hermosillo, Sonora, México. [In Spanish]

Segundo Galán, J. M., E. J. Torres-Romero, and L. C. Bender. 2024. Mapping core habitats, landscape permeability, and movement corridors for conservation of desert bighorn sheep in Sonora, México. *Therya* 15:279–288.

Segundo Galán, J. M., E. J. Torres-Romero, and L. C. Bender. 2025. Habitat associations of desert bighorn sheep in Sonora, Mexico. *Journal of Arid Environments* 229:105386.

Servicio de Información Alimentaria y Pesquera. 2023. Sistema de información agroalimentaria de consulta. Secretaria de agricultura y desarrollo rural. https://www.gob.mx/siap/docu-mentos/siacon-ng-161430. Accessed 5 Mar 2025. [In Spanish]

Servín, J., D. E. Carreón-González, F. Castro-Campos, A. Huerta, and M. Garza. 2018. *Las Unidades de Manejo para la Conservación de la Vida Silvestre (UMA) en el noroeste de México: Análisis de 10 años*. Universidad Autónoma Metropolitana, Ciudad de México, México. [In Spanish]

Shreve, F., and I. L. Wiggins. 1964. *Vegetation and Flora of the Sonoran Desert*. Stanford University Press, Redwood City, California, USA.

Sotelo Gallardo, H., J. A. García Salas, and A. J. Contreras Balderas. 2018a. Depredación del borrego *Ovis canadensis* (Artiodactyla: Bovidae) y el venado bura *Odocoileus hemionus* (Artiodactyla: Cervidae) por *Puma concolor* (Carnivora: Felidae) en Coahuila, México. *Revista de Biología Tropical* 66:1496–1503. [In Spanish]

Sotelo Gallardo, H., J. A. García-Salas, and A. J. Contreras-Balderas. 2018b. Influencia de la densidad de *Odocoileus hemionus crooki* Mearns, 1897, (Artiodactyla: Cervidae) en la abundancia relativa de *Puma concolor stanleyana* Goldman, 1938, (Carnivora: Felidae) y la relación en la depredación en una población reintroducida de *Ovis canadensis mexicana* Merriam, 1901, (Artiodactyla: Bovidae) en Coahuila, México. *Acta Zoológica Mexicana (nueva serie)* 34:1–11. [In Spanish]

Stronza, A. L., C. A. Hunt, and L. A. Fitzgerald. 2019. Ecotourism for conservation? *Annual Review of the Environment and Resources* 44:229–253.

Tapia Landeros, A. 2007. *Homo-Ovis: el Borrego Cimarrón en México*. Universidad Autónoma de Baja California, Mexicali, Sonora, México. [In Spanish]

Tarango, L. A., and P. R. Krausman. 1997. Desert bighorn sheep in Mexico. *Desert Bighorn Council Transactions* 41:1–7.

Taylor, W. A., M. F. Child, P. A. Lindsey, S. K. Nicholson, C. Relton, and H. T. Davies-Mostert. 2021. South Africa's private wildlife ranches protect globally significant populations of wild ungulates. *Biodiversity and Conservation* 30:4111–4135.

Teer, J. G. 2003. Nonnative large mammals in North America. Pages 1180–1187 *in* G. A. Feldhamer, B. C. Thompson, and J. A. Chapman, editors. *Wild Mammals of North America: Biology, Management and Conservation*. Second edition. Johns Hopkins University Press, Baltimore, Maryland, USA.

Tinker, B. 1978. *Mexican Wilderness and Wildlife*. University of Texas Press, Austin, Texas, USA.

Torres-Romero, E. J., A. J. Giordano, G. Ceballos, and J. V. López-Bao. 2020. Reducing the sixth mass extinction: understanding the value of human-altered landscapes to the conservation of the world's largest terrestrial mammals. *Biological Conservation* 249:108706.

Toweill, D. E., and V. Geist. 1999. *Return to Royalty: Wild Sheep of North America*. Boone and Crockett Club, Missoula, Montana, USA, and Foundation for North American Wild Sheep, Cody, Wyoming, USA.

Turner, R. M., J. E. Bowers, and T. L. Burgess. 2005. *Sonoran Desert Plants: An Ecological Atlas*. University of Arizona Press, Tucson, Arizona, USA.

United Nations Educational, Scientific and Cultural Organization. 2024. El Pinacate and Gran Desierto de Altar Biosphere Reserve. List of world heritage sites. https://www.whc.unesco.org/en/list/1410. Accessed 15 Oct 2024.

Uranga-Thomas, R., and R. Valdez. 2011. Reintroduction of desert bighorn sheep in Chihuahua, Mexico. *Desert Bighorn Council Transactions* 51:32–38.

Valdez, R. 1997. Mexico. Pages 303–307 *in* D. M. Shackleton, editor. *Wild Sheep and Goats and Their Relatives: Status Survey and Conservation Action Plan for Caprinae*. International Union for the Conservation of Nature and Natural Resources, Gland, Switzerland.

Valdez, R. 2011. Aoudad (*Ammotragus lervia*). Pages 714–715 *in* R. E. Wilson and R. A. Mittermeier, editors. *Handbook of the Mammals of the World. Volume 2. Hoofed Mammals*. Lynx Edicions, Barcelona, Spain.

Valdez, R. 2019. Historical and conservation perspectives of wildlife in Mexico. Pages 1–18 *in* R. Valdez and J. A. Ortega-S., editors. *Wildlife Ecology and Management in Mexico*. Texas A&M University Press, College Station, Texas, USA.

Valdez, R., J. C. Guzmán-Aranda, F. J. Abarca, L. A. Tarango-Arámbula, and F. Clemente-Sánchez. 2006. Wildlife conservation and management in Mexico. *Wildlife Society Bulletin* 34:1480–1488.

Villarreal González, J. G. 2006. *Venado Cola Blanca, Manejo y Aprovechamiento Cinegético*. Second edition. Unión Ganadera de Nuevo León, Fundación PRODUCE, y Confederación Nacional de Organizaciones Ganaderas, Monterrey, Nuevo Leon, México. [In Spanish]

Villarreal González, J. G. 2009. *Vida Silvestre de la Cuenca "Palo Blanco," Nuevo León, México*. Consejo Estatal de Flora y Fauna Silvestre de Nuevo León, Parques y Vida Silvestre de Nuevo León, y UANL, Monterrey, México. [In Spanish]

Villarreal González, J. G. 2013. *Ganadería Diversificada: Importancia, Ecológica, Cinegética, y Económica de los Venados Cola Blanca mexicano*. Secretaría de Agricultura, Ganadería, Desarrollo Rural, Pesca y Alimentación, Ciudad de México, México. [In Spanish]

Wakild, E. 2015. Parks, people, and perspectives: historicizing conservation in Latin America. Pages 41–52 *in* G. Wuerthner, E. Crist, and T. Butler, editors. *Protecting the Wild: Parks and Wilderness, the Foundation for Conservation*. Island Press, Washington, D.C., USA.

Wall, B., and B. Child. 2009. When does hunting contribute to conservation and rural development? Pages 255–265 *in* B. Dickson, J. Hutton, and W. M. Adams, editors. *Recreational Hunting, Conservation and Rural Livelihoods*. John Wiley & Sons, Chichester, West Sussex, United Kingdom.

Watson, J. E. M., N. Dudley, D. B. Segan, and M. Hockings. 2014. The performance and potential of protected areas. *Nature* 515:67–73.

White, R. J. 2000. Big game ranching. Pages 260–276 *in* S. Demarais and P. R. Krausman, editors. *Ecology and Management of Large Mammals in North America*. Prentice Hall, Upper Saddle River, New Jersey, USA.

Whitford, W. G., with B. D. Duval. 2020. *Ecology of Deserts*. Second edition. Academic Press, London, United Kingdom.

Wiggins, I. L. 1980. *Flora of Baja California*. Stanford University Press, Stanford, California, USA.

Wilson-Holt, O., D. Roe, and D. Natusch. 2022. *CITES and Livelihoods: Guidance on Maximizing Benefits to Indigenous Peoples and Local Communities for Trade in CITES-Listed Species*. CITES Secretariat, Geneva, Switzerland.

Wright, E. A., G. G. Brugette, K. R. Buckety, et al. 2024. Multi-locus sequence typing indicates multiple strains of Mycoplasma in bighorn sheep and aoudad in Texas. *Journal of Wildlife Management* 213:88e22594.

# 18 Human Dimensions and Mountain Sheep

*Kathryn A. Schoenecker, Nicholas W. Cole, and Saeideh Esmaeili*

## 18.1 INTRODUCTION

The way intrinsic linkages between humans and nature are perceived within various societies has played a significant role in shaping cultural norms and behaviors regarding wildlife (Berkes et al. 1998). Indigenous cultures that coexisted with wildlife for over 10,000 years valued wildlife for their use and their spiritual connections. Mountain sheep (bighorn sheep [*Ovis canadensis*], thinhorn sheep [*Ovis dalli*]) have held a sacred place in many Native American cultures, where they are believed to serve as the chief of all large game animals. Their ability to navigate mountainous, rugged terrain was admired and considered a metaphor for spiritual resilience and endurance (Scott 2015). Their deep connection to the species is reflected in their spiritual practices, which involve honoring the spirit of the mountain sheep for guidance and protection (Loendorf and Stone 2006). First Nations, Indigenous Peoples, and Tribal Nations in Canada, the United States, and Mexico all hold mountain sheep in high esteem, such that in these cultures, there exists a respect for mountain sheep that exceeds their consumptive value.

In Western societies, views have shifted widely over time from colonialism to current-day modern society (Manfredo et al. 2003). Relationships with wildlife during the European colonial era emphasized dominating nature as an external force, altering it to fit human needs, and extracting as much value as possible to benefit people. Following the period of overharvesting and resource depletion, an era emerged in human attitudes that emphasized the intrinsic value of nature beyond its extractive benefits to people and a recognition that humans exist within the same ecosystem. "The land is a community. It has a structure, it grows, it evolves, and it has life. It includes human beings as members, but it also includes non-human members to the community" (Leopold 1968:204). The resulting shift in perspective on wildlife management took steps forward toward reengaging with views that Indigenous groups around the globe have held for generations, where their communities maintain mutually beneficial or even familial relationships with the ecosystems they inhabit.

Structured policymaking, regulation, and funding of wildlife management are generally indicative of the complex relationships that influential groups have with wildlife but are recognized to fall short in some cases (Peterson and Nelson 2017). In North America, the important processes and tenets that describe how wildlife should be managed and what interactions or uses are appropriate have come to be called the North American Model of Wildlife Conservation (NAM; Organ et al. 2012, Mahoney and Geist 2019). The NAM emerged in response to shifting perspectives on wildlife in North America, successfully establishing a new paradigm that limits overharvest and provides funding sources for conserving game and non-game species (Organ et al. 2012). While providing a foundation for the governance of wildlife resources, the NAM has also recognized shortcomings and remains grounded in an anthropocentric lens that does not effectively incorporate Indigenous views and knowledge (Hessami et al. 2021).

Wildlife management has commonly been subdivided into three primary dimensions: humans, wildlife, and habitats (Giles 1978, Decker et al. 2012). Historically, the wildlife and habitat dimensions have composed the majority of wildlife management capacity around the world, but the expertise and methods that focus on the human dimension of wildlife have been increasingly integrated into the culture of wildlife management (Decker et al. 2012). Although complex relationships between humans and wildlife have always existed, the recognition of their importance in determining the outcomes of wildlife management actions has grown.

Human dimensions of wildlife are generally investigated and visualized through the social sciences. Often referred to as conservation social sciences, these disciplines comprise a broad set, all with distinct origins and philosophical or methodological foundations (Bennett et al. 2017). Human dimensions of wildlife research have historically been studied through questionnaire-based quantitative research rooted in economic and social-psychological theory and methods (Jacobson et al. 2010). Questionnaires are highly effective at describing the representativeness of a series of attitudes, values, or perceptions, comparing the frequencies of those results between disparate groups, and testing hypotheses that approximate reality. Qualitative and narrative-based approaches derived from humanities, anthropology, and consumer marketing are also used in areas where questionnaires are less effective at identifying nuanced viewpoints (Ebeling-Schuld and Darimont 2017, Stinchcomb et al. 2022, Cole et al. 2024). Describing the complex interactions between human and biological dimensions, and understanding the diverse cultural relationships people have with nature can be difficult to reduce to narrower concepts that can be effectively captured in a questionnaire.

DOI: 10.1201/9781003518686-20

Mountain sheep are commonly regarded as an iconic symbol of wildness and a vital part of the cultural heritage in North America for various communities and cultures (Brewer et al. 2014, Hurley et al. 2015, Jex et al. 2016). Bighorn and thinhorn sheep are highly sought after by recreational hunters, wildlife viewers, and photographers, and are culturally and religiously important to many Indigenous Peoples across North America. The human relationships with mountain sheep are understudied in comparison to other North American ungulate species, such as deer (*Odocoileus* spp.). Published studies investigating the relationship between humans and mountain sheep have largely focused on the implications of human activities on mountain sheep, including conflict and disturbance caused by human development (Schoenecker and Krausman 2002, Polfus and Krausman 2012, Lowrey and Longshore 2017, Brushett et al. 2023). Attitudes toward the management of mountain sheep and relevant predator species (Chase 2013, Harris et al. 1995), and human-wildlife conflict resulting from livestock disease spillover (Heinse et al. 2016). Culturally based views, which examine human dimensions and Indigenous relationships and perspectives, are less mainstream but contribute meaningfully to this topic (Hessami et al. 2021). Our objective in this chapter is to examine the relationships between humans and mountain sheep.

## 18.2 HUMAN DIMENSION CONSIDERATIONS OF MANAGING MOUNTAIN SHEEP

Prior to the arrival of early Europeans, Native North Americans held a subsistence and spiritual connection with mountain sheep. Dall's sheep (*O. dalli dalli*) were an important resource for many Indigenous People across northwestern North America (Demarchi and Hartwig 2004). Indigenous groups, such as the Yukon Indigenous People, Upper Tanana, Mountain (Sahtu), Koyukon, and Ahtna, relied on Dall's sheep as a key food source, as did the Tlingit, Tagish, and Tutchone, before moose (*Alces alces*) became more abundant. The Kutchin (Gwich'in), Tanaina, and other Alaskan Indigenous people also consumed Dall's sheep (Demarchi and Hartwig 2004). Stone's sheep (*O. d. stonei*) were primarily consumed by the Tahltan of the Stikine Plateau, where they were in alpine meadows above the timberline (Kuhnlein and Humphries 2017). For many Indigenous communities, hunting thinhorn sheep was a means of subsistence and held deep spiritual and cultural significance. Sheep hunting required patience, skill, and respect for the land, reinforcing values of stewardship and sustainable harvest. The animals were often honored through rituals and offerings, ensuring continued success in future hunts (Kuhnlein and Humphries 2017). Beyond their material value, thinhorn sheep played a role in oral traditions, stories, and ceremonies, symbolizing resilience, agility, and survival in harsh mountain environments (Kuhnlein and Humphries 2017).

In Mexico, desert bighorn sheep (*O. c.* spp.) were widely distributed throughout the northern states of Chihuahua, Coahuila, Nuevo Leon, Sonora, Baja California, and Baja California Sur (Espinosa-T 2006), where Indigenous Peoples traditionally hunted desert bighorn sheep (Sandoval et al. 2019). The La Proveedora archaeological site in Sonora, Mexico, features extensive petroglyphs, including numerous depictions of bighorn sheep, providing insights into the cultural significance of the species (Martínez-Pabello et al. 2022).

Early settlers, including frontiersmen (i.e., fur traders, explorers, and missionaries), and miners, relied on mountain sheep as a reliable source of food and income throughout most western mountain ranges in North America (Buechner 1960, Chapter 2). The high subsistence and eventual commercial value led to widespread overhunting, such that forage competition with domestic livestock, introduced diseases, and human encroachment had reduced bighorn sheep numbers and distribution in the western United States, southern Canada, and mainland and Baja Peninsula Mexico (Hurley et al. 2015). Furthermore, these early settlers viewed mountain sheep as competitors for grazing resources with their domestic livestock, so their removal was seen to enhance local agricultural economies (Buechner 1960), or they were seen in and of themselves as opportunities to create new economies, such as in attempts to hybridize Dall's sheep with domestic sheep for commercial meat production (Husby et al. 1998). So, mountain sheep remained popular game animals due to the quality of their meat and their value as recreational trophies. The significant decline in many wildlife populations and ranges eventually shifted public perception, leading to a rise of naturalism and wildlife management led by new thinkers such as Allen, Stoddard, and Leopold (Swanson 1987). Initially, restoration efforts were influenced by commercial and recreational hunting interests; however, over time, the intrinsic and cultural significance of these animals has become increasingly recognized (Whiting et al. 2023). These measurable shifts in perception underscore their importance to hunters and to broader communities, including ecotourism and private landowners (Manfredo et al. 2003).

In modern North America, the importance of mountain sheep remains multidimensional, encompassing a wide range of interests and support for continued restoration (Whiting et al. 2023). The presence of mountain sheep enriches the ecological diversity of an area and enhances its recreational appeal to outdoor enthusiasts (Purdy 1981, Harris et al. 1995). Most people visiting bighorn sheep reintroduction sites in Utah expect a positive experience with the species, which fosters a sense of connection to nature, provides educational moments for children, and offers shared wildlife experiences (Chase 2013). When negative interactions occur, like vehicle collisions, people raise concerns about sharing landscapes with bighorn sheep (Chase 2013). Interestingly, people's attitudes toward bighorn sheep varied depending on their self-perceived connection with nature: those who see themselves as separate from nature tend to find bighorn sheep intrusive, while those who view human and ecological systems as interconnected perceive bighorns as a natural part of their environment (Chase 2013).

Bighorn and thinhorn sheep represent an integral part of wildness and the natural heritage of western North America (Harris et al. 1995). These majestic animals have captivated human fascination, symbolizing resilience and the ruggedness of wilderness, as is depicted in numerous historical rock art sites across the west, and seen in modern culture through branding of sports teams, trucks, and other consumer products (Chapter 2). Economically, they play a vital role in local economies through ecotourism and regulated hunting, both of which support conservation efforts and contribute to local livelihoods. The economic influences that local communities receive from mountain sheep hunting, viewing, and general popular culture are most obvious in the direct purchase of permits, guide services, or lodging (Loomis et al. 1985). Visiting recreationalists spend money in remote and rural local communities on food, entertainment, and cross-recreational activities due to their interest in mountain sheep and the ranges they inhabit. The perception of increasing value of mountain sheep-related recreation would be expected to lead to increasing investments in services and hospitality that attract new and diverse audiences and expand economic influences of mountain sheep-related recreation. Uncertainty over the long-term viability of mountain sheep populations and perceived risk of diminishing access and opportunities would be expected to decrease value for recreationalists, local businesses, and the public (Brookshire et al. 1983). This connection also means that this segment of the population is more likely to care about the habitats that wild sheep need, often adding their voices to public processes or contributing financial support to conservation initiatives, thus influencing elected governments and land-use decision-making.

There are other forms of economic value besides direct use value. Instrumental value refers specifically to the benefits received by people and societies, and when applied to wildlife, it is usually presented as use value through recreational hunting or viewing. Hunters and wildlife viewers can derive positive emotional benefits from participating in their respective activities, but many people also experience similar benefits from simply being aware that the species is present and thriving. Intrinsic value refers to the value inherently associated with something that exists, independent of any direct benefit to people. Alternate valuations like intrinsic value are important to consider when evaluating the economic influences of mountain sheep on a community (Vucetich et al. 2015). This is especially the case for highly charismatic and easily recognizable species, such as mountain sheep, which commonly appear in popular media and represent conservation success stories that local communities develop a personal connection with.

The beliefs and values people hold toward wildlife have an influence on how they interact with wildlife (Manfredo et al. 2003). A person's belief system regarding wildlife is commonly represented as a wildlife value orientation, a theoretical framework used to describe patterns of behavior and the likelihood of expressing certain attitudes (Fulton et al. 1996, Teel et al. 2007). Within this framework, people are categorized based on their expression of traditional views (an emphasis on utilitarian views of wildlife and prioritizing people's needs over wildlife) and mutualist views (an emphasis on the belief that humans and wildlife exist together, and that wildlife deserve similar considerations to humans; Teel and Manfredo 2010). People's wildlife value orientations have continued to broaden from traditional utilitarian orientations to include more mutualist orientations since the emergence of animal rights movements in the 1970s (Singer 1975). A person with a mutualist orientation is more likely to apply an egalitarian view of wildlife where moral rights and care are extended to wildlife. North American Indigenous and Buddhist views are accepting of the utility aspects of wildlife, but spiritual values are held equally. A person with a more traditional orientation tends to have a utilitarian view of wildlife, emphasizing the use and economic benefits of wildlife to society. With increasing proportions of the population holding mutualistic views, it is expected that feelings of reciprocity and fairness will be increasingly applied to species like mountain sheep, which have been affected by land-use change, disease spread from livestock, and water diversion (Brewer et al. 2014).

People interact with wildlife in different ways, directly influencing public interests and perceptions of wildlife (Manfredo et al. 2003). These interactions can be beneficial, as many ungulate species provide recreational and commercial benefits through ecotourism and hunting (Loomis and White 1996). Half the United States population (148.3 million) is reported to engage in watching wildlife and 14.4 million people aged 16 or older hunt (U.S. Fish and Wildlife Service 2022). Mountain sheep are a vital component of this dynamic ecosystem and provide a wide range of benefits to people and society through outdoor recreation. Bighorn and thinhorn sheep are highly prized as game species, creating opportunities for wildlife viewing and photography. Their presence on public lands facilitates opportunity for humans to experience mountain sheep and mountain sheep habitats through related recreational opportunities. National parks and monuments serve as hotspots for ecotourism, offering unique viewing experiences to visitors from across the country and around the world.

Dall's sheep are considered one of the main species contributing to visitation satisfaction within Denali National Park (Skibins et al. 2012). Although Dall's sheep do not contribute to overall visitor satisfaction nearly as much as brown bears (*Ursus arctos*) or caribou (*Rangifer tarandus*), this is likely due to the increased likelihood of bears and caribou occurring near roadsides, which facilitate regular viewing opportunities (Skibins et al. 2012). The emotional benefits of encountering different wildlife species depend on the likelihood of their occurrence, the proximity of the wildlife, the authenticity of the setting, and the memorability of the situation (McIntosh and Wright 2017). A notable characteristic of mountain sheep is their distinctive behavior, which contributes to their intrinsic value and enhances the enjoyment of observing them. Visitors can delight in spectacular displays of head-butting males, adept hooves navigating impossibly steep terrain, and watch females with frolicking nursery groups of lambs on formidably steep slopes. Sometimes they will tolerate being close to humans, making viewing even more memorable. Rocky mountain bighorn sheep (*O. c. canadensis*) populations adjacent to human population centers in New Mexico (e.g., Rio Grande Gorge, Red River) or those habituated to hikers in alpine wilderness (Pecos) may allow humans to approach relatively short distances without fleeing (<50 m). These closer viewing opportunities can be more exciting and memorable for recreational viewers.

The rarity of hunting opportunities for bighorn and thinhorn sheep elevates demand. In 2016 approximately 12,000 hunting licenses were issued for mountain sheep in North America (Wild Sheep Initiative 2019). This equates to less than 0.1% of all hunters, whereas a majority of hunters (88%) had the opportunity to hunt deer in the same year (U.S. Department of the Interior 2016). This high demand, coupled with relatively low population densities, has motivated multiple research studies on how recreational hunting affects horn size in bighorn and thinhorn sheep (Douhard et al. 2016, Pigeon et al. 2016, LaSharr et al. 2019, Lassis et al. 2022). Increased harvest pressure and evolutionary response may lead to decreasing relative horn size, because hunters are likely to select for animals with larger horn sizes as the primary defining characteristic for determining trophy classification. Response to selective pressure may be confounded by factors that could lead to similar decreases in horn size, like nutritional limitations, disease, biased sampling of younger sheep, or habitat fragmentation (Festa-Bianchet et al. 2014, Heffelfinger 2018, Monteith et al. 2018). Even in areas where the average relative horn size has decreased, if harvest pressure is regulated by opportunity, participation, or low harvest intensity at local scales, selective pressure on horn size may be reduced (LaSharr et al. 2019). Antler or horn size is a common measure of quality and trophy status among many ungulate species (Messner 2011). In some cases, the demand for harvesting these trophy animals has led to commercialization and investment in captive production (Nox 2011). Antler or horn size is also a common threshold used by managers to set harvest restrictions that maintain high hunting opportunity with lower harvest success, while also discouraging harvest of younger males (Koenig 2024, Wallingford et al. 2017). For most game species, trophy animals are designated by proxy morphological characteristics that represent rarity, high-quality genetics, or advanced ages (Palazy et al. 2012). Chosen characteristics can vary widely across species, like body length or body mass for fish and predators,

beard and spur length for turkeys, and horn or antler mass for ungulates.

Participation in recreational harvest of bighorn and thinhorn sheep is commonly discussed as trophy-oriented, where harvesting a trophy animal is the primary cause for participation (Coltman et al. 2003, Monteith et al. 2013, Pigeon et al. 2016). There are often large expenditures on auctioned special permits (e.g., a governor's permit), where the proceeds of the auctions are earmarked for conservation activities (Coltman et al. 2003, Wild Sheep Foundation 2024). The willingness to spend hundreds of thousands, and in some cases, nearly a million United States dollars for auctioned hunting opportunities demonstrates clear evidence of high demand and value. That said, auctioned special permits make up a small fraction of the available permits each year. Lottery-drawn permits make up a much larger portion of permit available to the public and these permits are much more affordable. With the low population densities, demand for lottery-drawn hunting permits is much higher than the available permits, and drawing a tag for each mountain sheep species is considered a once-in-a-lifetime opportunity. Mountain sheep hunters may put in for a lottery-drawn permit for many years before being successful.

North American bighorn and thinhorn sheep populations tend to occupy isolated and rugged habitats in low densities due to stochastic predation, habitat fragmentation, competition with agricultural practices, disease, and water diversion (Dolan 2006, Festa-Bianchet et al. 2006, Bleich 2018, Rominger 2018, Whiting et al. 2023). Although great strides have been made in restoring mountain sheep populations and habitat, mountain sheep are substantially reduced relative to historical population estimates and distribution in the lower 48 states (Buechner 1960), and they are no more abundant than they were 25 years ago despite tremendous conservation effort (Western Association of Fish and Wildlife Agencies, Wild Sheep Initiative, Range-wide Status of Bighorn and Thinhorn Wild Sheep~2023 2024). The rarity of hunting opportunities, high cost of entry, limited supply of affordable permits, difficulty navigating the terrain and finding stalking opportunities, and low densities of mountain sheep create an inherently challenging environment that may increase the recreational value and prestige associated with successfully harvesting mountain sheep.

Recreational hunting motivations range widely, from harvesting a trophy animal to obtaining food and supplementing diets, experiencing and appreciating nature, relaxing and enjoying solitude, or socializing with friends and family (Kennedy 1973, Boulanger et al. 2006, Hinrichs et al. 2021). Hunter expectations and desires tend to be diverse and multidimensional, even when harvesting a trophy animal is emphasized as a preferred outcome. It is unlikely that any individual recreational motivation will be mutually exclusive; satisfaction can be achieved through a wide range of attributes (Whitehead et al. 2008, Schroeder et al. 2019, Sainsbury et al. 2024). As with most things that involve a diverse group of people, describing reality is likely complicated and requires engaging with complex relationships among ethics, culture, and leisure participation (Stedman and Heberlein 2001, Schroeder et al. 2006, 2018).

The decision to participate in any recreational activity begins with a set of motivations a person holds and the expectation that by participating, they will be able to fulfill some proportion of those motivations (Ryan and Deci 2000). A person may be looking to spend more time with their children, connect with nature and relax, get physical exercise, or challenge themselves with an activity that requires skill and practice. There are many possible recreational activities that provide these benefits to varying degrees, and individuals will rely on past experiences, support from their social network, and self-perceptions to determine which activity to pursue (Jun and Kyle 2015). All participants must continually evaluate their level of satisfaction based on the outcomes they expected and what they actually experienced. The actualized level of satisfaction or frustration provides information about future motivations and behavior, recalibrating expectations for future decisions (Ryan and Deci 2019). The motivations that influence participation in mountain sheep hunting, much like other recreational hunted species, are likely to be multi-faceted and heterogenous (Hendee 1973, Decker et al. 1980).

Recreational hunting motivations have been represented using many different frameworks (Manfredo et al. 1996, Gigliotti 2000, Schroeder et al. 2018). Decker et al. (1984) introduced a common and longstanding framework that reduces the dimensionality of recreation motivations to one of three goal orientations: achievement (goals associated with the activity), affiliation (goals associated with social relationships), and appreciation (goals associated with experiencing natural settings). When this framework was applied to hunters targeting ungulates (including several mountain sheep races), achievement goals were the most expressed goal orientation, and when multiple goal orientations were referenced, it tended to be the dominant orientation (Ebeling-Schuld and Darimont 2017). Notably, achievement goals do not specifically imply an emphasis on trophy hunting, such as horn size, but instead achievement goals can also be applied to any aspects of recreational hunting that can be gamified and present an activity-specific accomplishment. For example, obtaining food, using more challenging forms of equipment, or hunting in physically and mentally demanding terrains are all challenges that increase difficulty and, therefore, increase the perceived value of harvesting an animal (von Essen et al. 2019). The substantial interest in hunting mountain sheep may have less to do with the trophy orientation of participating hunters and more to do with pursuing a mountain sheep and the accomplishment of overcoming the many challenges associated with successfully harvesting one.

The conservation heritage of hunting in North America can also impart a sense of responsibility and stewardship, motivating participation (Decker et al. 2016). The Wild Sheep Foundation highlights its mission as including conservation, education, and the interests of hunters. Increasing involvement in recreational hunting and angling is likely to foster emotional relationships that evoke feelings of responsibility and stewardship (Shephard et al. 2024). Further, participation in outdoor recreation activities like hunting and wildlife viewing increases the likelihood of expressing pro-environmental behaviors and supporting conservation and efforts of conservation organizations (Cooper et al. 2015, Larson et al. 2015).

## 18.3 INDIGENOUS VALUES AND VIEWS OF MOUNTAIN SHEEP MANAGEMENT

The relationship between Indigenous Peoples in North America and mountain sheep is based on utilitarian purposes and a spiritual and cultural connection that has existed since time immemorial. Mountain sheep hold a sacred place in many Native American cultures, where they are believed to belong to the highest realm of the world, serve as the chief of all large game animals, and act as a bringer of rain and fertility (Garfinkel and Austin 2011). The Saulteau People believe that full-curl males have a social role as knowledge keepers and, biologically, are the most successful breeders, passing along ecological knowledge and genetics to the herd in a manner similar to how elders share knowledge within their community. By capturing the essence of the full circle of the horns, we are reminded to incorporate all ways of knowing and types of knowledge in our stewardship decisions, highlighting the interconnectedness of all, with no beginning, middle, or end (Saulteau First Nation 2022). In addition to meat, Indigenous Peoples also used harvested mountain sheep for clothing; their bones for tools and ribs for armor; their horns for vessels to carry liquids and fine materials, and when heated and shaped, they would make spoons, ladles, forks, knife handles, and fishhooks (Gwich'in Elders 1997, Kuhnlein and Humphries 2010, Chapter 2). Nations would also honor the species through carvings, songs, dances, and stories, particularly in the western mountainous regions where mountain sheep were a vital part of their livelihood and survival. In these cultures, social relationships with other-than-human beings are considered possible, leading to a respect for mountain sheep that exceeds their consumptive value. In some instances, they were even seen as superior to humans (Cross 1996) or as protectors who shared their abilities for climbing rugged terrain without falling, much like hunters who dreamed of the *divii* (Dall's sheep; Gwich'in Elders 1997).

The desert bighorn sheep (*O. c. nelsoni*) is known as Nah'gah among Southern Nevada tribes and is regarded as a protector of the tribes, believed to have sacrificed itself to provide food, medicine, shelter, tools, and clothing during hardship. The animal's physical features, such as its horns and hide, were also believed to carry protective and healing powers and were used in rituals to safeguard the community from harm. The Mountain Shoshone, also known as Sheepeaters, were semi-nomadic and followed the migrations of bighorn sheep, relying heavily on the species for food and materials (Price 2018). The bones of the bighorn sheep were fashioned into tools, while their hides were used for clothing and as a source of shelter. The Sheepeaters' intimate knowledge of bighorn sheep behavior and habitats allowed their culture and people to thrive in the harsh environments of what is now northwestern Wyoming. Their deep connection to the species is reflected in their spiritual practices, which often involved invoking the spirit of the bighorn sheep for guidance and protection (Loendorf and Stone 2006).

In Pueblo cultures, the desert bighorn sheep was seen as a spiritual guide, leading hunters to game and providing knowledge of the land. Pueblo art and pottery often feature depictions of the bighorn, symbolizing its role as a mediator between humans and the natural world (Scott 2015, Garfinkel et al. 2019). The bighorn's ability to navigate the rugged terrain of the mountains was admired and considered a metaphor for spiritual resilience and endurance (Scott 2015). The Hopi people, who inhabit the mesas of northeastern Arizona, also hold the bighorn sheep in high regard. The animal is featured in their kachina ceremonies, where masked dancers embody the spirit of the sheep to bring blessings of fertility and abundance to the community (Garfinkel et al. 2019). Among the Navajo, in the southwestern United States, the bighorn sheep has a special place in their mythology and ritual. The hunchback god, Ganskidi, shares characteristics with bighorns, has superpowers, and carries seeds of all vegetation in its hump (Grant 1980). Navajo legends often feature the bighorn sheep as a wise and noble creature with healing powers and control over hunting (Ami 2016). Rocky Mountain bighorn sheep have historically been the primary game species sustaining the Nez Perce people and are culturally and economically important for the current tribe along the Salmon River in Idaho (Mack et al. 2017).

Various Native American tribes, such as the Ute and Southern Paiute of Utah and northern Arizona, the Western Shoshone of eastern Idaho and northern Utah, and the Northern Paiute of Nevada and adjacent regions of Idaho, Oregon, and California, hunted bighorn sheep for food and other resources (Matheny et al. 1997). Their hunting techniques included surrounding the animals, constructing pitfalls, ambushing them near salt licks, driving them toward peaks or into enclosures and V-shaped fences, using fire or dogs to guide them, and running them past concealed hunters (Matheny et al. 1997). Additionally, the meat, bones, horns, and skin of mountain sheep were used to craft various tools and clothing (Matheny et al. 1997). In some Native American cultures, hunting bighorns was often accompanied by ceremonies intended to honor the animals and ensure their populations were not overexploited (Mykhailova and Garfinkel 2018). The legend of the Mountain Sheep Boy, shared by the Confederated Salish and Kootenai Tribes of the Flathead Reservation, ancestors of the people on the Flathead Reservation in Montana, illustrates the social and psychological principles of bighorn hunting, emphasizing

that breaking hunting rules could lead to the disappearance of the herds and the vital resources they provided (Cross 1996). Some Indigenous tribes, including the Nez Perce of the Pacific Northwest, practiced sustainable hunting methods. These included seasonal restrictions on hunting certain species to allow populations to recover, and rotational use of hunting grounds to prevent overexploitation (Hunn and Selam 1990). Similarly, the Shoshone had strict guidelines for bighorn hunting that involved the entire community. Elders and spiritual leaders were responsible for imparting knowledge about the appropriate times and methods for hunting and overseeing adherence to these practices (Loendorf and Stone 2006). These rules and taboos were designed to restrict rather than increase hunting, reflecting a deep cultural ethos of conservation and respect for the natural world (Cross 1996). Archaeological studies, however, occasionally reveal a decline in bighorn sheep remains at specific Native American hunting sites, potentially indicating overharvesting and the influence of other factors, such as climate change and evolving hunting practices (Speth and Spielmann 1983). Additionally, with the adoption of new hunting equipment, there are documented cases of overhunting and the subsequent depletion of game animals, including bighorn sheep, on reservation lands (Anderson 1996).

Peyton (2011) describes the evolution of cultural and commercial relationships involving northern thinhorn sheep populations and Indigenous Peoples in that region. With the commodification of wildlife resources, traditional First Nations relationships changed to resemble non-Indigenous adventurers, prospectors, ethnographers, and settlers who collectively pursued mountain sheep for a wide range of interests from subsistence use, to provisioning industrial camps, to servicing growing commercial and international trophy hunting markets. Hunting bighorn sheep by early European settlers also differed significantly from Native American cultural practices, particularly in terms of intensity and equipment. Settlers relied heavily on firearms, which increased the efficiency and range of their hunts compared to early traditional Indigenous methods (Geist 1971). Bighorn sheep were primarily hunted for their meat, and their populations began to decline most noticeably in areas that were easily accessible to humans, such as low-elevation habitats and regions near water sources (Buechner 1960). The hunting culture among early settlers emphasized immediate exploitation over sustainable practices, leading to widespread overharvesting and significant population declines in areas where bighorn sheep once thrived. The most important factors influencing bighorn sheep declines, however, were overhunting and the introduction of diseases transmitted by domestic sheep and goats that were brought by Europeans, and the expansion of the domestic sheep meat, dairy, and wool industry (Singer and Gudorf 1999, Gross et al. 2000). These pathogens and diseases, previously unknown to bighorn sheep, caused catastrophic mortality across their range (Hurley et al. 2015) and remain an obstacle to their continued restoration and management (Cassirer et al. 2018).

## 18.4  GOVERNANCE OF MOUNTAIN SHEEP RESOURCES

Multiple entities are engaged in mountain sheep conservation and management, including states, provinces, tribal governments, federal agencies, and non-governmental organizations. All share similar goals of promoting persistence and potential expansion of mountain sheep populations within landscapes that vary by private or government land ownership jurisdictions. The current governance system evolved from conservation concerns because of unregulated market hunting during the 19th century, amid resource exploitation, in which bighorn sheep ranges and populations declined to near extirpation, like other North American ungulates at the time (Chapter 2). The distribution of thinhorn sheep, according to all accounts, has not changed appreciably since historic times, as their habitat has remained mostly intact compared to that encountered by early European explorers (Demarchi and Hartwig 2004). As a result of overexploitation and possibly disease from domestic animals that were unknown at the time, the American Game Policy was enacted in 1930 (American Game Protection Association 1930), followed by the first federal excise tax to fund wildlife conservation, the Pittman-Robertson Act of 1937. Jurisdictions in Canada were also creating regulations at this same time that were focused on restoration and conservation of reduced numbers of mountain sheep. In the early 1970s, the development of the North American Wildlife Policy and the passage of the Endangered Species Act in the United States (Public Law 93–205) came. Each of these policy advances established the framework for developing a conceptual and historical narrative known as the NAM (Geist et al. 2001) and the Public Trust Doctrine. The NAM is rooted in a foundation of regulated, science-based take or harvest, to promote population and species sustainability. The seven general principles are: (1) wildlife resources are a public trust; (2) markets for game, shorebirds, and songbirds are eliminated; (3) allocation of wildlife is by law; (4) wildlife can only be killed for a legitimate purpose; (5) wildlife is considered an international resource; (6) science is the proper tool to discharge wildlife policy; and (7) hunting is democratic. The importance of ethics and stewardship in recreational hunting has its foundation in the social responses to overhunting during the colonial era (Peterson 2004). The NAM has served as the framework for continuing successful wildlife management, but it is changing. Since the adoption of the Declaration of the Rights of Indigenous Peoples by member countries of the United Nations in 2007, there has been a shift in the application of the seven principles listed in the NAM, to also include Traditional Indigenous Knowledge and Local Ecological Knowledge. Traditional Indigenous Knowledge has been defined as "a cumulative body of knowledge, practice, and belief, evolving through adaptive processes and handed down through generations by cultural transmission, about the relationship between living beings (including humans) and their environment" (Berkes 2000:1252). Local Ecological Knowledge is described as knowledge, practices, and beliefs regarding ecological relationships that are gained through shared experiences and interactions with local ecosystems and local resource users (Joa et al. 2018). Both types of knowledge can be incorporated into and improve the application of NAM principles. Indigenizing the NAM and Management interweaves various Indigenous worldviews and conservation practices (Hessami et al. 2021) across jurisdictions, as aspects of Indigenous governance are incorporated into government and agency policies, procedures, and regulatory frameworks.

### 18.4.1  STATE, PROVINCIAL, TERRITORIAL, FEDERAL, AND INDIGENOUS WILDLIFE MANAGEMENT AGENCIES

Land ownership and jurisdiction largely determine the lead management agency responsible for mountain sheep resources. State, provincial, and territorial wildlife agencies serve as the primary stewards and managers of wildlife resources within their administrative jurisdiction. As such, they are responsible for monitoring populations through research and surveys, regulating hunting activities through license sales and season limits, protecting critical habitats, mitigating human-wildlife conflicts, ensuring public safety, and educating the public on wildlife conservation practices. Agencies employ methods of Western science, seeking to use scientific data for decision-making to ensure sustainability of mountain sheep resources, and to strategize restoration actions accordingly, including translocations, habitat restoration, and predator management. Most managers are responsible for seeking out research studies and literature on mountain sheep to use in their management. Additionally, local, regional, and national conferences such as the Northern Wild Sheep and Goat Council Symposium, the Desert Bighorn Sheep Council meetings, and the North American Thinhorn Summit provide science resources, networking for mountain sheep managers and enthusiasts, and scientific foundation for management decisions. Funding for state, provincial, and territorial wildlife agencies comes from several sources, including federal dollars. Federal funding is distributed based on established criteria and a formula that is considerate of each state, province, or territory. In the United States funds are distributed as reimbursements, whereby a state covers a project cost first and then applies for reimbursement up to 75%. Traditionally, the sale of hunting and fishing licenses was thought to be the most important source of revenue for most wildlife agencies, followed by federal grants (e.g., the Pittman-Robertson and Dingell-Johnson Acts in the United States; Chapter 2). Due to the rarity of mountain sheep in some states and territories and some cases a once-in-a-lifetime opportunity to harvest each species of mountain sheep, permits can cost significantly more than other ungulate permits and, when offered at auction, can generate significant revenues that support program delivery. State wildlife agencies are advised by state wildlife commissions or boards, which are composed of elected and sometimes appointed officials. These officials do not have authority over agencies but serve in an advisory role. In practice, commissioners have considerable latitude to direct the business, actions, and funding allocations of state wildlife agencies, due to the inherently political nature of their positions. In Canada, formal stakeholder advisory committees, such as Renewal Resource Councils, function in a manner similar to state commissions and boards, with their roles and governance supported by provincial or territorial regulations.

First Nations people play an active role in conserving thinhorn and bighorn sheep populations, recognizing their ecological and cultural significance on landscapes their ancestors have long inhabited. They are also involved in the tourism industry, where mountain sheep are valued for hunting and wildlife viewing (Demarchi and Hartwig 2004). Almost all First Nation Tribes today have established natural resource agencies or wildlife agencies within their tribal governments who promote and manage wildlife resources on tribal lands. Indigenous governments often work with state, provincial, and territorial wildlife agencies on program deliverables such as mountain sheep population surveys, translocations, and habitat improvements. The Nez Perce of Idaho recognized bighorn sheep as a critically imperiled species and worked with local partners to institute bighorn sheep restoration programs (Nez Perce Tribe 2016). Taos Pueblo in New Mexico initiated the restoration of bighorn sheep to the Rio Grande Gorge, leading the initial translocation effort and continuing to work with New Mexico Game and Fish to conserve the population. These bighorns offer abundant recreational opportunities as they are easily viewed and hunted by the Pueblo and the state.

The rural community of Baja California, Mexico, has collaborated with conservation organizations, including the Wildlife Conservation Society and the Mexican Government, to manage desert bighorn sheep (Ruiz-Mondragón et al. 2023). In Mexico, hunting of bighorn sheep is permitted in the states of Sonora, Baja California Sur, Chihuahua, and Coahuila (Valdez 2019). Bighorn sheep hunting is legal only within Conservation Wildlife Management Units, which encompass most bighorn habitats. Local communal farmers manage about half of these units. A notable unit is on Tiburon Island, off the Sonoran coast, where the Indigenous Seri Indians oversee conservation efforts. Both community and Indigenous groups manage hunting expeditions, retain revenues, and conduct monitoring, anti-poaching, and wildlife management activities (Wild Sheep Foundation 2020).

In the Yukon, the Gwich'in First Nation has been involved in the conservation of thinhorn sheep through co-management agreements with the Yukon government and conservation organizations. Indigenous knowledge is utilized to monitor sheep populations, with the Gwich'in offering insights into traditional harvesting practices that prevent overexploitation of populations. This collaboration also includes efforts to mitigate the effects of climate change on sheep habitats and migration patterns (Environment Yukon 2019, Benson 2023). In 2019, the Province of British Columbia began drafting a management plan for thinhorn

sheep, which involved First Nations whose territories overlap with the sheep's range. Several First Nations, including Saulteau, Blueberry River, Taku River Tlingit, and Fort Nelson First Nations, joined a working group to develop a thinhorn Sheep Stewardship Framework that incorporates Indigenous knowledge and perspectives alongside scientific research. The framework aimed to guide the stewardship of thinhorn sheep in British Columbia and promote collaboration between Indigenous communities and the province (First Nations BC Wildlife Forum 2019).

Federal agencies are responsible for mountain sheep resources on federally managed landscapes (e.g., Department of Interior lands including national parks, fish and wildlife refuges, Bureau of Land Management rangelands, and national forests and grasslands; Zellmer et al. 2017), particularly in cases where wildlife preservation was the focus of enabling legislation and establishment of that land designation via statute. The reality, however, is that constructive intergovernmental cooperation among agencies and Indigenous governments is important for managing mountain sheep resources and is generally a reciprocal process. Agencies at all levels of governance need to participate collaboratively in decision-making that affects mountain sheep on state, provincial, territorial, federal, or tribal lands.

### 18.4.2 Non-Governmental Groups

The recovery of mountain sheep in North America is closely tied to contributions from non-governmental organizations, including recreational hunting organizations and non-hunting organizations. These groups have played an important role in the successful restoration and ongoing conservation of mountain sheep species. Historically, hunting communities have been instrumental in early efforts to restore wild sheep populations, supporting these goals both financially and through their participation in working groups associated with initial restoration programs (Hurley et al. 2015). These restoration efforts were sustained and expanded through various programs and conservation organizations, like the Wild Sheep Foundation. The Wild Sheep Foundation was originally established in 1977 by a group of wild sheep hunters, then known as the Foundation for North American Wild Sheep, and continues to collaborate with the Wild Sheep Initiative within the Western Association of Fish and Wildlife Agencies. The Wild Sheep Initiative is a collaborative working group of jurisdictional managers and researchers, who as science experts facilitate training and create scientific standards, guidelines and reference literature for wildlife managers, host meetings for wild sheep professionals to share research and management updates, and provide information to the public and science guidance for governments (https://wafwa.org/initiatives/wsi/, accessed Oct 2024). Non-governmental organizations support the management and growth of wild sheep populations and play an important role in educating the public and youth about the benefits of conservation and sustainable hunting practices. The Wild Sheep Foundation defines its mission as enhancing wild sheep populations, promoting scientific wildlife management, educating the public and youth on the conservation benefits of hunting, and advocating for the interests of hunters. This broad approach has contributed to the long-term success of wild sheep recovery efforts.

Many non-governmental organizations fund, support, and conduct habitat improvement projects for mountain sheep, thereby further supporting conservation efforts. In addition, 74% of the funding for wild sheep conservation by wildlife agencies comes from auction or raffle of conservation permits (Wild Sheep Foundation 2024). Non-governmental organizations facilitate most auctions and raffles. Some hunters support the welfare of rural communities by contributing to schools and medical facilities through winning bids on permits exceeding $100,000. In some cases, bidders donate their tags to raise additional funds or use their permits to capture animals for transplant programs (Lee 2011). In 2024, Colorado and New Mexico sold mountain sheep hunting permits for $600,000. In 2023, the Wild Sheep Foundation raised $2,730,000 by selling nine hunting permits (Wild Sheep Foundation 2024), and in 2025, a single state tag sold for 1.3 million USD. This infusion of non-government funding support, all of which was recirculated back to mountain sheep conservation, is an unparalleled investment in wildlife and ecosystem conservation.

### 18.4.3 Public Stakeholders

Others with a stake in management of mountain sheep include landowners with mountain sheep habitat, businesses such as hunting guides and outfitters, recreationists, the public, and residents of rural communities. Mountain sheep are considered watchable wildlife, and they generate increases in visitation rates to national parks and wildlife refuges. Recreation activities, such as hunting mountain sheep, hiking in mountain sheep ranges, or wildlife photography and viewing, generate financial resources for remote and rural communities where mountain sheep are found. Private landowners also often benefit from the stewardship of mountain sheep habitat, and in some states, they may receive private land hunting permits in exchange for habitat improvements or by granting public hunters access to their property. In New Mexico, private landowners can be provided private land bighorn hunting permits in exchange for opening their property to public, lottery-drawn bighorn sheep hunters. In 1986, Colorado implemented a pilot program called Ranching for Wildlife. Through this program, hunters play a role in managing bighorn sheep on private land. Participating landowners are required to improve habitat on their ranches for game and non-game animals in exchange for ranch-specific hunting permits issued to the landowner, which allow applicants to hunt on private ranches that are normally closed to the public. These programs seek to preserve and enhance open spaces for mountain sheep and other wildlife to live and thrive.

*Decision-making, public input, and information dissemination.* Obtaining and incorporating public and Indigenous input into the management decision-making process for mountain sheep is challenging but important for the long-term success of mountain sheep conservation. A wide range of connections, perceptions, and relationships with wildlife, their habitat, and competing societal interests will exist. Some segments of society may hold economic, recreational, and cultural viewpoints that directly conflict with the views of others, and the way the public evaluates and communicates about wildlife-related issues can differ widely (Byrd et al. 2017, Cole et al. 2024). To encourage input and participation in wildlife management, the public must be educated about wildlife management issues, the available and feasible actions, and the consequences of inaction. The public must also trust that the management and regulatory agencies involved understand and value their opinions.

Efforts to incorporate public input and participation would ideally be conducted through a structured, long-term public engagement process (Leshner 2003). This public process occurs for federal jurisdictions through the agency delivery of, for example, the United States National Environmental Policy Act, which requires solicitation from the public (public scoping) of new decisions and actions relative to wildlife and habitat resources on United States federal public lands. Typically, it requires the responsible federal land management agency to host public meetings and provide a period of time for written public input, which is considered and responded to before a management action decision of record is finalized. At the state level, solicitation and consideration of public input occur during state wildlife agency rule-making processes and through the establishment of state advisory boards. States are required to provide opportunities for public input on state wildlife action plans and comprehensive conservation plans. Comprehensive conservation plans are developed for national wildlife refuges and state wildlife action plans and aim to present a comprehensive overview of the state's wildlife conservation needs, proposed actions to address them, and a plan for monitoring progress and engaging the public. These plans are scheduled to be revised every 10 years. State wildlife agencies (e.g., ID, CO, NE) regularly develop and publish management action plans for mountain sheep. These documents highlight a need to conduct outreach activities, particularly with agricultural interests related to domestic sheep production, to minimize the spread of disease between domestic and wild populations. The importance of engaging these groups and working with them to identify mutually beneficial management approaches is consistently highlighted. These activities include promoting educational programs with 4-H and Future Farmers of America groups, working with domestic sheep industry groups to reach landowners directly, supporting and attending trade conventions and conferences, and developing educational materials that could be provided to domestic sheep producers (Idaho

Department of Fish and Game 2022, Nebraska Game and Parks Commission 2019).

In similar ways in Canada, engagement with the public and First Nations either through established Treaty and legal process, or through collaboration at the national or regional scales can be effective ways to achieve shared visions. Harvest regulations, including hunting seasons, quotas, and specific regulations for different subspecies (e.g., Dall's sheep, Stone's sheep), are implemented to ensure sustainable harvest and co-management with First Nations is used to ensure that management decisions reflect local knowledge and values. In this way, Traditional Indigenous Knowledge is merged with Local Ecological Knowledge for the benefit of thinhorn sheep habitat management and planning.

Many organizations advertise and hold webinars or presentations to get information out to the public. They also disseminate information through social media or email to engage stakeholders directly and post signs, brochures, or bulletins on public or private lands, and in publicly accessible office spaces. Bottom-up communication often occurs through formal institutional information collection efforts, such as public surveys or questionnaires, listening sessions, town-hall-style meetings, or open comment methods available over the phone or online. Colorado Parks and Wildlife conducts surveys of game management units for each managed bighorn sheep herd in the state (Colorado Parks and Wildlife 2024). Survey respondents were asked about their attitudes toward bighorn sheep conservation and management, the recreational activities they participate in related to bighorn sheep, preferences for specific management decisions and hunting opportunities. Across all surveys, respondents consistently rated the persistence of bighorn sheep populations in their game management unit as very important to them. Respondents also stated that management decisions that best support bighorn sheep should be prioritized over other interests, like agricultural or industrial priorities. Respondents also consistently preferred maintaining current hunting opportunities and sex ratios, even if it decreases the likelihood of drawing a license (Colorado Parks and Wildlife 2024).

## 18.5  SUMMARY

Human dimensions of wildlife are increasingly being integrated into wildlife management, and recognition of their importance in determining outcomes of wildlife management actions has grown substantially. The complex relationships people have with nature are incorporated into policymaking, regulation, and funding for wildlife management, guided by the principles outlined in the NAM. The NAM is rooted in the Public Trust Doctrine, which states that all wildlife should be managed as a public good. Institutions charged with managing wildlife at state, provincial, territorial, or federal levels operate as trustees, preserving wildlife for future generations and ensuring accessibility for future generations.

As iconic game species in North America, bighorn and thinhorn sheep have a long and dynamic history of interaction with humans. Mountain sheep are an important part of cultural heritage in North America for many different communities and Native American cultures. These majestic animals have long captivated human fascination. Economically, they have played a vital role in shaping how we manage our environments, and they continue to support local economies through ecotourism and regulated hunting, both of which contribute to conservation efforts and enhance local livelihoods. Bighorn sheep experienced a massive decline in range and population size during the 19th century, primarily due to human occupation of foothills and overgrazing by domestic sheep (Buechner 1960). This encroachment resulted in significant habitat loss, unregulated harvesting, and the spread of fatal diseases to bighorns (Toweill and Geist 1999, Chapter 2). Efforts to restore bighorn populations to their former ranges were largely influenced by the value of the species as recreational game species and for wildlife viewing (Hurley et al. 2015, Krausman 2000). This restoration brought many bighorn populations into direct contact with people because the remaining habitat was subject to urban expansions (Chase 2013). The dramatic population declines, fragmentation of habitat, and narrowing of bighorn ranges have had a profound effect on public opinion toward the species, shifting the animal from an accessible source of food through subsistence hunting to a species of conservation concern with limited harvest opportunities (Krausman and Bleich 2013). The role of the human dimension of wildlife and social science research will continue to be imperative to overcoming conflicts, maintaining support for investment in restoration efforts, and preserving bighorn and thinhorn sheep into the future.

Studies that investigate the relationship between humans and mountain sheep have largely focused on the implications of humans on mountain sheep including conflict and disturbance from human activity or development (Schoenecker and Krausman 2002, Polfus and Krausman 2012, Lowrey and Longshore 2017, Brushett et al. 2023), evolutionary pressure from selective hunting (Monteith et al. 2013, 2018; Douhard et al. 2016, Pigeon et al. 2016, Lassis et al. 2022), attitudes toward the management of mountain sheep and relevant predator species (Harris et al. 1995, Chase 2013), and human-wildlife conflict resulting from livestock disease spillover (Heinse et al. 2016). Human attitudes toward mountain sheep are understudied; however, hunters and wildlife viewers receive substantial positive emotional benefits from participating in their respective activities; others experience intrinsic benefit, which is inherently attributable to something existing, independent of any direct benefit to people; and still others continue to practice long-held spiritual and cultural practices closely tied to mountain sheep, even when hunting in some populations no longer occurs. Both economic and intrinsic values are important, especially for highly charismatic species, such as mountain sheep, which commonly appear in popular media and represent conservation success stories.

Although great strides have been made in restoring mountain sheep populations and habitat, the species is substantially reduced relative to historical population estimates and distribution (Buechner 1960). As a result, hunting for each species of mountain sheep in some jurisdictions is provided as an once-in-a-lifetime opportunity through lottery draws, special auctions, or raffles that tend to be more costly and harder to obtain than other ungulate species. The high cost of entry, the difficulty navigating the terrain and finding stalking opportunities, and low densities of the sheep create an inherently challenging environment that may increase the recreational value and prestige associated with successfully harvesting a sheep.

Mountain sheep populations have increased from approximately 25,000 in the 1950s to around 85,000 in 2023, due to national and international partnerships between provincial, state, territorial, and federal governments, and efforts by non-governmental organizations providing funding for translocations, research, habitat restoration, disease control, predator management, and educational initiatives. Tribal Nations and Indigenous governments across Canada, Mexico, and the United States have played a large and significant role in bighorn and thinhorn sheep restoration and their stewardship has been critical to stabilizing mountain sheep populations. The role of the human dimension of wildlife and social science research will continue to be imperative to overcoming conflicts, maintaining support for investment in restoration efforts, and preserving bighorn and thinhorn sheep into the future.

## REFERENCES

American Game Protection Association. 1930. Report to the National Game Conference on American Game Policy; A Proposed American Game Policy to be Discussed at the Seventeenth Annual American Game Conference, December 1–2, 1930, New York City, New York, USA.

Ami, C. M. W. 2016. *Dííjí Nída'iil'ah (Today, We Butcher): A Study of Navajo Traditional Sheep Butchering*. University of California, Davis, California, USA.

Anderson, T. L. 1996. Conservation—Native American Style PERC Policy Series Issue Number PS-6 July 1996.

Bennett, N. J., R. Roth, S. C. Klain, K. Chan, P. Christie, D. A. Clark, G. Cullman, D. Curran, T. J. Durbin, G. Epstein, A. Greenberg, M. P. Nelson, J. Sandlos, R. Stedman, T. L. Teel, R. Thomas, D. Veríssimo, and C. Wyborn. 2017. Conservation social science: understanding and integrating human dimensions to improve conservation. *Biological Conservation* 205:93–108.

Benson, K. 2023. Gwich'in Knowledge of Divii (Dall's Sheep). Gwich'in Tribal Council, Department of Cultural Heritage. Inuvik, Northwest Territories, Canada.

Berkes, F., J. Colding, and C. Folke. 2000. Rediscovery of traditional ecological knowledge as adaptive management. *Ecological Applications* 10:1251–1262.

Berkes, F., M. Kislalioglu, C. Folke, and M. Gadgil. 1998. Exploring the basic ecological unit: ecosystem-like concepts in traditional societies. *Ecosystems* 1:409–415.

Bleich, V. C. 2018. Maintaining momentum for conservation: bighorn sheep as an example. *Wildlife Society Bulletin* 42(3):540–546.

Boulanger, J. R., D. E. Hubbard, J. A. Jenks, and L. M. Gigliotti. 2006. A typology of South Dakota Muzzleloader Deer Hunters. *Wildlife Society Bulletin* 34:691–697.

Brewer, C. E., V. C. Bleich, J. A. Foster, T. Hosch-Hebdon, D. E. McWhirter, E. M. Rominger, M. W. Wagner, and B. P. Wiedmann. 2014. *Bighorn Sheep: Conservation Challenges and Management Strategies for the 21st Century*. Wild Sheep Working Group, Western Association of Fish and Wildlife Agencies, Cheyenne, Wyoming, USA.

Brookshire, D. S., L. S. Eubanks, and A. Randall. 1983. Estimating option prices and existence values for wildlife resources. *Land Economics* 59:1–15.

Brushett, A., J. Whittington, B. Macbeth, and J. M. Fryxell. 2023. Changes in movement, habitat use, and response to human disturbance accompany parturition events in bighorn sheep (*Ovis canadensis*). *Movement Ecology* 11:36.

Buechner, H. K. 1960. The bighorn sheep in the United States, its past, present, and future. *Wildlife Monographs* 4:3–174.

Byrd, E., J. G. Lee, and N. J. Olynk. 2017. Perceptions of hunting and hunters by U.S. respondents. *Animals* 7:83.

Cassirer, E.F., K. R. Manlove, E. S. Almberg, , P. L, Kamath, M. Cox, P. L. Wolff, A. Roug, J. Shannon, R. Robinson, R. B. Harris, and B. J. Gonzales. 2018. Pneumonia in bighorn sheep: Risk and resilience. *The Journal of Wildlife Management* 82:32–45.

Chase, L. D. 2013. Human dimensions of reintroduced bighorn sheep and an associated increased mountain lion harvest along an urban interface. *Desert Bighorn Council Transactions* 52:16–28.

Cole, N. W., E. J. Wilkins, K. R. Clements, R. M. Schuster, A. A. Dayer, H. W. Harshaw, D. C. Fulton, J. N. Duberstein, and A. H. Raedeke. 2024. Perceived constraints to participating in wildlife-related recreation. *Journal of Outdoor Recreation and Tourism* 45. https://doi.org/10.1016/j.jort.2023.100712

Colorado Parks and Wildlife. 2024. Bighorn Sheep – Conservation and Management. Colorado Parks and Wildlife. (Accessed Oct 2024).

Coltman, D. W., P. O'Donoghue, J. T. Jorgenson, J. T. Hogg, C. Strobeck, and M. Festa-Bianchet. 2003. Undesirable evolutionary consequences of trophy hunting. *Nature* 426(6967):655–658.

Cooper, C., L. Larson, A. Dayer, R. Stedman, and D. Decker. 2015. Are wildlife recreationists conservationists? Linking hunting, birdwatching, and pro-environmental behavior. *Journal of Wildlife Management* 79:446–457.

Cross, M. P. 1996. Bighorn sheep and the Salish world view: a cultural approach to the landscape. Graduate Student Theses, Dissertations, and Professional Papers. 1674.

Decker, D. J., R. W. Provencher, and T. L. Brown.1984. Antecedentes to hunting participation: an exploratory study of the social-psychological determinants of initiation, continuation, and desertion in hunting. Federal Aid in Wildlife Restoration, Project 1–146-R-8. Ithica, New Jersey, USA.

Decker, D., C. Smith, A. Forstchen, D. Hare, E. Pomeranz, C. Doyle-Capitman, K. Schuler, and J. Organ. 2016. Governance principles for wildlife conservation in the 21st century. *Conservation Letters* 9:290–295.

Decker, D. J., S. J. Riley, and W. F. Siemer. (Eds.). 2012. *The Human Dimensions of Wildlife Management* (2nd ed.). The John Hopkins University Press, Baltimore, Maryland, USA.

Demarchi, R. A., and C. L. Hartwig. 2004. Status of Thinhorn Sheep in British Columbia. B.C. Ministry of Water, Land and Air Protection, Biodiversity Branch, Victoria, British Columbia, Canada. Wildlife Bulletin No. B-119. 96pp.

Dolan, B. F. 2006. Water developments and desert bighorn sheep: "implications for conservation." *Wildlife Society Bulletin* 34:642–646.

Douhard, M., M. Festa-Bianchet, F. Pelletier, J. Gaillard, and C. Bonenfant. 2016. Changes in horn size of Stone's sheep over four decades correlate with trophy hunting pressure. *Ecological Applications* 26:309–321.

Ebeling-Schuld, A. M., and C. T. Darimont. 2017. Online hunting forums identify achievement as prominent among multiple satisfactions. *Wildlife Society Bulletin* 41:523–529.

Environment Yukon. 2019. Science-based guidelines for management of Thinhorn Sheep in Yukon (MR19-01). Whitehorse, Yukon, Canada.

Espinosa-T, A., A. V. Sandoval, and A. J. Contreras-B. 2006. Historical distribution of desert bighorn sheep (*Ovis canadensis mexicana*) in Coahuila, Mexico. *The Southwestern Naturalist* 51:282–288.

Festa-Bianchet, M., F. Pelletier, J. T. Jorgenson, C. Feder, and A. Hubbs. 2014. Decrease in horn size and increase in age of trophy sheep in Alberta over 37 years. *Journal of Wildlife Management* 78:133–141.

Festa-Bianchet, M., T. Coulson, J. M. Gaillard, J. T. Hogg, and F. Pelletier. 2006. Stochastic predation events and population persistence in bighorn sheep. *Proceedings of the Royal Society B* 273:1537–1543.

First Nations BC Wildlife Forum. 2019. Communication and education working group. Retrieved March 2025, from https://www.firstnationsbcwildlifeforum.ca/working-groups/communication-and-education-working-group/?utm_source=chatgpt.com. (Accessed Oct 2024).

Fulton, D., M. Manfredo, and J. Lipscomb. 1996. Wildlife value orientations: a conceptual and measurement approach. *Human Dimensions of Wildlife* 1:24–27.

Garfinkel, A. P., and D. R. Austin. 2011. Reproductive symbolism in Great Basin rock art: bighorn sheep hunting, fertility and forager ideology. *Cambridge Archaeological Journal* 21:453–471.

Garfinkel, A. P., T. Riley, R. Barlow, C. King, A. Rogers, R. Yohe II, … and R. Gerstner. 2019. The Green River Bighorn Sheep horned headdress, San Rafael Swell, Utah. *Rock Art Research: The Journal of the Australian Rock Art Research Association (AURA)* 36:29–42.

Geist, V. 1971. *Mountain Sheep: A Study in Behavior and Evolution* (pp. xv+-383pp). The Chicago University Press, Chicago, Illinois, USA.

Geist, V., S. A. Mahoney, and J. F. Organ. 2001. Why hunting has defined the North American Model of Wildlife Conservation. *Transactions of the North American Wildlife and Natural Resources Conferences* 66:175–185.

Gigliotti, L. M. 2000. A classification scheme to better understand satisfaction of black hills deer hunters: the role of harvest success. *Human Dimensions of Wildlife* 5:32–51.

Giles, R. H. 1978. *Wildlife Management*. W. H. Freeman and Company, New York, New York, USA.

Grant, C. 1980. The desert bighorn and aboriginal man. In: Monson, G., L. Sumner, and R. W. Allen (Eds.), *The Desert Bighorn: Its Life History, Ecology, and Management* (pp. 7–39). University of Arizona Press, Tucson, Arizona, USA.

Gross, J. E., Singer, F. J., and Moses, M. E. 2000. Effects of disease, dispersal, and area on bighorn sheep restoration. *Restoration Ecology* 8:25–37.

Gwich'in Elders. 1997. Nành' Kak Geenjit Gwich'in Ginjik - Gwich'in words about the land. Gwich'in Renewable Resource Board, Inuvik, Northwest Territories, Canada.

Harris, L. K., P. R. Krausman, and W. W. Shaw. 1995. Human attitudes and mountain sheep in a wilderness setting. *Wildlife Society Bulletin* 23:66–72.

Heffelfinger, J. R. 2018. Inefficiency of evolutionarily relevant selection in ungulate trophy hunting. *Journal of Wildlife Management* 82:57–66.

Heinse, L. M., L. H. Hardesty, and R. B. Harris. 2016. Risk of pathogen spillover to bighorn sheep from domestic sheep and goat flocks on private land. *Wildlife Society Bulletin* 40:625–633.

Hendee, J. C. 1973. A multiple-satisfaction approach to game management. *Wildlife Society Bulletin* 2(3):104–113. https://about.jstor.org/terms. (Accessed Oct 2024).

Hessami, M. A., E. Bowles, J. N. Popp, and A. T. Ford. 2021. Indigenizing the North American Model of Wildlife Conservation. *FACETS* 6:1285–1306.

Hinrichs, M. P., M. P. Vrtiska, M. A. Pegg, and C. J. Chizinski. 2021. Motivations to participate in hunting and angling: a comparison among preferred activities and state of residence. *Human Dimensions of Wildlife* 26:576–595.

Hunn, E. S., and J. Selam. 1990. *Nch'i-wana, "The Big River": Mid-Columbia Indians and Their Land*. University of Washington Press, Seattle, Washington, USA.

Hurley, K., C. Brewer, and G. N. Thornton. 2015. The role of hunters in conservation, restoration, and management of North American wild sheep. *International Journal of Environmental Studies* 72:784–796.

Husby, F. M., G. A. Mitchel, D. Grindle, and J. Hanscom, editors. 1998. Agroborealis 30:12, 22 (Spring). Agricultural and Forestry Experiment Station, University of Alaska, Fairbanks, USA. https://scholarworks.alaska.edu/handle/11122/1595. Accessed 27 October 2024.

Idaho Department of Fish and Game. 2022. Idaho Bighorn Sheep Management Plan.

Jacobson, C. A., J. F. Organ, D. J. Decker, G. R. Batcheller, and L. Carpenter. 2010. A conservation institution for the 21st century: implications for State Wildlife Agencies. *Journal of Wildlife Management* 74:203–209.

Jex, B. A., J. B. Ayotte, V. C. Bleich, C. E. Brewer, D. L. Bruning, T. M. Hegel, N. C. Larter, R. A. Schwanke, H. M. Schwantje, and M. W. Wagner. 2016. Thinhorn sheep: conservation challenges and management strategies for the 21st century. Wild Sheep Working Group, Western Association of Fish and Wildlife Agencies, Boise, Idaho, USA.

Joa, B., G. Winkel, and E. Primmer. 2018. The unknown known: a review of local ecological knowledge in relation to forest biodiversity conservation. *Land Use Policy* 79:520–530.

Jun, J., and G. T. Kyle. 2015. An identity-based conceptualization of recreation specialization. *Journal of Leisure Research* 47(4):425–443. https://www.researchgate.net/publication/280006712. (Accessed Nov 2024).

Kennedy, J. J. 1973. Motivation and rewards of hunting in a group versus alone. *Wildlife Society Bulletin* 2(1):3–7. https://about.jstor.org/terms. (Accessed Nov 2024).

Koenig, M. 2024. 2024 sheep and goat regulation brochure. Colorado Parks and Wildlife. https://spl.cde.state.co.us/artemis/nrserials/nr1430internet/nr14302024internet.pdf. (Accessed Dec 2024).

Krausman, P. R. 2000. An introduction to the restoration of bighorn sheep. *Restoration Ecology* 8(3):3–5.

Krausman, P. R., and V. Bleich. 2013. Conservation and management of ungulates in North America. *International Journal of Environmental Studies* 70:372–382.

Kuhnlein, H. V., and M. M. Humphries. 2010. Traditional animal foods of Indigenous Peoples of Northern North America. McGill University, Montreal, Quebec, Canada. https://traditionalanimalfoods.org/mammals/hoofed/page.aspx?id=6135. Accessed February 4, 2025.

Kuhnlein, H. V., and M. M. Humphries. 2017. Traditional animal foods of Indigenous Peoples of Northern North America: https://traditionalanimalfoods.org/. Centre for Indigenous Peoples' Nutrition and Environment, McGill University, Montreal, Canada.

Larson, L. R., R. C. Stedman, C. B. Cooper, and D. J. Decker. 2015. Understanding the multi-dimensional structure of pro-environmental behavior. *Journal of Environmental Psychology* 43:112–124.

LaSharr, T. N., R. A. Long, J. R. Heffelfinger, V. C. Bleich, P. R. Krausman, R. T. Bowyer, J. M. Shannon, R. W. Klaver, C. E. Brewer, M. Cox, and A. A. Holland. 2019. Hunting and mountain sheep: do current harvest practices affect horn growth? *Evolutionary Applications* 12:1823–1836.

Lassis, R., M. Festa-Bianchet, and F. Pelletier. 2022. Effects of hunting pressure and timing of harvest on bighorn sheep (*Ovis canadensis*) horn size. *Canadian Journal of Zoology* 100:507–516.

Lee, R. 2011. Economic aspects of and the market for desert bighorn sheep. *Desert Bighorn Council Transactions* 51:46–49.

Leopold, A. 1968. *A Sand County Almanac and Sketches Here and There*. Oxford University Press, New York, New York, USA.

Leshner, A. 2003. Public engagement with science. *Science* 299:977.

Loendorf, L. L., and N. M. Stone. 2006. *Mountain Spirit: The Sheep Eater Indians of Yellowstone*. University of Utah Press, Salt Lake City, Utah, USA.

Loomis, J., D. M. Donnelly, C. F. Sorg, and L. Oldenburg. 1985. Net economic value of hunting unique species in Idaho: bighorn sheep, mountain goat, moose, and antelope. U.S. Forest Service Resource Bulletin. RM-10.

Loomis, J. B., and D. S. White. 1996. Economic benefits of rare and endangered species: summary and meta-analysis. *Ecological Economics* 18:197–206.

Lowrey, C., and K. M. Longshore. 2017. Tolerance to disturbance regulated by attractiveness of resources: a case study of desert bighorn sheep within the River Mountains, Nevada. *Western North American Naturalist* 77:82–98.

Mack, C. M., M. R. Kasprzak, and K. Luiz. 2017. *Salmon River Bighorn Sheep Project Final Report 2007–2015*. Nez Perce Tribe, Lapwai, Idaho, USA.

Mahoney, S. P., and V. Geist. (Eds.). 2019. *The North American Model of Wildlife Conservation*. Johns Hopkins University Press, Baltimore, Maryland, USA.

Manfredo, M. J., B. L. Driver, and M. A. Tarrant. 1996. Measuring leisure motivation: a meta-analysis of the recreation experience preference scales. *Journal of Leisure Research* 28:188–213.

Manfredo, M. J., T. L. Teel, and A. D. Bright. 2003. Why are public values toward wildlife changing? *Human Dimensions of Wildlife* 8:287–306.

Martínez-Pabello, P. U., B. Menéndez Iglesias, R. López Martínez, T. Pi-Puig, J. Solé, A. Izaguirre Pompa, and S. Sedov. 2022. Lithodiversity and cultural use of desert varnish in the Northern Desert of Mexico. *Boletín de la Sociedad Geológica Mexicana* 74(3):A100622.

Matheny, R. T., T. S. Smith, and D. G. Matheny. 1997. Animal ethology reflected in the rock art of Nine Mile Canyon, Utah. *Journal of California and Great Basin Anthropology* 19:70–103.

McIntosh, D., and P. A. Wright. 2017. Emotional processing as an important part of the wildlife viewing experience. *Journal of Outdoor Recreation and Tourism* 18:1–9.

Messner, T. C. 2011. White-tailed deer management strategies and domestication processes. *Human Ecology* 39:165–178.

Monteith, K. L., R. A. Long, T. R. Stephenson, V. C. Bleich, R. T. Bowyer, and T. N. Lasharr. 2018. Horn size and nutrition in mountain sheep: can we handle the truth? *Journal of Wildlife Management* 82:67–84.

Monteith, K. L., R. A. Long, V. C. Bleich, J. R. Heffelfinger, P. R. Krausman, and R. T. Bowyer. 2013. Effects of harvest, culture, and climate on trends in size of horn-like structures in trophy ungulates. *Wildlife Monographs* 183:1–28.

Mykhailova, N., and A. P. Garfinkel. 2018. Horned hunter: shaman, ancestor, and deity. *Origin of Language and Culture: Ancient History of Mankind* 5:5–26.

Nebraska Game and Parks Commission. 2019. Nebraska Bighorn Sheep Management Plan. Outdoornebraska.gov,

Nez Perce Tribe. 2016. Plant and Wildlife Conservation Strategy of the Nimiipuu. Nez Perce Tribe, Department of Natural Resources, Wildlife Management Division, Lapwai, Idaho, USA.

Nox, M. W. 2011. The antler religion. *Wildlife Society Bulletin* 35:45–48.

Organ, J. F., V. Geist, S. P. Mahoney, S. Williams, P. R. Krausman, T. A. Batcheller, R. Decker, R. Carmichael, P. Nanjappa, R. Regan, R. A. Medellin, R. Cantu, R. E. McCabe, S. Craven, G. M. Vecellio, and D. J. Decker. 2012. The North American model of wildlife conservation. The Wildlife Society Technical Review 1–47.

Palazy, L., C. Bonenfant, J. Gaillard, and F. Courchamp. 2012. Rarity, trophy hunting, and ungulates. *Animal Conservation* 15:4–11.

Peterson, M. N. 2004. An approach for demonstrating the social legitimacy of hunting. *Wildlife Society Bulletin* 32:310–321.

Peterson, M. N., and M. P. Nelson. 2017. Why the North American model of wildlife conservation is problematic for modern wildlife management. *Human Dimensions of Wildlife* 22:43–54.

Peyton, J. 2011. Imbricated geographies of conservation and consumption in the Stikine Plateau. *Environment and* History 17:555–581.

Pigeon, G., M. Festa-Bianchet, D. W. Coltman, and F. Pelletier. 2016. Intense selective hunting leads to artificial evolution in horn size. *Evolutionary Applications* 9:521–530.

Polfus, J. L., and P. R. Krausman. 2012. Impacts of residential development on ungulates in the Rocky Mountain West. *Wildlife Society Bulletin* 36:647–657.

Price, C. R. 2018. What's for dinner?: A faunal analysis of the bison, elk, and bighorn sheep bones from the Windy Bison Site (48YE697), Yellowstone National Park, Wyoming, USA.

Purdy, K. G. 1981. Recreational use of desert bighorn sheep habitat in Pusch Ridge Wilderness. M.S. Thesis, University of Arizona, Tucson, Arizona, USA.

Rominger, E. M. 2018. The Gordian knot of mountain lion predation and bighorn sheep. *The Journal of Wildlife Management* 82:19–31.

Ruiz-Mondragón, E. D. J., G. Romero-Figueroa, R. Paredes-Montesinos, L. A. Tapia-Cabazos, L. A. Méndez-Rosas, C. S. Venegas-Barrera, … and E. A. Lozano-Cavazos. 2023. Community-based workshops to involve rural communities in wildlife management case study: bighorn sheep in Baja California, Mexico. *Animals* 13:3171.

Ryan, R. M., and E. L. Deci. 2019. Brick by brick: the origins, development, and future of self-determination theory. *Advances in Motivation Science* 6:111–156.

Ryan, R. M., and E. L. Deci. 2000. Intrinsic and extrinsic motivations: classic definitions and new directions. *Contemporary Educational Psychology* 25:54–67.

Sainsbury, K. A., H. W. Harshaw, D. C. Fulton, N. W. Cole, A. A. Dayer, J. N. Duberstein, A. H. Raedeke, R. M. Schuster, and M. P. Vrtiska. 2024. What waterfowl hunters want: exploring heterogeneity in hunting trip preferences. *Wetlands* 44:35.

Sandoval, A. V., R. Valdez, and A. Espinosa-T. 2019. Desert bighorn sheep in Mexico. In R. Valdez, and A. Ortega (Eds.), *Wildlife Ecology and Management in Mexico* (pp. 350–365). Texas A & M University Press, College Station, TX.

Saulteau First Nation. 2022. Traditional use study overview report: Treaty 8 BC thinhorn sheep. Saulteau First Nations Band, Moberly Lake, British Columbia, Canada. Unpublished report.

Schoenecker, K. A., and P. R. Krausman. 2002. Human disturbance in bighorn sheep habitat, pusch ridge wilderness, Arizona. *Journal of the Arizona-Nevada Academy of Science* 34(1):63.

Schroeder, S. A., D. C. Fulton, and J. S. Lawrence. 2006. Managing for preferred hunting experiences: a typology of Minnesota Waterfowl. *Wildlife Society Bulletin* 34:380–387.

Schroeder, S. A., L. Cornicelli, D. C. Fulton, and S. S. Merchant. 2018. Explicit versus implicit motivations: clarifying how experiences affect turkey hunter satisfaction using revised importance-performance, importance grid, and penalty-reward-contrast analyses. *Human Dimensions of Wildlife* 23:1–20.

Schroeder, S. A., L. Cornicelli, D. C. Fulton, and S. S. Merchant. 2019. The influence of motivation versus experience on recreation satisfaction: how appreciative-versus achievement-oriented recreation experience preferences relate to hunter satisfaction. *Journal of Leisure Research* 50:107–131.

Scott, S. T. 2015. *A Strange Mixture: The Art and Politics of Painting Pueblo Indians* (Vol. 16). University of Oklahoma Press, Norman, Oklahoma, USA.

Shephard, S., E. von Essen, T. Gieser, C. J. List, and R. Arlinghaus. 2024. Recreational killing of wild animals can foster environmental stewardship. *Nature Sustainability* 7:956–963.

Singer, F. J., and M. A. Gudorf. 1999. Restoration of bighorn sheep metapopulations in and near 15 national parks: conservation of a severely fragmented species; Volume I, Planning, problem definition, findings, and restoration (No. 99–102). US Geological Survey.

Singer, P. 1975. *Animal Liberation Now*. Harper Collins Publishers LLC, New York and London.

Skibins, J. C., J. C. Hallo, J. L. Sharp, and R. E. Manning. 2012. Quantifying the role of viewing the Denali "Big 5" in visitor satisfaction and awareness: conservation implications for flagship recognition and resource management. *Human Dimensions of Wildlife* 17:112–128.

Speth, J. D., and K. A. Spielmann. 1983. Energy source, protein metabolism, and hunter-gatherer subsistence strategies. *Journal of Anthropological Archaeology* 2: 1–31.

Stedman, R. and T. Heberlein. 2001. Hunting and rural socialization: contingent effects of the rural setting on hunting participation. *Rural Sociology* 66:599–617.

Stinchcomb, T. R., Z. Ma, and Z. Nyssa. 2022. Complex human-deer interactions challenge conventional management approaches: the need to consider power, trust, and emotion. *Ecology and Society* 27(1):13.

Swanson, G. A. 1987. Creation and early history. *Wildlife Society Bulletin* 15:9–14.

Teel, T., and M. Manfredo. 2010. Understanding the diversity of public interests in wildlife conservation. *Conservation Biology* 24:128–139.

Teel, T., M. Manfredo, and H. Stinchfield. 2007. The need and theoretical basis for exploring wildlife value orientations cross-culturally. *Human Dimensions of Wildlife* 12:297–305.

Toweill, D. E., and V. Geist. 1999. *Return of Royalty: Wild Sheep of North America*. Boone and Crockett Club and Foundation for North American Wild Sheep, Missoula, Montana, USA.

U.S. Department of the Interior, U.S. Fish and Wildlife Service. 2022. National Survey of Fishing, Hunting, and Wildlife-Associated Recreation. https://www.doi.gov. https://www.fws.gov.

U.S. Department of the Interior, U.S. Fish and Wildlife Service, and U.S. Department of Commerce, U.S. Census Bureau. 2016. National Survey of Fishing, Hunting, and Wildlife-Associated Recreation. www.commerce.gov. www.census.gov.

Valdez, R. 2019. *Wildlife Ecology and Management in Mexico* (1st ed.). Texas A&M University Press, College Station, Texas, USA. https://muse.jhu.edu/book/66869 (Accessed Oct 2024).

von Essen, E., E. van Heijgen, and T. Gieser. 2019. Hunting communities of practice: factors behind the social differentiation of hunters in modernity. *Journal of Rural Studies* 68:13–21.

Vucetich, J. A., J. T. Bruskotter, and M. P. Nelson. 2015. Evaluating whether nature's intrinsic value is an axiom of or anathema to conservation. *Conservation Biology* 29:321–332.

Wallingford, B. D., D. R. Diefenbach, E. S. Long, C. S. Rosenberry, and G. L. Alt. 2017. Biological and social outcomes of antler point restriction harvest regulations for white-tailed deer. *Wildlife Monographs* 196:1–26.

Whitehead, J. C., S. K. Pattanayak, G. L. Van Houtven, and B. R. Gelso. 2008. Combining revealed and stated preference data to estimate the nonmarket value of ecological services: an assessment of the state of the science. *Journal of Economic Surveys* 22:872–908.

Whiting, J. C., V. C. Bleich, R. T. Bowyer, and C. W. Epps. 2023. Restoration of bighorn sheep: history, successes, and remaining conservation issues. *Frontiers in Ecology and Evolution* 11:1083350. Frontiers Media S.A. https://doi.org/10.3389/fevo.2023.1083350

Wild Sheep Foundation. 2020. Community-based trophy hunting of Bighorn Sheep in Mexico. https://www.wildsheepfoundation.org

Wild Sheep Foundation. 2024. Wild Sheep Foundation conservation impact. Wild Sheep Foundation, Bozeman, Montana, USA. https://www.wildsheepfoundation.org/about-wsf. Accessed Oct 2024.

Wild Sheep Initiative. 2019. Wild sheep population estimate license harvest 1990–2018. https://wafwa.org/initiatives/wsi/.. Accessed March 20, 2025.

Zellmer, S. B., M. Nie, C. Barnes, J. Haber, J. Joly, and K. Pitt. 2017. Fish and wildlife management on federal lands: debunking state supremacy. *Faculty Law Review Articles* 182. https://scholarworks.umt.edu/faculty_lawreviews/182/. (Accessed Dec 2024).

# Section 3

*Habitat, Climate Change, and Conflicts*

# 19 Characteristics of Habitat for Mountain Sheep

*Paul R. Krausman*

## 19.1 INTRODUCTION

The simple definition of habitat for an organism is the place where it lives (Odum 1971:234). The definition is clear, but biologists must go further when discussing habitat in relation to wildlife management. Unfortunately, "…science had accumulated more knowledge of how to distinguish one species from another than of the habits, requirements, and inter-relationships of living populations" (Leopold 1933:20). Since Leopold's writings, habitat, biology, and human dimensions have become the cornerstones of contemporary wildlife management (Giles 1978). Most people understand basic habitat and recognize that it must include food, water, cover, and all necessary special factors required for life. Regardless, habitat use and selection by animals is complex and descriptive terms are often misused in the literature distorting communication among scientists and the public (Hall et al. 1997, Krausman and Morrison 2016). Thus, for clarity, my first objective is to define habitat, habitat use, habitat selection and selection processes, habitat preference, habitat availability, habitat quality, habitat type, and critical habitat as used in descriptions of mountain sheep (bighorn sheep [*Ovis canadensis*], thinhorn sheep [*Ovis dalli*]) habitat.

The other objective of this chapter is to discuss the important habitat components for bighorn and thinhorn sheep in North America: the distributions and elevations of habitats; topographic features; habitats used for thermal cover, escape (i.e., topography where sheep have a locomotive advantage over predators; Krausman and Bowyer 2003), lambing, watering, foraging, and mineral sources (i.e., mineral licks). Special factors that influence the habitats of mountain sheep are covered in Chapter 12. My goal is for the synthesis to be available as source material to better understand habitats for mountain sheep and to assist with future research.

## 19.2 COMMON HABITAT TERMS

### 19.2.1 HABITAT

Habitat includes resources and conditions in an area that facilitate occupancy, survival, and reproduction by a given organism. It is the sum of the specific resources (i.e., food, water, cover, special factors; Leopold 1933) needed by a species for survival and reproduction.

### 19.2.2 HABITAT USE

The way an animal uses the physical and biological resources in a habitat is habitat use (e.g., foraging, cover, lambing, escape, other life history traits).

### 19.2.3 HABITAT SELECTION AND SELECTION PROCESSES

Habitat selection is a hierarchical process involving a series of innate and learned behavioral decisions made by an animal about what habitat it would use at different scales within the environment (Hutto 1985) and whether they are active or inactive (Bose et al. 2018). Habitat selection is an active behavioral process by an animal. Selection has been defined as first-order selection (i.e., physical or geographical range of a species), second-order selection (i.e., the home range of an individual or group within their geographical range), third-order selection (i.e., how habitat components are used [foraging, bedding, breeding, parturition,…] within the home range), and fourth-order selection (i.e., how third-order selection is carried out [e.g., if third-order selection determines a foraging site, the fourth-order selection would be the actual procurement of forage items from those available at that site]; Johnson 1980).

### 19.2.4 HABITAT PREFERENCE

Habitat preference is the disproportional use of some resources over others. If components of habitat are used more than expected by chance, they are considered preferred.

### 19.2.5 HABITAT AVAILABILITY

True habitat availability of habitat to an animal is difficult to measure and often consists of *a priori* or *a posteriori* measure of abundance of resources in an area used by an animal instead of true availability (i.e., the accessibility and ability of an animal to obtain physical and biological components of a habitat by animals; Morrison et al. 1992, Wiens 1984, Morrison and Mathewson 2015).

### 19.2.6 HABITAT QUALITY

Habitat quality (i.e., the ability of the environment to provide conditions appropriate for individual and population persistence; Hall et al. 1997) is a continuous variable ranging from low (based on resources only available for survival), medium (based on resources available for reproduction), and high (based on resources available for population persistence).

### 19.2.7 HABITAT TYPE

Habitat type includes "all the areas (sum of discrete units) that now support, or within recent time has supported, and presumably is still capable of support in one plant association..." (Daubenmire 1968:32). Habitat type is often misused in literature (Dyksterhuls 1983, Hall et al. 1997) as a term describing wildlife habitat.

### 19.2.8 CRITICAL HABITAT

Critical habitat includes specific geographic areas that contain features essential for the conservation of listed species as designated by federal law (Salzman 1990); critical habitat is a legal term and not biological. Critical habitat should be linked to high-quality habitat, which would make it an operational and ecological term, not political (Murphy and Noon 1991).

### 19.3 HABITAT OF MOUNTAIN SHEEP

Bighorn and thinhorn sheep use an extensive array of landscapes from elevations 78 m below sea level in Death Valley National Park to peaks over 4,000 m in the White Mountains, California. They also use habitats above and below timberline, dry canyons and mesas, and river canyons crossing the Great Plains. The diversity of habitats is extensive. For example, Rocky Mountain bighorn sheep in Montana use as few as four habitats (Erickson 1972), other populations use six seasonal habitats (Geist 1971), and some use ≥15 different habitats (Hansen 1982, Kornet 1978). Many of the habitat descriptions are based primarily on vegetation associations without consideration of other habitat components (Van Dyke 1978). Mountain sheep are, however, generally dependent upon a combination of vegetative and topographic attributes such as steep, rugged cliffs and outcrops that provide natal range and escape terrain from predators, nearby areas suitable for foraging (Figure 19.1), watering areas, and numerous other habitat components (e.g., mineral licks) depending on local conditions.

Bighorn sheep habitats include grasslands, alpine, subalpine, shrub-steppes, rock outcrops, caves, cliffs, meadows, moist draws, stream sides, talus sides, plateaus, deciduous and coniferous forests, clear-cuts or burned forests, and conifer forests, among other landscape features (Valdez and Krausman 1999). When available, these different habitats meet different needs of mountain sheep, so the specific use bighorn make of different habitats varies daily and seasonally as their requirements for food, security, cover, mating, lambing (i.e., parturition and ~1 month post-parturition), and thermal regulation change (Valdez and Krausman 1999).

**FIGURE 19.1**   The author surveyed mountain sheep in Pusch Ridge Wilderness, Arizona, in 1985, in a habitat with rugged cliffs and slopes interspersed with foraging areas. (Photo by J. Hendee (deceased); used with permission.)

Their major needs include forage, water, thermal protection, and areas for escape, rutting, and lambing (Geist 1971, Kornet 1978, Valdez and Krausman 1999); all areas that include these habitat components should be protected for efficient management (discussed in Chapter 20). Some areas may only be used for 1 month or less out of the year (e.g., lambing habitat) but they are important for the persistence of herds and cannot be overlooked when managing populations. Ideally, mountain sheep habitat should include freedom from negative and cumulative human disturbance (Leslie and Douglas 1980), and proper juxtaposition of forage, escape terrain, and water. There are, however, populations of mountain sheep that do well sharing habitat with humans.

Generalizing about specific habitat requirements for mountain sheep is challenging because of the different factors involved including amount of use by humans, different weather patterns and a changing climate, hunting pressure, methods used in different studies, the presence of competitors (e.g., livestock) and predators, time periods of studies, variables measured, amount of disturbance (e.g., fire [Clapp and Beck 2016, Donovan et al. 2021], mining [Poole et al. 2016]), fixed wing aircraft, and other factors. These explanations seldom correctly consider use and behaviors in the sense of tolerance versus habituation (Harris et al. 2023) and trade-offs made by individual animals (i.e., male vs. female, young vs. old, dominant vs. subordinate, experienced vs. naive) in the context of risk-reward. For example, some populations of bighorn sheep avoid roads (Krausman et al. 1989, Smith et al. 1991, Papouchis et al. 2001), other populations show a preference for trails (Robinson et al. 2020), and the proximity to roads was not a consistent factor in habitat selection (DeCesare and Pletscher 2006). The difference is likely the amount of human use of roads and trails. When a preference for trails occurs, use by humans is low (Robinson et al. 2020) compared to higher human use in other populations. Also, consider cover, visibility, and distance to escape terrain within mountain sheep habitat. Most populations prefer areas where visibility is not impaired by vegetation (Risenhoover et al. 1988) and in only a few populations, visibility was not identified as an important habitat component (Dicus 2002, DeCesare and Pletscher 2006). Water is another important habitat component but is not universally a selection factor for mountain sheep. Many populations select areas with water but when water is abundant throughout the range, there is no consistent use between sheep and the proximity to water likely due to high abundance of water (DeCesare and Pletscher 2006) or moisture in vegetation (Ober 1931). Examining the individual races of bighorn and thinhorn reveals the complexity of their habitats based on numerous field studies (Ober 1931, Geist 1971, DeCesare and Pletscher 2006, Lula et al. 2020), modeling exercises (Smith et al. 1991, Bleich et al. 1992, Dicus 2002, Wockner et al. 2003), and habitat evaluation procedures (Smith et al. 1991, Dunn 1996, Johnson and Swift 2000, Zeigenfuss et al. 2000, Sappington et al. 2007, Thompson and Cain 2024).

## 19.4 BIGHORN SHEEP HABITAT

### 19.4.1 ROCKY MOUNTAIN BIGHORN SHEEP

*Distribution.* The historical distributions of mountain sheep (Chapter 1) have been reduced due to hunting, parasites and disease, competition with domestic livestock, and competition with humans for space (Smith 1954, Sudgen 1961, Van Dyke et al. 1983, Krausman 2017); many populations have been extirpated (Buechner 1960, Wagner 1978) due to alterations of habitat. Rocky Mountain bighorn sheep historically even used habitats far from rugged mountain terrain (Cowan 1940, Smith 1954, Wishart 1958, Buechner 1960, Shackleton 1985). For example, "…under conditions of abundance sheep habitually crossed extensive plains to any other range within their circle of vision" (Cowan 1940:579). Some populations once inhabited the river valleys and surrounding prairies east of the Rocky Mountains in Canada and the United States (Cowan 1940, Smith 1954, McCann 1956, Buechner 1960, Stelfox 1971), and occupied low-elevation ranges for most, or all, of the year (Smith 1954, Sudgen 1961, Drewek 1970, Morrison 1972, Ebert 1978, Whittaker et al. 2004). Rocky Mountain bighorn sheep are currently distributed from about 55°N latitude in Alberta and British Columbia, Canada, through Washington, Oregon, Montana, Idaho, to the Sierra Nevada in California (37°N), Utah, Wyoming, Colorado, and northern New Mexico, USA, at 36°N latitude. Chapter 1 provides detailed distributions of mountain sheep.

*Elevation.* Because of the wide distribution of Rocky Mountain bighorn sheep, their habitat ranges from 450 to >3,300 m (Buechner 1960, Berwick 1968, Riggs 1977, Morrison 1972, Valdez and Krausman 1999) with variable differences in individual populations. Some are resident (Smith 1954, Sudgen 1961, Drewek 1970, Spalding and Bone 1970) and others exhibit long annual elevational migrations (Smith 1954, Wishart 1958, Berwick 1968; additional details about migrations are presented in Chapter 11).

*Topographic features.* Bighorn commonly use river canyons, foothills, and mountains (Geist 1971, Adams et al. 1982, Shackleton 1985) that include rugged terrain (e.g., bluffs, canyons, gulches, mountain tops, outcroppings, river benches, steep slopes, talus cliffs; Buechner 1960, Stelfox 1975, Kornet 1978, Valdez and Krausman 1999, Robinson et al. 2020). Rugged terrain serves as important escape terrain (i.e., security cover) and is common to most mountain sheep populations (Geist 1971, Van Dyke et al 1983). Rugged terrain provides sheep with bedding areas and protection from predators (Frisina 1974, Geist 1971, Adams et al. 1982), especially during lambing (Blood 1961, Drewek 1970, Kornet 1978, Shank 1979, Akeson and Akeson 1992).

*Thermal cover.* Protection from weather elements is important for survival for bighorn sheep. Thus, they select habitats during adverse weather that benefit survival. For Rocky Mountain bighorn sheep, winter is an important season when sheep seek thermal cover. The best winter range indicated by modeling included solar radiation as an important habitat component (Dicus 2002). Increased solar radiation assists with the ability of sheep to obtain forage, thus they select windswept, southwest-facing grasslands at high elevations, close to escape terrain on warmer aspects (Poole et al. 2016). Coniferous and deciduous forests are used when strong winds and cold temperatures force them to seek shelter (Schallenberger 1966); landscapes are sparingly used during other times. Open forest stands can be important for thermal cover for shade and foraging in summer (McCann 1956, Sudgen 1961, Spalding and Bone 1970, Demarchi and Mitchell 1973, Pallister 1974, Wikeem 1984).

Because of warming climates Rocky Mountain bighorn sheep will be exposed to more days of extreme heat and fewer days of extreme cold (Thomas-Kuzilik 2020) but are able to change behaviors by modifying habitat selection and movement characteristics with increasing temperatures (Thomas-Kuzilik 2020; Chapter 21). More information will be needed as to how thermal refugia is selected as temperatures increase (Hetem et al. 2014).

*Escape terrain.* Numerous types of escape terrain have been described (Van Dyke et al. 1983) and the size of escape areas relates to use. For example, cliffs that provide bed sites and escape terrain may not be suitable for lambing areas, which should be ≥2 ha. Cliffs <8 m high by 200 m long provide bedding and thermal areas by providing shade in summer and reflect the sun or dissipate absorbed heat in winter. In winter, cliffs have less snow compared to flatter areas and cliffs can radiate sunrays that melt snow and keep it soft, making it easier for bighorn sheep to remove snow to obtain the limited forage available (Van Dyke et al. 1983).

Escape terrain or escape cover may be more important for females than males because males can defend themselves against predators better than females (Shank 1979) and females must make trade-offs between forage benefits and predation risk in selecting habitats related to escape terrain (Festa-Bianchet 1989). Females are generally within 100 m of escape terrain during lambing in spring, but males generally ranged >200 m from escape terrain (Hansen 1982, Leslie and Douglas 1979). The distance sheep are separated from escape terrain is generally <500 m (Gionfriddo and Krausman 1986, Smith et al. 1991, Discus 2002) but distance to escape terrain varies. Most (95%) bighorns were within 300 m of escape terrain in Utah, and in other assessments of habitat of Rocky Mountain bighorn sheep, animals were within 400 m of escape terrain (Beecham et al. 2007). In other populations, Rocky Mountain bighorn sheep generally remain within 800 m from escape terrain during all seasons (Pallister 1974) and spend ≥86% of their time within 100 m of rocky escape terrain and 75% of their time while foraging within 100 m of rocky escape terrain (Oldemeyer et al. 1971, Erickson 1972).

The best winter range indicated by modeling in Montana included slope, distance to escape terrain, solar radiation, snow cover, and cover types (Dicus 2002). Similar results occurred in Colorado where escape terrain, visibility, and continuous forage were important components of bighorn sheep habitat (Risenhoover et al. 1988). In southwest

Montana female bighorn sheep used rugged slopes and steep terrain, areas with limited canopy cover, ridgelines, and southwest aspects in summer and winter. In winter females used lower elevations than in summer and areas with a higher normalized digital vegetation index (i.e., a measure of vegetation abundance; Lula et al. 2020). Similar use of habitats by bighorn sheep (i.e., escape terrain, slope, aspect, ground cover) occurs in the Highland-Pioneer Mountains, Montana (Semmens 1996), and except in winter bighorn sheep selected for south-, east-, and west-facing slopes. In northwestern Montana, bighorn sheep preferred cliffs with >80% slopes and were within 322 m of steep terrain including rockland-scree. They preferred shrubland-grassland associations and open forests (Tilton 1977, Tilton and Willard 1982).

Although bighorn sheep use various habitat components, slope and distance to escape terrain are universally important habitat variables to bighorn sheep based on field studies (Geist 1971, Tilton and Willard 1982, McCarthy and Bailey 1994) and modeling exercises (Sweanor et al. 1996, Hughes 1997, Johnson and Swift 2000, Forbes 2001, Dicus 2002, DeCesare and Pletscher 2006), although other factors are important locally. As with escape terrain, the use of slopes varies between study areas. Adequate slopes were defined as areas with 27% slopes (Geist 1971, Zeignefuss et al. 2000, DeCesare and Pletscher 2006). Slopes of sheer cliffs and rocks were rarely used because they could not be traversed by sheep (Lula et al. 2020, Thompson and Cain 2024), but sheep in northwest Montana used slopes >80% (Simmens 1996). The maintenance of habitat components should be a primary activity for management to ensure productive bighorn sheep herds (Risenhoover et al. 1988, Chapter 20).

*Lambing habitat.* Defining lambing habitat is important for the conservation, reproductive biology, and perpetuation of bighorn populations (Etchberger and Krausman 1999, Wiedmann and Sargeant 2014, Smith et al. 2015, Brushett et al. 2023). Thus, predator avoidance influences the selection of neonate lambing habitat. For example, female Sierra Nevada bighorn sheep (*O. c. sierrae*) selected lambing habitat where the relative probability of encountering mountain lions (*Puma concolor*) was low and near escape terrain. When the risk of encounters with predators was low, females selected shrub cover suggesting they may have selected areas with dense shrubs to reduce risk from other predators (e.g., coyotes [*Canis latrans*], golden eagles [*Aquila chrysaetos*]). As the risk of mountain lions increased, females selected barren ground cover where they could detect and avoid mountain lions, thus avoiding mortality (Forshee et al. 2022). In South Dakota, lambing habitats were relatively flat spaces within rugged terrain, close to water, on south- and west-facing slopes, with minimal anthropogenic disturbance (Smith et al. 2015). In southwest Montana, females and lambs used steeper slopes and sites that are closer to escape terrain than during other periods of the year (Semmens 1996). In Green River, Utah, lambing habitat consisted of barren vegetation, slopes, ruggedness, north-facing slopes, and a preference for areas away from roads. In the intermountain West, lambing habitat included steep, north-facing slopes, rugged terrain, and low elevations, with an avoidance to roads (Robinson et al. 2020). Other populations also included steep slopes and short distances to escape terrain as important habitat explanatory variables but aspect was not identified as important (DeCesare and Pletscher 2006). In Banff National Park, Alberta, Canada, parturient females had a stronger selection for low snow depth, sites closer to barren ground, and further from trails. Immediately following lambing females selected high elevation sites on steeper aspects, with more rugged terrain, closer to escape terrain, and further from anthropogenic disturbances (e.g., trails; Brushett et al. 2023).

The importance of lambing habitat is emphasized in translocations. Translocations were more successful when Rocky Mountain bighorn sheep were placed in habitat that contained >10% of suitable lambing habitat. Further, the rate of population growth was best predicted by the amount of lambing habitat, winter range, and distance to domestic livestock (Zeigenfuss et al. 2001).

By understanding where and when bighorn sheep give birth managers can determine ways to avoid disturbance (e.g., camping near rivers, canyoneering, hiking, mountain biking, off-highway vehicles, rafting, rock climbing; Papouchis et al. 2001, Longshore et al. 2013, Sproat et al. 2019, Brushett et al. 2023) in land-use planning (Robinson et al. 2020). In many lambing areas, outdoor recreation and energy development have continued to increase since 1980 (Smith and Burr 2011, Sproat et al. 2019). Hiking has a negative effect because of the unpredictable locations of hikers approaching bighorn sheep, especially in spring when females are giving birth (MacArthur et al. 1979, Papouchis et al. 2001, Wiedmann and Bleich 2014).

When bighorn sheep are disturbed during lambing, population survival may decrease and females may move into less suitable areas (Papouchis et al. 2001, Longshore et al. 2013, Wiedmann and Bleich 2014, Smith et al. 2015) resulting in abandonment of previously used lambing habitat, which increases predation risk for lambs (Papouchis et al. 2001), and lower recruitment of young, resulting in a declining population (Papouchis et al. 2001, Wiedmann and Bleich 2014).

*Water.* Rocky Mountain bighorn sheep are well adapted to arid conditions and can subsist for long periods without free-standing water because they can meet their water requirements from succulent vegetation in summer and from snow and ice in winter (McCann 1956, Kornet 1978, Van Dyke 1978, Shackleton et al. 1999). Water did not limit Rocky Mountain bighorn sheep itself (McCann 1956, Hansen 1982, Kornet 1978) but investigations of water use and Rock Mountain bighorn persistence have not been examined as much as other habitat components (Shackleton et al. 1999). Since 1999, however, more research has been conducted on Rocky Mountain bighorn sheep in arid areas (Whiting et al. 2009a,b; Whiting et al. 2011). When Rocky

Mountain bighorn occur in arid areas, they spent more time around water in summer but not during droughts (Whiting et al. 2009a). Following drought, water use did increase following times of similar precipitation, indicating a time-lag in water use. The time-lag could be due to increased water in forage used by sheep, underscoring the importance of water to Rocky Mountain sheep in these arid habitats.

The mean time for sheep to use water in arid areas was 14:22 for all sheep regardless of drought or non-drought periods (Whiting et al. 2009b). On Antelope Island, Utah, there was no difference between sex and age groups drinking at catchments. There is still much to be learned about relationships between water and Rocky Mountain bighorn sheep but in the Great Basin the importance of water has been established (Whiting et al. 2011). In addition, there have been numerous advances in the development of waters for bighorn sheep (Bleich et al. 2005) that can assist managers in ensuring proper water placement for Indigenous and translocated herds. Water and bighorn sheep relationships are discussed in more detail in Chapters 7 and 12.

*Foraging habitat.* Foraging areas used by bighorn sheep are generally within 800–1,600 m of escape terrain (McQuivey 1978, Van Dyke 1978, Leslie and Douglas 1979, Bailey 1980) but are not usually key determinants in describing Rocky Mountain bighorn habitat. Many studies concentrate more on other factors than vegetation and when vegetation is mentioned it is often related to cover and visibility of sheep or distance to different vegetation associations (e.g., distance to open canopy forests, burnt grasslands, burned forests; DeCesare and Pletscher 2006). Horizontal visibility of Rocky Mountain bighorn sheep can be limited by vegetation and is often a key component measured in habitat studies (Risenhoover and Bailey 1980, 1985; Fairbanks et al. 1987, Smith et al. 1991, Johnson and Swift 2000, Zeigenfuss et al. 2000) that influence sheep use of areas in some populations but not in others (DeCesare and Pletscher 2004, 2006).

Vegetation structure is often more important to Rocky Mountain bighorn sheep than the specific plant species composition. Open habitats that yield high visibility were used most by bighorn sheep regardless of the vegetation association (Risenhoover and Bailey 1985, Wakelyn 1987). If fire or other disturbances are prevented, vegetation succession that reduces visibility can cause Rocky Mountain bighorn sheep to abandon habitats (Wakelyn 1987) in some populations but not others (DeCesare and Pletscher 2004). Juxtaposition of landscapes is also important. Even if bighorn sheep can detect predators in open habitat, they rarely use them if escape cover is not available (Shackleton et al. 1999).

In general, open grasslands of wheatgrass (*Agropyron* spp.), fescues (*Festuca* spp.), bluegrasses (*Poa* spp.), mesquite (*Prosopis* spp.), grasses (*Bouteloua* spp.), muhlys (*Muhlenbergia* spp.), needle grasses (*Stipa* spp.) and ridegrasses (*Oryzopsis* spp.), forbs, and shrubs are used for winter ranges based on an array of studies (Todd 1972, Shannon et al. 1975, Shackleton et al. 1999, Donovan et al. 2021).

Winter foraging areas are often on alpine areas that are dominated with grasses, kobresia (*Kobresia* spp.), and sedges (*Carex* spp.; Shackleton et al. 1999). Subalpine areas used by Rocky Mountain bighorn often include sedges, grasses, rushes (*Juncus* spp.), a variety of forbs, and some shrubs (Pallister 1974, Shackleton et al. 1999). On intermediate elevations bighorn sheep used moderate stands of bunchgrasses interrupted with scattered trees, shrubs, rock outcrops, and non-caespitose grasses (Hudson et al. 1976).

Because of the variety of landscapes used by Rocky Mountain bighorn sheep, the vegetation they use is also varied. In general bighorn sheep are opportunistic and eat most plants that are available to them. The diets and nutrition of mountain sheep are covered in Chapter 7.

## 19.4.2 DESERT BIGHORN SHEEP

*Distribution.* Desert bighorn sheep occur across several ecologically distinct life zones and are adapted to rugged, arid, and sparsely vegetated desert environments. Historically they ranged from Nevada south to Coahuila, Mexico, and from western Texas, southern New Mexico, and Arizona, western Chihuahuan, Colorado, and Utah west to California (Monson 1980). Since 2000, genetic information has been used to obtain a more detailed understanding of historical ranges (Espenosa-T. et al. 2006, Epps et al. 2010, Buchalski et al. 2016). Desert bighorn sheep currently inhabit the Chihuahuan, Colorado, Great Basin, Mojave, Painted, Sonoran, and Vizcaino-Magdalena deserts

(Hansen 1980). Throughout this vast area, desert bighorn sheep are in isolated areas characterized by landscapes that provide a high degree of visibility (e.g., cliffs, deep canyons, rock outcrops; Risenhoover and Bailey 1985). More detailed distribution information is in Chapter 1.

*Elevation.* The elevational preference of desert bighorn sheep is influenced by numerous variables that reflect the animal's environmental, physiological, and behavioral preferences (Krausman et al. 1999, Krausman and Bowyer 2003). Desert bighorn sheep use elevations from 78 m below sea level in Death Valley, California (Welles and Welles 1961), to >4,267 m above sea level in the White Mountains, California (Kovach 1979, Shackleton et al. 1999). Few populations remain at the same general elevation year-round (Sudgen 1961, Spalding and Bone 1970, Figure 19.2). The location of water was the major factor influencing elevational distribution in Arizona, especially in summer. As temperatures decreased into fall, and sheep were less dependent on surface water, sheep would disperse to cooler ranges at higher elevations (Simmons 1969). In other populations in Arizona (i.e., Kofa National Wildlife Refuge) females (84%) and lambs (93%) used the upper third of mountain ranges while males used the middle (62%) and lower slopes of mountains (9%). Conversely, males used higher elevations (1,830–1,981 m) in the San Andres Mountains, New Mexico, than females and lambs (1,525–1,667 m), reflecting local availability and use of various elevations. Elevation was an important factor in defining space use in relation to escape terrain in Arizona (Hoglander et al. 2015).

**FIGURE 19.2** Typical mountain used by desert bighorn sheep with various elevations, slopes, drainages, and escape terrain interspersed with foraging areas, Big Hatchet Mountains, New Mexico, 2015. (Photo by E. Rominger, New Mexico Department of Game and Fish, used with permission.)

*Topographic features.* Desert bighorn sheep prefer topographic landscapes with precipitous cliffs, rock outcrops, and steep and rugged topography across their range; landscapes vary from steep or gentle slopes, broken cliffs, rock outcrops, canyons, river benches to the tops of mesas that are often within 200 m of escape terrain (Buechner 1960, Sudgen 1961, Wilson 1968, Welch 1969, Drewek 1970, Merritt 1974, Stelfox 1975, McQuivey 1978, Leslie and Douglas 1979, Sandoval 1979, DeForge 1980, Holl and Bleich 1983, Etchberger et al. 1989, Krausman et al. 1989, Wakeling and Miller 1989, Alvarez-Cárdenas et al. 2001). Although use of topographic features varies among seasons and diel periods (Gedir et al. 2020), desert bighorn sheep commonly use slopes >80% (Guerrero-Cárdenas et al. 2003). Some of the earliest descriptions of desert bighorns stated they were "Bold and somewhat reckless climbers, that are capable of feats that really are astonishing" (Ober 1931:35). Desert bighorns are "…miraculous climbers, jumping here and there as they traverse the rocky ledges" (Ober 1931:35). Ruggedness in habitat selection by bighorn sheep was consistent across mountain ranges in Arizona (independent of slopes; Sappington et al. 2007) and New Mexico (close to topography with 60% slopes; Bangs et al. 2005b). During periods of average precipitation in Arizona, desert bighorn sheep selected topographic features beneficial for predator avoidance (i.e., escape terrain) and locations near water. During drought, sheep moved farther from those sites, likely in search of nutrients that would provide moisture (Gedir et al. 2020). Desert bighorn sheep use of habitats was not associated with gentle slopes, large

boulders, and sheetrock compared to areas that were not used (Krausman and Leopold 1986a). Slopes and river benches are commonly used for foraging and the precipitous landscapes are used as escape terrain. Desert bighorn sheep (especially females when giving birth; Blood 1961, Drewek 1970, Kornet 1978, Holl 1982, Akeson and Akeson 1992) are rarely far from escape terrain because it provides them the best defense against predators (Geist 1971, Oldemeyer et al. 1971, Erickson 1972, Pallister 1974, Shank 1979, Hansen 1982, Gionfriddo and Krausman 1986, Krausman and Leopold 1986a, Valdez and Krausman 1999).

Seasonal selection of habitats was documented in Arizona (Jones et al. 2022). In winter, females and males generally selected areas where visibility was reduced, and there were moderate slopes of moderate ruggedness, increased distance to water, with greater solar radiation. In winter, females selected ridgetops over all other topographic positions. In summer, males and females selected areas with less vegetation obstruction that were closer to desert scrub and grasslands, and moderately rugged and steep terrain. Females also selected canyon bottoms, steep slopes, and ridgetops with high values of normalized digital vegetation indices and increased distance to water (Jones et al. 2022, Figure 19.3).

*Thermal cover.* The wide range of temperatures desert bighorn are exposed to (i.e., −29°C in winter, ≥49°C in summer; Hansen 1980, Sandoval 1979, Cain et al. 2008) influence their activities and make thermal cover a serious consideration (Ober 1931, Chilelli and Krausman 1981, Gedir et al. 2020). Desert bighorn sheep need to

**FIGURE 19.3** Rugged terrain typical of Cabeza Prieta Mountains, Cabeza Prieta National Wildlife Refuge, Arizona. (Photo by J. W. Cain, used with permission.)

maintain their body temperature within physiologically acceptable limits and minimize water loss (Feldhamer et al. 1999). Desert bighorn sheep cope with high temperatures by using a suite of mechanisms (e.g., physiological, morphological, behavioral) to minimize heat stress and dehydration (Schmidt-Nielsen 1979, Cain et al. 2006). Selective use of habitats is a major contributor to reduction of heat stress and dehydration when temperatures are high. During hot, dry summers, heat stress is minimized by bedding in shade for long periods (Simmons 1969), limiting activities to times that reduce daily heat loads and water loss, diet selection (for succulents), body orientation, foraging and watering when slopes are shaded, use of north-facing slopes with lower solar radiation (Gedir et al. 2020), and use of microhabitats (i.e., caves, vegetative cover; Krausman 1979, Sargeant et al. 1994, Cain et al. 2006, Figures 19.4–19.7). Bighorn sheep bedded in shade 43%–85% of the time during cool and hot seasons, respectively (Simmons 1969, Cain et al. 2008). For example, in Arizona and other arid areas where caves are available, the thermal benefits of shade from caves and trees cannot be overlooked (Cain et al. 2008). Carmen Mountain white-tailed deer (*Odocoileus virginianus carminis*) used caves in Big Bend National Park (Krausman 1979) and they may have also been used by desert bighorn sheep. In some caves fecal pellets from ungulates were >23 cm deep and some may have been from bighorn sheep before they were extirpated by 1960 (Cook 1994). Bighorn sheep were in the Chisos Mountains in 1905 (Bailey 1905) but were extirpated by the 1960s (Cook 1994) and have recently returned (Brewer and Hernandez 2011). The thermal load for desert bighorn sheep was lower in caves and under cover than control sites; caves provided more shade than canopies. The thermal benefits of shade use in caves and under trees increased with higher daily temperatures (Cain et al. 2008). In addition, the surface of shaded areas in summer were on average 7.8°C lower than unshaded areas (Simmons 1969, Cain et al. 2008).

*Escape terrain*. Escape terrain is a universal topographic habitat component for desert bighorn sheep and sheep rarely occur far from escape terrain (Oldemeyer et al. 1971, Erickson 1972, Pallister 1979, Shank 1979, Hansen 1982, Gionfriddo and Krausman 1986, Krausman and Leopold 1986a, Valdez and Krausman 1999, Tarango et al. 2002, Sappington et al. 2007). Escape terrain is used for lambing, bedding, and other foraging because of protection offered by ruggedness. Forage quality did not influence the use of escape cover, however (Steel and Workman 1990). Bed sites were five times greater in escape terrain when compared to areas without cliffs (Welch 1969).

Escape terrain consists of rugged landscapes, talus slopes, cliffs, rocky outcrops, and about any rugged terrain where desert bighorn sheep have an advantage in escaping and observing potential predators. The degree of slope, amount of escape terrain, amount of rock cover, abundance of cliffs, group size, slope orientation, and juxtaposition of components all play a part in the effectiveness of escape terrain. Desert bighorn sheep commonly use slopes ≥60% (Tilton and Willard 1982, Holl 1982, McKinney et al. 2003). Escape terrain is especially important for

**FIGURE 19.4** Desert bighorn sheep seeking shelter from an overhang, Pecos Wilderness, New Mexico, 2000. (Photo by E. Rominger, New Mexico Department of Game and Fish, used with permission.)

FIGURE 19.5   Desert bighorn sheep hiding in caves during aerial surveys in the Sierra Pinta Mountains, Cabeza Prieta National Wildlife Refuge, Arizona. (Photo in the public domain from the Arizona Game and Fish Department.)

FIGURE 19.6   Caves commonly used as thermal cover by desert bighorn sheep, Cabeza Prieta Mountains, Cabeza Prieta National Wildlife Refuge, Arizona. (Photo by J. W. Cain, used with permission.)

females and young. Descriptions of escape terrain vary depending on location, but most include the importance of rugged slopes that can range from 21% (Sandoval 1979) in New Mexico to ≥80% (Alvarez-Cárdeas et al. 2001) in Mexico. The distribution of female-lamb groups on slope gradients between 21% and 40% was higher than on slope gradients between 5% and 20%. Barren females preferred slopes between 21% and 40% and males preferred slopes between 5% and 40% (Sandoval 1979). In California and Arizona there are significant relationships between the amount of escape terrain and the number of female desert bighorn sheep; the size of the female segment of the population is directly proportional to the amount of escape cover (Holl 1982, McKinney et al. 2003).

Additional habitat components are also important if escape terrain is an effective habitat to avoid predation.

**FIGURE 19.7** Caves commonly used as thermal cover by desert bighorn sheep, Sierra Pinta Mountains, Cabeza Prieta National Wildlife Refuge, Arizona. (Photo by J. W. Cain, used with permission.)

There was a direct correlation between the amount of escape terrain and bighorn sheep use in Utah. Over 60% of sightings of bighorn sheep were in steep terrain where rock cover varied between 71% and 100% (Wilson 1968). Cliffs accounted for 23% of available habitat in the San Andres Mountains, New Mexico, but were used 70% of the time by desert bighorn sheep (Sandoval 1979).

The area and configuration of escape terrain were important in determining population size in Arizona. There should be >15 km² of escape terrain (McKinney et al. 2003) within habitats with a continuous buffer around them of 150 m (Locke et al. 2005b). Group size is also important as escape terrain is used for predator avoidance (Mooring et al. 2004). Increased group size decreased the risk of mortality but was only marginally reduced at ten sheep/group (Jones et al. 2022).

Juxtaposition of escape terrain with other habitat features is also important. For example, in the Superstition Mountains in Arizona, escape terrain close to water and canyon terraces were preferred (Wakeling and Miller 1989). The location of adjacent patches of escape terrain and the distances between patches was also an important determination of use (Dunn 1996). In New Mexico, bighorn sheep locations were on steep, rugged terrain, close to topography with 60% slopes, that provided lower visibility than random sites (Bangs et al. 2005b). In Arizona, desert bighorn sheep selected areas with lower horizontal obstruction, and intermediate rugged slopes, which decreased the risk of mountain lion predation (Jones et al. 2022). There was, however, no evident relationship between the amount of horizontal obstruction and mountain lion predation on desert bighorn sheep (Jones et al. 2022). Models incorporating distance to water, escape terrain, and a normalized digital vegetation

index as metrics related to desert bighorn sheep use of habitats indicated that escape terrain was the dominant predictor for intensity of space use year-round (Hoglander et al. 2015). The importance of escape terrain to desert bighorn sheep has been demonstrated from early observational studies (Oldemeyer et al. 1971, Erickson 1972) to modeling exercises where sophisticated methods of measuring metrics have been used (Locke et al. 2005b, Blum et al. 2018, Hoglander et al. 2015, Jones et al. 2022); ruggedness is key.

*Lambing habitat.* Early descriptions of lambing habitat were primarily observational (Eustis 1962), and because parturition sites are used for only 2–4 days, biologists may have missed the actual parturition period. With the advent of technology (e.g., global positioning system-collars, vaginal implant transmitters), biologists in this century have the advantage of tools to precisely determine parturition periods. Yet, field biologists were accurate in describing lambing habitat as the "…most dangerous and inaccessible crags and precipices imaginable" (Ober 1931:36).

The time surrounding lambing is associated with habitat selection trade-offs between the risk of predation and obtainment of nutrition, both of which influence fitness (Berger 1991, Karsch et al. 2016, Blum 2021, Forshee et al. 2022, Blum et al. 2023a). Selection for habitats that reduce risk of predation often overrides the obtainment of nutrients in the late stages of pregnancy (Berger 1991). Descriptions of lambing sites vary depending on habitat features and methods used to define habitat. Many earlier studies based on observations concluded that parturition and nursery sites were in areas characterized by steep, rugged terrain at high elevations, with high horizontal visibility (Hansen 1980, DeForge and Scott 1982, Shackleton et al. 1999). In other populations, lambing habitat was not different from other sites used throughout the year (Locke 2003) and differences reported between use of pre- and post-partum sites in the Peninsular Ranges were related to different use by different female groups (Hines 2019). In the Piloncillo Mountains in New Mexico, parturition sites were at mid-elevations with intermediate slopes (Karsch et al. 2016).

Lambing occurs coincident with the period of vegetation development from Canadian National Parks to west Texas based on 22 (Thompson and Turner 1982) and 30 (Bunnell 1982) populations. Lambing in northern populations is restricted to brief, relatively predictable periods of vegetation growth (Robinson and Turner 1982). In southern populations lambing occurs over longer periods to coincide with unpredictable plant growth that is influenced by erratic precipitation (Bunnell 1982, Robinson and Turner 1982). For example, lambing in northern populations extends from 14 May±2.8 days to 17 June±3.9 days for a duration of 33.7±4.6 days. In desert bighorn sheep populations, lambing begins 24 January±16.5 days and ends 24 June±14.0 days with a duration of 151.5±24.4 days (Bunnell 1982). In some southern areas lambing is even longer; lambing in the Sonoran Desert, Arizona, occurred in every month except October (Witham 1983). Peak lambing was 17 May±6.8 days for all populations and departures are related to the availability of vegetation (Bunnell 1982, Douglas 2001). Most (75%) births in the Sonoran Desert occurred from January to March (Witham 1983).

Vegetation is certainly an important component of lambing as is the selection of sites that minimize predation. Predation sites were more common at low elevations with high vertical visibility and at high elevations with low visibility due to different predators (Karsch et al. 2016) in New Mexico. In the Peloncillo Mountains 12 of 14 lamb deaths were due to predation (i.e., coyotes, gray fox [*Urocyon cinereoargenteus*], mountain lion). Five of 14 lambs were killed before they were 1 week old (Cain et al. 2019). In the Fra Cristobal Mountains, New Mexico, the vertical visibility for bighorn sheep was lower at kill sites compared to control sites, and slope, elevation, ruggedness were lower at kill sites. Kill sites were closer to water and anthropogenic features (e.g., roads) than random sites (Parsons 2007). In the Peninsular Ranges, desert bighorn sheep used areas farther from urban areas during the peak months of parturition (Rubin et al. 2002). It is thus important for desert bighorn sheep to carefully select habitats where the risk of predation is minimized.

Prior to parturition (last trimester) females selected areas with less risk of predation but lower quality vegetation than those that were not pregnant (Blum et al. 2023a). Females used relatively flat areas with rugged terrain close to water on south- and west-facing slopes that were undisturbed by humans (Smith et al. 2015). Predation risk is a key determinant of lambing habitat. When the risk of predation by mountain lions was high, females selected pre-parturient sites near escape terrain. When the risk of mountain lion predation was low, they selected areas with shrub cover as protection against coyotes (Forshee et al. 2022).

Actual use of parturition sites is limited and often 2–4 days (Bangs et al. 2005a, Karsch et al. 2016). Parturition sites for desert bighorn sheep were higher in elevation, more rugged than pre-parturition sites, steeper, with less visibility than random sites and more hiding cover when lambs were less mobile and subject to higher rates of predation in New Mexico (Bangs et al. 2005a). In more southern New Mexican ranges, parturition sites included intermediate slopes, at lower elevations with increased visibility when compared to random sites (Karsch et al. 2016). Parturition sites in Arizona were similar to other sites used throughout the year with similar slopes, topographic positions, vegetation composition, but did have more thermal cover during the short parturition period (Etchberger and Krausman 1999).

During post-parturition, females and lambs used similar habitats to parturition sites (Bangs et al. 2005a) in New Mexico. In other ranges, post-parturition sites were secure from predators but had lower quality vegetation (Blum et al. 2023a). Nursery groups in southern New Mexico were intermediate in elevation and slope and parturition sites were <2 km from nursery areas (Karsch et al. 2016). Earlier descriptions of parturition sites (Hansen 1980, DeForge and Scott 1982, Shackleton et al. 1999) may have been post-parturition sites because females and lambs only stayed in parturition sites 2–4 days (Karsch et al. 2016). Lambs selected areas with greater cover than random sites and avoided north-facing slopes to avoid predation and enhance solar radiation (Smith et al. 2015). Following the death of a lamb, females selected lower elevations, in rugged terrain, with steep slopes, and close to water (Blum et al. 2023a).

There is limited evidence of fidelity to parturition sites. Only one of 38 females in the Fra Cristobal Mountains, New Mexico, returned to a parturition site the following year (Bangs et al. 2005a). In the Harquahala Mountains, Arizona, birth sites were within 300 m of those used by the same female in previous years (Etchberger and Krausman 1999). There was no evidence of parturition site fidelity in southern New Mexico (Karsch et al. 2016). There was also no evidence that >1 female used the same parturition site (Etchberger and Krausman 1999, Bangs et al. 2005a, Karsch et al. 2016).

The time spent at pre-parturition, parturition, and post-parturition sites was relatively short and the mean time to return to pre-parturition levels (e.g., without dependent young) varied for different habitat components but was from approximately 4 to 108 days (Blum 2021, Blum et al. 2023a). Returning to habitats with similar habitat components ranged from 2 to 15 days for the selection of tree cover and slope, 84–120 days for selection of grass and forb cover, and 41–75 days before they returned to similar distances to water, rugged terrain, perennial grass and forb cover, and shrub cover (Blum et al. 2023a).

*Water.* Because water is a limiting factor to many desert bighorn sheep populations, it has been researched more than in other populations of mountain sheep. Water is often considered a primary factor limiting the distribution and productivity of desert bighorn sheep (Leslie and Douglas 1979). Because water is often limited within desert bighorn sheep habitat, the importance of surface water has been a major research emphasis across their range (Blong and Pollard 1968, Turner 1973, 1979; Campbell and Remington 1979, Douglas and White 1979, Leslie and Douglas 1979, Sandoval 1979, Witham and Smith 1979, Turner and Weaver 1980, Valdez and Krausman 1999, Etchart 2021, Glass et al. 2022). Water in the form of precipitation can determine the distribution and production of forage, which is an important requirement for desert bighorn sheep survival in the Southwest (Russo 1956, Dolan 2006). Desert bighorn sheep can survive on preformed water from their food, metabolic water formed from oxidative metabolism, and free sources of water on the landscape. Thus, wildlife management agencies and sportsmen's organizations have invested substantial time and resources in the construction and maintenance of water catchments (Halloran and Deming 1958, Rosenstock et al. 1999, Cain et al. 2008) and most Western states have programs to enhance water for desert bighorn sheep and other wildlife. Although the availability of free-standing water sources influences many aspects of the ecology of desert bighorn sheep, the efficacy of these water sources has been questioned (Broyles 1995, Broyles and Cutler 1999, 2001; Dolan 2006, Rosenstock et al. 2001, Longshore et al. 2009, Whiting et al. 2011) and their continued construction and maintenance has become controversial (Broyles and Cutler 1999, Czech and Krausman 1999, Rosenstock et al. 2001), creating wicked issues in areas managed for desert bighorn sheep and naturalness (e.g., national monuments, parks; Longshore et al. 2009). Much of this controversy is because of the lack of experimental studies specifically designed to assess the influence of these water sources on ungulate populations and the conflicting results of observational studies involving use of water catchments, animal distribution, and habitat use relative to these water sources (Leslie and Douglas 1979, Krausman and Leopold 1986a,b; Krausman and Etchberger 1995). In addition, some populations of desert bighorn sheep occupy areas without perennial water sources, and it has been suggested that these animals can obtain sufficient water from forage (Ober 1931, Mendoza 1976, Watts 1979, Alderman et al. 1989, Warrick and Krausman 1989). The controversy about the importance of water in desert bighorn sheep ranges was revitalized in the 1990s (Broyles 1995, Broyles and Cutler 1999) by questioning the usefulness of maintaining artificial water sources for wildlife leading to a suite of research demonstrating the value of water catchments to desert bighorn sheep and other wildlife. For example, critics of supplementing water for desert bighorn sheep claimed that because there were populations in areas without free-standing water, it was not needed and water quality in catchments may be detrimental to wildlife, catchments are designed for single species, there is the possibility of mortality in catchments, enhancement of predator populations increases in predator populations, and competition would be increased (Broyles 1995, Rosenstock et al. 1999,

Simpson et al. 2011). None of the detrimental aspects of water developments have been supported (Simpson et al. 2011). For example, water quality in catchments was not detrimental to the health of wildlife populations based on numerous water catchments where water quality was tested (Bleich et al. 2006). In the Cabeza Prieta National Wildlife Refuge, Arizona, the availability of perennial surface water compared in areas with and without water was not different in relative abundance, lamb:female ratios, yearling:female ratios, or male:female ratios between 1986 and 1996 (Broyles and Cutler 1999). In a later study of the effects of removal of water in the range of desert bighorn sheep in the Cabeza Prieta, removal of water catchments in the treatment range did not result in predicted changes in diet, foraging area selection, home-range size, movement rates, mortality, productivity, or recruitment. Female desert bighorn sheep did use areas with more thermal cover during summer after removal of water catchments, but other characteristics of foraging areas used by bighorn sheep and their diet did not change appreciably with removal of water catchments (Cain et al. 2008). There were no recorded changes in home-range area, movement rates, or distance sheep were from water during hotter months; there were documented changes in home-range area, movement rates, and distance to water catchments during winter and autumn.

There was no increased mortality or a change in lamb:female or yearling:female ratios after removal of water catchments. Home-range area and movement rates declined with increasing precipitation. Annual survival rates increased with increases in the current year's total precipitation and total precipitation during the previous year; annual survival rates declined with increases in average daily temperature during winter.

There was a severe drought during pretreatment and abnormally wet conditions during posttreatment. The increase in precipitation that coincided with removal of water sources improved forage conditions during posttreatment and may have provided adequate water for female desert bighorn sheep. The lack of change in home-range size, movement rates, and distance to the nearest water catchment during hot, dry seasons after removal of water sources suggests that forage conditions played a greater role in determining home-range area and movement rates than did the presence of water catchments. Higher mortality rates during the drought of the pretreatment period indicates that during droughts as severe as that of 2002, presence of water catchments was not sufficient to prevent mortalities of desert bighorn sheep and a lack of forage quality and quantity was likely the primary limiting factor of the population during this time. Improving forage conditions during posttreatment, increases in forage moisture content, and availability of naturally occurring sources of free water in the treatment range likely minimized any effect of removing water catchments on survival rates and lamb:female and yearling:female ratios. Due to the climatic conditions during the study researchers could not assess how the treatment population would have responded to a lack of water sources during a drought period. The influence of anthropogenic water catchments, if any, on desert bighorn sheep populations may be strongest during years with weather conditions that are neither drought nor wet periods. Given the high interannual variability in precipitation in many areas of the arid Southwest, it is important for wildlife managers to determine when water is the primary limiting factor for populations of desert bighorn sheep to make the most efficient use of agency time and resources (Cain et al. 2008).

Regardless of the controversy surrounding the use of surface water by desert bighorn sheep there are some commonalities that have been uncovered over the years related to water and desert bighorn sheep. First, local meteorological conditions are important factors indicating when water use by desert bighorn sheep occurs (Waddell et al. 2007). Desert bighorn used waters during high temperatures and low precipitation (Foster et al. 2005, Harris et al. 2020, Terry et al. 2022). When moisture was high in desert bighorn sheep forage, sheep did not use catchments. Sheep avoided waters when used by predators or other competitors and escape terrain was needed near waters (Terry et al. 2022). The maximum daily temperature did closely predict the percent of bighorn sheep using water in the Mojave Desert (Harris et al. 2020, Glass 2022). Visitation by desert bighorn sheep increased 30% between 30°C and 40°C and shifted tortuous foraging movement to directed movement to obtain water. Maximum daily temperature, day length, minimum daily temperature, solar radiation, and the interaction between day length and minimum temperature were the best predictors of the distance sheep were from water (Glass et al. 2022).

Second, water catchments alone are not adequate to enhance the habitat of desert bighorn sheep. In the San Andreas National Wildlife Refuge, population size and trends in desert bighorn sheep were best described by models that only included total annual precipitation (Bender and Wisenberger 2005). The work of Bender and Wisenberger (2005) was criticized, however by Rominger et al. (2008) because Bender and Wisenberger (2005) did not consider the long-term predator control used to enhance the population, they reported lamb-female ratios year-round instead of with the dates associated with recruitment so the influence of precipitation on lambs was unknown. Further, the model predictions were not supported with data and the interpretation of model results was not correct, indicating that precipitation was a weak prediction of population increases. Bender and Weisenberger (2009) responded to the criticism that Rominger et al. (2008) oversimplified desert bighorn sheep population dynamics and exaggerated the effects of predator control and went on to justify their model. Exchanges like this demonstrate the importance of long-term research with experimental controls (Chapter 23). Other researchers included more than a single factor as important to populations.

In southwest Arizona, escape cover, forage, and water were all important in hot, dry seasons (Locke et al. 2005a,

Hoglander et al. 2015). In some cases, forage with adequate moisture may supply all the necessary water needed for sheep (Gedir et al. 2016). For example, the percent moisture content of forage in the Little Harquahala Mountains, Arizona, was >32% throughout the year and for any given hour (Alderman et al. 1989). Also, in Arizona, cacti such as barrel cactus (*Ferocactus* spp.) supplemented the water in the diet (Warrick and Krausman 1989). In the Cabeza Prieta National Wildlife Refuge, when water was removed in a treatment area, researchers examined water and nutrient intake and metabolic water production from dry matter intake and forage moisture and nitrogen content of forage to see if sheep could meet their nutritional and water requirements for forage alone. Water removal did not influence diet and under drought conditions without surface water, males and females could only meet their daily water requirements in winter (Gedir et al. 2016). For sheep to achieve water and nutrient balances year-round, they would need to shift their diet proportions by 8%–55% from lower to higher moisture and nitrogen forage species. In these cases, supplemental water may not be necessary (Gedir et al. 2016).

Third, habitat use can be enhanced with the addition of surface water. The addition of water to desert bighorn sheep habitats in the Sonoran Desert can increase habitat from 1.16 to 5.08 times (Bleich et al. 2010). In Johsua Tree National Monument, California, there was 583 km² of summer habitat for desert bighorn sheep. Due to droughts and anthropogenic influences the available water has been reduced and the summer habitat was reduced by 51.7%. Without supplemental water, nearly half of summer habitat will be lost (Longshore et al. 2009). In addition, surface water benefited sheep in the Mojave Desert. Their distribution was strongly influenced by the presence of water, more so than precipitation, availability of rugged slopes, and almost as much as temperature (Rich et al. 2019).

Fourth, water use may or may not be disrupted by anthropogenic influences. In the River Mountains, Nevada, construction near primary watering sites caused nine of 17 females to alter their watering patterns. In contrast, in the Little Harquahala Mountains, Arizona, desert bighorns did not alter their habitat use or use of water prior to, during, or after the construction of the Hayden-Rhodes Aqueduct (Krausman and Etchberger 1995).

Fifth, desert bighorn sheep make trade-offs between the use of water and other habitat components. For example, with average precipitation in the Cabeza Prieta National Wildlife Refuge, Arizona, desert bighorn sheep selected topographic features beneficial for predator avoidance and near water. During drought, sheep moved farther from escape terrain and water to obtain forage with higher moisture contents (Gedir et al. 2020).

Sixth, when considering the use of supplemental waters, managers interested in efficiency should consider the different sex and age groups of sheep using them. When sheep are segregated, males and females used different sources of water. Males visited water sources used by females more during aggregation (Whiting et al. 2010). Both sexes used water more in summer and males used water more during rut. Both sexes did not use water differently between drought and non-drought seasons but used water more in seasons following drought. Males and females should be considered separately in determining placements of waters (Whiting et al. 2010).

Finally, water catchments are rarely detrimental to desert bighorn sheep except in unusual situations. For example, desert bighorn sheep fell into a 6,500 L catchment, drowned and decomposed. Thirteen drowned and the water was contaminated with decomposing carcasses causing the death of 32 others from *Clostridium botulinum* poisoning in the Mojave Desert, California (Swift et al. 2000).

Although some bighorn sheep populations occupy ranges with free-standing water, and others occupy ranges without free-standing water, most biologists agree that more and longer research across the range of habitats used by bighorn sheep will be important to ferret out the specific advantages and disadvantages of supplying water to wildlife (Broyles 1995, Rosenstock et al. 1999, Cain et al. 2008, Simpson et al. 2011, Whiting et al. 2011). Given the current controversy over building water catchments to provide water for desert ungulates in the arid southwestern United States (Broyles and Cutler 1999, Rosenstock et al. 2001), more information is needed regarding the efficacy of water catchments, the physiological responses of desert bighorn sheep to dehydration, and the population-level consequences resulting from the presence or absence of these water sources. Further studies of the water turnover, minimum water requirements, and physiological responses of desert ungulates to dehydration will be needed, especially in the face of warming climates.

*Foraging habitat.* Foraging habitat and vegetation associations used by desert bighorn sheep are used differently depending on season, sex, year (Krausman et al. 1989, Tarango et al. 2002), precipitation, group size, distance from escape terrain, behavior, visibility (Risenhoover and Bailey 1985, Bender and Wisenberger 2005, McKinney et al. 2006a,b; Sesnie et al. 2012), presence of livestock (Garrison et al. 2016), predation risk and trade-offs (Berger 1991, Gedir et al. 2020, Blum et al. 2023b), availability of surface water (Bleich et al. 2010), presence of urban parks (Longshore et al. 2016), mining activity (Jansen et al. 2006), other anthropogenic disturbances (Lowrey and Longshore 2017), and quality of forage (e.g., %N; DeYoung et al. 2000, McKinney et al. 2006a,b) among other factors (O'Farrill et al. 2019). Desert bighorns forage on nearly every plant within their domain and although not mentioned by name, desert bighorns were observed to ride the green wave as vegetation greens up (Ober 1931). The amount of precipitation alone in the San Andres Mountains, New Mexico, was the best predictor of the size of the desert bighorn sheep population and trend because precipitation regulated vegetation growth that desert bighorn sheep depend on (Bender and Wisenberger 2005). In the River Mountains, Nevada, desert bighorn sheep lamb recruitment was directly influenced by November precipitation. When >30% of the population

began using a municipal park along the urban wildland interface in 1985 when the park was developed, desert bighorn sheep obtained water and forage from the park and November precipitation was not as important. Weather is clearly an important component of foraging habitat along with the required physical features of habitat (Bender and Wisenberger 2005).

When foraging habitat is compared between areas with cattle and without cattle, biomass of browse was 4.7 times lower on cattle grazed ranges. Sheep also had to travel longer distances to obtain forage in cattle-dominated ranges (Garrison et al. 2016). Only the most rugged areas (areas cattle did not use) had adequate forage for desert bighorn in New Mexico. Cattle forced bighorn to increase foraging efforts by feeding in the few areas where adequate forage was available (Garrison et al. 2016).

Because of the numerous variables influencing desert bighorn sheep foraging areas (e.g., latitude, precipitation, elevation, exposure, land-use practices; Valdez and Krausman 1999) it is not surprising that the most common vegetation in the diet of desert bighorn sheep varies across their range. In the Harquahala Mountains, Chihuahuan Desert, and Tiburón Island, Mexico, and other parts of their range, browse was the dominant vegetation in the diet followed by grass and forbs, and succulents (Krausman et al. 1989, Tarango et al. 2002, Brewer and Harveson 2007, O'Farrill et al. 2019). In central Arizona, however, forbs dominated the diet (McKinney et al. 2006b).

For foraging, desert bighorn sheep generally use open grasslands and shrub-steppe communities at different elevations and various slopes. Openness is important for the detection of predators and while foraging, sheep are usually near escape terrain. Open grasslands contain forbs, shrubs, sedges, and grasses such as fescues (*Festuca* spp.), wheatgrasses (*Agropyron* spp.), and rice grasses (*Oryzopsis* spp.) that sheep forage on during different seasons (Jaeger 1957, Jorgensen and Turner 1975, Kovach 1979, Shackleton et al. 1999).

In deserts, vegetation is sparse and plants are widely spaced because of the demands of their root systems in shallow soils and their necessity of being able to withstand droughts that can last for years (Douglas and White 1979, Kelly 1979, Leslie and Douglas 1979, Watts 1979, Krausman et al. 1989). Foraging habitat is naturally varied from the lowland browse types in lower elevations of California, cacti in Baja California and Sonora, Mexico (Flores et al. 1972, Alvarez 1976), through pine, subalpine, and alpine associations (McQuivey 1978, DeForge 1980) at higher elevations in their range. Fortunately, desert bighorn sheep are opportunistic feeders. Specific dietary components of sheep diets are covered in Chapter 7.

## 19.5 THINHORN SHEEP HABITAT

The habitat components of Dall's and Stone's sheep across their range include slope, aspect, terrain ruggedness, landcover, average snow depth, distance to escape terrain and glacial ice, mineral licks, and various measures of vegetation. While the variables used in studies differ, the habitat used by sheep can be summarized by open, rugged terrain with low growing vegetation and close to escape terrain (Schoenhardt 2023). In addition, "Sheep habitats must contain adequate amounts of forage, escape terrain, birthing, loafing or respite areas, and unobstructed corridors through which they can move during their annual cycles. The value and juxtaposition of these specific habitat attributes cannot be overemphasized" (Jex et al. 2016:3).

### 19.5.1 Dall's Sheep

*Distribution.* Dall's sheep populations represent the northernmost subspecies of mountain sheep in North America (Bowyer et al. 2000, Sim et al. 2019) and are unique from most other mountain sheep populations because their populations may have not been markedly reduced and they still occupy much of their original range (i.e., rugged and steep mountains in Alaska, USA; Northwest Territories, the Yukon, and British Columbia, Canada [69° 40′–59° 30′]; Krausman and Bowyer 2003). Their continued use of historical ranges is attributed to a lack of widespread anthropogenic influences (including the absence of livestock or management of livestock so they do not encounter mountain sheep), and sheep use of rugged and remote habitats (Bowyer et al. 2000). Horses, however, have been implicated in some Dall's sheep declines (Hoefs and Bayer 1983).

Dall's sheep populations fluctuate. Periodic die-offs and rebounds are well documented (Murie 1944, Murphy and Whitten 1996, Whitten 1997). Increases and decreases in sheep numbers of 28%–35% have been reported over 1–3 year periods in some areas in the Alaska Range (Arthur 2003) and in Canada (Hoefs and Bayer 1983). Aerial trend surveys suggest that sheep populations in southcentral Alaska have been experiencing a continual decline since approximately 1990 (Alaska Department of Game and Fish sheep management reports 2007). Definitive cause and effect relationships generating this decline have not been established (Lohuls 2015).

The distribution of habitat used by Dall's sheep consists of alpine habitats with steep rugged cliffs and rock outcrops that provide escape terrain, open grass and sedge meadows, and winter ranges that are swept clean of snow by high winds (>160 km/h; Nichols 1973, Hoefs and Cowan 1979) so they can forage (Geist 1971, Hoefs 1984, Rachlow and Bowyer 1998). Snow is relatively soft and shallow so sheep can move freely and paw through the snow for forage (Simmons 1982); however, because snow cover in mountain areas is highly variable in space and time, due to the interplay of temperature, precipitation, solar radiation, and stratigraphy (Cosgrove et al. 2021), snow cover can be a barrier to foraging.

In the southern part of their range where weather is influenced by maritime conditions, occasional thaws and deep snow can restrict movement and sheep are limited to small patches of forage (Demarchi and Hartwig 2004).

Temperatures in winter are rarely above freezing (Nichols and Bunnell 1999). For example, in 1983 in the western Brooks Range, Alaska, temperatures ranged from −20.6°C to 60°C in January 1983 (Ayers 1986). In summer weather was generally cool and moist in the south and warmer and dry in the interior (Nichols and Bunnell 1999). Climate change has, however, increased the temperatures by several degrees in the past 60 years (https://2017snapshot.epa. Gov>documents. Accessed 20 Sep 2024).

Some sheep herds make use of somewhat atypical habitat. For example, small populations along the Yukon River from Minto to Fort Selkirk and on Mount Hansen occupy river bluffs within their limited range. Other populations occupy lower areas within trees (i.e., tree sheep), which have darker horns due to rubbing on bark (M. Larivee, Yukon Department of Environment, Canada, personal communication).

Dall's sheep are generally migratory, but some are sedentary (Geist 1971, Hoefs and Cowan 1979). Throughout the year they occupy steep rugged areas for escape terrain. Habitat use varies seasonally, with different patterns of use by males and females (Koizumi and Derocher 2019). They occupy different summer and winter ranges related to plant phenology, snow depth, and temperatures (Aycrigg et al. 2021) and specific demographic groups (e.g., nursery groups) may form in specific areas in summer to access important habitats and minimize vulnerability to predators (Deane et al. 2022). Most of the year they are on winter ranges (271 days for males, 303 days for females; Geist 1971) with increased forage and security and moderate thermal conditions to minimize energetic costs (Aycrigg et al. 2021) but they occupy other ranges throughout the year. Males have ranges for pre-rut, rut, early to midwinter, late winter, spring, summer, and areas with salt licks (Geist 1971). Females segregate around parturition in spring (Bleich et al. 1997, Rachlow and Bowyer 1998, Weckerly 1998) where they select areas with lower snow depth, less snow, warmer temperatures (Aycrigg et al. 2021, Van de Kerk et al. 2018), access to habitat features such as mineral licks (Heimer 1974), with different ranges in summer and winter. Windswept areas with adequate forage and suitable escape terrain are key elements of winter habitat (Bowyer et al. 2000). Seasonal habitat use is influenced by multiple ecological requirements simultaneously (Aycrigg et al. 2021). Additional information on the distribution of Dall's sheep is in Chapter 1.

*Elevation.* Dall's sheep use different elevations depending on their life stages, predation risk, and seasonal requirements for different resources (Koizumi and Derocher 2019) but usually occupy elevations >1,000 m. For example, on Igloo Mountain in Denali National Park and Preserve, Alaska, where the elevation ranges from 855 to 1,480 m, sites used by sheep were 1,187 ± 101 (SD) m (Rachlow and Bowyer 1998). In the Chugach Mountains, Alaska, males, and females preferred mid-elevations of 800–1,700 m in summer and 700–1,500 m in winter (Smith 2020), and in ranges across Alaska where elevations exceeded 2,000 m, the mean elevation used was 1,008 (range = 775–1,348 m; Van de Kerk et al. 2020). On Wrangell-St. Elias National Park and Preserve, Alaska, the available elevation was 422–4,843 m but Dall's sheep only used elevations from 500 to 3,247 m (Roffler et al. 2017). The highest degree of sheep use in Kluane National Park and Reserve, Yukon, Canada, occurred from 1,500 to 2,000 m (Schoenhardt 2023). Other populations used elevations from 1,000 to 1,900 m (Dertien et al. 2017) in various combinations on ranges (Koizumi et al. 2011).

Elevation alone is not a determining factor in sheep distribution within their habitat. Distance to escape terrain, distance to persistent snow and ice, percent slope, snow depth, graminoid cover, strong winds that blow snow off available vegetation, and other factors all play a role (Nichols and Bunnell 1999, Dertien et al. 2017, Aycrigg et al. 2021, Schoenhardt 2023). Elevation was, however, one of eight habitat covariates consistently retained in models describing habitat selection for Dall's sheep (Roffler et al. 2017). As. with most habitat features, no single one operates independently of other biotic and abiotic influences.

*Topography.* Dall's sheep generally inhabit steep, rugged cliffs and rock outcrops that provide terrain to escape from predators, open areas with sedges and grass for foraging, and windswept and dry areas that have light snowfall in winter that can be removed with strong winds to expose forage (Murie 1944, Hoefs 1984, Nichols and Bunnell 1999, Koizumi et al. 2011, Krausman and Bowyer 2003, Figure 19.8). As with other habitat features, topography is not used in a uniform manner by males and females. For example, in the Richardson Mountains in the Canadian Arctic in the Northwest and Yukon territories, Dall's sheep select the topographic features of slope, elevation, ruggedness, and aspect differently. During winter, males selected rugged terrain, steep slopes, and barren lands. Females selected eastern and western slopes exposing them to higher risks of predation by wolves (*Canis lupus*; Koizumi and Derocher 2019). In spring, males avoided northern slopes and females selected steep slopes. During lambing, males selected southern slopes and landscapes covered with vegetation associated with forests, herbs, shrubs, and barrens. Females selected rugged and steep slopes and avoided northern slopes but left the security of this topography occasionally to access mineral licks (Heimer 1974, Nichols and Bunnell 1999). In summer males used rugged terrain, steep slopes, barrens, and bryoids, while females used higher elevations and rugged slopes. In fall, males used lower elevations, southeast slopes, barrens, forests, herbs, shrubs, and bryoids. There was no discernable pattern for females (Koizumi et al. 2011). How topography is selected depends on the life stage of an individual and requirements for protection from predators and obtainment of forage and mineral resources. When vegetation characteristics are included with topography to define habitat use, topographic variables are more important predictors (Rachlow and Bowyer 1998, Terwilliger 2005, Dertien et al. 2017, Koizumi and Derocher 2019).

*Thermal cover.* The severity of winters in the Arctic and subarctic (e.g., −20.6°C to 60°C, winds >160 km/h; Nichols 1973, Ayers 1986, Mehlam 2019) causes thermal

FIGURE 19.8   Female Dall's sheep and lambs on windswept winter range adjacent to escape cover, Golden Mountain, British Columbia, Canada. (Photo by Bill A. Jex, Province of British Columbia, used with permission.)

cover to be important for growth and development of young lambs, replenishment of female body reserves (Bunnell 1982, Rachlow and Bowyer 1991, 1994), and protection of young lambs from hypothermia (Frid 1997, Rachlow and Bowyer 1998). Thus, in winter and spring Dall's sheep selected habitats with more exposed forage, steep slopes, high wind speeds that remove snow from vegetation, and high solar radiation (Aycrigg et al. 2021). Climatic factors are important in habitat selection for thermal cover. In areas where winds and low temperatures are excessive, sheep may need to search for microsites for protection (Heimer et al. 1994) compared to areas with warmer temperatures. In all seasons, escape cover was important. During spring, forage for females and lambs and warmer temperatures were important for thermal regulation (Aycrigg et al. 2021). Production of lambs was reduced in years with deep or persistent late-spring snow cover, and colder fall temperatures (Rattenbury et al. 2018, Cosgrove et al. 2021). Across the range of Dall's sheep, higher spring snowline elevations, earlier snow disappearance dates, and fewer snow-covered days/year enhanced lamb recruitment (Van de Kerk et al. 2018).

*Escape terrain.* Escape terrain for Dall's sheep is directly tied to topographic features that allow sheep to avoid predators including steep, rugged terrain intermixed with forage and usually at elevations free of snow (Murie 1944, Hoefs 1984, Rachlow and Bowyer 1998, Koizumi et al. 2011, Aycrigg et al. 2021, Nichols and Bunnell 1999, Krausman and Bowyer 2003, Figures 19.9 and 19.10). Dall's sheep are rarely far from escape terrain. For example, the mean distance to escape terrain in Alaska was 20.5 m±46.07 m (SD) when compared to random sites. Adequate escape terrain is important in all seasons, and especially during lambing.

At most scales occupancy models include slope, snow depth, and distance to escape terrain as the most important factors influencing habitat use (Dertien et al. 2017). The importance of escape terrain to mountain sheep cannot be understated because it is tied to most other uses of the landscapes they inhabit.

*Lambing habitat.* Lambing habitat for Dall's sheep in interior Alaska is steep, rugged terrain intermixed with forage (Rachlow and Bowyer 1998). Maternal females usually select level (Rachlow and Bowyer 1991), warmer, snow-free microsites within the most rugged areas for lambing (Nichols and Bunnell 1999).

During pre-lambing, females selected sites with forage and avoided snow-covered areas. During lambing, forage and steep slopes continued to be used and distance to escape terrain was important all year long (Rachlow and Bowyer 1997). When females ventured farther from escape terrain to access forage and mineral resources, they usually were in larger groups (Rachlow and Bowyer 1997).

During lambing, females constantly make choices between foraging and security from predators (Rachlow and Bowyer 1997). Sites selected for lambing varied depending on environmental conditions (e.g., snow, snow depth) and chronology of lambing (Rachlow and Bowyer 1998). Dall's sheep balance resource selection to meet multiple needs across seasons that are closely related to climate and vegetation (Aycrigg et al. 2021). In some populations, elevation and distance to glacial ice were the strongest predictors of Dall's sheep habitat use along with steep and rugged habitat (Schoenhardt 2023).

While there were minimal differences between habitat characteristics used by males and nursery groups, the two groups appear to use nearby, but different areas consistent

**FIGURE 19.9** Dall's sheep above winter escape terrain near Carcross, Yukon Territories, Canada, March 2013. (Photo by Bill A. Jex, Province of British Columbia, used with permission.)

**FIGURE 19.10**  Dall's sheep use of escape terrain and day bed sites, Atlin, British Columbia, Canada. (Photo by Bill A. Jex, Province of British Columbia, used with permission.)

with the findings of previous researchers (Corti 2001). Because males and nursery groups use different areas, but with similar habitat characteristics, selection of habitats by males and nursery groups is either more influenced by fine scale differences in habitat, or their sexual segregation is a result of behavior. Females used steeper, more rugged terrain, and higher elevations during summer (Koizumi and Derocher 2019).

*Water.* Within the habitat of thinhorn sheep, water is available as snow, melting snow, springs, and creeks; surface water is plentiful throughout much of the year (Nichols and Bunnell 1999). There are areas within Dall's sheep habitat where water is limited, and sheep have to move to find water or to access north aspects where forage is more lush and contains more moisture (Schoenhardt 2023). Lack of water has not been identified as an issue for thinhorn sheep (Nichols and Bunnell 1999).

*Foraging habitat.* Habitat characteristics required by Dall's sheep vary by season (Rachlow and Bowyer 1998, Nichols and Bunnell 1999) and are similar to those of Stone's sheep (Seip and Bunnell 1985). Dall's sheep select habitat as a trade-off between areas that provide for predator avoidance and those which provide adequate nutritional forage (Rachlow and Bowyer 1998). The open, rocky slopes that sheep rely on for predator avoidance are often sparsely vegetated, requiring sheep to travel out of these terrains to find adequate forage. In some populations sheep choose habitat that offers preferred vegetation and is close to escape terrain (Roffler et al. 2017, Aycrigg et al. 2021).

Vegetation (e.g., species, phenology, nutritional quality and quantity) and topography are important factors in habitat selection (Figure 19.11).

Dall's sheep select habitats with bunchgrasses including fescues (*Festuca* spp.) and bluegrass (*Poa* spp.), forbs, sedges (*Carex* spp.), rushes (*Juncus* spp.), low shrubs including dwarf willows (*Salix* spp.), huckleberry (*Vaccinium* spp.), mountain avens (*Dryas* spp.), crowberry (*Empetrum* spp.), heather (*Cassiope tetragona*), lichens, and mosses (Hoefs and Cowan 1979, Hansen 1996, Rachlow and Bowyer 1998, Nichols and Bunnell 1999, Roffler et al. 2017, Aycrigg et al. 2021). On lower slopes used in winter, dwarf birch (*Betula* spp.), alder (*Alnus* spp.), and mountain hemlock (*Tsuga mertensiana*) often form dense stands interspersed with larger trees (Nichols and Bunnell 1999). In summer Dall's sheep use relatively dry mountain meadows where the vegetation is abundant and varied in contrast to winter when forage is reduced in availability and variety (Nichols and Bunnell 1999). Patterns of Dall's sheep habitat use vary seasonally between males and females (Koizumi and Derocher 2019), and from a niche occupancy perspective, they have been described as unique species (Roffler et al. 2017). Contrary to this, there is overlap between the diets of males and females, and range overlap between sexes is especially common when forage is sparse (Seip and Bunnel 1985, Barichello and Carey 1989, Nichols and Bunnell 1999, Koizumi and Derocher 2019), both suggesting little difference in habitat needs. Chapter 7 includes information on diets of Dall's sheep.

**FIGURE 19.11** Male Dall's sheep on summer range adjacent to escape terrain, near Carcross, Yukon Territories, Canada, July 2012. (Photo by Bill A. Jex, Province of British Columbia, used with permission.)

### 19.5.2 STONE'S SHEEP

Researchers have generally described Stone's sheep movement and behavioral ecology within a six-season calendar, with early winter (1 January to 28 February), late winter (1 March to 14 May), lambing (15 May to 14 June), summer (15 June to 14 August), fall (15 August to 31 October), and rut (1 November to 31 December; Walker et al. 2007, Hengeveld and Cubberley 2011). Seasonal movement behaviors in a population of Stone's sheep in northwestern British Columbia, had median start and end dates of spring migration (12–17 June) and fall migration (30 August to 22 September; Enns et al. 2023, 2024). Like Dall's sheep, Stone's sheep habitat use of seasonal ranges can be greatly affected by climate.

*Distribution.* Until recently the habitat of Stone's sheep has been reported to occur from Alaska, Northwest Territories, the Yukon, and British Columbia from 69° 40′ to 59° 30′ N latitude in subalpine brushlands, glaciers, and isolated areas far from the nearest mountains, and even in lower forested areas (Nichols and Bunnell 1999, Bowyer et al. 2000, Krausman and Bowyer 2003). These previous descriptions of the distribution of Stone's sheep are controversial. Recent genetic work has indicted that the previous phenotypic subspecies classifications based on geographic and pelage characteristics are not accurate for taxonomic classifications of thinhorn sheep (Sim et al. 2019) and their range is primarily limited to British Columbia, Canada. Additional genetic information is provided in Chapter 6.

Stone's sheep use rugged escape terrain with adjacent foraging areas and snow-free winter foraging sites, but they are not restricted to alpine areas in the same way as Dall's sheep (Nichols and Bunnell 1999), particularly in summers. Still, weather and climate patterns can influence seasonal habitat suitability for thinhorn sheep on various temporal scales (Hik and Carey 2000, Loehr et al. 2010). For example, currents in the Pacific Ocean create regular influences on climate cycles (i.e., Pacific Decal Oscillation); to a lesser degree Arctic Decal Oscillation also affects climate cycling in northern mountain sheep environs, influencing precipitation patterns, temperature, and wind at approximately 10-year intervals. Within this 10-year cycle, events such as El Niño and La Niña can also occur, with these events typically affecting weather patterns on 6- to 12-month timelines, primarily influenced by winds and ocean surface temperatures. Finally, localized events such as temperature inversions and chinook winds happen infrequently, lasting only hours, to a day and up to a week, and are often unpredictable in their effect on habitat suitability and mountain sheep use because they can bring with them rain-on-snow that creates icing effects that significantly restricts access to forage, or can create hardened snow that improves movement of predators such as wolves and wolverines (*Gulo gulo*) into winter sheep habitats. Both outcomes can result in population effects through direct mortality (e.g., starvation, predation; Chapter 21). "Wind speeds on these ranges can exceed 160 km/hr" (Luckhurst 1973, Nichols and Bunnell 1999:25). Stone's sheep are distributed the farthest southward of thinhorn sheep and despite local extirpations and range contractions still occur throughout much of their original range (Krausman and Bowyer 2003, Wild Sheep Foundation 2023). Additional information on distribution of Stone's sheep is available in Chapter 1.

*Elevation.* In northern British Columbia, the median elevation used by female Stone's sheep was 1,679 m (range = 702–2,282 m). In summer the median range was 1,709 m (range = 1,563–1,827 m). In winter, the median elevation was 1,673 m (range = 1,478–1,757 m; Enns et al. 2023). Elevations were selected to avoid deep snows and movements aligned with the forage nutrition hypothesis, which predicts herbivores exhibit migration during the growing season to maximize their intake of new, high-quality forage by moving in concert with plant phenology (Seip 1983, Seip and Bunnell 1985, Albon and Langvatn 1992, Walker et al. 2006, Hebblewhite et al. 2008) as they surf the green wave (Merkle et al. 2016). Similar patterns were observed in another range in British Columbia where the upper elevations (1,600–2,200 m) were used in summer, fall, rut, and early winter. In late winter Stone's sheep selected upper and lower elevations (1,200–2,200 m). Lambing season was the only season in which Stone's sheep did not select for the highest elevations and used elevations around 1,400 m (Hengeveld and Cubberly 2011) and 1,700 m (Walker et al. 2007). In the northern Rocky Mountains of British Columbia, Stone's sheep wintered from 1,500 to 2,200 m but others used lower elevations (1,000–1,500 m; Seip and Bunnell 1985).

*Topography.* In northern British Columbia, habitat models included topographic features and risk of predation to best explain resource selection by Stone's sheep (Walker et al. 2007). There was variability in the selection of topographic features by Stone's sheep, but topography that was preferred by Stone's sheep included steep slopes, ridge-like topography, southern aspects, and upper elevations (Walker et al. 2007). In other populations, slopes averaging 29°–37° with westerly aspects were favored over easterly aspects (Hengeveld and Cubberly 2011), and females selected southwestern slopes in rugged terrain at mid-elevations (Enns et al. 2024) during lambing. These sites provided warmer micro-climates and topographic features that facilitated predator avoidance (Enns et al. 2024). Stone's sheep also used caves (Figure 19.12) most likely for security cover and thermal refugia. In burned areas ruggedness was still an important component of habitat. Stone's sheep selected and used slopes that were steep, rugged, and often in rocky areas (Sittler et al. 2015). In other ranges in British Columbia, females used locations that were in rocky terrain in dry alpine areas in all seasons. They selected steep slopes, ridgelines, and mountain tops, and south-facing slopes (Parker and Walker 2007). As with Dall's sheep, Stone's sheep select rugged topography throughout their range.

*Thermal cover.* Stone's sheep demonstrate behavioral plasticity in relation to selection of habitat attributes (Walker et al. 2007, Enns et al. 2023). For example, in late winter Stone's sheep had two strategies for survival. They selected lower elevations that had warmer temperatures on south and west aspects. They also selected higher elevations that had increased exposure to wind that exposed vegetation (Walker et al. 2007). Snow depth and the severity of the

**FIGURE 19.12**    Male Stone's sheep using a cave in summer near Dease Lake, British Columbia, Canada, July 2010. (Photo by Bill A. Jex, Province of British Columbia, used with permission.)

weather are important attributes of site selection in winter (Geist 1971, Walker et al. 2007). The interaction between slope and aspect can contribute to the amount of solar radiation and subsequent air temperature of sites (Walker et al. 2007), with thermal properties of southerly slopes further enhanced by patches of conifers at tree-line that act to block winds (Hengeveld and Cubberley 2011). During warm weather Stone's sheep can gain a thermal advantage by resting with their legs extended and draw their legs tight against their bodies in colder weather in winter (Geist 1971). While visiting with me in the Harquahala Mountains, Arizona, Geist indicated this was an untested hypothesis (V. Geist, deceased, personal communication).

*Escape terrain.* Many features of habitat for Stone's sheep, including escape terrain, had not been examined in as much detail as for other races of mountain sheep and researchers called for more research to better understand habitat (Nichols and Bunnell 1999, Krausman and Bowyer 2003). Since then, with new technology there have been several researchers that have expanded the knowledge base of the ecology of Stone's sheep through field work and models. One of the first comprehensive studies of habitat use by Stone's sheep reported that habitat features were directly tied to topographic features but varied with different groups of sheep and between years. Regardless, steep slopes, ridge-like topography, south aspects, and upper elevations were common for the population all year (Demarchi and Hartwig 2004, Walker et al. 2007, Hengeveld and Cubberley 2011). In early winter, Stone's sheep avoided densely timbered areas with limited graminoids where visibility was limited to avoid predators (Walker et al. 2007).

*Lambing and nursery habitat.* In northern British Columbia, the lambing season for Stone's sheep has been defined as 15 May to 14 June (Walker et al. 2007, Hengeveld and Cubberley 2011). These dates were refined recently; the lambing period is the period that began with the parturition date and included the length of time that the female remained within her lambing habitat (1 May to 30 June), and the nursery period is the 30-day period following the lambing period (1 to 30 July; Enns et al. 2023, 2024). During the lambing period, female Stone's sheep initially selected for escape terrain (Enns et al. 2024), with grizzly bear (*Ursus arctos*) and wolf predation risk continuing to influence habitat selection throughout the lambing season (Walker et al. 2007) and nursery period (Enns et al. 2023, 2024). The lambing season was the only season in which Stone's sheep did not select the highest elevations; they selected elevations between 1,600 and 1,700 m with rocks, high-quality vegetation, and steep slopes (Walker et al. 2007, Enns et al. 2023, 2024). There is a wide range in the number of days that a female remains in her lambing habitat after giving birth: 1–2 days (Shackelton and Haywood 1985), 1.5–11.3 days (Enns et al. 2024), and 5–7 days (Geist 1971), but none of these authors were specifically able to identify the reason for this variance. It is likely given the topography in which Stone's sheep are found, that weather patterns, predator avoidance, body condition of females and young, and forage availability may play a role (Enns et al. 2024). Following the lambing period, females moved upslope and predominantly occupied higher-elevation summer range habitats, making movements during the lambing season to lower elevations to access mineral licks (Ayotte et al. 2008, Enns et al. 2023, 2024). In summer Stone's sheep selected dry alpine, burn-grass (i.e., recently burned and open disturbed sites dominated by ryegrass [*Elymus* spp.]), shrub, and rock areas increasing their use of higher quality plant growth as it became more seasonally available (with an accompanying risk of predation; Walker et al. 2007).

Unfortunately, Walker et al. (2007) could not determine if females in their study in the Besa-Prophet watersheds in the southeast portions of the Muskwa-Kechika Management Area in northern British Columbia had lambs, but described activity and habitat use by nursery groups as more active in solid rock escape habitat than females without lambs. The best predictive model describing intersexual differences incorporated distance to nearest escape feature and the size of the nearest escape feature (Walker et al. 2006). Females with lambs were closer to the nearest escape feature (8.6 m vs. 14 m) and associated with larger escape features (1.7 ha vs. 0.41 ha) and larger broken rocks acting as escape features (1.6 ha vs. 0.47 ha) compared to females without lambs. Escape features were three to four times larger for groups with lambs (Walker et al. 2006). When lambs were present in groups the groups were ≤69 m from escape terrain versus 150 m for a group without lambs. Regardless of the maternal status of female Stone's sheep, they selected areas with increased cover. In another population, females selected southwest slopes in rugged terrain at mid-elevations, suggesting selection for features that facilitate predator avoidance (Enns et al. 2024). These habitat selection patterns were explained by the predation-risk hypothesis (i.e., segregation in ungulates is related to anti-predator strategies; Festa-Bianchet 1988) and the forage-selection hypothesis (i.e., segregation in ungulates is related to nutritional demands; Ruckstuhl and Neuhaus 2002).

Habitat selection for lambing and nursery sites includes habitats to avoid predators and enhance the fitness of females and survival of young (Bunnell 1982, Festa-Bianchet 1988). Thus, they normally give birth when isolated in steep terrain at high elevations (Geist 1971, Bunnell 1982, Festa-Bianchet 1988). This was also the case in the Cassiar Mountains, British Columbia. Females selected southwest slopes in rugged terrain and mid-elevations, suggesting selection for warmer micro-climates and features that facilitated predator avoidance. Females also avoided roads during lambing (Enns et al. 2024). During nursing periods, females and young left lambing habitat and joined other females and young in nursing groups (Geist 1971, Karsch et al. 2016). Due to their high nutritional demands, females with lambs favored habitats at elevations and aspects that facilitate early growth of forage, access to mineral licks (Ayotte et al. 2008), with visual openness and adequate escape terrain for detection and evasion of predators (Walker et al. 2006). In the Cassiar Mountains, lambing and nursery habitat was

**FIGURE 19.13**   Male Stone's sheep on late summer range with escape terrain in the background, near Cassiar, British Columbia, Canada. (Photo by Bill A. Jex, Province of British Columbia, used with permission.)

similar with mid-elevations, steep slopes, rugged terrain, on southwest slopes with few exceptions. During lambing females were at elevations around 1,635 m in the alpine but were slightly higher during nursing (1,674 m). Also, during lambing, females avoided roads but were closer to roads during the nursery period and this finding may have been a function of the use of mineral licks that occurred nearer valley-bottoms where road networks were often constructed, more so than an actual selection in favor of roaded areas (Enns et al. 2024).

*Water.* A lack of water has not been identified as an issue for thinhorn sheep (Seip and Bunnell 1985, Nichols and Bunnell 1999). Water is available throughout most of their ranges as snow and surface water.

*Foraging habitat.* The vegetation composition on Stone's sheep habitat is similar to that of Dall's sheep habitat with graminoids, sedges, ryegrass (*Elymus* spp.), fescue, and bluegrass common on alpine and subalpine areas including a wide variety of forbs and shrubs (Seip 1983, Seip and Bunnell 1985, Nichols and Bunnell 1999, Figure 19.13). Foraging habitat also includes riparian, shrub, conifer, subalpine spruce, dry alpine, wet alpine, and burnt areas, which all provide forage for Stone's sheep (Walker et al. 2007). The vegetative types in Stone's sheep habitat along with aspect, curvature (i.e., convex or concave topography), elevation, and risk of wolf predation were consistent seasonal components of habitat selection. Generally, foraging habitat selection in Stone's sheep habitat along with aspect, curvature (i.e., convex or concave topography), elevation, and risk of wolf predation were consistent seasonal components

of habitat selection (Walker et al. 2007, Hengeveld and Cubberley 2011, Enns et al. 2024). Generally, foraging habitat selection can best be explained by incorporating vegetation, topography, and the risk of predation into habitat selection models (Walker et al. 2007).

At small scales there is extensive variability in the use of vegetation by groups of sheep, season, and year likely as a result of topographic influences on moisture and temperature that affect forage growth and nutrition (Hengeveld and Cubberley 2011). For example, in late winter in the Besa-Prophet watersheds, more groups of Stone's sheep selected dry alpine (i.e., dominated by Arctic avens [*Dryas integrifolia*] and altai fescue [*Festuca altaica*] on well drained soils) more than any other. In summer Stone's sheep selected rock (i.e., rocky sites dominated by steep outcrops, talus slopes, scree slides, non-vegetated bedrock) habitat more than other habitat features (Walker et al. 2007). In fall, Stone's sheep selected for or were indifferent to the use of burns and in late winter, they selected burn areas more than in other seasons. In other populations, Stone's sheep used burned areas more than unburned control areas (Sitter et al. 2015, 2019). Stone's sheep selection of younger burns is further influenced by terrain steepness and ruggedness, attributes that help to partition their use of burned areas from that of other species such as elk (*Cervus canadensis*; Sittler et al. 2015, 2019). In addition, during years of high snowfall, there is a relationship between the use of vegetation and wind in determining habitat use (Seip and Bunnell 1985, Figure 19.14). Chapter 7 outlines additional information related to nutrition of Stone's sheep.

**FIGURE 19.14**  Windswept foraging habitat used by Stone's sheep near Cassiar, British Columbia, Canada. (Photo by Bill A. Jex, Province of British Columbia, used with permission.)

## 19.6  SUMMARY

In this chapter I describe the important habitat components for mountain sheep in North America (i.e., distribution, elevation, topographic features, thermal cover, escape terrain, lambing, water, forage [special habitat factors are discussed in Chapter 12]). Habitat characteristics for mountain sheep vary between subspecies, sex and age groups, life stages, occupied habitat, and in relation to the availability or predators, livestock, anthropogenic activity, and a host of other factors. Many of the differences can be attributed to the methods used to describe habitats by different researchers, but commonalities exist in relation to habitat across the range of mountain sheep.

Because mountain sheep range from montane areas in Canada into Mexico, they use a wide range of elevations from 78 m below sea level to >4,267 m for desert bighorn sheep, the mountain sheep that occur the furthest south of all races, to >1,000 m (range=500–3,247 m) for Dall's sheep, the northernmost race. Rocky Mountain bighorn sheep and Stone's thinhorn sheep use elevations from 450–3,300 and 702–2,282 m, respectively.

Mountain sheep use all parts of montane areas from canyon bottoms and foothills, to tops of mesas, peaks, and the most rugged terrain with steep slopes (>27%). The use of rugged terrain is a universal characteristic of mountain sheep habitat; sheep are rarely far from rugged terrain. Thinhorn sheep select rugged terrain with windswept surfaces where snow has been removed and exposes forage or reduces snow depth so sheep can paw for forage.

Thermal cover is important as shade in summer and increased solar radiation in winter for bighorn sheep. Desert bighorn also use caves, overhangs, microsites, and behavior to adjust to weather. Thinhorn sheep are in much colder areas than bighorns and rely on areas with increased solar radiation, sheltering terrain, vegetative cover, and high wind speeds to remove snow from forage. They also use microsites for protection from the elements including caves.

Escape terrain is an important commonality for mountain sheep and is intertwined with all aspects of mountain sheep habitat. Escape terrain is rugged habitat where sheep can avoid predation, and sheep are rarely far from it (usually within 100–500 m). Escape terrain for thinhorn sheep usually includes rock complexes and bluffs and has limited snow.

Lambing areas are also related to escape terrain and occur in rugged areas intermixed with forage, but with desirable southerly aspects. Females often make decisions (i.e., trade-offs) related to predator avoidance or nutrition to enhance the survival of lambs. Lambing habitat is only used during a short period of the year, but the value of the areas cannot be understated with respect to lamb production.

Water is an important habitat feature for all mountain sheep but is readily available within the habitats of thinhorn sheep and is not a limiting factor in most seasons or years. In arid areas of bighorn habitat water can be obtained from surface water or vegetation but for many populations of desert bighorn sheep, water is a limiting factor, and agencies have spent considerable time, effort, and resources to provide surface water for sheep. As temperatures increase and affect patterns of precipitation, more research is needed to better understand the relationships between sheep and water.

Forage and foraging areas are also important for mountain sheep. The types of plants used and available may not be as important as the horizontal visibility. Mountain sheep generally avoid areas where their vision is restricted. Because of the range of distribution, forages used are related to season, sex, year, precipitation, group size, distance from escape terrain, visibility, presence of livestock, predation risk and trade-offs, availability of water, urban parks, mining activity, roads and highways, disturbance, and quality of forage among other influences.

There is still more to be learned about the habitat of mountain sheep, especially in isolated populations and local ranges. Biologists know, however, that rugged topography with adequate escape terrain adjacent to foraging areas is intertwined with all other aspects of mountain sheep habitat in North America.

## REFERENCES

Adams, L. G., K. L. Risenhoover, and J. A. Bailey. 1982. Ecological relationship of mountain goat and Rocky Mountain bighorn sheep. *Proceedings of the Biennial Symposium of the Northern Wild Sheep and Goat Council* 3:9–22.

Akeson, J. J., and H. A. Akeson. 1992. Bighorn sheep movements and summer lamb mortality in central Idaho. *Proceedings of the Biennial Symposium of the Northern Wild Sheep and Goat Council* 8:14–27.

Albon, S. D., and R. Langvatn. 1992. Plant phenology and the benefits of migration in a temperate ungulate. *Oikos* 65:502–513.

Alderman, J. A., P. R. Krausman, and B. D. Leopold. 1989. Diel activity of female desert bighorn sheep in western Arizona. *Journal of Wildlife Management* 53:264–271.

Alvarez, T. 1976. Status of bighorns in Baja, California. *Desert Bighorn Council Transactions* 20:18–21.

Alvarez-Cárdenas, S., I. Guerrero-Cárdenas, S. Díaz, P. Galina-Tessaro, and S. Gallina. 2001. The variables of physical habitat selection by the desert bighorn sheep (*Ovis canadensis weemsi*) in the Sierra del Mechudo, Baja California Sur, Mexico. *Journal of Arid Environments* 49:357–374.

Arthur, S. M. 2003. Interrelationships of Dall sheep and predators in the central Alaska Range. Federal Aid in Wildlife Restoration Research Final Report. Project 6.13. 1 July 1998 to 30 June 2003. Alaska Department of Fish and Game publication. Juneau, Alaska, USA.

Aycrigg, J. L., A. G. Wells, E. O. Garton, B. Magipane, G. E. Liston, L. R. Prugh, and J. L. Rachlow 2021. Habitat selection by Dall's sheep is influenced by multiple factors including direct and indirect climate effects. *PLoS ONE* 16:e0248763.

Ayers, L. A. 1986. The movement patterns and foraging ecology of Dall sheep (*Ovis dalli dalli*) in the Noatak National Preserve, Alaska. Thesis, University of California, Berkeley, California, USA.

Ayotte, J. B., K. L. Parker, and M. P. Gillingham. 2008. Use of natural licks by four species of ungulates in northern British Columbia. *Journal of Mammalogy* 89:1041–1050.

Bailey, J. A. 1980. Trickle Mountain forage allocation research. United States Bureau of Land Management, Denver Service Center, Colorado, USA. Report 5:1–30.

Bailey, V. 1905. Biological survey of Texas. United States Department of Agriculture Bureau of Biological Survey. *North American Fauna* 25:1–222.

Bangs, P. D., P. R. Krausman, K. E. Kunkel, and Z. D. Parsons. 2005a. Habitat use by desert bighorn sheep during lambing. *European Journal of Wildlife Research* 51:178–184.

Bangs, P. D., P. R. Krausman, K. E. Kunkel, and Z. D. Parsons. 2005b. Habitat use by female desert bighorn sheep in the Fra Cristobal Mountains, New Mexico, USA. *European Journal of Wildlife Research* 51:77–83.

Beecham, J. J., C. P. Collins, and T. D. Reynolds. 2007. Rocky Mountain bighorn sheep: a technical conservation assessment. U. S. Department of Agriculture Forest Service, Rocky Mountain Region Species Conservation Project, Lakewood, Colorado, USA.

Bender, L. C., and M. E. Weisenberger. 2005. Precipitation, density, and population dynamics of desert bighorn sheep on San Andres National Wildlife Refuge, New Mexico. *Wildlife Society Bulletin* 33:956–964.

Bender, L. C., and M. E. Weisenberger. 2009. Criticisms biologically unwarranted and analytically irrelevant: reply to Rominger et al. *Journal of Wildlife Management* 73:806–810.

Berger, J. 1991. Pregnancy incentives, predation constraints and habitat shifts: experimental and field evidence for wild bighorn sheep. *Animal Behaviour* 41:61–77.

Berwick, S. H. 1968. Observations on the decline of the Rock Creek, Montana, population of bighorn sheep. Thesis, University of Montana, Missoula, Montana, USA.

Bleich, V. C., J. G. Kie, E. R. Loft, T. R. Stephenson, M. W Oehler Sr., and A. L. Midian. 2005. Managing rangelands for wildlife. Pages 873–897 in C. E. Braun, editor. *Techniques for Wildlife Investigation and Management*. The Wildlife Society, Bethesda, Maryland, USA.

Bleich, V. C., J. P. Marshal, and N. G. Andrew. 2010. Habitat use by a desert ungulate: predicting effects of water availability on mountain sheep. *Journal of Arid Environments* 74:638–645.

Bleich, V. C., M. C. Nicholson, A. T. Lombard, and P. V. August. 1992. Preliminary tests of a mountain sheep habitat model using a geographic information system. *Proceedings of the Biennial Symposium of the Northern Wild Sheep and Goat Council* 8:256–263.

Bleich, V. C., N. G. Andrew, M. J. Martin, G. P. Mulcahy, A. M. Pauli, and S. S. Rosenstock. 2006. Quality of water available to wildlife in desert environments: comparisons among anthropogenic and natural sources. *Wildlife Society Bulletin* 34:627–632.

Bleich, V. C., R. T. Bowyer, and J. D. Weyhausen. 1997. Sexual segregation in mountain sheep. *Resources or predation? Wildlife Monographs* 134:1–50.

Blong, B., and W. Pollard. 1968. Summer water requirements of desert bighorn in the Santa Rosa Mountains. *California Fish and Game* 54:289–289.

Blood, D. A. 1961. An ecological study of California bighorn sheep (*Ovis canadensis californiana* Douglas) in southern British Columbia. Thesis, University of British Columbia, Vancouver, Canada.

Blum, M. E. 2021. Habitat selection of female desert bighorn sheep: tradeoffs associated with reproduction. Dissertation, University of Nevada, Reno, Nevada, USA.

Blum, M. E., K. M. Stewart, K. T. Shoemaker, T. E. Dilts, V. C. Bleich, M. Cox, J. R. Bennett, and B. F. Wakeling. 2023a. Changes in selection of resources with reproductive state in a montane ungulate. *Movement Ecology* 11:20.

Blum, M. E., K. M. Stewart, M. Cox, K. T. Shoemaker, J. R. Bennett, B. W. Sullivan, B. F. Wakeling, and V. C. Bleich. 2023b. Variation in diet of desert bighorn sheep (*Ovis canadensis nelsoni*): tradeoffs associated with parturition. *Frontiers in Ecology and Evolution* 10:1071771.

Blum, M. E., T. Dilts, and K. Stewart. 2018. Next generation ruggedness indices for modeling escape terrain of desert bighorn sheep at Lone Mountain, Nevada. *Wild Sheep and Goat Council Symposium* 21:1.

Bose, S., T. D. Forrester, D. S. Casady, and H. U. Wittmer. 2018. Effect of activity states on habitat selection by black-tailed deer. *Journal of Wildlife Management* 82:1711–1724.

Bowyer, R. T., D. M. Leslie Jr, and J. L. Rachlow. 2000. Dall's and Stone's sheep. Pages 491–516 *in* S. Demarais and P. R. Krausman, editors. *Ecology and Management of Large Mammals in North America*. Prentice Hall, Upper Saddle River, New Jersey, USA.

Brewer, C. E., and F. Hernandez. 2011. Status of desert bighorn sheep in Texas, 2009–2010. *Desert Bighorn Council Transactions* 51:76–79.

Brewer, C. E., and L. A. Harveson. 2007. Diets of bighorn sheep in the Chihuahuan Desert, Texas. *Southwestern Naturalist* 52:97–103.

Broyles, B. 1995. Desert wildlife water developments: questioning use in the Southwest. *Wildlife Society Bulletin* 23:663–675.

Broyles, B., and T. L. Cutler. 1999. Effect of surface water on desert bighorn sheep in the Cabeza Prieta National Wildlife Refuge, southwestern Arizona. *Wildlife Society Bulletin* 27:1082–1088.

Broyles, B., and T. L. Cutler. 2001. Reply to Rosenstock et al. (2001) regarding effects of water on desert bighorn sheep at Cabeza Prieta National Wildlife Refuge, Arizona. *Wildlife Society Bulletin* 29:738–743.

Brushett, A., J. Whittington, B. Macbeth, and J. M. Fryxel. 2023. Changes in movement, habitat use, and response to human disturbance accompany parturition events in bighorn sheep (*Ovis canadensis*). *Movement Ecology* 11:36.

Buchalski, M. R., B. N. Sacks, D. A. Gille, M. Cecilia, T. Penedo, H. B. Ernest, S. A. Morrison, and W. M. Boyce. 2016. Phylogeographic and population genetic structure of bighorn sheep (*Ovis canadensis*) in North American deserts. *Journal of Mammalogy* 97:823–838.

Buechner, H. K. 1960. The bighorn sheep in the United States, its past, present, and future. *Wildlife Monographs* 4:1–174.

Bunnell, F. L. 1982. The lambing period of mountain sheep: synthesis, hypothesis, and tests. *Canadian Journal of Zoology* 60:1–14.

Cain, J. W., B. D. Jansen, R. R. Wilson, and P. R. Krausman. 2008. Potential thermoregulatory advantages of shade use by desert bighorn sheep. *Journal of Arid Environments* 72:1518–1525.

Cain III, J. W., P. R. Krausman, J. R. Morgart, B. D. Jansen, and M. P. Pepper. 2008. Responses of desert bighorn sheep to removal of water sources. *Wildlife Monographs* 171:1–32.

Cain III, J. W., P. R. Krausman, S. S. Rosenstock, and J. C. Turner. 2006. Mechanisms of thermoregulation and water balance in desert ungulates. *Wildlife Society Bulletin* 34:570–581.

Cain III, J. W., R. C. Karsch, E. J. Goldstein, E. M. Rominger, and W. R. Gould. 2019. Survival and cause-specific mortality of desert bighorn sheep lambs. *Journal of Wildlife Management* 83:251–259.

Campbell, B. H., and R. R. Remington. 1979. Bighorn use of artificial water sources in the Buckskin Mountains, Arizona. *Desert Bighorn Council Transactions* 23:50–56.

Chilelli, M. E., and P. R. Krausman. 1981. Group organization and activity patterns of desert bighorn sheep. *Desert Bighorn Council Transactions* 25:17–24.

Clapp, J. G., and J. L. Beck. 2016. Short-term impacts of fire-mediated habitat alterations on an isolated bighorn sheep population. *Fire Ecology* 12:80–98.

Cook, R. L. 1994. A historical review of reports, field notes and correspondence on the desert bighorn sheep in Texas. Special Report to the Desert Bighorn Sheep Advisory Committee. Contribution to Federal Aid Project Number W-127-R and W-123-D. Texas Parks and Wildlife Department, Austin, Texas, USA.

Corti, P. 2001. Dall's sheep (Ovis dalli dalli Nelson, 1884) sexual segregation: interactions between two hypotheses. Dissertation, University of British Columbia, Canada.

Cosgrove, C. L., J. Wells, A. W. Nolin, J. Putera, and L. R. Prugh. 2021. Seasonal influence of snow conditions on Dall's sheep productivity in Wrangell-St Elias National Park and Preserve. *PLoS ONE* 16:e0244787.

Cowan, I. M. 1940. Distribution and variation in the native sheep of North America. *American Midland Naturalist* 24:505–580.

Czech, B., and P. R. Krausman. 1999. Controversial wildlife management issues in southwestern U. S. wilderness. *International Journal of Wilderness* 5:22–28.

Daubenmire, R. 1968. Plant communities. A textbook of plant synecology. Harper and Row, New Jersey, USA.

Deane, C. E., B. A. Flynn, D. L. Bruning, G. A. Breed, and K. A. Jochum. 2022. Daily abundance of Dall's sheep peaks during late summer in a seasonal habitat of high-management interest. *Ecosphere* 13(1):e03892.

DeCesare, N. J., and D. H. Pletscher. 2004. Bighorn sheep, horizontal visibility, and GIS. *Proceedings of the Biennial Symposium of the Northern Wild Sheep and Goat Council* 14:181–190.

DeCesare, N. J., and D. H. Pletscher. 2006. Movements, connectivity, and resource selection of Rocky Mountain bighorn sheep. *Journal of Mammalogy* 87:531–538.

DeForge, J. R. 1980. Population biology of desert bighorn sheep in the San Gabriel Mountains of California. *Desert Bighorn Council Transactions* 24:29–32.

DeForge, J. R., and J. E. Scott. 1982. Ecological investigations into high lamb mortality. *Desert Bighorn Council Transactions* 26:65–76.

Demarchi, D. A., and H. B. Mitchell 1973. The Chilcotin River bighorn population. *Canadian-Field Naturalist* 87:433–454.

Demarchi, R. A., and C. L. Hartwig. 2004. Status of thinhorn sheep in British Columbia. Wildlife Bulletin B-119 96. British Columbia Minister of Water, Land and Air Protection, Biodiversity Branch, Victoria, British Columbia, Canada.

Dertien, J. S., P. F. Doherty Jr, C. F. Bagley, J. A. Haddix, A. R. Brinkman, and E. S. Neipert. 2017. Evaluating Dall's sheep habitat use via camera traps. *Journal of Wildlife Management* 81:1457–1467.

DeYoung, R. W., E. C. Hellgren, T. E. Fulbright, W. F. Robbins Jr, and I. D. Humphreys. 2000. Modeling nutritional carrying capacity for translocated desert bighorn sheep in western Texas. *Restoration Ecology* 84S:57–65.

Dicus, G. H. 2002. An evaluation of GIS-based habitat models for bighorn sheep winter ranges in Glacier National Park, Montana. Thesis, University of Montana, Missoula, Montana, USA.

Dolan, B. F. 2006. Water developments and desert bighorn sheep: implications for conservation. *Wildlife Society Bulletin* 34:642–646.

Donovan, V. M., S. P. H. Dwinnell, J. L. Beck, C. P. Roberts, J. G. Clapp, G. S. Hiatt, K. L. Monteith, and D. Twidwell. 2021. Fire-driven landscape heterogeneity shapes habitat selection of bighorn sheep. *Journal of Mammalogy* 102:757–771.

Douglas, C. L. 2001. Weather, disease, and bighorn lamb survival during 23 years in Canyonlands National Park. *Wildlife Society Bulletin* 29:297–305.

Douglas, C. L., and L. D. White. 1979. Movements of desert bighorn sheep in the Stubbe Spring area, Joshua Tree National Monument. *Desert Bighorn Council Transactions* 23:71–77.

Drewek, J. R. 1970. Population characteristics and behavior of introduced bighorn sheep in Owyhee County, Idaho. Thesis, University of Idaho, Moscow, Idaho, USA.

Dunn, W. C. 1996. Evaluating bighorn habitat: a landscape approach. Technical Note 395:1–41. United States Bureau of Land Management, Washington, D.C., USA.

Dyksterhuls, E. J. 1983. Habitat-type: a review. *Rangelands* 5:270–271.

Ebert, P. 1978. Bighorn sheep in Oregon. *Oregon Wildlife* 33:3–6.

Enns, G. E., B. A. Jex, and M. S. Boyce. 2023. Diverse migration patterns and seasonal habitat use of Stone's sheep (*Ovis dalli stonei*). *PeerJ* 11:e15215.

Enns, G. E., B. A. Jex, and M. S. Boyce. 2024. Stone's sheep (*Ovis dalli stonei*) lambing and nursery habitat selection. *Canadian Journal of Zoology.* https://doi.org/10.1139/cjz-2023-0028

Epps, C. W., J. D. Wehausen, P. J. Palsbøll, and D. R. McCullough. 2010. Using genetic tools to track desert bighorn sheep colonizations. *Journal of Wildlife Management* 74:522–531.

Erickson, G. L. 1972. The ecology of Rocky Mountain bighorn sheep in the Sun River area of Montana with special reference to summer food habits and range movements. Federal Aid in Wildlife Restoration Project W-120-R-2 and R-3. Montana Fish and Game Department, Bozeman, Montana, USA.

Espenosa-T, A., A. V. Sandoval, and A. J. Contreras-B. 2006. Historical distribution of desert bighorn sheep (*Ovis canadensis mexicana*) in Coahuila, Mexico. *Southwestern Naturalist* 51:282–288.

Etchart, J. L. 2021. Evaluating water use and seasonal ranges of desert Bighorn sheep and Aoudad in the Sierra Vieja Mountains, Texas. Dissertation, Sul Ross State University, Alpine, Texas, USA.

Etchberger, R. C., and P. R. Krausman. 1999. Frequency of birth and lambing sites of a small population of mountain sheep. *Southwestern Naturalist* 44:354–360.

Etchberger, R. C., Krausman, P. R., and Mazaika, R. 1989. Mountain sheep habitat characteristics in the Pusch Ridge Wilderness, Arizona. *Journal of Wildlife Management* 53:902–907.

Eustis, G. P. 1962. Winter lamb surveys on the Kofa game Range. *Desert Bighorn Council Transactions* 6:83–86.

Fairbanks, W. S, J. A. Bailey, and R. S. Cook. 1987. Habitat use by a low-elevation, semicaptive bighorn sheep population. *Journal of Wildlife Management* 51:912–915.

Feldhamer, G. A., L. C. Drickamer, S. H. Vessey, and J. F. Merritt. 1999. *Mammalogy: Adaptation, Diversity, and Ecology.* McGraw Hill, Boston, Massachusetts, USA.

Festa-Bianchet, M. 1988. Seasonal range selection in bighorn sheep: conflicts between forage quality, forage quantity, and predator avoidance. *Oceologia (Berlin)* 75:580–586.

Festa-Bianchet, M. 1989. Individual differences, parasites, and the costs of reproduction for bighorn ewes (*Ovis canadensis*). *Journal of Animal Ecology* 58:785–795.

Flores, M. G., L. J. Moncayo, and R. F. Takaki. 1972. Tipos de vegetacion de la Republica Mexicana. Subsecretia di Planeacion, Dirección General de Estudios, Dirección Agricola, Mexico, D.F.

Forbes, K. M. 2001. Modeling bighorn sheep habitat in northwest Nebraska. Thesis, University of Nebraska, Lincoln, Nebraska, USA.

Forshee, S. C., M. S. Mitchell, and T. R. Stephenson. 2022. Predator avoidance influences selection of neonatal lambing habitat by Sierra Nevada bighorn sheep. *Journal of Wildlife Management* 8:e22311.

Foster, J. A., L. A. Harveson, and M. T. Pittman. 2005. Use of guzzlers by bighorn sheep in the Chihuahuan Desert. *Desert Bighorn Council Transactions* 48:12–22.

Frid, A. 1997. Vigilance by female Dall's sheep: interactions between predation risk factors. *Animal Behavior* 53:799–808.

Frisina, M. R. 1974. Ecology of bighorn sheep in the Sun River Area of Montana during fall and spring. Federal Aid in Wildlife Restoration Project W-120-R-4, 5. Montana Fish and Game Department, Helena, Montana, USA.

Garrison, K. R., J. W. Cain III, E. M. Rominger, and E. J. Goldstein. 2016. Sympatric cattle grazing and desert bighorn sheep foraging. *Journal of Wildlife Management* 80:197–207.

Gedir, J. V., J. W. Cain III, P. R. Krausman, J. D. Allen, G. C. Duff, and J. R. Morgart. 2016. Potential foraging decisions by a desert ungulate to balance water and nutrient intake in a water-stressed environment. *PloS ONE* 11:e0148795.

Gedir, J. V., J. W. Cain III, T. L. Swetnam, P. R. Krausman, and J. R. Morgart. 2020. Extreme drought and adaptive resource selection by a desert mammal. *Ecosphere* 11:e03175.

Geist, V. 1971. *Mountain Sheep, A Study in Behavior and Evolution.* University of Chicago Press, Chicago, USA, and London, United Kingdom.

Giles Jr., R. H. 1978. *Wildlife Management.* W. H. Freeman, San Francisco, California, USA.

Gionfriddo, J. P., and P. R. Krausman. 1986. Summer habitat use by mountain sheep. *Journal of Wildlife Management* 50:331–336.

Glass, D. M., P. R. Prentice, A. D. Evans, and O. J. Schmitz. 2022. Local differences in maximum temperature determine water use among desert bighorn sheep populations. *Journal of Wildlife Management* 86:e22313.

Guerrero-Cárdenas, I., I. Tovar-Zamora, and S. Alvarez-Cárdenas. 2003. Factores que afectan la distribución espacial del borrego cimarrón *Ovis canadensis weemsi* en la Sierra del Mechudo, BCS, México. *Anales del Instituto de Biología, Universidad Nacional Autónoma de México, Serie Zoológica* 74:83–98.

Hall, L. S., P. R. Krausman, and M. L. Morrison. 1997. The habitat concept and a plea for standard terminology. *Wildlife Society Bulletin* 25:173–182.

Halloran, A. F., and O. V. Deming. 1958. Water development for desert bighorn sheep. *Journal of Wildlife Management* 22:1–9.

Hansen, C. G. 1980. Habitat. Pages 64–79 *in* G. Monson, and L. Sumner, editors. *The Desert Bighorn: Its Life History, Ecology, and Management.* University of Arizona Press, Tucson, Arizona, USA.

Hansen, M. C. 1982. Status and habitat preference of California bighorn sheep on Sheldon National Wildlife Refuge, Nevada. Thesis, Oregon State University, Corvallis, Oregon, USA.

Hansen, M. C. 1996. Foraging ecology of female Dall's sheep in the Brooks Range, Alaska. Dissertation, University of Alaska, Fairbanks, Alaska, USA.

Harris, G. M., D. R. Stewart, D. Brown, L. Johnson, J. Sanderson, A. Alvidrez, T. Waddell, and R. Thompson. 2020. Year-round water management for desert bighorn sheep corresponds with visits by predators not bighorn sheep. *PLoS ONE* 15:e0241131.

Harris, R. B., K. Aluzas, L. Balyx, J. Belt, J. Berger, M. Biel, T. Chilton-Radandt, S. D. Côté, J. Cunningham, A. Ford et al. 2023. Habituated, tolerant, or salt-conditioned mountain goats and human safety. *Human-Wildlife Interactions* 17:11.

Hebblewhite, M., E. Merrill, and G. McDermid. 2008. A multi-scale test of the forage maturation hypothesis in a partially migratory ungulate population. *Ecological Monographs* 78:141–166.

Heimer, W. E. 1974. The importance of mineral licks to Dall's sheep in Interior Alaska and its significance to sheep management. *Biennial Symposium of the Northern Wild Sheep and Goat Council* 3:49–63.

Heimer, W. E., F. W. Mauer, and S. W. Keller. 1994. The effects of physical geography on Dall sheep habitat quality and home range size. *Biennial Symposium of the Northern Wild Sheep and Goat Council* 9:144–148.

Hengeveld, P. E., and J. C. Cubberley. 2011. Stone's sheep population dynamics and habitat use in the Sulphur/8 Mile oil and gas pre-tenure plan area, northern British Columbia, 2005–2010. Synergy Applied Ecology, Mackenzie, British Columbia, Canada.

Hetem, R. S., A. Fuller, S. K. Maloney, and D. Mitchell. 2014. Responses of large mammals to climate change. *Temperature* 1:15–127.

Hik, D. S., and J. Carey. 2000. Cohort variation in horn growth of Dall's sheep rams in the southwestern Yukon, 1969–1999. *Biennial Symposium of the Northern Wild Sheep and Goat Council* 12:88–100.

Hines, K. 2019. Post-partum habitat use for Peninsular bighorn sheep (*Ovis canadensis nelsoni*) in Southern California. Thesis, California State University San Marcos, California, USA.

Hoefs, M. 1984. Productivity and carrying capacity of a subarctic sheep winter range. *Arctic* 37:141–147.

Hoefs, M., and I. M. Cowan. 1979. Ecological investigation of a population of Dall sheep (*Ovis dalli dalli*). *Syesis* 12(Supplement 1):1–81.

Hoefs, M., and M. Bayer. 1983. Demographic characteristics of an unhunted Dall sheep (*Ovis dalli dalli*) population in southwest Yukon, Canada. *Canadian Journal of Zoology* 61:1346–1357.

Hoglander, C., B. G. Dickson, S. S. Rosenstock, and J. J. Anderson. 2015. Landscape models of space use by desert bighorn sheep in the Sonoran Desert of southwestern Arizona. *Journal of Wildlife Management* 79:77–91.

Holl, S. A. 1982. Evaluation of bighorn sheep habitat. *Desert Bighorn Sheep Council Transactions* 26:47–49.

Holl, S. A., and V. C. Bleich. 1983. San Gabriel Mountain sheep: biological and management considerations. No report number:1–134. U. S. Forest Service, San Bernardino National Forest Administration Report, California, USA.

Hudson, R. J., D. M. Hebert, and V. C. Brink. 1976. Occupational patterns of wildlife on a major East Kootenay winter-spring range. *Rangeland Ecology and Management* 29:38–43.

Hughes, L. G. 1997. A GIS-based evaluation of the Big Horn Mountains for reintroduction of Rocky Mountain bighorn sheep. Thesis, University of Wyoming, Laramie, Wyoming, USA.

Hutto, R. L. 1985. Habitat selection by nonbreeding migratory land birds. Pages 455–476 in M. L. Cody, editor. *Habitat Selection in Birds*. Academic Press, Orlando, Florida, USA.

Jaeger, E. C. 1957. *North American Deserts*. Stanford University Press, Palo Alto, California, USA.

Jansen, B. D., P. R. Krausman, J. R. Heffelfinger, and J. C. Devos Jr. 2006. Bighorn sheep selection of landscape features in an active copper mine. *Wildlife Society Bulletin* 34:1121–1126.

Jex, B. A., J. B. Ayotte, V. C. Bleich, C. E. Brewer, D. L. Bruning, T. M. Hegel, N. C. Larter, R. A. Schwanke, H. M. Schwantje, and M. W. Wagner. 2016. Thinhorn Sheep: conservation challenges and management strategies for the 21st century. Wild Sheep Working Group, Western Association of Fish and Wildlife Agencies, Boise, Idaho, USA.

Johnson, D. H. 1980. The comparison of usage and availability measurements for evaluation resource preferences. *Ecology* 61:65–71.

Johnson, T. L., and D. M. Swift. 2000. A test of a habitat evaluation procedure for Rocky Mountain bighorn sheep. *Restoration Ecology* 8(4S):47–56.

Jones, A. S., E. S. Rubin, M. J. Clement, L. E. Harding, and J. I. Mesler. 2022. Desert bighorn sheep habitat selection, group size, and mountain lion predation risk. *Journal of Wildlife Management* 86:e22173.

Jorgensen, M. C., and R. E. Turner. 1975. Desert bighorn of the Anza-Borrego Desert State Park. *Desert Bighorn Council Transactions* 19:51–53.

Karsch, R. C., J. W. Cain III, E. M. Rominger, and E. J. Goldstein. 2016. Desert bighorn sheep lambing habitat: parturition, nursery, and predation sites. *Journal of Wildlife Management* 80:1069–1080.

Kelly, W. E. 1979. A comparison of 3 bighorn areas on the Humboldt National Forest. *Desert Bighorn Council Transactions* 23:37–39.

Koizumi, C. L., and A. E. Derocher. 2019. Predation risk and space use of a declining Dall's sheep (*Ovis dalli dalli*) population. *PLoS ONE* 14:e0215519.

Koizumi, C. L., J. Carey, M. Branigan, and K. Callaghan. 2011. Status of Dall's sheep (*Ovis dalli dalli*) in the Northern Richardson Mountains. Yukon Fish and Wildlife Branch Report, Canada.

Kornet, C. A. 1978. Status and habitat use of California bighorn sheep on Hart Mountain, Oregon. Thesis, Oregon State University, Corvallis, Oregon, USA.

Kovach, S. D. 1979. An ecological survey of the White Mountain Peak bighorn. *Desert Bighorn Council Transactions* 23:57–61.

Krausman, P. R. 1979. Use of caves by white-tailed deer. *Southwestern Naturalist* 24:203–203.

Krausman, P. R. 2017. *And Then There Were None: The Demise of Desert Bighorn Sheep in the Pusch Ridge Wilderness*. University of New Mexico Press, Albuquerque, New Mexico, USA.

Krausman, P. R., A. V. Sandoval, and R. C. Etchberger. 1999. Natural history of desert bighorn sheep. Pages 139–191 in R. Valdez, and P. R. Krausman. *Mountain Sheep of North America*. University of Arizona Press, Tucson, Arizona, USA.

Krausman, P. R., and B. D. Leopold. 1986a. Habitat components for desert bighorn sheep in the Harquahala Mountains, Arizona. *Journal of Wildlife Management* 50:504–508.

Krausman, P. R., and B. D. Leopold. 1986b. The importance of small populations of desert bighorn sheep. *Transactions of the North American Wildlife and Natural Resource Conference* 51:52–61.

Krausman, P. R., and M. L. Morrison. 2016. Another plea for standard terminology. *Journal of Wildlife Management* 80:1143–1144.

Krausman, P. R., and R. C. Etchberger. 1995. Response of desert ungulates to a water project in Arizona. *Journal of Wildlife Management* 59:292–300.

Krausman, P. R., and R. T. Bowyer. 2003. Mountain sheep. Pages 1095–1115 in G. A. Feldhamer, B. C. Thompson, and J. A. Chapman, editors. *Wild Mammals of North America*, Second edition. Johns Hopkins University Press, Baltimore, Maryland, USA.

Krausman, P. R., B. D. Leopold, R. F. Seegmiller, and S. G. Torres. 1989. Relationships between desert bighorn sheep and habitat in western Arizona. *Wildlife Monographs* 53:1–66.

Leopold, A. 1933. *Game Management*. Charles Scribner's Sons, New York, New York, USA.

Leslie Jr, D. M., and C. L. Douglas. 1980. Human disturbance at water sources of desert bighorn sheep. *Wildlife Society Bulletin* 8:284–290.

Leslie, D. M., and C. L. Douglas. 1979. Desert bighorn sheep of the River Mountains, Nevada. *Wildlife Monographs* 66:1–56.

Locke, S. L. 2003. Habitat use and movements of desert bighorn sheep at Elephant Mountain Wildlife Management area, Texas. Thesis, Sul Ross State University, Alpine, Texas, USA.

Locke, S. L., C. E. Brewer, and L. A. Harveson. 2005a. Habitat use and movements of bighorn sheep in west Texas. *Desert Bighorn Council Transactions* 48:1–11.

Locke, S. L., C. E. Brewer, and L. A. Harveson. 2005b. Identifying landscapes for desert bighorn sheep translocations in Texas. *The Texas Journal of Science* 57:25–35.

Loehr, J., J. Carey, R. B. O'Hara, and D. S. Hik. 2010. The role of phenotypic plasticity in responses of hunted thinhorn sheep ram horn growth to changing climate conditions. *Journal of Evolutionary Biology* 23:783–790.

Lohuis, T. 2015. Ewe Dall's sheep survival, pregnancy and parturition rates, and lamb recruitment in GMU 13D, Chugach Mountains, Alaska. Federal Aid in Wildlife Restoration Project AKW-4. Alaska Department of Fish and Game, Juneau, Alaska, USA.

Longshore, K. M., C. Lowrey, and D. B. Thompson. 2009. Compensating for diminishing natural water: predicting the impacts of water development on summer habitat of desert bighorn sheep. *Journal of Arid Environments* 73:280–286.

Longshore, K., C. Lowrey, and P. Cummings. 2016. Foraging at the wildland–urban interface decouples weather as a driver of recruitment for desert bighorn sheep. *Wildlife Society Bulletin* 40:494–499.

Longshore, K., Lowrey, C., and Thompson, D. B. 2013. Detecting short-term responses to weekend recreation activity: desert bighorn sheep avoidance of hiking trails. *Wildlife Society Bulletin* 37:698–706.

Lowrey, C., and K. M. Longshore. 2017. Tolerance to disturbance regulated by attractiveness of resources: a case study of desert bighorn sheep within the River Mountains, Nevada. *Western North American Naturalist* 77:82–98.

Luckhurst, A. J. 1973. Stone sheep and their habitat in the northern Rocky mountain foothills of British Columbia. Thesis, University of British Columbia, Vancouver, Canada.

Lula, E. S., B. Lowrey, K. M. Proffitt, A. R. Litt, J. A. Cunningham, C. J. Butler, and R. A. Garrott. 2020. Is habitat constraining bighorn sheep restoration? A case study. *Journal of Wildlife Management* 84:588–600.

MacArthur, R. A., Johnston, R. H., and Geist, V. 1979. Factors influencing heart-rate in free-ranging bighorn sheep-physiological approach to the study of wildlife harassment. *Canadian Journal of Zoology* 57:2010–2021.

McCann, J. L. 1956. Ecology of mountain sheep. *American Midland Naturalist* 56:297–324.

McCarthy, C. W., and J. A. Bailey. 1994. Habitat requirements of desert bighorn sheep. Special Report 69. Colorado Division of Wildlife, Denver, Colorado, USA.

McKinney, T., J. C. Devos Jr, W. B. Ballard, and S. R. Boe. 2006a. Mountain lion predation of translocated desert bighorn sheep in Arizona. *Wildlife Society Bulletin* 34:1255–1263.

McKinney, T., S. R. Boe, and J. C. deVos Jr. 2003. GIS-based evaluation of escape terrain and desert bighorn sheep populations in Arizona. *Wildlife Society Bulletin* 31:1229–1236.

McKinney, T., T. W. Smith, and J. C. deVos, Jr. 2006b. Evaluation of factors potentially influencing a desert bighorn sheep population. *Wildlife Monographs* 164:1–36.

McQuivey, R. P. 1978. The bighorn sheep of Nevada. Nevada Department of Fish and Game, Reno, USA. *Biological Bulletin* 6:1–81.

Mehlam, M. 2019. Pairwise comparisons of shrub change across alpine climates show heterogeneous response to temperature in Dall's sheep range. Thesis, University of Alaska, Fairbanks, Alaska, USA.

Mendoza, V. J. 1976. The bighorn sheep of the state of Sonora. *Desert Bighorn Council Transactions* 20:25–26.

Merkle, J. A., K. L. M. Monteith, E. O. Hayes, K. R. Hersey, A. D. Middleton, B. A. Oates, H. Sawyer, B. M. Scurlock, and M. J. Kauffman. 2016. Large herbivores surf waves of green-up during spring. *Proceedings of the Royal Society B* 283:20160456.

Merritt, M. F. 1974. Measurement of utilization of bighorn sheep habitat in the Snata Rosa Mountains. *Desert Bighorn Council Transactions* 18:4–17.

Monson, G. 1980. Distribution and abundance. Pages 40–51 in G. Monson, and L. Sumner, editors. *The Desert Bighorn: Its Life History, Ecology, and Management*. University of Arizona Press, Tucson, Arizona, USA.

Mooring, M. S., T. A. Fitzpatrick, T. T. Nishihira, and D. D. Reisig. 2004. Vigilance, predation risk, and the Allee effect in desert bighorn sheep. *Journal of Wildlife Management* 68:519–532.

Morrison, D. C. 1972. Habitat utilization by mule deer in relation to cattle and California bighorn sheep in the Ashnola River Valley, British Columbia, Thesis, University of British Columbia, Vancouver, Canada.

Morrison, M. L., and H. A. Mathewson. 2015. *Wildlife Habitat Conservation*. Johns Hopkins University Press, Baltimore, Maryland, USA.

Morrison, M. L., B. G. Marcot, and R. W. Mannan. 1992. *Wildlife-Habitat Relationships: Concepts and Applications*. University of Wisconsin Press, Madison, Wisconsin, USA.

Murie, A. 1944. The wolves of Mount McKinley. *U. S. National Park Service Fauna Series* 5:1–238.

Murphy, D. D., and B. D. Noon. 1991. Coping with uncertainty in wildlife biology. *Journal of Wildlife Management* 55:773–782.

Murphy, E. C., and K. R. Whitten. 1996. Dall sheep demography in McKinley Park and a reevaluation of Murie's data. *Journal of Wildlife Management* 60:597–609.

Nichols, L. 1973. Sheep report. Alaska Department of Fish and Game, federal Aid in Wildlife Restoration Annual Project Report 14. Alaska Department of Fish and Game, Juneau, Alaska, USA.

Nichols, L., and F. L. Bunnell. 1999. Natural history of thinhorn sheep. Pages 23–77 *in* R. Valdez, and P. R. Krausman, editors. *Mountain Sheep of North America*. University of Arizona Press, Tucson, Arizona, USA.

O'Farrill, G., R. A. Medellín, S. F. Matter, and G. N. Cameron. 2019. Habitat use and diet of desert bighorn sheep (*Ovis canadensis mexicana*) and endemic mule deer (*Odocoileus hemionus nelsoni*) on Tiburón Island, Mexico. *Southwestern Naturalist* 63:225–234.

Ober, E. H. 1931. The mountain sheep of California. *California Fish and Game* 17:27–39.

Odum, E. P. 1971. *Fundamentals of Ecology*. Third edition. W. E. Saunders, Philadelphia, Pennsylvania, USA.

Oldemeyer, J. L., W. J. Barmore, and D. L. Gilbert. 1971. Winter ecology of bighorn sheep in Yellowstone National Park. *Journal of Wildlife Management* 35:257–269.

Pallister, G. L. 1974. The seasonal distribution and range use of bighorn sheep in the Beartooth Mountains, with special reference to the West Rosebud and Stillwater herds. Federal Aid in Wildlife Restoration Project W-120-R-5, Montana Fish and Game Department, Bozeman, Montana, USA.

Papouchis, C. M., Singer, F. J., and Sloan, W. B. (2001). Responses of desert bighorn sheep to increased human recreation. *Journal of Wildlife Management* 65:573–582.

Parker, K. L., and A. B. D. Walker. 2007. Habitat selection and behavioural strategies of Stone's sheep in the Besa-Prophet. Muskwa-Kechika Advisory Board report, British Columbia, Canada.

Parsons, Z. D. 2007. Cause specific mortality of desert bighorn sheep lambs in the Fra Cristobal Mountains, New Mexico, USA. Thesis, University of Montana, Missoula, Montana, USA.

Poole, K. G., R. Serrouya, I. E. Teske, and K. Podrasky. 2016. Bighorn sheep winter habitat selection and seasonal movements in an area of active coal mining. *Canadian Journal of Zoology* 94:733–745.

Rachlow, J. L., and R. T. Bowyer. 1991. Interannual variation in timing and synchrony of parturition in Dall's sheep. *Journal of Mammalogy* 72:487–492.

Rachlow, J. L., and R. T. Bowyer. 1994. Variability in maternal behavior by Dall's sheep: environmental tracking or adaptive strategy? *Journal of Mammalogy* 75:328–337.

Rachlow, J. L., and R. T. Bowyer. 1998. Habitat selection by Dall's sheep (Ovis dalli): maternal trade-offs. *Journal of Zoology (London)* 245:457–465.

Rattenbury, K. L., J. H. Schmidt, D. K. Swanson, B. L. Borg, B. A. Mangipane, and P. J. Sousanes. 2018. Delayed spring onset drives declines in abundance and recruitment in a mountain ungulate. *Ecosphere* 9:e02513.

Rich, L. N., S. R. Beissinger, J. S. Brashares, and B. J. Furnas. 2019. Artificial water catchments influence wildlife distribution in the Mojave Desert. *Journal of Wildlife Management* 83:855–865.

Riggs, R. A. 1977. Winter habitat use patterns and population of bighorn sheep in Glacier national Park. Thesis, University of Idaho, Moscow, Idaho, USA.

Risenhoover, K. L., and J. A. Bailey. 1980. Visibility: an important habitat factor for and indigenous, low-elevation, bighorn herd in Colorado. *Biennial Symposium of the Northern Wild Sheep and Goat Council* 2:18–28.

Risenhoover, K. L., and J. A. Bailey. 1985. Foraging ecology of mountain sheep: implications for habitat management. *Journal of Wildlife Management* 49:797–804.

Risenhoover, K. L., J. A. Bailey, and L. A. Wakelyn. 1988. Assessing the Rocky Mountain bighorn sheep management problem. *Wildlife Society Bulletin* 16:346–357.

Robinson, R. W., T. S. Smith, J. C. Whiting, R. T. Larsen, and J. M. Shannon. 2020. Determining timing of births and habitat selection to identify lambing period habitat for bighorn sheep. *Frontiers in Ecology and Evolution* 8:97.

Roffler, G. H., L. G. Adams, and M. Hebblewhite. 2017. Summer habitat selection by Dall's sheep in Wrangell-St. Elias National Park and Preserve, Alaska. *Journal of Mammalogy* 98:94–105.

Rominger, E. M., E. J. Goldstein, and M. A. Evans. 2008. Biological and statistical errors make inferences circumspect: response to Bender and Weisenberger. *Journal of Wildlife Management* 72:580–582.

Rosenstock, S. S., J. J. Hervert, V. C. Bleich, and P. R. Krausman. 2001. Muddying the water with poor science: a reply to Broyles and Cutler. *Wildlife Society Bulletin* 29:734–738.

Rosenstock, S. S., W. B. Ballard, and J. C. DeVos. 1999. Benefits and impacts of wildlife water developments. *Journal of Range Management* 52:302–311.

Rubin, E. S., W. M. Boyce, C. J. Stermer, and S. G. Torres. 2002. Bighorn sheep habitat use and selection near an urban environment. *Biological Conservation* 104:251–263.

Ruckstuhl, K. E., and P. Neuhaus. 2002. Sexual segregation in ungulates: a comparative test of three hypothesis. *Biological Review* 77:77–96.

Russo, J. 1956. The desert bighorn in Arizona. Arizona Game and Fish Department, Phoenix, USA. *Bulletin* 1:1–153.

Salzman, J. 1990. Evolution and application of critical habitat under the Endangered Species Act. *Harvard Environmental Law Review* 14:311–342.

Sandoval, A. V. 1979. Preferred habitat of desert bighorn sheep in the San Andres Mountains, New Mexico. Thesis, Colorado State University, Fort Collins, Colorado, USA.

Sappington, J. M., K. M. Longshore, and D. B. Thompson. 2007. Quantifying landscape ruggedness for animal habitat analysis: a case study using bighorn sheep in the Mojave Desert. *Journal of Wildlife Management* 71:1419–1426.

Sargeant, G. A., L. E. Eberhardt, and J. M. Peek. 1994. Thermoregulation by mule deer (*Odocoileus hemionus*) in arid rangelands of southcentral Washington. *Journal of Mammalogy* 75:536–544.

Schallenberger, A. D. 1966. Food habits, range use and interspecific relationships of bighorn sheep in the Sun River area, west-central Montana. Thesis, Montana State University, Bozeman, Montana, USA.

Schmidt-Nielsen, K. 1979. *Desert Animals: Physiological Problems of Heat and Water*. Dover, New York, New York, USA.

Schoenhardt, M. A. M. 2023. Post-lambing spatial distribution of Dall's Sheep in southwest Yukon. Thesis, Queen's University, Kingston, Ontario, Canada.

Seip, D. R. 1983. Foraging ecology and nutrition of Stone's sheep. Dissertation, University of British Columbia, Vancouver, Canada.

Seip, D. R., and F. L. Bunnell. 1985. Foraging behaviour and food habits of Stone's sheep. *Canadian Journal of Zoology* 63:1638–1646.

Semmens, W. J. 1996. Seasonal movements and habitat use of the Highland/Pioneer Mountain bighorn sheep herd of southwest Montana. Thesis, University of Montana, Missoula, Montana, USA.

Sesnie, S. E., B. G. Dickson, S. S. Rosenstock, and J. M. Rundall. 2012. A comparison of Landsat TM and MODIS vegetation indices for estimating forage phenology in desert bighorn sheep (*Ovis canadensis nelsoni*) habitat in the Sonoran Desert, USA. *International Journal of Remote Sensing* 33:276–286.

Shackleton, D. M. 1985. *Ovis canadensis. Mammalian Species* 230:1–9.

Shackleton, D. M., and Haywood, J. 1985. Early mother–young interactions in California bighorn sheep, *Ovis canadensis californiana. Canadian Journal of Zoology* 63:868–875.

Shackleton, D. M., C. C. Shank, and B. M. Wikeem. 1999. Natural history of Rocky Mountain and California bighorn sheep. Pages 78–138 *in* R. Valdez, and P. R. Krausman, editors. *Mountain Sheep of North America.* University of Arizona Press, Tucson, Arizona, USA.

Shank, C. C. 1979. Sexual dimorphism and the ecological niche of wintering Rocky Mountain bighorn sheep. Dissertation, University of Calgary, Alberta, Canada.

Shannon, N. H., R. J. Hudson, V. C. Brink, and W. D. Kitts. 1975. Determinants of spatial distribution of Rocky Mountain bighorn sheep. *Journal of Wildlife Management* 39:387–401.

Sim, Z., C. S. Davis, B. A. Jex, T. Hegel, and D. W. Coltman. 2019. Management implications of highly resolved hierarchical population genetic structure in thinhorn sheep. *Conservation Genetics* 20:185–201.

Simmons, N. M. 1969. Heat stress and bighorn behavior in the Cabeza Prieta game Range, Arizona. *Desert Bighorn Council Transactions* 13:56–63.

Simmons, N. M. 1982. Seasonal ranges of Dall's sheep, Mackenzie Mountains, Northwest Territories. *Arctic* 35:512–518.

Simpson, N. O., K. M. Stewart, and V. C. Bleich. 2011. What have we learned about water developments for wildlife? Not enough! *California Fish and Game* 97:190–209.

Sittler, K. L., K. L. Parker, and M. P. Gillingham. 2015. Resource separation by mountain ungulates on a landscape modified by fire. *Journal of Wildlife Management* 79:591–604.

Sittler, K. L., K. L. Parker, and M. P. Gillingham. 2019. Vegetation and prescribed fire: implications for Stone's sheep and elk. *Journal of Wildlife Management* 83:393–409.

Smith, D. R. 1954. The bighorn sheep in Idaho: its status, life history and management. Wildlife Bulletin 1. Idaho Game and Fish Department, Boise, Idaho, USA.

Smith, J., and Burr, S. 2011. Environmental attitudes and desired social-psychological benefits of off-highway vehicle users. *Forests* 2:875–893.

Smith, J. B., Grovenburg, T. W., and Jenks, J. A. 2015. Parturition and bed site selection of bighorn sheep at local and landscape scales. *Journal of Wildlife Management* 79:393–401.

Smith, K. R. 2020. Habitat selection of Dall's Sheep in the Chugach Mountains using step-selection function analysis. Thesis, Alaska Pacific University, Anchorage, Alaska, USA.

Smith, T. S., J. T. Flinders, and D. S. Winn. 1991. A habitat evaluation procedure for Rocky Mountain bighorn sheep in the Intermountain West. *Great Basin Naturalist* 51:205–225.

Spalding, D. J., and J. N. Bone. 1970. The California bighorn sheep of the south Okanagan Valley, British Columbia. Fish and Wildlife Branch, Victoria, British Columbia, Canada. *Wildlife Management Publication* 3:1–45.

Sproat, K. K., N. R. Martinez, T. S. Smith, W. B. Sloan, J. T. Flinders, J. W. Bates, J. G. Cresto, and V. C. Bleich. 2019. Desert bighorn sheep responses to human activity in south-eastern Utah. *Wildlife Research* 47:16–24.

Steel, T. L., and G. W. Workman. 1990. Influence of forage quantity on microhabitat use by desert bighorn sheep. *Desert Bighorn Council Transactions* 34:1–5.

Stelfox, J. G. 1971. Bighorn sheep in the Canadian Rockies: a history, 1800–1970. *Canadian Field Naturalist* 85:101–122.

Stelfox, J. G. 1975. Range ecology of Rocky Mountain bighorn sheep in Canadian National Parks. Dissertation, University of Montana, Missoula, Montana, USA.

Sudgen, L. G. 1961. The California bighorn in British Columbia with particular reference to the Churn Creek herd. British Columbia Department of Recreation and Conservation, Victoria, British Columbia, Canada.

Sweanor, P. Y., M. Gudorf, and F. J. Singer. 1996. Application of a GIS-based bighorn sheep habitat model in Rocky Mountain region of national parks. *Proceedings of the Biennial Symposium of the Northern Wild Sheep and Goat Council* 10:118–125.

Swift, P. K., J. D. Wehausen, H. B. Ernest, R. S. Singer, A. M. Pauli, H. Kinde, T. E. Rocke, and V. C. Bleich. 2000. Desert bighorn sheep mortality due to presumptive type C botulism in California. *Journal of Wildlife Diseases* 36:184–189.

Tarango, L. A., P. R. Krausman, and R. Valdez. 2002. Habitat use by desert bighorn sheep in Sonora, Mexico. *Pirineos* 157:219–226.

Terry, P. J., A. C. Alvidrez, and C. W. Black. 2022. Factors affecting bighorn sheep activity at water developments in southwestern Arizona. *Journal of Wildlife Management* 86:e22134.

Terwilliger, M. L. N. 2005. Population and habitat analyses for Dall's Sheep (*Ovis dalli*) in Wrangell - St. Elias National Park and Preserve. Thesis, University of Alaska, Fairbanks, Alaska, USA.

Thomas-Kuzilik, R., J. A. Becker, and J. A. Merkle. 2020. Behavioral plasticity of large mammals in the Rocky Mountain to variation in temperature. *University of Wyoming-National Park Service Research Station Annual Reports* 43:48–55.

Thompson, C. J., and J. W. Cain III. 2024. Biological feasibility of introducing bighorn sheep to the Jicarilla Apache Nation. Report to the Jicarilla, Apache Nation, New Mexico, USA.

Thompson, R. W., and J. C. Turner. 1982. Temporal geographic variation in the lambing season of bighorn sheep. *Canadian Journal of Zoology* 60:1781–1793.

Tilton, M. E. 1977. Habitat selection and use by bighorn sheep (*Ovis canadensis*) on a northwestern Montana winter range. University of Montana, Missoula, Montana, USA.

Tilton, M. E., and E. E. Willard. 1982. Winter habitat selection by mountain sheep. *Journal of Wildlife Management* 46:359–366.

Todd, J. W. 1972. Foods of Rocky Mountain bighorn sheep in southern Colorado. Thesis, Colorado State University, Fort Collins, Colorado, USA.

Turner, J. C. 1973. Water energy and electrolyte balance in the bighorn sheep (*Ovis canadensis*). Dissertation, University of California, Riverside, California, USA.

Turner, J. C. 1979. Osmotic fragility of desert bighorn sheep red blood cells. *Comparative Biochemistry Physiology A* 64:167–175.

Turner, J. C., and R. A. Weaver. 1980. Water. Pages 100–112 *in* G. Monson, and L. Sumner, editors. *The Desert Bighorn: Its Life History, Ecology, and Management.* University of Arizona Press, Tucson, Arizona, USA.

Valdez, R., and P. R. Krausman. 1999. *Mountain Sheep of North America.* University of Arizona Press, Tucson, Arizona, USA.

Van de Kerk, M., D. Verbyla, A. W. Nolin, K. J. Sivy, and L. R. Prugh. 2018. Range-wide variation in the effect of spring snow phenology on Dall sheep population dynamics. *Environmental Research Letters* 13:075008.

Van de Kerk, M., S. Arthur, M. Bertram, B. Borg, J. Herriges, J. Lawler, B. Mangipane, C. L. Koizumi, C. Wendling, and L. Prugh. 2020. Environmental influences on Dall's sheep survival. *Journal of Wildlife Management* 84:1127–1138.

Van Dyke, W. A. 1978. Population characteristics and habitat utilization of bighorn sheep, Steens Mountain, Oregon. Thesis, Oregon State University, Corvallis, Oregon, USA.

Van Dyke, W. A., A. Sands, J. Yoakum, A. Polenz, and J. Blasdell. 1983. Wildlife habitats in managed rangelands: the Great Basin of southeastern Oregon: bighorn sheep. *Pacific Northwest Forest and Range Experiment Station, United States Department of Agriculture Forest Service, General Technical Review* 159:1–37.

Waddell, R. B., C. S. O'Brien, and S. S. Rosenstock. 2007. Bighorn sheep use of a developed water in southwestern Arizona. *Desert Bighorn Council Transactions* 49:8–17.

Wagner, F. H. 1978. Livestock grazing and the livestock industry. Pages 121–145 *in* H. R. Brokaw, editor. Wildlife and America: Contributions to an Understanding of American Wildlife and Its Conservation. Council on Environmental Quality, Washington, D.C., USA.

Wakeling, B. F., and W. H. Miller. 1989. Bedside characteristics of desert bighorn sheep in the Superstition Mountains, Arizona. *Desert Bighorn Council Transactions* 33:6–8.

Walker, A. B. D., K. L. Parker, and M. P. Gillingham. 2006. Behaviour, habitat associations, and intrasexual differences of female Stone's sheep. *Canadian Journal of Zoology* 84:187–1201.

Walker, A. B. D., K. L. Parker, M. P. Gillingham, D. D. Gustine, and R. J. Lay. 2007. Habitat selection by female Stone's sheep in relation to vegetation, topography, and risk of predation. *Ecoscience* 14:55–70.

Warrick, G. D., and P. R. Krausman. 1989. Barrel cacti consumption by desert bighorn sheep. *Southwestern Naturalist* 34:483–486.

Watts, T. J. 1979. Status of the Big Hatchet desert sheep population, New Mexico. *Desert Bighorn Council Transactions* 23:92–94.

Weckerly, F. L. 1998. Sexual size dimorphism: influences of mass and mating systems in the most dimorphic mammals. *Journal of Mammalogy* 79:33–52.

Welch, R. D. 1969. Behavioral patterns of desert bighorn sheep in south-central New Mexico. *Desert Bighorn Council Transactions* 13:114–129.

Welles, R. E., and F. B. Welles. 1961. The bighorn of Death Valley. *United States National Park Service, Fauna Series* 6:1–242.

Whiting, J. C., D. D. Olson, J. M. Shannon, R. T. Bowyer, R. W. Klaver, and J. T. Flinders. 2012. Timing and synchrony of births in bighorn sheep: implications for reintroduction and conservation. *Wildlife Research* 39:565–572.

Whiting, J. C., R. T. Bowyer, and J. T. Flinders. 2009a. Annual use of water sources by reintroduced Rocky Mountain bighorn sheep (*Ovis canadensis canadensis*): effects of season and drought. *Acta Theriologica* 54:127–136.

Whiting, J. C., R. T. Bowyer, and J. T. Flinders. 2009b. Diel use of water by reintroduced bighorn sheep. *Western North American Naturalist* 69:407–412.

Whiting, J. C., R. T. Bowyer, J. T. Flinders, V. C. Bleich, and J. G. Kie. 2010. Sexual segregation and use of water by bighorn sheep: implications for conservation. *Animal Conservation* 13:541–548.

Whiting, J. C., V. C. Bleich, R. T. Bowyer, and R. T. Larsen. 2011. Water availability and bighorn sheep: life-history characteristics and persistence of populations. *Advances in Environmental Research* 21:131–163.

Whittaker, D. G., S. D. Ostermann, and W. M. Boyce. 2004. Genetic variability of reintroduced California bighorn sheep in Oregon. *Journal of Wildlife Management* 68:850–859.

Whitten, K. R. 1997. Estimating population size and composition of Dall's sheep in Alaska: assessment of previously used methods and experimental implementation of new techniques. Federal Aid in Wildlife Restoration Research Report. Project 6.11. Alaska Department of Fish and Game, Juneau, Alaska, USA.

Wiedmann, B. P., and V. C. Bleich. 2014. Demographic responses of bighorn sheep to recreational activities: a trial of a trail. *Wildlife Society Bulletin* 38:773–782.

Wiedmann, B. P., and G. A. Sargeant. 2014. Ecotypic variation in recruitment of reintroduced bighorn sheep: implications for translocation. *Journal of Wildlife Management* 78:394–401.

Wiens, J. A. 1984. Resource systems, populations, and communities. Pages 397–436 *in* P. W. Price, C. N. Slobodchikoff, and W. S. Gaud, editors. *A New Ecology: Novel Approaches to Interactive Systems*. John Wiley & Sons, New York, New York, USA.

Wikeem, B. M. 1984. Forage selection by California bighorn sheep and the effects of grazing on an Artemisia-Agropyron community in southern British Columbia. Dissertation, University of British Columbia, Vancouver, Canada.

Wild Sheep Foundation. 2023. Jurisdictional Western Association of Fish and Wildlife Agencies (WAFWA) summaries providing in-depth, statistical sheep information. WAFWA Update - British Columbia: map of wild mountain sheep populations. Wild Sheep Foundation, Bozeman, Montana, USA.

Wilson, L. O. 1968. Distribution and ecology in desert bighorn sheep in southeastern Utah. Thesis, Utah State University, Logan, Utah, USA.

Wishart, W. D. 1958. The bighorn sheep of the Sheep River Valley. Thesis, University of Alberta, Edmonton, Canada.

Witham, J., and E. Smith. 1979. Desert bighorn movements in a southwestern Arizona mountain complex. *Desert Bighorn Council Transactions* 23:20–24.

Witham, J. H. 1983. Desert bighorn sheep in southwestern Arizona (reproduction, survival, telemetry). Dissertation, Colorado State University, Fort Collins, Colorado, USA.

Wockner, G., F. Singer, and K. Schoenecker. 2003. Habitat suitability model for bighorn sheep and wild horses in Bighorn Canyon and the Pryor Mountain wild horse range. Natural Resource Ecology Lab, Colorado State University, Fort Collins, Colorado, USA: 1–42.

Zeigenfuss, L., F. J. Singer, and M. A. Gudorf. 2000. Test of a modified habitat suitability model for bighorn sheep. *Restoration Ecology* 8:38–46.

# 20 Habitat Management
## The Key to Mountain Sheep Conservation

*James C. deVos, Jr. and Raymond M. Lee*

## 20.1 INTRODUCTION

Ancestors of North America's mountain sheep evolved into the species of bighorn sheep (*Ovis canadensis*) and thinhorn sheep (*Ovis dalli*), following their migration across the Bering land bridge from Eurasia during the Pleistocene, more than 750,000 years ago (McTaggart-Cowan 1940, Chapter 1). Following the migration, continental glaciations separated mountain sheep populations, resulting in genetic variance. For thinhorn sheep as we know them today, the subspecies origins date back prior to their survival through the Last Ice Age, more than 18,000 years ago (Sim et al. 2018, Chapter 1). The earliest known fossil records of bighorn sheep were found in Sangamonian interglacial deposits near Medicine Hat, Alberta, Canada, dating to 100,000 years before present (Stalker and Churcher 1982). Bighorn sheep spread through mountainous regions of western North America, arriving in the southern extent of their range to the tip of Baja California and northern Sierra Madre between 9,000 and 12,000 years before present with the southern extent likely being impeded due to the presence of dense vegetation (Brown 1989).

While thinhorns remained in the remote and northern regions in Alaska, Yukon, Northwest Territories, and British Columbia, historical bighorn distribution varied and was correlated with low precipitation levels in winter and spring and ranged from the high-elevation alpine meadows of the Rocky Mountains and Sierra Nevada, southward to the low-elevation desert mountain ranges in the southwestern United States and northern Mexico (Shackleton 1985, Chapter 1). Bighorn sheep were historically a wilderness animal with low tolerance for interactions with Euro-American settlement (Buechner 1960). In the 1800s and as settlement occurred, the once broader occupied habitats were contracted to the most rugged and isolated areas where interactions with humans and their associated livestock were infrequent. At the widest point in occupied range, most mountain ranges in western North America were occupied by bighorn sheep and many of the larger river drainages and badlands allowed eastern expansion away from mountains (Buechner 1960). Thinhorns appear to have escaped much of the settlement pressures faced by bighorns 200 years ago; however, based on archival reports (Chapter 2), they faced their range contractions in more front-country areas closer to gold rush era settlements. Despite the selection of remote and rugged terrain in the face of anthropogenic pressures, neither species escaped the pressures of hunting or the interest of scientists, trophy hunters, or subsistence-supported lifestyles (Chapter 2).

Reliable mountain sheep population estimates before the 1800s are lacking but estimates range as high as hundreds of thousands with distribution ranging from central Alaska southward to Mexico and eastward to Nebraska and Texas (Buechner 1960, Brewer et al. 2014, Chapter 2). As the human footprint increased within historical mountain sheep habitat, the genus retreated to the most inaccessible and isolated areas; however, adverse influences to mountain sheep habitat are complex, substantial, and increasing (Brewer et al. 2014, Chapters 19 and 21). Regardless of the maximum historical number of mountain sheep in North America, there has been decline in numbers and distribution, with much of this decline tied to deteriorated habitat and diminished habitat availability. Active habitat conservation and management are key to continuing the recent trend of increasing population levels and distribution (Brewer et al. 2014, Chapter 2).

We consider the current primary threats to bighorn sheep habitat to be population isolation due to human activities such as highway corridor expansion, introduction of exotic equids, fire suppression leading to impermeable vegetative conditions, lack of dependable free-standing water sources, introduction of novel disease organisms from domestic livestock, and invasive vegetation limiting diet quality and availability. Other recognized limiting factors currently experienced by both species include mining; high levels of recreation, particularly in sensitive areas such as lambing locations and winter ranges; and urbanization (Brewer et al. 2014, Jex et al. 2016, Chapter 21).

Many habitat specialist species that occur over large geographic areas are comprised of smaller, but interconnected populations (Levins 1970). When applied to mountain sheep, it establishes an important baseline understanding for habitat management (Schwartz et al. 1986). Therefore, important actions for habitat management are maintaining and enhancing the habitat patches occupied by mountain sheep; to ensure connectivity between key habitat patches within the metapopulation; and ensuring that sufficient high-quality resources are available to avoid localized extirpation of these patches that would reduce fitness of the larger metapopulations.

The relationship between Rocky Mountain bighorn sheep (*Ovis c. canadensis*) and the habitat they occupy in the Madison Range in Montana is the result of a strategy of restoration actions that focused on developing or enhancing

DOI: 10.1201/9781003518686-23

a metapopulation of smaller, local subpopulations, which would support successful repatriation or maintenance of larger, robust bighorn sheep populations (Lula et al. 2020). Mountain sheep fit this model of metapopulation existence but have failed to experience substantial rebounds from the lows that occurred in the 19th century as have other large herbivores in North America (DeCesare and Pletscher 2006). Exacerbating this trend of slow population growth and recolonization is habitat specialists like mountain sheep are typically slow to recolonize through natural dispersion (Geist 1970).

### 20.1.1 IMPORTANCE OF HABITAT

Understanding the importance of habitat to the persistence, or even expansion, of mountain sheep is key to their management. At a basic level, mountain sheep habitat exists within a landscape that is a functioning, suitable, or capable habitat; not all capable habitat can become suitable and not all suitable habitat is functioning. Hence the management of habitat generally requires an understanding of both historical and current mountain sheep distributions, and their biogeography (Chapter 19). Habitat is the ability of the environment to provide suitable conditions and life requisites (e.g., food, water, cover) required for population persistence and habitat quality relates to how effectively the environment provides these. Given the diminished abundance and distribution of mountain sheep in most areas, habitat and the quality of that habitat must be included in any effective management program. While some studies on the effectiveness of habitat quality look at single components, it is important to understand that often, multiple influences on habitat quality act in synergy, and this interaction may result in the expansion of a population, or extirpation of a population, as in the Santa Catalina Mountains in Arizona (Cain III et al. 2005, Krausman 2017). In the timeframe from the 1920s to the 1990s, these authors suggest that urban expansion, increased outdoor recreation, and fire suppression, which increased vegetative density, resulted in the extirpation of what was once a thriving bighorn sheep population.

Availability of suitable habitat ultimately determines the distribution and numbers of mountain sheep that a given area can support (Brewer et al. 2014, Jex et al. 2016). To meet the social and biological requirements of suitable mountain sheep habitat, the area must contain adequate amounts of forage, escape terrain to evade predators, lambing and loafing areas, water, and unobstructed movement corridors (McKinney et al. 2003). These habitat components must be juxtaposed and well-distributed throughout mountain sheep ranges to meet annual and seasonal needs (Brewer et al. 2014, Jex et al. 2016). Within the context of what was historically occupied, there remain landscapes that are not currently functioning as bighorn habitat, because they are not occupied by bighorn sheep. Hence, effective management of mountain sheep should ensure that functioning habitats remain connected to those vacant suitable habitats,

where population growth and expansion are a conservation goal.

Managing to improve adequate quality and quantity of mountain sheep habitat is an ongoing challenge for resource managers now and it will continue into the future. There are alternative views on the relationship between mountain sheep and the habitat they depend on, with Bailey (1980) suggesting that this genus is poorly adapted to more xeric environments and should be considered a relict species. In contrast, Hansen (1982) indicates that mountain sheep are well-adapted to the environment which they evolved from with respect to dealing with the physiological conditions associated with the environment in which they exist. Rather, habitat limitations are imposed by human-related influences including loss of forage and water resources, introduction of exotic competitors, and diseases associated with these introductions. Regardless of this debate, it is important to recognize that given the habitat conditions that mountain sheep rely on, improving habitat conditions is the cornerstone of mountain sheep management and repatriation into now vacant, historical ranges.

Acting alone or in combination, direct and indirect human effects and vegetative changes lead to degradation of habitat quality, fragmentation or loss of habitat, and, ultimately, limit viability and distribution of populations. Addressing the effects of human encroachment leading to habitat loss is one of the greatest challenges faced by mountain sheep and their managers (Brewer et al. 2014, Jex et al. 2016).

Our objective is to identify management actions that can be taken to reduce or eliminate the primary threats to habitat features in specific instances that have been identified as posing a risk to the conservation of mountain sheep in North America. We have not reviewed the potential influence of climate change because this threat to mountain sheep habitat is beyond the scope of what most individuals working on mountain sheep conservation can address, and the topic is broadly covered in Chapter 21.

## 20.2 MANAGEMENT ACTIONS

### 20.2.1 HIGHWAY CORRIDOR MANAGEMENT

The expansion of roadways is increasing in most landscapes in North America. Key adverse effects associated with expanding roadways include direct mortality and alteration of habitat use patterns (Trombulak and Frissell 2000). As highway infrastructure expands to accommodate a growing human population, genetic isolation and reduced fitness can occur in as little as a few decades, and left unremedied can lead to the loss of populations (Epps et al. 2005). Further, vehicle collisions can be a major source of direct mortality of many wildlife species. Roadways are often thought of as direct influences on populations due to collisions and bisecting movement corridors, but Trombulak and Frissell (2000) point to a much broader suite of influences including changes in water flow and sedimentation patterns, increased spread of noxious weeds, and the presence of roads is highly correlated with changes in species composition and population size.

There are also multiple Stone's sheep (*Ovis dalli stonei*) mortalities along a stretch of the Alaska highway in northern British Columbia caused by vehicle strikes, and the losses were additive to natural mortality and the direct result of Stone's sheep moving to mineral salt licks (Hengeveld and Cubberley 2012). Even within Rocky Mountain National Park, which is managed for its natural setting, roadways cause bighorn sheep to lose access to key resources with negative consequences where traffic reduced access to a mineral lick and reduced access to escape terrain (Keller and Bender 2006). Stress and disruption of seasonal migrations of female Stone's sheep to a mineral lick has also occurred (Tahltan Central Government 2021). Use by as few as two vehicles per day increased the number of crossing attempts made by Stone's sheep as a result of movement deflection (i.e., the individual sheep turned back from the route and failed to cross) and increased the time it took for cross-valley migrations. The increased stress (measured through fecal cortisol concentrations) and predation risk increased for females and newborn lambs.

In northwestern Arizona, a proposal to expand the U.S. Highway 93 travel corridor prompted research on highway-related mortality and subpopulation isolation before highway redesign and expansion (Cunningham and deVos Jr. 1992). Half of the mortalities in the Black Mountains were the result of vehicle collisions; as many as 12 mountain sheep were killed per year (Cunningham and Hanna 1992). Without establishing usable corridors to permit successful crossings, this population was at risk of being eliminated (Cuningham and Hanna 1992, Bristow and Crabb 2008). A key to incorporating crossing features into highway design was the need for balancing construction cost with crossing efficiencies.

Studies were initiated in the region to determine what types of highway crossing structures were suitable for mountain sheep crossings with evaluations conducted to test the permeability of highway underpasses and overpasses. Underpasses were frequently employed in highway development in the area to facilitate water passage below the road grade. In a nearby area (Highway 68, Arizona), Bristow and Crabb (2008) used remote sensing cameras to examine underpass designs that were constructed to facilitate wildlife movements as vehicle use on this highway due to increased tourism as the Laughlin, Nevada, area expanded. Only 12% of the population used these underpass crossings and the female segment of the population was being isolated and not using these features. The crossings evaluated were too narrow and low to facilitate mountain sheep crossings, and more open structures were required to maintain population connectivity. Similar work in British Columbia has also identified that the sizes of constructed overpasses do not effectively support use of the structure by mountain sheep (although other species such as deer [*Odocoileus* spp.] use the crossing structure (Brennan et al. 2022). Overpasses 34–50m wide on average would ensure optimal wildlife use (Brennan et al. 2022).

Highway overpasses on U.S. 93, near Hoover Dam, Arizona (Figure 20.1), connect bighorn sheep where subpopulations were isolated. Prior to the implementation of these crossing features vehicle-related bighorn sheep mortalities averaged 12 per year, crossings were sporadic, and

**FIGURE 20.1**   An overpass near Hoover Dam, Arizona, was designed to take advantage of a ridgeline used by bighorn sheep for crossing. (Photo by George Andrejko; used with permission.)

crossings were primarily from the male segment of the population (Cunningham and Hanna 1992). Under these conditions, this population of bighorn sheep was not likely to persist in the future due to high mortality and genetic drift. With the implementation of crossings, mortality was reduced by 93.3% (from 12 sheep/year pre-crossing implementation, to 0.8 sheep/year post-development). Successful crossings increased from 1.8 in year one to 11.4 in the last year of the study (Gagnon et al. 2022). With this rate of reduced mortality and higher numbers of crossings, the concern that genetic isolation would occur was greatly reduced and perhaps eliminated.

Although highway overpasses are considered the optimally effective crossing structure, underpasses can also be effective when they are wide and tall enough to offer mountain sheep visual security to avoid predators. McKinney and Smith (2007) used radio-collared mountain sheep movements to evaluate underpass and overpass locations. They identified that ridgelines were most used as crossing locations of Highway 93, with 83% of the crossings occurring on five prominent ridgelines. Using expert opinion, three of these ridgelines were selected for overpass construction. These construction sites, in combination with water flow bridges and underpasses, were studied by Gagnon et al. (2022) using global positioning system collar locations and remote sensing cameras. In total, eight separate features were monitored: two water transport bridges, three water passing culverts, and three wildlife overpasses. In addition to these structures, impermeable fencing was placed on both sides of the highway to funnel mountain sheep to potential crossing areas, a factor that was not present prior to the implementation of this phase of the study (Figure 20.2).

While there are several road crossing designs that have proven effective, there are at least five key elements that will enhance the use of crossings:

1. Predesign monitoring is key to identifying areas and topographic features that wildlife is predisposed to use; this will avoid making costly design errors and minimize the time required for wildlife to use crossings.
2. Impervious fencing to funnel wildlife to crossing locations is important in reducing random crossing locations.
3. For mountain sheep, animals that rely on high visibility to reduce danger, overpasses are preferred to underpasses and to optimize use, most overpasses should use a width to length ratio of 0.8, and widths of approximately 50m will support the highest level of large mammal use (Brennan et al. 2022); in some locations such as those that facilitate access to water or other resources, widths between 15 and 30m are effective.
4. Implement a long-term monitoring program to document the use of these crossing features and base adaptive management changes as needed to maintain connectivity. This is particularly important as an increased human footprint will occur that may influence the suitability of connectivity features such as overpasses.
5. It is important for wildlife managers to work closely with departments of transportation to identify potential conflicts between wildlife crossings and roadway design as early in highway construction planning as possible. While transportation

**FIGURE 20.2** Mountain sheep and other wildlife can be funneled to crossing features with impermeable fences that reduce spontaneous and sometimes fatal crossings. (Photo by George Andrejko; used with permission.)

corridors and roadway expansion will continue to influence wildlife habitat in North America and elsewhere, it is important to recognize that without mitigation road construction can compromise metapopulation structure. *Post-hoc* remediation will always be less effective, and more costly than early involvement in pre-construction planning.

## 20.2.2  FERAL EQUIDS AND THEIR EFFECT ON HABITAT

While feral equids and other non-native species are not a component of what is typically thought of as mountain sheep habitat, any evaluation of habitat improvements for this genus would be remiss without mention of non-native species. Although there were equids in North America, this group of animals was lost to the faunal component during the megafaunal extinction (Geist 1970, Chapter 10). Feral burros are descendants of animals released or that escaped during the settlement of the American West by Spanish explorers and subsequently released or escaped during the era of exploration and mining in the West (Knight 1958). Feral horses remaining in extant populations in North America are not relatives of those animals that became extinct in North America, along with the demise of many large animals during the last era of mass extinction that occurred approximately 15,000 years before present (Martin 1970).

Populations of feral equids were established as western North America was settled by European explorers and settlers. Horses that were traded as a commodity or that later become feral contributed greatly to Native American cultures in both the United States and Canada as early as the 1700s (Chapter 2). It was probably not until the latter half of the 19th century that burros became widely dispersed in southwestern United States, primarily through their use as pack animals by prospectors. Following the decline in the mining boom, and with the advent of the railroad, the need for pack animals declined. Many burros were traded with Indigenous communities, abandoned, became feral, and established themselves principally in the desert mountains and valleys in Arizona, California, and Nevada (Seegmiller and Ohmart 1989).

Management of feral equids is an ongoing social and political controversy in the western United States, with advocates supporting the importance of these species as a component of American western history (Beever 2003). Conversely, resource managers and wildlife conservation organizations are concerned that the number of burros in several states, and horses in others, exceed levels that maintain "a thriving natural ecological balance" as required in the 1971 Wild Free-Roaming Horses and Burros Act (WHBA: Section 1333 [a]). The management of free-roaming feral horses; however, is often challenged and criticized by special interest groups (Symanski 1996) and horse removal is always controversial. Before the passage of the WHBA, feral equids received little protection and free-roaming populations were seldom managed. Free-roaming horses

and burros are a serious conservation concern because of their effect on environmental resources, which results in significant habitat degradation (Beever 2003, Figure 20.3).

Although other non-native species have been brought to the western United States (particularly domestic cattle), there are substantial physical differences that make feral equids more influential to bighorn sheep habitat. In comparison to cattle, equids have a cecal digestive system that allows them to thrive even on poor quality forages; their elongated head, flexible lips, and presence of upper and lower incisors enhance forage consumption (Beever 2003). In the aggregate, large, non-native herbivores have changed the structure of rangelands to the detriment of native ungulates and mountain sheep, and high numbers of these feral species may overuse resources that otherwise would be available to mountain sheep (Brewer et al. 2014).

Horses and burros can be more aggressive, and can behaviorally displace mountain sheep, particularly in arid environments and reduce access to important resources such as water (Ostermann-Kelm et al. 2009, Chapter 10). Due to their large and growing populations, burros also outcompete bighorn sheep for water in areas where water is limited, also polluting sources of fresh water in desert

**FIGURE 20.3** Feral burros, a large non-native ungulate, can damage vegetation that bighorn sheep use (the damage in this figure is extreme). (Photo by George Andrejko; used with permission.)

environments with feces and urine. Further, unlike domestic livestock that are managed on a seasonal and numeric basis, feral equids continuously graze preferred plants and sensitive locations such as riparian areas and can threaten the sustainability of ecosystems via reduced hydrological flow and forage production that then adversely influence wildlife habitats (Davies and Boyd 2019). In extreme drought conditions, burros fare well relative to mountain sheep, because they have a broader forage niche and can make better use of poor quality or coarse forages (Brewer et al. 2014).

Two primary issues associated with the number and distribution of feral horses and burros are the lack of predation and a high reproduction rate. Predation, other than by humans, appears to be an insignificant factor in limiting feral burro populations. Knight (1958) suggested that the only predators capable of killing a mature burro were cougars (*Puma concolor*), wolves (*Canis lupus*), and jaguars (*Felis onca*) and that these predators were largely absent from the landscape at that time. There are few documented instances of feral equid depredations in the published literature that we could find, with Messler and Jones (2021) identifying burro remains at 39 of 279 investigated cougar kill sites being the first documentation in the primary literature of mountain lion preying on burros in Arizona.

A second broad concern with the presence of feral equids is their high reproductive rates. Of 553 feral horses tested from Nevada, Oregon, and Wyoming, and 173 feral burros from California to determine pregnancy rates, 57% (NV), 61% (OR), and 81% (WY) of the adult females two years or greater in age were pregnant. Additionally, yearling females were documented to be pregnant as much as 22% of the time (NV). Reproductive rates in burros were reported to be about 73% for females greater than two years old, with 25% of the yearling burros also being pregnant (Wolfe et al. 1989).

Although the WHBA requires balancing the number of feral equids with native wildlife and the available habitat that they depend on, this has not occurred in most areas, and control measures such as contraception, gathers, and adoptions have failed to control feral equid populations at levels that meet the legislative requirements of the WHBA (Stoner et al. 2021). The target numbers set in each unit are estimated to be the maximum number of each equid species to meet the elusive target of maintaining a thriving ecological balance. This value is referred to as the appropriate management level and is being exceeded in almost all herd management areas. Appropriate management levels are not static and when numbers exceed levels for protracted times, the effective appropriate management level is reduced by excessive herbivory (Rubin 2021, Rubin et al. 2024). Of concern is that the numbers reported by the Bureau of Land Management are only for the established herd management areas and do not reflect burros that occur outside the herd management areas. The number of burros on the landscape is unknown, and likely much greater than reported by the Bureau of Land Management; thus, the ecological disturbance from the feral animals is more widespread and harmful than if one considers only the actual established herd areas. The ecological influence of expanding burro populations has demonstrated adverse effects to key Sonoran Desert plants, including the desert's iconic species, the saguaro cactus (*Carnegiea gigantea*), a plant that is extremely long lived and slow growing.

It is important to recognize that although there are differing views on the presence of feral horses and burros on public lands, the goal of their management is not elimination but to find population levels that are consistent with the congressional intent of the WHBA, which is that these species are managed in a fashion that exists within the context of a thriving ecological balance. The following seven recommendations are for managing feral equids where they overlap with mountain sheep populations.

1. There are a variety of jurisdictional agencies that are responsible for management of feral equids within their respective lands, and wildlife managers need to be actively involved in all land-use plan development addressing equid distribution and densities. This should include planning to restore key resource sites such as streamside vegetation or key water resources important to mountain sheep.

2. Work cooperatively with the jurisdictional agencies to develop specific management actions to reduce and maintain feral equid populations within the agencies' established herd management areas as prescribed by the WHBA.

3. Work cooperatively with land resource and wildlife management agencies, and involved nongovernmental organizations, to develop funding mechanisms to provide the protection (fencing and other exclusionary mechanisms) required to preclude equid use in sensitive areas for mountain sheep.

4. As appropriate, promote revisions to the WHBA, and other relevant guidelines that will facilitate funding and approve management of these two species to benefit the many wildlife species that occur in common use areas.

5. There continues to be a need for multi-disciplinary support for research and development of fertility control methods to reduce the need for costly gathers and adoptions, both of which only defer the problem to a future date.

6. The use of gathers to control feral equid populations and reduce adverse effects from these animals is a failed paradigm given the high reproductive rates for these species. The Bureau of Land Management has, in the past, removed large numbers of animals from the range. Between 1980 and 1990, over 90,000 animals were removed from the range, and placed in holding facilities, which aided meeting target populations. By 2000, an additional 80,000 animals were removed from the range, pointing to the necessity of finding an acceptable control method that eliminates the need for large removal programs.

7. Many jurisdictional agencies have guidelines regarding the commingling of domestic goats and sheep with mountain sheep; wildlife managers should become aware of these guidelines and work with appropriate entities to eliminate the feral species as soon as practicable.

The Wildlife Society (2020) issued a statement on the management of feral equids, noting the increasingly negative ecological effects caused by the inability of wildlife professionals to meet their mandate of conserving native wildlife species on public lands. It was further stated that it is important for the federal agencies to enhance their collaboration with state wildlife agencies when making management decisions for wild horses and burros.

Feral equids are not the only species of concern to mountain sheep managers as both domestic goats and domestic sheep have been an important source of disease exposure to the native species that resulted in bighorn sheep die-offs (Heinse et al. 2016, Chapters 2, 8, and 10). Based on nasal sampling for *Mycoplasma ovipneumoniae*, the primary pathogen in many bighorn sheep disease-related die-off events, this pathogen was detected in 37.5% of domestic sheep or goat flocks sampled. A major concern for mountain sheep managers is that in the 24 flocks of domestic animals examined, 78% of the flocks had animals escape their enclosures and occupy areas with or proximate to mountain sheep populations (Heinse et al. 2016).

### 20.2.3   Habitat Alteration and Invasive Plant Communities

At least seasonally, and in many regions yearly, mountain sheep inhabit some of North America's most forage-depauperate and climatically harsh landscapes (Valdez and Krausman 1999). Adequate vegetative conditions in mountain sheep habitat are important to the management of this genus from a dietary standpoint (Dodd 1989). Adequate nutritional condition was a key mechanism that determined demographic performance, and population-level performance reflects the nutritional value of the occupied habitat (Stephenson et al. 2020). The top model fit in a survival analysis to further recovery of Sierra bighorn (*Ovis canadensis sierrae*) was related to the date of peak normalized difference vegetation index from the previous summer, and this value had a positive relationship for survival in the following year (Conner et al. 2018).

Dense vegetation can also curtail access to traditional use areas such as lambing grounds, water sources, and movement corridors (McKinney et al. 2003) and restrict visual sight distance needed to avoid predators (Jones et al. 2022). Even before European colonization of North America, considerable landscape change has adversely affected vegetative conditions that mountain sheep rely upon (Holl 1982, Belsky 1996, Cain III et al. 2005).

*Habitat alteration.* Cultural burning has been practiced in varying degrees by Indigenous peoples in North America for thousands of years. Based upon the tradition of using this treatment, these early habitat managers used fire near their communities to manage their environment for several benefits that supported domestic livestock forage, and improved sightlines that enhanced community security and communal protection, wildlife and their habitats. Fire, whether natural wildfire, human-caused, or managed or prescribed fire, can have beneficial and catastrophic effects on habitat value, and beneficial outcomes and use by wildlife of burned areas is not a foregone conclusion. Use of burned areas by bighorn sheep in Wyoming varied in response to burn severity (Donovan et al. 2021); the preferential use of burned areas by Stone's sheep and Rocky Mountain elk (*Cervus canadensis*) in northern British Columbia varied as a result of forage response and landscape topography (Sittler et al. 2019); the role of Indigenous use of fire in California's coastal ranges was a common practice to reduce chapparal plant density to enhance hunted resources (Keeley 2002). Therefore, despite the consensus that there is a relationship between First Nations' use of cultural fire and wildlife responses to that human-caused fire, it is likely that at least in some instances, the successes in achieving the desired environmental responses would have varied.

A prescribed burn uses a specific plan to meet a specific objective. Prescribed burns are typically used to reduce fuel loads (decadent vegetation) to preclude catastrophic fires. The use of fire can produce many benefits: controlling invasive plant and noxious weed species, increasing vegetative diversity and encouraging growth of native vegetation, and reducing shrub and tree overstory thereby promoting desirable forage species abundance and improving nutritional quality. While properly planned and managed prescribed burns can be a valuable tool for habitat improvement, prescribed fire has resulted in catastrophic fires causing stand conversions when overstory density is high (Veblen et al. 2000).

The effects of fire have been studied across most of western North America. Most studies identify decreased biomass of dead vegetation, increased live vegetation biomass, and short-term positive effects on forage quality (Ruckstuhl et al. 2000). Fire serves to enhance forage value (e.g., quality, quantity, digestible dry matter) in the short-term, by returning vegetative communities to an earlier successional stage (Sittler 2019). Fire also rejuvenates the soil by releasing nitrogen and carbon, nutrients that are stored in vegetation and subsequently released through the fire (Leverkus et al. 2018).

This general perception of the outcomes of prescribed fire and its ability to restore mountain sheep habitat has been broadly supported by advocacy groups interested in growing mountain sheep populations through habitat restoration activities. For example, the Wild Sheep Society of British Columbia has been actively using prescribed burns to enhance the habitat of bighorn and thinhorn sheep. Over the past two years, the Society has contributed $1.25 million Canadian dollars to mountain sheep projects in British Columbia. In 2022, the Wild Sheep Foundation (formerly the Foundation for North American Wild Sheep) contributed $250,000 toward the province's Northern Burn projects, which focused on benefiting Stone's sheep and developing science-supported conclusions about the responses of Stone's sheep populations and their habitats to prescribed fire interventions. These burns reduce the density of subalpine conifers and shrub layers, decreasing cover types on mountain sheep ranges, which improves visibility and reduces predation by terrestrial species such as wolves (*Canis lupus*), bears, and cougars.

When considering habitat treatments to reduce vegetative density and improve forage conditions for bighorn sheep, fire is an effective tool to improve habitat and sustain or restore bighorn sheep populations (Hobbs and Spowart 1984, Bleich et al. 2008, Donovan et al. 2021), when it is applied in ecosystems that have natural fire associations. In chaparral habitats in California, models were used to evaluate the uses of fire on the habitat of mountain sheep. The model's best fit was for areas that had been burned within 15 years, as bighorn sheep preferred burned areas and avoided those that remained unburned for longer than 15 years (Bleich et al. 2008). These results notwithstanding, it is important to consider the historical fire regime in an area being considered for introduction of fire as a habitat management tool, as the introduction of unnaturally frequent fires can increase the spread of undesired exotic plants and result in a monoculture of undesired plant species (Bleich et al. 2008). These authors successfully applied the model they developed to identify potential mountain

sheep reestablishment areas, where long-term fire suppression had reduced habitat suitability for bighorn sheep. They concluded that maintaining the distribution of this ungulate would depend on more frequent fires within its historical range. In more northern environments, relatively short-lived (1–7 years) responses from fire were observed (Ruckstuhl et al. 2000, Sittler 2019, Sittler et al. 2019); the short-term positive effect on forage quantity and quality improved mountain sheep nutrition.

Mechanical treatments have become important in restoring suitable habitats that have been heavily invaded by tree species or non-native invasive plants. This treatment is less affected by weather conditions, can be completed relatively quickly, and prescriptions can be tailored to specific local conditions (Peterson et al. 2007, Figure 20.4). Mastication, where woody species are ground by mulching machines, is effective at reducing both understory and overstory density as it is selective and can target a single tree at a time. In western forests, mastication removed the highest amount of woody plant cover over the largest treatment area in one study, but at much greater cost than harrowing and chaining. Mastication treatments typically have a higher cost per area because they selectively remove individual woody plants and require the use of heavy equipment (Munson et al. 2020).

Another effective treatment to reduce the density of undesired plant species is the use of herbicides. The use of herbicides can be effective in managing invasive species; in particular, inexpensive herbicide application has a large positive effect on seeded perennial grass cover (Munson et al. 2020).

*Invasive plant communities.* Across western North America, woody species have altered the plant communities important to mountain sheep. Pinyon pine (*Pinus* spp.)

**FIGURE 20.4** Mechanical thinning has the advantage of being able to select specific defined areas for treatment and eliminates the risk of fire escapement. (Photo by George Andrejko; used with permission.)

and juniper (*Juniperus* spp.) woodlands comprise the dominant overstory vegetation on approximately 40 million ha in the United States, and there is concern that the ecological dynamics of this woodland community have been altered by Euro-American settlement. The primary concern is that this vegetative community is becoming unnaturally dense, impacting landscapes that were formerly dominated by grassland and shrubland communities (Belsky 1996, Romme et al. 2009). The causes of pinyon-juniper expansion are largely attributed to livestock grazing, alterations in the natural fire regime, and recovery from past disturbances (Romme et al. 2009). Dense stands of pinyon-juniper reduce openings in the forest that are most conducive to forage production used by bighorn sheep (Figure 20.5). Additionally, the sight distance is reduced at this tree density, which increases the potential for predation and disrupts movement corridors.

There are two elements to increasing pinyon-juniper densities, including infill, where naturally occurring gaps in the tree canopy are filled with newly recruited plants, and pioneering expansion into new areas. Unless treated to reduce the density of these species, these stands remain in transition, with the endpoint being a dense overstory that adversely influences local, native plant communities and, in turn, mountain sheep habitat.

Increasing density of shrub species not used as forage by mountain sheep can adversely influence food resources by reducing the density of shrubs and forbs that constitute preferred forage for mountain sheep (Huisinga et al. 2005, Miller et al. 2000). In addition to reduced forage abundance, dense vegetation can reduce the viewscape that female mountain sheep need for predator detection and eliminates some areas used as parturition sites (Karsch

et al. 2016). Traveling through areas with dense brush can increase predation (deVos et al. 1981). When woody species density approaches 35% cover, mountain sheep populations decline (Holl 1982).

In addition to conifers, there are other invasive plant species that can adversely influence habitat quality for mountain sheep. It is estimated that honey mesquite (*Prosopis glandulosa*) infests more than 23 million ha of natural grasslands in Texas (Scifres et al. 1971). In southern New Mexico, conversion from grasslands and low-density shrublands to dense mesquite-dominated shrublands has occurred over the past 100 years. This invasion has led to the conversion of ecologically diverse grasslands and shrublands into near-monocultures of mesquite and creosote bush (*Larrea tridentata*), resulting in reduced forage quality and abundance.

Bighorn sheep diets have been studied throughout much of their range, including Mexico. In general, their diets are complex and primarily comprised of browse, succulents, grasses, and forbs. In northern Mexico, bighorn sheep consumed 41 plant species including 22 browse, ten forbs, five grasses, and four succulents (Tarango et al. 2002). In Arizona, during the winter of 1965, female bighorn sheep ate 61% grass, 17% forbs, and 22% browse, and 1.5% unknown (Krausman et al. 1989). In a southwestern U.S. study, fecal pellet groups (209 males, 209 females, 14 lambs) were collected every two weeks between September 1998 and August 2000. Ninety-four plant species were identified; bighorn sheep diets consisted of 50% browse, 35% forbs, 11% grasses, and 4% succulents (Brewer and Harverson 2007).

Although there were compositional differences between desert bighorn sheep and those in more northern climes, diet

**FIGURE 20.5**   Post forest treatment, growth of grasses, forbs, and palatable shrubs increase as competition for soil moisture is reduced and increased sunlight spurs growth of desired food items for bighorn sheep. (Photo by Jim deVos.)

studies for this genus revealed a strong preference for plant species commonly found in more open vegetation classes. For example, in Yellowstone National Park, bighorn sheep consumed 61% grass, 17% forbs, and 22% browse (Oldemeyer et al. 1971). In Colorado, Rocky Mountain bighorn sheep in the Sheep Creek-Trickle Mountain area of Saguache County consumed a yearly diet of grasses and sedges (*Carex* spp.) (46%), followed closely by shrubs (45%) and forbs (9%) (Todd 1975). The importance of available grasses and forbs cannot be overstated when describing and defining higher value forage communities for mountain sheep across their range in North America. Thus, the invasive nature and rapid spread of annual grass species, including but not limited to downy brome (*Bromus tectorum*), more commonly known as cheatgrass, has high potential to reduce forage quantity, quality, and ultimately the carrying capacity of the mountain sheep range. In addition to its ability to outcompete and replace native perennial grasses, its fire-prone nature means it can benefit from changing climate conditions that result in hotter and drier wildfire conditions, but also because of the way its seasonal life cycle occurs; areas dominated by cheatgrass create large horizontal masses of dead, fire-prone materials that can alter natural fire cycles (Colorado State University 2012).

Overstory disturbances that reduce canopy closure allowed understory vegetation to benefit from increased availability of sunlight, water, and soil nutrients to grow and reproduce at higher rates than in undisturbed forests (Stone and Wolfe 1996). These researchers measured understory density of grasses, forbs, and shrubs in a beetle-killed forest in northern Utah. Understory vegetation was at its lowest density under an undisturbed forest canopy with high canopy closure and increased across all canopy disturbance gradients. This is an important finding when considering the reliance on a diverse food base available to mountain sheep (Dodd 1989). Forests treated to increase canopy gaps and spatial variability had greater overall species richness in the understory, with this richness comprised primarily of early seral stage herbs and shrubs (Ares et al. 2009).

In any program to use fire or mechanical restoration tools to improve vegetative conditions for mountain sheep, managers should consider at least four guidelines:

1. Wildlife managers must recognize that mountain sheep habitat in most historical and occupied areas has been adversely affected by human-caused changes, including fire suppression, the introduction of non-native herbivores, infrastructure construction-induced fragmentation, and outdoor recreation, and that these factors often act in synergy. Maintaining or restoring mountain sheep populations requires active habitat management programs, and engaging with federal, state, provincial, and territorial land management agencies to undertake collaborative planning is important.

2. Adaptive approaches to site-specific restoration needs are paramount because fire, mechanical, and in some areas chemical restoration actions are not a one-size-fits-all solution. Tailoring restoration has to be site-specific to be effective and include measurable and achievable goals.

3. The use of prescribed fire must be carefully considered and given the changing fire regime over the last century, knowledge of the historical fire regime and the Pacific Decal Oscillation and recent El Nino-Southern Oscillation patterns (Chapter 21) are important in designing fire programs that benefit mountain sheep and other wildlife (Veblen et al. 2000). An effective fire plan needs to consider historical fire intervals and fuel loads to reduce the potential of catastrophic fires resulting in monotypic communities.

4. In addition to partnering with federal, state, provincial, and territorial land management agencies, there are several key non-governmental organizations such as the Wild Sheep Foundation and its chapter and affiliate state organizations (e.g., Arizona Desert Bighorn Sheep Society) that provide political and financial support for well-planned habitat and population programs. Developing effective working relationships with these organizations can amplify the effectiveness of habitat management programs.

## 20.2.4 The Role of Water Development in Maintaining or Enhancing Bighorn Sheep Habitat

Most researchers believe that robust mountain sheep populations require access to free water, not just ephemeral waters, dew, or metabolic water. There is a close correlation between free-standing water and the presence of desert bighorn sheep (Bates and Workman 1983, Cunningham and Ohmart 1986, Figure 20.6). These correlations increase during hot, dry periods and particularly during breeding and lactation periods (Hansen 1980, Turner and Weaver 1980). There have been very few reports of desert bighorn sheep living without access to free water, with these instances typically occurring in small mountain ranges and small populations (Krausman and Leopold 1986). Considering the large intermountain movements made by some desert bighorn sheep, it is possible that even in these few cases, the bighorn sheep were able to access free water found outside of their typical home range (Witham and Smith 1979, Ough and deVos Jr 1984).

Habitat models always include food, water, and shelter, indicating that access to water is a crucial component of a suitable habitat. Water availability is the most important parameter for predicting the success of transplants (Hansen 1980, Brown 1983). Models that incorporate water as a key criterion for successful transplants have been effective in directing transplant efforts for many bighorn sheep populations into their historical habitat (Cunningham 1989).

**FIGURE 20.6** Mountain sheep of all ages and both sexes use available water resources, primarily in hot and dry periods, with broad research results supporting the importance of this resource. (Photo by George Andrejko; used with permission.)

Early projects were typically small, often merely modifications to existing natural waters, such as mucking out a natural catchment, sealing cracks, and possibly installing a shade to reduce evaporation (Werner 1989). Later, catchment rims would be raised, gabions constructed, and steps cut into the rock. Then fences, or grates, were built to exclude use by feral and domestic livestock, and in some cases, use by predators. Eventually, modern projects became quite large (Figure 20.7), requiring dozens of laborers and heavy equipment (except in wilderness areas), with multiple 2,500-gallon storage tanks, and remote sensing to determine water storage levels and check for leaks.

Although a few water developments for mountain sheep were built before 1950, most arid states with desert bighorn sheep have implemented active water development programs since this period. While initially used to distribute mountain sheep into otherwise suitable habitats, developments have also been used to provide water to bighorn sheep that have lost access to historical waters due to human activities (e.g., mining, ground water pumping, housing, roads). Conversely, mines, canals, and dams have sometimes provided new water sources through mitigation for adverse project effects (Dolan 2006).

In 1946, the Arizona Game and Fish Department's (AZGFD) Development and Maintenance crew built the first Arizona guzzler. Water development continued in the 1950s, with the AZGFD primarily developing waters for quail and supporting the introduction of upland game birds. Switching more to big game in the 1960s, the AZGFD has now developed over 700 waters statewide, many of which benefit bighorn sheep and a wide variety of other wildlife species.

Water development is the *raison d'être* for the Arizona Desert Bighorn Sheep Society and many similar organizations. Their motto is *sin agua mortis* (without water death). In a climate where 2.4m of water can evaporate vertically during summer, providing artificial water sources is a reasonable idea. In the late 1960s, the Arizona Desert Bighorn Sheep Society began collaborating with the AZGFD and various land management agencies to develop waters primarily for the benefit of desert bighorn sheep. This program expanded to major development projects involving as many as 100 volunteers, many of whom had construction experience and had completed dozens of projects. The camaraderie established at these work projects is often the glue that holds these groups together. The Arizona Desert Bighorn Sheep Society typically builds or redevelops five to six waters per year, with support from the state and federal agencies responsible for the management of desert bighorn sheep habitat, and has now developed more than 80 wildlife waters.

The Wild Sheep Foundation's Grant-in-Aid Program has funded hundreds of similar improvement projects. In the southern states, most of these projects were water developments in Arizona, Nevada, California, Utah, New Mexico, and Texas. Although water developments targeting bighorn sheep population management are most common in the American Southwest, in more mesic states, including Oregon, Wyoming, Idaho, Colorado, and Montana, water developments have been funded by the Wild Sheep Foundation.

A number of mountain sheep advocacy groups, including the Wild Sheep Foundation and its chapters; Nevada Bighorns Unlimited and its chapters; Fraternity of the Desert Bighorn; Texas Bighorn Society; Society for the Conservation of Bighorn Sheep; Rocky Mountain Bighorn Society; Wild Sheep Society of British Columbia; and other affiliates, all raise funds for mountain sheep-related projects that often include water developments. Perhaps no other animal species has garnered as much funding from so many devotees.

The AZGFD has developed Wildlife Water Development Standards (2014, 2024). The 2024 version of the standards included 23 pages of text, three pages of engineering standards, and 25 pages of technical drawings and photos. These standards address the following items: tanks, troughs, precipitation collection aprons, tinaja sealants, shade structures, gabions, dams, fencing, escape passages, membranes (liners), evaporation covers, and camouflage finishes (artificial rocks).

FIGURE 20.7 Expansive water developments offer advantages over many previous designs by reducing evaporation losses and providing sufficient storage capacity to significantly reduce the need for water hauling, thereby ensuring against drying. (Photo by Mike Rice; used with permission.)

They also note an important aspect of water development: since the mid-1970s, the design, site selection, materials, and construction methods for water development have evolved, resulting in a range of catchment types that contribute to a growing maintenance load. Once a water site has been established, it is nearly impossible to close it. Research projects on water development removal are challenging to accomplish, as no one wants to be responsible for social opposition to recommending the closure of a water source after it has been used by bighorn sheep. In an experimental approach to removing water in desert bighorn sheep range, there were few demographic differences between the treatment and control areas (Cain III et al. 2008). These authors recommended that, given the high interannual variability in precipitation in many areas of the arid Southwest, it is important for wildlife managers to determine whether water is the primary limiting factor for populations of desert bighorn sheep.

Of course, wildlife water serves far more than just the target species and bats, bees, birds, small game and other wildlife (Figure 20.8) can access the water (Rosenstock et al. 1999). Usually, steps or ramps are built to allow easy egress for an animal that has the misfortune of miscalculating the water level.

We suggest that adequate water sources, free of shrubbery and competition, distributed through their habitat are important for supporting healthy and abundant mountain sheep populations in arid and desert ecosystems. These can be accomplished with at least four guidelines:

FIGURE 20.8 Wildlife water is crucial for mountain sheep and other wildlife species that rely on free-standing water, underscoring the importance of developing this habitat resource. (Photo by: George Andrejko; used with permission.)

1. Incorporating well-designed and placed water developments in all land-use planning activities at the earliest possible time.
2. Developing close working relationships with mountain sheep advocacy groups to help secure funding for water development projects. It is also important to tap into the large labor pool that these non-governmental organizations provide.

3. For new projects that encroach on mountain sheep habitat, such as mines, roadways, or electric transmission lines, wildlife managers should collaborate with project proponents, land stewards, and non-governmental organizations to seek adequate mitigation for these influences.

4. Coordinate with conservation organizations to secure funding for maintaining and developing water sources in historical mountain sheep habitats where water availability is limited.

### 20.2.5 EFFECTS OF MINING

There are few studies on the effects of mining activities on mountain sheep in the published literature. Nishi (2020) conducted a literature review focused on summarizing potential effects of mining projects and mitigations for thinhorn sheep in Yukon, Canada, finding only three of 461 publications were based on empirical field study; K. Poole, British Columbia (unpublished data) noted that apart from the creation of escape terrain and use of reclamation prescriptions designed for bighorn sheep on reclaimed mine areas in British Columbia, active mitigation during mining operations is not generally observed. Wickham et al. (2013) suggested that broad categories of effects arise from the removal of mountaintops and the forests that cover them. Brock and Green (2003) stated that environmental disturbance varies among different mining practices. The degree to which any effect or environmental change occurs as a result of mining is somewhat dependent on the mineral resource being extracted and the type of mine built (e.g., open-pit, underground). Generally, though, above-ground or open-pit mining

can result in the loss of topographic complexity, grassland community, forest loss and fragmentation, soil erosion, changes to watershed flows, and water quality. Thus, how an effect is quantified, as inherently positive or negative, will be variable and dependent on the interactions of many biotic and abiotic factors (Chapter 21).

In Arizona, bighorn sheep have been successful in occupying copper mine sites, including the Silverbell Mine near Tucson (Jansen et al. 2006), where habitat use patterns and primary behaviors were similar within and outside the mine perimeter. Similarly, at a copper mine in Morenci, Arizona, bighorn sheep have successfully established extensive use of the mine area, using many of the same features, such as steep rock faces, as they do beyond the mine (AZGFD unpublished data). In both locations, multiple captures and transplants have occurred, leading to successful translocations throughout the southwestern United States (AZGFD, unpublished data; Figure 20.9). The opposite effect has been shown in Alberta, Canada, where excessively high levels of selenium (a trace mineral known to effect reproduction and health) resulted in bighorn sheep on coal mine sites to have the lowest reproductive potential and reproductive output (i.e., resilience) of any populations that had been measured in Canada (Kneteman 2016). Despite the potential for negative effects on resilience, the landform transformation created by surface mining activities converted historically forested land into simplified agronomic plant communities, attracting females, likely due to higher nutritional forage and landscape features that provided protection from predators (Karabatsos 2020). Whether or not there will be negative, long-term implications to bighorn sheep as a result of exposure to extremely high

**FIGURE 20.9** Bighorn sheep have been successful in using copper mines in Arizona to the point that several translocations have occurred from these mines, and several historical populations have been restored from these sources. (Photo by Jim Heffelfinger; used with permission.)

levels of trace minerals, or as a result of the range shift abandonment of historical habitat areas, and onto the artificial mine environments, remains to be understood.

Although environmental disturbance occurs in varying degrees, mining always alters abiotic factors and biotic communities immediately within the project footprint, and in adjacent portions of the watershed and downstream receiving areas. With carefully planned ecological restoration, mined lands can be remediated into functional environments; however, most of these sites will have lasting effects on watersheds and habitats. While the approach to recovering mining sites is variable depending on the mine type and location, the Intergovernmental Forum on Mining, Minerals, Metals, and Sustainable Development (2019) does provide general guidance on best management practices that can be used in planning and implementing restoration of spent mine sites. The following considerations should be implemented to ensure successful closure and recovery of any mining operation (see also Chapter 21):

1. The mine owner or operator should ensure mine project closure planning is an integral component of mine project planning.
2. The mine owner or operator should consult and involve all levels of government, including federal, state, provincial, and territorial wildlife and land management agencies, other nearby landholders, affected communities, and other relevant stakeholders in the mine project rehabilitation and closure planning.
3. The mine owner or operator should be responsible for maintaining adequate financial resources to ensure the ability to meet the mining project closure obligations.
4. In many habitats occupied by mountain sheep, full restoration of the mine site and associated

developments is difficult to achieve, and therefore, pre-operation mining infrastructure should be designed to minimize unnecessary adverse influences to slope and vegetation features. It is also important to design and implement mitigative measures to offset the influences of mine development and operations.
5. Wildlife management agencies and nongovernmental organizations need to establish and maintain positive working relationships with mine owners and operators as experience has demonstrated that these mining operators can be good conservation partners helping to build or maintain water sources, physical habitat features, and afford protection of mountain sheep resources.

### 20.2.6 Effects from Nonconsumptive Outdoor Recreation

Carrying capacity is a concept that has been in place in the field of wildlife management and is defined as the number of animals that can exist in an area without undue resource damage. This concept has been expanded into assessing the appropriate level of human recreation in outdoor settings and carrying capacity has resource and social dimensions (Manning et al. 2005). There is a strong relationship between the natural resources of parks and the quality of the visitor experience, and these aesthetic influences offer important justification and guidance for setting standards of quality. Without balancing resource needs and recreational access, conflicts are sure to occur when plans are developed to limit visitor access.

Nonconsumptive recreational use has been thought to be benign to wildlife; however, as outdoor recreation such as hiking, mountain biking, and camping continue to increase and expand its footprint (Figure 20.10), researchers such

**FIGURE 20.10** Outdoor recreation has experienced significant growth with the advent of high-quality equipment, and many previously lightly used areas are now experiencing increased recreational influences. (Photo by George Andrejko; used with permission.)

as Kuwaczka et al. (2023), Courtemanch (2014), and Boyle and Samson (1985) suggest that these users are affecting wildlife and their habitats (Chapter 21). Managing this activity can be difficult for resource managers as there is often insufficient pre-disturbance information upon which to measure an effect that could justify limiting outdoor use of public lands (Boyle and Samson 1985). There is also resistance to modifying use levels as evidenced by research on outdoor recreationists in Yellowstone National Park (Borrie et al. 2001). Park users expressed that restricting recreational density to protect bison was inappropriate given their perception that the park's primary role is that of a place for recreation, and that people have a right to visit and experience the park as they wish. This contradiction between land protections established for natural values and the perceptions of the users that the land is theirs often results in conflicted approaches to management.

There are other research results that suggest that in some instances, recreational influences may not be benign. The bighorn sheep's use of a mineral lick in Rocky Mountain National Park was reduced by a combination of highway vehicle use and tourist visitation, which decreased the species' ability to use the lick (Keller and Bender 2006). The negative influence from this reduced use may compromise the health and productivity of this bighorn sheep herd.

The influence of recreational uses of the Santa Catalina Mountains, particularly the Pusch Ridge Wilderness Area, indicates hiking activities in currently or previously occupied areas by mountain sheep were, at least in part, responsible for the near extirpation of this once thriving bighorn sheep herd (Schoenecker and Krausman 2002, Krausman 2017). Although no single factor was the cause of the near extirpation of the bighorn sheep, adverse effects were evident from recreation use, and without control of these effects, it would be unlikely that the population could be restored to a viable level.

At least two considerations should be implemented to balance the need for adequate outdoor recreation and those of the mountain sheep in the area:

1. The effects of human recreation, disturbance, and habitat alienation can occur in a single season or include multiple seasons. It is important to document the level of recreational activity and then assess the effects on mountain sheep populations, and if recreational activity is a primary factor to declines, developing a land-use plan that incorporates input from land management agencies, recreational user groups, and the appropriate wildlife management agencies is essential to addressing and potentially reversing the adverse effects influencing the mountain sheep population.
2. Habitats, such as natal ranges or escape terrain, exist because they are linked to the vegetative-topographic characteristics of the landscape and the ruggedness of the terrain. Early involvement in land-use plan development is the best approach

to developing the actions needed to avoid disturbance effects to key habitat locations and resource features, such as water sites, mineral licks, winter ranges, or lambing areas for mountain sheep. Developing mitigation measures, such as establishing alternative trail routes or creating water sources away from main human traffic areas, can help mitigate the effects of human use and activities.

### 20.2.7 Adverse Impacts of Urban Development

Unlike some influences on wildlife habitats, the presence of urban development is readily apparent, and many urban and suburban developments are extensive, capable of having a significant influence (Figure 20.11; Chapter 21). Ongoing human population growth will exacerbate the influence of anthropogenic stressors that challenge efforts to reverse biodiversity loss (Mora 2014). The western states are projected to be an area of human population expansion at the highest rates in the United States, with most states with mountain sheep populations growing at rates that exceed the national average (University of Virginia, Weldon Cooper Center for Public Service 2024); similar examples exist in Canada (Chapter 21). With this projection, urban and suburban development will be an important factor for land and wildlife management agencies, and maintaining mountain sheep habitat will require close coordination with land-use planning and zoning organizations.

One of the key problems associated with suburban development is that there is seldom a cohesive approach at a landscape scale, where the cumulative effects of urban development can be coordinated to reduce influences on wildlife habitats (Hansen et al. 2005) and maintain the region's biodiversity. Polfus and Krausman (2012) reviewed the available scientific literature to compile and assess the effects of urban development on ungulates in the western United States. In earlier eras of westward expansion, most development was concentrated in valley bottoms and lower foothills, creating conflicts in winter ranges. As many of these locations were developed, there was expansion upslope into higher mountain regions on largely private lands, and herein lies the greatest potential for new negative effects on mountain sheep habitat (Polfus and Krausman 2012).

While urban and suburban expansion will continue, the following recommendations can be implemented to reduce the influence on mountain sheep habitat:

1. Establish a working relationship with local and regional land-use planning entities and incorporate, to the extent possible, measures to include wildlife-friendly designs in urban and exurban development and maintain effective movement corridors between metapopulation patches.
2. Where feasible, acquire and manage open spaces as reserve lands for mountain sheep use. This is

**FIGURE 20.11** Suburban developments, which often lack cohesive land-use plans, can have a greater influence on mountain sheep habitat than urban locations. (Photo by George Andrejko; used with permission.)

particularly important when the opportunity occurs to acquire core use areas.

3. Seek input in all federal, territorial, and provincial land-use planning projects to incorporate effective measures to plan land exchanges to optimize the spatial integrity of mountain sheep habitat.

4. Coordinate with conservation organizations that focus on mountain sheep management to solicit grants and other funding for core use areas, and integrate these acquisitions into long-term conservation programs.

## 20.3 SUMMARY

Most lands held in public trust are managed by federal, state, provincial, and territorial entities, and the unique values and priorities reflected within these authorities often lead to differences in actions that may not benefit mountain sheep. Managers should work cooperatively to protect important habitats that are currently in good condition; to improve habitats that are not but were historically; and to address other factors limiting the potential for mountain sheep populations to thrive (Brewer et al. 2014, Jex et al. 2016). There are many influences to habitats discussed herein and in other chapters, and elsewhere in the literature, that advise that the expanded linear corridors (e.g., roadway and electric distribution networks), unregulated populations of feral equids, changes in habitat composition (i.e., expanded woody and invasive plant communities), and limited surface water availability are primary threats to the sustainable management of resilient mountain sheep populations.

We counsel new resource managers to consider historical and future ecological effects over long periods. Robinett

(1990) recounts the history of the Tohono O'odham range in southern Arizona over several centuries, detailing the ecological conditions that changed during this time. Before the arrival of Spanish missionaries in the late 1600s, this native culture lived in scattered groups, moving as available water conditions dictated. Primarily through hunting of native species and farming, the influence on the landscape was minimal. The subsequent arrival of missionaries, who gifted the people cattle and horses, created a new era of culturally influenced landscape change, as these herds of domestic livestock grew and some became large, feral herds. Deep wells were dug, and by 1919, at one time, approximately 60,000 non-native livestock occupied one of the most arid regions in the West. Eventually, livestock-related defoliation led to a significant shift in the type of forage species, from more palatable native species to ones that are largely unpalatable. Environmental degradation included the drying of many rivers and other water sources, deep erosion of watersheds, and a decline in native fauna. Following the Euro-American settlement of North America, ecological changes have been relentless, and understanding the history of these changes is an important component of managing mountain sheep and their habitats, now and into the future.

## REFERENCES

Ares, A., S. D. Berryman, and K. J. Puettmann. 2009. Understory vegetation response to thinning disturbances of varying complexity in coniferous forests. *Applied Vegetation Science* 12:472–487.

Arizona Game and Fish Department. 2014. *Wildlife Water Development Standards*. Arizona Game and Fish Department, Phoenix, Arizona, USA.

Arizona Game and Fish Department. 2024. *Wildlife Water Development Standards*. Arizona Game and Fish Department, Phoenix, Arizona, USA.

Bailey, J. A. 1980. Desert bighorn forage competition and zoogeography. *Wildlife Society Bulletin* 8:208–216.

Bates Jr., J. W., and G. Workman. 1983. Desert bighorn sheep habitat utilization in Canyonlands National Park. *Desert Bighorn Council Transactions* 27:25–28.

Beever, E. 2003. Management implications of the ecology of free-roaming horses in semi-desert ecosystems of the western United States. *Wildlife Society Bulletin* 31:887–895.

Belsky, A. J. 1996. Viewpoint: Western juniper expansion: is it a threat to northwestern ecosystems? *Journal of Range Management* 49:53–59.

Bleich, V. C., H. E. Johnson, S. A. Holl, L. Konde, S. G. Torres, and P. R. Krausman. 2008. Fire history in a chaparral ecosystem. Implications for conservation of a native ungulate. *Rangeland Ecology and Management* 61:571–578.

Borrie, W., W. Freimund, M. Davenport, and R. Manning. 2001. Crossing methodological boundaries: assessing visitor motivations and support for management actions at Yellowstone National Park using quantitative and qualitative research approaches. *The George Wright Forum* 18:72–84.

Boyle, S. A., and F. R. Samson. 1985. Effects of nonconsumptive recreation on wildlife: a review. *Wildlife Society Bulletin* 13:110–116.

Brennan, L., E. Chow, and C. Lamb. 2022. Wildlife overpass structure size, distribution, effectiveness in Arizona. Arizona Game and Fish Department, Phoenix, Arizona, USA.

Brewer, C. E., and L. A. Harverson. 2007. Diets of bighorn sheep in the Chihuahuan Desert. *Southwestern Naturalist* 57:97–103.

Brewer, C. E., V. C. Bleich, J. A. Foster, T. Hosch-Hebdon, D. E. McWhirter, E. M. Rominger, M. W. Wagner, and B. P. Wiedmann. 2014. *Bighorn Sheep: Conservation Challenges and Management Strategies for the 21st Century*. Wild Sheep Working Group, Western Association of Fish and Wildlife Agencies, Cheyenne, Wyoming, USA.

Bristow, K., and M. Crabb. 2008. Evaluation of distribution and trans-highway movement of desert bighorn sheep: Arizona Highway 68, Arizona, USA. Final Report 588. Arizona Department of Transportation, Research Center, Phoenix, Arizona, USA.

Brock, J. H., and D. M. Green. 2003. Impacts of livestock grazing, mining, recreation, roads, and other land uses on watershed resources. *Journal of the Arizona-Nevada Academy of Science* 35:11–22.

Brown, D. E. 1983. Guide to formulating desert bighorn sheep transplant priorities. Federal Aid Report W53R. Arizona Game and Fish Department, Phoenix, Arizona, USA.

Brown, D. E. 1989. Early history. Pages 1–11 *in* R. M. Lee, editor. *The Desert Bighorn Sheep in Arizona*. Arizona Game and Fish Department, Phoenix, Arizona, USA.

Buechner, H. K. 1960. The bighorn sheep in the United States, its past, present, and future. *Wildlife Monographs* 4:3–174.

Cain III, J. W., H. E. Johnson, and P. R. Krausman. 2005. Wildfire and bighorn sheep habitat. *Southwestern Naturalist* 50:506–513.

Cain III, J. W., P. R. Krausman, J. R. Morgart, B. D. Jansen, and M. P. Pepper. 2008. Responses of desert bighorn sheep to removal of water sources. Wildlife Monographs No.171.

Colorado State University. 2012. Cheatgrass and wildfire. Natural Resources Series, Forestry, Extension Note 6.310. Colorado State University, Fort Collins, Colorado, USA.

Conner, M. M., T. R. Stephenson, D. W. German, K. L. Monteith, A. P. Few, and E. H. Bair. 2018. Survival analysis: informing recovery of Sierra Nevada bighorn sheep. *Journal of Wildlife Management* 87:1442–1458

Courtemanch, A. B. 2014. Seasonal habitat selection and impacts of backcountry recreation on a formerly migratory bighorn sheep population in northwest Wyoming. Thesis, Department of Zoology and Physiology, University of Wyoming, Laramie, Wyoming, USA.

Cunningham, S. C. 1989. Evaluation of bighorn sheep habitat. Pages 135–160 *in* R. M. Lee, editor. *The Desert Bighorn Sheep in Arizona*. Arizona Game and Fish Department, Phoenix, Arizona, USA.

Cunningham, S. C., and J. C. deVos. 1992. Mortality of mountain bighorn sheep in the Black Canyon area of northwestern Arizona. *Desert Bighorn Council Transactions* 36:27–29.

Cunningham, S. C., and L. Hanna. 1992. *Movements and Habitat Use of Desert Bighorn in the Black Canyon Area: Final Report*. U. S. Bureau of Reclamation Lower Colorado River Office, Environmental Division, Boulder City, Nevada, USA.

Cunningham, S. C., and R. D. Ohmart. 1986. Aspects of the ecology of desert bighorn sheep in Carrizo Canyon, California. *Desert Bighorn Council Transactions* 30:14–19.

Davies, K. W., and C. S. Boyd. 2019. Ecological effects of free-roaming horses in North America rangelands. *BioScience* 69:558–565.

DeCesare, N. J., and D. H. Pletscher. 2006. Movements, connectivity, and resource selection of Rocky Mountain bighorn sheep. *Journal of Mammalogy* 87:531–538.

deVos, J. C., W. Ough, D. Taylor, R. Miller, S. Walchuk, and R. Remington. 1981. Evaluation of a desert bighorn release. *Desert Bighorn Council Transactions* 25:29–30.

Dodd, N. 1989. Dietary considerations. Pages 109–134 *in* R. M. Lee, editor. The desert bighorn sheep in Arizona. Arizona Game and Fish Department, Phoenix, Arizona, USA.

Dolan, B. F. 2006. Water developments and desert bighorn sheep: Implications for conservation. Wildlife Society Bulletin 34:642–646.

Donovan, V. M., S. P. H. Dwinnell, J. L. Beck, C. P. Roberts, J. G. Clapp, G. S. Hiatt, K. L. Monteith, and D. Tidwell. 2021. Fire-driven landscape heterogeneity shapes habitat selection of bighorn sheep. *Journal of Mammalogy* 102:757–771.

Epps, C. W., P. J. Palsboll, J. D. Wehhausen, G. K. Roderick, R. R. Ramey, and D. R. McCullough. 2005. Highways block gene flow and cause a rapid decline in genetic diversity of desert bighorn sheep. *Ecology Letters* 8:1029–1038.

Gagnon, J. W., C. D. Loberger, K. S. Ogren, S. C. Sprague, S. R. Boe, and R. E. Schweinsburg. 2022. Mitigating bighorn sheep-vehicle collisions and habitat fragmentation with overpasses and adaptive mitigation. *Human-Wildlife Interactions* 16:353–372.

Geist, V. 1970. On the home range fidelity of bighorn rams. *Desert Bighorn Council Transactions* 14:51–53.

Hansen, A. J., R. L. Knight, J. M. Marzluff, S. Powell, K. Brown, P. H. Gude, and A. Jones. 2005. Effects of exurban development on biodiversity: patterns, mechanisms, and research needs. *Ecological Applications* 15:1893–1905.

Hansen, C. G. 1980. Habitat. Pages 64–99 *in* G. Monson and L. Sumner, editors. *The Desert Bighorn: Its Life History, Ecology and Management*. University of Arizona Press, Tucson, Arizona, USA.

Hansen, M. C. 1982. Desert bighorn sheep: another view. *Wildlife Society Bulletin* 10:133–140.

Heinse, L. M., L. H. Hardesty, and R. B. Harris. 2016. Risk of pathogen spillover to bighorn sheep from domestic sheep and goat flocks on private land. *Wildlife Society Bulletin* 40:625–633.

Hengeveld, P. E., and J. C. Cubberley. 2012. Sulphur/8 Mile Stone's sheep project: research summary and management considerations. Synergy Applied Ecology, Mackenzie British Columbia, Canada.

Hobbs, N. T., and R. A. Spowart. 1984. Effects of prescribed fire on nutrition of mountain sheep and mule deer during winter and spring. *Journal of Wildlife Management* 48:551–556.

Holl, S. A. 1982. Evaluation of bighorn sheep habitat. *Desert Bighorn Council Transactions* 26:47–49.

Huisinga, K. D., D. C. Laughlin, P. Z. Fule, J. D. Springer, and C. M. McGlone. 2005. Effects of intense prescribed fire on understory vegetation in a mixed-conifer forest. *Journal of the Torrey Botanical Society* 132:590–601.

Intergovernmental Forum on Mining, Minerals, Metals and Sustainable Development. 2019. Mining project rehabilitation and closure guidelines. International Institute for Sustainable Development. https://www.igfmining.org. Accessed 26 Mar 2025.

Jansen, B. D., P. R. Krausman, J. R. Heffelfinger, and J. C. deVos Jr. 2006. Bighorn sheep selection of landscape features in an active copper mine. *Wildlife Society Bulletin* 34:1121–1126.

Jex, B. A., J. B. Ayotte, V. C. Bleich, C. E. Brewer, D. L. Bruning, T. M. Hegel, N. C. Larter, R. A. Schwanke, H. M. Schwantje, and M. W. Wagner. 2016. *Thinhorn Sheep: Conservation Challenges and Management Strategies for the 21st Century.* Wild Sheep Working Group, Western Association of Fish and Wildlife Agencies, Boise, Idaho, USA.

Jones, A. J., E. S. Rubin, M. J. Clement, L. E. Harding, and J. I. Mesler. 2022. Desert bighorn sheep habitat selection, group size, and mountain lion predation risk. *Journal of Wildlife Management* 86:1–28.

Karabatsos, S. 2020. Horns and hotspots: detecting change in mountain sheep populations over large spatiotemporal scales. Thesis, University of Alberta, Edmonton Alberta, Canada.

Karsch, R. C., J. W. Cain, III, E. M. Rominger, and E. J. Goldstein. 2016. Desert bighorn sheep lambing habitat: parturition, nursery, and predation sites. *Journal of Wildlife Management* 80:1069–1080.

Keeley, J. E. 2002. Native American impacts on fire regimes of the California coastal ranges. *Journal of Biogeography* 29:303–320.

Keller, B. J., and L. C. Bender. 2006. Bighorn sheep response to road-related disturbance in Rocky Mountain National Park, Colorado. *Journal of Wildlife Management* 71:2329–2337.

Kneteman, J. G. 2016. Resilient space: Bighorn sheep (*Ovis canadensis*) ecological resilience in the northern Rocky Mountains. Thesis, University of Alberta, Edmonton Alberta, Canada.

Knight, T. L. 1958. The feral burro in the United States: distribution and problems. *Journal of Wildlife Management* 22:163–179.

Krausman, P. R. 2017. *And Then There Were None: The Demise of Desert Bighorn Sheep in the Pusch Ridge Wilderness.* University of New Mexico Press, Albuquerque, New Mexico, USA.

Krausman, P. R., and B. D. Leopold. 1986. Habitat components for desert bighorn sheep in the Harquahala Mountains, Arizona. *Journal of Wildlife Management* 50:504–508.

Krausman, P. R., B. D. Leopold, R. F. Seegmiller, and S. G. Torres. 1989. Relationships between desert bighorn sheep and habitat in western Arizona. *Wildlife Monographs* 102:3–66.

Kuwaczka, L. F., V. Mitterwallner, V. Audorff, and M. J. Steinbauer. 2023. Ecological impacts of (electrically assisted) mountain biking. *Global Ecology and Conservation* 44:e02475.

Leverkus, S. E. R., J. D. Scasta, R. L. Concepcion, M. Lavallée, and K. White. 2018. *Towards a Peace-Liard Prescribed Fire Program: Part A – Rationale.* Ministry of Forests, Lands, Natural Resource Operations, and Rural Development, Fort St. John, British Columbia, Canada.

Levins, R. 1970. Extinction. Pages 77–107 *in* M. Gesternhaber, editor. *Some Mathematical Questions in Biology.* American Mathematical Society, Providence, Rhode Island.

Lula, E. S., B. Lowrey, K. M. Proffitt, A. R. Litt, J. A. Cunningham, C. J. Butler, and R. A. Garrott. 2020. Is habitat constraining bighorn sheep restoration? A case study. *Journal of Wildlife Management* 84:588–600.

Manning, R., Y. Leung, and M. Dudruk. 2005. Research to support management of visitor carrying capacity of Boston Harbor Islands. *Northeastern Naturalist* 12:201–220.

Martin, P. S. 1970. Pleistocene niches for alien animals. *Bioscience* 20:218–221.

McKinney, T., S. R. Boe, and J. C. deVos Jr. 2003. GIS-based evaluation of escape terrain and desert bighorn sheep populations in Arizona. *Wildlife Society Bulletin* 31:1229–1236.

McKinney, T., and T. Smith. 2007. *US 93 Bighorn Study: Distribution and Trans-Highway Movement of Desert Bighorn Sheep in Northwestern Arizona.* FHWA-AZ-07–576. Arizona Department of Transportation, Phoenix, Arizona, USA.

McTaggart-Cowan, I. 1940. Distribution and variation in the native sheep of North America. *American Midland Naturalist* 24:505–580.

Messler, J. I., and A. S. Jones. 2021. Feral burros as a mountain lion prey item in west central Arizona. *Southwestern Naturalist* 66:338–342.

Miller, R. F., T. J. Svejcar, and J. A. Rose. 2000. Impacts of western juniper on plant community and structure. *Journal of Range Management* 53:574–585.

Mora, C. 2014. Revisiting the environmental and socioeconomic effects of population growth: a fundamental but fading issue in modern scientific, public, and political circles. *Ecology and Society* 19:38.

Munson, S. M., E. O. Yackulic, L. S. Bair, S. M. Copeland, and K. L. Gunnell. 2020. The biggest bang for the buck: cost-effective vegetation treatment outcomes across drylands of the western United States. *Ecological Applications* 30:1–14.

Nishi, J. 2020. A literature review of mining project impacts and mitigations for thinhorn sheep. Technical report, EcoBorealis Consulting Inc., Millerville, Alberta, Canada.

Oldemeyer, J. L., W. J. Barmore, and D. L. Gilbert. 1971. Winter ecology of bighorn sheep in Yellowstone National Park. *Journal of Wildlife Management* 35:257–269.

Ostermann-Kelm, S. D., E. A. Atwill, E. S. Ruben, I. E. Hendrickson, and W. M. Boyce. 2009. Impacts of feral horses on a desert environment. *BMC Ecology* 9:22–31.

Ough, W. D., and J. C. deVos Jr. 1984. Intermountain travel corridors and their management implications for bighorn sheep. *Desert Bighorn Council Transactions* 28:32–36.

Peterson, D. W., P. B. Reich, and K. J. Wrage. 2007. Plant functional group response to fire frequency and tree canopy cover gradients in oak savannas and woodlands. *Journal of Vegetation Science* 18:3–13.

Polfus, J. L., and P. R. Krausman. 2012. Impacts of residential development on ungulates of the Rocky Mountain West. *Wildlife Society Bulletin* 36:647–657.

Robinett, D. 1990. Tohono O'odham range history. *Society for Range Management* 12:296–300.

Romme, W. H., C. D. Allen, J. D. Bailey, W. L. Baker, B. T. Bestelmeyer, P. M. Brown, K. S. Wisenhart, M. L. Floyd, D. W. Huffman, B. F. Jacobs, R. F. Miller, E. H., Muldavin, T. W. Swetman, R. J. Tausch, and P. J. Weisberg. 2009. Historical and modern disturbance regimes, stand structures and landscape dynamics in Pinon-juniper vegetation of the western United States. *Rangeland Ecology and Management* 62:203–222.

Rosenstock, S. S., W. B. Ballard, and J. C. deVos Jr. 1999. The benefits of and impacts of wildlife water development. *Journal of Range Management* 52:302–311.

Rubin, E. S. 2021. Feral burro report. *Desert Bighorn Council Transactions* 56:75–81.

Rubin, E. S., D. Conrad, L. E. Harding, and B. M. Russo. 2024. Associations between a feral equid and the Sonoran Desert ecosystem. *Wildlife Monographs* 215:1–72.

Ruckstuhl, K. E., M. Fiesta-Bianchet, and J. T. Jorgenson. 2000. Effects of prescribed grassland burns on forage availability, quality and bighorn sheep use. *Biennial Symposium Northern Wild Sheep and Goat Council* 12:11–25.

Schoenecker, K. A., and P. R. Krausman. 2002. Human disturbance in bighorn sheep habitat, Pusch Ridge Wilderness, Arizona. *Journal of the Arizona-Nevada Academy of Science* 34:63–68.

Schwartz, O. A., V. C. Bleich, and S. A. Holl. 1986. Genetics and the conservation of mountain sheep *Ovis canadensis nelsoni*. *Biological Conservation* 37:179–190.

Scifres, C. J., J. H. Brock, and R. R. Hahn. 1971. Influence of secondary succession on honey mesquite invasion in north Texas. *Journal of Range Management* 24:206–210.

Seegmiller, R. F., and R. D. Ohmart. 1989. Ecological relationships of feral burros and desert bighorn sheep. Wildlife Monographs No. 78.

Shackleton, D. M. 1985. Bighorn sheep. *American Society of Mammalogists. Mammalian Species* 230:1–9.

Sim, Z., C. S. Davis, B. A. Jex, T. Hegel, and D. W. Coltman. 2018. Management implications of highly resolved hierarchical population genetic structure in thinhorn sheep. *Conservation Genetics* 20:185–201.

Sittler, K. L. 2019. *Stone's Sheep and Elk Forage Characteristics – 7-Years Post Fire.* Report 663:1–33. Wildlife Infometrics, Mackenzie, British Columbia, Canada.

Sittler, K. L., K. L. Parker, and M. P. Gillingham. 2019. Vegetation and prescribed fire: implications for Stone's sheep and elk. *Journal of Wildlife Management* 83:393–409.

Stalker, A., and C. S. Churcher. 1982. Ice age deposits and animals from the southwestern part of the great plains of Canada, Geological Survey Canada, Miscellaneous Report 31.

Stephenson, T. R., D. W. German, E. F. Cassierer, D. P. Walsh, M. E. Blum, M. Cos, K. M. Stewart, and K. L. Monteith. 2020. Linking population performance to nutritional condition in an alpine ungulate. *Journal of Mammalogy* 101:1244–1256.

Stone, W. E., and M. L. Wolfe. 1996. Response of understory vegetation to variable tree mortality following a mountain pine beetle epidemic in lodgepole pine stands in northern Utah. *Vegetation* 122:1–12.

Stoner, D. C., M. T. Anderson, C. A. Schroeder, C. A. Bleke, and E. T. Thacker. 2021. Distribution of competition potential between native ungulates and free-roaming equids on western rangelands. *Journal of Wildlife Management* 85:1062–1073.

Symanski, R. 1996. Dances with horses: lessons learned from the environmental fringe. *Conservation Biology* 10:708–712.

Tahltan Central Government. 2021. Jade-Boulder sheep project. Fish & Wildlife Newsletter. Tahltan Central Government, Dease Lake, British Columbia, Canada.

Tarango, L. A., P. R. Krausman, R. Valdez, and R. M. Kattnig. 2002. Research observation: desert bighorn sheep diets in northwestern Sonora, Mexico. *Journal of Range Management* 55:530–534.

The Wildlife Society. 2020. The Wildlife Society calls to reduce wild horse and burro populations. 7 October 2020 position statement Nashville, Tennessee, USA.

Todd, J. W. 1975. Mountain bighorn sheep in southern Colorado. *Journal of Wildlife Management* 39:108–111.

Trombulak, S. C., and C. Frissell. 2000. Review of ecological effects of roads on terrestrial and aquatic communities. *Conservation Biology* 14:18–30.

Turner, J. C., and R. A. Weaver. 1980. Water. Pages 100–112 *in* G. Monson and L. Sumner, editors. *The Desert Bighorn: Its Life History, Ecology and Management.* University of Arizona Press, Tucson, Arizona, USA.

University of Virginia, Weldon Cooper Center for Public Service. 2024. National 50-state population projections: 2030, 2040, 2050: July 2024. University of Virginia, Charlottesville, Virginia, USA.

Valdez, R., and P. R. Krausman. 1999. *Mountain Sheep of North America.* University of Arizona Press, Tucson, Arizona, USA.

Veblen, T. T., T. Kitzberger, and J. Donnegan. 2000. Climatic and human influences on fire regimes in ponderosa pine forests in the Colorado front range. *Ecological Applications* 10:1178–1195.

Werner, W. 1989. Water development. Pages 161–175 *in* R. M. Lee, editor. *The Desert Bighorn Sheep in Arizona.* Arizona Game and Fish Dept., Phoenix, Arizona, USA.

Wickham, J., P. B. Wood, M. C. Nicholson, W. Jenkins, D. Druckenbrod, G. W. Suter, M. P. Strager, C. Mazzarella, W. Galloway, and J. Amos. 2013. The overlooked terrestrial impacts of mountaintop mining. *BioScience* 63:335–348.

Witham, J. H., and L. Smith. 1979. Desert bighorn movements in a southwestern Arizona mountain complex. *Desert Bighorn Council Transactions* 23:20–24.

Wolfe, M. L., L. C. Ellis, and R. MacMullen. 1989. Reproductive rates of feral horses and burros. *Journal of Wildlife Management* 53:916–924.

# 21 Threats to Mountain Sheep Habitat

*Bill A. Jex*

## 21.1 INTRODUCTION

Mountain sheep, comprising various subspecies of bighorn (*Ovis canadensis*) and thinhorn sheep (*Ovis dalli*), are iconic inhabitants of the rugged landscapes of North America. Renowned for their agility, strength, and majestic appearance, these creatures epitomize the resilience required to thrive in harsh environments. Their continued existence in those unique landscapes, however, is increasingly threatened by a myriad of factors, ranging from the effects of changing climatic patterns, shifting biomes and species assemblages, and human activities that continue to affect and encroach into historical habitats.

By exploring the unique ecotypic adaptations of these mountain sheep populations across western North America, to their preferred habitats and the emerging challenges they face, managers can gain a deeper understanding of their conservation needs and the actions necessary to ensure their long-term survival.

## 21.2 FACTORS AFFECTING HABITAT AND HABITAT FUNCTION

Mountain sheep inhabit diverse ecosystems that range from southern deserts to canyon lands, open grasslands and montane forests, to high elevation alpine plateaus and mountains in the Arctic, even persisting in landscapes that are now interface areas between remaining backcountry environs and front-country human developments (Chapter 19). Each of these habitats is facing distinct threats and cumulative effects that alter habitat function and subsequently mountain sheep habitat use. Understanding these challenges is central to devising effective conservation strategies as fundamentally, populations of wild sheep are limited by the availability and function of suitable habitats, with those habitats limited by the often-compounding influences of anthropogenic disturbance. The following sections detail the primary challenges confronting mountain sheep habitats.

### 21.2.1 CLIMATE CHANGE

There is ongoing debate about defining what climate change is. Here I describe it simply as: climate change occurs when evidence shows the historical and relatively regular patterns of seasonality within annual and multi-year cycles become less stable and more variable than humans have come to understand, or wildlife have adapted to. Changing climate poses a significant threat to mountain sheep populations

where changes to habitat condition on a temporal scale occur more rapidly than wild sheep can adapt to (this ability to adapt is often described as phenotypic plasticity; Pojar 2009).

In western North America, annual climate cycles traditionally store winter precipitation in the form of mountain snowpacks, with subsequent snowmelt in spring and summer months, providing water to support ecological function and the natural processes that occur between ecosystems and landscapes (Hale et al. 2023, Reyes and Kramer 2023), and healthy populations of mountain sheep across their range from Mexico, the United States, and Canada. Studies from across western North America document the effects of rising temperatures, more stochastic patterning of storm events, more frequent rain-on-snow precipitation, decreased snow-pack volumes and melt duration, causing changes in snow melt timing and annually fewer days with snow on the ground and in others, more persistent snow cover and cold temperatures into spring (Delbart et al. 2006, Shabbar 2006, Bevington et al. 2019, 2020; Xie et al. 2024). Increased lamb recruitment and survival in Dall's sheep (*O. d. dalli*) in the Alaskan Brooks Range is linked with the occurrence of higher spring snowline elevations and earlier spring snow disappearance dates and higher normalized difference vegetation index (NDVI), and in years with late snow disappearance dates, population growth could be reduced by 2% at intermediate northern latitudes, and by 5% at higher latitudes (Van de Kerk et al. 2018, 2020).

Responses in vegetation communities, plant distributions, and phenology have also been observed, with these changes affecting the availability and nutritional quality of forage, particularly in summer and autumn months when native perennial grasses provide significantly higher forage quality than annual grasses (e.g., cheatgrass [*Bromus tectorum*]; Ganskopp and Bohnert 2001, Wagler et al. 2023). For populations of thinhorn sheep in Alaska and Canada, the earlier arrival of spring green-up dates may improve female fitness as a result of more available new forage pre-parturition, also enhancing milk quality improving lamb growth and fitness. Improved growing conditions, however, are also likely to result in changes in vegetative cover and structure, expanding the degree of use and area of range overlap with other species of ungulates (i.e., interference and exploitative competition; Brewer et al. 2014, Jex et al. 2016 [competition definitions described in Table 21.1]), and benefiting some species of predators (apparent competition; Brewer et al. 2014, Jex et al. 2016). In these instances, the potential positive effects of climate-induced accessibility and increases in forage production are diminished

DOI: 10.1201/9781003518686-24

## TABLE 21.1

### Types of Competition Affecting Wild Sheep in North America

| Type of Competition | Definition |
| --- | --- |
| Apparent | Apparent competition occurs when one species indirectly competes with one or more other species, but each serves as prey of a common predator. This situation occurs when one (or more) species increases in number(s), and results in an increase in predator numbers in a particular area. As a result, there are more predators hunting for individuals belonging to the initial group occupying that area. |
| Exploitative | Exploitative competition occurs when two species utilize a resource that is in short supply (e.g., food, water, mineral licks, or cover) to the extent that occupation and use of the site or resource benefit one of those species at the expense of the other. |
| Interference | Interference competition occurs when one species excludes another from, or limits access to, a particular resource, and thus inhibits survival, reproduction, or other parameters as a result of behavioral interactions. |
| Manufactured (facilitated) | Manufactured competition occurs where a human intervention or habitat manipulation improves conditions or enhances a target species (e.g., prescribed fire habitat enhancement or desert water guzzler installations for mountain sheep), which also alters the distribution, density, or behavior of other sympatric species, leading to increased predation rates or other negative pressures on the target species. |

through the effects of competition and longer exposure to predation through earlier arrival of migratory predators such as golden eagles (Davidson et al. 2020). Pojar (2009) concluded that thinhorn sheep numbers in northwestern British Columbia will most likely decline in response to shifting climates, while elk (*Cervus canadensis*), bison (*Bison bison*), and mountain goat (*Oreamnos americanus*) populations are expected to increase. Because climate patterns in northern thinhorn sheep ranges are similar across jurisdictions due to influences from the Arctic Oscillation (Lindsey 2009) and Pacific Decadal Oscillation (Hamlington et al. 2019), negative effects from changing climates observed in Alaska could also be expected to occur in Yukon, Northwest Territories, and British Columbia. For populations of desert bighorn sheep (*Ovis canadensis nelsoni, O. c. cremnobates, O. c. mexicana,* and *O. c. weemsi*), California bighorn sheep (*O. c. californiana*), and Rocky Mountain bighorn sheep (*O. c. canadensis and O. c. sierra*; Chapters 1 and 6) already experiencing chronic shortages of water during hot summer months, as a result of reduced snowpacks and available surface water, earlier snow-free dates could affect forage quality and quantity, reducing landscape carrying capacity (Brewer et al. 2014). Without access to natural sources of water either because of increased precipitation or prolonged snow melt, population reductions, localized extirpations, or range contractions could be expected to occur (Weaver and Mensch 1971, Epps et al. 2004, Whiting et al. 2012, Cain et al. 2018).

Changes in precipitation patterns (i.e., wet vs. dry seasons) and the way in which precipitation falls (i.e., rain, snow, rain-on-snow) can disrupt water sources important for sheep survival, or it can reduce the quality of and access to autumn and winter forage, while also increasing the risk of accidental injury or death associated with landform instability (e.g., increased erosion exacerbating landslides) or the frequency of avalanche activity as has been observed in Alaska (Lohuis 2017, Lohuis et al. 2018).

Moreover, increased frequency and intensity of weather events such as extreme heat events that create prolonged droughts can fuel destructive wildfires, further affecting habitat suitability (Clapp and Beck 2016). This may lead to degraded habitat, or fragment contiguous habitats, when catastrophic wildfire is widespread and consumes most of the shrub and grass communities, resulting in little post-fire vegetative response (Clapp and Beck 2016). These outcomes can ultimately transform ecosystem biodiversity and shift biomes (Pojar 2009). Where biome shift is significant and results in less favorable habitat condition, phenotypic plasticity might reach its limits and result in population declines. If environmental change trends to more preferred habitat however, ecotypic differentiation may enable the populations to adapt to new environmental conditions (Franks et al. 2014). For example, in the case of an abomasal nematode (*Marshallagia marshalli*) common to the stomach of Dall's sheep, it negatively affects fitness (Aleuy et al. 2018) and changes in environmental conditions trending toward warmer and drier climate, and may favor the nematode's reproductive lifecycle. In this case, parasitized Dall's sheep could be expected to have higher parasite loads that cause poorer body condition resulting in population declines, whereas if changed climates have a less favorable effect on life requisites for this parasite, corresponding improvement in the fitness of Dall's sheep could be observed, where lower parasite loads result. Similarly, warmer temperatures may create an environment that allows infected Dall's sheep to survive winters in larger numbers, increasing the prevalence and geographic distribution of the parasites or pathogens those individual sheep carry (Hueffer et al. 2013).

Change resulting from weather will produce some widespread outcomes (Xie et al. 2024); however, there will also be uniquely regionalized effects (Table 21.2). The effects of changing climates are anticipated to influence future temperature trends and patterns of precipitation, reshaping ecosystems across North America (Figure 21.1a and b). The ability to reliably predict and forecast those changes in key

**TABLE 21.2**

**Summary of Generalized National and Jurisdictional Temperature Trends and Precipitation Patterns Sourced from Climate Data Websites (August 2024), for Regions Inhabited by Bighorn and Thinhorn Sheep in North America**

| Region | Temperature Trends | Precipitation Patterns | Jurisdiction and Species |
|---|---|---|---|
| Rocky Mountains United States | Increasingly warm; warming trend noted | Variable; some areas show increase | • MT, WY, CO, NE, NV, NM and AZ: Rocky Mountain bighorn sheep<br>• UT: Rocky Mountain and desert bighorn sheep<br>• ID: California and Rocky Mountain bighorn sheep range |
| Rocky Mountains Canada | Warming trend; winter temperatures rising | Variable; influenced by Pacific systems | • BC: Stone's sheep, Dall's sheep in northern portions<br>• AB and BC: includes most northerly Rocky Mountain bighorn sheep population in North America; California and Rocky Mountain bighorn sheep range throughout the southern portions |
| Badlands (US) | Increasingly warm; warming trend noted | Variable; some areas show increase | • ND, SD, NE: Rocky Mountain bighorn sheep |
| Sierra Nevada (US) | Warming trend; significant in summer | Declining; notable drought periods and stochastic severe winter storm events | • CA: Endangered Sierra Nevada bighorn sheep (*O. c. Sierra*)<br>• NV: California, Rocky Mountain, and desert bighorn sheep |
| Coast Mountains (US) | Generally warming; pronounced in lower elevations, cooler spring periods with more frequent storm events | Variable; influenced by coastal systems, increased rain-on-snow and winter/spring ground freezing and persistent snow cover | • WA, OR, ID: California and Rocky Mountain bighorn sheep |
| Coast Mountains (Can) | Generally warming; pronounced in lower elevations, cooler spring periods with more frequent storm events | Variable; influenced by coastal systems, increased rain-on-snow and winter/spring ground freezing and persistent snow cover | • BC and YT: Dall's sheep (thinhorn sheep) primarily inhabit northern portions of this mountain range, with north-central areas in BC dominated by Stone's sheep whose range extends slightly into south-central YT. |
| Alaska and Brooks Ranges (US) | | | • AK and YT: Dall's sheep |
| Mackenzie and Ogilvie Mountains (Can) | | | • YT and NT: Dall's sheep |
| Desert Southwest (US) | Significant warming; extreme heat events | Variable; some areas drying out | • AZ, CA, NV, NM, TX, UT: desert bighorn sheep |
| Sierra Madre and Sonoran Desert (Mexico) | Increasingly warm; extreme heat events | Highly variable; sporadic rains | • Desert bighorn sheep in northern mountains |
| Chihuahuan Desert (Mexico) | Warming trend; hot summers | Variable; monsoon dependent | • Home to bighorn sheep populations |

regions inhabited by bighorn and thinhorn sheep populations will become an important aspect of habitat management and habitat enhancement activities. Being able to identify emerging trends in regional precipitation may help prioritize areas for habitat interventions (e.g., prescribed fire) to address shrub encroachment into alpine grassland communities. In dry climates in southern and desert jurisdictions, interventions such as artificial water sources (e.g., guzzlers) may be required to mitigate negative effects created by prolonged drought conditions (Whiting et al. 2012), or where weather patterns shift and improve water availability, the increased precipitation could address forage quantity and nutritional limitations, improving landscape carrying

capacity and benefiting local sheep herds (Chapter 12). Trends and events, however, may also result in significant precipitation that causes landslides or creates avalanche conditions in mountain sheep habitat; contribute to encroachment of woody plants into native grasslands, subalpine and alpine grassland areas; convert savannas into forested areas; or change plant phenology (Lohuis et al. 2018, Meyers-Smith and Hik 2018, Hicke et al. 2022).

Mountain ecosystems are vegetatively and topographically complex environments that influence lower elevation receiving environments such as canyon lands and desert badland terrains (Acreman et al. 2012). These environments typically provide the most suitable habitats for

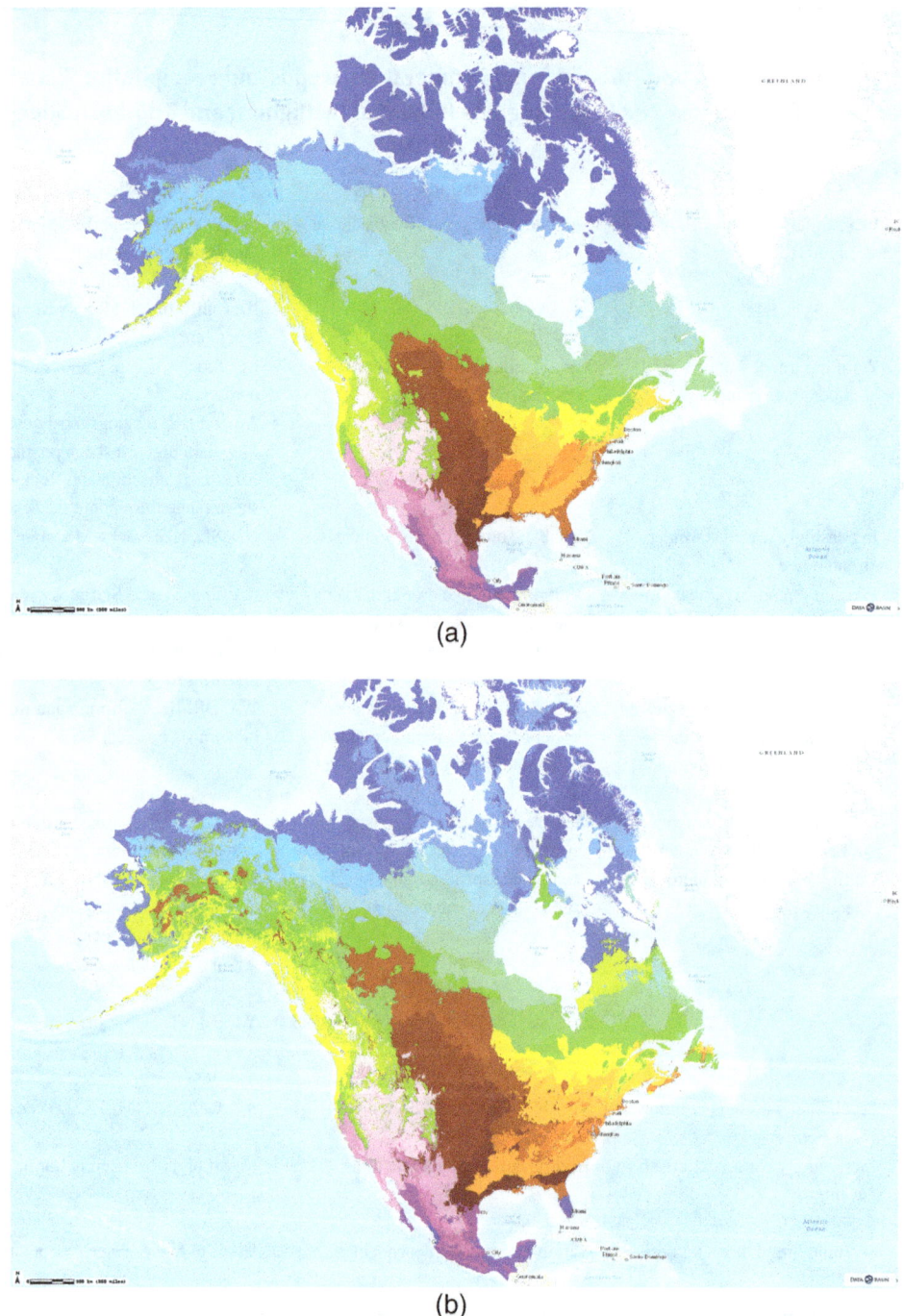

(a)

(b)

**FIGURE 21.1**   (a) Model-predicted distribution of current (2018) North American ecoregions, from AdaptWest: a climate adaptation and conservation planning database for North America (map adapted from Stralberg 2018). (b) Model-predicted distribution of future (2070) North American ecoregions, from AdaptWest: a climate adaptation and conservation planning database for North America (map adapted from Stralberg 2018).

wild sheep (Chapter 19). It is reasonable to surmise that where changing climates create altered or new transitional habitat at any geographic scale, the resulting changes in functional suitability of those habitats will change along with the inherent capability of those areas to provide the necessary life requisites for mountain sheep. Such climate and environmental change could then stimulate a positive or negative effect on population trajectory and resilience. This highlights the

challenge in classifying climate change as inherently positive or negative. As global temperatures warm and new climate trends manifest, annual climate cycles may create negative effects for bighorn and thinhorn sheep because of the severity or stochastic nature of events (Whiting et al. 2012, Hik and Carey 2000) that ultimately lead to population decline (Bleich 2009). Instead, environmental responses to the new climate norms may be positive on some levels for mountain sheep, where for example, later arrival of snow and earlier spring melts improve access to better habitat conditions (Goodson and Stevens 1992, Hik and Carey 2000, Cosgrove et al. 2021). Regardless, the projected shifts in climate patterns will reshape distribution of habitats used by mountain sheep (Pojar 2009). Some subspecies of mountain sheep may be able to adapt to the effects of changing climates where those changes result in more favorable environments, but may be unable to overcome the effects of severe weather events occurring irregularly within annual seasonal weather cycles. These irregular and extreme weather events have been associated with population decline in some areas (Bleich 2009), but due to their temporary nature, may go undetected as a factor affecting population trajectory. Conversely, gradual but unfavorable changes in habitat suitability may be easy to observe and offer opportunities to implement habitat restoration projects (e.g., using prescribed fire or mechanical interventions, or water source management and the creation of supplemental water sources; Chapters 12 and 20) that improve population trend and resilience, prior to significant declines occurring. Population resilience, being inherently

limited by species fecundity, biology, and environmental suitability, could be positively or negatively influenced by the rate and type of changes in climate that occur. Considering trends in temperature and precipitation, and planning for population-changing events when undertaking species and habitat management, will be imperative for sustaining some smaller and more localized and isolated populations of mountain sheep. Where shrubs invade alpine grasslands due to gradually warming temperatures and increased precipitation (Meyers-Smith and Hik 2018; the encroachment of shrubs into alpine grasses is often referred to as shrubification), conventional practice is to apply a prescribed fire intervention to restore their suitability for mountain sheep as grassland (Brewer et al. 2014, Sittler 2019). With shifts in forested areas upslope, however, new forested stands may be more prone to lightning-caused wildfire, converting the climate-altered habitat structure back into grassland (Figure 21.2; Cansler et al. 2018), resulting in positive effects on habitat structure and suitability for mountain sheep.

Outcomes resulting from climate-affected ecosystems may create additional habitat and management challenges.

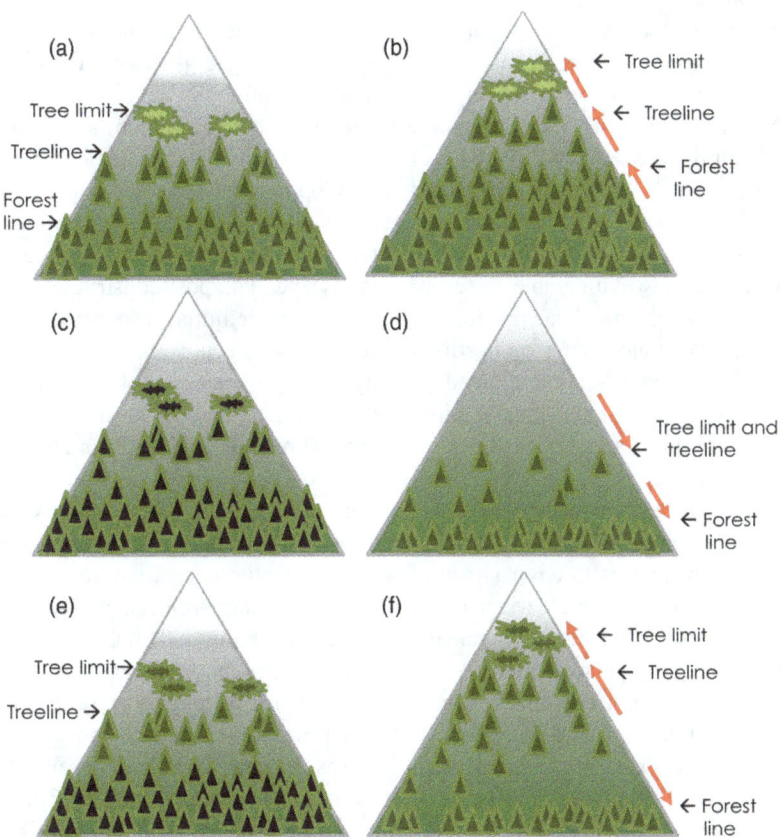

**FIGURE 21.2**  Conceptual model of vegetation changes in the alpine treeline ecotone (ATE) due to climate warming (top row) and the combined effects of climate warming and fire (bottom two rows), in the western United States. The current condition is represented by (a). With climate change, tree limit, treeline, and forest line are expected to rise in elevation (grey arrows, b). (c) represents post-fire conditions if the entire ATE burned. Fire results in a long-term suppression of tree limit, treeline, and the forest line (grey arrows, d). Post-fire recovery is slow but is likely to be faster near the forest line than the tree limit. If fire were to burn only to the forest line (e), with climate warming the upper tree limit would ascend but the lower edge of the ecotone would descend (f). In sum, the change due to climate warming likely differs from that due to fire (increase vs. decrease in tree cover, respectively), resulting in combined effects that differ from those caused by either factor alone (open access source: Cansler et al. 2018; https://doi.org/10.1002/ecs2.2091).

For example, conventional practices for enhancing mountain sheep habitats encourage the use of prescribed fire interventions to restore open canopies, understory and perennial grasses, also allowing better evasion of predators, and improved access to high-quality forage. Interference, exploitation, and apparent competition from other ungulates and their predators, or the expansion of invasive plant distributions that compete with native grasses, are unintentional outcomes of applying prescribed fire to the landscape and this may require careful consideration and pre-planning, to ensure the highest return from the enhancement activity (Ministry of Forests, Lands, Natural Resource Operations and Rural Development 2022, Whiting et al. 2023).

In northern environs, climate change is occurring even faster than in southern locales. Annual warming and thawing cycles, increases in the frequency and severity of climate-driven disturbances (e.g., wildfire, forest and animal health pathogens, insects), and widespread changes to precipitation, surface water extent, soil moisture, vegetation structure and function (Youngwook et al. 2024, Arctic-Boreal Vulnerability Experiment 2024) are changing wildlife habitats and affecting wildlife populations and mountain sheep population resilience (Kneteman 2016). With changed habitats and more moderate seasonal effects, expanded range distributions of more southern species (e.g., deer [*Odocoileus* spp.], elk, and cougar [*Puma concolor*]) are occurring (Ministry of Forests, Lands, Natural Resource Operations and Rural Development 2019, Mowat et al. 2023). Unlike native northern wildlife species adapted to historical climate patterns, these new arrivals are able to increase their distribution and numbers rapidly when conditions are conducive but are also vulnerable to severe weather events more typical of historical norms for the north. Mitigating the effects of climate change on northern thinhorn sheep habitats will require application of an adaptive management framework that explores longstanding mountain sheep habitat relationships, and the dynamics of new climate-altered relationships. Creating positive effects for thinhorn sheep may center around projects that include habitat restoration, managing predator densities, collaborative monitoring of population responses to environmental changes at localized scales across multiple jurisdictions, or a combination of activities, to effectively manage cross-jurisdictional habitat and species management.

Climate change in northern environments is also enabling expansion of financial and infrastructure investments into indigenous agricultural systems to assist with food security at a community level. Shifting abundance of fisheries, wildlife, and traditional plant-based foods is creating a community need to find new areas to hunt with new modes of transport required to access those areas, but also to substitute traditionally harvested species with other wild foods (Brown et al. 2021) and implement new opportunities to address quantity and quality of available, local foods (New Relationship Trust 2024). These climate-facilitated

changes are likely to result in changes to natural environments by shifting forested and natural landscapes into hot-house and field-crop purposed landscapes (sensu Inuvik Community Greenhouse 2024), changing the traditional hunter-gatherer culture centered around fisheries and wildlife food acquisition toward an on-farm livestock-centric one. This in turn will lead to higher numbers and densities of livestock in northern landscapes, with increased risk to immunologically naive wildlife species such as thinhorn sheep (Jex et al. 2016). With the introduction of new and novel parasites and pathogens, negative population effects are expected to occur (Jex et al. 2016).

### 21.2.2 Natural Resource Extraction

The extraction of natural resources, such as minerals, oil and gas, and vegetation (e.g., livestock grazing, forest harvesting) poses a significant threat to mountain sheep habitats where the activity occurs without implementing active and adaptive mitigations. Mining activities often involve habitat alteration at very large geographic scales that may lead to destruction or loss, fragmentation of habitat connectivity and movement corridors, changes in plant community species assemblages, introduction of non-native species, changed geology affecting the nutritional qualities and trace minerals concentrations in available forage, and alteration, contamination, and poisoning of water sources (Poole et al. 2020, Anderson et al. 2022). Infrastructure development associated with resource extraction, such as roads and pipelines, further fragments habitat and expands human disturbance by increasing the permeability of the landscape. The post-construction use of an area is seldom limited to the initial resource extraction activity leading to secondary effects from concurrent increases in accessibility and subsequent use of other resources such as timber, or the wildlife resource itself by recreational or commercial users (Freeman and Maclean 2021).

Efforts to mitigate the effects of natural resource extraction on mountain sheep habitats may include habitat restoration projects, conservation easements, protections for specific features (e.g., natural mineral licks or known seasonal migration routes), and collaborative planning processes involving stakeholders from the conservation and resource extraction sectors. Governments, responsible for the issuance of tenures or project approvals, create policy and procedural directives to guide these permitting processes that may include requirements for incorporating best management practices and other mitigations with the goal of minimizing negative effects (Ministry of Forests, Lands and Natural Resource Operations 2014, Mining Environment Research Group 2002, Table 21.3). Understanding the net outcomes, however, is often limited by an incomplete understanding of the pre-existing habitat values and patterns of sheep use that existed pre-development; seldom is there a requirement for adequate baseline wildlife data

**TABLE 21.3**

**Overview of Management Objectives, Mitigations, or Restorations Used in Natural Resource Extraction in British Columbia**

Mountain sheep management objectives:

- Maintain suitable, connected mountain sheep habitat and its functional integrity;
- Avoid human disturbance and access during critical times and to critical habitats;
- Maintain the long-term viability of mountain sheep populations; and
- Reduce and eliminate disease transmission and competition for forage.

Identify mountain sheep indicators within the project area. Key components and indicators include:

- location, amount and condition of habitat;
- abundance and distribution of suitable winter and seasonal habitat;
- population structure and dynamics;
- mortality risk factors;
- population size and trend;
- adult survival;
- lamb survival; and
- population growth rate (lambda).

**Guidance for mountain sheep management:**

| | |
|---|---|
| Identify mountain sheep habitat and historic and current use of those habitats within the proposed project footprint and area of influence. | 1. Review, update, and validate or refine habitat suitability modeling and mapping.<br>2. Identify movement corridors for use between critical ranges and features.<br>3. Plan all activities in and around bighorn and thinhorn sheep habitat to coordinate with least-risk timing windows.<br>4. Avoid physical destruction of critical mountain sheep habitats. Locate facilities, trails, and roads away from these habitats. Facility development must be greater than 2 km from confirmed natal areas.<br>5. Maintain visual screening (i.e., vegetated areas with cliffs and gullies) and a 500 m no-disturbance buffer around mineral lick sites and natal areas to provide security and escape cover to animals using the lick and during lambing. |
| Identify the impacts of proposed activities on mountain sheep and their habitat. | 1. Quantify the amount of habitat lost or adversely affected by project footprint or through sensory disturbance.<br>2. Identify duration and sources of sensory disturbance throughout the project.<br>3. Identify potential impacts of project footprint on predator-prey dynamics.<br>4. Quantify the increase in density of linear features within the regional study area. |
| Avoid or minimize new disturbance to important habitats. | 1. Avoid physical destruction of mountain sheep escape terrain and moderate and high-value habitats.<br>2. Identify mineral lick locations.<br>3. Maintain known and potential mineral licks in a natural state.<br>4. Avoid physical destruction of mineral lick sites and connecting wildlife trails.<br>5. Maintain visual screening and forested cover around mineral lick sites. |
| Avoid increasing access to or within mountain sheep habitat. | 1. Limit operating motorized vehicles within 500 m of natal areas during critical risk timing windows. Restrict off-road travel during critical winter periods.<br>2. Limit the amount of road access in close proximity to critical sheep habitats (within 1–2 km) to reduce human disturbance and the potential for illegal harvest.<br>3. Design temporary structures, including roads, when industrial activity cannot be avoided within 500 m of mountain sheep habitat.<br>4. Adopt the principle of adaptive management and employ appropriate methods to reduce development impacts to mountain sheep. |
| Avoid displacing mountain sheep and minimize direct and indirect mortality on mountain sheep populations. | 1. Follow risk timing windows when planning and conducting development activities.<br>2. Apply a precautionary approach to aerial operations for both helicopter and fixed-wing aircraft by applying both horizontal and vertical buffers, along with site-specific best management practices when operating in and around identified critical wild sheep habitats and habitat features (e.g., mineral licks).<br>3. Develop and implement access management plans and access control points near mountain sheep habitats.<br>4. Coordinate planning, development, and deactivation schedules for all development activities with other users if possible.<br>5. Within 16 km of known mountain sheep ranges, exclude domestic sheep and goats to reduce disease transmission risk and competition for forage.<br>6. Restrict dogs on important mountain sheep habitat when occupied. |
| Avoid contaminating mountain sheep habitat. | Manage all waste products, during all project phases, to ensure contamination of water and vegetation does not occur due to development activities. |

*(Continued)*

**TABLE 21.3** (*Continued*)

## Overview of Management Objectives, Mitigations, or Restorations Used in Natural Resource Extraction in British Columbia

| | |
|---|---|
| Restore habitats to a condition that provides a similar level of functionality as before industrial activity. | 1. Restore habitat as soon as possible following development.<br>2. Deactivate linear features (e.g., roads and trails) as soon as possible after cessation of development.<br>3. When clearing land, prevent the establishment and spread of invasive species by re-vegetating disturbed areas as soon as possible using native plants. Never plant invasive or non-native plants.<br>4. Implement ongoing monitoring for the occurrence of invasive species and undertake actions to manage and control occurrences when found. |
| Develop a monitoring and adaptive management plan to monitor the effectiveness of mitigation and restoration measures. | 1. The proponent is responsible for carrying out relevant monitoring to ensure that mitigation measures are implemented as planned and are effective at meeting the intended objectives to reduce impacts on mountain sheep and mountain sheep habitat. Monitoring must be done at an early phase of the project and revised as required. Monitoring objectives and commitments must be established prior to finalization of the mitigation plan.<br>2. The type and degree of monitoring should be proportional to the degree of risk to mountain sheep posed by the development.<br>3. Implementation and effectiveness monitoring of mitigation measures must be planned and undertaken by an appropriately qualified professional. All data must be shared with the province to facilitate current and future management and adaptive mitigations. |
| Risk timing windows for mountain sheep in northern British Columbia: | Low risk: 16 July–14 November<br>Critical: 15 January–15 July<br>Caution: 15 November–14 January<br><br>• *Low risk.* Restrictions would not normally apply. Where ground conditions permit, plan development activities within these timeframes. Best management practices and strategies apply.<br>• *Caution.* Operators should avoid development activities during these timeframes. Aerial activities should adhere to guidelines. In the event that working within a cautionary window is unavoidable, the proponent must contact an appropriate qualified professional to discuss alternatives, and potential mitigation and monitoring plans. A referral to work in this window must be accompanied by a rationale, mitigation and monitoring plan.<br>• *Critical.* Development activities are not appropriate during this timeframe. Aerial activities should adhere to guidelines. If working within a critical window is unavoidable, the proponent must contact an appropriate qualified professional to discuss alternatives, and potential mitigation and monitoring plans. A referral to work in this window must be accompanied by a rationale, mitigation and monitoring plan. |

*Source:*  Ministry of Forests, Lands and Natural Resource Operations (2014).

collection or habitat function assessment prior to site disturbance occurring or earlier in the project during initial exploration activities. The resulting knowledge gap created by activities occurring prior to initial baseline investigation and documentation of the environmental values that existed in the undisturbed, natural state often does not get incorporated into the quantification of negative-effects-accounting developed through permit issuance processes, with these effects often identified as out of scope of the assessment. This can occur where exploration activities occur over a long period of time, in advance of an environmental assessment linked to a resource extraction development tenure application (Environmental Law Centre 2010, British Columbia Environmental Assessment Office 2020).

In populations of mountain sheep naive to the activities of humans, the initial effects of resource extraction activities are more likely to lead to temporary abandonment of seasonal ranges and preferred foraging habitats (Frid and Dill 2002, Poole et al. 2020). While mountain sheep have developed tolerance to some predictable activities (Beale and Boyce 2020), this tolerance appears to exist as a function of site attractiveness (i.e., improved forage supply, modified terrain features [e.g., excavated escape terrain],

decreased predator occurrence) more so than as a result of habituation to the activity (Frid and Dill 2002). This association between activity, occurrence, and tolerance behavior is consistent with the risk-disturbance theory: animals perceive nonlethal human disturbance similar to nonlethal predation stimuli, temporally and spatially balancing predation risk with resource richness and the availability of alternative areas of suitable habitat with less human disturbance (Frid and Dill 2002). In this context, responses of mountain sheep to natural resource extraction activities have similar observed outcomes, often showing that the disturbed area associated with the mines has a moderate influence on sheep distribution, with sheep at times being associated with areas closer to the mine, using altered terrain features and vegetation structure (Bleich et al. 2008). Whether increases in sheep use or proximity to the disturbed area reflects a benefit depends on the demographic responses of those sheep to the resources and conditions available on mine sites; information about those specific responses is lacking (Bleich et al. 2008, Jansen et al. 2009, Poole et al. 2016).

Considerations of predation risk, a key factor affecting mountain sheep habitat selection (Frid and Dill 2002), may be more prominent in influencing sheep use and selection of disturbed habitats or areas close to humans. In areas where predation risk is not reduced by a disturbance or the presence of humans, anecdotal reports of some level of abandonment or alienation of habitats by thinhorn sheep have been observed (Bill A. Jex, Ministry of Water, Land and Resource Stewardship, British Columbia, unpublished data). To interpret behavioral responses to resource developments such as mining, biologists often use resource selection function modeling; however, the outputs of this modeling require careful interpretation and full consideration of other trade-offs being made (e.g., predation risk) by individual sheep in that specific environment based on their sex, age, social grouping, and seasonal timing (e.g., lambing, rutting), to ensure erroneous assumptions are not made with respect to preferred habitat associations (Chapter 19) and the net effect of habitat alterations on use and selection.

## 21.2.3 INVASIVE SPECIES: NON-NATIVE AND IMMIGRATING FLORA AND FAUNA

Changes in habitat function associated with altered climates, wildfire, and invasive plants can lead to changes in competition from other ungulates either native to the area (e.g., elk) or that have been introduced and become feral (e.g., horse [Equus caballus] and burro [Equus asinus]) and have expanded range distributions into traditional mountain sheep range previously devoid of such species (e.g., aoudad [Ammotragus lervia]; Brewer et al. 2014). Similarly, predation by new predator species colonizing expanded geographies (e.g., cougar range expansion into northern areas in British Columbia and the Yukon; Mowat et al. 2023, Environment Yukon 2024) can increase mortality rates and predation risk in mountain sheep populations.

The introduction of non-native flora and translocation of fauna and natural immigration of fauna pose a threat to mountain sheep habitats by altering ecosystems, changing species assemblages and vegetative communities, and potentially affecting populations because of apparent, exploitative, or interference competition (Brewer et al. 2014, Jex et al. 2016). An additional form of competition called manufactured or facilitated competition (Table 21.1) may also occur. Apparent, exploitative, and interference competition differ from manufactured competition in that they are the result of changes in ecosystem or the supply of resources through a somewhat natural evolution of habitat condition, whereas the latter is the result of a human intervention or perturbation to the environment, such as the translocation of aoudad into Texas and New Mexico, or the movement of elk into Stone's sheep habitats in northeast British Columbia following widespread application of prescribed fire to the landscape. In some manufactured competition situations, there may be a net-negative effect to the target species despite the original intent to improve outcomes through the intervention activity. The cost of predation coupled with effects of other types of competition may, for example, outweigh the anticipated benefits of open

habitat and improved forage quality for mountain sheep that could be created by such interventions as prescribed burns. In all cases, the degree of competition will depend on the resources available, and the number and species of competitors involved (Brewer et al. 2014, Jex et al. 2016).

Where these habitat restoration interventions lead to changes in the types of grasses recolonizing those treatment areas, or create a shift in vegetation species assemblages toward a higher proportion of invasive annual grasses (e.g., cheatgrass [Bromus tectorum], North Africa grass [Ventenata dubia], Medusahead [Taeniatherum caput-medusae]) or invasive weeds (e.g., spotted knapweed [Centaurea stoebe], common St. John's wort [Hypericum perforatum]), planning to include post-fire mitigation and remediation will be required to ensure net benefits to mountain sheep can be realized. Wildfire remediation and prescribed fire interventions can be costly; adding the considerations for post-project mitigation from competition and invasive species will increase such costs.

Animals are inherently tied to the nutritional resources of the landscape (Wagler et al. 2023), and where winter months create nutritional intake challenges, the availability and quality of forage on summer range and late season areas has an important influence on individual nutritional health and reproductive success (Pigeon et al. 2019, Wagler et al. 2023, Chapter 7). Because invasive grasses generally grow faster than native grasses (Frank et al. 2023), in areas where annual grasses and invasive weed abundance becomes widespread enough to reduce the availability of native perennial grasses, it changes vegetation composition and can significantly reduce the quantity and quality of summer forage (Gianella and Sanchez 2024) and late season forage (Ganskopp and Bohnert 2001). Changed vegetation communities, where native grass species are outcompeted by faster non-native annual grass growth, reduce mountain sheep range carrying capacity. Bighorn sheep show a forage selection preference toward native perennial grass species (Wikeem 1984), having evolved with native grasses providing the bulk of their forage intake in some areas. In regions where invasive annual grass communities are established there can be active selection against areas that have converted from annual grass communities (Figure 21.3; Ministry of Water, Land and Resource Stewardship 2023). This may have implications to the practice and planning of applying prescribed fire to resolve forest encroachment effects when attempting to improve or restore bighorn sheep habitat conditions (Ministry of Forests, Lands, Natural Resource Operations and Rural Development 2022).

Wildfire is different than prescribed fire in its origin, timing, severity, and geographic scale, and is considered a naturally occurring phenomenon in many ecosystems, associated with biogeoclimatic zone or biome (Forest Practices Code of British Columbia 1995, Nitschke and Innes 2013). Unmanaged wildfires often affect expansive areas of geography. Those wildfires that occur close to areas with invasive annual grasses can increase invasive annual grass distributions, subsequently creating a relatively rapid

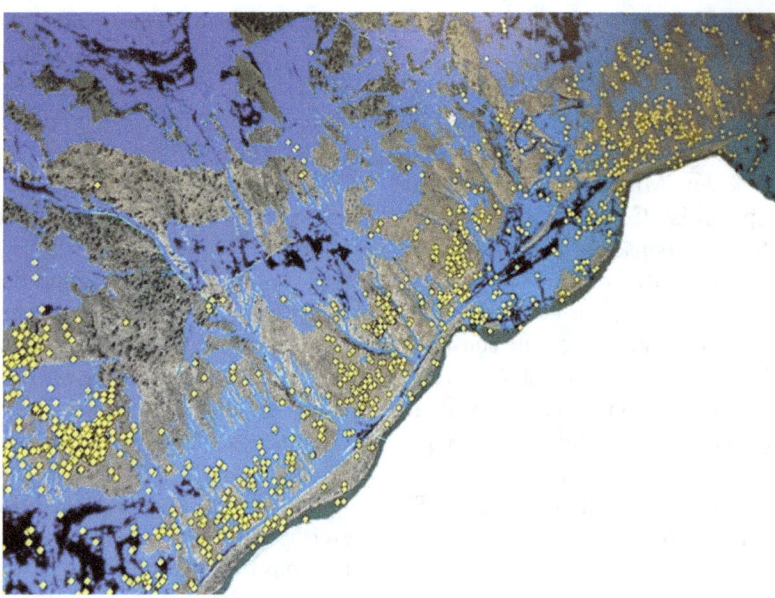

**FIGURE 21.3** Global positioning system-based point locations (yellow) of collared bighorn sheep in relation to native grasslands (gray) and mapped invasive cheatgrass communities (blue) in the Kamloops region of British Columbia, Canada (Ministry of Water, Land and Resource Stewardship 2023; unpublished data.)

cyclical and self-fueling relationship between invasive annual grass occurrence and abundance, and reoccurring frequency of wildfire (Fusco et al. 2019, Stanton et al. 2023). The resulting shortened fire return intervals (Fusco et al. 2019) can alter the ecosystem and habitat suitability for mountain sheep and other species of wildlife (Fish and Wildlife Compensation Program 2020, Grow Wild Montana 2024a,b), while changes in precipitation patterns can alter soil hydrology leading to shifting vegetation successional states. These changes may affect forage availability and forage nutrition, reducing the available habitat's carrying capacity for mountain sheep. Where the transformed habitats become unsuitable for perennial native grasses, a progression of habitat replacement will occur with annual grasses replacing perennial grassland communities resulting in a long-term reduction in the area's ability to support as many sheep (California Department of Fish and Wildlife 2020, Grow Wild Montana 2024a). Efforts to control invasive species and restore native vegetation are essential for maintaining healthy mountain sheep habitats. Such efforts may include targeted eradication programs, habitat restoration projects, and public education campaigns to prevent the introduction and spread of invasive species (Chapter 20).

### 21.2.4 Recreation

Motorized and non-motorized recreational activities, such as hiking, mountain biking and electrically assisted mountain biking, skiing and backcountry skiing, hunting and off-road vehicle use, can disrupt mountain sheep habitats by causing habitat degradation and a zone of disturbance

that results in changed seasonal migration patterns (Legg 1998, Courtemanch 2014), causing displacement of animals leading to temporary or permanent habitat alienation and range abandonment. The effects of human activity can mimic predation risk (Lowrey and Longshore 2017) and increased human presence in sensitive habitat areas can lead to heightened stress levels in mountain sheep, affecting habitat selection (Papouchis et al. 2001), behavior, and reproductive success (Wiedmann and Bleich 2014, Sloan 2022). Activities such as mountain biking can result in immediate responses of animals, changes in habitat use and diurnal activity patterns, reduction in reproductive success, increases in seed dispersal and trampling damage on flora leading to vegetation changes in areas adjacent to trails, and soil compaction, exposure, and erosion (Kuwaczka et al. 2023). Additional effects of human recreational activities include increased heart rates, displacement from preferred foraging areas, and increased energetic costs (MacArthur et al. 1982, Wiedmann and Bleich 2014, Brushett et al. 2023). Negative effects on mountain sheep from recreation are no longer limited to front-country habitats (Courtemanch 2014, Sproat et al. 2019, Papouchis et al. 2001, Brushett et al. 2023) and terrestrial-based activities now include the use of electrically assisted mountain bikes that increases the spatial scale of effects beyond those from unassisted mountain biking activity (Kuwaczka et al 2023). Similarly, conventional motorized recreation (motorcycles, off-road vehicles, all-terrain vehicles) and snowmobiles (including snow-bikes) have continued to advance in technology and design, and now permit a significantly expanded motorized terrestrial footprint into areas that previously served as refugia habitats either through distance isolation or geophysical limitations that are now no longer restrictive; and those geophysical limitations are becoming increasingly less restrictive as changing climate conditions reduce barriers to overland movement of humans (Segal 2024).

Helicopter supported recreation can also result in widespread consequences to mountain sheep habitat effectiveness (Legg 1998, Frid 2003), moving human recreational disturbances (e.g., heli-skiing, heli-mountain biking, alpine weddings) deeper into the backcountry and deeper into wildland sheep habitats. In response, most jurisdictional regulators have implemented policy and legislation, supported by monitoring programs of licensed or commercial activities and tenures. Similarly, the increased personal and commercial use of unmanned aerial vehicles or drones in support of photography, filming, and image capture for general and commercial interests elicit negative behavioral responses in mountain sheep and other wildlife (Mulero-Pázmány et al. 2017, Rebolo-Ifrán et al. 2019), mimicking aerial predator avoidance. This relatively novel activity is increasingly seeing jurisdictions and organizations creating policy and legislation to minimize disturbance effects (e.g., Wild Sheep Foundation Position Statement: Effects of Recreational and Commercial Use of Drones on Wild Sheep 2021).

Often, the individual people or groups seldom have the personal reference to understand the degree to which their activities are influencing mountain sheep behavior, habitat selection and use, and ultimately on individual animal fitness and population trajectory (Courtemanch 2014). As compared to direct causes of mortality that are easy to see and understand (e.g., hunter harvest; Chapter 16), the population effects of widespread and repeated human activity occurring in important habitats, with continued encroachment further into the backcountry, is not inherently understood by many recreationalists (Sterl et al. 2008). Most people carry the opinion that because of the scale of their activity (e.g., individuals participating in non-motorized, backcountry skiing, hiking, or mountain biking) the consequences of their recreation are benign and even justifiable, when considered in comparison to the more obvious hunter harvest or industrial scale effects (Peterson 2020). Increasingly, wildlife and mountain sheep managers are struggling to educate these users to the significance of the effects they have on wildlife (Wiedmann and Bleich 2014, Teton Range Bighorn Sheep Working Group 2021). Unlike hunting or resource extraction activities that are regulated through licensing, laws, and established engagement and communication pathways, recreation has few regulatory restrictions in most jurisdictions and is largely unorganized and unlicensed. Given the lack of regulatory options in many jurisdictions, there is often no direct means for managers to influence or manage many individual recreationalists. Communications about access restrictions or prohibited activities often occur through signage or passive communications, both of which do not strongly elicit changes in behaviors within this demographic (Sterl et al. 2008, Wiedmann and Bleich 2014, Kopp and Coppes 2020). In some areas land management agencies have successfully implemented measures to minimize the effects of individual and commercial recreational activities on mountain sheep habitats by implementing actions such as targeted education programs (i.e., rather than passive communications), area activity, and geographic closures based on clear science-supported rationales, and establishing designated use areas with marked trails and improved accessibility to areas where there will be less disturbance effect on mountain sheep (Teton Range Bighorn Sheep Working Group 2021, Chapter 20). This can help minimize and mitigate negative effects some sectors or activities may have on wildlife and the suitability and function of important habitats. Other strategies include fostering self-regulating opportunities by non-government and stakeholder-led organizations as an alternative to broader legislated access restrictions.

### 21.2.5 Urbanization and Human Population Growth

Urbanization and infrastructure development encroach upon mountain sheep habitats, leading to habitat loss, fragmentation, and increased human-wildlife conflicts (Radium 2022, Vail 2024). Expansion of urban areas into natural habitats restricts the movement of mountain sheep populations, alienates habitats from use by sheep, and often isolates fragmented patches of suitable or seasonally important habitat (e.g., winter range). There is often more landscape with suitable seasonal habitat attributes than there is habitat that is occupied and used by mountain sheep during different seasons. Wild sheep display a high degree of fidelity to seasonal habitats (Brushett et al. 2023, Enns et al. 2023). A population of mountain sheep that has learned to use an area of suitable seasonal habitat (e.g., winter range) can be disproportionally affected by anthropogenic disturbance or development where it occurs in or adjacent to traditionally used seasonal ranges, even though within the context of the larger landscape there appear to be alternative areas with suitable seasonal habitat attributes (e.g., a highway constructed through an important area of winter range habitat; Wild Wise Yukon 2023). While examples of this can most commonly be observed near ski resorts (Teton Range Bighorn Sheep Working Group 2021, Vail 2024), the effects of human population growth and human activity also play out in direct habitat loss associated with our roadways and communities (Ament et al. 2008, Radium 2022). Wild sheep conservation advocates and land managers have been struggling to convince cultural and commercial interests to fully consider the needs of species such as bighorn sheep, even when the risk of local extirpation has been proven, financial compensations provided, or legal decisions being adjudicated. In some communities the commercial development sector that facilitates urban growth has not yet implemented alternative solutions to removing habitat and mitigating negative effects and often, this lack of consideration is in direct conflict with the more public-facing corporate social license pledges (Vail 2024). For other municipalities, mountain sheep generate significant tourism income and, in these communities, residents and sheep often occupy space in close association (Radium 2022). In these situations, however, there tends to remain a misconception by many that continued use of a now urban area by wild sheep implies a level of habituation and a symbiotic state. This is more often the result of individual animal tolerances and a complex array of trade-offs that an individual animal is making, manifesting in a risk-reward tolerance-based behavior (Frid and Dill 2002, Lowrey and Longshore 2017, Harris et al. 2023), that place mountain sheep in high-risk situations such as eating salt residue along highways (Wildlife Collision Prevention Program 2024, Radium 2022).

The secondary effects of urbanization and increased resident and tourist activity in sheep habitat include increasing highway footprints and road network densities, coupled with a higher number of vehicles traveling at higher speeds, through more of the hours in the day (Langley 2020, Gagnon et al. 2022, Radium 2022). This has motivated some stakeholder groups to persuade governments to invest in infrastructure improvements with targeted mitigations (e.g., wildlife overpasses) and public education. The most effective mitigations are those created and implemented to fit specific situations; however, these solutions can be unfamiliar to travelers who, without an understanding of the local situation and expected behaviors that support sheep, can unknowingly act in ways that lead to negative outcomes (e.g., traveling at speeds higher than those posted in areas where mountain sheep intersect with highways; East Kootenay News Online Weekly 2022, Radium 2022).

The United Nations projects annual increases in human population in North America, albeit with a declining rate of population growth, through to the year 2100 (Macrotrends 2024). An increased North American human population will of course require space and places to live and at least some of these places will be in or close to habitats presently occupied by mountain sheep. Currently however, human population in some regions inhabited by mountain sheep is growing by double-digits (e.g., population of Canmore, Alberta, grew by 14.3% between 2016 and 2021 and 11.8% over the last 5 years; Government of Alberta 2024). This population growth is increasing demand for land, water, and recreational opportunities, exerting pressure on natural resources and mountain sheep habitats, leading to habitat alienation, fragmentation, and loss, exacerbating the challenges facing regional mountain sheep populations. To address the effects of urbanization and infrastructure on mountain sheep and their habitats, land use planning efforts have begun to proactively prioritize conservation and connectivity of mountain sheep habitats, with a focus on reducing the potential for negative effects on sheep populations (e.g., construction of wildlife highway overpasses, protection of important habitat areas and connectivity; Kopach et al. 2020). This approach has a far greater likelihood of success and is much more economical in the long term compared to reclaiming land or retrofitting constructed communities and highway infrastructure. This will however require collaborative efforts involving government agencies, conservation organizations, and local communities. Strategies may include land use planning, zoning regulations, construction incentives and community outreach programs to promote sustainable development practices and minimize the ecological footprint of human communities and activities.

## 21.3  SUMMARY

As evidenced through economic investment, society wants wildlife populations and wild places managed sustainably, conserving their inherent aesthetic and intrinsic values while also providing for recreation, harvest, and economic and scientific uses (Canadian Tourism Commission 2012, Association of Fish and Wildlife Agencies 2023, Outdoor Recreation Roundtable 2024). Accomplishing this while also managing and mitigating direct human effects on mountain sheep habitat (i.e., quality, availability) and the indirect effects from climate change, resource, and land use decisions and socio-political activism (Table 21.4) will continue to challenge wildlife managers and, by extension, mountain sheep in North America.

The issues affecting habitats of mountain sheep in western North America are complex. Land management agencies responsible for management of mountain sheep range must actively prioritize the stewardship of wildlife habitats in resource management and land use planning, directing those solutions using laws and policies, implemented on a scale large enough to be effective in sustaining healthy habitats and populations into the future. Recruiting biologists will need to recognize that the effects of change may not all be negative in nature. With this perspective, they will need to help shape effective strategies for mitigating negative outcomes, by building on the lessons learned from the past, while also incorporating new and relevant knowledge of the current day.

Mountain sheep populations in North America face a myriad of evolving threats. Effective conservation and management strategies are imperative to ensure the long-term survival of these globally unique and iconic species. By understanding the complex interactions between mountain sheep ecology and the evolving threats they face, we can work toward safeguarding these magnificent creatures by stewarding the ecosystems they inhabit, in consideration of their benefit to future generations.

## TABLE 21.4

### Examples of Direct and Indirect Human Effects on Mountain Sheep and Their Habitats in North America

| Direct Effects | Indirect Effects |
|---|---|
| • Oil, gas, mineral exploration, and extraction in important habitats or refugia resulting in direct habitat loss. | • Climate change: |
| • Expanded commercial recreation tenuring in both front and backcountry habitats. |   • Drought, changes in precipitation patterns, seasonality and plant phenology; |
| • Commercial and private use of new technologies (e.g., drones, electrically assisted mountain bikes, snow-bikes, etc.) that create widespread disturbances. |   • Forest ingrowth in lower elevations and alpine shrubification; |
| • Landscape fragmentation through privatization of public lands and increased urban growth (habitat loss) accompanied by higher use in adjacent areas, expanding the area of effect. |   • Winter rain-on-snow icing events and warmer, wetter winters; <br>   • Altered precipitation patterns leading to drought, flash flood, or landslides; |
| • Linear corridors, highway, railroad and fence-line development that create impediments to migration and seasonal habitat use and sources of injury and accidental mortality. |   • Expanded species distributions of both competitors and predators. <br> • Front and backcountry recreation that is largely unregulated (e.g., hiking, biking, skiing, camping) causing disturbance and alienating important habitats. |
| • Expanded backcountry industrial access corridors (e.g., roads, trails) increasing landscape permeability and ease of access for direct use (e.g., hunting, photography). | • Increased public perceptions associated with personal use of temporary access routes made permanent, thereby supporting permanent use and effect (most time negative). |
| • Fire suppression and advanced vegetative succession that changes plant communities. | • Off-road and four-season vehicle use over expanded areas resulting in habitat alienation and elevated stress hormone levels, affecting fitness and health. |
| • Introduction of invasive plants that effect native species and forage availability. | • Increased year-round access for predators along linear travel corridors. |
| • Range use and grazing by livestock, feral equids, and other exotics that create overlapping use of habitats, impacting forage availability. | • Socio-political activism that favors single species protections over holistic ecosystem management (e.g., "wild" horses, cougars or mountain lions, wolves, bears). |
| • Transmission of domestic disease and parasites resulting in mortality, chronic poor fitness, and population declines. | • Increased risk and transmission of domestic disease and parasites. |
| • Interventions that change habitat type or function, altering existing species assemblages and distributions. | • Reduced fitness and overall health from livestock-affected forage (i.e., quality and available quantity) with implications to mountain sheep population fitness. |
| • Reintroduction and augmentation of wildlife populations expanding management responsibilities and long-term financial investments. | • Enhancement activities that result in wildlife population over-abundance (i.e., wild sheep or competing ungulates). |

## REFERENCES

Acreman, M., J. Albertengo, T. Amado, M. Amis, A. Anderson, I. Bacchur, G. Basch, A. Calegari, N. Chappell, …, and E. Yakushina. 2012. Report of the work of the expert group on maintaining the ability of biodiversity to continue to support the water cycle. Report number: UNEP/CBD/COP/11/INF/2. doi:10.13140/RG.2.2.34164.40329

Aleuy, O. A., K. Ruckstuhl, E. P. Hoberg, A. Veitch, N. Simmons, and S. J. Kutz. 2018. Diversity of gastrointestinal helminths in Dall's sheep and the negative association of the abomasal nematode, *Marshallagia marshalli*, with fitness indicators. *PLoS ONE* 13:e0192825.

Ament, R., A. Clevenger, O. Yu, and A. Hardy. 2008. An assessment of road impacts on wildlife populations in U.S. National Parks. *Environmental Management* 42:480–496.

Anderson, D. J., V. C. Bleich, and J. T. Villepique. 2022. The bighorn habitat assessment tool: a method to quantify conservation value on landscapes impacted by mining. *Land* 11:552.

Arctic Boreal Vulnerability Experiment. 2024. ABoVE, National Aeronautics and Space Administration. NASA Earth Science Division Washington, D.C., USA. https://above.nasa.gov/about.html#:~:text=What%20is%20ABoVE%3F,implications%20for%20social%2Decological%20systems. Accessed 11 July 2024.

Association of Fish and Wildlife Agencies 2023. AFWA 2023 Annual Report. Washington, D.C., USA. https://www.fishwildlife.org/application/files/3317/1052/2834/2023-AFWA-AnnualReport-v01F.pdf. Accessed 9 July 2024.

Beale, M., and M. Boyce. 2020. Mine reclamation enhances habitats for wild ungulates in west-central Alberta. *Restoration Ecology* 28:828–840.

Bevington, A. R., H. E. Gleason, and V. N. Foord. 2020. Snow cover timing from satellite imagery: opportunities for near-real-time mapping and seasonal forecasting. Technical Report 129. Province of British Columbia, Victoria, British Columbia, Canada.

Bevington, A. R., H. E. Gleason, V. N. Foord, W. C. Floyd, and H. P. Griesbauer. 2019. Regional influence of ocean–atmosphere teleconnections on the timing and duration of MODIS-derived snow cover in British Columbia, Canada. *Cryosphere* 10:2693–2712.

Bleich, V., J. Davis, J. Marshal, S. Torres, and B. Gonzales. 2008. Mining activity and habitat use by mountain sheep (Ovis canadensis). *European Journal of Wildlife Research* 55:183–191.

Bleich, V. C. 2009. Factors to consider when reprovisioning water developments used by mountain sheep. *California Fish and Wildlife Journal* 95:153–159.

Brewer, C. E., V. C. Bleich, J. A. Foster, T. Hosch-Hebdon, D. E. McWhirter, E. M. Rominger, M. W. Wagner, and B. P. Wiedmann. 2014. *Bighorn Sheep: Conservation Challenges and Management Strategies for the 21st Century*. Wild Sheep Working Group, Western Association of Fish and Wildlife Agencies, Cheyenne, Wyoming, USA.

British Columbia Environmental Assessment Office. 2020. Process Planning Policy Version 1.0. Environmental Assessment Office, Victoria, British Columbia, Canada. https://www2.gov.bc.ca/assets/gov/environment/natural-resource-stewardship/environmental-assessments/guidance-documents/2018-act/process_planning_policy_v1_-_april_2020.pdf. Accessed 17 August 2024.

Brown, C. L., S. F. Trainor, C. N. Knapp, and N. P. Kettle. 2021. Alaskan wild food harvester information needs and climate adaptation strategies. *Ecology and Society* 26:2.

Brushett, A., J. Whittington, B. Macbeth, and J. M. Fryxell. 2023. Changes in movement, habitat use, and response to human disturbance accompany parturition events in bighorn sheep (*Ovis canadensis*). *Movement Ecology* 11:36.

Cain, J. W., J. V. Gedir, C. A. Caldwell, and S. Carleton. 2018. Final report: assessment of drought impacts on selected fish and wildlife species in the southwestern United States. Agreement number: G13AC00257. U.S. Geological Survey, National Climate Change and Wildlife Science Center, Las Cruces, New Mexico, USA.

California Department of Fish and Wildlife. 2020. Peninsular bighorn sheep 2019–2020 annual report. California Department of Fish and Wildlife, South Coast Region, San Diego, California, USA.

Canadian Tourism Commission. 2012. Sport fishing and game hunting in Canada: an assessment on the potential international tourism opportunity. Canadian Tourism Commission Research and Evaluation, Ottawa, Ontario, Canada. https://publications.gc.ca/collections/collection_2013/ic/Iu86-49-2012-eng.pdf. Accessed 15 August 2024.

Cansler, C., D. McKenzie, and C. Halpern. 2018. Fire enhances the complexity of forest structure in alpine treeline ecotones. *Ecosphere* 9:e02091.

Clapp, J. G., and J. L. Beck. 2016. Short-term impacts of fire-mediated habitat alterations on an isolated bighorn sheep population. *Fire Ecology* 12:80–98.

Cosgrove, C. L., J. Wells, A. W. Nolin, J. Putera, and L. R. Prugh. 2021. Seasonal influence of snow conditions on Dall's sheep productivity in Wrangell-St Elias National Park and Preserve. *PLoS ONE* 16:e0244787.

Courtemanch, A. B. 2014. Seasonal habitat selection and impacts of backcountry recreation on a formerly migratory bighorn sheep population in northwest Wyoming. Thesis, Department of Zoology and Physiology, University of Wyoming, Laramie, Wyoming, USA.

Davidson, S. C., G. Bohrer, E. Gurarie, S. LaPoint, P. J. Mahoney, N. T. Boelman, J. U. H. Eitel, L. R. Prugh, L. A. Vierling, …, and Mark Hebblewhite. 2020. Ecological insights from three decades of animal movement tracking across a changing Arctic. *Science* 370:712–715.

Delbart, N., T. Le Toan, L. Kergoat, and V. Fedotova. 2006. Remote sensing of spring phenology in boreal regions: a free of snow-effect method using NOAA-AVHRR and SPOT-VGT data (1982–2004). *Remote Sensing and Environment* 101:52–62.

East Kootenay News Online Weekly. 2022. Slow your roll campaign launched to protect sheep. e-KNOW.ca, East Kootenays, British Columbia, Canada. https://www.e-know.ca/regions/columbia-valley/radium/slow-your-roll-campaign-launched-to-protect-sheep/. Accessed 11 August 2024.

Enns, G. E., B. A. Jex, and M. S. Boyce. 2023. Diverse migration patterns and seasonal habitat use of Stone's sheep (*Ovis dalli stonei*). *PeerJ* 11:e15215.

Environment Yukon. 2024. Yukon wildlife: Cougar. https://www.env.gov.yk.ca/animals-habitat/mammals/cougar.php. Accessed 28 August 2024.

Environmental Law Centre. 2010. *Environmental Assessment in British Columbia*. University of Victoria, Victoria, British Columbia, Canada.

Epps, C. W., D. R. McCullough, J. D. Wehausen, V. C. Bleich, and J. L. Rechel. 2004. Effects of climate change on population persistence of desert-dwelling mountain sheep in California. *Conservation Biology* 18:102–113.

Fish and Wildlife Compensation Program. 2020. Invasive plant management on bighorn sheep winter ranges: Bull River and Wigwam Flats. Project No COL-F20-W-3041. Technical report prepared by the Ministry of Forests, Lands, Natural Resource Operations and Rural Development, Cranbrook, British Columbia, Canada.

Forest Practices Code of British Columbia. 1995. Biodiversity guidebook – September 1995. Forest Practices Code of British Columbia, BC Environment, Victoria, British Columbia, Canada.

Frank, D. A., K. M. Becklin, J. F. Penner, K. A. Lindsay, and C. J. Geremia. 2023. Feast or famine: how is global change affecting forage supply for Yellowstone's ungulate herds? *Ecological Applications* 33:e2735.

Franks, S. J., J. J. Weber, and S. N. Aitken. 2014. Evolutionary and plastic responses to climate change in terrestrial plant populations. *Evolutionary Applications* 7:123–139.

Freeman, S. D. and N. Maclean. 2021. Habitat use and movement of the Dome Mountain Stone's sheep. Tahltan Guide Outfitters Association, Dease Lake, British Columbia, Canada.

Frid, A. 2003. Dall's sheep responses to overflights by helicopter and fixed-wing aircraft. *Biological Conservation* 110:387–399.

Frid, A., and L. M. Dill. 2002. Human-caused disturbance stimuli as a form of predation risk. *Conservation Ecology* 6:11.

Fusco, E. J., J. T. Finn, J. K. Balch, and B. A. Bradley. 2019. Invasive grasses increase fire occurrence and frequency across US ecoregions. *Proceeding of the National Academy of Sciences* 116:23594–23599.

Gagnon, J. W., C. D. Loberger, K. S. Ogren, S. C. Sprague, S. R. Boe, and R. E. Schweinberger. 2022. Mitigating bighorn sheep-vehicle collisions and habitat fragmentation with overpasses and adaptive management. *Human-Wildlife Interactions* 16:353–372.

Ganskopp, D., and D. Bohnert. 2001. Nutritional dynamics of 7 northern Great Basin grasses. *Journal of Range Management* 54:640–647.

Gianella, T. D., and D. M. Sanchez. 2024. Summer mule deer use of annual grass herbicide-treated plots. *Rangeland Ecology and Management* 94:178–183.

Goodson, N. J., and D. R. Stevens. 1992. Management implications of an intensive study of winter foraging ecology of bighorn sheep. *Biennial Symposium of the Northern Wild Sheep and Goat Council* 8:58–67.

Government of Alberta. 2024. Canmore – Population. Office of Statistics and Information, Alberta Treasury Board and Finance, Edmonton, Alberta, Canada. https://regionaldashboard.alberta.ca/region/canmore/population/#/?from=2019&to=2023. Accessed 8 August 2024.

Grow Wild Montana. 2024a. Bighorn sheep winter range project report. https://www.growwildmt.org/bighorn-sheep-habitat-improvement. Accessed 17 August 2024.

Grow Wild Montana. 2024b. Impacts of invasive plant species – aka noxious weeds. https://www.growwildmt.org/invasive-species-impacts. Accessed 17 August 2024.

Hale, K. E., K. S. Jennings, K. N. Musselman, B. Livneh, and N. P. Molotch. 2023. Recent decreases in snow water storage in western North America. *Communications Earth and Environment* 4:170.

Hamlington, B. D., S. H. Cheon, C. G. Piecuch, K. B. Karnauskas, P. R. Thompson, K. Y. Kim, J. T. Reager, F. W. Landerer, and T. Frederikse. 2019. The dominant global modes of recent internal sea level variability. *Journal of Geophysical Research: Oceans* 124:2750–2768.

Harris, R. B., K. Aluzas, L. Balyx, J. Belt, J. Berger, M. Biel, T. Chilton-Radandt, S. D. Côté, J. Cunningham, A. Ford, P. Happe, C. P. Lehman, K. Poole, C. G. Rice, K. Safford, W. Sarmento, and L. Wolf. 2023. Habituated, tolerant, or salt-conditioned mountain goats and human safety. *Human-Wildlife Interactions* 17(1):1–24.

Hicke, J. A., S. Lucatello, L. D. Mortsch, J. Dawson, M. Domínguez Aguilar, C. A. F. Enquist, E. A. Gilmore, D. S. Gutzler, S. Harper, K. Holsman, E. B. Jewett, T. A. Kohler, and K. A. Miller. 2022. *Climate Change 2022: Impacts, Adaptation and Vulnerability*. Contribution of Working Group II to the Sixth Assessment Report of the Intergovernmental Panel on Climate Change. Cambridge University Press, New York, New York, USA.

Hik, D. S., and J. Carey. 2000. Cohort variation in horn growth of Dall sheep rams in the southwest Yukon, 1969–1999. *Biennial Symposium of the Northern Wild Sheep and Goat Council* 12:88–100.

Hueffer, K., A. J. Parkinson, R. Gerlach, and J. Berner. 2013. Zoonotic infections in Alaska: disease prevalence, potential impact of climate change and recommended actions for earlier disease detection, research, prevention and control. *International Journal of Circumpolar Health* 2013:72.

Inuvik Community Greenhouse. 2024. The most northern commercial and community greenhouse in North America. https://spectacularnwt.com/attractions/inuvik-community-greenhouse/. Accessed 3 December 2024.

Jansen, B. D., P. R. Krausman, K. D. Bristow, J. R. Heffelfinger, and J. C. deVos, Jr. 2009. Surface mining and ecology of desert bighorn sheep. *Southwest Naturalist* 54:430–438.

Jex, B. A., J. B. Ayotte, V. C. Bleich, C. E. Brewer, D. L. Bruning, T. M. Hegel, N. C. Larter, R. A. Schwanke, H. M. Schwantje, and M. W. Wagner. 2016. *Thinhorn Sheep: Conservation Challenges and Management Strategies for the 21st Century*. Wild Sheep Working Group, Western Association of Fish and Wildlife Agencies, Boise, Idaho, USA.

Kneteman, J. G. 2016. Resilient space: bighorn sheep (*Ovis canadensis*) ecological resilience in the northern Rocky Mountains. Thesis, Department of Biological Sciences, University of Alberta, Edmonton, Alberta, Canada.

Kopach, B., S. Gavin, S. Gutsell, K. Simpson, A. Stewart, and K. Gargus. 2020. Technical review of the environmental impacts statement: Smith Creek area structure plan. Prepared for the Town of Canmore. Unpublished Report. Management and Solutions in Environmental Science, Calgary, Alberta, Canada. https://www.tsmv.ca/wp-content/uploads/2021/01/Third-Party-Review-Smith-Creek-Environmental-Impact-Statement_20210106.pdf. Accessed 18 October 2024.

Kopp, V., and J. Coppes. 2020. Why do people leave marked trails? Implications for managing outdoor recreationists. *Eco.mont – Journal on Protected Mountain Areas Research and Management* 12:2.

Kuwaczka, L. F., V. Mitterwallner, V. Audorff, and M. J. Steinbauer. 2023. Ecological impacts of (electrically assisted) mountain biking. *Global Ecology and Conservation* 44:e02475.

Langley, M. 2020. Limiting factors on a small herd of Rocky Mountain bighorn sheep residing in the Kicking Horse Canyon, near Golden, British Columbia. *Biennial Symposium of the Northern Wild Sheep and Goat Council* 22:106–112.

Legg, K. 1998. A review of the potential effects of winter recreation on bighorn sheep. Yellowstone National Park, National Park Service. *Biennial Symposium of the Northern Wild Sheep and Goat Council* 11:14–19.

Lindsey, R. 2009. Climate variability: arctic oscillation. National Oceanic and Atmospheric Administration, Springfield, Maryland, USA. https://www.climate.gov/news-features/understanding-climate/climate-variability-arctic-oscillation. Accessed 29 November 2024.

Lohuis, T. 2017. Dall's sheep research in the Chugach Range, GMUs 13D and 14C. Alaska Department of Fish and Game, Anchorage, Alaska, USA. https://www.adfg.alaska.gov/static/applications/web/nocache/species/speciesinfo/dallsheep/pdfs/chugach_range_dalls_sheep_research.pdf0EFD2D9EE22249846CB36BDA8F2117D1/chugach_range_dalls_sheep_research.pdf. Accessed 12 October 2024.

Lohuis, T., K. Smith, L. Metherell, and D. Roman. 2018. Dall's sheep population declines in Alaska's Chugach Range may be related to climate and weather patterns. *Biennial Symposium of the Northern Wild Sheep and Goat Council* 21:76.

Lowrey, C., and K. M. Longshore. 2017. Tolerance to disturbance regulated by attractiveness of resources: a case study of desert bighorn sheep within the River Mountains, Nevada, USA. *Western North American Naturalist* 77:82–98.

MacArthur, R. A., V. Geist, and, R. H. Johnston. 1982. Cardiac and behavioural responses of mountain sheep to human disturbance. *Journal of Wildlife Management* 46:351–358.

Macrotrends. 2024. North American population growth rate and United Nations projections. https://www.macrotrends.net/global-metrics/countries/NAC/north-america/population-growth-rate. Accessed 5 September 2024.

Meyers-Smith, I. H., and D. S. Hik. 2018. Climate warming as a driver of tundra shrubline advance. *Journal of Ecology* 106:547–560.

Mining Environment Research Group. 2002. Flying in sheep country: how to minimize disturbance from aircraft. Mining Environment Research Group Report 2002–6. Unpublished Report. Geoscience Information and Sales, Yukon Government, Whitehorse, Yukon, Canada.

Ministry of Forests, Lands and Natural Resource Operations. 2014. A compendium of wildlife guidelines for industrial development projects in the North Area, British Columbia – Interim Guidance. Unpublished Report. Skeena Region, Smithers, British Columbia, Canada.

Ministry of Forests, Lands, Natural Resource Operations and Rural Development. 2019. Management framework for Rocky Mountain elk (*Cervus canadensis nelsoni*) in British Columbia (Draft). Victoria, British Columbia, Canada.

Ministry of Forests, Lands, Natural Resource Operations and Rural Development. 2022. Prescribed fire and invasive plants – a reference guide and manual of best practices. Unpublished Report. Invasive Plant Program, Victoria, British Columbia, Canada.

Ministry of Water, Land and Resource Stewardship. 2023. Bighorn sheep habitat use in relation to mapped cheatgrass communities in the Kamloops Lake area of British Columbia (draft). Wildlife Branch, Kamloops, British Columbia, Canada.

Mowat, G., S. Darlington, S. Wilson, L. Vander Vennen, T. J. Gooliaff, and S. MacIver. 2023. A review of cougar biology and management in British Columbia. Technical Report 150. Victoria, British Columbia, Canada.

Mulero-Pázmány, M., S. Jenni-Eiermann, N. Strebel, T. Sattler, J. J. Negro, and Z. Tablado. 2017. Unmanned aircraft systems as a new source of disturbance for wildlife: a systematic review. *PLoS ONE* 12:e0178448.

New Relationship Trust. 2024. Indigenous Food Security and Sovereignty program. First Nation Leadership Council, British Columbia and Canada. https://newrelationshiptrust.ca/apply-for-funding/sustainability-development-goals-sdg-initiatives/food-security-grants/. Accessed 7 July 2024.

Nitschke, C. R., and J. L. Innes. 2013. Potential effect of climate change on observed fire regimes in the Cordilleran forests of South-Central Interior, British Columbia. *Climatic Change* 116:579–591.

Outdoor Recreation Roundtable. 2024. Outdoor Recreation Roundtable Association: Together Outdoors Initiative. https://recreationroundtable.org/programs/together-outdoors/. Accessed 5 August 2024.

Papouchis, C. M., F. J. Singer, and W. B. Sloan. 2001. Responses of desert bighorn sheep to increased human recreation. *Journal Wildlife Management* 65:573–582.

Peterson, C. 2020. When the love of skiing endangers wildlife. High Country News: Recreation March 16, 2020. https://www.hcn.org/issues/52-3s/special-recreation-when-the-love-of-skiing-endangers-wildlife. Accessed 7 August 2024.

Pigeon, G., L. E. Loe, R. Bischof, C. Bonenfant, M. Forchhammer, R. J. Irvine, E. Ropstad, A. Stien, V. Veiberg, and S. Albon. 2019. Silver spoon effects are constrained under extreme adult environmental conditions. *Ecology* 100:e02886.

Pojar, J. 2009. Climate change and land use planning in the Atlin-Taku area. Report prepared for the Taku River Tlingit First Nation and for the Integrated Land Management Bureau. Unpublished Report. Ministry of Agriculture and Lands, Victoria, British Columbia, Canada.

Poole, K., I. Teske, K. Podrasky, C. Berdusco, C. Conroy, R. MacDonald, R. Davies, H. Schwantje, E. Chow, C. van Rensen, and T. Ayele. 2020. Bighorn sheep cumulative effects assessment report Elk Valley, Kootenay-Boundary Region, Government of British Columbia. Cumulative Effects Management Framework. Unpublished Draft Report.

Poole, K. G., R. Serrouya, I. E. Teske, and K. Podrasky. 2016. Rocky Mountain bighorn sheep (*Ovis canadensis canadensis*) winter habitat selection and seasonal movements in an area of active coal mining. *Canadian Journal of Zoology* 94:733–745.

Radium. 2022. 5 ways to save the sheep: Tourism Radium. Village of Radium, British Columbia, Canada. https://www.radiumhotsprings.com/5-ways-to-save-the-sheep/. Accessed 12 August 2024.

Rebolo-Ifrán, N., M. Graña Grilli, and S. Lambertucci. 2019. Drones as a threat to wildlife: YouTube complements science in providing evidence about their effect. *Environmental Conservation* 46:205–210.

Reyes, L., and M. G. Kramer. 2023. High-elevation snowpack loss during the 2021 Pacific Northwest heat dome amplified by successive spring heatwaves. *NPJ Climate and Atmospheric Science* 6:208.

Segal, M. 2024. Following the signs: what mountain guides are seeing up in the Rockies can tell us a lot about the pace of global warming. CBC News Radio's What on Earth, April 14, 2024. https://www.cbc.ca/newsinteractives/features/mountain-guides-climate-change. Accessed 12 August 2024.

Shabbar, A. 2006. The impact of El Niño-Southern Oscillation on the Canadian climate. *Advances in Geosciences* 6:149–153.

Sittler, K. L. 2019. Stone's sheep and elk forage characteristics – 7-years post fire. Wildlife Infometrics Report No. 663. Wildlife Infometrics Inc., Mackenzie, British Columbia, Canada.

Sloan, W. B. 2022. Bighorn sheep reclaimed abandoned habitat during the pandemic and had lots of young. Now what? Canyonlands National Park, National Park Service Perspectives, Park Science Magazine 36:1, Summer.

Sproat, K. K., N. R. Martinez, T. S. Smith, W. B. Sloan, J. T. Flinders, J. W. Bates, J. G. Cresto, and V. C. Bleich. 2019. Desert bighorn sheep responses to human activity in south-eastern Utah. *Wildlife Research* 47:16–24.

Stanton, R. L., B. C. Nusink, K. L. Cass, T. B. B. Bishop, B. M. Woodbury, D. N. Armond, and S. B. Clair. 2023. Fire frequency effects on plant community characteristics in the Great Basin and Mojave deserts of North America. *Fire Ecology* 19:60.

Sterl, P., C. Brandenburg, and A. Arnberger. 2008. Visitor's awareness and assessment of recreational disturbance of wildlife in that Donau-Auen National Park. *Journal For Nature Conservation* 16:135–145.

Stralberg, D. 2018. Climate-projected distributional shifts in North American ecoregions. In AdaptWest: A Climate Adaptation and Conservation Planning Database for North America. Conservation Biology Institute, Corvallis, Oregon, USA. https://adaptwest.databasin.org/pages/ecoregion-displacement-and-refugia/. Accessed 4 July 2024.

Teton Range Bighorn Sheep Working Group. 2021. Recommendations from the Teton Bighorn Sheep and Winter Recreation Community Collaborative Learning Process – Teton Range Bighorn Sheep and Winter Recreation Strategy. https://static1.squarespace.com/static/622641bdfa0b7a00f8f78c88/t/63c76cdb826d6e442af83405/1674013921810/FINAL+BHS+Winter+Rec+Strategy_Oct+2021.pdf. Accessed 5 September 2024.

Vail. 2024. Vail Bighorn Sheep Initiative. Town of Vail, Colorado, USA. https://www.vailbighorn.com/. Accessed 15 August 2024.

Van de Kerk, M., D. Verbyla, A. W. Nolin, K. J. Sivy, and L. R. Prugh. 2018. Range-wide variation in the effect of spring snow phenology on Dall sheep population dynamics. *Environmental Research Letters* 13:7.

Van de Kerk, M., S. Arthur, M. Bertram, B. Borg, J. Herriges, J. Lawler, B. Mangipane, C. Lambert Koizumi, B. Wendling, and L. Prugh. 2020. Environmental influences on Dall's sheep survival. *Journal of Wildlife Management* 84:1127–1138.

Wagler, B. L., R. A. Smiley, A. B. Courtemanch, D. Lutz, D. McWhirter, D. Brimeyer, P. Hnilicka, T. J. Robinson, and K. L. Monteith. 2023. Implications of forage quality for population recovery of bighorn sheep following a pneumonia epizootic. *Journal of Wildlife Management* 87:e22452.

Weaver, R. A., and J. L. Mensch. 1971. Bighorn sheep in northeastern Riverside County. Wildlife management administrative report 71-1. California Department of Fish and Game, Sacramento, California, USA.

Whiting, J. C., V. C. Bleich, R. T. Bowyer, and C. W. Epps. 2023. Restoration of bighorn sheep: history, successes, and remaining conservation issues. *Frontiers in Ecology and Evolution* 11:1083350.

Whiting, J. C., V. C. Bleich, R. T. Bowyer, and R. T. Larsen. 2012. Water availability and bighorn sheep: life history characteristics and persistence of populations. Pages 127–158 in J. A. Daniels, editor. *Advances in Environmental Research*. Nova Science Publishers, Hauppauge, New York, USA.

Wiedmann, B., and V. C. Bleich. 2014. Demographic responses of bighorn sheep to recreational activities: a trial of a trail. *Wildlife Society Bulletin* 38:773–782.

Wikeem, B. M. 1984. Forage selection by California bighorn sheep and the effects of grazing on an Artemisia-Agropyron community in southern British Columbia. Dissertation, University of British Columbia, Vancouver, British Columbia, Canada.

Wild Sheep Foundation. 2021. Position statement: effects of recreational and commercial use of drones on wild sheep. Wild Sheep Foundation, Bozeman, Montana, USA.

Wild Wise Yukon. 2023. Dan Keyi sheep brochure. Wild Wise Yukon, Whitehorse, Yukon Territory, Canada. Public Information Pamphlet: https://wildwise.ca/wp-content/uploads/2020/08/Dan-Keyi-sheep-brochure.pdf. Accessed 21 August 2024.

Wildlife Collision Prevention Program. 2024. When collisions occur: bighorn sheep. https://www.wildlifecollisions.ca/collision/when-collisions-occur.htm. Accessed 4 August 2024.

Xie, J., Q. Tang, J. Golaz, and W. Lin. 2024. Record high 2022 September-mean temperature in western North America. *Bulletin of the American Meteorological Society* 105:E306–E312.

Youngwook, K., J. S. Kimball, N. Parazoo, X. Xu, A. Colliander, R. Reichle, J. Xiao, and X. Li. 2024. Diagnosing spring onset across North American Arctic-boreal region using complementary satellite environmental data records. *Journal of Geophysical Research, Biogeosciences* 129:e2023JG007977.

# 22 Conflict Management

*Brian F. Wakeling, Stephen L. Webb, Jeffrey W. Gagnon, Emily S. Almberg, and Jared T. Beaver*

## 22.1 INTRODUCTION

Conflict in wildlife management is often poorly defined (Wakeling et al. 2023), frequently referring to adverse interactions between wildlife and people. Yet specialists who study human dimensions and wildlife conflict draw a clear distinction between the effects of wildlife on humans and disagreement among people regarding wildlife and its management. Specialists define a conflict as a situation that occurs when two or more parties with intensely held opinions disagree over objectives and one party is perceived to assert its interests at the expense of another (Redpath et al. 2013). Conflict occurs among people, but the effects of wildlife on people (or effects of people on wildlife) may be catalysts for conflict if management objectives and approaches are not mutually acceptable.

Managing the effects of wildlife-human interactions requires addressing the occurrence and severity of interactions between wildlife and people (and their interests that could include recreational pursuits, livestock husbandry, or disease transmission), which may alter the activities of wildlife, humans, or both. Equally important are human actions that have unacceptable effects on wildlife, such as interruptions to migratory patterns (e.g., due to exclusionary fencing or public transportation networks), population health and resilience (e.g., transmission of domestic livestock pathogens), or loss of important seasonal habitat (e.g., because of commercial, industrial, or recreational development). Some actions, such as highway collisions, may influence both humans and wildlife, underscoring the complexity in trying to define the source of conflict as either wildlife or human-caused. Effective management requires an understanding of the root cause of the conflict situation and the options to create solutions to address the effects wildlife are having on humans or the degree to which human activities are detrimental to wildlife (Decker et al. 2019).

Equally complex is addressing conflict among humans, requiring those leading public stakeholder engagements to facilitate information sharing among parties, while retaining the focus of the engagement on solutions for wildlife and reducing conflict. This delivery can be about managing expectations and broadening understanding among participants (Decker et al. 2019) as initial steps, and requires inclusion of management agencies, rights and title holders (e.g., First Nations), and other stakeholders (e.g., non-government organizations, private landowners, tenure holders, members of the public [often recreationalists]).

Mountain sheep (bighorn [*Ovis canadensis*], thinhorn [*Ovis dalli*]) can generate conflict among people in many of the same ways as other wildlife. To date, most conflicts associated with the management of mountain sheep involve bighorn sheep, largely due to the high degree of geographical overlap with humans and private land interests. Trends observed with bighorn sheep may soon be replicated in thinhorn sheep range as people become more widespread and expand their recreational, agricultural, or resource development footprints into remote and northern environments (Chapters 20 and 21).

In this chapter, we review the social and ecological context of mountain sheep management, identify areas of conflict, and discuss tools to mitigate such conflict. Chapters 8, 11, and 18 provide valuable context for the biology, ecology, and sociology that influence some of the key conflicts surrounding mountain sheep that we discuss here.

## 22.2 BIOLOGICAL, ECOLOGICAL, AND SOCIAL CARRYING CAPACITY

A primary factor that influences the likelihood of conflict developing among stakeholders is the distribution and abundance of the species from which conflict may arise. Human interactions with a species are greater when the animal is more abundant or when specific behaviors are considered unfavorable. For instance, observation of a single mountain sheep within a municipality feeding on ornamental vegetation is often interesting, whereas large numbers of mountain sheep engaged in the same activity may cause economic damage and traffic congestion. The municipality may decide to address the interaction in several ways that may include hazing, altering vegetation, or requesting removal of the animals by the state, provincial, or territorial wildlife conservation agency; any action undertaken could then create conflict among the residents and interested stakeholders.

Determining whether wildlife is overabundant is based on biological, ecological, and social considerations (Krausman et al. 2014). Carrying capacity is defined by the size or density of a population that can be sustained or tolerated within a specified area, which may be determined based on three different, but interconnected, dimensions of management contexts: biological, ecological, and social.

Biological carrying capacity is the maximum number of individuals of a species that the habitat can support for a defined period. At the biological carrying capacity, the

DOI: 10.1201/9781003518686-25

area can continuously supply life requisites necessary to meet energy and physiological requirements. Biological carrying capacity can fluctuate with changing conditions on the landscape as the result of environmental variability and direct management (or lack thereof) or based on the season and time of year. When forage quality and quantity improve, the biological carrying capacity increases; conversely, if an environmental variable diminishes resources (e.g., severe weather causing drought), then the biological carrying capacity may be reduced. In human-dominated landscapes, biological carrying capacity may be enhanced by greater access to artificial or augmented food sources (Furnas et al. 2020).

Ecological carrying capacity is the maximum number of individuals of a species at which the population does not negatively influence native plants and animals. When domestic or wild ungulate populations are greater than ecological carrying capacity, they can affect the regeneration of some plant species and degrade habitat (Allombert et al. 2005, Adams et al. 2006). In turn, this can affect species such as mountain sheep that also depend on those plant species or habitat characteristics (DeCalesta 1994, Tilghman 1989). When this occurs, the effect can also be described as either exploitative or interference competition. This type of competition is the result of one species causing overbrowsing, which can influence the distribution and abundance of another species (Jex et al. 2016). Reversing the effects of extended overbrowsing can take years to decades.

Social carrying capacity is the population level of a wildlife species at which the local human population can tolerate or accept the problems associated with that species. The social carrying capacity for wildlife abundance is determined by stakeholders (often local) and the values they place on wildlife (Manfredo et al. 2017). Because the tolerances by stakeholder group differ, social carrying capacity can vary within an area (Wingard and Krausman 2019). Social carrying capacity is often contextualized by concerns about public safety and financial considerations, but it can also consider conflicts among recreationists, advocates, and producers (Hendee et al. 1978). Other related concepts include wildlife acceptance capacity (the largest wildlife population acceptable to people; Decker and Purdy 1988) and stakeholder acceptance capacity (the capacity of society to tolerate or accept effects of wildlife in particular situations; Carpenter 2000). Although each of these views about human tolerance of wildlife differs slightly, we treat them in a conglomerate within this chapter.

## 22.2.1 Social Values and Normative Ethics

Most people believe wildlife, including mountain sheep, adds value to their lives, but the values they ascribe to various species of wildlife may differ markedly. When values associated with wildlife and wildlife management differ markedly among members of society, especially in situations where human-wildlife interactions result in some people experiencing tangible effects (e.g., economic loss,

health and safety threats), conflicts may arise. These conflicts could be due to ethical considerations regarding management objectives and actions to reduce conflicts (Smith et al. 2023). For instance, some people may find lethal removal through hunting or other means to be an acceptable method to reduce abundance of wildlife populations responsible for creating negative effects in an agricultural setting. Conversely, other people may have greater tolerance for wildlife behavior (and the effects of such behavior) and the behavior of humans that lead to conflicts (e.g., maintaining nutritious ornamental plants that attract wildlife close to humans and residences). People's core values (often expressed as ethical stances) frequently contribute to conflict. A familiar example in wildlife management is killing animals, where some people support hunting or culling as a technique for population control, whereas others are opposed to hunting or any lethal actions. This tension exists in mountain sheep management. Livestock producers and vocal mountain sheep advocates may share utilitarian values toward animals, yet they may find themselves on opposing sides of proposed management actions (e.g., exclusion of domestic sheep grazing, prohibiting reintroduction of mountain sheep). Nevertheless, ethically acceptable management actions intended to resolve wildlife conflicts are increasing.

Ethics represents the disciplined effort to examine and understand how societal values underlie norms of behavior, and to offer sustained analysis supporting or opposing those norms (Smith et al. 2023). Normative ethics involves the search for specific criteria that justify certain actions or policies. Applied normative ethics generally refers to the moral actions or courses of action achieved through discussion of a particular topic (e.g., environmental ethics, biomedical ethics; Smith et al. 2023).

Historically, management actions occurred without conscious consideration of social values or ethics (also referred to by some as social license). Humans established towns and cities in places where natural resources were relatively easy to acquire (e.g., access to water, productive soils, grazeable land), resulting in wildlife, humans, and livestock using areas proximate to each other. In these situations, wildlife conflicts are inevitable, so the most common strategy has been to preclude or reduce access by wildlife (e.g., with the use of fences); many of these strategies are cost-prohibitive or impractical across large areas. Still, the primary recommendation for managing conflict between domestic and mountain sheep has been temporal and geographic separation, which has been difficult to achieve and maintain.

In these conflict situations, stakeholders may share mutual interests. For example, livestock producers that graze livestock on or near public land, land management agencies that regulate livestock use on public lands and derive income from grazing leases, and wildlife management agencies that manage mountain sheep populations for viewing and hunting on the same lands all wish to maintain sustainable range conditions that permit continued landscape uses. Many wildlife conservation organizations have

ative influence decision-making outcomes. A focus on all potential stakeholders, not just organized interest groups, recognizes the existence of a larger set of beneficiaries of public wildlife management, including future generations (Decker et al. 2019). Each user of public resources may view or envision purposes for a portion of the landscape that differ from another user of the same public resource.

Wildlife management agencies sometimes disproportionately consider the hunting public when formulating a strategy to address a specific conflict. Similarly, a municipality or department of transportation that is being asked to reduce conflicts involving wildlife may only consider the segment of the public voicing a complaint. Reaching suitable resolutions to conflicts between humans and mountain sheep requires stakeholder involvement during decision-making (see Section 22.5); this ensures that anthropogenic influences giving rise to conflict are addressed sustainably and acceptably (Westerfield et al. 2019) by looking at all issues and not just those related to a single species.

Identifying stakeholders is important to address conflict management and conflict resolution. Stakeholders generally have varying interests in aspects surrounding potential conflicts. Important considerations include preventing conflict and avoiding situations where the resolution creates or perpetuates another conflict through collateral effects.

## 22.4 MANAGEMENT AUTHORITY AND RESPONSIBILITY

Many challenges must be overcome when addressing and attempting to reduce conflicts. A primary challenge for agencies is establishing the scale and scope for engagement to ensure that the diversity of stakeholders involved or potentially affected by the final actions to resolve the conflict has a vested interest in the effort. Rights and title holders, individuals, groups, and organizations value wildlife differently, and they may have cultural and ethical beliefs that are not aligned, especially when it comes to potential actions intended to address management problems. These differences may make consensus difficult. The lack of consensus among the public is a source of controversy for statutory decision-makers and elected officials because polarized constituents may propose fundamentally different solutions to address perceived challenges. Other stakeholders may remain silent because they do not perceive that a problem exists, yet they may be concerned about the type of management response proposed.

Addressing conflicts involving humans and wildlife can be made easier where relevant legislation, policies, or procedures exist, but solutions still commonly require considerable time, human resources, and financial support. Entities involved with wildlife conflict can include wildlife and land management agencies, landowners, organizations, industry, tribal nations, cities, and municipalities, which typically have multiple priorities and competing needs and limited budgets. In most situations, no single authority exists that can influence all aspects of conflict resolution. Management authorities for wildlife generally rest with the state, provincial, or territorial wildlife management agencies. Both federal land management agencies and local governments influence land and commercial uses, which can influence habitat suitability. Anyone who may be affected by or may influence wildlife or a wildlife management decision must work within the engagement framework to define objectives and identify acceptable management actions. In the end, communication, coordination, and collaboration are important to success.

### 22.4.1 WILDLIFE AGENCIES

State, provincial, territorial, and tribal wildlife agencies are constrained by regulatory authority, policy directives, and program mandates. A wildlife agency's operational funding is limited, and many do not have a dedicated budget to address conflicts involving wildlife. Several jurisdictions and agencies do have access to funding derived from special auctions or raffles of limited hunting permits, marketed through non-government wildlife conservation organizations. Funds generated through these opportunities are maximized when they are directed into tangible projects and outcomes that ultimately provide direct benefit to mountain sheep populations or hunting opportunities, so funding for conflict management (an intangible benefit) may not be available.

When mountain sheep occupy public or private landscapes, they often share these habitats with domestic livestock. Wildlife management agencies generally do not administer livestock grazing permits or the distribution of livestock on either public or private lands yet contact with domestic livestock can lead to pathogen spillovers. Of primary concern are respiratory pathogens because they can have debilitating and chronic effects on mountain sheep populations (Cassier et al. 2018). Communicating, coordinating, and collaborating with land management agencies and private producers is an important aspect of mitigating risk and conflict associated with disease. Wildlife management agencies have management authority over wildlife and can implement some management strategies to reduce conflict. For instance, most wildlife management agencies will seek to remove individual mountain sheep after known or suspected contact with domestic sheep or goats. When mountain sheep are translocated, plans and protocols are implemented to ensure connectivity of mountain sheep populations while reducing proximity to domestic sheep and goats. Where respiratory infection is prevalent and chronic in a population of mountain sheep, wildlife management agencies may choose to employ a test-and-remove approach; some respiratory pathogens require capture and testing for respiratory bacteria such as *Mycoplasma ovipneumoniae* (*M. ovi*), and culling of sick and chronic carriers of the bacteria has shown success in supporting lamb survival, recruitment, and population recovery (Wild Sheep Initiative 2023). These activities are within the purview of wildlife management agencies, but communication, coordination, and collaboration among stakeholders remain important to ensure durable outcomes.

When wildlife agencies need to address conflicts that arise in urban settings (including suburban and exurban), prioritizing both public and wildlife safety is important. This prioritization may create issues regarding budget reallocation or the final action taken to reduce the conflict. Some wildlife agencies have well-defined plans or policies outlining the processes they can take to help communities manage urban conflicts. Addressing these urban-associated conflicts may include tackling broader issues, such as those related to highway crossing and wildlife-vehicle collisions (Wakeling et al. 2007). Wildlife agencies have little regulatory influence over municipal or roadway development, but agency biologists can engage during early stages of land use planning so that potential concerns can be considered and potentially mitigated.

Wildlife agencies have limited tools to address overabundant mountain sheep populations where they may exist; typical solutions include lethal removal (of sick individuals) and relocation (Westerfield et al. 2019). Lethal removal with firearms is generally not used as a tool to manage mountain sheep, and where it might be considered, it may not be feasible in many circumstances. Regardless, public acceptance of lethal removal is often difficult to obtain for any wildlife species. Trap and removal measures may be perceived as safer, but substantial expenses, equipment, and expertise are required to undertake this effort humanely and effectively. Furthermore, relying primarily on removing animals provides only temporary relief if attractants and access are not mitigated or eliminated.

## 22.4.2 Land Management Agencies

Federal, state, provincial, territorial, and tribal land management agencies vary in their mandates, but all generally include some commitment to sustainable management of natural resources and ecosystems on multiple-use landscapes. Although there is little in their mandates that specifically address wildlife conflict, some federal land management agencies must balance management of socio-economic values across multiple-use landscapes, which allow for livestock grazing, energy development, recreation, and timber harvesting. These uses can be compatible with goals for conserving natural, economic, and social resources, but each of these multiple uses may generate conflict, so communication, coordination, and collaboration with and among stakeholders are necessary to avoid conflict. Land management agencies primarily provide public grazing allotments and permit the number and class of livestock that may be grazed on public lands. Federal grazing permits in the United States generally involve environmental evaluations consistent with the National Environmental Policy Act, and some states, provinces, and territories also include environmental evaluations. Part of any evaluation of the affected environment includes effects on existing wildlife populations. For mountain sheep, evaluations include the potential influence and transmission of pathogens from domestic sheep or goats, along with effects from various habitat manipulations, resource development, and construction along linear corridors (e.g., highways, row-housing).

## 22.4.3 Federal, State, Provincial, and Territorial Departments of Transportation

Construction of roadways is a lengthy process that often requires decades of environmental and pre-construction planning, assessments, and siting decisions, followed by contracting, construction, and post-construction assessments. The process can be almost as lengthy during improvement projects related to resurfacing or widening of roadways. Incorporating features such as overpasses, underpasses, fencing, and crosswalks requires additional foresight to ensure public safety once built and to limit wildlife mortalities because of collisions with vehicles. Identifying the location of mitigation structures requires wildlife and landscape data to support structure placement. Incorporating this type of information early in the planning process is fundamental to successful mitigation and reducing negative effects. This necessitates communication, coordination, and collaboration across agencies and with stakeholders that have data or interests regarding these features. Roadway mitigation contributes to protection of human life and reduces property damage, improves wildlife population connectivity by maintaining movement corridors, and enhances genetic exchange (Gagnon et al. 2017, 2022).

---

### Case Study 22.1: Effectiveness of US Highway 93 Overpasses for Maintaining Mountain Sheep Habitat Connectivity

#### CONFLICT ISSUE OR PROBLEM

Arizona Game and Fish Department (AZGFD) was concerned about the effects that the newly expanded highway (US 93) would have on the local desert bighorn sheep (*O. c. nelsoni*) population because of the risk for increased desert bighorn sheep-vehicle collisions and the associated habitat fragmentation that would follow.

#### GOAL

To maintain or increase habitat connectivity and reduce desert bighorn sheep-vehicle collisions across the newly expanded US 93.

#### BACKGROUND AND CONTEXT

The AZGFD and partners were concerned about increases in wildlife-vehicle collisions and habitat fragmentation for a desert bighorn sheep population in the Black Mountains, Arizona, during the upgrade of US 93 from a two-lane roadway to a four-lane divided highway to accommodate increasing traffic volumes. Stakeholder coordination was important in supporting the development and implementation of a comprehensive strategy to ensure sheep and motorists could coexist with minimal human-wildlife conflicts.

#### STAKEHOLDERS

- *Who they are.* The primary stakeholders for US 93 included the Arizona Department of Transportation (ADOT), AZGFD, National Park Service, Bureau of Land Management, and Bureau of Reclamation. Other stakeholders included conservationists interested in the safety and connectivity of bighorn sheep populations and the traveling public.
- *Who they should be.* Ideally, stakeholders should include departments of transportation, public and private landowners, wildlife agencies, conservation organizations,

other state or federal agencies as needed (e.g., US Forest Service (USFS), United States Fish and Wildlife Service, Bureau of Land Management (BLM)), and support from the traveling public.

#### MANAGEMENT AUTHORITIES

- *Entities.* The AZGFD, a public agency responsible for wildlife management and conservation; ADOT, a public agency responsible for a safe transportation system; and National Park Service, BLM, and Bureau of Reclamation, the federal landowners in the project area.
- *Scale and dynamics.* The AZGFD, working with ADOT and adjacent state's departments of transportation and wildlife, manages diverse wildlife populations and habitats in consideration of urban expansion and human population growth. Agency collaboration includes public input and landscape planning to identify local placement of mitigation structures designed to benefit wildlife.

#### METHOD, TOOLS, RISK ASSESSMENT, AND APPROACH

- *Previous research.* The AZGFD relied on the performance of prior wildlife crossing projects in Arizona and elsewhere.
- *Global positioning system telemetry (GPS) data.* Data from 30 desert bighorn sheep fitted with GPS collars provided the information needed to pinpoint the location of five major highway crossing areas, of which three were selected for the installation of wildlife overpasses. Continuous exclusionary fencing would also be constructed to guide sheep to the overpasses, under bridges and through culverts, in the construction zone and study area. Global positioning system data provided baseline connectivity information for future comparisons.
- *Post-construction monitoring.* The AZGFD and ADOT used GPS movement data and records of sheep-vehicle collisions before, during, and after US Highway 93 construction upgrade and included motion-sensitive camera data after the construction to evaluate success.
- *Adaptive mitigation.* The AZGFD and ADOT used the data collected after construction to identify areas of improvement, which included addressing gaps in the fence, modifying escape ramps, and widening cattle guards; monitoring continued to provide additional information for adapting to changing conditions. This continuous feedback loop (i.e., the essence of adaptive management) allowed for maximum project effectiveness.

#### OUTCOME

The culmination of information collected through desert bighorn sheep movement studies, lessons learned from prior and ongoing projects, and adaptive mitigation guided by post-construction monitoring led to successful mitigation of habitat fragmentation and sheep-vehicle collisions.

These highway structures are the first of their kind because they wee specifically designed for desert bighorn sheep. Post-construction monitoring verified the preference for use of overpasses by mountain sheep. Vehicular collisions with sheep were reduced by 68% during the first 2 years of monitoring; once ADOT maintenance personnel addressed desert bighorn sheep breach points, only one vehicle-killed desert bighorn sheep was documented. Since 2014, to the writing of this book, no desert bighorn sheep-vehicle collisions have occurred along this stretch of US Highway 93. The success of this project led to a partnership between AZGFD, Nevada Department of Transportation, and Nevada Department of Wildlife. The Nevada Department of Transportation contracted AZGFD to assist in design, implementation, and post-construction monitoring of a new highway in Nevada (i.e., Interstate 11) and the efficacy of wildlife crossings. As of this writing, >22,000 sheep crossings have occurred at designed wildlife crossing locations and no collisions have been documented.

### 22.4.4 NONPROFIT WILDLIFE CONSERVATION ORGANIZATIONS

Mountain sheep, perhaps to an extent greater than any other single taxonomic group of wildlife, are the focus of influential nonprofit wildlife conservation organizations (e.g., Wild Sheep Foundation and its Chapters and Affiliates, Grand Slam Club – Ovis, Safari Club International) that raise substantial funds and coordinate volunteer efforts to further mountain sheep conservation (e.g., water source development, translocations, wild sheep health investigations). Despite their many benefits, these groups have limited to no official authority in most decision-making processes, albeit they are usually included as stakeholders during decision-making. These organizations have a large effect on the outcome of most management proposals because they provide large volunteer resources, contribute substantial funding for projects and proposals, and influence views and perceptions using social media platforms. With this influence, nonprofit wildlife conservation organizations can have an important effect on the likelihood of resolving conflict because of their willingness to collaborate or communicate messages either in support or opposition to proposed activities (Hawkins 2017).

### Case Study 22.2: Reintroducing Desert Bighorn Sheep into an Exurban Mountain Range

#### CONFLICT ISSUE OR PROBLEM

Desert bighorn sheep had historically occupied the Catalina Mountains near Tucson, Arizona, although lack of fires and brush encroachment, high recreational use by humans, mountain lion (*Puma concolor*) predation, and possible

disease contributed to the extirpation of this population (Krausman 2017). Reintroducing desert bighorn sheep into this isolated mountain range, while popular among the public, could require substantial management intervention to establish a self-sustaining herd of desert bighorn sheep. With the large number of conservation organizations within Tucson (more than 50) and with their differing levels of acceptance for a variety of management activities, gaining public support for this management action was essential and challenging.

#### GOAL

Engage an *ad hoc* stakeholder group that represents the diverse interests and viewpoints of the conservation-minded public in and around Tucson, Arizona, to develop management options informed by scientific knowledge and supported by the stakeholders involved.

#### BACKGROUND AND CONTEXT

Restoring extirpated wildlife to historically occupied range may require management intervention that extends beyond simply returning wild animals to the environment. In the Catalina Mountains, the most likely impediment to successful reestablishment of desert bighorn sheep following a translocation was the likelihood of predation to limit or extirpate the released herd. Various stakeholders may not support hunting or lethal removal of mountain lions, captures of desert bighorn sheep that may result in related mortality, management actions within an established wilderness area, limiting recreational activity like hiking on public land, or active fire management, a consensus-influenced stakeholder group was needed to publicly advocate for actions by management agencies.

#### STAKEHOLDERS

- *Who they are.* The primary stakeholders that participated in a self-formed advisory committee included the Arizona Antelope Foundation, Arizona Desert Bighorn Sheep Society, Arizona Wilderness Coalition, Arizona Wildlife Federation, Center for Biological Diversity, Sky Island Alliance, and The Wilderness Society. Secondary stakeholders that attended meetings included the AZGFD and the USFS.
- *Who they should be.* Getting all groups to come to consensus on management actions such as landing helicopters in wilderness areas or lethally removing mountain lions may be impossible. The Sierra Club and Friends of Wild Animals chose not to serve on the advisory committee, but engaged in the project, sometimes in protest, primarily about lethal removal of mountain lions. The groups that chose not to participate were unwilling to accept certain actions (e.g., mountain lion killing or desert bighorn sheep mortality during capture events) to achieve a stated goal (e.g., reestablishing desert bighorn sheep in historically occupied habitat).

## MANAGEMENT AUTHORITIES

- *Entities.* The AZGFD is a public agency with authority over wildlife management and conservation in Arizona; the USFS is a public agency with regulatory authority involving land and recreational use on lands under their jurisdiction, including the Pusch Ridge Wilderness Area in the Catalina Mountains; the *ad hoc* advisory committee was self-formed with no formal authority (as they were not a public agency, they were not subject to the public open meeting laws of Arizona).
- *Scale and dynamics.* Mountain sheep reintroduction projects in Arizona typically involved wildlife and land management agencies and supporting wildlife conservation organizations. Wildlife conservation organizations formed an advisory group because they recognized the need to address likely conflict arising from the environmental focus of this proposed reintroduction and the likelihood for the need to mitigate mountain predation on translocated mountain sheep.

## METHOD, TOOLS, RISK ASSESSMENT, AND APPROACH

- The *ad hoc* advisory committee was formed to find common ground concerning the acceptable management activities that could be used and publicly supported to achieve the reestablishment of desert bighorn sheep within the Catalina Mountains, Arizona.
- No formal tools were used other than facilitated discussions among the members to reach consensus on the types of management actions (e.g., limited lethal removal of mountain lions that preyed on desert bighorn sheep, net-gun captures of desert bighorn sheep within designated wilderness areas). A recommendation to proceed with a translocation was only adopted by the *ad hoc* advisory committee when the entire project was outlined to their satisfaction. The AZGFD and the USFS were not bound by any recommendations offered by the *ad hoc* committee.

## OUTCOME

One hundred ten desert bighorn sheep were released over 4 years, with 43 documented mortalities of radio-marked desert bighorn sheep and eight mountain lions removed following known predation events. There were an estimated 85 desert bighorn sheep occupying the Catalina Mountains in 2017, 4 years following first release (reproduction and recruitment were documented annually). The project enjoyed high support by the public in general, although Friends of Wild Animals frequently and publicly protested at agency offices, commission meetings, and other public events over culling of mountain lions. The project was not supported by Sierra Club. All participating stakeholders on the *ad hoc* advisory committee publicly supported the project and noted that while their own values did not change, their attitudes did, especially toward one another.

### 22.4.5  Livestock Producers

Like nonprofit wildlife conservation organizations, livestock producers have little to no official authority in decision-making processes. Nevertheless, the producers, primarily of domestic sheep, typically have a large influence on the outcome of management proposals that potentially influence their operations because they are well organized, resulting in a substantive influence in many political environments. As an example, these organized groups include domestic sheep and goat associations and wool growers' associations, which can be found at state, provincial, territorial, and national levels. Livestock products are an important commodity, and many livestock producers own private land, which wildlife depend on for habitat and resources that contribute to survival and reproduction. Therefore, the support of producers and their input when managing populations of mountain sheep across landscapes used by domestic sheep and goats on public land helps to achieve positive conservation outcomes.

### 22.4.6  Private Landowners

In most jurisdictions, private lands often comprise a substantial proportion of habitats used by wildlife populations. Although much of the occupied mountain sheep habitat is located on public land, private landowners often own and manage portions of seasonal ranges that can have disproportionate influence on some populations. Frequently, bighorn sheep will move to lower elevations and valleys during winter, and these areas are commonly owned by individuals or organizations that may have agricultural interests involving farming or animal husbandry. Although some private landowners tolerate wildlife and mountain sheep on their land, mountain sheep may have a negative financial influence on their businesses. Recognizing the interests of private landowners and the effect of mountain sheep on their properties is important when making management decisions.

### 22.4.7  Cities and Municipalities

City officials often are expected to solve urban issues, but they lack authority when conflicts extend beyond their jurisdictional boundaries. Wildlife management agencies primarily manage mountain sheep population size outside of municipalities through regulated public hunting. Cities usually have ordinances and bylaws that protect public safety and prohibit the discharge of firearms within city limits. Mountain sheep, however, can be abundant near urban environments, or even found within city limits, yet cities do not have the same tools to manage conflicts as do wildlife management agencies. For example, if hunting is not allowed (or is deemed unsafe) within city limits, then hunters cannot be used to regulate sheep populations, effectively eliminating a primary tool used by wildlife agencies to reduce herd size. People living in urban areas may not consider hunting acceptable (Manfredo et al. 2018), although some stakeholders (e.g., locavores) in urban areas may accept hunting as a tool to reduce wildlife densities while obtaining locally sourced protein. Contributing to problems with wildlife in urban areas is the prevalence of artificial feeding, which attracts animals close to humans, houses, and roadways (Westerfield et al. 2019), and can result in human food conditioning by individuals or groups of wildlife. Often, municipalities lack ordinances that prohibit the feeding of wildlife, even though a strategy to reduce conflict is to restrict or prohibit artificial feeding.

### 22.4.8 WORKING TOGETHER TO MANAGE MOUNTAIN SHEEP

Communication, coordination, and collaboration among state, provincial, territorial, and federal wildlife management agencies, land management agencies, departments of transportation, stakeholders, producers, and municipalities are needed for effective management of mountain sheep conflicts. Mountain sheep management is not simply finding a way to harvest animals in situations where conflicts arise; rather, it is a strategic plan to avoid, eliminate, mitigate, or reduce conflict. Most challenges arise because issues are poorly defined, tenuously understood, or viewed or valued differently by stakeholders. Management decisions are value-based and determined from social science, ethics assessment, business efficacy, biological science, and expert opinion (Decker et al. 1996, Mangel et al. 1996, Organ and Ellingwood 2000, Riley et al. 2002, Raik et al. 2003, Smith et al. 2023).

## 22.5 DECISION-MAKING

Managing natural resources is complex, but it is even more complex when conflict arises among various stakeholders that have deeply entrenched values and different visions and priorities for shared public resources. Decisions must be made to address or prevent conflict. Several formal decision-making frameworks exist to assist with the task, but basic principles and components guide decision-making regarding conflict (Table 22.1).

Decision analysis is a process commonly known as structured decision-making and widely used in natural resource fields (Gregory et al. 2012a). Structured decision-making has five principal elements: **pr**oblem framing, **o**bjectives, **a**lternatives, **c**onsequences, and **t**rade-offs (PrOACT; Runge and Bean 2020). During problem framing, the context for the decision is articulated, including spatial and temporal aspects and identification of the stakeholders, decision-makers, and management authorities. Objectives are articulated to explicitly state what future goal or condition is targeted. Alternatives, developed by subject matter

experts, are potential actions that could be used to achieve an explicit objective. Each alternative will be subject to evaluation, modeling, or prediction of potential outcomes. The use of the predictions depends on systems knowledge and thinking, available information and data, and forecasts of future conditions (Runge and Bean 2020). Finally, trade-offs will occur to balance value, policy, and science when choosing one alternative over another.

In addition to the five principal elements of decision-making, others need to be included in the decision analysis (Smith 2020, Table 2.1). It is important to identify who is involved with, or potentially influenced by, the decision and what rights and title encumbrances they may have or exert when developing options to achieve an outcome. These stakeholders may consist of decision-makers, legal authorities, First Nations and Tribal Agencies, interested audiences, and technical experts (see Section 2.3). Next, decisions must consider legal and regulatory authorities (see Section 22.4), which may include local, state, provincial, territorial, or federal agencies or land management authorities that regulate activities on public land. There are also key considerations to include as part of PrOACT, which include how frequently a decision is made, the timing of the decision or implementation, and the scope, scale, and legal designation of the project. Still other considerations (e.g., objectives, actions, constraints, and uncertainties; Smith 2020, Table 2.1) are formalized parts of PrOACT. Last, analyses will be required to evaluate alternatives and their potential effects if chosen. Because natural systems are always changing, so too will the conflicts that arise, so plans will need to be revised to ensure social license and broad public support for ongoing management actions are maintained.

A more general decision analysis framework is known as adaptive management (Walters 1986) and can use any formal decision analysis process. Adaptive management recognizes there is uncertainty surrounding decisions, so as more information is acquired, management directions may need to be changed (i.e., adapted to the current situation, condition, or state). Uncertainty exists for human influences on natural systems, and when making any decision or applying any management action. Therefore, effective monitoring is important to gain information about the state and trajectory of the system under management and to determine the extent to which change in the system can be attributed to the management actions deployed, creating a feedback loop of ongoing monitoring, data assessment, and revised action (Sells et al. 2016).

Another approach is adaptive impact management (Riley et al. 2003), which was developed to address some of the shortcomings associated with adaptive management. Models and approaches to conventional adaptive management are primarily based on objectives for the condition of wildlife populations, species, or habitat rather than effects on society defined by stakeholders (Walters 1986). This shortcoming, unfortunately, leads to a focus on means or

**TABLE 22.1**

**Glossary of Terms Commonly Used When Managing Conflict Arising from Wildlife Interactions**

| Term | Definition | Source |
|---|---|---|
| Action | Management or other activity by humans that may influence a resource state. | Conroy and Peterson (2013) |
| Adaptive management | An approach for making ongoing management decisions in the face of uncertainty, with feedback monitoring that allows decisions to improve as uncertainty is reduced. | Walters (1986) |
| Alternatives | The potential management solutions designed to meet the objectives and solve the management issue. | Runge et al. (2013) |
| Conflict | A situation that occurs when two or more parties with intensely held opinions strongly disagree over conservation objectives and one party is perceived to assert its interests at the expense of another. | Redpath et al. (2013) |
| Constraint | Factor(s) that limit the range of decisions that can be made such as costs or logistical issues. | Conroy and Peterson (2013) |
| Cost-benefit analysis | The process of comparing the projected or estimated costs and benefits or opportunities associated with a project decision. | Stobierski (2019) |
| Decision | Choice(s) between/among options or alternative actions directed at achieving an objective. | Conroy and Peterson (2013) |
| Facilitation | The act of making, helping, or bringing about something to make it easier or to be more effective. | Merriam-Webster (2024) |
| Implementation | The process of making something active or effective. | Merriam-Webster (2024) |
| Judgment | The process of forming an opinion or evaluation by discerning or comparing; an opinion or estimate. | Merriam-Webster (2024) |
| Mitigation | The process or result of making something less severe, dangerous, harsh, or damaging. | Merriam-Webster (2024) |
| Monitoring | Tracking of a natural resource system or its components through time to determine its status and trajectory, inform decision-making, evaluate the outcomes of management, and reduce uncertainty | Conroy and Peterson (2013) |
| Objective | Quantifiable outcomes that reflect the values of decision-makers or stakeholders, and are influenced by management decisions. | Conroy and Peterson (2013) |
| Optimal | The decision or combination of decisions culminating in the best result(s). | Conroy and Peterson (2013) |
| Proactive | Acting in anticipation of future problems, needs, or changes. | Merriam-Webster (2024) |
| Reactive | Occurring as the result of something; responding to something that has already happened. | Merriam-Webster (2024) |
| Resolution | The solution, albeit not necessarily optimal, identified through negotiation, mediation, and/or litigation. | Smith (2020) |
| Risk | The likelihood that a hazard or decision results in an undesirable outcome. | Conroy and Peterson (2013) |
| Stakeholder | Any citizen potentially affected by or having a vested interest or influence in an issue, decision, or action. | Decker et al. (1996) |
| Structured decision-making | A method of decision analysis used to identify the most effective, efficient, and realistic decisions while accounting for values and priorities of the decision-maker(s). | Sells et al. (2016) |
| Structured decision-making | A decision analytic method that breaks decision-making into logical components of identifying the management problem, defining objectives for solving the problem, developing alternative approaches to achieve the objectives, and formally evaluating which of the alternatives is most likely to accomplish the objective(s). | Runge et al. (2013) |
| Tolerance | The ability of the decision-makers and stakeholders to accept or absorb risk. | Conroy and Peterson (2013) |
| Transparency | The quality or ability to be seen, known, or understood often through proper communication. | Merriam-Webster (2024) |
| Uncertainty | Indetermination of an attribute, system state, or other feature; associated with limited knowledge of the state of a system due to limitations, insufficient data, or variability. | Conroy and Peterson (2013); Groves and Game (2016) |
| Value | The meaning, goodness, or worth that humans place on something; relative importance or worth of an object to an individual or group (often context dependent). | Brown (1984) |

enabling objectives rather than fundamental objectives that were defined by stakeholders; therefore, adaptive impact management incorporates a value-based, decision-making philosophy. The key difference between adaptive impact management and conventional adaptive management is that adaptive impact management seeks to define objectives in terms of desired stakeholder-identified outcomes. These objectives become performance measures for subsequent management actions. Regardless of the approach used, formal stakeholder engagement must match the individuals with the duration of effects, and ensure they are aligned with the scale of management actions and with the biological, political, economic, and technological limitations of management (Riley et al. 2003).

Decision-making, socio-economic or otherwise, often employs community-based processes. Applying a community-based process ensures social license support for the solutions implemented and that those solutions will address the biological and social issues at hand (Raik et al. 2003). If the conflict is more widespread than the local community, there may be value in developing a formal process such as structured decision-making (Gregory et al. 2012b), appreciative inquiry (Coghlan et al. 2003), or some other decision analysis. Early on, public surveys and public engagement processes may be useful in defining the depth and breadth of a specific concern (Wingard and Krausman 2019). These approaches are adaptive and allow wildlife managers and stakeholders to refocus priorities as more information is acquired. A proactive approach to conflict management may preemptively keep a problem from developing into a public controversy (Decker et al. 2019).

## 22.6  SOURCES OF CONFLICT

There are two important considerations for conflict resolution: the type of conflict and the context in which the conflict arises. Considering the context in which a conflict is occurring will guide managers toward solutions that may best resolve the conflict. There are many tools available for mitigating conflict. Most tools can be used for more than one source, type, or context of conflict.

### 22.6.1  Types of Conflict

A variety of perspectives may bring about conflict among individuals interested in wildlife. We provide several examples below and explain how conflict may occur:

1. Nuisance (wildlife effect on people [e.g., deer in yard causing pet dog to bark])
2. Economic (wildlife effect on people [e.g., ornamental plant damage around home, crop damage, changes to public grazing allotments], or industrial or commercial effect on wildlife [e.g., a housing development being constructed in bighorn sheep winter range that also supports a guide-outfitting operation for which bighorn sheep are a harvested species; Chapters 20 and 21])
3. Health (wildlife may influence people or domesticated animals or humans may have this effect on wildlife, possibly through domesticated animals [e.g., bovine tuberculosis, rabies, respiratory disease])
4. Safety (wildlife effect on people [e.g., mountain sheep-vehicle collisions, head butting])
5. Psychological (wildlife effect on people [e.g., fear of car collisions or physical attack], people effect on people [e.g., concern about well-being of wildlife including endangerment of populations of specific species, biodiversity generally being threatened by human activities])
6. Ecological (wildlife effect on ecosystem [e.g., mountain sheep effects on plant species diversity, species composition, structure of forest], or domestic animal or feral livestock effect on ecosystem [e.g., cattle grazing or feral equid effects on forage availability, plant species diversity, species composition, structure of forest; Chapter 20])

### 22.6.2  Contexts for Conflict

*Agriculture.* Mountain sheep will feed on crops such as alfalfa when the opportunity exists. Damage to standing crops by mountain sheep is usually minimal and literature documenting damage is sparse. When mountain sheep are present in agricultural settings, conflict is usually related to commingling with domestic sheep or goats and concerns over disease transmission. Although respiratory disease transmission is considered more detrimental to mountain sheep populations (Cassier et al. 2018), which may result in all-age die-offs or chronic disease within populations, domestic sheep reproduction and recruitment are also affected by respiratory pathogens (Manlove et al. 2019). Independent of disease transmission risk, mountain sheep may cause local economic losses to producers when they bed in or consume agricultural forages.

Grazing of domestic sheep on mountain meadows in wildland conditions is not uncommon. Damage to the quantity and species of forage can occur because of overgrazing by domestic livestock (Krahulec et al. 2001), which can have negative implications on wildlife (Chapter 20). The primary concern for mountain sheep when grazing occurs in wildland habitat is that commingling between mountain and domestic sheep and goats increases disease transmission risk among the species, which is the main source of conflict between domestic sheep producers and wild sheep conservation advocates.

*Urban or exurban interface.* Like other species, mountain sheep may be attracted to the interface between wildlands and human development areas. Attraction to areas occupied by humans occurs when the benefits to mountain sheep of using these areas outweigh the potential risks (Dzialak et al. 2011). For instance, desert bighorn sheep use of Hemenway Park, a municipal park in proximity to Boulder City and to the River Mountains, Nevada, occurs during summer when water and succulent herbaceous vegetation were scarce in other portions of their range; this pattern was not observed during winter (Lowery and Longshore 2017). Of concern, related to the well-being of mountain sheep within and near urban areas, is mountain biking. Mountain biking has been deemed to be disruptive and unpredictable, causing desert bighorn sheep to avoid locations with mountain biking activity (Lowery and Longshore 2017). In Arizona, translocated desert bighorn sheep moved through municipalities like Sedona, browsed on cultivated rose bushes, and head-butted sliding glass doors with destructive results before bedding in the residential yard (B. F. Wakeling, Montana Fish, Wildlife and Parks, personal observation).

Effects of resource extraction activities (e.g., oil, natural gas) are more likely to lead to temporary abandonment of seasonal ranges and preferred foraging habitats (Frid and Dill 2002, Poole et al. 2016). While mountain sheep have developed tolerances to some predictable activities (Beale and Boyce 2020), this tolerance appears to exist as a function of site attractiveness (i.e., improved forage supply, modified terrain features [e.g., excavated escape terrain], decreased predator occurrence) more so than because of habituation to the activity (Frid and Dill 2002, Dzialak et al. 2011, Chapter 21). Rocky Mountain bighorn sheep (*O. c. canadensis*) in an active mining area near Clifton-Morenci, Arizona, have been observed crawling under moving ore belts to access water and lawns (B. F. Wakeling, personal observation). The level of disturbance tolerated by mountain sheep is apparently situational and site specific.

*Human safety.* Mountain sheep rarely present a true risk to human safety; however, humans may be at risk due to biological and physiological changes that occur in mountain sheep, especially male sheep during the breeding season. Males may be more aggressive during the breeding season because of increased testosterone levels, so males may not be as tolerant of the presence of humans or their pets. This situation can be compounded if mountain sheep have become habituated with human activity, such as in urban areas, and their avoidance of humans is diminished.

*Roadways.* One of the most frequent types of conflict relates to mountain sheep-vehicle collisions. Wildlife-vehicle collisions are an almost ubiquitous challenge, resulting in the expenditure of more than US$8 billion in the United States (Huijser et al. 2008) and account for the highest source of human injury or fatality associated with wildlife conflicts in the United States (Conover 2019). Although collisions with mountain sheep are relatively rare and account for only a small percentage of collisions, mountain sheep-vehicle collisions can result in motorist injury, fatality, and property damage.

Vehicle collisions with mountain sheep occur throughout their range and are often associated with the presence of preferred resources or vegetative associations that are either naturally occurring or created by humans adjacent to roads. A common cause of mountain sheep-vehicle incidents within higher elevations and northern latitudes is the use of chloride-based road deicers, or road salt, which entices mountain sheep to lick the deicer from the road, increasing the potential for mountain sheep-vehicle collisions (Tiwari and Rachlin 2018, Huijser et al. 2021). The conflict can be further exacerbated when roadside safety grooves (rumble strips) have been installed along the centerline and shoulders sections of the paved surface; these grooves collect and concentrate the brine that is sprayed onto the road surface, increasing the attractiveness of those specific portions of the traveled roadway. Road salt is anecdotally referenced as among the most prominent sources of mountain sheep-vehicle collisions in northern parts of the United States and Canada. At Thompson Falls, Montana, about 15 mountain sheep are killed annually, and a large proportion involved mountain sheep licking road deicer on Highway 200 (Huijser and Clevenger 2013). Road salt is a leading cause of Rocky Mountain bighorn sheep-vehicle collisions along several roads in South Dakota (Figure 22.1; M. Rippentrop, South Dakota Department of Transportation, unpublished data). Dall's sheep (*O. dalli*) in the Thechàl Dhâl' in Kluane National Park and Reserve in the Yukon along the Alaska Highway are struck regularly when licking road

salt, including a single accident in 2018 that killed eight in a single collision (Cumming and Taveres 2022).

Traffic volume on roadways influences mountain sheep movements and mortality in various ways. Mountain sheep populations that attempt to move across roadways with low to moderate traffic volumes experience greater mountain sheep-vehicle collisions when compared to areas with higher traffic volumes. In areas with higher traffic volumes, such as interstate highways (e.g., more than 10,000 vehicles average daily traffic), roadways may create a substantive barrier to mountain sheep movements, even to the point of blocking gene flow (Epps et al. 2005). The presence of existing drainage structures may allow occasional crossings under high traffic volume interstates, but the characteristics of those drainage structures and roadway fencing influence the likelihood of use (Gagnon et al. 2017, Aiello et al. 2024). If traffic volumes are high and suitable drainage structures do not exist, then the addition of wildlife crossings may be necessary to restore or maintain connectivity (Chapter 11). Wildlife crossings reduce the effect of high traffic volumes on wildlife movements (Gagnon et al. 2007a,b; Dodd and Gagnon 2011, Gagnon et al. 2022).

*Wildlife and livestock disease transmission.* Another major source of human conflict involves disease transmission. While contact between wild and domestic sheep may facilitate transmission of pathogens in both directions, most of the conflict has centered on the perceived or documented transmission of pathogens from domestics to wild sheep. Understanding the origins of conflict regarding mountain sheep, domestic sheep and goats, and disease is important for recognizing challenges to resolutions. Perhaps more so than any other conflict involving mountain sheep, economic and psychological factors of conflict may play an important role in understanding and addressing disease transmission concerns.

Throughout the 1980s and 1990s, the theory that domestic sheep could transmit disease in mountain sheep was examined through captive studies where mountain sheep were mixed with domestic sheep and other livestock species (Schommer and Woolever 2008, Wehausen et al. 2011). These studies demonstrated a consistent pattern of severe respiratory disease and death of bighorn sheep when commingled with apparently healthy domestic sheep (Foreyt and Jessup 1982, Onderka and Wishart 1988, Foreyt 1990, Callan et al. 1991, Foreyt 1994). By contrast, limited experiments in which mountain sheep were commingled with cattle, elk (*Cervus canadensis*), white-tailed deer (*Odocoileus virginianus*), mule deer (*O. hemionus*), domestic goats, mountain goats (*Oreamnos americanus*), llamas, horses, or steers demonstrated no consistent disease transmission (Foreyt 1992, 1994; Foreyt et al. 1996), and suggested that the focus should remain on domestic sheep. At the time, the causative agents of respiratory disease were not identified, nor was it understood that all members of the Caprinae family could serve as disease hosts for *M. ovi* (Cassirer et al. 2018). Yet, captive studies and a large volume of field observations led to the conclusion that mountain and domestic sheep were incompatible on shared landscapes (Schommer

**FIGURE 22.1** Many chemical deicers used on roadways contain salt, which may attract mountain sheep onto roadways, creating a risk for vehicle collisions. (Photo by Brian Wakeling.)

and Woolever 2001). This conclusion focused attention on spatial and temporal separation between mountain and domestic sheep as a tool for mountain sheep recovery.

Early efforts to establish federal and state guidelines, best management practices, and practical guidance were important in their formal acknowledgment of the conflict and in their description of tools to increase separation between mountain sheep and domestic sheep. In 1992, the Bureau of Land Management used recommendations from the Desert Bighorn Council (1990) and input from a diverse stakeholder committee to prepare grazing guidelines for the management of domestic sheep in bighorn sheep habitats (Bureau of Land Management 1992), revised to include domestic goats in 1998 (Bureau of Land Management 1998). These polices acknowledged the growing conflict between mountain and domestic sheep interests and the need for cooperation and coordination across state and federal agencies, interest groups, and the domestic sheep industry to maintain or expand mountain sheep numbers. The guidelines included holding agencies and the domestic sheep industry harmless in the event of disease affecting mountain sheep or domestic sheep or goats, reducing contact through discouraging domestic sheep and goat grazing and trailing in mountain sheep ranges, close management and careful herding of domestics adjacent to mountain sheep ranges, removal of stray domestics or mountain sheep in areas that would allow close contact between the species, implementation of 13.5 km buffers around any new domestic sheep or goat grazing permits or conversion of cattle to sheep permits, no new reintroductions of mountain sheep into areas with existing domestic sheep or goat grazing, and extraordinary precautions to protect special status subspecies such as desert bighorn sheep.

The USFS made internal recommendations to regional foresters to consider separation in Forest Management Plans as early as 1981 (Goodson 1982). Over the next few decades, the USFS invested substantial effort in supporting mountain sheep restoration and dedicated staff to work with local National Forest Districts on implementing mountain and domestic sheep separation throughout the West. In 2001, the USFS published guidelines for finding management solutions to the incompatibility between domestic and bighorn sheep (Schommer and Woolever 2001) in which they advocated for a collaborative, consensus-based approach to find solutions that benefited mountain sheep and domestic sheep producers. Implementing spatial separation on public land by closing, retiring, or buying out domestic sheep allotments, and the risk of further closure with continued mountain sheep restoration were viewed as threats to domestic sheep producers and industry, generating conflict, distrust, and frustration. The USFS issued three internal guidance memorandums encouraging work toward separation (Holtrop 2011, Weldon 2012, 2014), but a formal USFS policy was not adopted, and local implementation of separation varied as land managers grappled with the contentious trade-offs between mountain sheep recovery and domestic sheep grazing interests.

Legal challenges to grazing management decisions in Hells Canyon and the Payette National Forest in Idaho inflamed west-wide tensions between mountain sheep and domestic sheep interests. In 1993, after 20 years of struggling to maintain separation while balancing domestic and mountain sheep use in Hells Canyon, Idaho, the USFS revised their Wallowa-Whitman National Forest Plan and made the difficult decision to close three long-term domestic sheep allotments in Hells Canyon to prioritize mountain sheep restoration (Schommer 2002, Schommer and Woolever 2001). The decision was challenged but upheld in court. Short-notice, court-ordered closures of high-risk domestic sheep allotments were concerning because of the economic and logistical difficulties they imposed, but also because of permittees' long-term, multi-generational attachments to use of public land.

In 2003, the USFS published their Final Environmental Impact Statement for their 20-year update to the Payette National Forest Land and Resource Management Plan, which included forest management relevant to the mountain sheep metapopulation in Idaho (USFS 2010). The proposed revisions were challenged over concerns that mountain sheep viability was not adequately addressed with respect to domestic sheep grazing. In 2007, The Wilderness Society, Western Watersheds Project, and Hells Canyon Preservation Council filed a lawsuit against the USFS for failing to protect a viable population of mountain sheep. This lawsuit resulted in a court order removing domestic sheep from five allotments on the Payette National Forest and a requirement that the USFS complete an amended environmental impact statement that accounted for the effects of domestic sheep grazing on mountain sheep population viability (USFS 2010). In 2010, the USFS published their Final Supplemental Environmental Impact Statement and Record of Decision, which included the use of refined predictive tools to assess contact and disease risk, which led to the decision to reduce domestic sheep grazing on the Payette National Forest by about 70%. In 2012, sheep producers in Idaho sued the USFS over the reduction in domestic sheep grazing on the Payette, but in 2014, the Payette decision was upheld by a federal court order (Idaho Wool Growers Association vs. Tom Vilsack 2014).

The Hells Canyon lawsuit and the Payette decision were closely watched across the West. As tensions mounted, fueled by ongoing litigation, many local working groups struggled to make progress due to polarization, frustration, and lack of trust. State and national woolgrower and stockmen associations developed policy stances against restrictions to or phaseout of public grazing allotments, questioning the basic premise that domestic sheep posed a threat to mountain sheep population viability (Wyoming Wool Growers Association 2022, American Sheep Industry Association 2024). Continued disease outbreaks among mountain sheep were frequently blamed on the presence of domestic sheep even after it became clear that chronically infected mountain sheep could act as a spillover source to uninfected mountain sheep or to herd reinfections.

Additional challenges occasionally must be addressed that result from overlapping wild and domestic sheep ranges. Interbreeding among male mountain sheep and domestic females can occur, and lambs rarely survive for long (Figure 22.2). Producers may complain that interbreeding limits recruitment in their herds. These conflicts are not common but can be real for affected producers.

The heightened conflict concerning mountain sheep and domestic sheep led the Western Association of Fish and

**FIGURE 22.2** Commingling between mountain sheep and domestic sheep carries the risk of respiratory disease transmission and the possibility of interbreeding between mountain sheep males and domestic sheep females. Hybrid lambs rarely survive to adulthood and may limit domestic sheep recruitment. (Photo by Jami Murdoch; used with permission.)

Wildlife Agencies to call for standardized North American mountain-domestic sheep policies and their proposal of formal, and exhaustive, recommendations for effective separation aimed at federal, state, provincial, and territorial wildlife agencies, land management agencies, mountain sheep conservation organizations, domestic sheep and goat permittees and owners, and private landowners (Wild Sheep Working Group 2007, 2012). The Western Association of Fish and Wildlife Agencies guidelines set the standard for separation protocols aimed at conserving mountain sheep.

### 22.6.3 SPECIFIC TOOLS TO MITIGATE CONFLICT

Although conflict occurs among stakeholders, conflict arises in the context of specific conditions. Conflict is best addressed through prevention. Once a conflict exists, managers must consider the perspectives of those immediately involved and the perspectives of those that may be affected by the proposed actions to mitigate conflict (Wakeling et al. 2023). For instance, people who have no interest in mountain sheep may become interested and strongly opinionated once they recognize that mountain sheep are being captured and relocated. The methods for mitigating conflict may be used in a variety of contexts (Table 22.2), but not all methods are appropriate in every context. For instance, a helicopter net-gun capture to translocate sheep may not be an appropriate capture method within a densely populated municipal setting. Most tools can be used to mitigate multiple types of conflict, and multiple tools may be necessary to successfully mitigate the conflict. The following is a list of tools that can be used for conflict resolution, many of which can be used for multiple conflicts or those that are especially complex (e.g., sheep-vehicle collisions, disease spread).

*Fencing.* Fencing can be used as a permanent or temporary physical barrier that excludes wildlife from

**TABLE 22.2**

**Situations or Context in Which Conflict May Arise and Suggested Tools for Conflict Mitigation (Mitigation Approaches)**

| Context | Mitigation Approaches |
| --- | --- |
| Agriculture | Fencing, harassment, community-based processes |
| Urban-exurban | Fencing, harassment, guard dogs, community-based processes, local ordinances, capture and removal |
| Human safety | Fencing, harassment, guard dogs, community-based processes, local ordinances |
| Roadways | Fencing, lethal removal (including hunting), capture and removal, wildlife crossings, design, sensory devices |
| Wildlife and livestock disease transmission | Fencing, harassment, guard dogs, lethal removal (including hunting), sharpshooters, separation |

accessing areas where they can cause damage or are otherwise unwanted. When properly constructed and maintained, fencing can effectively reduce or prevent damage, reduce wildlife-vehicle collisions, and reduce transmission of disease (Conover 2001, Gagnon et al. 2022). There are, however, logistical and financial constraints if mountain sheep are to be excluded from agricultural fields or other large areas. Generally, fencing is justified financially when protecting high-value items such as haystacks, small fields, or yards and landscaping (e.g., orchards of high-yielding and browse-vulnerable dwarf and semi-dwarf apple trees).

Fence design varies based on need and the type of wildlife species most likely to be involved with the conflict. Fencing can be categorized into two main types: electrified and non-electrified. Non-electric fencing must be tall enough (3 m or taller) to prevent mountain sheep from jumping over it and tight to the ground so animals cannot crawl under it. Depending on the type of non-electrified fence, the strands of a barbed wire (or high-tensile wire) fence should be close enough together (20–25 cm apart) and taut enough (90 kg or more of tension) to prevent wildlife from passing between strands (DeNicola et al. 2000). Woven wire fences (or similarly, chain-link fences) have strands running horizontally and vertically, creating a mesh too small for an animal to pass through, but this wire material tends to be more expensive than high-tensile or barbed wire. Fencing that is not electrified can create effective physical barrier when constructed and maintained properly (Stewart et al. 2025). Woven wire, barbed wire, high-tensile, and chain-link fences, when constructed with quality components, have a life expectancy of two to three decades. Heavy plastic mesh is also an option that is more economical and can be effective under certain circumstances. Generally, plastic mesh fencing is used to solve temporary conflicts on restricted scales (i.e., smaller areas). As part of the decision-making process to find suitable solutions, cost-benefit analyses are usually conducted; considering the initial cost of permanent fencing versus its long-term benefits and lower maintenance costs, it can be an important trade-off with temporary fencing.

Electric fencing provides a physical and psychological barrier to wildlife (VerCauteren et al. 2006, Webb et al. 2009). Most often, electric fences are constructed using high-tensile strands of wire and are ≥3 m in height to prevent animals from crossing. Electric fences can be effective on large or small scales, but the efficacy of the fence excluding wildlife is dependent on a properly maintained fence, a continuous supply of power, and control of vegetation near the fence (Webb et al. 2009).

*Designed structures.* Designed structures, such as culverts, underpasses, overpasses, or similar structures (more generally, wildlife crossings), can be used on their own or in association with fencing (Clevenger et al. 2001, Grilo et al. 2008, Gagnon et al. 2015, Donaldson and Elliott 2021). Designed structures, when coupled with fencing, can reduce conflict, especially along roadways experiencing frequent wildlife-vehicle collisions (Figure 22.3). When using fencing along roads to reduce mountain sheep-vehicle collisions, provision for wildlife escape structures, to prevent entrapment in the road right-of-way, is an important consideration. Escape ramps, also referred to as jump-outs, allow mountain sheep to exit the right-of-way while preventing them from jumping into the right-of-way (Gagnon et al. 2017, 2020). Wildlife or cattle guards also are an option for reducing wildlife passage, especially along roadways (Gagnon et al. 2020).

Whether to design and construct an overpass or underpass will depend on the type of species targeted for reducing conflicts, especially in instances of wildlife-vehicle collisions. For instance, longer and broader sight distances and open drainages may be required for desert bighorn sheep to use a designed wildlife crossing (Gagnon et al. 2017). Designed underpasses have been successful on some sites at mitigating roadway conflicts for several wildlife species but show variable use and success for mountain sheep (Bristow and Crabb 2008). Therefore, at least for mountain sheep, properly designed overpasses may receive greater use (McKinney and Smith 2007) and may reduce collisions substantially (Gagnon et al. 2022).

The efficacy of a wildlife crossing structure will be influenced by the type of habitat where the structure is built and the habitat needs of individual species. For example, desert bighorn sheep generally inhabit more open spaces and rely heavily on eyesight, making overpasses more suitable structures for reducing roadway conflicts. On the other hand, Rocky Mountain bighorn sheep regularly travel through forests and cover, possibly reducing natural avoidance behaviors just enough to support at least intermittent use of an underpass.

**FIGURE 22.3**    A highway overpass constructed over US Highway 93 in Arizona to facilitate desert bighorn sheep crossing and reduce roadway collisions. (Photo by G. Andrejko; used with permission.)

*Sensory.* Most mitigating approaches used to avoid vehicle collisions with mountain sheep were developed primarily for other species like deer, elk, and moose (*Alces alces*). These approaches can be broadly classified into three categories: options that influence wildlife behavior, options that influence motorist behavior, and options that influence both motorist and wildlife behavior simultaneously. To influence wildlife behavior, tools can be designed to affect an animal's sensory system (olfactory, auditory, and visual). Sensory tools cause wildlife to avoid or lure them away from the road. Attempts to modify ungulate behavior through visual senses have included enhancing the perceived threat of oncoming cars by using flashing lights, lasers, and installing roadside reflectors or fence tags, which have mixed or inconclusive results (Waring et al. 1991, Reeve and Anderson 1993, Ujvári et al. 1998, D'Angelo et al. 2006, Vercauteren et al. 2006, Blackwell and Seamans 2009, Brieger et al. 2016, Gagnon et al. 2023). Auditory signals, such as sirens or horns, also have been used to alert wildlife of danger, but with variable efficacy (Westerfield et al. 2019, Wakeling et al. 2023).

To reduce wildlife-vehicle collisions, several tools or approaches have been developed to influence human behavior such as decoy deterrents, nighttime or seasonal speed limits, and static or dynamic signage, including animal-activated detection systems. Decoy deterrents are intended to make motorists react to the visual cue of seeing a simulated wildlife species and respond by slowing down but have received minimal use and may pose a potential public safety liability (Wakeling et al. 2015). Adjusting speed limits in areas with greater wildlife activity is common and can be easy to implement, particularly where highway managers can associate a direct risk to public safety.

Nighttime and seasonal speed limit reductions are used in some jurisdictions to reduce motorist speed during periods of greater wildlife activity (e.g., dawn, dusk), giving drivers more time to identify and react to potential wildlife encounters. Research indicates mixed results on lower posted speed limits, and it may not be an effective option to reduce mountain sheep-vehicle collisions (Bissonette and Kassar 2008, Meisingset et al. 2014, Riginos et al. 2022). Like speed limit signs, motorist warning signs are designed to reduce motorist speed or increase alertness of drivers to potential hazards with wildlife along the roadway (Figure 22.4). Signage designs range from static text or silhouette signs to advanced animal-activated detection systems. Permanent signs are generally ineffective for long-term mitigation of wildlife-vehicle collisions because motorists largely ignore them (Wakeling et al. 2015). Temporary and seasonal signs appear to be more effective than permanent signs because signs are in place for a shorter time, the novelty of the signs apparently increasing the likelihood for motorists to note and react to the new signage. Variable or dynamic message signs appear to be most effective at garnering motorist attention, in turn reducing collisions. For example, studies in Virginia and Montana noted a reduction in driver speed when using variable message signs (Hardy et al. 2006, Stanley et al. 2006, Donaldson and Kweon 2019).

The most advanced warning devices are animal-activated detection systems, which can include flashing or dynamic signs and may be the most effective at reducing motorist speed because there is limited opportunity for motorists to become accustomed to and ignore them. Animal-activated detection systems are activated when animals are present; animals can be detected by radar, light detection and ranging, thermal, visible light cameras, buried cables, and

**FIGURE 22.4**   Highway crossing signs like this one in South Dakota are not generally effective at alerting drivers to the risk of collision with mountain sheep because they are static and unchanging. (Photo by M. Rippentrop; used with permission.)

break-beam technologies (Huijser et al. 2005, 2009, 2012; Druta and Alden 2015, Grace et al. 2015, 2017; Huijser et al. 2017, Gagnon et al. 2019). Many of the current detection technologies have a limited range but they can be linked with additional sensors to increase coverage distance. Motorist warning devices are relatively inexpensive to deploy when compared to wildlife crossings. More sophisticated warning devices (e.g., animal-activated detection systems) are expensive and can require greater maintenance. Animal-activated detection systems may be a reasonable option to consider because they are well suited for low to moderate traffic volumes with lower speeds, especially in areas where mountain sheep-vehicle collisions predominately occur.

*Harassment.* Harassment is an approach to reduce conflict by discouraging wildlife from using specific areas. The types of harassment, often referred to as frightening devices, used as a deterrent are variable but include noise makers (e.g., gunshots, cannons, playbacks, fireworks) and visual deterrents. The goal of frightening devices is to disperse wildlife from an area to reduce conflict. Many wildlife species become habituated to loud noises and may not be deterred unless there is an escalated threat that follows the noise (Westerfield et al. 2019), so these devices should be used conservatively to maintain their efficacy.

*Guard dogs.* Guard dogs (also known as guardian dogs when used in livestock settings) are a specialized type of harassment technique and may be effective in frightening mountain sheep from an area. The movement of guard dogs should be restricted or fenced to focus their deterrence efforts (Figure 22.5). Sometimes a single dog may not be capable of effectively patrolling a large perimeter,

and substantial training may be necessary to teach the dog how to protect specific resources (Westerfield et al. 2019). However, documented effectiveness in mountain sheep systems is limited and largely anecdotal (producer reports, case examples); outcomes appear highly dependent on fencing, pasture size, vegetative structure, dog training, and handler presence. Use can also create new risks (e.g., stress-chase responses, non-target wildlife disturbance, liability-ordinance conflicts). Accordingly, this practice remains experimental and warrants structured field evaluation before being recommended broadly.

*Supplemental feeding.* Supplemental feeding or diversionary feeding to draw mountain sheep away from areas where they are not wanted (Inslerman et al. 2006, Sorensen et al. 2014) is not a good practice because feeding generally exacerbates existing problems or creates new ones (Westerfield et al. 2019). Increasing access to anthropogenic foods may attract more wildlife to an area possibly creating an overabundance situation and attendant conflicts (DeNicola et al. 2000). Concentrating mountain sheep at a single feeding area can increase the risk of disease transmission, disrupt migration, and damage habitat, which may have longer term consequences than reduction of infrequent conflicts. Placing salt or food sources to supplement, intercept, or divert wildlife may be unlawful within some jurisdictions (e.g., Montana).

The influence of road salt as an attractant for wildlife is most studied for moose (Fraser and Thomas 1982, Grosman et al. 2011). Avoiding the use of road salt as a deicing agent will reduce the desire of wildlife to use roadways for acquiring salt. Despite this known attraction for wildlife, deicing

**FIGURE 22.5** Livestock protection dogs may be trained to reduce commingling between mountain sheep and domestic sheep, although factors including size of pasture, fencing, and vegetative associations may influence their efficacy. (Photo by Montana State University, used with permission.)

is essential on roads during winter, and salt is still the primary deicing agent. Other options to reduce wildlife reliance on salt near roads is to place supplemental salt away from the road (Grosman et al. 2009; but see other potential conflicts above), removing salt pools that form following ice melting (Grosman et al. 2011), and adding aversive olfactory or taste components to road salt (Fraser and Hristienko 1982, Brown et al. 2000).

*Capture and relocation.* Capture and relocation are options that may be used to reduce the number and density of animals in a specific area, or to remove specific individuals involved in unwanted behavior from the area. There are a variety of methods used to catch mountain sheep like drop nets, chemical immobilization, and shoulder-fired net guns from the ground or air (Krausman et al. 1984, 1985), but the method of choice will depend on the situation (e.g., individual behavior, topographic features, proximity to humans and developments; Figure 22.6). It is important to assess the release site for the relocated individuals to ensure that they are not spreading disease, being introduced into unsuitable habitat or unnatural and unsafe conditions, and increasing the likelihood for conflict in the new location. The success of a relocation effort is largely dependent on release site attributes and local populations of sheep (Robinson et al. 2019). Trap and translocation efforts may only be a short-term solution to the conflict if mountain sheep populations quickly reestablish. Although this method can be successful, translocation can result in post-release mortality or predispose animals to vehicle collisions near the release site (Beringer et al. 2002, Massei et al. 2010).

*Lethal removal.* Lethal removal of mountain sheep is another method of addressing sources of conflict, yet this approach has a substantial likelihood for creating new conflict among stakeholders who do not support this type of conflict resolution (Decker et al. 2019). Nevertheless although, reducing wildlife numbers generally reduces wildlife-vehicle collisions in specific situations (Schwabe et al. 2002, Muller et al. 2014). There are multiple lethal removal techniques, including regulated hunting or the use of qualified sharpshooters. Wildlife agencies can create hunts to increase harvest in certain areas if populations are creating conflict, but population reductions may be unpopular among the hunting public and hunts within some areas may not be possible (e.g., mines or municipalities). Sharpshooting is a specific type of lethal removal that uses trained personnel to systematically remove wildlife (Stradtman et al. 1995). Sharpshooting can be effective in areas where other hunting options are limited. Because sharpshooting is highly controlled, its efficacy can be high if the effort is sufficient to ensure the appropriate number of animals are removed over a short period of time and there is access to affected properties.

Wildlife agencies can lethally remove mountain sheep, or the agency may delegate this authority to landowners or Indigenous hunters (e.g., Parks Canada, Stoney and Simpcw First Nations). Regardless of which approach is used, it is likely that there will be implications for other stakeholders or the public at large. Typically, 40%–60% of an ungulate population, predominately females, must be removed annually to curb population growth (Etter et al. 1999). Not all property owners will allow access to their lands to facilitate

**FIGURE 22.6** Capture and relocation may be used to reduce the abundance of mountain sheep in some areas, but the method must be tailored to the location and the available financial and human resources. Attempts to reduce abundance through removal may only be effective temporarily because population numbers may increase following any management action. (Photo by G. Andrejko; used with permission.)

lethal removal and some may harbor concerns that hunters may not respect rules or stipulations for using private land. There are real and perceived safety concerns when hunting near roadways and areas of concentrated human occupation, so these concerns need to be incorporated into decisions about using lethal means to remove mountain sheep.

*Ordinances and bylaws.* In some jurisdictions, local ordinances or bylaws can be useful in addressing particular challenges in a specific context, where statutory authority at the local level has been established by federal, state, provincial, or territorial legislation. Ordinances can be tailored to address local situations and may be better supported than statewide statutory regulation. For example, artificial feeding of wildlife may not be prohibited by a state authority, but community-supported ordinances to restrict feeding may resolve issues (Westerfield et al. 2019).

In another example from British Columbia, the Regional District of Okanagan-Similkameen established a local bylaw prescribing specific actions within its boundaries related to the keeping of domestic sheep as part of a separation strategy for mountain and domestic sheep. This was to protect local mountain sheep herds from being in close contact with domestic sheep and potentially becoming infected with respiratory pathogens that could result in die-off events. That bylaw was later rescinded and never implemented since the provincial Ministry of Agriculture (the statutory authority) did not establish authority for that type of action to occur, making the bylaw non-compliant with the Provincial Farm Practices Protection Act (Porter 2014).

A third example has been implemented by the Yukon Territory, established under the Animal Health Act, protecting wild sheep and goat populations from respiratory diseases found in domestic sheep and goats. This legislation, known as the Sheep and Goat Control Order, requires that all owners or transporters of domestic sheep or goats in the Yukon have animals in their care tested for respiratory pathogens, with animals testing positive for listed pathogens being destroyed or immediately removed from the Territory, and keeping all domestic sheep and goats below 1,000 m elevation (i.e., below wild sheep habitat). The Northwest Territories, Canada also has legislation prescribing geographic exclusion zones for domestic sheep and goats.

*Fertility control.* Fertility control has been proposed to control wildlife numbers for some species, especially when conflicts result because of high population density (Westerfield et al. 2019). The efficacy of contraceptives is limited because they require repeated, annual doses and a large proportion (85%–90%) of the population must be treated to be effective (Boulanger et al. 2012), which is impractical in free-ranging conditions. Two contraceptives have been approved for deer (Wakeling et al. 2023), but none are currently available for mountain sheep. The need for approved contraceptives for mountain sheep currently remains to be seen.

Surgical procedures exist for permanent sterilization of females, but implementation is expensive ($1,500/animal) and labor intensive (Wakeling et al. 2023). Male sterilization is generally recommended to occur in conjunction with control of females (Locke et al. 2007, Miller et al. 2008,

Gionfriddo et al. 2009, 2011). Even when fertility control is applied effectively, population reductions may not be observed for 5–10 years. The use of fertility control can increase the lifespan of individuals of certain species such as deer, further hampering efficacy (Swihart and DeNicola 1995, Warren et al. 1995, Boulanger et al. 2012). Because of these limitations, fertility control is not currently an effective means to reduce overabundant populations and has yet to be used for wild populations of mountain sheep.

*Mitigating disease and parasite transmission.* Many of the tools listed above can be used to reduce contact among animals (direct transmission) or shared space (indirect transmission), and thereby decrease likelihood of pathogen transmission (Chapter 8). Maintaining physical separation of individuals, herds, or populations using fences, guard dogs, geographic separation, and removal of known or suspected individuals with infection is the primary means for mitigating pathogen spread. Given the complexity related to disease transmission and associated conflicts, however, this section integrates tools with context-dependent information to highlight methods to prevent, or reduce, disease transmission.

Efforts to prevent disease transmission have largely focused on preventing contact to reduce the risk of transmission between domestic sheep and goats and mountain sheep (Wild Sheep Working Group 2012). Most disease prevention efforts try to maintain physical and temporal separation between mountain and domestic sheep through distance and activity time. Areas at risk for wild and domestic sheep contact have typically been identified by simple spatial overlap of the species, proximity, or through the use of the risk-of-contact tool (O'Brien et al. 2021). Other models of risk have considered more nuanced behavior and habitat use (Anderson et al. 2022). Within these areas of identified risk, managers can employ a range of tools including the prioritized closure of high-risk domestic sheep grazing allotments, voluntary range vacancy, seasonal use of specific areas, or the conversion of grazing allotments from domestic sheep or goats to a livestock species of lower transmission risk (e.g. cattle).

Maintaining physical separation through double-fencing, use of livestock guardian and herding dogs, attentive herding, and trucking domestic sheep between ranges instead of physical herding between seasonal ranges are alternative strategies where geographic distancing or seasonal timing cannot be achieved or relied upon. Lethal removal, as described above, is another strategy for stray domestic livestock or mountain sheep that are known or believed to have contacted domestic sheep or goats. Electronic livestock tags and geofencing could offer additional control over the movement and tracking of domestic sheep on large landscapes, helping to prevent or to identify when domestic sheep may have a greater probability of contacting wild sheep or occupying overlapping ranges. Manipulating distribution, abundance, age, and sex structure of mountain sheep populations may reduce foray behaviors and the risk

of contact with domestic sheep. The reported relationship between herd size and epizootic risk (Monello et al. 2001, Sells et al. 2015) has assisted some state decisions to reduce mountain sheep population size to limit the risk of contracting disease or resulting in an outbreak.

Other approaches may reduce the consequences of commingling between mountain and domestic sheep. Efforts to create domestic flocks free of *M. ovi* using test-and-remove (removal of domestic sheep that test positive for *M. ovi*) or the early separation of domestic lambs from their dams have only been feasible since the discovery of the primary disease agent (Besser et al. 2012). High prevalence of *M. ovi* infection in domestic sheep, high levels of biosecurity needed to maintain freedom from infection, and the ease with which domestic sheep can become reinfected make these practices challenging. Vaccinations or treatments to reduce infection are being actively pursued (Besser et al. 2019, Manlove et al. 2019), yet disease management options for mountain sheep remain limited and do not appear practical or feasible for wild populations. Nevertheless, some treatments in captive mountain sheep may provide insight for implementation in free-ranging situations at some point in the future (Wood et al. 2023). While new tools may hold promise for treatment or preventing *M. ovi* spillover, separation provides ancillary benefits such as preventing interbreeding and competition for forage and range resources.

While litigation is sometimes used to clarify legal boundaries, it tends to be divisive and generally heightens conflict, underscoring the need for collaborative approaches. Intensive litigation has defined winners and losers, caused increased polarization, and detracted from the important need to work together to find mutually beneficial, long-term solutions. Finding ways to build trust and collaboration is paramount for success, as are tools to offset financial and logistical burdens associated with resolution actions. One example is cooperatively negotiated, voluntary agreements to remove domestic sheep from high-risk allotments in exchange for funds, which are often offered by mountain sheep interest groups. Other solutions include economic incentives to producers through the branding and marketing of mountain sheep friendly wool, double-fencing (on small scales), and experimental attempts to clear *M. ovi* from small domestic flocks using repeated antibiotic nasal washes.

## 22.7 SUMMARY

Conflict occurs when two or more parties with intensely held opinions strongly disagree on conservation objectives and one party is perceived to assert its interests at the expense of another. Conflict among humans may involve stakeholders who had no prior concern regarding mountain sheep management until an objective or tool used to meet that objective initiated controversy. Resolving conflicts requires stakeholder engagement, and wildlife managers will make better progress resolving these issues when they take time to listen and understand the perspectives of those affected by mountain sheep. Mountain sheep may serve as the catalyst for conflict in many of the same ways that other species do, yet the two scenarios with the greatest financial implications are when they are involved in roadway collisions and when perceived risk of respiratory disease transmission between domestic and mountain sheep becomes unacceptable.

Managers should actively engage stakeholders using tools that have demonstrated efficacy, such as structured decision-making or other decision-making processes, to ensure that concerns are heard and broadly understood among stakeholders, and acceptable solutions are crafted to proactively address potential conflicts. Many processes routinely used by agency decision-makers, such as structured decision-making, are perceived as being too academic or insufficiently transparent. Although this perception may be fading among some, many stakeholders remain skeptical and may prefer to use old-fashioned political influence to have policy preferences prevail, rather than using civic engagement. Wildlife managers need to be realistic about the challenges ahead. Wildlife management agencies have continued to progress when considering affected stakeholders, but each situation merits careful consideration and inclusion. Agencies that hold the appropriate decision-making authority must be involved in the process as well as those affected by the decisions. Patience, sincere understanding, and a commitment to a fair and transparent process are critical to selecting the right approaches to prevent and mitigate conflict.

## REFERENCES

Adams, C. E., K. J. Lindsey, and S. J. Ash. 2006. *Urban Wildlife Management*. CRC Press, Boca Raton, Florida, USA.

Aiello, C. M., N. L. Galloway, K. Freatella, P. R. Prentice, N. W. Darby, and C. W. Epps. 2024. Highway underpasses offer little fragmentation relief for desert bighorn sheep near Mojave National Preserve, CA. *California Fish and Wildlife Journal* 110:e5.

Allombert, S., A. J. Gaston, and J. Martin. 2005. A natural experiment on the impact of overabundant deer on songbird populations. *Biological Conservation* 126:1–13.

American Sheep Industry Association. 2024. Bighorn sheep in domestic sheep grazing allotments. American Sheep Industry Association Policy Book. https://www.sheepusa.org/wp-content/uploads/2024/04/2024-ASI-Policy-Book.pdf. Accessed 05 Dec 2024.

Anderson, K., M. L. Cahn, T. R. Stephenson, A. P. Few, B. E. Hatfield, D. W. German, J. Weissman, and B. Croft. 2022. Cost distance models to predict contact between bighorn sheep and domestic sheep. *Wildlife Society Bulletin* 46:e1329.

Beale, M. and M. Boyce. 2020. Mine reclamation enhances habitats for wild ungulates in west-central Alberta. *Restoration Ecology* 28:828–840.

Beringer, J., L. P. Hansen, J. A. Demand, J. Sartwell, M. Wallendorf, and R. Mange. 2002. Efficacy of translocation to control urban deer in Missouri: costs, efficiency, and outcome. *Wildlife Society Bulletin* 30:767–774.

Besser, T. E., J. Levy, M. Ackerman, D. Nelson, K. Manlove, K. A. Potter, J. Busboom, and M. Benson. 2019. A pilot study of the effects of *Mycoplasma ovipneumoniae* exposure on domestic lamb growth and performance. *Plos One* 14:e0207420.

Besser, T. E., M. A. Highland, K. Baker, E. F. Cassirer, N. J. Anderson, J. M. Ramsey, K. Mansfield, D. L. Bruning, P. Wolff, J. B. Smith, and J. A. Jenks. 2012. Causes of pneumonia epizootics among bighorn sheep, western United States, 2008–2010. *Emerging Infectious Diseases* 18:406.

Bissonette, J. A., and C. Kassar. 2008. Locations of deer-vehicle collisions are unrelated to traffic volume or posted speed limit. *Human–Wildlife Conflicts* 2:122–130.

Blackwell, B. F., and T. W. Seamans. 2009. Enhancing the perceived threat of vehicle approach to deer. *The Journal of Wildlife Management* 73:128–135.

Boulanger, J. R., P. D. Curtis, E. G. Cooch, and A. J. DeNicola. 2012. Sterilization as an alternative deer control technique: a review. *Human–Wildlife Interactions* 6:article 9.

Brieger, F., R. Hagen, D. Vetter, C. F. Dormann, and I. Storch. 2016. Effectiveness of light-reflecting devices: a systematic reanalysis of animal-vehicle collision data. *Accident Analysis and Prevention* 97:242–260.

Bristow, K. D., and M. L. Crabb. 2008. Evaluation of distribution and trans-highway movement of desert bighorn sheep: Arizona Highway 68. Final Report 588, Arizona Department of Transportation, Phoenix, Arizona, USA.

Brown, T. C. 1984. The concept of value in resource allocation. *Land Economics* 60:231–246.

Brown, W. K., W. K. Hall, L. R. Linton, R. E. Huenefeld, and L. A. Shipley. 2000. Repellency of three compounds to caribou. *Wildlife Society Bulletin* 28:365–371.

Bureau of Land Management. 1992. Grazing Guidelines for Management of Domestic Sheep in Bighorn Habitats. Instruction Memorandum No. 92–264. United States Department of the Interior, Washington D.C., USA.

Bureau of Land Management. 1998. Grazing Guidelines for Management of Domestic Sheep in Bighorn Habitats, Instruction Memorandum No. 98–140. United States Department of the Interior, Washington, D.C., USA.

Callan, R. J., T. D. Bunch, G. W. Workman, and R. E. Mock. 1991. Development of pneumonia in desert bighorn sheep after exposure to a flock of exotic wild and domestic sheep. *Journal of the American Veterinary Medical Association* 198:1052–1056.

Carpenter, L. H. 2000. Harvest management goals. Pages 192–213 *in* S. Damarais and P. R. Krausman, editors. *Ecology and Management of Large Mammals in North America*. Prentice-Hall, Upper Saddle River, New Jersey, USA.

Cassirer, E. F., K. R. Manlove, E. S. Almberg, P. L. Kamath, M. Cox, P. Wolff, A. Roug, J. Shannon, R. Robinson, R. B. Harris, B. J. Gonzalez, R. K. Plowright, P. J. Hudson, P. C. Cross, A. Dobson, and T. E. Besser. 2018. Pneumonia in bighorn sheep: risk and resilience. *Journal of Wildlife Management* 82:32–45.

Clevenger, A. P., B. Chruszcz, and K. Gunson. 2001. Drainage culverts as habitat linkages and factors affecting passage by mammals. *Journal of Applied Ecology* 38:1340–1349.

Coghlan, A. T., H. Preskill, and T. Tzavaras Catsambas. 2003. An overview of appreciative inquiry in evaluation. *New Directions for Evaluation* 2003:5–22.

Conover, M. R. 2001. *Resolving Human-Wildlife Conflicts: The Science of Wildlife Damage Management*. CRC Press, Boca Raton, Florida, USA.

Conover, M. R. 2019. Number of human fatalities, injuries, and illnesses in the United States due to wildlife. *Human–Wildlife Interactions* 13:264–276.

Conroy, M. J., and J. T. Peterson. 2013. *Decision Making in Natural Resource Management: A Structured, Adaptive Approach.* John Wiley & Sons, Oxford, United Kingdom.

Cumming, K., and D. Tavares. 2022. Using strategic environmental assessment and project environmental impact assessment to assess ecological connectivity at multiple scales in a national park context. *Impact Assessment and Project Appraisal* 40:507–516.

D'Angelo, G., J. G. D'Angelo, G. R. Gallagher, D. A. Osborn, K. V. Miller, and R. J. Warren. 2006. Evaluation of wildlife warning reflectors for altering white-tailed deer behavior along roadways. *Wildlife Society Bulletin* 34:1175–1183.

DeCalesta, D. S. 1994. Effect of white-tailed deer on songbirds within managed forests in Pennsylvania. *Journal of Wildlife Management* 58:711–718.

Decker, D. J., A. B. Forstchen, W. F. Siemer, C. A. Smith, R. K. Frohlich, M. V. Schiavone, P. E. Lederle, and E. F. Pomeranz. 2019. Moving the paradigm from stakeholders to beneficiaries in wildlife management. *Journal of Wildlife Management* 83:513–518.

Decker, D. J., C. C. Kreuger, R. A. Baer, Jr., B. A. Knuth, and M. E. Richmond. 1996. From clients to stakeholders: a philosophical shift for fish and wildlife management. *Human Dimensions of Wildlife* 1:70–82.

Decker, D. J., and K. G. Purdy. 1988. Toward a concept of wildlife acceptance capacity in wildlife management. *Wildlife Society Bulletin* 16:53–57.

DeNicola, A. J., K. C. VerCauteran, P. D. Curtis, and S. E. Hygnstrom. 2000. *Managing White-Tailed Deer in Suburban Environments: A Technical Guide.* Cornell Cooperative Extension Service, Ithaca, New York, USA.

Desert Bighorn Council. 1990. Guidelines for management of domestic sheep in the vicinity of desert bighorn habitat. *Desert Bighorn Council Transactions* 34:33–35.

Dodd, N. L., and J. W. Gagnon. 2011. Influence of underpasses and traffic on white-tailed deer highway permeability. *Wildlife Society Bulletin* 35:270–281.

Donaldson, B. M., and K. E. M. Elliott. 2021. Enhancing existing isolated underpasses with fencing reduces wildlife crashes and connects habitat. *Human–Wildlife Conflicts* 15:148–161.

Donaldson, B. M., and Y. J. Kweon. 2019. Effectiveness of seasonal deer advisories on changeable message signs as a deer crash reduction tool. Final Report 19-R8, Virginia Transportation Research Council. https://www.virginiadot.org/vtrc/main/online_reports/pdf/19-R8.pdf.

Druta, C., and A. S. Alden. 2015. Evaluation of a buried cable roadside animal detection system. Final Report VCTIR 15-R25. Virginia Center for Transportation Innovation and Research.

Dzialak, M. R., S. L. Webb, S. M. Harju, J. B. Winstead, J. Wondzell, J. P. Mudd, and L. D. Hayden-Wing. 2011. The spatial pattern of demographic performance as a component of sustainable landscape management and planning. *Landscape Ecology* 26:775–790.

Epps, C. W., P. J. Palsbøll, J. D. Wehausen, G. K. Roderick, R. R. Ramey, and D. R. McCullough. 2005. Highways block gene flow and cause a rapid decline in genetic diversity of desert bighorn sheep. *Ecology Letters* 8:1029–1038.

Etter, D. R., T. R. Van Deelen, R. E. Warner, and B. M. Hannon. 1999. An empirical model for predicting deer population trends in suburban Chicago, Illinois. *Proceedings of the Fifth Annual Wildlife Society Conference*, Bethesda, Maryland, USA.

Foreyt, W. J. 1990. Pneumonia in bighorn sheep: effects of *Pasteurella haemolytica* from domestic sheep and effects on survival and long-term reproduction. *Proceedings of the Biennial Symposium of the Northern Wild Sheep and Goat Council* 7:92–101.

Foreyt, W. J. 1992. Experimental contact association between bighorn sheep, elk, and deer with known *Pasteurella haemolytica* infections. *Proceedings of the Biennial Symposium of the Northern Wild Sheep and Goat Council* 8:213–218.

Foreyt, W. J. 1994. Effects of controlled contact exposure between healthy bighorn sheep and llamas, domestic goats, mountain goats, cattle, domestic sheep, or mouflon sheep. *Proceedings of the Biennial Symposium of the Northern Wild Sheep and Goat Council* 9:7–14.

Foreyt, W. J., and D. A. Jessup. 1982. Fatal pneumonia of bighorn sheep following association with domestic sheep. *Journal of Wildlife Diseases* 18:163–168.

Foreyt, W. J., R. M. Silflow, and J. E. Lagerquist. 1996. Susceptibility of Dall sheep (*Ovis dalli dalli*) to pneumonia caused by *Pasteurella haemolytica*. *Journal of Wildlife Diseases* 32:586–593.

Fraser, D., and E. R. Thomas. 1982. Moose-vehicle accidents in Ontario: relation to highway salt. *Wildlife Society Bulletin* 10:261–265.

Fraser, D., and H. Hristienko. 1982. Moose-vehicle accidents in Ontario: a repugnant solution? *Wildlife Society Bulletin* 10:266–270.

Frid, A., and L. M. Dill. 2002. Human-caused disturbance stimuli as a form of predation risk. *Conservation Ecology* 6:11.

Furnas, B. J., R. H. Landers, R. G. Paiste, and B. N. Sacks. 2020. Overabundance of black-tailed deer in urbanized coastal California. *Journal of Wildlife Management* 84:979–988.

Gagnon, J. W., C. A. Beach, T. P. McCarthy, C. D. Loberger, H. P. Nelson, and S. C. Sprague. 2023. Evaluation of right-of-way fence tags to reduce animal-vehicle collisions. Final project report 758, Arizona Department of Transportation Research Center, Phoenix, Arizona, USA.

Gagnon, J. W., C. D. Loberger, K. Ogren, C. A. Beach, H. D. Nelson, and S. C. Sprague. 2020. Evaluation of effectiveness of wildlife guards and right of way escape mechanisms for large ungulates in Arizona. Final Report SPR-729. Arizona Department of Transportation Research Center, Phoenix, Arizona, USA.

Gagnon, J. W., C. D. Loberger, K. Ogren, S. C. Sprague, S. Boe, and R. E. Schweinsburg. 2017. Evaluation of desert bighorn sheep movement along Interstate-15; Virgin River Gorge. Final Project Report Presented to Arizona Department of Transportation, Phoenix, Arizona, USA.

Gagnon, J. W., C. D. Loberger, K. S. Ogren, S. C. Sprague, S. R. Boe, and R. E. Schweinsburg. 2022. Mitigating bighorn sheep-vehicle collisions and habitat fragmentation with overpasses and adaptive mitigation. *Human–Wildlife Interactions* 16:353–372.

Gagnon, J. W., C. D. Loberger, S. C. Sprague, K. S. Ogren, S. L. Boe, and R. E. Schweinsburg. 2015. Cost-effective approach to reducing collisions with elk by fencing between existing highway structures. *Human–Wildlife Interactions* 9:248–264.

Gagnon, J. W., N. L. Dodd, S. C. Sprague, K. S. Ogren, C. D. Loberger, and R. E. Schweinsburg. 2019. Animal-activated highway crosswalk: long-term impact on elk-vehicle collisions, vehicle speeds, and motorist braking response. *Human Dimensions of Wildlife* 24:132–147.

Gagnon, J. W., T. C. Theimer, N. L. Dodd, A. L. Manzo, and R. E. Schweinsburg. 2007b. Effects of traffic on elk use of wildlife underpasses in Arizona. *Journal of Wildlife Management* 71:2324–2328.

Gagnon, J. W., T. C. Theimer, N. L. Dodd, S. R. Boe, and R. E. Schweinsburg. 2007a. Traffic volume alters elk distribution and highway crossings in Arizona. *Journal of Wildlife Management* 71:2318–2323.

Gionfriddo, J. P., A. J. DeNicola, L. A. Miller, and K. A. Fagerstone. 2011. Efficacy of GnRH immunocontraception of white-tailed deer in New Jersey. *Wildlife Society Bulletin* 35:142–148.

Gionfriddo, J. P., J. D. Eisemann, K. J. Sullivan, R. S. Healy, L. A. Miller, K. A. Fagerstone, R. M. Engeman, and C. A. Yoder. 2009. Field test of a gonadotrophin-releasing hormone immunocontraceptive vaccine in female white-tailed deer. *Wildlife Research* 36:177–184.

Goodson, N. J. 1982. Effects of domestic sheep grazing on bighorn sheep populations: a review. *Biennial Symposium of the Northern Wild Sheep and Goat Council* 3:287–313.

Grace, M. K., D. J. Smith, and R. F. Noss. 2015. Testing alternative designs for a roadside animal detection system using a driving simulator. *Nature Conservation* 11:61–77.

Grace, M. K., D. J. Smith, and R. F. Noss. 2017. Reducing the threat of wildlife-vehicle collisions during peak tourism periods using a Roadside Animal Detection System. *Accident Analysis & Prevention* 109:55–61.

Gregory, R., G. Long, M. Colligan, J. G. Geiger, and M. Laser. 2012b. When experts disagree (and better science won't help much): using structured deliberations to support endangered species recovery planning. *Journal of Environmental Management* 105:30–43.

Gregory, R., L. Failing, M. Harstone, G. Long, T. McDaniels, and D. Ohlson. 2012a. *Structured Decision Making: A Practical Guide for Environmental Management Choices.* Wiley-Blackwell, West Sussex.

Grilo, C., J. A. Bissonette, and M. Santos-Reis. 2008. Response of carnivores to existing highway culverts and underpasses: implications for road planning and mitigation. *Biodiversity and Conservation* 17:1685–1699.

Grosman, P. D., J. A. G. Jaeger, P. M. Biron, C. Dussault, and J.-P. Ouellet. 2009. Reducing moose-vehicle collisions through salt pool removal and displacement: an agent-based modeling approach. *Ecology and Society* 14:17.

Grosman, P. D., J. A. G. Jaeger, P. M. Biron, C. Dussault, and J.-P. Ouellet. 2011. Trade-off between road avoidance and attraction by roadside salt pools in moose: an agent-based model to assess measures for reducing moose-vehicle collisions. *Ecological Modelling* 222:1423–1435.

Groves, C. R., and E. T. Game. 2016. *Conservation Planning: Informed Decisions for a Healthier Planet.* Roberts and Company Publishers, Greenwood Village, Colorado, USA.

Hammond, J. S., R. L. Keeney, and H. Raiffa. 1999. *Smart Choices: A Practical Guide to Making Better Life Decisions.* Broadway Books, New York, New York, USA.

Hardy, A., S. Lee, and A. Al-Kaisy. 2006. Effectiveness of animal advisory messages on dynamic message signs as a speed reduction tool: case study in rural Montana. *Transportation Research Record: Journal of the Transportation Research Board* 1973:64–72.

Hawkins, T. O. 2017. A case study analysis of collaborative conservation: restoring bighorn sheep to the Santa Catalina Mountains. Thesis, University of Arizona, Tucson, Arizona, USA.

Hendee, J. C., J. H. Stankey, and R. C. Lucas. 1978. *Wilderness Management.* USDA Forest Service Miscellaneous Publication 1365. USDA Forest Service, Washington, D.C., USA.

Holtrop, J. D. 2011. *Bighorn Sheep Analysis for NEPA Documents.* U.S. Forest Service, Washington D.C., USA.

Huijser, M. P., and A. P. Clevenger. 2013. Review of proposed bighorn sheep mitigation measures along Montana Hwy 200, East of Thompson Falls, Montana. Western Transportation Institute, College of Engineering, Montana State University, Bozeman, Montana, USA.

Huijser, M. P., C. Haas, and K. R. Crooks. 2012. The reliability and effectiveness of an electromagnetic animal detection and driver warning system. Final Report No. CDOT-2012-2. Colorado Department of Transportation Applied Research and Innovation Branch.

Huijser, M. P., E. R. Fairbanks, and F. D. Abra. 2017. The reliability and effectiveness of a radar-based animal detection system. RP 247 - Idaho Department of Transportation Research Program.

Huijser, M. P., P. McGowen, J. Fuller, A. Hardy, A. Kociolek, A. P. Clevenger, D. Smith, and R. Ament. 2008. Wildlife-vehicle collision reduction study. Report to Congress. U.S. Department of Transportation, Federal Highway Administration, Washington, D.C., USA.

Huijser, M. P., T. D. Holland, B. Matt, M. C. Greenwood, P. T. McGowen, B. Hubbard, and S. Wang. 2009. The comparison of animal detection systems in a test-bed: a quantitative comparison of system reliability and experience with operation and maintenance: final report. Prepared by Western Transportation Institute for Federal Highway Administration and Montana Department of Transportation.

Huijser, M. P., W. Camel, and A. Hardy. 2005. Reliability of the animal detection system along US Hwy 191 in Yellowstone National Park, Montana, USA. Pages 509–523 *in* C. L. Irwin, P. Garrett, and K. P. McDermott, editors. *Proceedings of the International Conference on Ecology and Transportation.* Center for Transportation and the Environment, North Carolina State University, Raleigh, North Carolina, USA.

Inslerman, R. A., J. E. Miller, D. L. Baker, J. E. Kennamer, R. Cumberland, E. R. Stinson, P. Doerr, and S. J. Williamson. 2006. Baiting and supplemental feeding of game wildlife species. Technical Review 06-01. The Wildlife Society, Bethesda, Maryland, USA.

Jex, B. A., J. B. Ayotte, V. C. Bleich, C. E. Brewer, D. L. Bruning, T. M. Hegel, N. C. Larter, R. A. Schwanke, H. M. Schwantje, and M. W. Wagner. 2016. *Thinhorn Sheep: Conservation Challenges and Management Strategies for the 21st Century.* Wild Sheep Working Group, Western Association of Fish and Wildlife Agencies, Cheyenne, Wyoming, USA.

Krahulec, F., H. Skálová, T. Herben, V. Hadincová, R Wildová, and S. Pecháčková. 2001. Vegetation changes following sheep grazing in abandoned mountain meadows. *Applied Vegetation Science* 4:97–102.

Krausman, P. R. 2017. *And Then There Were None: The Demise of Desert Bighorn Sheep in the Pusch Ridge Wilderness.* University of New Mexico Press, Albuquerque, New Mexico, USA.

Krausman, P. R., J. J. Hervert, and L. L. Ordway. 1985. Capturing deer and mountain sheep with a net-gun. *Wildlife Society Bulletin* 13:71–73.

Krausman, P. R., J. J. Hervert, L. L. Ordway, K. Rautenstrauch, and R. Remington. 1984. Immobilization of desert mule deer with etorphine plus azaperone. Pages 103–105 *in* P. R. Krausman and N. Smith, editors. *Deer in the Southwest: A Workshop*. School of Renewable Natural Resources, University of Arizona, Tucson, Arizona, USA.

Krausman, P. R., S. A. Christensen, J. E. McDonald, and B. D. Leopold. 2014. Dynamics and social issues of overpopulated deer ranges in the United States: a long-term assessment. *California Fish and Game* 100:436–450.

Lackey, C. W., S. W. Breck, B. F. Wakeling, and B. White. 2018. Human–black bear conflicts: a review of common management practices. *Human–Wildlife Interactions Monograph* 2:1–68.

Locke, S. L., M. W. Cook, L. A. Haverson, D. S. Davis, R. R. Lopez, N. J. Silvy, and M. A. Fraker. 2007. Effectiveness of Spayvac for reducing white-tailed deer fertility. *Journal of Wildlife Diseases* 4:726–730.

Lowery, C., and K. M. Longshore. 2017. Tolerance to disturbance regulated by attractiveness of resources: a case study of desert bighorn sheep within the River Mountains, Nevada. *Western North American Naturalist* 77:82–98.

Manfredo, M. J., L. Sullivan, A. W. Don Carlos, A. M. Dietsch, T. L. Teel, A. D. Bright, and J. Bruskotter. 2018. America's wildlife values: the social context of wildlife management in the U. S. National report from the research project entitled "America's Wildlife Values." Colorado State University, Department of Human Dimensions of Natural Resources, Fort Collins, Colorado, USA.

Manfredo, M. J., T. L. Teel, L. Sullivan, and A. M. Dietsch. 2017. Values, trust, and cultural backlash in conservation governance: the case of wildlife management in the United States. *Biological Conservation* 214:303–311.

Mangel, M., L. M. Talbot, M. Meffe, G. K. Agardy, M. Tundi, D. L. Alverson, J. Barlow, D. B. Botkin, G. Budowski, T. Clark, et al. 1996. Principles for the conservation of wild living resources. *Ecological Applications* 6:338–362.

Manlove, K., M. Branan, K. Baker, D. Bradway, E. F. Cassirer, K. L. Marshall, R. S. Miller, S. Sweeney, P. C. Cross, and T. E. Besser. 2019. Risk factors and productivity losses associated with *Mycoplasma ovipneumoniae* infection in United States domestic sheep operations. *Preventive Veterinary Medicine* 168:30–38.

Massei, G., R. Quy, J. Gurney, and D. Cowan. 2010. Can translocations be used to mitigate human-wildlife conflict? *CSIRO Wildlife Research* 37:428–439.

McKinney, T., and T. Smith. 2007. US93 bighorn sheep study: distribution and trans-highway movements of desert bighorn sheep in northwestern Arizona. FHWA-AZ-07-576. Arizona Department of Transportation, Phoenix, Arizona, USA.

Meisingset, E. L., L. E. Loe, Ø. Brekkum, and A. Mysterud. 2014. Targeting mitigation efforts: the role of speed limit and road edge clearance for deer–vehicle collisions. *The Journal of Wildlife Management* 78:679–688.

Merriam-Webster Dictionary. 2024. https://www.merriam-webster.com/dictionary/. Accessed on 6 December 2024.

Miller, L. A., J. P. Gionnfriddo, J. C. Rhyan, K. A. Fagerstone, D. C. Wagner, and G. J. Killian. 2008. GnRH contraception of male and female white-tailed deer fawns. *Human-Wildlife Conflicts* 2:93–101.

Monello, R. J., D. L. Murray, and E. F. Cassirer. 2001. Ecological correlates of pneumonia epizootics in bighorn sheep herds. *Canadian Journal of Zoology* 79:1423–1432.

Muller, L. I., A. M. Hackworth, N. R. Giffen, J. W. Evans, J. Henning, G. J. Hickling, and P. Allen. 2014. Spatial and temporal relationships between deer harvest and deer–vehicle collisions at Oak Ridge Reservation, Tennessee. *Wildlife Society Bulletin* 38:812–820.

O'Brien, J., A. Titolo, P. Cross, F. Quamen, and M. Woolever. 2021. Bighorn sheep risk of contact tool. United States Geological Survey, Northern Rocky Mountain Science Center (NOROCK) Headquarters, Bozeman, Montana, USA.

Onderka, D. K. and W. D. Wishart. 1988. Experimental contact transmission of *Pasteurella haemolytica* from clinically normal domestic sheep causing pneumonia in Rocky Mountain bighorn sheep. *Journal of Wildlife Diseases* 24:663–667.

Organ, J. E., and M. R. Ellingwood. 2000. Wildlife stakeholder acceptance capacity for black bears, beavers, and other bests in the east. *Human Dimensions of Wildlife* 5:63–75.

Poole, K. G., R. Serrouya, I. E. Teske, and K. Podrasky. 2016. Rocky Mountain bighorn sheep (*Ovis canadensis canadensis*) winter habitat selection and seasonal movements in an area of active coal mining. *Canadian Journal of Zoology* 94:733–745.

Porter, H. 2014. Bighorn sheep herd die-offs in British Columbia: the need for a provincial wild/domestic sheep separation strategy. Environmental Law Center, University of Victoria, Victoria, British Columbia, Canada.

Raik, D., D. Decker, and W. Siemer. 2003. Dimensions of capacity in community-based suburban deer management: the managers' perspective. *Wildlife Society Bulletin* 31:854–864.

Redpath, S. M., J. Young, A. Evely, W. M. Adams, W. J. Sutherland, A. Whitehouse, A. Amar, R. A. Lambert, J. D. C. Linnell, A. Watt, and R. J. Gutiérrez. 2013. Understanding and managing conservation conflicts. *Trends in Ecology and Evolution* 28:100–109.

Reeve, A. F., and S. H. Anderson. 1993. Ineffectiveness of Swareflex reflectors at reducing deer-vehicle collisions. *Wildlife Society Bulletin* 21:127–132.

Riginos, C., E. Fairbank, E. Hansen, J. Kolek, and M. P. Huijser. 2022. Reduced speed limit is ineffective for mitigating the effects of roads on ungulates. *Conservation Science and Practice* 4:e618.

Riley, S. J., D. J. Decker, L. H. Carpenter, J. F. Organ, W. F. Siemer, G. F. Mattfeld, and G. Parsons. 2002. The essence of wildlife management. *Wildlife Society Bulletin* 30:585–593.

Riley, S. J., W. F. Siemer, D. J. Decker, L. H. Carpenter, J. F. Organ, and L. T. Berchielli. 2003. Adaptive impact management: an integrative approach to wildlife management. *Human Dimensions of Wildlife* 8:81–95.

Robinson, R. W., J. C. Whiting, J. M. Shannon, D. D. Olson, J. T. Flinders, T. S. Smith, and R. T. Bowyer. 2019. Habitat use and social mixing between groups of resident and augmented bighorn sheep. *Scientific Reports* 9:14984.

Runge, M. C., and E. A. Bean. 2020. Decision analysis for managing public natural resources. Pages 3–11 *in* M. C. Runge, S. J. Converse, J. E. Lyons, and D. R. Smith, editors. *Structured Decision Making: Case Studies in Natural Resource Management*. Johns Hopkins University Press, Baltimore, Maryland, USA.

Runge, M. C., J. B. Grand, and M. S. Mitchell. 2013. Structured decision making. Pages 51–72 *in* P. R. Krausman and J. W. Cain, editors. *Wildlife Management and Conservation: Contemporary Principles and Practices*. Johns Hopkins University Press, Baltimore, Maryland, USA.

Schommer, T. 2002. Bighorn Sheep in Hells Canyon: Historical Background and the Hells Canyon Bighorn Sheep Restoration Project. U.S. Forest Service, Wallowa-Whitman National Forest. https://www.fs.usda.gov/detail/wallowa-whitman/landmanagement/resourcemanagement/?cid=stelprdb5287260. Accessed 04 December 2024.

Schommer, T., and M. Woolever. 2001. A process for finding management solutions to the incompatibility between domestic and bighorn sheep. US Department of Agriculture, Forest Service, Washington, D.C., USA.

Schommer, T. J., and M. M. Woolever. 2008. A review of disease related conflicts between domestic sheep and goats and bighorn sheep. US Department of Agriculture, Forest Service, Washington, D.C., USA.

Schwabe, K. A., P. W. Schuhmann, and M. Tonkovich. 2002. A dynamic exercise in reducing deer-vehicle collisions: management through vehicle mitigation techniques and hunting. *Journal of Agricultural and Resource Economics* 27:261–280.

Sells, S. N., M. S. Mitchell, J. J. Nowak, P. M. Lukacs, N. J. Anderson, J. M. Ramsey, J. A. Gude, and P. R. Krausman. 2015. Modeling risk of pneumonia epizootics in bighorn sheep. *The Journal of Wildlife Management* 79:195–210.

Sells, S. N., M. S. Mitchell, V. L. Edwards, J. A. Gude, and N. J. Anderson. 2016. Structured decision making for managing pneumonia epizootics in bighorn sheep. *Journal of Wildlife Management* 80:957–969.

Smith, C. A., J. A. Tantillo, B. Hale, D. J. Decker, A. B. Forstchen, E. F. Pomeranz, T. B. Lauber, M. V. Schiavone, K. Frohlich, P. E. Lederle, R. J. Benedict, J. Hurst, R. King, W. F. Siemer, and M. S. Baumer. 2023. A practical framework for ethics assessment in wildlife management decision-making. *Journal of Wildlife Management.* doi:10.1002/jwmg.22502

Smith, D. R. 2020. Introduction to structuring decisions. Pages 15–22 *in* M. C. Runge, S. J. Converse, J. E. Lyons, and D. R. Smith, editors. *Structured Decision Making: Case Studies in Natural Resource Management.* Johns Hopkins University Press, Baltimore, Maryland, USA.

Sorensen, A., F. M. Beest, and R. K. Brook. 2014. Impacts of wildlife baiting and supplemental feeding on infectious disease transmission risk: a synthesis of knowledge. *Preventive Veterinary Medicine* 113:356–363.

Stanley, L., A. Hardy, and S. Lassacher. 2006. Responses to enhanced wildlife advisories in a simulated environment. *Transportation Research Record: Journal of the Transportation Research Board* 1980:126–133.

Stewart, D. G., J. T. Beaver, M. L. Cooksey, C. Grantham, B. L. Pierce, R. R. Lopez, and S. L. Webb. 2025. Water gaps and standard fencing facilitate white-tailed deer movement across high fences while maintaining property fidelity. *Southwestern Naturalist* 69:1–13. doi:10.1894/0038-4909-69.2.9

Stobierski, T. 2019. How to do a cost-benefit analysis and why it's important. Harvard Business School Online, Business Insights. https://online.hbs.edu/blog/post/cost-benefit-analysis. Accessed on 6 December 2024.

Stradtman, M. L., J. B. McAninch, E. P. Wiggers, and J. M. Parker. 1995. Police sharpshooting as a method to reduce urban deer populations. Pages 117–122 *in* J. B. McAninch, editor. *Urban deer: a manageable resource? Proceedings of the Symposium: 55th Midwest Fish and Wildlife Conference, 12–14 December 1993,* North Central Section of The Wildlife Society, St. Louis, Missouri, USA.

Swihart, R. K., and A. J. DeNicola. 1995. Modeling the impacts of contraception on populations of white-tailed deer. Pages 151–163 *in* J. B. McAninch, editor. *Urban Deer: A Manageable Resource? Proceedings of the 1993 Symposium of the North Central Section.* The Wildlife Society, Bethesda, Maryland, USA.

Tilghman, N. G. 1989. Impacts of white-tailed deer on forest regeneration in northwestern Pennsylvania. *Journal of Wildlife Management* 53:524–532.

Tiwari, A., and J. W. Rachlin. 2018. A review of road salt ecological impacts. *Northeastern Naturalist* 25:123–142.

Ujvári, M., H. J. Baagøe, and A. B. Madsen. 1998. Effectiveness of wildlife warning reflectors in reducing deer-vehicle collisions: a behavioral study. *Journal of Wildlife Management* 62:1094–1099.

United States Forest Service. 2010. *Payette National Forest Record of Decision for the Final Supplemental Environmental Impact Statement and Forest Plan Amendment Identifying Suitable Rangeland for Domestic Sheep and Goat Grazing to Maintain Habitat for Viable Bighorn Sheep Populations.* USDA Forest Service, Payette National Forest, Idaho, USA.

VerCauteren, K. C., M. G. Jason, S. E. Hygnstrom, P. B. Fioranelli, J. A. Wilson, and S. Barras. 2006. Green and blue lasers are ineffective for dispersing deer at night. *Wildlife Society Bulletin* 34:371–374.

Wakeling, B. F., H. S. Najar, and J. C. Odell. 2007. Mortality of bighorn sheep along US Highway 191 in Arizona. *Desert Bighorn Council Transactions* 49:18–22.

Wakeling, B. F., J. W. Gagnon, D. Olson, D. W. Lutz, T. W. Keegan, J. Shannon, A. Holland, A. Lindbloom, and C. Schroeder. 2015. *Mule Deer and Movement Barriers.* Mule Deer Working Group, Western Association of Fish and Wildlife Agencies, Boise, Idaho, USA.

Wakeling, B. F., O. V. Duvuvuei, J. M. Shannon, A. Roug, C. Wilson, and S. J. K. Hansen. 2023. Conflict management. Pages 333–348 *in* J. R. Heffelfinger and P. R. Krausman, editors. *Ecology and Management of Black-tailed and Mule Deer of North America.* CRC Press, Boca Raton, Florida, USA.

Walters, C. J. 1986. *Adaptive Management of Renewable Resources.* Macmillan, New York, New York, USA.

Waring, G. H., J. L. Griffis, and M. E. Vaughn. 1991. White-tailed deer roadside behavior, wildlife warning reflectors, and highway mortality. *Applied Animal Behaviour Science* 29:215–223.

Warren, R. J., L. M. White, and W. R. Lance. 1995. Management of urban dee populations with contraceptives: practicality and agency concerns. Pages 164–170 *in* J. B. McAninch, editor. *Urban Deer: A Manageable Resource? Proceedings of the 1993 Symposium of the Northcentral Section.* The Wildlife Society, St. Louis, Michigan, USA.

Webb, S. L., K. L. Gee, S. Demarais, B. K. Strickland, and R. W. DeYoung. 2009. Efficacy of a 15-strand high-tensile electric fence to control white-tailed deer movements. *Wildlife Biology in Practice* 5:45–57.

Wehausen, J. D., S. T. Kelley, and R. R. Ramey. 2011. Domestic sheep, bighorn sheep, and respiratory disease: a review of the experimental evidence. *California Fish and Game* 97:7–24.

Weldon, L. A. 2012. *Bighorn Sheep Analysis for NEPA Documents.* U.S. Forest Service, Washington D.., USA.

Weldon, L. A. 2014. *Bighorn Sheep Analysis for NEPA Documents.* U.S. Forest Service, Washington D.C., USA.

Westerfield, G. D., J. M. Shannon, O. V. Duvuvuei, T. A. Decker, N. P. Snow, E. D. Shank, B. F. Wakeling, and H. B. White. 2019. Methods for managing human-deer conflicts in urban, suburban, and exurban areas. *Human–Wildlife Interactions Monograph* 3:1–99.

Wild Sheep Initiative. 2023. Pneumonia and bighorn sheep: test and remove factsheet 7–14–2023. Western Association of Fish and Wildlife Agencies, Boise, Idaho, USA.

Wild Sheep Working Group. 2007. Recommendations for Domestic Sheep and Goat Management in Wild Sheep Habitat. Western Association of Fish and Wildlife Agencies, Boise, Idaho, USA.

Wild Sheep Working Group. 2012. Recommendations for Domestic Sheep and Goat Management in Wild Sheep Habitat. Western Association of Fish and Wildlife Agencies, Boise, Idaho, USA.

Wingard, R. P., and P. R. Krausman. 2019. Hunter and public opinions of a Columbian black-tailed deer population in a Pacific Northwest island landscape. *Human–Wildlife Interactions* 13:474–488.

Wood, M. E., W. H. Edwards, J. E. Jennings-Gaines, M. Gaston, P. Van Wick, S. Amundson, S. E. Allen, and L. L. Wolfe. 2023. Clearance of *Mycoplasma ovipneumoniae* in captive bighorn sheep (*Ovis canadensis*) following extended oral doxycycline treatment. *Journal of Wildlife Diseases* 59:753–758.

Wyoming Wool Growers Association. 2022. Wyoming Wool Growers Association Policy Manual. https://www.wyo-wool.com/_files/ugd/7a82ea_c206549de7f34eb085e5b-5f6183e097a.pdf. Accessed 05 December 2024.

# Section 4

## Science-Related Management Opportunities

# 23 Long-Term Research on Individual Sheep

*Marco Festa-Bianchet and Fanie Pelletier*

## 23.1 INTRODUCTION

Long-term studies of marked individuals have made substantial contributions to ecology, evolution, and conservation. Their value increases as years of monitoring are added, because with more information new questions can be addressed (Clutton-Brock and Sheldon 2010, Festa-Bianchet et al. 2017). Many ecological and evolutionary processes take place over decades or longer and cannot be understood without long-term monitoring. For mountain sheep (bighorn [*Ovis canadensis*], thinhorn [*O. dalli*]), events happening over different life stages, but particularly early in life, can have important long-lasting consequences on growth, reproduction, and survival (Marcil-Ferland et al. 2013, Pigeon et al. 2017). Monitoring known individuals from birth to death can quantify differences in performance and identify some of the causes of those differences that could not be understood by a cross-sectional sample. For example, in bighorn sheep, 42 years of monitoring revealed that females born at high density had lower lifetime reproductive success and shorter life expectancy than those born at low density, even though those born at high density were heavier as adults (Pigeon and Pelletier 2018). Without knowledge about early-life environmental conditions, the apparent negative effect of large mass on fitness would have been puzzling. Important variables that affect individual fitness and population dynamics can vary substantially over time (e.g., local weather, density, predation, disease, vegetative cover and composition, parasites, and competitors). Long-term studies are necessary to quantify that temporal variation and measure the relative importance of different variables in affecting population dynamics over the long run (Turgeon et al. 2024). For bighorn sheep, the population dynamics influences of rare but influential events such as pneumonia epizootics and episodes of intense cougar (*Puma concolor*) predation can only be assessed with many years of monitoring (Cloutier et al. 2024, Turgeon et al. 2024). Similarly, the consequences of management strategies such as changes in hunting regulations, translocations, and habitat manipulations often require many years of monitoring to be adequately understood (Hogg et al. 2006). For example, to avoid undesirable evolutionary consequences of selective harvests, management of mountain sheep as a trophy species should direct the harvest to males aged 8 years

or older (Karabatsos et al. 2024). Yet, the number and horn size of 8-year-old males available 1 year depends on events that affect survival and growth over the previous 8 years, or even longer if maternal effects are considered. Finally, the effects of climate change on environmental variables and population ecology can only be measured through long-term monitoring.

In this chapter, we will first consider a few contributions of long-term monitoring of marked individuals to our knowledge of mountain sheep ecology, evolution, and conservation. Then we examine how analyses of long-term horn measurements of harvested males could assist management. The annual horn increments of mountain sheep provide a valuable source of long-term information on the effects of changes in environmental variables and management schemes. We briefly present four long-term studies of marked bighorn sheep, with emphasis on how and why the studies started and persisted, including funding sources and the involvement of different people and agencies. After an overview of knowledge generated by long-term studies, we will explore why long-term studies are rare, despite their acknowledged disproportionate contribution to science. Most long-term studies do not begin with a long-term vision. For example, the Ram Mountain study began as an experimental test of a hunting season on female sheep, while at Sheep River the initial objective was to test if an early return to the winter range increased lungworm infection. As long-term data accumulate, new questions can be investigated, some motivated by advances in ecological theory or techniques that did not exist when the studies were initiated. Using data from Ram Mountain, we will illustrate how short-term studies may provide unreliable estimates of the importance of ecological variables by examining the relationship between population density and lamb survival over 5–49 years. We will then present how important contributions to our knowledge of mountain sheep biology emerged by combining results from multiple long-term studies. That will be followed by a consideration of what factors facilitate or hinder long-term studies, and what their possible drawbacks are. We will conclude with an examination of advances in our understanding of mountain sheep ecology, evolution, and conservation that were brought about by combining monitoring of marked individuals with analyses of data on harvested males.

DOI: 10.1201/9781003518686-27

## 23.2 LONG-TERM PROGRAMS ON MARKED MOUNTAIN SHEEP

### 23.2.1 RAM MOUNTAIN

Ram Mountain is an isolated outcrop in west-central Alberta, Canada (Figure 23.1), about 30 km east of the main Rocky Mountain range. Research on this bighorn sheep population was initiated by W.D. Wishart, wildlife biologist with the Alberta provincial government, in 1971 to test the feasibility of a hunting season on females. At the time, there were concerns that high density would increase the risk of pneumonia epizootics, and female hunting was thought to maintain moderate population density. From 1972 to 1980, female removals simulated a hunting season to determine a sustainable female harvest rate and monitor growth and survival of orphaned lambs. In its 55th field season in 2025, this research program uses a corral trap (Figure 23.2) baited with salt where sheep are repeatedly captured from late May to late September. J.T. Jorgenson contributed fieldwork, expertise, and advice for 48 years, well past his retirement as a biologist from the Alberta government. This research was mostly financed by the Alberta government

until 1988. In 1988–1990, the study was headed by Dr. J.T. Hogg, then with the Craighead Wildlife-Wildland Institute, mostly supported by grants from the Charles Engelhard Foundation. Hogg initiated the tissue sampling program that was instrumental in assessing paternities and building a pedigree. Since 1991, the program has been financed mostly by Discovery Grants from the Natural Sciences and Engineering Research Council of Canada to M. Festa-Bianchet, F. Pelletier, and D.W. Coltman, in addition to funding from the Canada Research Chairs program, the Alberta Conservation Association, and continued logistic support by the Alberta government. Additional support was obtained from the Alberta Chapter of the Wild Sheep Foundation for transplants of bighorn sheep, the Fonds de recherche du Québec – Nature et technologies, NSERC scholarships, and various other sources. This program has trained over 40 graduate students or postdocs and 60 field assistants from eight countries.

The female removal provided a rare experimental manipulation of population density. There was no detectable effect of female removal on orphan lamb survival, but as young adults, orphaned males had smaller horns than

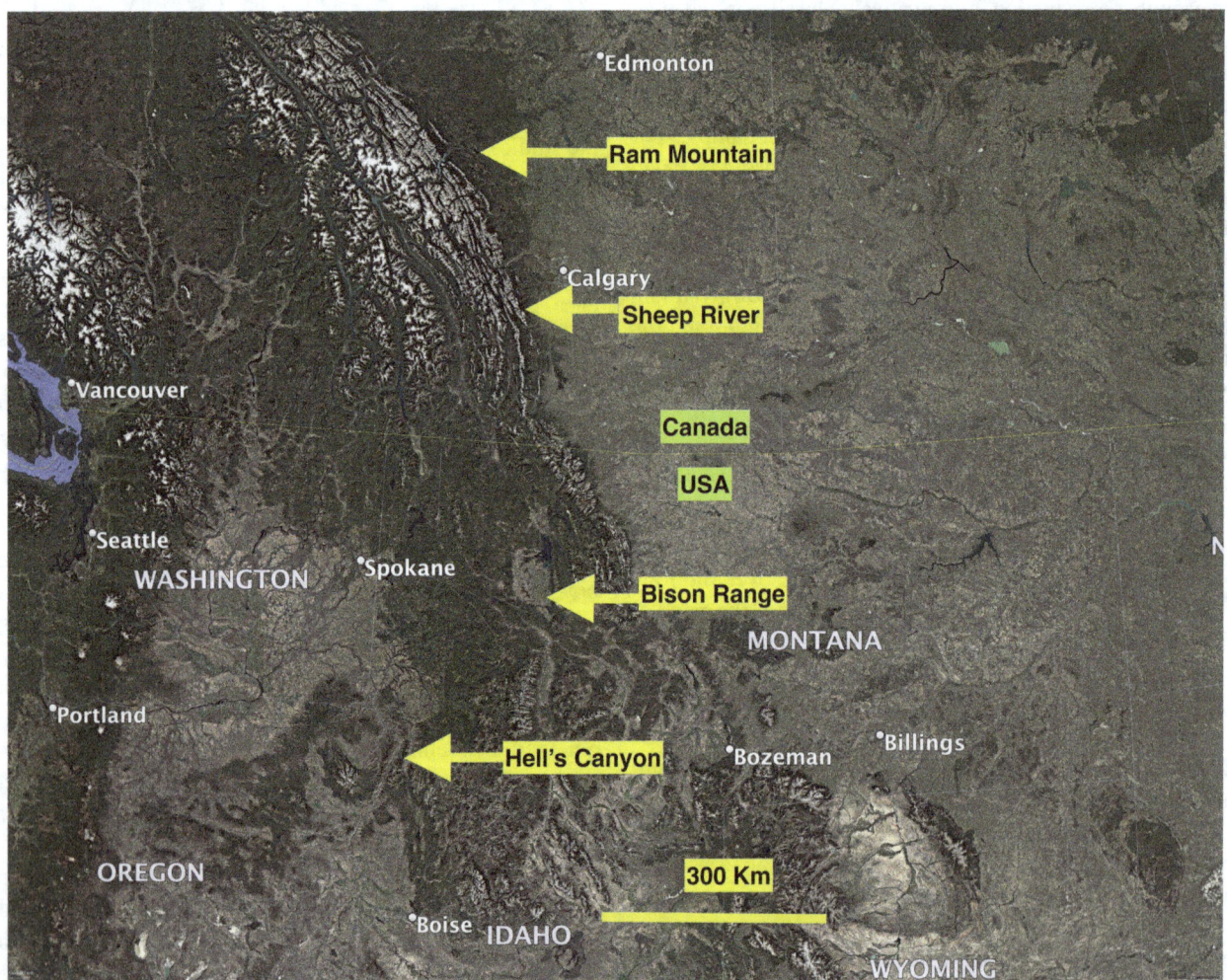

**FIGURE 23.1** Google Earth screenshot of four long-term study areas of individually marked bighorn sheep in Canada and the United States.

**FIGURE 23.2** The corral trap at Ram Mountain, Alberta, Canada. (Photo by F. Pelletier.)

males whose mother was not removed (Festa-Bianchet et al. 1994). The experiment suggested that up to 12% of females could be taken without causing a decline in population size because of compensatory recruitment (Jorgenson et al. 1993). Over the first 22 years of monitoring there was strong evidence of density dependence; as the population increased the minimum age of primiparity went from 2 to 4 years, while lamb survival and male horn growth declined (Jorgenson et al. 1998). It was the subsequent 32 years, however, that revealed the importance of long-term studies. After an 80% population decline likely due to combined density-dependent effects, additional female removals in 1997, and cougar predation, the population showed signs of inbreeding (Rioux-Paquette et al. 2011) and density dependence mostly disappeared (Turgeon et al. 2024). To counter the drastic decrease in population size and the increase in inbreeding, 35 sheep were transplanted to Ram Mountain in 2002–2015, providing an experimental test of genetic rescue, with beneficial short-term effects on reproduction and survival (Poirier et al. 2019).

Since 1975, all sheep on Ram Mountain have been marked, except for a few lambs that were caught as yearlings and very rare immigrants of both sexes. Because after 1972 most sheep were first caught as lamb, this study assembled a database on age-specific reproduction and survival, providing fundamental contributions to our knowledge of reproductive tactics of both sexes (Festa-Bianchet et al. 2019). Individual monitoring also allowed researchers to assess cohort effects (Pigeon and Pelletier 2018) and quantify the positive effects of a favorable early-life environment on subsequent fitness (Marcil-Ferland et al. 2013, Pigeon et al. 2017). The deep pedigree was used to characterize

heritability of physical and life-history traits (Poissant et al. 2008, Poissant et al. 2011) and provided evidence of evolutionary shrinking of horn size induced by intense selective hunting (Pigeon et al. 2016). The latter was assisted by experimental changes in hunting regulations. Initially, the population was subject to a trophy hunt based on a four-fifth curl definition of legal male. That definition was changed to full-curl in 1996 and the hunt was closed in 2011.

### 23.2.2 National Bison Range

Research at what was until 2020 the National Bison Range (NBR) and is now the Confederated Salish and Kootenai Tribes Bison Range, Montana, was initiated in 1979 by J.T. Hogg, then a Ph.D. student at the University of Montana, who led this research through to the present. The study was funded until 1986 by grants to the University of Montana and Montana Cooperative Wildlife Research Unit. Over four decades, the main funders have been the Charles Engelhard Foundation, the National Geographic Society, and the Eppley Foundation. From 1987 to 2001, the study was based at the Craighead Wildlife-Wildland Institute. Since 2001, the research has been based at the Montana Conservation Science Institute. All individuals in this population have been recognizable since 1979. One advantage of this study is that the open, gently sloping terrain allows researchers to locate, sample, and mark newborn lambs. Intense searches and observations during the rut and at lambing have provided valuable data on parturition date, gestation length, and birth mass (Hogg et al. 2017). The open terrain and relative ease of access are also ideal for behavioral observations year-round (Hogg 1988).

Bighorn sheep were introduced to the NBR from Banff, Alberta, in 1922. The range is fenced but sheep can move in and out. There is no sport hunting of sheep at the NBR. It was from this population that the first molecular-based data on paternity became available, at a time when paternity data for wild ungulates were only available for red deer (*Cervus elaphus*). Unlike in red deer, alternative mating tactics frequently led to fertilizations, so that mating was not monopolized by the most dominant males (Hogg 1984, Hogg and Forbes 1997). Three other contributions from this program illustrate the value of long-term monitoring of known individuals. First, an experimental genetic rescue led to strong positive individual and population-level effects. These effects could only be documented through comprehensive long-term monitoring of all individuals over multiple sheep generations, including several years before and after admixture. Following the introduction of ten sheep from other populations, outbred sheep of both sexes showed 28% increases in lifespan and doubling of reproductive success. Population growth rate also increased with the average level of individual genetic admixture (Hogg et al. 2006, Miller et al. 2012).

Second, by combining long-term data on reproductive phenology with an ongoing study of pronghorn (*Antilocapra americana*) at the NBR, this study revealed that plasticity of gestation duration is substantial in pronghorn and minimal in bighorn sheep (Hogg et al. 2017). A pneumonia epizootic in 2016 provided an unwelcome but important opportunity to document the effects of the disease on individuals and on the population (Besser et al. 2021), including a substantial negative effect on horn growth of surviving males that was also documented in other populations affected by pneumonia (Martin et al. 2022).

### 23.2.3 Sheep River

The Sheep River, Alberta, research program was initiated by W.D. Wishart. In 1978, the population experienced a pneumonia die-off (Wishart et al. 1980). At that time,

many biologists thought that lungworms (*Protostrongylus* spp.) and high population density increased the risk of pneumonia. Wishart hypothesized that an early return to the low-elevation winter range, possibly due to a population increase, led to higher lungworm infection. M. Festa-Bianchet worked on the program for the Alberta government from 1981 to 1983, then as a Ph.D. student at the University of Calgary until 1987, as a postdoctoral fellow at the University of Cambridge until 1990, and as a professor at the Université de Sherbrooke until 2010. F. Pelletier was involved as a graduate student from 2001 to 2005 and as a collaborator until 2010. The study is continued by K.E. Ruckstuhl at the University of Calgary (Wang et al. 2018), after her Ph.D. at Sheep River from 1994 to 1998. Financing was provided by the government of Alberta from 1981 to 1983, then by Natural Sciences and Engineering Research Council of Canada scholarships and fellowships and funds from the Alberta Recreation, Parks and Wildlife Foundation from 1984 to 1990. Since 1990, the study has been financed by Natural Sciences and Engineering Research Council of Canada Discovery Grants, with continued logistic support from the Alberta government. Research by J.T. Hogg on male mating behavior and breeding migrations in 1987–2009 was supported by the same founders who financed research at the NBR (Hogg and Forbes 1997).

Sheep River bighorns are part of a metapopulation, with frequent contacts with other groups of sheep and seasonal migrations between a low-elevation winter range in the foothills, easily accessible by vehicle, and a high-elevation summer range in the Rockies 10–15 km to the west (Festa-Bianchet 1986a,b). Nearly all sheep were captured using a dart gun (Jorgenson et al. 1990) and most were only caught once, usually as lambs. F. Pelletier used a platform scale baited with salt to repeatedly weigh free-ranging sheep (Figure 23.3), revealing that body mass had an increasingly important effect on social rank for males of 120 kg and more (Pelletier et al. 2006). The very open winter range at Sheep River is ideal for behavioral observations, allowing the construction of detailed time budgets (Festa-Bianchet

(a)                                                                                                          (b)

**FIGURE 23.3**  Platform scale used to weigh bighorn sheep at Sheep River. (a) The scale equipped with a salt block and two side walls to ensure that one sheep stands on the platform.(b) A ram weighing himself. The weight is read on a display in a truck parked 20 m away. (Photo by F. Pelletier.)

1991, Pelletier et al. 2005). Ability to observe sheep for up to 14 consecutive hours was key to research on sexual segregation, suggesting that sexual differences in rumination and foraging may explain why males and females form separate groups (Ruckstuhl 1998). Sheep from this population are hunted outside the protected winter range. Males are hunted under a four-fifth-curl regulation with no quota, while a few permits for females are issued on a draw.

The initial hypothesis of a relationship between late-summer use of the winter range and lungworm infection was not supported, but fecal samples from known individuals revealed that lungworm larval counts are affected by reproductive effort in both sexes (Festa-Bianchet 1991, Pelletier et al. 2005), suggesting a trade-off between allocation to reproduction and defense against parasites. Research elsewhere has largely discounted the role of lungworms in pneumonia: it is now clear that to prevent pneumonia epizootics, contact with domestic sheep must be avoided (Cassirer et al. 2018). Research at Sheep River contributed substantially to our knowledge of male mating tactics, including breeding migrations, and male reproductive success (Hogg and Forbes 1997, Hogg 2000, Pelletier and Festa-Bianchet 2004, Pelletier et al. 2006, Lassis et al. 2022). These results are important for management because they document how some males present in an area late in the hunting season likely originate from populations in protected areas, how breeding migrations decrease the risk of inbreeding, and how alternative mating tactics limit the ability of dominant males to monopolize matings. Because the study population was affected by pneumonia epizootics in 1985 and 2002, this research also documented the individual and population effects of the disease (Turgeon et al. 2024).

### 23.2.4 HELLS CANYON

Bighorn sheep were extirpated in 1945 from Hells Canyon on the Snake River (Figure 23.1). They were reintroduced beginning in 1971 (Cassirer et al. 2013). This study illustrates the complexities of successful long-term research with many collaborators and funding from a succession of multiple funders. A multi-agency, multidisciplinary restoration program began in 1997 involving the wildlife agencies of Idaho, Oregon, and Washington, the Bureau of Land Management, and the United States Forest Service, in collaboration with the Foundation for North American Wild Sheep (now the Wild Sheep Foundation), after several years of informal collaboration. This program aimed to accelerate the restoration of bighorn sheep in Hells Canyon and guide bighorn sheep restoration and management elsewhere. Reintroduced bighorn sheep in this area were affected by a series of pneumonia-related die-offs, following contact with domestic sheep. Consequently, the major research emphasis of this program has been understanding and preventing pneumonia epizootics. The program also involved buying out nearby public land allotments where domestic sheep grazed. The primary players were V. Coggins (OR), L. Oldenburg (ID), P. Fowler (WA), K. Martin and

T. Schommer (U.S. Forest Service), P. Cimellaro and D. Gilchrist (Wild Sheep Foundation) and several Chapters of the Wild Sheep Foundation. The program was initially funded by a combination of state wildlife agencies, the Wild Sheep Foundation, the Oregon Hunters' Association, the Bureau of Land Management, and the United States Forest Service. Over time, many other contributors and collaborators took part in this program, including the Shikar-Safari Club, Morris Animal Foundation, and the Smith Fellowship program. Researchers from several universities provided important contributions, including the Pennsylvania State University Center for Infectious Disease Dynamics, Washington State University Animal Disease Diagnostic Laboratory, and University of Idaho Wildlife Health Laboratory. This program has substantially advanced our knowledge of pneumonia, the main threat to bighorn sheep in nearly all their range. It identified *Mycoplasma ovipneumoniae* as the main pathogen responsible for epizootics (Besser et al. 2012). Research here revealed the presence of multiple pathogen clades in domestic sheep and goats, that can infect bighorns, with little cross-strain immunity (Cassirer et al. 2017), and identified both chronic and intermittent carriers among sheep that survived the initial die-off. These carriers continue to transmit pneumonia, especially to lambs, years after the epizootic (Plowright et al. 2017). Modeling based on this research suggested that removal of carriers may be necessary to clear pneumonia from populations (Plowright et al. 2017, Almberg et al. 2022). Initial applications of the test-and-remove strategy are encouraging (Garwood et al. 2020). Research by K.R. Manlove provided important insights into how different sex-age classes vary in potential to maintain and transmit the disease (Manlove et al. 2017) and how disease transmission is frequency-dependent rather than density-dependent (Manlove et al. 2014). Both findings are key to reducing the effects of pneumonia on bighorn sheep populations. They required long-term monitoring of individuals in different subpopulations with different disease status. That monitoring also revealed that pneumonia can reduce population growth for decades after initial exposure (Manlove et al. 2016), further emphasizing the need to prevent contact with domestic sheep and to remove disease carriers (Chapter 8).

### 23.2.5 CONTRASTING SHORT- AND LONG-TERM TIME SERIES: POPULATION DENSITY AND LAMB SURVIVAL

Consumptive wildlife management is often based upon the assumption of density dependence, where harvest is compensated by greater recruitment (Caughley and Sinclair 1994). The strength and form of the density-dependent response are therefore important, as is the identification of which vital rates are most sensitive to changes in density. There is little question that at some point a growing herbivore population will outstrip its resources and some density-dependent mechanism will slow or reverse population growth (Bonenfant et al. 2009a). Yet, many density-independent factors can act on demographic rates and reduce the chances that a population will attain the level where density dependence is expressed. These include weather and some types of disease, parasitism, and predation. Researchers analyzed time series for 217 populations of bighorn sheep, and reported that about a third showed no clear evidence of density dependence. Although negative density dependence was evident in many populations, in some the relationship was positive (Donovan et al. 2020). In ungulates, as populations increase density dependence typically first manifests itself as reduced juvenile survival (Eberhardt 2002). What can the Ram Mountain study tell us about density dependence in lamb survival to 1 year over

different time periods? Short-term studies of 5 or 10 years would have produced widely different results, from very negative to weak or even positive density dependence (Figure 23.4). Longer time series suggest a negative but weak effect of population density on lamb survival (Turgeon et al. 2024). Some of the variability in the 10-year slopes is likely real, due to changes in the importance of density dependence in this population as discussed above. Early publications from the Ram Mountain study suggested that managers could safely harvest over 12% of females because harvest would be compensated by increased recruitment (Jorgenson et al. 1993), but a longer-term view (Figure 23.4) shows the dangers of assuming strong and constant density dependence when planning female harvests.

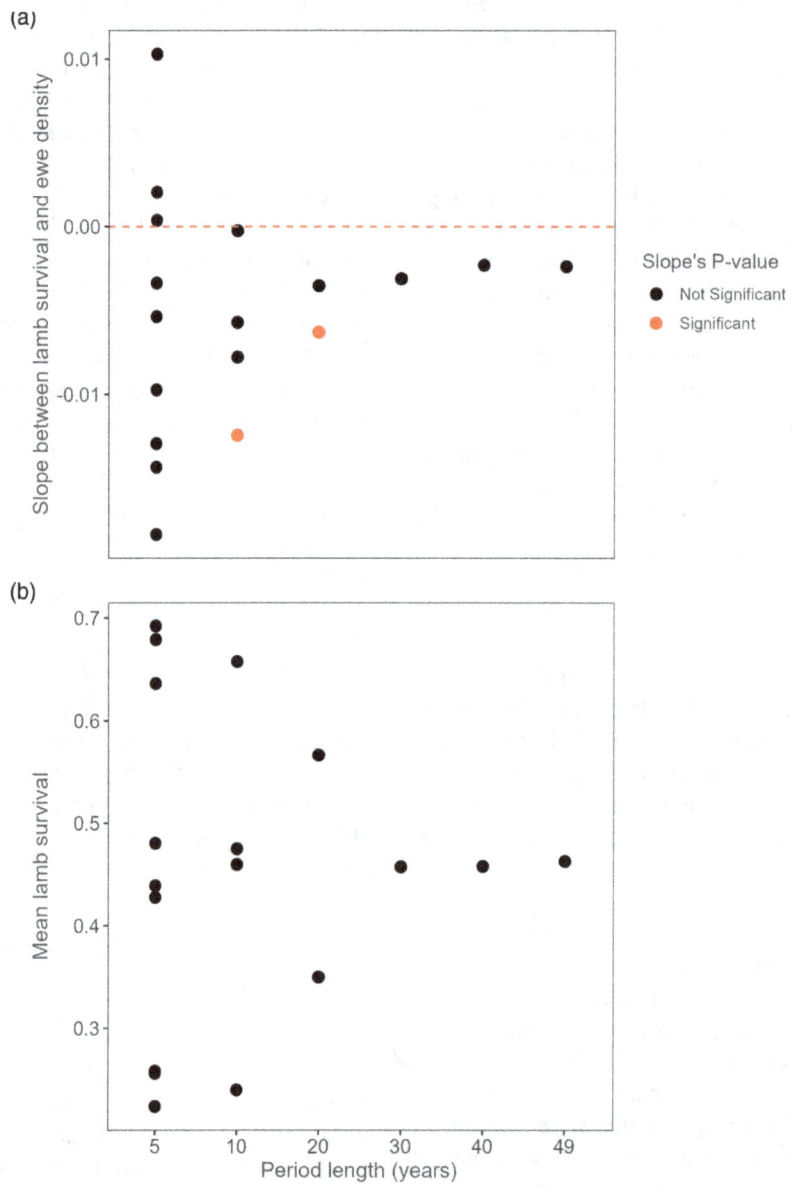

**FIGURE 23.4** (a) Slopes of the relationship between population density, estimated by the number of mature females, and lamb survival to 1 year of age at Ram Mountain, Alberta, Canada, 1972–2023. The slope was calculated for each available period of 5, 10, 20, 30, 40, and 49 years. Note that in three of nine 5-year periods and one of four 10-year periods, the slope is positive or near-zero, indicated by the dotted red line. (b) Estimates of mean lamb survival at Ram Mountain obtained from monitoring periods ranging from 5 to 49 years.

## 23.3  COMBINING LONG-TERM STUDIES OF MARKED INDIVIDUALS

Long-term studies of marked ungulates are rare, and typically only one study per species is available. Given the strong effects of local conditions on population ecology, the external validity of these studies is uncertain. Most have been conducted in areas without large predators, others on islands or inside fences, limiting dispersal and connectivity (Festa-Bianchet et al. 2017). Analyses combining results from long-term research on multiple species have led to major scientific advances (Nussey et al. 2008, de Villemereuil et al. 2020, Bonnet et al. 2022). We present two examples of how combining results from several studies has advanced our knowledge of bighorn sheep ecology.

### 23.3.1  Population Dynamics

A recent analysis examined the factors influencing population dynamics in three populations of mountain ungulates in Alberta (Turgeon et al. 2024). In addition to bighorn sheep at Ram Mountain and Sheep River, this research included mountain goats (*Oreamnos americanus*) at Caw Ridge (Festa-Bianchet et al. 2019). All three populations were monitored for decades with similar methods, attempting to mark all animals as young as possible and then following their growth, survival, and reproduction over their lifetime. This study considered four major variables: population density, predation, disease, and weather. Negative density dependence was evident in some years at Ram Mountain (Portier et al. 1998, Figure 23.4) and Caw Ridge (Panagakis et al. 2017). Yet, over the long term, changes in population size were mostly influenced by sporadic events: intense cougar predation in all three populations and pneumonia epizootics at Sheep River. Density effects were only detected on juvenile survival, but pneumonia and predation affected all age classes. Weather, often thought to be a major factor in ungulate population dynamics, played a minor role, mostly affecting juvenile survival (Turgeon et al. 2024). Combining over 100 population-years of data, this research confirmed that changes in adult female survival have drastic effects on population dynamics (Gaillard et al. 1998) and suggested that in most years density-independent factors keep populations below the level where competition for resources may lead to density-dependent responses, such as the decreased lamb survival and later primiparity seen over the first 20 years of the Ram Mountain research. While density dependence is often a major factor in the population dynamics of ungulates (Bonenfant et al. 2009a), in small and demographically isolated populations, such as mountain sheep populations that include fewer than 100 adult females, it may be less important than density-independent factors.

### 23.3.2  Cougar Predation

Cougar predation is a concern for many populations of bighorn sheep (Rominger 2018). All long-term bighorn sheep study areas have resident cougar populations. For the first

26 years of research at Ram Mountain, 12 at Sheep River, and 17 years at the NBR, however, cougar predation on sheep was rare, presumably because cougars mostly preyed on cervids in forested habitat (Elbroch and Quigley 2019). All three study areas then experienced episodes of intense cougar predation on sheep, apparently due mostly to specialist individuals (Festa-Bianchet et al. 2006). At Sheep River, the first episode was mostly caused by one radio-collared female, which had been monitored from when she was a kitten. After several years of preying on deer (*Odocoileus* spp.), in 1993–1994 she killed 11 sheep, or 9% of the pre-winter population (Ross et al. 1997). All six episodes of intense cougar predation in the three study areas followed the same temporal pattern of sudden initiation and ending. We speculate that each episode began when a cougar started ambushing sheep and ended when that cougar died. All populations declined under intense cougar predation. Simulations suggested that intense cougar predation could lead to local extirpation if it lasted more than a few years or if the frequency of specialist cougars was greater than about one every 10 years (Festa-Bianchet et al. 2006). Cougar predation on bighorn sheep is a case of predator pit; cougars could cause extirpation of a local herd, then go back to preying upon cervids (Chapter 9). Although not all cases of cougar predation on bighorn sheep involve specialist individuals (Rominger 2018), the pattern observed in long-term studies suggests the effects of individual specialists. Onset of predation events was unrelated to population density and cougars preyed upon all sex-age classes.

How did long-term studies of marked bighorn sheep increase our knowledge of cougar predation? First, they documented that predation was sporadic, leading to much uncertainty in predicting population growth (Turgeon et al. 2024). Second, they documented what sex-age classes were killed, and how predation affected population dynamics (Festa-Bianchet et al. 2006, Bourbeau-Lemieux et al. 2011, Cloutier et al. 2024). Predation by cougars contributed to Allee effects, or positive density dependence, causing greater declines in lamb survival when population size was small (Bourbeau-Lemieux et al. 2011). Third, monitoring of individuals suggested that, in addition to causing direct mortality, intense cougar predation had important indirect effects, reducing growth and fitness of surviving sheep. The indirect effects of predation, where stressed prey change behaviors, reduce foraging, and as a result decrease reproduction or suffer higher mortality without being killed by predators, are a controversial topic of substantial interest. Part of the controversy stems from criticism of experimental evidence, exposing prey to predator stimuli at rates far higher than expected in a natural setting (Zanette et al. 2011). For ungulates, reports of strong indirect effects on elk (*Cervus canadensis*) following wolf (*Canis lupus*) reintroduction in Yellowstone (Creel et al. 2007) were refuted by evidence of low frequency of wolf-elk encounters (Middleton et al. 2013). The key influence of indirect effects is how often individual prey experience stressful predator attacks. If a cougar killed a sheep every 10 days, and if, say, one in three attacks were successful, there may be an attack every 2–4 days. Given that at small population size we often saw all or nearly all nursery sheep in a single group, every female in a population may experience an attack every 2–4 days. That could be enough to cause changes in behavior, increase stress, and reduce allocation to reproduction. Data from two populations support that expectation. At Sheep River, lambs were smaller in years of intense cougar predation than in other years. Given that smaller lambs have reduced survival, almost a third of the increase in lamb mortality seen during years of intense predation could be due to indirect effects (Bourbeau-Lemieux et al. 2011). At Ram Mountain, in years of high cougar predation, lamb production declined by 14%, female summer mass gain by 16%, and lamb weaning mass by 8% (Cloutier et al. 2024).

## 23.4   LONG-TERM STUDIES: ASSEMBLY INSTRUCTIONS

We have participated in 12 long-term studies of terrestrial and marine mammals and birds. Here we present our views on what makes a long-term study successful, and what some of the main obstacles to long-term research are, with emphasis on studies of mountain sheep in North America.

### 23.4.1   RESEARCH SITE

The most important requirements for a successful long-term study are an accessible site and good field accommodation. Both access and accommodation should be under the control of the researchers, and not some external entity that may block one or the other following changes in personnel, policy, or priorities. The Sheep River and NBR camps are accessible by road, but Ram Mountain is not. Since the lookout road was reclaimed in 1985, getting to camp involves 4 km by quad and 1 km on foot. That presents some logistical challenges and increases the cost of research. Electricity is provided by a gasoline generator and a solar panel. For propane tanks and drinking water, a helicopter is required. Every decade or so, replacing the propane fridge causes a crisis. Over time, fieldwork has been hampered by floods that washed away bridges or the quad access trail. Recently, climate change (Chapter 21) has led to an increase in fires in the boreal forest near Sheep River and Ram Mountain. On some days, the smoke makes it impossible to work outside, and the risk of emergency evacuations is increasing, presenting additional challenges for the field crew. Adequate field accommodation is important for several reasons. First, people who work under difficult field conditions for several months need a place where they can safely sleep, cook, eat, stay warm, and relax. A fridge is necessary to store perishable food and to keep samples cool or frozen. Electricity to recharge phones and computers is essential for data management and communication, which in turn is important for both mental health and safety. Cell services and internet connections have made fieldwork more efficient and safer

in recent years, particularly in the face of increasing threats from forest fires. Good accommodation is also important for fostering collaborations, which typically involve larger field crews.

### 23.4.2   STUDENTS AND ASSISTANTS

Most field research is done by graduate students and assistants, who need to be looked after and adequately compensated. Reliance on unpaid volunteers is a form of exploitation and limits opportunities to those with adequate financial resources. Increasingly, students come from diverse backgrounds, and long-term programs must consider how to include underrepresented groups. Field research requires knowledge of multiple aspects of data collection, camp and equipment maintenance, backcountry safety, managing social relationships within the team, communication with research partners, and where to find supplies. It is therefore important that principal investigators (PIs) spend time in the field to show all procedures, ensure roles and responsibilities are clear and transparent, provide emergency preparedness information, and discuss possible issues. Spending time in the field, getting dirty, wet, cold, hot, tired, bitten by ticks, or dealing with stranded vehicles reminds all involved that data do not magically appear in computer files. Some students must return for more than 1 year, so that knowledge is passed on to new field crews. Some programs can hire research coordinators, but many others rely on students and field assistants, who are in the field for one to three seasons. Overlap of personnel from year to year and presence of PIs is important to ensure that data collection follows standardized procedures. Graduate students benefit greatly from a long-term database: our students now analyze data collected before they were born. Students can bring new ideas, research questions, techniques, and analytical expertise that allow scientific progress in new directions. It is essential that students and PIs mutually agree from the beginning on what data will be available, how the research will be financed, what is expected of each student in terms of data collection not strictly related to their research, and what authorship and data management practices apply, including contributions after students have left. Poor communication about expectations, compensation, and authorship are sources of frustration for students in long-term studies (Wishart et al. 2024). Increasingly, students must also be aware of requirements for data and code-sharing after publication, including restrictions placed on data obtained in partnership with government agencies.

### 23.4.3   CAPTURES

Many major contributions of long-term studies required monitoring of known individuals. Consequently, another essential attribute of these programs is the ability to capture and mark animals. Regular recaptures provide data on growth, disease status, and other variables that elude studies where recaptures are difficult. Recaptures also allow repeated sampling of blood, hair, milk, and swabs to test for infections. It is important that researchers directly control capture operation without need for external assistance. For example, some national parks may insist that only their staff perform captures. Restrictions on immobilizing drugs, and legislations limiting chemical captures to veterinarians are increasing obstacles to long-term studies of large mammals. Mountain sheep are particularly suited to long-term studies because they are attracted to salt, can become tolerant to human observers, and use open habitats where they are easy to observe. Attraction to salt has allowed us to recapture some sheep at Ram Mountain 50 times over their lifetime, and salt can attract mountain ungulates to platform scales where they can be weighed without handling (Bassano et al. 2003, Figure 23.3).

### 23.4.4 AGENCY COOPERATION

Knowledge co-production (Wisdom et al. 2020) is a key element of long-term studies of mountain sheep. All long-term research programs presented here were initiated either by or in collaboration with wildlife agencies. Government agencies facilitate field logistics, provide in-kind support, and control issuance of research and collection permits. Field camps on public land require government permits. Government biologists contribute fieldwork, expertise, and local knowledge to long-term studies of mountain sheep. Collaboration with government biologists facilitates and expedites inclusion of new research findings into policy and management practices. Interactions with government agencies are an important component of the training of graduate students, many of whom will eventually work for or with those agencies.

### 23.4.5 FUNDING

Field research is expensive, and long-term programs require long-term funding. In many cases, however, these programs were financed by short-term funding, which had to be renewed every year, or switched from one funding source to another. Some funders may challenge the justification for continued support of long-term studies, which must constantly develop new objectives to justify continued data collection (Sheldon et al. 2022). In Canada, the Natural Sciences and Engineering Research Council supports both long-term programs and the training of graduate students through its Discovery Grant program.

### 23.4.6 COLLABORATIONS

Long-term programs that monitor marked individuals, especially if they can recapture those individuals, or collect samples such as tissue, feces, milk, hair, or blood, foster interdisciplinary collaborations. Some collaborations are based on advances in technology or discoveries in other fields of science. For example, when the long-term studies

presented above were started, there was no DNA-based technology that allowed paternity assignment. Now, stored tissue samples from sheep captured years ago are being used for genomic analyses to identify genes responsible for specific traits, such as horn growth. Data on bighorn sheep have been included in recent studies combining many long-term studies of wild animals to examine selection on breeding date (de Villemereuil et al. 2020) and quantify genetic variance in fitness, revealing the potential for rapid evolution (Bonnet et al. 2022). Collaborations can be limited by the availability of space in field camps, again underlining the importance of accommodation and accessibility. A prominent issue in collaboration is communicating to laboratory-based scientists the challenges of collecting data in the wild, including unpredictability and difficulties in maintaining sampling schedules.

### 23.4.7 LUCK

Long-term programs need long-term luck. Wildfires must not burn down research facilities, natural disasters must not destroy study areas, changes in agency personnel or government policies must not lead to withdrawal of funds, access rights, or collaborations. The PIs must stay alive and healthy, or find successors. The study population must remain at a level where it provides an adequate sample size for the study objectives. For bighorn sheep, catastrophic population declines can be induced by pneumonia or intense cougar predation. The Ram Mountain population declined from 66 sheep in 2019 to 35 in 2020, likely because of cougar predation. It recovered to 52 the following year, apparently after the sheep-killing cougar died. Had the high level of predation continued, there may have been fewer than ten sheep in 2021, and hard choices would have had to be made about whether to continue the study. Availability of long-term data with a large existing sample size makes these studies resilient to unexpected declines, such as those caused at both Sheep River and NBR by pneumonia epizootics.

### 23.4.8 DRAWBACKS

There are two main drawbacks to long-term research on marked individuals. First, they require substantial commitments from PIs every year, including field work at key times, maintenance of funding, collaborations, and recruitment of students. Those requirements limit the opportunities to pursue other research goals and may clash with personal commitments. Second, these studies can generally be conducted on only one population, which can limit their external validity. Bighorn sheep are unique among ungulates because they are the subject of several long-term studies, including three (Ram Mountain, NBR, Sheep River), based on multi-generational monitoring of all individuals in a population. Consistent results from multiple study areas, such as the strong effects of cougar predation and pneumonia epizootics, the variable effects of density, and the main factors affecting male and female reproductive success, provide greater confidence in their validity and usefulness for conservation. Unfortunately, no comparable research exists for desert or thinhorn sheep.

## 23.5 COMBINING INDIVIDUAL MONITORING AND LONG-TERM HARVEST RECORDS IMPROVES CONSERVATION

All North American jurisdictions require compulsory inspection of harvested males, and many have collected these data for decades (LaSharr et al. 2019) so that long-term records of horn size are available for large areas. Horn measurements represent a valuable long-term record, especially when annual growth increments are measured. A 10-year-old male has 11 years of data in his horns. Annual increment and base circumference measurements can be analyzed according to weather, population density, other ecological variables, and management regimes. Unfortunately, some jurisdictions only measure total length and base circumference of the horns, or calculate a trophy score whose relationship to biology is uncertain. Measurements of annual increments take time and require training, but are extremely valuable in guiding policy. A training guide is available at https://figshare.com/articles/ online_resource/Guide_to_ageing_mountain_sheep_with_ emphasis_on_methods_for_measuring_yearly_horn_incr ements/19398710?file=34656401, accessed 27 April 2025.

Horn measurements are also an excellent source of long-term data for those interested in ecology and wildlife conservation. Measuring length and base circumference of each increment allows researchers to estimate horn volume, and if basic shape measurements are collected, such as tip-to-tip spread or curl diameter, one could also look at possible changes in shape with changes in environmental conditions or hunting regulations (Garel et al. 2007). As technology advances, it may become possible to create digital 3D images of the horns, which would also allow us to estimate their volume.

Harvested males are not a random sample of the population. Males with small horns are less likely to be shot, and those with rapid horn growth are shot at a younger age (Hengeveld and Festa-Bianchet 2011, Pelletier et al. 2012). At Ram Mountain, records of harvested males underestimated the decline in horn size by almost half, because it was illegal to harvest small-horned males (Pelletier et al. 2012). Once those biases are considered, however, long-term data on horn growth can provide important advances in our understanding of mountain sheep ecology and conservation. For example, horn measurements collected by government agencies have been used to examine the effects of weather (Loehr et al. 2010, LaSharr et al. 2019), harvest strategies (Morrissey et al. 2021, Karabatsos et al. 2024), hunter density and hunting pressure (Douhard et al. 2016), and disease (Martin et al. 2022). A promising line of

investigation based on horn measurements is the effects of climate change (Chapter 21), especially given the variety of climatic conditions experienced by North American mountain sheep over their vast geographical range.

### 23.5.1 COMBINING LONG-TERM INDIVIDUAL MONITORING AND HORN MEASUREMENTS: IMPLICATIONS FOR TROPHY MANAGEMENT

The complex relationship between horn size, age, and mating success in males (Chapter 5) has many implications for the trophy management of mountain sheep. For males aged 7 years or older, larger horns increase the ability to attain a very high rank and adopt the highly successful tending tactic (Coltman et al. 2002). Among younger males, however, the correlation between horn size and mating success is weaker (Martin et al. 2016), possibly because even the largest males aged 4–6 years are likely subordinate to older males (Pelletier and Festa-Bianchet 2006). Yet, most horn growth takes place in the first 6 years of life (Bonenfant et al. 2009b). Males with rapidly growing horns reach trophy status at a younger age. For example, in British Columbia, under a full-curl minimum legal size for males of the California ecotype, average cumulative horn length grown over the second and third year decreased from 57 cm for males shot at age 4 to 37 cm for those killed at 8 years, a difference of 18%. For the Rocky Mountain ecotype, also in British Columbia, over the same range in age at harvest, early horn growth decreased by 20% (Hengeveld and Festa-Bianchet 2011). Similarly, for Stone's sheep (*O. d. stonei*) under regulations limiting harvest to males with full-curl horns or a minimum age of 8 years, Douhard et al. (2016) reported that horn growth over the second and third year of life in a heavily hunted region was 45 cm for males shot at age 7 and 24 cm for those killed at age 9, a decrease of 47%. In a region with lower hunting pressure, the corresponding decrease in early horn growth was slightly lower at 41%.

The mating system of mountain sheep selects for rapid horn growth, because males whose horns grow rapidly attain high dominance rank at an earlier age. Hunting management where males with rapidly growing horns are harvested at 4 or 5 years of age, as in many bighorn sheep populations, however, selects for slow growth because males with slow-growing horns attain trophy status at a later age, or never. Horn size is inheritable; although much of its variation is environmental, about 30%–40% is genetic (Pigeon et al. 2016). The negative selective pressure on rapid horn growth from trophy hunting starts 2–3 years earlier than the positive selective pressure from high mating success of large-horned mature males. If a large proportion of males with rapidly growing horns are shot before they can attain high mating success, hunting could induce evolutionary shrinkage of horns.

Trophy hunting is selective, but selection will not necessarily lead to evolution. Therefore, the key question for both biology and management of mountain sheep is whether trophy hunting leads to smaller horns over sheep generations because of genetic changes. If that were the case, managers and hunters interested in evolutionary sustainable harvests should reconsider harvest strategies. Multiple lines of evidence suggest that when hunting removes a large proportion of trophy males at a young age, with regulations that do not change over decades and a large area, and there is no effective genetic rescue from protected areas (Lassis et al. 2023), horns will evolve to be smaller. The pedigree-based long-term study at Ram Mountain measured a genetic decline in horn size of about 2.5 cm over 23 years when the yearly harvest of males with horns of four-fifths curl or greater was about 40%. After the regulation changed to full-curl, the decline stopped and may have been reversed (Pigeon et al. 2016). Researchers who examined horn size of harvested males generally reported declines where trophy hunting regulations were very liberal, but not where harvest was restricted (Hengeveld and Festa-Bianchet 2011, Festa-Bianchet et al. 2014, Douhard et al. 2016, LaSharr et al. 2019, Morrissey et al. 2021). A 16-year study of Dall sheep (*O. d. dalli*) in the Northwest Territories, Canada, reported no change in horn size with a harvest pressure much lower than in other jurisdictions and average age at harvest of 9.7 years (Karabatsos et al. 2024). In contrast, in Alberta, Canada, between 1980 and 1995, about 20%–25% of the harvest was made up of males aged 4 or 5 years. Between 1980 and 2010, horn length of harvested 6-year-olds in Alberta declined by about 3 cm, consistent with the genetic decline measured at Ram Mountain (Festa-Bianchet et al. 2014).

Evolutionary enlightened management of mountain sheep can avoid undesirable evolutionary change by allowing some large-horned males to reach 8 years of age (Festa-Bianchet 2017, Festa-Bianchet and Mysterud 2018). Although evolutionary biologists may find it trivial that persistent, strong, directional selective pressure on a heritable trait will lead to evolution, the result is controversial among some hunters and wildlife biologists (Boyce and Krausman 2018), possibly because it was taken up by anti-use groups to incorrectly argue that any trophy hunting will lead to evolution. One paper analyzed data on body mass from Ram Mountain with an integral projection model and claimed that observed changes in horn size were demographic and not genetic (Traill et al. 2014). Traill et al. (2014) used a technique that cannot quantify evolution and an inappropriate measure of inheritance, as demonstrated by multiple critiques (Hedrick et al. 2014, Chevin 2015, Janeiro et al. 2017, vanBenthem et al. 2017). Another model claiming that evolutionary change was irrelevant to mountain sheep management (Coulson et al. 2018) used very small values for additive and phenotypic variance, forcibly reducing the potential for evolution. When variance values empirically estimated for bighorn sheep are used, the model predicts substantial evolutionary change, comparable to that seen at Ram Mountain (Boyce et al. 2019).

If large-horned males have higher reproductive success, why do horns not grow larger and larger in unhunted populations? Presumably, there is some countervailing natural selection against very large horns, but so far that has not been detected empirically. Earlier suggestions that large-horned males had shorter life expectancy (Geist 1966) were based on a small sample ($n=40$) of males that were found dead, a technique subject to multiple biases (Murphy and Whitten 1976). A comparison of early horn growth and natural survival for 172 bighorn males found no evidence of a survival cost of rapidly growing horns for adults, and a positive effect on survival of yearlings (Bonenfant et al. 2009b). Perhaps, very large horns may reduce the natural survival of mature males, but that has not been explicitly investigated. A negative effect of large horns on survival was reported for Pyrenean chamois (*Rupicapra pyrenaica*; Douhard et al. 2020).

## 23.6 SUMMARY

Long-term studies have made substantial contributions to our knowledge of the ecology, evolution, and conservation of mountain sheep. They have measured and shown the importance of temporal changes in ecological and management variables and revealed the strong influence of early-life environment on lifetime performance. They are, however, limited in number and do not reflect the taxonomic and ecological variability of mountain sheep in North America. Future research progress will require the continuation of monitoring programs that started in the last few years, and more collaboration among university and government researchers, particularly in securing long-term funding and access to study sites. Both individual-based monitoring and accurate measurement of annual horn increments of harvested males will be required to investigate the effects of climate change.

## REFERENCES

Almberg, E. S., K. R. Manlove, E. F. Cassirer, C. B. Ramsey, K. Carson, J. A. Gude, and R. K. Plowright. 2022. Modelling management strategies for chronic disease in wildlife: predictions for the control of respiratory disease in bighorn sheep. *Journal of Applied Ecology* 59:693–703.

Bassano, B., A. V. Hardenberg, F. Pelletier, and G. Gobbi. 2003. A method to weigh free-ranging ungulates without handling. *Wildlife Society Bulletin* 31:1205–1209.

Besser, T. E., E. F. Cassirer, A. Lisk, D. Nelson, K. R. Manlove, P. C. Cross, and J. T. Hogg. 2021. Natural history of a bighorn sheep pneumonia epizootic: source of infection, course of disease, and pathogen clearance. *Ecology and Evolution* 11:14366–14382.

Besser, T. E., M. A. Highland, K. Baker, E. F. Cassirer, N. J. Anderson, J. M. Ramsey, K. Mansfield, D. L. Bruning, P. Wolff, J. B. Smith, and A. Jenks. 2012. Causes of pneumonia epizootics among bighorn sheep, western United States, 2008–2010. *Emerging Infectious Diseases* 18:406–414.

Bonenfant, C., F. Pelletier, M. Garel, and P. Bergeron. 2009b. Age-dependent relationship between horn growth and survival in wild sheep. *Journal of Animal Ecology* 78:161–171.

Bonenfant, C., J. M. Gaillard, T. Coulson, M. Festa-Bianchet, A. Loison, M. Garel, L. E. Loe, P. Blanchard, N. Pettorelli, N. Owen-Smith, J. Du Toit, and P. Duncan. 2009a. Empirical evidence of density-dependence in populations of large herbivores. *Advances in Ecological Research* 41:313–357.

Bonnet, T., M. B. Morrissey, P. de Villemereuil, S. C. Alberts, P. Arcese, L. Bailey, S. Boutin, P. Brekke, L. J. N. Brent, G. Camenisch, A. Charmantier, et al. 2022. Genetic variance in fitness indicates rapid contemporary adaptive evolution in wild animals. *Science* 376:1012–1016.

Bourbeau-Lemieux, A., M. Festa-Bianchet, J. M. Gaillard, and F. Pelletier. 2011. Predator-driven component Allee effects in a wild ungulate. *Ecology Letters* 14:358–363.

Boyce, M. S., and P. R. Krausman. 2018. Special section: controversies in mountain sheep management. *Journal of Wildlife Management* 82:5–7.

Boyce, M. S., T. Coulson, J. R. Heffelfinger, and P. R. Krausman. 2019. Mountain sheep management must use representative data: a reply to Festa-Bianchet (2019). *Journal of Wildlife Management* 83:9–11.

Cassirer, E. F., K. R. Manlove, E. S. Almberg, P. L. Kamath, M. Cox, P. Wolff, A. Roug, R. Shannon, R. B. Harris, B. J. Gonzales, et al. 2018. Pneumonia in bighorn sheep: risks and resilience. *Journal of Wildlife Management* 82:32–45.

Cassirer, E. F., K. R. Manlove, R. K. Plowright, and T. E. Besser. 2017. Evidence for strain-specific immunity to pneumonia in bighorn sheep. *Journal of Wildlife Management* 81:133–143.

Cassirer, E. F., R. K. Plowright, K. R. Manlove, P. C. Cross, A. P. Dobson, K. A. Potter, and P. J. Hudson. 2013. Spatio-temporal dynamics of pneumonia in bighorn sheep. *Journal of Animal Ecology* 82:518–528.

Caughley, G., and A. R. E. Sinclair. 1994. *Wildlife Ecology and Management.* Blackwell Scientific Publications, Boston, Massachusetts, USA.

Chevin, L.-M. 2015. Evolution of adult size depends on genetic variance in growth trajectories: a comment on analyses of evolutionary dynamics using integral projection models. *Methods in Ecology and Evolution* 6:981–986.

Cloutier, Z., M. Festa-Bianchet, and F. Pelletier. 2024. Direct and indirect effects of cougar predation on bighorn sheep fitness. *Ecology* 105:e4374.

Clutton-Brock, T. H., and B. C. Sheldon. 2010. Individuals and populations: the role of long-term, individual-based studies of animals in ecology and evolutionary biology. *Trends in Ecology and Evolution* 25:562–573.

Coltman, D. W., M. Festa-Bianchet, J. T. Jorgenson, and C. Strobeck. 2002. Age-dependent sexual selection in bighorn rams. *Proceedings of the Royal Society B* 269:165–172.

Coulson, T., S. Schindler, L. W. Traill, and B. E. Kendall. 2018. Predicting the evolutionary consequences of trophy hunting on a quantitative trait. *Journal of Wildlife Management* 82:46–56.

Creel, S., D. Christianson, S. Liley, and J. A. Winnie. 2007. Predation risk affects reproductive physiology and demography of elk. *Science* 315:960.

de Villemereuil, P., A. Charmantier, D. Arlt, P. Bize, P. Brekke, L. Brouwer, A. Cockburn, S. D. Côté, F. S. Dobson, S. R. Evans, et al. 2020. Fluctuating optimum and temporally variable selection on breeding date in birds and mammals. *Proceedings of the National Academy of Sciences* 117:31969–31978.

Donovan, V. M., C. P. Roberts, C. L. Wonkka, J. L. Beck, J. N. Popp, C. R. Allen, and D. Twidwell. 2020. Range-wide monitoring of population trends for Rocky Mountain bighorn sheep. *Biological Conservation* 248:108639.

Douhard, M., J. P. Crampe, A. Loison, and C. Bonenfant. 2020. A negative association between horn length and survival in a weakly dimorphic ungulate. *Ecology and Evolution* 10:2793–2802.

Douhard, M., M. Festa-Bianchet, F. Pelletier, J. M. Gaillard, and C. Bonenfant. 2016. Changes in horn size of Stone's sheep over four decades correlate with trophy hunting pressure. *Ecological Applications* 26:309–321.

Eberhardt, L. L. 2002. A paradigm for population analysis of long-lived vertebrates. *Ecology* 83:2841–2854.

Elbroch, L. M., and H. B. Quigley. 2019. Age-specific foraging strategies among pumas, and its implications for aiding ungulate populations through carnivore control. *Conservation Science and Practice* 1:e23.

Festa-Bianchet, M. 1986a. Seasonal dispersion of overlapping mountain sheep ewe groups. *Journal of Wildlife Management* 50:325–330.

Festa-Bianchet, M. 1986b. Site fidelity and seasonal range use by bighorn rams. *Canadian Journal of Zoology* 64:2126–2132.

Festa-Bianchet, M. 1991. Numbers of lungworm larvae in feces of bighorn sheep: yearly changes, influence of host sex, and effects on host survival. *Canadian Journal of Zoology* 69:547–554.

Festa-Bianchet, M. 2017. When does selective hunting select, how can we tell, and what should we do about it? *Mammal Review* 47:76–81.

Festa-Bianchet, M., and A. Mysterud. 2018. Hunting and evolution: theory, evidence, and unknowns. *Journal of Mammalogy* 99:1281–1292.

Festa-Bianchet, M., F. Pelletier, J. T. Jorgenson, C. Feder, and A. Hubbs. 2014. Decrease in horn size and increase in age of trophy sheep in Alberta over 37 years. *Journal of Wildlife Management* 78:133–141.

Festa-Bianchet, M., J. T. Jorgenson, and W. D. Wishart. 1994. Early weaning in bighorn sheep, *Ovis canadensis*, affects growth of males but not females. *Behavioral Ecology* 6:21–27.

Festa-Bianchet, M., M. Douhard, J. M. Gaillard, and F. Pelletier. 2017. Successes and challenges of long-term field studies of marked ungulates. *Journal of Mammalogy* 98:612–620.

Festa-Bianchet, M., S. D. Côté, S. Hamel, and F. Pelletier. 2019. Long-term studies of bighorn sheep and mountain goats reveal fitness costs of reproduction. *Journal of Animal Ecology* 88:1118–1133.

Festa-Bianchet, M., T. Coulson, J. M. Gaillard, J. T. Hogg, and F. Pelletier. 2006. Stochastic predation and population persistence in bighorn sheep. *Proceedings of the Royal Society B-Biological Sciences* 273:1537–1543.

Gaillard, J.-M., M. Festa-Bianchet, and N. G. Yoccoz. 1998. Population dynamics of large herbivores: variable recruitment with constant adult survival. *Trends in Ecology and Evolution* 13:58–63.

Garel, M., J.-M. Cugnasse, D. Maillard, J.-M. Gaillard, A. J. M. Hewison, and D. Dubray. 2007. Selective harvesting and habitat loss produce long-term life history changes in a mouflon population. *Ecological Applications* 17:1607–1618.

Garwood, T. J., C. P. Lehman, D. P. Walsh, E. F. Cassirer, T. E. Besser, and J. A. Jenks. 2020. Removal of chronic *Mycoplasma ovipneumoniae* carrier ewes eliminates pneumonia in a bighorn sheep population. *Ecology and Evolution* 10:3491–3502.

Geist, V. 1966. The evolutionary significance of mountain sheep horns. *Evolution* 20:558–566.

Hedrick, P. W., D. W. Coltman, M. Festa-Bianchet, and F. Pelletier. 2014. Not surprisingly, no inheritance of a trait results in no evolution. *Proceedings of the National Academy of Sciences* 111:E4810.

Hengeveld, P. E., and M. Festa-Bianchet. 2011. Harvest regulations and artificial selection on horn size in male bighorn sheep. *Journal of Wildlife Management* 75:189–197.

Hogg, J. T. 1984. Mating in bighorn sheep: multiple creative male strategies. *Science* 225:526–529.

Hogg, J. T. 1988. Copulatory tactics in relation to sperm competition in Rocky Mountain bighorn sheep. *Behavioral Ecology and Sociobiology* 22:49–59.

Hogg, J. T. 2000. Mating systems and conservation at large spatial scales. Pages 214–252 *in* M. Apollonio, M. Festa-Bianchet, and D. Mainardi, editors. *Vertebrate mating systems*. World Scientific, Singapore.

Hogg, J. T., and S. H. Forbes. 1997. Mating in bighorn sheep: frequent male reproduction via a high-risk "unconventional" tactic. *Behavioral Ecology and Sociobiology* 41:33–48.

Hogg, J. T., S. H. Forbes, B. M. Steele, and G. Luikart. 2006. Genetic rescue of an insular population of large mammals. *Proceedings of the Royal Society B-Biological Sciences* 273:1491–1499.

Hogg, J. T., S. J. Dunn, J. Poissant, F. Pelletier, and J. A. Byers. 2017. Capital vs. income-dependent optimal birth date in two North American ungulates. *Ecosphere* 84:e01766.

Janeiro, M. J., D. W. Coltman, M. Festa-Bianchet, F. Pelletier, and M. B. Morrissey. 2017. Towards robust evolutionary inference with integral projection models. *Journal of Evolutionary Biology* 30:270–288.

Jorgenson, J. T., J. Samson, and M. Festa-Bianchet. 1990. Field immobilization of bighorn sheep with xylazine hydrochloride and antagonism with idazoxan. *Journal of Wildlife Diseases* 26:522–527.

Jorgenson, J. T., M. Festa-Bianchet, and W. D. Wishart. 1993. Harvesting bighorn ewes: consequences for population size and trophy ram production. *Journal of Wildlife Management* 57:429–435.

Jorgenson, J. T., M. Festa-Bianchet, and W. D. Wishart. 1998. Effects of population density on horn development in bighorn rams. *Journal of Wildlife Management* 62:1011–1020.

Karabatsos, S., N. C. Larter, D. Allaire, K. Eykelboom, C. Estevo, M. Iravani, I. C. Barrio, and D. S. Hik. 2024. Dall's sheep horn growth and harvest management in the Mackenzie Mountains, Northwest Territories, Canada. *Journal of Wildlife Management* 88:e22536.

LaSharr, T. N., R. A. Long, J. R. Heffelfinger, V. C. Bleich, P. R. Krausman, R. T. Bowyer, J. M. Shannon, R. W. Klaver, C. Brewer, M. Cox, et al. 2019. Hunting and mountain sheep: Do current harvest practices affect horn growth? *Evolutionary Applications* 12:1823–1836.

Lassis, R., M. Festa-Bianchet, and F. Pelletier. 2022. Breeding migrations by bighorn sheep males are driven by mating opportunities. *Ecology and Evolution* 12:e8692.

Lassis, R., M. Festa-Bianchet, J. Van de Walle, and F. Pelletier. 2023. Genetic rescue from protected areas is modulated by migration, hunting rate, and timing of harvest. *Evolutionary Applications* 16:1105–1118.

Loehr, J., J. Carey, R. B. O'Hara, and D. S. Hik. 2010. The role of phenotypic plasticity in responses of hunted thinhorn sheep ram horn growth to changing climate conditions. *Journal of Evolutionary Biology* 23:783–790.

Manlove, K. R., E. F. Cassirer, P. C. Cross, R. K. Plowright, and P. J. Hudson. 2014. Costs and benefits of group living with disease: a case study of pneumonia in bighorn lambs (*Ovis canadensis*). *Proceedings of the Royal Society B* 281:20142331.

Manlove, K. R., E. F. Cassirer, P. C. Cross, R. K. Plowright, and P. J. Hudson. 2016. Disease introduction is associated with a phase transition in bighorn sheep demographics. *Ecology* 97:2593–2602.

Manlove, K. R., E. F. Cassirer, R. K. Plowright, P. C. Cross, and P. J. Hudson. 2017. Contact and contagion: probability of transmission given contact varies with demographic state in bighorn sheep. *Journal of Animal Ecology* 86:908–920.

Marcil-Ferland, D., M. Festa-Bianchet, A. M. Martin, and F. Pelletier. 2013. Despite catch-up, prolonged growth has detrimental fitness consequences in a long-lived vertebrate. *American Naturalist* 182:775–785.

Martin, A. M., J. T. Hogg, K. R. Manlove, T. N. Lasharr, J. M. Shannon, D. E. McWhirter, H. M. Miyasaki, K. B. Monteith, and P. C. Cross. 2022. Disease and secondary sexual traits: effects of pneumonia on horn size of bighorn sheep. *Journal of Wildlife Management* 86:e22154.

Martin, A. M., M. Festa-Bianchet, D. W. Coltman, and F. Pelletier. 2016. Demographic drivers of age-dependent sexual selection. *Journal of Evolutionary Biology* 29:1437–1446.

Middleton, A. D., M. J. Kauffman, D. E. McWhirter, M. D. Jimenez, R. C. Cook, J. G. Cook, S. E. Albeke, H. Sawyer, and P. J. White. 2013. Linking anti-predator behaviour to prey demography reveals limited risk effects of an actively hunting large carnivore. *Ecology Letters* 16:1023–1030.

Miller, J. M., J. Poissant, J. T. Hogg, and D. W. Coltman. 2012. Genomic consequences of genetic rescue in an insular population of bighorn sheep (*Ovis canadensis*). *Molecular Ecology* 21:1583–1596.

Morrissey, M. B., A. Hubbs, and M. Festa-Bianchet. 2021. Horn growth appears to decline under intense trophy hunting, but biases in hunt data challenge the interpretation of the evolutionary basis of trends. *Evolutionary Applications* 14:1519–1527.

Murphy, E. C., and K. R. Whitten. 1976. Dall sheep demography in McKinley Park and a reevaluation of Murie's data. *Journal of Wildlife Management* 40:597–609.

Nussey, D. H., T. Coulson, M. Festa-Bianchet, and J. M. Gaillard. 2008. Measuring senescence in wild animal populations: towards a longitudinal approach. *Functional Ecology* 22:393–406.

Panagakis, A., S. Hamel, and S. D. Côté. 2017. Influence of early reproductive success on longevity and late reproductive success in an alpine ungulate. *American Naturalist* 189:667–683.

Pelletier, F., J. T. Hogg, and M. Festa-Bianchet. 2006. Male mating effort in a polygynous ungulate. *Behavioral Ecology and Sociobiology* 60:645–654.

Pelletier, F., K. A. Page, T. Ostiguy, and M. Festa-Bianchet. 2005. Fecal counts of lungworm larvae and reproductive effort in bighorn sheep, *Ovis canadensis. Oikos* 110:473–480.

Pelletier, F., and M. Festa-Bianchet. 2004. Effects of body mass, age, dominance and parasite load on foraging time of bighorn rams, *Ovis canadensis. Behavioral Ecology and Sociobiology* 56:546–551.

Pelletier, F., and M. Festa-Bianchet. 2006. Sexual selection and social rank in bighorn rams. *Animal Behaviour* 71:649–655.

Pelletier, F., M. Festa-Bianchet, and J. T. Jorgenson. 2012. Data from selective harvests underestimate temporal trends in quantitative traits. *Biology Letters* 8:878–881.

Pigeon, G., and F. Pelletier. 2018. Direct and indirect effects of early-life environment on lifetime fitness of bighorn ewes. *Proceedings of the Royal Society B* 285:20171935.

Pigeon, G., M. Festa-Bianchet, D. W. Coltman, and F. Pelletier. 2016. Intense selective hunting leads to artificial evolution in horn size. *Evolutionary Applications* 9:521–530.

Pigeon, G., M. Festa-Bianchet, and F. Pelletier. 2017. Long-term fitness consequences of early environment in a long-lived ungulate. *Proceedings of the Royal Society B* 284:20170222.

Plowright, R. K., K. R. Manlove, T. E. Besser, D. Paez, K. R. Andrews, P. E. Matthews, L. P. Waits, P. J. Hudson, and E. F. Cassirer. 2017. Age-specific infectious period shapes dynamics of pneumonia in bighorn sheep. *Ecology Letters* 20:1325–1336.

Poirier, M.-A., D. W. Coltman, F. Pelletier, J. T. Jorgenson, and M. Festa-Bianchet. 2019. Genetic decline, restoration and rescue of an isolated ungulate population. *Evolutionary Applications* 12:1318–1328.

Poissant, J., A. J. Wilson, M. Festa-Bianchet, J. T. Hogg, and D. W. Coltman. 2008. Quantitative genetics and sex-specific selection on sexually dimorphic traits in bighorn sheep. *Proceedings of the Royal Society of London B* 275:623–628.

Poissant, J., C. S. Davis, R. M. Malenfant, J. T. Hogg, and D. W. Coltman. 2011. QTL mapping for sexually dimorphic fitness-related traits in wild bighorn sheep. *Heredity* 1908:253–263.

Portier, C., M. Festa-Bianchet, J.-M. Gaillard, J. T. Jorgenson, and N. G. Yoccoz. 1998. Effects of density and weather on survival of bighorn sheep lambs (*Ovis canadensis*). *Journal of Zoology* 245:271–278.

Rioux-Paquette, E., D. W. Coltman, and M. Festa-Bianchet. 2011. Sex-differential effects of inbreeding on overwinter survival, birth date and mass of bighorn lambs. *Journal of Evolutionary Biology* 24:121–131.

Rominger, E. M. 2018. The Gordian knot of mountain lion predation and bighorn sheep. *Journal of Wildlife Management* 82:19–31.

Ross, P. I., M. G. Jalkotzy, and M. Festa-Bianchet. 1997. Cougar predation on bighorn sheep in southwestern Alberta during winter. *Canadian Journal of Zoology* 75:771–775.

Ruckstuhl, K. E. 1998. Foraging behaviour and sexual segregation in bighorn sheep. *Animal Behaviour* 56:99–106.

Sheldon, B. C., L. E. B. Kruuk, and S. C. Alberts. 2022. The expanding value of long-term studies of individuals in the wild. *Nature Ecology and Evolution* 6:1799–1801.

Traill, L. W., S. Schindler, and T. Coulson. 2014. Demography, not inheritance, drives phenotypic change in hunted bighorn sheep. *Proceedings of the National Academy of Sciences* 111:13223–13228.

Turgeon, R., F. Pelletier, S. D. Côté, M. Festa-Bianchet, and S. Hamel. 2024. Sporadic events have a greater influence on the dynamics of small isolated populations than density-dependence and environmental conditions. *American Naturalist* 204:574–588.

vanBenthem, K. J., M. Bruijning, T. Bonnet, E. Jongejans, E. Postma, and A. Ozgul. 2017. Disentangling evolutionary, plastic and demographic processes underlying trait dynamics: a review of four frameworks. *Methods in Ecology and Evolution* 8:75–85.

Wang, M., J. Alves, M. Tucker, W. Yang, and K. E. Ruckstuhl. 2018. Effects of intrinsic and extrinsic factors on ruminating, grazing, and bedding time in bighorn sheep (*Ovis canadensis*). *PLoS One* 13:e0206664.

Wisdom, M. J., R. M. Nielson, M. M. Rowland, and K. M. Proffitt. 2020. Modeling landscape use for ungulates: tenets of ecology, management and inference. *Frontiers in Ecology and Evolution* 8:211.

Wishart, A., M. R. Boudreau, and A. K. Menzies. 2024. Graduate student experiences and perspectives related to conducting thesis research within long-term ecological projects. *Facets* 9:1–12.

Wishart, W. D., J. T. Jorgenson, and M. Hilton. 1980. A minor die-off of bighorns from pneumonia in southern Alberta (1978). Pages 229–245 *in Northern Wild Sheep & Goat Council*. Idaho Fish & Game, Salmon, Idaho, USA.

Zanette, L. Y., A. F. White, M. C. Allen, and M. Clinchy. 2011. Perceived predation risk reduces the number of offspring songbirds produce per year. *Science* 334:1398–1401.

# 24 Challenges for Mountain Sheep Management and Conservation in North America

*Paul R. Krausman and Bill A. Jex*

## 24.1 INTRODUCTION

Historical accounts of mountain sheep populations (bighorn [*Ovis canadensis*], thinhorn [*Ovis dalli*]) in North America, dating back to the late 1600s, indicate robust numbers in remote wilderness areas. Early explorers and settlers reported abundant populations in Alaska, Canada, the western mountain ranges, and the deserts in the United States and Mexico. Perspectives on wildlife and mountain sheep abundance have been, and continue to be, shaped by conflicting views: the Old-World view that wildlife exists to be exploited and controlled, in contrast to Indigenous Peoples' holistic connections to these animals and the habitats that support them. Despite a long-standing appreciation for mountain sheep, their populations and distribution have declined since the mid-1800s, particularly in some areas. Understanding the challenges faced by mountain sheep today is essential for effective conservation, management, and restoration of resilient populations and their environments.

As Western settlement expanded, mountain sheep populations dwindled due to the introduction of livestock (i.e., disease, grazing), industrial-scale resource use (e.g., mining, logging, agriculture), unregulated market and subsistence hunting, recreation, and adverse weather conditions. While these factors contributed locally, the previously unknown effects of domestic livestock pathogens likely played the largest role in the declines and extirpations of mountain sheep. Early explanations for population losses often focused on more apparent causes, such as illegal hunting and predation. By the late 1800s, mountain sheep had become rare in most of the United States and the southern parts of Canada, prompting management changes aimed at restoring their populations in the mid-20th century. Bighorn sheep thrive in undisturbed montane habitats, while thinhorns are typically found in climax alpine and northern grassland ecosystems. Fire suppression and shrub and forest encroachment into these areas can lead to overgrowth, reducing desirable forage and affecting visibility, which is essential for predator avoidance.

Land-use changes, harvest regulation, and advances in wildlife management facilitated a rebound in bighorn populations during the 1960s, peaking in various regions in the late 1970s and early 1980s, partly due in some areas to reintroduction efforts and in others through habitat interventions (protected areas and enhancement actions) and predator removals (e.g., poisoning). By the mid-1990s, however, trends shifted again, with some jurisdictions experiencing declines. This shift may be somewhat related to the evolution of government and agency 1960s–1980s management policies, and changes in rural economies that saw curtailment of some direct wildland interventions (e.g., aggressive wildfire attack programs, prohibition on the use of poisons, increased industrial demands for new resources, dramatic declines in demand for wild fur). Currently, North American mountain sheep populations number less than half a million, while most other big game species exceed a million. Nonetheless, mountain sheep conservation has seen successes, particularly through innovative management techniques and widespread conservation efforts initiated by conservation leaders such as Leopold (1933).

Despite these successes, ongoing monitoring and evaluation are essential. There is no expectation that mountain sheep and their habitats can be sustained without ongoing management efforts. While we may never know the pre-colonial population sizes, solid scientific data on current populations and trends are vital for future conservation. Ongoing monitoring must address challenges stemming from human population growth's influence on the environment, the expansion of energy and natural resource extraction, climate change, and the introduction of invasive species. Additionally, population management must consider taxonomy and genetics, changing public perceptions, predation, disease, and competition. Long-term research is crucial for ensuring the persistence, enhancement, and resilience of mountain sheep populations.

Our objective is to present challenges to the conservation and management of mountain sheep in North America. None of the challenges are isolated, and they all interact with one another. Consider competition as an example. When mountain sheep inhabit areas also used by domestic goats and sheep, there is competition for forage resources, the spread of disease from domestic animals to mountain sheep, habitat alteration, and human conflicts between livestock producers and wildlife managers. Thus, challenges cannot be isolated, solved, and then forgotten. Each challenge must be continually monitored and evaluated for effectiveness so the best scientific conservation and management actions can be initiated and sustained.

DOI: 10.1201/9781003518686-28

## 24.2 CHALLENGES

### 24.2.1 NUTRITION

Nutrition is a key determinant of mountain sheep biology and physiology, influencing productivity, survival, and nearly every aspect of their biology (Smiley et al. 2022, Chapters 4 and 8). Limitations can stem from environmental changes or resource competition, leading to declines in reproductive success and overall health. Evaluating the nutritional status of mountain sheep is currently best accomplished by directly assessing their body fat levels through validated ultrasonography and manual palpation techniques; this information can be correlated with demographic rates to determine how nutrition influences population dynamics. Managing nutrition for mountain sheep across landscapes and jurisdictions can be challenging, but success is most achievable with a strong understanding of the nutritional ecology of mountain sheep.

### 24.2.2 WATER REQUIREMENTS

Mountain sheep obtain water from environmental sources, forage, and metabolic processes. When preformed water and metabolic water are insufficient to meet water demands, mountain sheep require access to sources of drinking water. Natural perennial sources of drinking water can be limited seasonally or in some years due to extreme drought, but in desert environments, they are always scarce; thus, wildlife managers have constructed and maintained sources of water for desert bighorn sheep (*Ovis canadensis* spp.) and some populations of Rocky Mountain bighorn sheep (*O. c. canadensis*) inhabiting arid ranges. These additions to the habitat that support the life requirement of water have been instrumental in establishing and maintaining many populations of mountain sheep. Climatic conditions also influence the survival, reproduction, and population dynamics of mountain sheep. Extremely high summer temperatures in desert environments can lead to dehydration and heat stress, potentially resulting in higher mortality rates and poorer body condition. In contrast, northern species of mountain sheep must contend with extreme winter conditions and are exposed to freezing temperatures, stochastic weather events, and heavy snowfall, which can increase energy expenditure to maintain body heat and increased snow depth can increase costs of locomotion and impede access to forage. These conditions can be particularly challenging for young-of-the-year. Extreme weather events, such as droughts, heavy snowfall, and storms, can have a profound and severe influences on mountain sheep populations. These events can exacerbate existing stressors and cause new ones, resulting in significant demographic changes and increases of the magnitude of effect from stressors within their ecosystem (e.g., predation, competition with other ungulates).

Availability and access to forage is directly affected by precipitation patterns. Forage quantity and quality are heavily determined by the timing and amount of rainfall,

influencing the body condition, fecundity, and reproductive success of mountain sheep. Droughts can influence desert bighorn sheep, leading to decreased forage quality and quantity, and limited water availability. Prolonged droughts can further reduce individual fitness, lamb survival, recruitment, and adult survival. Heavy winter and spring snow, or rain-on-snow events that lead to ground icing, can restrict access to forage for northern bighorn and thinhorn sheep, affecting fitness, survival, and recruitment rates. Maintaining water balance has been, and will continue to be, an ongoing challenge for mountain sheep managers, especially when considering the role that changing climate and weather patterns will play.

### 24.2.3 BODY AND HORN SIZE

Robust mountain sheep populations are desirable for managers and the public; however, achieving large body and horn sizes is a multifactorial process influenced by age, genetics, health, and nutrition. There is no single strategy that will produce large-horned and large-bodied mountain sheep. The growth and size of horns and bodies in mountain sheep result from a complex interplay of multiple factors, including age, genetics, and nutrition (Chapters 4 and 6). While all three pieces influence the size, conformation, and growth of horns and bodies in mountain sheep, nutrition often plays an important role, as it affects growth and reproductive success. Moreover, nutrition is one of the pieces that managers may be able to influence through habitat intervention, conservation, and management actions.

Managers must consider how factors like species management policies, harvest practices, and habitat management influence individual growth and population dynamics. For example, scarce resources can lead females to prioritize their own survival over that of their lambs, influencing lamb survival and, in turn, the overall health and resilience of the population. This is a topic that continues to require improvement and a deeper understanding, as nutrition is important for all life stages and is intricately linked to life history characteristics from birth to death. Nutritional effects will influence population numbers, and considering disease and harvest, robust and healthy populations will cease to exist without adequate nutrition. Biologists need to have a firm understanding of how resources influence population dynamics. In addition, large-horned mountain sheep have reproductive advantages over smaller-horned mountain sheep as they search for mates. These searches are important for maintaining the genetic diversity of mountain sheep populations. Behavior and nutrition are closely tied together and managing mountain sheep is not just a numbers game. Biologists must also be aware of the changing environmental factors influencing populations. Understanding the interplay between nutrition and population dynamics is essential for effective management. Finally, wild sheep managers and biologists will need to ensure that stakeholders and non-governmental organizations fully understand the outcomes of any directives on population dynamics, genetic heterozygosity, and phenotypic expression when implementing policy and creating regulations to achieve specific objectives. Having competing objectives can complicate management actions and negatively affect any success achieved from those actions.

### 24.2.4 HABITAT

Protecting and conserving mountain sheep habitats is increasingly challenging in the face of human effects and climate change. Managers must prioritize habitat availability and function, recognizing that each population has unique needs. Collaborative approaches among land management agencies are important to ensuring the sustainability of mountain sheep habitats (Chapters 7, 20, and 21).

Habitat requirements for mountain sheep vary between subspecies, sex and age groups, life stages, occupied and seasonal habitats, predators and predator assemblages, livestock competition, disease, anthropogenic activities, and a host of other factors which also peripherally influence habitat suitability and function. One size does not fit all. Human activities, such as development, recreation, and land-use changes, can lead to habitat alienation and loss, and degradation in suitability, influencing mountain sheep populations.

Managers must continue to monitor habitats for effectiveness and better understand the important components that comprise habitat suitability and function for mountain sheep. This will be a continuing challenge because each population and subspecies are somewhat unique. For example, thinhorn sheep are particularly vulnerable to stochastic weather events and extreme winter conditions, which can result in high mortality rates. In contrast, desert bighorn sheep are more susceptible to the hot, arid summer climates and the limited availability of water, leading to increased mortality. In addition, wildfire suppression and subsequent forest encroachment can negatively influence the habitat of Rocky Mountain bighorn sheep, whereas desert bighorn sheep are not typically found in forests. Thus, there is still more to be learned about the habitats of mountain sheep, especially in isolated populations and local ranges. Managing habitats is crucial for maintaining healthy populations of mountain sheep. Biologists are aware of this, and they recognize that rugged topography with adequate escape terrain adjacent to foraging areas is intricately linked to all other aspects of mountain sheep habitat in North America. A significant challenge remains, however; there is no single entity that manages mountain sheep.

Most lands held in public trust are managed by federal, state, provincial, or territorial entities, and the unique values and priorities reflected within these authorities often lead to differences in actions that may not be favorable to mountain sheep. Managers should work cooperatively to protect all habitats that are currently in good condition, improve those that are not, and address other factors limiting the potential for mountain sheep populations to thrive (Brewer et al. 2014). There are many influences to

mountain sheep habitats discussed herein, and elsewhere in the literature, discussing expanded road networks, linearly designed urban developments, unregulated populations of non-native ungulates and feral equids, woody and invasive plant distribution, and limited surface water availability as the primary threats to the management of healthy mountain sheep populations. Mountain sheep populations today are the result of anthropogenic disturbances and actions; they are being forced to coexist with humans in the landscape. New resource managers must understand that the ecological effects across landscapes have been evolving for a very long time and thus require consideration of past and future impacts in the context of generational timescales, rather than from the date of their arrival on the job. Robinett (1990) recounts the history of the Tohono O'odham range in southern Arizona over several centuries, detailing the ecological conditions that changed during this time. Prior to the arrival of Spanish missionaries in the late 1600s, this native culture lived in scattered groups, moving seasonally between summer and winter settlements as dictated by available water conditions. They subsisted primarily through hunting of native species and the cultivation and farming of edible, native plant species, so their influence on the landscape was minimal. Cultural changes concurrent with the arrival of Spanish missionaries led to the gifting of cattle and horses to various communities, marking the beginning of a new era in landscape use (and damage) as these herds expanded; over the years, many escaped, forming large, feral herds. Deep wells were dug, and by 1919, approximately 60,000 non-native livestock once occupied one of the most arid regions in the West (Chapter 2). Eventually, livestock-related overgrazing resulted in a significant shift in vegetation assemblages from native species more palatable to mountain sheep to one dominated by largely unpalatable ones. Environmental degradation included streambank destabilization, the drying of many rivers and other water sources, deep erosional headcuts, and a decline in native fauna. The point we are making is that following the Euro-American settlement of the West, ecological change has been ever-present and relentless, and understanding the history of these changes and their continued effect today is an important component of managing mountain sheep and their habitat, now and into the future. We must understand and learn from the past to advance science and management of mountain sheep and their environments successfully into the future.

As evidenced through economic investment, society wants wildlife populations and wild places managed sustainably, conserving their inherent aesthetic and intrinsic values while also providing for recreation, harvest, economic, and scientific uses (Canadian Tourism Commission 2012, Association of Fish and Wildlife Agencies 2023, Outdoor Recreation Roundtable 2024, Chapter 21). Accomplishing this while also managing and mitigating direct human effects on mountain sheep habitat (i.e., quality, availability) and the indirect effects from climate change (i.e., climate change, including altered precipitation patterns and increased temperatures, can negatively affect habitat quality and availability), resource and land-use decisions, and socio-political activism will continue to challenge wildlife managers and, by extension, mountain sheep in North America.

The issues affecting the habitats of mountain sheep in western North America are complex. Land management agencies responsible for management of mountain sheep range must actively prioritize the stewardship of wildlife habitats in resource management and land-use planning, directing those solutions using laws and policies, implemented on a scale large enough to be effective in sustaining healthy habitats and populations into the future. Recruiting biologists will need to recognize that the effects of change may not all be negative in nature. With this perspective, they will need to help shape effective strategies for mitigating negative outcomes by building on the lessons learned from the past while also incorporating new and relevant knowledge of the current day.

Mountain sheep populations in North America face a myriad of evolving threats. Effective conservation and management strategies are imperative to ensure the long-term survival of mountain sheep. By understanding the complex interactions between mountain sheep ecology and the evolving threats they face, we can work toward safeguarding these magnificent creatures by stewarding the ecosystems they inhabit, considering their benefits to future generations.

### 24.2.5 POPULATION MODELING

Modeling of populations is another field that is rapidly advancing, and biologists will be well advised to understand the advantages and limitations inherent in the latest models used to manage and conserve mountain sheep. Model descriptions, development, and use can be challenging for biologists who are not well-versed in statistics or who may not fully understand the unique strategies of locally adapted habitat use by small or remote mountain sheep populations. In such situations, we strongly encourage biologists to seek the assistance of modelers to better understand how the parameters incorporated into models can be effectively used. Some models are more useful than others when developing specific answers to behavioral or resource use questions, and they can be a valid component of conservation when locally relevant resource data are applied at a geographically appropriate scale.

### 24.2.6 TAXONOMY

Since the 1980s, genetic data have been important to our collective understanding of intraspecific variation in mountain sheep. The subspecies taxonomic framework, however, remains unresolved due to poorly defined boundaries. Although initially intended to illustrate variation within species, the subjective nature of these divisions obscured true intraspecific differences. Genetic methodologies,

such as karyotyping and mitochondrial DNA sequencing, have facilitated the identification of unique populations of mountain sheep, aiding conservation efforts. New genetic insights enable the reevaluation of historical ranges, enhancing our understanding of population structure, connectivity, and demographics. Ongoing research into adaptive variation, disease susceptibility, and dietary needs will further enhance the conservation of mountain sheep. Still, some entities remain uncomfortable with the evolution of subspecies assemblages and assignments dictated solely through genetic analyses. The need will be to acknowledge the viewpoints of those who are more comfortable with the historical assignment of subspecies, whether they be academics or trophy scoring organizations, while advancing a more current and integrated approach to species conservation and management in the future.

### 24.2.7 DISEASE

Disease directly influences mountain sheep populations. An array of diseases can affect mountain sheep in North America (Chapter 8). The effects of most diseases are limited to impairing individual or local (i.e., herd-level) productivity, survival, or resilience. Some diseases, particularly those from non-native sources, have had severe implications for bighorn sheep conservation. Diseases in their broadest context, particularly those involving apparently non-native (i.e., introduced) agents, may have had more cumulative, range-wide consequences for bighorn sheep than for any other native ungulate species (and perhaps bison [*Bison bison*]). Some diseases continue to affect mountain sheep populations, undermining the conservation of bighorn and thinhorn subspecies. The challenge is to be able to monitor populations, determine the initial influences of disease, and create socio-political support for solutions that foster recovery and future viability. This can be a burden on time and be expensive, but it remains a much cheaper approach than widespread die-offs that lead to populations being considered as an endangered species, then requiring a suite of mandated outcomes that are much less flexible and unaccepting of local realities than those that could be created through a proactive social engagement process. Monitoring disease prevalence and understanding its effects are important for the effective management of mountain sheep.

### 24.2.8 PREDATION

Understanding predator-prey dynamics is complex, particularly in light of altered ecological communities (Bleich 2020, 2021, 2022). Another challenge to understanding predator-prey relationships is the complexity of past management actions that could potentially affect the relationships among multiple prey and multiple predators inhabiting the various landscapes where mountain sheep reside. Much remains to be learned about predator-prey interactions; however, the investment in designing future experiments may be substantial unless they are carefully constructed.

There are at least five areas that need more attention: (1) is there a specific set of circumstances (e.g., mountain sheep population size, relation to carrying capacity, proportional reduction of predators or species of predator) when reducing predator numbers will result in an increase in mountain sheep numbers; (2) is there a set of habitat conditions within which predators do not limit mountain sheep populations; (3) are there anthropogenic influences that change regulating factors into limiting factors for mountain sheep, and are there thresholds within which mountain sheep may thrive with human influence present; (4) are there optimum densities of predators or predator assemblages (wolves [*Canis lupus*], mountain lions [*Puma concolor*], coyotes [*Canis latrans*], bears [*Ursus* spp.], eagles [*Aquila* spp. and *Haliaeetus* spp.]) that would mitigate predation rates on mountain sheep; and (5) are there optimum densities of prey (mule deer [*Odocoileus hemionus*], elk [*Cervus canadensis*], free-ranging feral equids) that mitigate predation rates among all prey species? These and other questions need to be addressed as our understanding of predator-prey dynamics continues to develop. Effective management of predator-prey interactions should consider the interplay of predators, prey, and habitat (Chapter 9).

## 24.2.9 COMPETITION

Competition for resources with native, non-native, feral, and domestic ungulates can have a negative influence on mountain sheep. Typically, competition among these ungulates is inversely related to forage or water availability, and competition can intensify if the availability of these resources is reduced by common use. Feral equids, in particular, can have a significant influence on the availability of forage, water, cover, and areas used by Rocky Mountain and desert bighorn sheep (McKnight 1958, Seegmiller and Simpson 1979, Seegmiller and Ohmart 1981). Similar competition may occur for Dall's and Stone's sheep (Hoefs and Brink 1978, Bleich 2009, Jex et al. 2016). As anthropogenic pressures and climate change intensify, competition across races is likely to increase. Future research should explore how these changes influence resource availability, habitat suitability and function, and competition dynamics. Global positioning system technology can aid in understanding how species may compete temporally and spatially (Tomkiewicz et al. 2010). Proper conservation of mountain sheep populations and habitats will require biologists to understand and manage competition among mountain sheep and naturalized, feral, and domestic ungulates (Chapter 10).

## 24.2.10 MIGRATION

Mountain sheep exhibit diverse migratory behaviors, which are important for accessing seasonal resources (Chapter 11), particularly during years when environmental stressors (e.g., drought, severe weather, habitat change) result in conditions outside the natural range of variability and into extremes. Native populations maintain strong migratory traditions, while restored populations often lack this knowledge. Managing migratory diversity can help sustain migratory behavior in the face of climate change and other extremes, and expanded anthropogenic pressures, which can limit landscape connectivity between seasonal ranges or alter the spatiotemporal dynamics of regional phenology. These can have cascading effects on other biotic and abiotic interactions, such as the need to balance access to quality forage resources and risk of predation, or seeking security provided by spatial refugia in the face of managing thermal stress. In each population scenario (i.e., native, augmented, rest ored), wildlife managers will need to consider how specific herds may respond to stressors and how this may affect patterns of seasonal habitat use, so that they can focus management actions on the most significant and immediate issues at hand.

## 24.2.11 TRANSLOCATIONS

Translocations have been vital for establishing and augmenting mountain sheep populations, particularly in the United States. Keeping abreast of the latest developments in capturing and translocating has continually advanced, and biologists must keep pace with current advances and developments, applying information appropriately in the best interest of conservation (Chapter 14). While translocation has been the primary tool used to reestablish bighorn sheep across western North America, its future use may be limited (Bleich 2021). Reducing the risk of disease spillover from domestic livestock to bighorn sheep is important for the effective restoration of the species, and future translocation efforts must address ways to minimize contact with domestic sheep or goats. In some areas where the risk of pathogen spillover from domestic livestock may be too great to overcome, managers may need to shift from a paradigm where individuals from successful populations were used as translocation stock, to new approaches where both males and females are subject to hunter-harvest in order to manage either population growth or density. Harvest of female bighorn sheep remains uncommon, but it is practiced in several states and provinces (Monteith et al. 2013, 2018; Whiting et al. 2023, LaSharr et al. 2019). A male-biased harvest, however, can have limited effect on the population dynamics of ungulates (where harvest pressure is sustainably managed at harvest rates similar to compensatory mortality), because the abundance of males has little influence on the nutrition of females and, consequently, recruitment of young into the population (McCullough 1979, 2001; Bowyer et al. 2014, 2020; Monteith et al. 2018). Thus, female harvest offers a beneficial, but underused, management option for regulating density-dependent processes for many ungulate populations by holding populations below the ecological carrying capacity (Monteith et al. 2018), allowing managers to manipulate population size to decrease nutritional limitations and competition for resources (McCullough 1979, 2001), or to decrease the potential for pathogen transmission by reducing population density. The decision to implement female harvests, however, is contingent upon several biological and sociological considerations.

Translocation is subject to other constraints, and there is growing recognition that moving animals always includes the risk of moving pathogens. In some instances, a lack of public support for efforts that may increase the risk of death of translocated individuals may require the removal of apex or specialized predators or otherwise raise concerns about animal welfare or influences to human interests. Managers also face continued uncertainty about the appropriateness of allowing the mixing of mountain sheep from different evolutionary lineages, and ongoing questions about the most suitable source of bighorn sheep for use in restoration efforts (Epps et al. 2019). Finally, due to the lingering effects of founder effects and genetic bottlenecks, managers face questions about how to best manage genetic diversity while minimizing disease transmission risk among bighorn sheep resulting from natural movement and translocation. Despite recognizing the evolutionary benefits of genetic adaptations across mountain sheep distributions, little is yet known about how variation in genetic diversity contributes to the success of both natural and restored populations.

## 24.2.12  HUNTING

Another challenge to the management and conservation of mountain sheep involves hunting and social license as it relates to the evolutionary development of harvest-related regulations. As managers face challenges associated with expanded anthropogenic effects, disease epizootics, and habitat loss, it is incumbent upon resource professionals to recognize the inherent rights of Indigenous Peoples and develop information of public consumption that includes sound rationales in support of innovative harvest strategies that improve our ability to manage mountain sheep sustainably into the future.

For species such as the thinhorn sheep, future challenges will be largely influenced by climate-related changes in range conditions and alterations to species assemblages. The current harvest strategy focused on mature males is a conservative approach that will likely continue given public interest and current levels of support for this regulated approach; however, as harvest pressure increases to levels where effects on mature male representation, male rut behaviors, or population effects are observed, a departure from the general-open-season or over-the-counter regulations that maximize hunter opportunity may be replaced with more restrictive hunting seasons or move to a draw-only hunting opportunity. Alternative harvest strategies will require more research on abiotic and biotic factors affecting productivity and recruitment, but also include a deeper understanding of hunter motivations, human dimensions, and societal expectations for hunted thinhorn sheep populations. Indigenous and public interests will need to be researched and monitored to assist with defining allocations of allowable harvests and hunting opportunity, and to assess social license supporting new regulatory actions.

Because thinhorn population trajectories are primarily influenced by factors outside management control (e.g., weather), focusing on the resilience of population dynamics in an uncertain future will aid in developing adaptive and innovative harvest management options.

## 24.2.13  HUMAN DIMENSIONS

The human dimensions of wildlife management play a vital role in conservation outcomes. The complex relationships that Indigenous and non-Indigenous Peoples have with nature are incorporated into the policymaking, regulation, and funding of wildlife management, in concert with the philosophy of the North American Model of Wildlife Conservation (NAM). The NAM is rooted in the public trust doctrine, which requires that resources (i.e., wildlife) be managed in the interest of both present and future generations (i.e., in the public's best interest). Agencies charged with managing wildlife at state, provincial, territorial, or federal levels operate as trustees, within a complex suite of regulatory frameworks, employing the principles of the NAM in preserving wildlife and access to it for future generations.

Mountain sheep hold cultural and economic value; they influence local economies through non-consumptive use (i.e., recreation, ecotourism) and consumptive use (e.g., subsistence, ceremonial, regulated hunting). As iconic game species in North America, bighorn and thinhorn sheep have a long and dynamic history of interaction with humans, playing a vital role in local economies that support conservation efforts and contribute to the livelihoods of local communities. Bighorn sheep experienced a massive decline in range and population size during the 19th century, primarily due to human occupation of foothills, overgrazing by domestic livestock, and the introduction of pathogens (Buechner 1960, Chapters 2 and 8). At local scales, this encroachment led to significant loss of habitat through displacement and competition for forage resources, unregulated harvest, and the spread of fatal diseases to the bighorns (Toweill and Geist 1999). Efforts to restore bighorn populations to their former ranges were largely influenced by the value of the species as recreational game species and for wildlife viewing (Krausman 2000, Hurley et al. 2015). Restoration, however, subsequently led to many bighorn populations being in direct contact with people because urban expansions overlapped remaining habitats (Chase 2013, Chapters 20 and 21).

The population declines, fragmentation of habitat, and narrowing of bighorn ranges have had a profound effect on cultural relationships held by Native American peoples and broader public opinion toward the species. The loss of direct connections shifted the animal from being a spiritual guide that fostered close connections to a healthy environment, while also supporting an accessible source of food through subsistence hunting, to a species of conservation concern with limited opportunities for firsthand connection and consumptive use (Krausman and Bleich 2013,

Hessami et al. 2021). Recognition of long-held, Indigenous beliefs and knowledge and the role of human dimensions in wildlife and social science research will continue to be imperative to overcoming conflicts, maintaining support for investment in restoration efforts, and preserving bighorn and thinhorn sheep and their habitats into the future.

Studies that investigate the relationship between humans and mountain sheep have largely focused on the implications of humans on mountain sheep including conflict and disturbance from human activity or development (Schoenecker and Krausman 2002, Polfus and Krausman 2012, Lowrey and Longshore 2017, Brushett et al. 2023), evolutionary pressure from selective hunting (Monteith et al. 2013, 2018; Pigeon et al. 2016, Douhard et al. 2016, Lassis et al. 2022), attitudes toward the management of mountain sheep and relevant predator species (Harris et al. 1995, Chase 2013), and human-wildlife conflict resulting from livestock disease spillover (Heinse et al. 2016). Human attitudes toward mountain sheep are understudied. Hunters (including Indigenous harvesters) and wildlife viewers, however, receive substantial positive emotional and spiritual benefits from participating in their respective activities. Others experience intrinsic benefit, which is valued inherently attributable to something existing, independent of any direct benefit to people. Both valuations are important, especially for highly charismatic species like mountain sheep, which frequently appear in popular media and represent success stories in conservation.

Although advances have been made in restoring mountain sheep populations and their habitats, the species' population has been substantially reduced relative to historical estimates and its original distribution (Buechner 1960). As a result, hunting for mountain sheep in some jurisdictions is provided as once-in-a-lifetime or limited opportunity through lottery draws, special auctions, or raffles that tend to be more costly and harder to obtain than similar opportunities for other ungulate species. The high cost of entry, the difficulty navigating the terrain and finding stalking opportunities, relatively low success rates, and low densities of the sheep create an inherently challenging environment that may increase the recreational value and prestige associated with successfully harvesting a sheep. Although likely not the intent at the outset of restoration efforts, the fact that hunting opportunities, although very limited, continue to be viewed positively by those who can participate suggests deeper meaning beyond the simple act of harvest that those opportunities provide; this may possibly parallel traditional and cultural pursuits of mountain sheep historically undertaken by Indigenous Peoples.

Bighorn sheep populations have grown from approximately 25,000 in the 1950s to an estimated 85,000 today, and thinhorns exceed 50,000, due to partnerships between state and federal governments and efforts by non-governmental and conservation organizations that provide funding for translocations, research, habitat restoration,

disease control, predator management, and educational initiatives. Tribal Nations have played a significant role in bighorn sheep restoration in the United States, including efforts by the Navajo Nation, the Confederated Tribes of the Colville Reservation, the Pyramid Lake Paiute Tribe, Eastern Shoshone and Northern Arapaho Tribes in the Wind River Reservation, and The Salmon River Bighorn Sheep Project on Nez Perce tribal lands. In Canada, First Nations governments have also supported focused management efforts that benefit mountain sheep and their habitats in British Columbia and Alberta. Relationships between government agencies, First Nations, and both consumptive and non-consumptive conservation-minded individuals and organizations will continue to be imperative to overcoming conflicts, maintaining support for investment in restoration efforts, and preserving bighorn and thinhorn sheep into the future.

## 24.2.14 Conflicts in Management

Conflicting conservation goals or strategies among agencies sharing management responsibilities can hinder effective habitat management for mountain sheep. Similarly, conflict between these agencies and the public over species management priorities can also influence success and damage relationships and trust. Conflict may involve stakeholders who had no prior concern regarding mountain sheep management until an objective or tool used to achieve a management outcome was initiated. Resolving conflicts requires wildlife managers to engage with rights and title holders and stakeholders to gain an understanding of their immediate concerns and the values they place on wildlife resources, and the perceptions they hold regarding the issues at hand and how to best achieve the identified stewardship goals. Mountain sheep may serve as the catalyst for conflict in many of the same ways that other species do (e.g., highway collisions, disease transmission between domestic livestock and wildlife). Agency managers who hold the appropriate decision-making authority should proactively engage on topics where conflict is anticipated, employing approaches such as structured decision-making to ensure that concerns are clearly heard, understood, and that acceptable solutions are collaboratively crafted. Wildlife management agencies have continued to make progress with respect to public engagement and understanding the interests of affected parties, stakeholders, and the public in general. Delivering an open and transparent engagement process is important when crafting collaborative solutions and making science-based decisions that prevent and mitigate conflict. Creating feedback loops and touchpoints with decision-makers is crucial for enhancing communication and collaboration among agencies, rights and title holders, researchers, stakeholders, and the public, thereby improving the addressing of complex conservation challenges faced by mountain sheep.

### 24.2.15  MANAGEMENT OF MOUNTAIN SHEEP IN MEXICO

Mountain sheep management varies across North America but is substantially different in Mexico, where desert bighorn sheep are the highest-priced sport-hunted big game mammal in the country. Mexico also hosts the largest population of desert bighorn sheep in North America. Due to the rarity of desert bighorn sheep hunting opportunities outside Mexico, there is high international demand from big game hunters willing to pay exorbitant prices for hunting permits (Chapter 17).

As a native, desert-adapted large herbivore, the desert bighorn sheep can be considered a potential flagship species for arid mountainous ecosystems within its distribution in northern Mexico and the Baja California Peninsula. Desert bighorn sheep are a charismatic and economically valuable species to communities (i.e., communal interests) and the state and federal governments. Management of the species and both non-consumptive and consumptive use is shared between the various levels of government. These direct connections to bighorn sheep promote public awareness and support for ecological and conservation values, and inspire the development of social responsibility and a social license for managing wildlife resources and their habitats sustainably. A change in environmental awareness among community members within protected and surrounding areas involves obviating social ineffectiveness and creating changes in the values, attitudes, and behaviors that are vital to Mexico's communal natural resource management programs.

Wildlife managers in Mexico have also pioneered extensive captive breeding programs for bighorn sheep in the country. Captive-bred sheep have been used to restore desert bighorn sheep populations in the Mexican states of Chihuahua and Coahuila, where they had been extirpated. The number of desert bighorn sheep has increased substantially in Mexico because of the establishment of additional wild populations, augmentation of existing populations, and successful captive breeding programs. Mexico is the only country in North America that prioritizes an integrated mountain sheep captive breeding program and free-ranging population management, a management policy that has had great success. Despite this success, Mexico's management remains somewhat controversial, and many harvested sheep cannot be listed in record books because they were raised behind wire, a practice that does not align with the ethics of fair chase hunting held by some international trophy scoring organizations.

### 24.2.16  LONG-TERM RESEARCH

Because many natural processes relevant to mountain sheep ecology and conservation occur over several years and even decades, long-term monitoring is necessary to characterize and quantify their variability and the significance of their effects. Events happening over different stages of life, particularly early in life, can have important consequences for the growth, reproduction, and survival of individuals. Four long-term programs, each monitoring individually marked bighorn sheep, have enhanced our understanding of their ecology, behavior, evolution, the effects of disease, and the conservation of mountain sheep as a whole. These programs require stable funding, adequate field facilities, the ability to mark and recapture individuals, support from provincial, state, territorial, and federal government agencies, and strong contributions from students and academia. Analyses combining results from multiple long-term studies have revealed similarities and differences in the effects of cougar predation, implications from disease, variability of population density, and responses to various management regimes. Measurements of the horn size of harvested males, particularly annual horn growth increments, provide an invaluable source of data for testing multiple hypotheses on the ecology, evolution, and conservation of mountain sheep. Long-term research across large landscapes, varied ecosystems, and diverse species assemblages will significantly contribute to a deeper understanding of the biology and ecology of mountain sheep, thereby supporting the stewardship, conservation, and management of these animals throughout North America.

## 24.3  SUMMARY

Managing any wild animal is challenging; it becomes even more challenging when conflicting opinions arise regarding the use of landscapes shared by that animal. As humans encroach on the habitats of mountain sheep across North America, the challenges sustainable management must address become increasingly important. Throughout this book, the authors of chapters have emphasized the importance of cooperation between the various managers of mountain sheep and the public. In this chapter we highlight the importance of understanding nutrition, water requirements, body and horn size, habitat, population modeling, taxonomy, disease, predation, competition, migration, translocation, hunting, human dimensions, cultural and economic relationships, conflicts in management, the unique management of mountain sheep in Mexico, and the importance of long-term research. If humans and mountain sheep are to continue coexisting, we will need more sophisticated information and a deeper understanding of these topics for the successful conservation and management of mountain sheep into the future.

## REFERENCES

Association of Fish and Wildlife Agencies. 2023. AFWA 2023 Annual Report. Washington, D.C., USA. https://www.fish-wildlife.org/application/files/3317/1052/2834/2023-AFWA -AnnualReport-v01F.pdf. Accessed 9 Jul 2024.

Bleich, V. C. 2009. Perceived threats to wild sheep: levels of concordance among states, provinces, and territories. *Desert Bighorn Council Transactions* 50:32–39.

Bleich, V. C. 2020. Mountain lions, minimum viable populations, and intact ecosystems: a cautionary note. *Ecological Applications* 30:e01990.

Bleich, V. C. 2021. Fire on the mountain—run boys, run! *California Fish and Wildlife Journal* 107:33–43.

Bleich, V. C. 2022. Feral horses, feral asses, and professional politicians: broodings from a beleaguered biologist. *Human-Wildlife Interactions* 16:337–342.

Bowyer, R. T., K. M. Stewart, V. C. Bleich, J. C. Whiting, K. L. Monteith, M. E. Blum, and T. N. Lasharr. 2020. Metrics of harvest for ungulate populations: misconceptions, lurking variables, and prudent management. *Alces* 56:15–38.

Bowyer, R. T., V. C. Bleich, K. M. Stewart, J. C. Whiting, and K. L. Monteith. 2014. Density dependence in ungulates: a review of causes, and concepts with some clarifications. *California Fish Game Journal* 100:550–572.

Brewer, C. E., V. C. Bleich, J. A. Foster, T. Hosch-Hebdon, D. E. McWhirter, E. M. Rominger, M. W. Wagner, and B. P. Wiedmann. 2014. *Bighorn Sheep: Conservation Challenges and Management Strategies for the 21st Century*. Wild Sheep Working Group, Western Association of Fish and Wildlife Agencies, Cheyenne, Wyoming, USA.

Brushett, A., J. Whittington, B. Macbeth, and J. M. Fryxell. 2023. Changes in movement, habitat use, and response to human disturbance accompany parturition events in bighorn sheep (*Ovis canadensis*). *Movement Ecology* 11:36.

Buechner, H. K. 1960. The bighorn sheep in the United States, its past, present, and future. *Wildlife Monographs* 4:3–174.

Canadian Tourism Commission. 2012. Sport fishing and game hunting in Canada: an assessment on the potential international tourism opportunity. Canadian Tourism Commission Research and Evaluation, Ottawa, Ontario, Canada. https://publications.gc.ca/collections/collection_2013/ic/Iu86-49-2012-eng.pdf. Accessed 15 Aug 2024.

Chase, L. D. 2013. Human dimensions of reintroduced bighorn sheep and an associated increased mountain lion harvest along an urban interface. *Desert Bighorn Council Transactions* 52:16–28.

Douhard, M., M. Festa-Bianchet, F. Pelletier, J. Gaillard, and C. Bonenfant. 2016. Changes in horn size of Stone's sheep over four decades correlate with trophy hunting pressure. *Ecological Applications* 26:309–321.

Epps, C. W., E. Bowen, M. R. Buchalski, F. Cassirer, D. W. Coltman, W. C. Conway, M. Cox, R. Harris, J. Jahner, M. Matocq, and T. Parchman. 2019. Frequently-asked questions about wild sheep genetics and genomics. *Proceedings of 2018 North American Wild Sheep Goat Symposium* 21:45–75.

Harris, L. K., P. R. Krausman, and W. W. Shaw. 1995. Human attitudes and mountain sheep in a wilderness setting. *Wildlife Society Bulletin* 23:66–72.

Heinse, L, M., L. H. Hardesty, and R. B. Harris. 2016. Risk of pathogen spillover to bighorn sheep from domestic sheep and goat flocks on private land. *Wildlife Society Bulletin* 40:625–633.

Hessami, M. A., E. Bowles, J. N. Popp, and A. T. Ford. 2021. Indigenizing the North American model of wildlife conservation. *FACETS* 6:1285–1306.

Hoefs, M., and V. C. Brink. 1978. Forage production and utilization of a Dall sheep winter range, southwest Yukon Territory. *Northern Wild Sheep and Goat Council Proceedings* 1:87–105.

Hurley, K., C. Brewer, and G. N. Thornton. 2015. The role of hunters in conservation, restoration, and management of North American wild sheep. *International Journal of Environmental Studies* 72:784–796.

Jex, B. A., J. B. Ayotte, V. C. Bleich, C. E. Brewer, D. L. Bruning, T. M. Hegel, N. C. Larter, R. A. Schwanke, H. M. Schwantje, and M. W. Wagner. 2016. *Thinhorn Sheep: Conservation Challenges and Management Strategies for the 21st Century*. Wild Sheep Working Group, Western Association of Fish and Wildlife Agencies, Cheyenne, Wyoming, USA.

Krausman, P. R. 2000. An introduction to the restoration of bighorn sheep. *Restoration Ecology* 8:3–5.

Krausman, P. R., and V. C. Bleich. 2013. Conservation and management of ungulates in North America. *International Journal of Environmental Studies* 70:372–382.

LaSharr, T. N., R. A. Long, J. R. Heffelfinger, V. C. Bleich, P. R. Krausman, T. R. Bowyer, J. M. Shannon, R. W. Klaver, C. E. Brewer, M. Cox, A. Andrew Holland, A. Hubbs, C. P. Lehman, J. D. Muir, B. Sterling, and K. L. Monteith. 2019. Hunting and mountain sheep: do current harvest practices affect horn growth? *Evolutionary Applications* 12:1823–1836.

Lassis, R., M. Festa-Bianchet, and F. Pelletier. 2022. Effects of hunting pressure and timing of harvest on bighorn sheep (*Ovis canadensis*) horn size. *Canadian Journal of Zoology* 100:507–516.

Leopold, A. 1933. *Game Management*. Charles Scribner's Sons, New York, New York, USA.

Lowrey, C., and K. M. Longshore. 2017. Tolerance to disturbance regulated by attractiveness of resources: a case study of desert bighorn sheep within the River Mountains, Nevada, USA. *Western North American Naturalist* 77:82–98.

McCullough, D. R. 1979. *The George Reserve Deer Herd: Population Ecology of a K-Selected Species*. The University of Michigan Press, Ann Arbor, Michigan, USA.

McCullough, D. R. 2001. Male harvest in relation to female removals in a black tailed deer population. *Journal of Wildlife Management* 65:46–58.

McKnight, T. L. 1958. The feral burro in the United States: distribution and problems. *Journal of Wildlife Management* 22:163–179.

Monteith, K. L., R. A. Long, T. R. Stephenson, V. C. Bleich, R. T. Bowyer, and T. N. Lasharr. 2018. Horn size and nutrition in mountain sheep: can ewe handle the truth? *Journal of Wildlife Management* 82:67–84.

Monteith, K. L., R. A. Long, V. C. Bleich, J. R. Heffelfinger, P. R. Krausman, and R. T. Bowyer. 2013. Effects of harvest, culture, and climate on trends in size of horn-like structures in trophy ungulates. *Wildlife Monographs* 183:1–28.

Outdoor Recreation Roundtable. 2024. Outdoor Recreation Roundtable Association: Together Outdoors Initiative. https://recreationroundtable.org/programs/together-outdoors/. Accessed 5 Aug 2024.

Pigeon, G., M. Festa-Bianchet, D. W. Coltman, and F. Pelletier. 2016. Intense selective hunting leads to artificial evolution in horn size. *Evolutionary Applications* 9:521–530.

Polfus, J. L., and P. R. Krausman. 2012. Impacts of residential development on ungulates in the Rocky Mountain West. *Wildlife Society Bulletin* 36:647–657.

Robinett, D. 1990. Tohono O'odham range history. *Society for Range Management* 12:296–300.

Schoenecker, K. A., and P. R. Krausman. 2002. Human disturbance in bighorn sheep habitat, Pusch Ridge Wilderness, Arizona. *Journal of the Arizona-Nevada Academy of Science* 34:63–68.

Seegmiller, R. F., and C. D. Simpson. 1979. The Barbary sheep: some conceptual implications of competition with desert bighorn. *Desert Bighorn Council Transactions* 23:47–49.

Seegmiller, R. F., and R. D. Ohmart. 1981. Ecological relationships of feral burros and desert bighorn sheep. *Wildlife Monographs* 78:1–58.

Smiley, R. A., B. L. Wagler, T. N. Lasharr, K. Denryter, T. R. Stephenson, A. B. Courtemanch, T. W. Mong, D. Lutz, D. E. McWhirter, D. Brimeyer, P. Hnilicka, B. Lowrey, and K.

L. Monteith. 2022. Heterogeneity in risk-sensitive allocation of somatic reserves in a long-lived mammal. *Ecosphere* 13:e4161.

Tomkiewicz, S. M., M. R. Fuller, J. G. Kie, and K. K. Bates. 2010. Global positioning system and associated technologies in animal behaviour and ecological research. *Philosophical Transactions of the Royal Society B: Biological Sciences* 365:2163–2176.

Toweill, D. E., and V. Geist. 1999. *Return of Royalty: Wild Sheep of North America.* Boone and Crockett Club, Missoula, Montana, USA.

Whiting, J. C., V. C. Bleich, R. T. Bowyer, and C. W. Epps. 2023. Restoration of bighorn sheep: history, successes, and remaining conservation issues. *Frontiers in Ecology and Evolution* 11:1083350.

# Index

Note: Page numbers followed by "f", as well as *italicized* page numbers, denote figures; page numbers followed by "t" denote tables.

4-H, 338

## A

Abortion, 114, 221
Abscess, 140, 142–143, 152
Abundance, 28, 48, 50, 52, 137, 139, 148, 151, 160–161, 163, 167, 169–170, 175–176, 184, 196, 201, 213–215, 218–219, 223, 233, 235–238, 241–244, 246–248, 280, 282, 284–285, 290, 296, 298, 335, 347, 349, 354, 358, 379, 384, 402–403t, 405, 409t, 414–415, *431*–432, 455, 459
  population, 48–49, 136, 160, 175, 199, 223, 281, 296
  prey, 162, 167
  resource, 176, 347
  vegetation, 282, 350
Acacia, 119
*Acacia* spp., 119
*Achillea* sp., 118
Acidosis, 118, 152–153
Activity, 56, 116, 122, 140, 166, 175, 209–210, 216, 298, 353–354, 367, 406, 414, 428, 432
Adaptive capacity, 97, 100–101
  adaptive mitigation, 402, 404t, 418
  impact management, 421–422
  management, 141, 169, 247, 273, 285, 301, 324, 381, 402, 403t–404t, 418, 421t–422
  variation, 84, 100–101, 104, 458
Adits, 213
Adenoviral hemorrhagic disease, 146
*Aedes* spp., 220
Age, 14, 55–64, 66, 72–74, 74f–75, 76f–77, 77f–81, 77f, 89, 92, 100, 104–105, 114–115, 121, 126, 139–140, 143, 145, 152, 167, 179f, 196, 202, 209, 240, 245, 266, 280–281, 291–294, 296–298, 300, 311, 317, 333, 351, 359, 369, 383, 388f, 405, 432, 443, 446f, 450–451, 456–457
  Class, 60, 77f, 167, 222, 244–245, 280, 283, 423, 447
  distribution, 90, 280, 300
  ratio, 240, 244–245
  -specific survival, 72, 280
  Structure, 64, 77–78, 160, 234t, 238, 246, 265, 280–281, 293, 297, 299, 323
Agriculture, 7, 34, 423, 427t, 455
*Agropyron* spp., 119, 351, 360
Alaska Board of Game, 298, 300–301
  Fish and Game, 220
  National Interest Lands Conservation Act, 298–299
Alberta Chapter of the Wild Sheep Foundation, 442

Alberta Recreation, Parks and Wildlife Foundation, 444
  Conservation Association, 442
*Alces alces,* 26, 150, 160, 331, 428
Alder, 364
Alkaloids, 120–121
*Alnus* spp., 364
American Game Policy, 336
  Museum of Natural History, 25
Animal-activated detection system, 428–429
  indicator concept, 125
*Ammotragus lervia,* 9, 36, 51, 181, 216, 312, 405
Anemia, 120, 150–151
*Anabaena, 216*
Animal Health Act, 431
*Anopheles* spp., 220
Anthrax, 151
Anthropogenic disturbance, 166–167, 266, 350–351, 360, 397, 407, 457
Anthropogenic
  activity, 369, 457
  anthropogenic pressure 175
  anthropogenic mineral source 221
  changes, *See* anthropogenic influence
  development, *See* anthropogenic influence
  disturbance, 166–167, 266, 350–351, 360, 397, 407, 457
  effects, *See* anthropogenic influence
  factors, *See* anthropogenic influence
  features, 167, 201, 357
  foods, 429
  habitat degradation/fragmentation/ modification/perturbation, 3, 7, 89, 180
  influence, 7, 9, 136–137, 162, 169, 219, 258, 359, 361, 416, 459–460
  land use, 312–313
  lick, *See* anthropogenic mineral source
  mineral sources, 198–199, 201, 221
  mineral supplementation, 201
  movement barriers, 305
  pressures, 175, 177, 185, 201, 378, 392, 459
  stressors, 392
  water, 213, 215, 239, 241, 359
*Antilocapra americana,* 8, 27, 142, 160, 176, 213, 311, 322, 444
  *peninsularis,* 322
  *sonoriensis,* 213
Antinutritional factors, 119–121
Aoudad, 8–10, 36, 51, 66, 181–182, 312, 405
  population, 10
  range, 10
*Aquila chrysaetos,* 36, 162, 312, 350
  spp., 459
Argali, 4, *10*, 13
Argaliform, 4–5
Arizona Antelope Foundation, 419
  Department of Transportation (ADOT), 418–419

Desert Bighorn Sheep Society, 387–388, 419
  Game and Fish Department (AZGFD), *267,* 388, 418–420
  Wilderness Coalition, 419
  Wildlife Federation, 419
*Artemisia* spp., 118–119, 121
  *tridentate,* 180
Arthropod parasites, 144–145, 149, 153
Artiodactyla, 8, 19
Asociación Nacional de Ganaderos Diversificados, 319
Assignment tests. *See* genetic assignment
*Atriplex* spp., 119
Auction, 49t, 290, 296, 301, 320, 333, 338, 340, 417, 461
Avens
  Arctic, 368
  mountain, 364

## B

Bacterial disease, 151–152
Bag limit, 35–36, 297
Bear, 332–333, 385, 409t, 459
  black, 162, 164t
  brown, 104, 162, 164t, 332
    predation, 367
  grizzly, *See* brown bear
  polar, 212
Bed, 423
  site, 350, 354, 364
Bedding, 199, 210–211, 245, 347, 349–350, 354, 423
Breeding, 61, 72, 76, 79, 103, 145, 196, 200, 217, 270, 293, 297, 311, 322, 347, 387, 449
  group, 76, 78
  migrations, 78–79, 81, 200, 444–445
  opportunity, 78, 200, 297–298
  season, 76, 79, 164, 192, 424
  suppression, 168
Behavior, 27, 30, 38, 72–73, 75, 77, 81, 98, 114, 117, 128, 137, 143, 150, 163, 166–168, 175, 183, 191, 193–195, 198, 213, 217, 234, 242–243, 247, 268, 272–273, 282, 285, 290, 332–335, 360, 364, 369, 398t, 404, 406–408, 415, 428, 430, 432, 448, 456, 462
  avoidance, 40, 72, 81, 98–99, 162–163, 166, 197, 219, 283, 350–351, 353, 356, 359, 364, 366–367, 369, 407, 424, 427, 455
  foray, 79, 99, 145, 191, 200, 202–203, 268, 273, 432
  grazing, 35
  grouping, 272
  rut, *See* rut
  social, 72–81, 145, 323
  suckling, 56

Behavioral adaptations, 81, 122–123, 176, 192, 212, 219, 223
  alteration, 166
  awareness, 324
  change, 166, 176
  characteristics, 8, 266
  decisions, 347
  differences, 149
  displacement, 183
  ecology, 365
  factors, 280
  grooming, 150
  interactions, 398t
  mechanism, 191, 208–209, 354
  observations, 443–444
  plasticity, 366
  preference, 352
  process, 347
  regulation of body temperature, 210–211
  relationships, 26, 38
  response, 162, 191, 405, 407
  shifts, 93
  tactic, 72
  trade-off, 72
*Betula* spp., 364
Bharal, 9–10
*Bibersteinia* spp., 142
  *trehalose*, 142
Birth. *See* parturition
Bite (food), 114–128
  count technique, 123
  mass, 123
  rate, 123, 184
  size, 283
Bison, 9, 36, 142, 147, 160, 180, 398, 458
  population, 35
*Bison bison,* 9, 35, 142, 160, 180, 398, 458
Bitterbrush, 119
Blackflies, 220
Blueberry, 119
Bluegrass, 118–119, 351, 364, 368
Bluetongue virus (BTV), 146, 215, 270
Board of Game Commissioners 36
Bobcat, 161–164t, 216, 312
Body characteristics. *See* morphology
  color, *See* morphology
  condition, 125, 146, 149–150, 152, 166, 194, 197, 217–220, 223–224, 367, 398, 456
  conformation, *See* morphology
  fat, 28, 61, *63*, 114–*117*, 122, 125–128, 197, 218, 282–283, 456
  fluid volume, 120
  growth, 62–63, 322
  heat, 209–211, 217, 224, 456
  length, *See* morphology
  maintenance, 64
  mass, 3, 55–56, 60–*63*, 73, 75–78, 100, 114, 122, 126, 164, 192, 211–212, 220, 333, 444, 451
  measurements, *See* morphology
  pH, 120
  reserves, 114–115, 283, 363
  shape, *See* morphology
  size, 55t, 57, 59–64, 72–73, 77–78, 99, 122, 126, 208, 211–*212*, 218, 247, 293, 456
  temperature, 209–212, 217, 221, 354
  water, 208–210
Boone and Crockett Club, 5, 56, 58t, 320
Border disease virus, 147

*Bos grunniens,* 9
Bot fly, 145, 150, 220
  nasal botfly, 145, 150
Bottleneck, 84, 87, 92–93, 97–99, 103–104, 114, 283, 458, 460
*Bouteloua* spp., 351
Bovidae, 8–9
Bovine viral diarrhea (BVD), 147–148
Bovine viral diarrhea (BVD), tuberculosis, 148, 423
*Branhamella* spp., 147
Breeding, 61, 72, 76, 79, 103, 145, 196, 200, 217, 270, 293, 297, 311, 322, 347, 387, 449
  group, 76, 78
  migrations, 78–79, 81, 200, 444–445
  opportunity, 78, 200, 297–298
  season, 76, 79, 164, 192, 424
  suppression, 168
Bromegrass, 119
*Bromus* spp., 119
  *tectorum,* 65, 387, 397, 405
Bronchopneumonia, 140, 144, 146
Brooming. *See* horn loss
Buckthorn, 119
Buckwheat, 119
Bureau of Biological Survey, 26
  of Land Management (BLM), 182, *264*–265t, 337, 383, 418, 425, 445
  Reclamation, 418–419
Burro, 34, 52, 150, 182–184, 216, 312–313, 382–384, 405
  distribution, 182, 383, 405
  population, 383
Bylaws, 420, 431

C

Cactus, 74, 208, 210, 359–360
  barrel, 212, 359
  saguaro, 383
Camel, 8
Camelids, 117
*Camelus* spp., 8
*Camptorhynchus labradorius,* 160
Canada Research Chairs program, 442
*Canis latrans,* 36, 72, 162, 216, 312, 350, 459
  *lupus,* 72, 160, 220, 285, 312, 362, 383, 447, 459
*Capra,* 9, 10, 86
  *aegagrus,* 10
  *aegagrus aegagrus,* 181
  *cylindricornis,* 10
  *falconeri,* 10
  *ibex,* 60, 72
  *sibirica,* 10
*Capricornis* species, 9
Caprinae, 8–10, 424
Caprini, 3, 8–10
Captive breeding, 36, 99, 311, 314, 321–322, 324, 462
  populations, *18,* 62, 270, 306, 314, 316–317, 320–321
Capture, 84–85, 93, 97–98, 101, 120, 125, 152, 239, 258–274, 314, 338, 390, 417, 419–420, 426–427t, 430–*431,* 442, 444, 448–449
  -associated mortality, 120, 153, 261
  methods, 260–264
  myopathy, *See* capture-associated mortality
  -recapture, 97, 239–240, 247

*Carex* spp., 118, 352, 364, 387
Caribou, 9, 74, 141, 166, 168, 180, 332
*Carnegiea gigantea,* 383
Carotid worm, 149
Carrying capacity, 65, 127, 163, 167, 169, 184, 280, 282, 285, 297, 312, 387, 391, 405–406, 414, 459
  biological, 414–416
  ecological, 322, 414–416, 459
  landscape, 398–399
  nutritional, 65, 114–115, 127
  social, 414–416
*Cassiope tetragona,* 364
Cattle, 8, 9, 33–35, 97, 138, 142–143, 146–148, 177, 184–185, 312–313, 319, 321, 360, 382, 393, 418, 423–425, 427, 432, 457
Caucasian tur, 10
*Ceanothus* spp., 119
*Centaurea stoebe,* 405
Center for Biological Diversity, 419
*Cephenemyia trompe,* 220
*Cercocarpus* spp., 119
  *canadnesis nelson,* 150
  *Cervus canadnesis,* 26, 74, 149, 160, 176, 216, 236, 368, 384, 398, 424, 444, 447, 459
  *elaphus,* 74, 444
Cestodes, 149–150
Chaining, 385
Chamois, 8, 60
  Eurasian, 9
  Pyrenean, 451
Charles Engelhard Foundation, 442–443
Cheatgrass, 65, 387, 397, 405–*406*
Chemical immobilization, 260–261, 273, 430
Chevrotains, 8
*Chlamydia* spp., 147
Chronic sinusitis, 144–145
*Chrysothamnus* spp., 119
Classification, 3–19, 85, 92, 102, 305, 365
Climate change, 3, 39, 51, 65, 100, 103, 127, 151, 175, 177, 184–185, 201–203, 217, 219–222, 285, 301, 335, 337, 379, 397–402, 408–409t, 441, 448, 450–451, 455, 457, 459
*Clostridium botulinum,* 215, 360
Colorado Colorado Department of Forestry, Game and Fish, 36
  Parks and Wildlife, 339
Colorado Department of Forestry, Game and Fish, 36
Competition, 10, 29, 35, 63–65, 72, 77, 114–115, 123, 142, 145, 168, 175–185, 191, *198,* 215–216, 220, 243, 290, 296, 300, 319, 321, 349, 358, *386,* 389, 398, 405, 455–456, 459, 462
  apparent, 177–178, 180, 194, 397–398t, 402
  with bison, 180
  with caribou, 180
  consumptive, 30
  direct, 34, 176–177, 183
  with domestic ungulates, 175, 177, 184–185, 459
  with elk, with exotic (feral) ungulates, 178–184
  exploitation (exploitative), 177, 183, 397–398t
  interference, 169, 177, 180, 183, 216, 398t, 405, 415
  interspecific, 175, 177, 258, 312

manufactured (facilitated), 398t, 405
  with mountain goat, 180
  with mule deer, 178–180
  with pronghorn, 178–180
  spatial, 184
  sperm, 76
  types, 177–178, 398t
Competitive ability, 176
  advantage, 175, 182
  exclusion, 175, 177
  intensity, 175
  interactions, 166, 175, 178
Conflict, 34, 40, 58, 87, 176–177, 180, 183,
    204, 220, 301, 331, 336, 338–340,
    345–462
  management, 414–433
Connectivity, 85, 93, 95, 98–99, 102, 104, 201,
    203, 219, 222, 305–306, 322, 378,
    380–381, 402, 408, 417–419, 424,
    447, 458–459
Conservation actions, 52, 66, 102, 247,
    319, 323
  activities, 333
  advantages, 317–320
  advocates, 407, 423
  agency, 414
  benefits, 319, 337
  challenges, 461
  concerns, 3, 290, 301, 336, 339, 382, 460
  considerations, 66
  decisions, 241
  easements, 402
  efforts, 40, 136, 202, 222, 311, 313,
    319–320, 323, 332–333, 337–339,
    455, 458, 460
  framework, 323
  genetic metric, 103
  genetics, 105
  genetic studies, 105
  goal, 379, 461
  implications, 87
  initiatives, 332
  instrument, 313
  interest, 300
  investment, 320
  leaders, 455
  measures, 291
  monitoring, 244
  movement, 30
  need, 26, 52, 298, 301, 338
  objectives, 313, 422t, 432
  opportunities, 311
  organization, 290, 298, 301, 334, 337, 382,
    390, 393, 408, 415, 417–420,
    426, 461
  outcomes, 35, 420, 460
  partners, 391
  permit, 320, 338
  plan, 322, 338, 400
  policy, I, 311
  practices, 336
  priority, 202–203, 265
  program, 313, 320, 323–324, 393
  projects, 39, 320
  purposes, 248, 298
  regulations, 292, 301, 311
  social sciences, 330
  stakeholders, 300
  standpoint, 196
  status, 103–104

success, 290, 301, 332, 340
  thresholds, 299
  units, 19
  value, 179, 324, 462
Contagious ecthyma, 147, 270
Convention on International Trade in
    Endangered Species of Wild Fauna
    and Flora (CITES), 305, 316–318
Cooperative Fish and Wildlife Research
    Units, 26
Core area, 306, 313
Corral traps, 260, 266, 442–443
Corridor, 3, 19, 73, 86, 201, 266, 306, 312,
    361, 378–382, 384, 386, 392–393,
    402–403t, 409t, 417–418
Cougar, 36, 72, 161–162, 164t, 166–169, 194,
    216, 284–285, 312, 350, 357, 383,
    385, 402, 409t, 447, 449, 459
  abundance, 167
  density, 270
  distribution, 167, 169
  home range, 216
  hunting, 419
  kill site, 216, 383
  populations, 162, 167–169, 269, 284,
    312, 447
  predation, 52, 81, 128, 162, 166–169,
    177, 268, 270, 284, 356–357, 383,
    419–420, 441, 443, 447–449, 462
  range expansion, 405
  removals, 168–169, 177, 270, 419–420
Cover, 27, 64–65, 90, 125, 163, 175–176,
    178–179, 185, 194–195, 208, 211,
    218, 246–247, 266, 282, 347–351,
    353–359, 361–363, 366–367, 369,
    379, 385–386, 390, 397–399t, 401,
    403t, 427, 441, 459
Coyote, 72, 162–164t, 166–168, 216, 220, 312,
    350, 357, 459
  predation, 36, 357
  removal, 167
Craighead Wildlife-Wildland Institute,
    442–443
Creosote bush, 386
Crowberry, 364
Culex spp., 220
Culicoides, 146, 215
  sonorensis, 215
Culiseta spp., 220
Cull, 30, 66, 139t, 142, 415, 417, 420
Culverts, 381, 418, 427
Currant, 119
Cyanobacteria, 216

D

Decision, 39, 52, 81, 89–90, 102, 161, 194,
    197, 210, 212, 233, 241, 248, 272,
    290, 311, 334, 336, 338–339, 347,
    369, 384, 408, 416–417, 420–422t,
    425–426, 431–433, 458–459, 461
  analysis, 421–423
  makers, 416, 421, 422t
  making, 30, 87, 138, 305, 324, 332, 336–
    339, 416, 419–423, 427, 433, 461
Declaration of the Rights of Indigenous
    Peoples, 336
Deer, 8–9, 12–13, 26–28, 36, 38, 56, 93, 120,
    125, 145–146, 149, 331, 333, 380,
    423, 428, 431–432, 447

antlerless water, 9
antlers, 9
Columbian black-tailed, 38, 146
mule, 27, 56, 142, 146, 150, 160, 166–167,
    169, 176–180, 194, 216, 235–236,
    238–239, 282, 312–313, 424, 459
  abundance, 169
  body fat, 127
  distribution, 178
  habitat, 178–179
  hunting, 319–320
  migrations, 195
  population, 38, 160, 236, 320
  predation mortality, 128
  range expansion, 177
  recruitment, 128
  survival, 128
  trophy hunting, 131
  winter range, 178
musk, 8
range distributions, 402
red, 74, 444
season, 36
Sitka, 146
white-tailed, 38, 146, 149, 160, 312,
    354, 424
  hunting, 319
Deer atadenovirus A, 146
Dehydration, 151–152, 210, 217, 224, 354,
    360, 456
Delphinium spp., 121
Demographic, 104, 167, 217, 291, 300, 407,
    451, 458
  assessments, 233
  changes, 217, 224, 456
  characteristics, 233, 246, 266
  consequences, 169, 183
  data, 242, 244, 292, 299
  decline, 92, 99
  differences, 389
  dispersal, 78
  estimates, 234, 247
  factors, 149
  groups, 139–140, 361
  history, 84, 98
  information, 237, 241–242, 322
  models, 280, 285
  monitoring, 268
  numbers, 88
  objective, 237
  parameters, 241
  performance, 196, 384
  profiles, 241
  rate, 114–115, 128, 208, 223, 446, 456
  reconstruction, 92
  responses, 404
  simulations, 281
Demography, 85, 87, 91, 101, 104, 126–127,
    183, 213, 217–223, 243, 280–281,
    285, 291
Department of Interior, 337
Department of transportation, 416–419
Dermacentor albipictus, 150
  spp., 150
Desert Bighorn Council, 38, 425
Die-off. See disease
Diet, 61–62, 98, 101, 104, 116, 118–120, 124,
    152, 176–177, 212–213, 247, 354,
    358–360, 378, 386–387
  composition, 119, 123–126, 387

Dietary changes, 118
  choices, 123
  components, 360
  differences, 126, 177
  needs, 216, 458
  nitrogen, 124
  overlap, 180, 183–184
  preference, 176–177, 180–181, 183
  protein, 121
  similarities, 178
Digestible energy, 119, 121, 124, 184
  protein, 120, 124
Digestibility, 121, 124, 182, 192
Digestive capacity, 126
  constraints, 121
  efficiencies, 183
  issues, 152
  losses, 221
  morphology, 116–121
  processes, 114
  system, 149–150, 182, 382
  tract, 101, 116–118, 121, 208
Dingell-Johnson Act, 336
Discovery Grants, 442–443
Disease, 3, 10, 25, 29–30, 33–36, 38–39,
    50–52, 61, 64–66, 79, 88, 100–101,
    104, 114, 116, 126, 128, 136–153,
    160, 165–166, 168, 180, 182, 191,
    202, 218–219, 223, 258, 264–266,
    270, 273, 280, 284, 290–291, 293,
    296–297, 300–301, 312, 321–323,
    331, 333, 335–336, 338–339, 349,
    378–379, 409t, 417, 419, 423–425,
    427, 430, 432, 441, 444–447, 450,
    455–458, 460, 462
  agent, 136, 138, 432
  control, 136–139, 340, 461
  die-off, 10, 34, 36, 38, 51, 64, 79, 139, 146,
    166, 202, 269, 271–272, 297, 361,
    384, 423, 431, 444–445, 458
  dynamics, 138, 199, 215, 296
  ecology, 136–139
  enzootic, 51, 137
  epizootic, 81, 146, 271–272, 301, 432,
    441–442, 444–445, 447, 449, 460
  event, 50, 97, 218, 247, 284
  exposure, 146, 322, 384
  incidental, 152
  management, 7, 36, 64, 66, 138–139t,
    305, 432
  outbreak, 3, 37, 64, 103, 137–138, 140,
    143–144, 146–147, 223, 233, 244,
    269, 272, 426
  pillover, 139–140, 143, 146–148, 150, 297,
    331, 339, 417, 426, 432, 459, 461
  prevalence, 100, 284, 458
  risk, 200, 281, 284, 426
  spread, 93, 95, 200, 332, 426
  status, 100, 265, 445, 448
  surveillance, 93, 98, 271, 273
  susceptibility to predation, 164–166
  susceptibility, 218, 458
  testing, 101, 104
  transfer, *198–199*
  transmission, 39, 73, 93, 181, 200, 202,
    215, 273, 297, 305, 312, 403t, 414,
    423–427t, 429, 432–433, 445,
    460–461
Dispersal, 4–5, 29, 78–79, 81, 89, 93, 95,
    97–98, 145, 181, 191, 200–203, *268,*
    281, 297, 406, 447

Distribution, 3–19, 25–26, 28–*29,* 33, 35,
    37, 47–48, 52, 73, 87, 89–93, 98,
    101–102, 106, 115–116, 123, 136,
    139, 143–144, 146, 148, 150–151,
    161–162, 167–169, 178–180, 182–
    184, 191–192, 195–196, 199–202,
    208, 212–216, 233–234, 236–241,
    244, 247–248, 258, 265, *267*–268,
    280–*281,* 283–285, 290–291, 298,
    300, 305–*310,* 313, 316, 319–320,
    324, 331, 333, 335–336, 338, 340,
    347, 349, 352, 355, 357–359, 361–
    362, 365–366, 369–370, 378–379,
    383, 385, 393, 397–398, *400*–405,
    409t, 414–415, 417, 432, 455–457,
    460–462
  historical, 3, 19, *29,* 102, 161, 239, *307,* 349
Disturbance, 34, 152, 166–167, 201, 203,
    241, 243, 245–246, 258, 266, 282,
    331, 339, 349–351, 360, 370, 383,
    386–387, 390–392, 397, 402–407,
    409t, 424, 429, 457, 461
Domestic ungulates, 3, 29, 34, 148, 175–178,
    184–185, 312, 459
Dominance, 58, 73–78, 81, 101, 196,
    297–298, 450
Donkey. *See* burro
Downy brome, 387
Draw (hunt), *49–50,* 291–294, 296, 298–301,
    333, 338–340, 445, 460–461
Drive (trap), 27–28, 260
Drop-net, 260–*263,* 273
Drought, 52, 62, 65, 123, 127, 149, 151, 166,
    168, 175, 184, 202, 208, 210, 213,
    216–217, 219–222, 224, 284–285,
    351, 353, 358–360, 383, 398–399,
    409t, 415, 456, 459
*Dryas integrifolia,* 368
  spp., 364
Dwarf birch, 364

**E**

Eagle, 459
  golden, 36, 162, 164t, 167, 169, 312, 350,
    398
Economic activity, 312
  advantages, 264, 270, 317–319
  benefit, 181, 319, 323, 332
  damage, 414
  gain, 35
  importance, 148, 301
  incentives, 311, 319, 432
  investment, 319, 408, 457
  limitations, 422
  loss, 415, 423
  opportunities, 30
  resource, 301, 417
  returns, 319, 324
  uses, 408, 457
  value, 9, 311, 324, 332, 339, 417, 460
  viability, 311
Ectoparasites, 150
*Ectopistes migratorius,* 160
Ejido, 305–306, 311–312, 314–315, 319–320
Elk, 26, 28, 36, 149, 160, 166, 169, 176–180,
    216, 236, 236f, 368, 384, 398, 405,
    424, 428, 459
  antlers, *33*
  distribution, 179–180
  populations, 38

  range distributions, 402
  Rocky Mountain, 150, 384
  season, 36
  winter range, 179–180
*Elaeophora schneideri,* 149
Elevation, *16,* 98, 100, *126,* 192–194, 201,
    221, 245, 265t, 283, 312, 314,
    349, 351–352, 357, 360–363, 366,
    368–369, 397, 399, *401,* 431
*Elymus* spp., 118, 367–368
Emigration, 78, 126, 148, 285
*Empetrum* spp., 364
Encephalitis, 151
Endangered Species Act, 270, 336
Endotoxemia, 140
Energetic burden, 66
  costs, 62, 64, 78, 125, 209, 218–219,
    361, 406
  demands, 64, 266
Energetics modeling, 122, 126
Energy, 52, 58–59, 61–62, 64, 74, 78, 116–117,
    119, 121–127, 140, 152, 163, 176,
    218, 220–221
  balance, 126
  budget, 118
  development, 351, 415
  expenditure, 122, 125, 176, 211, 217–218,
    220, 224, 456
  extraction, 455
  intake, 121, 125, 192
  metabolism, 117–118, 120
  requirements, 121–122, 126, 415
  reserves, 122, 221
Environmental assessment, 404, 462
  awareness, 323–324
  change, 40, 52, 202, 390, 398, 400,
    402, 456
  conditions, 52, 61, 80, 103–104, 122, 126,
    137, 145, 175, 194, 196, 208–209,
    219–220, 222–223, 239–240, 246,
    282, 312, 322, 363, 398, 441, 450
  contamination, 147
  cues, 220
  degradation, 393, 457
  disturbance, 390–391
  DNA (eDNA), 39, 240
  education, 311, 313
  ethics, 415
  evaluations, 417
  factors, 26, 86, 100, 103, 136, 138, 141, 145,
    164, 175, 221, 245, 291, 456
  heat load, 210–212
  indicator, 52
  influences, 137, 281
  manipulations, 139t
  persistence, 143
  perturbations, 239, 282
  planning, 417
  policy, 338, 417
  responses, 384, 400
  stewardship, 313
  stochasticity, 283–284
  strategies, 311
  stressors, 145, 462
  variables, 125, 285, 415, 441
Environmental Impact Statement for their
    20-year update to the Payette
    National Forest Land and Resource
    Management Plan, 426
Ephedra, 119
*Ephedra* spp., 119

Epizootic hemorrhagic disease virus (EHDV), 146, 215
Eppley Foundation, 443
Equid. *See* horse
Equine. *See* horse
*Equus asinus*, 405
    *caballus*, 405
*Eremogone* spp., 119
*Eriogonum* spp., 119
Escape cover, 27, 163, 350–351, 354–355, 359, *362*–363, 403t
    terrain, 72, 81, 97, 116, 123, 128, 163–164, 166, 169, 180, 200, 211, 213, 216, 220, 265t, 268, 284, 348–357, 359–365, 367–370, 379–380, 390, 392, 403t–404, 423, 457
Ethics, 273, 334, 336, 415–416, 421, 462
Eurasian mouflon, 10
European, 26, 29–30, 32, 35, 40, 95, 139, 142, 237, 312, 330–331, 335
    colonization (colonizer), 7, 11, 29, 151, 167, 384
    culture, 30–33
    evolution, *See* origin
    exploration (explorer), 25, 29–33, 336, 382
    hunters, 32
    mammalogists, 11, 19
    settlement (settlers), 29, 47, 91, 160, 335
Evaporative cooling, 209–210
    heat loss, 209–211
    loss, 122
Extirpation, 3, 7, 19, 26, 34, 40, 52, *91*–92, 99, 102, 140, 167–168, 175, 179, 196, 201, 221, 258, 270–271, 281, 284, 290–292, 319, 336, 365, 378–379, 392, 398, 407, 419, 447, 455

**F**

*Fascioloides magna*, 149
Fat, 62, 81, 114, 117–118, 125–126, 152, 208, 218, 221
    accretion, 122, 125
    catabolism, 125
    costs of reproduction, 126
    dynamics, 125–126
    niche, 180, 383
    reserves, 125, 127, 164, 176, 192, 194, 202, 218, 220–221
    stores, 56, 152, 210
Fecundity, 81, 280, 401, 456
Feeding, 73, 78, 124, 148, 153, 283, 360, 414, 421, 429, 431
    activities, 151
    artificial, 152, 421, 431
    habits, 184
    site, 123
    supplemental, 429–430
    time, 73
*Ferocactus* spp., 212, 359
*Festuca altaica*, 368
    spp., 119, 351, 360, 364
Female choice, 79
Fencing, 33, 139t, 201, 216, 381, 383, 388, 414, 417, 418, 424, 426–427, 429–*430*, 432
Fertility, 11, 120, 126, 334–335
    control, 383, 431–432
Fescue, 119, 351, 360, 364
    altai, 368

Fire, 27, 40, 65, 78, *179*–180, 184, 335, 349, 351, 384–385, 387, 398, *401*, 405–406, 419, 448
    cycles, *See* regime
    management, 184, 419
    prescribed, 65, 81, 169, 384–385, 398t–399, 401–402, 405
    regime, 27, 169, 385–387
    suppression, 27, 40, 52, 65, 167, 169, 177, 378–379, 385, 387, 409t, 455, 457
    wildfire, 65, 169, 244, 384, 387, 398, 401–402, 405–406, 449, 455
First Nation people. *See* Native Americans
Flatworms. *See* trematodes
Fonds de recherche du Québec–Nature et technologies, NSERC scholarships, 442
Food, 10, 29–30, 35, 64, 74, 117–123, 125, 175, 185, 209, 217, 219–220, 259, 298, 321, 331–335, 347–348, 358, 359, 379, 387, 398t, 402, 421, 429, 448
    competition, 180
    intake, 122–124
    item, 117, 124, *386*
    processing, 116–121
    quality, 56, 64, 119–125, 322
    quantity, 123–125
    resources, 180, 282, 386
    -scape models, 286
    selection, 123
    source, 9, 27, 177, 331, 415, 429, 460
    supply, 35, 114, 127, 223
Foodscapes, 283
Forage, 13, 35, 40, 62, 64–65, 72–73, 80, 115–116, 118–121, 123–128, 139, 152, 163, 166, 175–184, 191–197, 199, 201–202, 208–213, 215–224, 242, 266, 269, 272, 282–284, 312, 347, 349–351, 358–365, 367–370, 379, 383–384, 386–387, 390, 393, 397, 402–403t, 405, 455–456, 459
    abundance, 121, 176, 184–185, 386
    availability, 98, 114, 136, 176, 178–180, 183, 208, 216–222, 305, 312, 367, 406, 409t, 423
    base, 115, 283
    competition, 145, 178, 216, 331, 403t, 432, 455, 460
    conditions, 192, 197, 212–213, 221–223, 282–284, 358, 385
    intake, 72, 123, 164, 197, 210, 405
    moisture content, 210, 212–213, 358–359
    nutrition hypothesis, 366
    production, 182, 216, 312, 358, 383, 386, 397
    quality, 12, 64–65, 114–116, 121, 123–124, 127, 152, 163–164, 166, 179, 182, 192, 201–202, 212, 217–222, 224, 266, 297, 354, 358, 360, 366, 382, 384–387, 397–398, 402, 405, 409t, 415, 456
    quantity, 12, 64, 115, 123, 176, 217–218, 220, 224, 266, 358, 385, 387, 398–399, 409t, 415, 456
    resources, 29, 34, 123, 223, 283, 455, 459–460
    selection, 119, 168, 405
    -selection hypothesis, 367
    spring forage, 121, 198, 217–218, 221, 363
    summer forage, 192, 220–221, 405

    supply, 25, 404, 423
    winter forage, 197, 220, 398
Foraging, 72–73, 78, 114, 122–123, 126, 150, 166, 196, 202, 210, 218, 220, 238, 245, 282, 347–348, 353–354, 359–363, 445, 447
    areas, 168, 180, 192, 211, *348*, 350–352, 358, 360, 365, 370, 406, 457
    conditions, 116, 194, 197
    decisions, 210, 212
    efficiency, 123, 127
    foraging behavior, 116–121, 125
    foraging habitat, 27, 351–352, 360, 364, 368–*369*, 404, 423
    site, 166, 347, 365
    strategy, 72, 167, 183
    time, 73, 168
Forbs, 65, 118–120, 177, 180, 208, 210, 265t, 351–352, 357, 360, 364, 368, 386–387
Fort Nelson First Nations, 337
Foundation for North American Wild Sheep, 337, 385, 445
Free-ranging, 10, 37, 101, 120, 123, 137
Fraternity of the Desert Bighorn, 388
Friends of Wild Animals, 419–420
Fungal disease, 151–152
*Fusobacterium necrophorum*, 143
Future Farmers of America, 338

**G**

Game Amendment Act, 1905, 30
Gas, 140, 402, 409t, 423
Gemsbok, 181
Genetic adaptations, 322, 460
    analysis, 17, 77, 79, 88, 92, 95, 141, 181, 238, 458
    ancestry, *96*, 104
    applications, 19, 84–85, 93, 100–102, 234t, 240, 248, 458
    approaches, *See* genetic applications
    artificial selection, 61, 99–100, 104, 123
    assignment, 56, 88, *94*–96
    basis of traits, 4, 100–101, 103–104
    clusters, 17, 95–*96*
    consequences, 99, 266
    data (datasets), 11, 19, 85–87, 89, *91*–93, 98–99, 102, 104, 458
    differentiation, 4–5, 7, 86, 90, 92–95
    dispersal, 78
    divergence, 92, 105
    diversity, 17, 64, 77, 84–85, 88, 92–93, 97–100, 102–105, 199–201, 219, 222, 305, 313, 322, 456, 460
    drift, 86, 89, 93, 98, 104–105, 322, 381
    heterogeneity, 219, 322, 457
    information, 12, 51, 98, 239, 300, 352, 365
    isolation, 11, 100, 379, 381
    load, 85, 103, 105
    management, 97–99, 103–104, 322
    markers, 84, 95, 97, 100, 103–105
    methods (methodology), *See* genetic applications
    monitoring, 98–99, 322
    mutations, 84
    populations, 92
    potential, 61, 322
    rescue, 79, 85, 99, 104, 443–444, 451
    sampling (samples), 93, 99, 240
    selection, 139t, 291

Genetic adaptations (*cont.*)
  structure, 17, 61, 90, 93–95, 97–98,
    103, 105
  study, 4–5, 7, 11, 17, 85, 87, 89, 93, 97, 100,
    105, 323
  techniques, *See* genetic applications
  tools, 95, 104
  variant, 105–106, 137–138, 141
  variation (variability), 4, 11, 14, 17, 19, 79,
    81, 84–86, 89–92, 97, 100, 103–104
  viability, 52, 281
  vicariance, 378
Genetics, 7, 12, 17, 59–64, 66, 78, 84–106,
    266, 290, 333–334, 455–456
  management, 322
  spatial, 95
Geological Survey of Canada, 26
Gestation, 58, 60, 62–63, 79–80, 114–115,
    121–122, 126–128, 138, 143, 149,
    197–198, 218, 220–223, 261, 283,
    297, 356, 383, 443–444
Giant liver fluke, 149–150
Giraffe, 8–9
*Giraffa* spp., 8
Gland, 9–11, 56, 117, 209
Goat, 8–10, 13, 33–34, 36, 66, 182,
    184–185, 266
  bezoar, 10
    population, 10
  domestic, 10, 34, 51, 87, 128, 138, 141,
    143, 145, 147–148, 182, 184, 268,
    284, 290, 292, 297, 305, 312, 321,
    335, 384, 403t, 417, 420, 423–426,
    431–432, 445, 455, 459
    population, 10
  feral, 182, 312
  mountain, 9, 36, 60, 104, 141, 147–148,
    166, 168, 176–177, *179*–180,
    *198*–199, 212, 239, 424, 447
    habitat, 177, 180
    niche, 180
    population, 39, 398
  populations, 431
  Rocky Mountain, 9, 32
Goral, 9
Governance, 311, 330, 336–337
Graminoids, 65, 118–119, 208, 210, 367–368
Grand Slam Club–Ovis, 419
Grass, 118–119, 123, 177, 180, 184, 208, 210,
    265t, 312, 351–352, 357, 360–362,
    367, 385–387, 397–398, 401–402,
    405–406
  abundance, 406
Gray fox, 162, 164t
Gregariousness, 72–74, 81
Green-up, 80, 192–193, 195, 197, 199, 202,
    283, 360, 397
  wave, 194–195, 197, 202, 360, 366
Grey kangaroo, 80
Growth, 34, 39–40, 55, 57–66, 72, 80–81, 101,
    114–115, 118, 120–122, 125–128,
    136, 144, 147, 153, 164, 167–168,
    182, 195, 198–199, 202, 217–219,
    221–223, 245, 269, 280–284, 291–
    293, 296–297, 312, 320, 322–323,
    337, 351, 357, 360, 363, 367–368,
    379, 384, *386, 391*–392, 397, 405,
    407–409t, 418, 430, 441–451,
    455–456, 459, 462
  models, 285
  patterns, 3, 11, 58

rate, 115, 120, 127, 164, 222–223, 270,
    281–282, 285, 397, 403t
  rings, 58
Guard dogs, 429
*Gulo gulo,* 162, 220, 365
Guzzler, 153, 200, 208, 213–214, 388,
    398t–399
Gwich'in First Nation, 337

**H**

Habitat, 3–5, 9–11, *15–16*, 19, 25–40, 52, 55,
    65–66, 72, 76, 78, 81, 88, 90–93,
    95, 98–100, 102, 104, 116, 122–123,
    125, 127–128, 136–139, 145–146,
    150, 153, 160, 162–163, 166–167,
    169, 175–177, 179–185, 191, 196,
    201–203, 208, 213, 215–219, 234t,
    237–238, 240–241, 243–247, 258,
    264–265, 265t–266, 268–271, 273,
    284, 286, 290–291, 300–301, 306,
    312–314, 319–323, 330, 332–333,
    335–340, 347–433, 441, 449,
    455–462
  abandonment, 34
  alienation, 392, 406–407, 409t, 457
  alteration, 34, 169, 233, 244, 384–387, 402,
    405, 455
  areas, 166, 201, 391, 406, 408
  associations, 241, 265, 265t, 405
  attributes, 162, 265, 361, 366, 407
  availability, 90, 201, 219, 347, 378, 408,
    457–458
  change, 5, 168, 293, 459
  characteristics, 103, 197, 246, 347–370, 415
  components, 213, 306, 347–351, 354–355,
    357, 359–360, 369, 379
  conditions, 37, 136, 169, 197, 280, 379,
    397–398, 401, 405, 459
  connectivity, 201, 219, 222, 402, 418–419
  conservation, 319, 323, 378
  corridors, 306
  critical, 347–348, 403t
  degradation, 3, 34–35, 38, 258, 305,
    382, 406
  enhancement (enhancing), 143, 169, 184,
    208, 387–390, 398t–399
  escape, 367
  evaluation, 266, 268, 349
  features, 265t, 356, 361–362, 367–369, 379,
    391, 403t
  fragmentation, 3, 26, 33–34, 89, 98, 136,
    168, 219, 222, 305, 312, 319, 333,
    379, 407, 419
  function, 33, 397–408, 409t
  historical, 25–40, 387, 391, 397
  improvement, 65, 208, 265t, 313, 337–338,
    382, 384
  interventions, 214, 300, 399, 455–456
  loss, 33–34, 65, 177, 213, 219, 258, 301,
    319, 339, 379, 407, 409t, 460
  management, 26, 38–39, 65–66, 169–170,
    233, 297, 339, 378–393, 399, 401,
    456, 461
  managers, 26–27, 202, 384
  manipulation, 48, 64–66, 208, 398t,
    417, 441
  in Mexico, 312–314
  microhabitats, 218, 354
  model, 268, 366, 387
  modifications, 7, 146

niche, 177–*178*, 181
  nursery, 367–368
  partitioning, 180
  patches, 192, 201, 378
  preferences, 178, 181, 183, 347
  protection, 319
  quality, 61, 65, 98, 125, 182, 184, 208,
    212–213, 215, 218, 347, 379, 386,
    408, 457–458
  record, 87, 450
  requirements, 25, 93, 162, 195, 349, 457
  restoration, 336, 340, 384, 401–402,
    405–406, 461
  risky, 166
  seasonal, 238, 245, 266, 348, 361, 403t,
    407, 414, 457, 459
  selection, 116, 166, 175, 265, 347, 349–350,
    353, 356, 362–364, 367–368, 404,
    406–407
  specialist, 180, 378–379
  suitability (suitable habitat), 51–52, 90, 95,
    136, 150, 153, 166–169, 264–265,
    265t, 266, 271, 365, 379, 385,
    387–388, 397–399, 401, 403t–404,
    406, 416, 430, 457, 459
  suitable, 90, 95, 136, 153, 264–266, 271,
    379, 385, 387–388, 397, 399, 404,
    416, 430
  summer, 359, 367
  threats, 378–379, 397–409
  treatments, 114, 238, 385
  type, 347–348, 409t
  unsuitable, 430
  use, 72, 114, 116, 146, 150, 166–167,
    175–178, 180, 183, 185, 202, 208,
    241, 243–244, *267*–268, 290, 322,
    347, 358–359, 361–365, 367–368,
    379, 390, 397, 406–407, 409t,
    432, 458
  winter, 180, 201, 245, 361, 407
*Haliaeetus* spp., 459
Hand-capture, 260
Harassment, 427t, 429
Hard release, 266–*267,* 272–273
Harvest, 26, 28, 30, 32, 35, 38, 49–*50*, 52,
    59–62, 64–65, 79, 81, 87–88,
    99–102, 104, 147, 149, 160, 202,
    213, 242, 247, 269, 284–286, 290–
    301, 311, 318, 333–334, 336–337,
    339–340, 402–403t, 407–408,
    417, 421, 423, 430, 441, 445–446,
    450–451, 456–457, 459–462
  age structure model, 299
  curl, 297, 301
  data, 296, 298–300
  evolutionary response to, 100–101
  female (ewe), 49t,-50t, 65, 297, 442, 446,
    459–460
  history and early regulation, 291–294,
    297–298
  hunter, 168, 265, 291, 294, 297, 407
  illegal, 311, 403t
  information, *See* harvest data
  intensity, 61, 291, 333
  legal, 100, 296
  limits, *See* harvest uotas
  male, 49t–50t, 79, 100, 291, 299
  management, 48, 64, 78, 290–301, 318,
    322, 460
  models, 284–285
  opportunity, 297–298, 301, 339

permit, 291, 298
practices, 61, 64, 66, 290–291, 456
pressure, 100, 291, 293, 301, 333, 451, 459–460
quotas, 30, 35, 281, 292, 299–300
rate, 284, 291, 293–294, 296, 299–300, 318, 442, 459
recommendations, 241, 245
records, 87, 450
recreational, 333
regulations, 35, 100, 238, 291–292, 298–301, 339, 455
reports, 296, 299–300
restrictions, 40, 333
season, 293, 299
selective, 61, 81, 300, 441
strategies, 247, 280, 291, 293, 298–299, 301, 450–451, 460
subsistence, 299
success, 294, 299, 333
survey, 299
sustainable, 52, 284, 290, 299, 301, 331, 339, 451
trophy, 81, 291
Heat absorption, 211
    acquisition, 212
    dissipation, 56
    exchange, 209–211
    gain, 210–212
    load, 209–212, 354
    loss, 209–212
    production, 124, 210
    stress, 217, 224, 354, 456
    transfer, 211
Heath hen, 160
Heather, 364
Hells Canyon, 425–426, 445
    Preservation Council, 426
Helminth, 137, 144–145, 149–150, 220
    population, 149
*Hemitragus jemlahicus*, 10
Hemorrhagic disease, 146
    septicemia, 142
Herbicide, 65, 385
Herpesvirus, 151
Highway. *See* road
Himalayan tahr, *10*
*Hippopotamus* spp., 8
Hippos, 8
Home range, 116, 122, 200, 213, 216, 221, 283, 347, 387
Horn, 12, 74–75, 98, 100, 273, *292*, 296, 299–300, 322, 333, 441, 450
    annuli, 56–58, 104, 293
    buds, 58
    characteristics (characters), 57–60, 89, 211
    circumference, 11, 61–*62*, 101
    core, 9, 58, 212
    curl, 61, 64, 100, 238, 291–293, 297–301, 334, 443, 445, 450–451
    development, 55–66
    growth, 58, 60–66, 81, 115, 122, 245, 291, 293, 296–297, 322, 443–444, 449–451, 462
    increments, 75, 441, 451
    length, 4, 9–10, 60–*62*, 64, 78, 100–101, 292–*293*, 297, 317, 450–451
    loss, 58, 60, 75
    measurement, 13–14, 293, 441, 450–451

morphology (morphometric), 26, *57*, 100, 245, 298, 300
    rings, *See* annuli
    shape, *8*, 10–11, 14, 57
    sheath, 9, 58, 152
    size, 55, *57*–66, 72, 74–75, 77–78, 81, 84, 99–101, 290–293, 323, 333–334, 441, 443, 450–451, 456–457, 462
    tip, 11, 58, 75, 297
    volume, 60–61, 101, 317, 450
Horse, 27, 29–30, 34, 52, 182–183, 313, 361, 378, 382, 384, 393, 405, 409t, 424, 457
    distribution, 182, 383, 405
    domestic, 177, 182, 312
    feral, 34, 52, 66, 122, 175, 177, 182–184, 216, 243, 382–384, 393, 409, 423, 457, 459
    population, 34
Host, 101, 136–138, 140–151, 424
    -agent coevolutionary paths, 145
    -agent dynamics, 138
    -agent environment, 136, 138
    -agent interaction, 136–137
    -agent relationship, 137
    factors, 145
    genetics, 101
    -parasite environment, 149
    -parasite relationship, 143
    population structure, 101, 137
    response, 137–138, 141–142, 144
Huckleberry, 364
Hudson's Bay Company, 34
Human, 26–27, 35, 39, 56, 58, 61, 81, 117, 136, 138–139, 145, 147–148, 150–151, 160–161, 166, 184, 201–202, 236, 238, 247–248, 258, 292, 330–336, 339, 349, 357, 378, 383–384, 387, 392, 397, 404–405, 407, 414–416, 418–419, 421t, 423–424, 430–432, 449, 457, 460–462
    activity, 52, 85, 88, 104, 153, 161, 166–167, 258, 283, 331, 339, 378, 388, 392, 397, 406–408, 414, 423–424, 457, 461
    attitudes, 330, 339, 461
    behavior, 428
    communities, 313, 319, 408
    development, 79, 81, 201, 213, 265t, 331, 339, 397, 423
    dimensions, 300, 330–340, 347, 414, 460–462
    disturbance, 34, 166, 243, 245, 349, 402–404
    effects, 72, 379, 408–409t, 457
    encroachment, 203, 290, 331, 379
    footprint, 26, 29, 378, 381
    health, 149, 151
    influence, 201, 379, 421, 459
    interactions, 27, 414–415
    intervention, 27, 93, 102, 203, 398t, 405
    population, 35, 39, 313, 333, 379, 392, 407–408, 415, 418, 455
    predation, 7, 236
    presence, 29, 52, 166, 406
    recreation, 201, 391–392, 406–407
    safety, 260, 424, 427t
    use, 201, 349, 392
    -wildlife conflict, 331, 336, 339, 407, 418, 424, 455, 461

Hunting, 25, 27–28, 30, 32–33, 35–37, 49t, 60–61, 102, 139t, 202, 220, 234, 284–285, 292, 296–301, 305, 313–314, 316, 319–320, 331–332, 334–340, 349, 393, 398t, 406–407, 409t, 415–416, 419–421, 427t, 431, 450–451, 457, 460–462
    activity, 32, 336
    architecture, *See* hunting structure
    area, 292, 299–300
    big game, *32–33,* 36–37, 305, 320
    blinds, *See* hunting structure
    efficiency, 167
    enclosures, *See* hunting structure
    expeditions, 30, 35, 337
    experience, 317
    female, 49t–50t, 65, 294, 297, 442, 446, 459
    grounds, 30, 320, 335
    guides, 36, 388
    hunting management, 72, 305, 315–320, 450
    illegal, 30, 37–38, 40, 101, 305, 313, 455
    Indigenous, 29–30
    -induced evolution, 81
    industry, 30, 320
    interest, 32, 321, 331
    legal, 293, 316–317
    licenses, *See* hunting licenses
    male, 49t–50t, 79, 100, 291, 299
    market, 25, 29, 30, 33, 35–36, 40, 160, 290–292, 297, 336
    method, 27–28, 335
    opportunity, 26, 290–292, 294, 296, 298, 301, 312, 316, 319, 320, 322, 324, 333, 339, 417, 460–462
    options, 292, 430
    organization, 320, 337
    outfitters, *See* hunting guide
    overhunting, 331, 335–336
    permit, 36–38, 168, 285, 296–297, 314, 316–*317,* 319–320, 324, 333, 336, 338, 417, 462
    practices, 61, 335, 337
    pressure, 79, 300–301, 349–350
    program, 318–320
    public, 416, 430
    recreational, 319, 331–334, 336–337
    regulated, 332, 339, 430, 460
    regulation, 292, 297, 311, 319–320, 335, 441, 443, 450–451
    rules, *See* hunting regulation
    season, 26, 35–37, 79, 147, 245, 285, 292–293, 296, 298–299, 314, 316–*318,* 339, 441–442, 445, 460
    selective, 79, 339, 443, 461
    sport, 10, 25, 30, 32–*33,* 37, 305, 311, 319, 321, 444
    structures, 27, 234
    subsistence, 30, 34, 298, 339, 455, 460
    tag, 320
    technique, 162, 335
    trip, 28, 313
    trophy, 30, 78–79, 285, 305, 313, 316–319, 334–335, 443, 450–451
    unregulated, 33, 258, 291
*Hydropotes inermis,* 9
*Hypericum perforatum,* 405
*Hypoderma tarandi,* 220
Hypothermia, 217, 221, 363

**I**

ibex, 13
    Alpine, 60, 72
    Himalayan, *See* Siberian ibex
    Persian, 181
    population, 10
    Siberian, 10
Immigration, 126, 167, 285, 405
Inbreeding, 13, 79, 84–85, 98–100, 103–106, 443, 445
Indigenous First Nations. *See* Native Americans
Indigenous people. *See* Native Americans
Individual identification, 97
Infectious keratoconjunctivitis, 147
Information dissemination, 338–339
Intraspecific battles/conflict, 58, 220
    relationships, 26, 38
    variation/differences/diversity, 11, 17, 458
Interspecific competition, 175, 177, 258, 312
    genetic variation, 14
    interactions, 175, *198*
    relationships, 26, 38, 170, 175
Invasive plant, 65, 177, 305, 384–387, 393, 402, 405, 409t, 457
    species, 216, 385, 404t–406, 455

**J**

Jaguar, 162, 383
Janusia, 119
*Janusia* spp., 119
Johne's disease, 148
*Juncus* spp., 352, 364
Junegrass, 119
Juniper, 386
*Juniperus* spp., 386

**K**

Kobresia, 352
*Kobresia* spp., 352
*Koeleria* spp., 119

**L**

Labrador duck, 160
Lactation (lactating), 63, 73, 77, 80–81, 114–116, 121–122, 125–127, 147, 163, 198–199, 209, 213, 217–218, 221, 367–368, 387
*Lama* spp., 8
Lamb, *4*, 51, 55t–56, 58, 60, 72–73, 77–81, 98–100, 115–116, 118, 120, 126, 139–141, 143–145, 147, 149, 152, 164–*165*, 167–168, *198*–199, 208, 213, 217–223, 241, 243–245, 269–270, 272, 280, 284, 296, 299, 333, 350–352, 357, 359, *362*–363, 367, 380, 386, 426, 432, 442–445, 448, 456
    birth rates, 115
    deaths, *See* lamb mortality
    groups, *4*, 145, 166, 355
    growth, 115, 397
    horn, 58
    lam ratios, 222, 223, 241, 247, 269–270, 293, 296, 358–359
    mass, 81, 448
    mortality, 80, 140, 199, 221, 271, 357, 448

population, 300
production, 168, 299, 369, 448
recruitment, 52, 115, 219, 221, 246, 271–272, 360, 363, 397
survival, 51–52, 80–81, 115, 126–127, 144, 197, 202, 217, 219–224, 247, 269, 271, *281*, 403t, 417, 441–443, 445–447, 456
Lambing, 80, 176, 196–197, 202, 218–223, 268–269, 347–351, 354, 356–357, 362–363, 365–369, 378–379, 403t, 405, 443
    areas, 197, 202, 350–351, 369, 392
    grounds, 384
    habitat, 265t, 349–351, 356–357, 363, 367, 369
    parturition synchrony, 222
    period, 197, 222, 269, 367
    ranges, 197
    season, 221–223, 366–367
    sites, 166–167, 356, 367
Land management, 78, 290, 324, 338, 387–388, 391–392, 407–408, 415–417, 420–421, 426, 457–458
Larkspur, 121
*Larrea tridentata*, 386
Leg snares, 260
*Leopardus pardalis*, 162
*Lepus americanus*, 160
Lethal removal, 139t, 168–169, 268, 415, 417, 419–420, 427t, 430–432
Lichens, 177, 180, 188, 364
Limiting factor, 37, 114–116, 128, 167–169, 195, 214, 247, 357–359, 369, 378, 389, 459
Livestock, 25, 29, 33–37, 39–40, 65–66, 123–124, 139, 141–151, 169, 177, *182*, 184, 213, 216, 258, 290–291, 297, 311–312, 319, 321, 323, 331–332, 349, 351, 360–361, 369, 370, 378, 383–384, 386, 388, 393, 402, 409t, 414–417, 420, 423–424, 429–*430*, 432, 455, 457, 459–461
    disease, 88, 270, 291, 331, 339, 424–427t, 461
    overgrazing, 33–35
Llamas, 8, 424
Loafing areas, 361, 379
Local adaptation, 92, 102–104
Local Ecological Knowledge, 336, 339
Locomotion, 122, 224, 456
Locoweed, 118
Logging, 33, 455
Long-term monitoring. *See* long-term research
    research, 26, 38–39, 80, 98–100, 104, 127, 197, 216, 219, 240, 244, 259, 381, 441–451, 455, 462
    studies, *See* long-term research
Lupine, 121
*Lupinus* spp., 121
Lumpy jaw, 220
Lungworm, 143–145, 149, 218, 220, 444–445
    abundance, 100
    distribution, 143–144
    infection, 142–144, 441, 444–445
    scabies, 38
    transmission, 143
    treatment, 143
Lynx, 160, 162, 164t, 220
*Lynx canadensis*, 160, 220
    *rufus*, 161, 216, 312

**M**

*Macropus giganteus*, 80
Maintenance, 61–64, 72, 81, 115, 120–122, 145, 197, 203, 212–213, 217, 259
Malignant catarrhal fever-associated viruses, 151
Malnutrition, 152, 218–222
*Mannheimia haemolytica*, 142–143
    spp., 142
Management action, 52, 60, 65–66, 88, 102, 128, 139, 152, 168–169, 195, 197, 200–202, 233, 265, 300, 321–322, 324, 330, 338–339, 379–393, 415–416, 419–422, *431*, 455–459
    active, 270, 419
    activities, 168, 319, 337, 419–420
    agencies, 26, 48, 213, 234, 246–247, 265, 281, 290, 321, 336–338, 358, 383, 387–388, 391–392, 407–408, 414–417, 419–421, 426, 433, 457–458, 461
    agreements, 337
    alternatives, 285
    area, *96*, 291, 298, 323, 367, 383
    authority, 305, 321, 416–421
    challenge, 34, 202, 401
    concerns, 64, 137, 149–150, 202
    consequence, 137, 139, 151
    decision, 30, 39, 52, 89, 233, 248, 290, 311, 336, 338–339, 384, 416, 420–422t, 425
    grazing, 35, 65, 425
    highway corridor, 379–382
    implications, 5, 64–66, 136, 153
    intervention, 27, 140, 151, 168, 285, 419
    in Mexico, 37, 305–324
    objectives, 265, 280, 311, 403t–404t, 414–415
    opportunities, 439–464
    plan, 311, 322–323, 337, 403t–404t, 425–426
    policy, 305–311, 313, 321, 324, 462
    practice, 27, 147, 184, 202, 319, 324, 391, 402, 403t–404t, 425, 448–449
    program, 26, 38–39, 52, 60, 301, 305, 311, 312, 316–324, 379, 387, 462
    resource, 234, 247, 408, 426, 458, 462
    strategy, 37, 48, 52, 65–66, 141, 143, 290–291, 298, 305, 324, 408, 417, 441, 458
    sustainable, 323, 393, 417, 462
    trophy management, 290–291, 450–451
    unit, *96*, 306, 311, 315–316, 337, 339
Mange, 138, 150–151
    psoroptic, 150–151
Markhor, 10
Mark-recapture, 97
    -resight, 97, 233, 243, 246
*Marshallagia marshalli*, 149, 398
Mass, 3, 9, 11, 55–56, 58, 60–63, 73, 75–77, 80, 81, 100, 114, 116, 122–123, 126, 144, 164, 168, 192, 211–212, 220, 222–223, 333, 382, 441, 443–444, 448, 451
Mastication, 385
Mating, 58, 79, 98, 104–105, 322, 348, 444
    behavior, 76, 78, 444
    competition, 79
    opportunities, 79
    season, 220

skew, 78
strategies, 81
success, 72, 81, 100, 293, 450
system, 72, 76–79, 81, 93, 450
tactic, 60, 72, 75–76, 81, 444–445
Mechanical (treatment), 65, 385, 387, 401
The Medical Repository, 13, 30
Medusahead, 405
Mesquite, 351, 386
honey, 386
Metabolic disturbances, 152
function, 120
heat, 210
needs, 115, 163
pathways, 209
processes, 119–120, 125, 456
rate, 209–210, 217
requirements, 122, 163
water, 122, 208, 212, 224, 358–359,
387, 456
Metabolism, 62, 103, 117–120, 122, 209, 358
Metapopulation, 78, 89, 93, 95, 97–100, 201,
213, 239, 242, 248, 271–272, 306,
322, 378–379, 382, 392,
426, 444
Microhabitat, 218, 354
Migration, 78–79, 81, 86, 98, 105, 122–123,
166, 176, 191–198, 200–203, 221,
266, 269, 335, 349, 376, 378, 380,
409t, 429, 444–445, 462
abbreviated, 192–194, 197, 201
altitudinal, See elevational migration
corridors, 266
distance, 192, 194–196
elevational, 192–194, 201, 203, 349
fall, 192, 194, 200, 203, 365
geographic, 192, 194–195, 203
partial, 197
patterns, 128, 196, 201, 337, 406
rates, 84
resident, 77, 79, 117, 168, 191, 192,
194–195, 197, 201, 213,
349, 447
routes, 27, 78, 193–196, 201, 402
seasonal, 116, 166, 191–203, 380, 402,
406, 444
spring, 192, 196, 199, 365
vacillating, 194, 202–203
Migratory behavior, 126, 128, 191–197,
202–203, 459
classifications, 192
diversity, 195–196, 201, 203, 459
movements, 27, 192, 196, 198–200,
202–203
patterns, 196, 414
plasticity, 197–198
populations, 195
predators, 398
propensity, 194, 196–197, 203
routes, 191, 196, 199
stop-over site, 166
subpopulations, 150
switching, 197–198
tactic, 115–117, 126–127
Milk, 56, 81, 114–115, 118, 122, 152, 218–219,
223, 269, 397, 431, 449
Mineral, 3, 52, 58, 119–123, 139t, 145, 147,
152, 180, 198–199, 201, 218, 221,
242, 347, 362–363, 390–391,
398t, 402

lick, 97, 116, 147, 166, 177, 191, 198–203,
218, 221, 234t, 239–240, 242, 266,
347–348, 361–362, 367–368, 380,
392, 402–403t
requirements, 116, 123
Mining, 33–34, 37, 40, 153, 166, 291, 312–313,
349, 360, 370, 378, 382, 388,
390–391, 402, 405, 424, 455
Ministry of Agriculture, 431
Mite, 150–151
infestation, 140
Mitigation, 258, 382, 388, 390, 392, 402–405,
408, 418, 422t, 427t–428
Monitoring, 26, 38, 52, 61, 78, 98–100, 140,
151, 197, 200, 216, 233, 240–241,
244, 246–248, 265–266, 268,
273, 280, 285, 311, 313, 319, 322,
336–338, 381, 402, 404t, 407,
418–419, 421–422t, 441, 443–451,
455, 458, 462
range-wide status, 52
post-release monitoring, 266–268
Montana Conservation Science Institute, 443
Moose, 26, 30, 160, 166, 168, 331, 428–429
season, 36
winter range, 150
Moraxella bovis, 147
Morphological adaptations for
thermoregulation and water
balance, 211–212
Morphology, 4–5, 11, 13, 26, 55–65, 88–89,
91–92, 100, 102, 114, 116–118, 211,
245, 323, 333
Mortality, 25, 34, 52, 63, 78, 80, 116, 123,
127–128, 139–140, 142, 144–146,
148, 151, 153, 160, 162–163, 167,
199, 201, 213, 215–221, 223–224,
246, 261, 261f, 263, 268–271, 273,
284, 297–298, 322, 335, 350, 356,
358, 365, 379–381, 403t, 405, 407,
409t, 418–420, 424, 430, 447–448,
456–457, 459
cause-specific mortality, 284
Morris Animal Foundation, 445
Moschidae, 9
Moschids, 8
Moschus spp., 8
Mosquitoes, 220
Mosses, 364
Mountain hemlock, 364
lion, See cougar
mahogany, 119
Movement, 27–28, 33, 79, 89, 93–95, 97, 99,
114, 116, 139t, 165–166, 191–203,
208, 213, 217, 219–221, 233, 240,
243, 245, 247, 264–266, 271–272,
281, 283, 306, 323, 350, 359, 361,
365–367, 380–381, 387, 405, 407,
418, 424, 429, 432, 460
barriers, 305
behaviors, 365
corridors, 73, 306, 312, 379, 384, 386, 392,
402–403t, 418
exploratory, 79, 192, 200
patterns, 243
rates, 123, 358
routes, 203
Muellerius capillaris, 143
Muhlenbergia spp., 119, 351
Muhly, 119, 351

Muskox, 9
Mycobacteriosis, 148–149
Mycobacterium, 148
distribution, 148
Mycoplasmas, 141–145
Mycoplasma arginine, 142
bovirhinis, 142
bovis, 142
conjunctivae, 147
ovipneumoniae, 10, 34, 50–51, 100–101,
103, 141–142, 147, 166, 181, 202,
264, 271–272, 284, 384, 417,
424, 432
sp., 142
spp., 142, 147

N

Nasal bot infestation, 144–145
Natal areas, 403t
group, 73, 81
population, 79
range, 268, 348, 392
National Bison Range (NBR), 77–78, 99,
443–444, 447–449
Environmental Policy Act, 338, 417
Geographic Society, 443
Park Service, 100, 265t, 418
National protected areas (Mexico), 313–314
Native Americans, 13, 25–30, 32, 35–36, 39,
151, 234–236, 258, 273, 300–301,
330–331, 334–337, 339, 382, 384,
414, 416, 421, 430, 455,
460–461
art, 28, 234–236, 248, 332, 335
hunting, 29–30, 335
values, 334–336
Natural Sciences and Engineering Research
Council of Canada, 442, 444
Needlegrass, 119, 351
Nematode, 149, 398
Nemorhedus species, 9
Neonatal losses, 149
deaths, 149
period, 152, 245
Nerium spp, 153
Net-gun, 260–264, 273, 420, 426
Nevada Bighorns Unlimited, 388
Department of Transportation, 419
Department of Wildlife, 419
New Mexico Department of Game
and Fish, 36
Nez Perce, 335, 337, 461
Niche, 28, 175–181, 185, 364, 383
fundamental, 175
realized, 175
Non-governmental group (organization), 318,
337–338, 414
Nonmigratory, 196, 203, 269
individuals, 191
movements, 191, 198–200, 202
Nonprofit wildlife conservation, 419–420
North Africa grass, 405
North American Model of Wildlife
Conservation (NAM), 330, 336,
339, 460
Thinhorn Summit, 336
Wildlife Policy, 336
Northern Wild Sheep and Goat Council
Symposium, 336

Nursery group, 72–73, 145, 333, 357, 361,
        363–364, 367, 448
    habitat, 367–368
    period, 367–368
    sites, 356, 367
Nursing. *See* lactation
Nutrient, 119–123, 126, 149–150, 163–164,
        208, 215–216, 258, 353, 356, 359,
        384, 387
    acquisition, 117, 166
    availability, 12, 163, 245, 283
    intake, 115, 119, 122, 163, 166, 359
Nutrition, 59–66, 101, 103–104, 114–117, 119,
        122, 125–128, 136, 145, 164, 196,
        201–202, 282, 284, 322, 352, 356,
        366, 368–369, 385, 406, 456,
        459, 462
Nutritional ecology, 114–128, 456
    influences on predation, 163–164
    requirements, 115, 119–123, 127, 283
    resources, 65–66, 127, 163

**O**

Ocelot, 162
*Odocoileus hemionus,* 27, 38, 56, 142, 146,
        160, 176, 194, 216, 236, 312, 424,
        459
    *hemionus columbianus,* 38, 146
    *hemionus sitkensis,* 146
    spp., 26, 38, 56, 93, 120, 145–146, 282, 331,
        380, 402, 447
    *virginianus,* 38, 146, 160, 312, 319, 424
    *virginianus carminis,* 354
*Oestrus ovis,* 145, 150
Oil, 402, 409t, 423
Okapi, 8–9
*Okapi* spp., 8
Oleander, 153
*Opuntia* spp., 119
Orbiviruses, 146
Ordinances, 420–421, 427t, 429, 431
*Oreamnos americanus,* 9, 36, 60, 104, 141,
        166, 176, 199, 212, 239,
        398, 447
Oregon Hunters' Association, 445
Origin, 3–19, 61, 84–90, 92–93, 98, 104, 150,
        266, 441, 451, 458, 460, 462
    in North America, 4–6
*Oryx gazella,* 181
*Oryzopsis* spp., 351, 360
*Otobius* spp., 150
Overgrazing, 33–38, 65, 183–184, 339, 423,
        457, 460
Overharvest, 29, 35, 285, 290–292, 297–298,
        330, 335
Overpass, 93, 380–381, 408, 417–419, 427–*428*
*Ovibos moschatus,* 9
    spp., 9
*Ovis,* 3–4, 8–11, 56, 60, 86, 319
    *ammon,* 5, *10,* 13
    *aries,* 60, 85, 99
    *californiana,* 13
    *canadensis auduboni,* 14, 16–17, 88, *91*
    *canadensis californiana,* 14, 16–17, 48, 60,
        89, 136, 209, 245, 270, 398
    *canadensis canadensis,* 14t, 48, 56, 85, 90,
        136, 175, 191, 209, 237, 270, 305,
        333, 378, 398, 424, 456
    *canadensis cremnobates,* 14, 16–17, 89–91,
        305, 398

*canadensis mexicana,* 14, 17, 25, 89–91,
        175, 305, 398
*canadensis nelsoni,* 14, 16–17, 56, 84,
        89–91, 136, 150, 153, 162, 175, 191,
        209, 284, 334, 398, 423
*canadensis sheldoni,* 11
*canadensis sierrae,* 14t, 17, 57, 89, 119, 191,
        240, 269, 350, 384
*canadensis* spp., 162
*canadensis texianus,* 92
*canadensis weemsi,* 14, 17, 89, 92, 98, 305
*canadensis,* 14t, 25, 47, 55, 84, 88–93, 114,
        136, 160, 175, 191, 208, 233, 258,
        280, 290, 305, 330, 347, 378, 397,
        414, 441, 455
*cervine,* 12
*dalli,* 3, 14t, 17, 25, 47, 55, 84, 86–88, 114,
        136, 160, 175, 191, 208, 233, 258,
        280, 290, 319, 330, 347, 378, 397,
        414, 455
*dalli dalli,* 3, 14t, 17, 26, 48, 56,
        84, 116, 136, 284, 290, 331,
        397, 451
*dalli fannini,* 14t, 17
*dalli kenaiensis,* 14
*dalli* spp., 162
*dalli stonei,* 7, 14, 17, 26, 48, 56,
        86, 118, 136, 167, 175, 191,
        211, 237, 290, 319, 331,
        380, 450
*gmelina,* 10
*mexicana,* 13
*montana,* 12–13
*nelsoni,* 13
*nivicola,* 3–4, 25–26, 86
    spp., 72
*stonei,* 17
*vignei,* 10
*Oxytropis* spp., 118

**P**

Pachyceriforms, 4, 86
*Panthera onca,* 162, 383
    *uncia,* 72
Parainfluenza 3 virus, 144
Paranasal sinus tumors, 144–145
Parapoxvirus, 146–147
Parasite, 25, 59, 101, 136–138, 141, 148–151,
        165–166, 218, 223, 349, 398, 402,
        409t, 441, 445
    endoparasites, 149
    microparasites, 137, 139t
    protozoal parasite, 148
    transmission, 432
Parasitic worms. *See* helminth
Paratuberculosis, 148
*Parelaphostrongylus odocoilei,* 143, 165
    *tenuis,* 165
Parks Canada, 430
Parturition, 58, 60, 72, 77–81, 114–116,
        120, 143–144, 163–164, 181, 197,
        218–219, 221–223, 265t, 269, 280,
        285, 347–348, 351, 353, 356–357,
        361, 367, 386, 397, 441,
        443, 456
Passenger pigeon, 160
*Pasteurella multocida,* 142
Pasteurellaceae, 141–145, 148
Pasteurellosis, 142–143, 145,
    *See also* disease

Pathogen, 10, 25, 29, 33–35, 40, 64, 66, 103,
        136–138, 140–142, 144, 146, 148,
        151, 153, 165, 181, 200, 202, 218,
        240, 264, 269, 273, 296, 335, 384,
        398, 402, 414, 431, 445,
        455, 460
    apillover, 417, 459
    carriers, 51
    clearance, 116
    detection, 101
    exposure, 145
    infections, 81
    introduction, 145, 147
    spread, 432
    transfer, 215, 271–272
    transmission, 145, 182, 272, 432, 459
*Pecari* spp., 8
    *tajacu,* 312
Peccary, 8
    collared, 312
Pecora, 8–9
Pelage, 3, 7, 11–12, 14, 17, 87, 211–212, 365
Pennsylvania State University Center for
        Infectious Disease Dynamics, 445
Pestiviruses, 147
Phenolics, 120–121
Phlox, 119
*Phlox* spp., 119
Physical characteristics. *See* morphology
Physiological adaptations, 3–4, 116, 122,
        208–209, 212, 219, 223
    mechanisms, 209, 354
    processes, 119, 208
    requirements, 415
    responses, 360
Pine, 360
    pinyon, 385
*Pinus* spp., 385
Pittman-Robertson Act, 26, 38, 336
Plant-herbivore interactions, 281–286
*Poa* spp., 118, 351, 364
Poaching, 177, 270, 337
Polioencephalomalacia, 152
Population-changing events, 401
    abundance, 48, 160, 199, 223,
        281, 296
    allele frequency, 102
    ancestry, 95
    assessments, 52, 233, 247–248
    assignment, 102
    augmentation, 49, 88, 99, 196, 242, 259,
        264–265, 269, 305, 324, 409t, 462
    change, 48, 52, 126, 245
    connection, 93
    connectivity, 380, 418
    control, 415
    core areas, 306
    crash, 64, 219, 282
    declines, 3, 34, 40, 48, 51–52, 99, 103,
        128, 151, 213, 218–221, 290–291,
        296–297, 299, 319, 335, 339, 398,
        400–401, 409t, 443, 449, 460
    demography, 101
    density, 63, 73, 80, 115, 123, 128, 176, 182,
        239, 285, 298, 333, 431, 441–442,
        444–447, 450, 459, 462
    dynamics, 29, 66, 72, 80–81, 115, 126–128,
        137, 163, 167, 169, 191, 196, 217,
        219, 221–224, 233, 242, 268,
        280–286, 301, 322, 359, 403t, 441,
        447, 456, 459, 460

ecology, 243, 441, 447
effects, 143, 365, 402, 407, 445, 460
equilibrium, 282
estimate, 36, 51, 97, 160, 233, 239–241, 245–247, 290, 296, 299, 315, 333, 340, 378
estimation, 97, 240
expansion, 35, 87, 99, 199, 392
extirpation, 3, 19, 34, 40, 48, 52, 99, 196, 290, 319
fluctuations, 48, 52, 90, 281–282, 284–285, 293, 297
fragmentation, 89
free-ranging, 10, *18*, 37, 151, 180–181, 305–*307*, 311, 314–317, 320–324, 462
genetics, 78, 84–85, 93–95, 105, 379
growth, 34, 37, 65, 115, 127, 136, 143, 167–168, 182, 202, 218–219, 221, 223, 269, 280–281, 284, 296–297, 312, 320, 351, 379, 392, 397, 403t, 407–408, 418, 430, 444–447, 455, 459
health, 300, 414
historical trends, 25–40
history, 84, 93, 104
home range, 283
increase, 49, 213, 314–315, 359, 443–444
indices, 168, 244
interconnectedness, 139
inventory, 299
isolation, 89, 378, 380
-level consequences, 360
-level effects, 120, 162, 167–168, 216, 268, 444
-level health effects, 143
-level influence, 213
-level knowledge, 203
-level linkage, 61
-level management, 197
-level performance, 384
-level vital rates, 197
-level vulnerability, 201
limitation, 167–168
-limiting effects, 149
losses, 163, 455
management, 101, 143, 153, 233–324, 388, 455, 462
in Mexico, 305–306, 314–316, 319–320, 322, 324
model, 223, 280–285, 458, 462
monitoring, 247–248, 319
numbers, 65, 114, 247, 282, 291, 301, 315, *431*, 456
over-abundance, 409t
performance, 101, 140, 148, 164, 223, 265
persistence, 98, 153, 167, 175, 177, 201–203, 247, 271, 347–348, 379
prey, 29–30, 160, 162–163
productivity, 114–115
projection, 280–281
recovery, 51, 322, 417
reduction, 65, 398, 430, 432
regulation, 167–168
resilience, 136, 139, 202, 219, 301, 401–402, 414
response, 247, 285, 402
restoration, 136
restored, 99, 195–196, 459–460

size, 19, 48–49, 52, 73, 78, 84–85, 90, 93, 95, 97–99, 103–104, 127, 163, 169, 213–214, 219, 221, 223, 233, 236, 239–240, 244–247, 281–282, 284–285, 293, 296, 323, 339, 356, 359, 403t, 420, 432, 443, 447–448, 455, 459–460
stability, 139, 220
stagnation, 51
status, 37–38, 48–52, 314–316
structure, 84, 89, 93–99, 101, 104–105, 219, 233, 239, 242, 281, 291, 382, 403t, 458
survey, 38, 79, 233, 240, 246–247, 296, 300, 311, 337
survival, 351
trajectories, 64, 167, 301, 400–401, 407, 460
translocated, 99, 104, 196, 223, 265, 270
trends, 38–40, 48, 50–52, 163, 167, 242, 244–246, 315, 359–360, 401, 403t, 455
variation, 196, 283
viability, 97, 220, 280, 285, 297, 426
Precipitation, 3, 103, 115, 208–224, 239, 312, 314, 351, 353, 357–361, 365, 370, 378, 388–389, 397–399, 401–402, 406, 409t, 456–457
Predation, 7, 36, 52, 72–73, 80–81, 116, 126–128, 140, 147, 151, 153, 160–169, 175–177, 180, 191, 194, 197, 201–202, 215–219, 221–223, 236, 265, 268–270, 273, 280, 282, 284–285, 312, 321, 333, 355–357, 362, 365–370, 383, 385–386, 398, 404–405, 419–420, 441, 443, 446–449, 455–456, 458–459, 462
predation management, 168–169
rates, 163–167, 169, 194, 398, 459
risk, 73, 116, 123, 126, 163, 166–168, 194, 197, 268, 350–351, 357, 360–361, 367, 380, 404–406
Predator, 3, 9, 27, 30, 65, 72–73, 81, 114, 116, 122, 151, 153, 160–170, 176–178, 194, 197, 216, 219–221, 243, 266, 268, 282–285, 301, 312, 322, 331, 333, 339, 347–351, 353–354, 357, 359–365, 367, 369, 379, 381, 383–384, 388, 390, 397–398, 402, 405, 409t, 447, 457–461
abundance, 166, 243, 284
assemblages, 161–162, 167, 457, 459
attack, 72–73, 81, 447
avoidance, 40, 72, 163, 166, 197, 219, 350, 353–354, 359, 364, 366–367, 369, 407, 455
behavior, 168
communities, 167–168
control, 30, 35, 284, 359
density, 169, 285, 402
detection, 40, 386
evasion, 163, 178–180
management, 36, 38, 238, 268–269, 300, 336, 340, 461
occurrence, 404, 424
populations, 168, 177, 358
-prey dynamics, 403t, 458–459
-prey interactions, 458–459
-prey models, 280, 284
-prey relationships, 39, 160–162, 169, 458
removal, 30, 128, 160, 168, 268–269, 455

Pregnancy. *See* gestation
Prickly pear, 119
Private landowner, 37, 268, 305, 311–312, 315, 319, 321, 331, 338, 414, 418, 420, 426
PrOACT, 421
Pronghorn, 8–9, 27, 36, 38, 142, 160, 176–180, 311, 444
peninsular population, 322
populations, 38, 168
Sonoran, 213
*Prosopis glandulosa*, 386
spp., 351, 386
Protein, 27, 62, 84, 103, 104, 106, 114, 117–122, 124–125, 142, 163, 192, 208, 212, 224, 258, 269, 273, 421
requirements, 121–122
*Protostrongylus, rushi*, 143
spp., 143, 444
*stilesi*, 143–144
*Protovis himalayensis*, 3
Protozoal disease, 151–152
Provincial Farm Practices Protection Act, 431
Game Warden for British Columbia, 35
Prune, 153
*Prunus* spp., 153
*Pseudois nayaur, 9–10*
*Psoroptes* cuniculi, 150
*ovis*, 150
spp., 38, 138, 150–151, 166
Public input, 338–339, 418
Public Trust Doctrine, 336, 339
Puget Sound Agricultural Company, 34
Puma. *See* mountain lion
*Puma concolor*, 36, 52, 72, 128, 161, 177, 194, 216, 268, 284, 312, 350, 383, 402, 419, 441, 459
*Purshia* spp., 119

R

Rabies virus, 138, 151, 423
Rabbitbrush, 119
Ram Mountain, 73–75, 77–78, 80–81, 98–100, 293, 320, 441–443, 446–451
Ranching for Wildlife, 338
abandonment, 406
assessments, 115
bighorn, 14t, 16, 19, 33–35, 48, 50, 78, 88, 90, 92, 102, 138, 145–146, 149, 151, 168, 177, 194–195, 213–214, 222–224, 241–242, 282–283, 290, 293, 306, 335–336, 339, 352–353, 357–358, 360, 378, 386, 389, 399t, 423, 444–445, 456, 460
conditions, 297, 415
contractions, 365, 378, 398
distribution, 7, 17, 19, 214
expansion, 86–87, 177, 191, 216, 405
geographic, 11, 17, 89, 347, 450
high-elevation, 192, 194–195, 197
historic(al), 50, 103–104, 161–162, 167, 179, 195–196, 270, 273, 290, 292, 301, 314, 352, 361, 379, 385, 458
loss, 296
low-elevation, 192, 245
map, 90, 102, 161
native, 90, 264, 280, 290
occupied, 127, 270, 290, 378, 419
over-exploitation, 177
overlap, 162, 176, 364, 397

Ranching for Wildlife (*cont.*)
  population, 97
  quality, 65, 120
  range, 3, 7, 25, 39, 48, 52, 55t, 59, 61,
    64–65, 72, 79, 85, 98, 115, 127–128,
    136, 138–139t, 143–144, 146, 148,
    153, 191–195, 197–199, 201–203,
    209, 217, 223, 237, 239, 245, 258,
    261, 270, 283, 285, 290, 297, 312,
    331–332, 338, 349, 369–370, 379,
    383–385, 387, 397, 402–403t, 405,
    408, 423, 425–426, 432, 457–458
  seasonal, 72–73, 78, 167, 176, 191–197,
    199–200, 202–203, 265t–266,
    282–283, 365, 404, 407, 420, 423,
    432, 459
  shift, 195, 219, 391
  summer, 191–195, 199, 203, 265t, 282–283,
    *365*, 367–*368*, 405, 444
  thinhorn, 3, 17, 50–51, 66, 87–88, 151,
    162–163, 167–168, 180, 193, 213,
    291, 298, 337, 360–363, 365–366,
    368, 378, 398–399t, 460
  use, 243, 283, 409t
  -wide abundance estimate, 48
  -wide consequences, 136, 153, 458
  -wide status, 47–52
  winter, 150, 178–180, 191–197, 201, 203,
    240–242, 245, 265t, 282, 349–351,
    361–*362*, 378, 392, 407, 423, 441,
    444–445
*Rangifer tarandus,* 9, 74, 141, 166, 180, 332
Regional District of
  Okanagan-Similkameen, 430
Renewal Resource Councils, 337
Recovery, 36–38, 141, 143, 145, 147, 163, 191,
  218, 221, 284, 293, 301, 311, 316,
  319–320, 324, 337, 384, 386, 391,
  *401,* 417, 425, 458
  efforts, 3, 38, 61, 338
  goals, 322
  plans, 322
  programs, 293–294, 316
Recreation, 40, 146, 166, 177, 182, 201,
  245, 301, 331–334, 337–340, 351,
  378–379, 387, 391–392, 402,
  406–409t, 414, 417, 419–420, 455,
  457, 460–461
Recruitment, 50–52, 80, 114–115, 128, 139,
  147, 166–168, 202, 213, 215, 219,
  221–224, 245–246, 269, 271–272,
  280–281, 284, 299, 301, 322, 351,
  358–361, 363, 397, 417, 420, 423,
  426, 443, 445–446, 449, 456,
  459–460
Reintroduction, 38, 40, 48–49, 78, 184,
  264–265, 269, 271, 290, 293, 297,
  331, 409t, 415, 420, 425, 447, 455
Relocation, 417, 430–*431*
Reproduction, 61–62, 64, 72, 76, 78, 80–81,
  98, 101, 103, 114–115, 120–122,
  125–126, 143–144, 149, 197–198,
  208, 217–222, 224, 280, 283–284,
  297, 322, 347–348, 383, 390, 398t,
  420, 423, 441, 443, 445, 447–448,
  456, 462
Reproductive activity, 144
  effort, 149, 222, 445
  isolation, 11
  outcomes, 220, 322
  output, 65, 390

  potential, 80, 322, 390
  rates, 120, 138, 247, 383
  skew, 78, 100
  status, 73, 77, 128, 209
  strategy, 72–81
  success, 60, 64, 72, 76–78, 80, 99–100, 114,
    125, 166, 197, 215, 217, 220–221,
    224, 405–406, 441, 444–445,
    450–451, 456
  tactics, 72, 76–79, 443
Resource abundance, 176
  allocation, 63–64
  availability, 61–62, 80, 163, 176, 217,
    222, 459
  competition, 176–184, 447, 455–456, 460
  damage, 391
  depletion, 330
  development, 405, 414, 417
  exploitation, 336
  extraction, 27, 35, 201, 313, 402–405, 407,
    423, 455
  management, 234, 247, 408, 458, 462
  partitioning, 176–177
  quality, 61
  scarcity, 57, 176
  selection, 363, 366, 405
  use, 52, 150, 176, 234, 458
Respiratory disease, 34, 66, 126, 139–149, 181,
  202, 264, 270–271, 321, 423–424,
  *426,* 431
  pathogen, 10, 34, 66, 101, 103, 141, 144,
    153, 417, 431
  syncytial virus, 144
  virus, 144
Restoration, 38–40, 93, 102, 136, 161, 169, 195,
  258, 264, 266, 270–272, 292, 324,
  331, 335–337, 339–340, 378, 384,
  387, 391, 401–406, 425–426, 445,
  455, 459–461
Retention dams, 213
*Ribes* spp., 119
Rice grasses, 360
Ridgegrasses, 351
Road, 33, 52, 79, 93–*94*, 147, 166–167,
  198–201, 221, 245, 306, 332, 349,
  351, 357, 367–368, 370, 378–382,
  390, 392–393, 403t–404t, 407–409,
  409t, 414, 417–421, 424, *425,*
  427–431, 433, 448, 457, 461
Rocket-nets, 260
Rocky Mountain Bighorn Society, 388
Roundworms. *See* nematode
Ruminantia, 8–9
*Rupicapra pyrenaica,* 451
  *rupicapra,* 60
  species, 9
Rushes, 118, 352, 364
Rut, 28, 72, 75–79, 81, 125–126, 145, 149, 192,
  196–197, 200, 220, 222, 245, 297,
  349, 360–361, 365–366, 405, 443
  behaviors, 460
Ryegrass, 118, 367–368

S

Safari Club International, 419
Sagebrush, 119, 121, 180
Sagewort, 118
*Salix* spp., 119, 364
Salmon River Bighorn Sheep Project, 461
*Salsola tragus,* 65

Saltbrush, 119
Sampling, 77, 85, 95, 97, 105, 115, 233,
  239–241, 243, 260, 333, 449
  area, 243–244
  design, 245
  distance, 233, 243–244
  double, 125
  effort, 240
  error, 240, 280
  group, 87
  localities, *96*
  method, 237, 239, 244
  nasal, 384
  non-invasive, 97
  periods, 239, 243
  plant, 125
  protocols, 247
  schedules, 449
  schemes, 238
  surveys, 246
  tissue, 77, 442
  units, 244
Sand tanks, 213–*214*
Sandwort, 119
*Sarcoptes scabiei,* 150
  distribution, 150
Saulteau, Blueberry River, 337
Scabies, 38, 150
Scrapie, 181
Security cover, 348–349, 366, 403t
Sedges, 118, 177, 352, 360–362, 364, 368, 387
Senescence, 80, 121, 297
Seri Indians, 337
Serow, 9
Sex, 14, 55t, 58t, 60, 73, 79, 104, 122, 167, 175,
  209, 238–240, 244–246, 266, 283,
  317, 351, 359–360, 369–370, 405,
  457
  categories, 245
  classifications, 245
  ratio, 72, 77–79, 240, 242, 244–245, 280,
    323, 339
  stages, 281
  structure, 265, 432
Sexual segregation, 72–73, 126, *163,* 233,
  242–243, 245, 364, 445
Sheep
  Alaskan, 13
  Audubon's bighorn, 14t–17, 19, 88, *91*
  Barbary, 9, 51, 181, 216
  blue, *See* bharal
  California bighorn, *6,* 14t, 16–17, 19, 36,
    38, 48–50, 60, 89–90, 92, 99, 102,
    136, 209–210, 213, 215, 217, 245,
    270, 305–306, 398–399t
    distribution, 17, 90
    population, 16–17, 49
  Canadian, 12
  Dall's, 3–5, 7, 11, 13–14, 17, 19, 34–36, 48,
    51, 56, 60, *62,* 86–*88,* 94–95, 101,
    116, 118–119, 123, 141, 149, 162,
    175, 183, 191, 199, *211*–212, 221,
    239, 242, 244, 247, 290, 294t–*295,*
    298–299, 301, 319, 331–332, 334,
    339, 361–369, 398–399t, 424, 459
    abundance, 50, 284
    body size, 55t, *212*
    classification, 17
    diet, 123, 364
    distribution, 7, 19, 361
    habitat, 35, 243, 360–364, 368

horn, 10, 57–58t, 61–62
  nutrition, 61
  origins, 7, 56, 84, 86, 88
  population, 17, 26, *50–52, 62,* 94, 136,
    168, 220, 284, 361
  predation, 166, 168
  range, 7, 19, 50–51, 167, 362–363, *365*
  recruitment, 52, 397
  survival, 397
  taxonomy, 17
deer-like, 12
desert bighorn, *6–7,* 13–14, 17, 19, 28,
    30–*31,* 34, 36–38, 48, 50, 56, 73,
    77, 79–80, 84, 89–93, 99–100,
    102, 116, 121, 123, 126–*127,* 145,
    147, 150, 153, 162, 167, 169, 175,
    181–184, 191, 194, 200, 209–214,
    216–217, 223–224, 234, 237, 240–
    243, 245, *264,* 270, 284, 305–306,
    311–314, 316, 319–324, 331, 334,
    337, 352–361, 369, 387–389, 399t,
    418–420, 423, 425–*428,* 450,
    456–457, 459, 462
  abundance, 136
  behavior, 243, 369
  body condition, 222
  body size, 55t
  captive-breeding, 315–317, 320–324
  classification, 16–17
  demography, 222–223
  diet, 34, 119, 126, 386
  distribution, 28, 90, 93, 136, 183, 200,
    212, 214–216, 241, 306–*310,* 316,
    324, 352, 357–359, 462
  habitat, 183, 212, 217, 240, 312, 353,
    356, 358–360, 419
  harvest, 295
  home range, 122
  horns, 57–58t, 212
  hunting, 293, 305, 313, *317–320,*
    324, 462
  management, 305, 320–324, 388
  migration, 200, 203
  mortality, 215, 358, 419
  movement, 200, 243, 387, 418
  parturition, 80
  population, 16–17, 25, 33, 36–38,
    49t–50, 79, 90, 92–95, 97–98,
    100, 168–169, 183–184, 195, 199,
    213–215, 222–223, 243, 272, 284,
    305, *307,* 312, 314–316, 320–322,
    324, 357–360, 369, 389, 398,
    418, 462
  predation, 167, 356, 420
  range, 19, 102, 222, 358, 360
  recruitment, 222–223, 360
  reproduction, 222–223
  reproductive success, 77
  survival, 222–223, 358
  taxonomy, 16–17
domestic, 10, 25, 29, *31,* 33–36, 40, 51, 56,
    66, 79, 85–86, 97, 99, 101, 103–104,
    128, 138, 141–148, 150–151, 153,
    *182,* 184, 200, 202, 210, 266, 268,
    270–272, 284, 292, 297, 312, 321,
    331, 335, 338–339, 384, 403t, 415,
    417, 420, 423–426, *430–*432,
    445, 459
Fannin, 7, 13–14, *16–17,* 36, 56, 87
  population, 14t, 17
Liard, 13

Mexican (desert) bighorn, 13, 89–91,
    305–360
Nelson's bighorn, 14t, 16–17, 91–92
peninsular bighorn, 16, 89–92, 95,
    305–306, 322
Rocky Mountain, *5,* 12–14, 16–17, 19,
    26, 30–32, 36, 38, 55–56, 60, 85,
    88–90, 92, 102, 118–119, 122, 145,
    147–148, 168, 175, 183–184, 191,
    197–*198,* 201, 209–210, *212–*213,
    215–219, 224, 237, 242, 243, 270,
    319, 335, 349–352, 369, 387, 399t,
    424, 427, 459
  abundance, 48, 136, 244
  behavior, *193*
  body size, 55t
  classification, 16–17
  distribution, 19, 136, 305, 349
  habitat, 34, 181, 183, 348–351, 378, 457
  harvest, 295
  horns, 10, 57–58t, 212
  migrations, 192
  mortality, 218
  population, 19, 38, 49–*50,* 77, 79, 89,
    95, 98, 102, 195, 213, 218–219, 240,
    333, 349–350, 398, 456
  range, 19, 196, 224, 245
  reproduction, 218–219
  reproductive success, 77
  survival, 217–218, 222
  taxonomy, 16–17
Soay, 60, 99–100
Sheldon's desert bighorn, 11
Sierra Nevada bighorn, 14t, 17, 36, 38, 55t,
    58t, *62–63,* 89–90, 92–93, 99, 102,
    *116–117,* 119, 126–127, 166, 191,
    197, 202, 217–219, 240–241, 269,
    350, 399t
  population, 17, 61–*63*
snow, 3–5, 7–8, 19, 25–26, 86–87, 103
Stone's, *5,* 7, 11, 13–14t, 17, 19, 36, 51, 56,
    60, 80, 86–*88,* 94, 118–119, 153,
    175, 191, 193, 211, 237, 241–242,
    247, *263,* 290, 301, 319, 331, 339,
    365–369, 384–385, 399t, 459
  abundance, 48
  behavior, 365
  body size, 55t
  classification, 17
  diet, 123
  distribution, 17, 365–366
  habitat, 180, 243, 360, 364–*369,* 405
  harvest, 294t–*295,* 450
  horns, 57–58t
  migration, 194, 380
  mortality, 380
  movement, 365
  nutrition, 368
  population, 17, 26, 50t–51, 87, 94, 136,
    150, 365, 385
  range, 51, 88, 167, 180, 194, 365
  survival, 366
  taxonomy, 17
Texas bighorn, 92
Weem's bighorn, 17, 89, 92
white sheep, 13
Sheep and Goat Control Order, 431
  River, 73–75, 77–78, 80, 441, 444–445,
    447–449
  Stewardship Framework, 337
Shikar-Safari Club, 445

Shrub, 28, 65, 81, 118–119, 177, 180, 208, 350–
    352, 357, 360, 362, 364, 367–368,
    384–387, 398–399, 401, 455
Sierra Club, 419–420
*Simulium vittatum,* 220
Sky Island Alliance, 419
Smith Fellowship, 445
Snow, 51, 72, 74, 78, 115–116, 122–123,
    166, 169, 175–176, 180, 192–195,
    199, 201–202, 208, 213, 217–224,
    245–247, 265t–266, 283, 299, 301,
    350–351, 360–366, 369, 397–399t,
    401, 409t, 456
Snow leopard, 72
Snowshoe hare, 160
Social structure, 72–76, 123
Society for the Conservation of Bighorn Sheep,
    388
Soft releases, 266, 272–273
Source stock, 89, 266, 269
Space use, 97, 175, 280, 284, 352, 356
Spatial behavior, 116
  competition, 184
  distribution, 35, 115, 123, 147, 298
  ecology, 116, 191
  genetics, 95
  overlap, 166, 183, 432
  patterns, 196
  segregation, 176–177, 180
  separation, 176–177, 183, 216, 425
  scale, 196, 283, 301, 406
Spotted knapweed, 405
St. John's wort, 405
Stakeholder, 26, 39, 261, 268, 296–297, 300,
    311, 337–339, 391, 402, 407–408,
    414–423, 425–426, 430, 432–433,
    456, 461
*Stipa* spp., 119, 351
Stopover, 191, 199
Stress, 56, 81, 103–104, 137, 142–143, 145,
    149, 166, 168, 199, 201–202, 208,
    217–218, 220–221, 224, 239, 266,
    272, 321, 354, 380, 392, 406, 409t,
    429, 447–448, 456, 459
Stoney and Simpcw First Nations, 430
Subpopulation, 104, 150, 191, 290, 379–380,
    445
Subspecies, 4–5, 7–9, 11–17, 19, 48, 51, 55t,
    57–58, 79, 84–93, 95–*96,* 98–99,
    101–104, 136, 139, 146, 151, 153,
    270, 293–295, 305–306, 339,
    361, 365, 369, 378, 397, 401, 425,
    457–458
Supplemental feeding, 429–430
Survey, 36–38, 48, 52, 79, 161, 221, 233–248,
    264, 296, 299–300, 311, 323,
    336–337, *348*
  aerial, 97, 233–234t, 239–242, 244–246,
    248, 269, *298,* 361, 403t
  cameras, *See* remote photography
  census, 95, 233, 241, 244, 311
  DNA, 16–17, 19, 39, 77, 84–87, 89–92,
    97–98, 101–106, 124, 233–234t,
    238–240, 242, 300, 449, 458
  direct evidence, 97, 240
  ground, 233–234t, 240–242,
    244–245, 248
  government, 30–32
  hiking, *See* ground survey
  indirect evidence, 237–238
  mineral lick, 242

Survey (*cont.*)
pellet counts (groups), 233, 238–240, 242, 386
rock art, *See* Native American art
remote photography, 199, 216, 233–234t, 240, 242–247, 380–381, 418, 428
sign, 238, 241
sightability, 244, 246, 296, 299
transect, 92, 97, 239–240
waterhole, 241–242
winter range, 242
Survival, 56, 61–62, 64–65, 72, 87, 92, 98–99, 114–116, 118, 121, 125–128, 136, 146, 149–151, 153, 166, 192, 196–197, 199, 202, 208, 213, 215, 217–224, *235*, 247, 260, 266–*267*, 272–273, 280–281, 283–284, 299, 331, 334, 347–349, 351, 358, 366–367, 369, 378, 384, 397–398, 403t, 408, 420, 441–443, 447–448, 451, 456, 458, 462
adult, 115, 126, 127, 163, 208, 219, 222–224, 270, 281, 403t, 447, 456
female, 60, 63, 73, 80–81, 115, 126, 223, 281, 447
juvenile, 223, 281, 283–284, 446–447
male, 60, *281*
offspring, 114, 163
winter, 98, *117*, 192, 202, 283, 298
Succulent, 119, 208, 210, 212, 351, 354, 360, 386, 423
Swine, 8

**T**

*Taeniatherum caput-medusae,* 405
Taku River Tlingit, 337
Taos Pueblo, 337
Tapeworms. *See* cestodes
Taxonomic classification, 85, 102, 365
history, 3, 12–15
Taxonomy, 3–4, 7–8, 11, 13–14, 16–17, 19, 84–106, 455, 462
Teeth, 9, 56–57, 119, 220
Temperature, 3, 56, 93, 102–103, 122, 192, 201–202, 208–210, 212, 215–218, 220, 223–224, 246–247, *264*, 299, 312, 314, 350, 352–354, 358–359, 361, 363, 365–369, 397–401, 456, 458
Terpenes, 120
Texas Bighorn Society, 388
Thermal cover, *211*, 347, 349–350, 353–358, 362–363, 366–367, 369
load, 354
neutral range, 56, 209
protection, 349
refugia, 202, 350, 366
regulation, 348, 363
stress, 56, 459
Thermoregulation, 122, 208–212
Thistle
Russian, 65
Throw-net, 260
Tick, 150
infestations, 140, 150
survival, 150
winter, 150
Time series, 445–446
Tinajas, 213–*214*, 234, 388

Topography (topographic features), 28, 246, 260, 283, 347, 349, 353, 356, 359, 362–364, 366–370, 381, 384, 430, 457
Toxin, 116, 121, 136–137, 140, 142, 145, 153, 215–216
*Toxoplasma gondii,* 148–149
Toxoplasmosis, 148–149
Traditional Indigenous Knowledge, 336, 339
Tragulidae, 9
Tragulids, 8
*Tragulus* spp., 8
Translocation, 26, 36, 38–40, 88, *91,* 93, 95, 97–99, 103, 139t, 143, 145, 148, 181, 195–196, 199, 202–203, 244–245, 247, 258–274, 305, 313, 336–337, 340, 351, 390, 405, 419–420, 430, 441, 459–462
Trap, 26–27, 52, 161, 234, 243–244, 246, 258–260, 417, 430, 442–*443*
Trematodes, 149–150
*Trueperella pyogenes,* 143
*Tsuga mertensiana,* 364
*Tympanuchus cupido cupido,* 160

**U**

Underpass, 380–381, 417, 427
Unidades para la Conservación, Manejo y Aprovechamiento Sustentable de la Vida Silvestre (UMAs), 306, 311, 313–316, 319
United States United States Fish and Wildlife Service, 418
Forest Service (USFS), 35, 182, 265t, 418–420, 425–426, 445
University of Idaho Wildlife Health Laboratory, 445
Urban area, 357, 407–408, 421, 423–424
conflicts, 417
development, 392–393, 457
environments, 420
expansion, 339, 379, 392, 418, 460
growth, 408–409t
interface, 360, 423–424
issues, 420
management, 238
parks, 360, 370
settings, 417
Urbanization, 378, 407–408, 453
*Urocyon cinereoargenteus,* 162, 357
*Ursus americanus,* 162
*arctos,* 104, 332, 367
*arctos horribilis,* 162
*maritimus,* 212
spp., 459

**V**

*Vaccinium* spp., 119, 364
Values, 52, 324, 330–332, 334–335, 339, 392–393, 404, 408, 415–416, 420–422t, 457, 461–462
Vehicle collision, 199, 201, 331, 379–380, 417–418, 423–430
*Ventenata dubia,* 405
Viral disease, 151–152
Virus, 137, 140, 144, 146–147, 151
Vital rate, 80, 197, 219, 223, 280–286, 445
Vitamin, 117, 119–121, 123
requirements, 123

**W**

Wallowa-Whitman National Forest Plan, 425
Washington State University Animal Disease Diagnostic Laboratory, 445
Water, 3, 85, 94, 97, 118–122, 143, 166, 175–177, 181–185, 191, 195, 200–201, 208–224, 234, 239–241, 243, 261, 265t–266, 321, 347, 349–353, 356–360, 364, 368–370, 378–382, 387–389, 393, 397–398, 402–403t, 408, 415, 423–424, 448, 456–457, 459
access, 139t, 150
areas, *See* water source
availability, 94, 166, 185, 192, 201, 208, 213, 215, 217, 224, 266, 322, 387, 390, 393, 399, 456–457, 459
balance, 122, 208–212, 223, 456
catchments, 213, 358–360
content, 125, 210
demands, 210, 212, 224, 456
developments, 36, 208, 213, 215–217, 239, 243, 323, 358, 387–390
diversion, 332–333
free water, 3, 122, 208–210, 212–213, 223, 351, 358, 360, 387, *389*
guzzler, 214, 398t
intake, 209–210
loss, 209, 212, 354
management, 208, 213–216
preformed water, 122, 208, 210, 212, 223–224, 358, 456
quality, 122, 215–216, 358, 390
removal, 213, 359
requirements, 122–123, 210, 212–213, 223, 242, 351, 359–360, 456, 462
resources, 34, 176, 379, 383, *388*
site, *See* water source
source, 52, 97, 122–123, 150, 153, 180, 183–184, 198, 200–202, 208, 212–217, 220, 234t, 239–244, 283, 305, 312, 320, 335, 348, 358–360, 378–379, 384, 388–393, 398–399, 401–402, 419, 456–457
stress, 208
tank, *214*
turnover, 208–210, 360
use, 212, 244, 351, 359
Weather, 40, 51–52, 62, 115, 122, 125, 127, 136–137, 145–146, 166, 168, 191–192, 194, 208–224, 241–242, 261, *263,* 266, 280, 282, 298–301, 349, 359–361, 365, 367, 369, 385, 398–399, 401–402, 415, 441, 446–447, 450, 455–457, 459–460
seasonal variation, 61, 217
Western Watersheds Project, 426
Association of Fish and Wildlife Agencies, 260, 337, 426
Wheatgrass, 119, 351, 360
White muscle disease, 120
The Wilderness Society, 419, 426
Wild Free-Roaming Horses and Burros Act (WHBA), 382–383
Grant-in-Aid Program, 388
Initiative, 49, 260, 333, 337
Position Statement: Effects of Recreational and Commercial Use of Drones on Wild Sheep 2021, 407

Sheep Foundation, 39, 214, 315, 320, 334, 337, 385, 387–388, 419, 442, 445
Society of British Columbia, 384, 388
Wildlife agency, 26, 208, 247, 260, 315, 321, 336–338, 384, 416–418, 421, 426, 430, 445, 449, 457
    crossing, 243, 381, 418–419, 424, 427, 429
Wildlife Conservation Society, 337

Water Development Standards, 388
Wild rye. *See* ryegrass
Willow, 119, 364
Wolf, 72, 160, 162, 164t, 166–169, 220, 312, 365, 383, 385, 409t, 447, 459
    predation, 162, 166, 168, 285, 362, 367–368
    gray, 162
Wolverine, 162, 164t, 169, 220, 365

Worms, 143–144
Wyoming Game and Fish Department, 191
*Wyominia tetoni*, 150

Y

Yak, 9
Yarrow, 118

For Product Safety Concerns and Information please contact our EU
representative  GPSR@taylorandfrancis.com
Taylor & Francis Verlag GmbH, Kaufingerstraße 24, 80331 München, Germany